Algebra and Trigonometry

SIXTH EDITION

Algebra and Trigonometry

SIXTH EDITION

MERVIN L. KEEDY

Purdue University

MARVIN L. BITTINGER

Indiana University—Purdue University at Indianapolis

JUDITH A. BEECHER

Indiana University—Purdue University at Indianapolis

ADDISON-WESLEY PUBLISHING COMPANY

Reading, Massachusetts • Menlo Park, California • New York
Don Mills, Ontario • Wokingham, England • Amsterdam
Bonn • Sydney • Singapore • Tokyo • Madrid
San Juan • Milan • Paris

Sponsoring Editor	Bill Poole
Development Editor	Lenore Parens
Managing Editor	Karen Guardino
Production Supervisor	Jack Casteel
Design, Editorial, and Production Services	Quadrata, Inc.
Illustrator	Scientific Illustrators
Manufacturing Supervisor	Roy Logan
Cover Designer	Geri Davis, Quadrata, Inc. and Hannus Design Associates
Cover Illustration	Masato Nishimura, The Pushpin Group

Photo credits: **347 (top)**, AP/Wide World Photos **347 (bottom)**, Rick Haston

Library of Congress Cataloging-in-Publication Data
Keedy, Mervin Laverne.
    Algebra and trigonometry : (unit circle) / Mervin L. Keedy, Marvin L.
Bittinger, Judith A. Beecher. — 6th ed.
      p.   cm.
    Includes index.
    ISBN 0-201-52517-8
    1. Algebra.   2. Trigonometry.   I. Bittinger, Marvin L.
II. Beecher, Judith A.   III. Title.
[QA154.2.K426   1993]
512′.13—dc20

                                       92-44348
                                          CIP

14 15 **CRK** 00 99 98

This volume is one of a series of texts for college algebra and trigonometry courses. Included in the series are the following.

KEEDY/BITTINGER/BEECHER

Algebra and Trigonometry (Unit Circle), Sixth Edition, covers precalculus algebra and trigonometry with an emphasis on functions. Trigonometric functions are presented first with the unit circle and then through a discussion of right-triangle trigonometry.

CONTENTS

Basic Concepts of Algebra. Equations, Inequalities, and Problem Solving. Functions, Graphs, and Transformations. Polynomial and Rational Functions. Exponential and Logarithmic Functions. The Trigonometric or Circular Functions. Trigonometric Identities, Inverse Functions, and Equations. Triangles, Vectors, and Applications. Systems and Matrices. Conic Sections. Sequences, Series, and Combinatorics.

BITTINGER/BEECHER

Algebra and Trigonometry (Right Triangle), Second Edition, covers precalculus algebra and trigonometry with an emphasis on functions. Trigonometric functions are presented first through a discussion of right-triangle trigonometry and then with the unit circle.

CONTENTS

Basic Concepts of Algebra. Equations, Inequalities, and Problem Solving. Functions, Graphs, and Transformations. Polynomial and Rational Functions. Exponential and Logarithmic Functions. The Trigonometric or Circular Functions. Trigonometric Identities, Inverse Functions, and Equations. Triangles, Vectors, and Applications. Systems and Matrices. Conic Sections. Sequences, Series, and Combinatorics.

College Algebra, Second Edition, covers the basic topics of college-level precalculus algebra.

CONTENTS

Basic Concepts of Algebra. Equations, Inequalities, and Problem Solving. Functions, Graphs, and Transformations. Polynomial and Rational Functions. Exponential and Logarithmic Functions. Systems and Matrices. Conic Sections. Sequences, Series, and Combinatorics.

Preface

This text covers college-level algebra and trigonometry and is appropriate for a one- or two-term course in precalculus mathematics. Its approach is more interactive than most precalculus texts, and is designed to help students achieve a greater understanding of sophisticated mathematical concepts than they might be able to achieve with other texts. Our goal is to provide as much support and help for students as we can, in order to ease the difficult transition into their first college-level mathematics course. At the same time, we want to give students solid preparation for calculus and maintain the appropriate topical coverage and level of presentation for a precalculus course.

CONTENT FEATURES

Algebra Review

- A lack of mastery of basic algebra skills is a common cause for student struggles in precalculus and calculus. This text contains a careful and detailed review of intermediate algebra topics in Chapters 1 and 2. (See pp. 1–128.)

Graphing and Functions

- Graphing and functions are both introduced in Chapter 3. We present graphing as a means of representing relations visually, and use it throughout the text to give students an intuitive and visual understanding of the material they are studying. For example, whenever a new type of function is introduced, we immediately show its graph. (See pp. 129, 153, 217, 296, 303, and 362.)

- Functions are presented in Chapter 3 after relations and graphs have been introduced. We include a detailed presentation of topics that will be important in calculus: the combination of functions, the composition of functions. (See pp. 144 and 189.)

Problem Solving and Applications

- Throughout the text, we try to motivate the material by providing meaningful real-life applications of the mathematics. We also include many problem-solving sections. (See pp. 161, 173, 325, and 507.)
- Since solving applied problems is one of the most difficult and important areas of algebra, we present a problem-solving algorithm in Section 2.3, and use this algorithm consistently throughout the text. The algorithm provides students with a starting point, and allows them to focus on the mathematics needed to solve the problem. To give students the background they need to translate problems themselves, we discuss extensively the "familiarization" and "translation" steps of the problem-solving process. (See pp. 78 and 563.)

Trigonometry

- We begin the trigonometry material with a discussion of distances on the unit circle in Section 6.1, to give the student a thorough understanding of where points like π and $-\pi/3$ are located. We then develop trigonometric functions of the real numbers on the unit circle in Sections 6.2 and 6.3, before introducing angles, radians, and degrees in Section 6.4. Trigonometric functions of the right triangle are introduced in Section 6.5 and applied in Section 8.1. (See pp. 353–408 and 503.)
- In Chapter 7, we devote an entire section to discussing and practicing techniques for proving trigonometric identities. We present two different methods for proving identities and discuss the merits of each one. (See pp. 470–475.)

Calculus Preparation

- Throughout the text, we give careful attention to the presentation of concepts that will become important in later calculus courses. For example, in the introduction to relations and functions, we emphasize the idea of domain, discussing it first in the context of relations, then with graphs of relations, and finally with functions. (See pp. 136, 144, 149, 150, 447–448.)

Exponential and Logarithmic Functions

- The material on exponential and logarithmic functions is introduced in the context of inverse functions, with careful emphasis placed on the relationship between exponents and logarithms. (See pp. 283 and 303.)

WHAT'S NEW IN THE SIXTH EDITION?

We have rewritten many key topics in response to user feedback, and have made significant improvements to several chapters. A detailed list of changes made to this material is available in the form of a *Conversion Guide*. Please ask your local Addison-Wesley sales representative for more information. The following is a list of the major changes in this revision.

New Design

- The new design is more open and readable. Pedagogical use of color makes it easier to tell where exercises, explanations, and examples begin and end. Highlighted boxes of information and the theorem and definition boxes stand out more.

- As always in a Keedy–Bittinger text, the objective headings within the exposition provide a concise outline of the material for both instructor and student, lessening instructor preparation time and student study time. The new design also allows for the presentation of material in a more organized, understandable manner.

New Art in Both Text and Answer Section

- The entire art program is new for this edition.

- We have maximized the accuracy of the graphs by using computer-generated art throughout the text and answers. Those who have used these texts in previous editions will immediately observe the vast improvements in the accuracy of the shape of the graphs of the polynomial, rational, and trigonometric functions.

- The use of color in the art is not just window dressing, but has been carried out in a methodical and precise manner to enhance the readability of the text for the student and the instructor. The added use of gray in the graphic art contributes to better visualization.

- The new design of the polar grid is clearer and more easily read.

- As in all Keedy–Bittinger texts, the art is placed in the most understandable location for the student. Students need not refer to figure numbers.

Technology Connections by Stuart Ball

- With this new edition, students have the added option of using graphing calculators or computer software to help visualize concepts.

- The *Technology Connection* package by Stuart Ball consists of 61 features throughout the text. These features integrate technology, motivate discovery learning, and increase understanding of the concepts through immediate visualization. Optional Technology Connection exercises are at the end of many exercise sets. The art for these features and answers has been computer-generated with a program to match what the student sees on a calculator or computer. A complete index for the topics covered in these lessons is found on p. xvii.

Thinking and Writing Exercises

- These exercises are included at the end of each chapter and aid in student comprehension, critical thinking, and conceptualization. They can be used for individual assignment or as a focus for classroom discussion. These exercises allow students to verbalize concepts, leading to better understanding, and they also help to develop self-confidence by having students write about mathematics.

Content Changes

- Chapters 3, 4, and 11 were reorganized into Chapters 3 and 4, placing the discussion of polynomial and rational functions in a more logical position, closer to that of linear and quadratic functions.

PEDAGOGICAL FEATURES

Interactive Presentation

The specialized design of this text promotes student interaction and immediate involvement with concepts. The objectives, which are stated in the margin and repeated in the textual material, and the related margin exercises, together with the annotated examples, provide a concise outline of concepts and procedures.

- **Section Objectives.**   As each new section begins, its objectives are clearly stated in the margin. These can be spotted easily by the student, and provide the answer to the typical question, "What material am I responsible for?" Symbols such as ▶ are used throughout the text to key specific objectives to the exposition, related exercises, and answers to reviews and tests at the back of the text. This allows students to easily identify appropriate material for review. (See p. 253.)

- **Margin Exercises.**   Each section contains margin exercises (over 1300 in the text), which offer practice, development, and exploration. As students work through the material, they are encouraged to work the related margin exercises. This actively involves them in the development of each topic, and gives them a deeper understanding of the material they are studying. The answers to all margin exercises are included at the back of the text. It is recommended that students work out all of these problems, stopping to do them when the text so indicates. (See pp. 192 and 374.)

Examples

The carefully chosen, detailed examples in each section create a smooth transition from the exposition to the exercise sets. Numerous annotations accompany the detailed steps in the over 900 examples.

Variety of Exercises

The more than 8000 exercises in this text excel in meeting the needs of varied skill levels and career interests.

- **Synthesis and Challenge Exercises.**   In addition to traditional computational and applied exercises, most exercise sets contain two levels of challenge exercises. Synthesis exercises require students to synthesize objectives from several sections and help to develop their critical thinking skills. Challenge exercises go beyond the section objectives and are designed to challenge the brightest students. (See pp. 143–144 and 302.)

- **Technology Connections.** See above.

- **Thinking and Writing Exercises.**   See above.

Tests and Review

- **Summary and Review.**   At the end of each chapter is a summary and review, which is designed to provide students with all the material they need for successful review. Answers are at the back of the book, together with section references so that students can easily find the correct material to restudy if they have difficulty with a particular exercise. (See p. A-4.)

- **Tests.**   Each chapter ends with a chapter test that includes synthesis questions. The answers, with appropriate section references, are given at the back of the book. Six additional forms of each of the chapter tests with answer keys appear, ready for classroom use, in the Instructor's Manual and Printed Test Bank.

ACKNOWLEDGMENTS

The authors wish to express their appreciation to the many people who helped with the development of this book, particularly Stuart Ball, Linda Collins, Barbara Johnson, and Judith Penna for their precise proofreading and checking of the manuscript.

In addition, we wish to thank the following professors for their thorough reviewing and feedback:

Helen Burrier, *Kirkwood Community College*

Marilyn Carlson, *University of Kansas*

Daniel Comenetz, *University of Massachusetts at Boston*

Chuck Cummins, *Salt Lake Community College*

Floyd L. Downs, *Arizona State University*

F. Wayne Edwards, *Clinch Valley College*

Nina R. Girard, *University of Pittsburgh at Johnstown*

Brian Gray, *Jamestown Community College*

Julie Guelich, *Normandale Community College*

Leon R. Jones, *St. Charles County Community College*

Stephen Mondy, *Normandale Community College*

Julia R. Monte, *Daytona Beach Community College*

Sarah K. Percy-Janes, *San Jacinto College–North*

Janice Recht, *University of Nebraska at Omaha*

Perry L. Weston, *Chattanooga State Technical Community College*

M.L.K.
M.L.B.
J.A.B.

SUPPLEMENTS FOR THE INSTRUCTOR

Addison-Wesley is committed to providing the best possible service and support for your classroom needs. If you have any questions about the extensive supplements package that accompanies this text for you or your students, please feel free to call 1-800-447-2226.

Instructor's Manual and Printed Test Bank by Donna DeSpain

This contains six forms of each chapter test and six final examinations with answers. Test Forms A, B, C, and D are equivalent in length and difficulty. Test Form E adds variety in style of questions and in the objectives tested. In most chapters, Test Form E is more difficult than Test Forms A, B, C, D, and F. With the exception of mathematical induction and proving identities, all questions on Test Form F are multiple choice. The manual also contains a correlation chart to the videotapes.

Instructor's Solutions Manual by Judith A. Penna

This contains worked-out solutions to every exercise in the exercise sets.

Videotapes

Using the chalkboard and manipulative aids, Professor John K. Baumgart of North Park College gives a careful, detailed, and polished presentation of the text material in 50 videotapes ranging from 30 minutes to 45 minutes in length. Professor Baumgart's use of the chalkboard and step-by-step discussion brings the concepts of algebra and trigonometry to life and provides ideal review or supplementary coverage for classroom lectures.

Transparency Masters

A set of 50 black-line masters of key examples and figures taken from the text is available for classroom use.

SUPPLEMENT FOR THE STUDENT

Student's Solutions Manual by Judith A. Penna

This contains completely worked-out solutions to the odd-numbered exercises and answers to the even-numbered exercises in the exercise sets.

SOFTWARE SUPPLEMENTS

Omnitest II. This is an extremely easy-to-use, state-of-the-art computerized testing system. As an algorithm-driven system, it can easily create up to 99 versions of the same test by automatically inserting random numbers into model problems. Although the numbers are random, they are constrained to result in reasonable answers. *Omnitest II* also allows instructors to add and edit questions with full graphics capability, as well as preview and edit items on the screen. It uses pull-down menus for fast access to the functions used most, and utilizes a WYSIWYG ("What You See Is What You Get") interface for perfect match from screen to printout. *Omnitest II* runs on IBM PC and compatible computers.

The Precalculus Explorer.   This is a practical and easy-to-use software tool that aids in the investigation of a variety of precalculus topics. Consisting of ten programs, the *Precalculus Explorer* provides a convenient way to see mathematics come alive. It gives students a useful tool for self-study, exploration, and problem solving. In addition, the *Precalculus Explorer* offers instructors a convenient way to construct dynamic exercises for classroom instruction. The *Precalculus Explorer* runs on IBM PC and compatible computers.

To the Student

THE USE OF CALCULATORS

We assume that you own a calculator and will use it while studying this book. A scientific calculator with a pi key, a power key, exponential and logarithmic keys, and trigonometric keys will be the most beneficial to you. You will note that certain exercises are designed for use of a calculator. They are indicated by the symbol ▦. Certain examples and discussions are similarly marked.

You should realize that there can be minor discrepancies due to different rounding procedures. Some calculations can be performed more conveniently on your calculator by not stopping to round. Answers will be found that way in all examples and in the exercise sets. You may note some variance in the last one or two decimal places if you round as you go. Please do not be concerned about a small variation in an answer that might be due only to a difference in rounding.

PRECISION

We should keep in mind that for most applications, three-digit accuracy is sufficient. In fact, many distances cannot even be measured to a precision greater than three digits. Thus in most examples, if we use more than three-digit precision, we are only fooling ourselves—those extra digits are actually meaningless. Thus values obtained from a calculator are too precise even to be realistic. Would it make much practical sense, for example, to calculate the distance from New York to Chicago to the nearest hundredth of an inch? In most cases, not.

The given data in the exercises in this text may be provided with greater precision than is warranted. We have adopted the convention of rounding answers to the same number of decimal places as appears in the exercise and not worrying about what is appropriate in an actual application. In a physics or engineering course, you will need to give attention to realistic rounding, but that is not the subject of this book.

INDEX OF TECHNOLOGY CONNECTIONS

*Indicates a special "discovery" feature that introduces concepts through a process of learning through discovery.

Contents

3

FUNCTIONS, GRAPHS, AND TRANSFORMATIONS 129

4

POLYNOMIAL AND RATIONAL FUNCTIONS 211

5

EXPONENTIAL AND LOGARITHMIC FUNCTIONS 283

6

THE TRIGONOMETRIC OR CIRCULAR FUNCTIONS   353

7

TRIGONOMETRIC IDENTITIES, INVERSE FUNCTIONS, AND EQUATIONS   443

8

TRIANGLES, VECTORS, AND APPLICATIONS   503

9

SYSTEMS AND MATRICES   557

10

CONIC SECTIONS   633

11

SEQUENCES, SERIES, AND COMBINATORICS 675

TABLES 739

ANSWERS A-1

INDEX I-1

This chapter is a study of the basic concepts of algebra. We assume that you have already studied intermediate algebra. That being the case, you should find most of this material merely a review. We cover the properties of the real-number system and various kinds of algebraic expressions and manipulations of them—for example, how to add them, multiply them, factor them, and so on. • If your study of algebra is recent, you might be able to skip this chapter, or at least most of it. To determine whether that is the case, you can work through the chapter test. If you answer 75% to 85% of the questions correctly, then it might be wise for you to go on to Chapter 2. •

Basic Concepts of Algebra

1.1

THE REAL-NUMBER SYSTEM

OBJECTIVES

You should be able to:

A Identify various kinds of real numbers.

B Add, subtract, multiply, and divide positive and negative real numbers.

 Real Numbers

There are various kinds of numbers. The set of numbers most used in algebra is the set of *real numbers*. Later, we will consider a more comprehensive set of numbers called the *complex numbers*. The real numbers are often shown in one-to-one correspondence with the points of a line, as follows.

$$
\begin{array}{c}
 -2.5 -\tfrac{1}{2} \tfrac{1}{2} \sqrt{2} \pi \\
\xleftarrow{} \underset{-5}{\mid} \underset{-4}{\mid} \underset{-3}{\overset{\bullet}{\mid}} \underset{-2}{\mid} \underset{-1}{\mid} \underset{0}{\mid} \underset{1}{\overset{\bullet}{\mid}} \underset{2}{\overset{\bullet}{\mid}} \underset{3}{\overset{\bullet}{\mid}} \underset{4}{\mid} \underset{5}{\mid} \xrightarrow{}
\end{array}
$$

The positive numbers are shown to the right of zero and the negative numbers to the left. Zero itself is neither positive nor negative.

There are many ways in which to name sets. For example, the set containing the numbers 2, 3, 4, and 5 can be denoted $\{2, 3, 4, 5\}$. This method of naming sets is called the **roster method.**

There are several subsets of the real-number system. Some important subsets are as follows.

All answers to the margin exercises throughout the book are located in the answer section at the back of the book.

Consider the numbers

$$1, \frac{3}{4}, -6, 0, 19, -\frac{8}{7}.$$

1. Which are natural numbers?

DEFINITION   **Sets of Numbers**

Natural Numbers.   Those numbers used for counting: $\{1, 2, 3, \ldots\}$.

Whole Numbers.   The natural numbers and 0: $\{0, 1, 2, 3, \ldots\}$.

Integers.   The whole numbers and their opposites:

$$\{\ldots, -3, -2, -1, 0, 1, 2, 3, \ldots\}.$$

Rational Numbers.   The integers and all quotients of integers (excluding division by 0): for example, $\frac{4}{5}, \frac{17}{3}, \frac{9}{1}, 6, 0, -4, \frac{78}{-5}, -\frac{2}{3}$ (can also be named $\frac{-2}{3}$ or $\frac{2}{-3}$).

We can describe the set of rational numbers precisely using what is known as **set-builder notation.** To do so, we abbreviate the sentence

2. Which are whole numbers?

"The set of all quotients $\dfrac{a}{b}$ such that a and b are integers and $b \neq 0$"

as

$$\left\{ \frac{a}{b} \,\middle|\, a \text{ and } b \text{ are integers and } b \neq 0 \right\}.$$

We will consider set-builder notation more extensively when we study equations and inequalities.

DO EXERCISES 1–4 (IN THE MARGIN).

Any real number that is not rational is called **irrational.** The rational numbers and the irrational numbers can be described in several ways.

3. Which are integers?

DEFINITION   **Rational Numbers**

The *rational numbers* are:

1. Those numbers that can be named with fractional notation a/b, where a and b are integers and $b \neq 0$ (definition).
2. Those numbers for which decimal notation either ends or repeats.

Examples   All of these are rational numbers.

1. $\frac{5}{16} = 0.3125$ Ending (terminating) decimal
2. $-\frac{8}{7} = -1.142857142857\ldots = -1.\overline{142857}$ Repeating decimal. The bar indicates the repeating part.
3. $\frac{3}{11} = 0.2727\ldots = 0.\overline{27}$ ◄

4. Which are rational numbers?

DEFINITION   **Irrational Numbers**

The *irrational numbers* are:

1. Those real numbers that are not rational (definition).
2. Those real numbers that cannot be named with fractional notation a/b, where a and b are integers and $b \neq 0$.

There are many irrational numbers. For example, $\sqrt{2}$ is irrational. We can find rational numbers a/b for which $(a/b)^2$ is close to 2, but we cannot find such a number a/b for which $(a/b)^2$ is *exactly* 2. Unless a whole number is a perfect square, its square root is irrational. For example, $\sqrt{9}$ and $\sqrt{25}$ are rational, but

$$\sqrt{3}, \qquad -\sqrt{14}, \quad \text{and} \quad \sqrt{45}$$

are irrational. There are also many irrational numbers that cannot be obtained by taking square roots. The number π is an example.* Decimal notation for π does not end and does not repeat.

Examples   All of these are irrational numbers.

4. $\pi = 3.1415926535\ldots$ There is no repeating block of digits.

5. $-1.101001000100001000001\ldots$ Though there is a pattern, there is no repeating block of digits.

6. $\sqrt{45} = 6.70820393\ldots$ There is no repeating block of digits.

7. $\sqrt[3]{2} = 1.25992105\ldots$ There is no repeating block of digits.   ◄

The following figure shows the relationship between the various kinds of numbers.

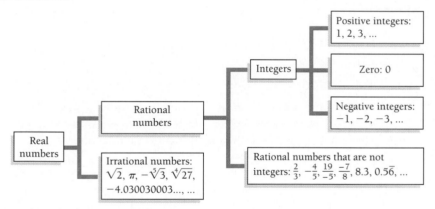

DO EXERCISES 5–10.

Ⓑ Operations on the Real Numbers

In arithmetic, we use numbers, performing calculations to obtain certain answers. In algebra, we use arithmetic symbolism, but we also use symbols to represent unknown numbers. We do calculations and manipulations of symbols on the basis of the properties of numbers, which we review now. Algebra is thus an extension of arithmetic and a more powerful tool for solving problems.

Addition

Let us review how the definition of addition is extended to include the negative numbers.

---

* $\frac{22}{7}$, 3.14, and $\frac{355}{113}$ are only rational approximations to the irrational number π.

Determine whether each of the following is rational or irrational.

5. $\dfrac{-4}{5}$

6. $-\sqrt{64}$

7. 7.42

8. 0.47474747...
   (There is a repeating block of digits.)

9. 2.57340046631...
   (There is no repeating block of digits.)

10. $\sqrt{27}$

Add.

11. $-5 + (-7)$

12. $-1.2 + (-3.5)$

13. $-\dfrac{6}{5} + \dfrac{2}{5}$

14. $0.5 + (-0.7)$

15. $8 + (-3)$

16. $\dfrac{14}{3} + \left(-\dfrac{14}{3}\right)$

To add two real numbers:

1. *Positive numbers:* Add the numbers. The result is positive.
2. *Negative numbers:* Add the absolute values. Make the answer negative.
3. *A positive number and a negative number:* Subtract the smaller absolute value from the larger. Then:
   a) If the positive number has the greater absolute value, make the answer positive.
   b) If the negative number has the greater absolute value, make the answer negative.
   c) If the numbers have the same absolute value, make the answer 0.
4. *One number is zero:* The sum is the other number.

Examples Add.

8. $-5 + (-6) = -11$

9. $8.6 + (-4.2) = 4.4$

10. $-5 + 3 = -2$

11. $\pi + (-\pi) = 0$

12. $8 + (-5) = 3$

13. $-\sqrt{3} + (-4\sqrt{3}) = -5\sqrt{3}$

14. $-\frac{9}{5} + \frac{3}{5} = -\frac{6}{5}$

15. $-\frac{5}{6} + (-\frac{7}{8}) = -\frac{20}{24} + (-\frac{21}{24})$
$= -\frac{41}{24}$ ◀

DO EXERCISES 11–16.

Properties of Real Numbers Under Addition In solving equations and doing other kinds of work in algebra, we manipulate algebraic symbols in various ways, such as collecting like terms. For example, instead of

$$4x + 7x,$$

we might write

$$11x,$$

knowing that the two expressions represent the same number no matter what x represents. In that sense, the expressions $4x + 7x$ and $11x$ are *equivalent*.

DEFINITION   **Equivalent Expressions**

If two expressions represent the same number for *any* meaningful replacement, then the expressions are said to be *equivalent*.

A replacement for an expression that is not meaningful is a number that when substituted for a variable in the expression does not give us a real number. For example, the number 2 is not a meaningful replacement for y in the expression

$$\frac{8 + y}{y - 2}$$

because it gives a denominator of 0, and we know that we cannot divide by 0

The following are properties of addition on which algebraic manipulations are based, especially when symbols for unknown numbers are used. These properties allow us to find equivalent expressions.

Commutative Law.   For any real numbers a and b,

$$a + b = b + a.$$

(The *order* in which numbers are added does not affect the sum.)

Associative Law.   For any real numbers a, b, and c,

$$a + (b + c) = (a + b) + c.$$

(When *only* additions are involved, parentheses for grouping purposes can be placed as we please without affecting the sum.)

Additive Identity.   There exists a unique real number 0 such that for any real number a,

$$a + 0 = 0 + a = a.$$

Additive Inverses.   For any real number a, there exists a unique real number, denoted $-a$, called the *additive inverse,* or *opposite,* for which

$$-a + a = a + (-a) = 0.$$

Opposites, or Additive Inverses   Concerning opposites, or additive inverses, we caution you about one of the most misunderstood, confusing ideas in elementary algebra: It is common to read an expression such as $-x$ as "negative x." This can be confusing, because $-x$ can be positive, negative, or zero, depending on the value of x. The symbol $-$, used in this way, indicates an *opposite,* or *additive inverse.* The same symbol can also indicate a negative number, as in -5, or it can indicate subtraction, as in $3 - x$.

CAUTION!   An initial $-$ sign, as in $-x$ or $-(x^2 + 3x - 2)$, should always be interpreted as meaning "the opposite of" or "the additive inverse of", *not* "the negative of"! The entire expression can be positive, negative, or zero, depending on the value of the part of the expression that follows the $-$ sign.

Taking the opposite is sometimes called "changing the sign."

Examples   Find each of the following.

16. $-x$, when $x = 5$ $-(5) = -5$ The opposite of 5 is negative 5.

17. $-x$, when $x = -3$ $-(-3) = 3$ The opposite of negative 3 is 3.

18. $-x$, when $x = 0$ $-(0) = 0$ The opposite of 0 is 0.

19. $-(x^2 + 4x + 2)$, when $x = 1$ $-(1^2 + 4 \cdot 1 + 2)$

$$= -(7) = -7$$

$-(x^2 + 4x + 2)$, when $x = -1$ $-[(-1)^2 + 4(-1) + 2]$

$$= -(-1) = 1 \qquad \blacktriangleleft$$

Find $-x$ and $-1 \cdot x$ when:

17. $x = 6$.

18. $x = -8$.

19. $x = -3.4$.

20. Find $-(x^2 - 3x)$ (a) when $x = 2$ and (b) when $x = -1$.

Multiply.

21. $4 \cdot (-6)$

22. $-\frac{7}{5} \cdot \left(-\frac{3}{5}\right)$

23. $(-2) \cdot (-3) \cdot (-4) \cdot (-6)$

It can be shown that $-1 \cdot x = -x$ and $-(-x) = x$ for any number x. That is, multiplying a number x by negative 1 produces the opposite of x, and the opposite of the opposite of a number is the number itself.

THEOREM 1 **Opposites and Multiplying by -1**

For any real number a,

$$-1 \cdot a = -a \quad \text{and} \quad -(-a) = a.$$

(Multiplying a number by -1 produces its opposite, and the opposite of the opposite of a number is the number itself.)

DO EXERCISES 17–20.

Multiplication

Assuming that multiplication of nonnegative real numbers poses no problem, let us review how the definition of multiplication is extended to include the negative numbers.

To multiply two real numbers:

1. Multiply the absolute values.
2. If the signs are the same, the answer is positive.
3. If the signs are different, the answer is negative.

Examples Multiply.

20. $3 \cdot (-4) = -12$

21. $1.5 \cdot (-3.8) = -5.7$

22. $-5 \cdot (-4) = 20$

23. $-\frac{2}{3} \cdot (-\frac{4}{5}) = \frac{8}{15}$

24. $-3 \cdot (-2) \cdot (-4) = -24$

DO EXERCISES 21–23.

Properties of Real Numbers Under Multiplication We now list the properties of the real numbers under multiplication. Again, these are properties that allow us to find equivalent expressions.

Commutative Law. For any real numbers a and b,

$$ab = ba.$$

(The *order* in which numbers are multiplied does not affect the product.)

Associative Law. For any real numbers a, b, and c,

$$a(bc) = (ab)c.$$

(When *only* multiplications are involved, parentheses for grouping purposes can be placed as we please without affecting the product.)

Multiplicative Identity.   **There exists a unique real number 1 such that for any real number a,**

$$a \cdot 1 = 1 \cdot a = a.$$

(Multiplying any number by 1 gives that same number.)

Multiplicative Inverses.   **For each nonzero real number a, there exists a unique number, denoted $1/a$ or a^{-1}, called the *multiplicative inverse*, or *reciprocal*, for which**

$$a\left(\frac{1}{a}\right) = \frac{1}{a}(a) = 1.$$

Examples

25. The multiplicative inverse, or reciprocal, of 2 is $\frac{1}{2}$, because $2(\frac{1}{2}) = 1$.
26. The multiplicative inverse of $-\frac{2}{3}$ is $-\frac{3}{2}$, because $(-\frac{2}{3})(-\frac{3}{2}) = 1$.
27. The reciprocal of 0.16 is 6.25, because $(0.16)(6.25) = 1$. ◄

There is a very special property that connects addition and multiplication as follows.

Distributive Law.   **For any real numbers a, b, and c,**

$$a(b + c) = ab + ac.$$

(This is also called the distributive law of multiplication over addition.)

The expression $ab + ac$ means $(a \cdot b) + (a \cdot c)$. By agreement, we can omit parentheses around multiplications. According to this agreement, multiplications are to be done before additions or subtractions.

Any number system having the preceding properties for addition and for multiplication is called a **field.** Thus we refer to these properties as the **field properties.** Many other properties important in algebraic manipulations can be proved from the field properties. We list some of these as theorems.

Subtraction

Subtraction is the operation opposite to addition.

DEFINITION   **Subtraction**

For any real numbers a and b,

$$a - b = c \quad \text{if and only if} \quad b + c = a.$$

In any field, we actually subtract by adding the opposite of the number being subtracted.

Subtract.

24. $2.5 - 1.2$

25. $12 - (-5)$

26. $-\dfrac{8}{5} - \dfrac{3}{5}$

27. $-20 - (-7)$

THEOREM 2*

For any real numbers a and b,

$$a - b = a + (-b).$$

To subtract a number, we can add its opposite.

Examples Subtract.

28. $8 - 5 = 8 + (-5) = 3$

29. $3 - 7 = 3 + (-7) = -4$

30. $8.6 - (-2.3) = 8.6 + 2.3$
$$= 10.9$$

31. $-15 - (-5) = -15 + 5$
$$= -10$$

32. $10 - (-4) = 10 + 4 = 14$

33. $\frac{2}{3} - \frac{5}{6} = \frac{4}{6} + (-\frac{5}{6}) = -\frac{1}{6}$ ◀

DO EXERCISES 24–27.

Multiplication is distributive over subtraction in the real-number system, as the following theorem states.

THEOREM 3   **The Distributive Law**

For any real numbers a, b, and c,

$$a(b - c) = ab - ac.$$

(This is the distributive law of multiplication over subtraction.)

Theorem 3 follows easily from Theorem 2 and the other distributive law. We will often use the term "distributive law" when we mean either law, since one is easily proven from the other and subtraction can always be expressed as an addition.

Division

Division is the operation opposite to multiplication.

DEFINITION   **Division**

For any number a and any nonzero number b,

$$a \div b = c \quad \text{if and only if} \quad b \cdot c = a.$$

In any field, we usually divide by multiplying by a reciprocal.

---

* Theorem 2 is often used as a *definition* of subtraction. The definition of subtraction used here is more general, since it does not depend on the existence of inverses. Our definition is valid in the system of natural numbers, for example, where Theorem 2 would not even make sense since opposites do not exist.

THEOREM 4

For any real number a and any nonzero number b,

$$a \div b = a \cdot \left(\frac{1}{b}\right).$$

The definition of division parallels the one for subtraction. Theorems 2 and 4 are also parallel.

Examples   Divide by multiplying by a reciprocal.

34. $\frac{3}{4} \div \left(-\frac{2}{3}\right) = \frac{3}{4}\left(-\frac{3}{2}\right) = -\frac{9}{8}$

35. $-\frac{6}{7} \div \left(-\frac{3}{5}\right) = -\frac{6}{7}\left(-\frac{5}{3}\right) = \frac{30}{21} = \frac{10}{7}$

36. $-\frac{24}{8} = -24 \cdot \frac{1}{8} = -3$ ◀

From Theorem 4, it follows easily that the quotient of two negative numbers is positive and the quotient of a positive number and a negative number is negative.

DO EXERCISES 28–31.

Order

The order of the real numbers is shown intuitively by a number line. If a number a is pictured to the left of a number b, then a is less than b ($a < b$).

In this case, if we subtract a from b, then the answer will be a positive number. For example,

$-4 < 9$ because $9 - (-4) = 13$ and 13 is positive, and

$-7 < -5$ because $-5 - (-7) = 2$ and 2 is positive.

We also say that $a > b$ is true if $b < a$ is true. The symbol $a \leq b$, read "a is less than or equal to b," is true when either $a < b$ is true or $a = b$ is true.

DO EXERCISES 32–35.

Divide.

28. $\dfrac{-20}{-5}$

29. $\dfrac{4.5}{-1.5}$

30. $-\dfrac{4}{5} \div \dfrac{3}{10}$

31. $-\dfrac{5}{6} \div \left(-\dfrac{5}{12}\right)$

Determine whether each of the following is true.

32. $-3.4 < -3.8$

33. $-3 \leq 5$

34. $234 > -56$

35. $\dfrac{2}{3} \leq 0.\overline{3}$

● EXERCISE SET   1.1

Review all the objectives at the beginning of the section before beginning the exercise set.

 Consider the numbers

$-6, 0, 3, -\frac{1}{2}, \sqrt{3}, -2, -\sqrt{7}, \sqrt[3]{2},$
$\frac{5}{8}, 14, -\frac{9}{4}, 8.53, 9\frac{1}{2}, \sqrt{16}, -\sqrt[3]{8}.$

1. Which are natural numbers?

2. Which are whole numbers?

3. Which are irrational numbers?

4. Which are rational numbers?

5. Which are integers?

6. Which are real numbers?

Determine whether each of the following is rational or irrational.

7. $-\frac{6}{5}$

8. $-\frac{3}{7}$

9. -9.032

10. 3.14

11. $4.\overline{516}$

12. $-7.\overline{32}$

13. 4.303003000300003 . . . (No repeating block of digits)

14. 6.414114111411114 . . . (No repeating block of digits)

15. $\sqrt{6}$ **16.** $\sqrt{7}$

17. $-\sqrt{14}$ **18.** $-\sqrt{12}$

19. $\sqrt{49}$ **20.** $-\sqrt{16}$

21. $\sqrt[3]{5}$ **22.** $\sqrt[4]{10}$

▶   Find $-x$ and $-1 \cdot x$, when:

23. $x = -7$.   **24.** $x = -\frac{10}{3}$.

25. $x = 57$.   **26.** $x = \frac{13}{14}$.

Find $-(x^2 - 5x + 3)$, when:

27. $x = 12$.   **28.** $x = -8$.

Find $-(7 - y)$, when:

29. $y = -9$.   **30.** $y = 19$.

Compute.

31. $-3.1 + (-7.2)$ **32.** $-735 + 319$

33. $\frac{9}{2} + \left(-\frac{3}{5}\right)$ **34.** $-6 + (-4) + (-10)$

35. $-7(-4)$ **36.** $-\frac{8}{3}\left(-\frac{9}{2}\right)$

37. $(-8.2) \times 6$ **38.** $-6(-2)(-4)$

39. $-7(-2)(-3)(-5)$ **40.** $(-7.1)(-2.3)$

41. $-\frac{14}{3}\left(-\frac{17}{5}\right)\left(-\frac{21}{2}\right)$ **42.** $-\frac{13}{4}\left(-\frac{16}{5}\right)\left(\frac{23}{2}\right)$

43. $\frac{-20}{-4}$ **44.** $\frac{49}{-7}$

45. $\frac{-10}{70}$ **46.** $\frac{-40}{8}$

47. $\frac{2}{7} \div \left(-\frac{14}{3}\right)$ **48.** $-\frac{3}{5} \div \left(-\frac{6}{7}\right)$

49. $-\frac{10}{3} \div \left(-\frac{2}{15}\right)$ **50.** $-\frac{12}{5} \div (-0.3)$

51. $11 - 15$ **52.** $-12 - 17$

53. $12 - (-6)$ **54.** $-13 - (-4)$

55. $15.8 - 27.4$ **56.** $-19.04 - 15.76$

57. $-\frac{21}{4} - \left(-\frac{7}{8}\right)$ **58.** $\frac{2}{3} - \left(-\frac{17}{4}\right)$

●   **SYNTHESIS** _____

Calculate. Round to six decimal places. The symbol ▦ indicates an exercise meant to be done with a calculator.

59. ▦ **a)** $(1.4)^2$
              $(1.41)^2$
              $(1.414)^2$
              $(1.4142)^2$
              $(1.41421)^2$

   b) What number does the sequence of numbers in part (a) seem to approach?

60. ▦ **a)** $(2.1)^3$
              $(2.15)^3$
              $(2.154)^3$
              $(2.1544)^3$
              $(2.15443)^3$

   b) What number does the sequence of numbers in part (a) seem to approach?

What property is illustrated by each sentence?

61. $k + 0 = k$

62. $ax = xa$

63. $-1(x + y) = (-1x) + (-1y)$

64. $4 + (t + 6) = (4 + t) + 6$

65. $c + d = d + c$

66. $-67 \cdot 1 = -67$

67. $4(xy) = (4x)y$

68. $5(a + t) = 5a + 5t$

69. $y\left(\dfrac{1}{y}\right) = 1, y \neq 0$

70. $-x + x = 0$

71. $a + (b + c) = a + (c + b)$

72. $a(b + c) = (b + c)a$

73. $a(b + c) = a(c + b)$

74. $ab + ac = a(b + c)$

75. Show that subtraction is not commutative. That is, find real numbers a and b such that $a - b \neq b - a$.

76. Show that division is not commutative.

77. Show that division is not associative.

78. Show that subtraction is not associative.

79. Which is a better approximation to π: 3.14, $\frac{22}{7}$, or $\frac{3927}{1250}$?

80. At what decimal place does $\frac{22}{7}$ differ from π?

To convert from repeating decimal notation to fractional notation, consider an example such as 8.97656565 . . . or 8.97$\overline{65}$. Let $n = 8.97\overline{65} = 8.976565 \ldots$. Then

$$10,000n = 89765.6565 \ldots$$
$$\underline{100n = 897.6565 \ldots}$$
$$9,900n = 88,868$$
$$n = \frac{88,868}{9900}.$$

Convert to fractional notation.

81. $0.\overline{9}$

82. $3.\overline{74}$

83. $18.3\overline{245}$

84. $12.34\overline{7652}$

●   **CHALLENGE** _____

85. Prove that for any real numbers
$$(b + c)a = ba + ca.$$

86. Prove that any positive number is greater than 0.

1.2
EXPONENTIAL, SCIENTIFIC, AND ABSOLUTE-VALUE NOTATION

A Integers as Exponents

When an integer greater than 1 is used as an **exponent,** the integer tells us the number of times that the base is used as a factor. For example, 5^3 means $5 \cdot 5 \cdot 5$. An exponent of 1 does not change the meaning of an expression. For example, $(-3)^1 = -3$. When 0 occurs as the exponent of a nonzero expression, we agree that the expression is equal to 1. For example, $37^0 = 1$.

DEFINITION	**Exponents of 1 and 0**

For any real number a, $a^1 = a$.
For any nonzero real number a, $a^0 = 1$.

We will see later why 0 is not allowed as an exponential base.

CAUTION! When a negative sign occurs in exponential notation, a certain caution is in order. For example, $(-4)^2$ means that -4 is to be raised to the second power. Hence, $(-4)^2 = (-4)(-4) = 16$. On the other hand, -4^2 represents the opposite of 4^2. Thus, $-4^2 = -16$. It may help to remember that $-x^2$ and $-1 \cdot x^2$ are equivalent, according to Theorem 1.

DO EXERCISES 1–10.

Negative integers as exponents are defined as follows.

DEFINITION	**Negative Exponents**

If n is any positive integer, then a^{-n} means $\dfrac{1}{a^n}$ for $a \neq 0$.

In other words, a^n and a^{-n} are reciprocals of each other.

Examples

1. $\dfrac{1}{5^2} = 5^{-2}$ 2. $7^{-3} = \dfrac{1}{7^3} = \dfrac{1}{343}$

3. $5^{-4} = \dfrac{1}{5^4} = \dfrac{1}{5 \cdot 5 \cdot 5 \cdot 5} = \dfrac{1}{625}$ ◀

DO EXERCISES 11–13.

Rename with exponents.

1. $8 \cdot 8 \cdot 8 \cdot 8$

2. xxx

3. $4y \cdot 4y \cdot 4y \cdot 4y$

Rename without exponents.

4. 3^4

5. $(5x)^4$

6. $(-5)^4$

7. -5^4

8. $(3x)^0$

Simplify.

9. $(5y)^2$

10. $(-2x)^3$

11. Rename $1/4^3$ using a negative exponent.

12. Rename 10^{-4} without using a negative exponent.

13. Write three other symbols for 4^{-3}.

Multiply and simplify.

14. $8^{-3} \cdot 8^7$

15. $y^7 y^{-2}$

16. $(9x^4)(-2x^7)$

17. $(-3x^{-4})(25x^{-10})$

18. $(5x^{-3}y^4)(-2x^{-9}y^{-2})$

19. $(4x^{-2}y^4)(15x^2y^{-3})$

Properties of Exponents

Multiplication   Let us consider an example involving multiplication:

$$b^5 \cdot b^{-2} = (b \cdot b \cdot b \cdot b \cdot b) \cdot \frac{1}{b \cdot b}$$

$$= \frac{b \cdot b}{b \cdot b} \cdot (b \cdot b \cdot b) = 1 \cdot (b \cdot b \cdot b) = b^3.$$

Note that the result can be obtained by adding the exponents.

THEOREM 5	**The Product Rule**

For any number a and any integers m and n,

$$a^m \cdot a^n = a^{m+n}.$$

We can use Theorem 5 to find equivalent expressions for products of exponential expressions with the same base. We will usually express our final answer with a positive exponent where possible.

Examples   Multiply and simplify.

4. $x^4 \cdot x^{-2} = x^{4+(-2)} = x^2$ 5. $a^4 \cdot a^3 = a^7$

6. $c^{-3} \cdot c^{-2} = c^{-5} = \dfrac{1}{c^5}$

7. $(-8a^{-5}b^4)(10a^7b^{-13}) = -80a^2b^{-9}$

$$= -\frac{80a^2}{b^9}$$

DO EXERCISES 14–19.

Division   Let us consider an example involving division:

$$\frac{a^5}{a^3} = a^5 \cdot \frac{1}{a^3} = a^5 \cdot a^{-3} = a^{5+(-3)} = a^{5-3} = a^2.$$

Note that we could also have obtained this result by subtracting the exponents. Here is another example:

$$\frac{b^{-2}}{b^3} = b^{-2} \cdot b^{-3} = b^{-2+(-3)} = b^{-2-3} = b^{-5} = \frac{1}{b^5}.$$

Again, the result could be obtained by subtracting the exponents.

THEOREM 6	**The Quotient Rule**

For any nonzero number a and any integers m and n,

$$\frac{a^m}{a^n} = a^{m-n}.$$

Examples   Divide and simplify.

8. $\dfrac{a^{-2}}{a^5} = a^{-2-5} = a^{-7} = \dfrac{1}{a^7}$

9. $\dfrac{b^{-4}}{b^{-5}} = b^{-4-(-5)} = b^1 = b$

10. $\dfrac{32x^{15}}{-16x^8} = \dfrac{32}{-16}x^{15-8} = -2x^7$

11. $\dfrac{-54x^{-3}y^5}{48x^{-7}y^{13}} = -\dfrac{9}{8}x^{-3-(-7)}y^{5-13} = -\dfrac{9}{8}x^4y^{-8} = -\dfrac{9x^4}{8y^8}$ ◄

We can use the quotient rule to show why a^0 is not defined when $a = 0$. Consider the following:

$$a^0 = a^{3-3} = \dfrac{a^3}{a^3}.$$

If a were 0, we would then have $\frac{0}{0}$, which is meaningless.

DO EXERCISES 20–26.

Raising Powers to Powers   Consider this example:

$$(a^2)^4 = a^2 \cdot a^2 \cdot a^2 \cdot a^2 = a^8.$$

We can obtain the same result by multiplying the exponents.

THEOREM 7	The Power Rule

For any number a and any integers m and n,

$$(a^m)^n = a^{mn}.$$

When an expression inside parentheses is raised to a power, the inside expression is the base. For example, $(3a)^2 = (3a)(3a) = 3^2 \cdot a^2 = 9a^2$. We can evaluate the power $(3a)^2$ by raising 3 to the power 2 and a to the power 2. A similar thing happens to quotients.

THEOREM 8	Raising a Product or Quotient to a Power

For any real numbers a and b and any integer n,

$$(ab)^n = a^n b^n.$$

For any real numbers a and b, $b \neq 0$, and any integer n,

$$\left(\dfrac{a}{b}\right)^n = \dfrac{a^n}{b^n}.$$

Examples   Simplify. Write answers with positive exponents.

12. $(8^{-2})^3 = 8^{-2\cdot3} = 8^{-6} = \dfrac{1}{8^6}$

13. $(x^{-5})^4 = x^{-20} = \dfrac{1}{x^{20}}$

Divide and simplify.

20. $\dfrac{4^8}{4^5}$

21. $\dfrac{5^4}{5^{-2}}$

22. $\dfrac{10^{-5}}{10^9}$

23. $\dfrac{9^{-8}}{9^{-2}}$

24. $\dfrac{y^6}{y^{-5}}$

25. $\dfrac{10y^2}{2y^3}$

26. $\dfrac{42x^7y^6}{-21y^{-3}x^{10}}$

Simplify.

27. $(3^7)^7$

28. $(8^2)^{-7}$

29. $(y^4)^{-7}$

30. $(2xy)^3$

31. $(4x^{-2}y^7)^2$

32. $\left(\dfrac{3x^4y^2}{z^5}\right)^{-3}$

33. $\left(\dfrac{10x^{-4}y^7z^{-2}}{5x^6y^{-8}z^{-3}}\right)^3$

14. $(4x^{-2})^3 = 4^3(x^{-2})^3 = 64x^{-6} = \dfrac{64}{x^6}$

15. $(5x^3y^{-5}z^2)^4 = 5^4x^{12}y^{-20}z^8 = 625x^{12}y^{-20}z^8 = \dfrac{625x^{12}z^8}{y^{20}}$

16. $\left(\dfrac{12x^2y^{-3}}{3y^8z^{-5}}\right)^{-4} = \left(\dfrac{4x^2}{y^{11}z^{-5}}\right)^{-4} = \dfrac{4^{-4}(x^2)^{-4}}{(y^{11})^{-4}(z^{-5})^{-4}}$

$$= \dfrac{x^{-8}}{256y^{-44}z^{20}} = \dfrac{y^{44}}{256x^8z^{20}}$$

> **CAUTION!** When raising a product such as $8x^2y^{-2}$ to a power, don't forget to raise *each* factor to the power. For example,
> $$(8x^2y^{-2})^3 = 8^3(x^2)^3(y^{-2})^3.$$

DO EXERCISES 27–33.

▶ B Scientific Notation

Scientific notation is particularly useful for naming very large or very small numbers. The following are examples of scientific notation:

$$7.8 \times 10^{13} \quad \text{and} \quad 5.64 \times 10^{-8}.$$

> **DEFINITION**   **Scientific Notation**
>
> *Scientific notation* for a number is an expression of the type
> $$N \times 10^n,$$
> where N is greater than or equal to 1 and less than 10 ($1 \le N < 10$) and N is expressed in decimal notation. When $N = 1$, the expression 10^n is considered to be scientific notation.

We keep in mind that positive exponents correspond to large numbers and negative exponents correspond to small numbers.

Example 17   The population of the United States is about 257,000,000. Convert this number to scientific notation.

Solution   We want the decimal point to be positioned between the 2 and the 5. We count the number of moves of the decimal point. It is 8. Since the number to be converted is large, the exponent is positive. Thus,

$$257,000,000 = 2.57 \times 10^8.$$

Example 18   The mass of a hydrogen atom is

0.00000000000000000000000017 gram.

Convert this number to scientific notation.

Solution We want the decimal point to be positioned between the 1 and the 7. We count the number of moves of the decimal point: It is 24. Since the number to be converted is small, the exponent is negative. Thus,

$$0.0000000000000000000000017 = 1.7 \times 10^{-24}.$$ ◄

Examples Convert to decimal notation.

19. $6.043 \times 10^5 = 604{,}300$
20. $4.7 \times 10^{-8} = 0.000000047$ ◄

DO EXERCISES 34–39. _____

On a calculator, a number like 370,000,000 might be expressed using notation like "3.7 E 8", or with a space, simply as "3.7 8". This is the way the calculator would show a very large or small number using scientific notation.

Examples Convert to decimal notation.

21. $4.23 \text{ E} -5 = 0.0000423$
22. $7.31 \text{ E } 12 = 7{,}310{,}000{,}000{,}000$ ◄

DO EXERCISES 40 AND 41. _____

Scientific notation is often used in problem solving.

Example 23 Alpha Centauri is the star, apart from the sun, that is closest to the earth. It is about 4.3 light-years from the earth. One **light-year** is the distance that light travels in one year and is about 5.88×10^{12} miles. How many miles is it from earth to Alpha Centauri? Express your answer in scientific notation.

Solution The distance from the earth to Alpha Centauri is the number of light-years times the number of miles that light travels in one year. It is given by

$$
\begin{aligned}
4.3 \times (5.88 \times 10^{12}) &= (4.3 \times 5.88) \times 10^{12} \\
&= 25.284 \times 10^{12} \\
&= (2.5284 \times 10^1) \times 10^{12} \\
&= 2.5284 \times 10^{13}.
\end{aligned}
$$ ◄

In Example 23, you may have been tempted to quit when you obtained 25.284×10^{12}. That would be a correct answer numerically, but it is not in scientific notation, since 25.284 is not a number between 1 and 10.

DO EXERCISES 42–44. _____

  Order of Operations

What does $3 + 2 \cdot 7^2$ mean? If we add 3 and 2, to get 5, and then multiply by 7^2, which is 49, we get 245. If we multiply 2 times 49 and then add 3,

---

Convert to scientific notation.

34. 465,000

35. 3789

36. 0.000145

37. 0.00000000067

Convert to decimal notation.

38. 4.67×10^{-5}

39. 7.894×10^{12}

Convert to decimal notation.

40. 8.166 E 9

41. 1.103 E −6

42. Find scientific notation for the number of seconds in one year. Use 365 days for a year.

Compute. Write scientific notation for the answer.

43. $(8.3 \times 10^{-3})(7.6 \times 10^{-8})$

44. $\dfrac{1.8 \times 10^{-16}}{2.5 \times 10^{-7}}$

Calculate.

45. a) $3 \cdot 5^2 + 4$

b) $3 \cdot (5^2 + 4)$

46. $\left(\dfrac{(3+2)^2 - 3 + 2^2 + 1}{2^3 + 5^0} \right)^3$

47. a) Calculate $20 \div 4 - 3 \cdot 6$ using the rules for order of operations.

b) Do the same calculation by first converting the division to a multiplication and the subtraction to an addition.

we get 101. Clearly, only one result can be correct. To determine which procedure to use, mathematicians have agreed on the following rules for *order of operations.*

Rules for Order of Operations

1. Do all calculations within grouping symbols before operations outside.
2. Evaluate all exponential expressions.
3. Do all multiplications and divisions in order from left to right.
4. Do all additions and subtractions in order from left to right.

Example 24 Calculate: $8 + 2(4 - 9)^2$.

Solution

$$
\begin{aligned}
8 + 2(4 - 9)^2 &= 8 + 2(-5)^2 && \text{Working within parentheses first} \\
&= 8 + 2(25) && \text{Simplifying } (-5)^2 \\
&= 8 + 50 && \text{Multiplying} \\
&= 58 && \text{Adding}
\end{aligned}
$$

In addition to the common grouping symbols such as parentheses ( ), brackets [ ], and braces { }, fraction bars can act as grouping symbols.

Example 25 Calculate: $\dfrac{14(11 - 2) + 8 \cdot 6}{5^2 + 2^3}$.

Solution An equivalent expression using brackets as grouping symbols is

$$[14(11 - 2) + 8 \cdot 6] \div [5^2 + 2^3].$$

What this shows, in effect, is that first we do the calculations in the numerator and in the denominator, and then we divide the results:

$$\frac{14(11 - 2) + 8 \cdot 6}{5^2 + 2^3} = \frac{14(9) + 8 \cdot 6}{25 + 8} = \frac{126 + 48}{33} = \frac{174}{33} = \frac{58}{11}.$$

DO EXERCISES 45–47.

D Absolute Value

Informally, we say that the **absolute value** of a number is its distance from 0 on a number line. The absolute value of a number a is denoted $|a|$.

Examples Simplify.

26. $|-7|$ The distance of -7 from 0 is 7, so $|-7| = 7$.

27. $|5|$ The distance of 5 from 0 is 5, so $|5| = 5$.

28. $|0|$ The distance of 0 from 0 is 0, so $|0| = 0$.

We can use the notation of opposite to give a formal definition of absolute value.

DEFINITION   **Absolute Value**

For any real number a,

$$|a| = a \quad \text{if } a \geq 0$$

(if a number is nonnegative, then its absolute value is the number itself) and

$$|a| = -a \quad \text{if } a < 0$$

(if a number is negative, then its absolute value is its opposite).

DO EXERCISES 48–51. _____

We now consider certain properties involving absolute value and use them to simplify certain expressions. In that way, we can find equivalent expressions. Consider, for example, the absolute value of the product $(-3)5$. Now

$$|-3 \cdot 5| = |-15| = 15 \quad \text{and} \quad |-3| \cdot |5| = 3 \cdot 5 = 15,$$

so

$$|-3 \cdot 5| = |-3| \cdot |5|.$$

Note that the absolute value of a product is the product of the absolute values.

Similarly, the absolute value of a quotient is the quotient of the absolute values. We can check this as follows:

$$\left|\frac{25}{-5}\right| = |-5| = 5 \quad \text{and} \quad \frac{|25|}{|-5|} = \frac{25}{5} = 5, \quad \text{so} \quad \left|\frac{25}{-5}\right| = \frac{|25|}{|-5|}.$$

Note too that

$$|(-3)^2| = |9| = 9 \quad \text{and} \quad (-3)^2 = 9, \quad \text{so} \quad |(-3)^2| = (-3)^2.$$

Also,

$$|-3| = 3 \quad \text{and} \quad |3| = 3, \quad \text{so} \quad |-3| = |3|.$$

Theorem 9 summarizes the properties of absolute value that we use. Each can be proven using the definition of absolute value.

THEOREM 9   **Properties of Absolute Value**

For any real numbers a and b and any nonzero number c:

1. $|ab| = |a| \cdot |b|$.
2. $\left|\dfrac{a}{c}\right| = \dfrac{|a|}{|c|}$.
3. $|a^n| = a^n$, if n is an even integer.
4. $|-a| = |a|$.

We can use Theorem 9 to find equivalent expressions.

Simplify.

48. $|2|$

49. $|\sqrt{3}|$

50. $|-11.3|$

51. $\left|-\dfrac{3}{4}\right|$

Simplify.

52. $|(-5)(-4)|$ **53.** $|-5| \cdot |-4|$

54. $\dfrac{|-20|}{|-5|}$ **55.** $\left|\dfrac{-20}{-5}\right|$

56. $|-6ab|$ **57.** $|x^8|$

58. $|10m^2n^3|$ **59.** $\left|\dfrac{-2x^3}{y^2}\right|$

Examples Simplify, leaving as little as possible inside the absolute-value signs.

29. $|3x| = |3| \cdot |x| = 3|x|$ **30.** $|x^2| = x^2$

31. $\left|\dfrac{x^2}{y}\right| = \dfrac{|x^2|}{|y|} = \dfrac{x^2}{|y|}$

32. $|x^2y^3| = |x^2 \cdot y^2 \cdot y| = |x^2| \cdot |y^2| \cdot |y| = x^2y^2|y|$

33. $|-3x| = |-3| \cdot |x| = 3|x|$

◀

DO EXERCISES 52–59.

● **EXERCISE SET** **1.2**

A Simplify.

1. $2^3 \cdot 2^{-4}$ **2.** $3^4 \cdot 3^{-5}$

3. $b^2 \cdot b^{-2}$ **4.** $c^3 \cdot c^{-3}$

5. $4^2 \cdot 4^{-5} \cdot 4^6$ **6.** $5^2 \cdot 5^{-4} \cdot 5^5$

7. $2x^3 \cdot 3x^2$ **8.** $3y^4 \cdot 4y^3$

9. $(5a^2b)(3a^{-3}b^4)$ **10.** $(4xy^2)(3x^{-4}y^5)$

11. $(2x)^3(3x)^2$ **12.** $(4y)^2(3y)^3$

13. $(6x^5y^{-2}z^3)(-3x^2y^3z^{-2})$

14. $(5x^4y^{-3}z^2)(-2x^2y^4z^{-1})$

15. $\dfrac{b^{40}}{b^{37}}$ **16.** $\dfrac{a^{39}}{a^{32}}$ **17.** $\dfrac{x^2y^{-2}}{x^{-1}y}$

18. $\dfrac{x^3y^{-3}}{x^{-1}y^2}$ **19.** $\dfrac{9a^2}{(-3a)^2}$ **20.** $\dfrac{16y^2}{(-4y)^2}$

21. $\dfrac{24a^5b^3}{8a^4b}$ **22.** $\dfrac{30x^6y^4}{5x^3y^2}$ **23.** $\dfrac{12x^2y^3z^{-2}}{21xy^2z^3}$

24. $\dfrac{15x^3y^4z^{-3}}{45xyz^5}$ **25.** $(2ab^2)^3$ **26.** $(4xy^3)^2$

27. $(-2x^3)^4$ **28.** $(-3x^2)^4$ **29.** $-(2x^3)^4$

30. $-(3x^2)^4$ **31.** $(6a^2b^3c)^2$ **32.** $(5x^3y^2z)^2$

33. $(-5c^{-1}d^{-2})^{-2}$ **34.** $(-4x^{-1}z^{-2})^{-2}$

35. $\dfrac{4^{-2} + 2^{-4}}{8^{-1}}$ **36.** $\dfrac{3^{-2} + 2^{-3}}{7^{-1}}$

37. $\dfrac{(-2)^4 + (-4)^2}{(-1)^8}$ **38.** $\dfrac{(-3)^2 + (-2)^4}{(-1)^6}$

39. $\dfrac{(3a^2b^{-2}c^4)^3}{(2a^{-1}b^2c^{-3})^2}$ **40.** $\dfrac{(2a^3b^{-3}c^3)^3}{(3a^{-1}b^{-3}c^{-5})^2}$

41. $\dfrac{6^{-2}x^{-3}y^2}{3^{-3}x^{-4}y}$ **42.** $\dfrac{5^{-2}x^{-4}y^3}{2^{-3}x^{-5}y}$

43. $\left(\dfrac{24a^{10}b^{-8}c^7}{3a^6b^{-3}c^5}\right)^5$ **44.** $\left(\dfrac{125p^{12}q^{-14}r^{22}}{25p^8q^6r^{-15}}\right)^{-4}$

Find $-x^2$ and $(-x)^2$, when:

45. $x = 5$. **46.** $x = -7$.

47. $x = -1.08$. **48.** $x = \sqrt{3}$.

B Convert to scientific notation.

49. 58,000,000 **50.** 27,000

51. 365,000 **52.** 3645

53. 0.0000027 **54.** 0.0000658

55. 0.027 **56.** 0.0038

57. The mass of an electron is
　0.00000000000000000000000000000911 gram.

58. The distance from the earth to the sun is 93,000,000 miles.

59. The distance from the sun to Pluto is 3,664,000,000 miles.

60. An oxygen atom is about 0.000000001 times the size of a drop of water.

Convert to decimal notation.

61. 4×10^5 **62.** 5×10^{-4}

63. 6.2×10^{-3} **64.** 7.8×10^6

65. 7.69×10^{12} **66.** 8.54×10^{-7}

67. 5.67 E −7 **68.** 1.314 E 12

69. Light travels 9.46×10^{12} kilometers in one year.

70. The wavelength of a certain red light is 6.6×10^{-5}.

71. 7.69 E −8 **72.** 8.603 E −10

73. 2.567 E 8 **74.** 1.113 E 11

Compute. Write scientific notation for the answer.

75. $(3.1 \times 10^5)(4.5 \times 10^{-3})$

76. $(9.1 \times 10^{-17})(8.2 \times 10^3)$

77. $\dfrac{6.4 \times 10^{-7}}{8.0 \times 10^{6}}$

78. $\dfrac{1.1 \times 10^{-40}}{2.0 \times 10^{-71}}$

Solve. Write the answers using scientific notation.

79. The average discharge of water from the mouth of the Amazon river is 4,200,000 cubic feet per second. How much water is discharged in one hour?

80. Americans drink 3 million gallons of orange juice in one day. How much orange juice is consumed by Americans in one year? Use 365 days for 1 year.

81. A *nanosecond* is one billionth of a second. Find scientific notation for 1 nanosecond.

82. The average distance from the earth to the sun is 9.3×10^{7} miles. About how far does the earth travel in a yearly orbit about the sun? (*Hint:* Assume a circular orbit.)

83. How far, in miles, does light travel in 13 weeks?

84. A certain thin plastic sheet is used in many applications of building and landscaping. The sheet is packaged in rolls that are 1 m wide and 30 m long. The thickness of the sheet is 0.8 mm. Find the volume of plastic in a roll.

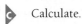 Calculate.

85. $3 \cdot 2 + 4 \cdot 2^{2} - 6(3 - 1)$

86. $3[(2 + 4 \cdot 2^{2}) - 6(3 - 1)]$

87. $\dfrac{4(8 - 6)^{2} + 4 \cdot 3 - 2 \cdot 8}{3^{1} + 19^{0}}$

88. $\dfrac{[4(8 - 6)^{2} + 4](3 - 2 \cdot 8)}{2^{2}(2^{3} + 5)}$

89. $16 \div 4 \cdot 4 \div 2 \cdot 256$

90. $2^{6} \cdot 2^{-3} \div 2^{10} \div 2^{-8}$

91. $\left[\dfrac{5^{2}}{8} + 5(5) - \dfrac{5^{3}}{12}\right] - \left[\dfrac{(-4)^{2}}{8} + 5(-4) - \dfrac{(-4)^{3}}{12}\right]$

92. $\left[\dfrac{2^{2}}{8} + \dfrac{2}{2} - \dfrac{2^{3}}{12}\right] - \left[\dfrac{(-1)^{2}}{8} - \dfrac{1}{2} - \dfrac{(-1)^{3}}{12}\right]$

Simplify.

93. $|12|$ **94.** $|-2.56|$ **95.** $|-47|$

96. $|0|$ **97.** $|-7a|$ **98.** $|-10mn|$

99. $|-8x^{6}|$ **100.** $|5x^{4}y^{8}|$ **101.** $|9xy|$

102. $|y^{4}|$ **103.** $|3a^{2}b|$ **104.** $\left|\dfrac{4a}{b^{2}}\right|$

● **SYNTHESIS**

Simplify. Assume that all exponents are integers.

105. $(x^{t} \cdot x^{3t})^{2}$ **106.** $(x^{y} \cdot x^{-y})^{3}$

107. $(t^{a+x} \cdot t^{x-a})^{4}$ **108.** $(m^{x-y} \cdot m^{3y-x})^{t}$

109. $(x^{a}y^{b} \cdot x^{b}y^{a})^{c}$

110. $(m^{x-b} \cdot n^{x+b})^{x}(m^{b}n^{-b})^{x}$

111. $\left[\dfrac{(3x^{a}y^{b})^{3}}{(-3x^{a}y^{b})^{2}}\right]^{2}$

112. $\left[\left(\dfrac{x^{r}}{y^{t}}\right)^{2}\left(\dfrac{x^{2r}}{y^{4t}}\right)^{-2}\right]^{-3}$

The formula

$$M = P\left[\dfrac{\dfrac{i}{12}\left(1 + \dfrac{i}{12}\right)^{n}}{\left(1 + \dfrac{i}{12}\right)^{n} - 1}\right]$$

gives the monthly mortgage payment M on a home loan of P dollars at interest rate i, where n is the total number of payments (12 times the number of years).

113. ▦ The cost of a house is \$92,000. The down payment is \$14,000, the interest rate is $10\frac{3}{4}\%$, and the loan period is 25 years. What is the monthly payment?

114. ▦ Repeat Exercise 113 for loan periods of 20 years and 30 years.

Find the error(s) in each of the following. Explain why each is an error. Then find the correct answer.

115. $x^{4}(x^{3})^{2} = x^{9}$

116. $\dfrac{x^{4}y^{-7}}{x^{-2}y^{5}} = \dfrac{x^{2}}{y^{2}}$

117. $(2x^{-4}y^{6}z^{3})^{3} = 6x^{-1}y^{3}z^{6}$

1.3

ADDITION AND SUBTRACTION OF ALGEBRAIC EXPRESSIONS

▲ Polynomials

Expressions like the following are called **polynomials in one variable**:

$$-7x + 5, \qquad 3y^{3} - 5y^{2} + 7y - 4, \qquad 0, \qquad -5t^{4}, \qquad x^{5} - 9.$$

A **variable** is a symbol that can represent different numbers. Letters are generally used for variables. For example, if a is used to represent your age,

OBJECTIVES

You should be able to:

▲ Determine the degree of each term of a polynomial and the degree of the polynomial.

� B Add polynomials and other algebraic expressions.

▶ C Find the opposite of an algebraic expression.

▶ D Subtract polynomials and other algebraic expressions.

then a is a variable. Letters used to represent numbers are not always variables, however. For example, if we choose to represent the distance to the moon by the letter d, then in that context d is not a variable, but a **constant**.

DEFINITION   **Polynomial in One Variable**

A *polynomial in one variable* is any expression of the type

$$a_n x^n + a_{n-1} x^{n-1} + \cdots + a_2 x^2 + a_1 x + a_0,$$

where n is a nonnegative integer and $a_n, \ldots, a_0$ are real numbers, called *coefficients*.

Some or all of the coefficients of a polynomial may be 0. Each of the parts separated by plus signs is called a **term.**

The following are *not* polynomials:

$$(1)\ \ x^2 + 3x + \frac{2}{x}; \qquad (2)\ \ 2x + \sqrt{x}; \qquad (3)\ \ \frac{x^3 + 4}{x - 7}.$$

Expression (1) is not a polynomial because $2/x = 2x^{-1}$ and -1 is a negative integer. Expression (2) is not a polynomial because $\sqrt{x} = x^{1/2}$ and $\frac{1}{2}$ is not an integer. Expression (3) is not a polynomial because it represents a division that cannot be expressed as an equivalent polynomial.

The question might arise whether an expression such as

$$8x^3 - 6x^2 + 7x - 5$$

is a polynomial. It is indeed, because it is equivalent to

$$8x^3 + (-6x^2) + 7x + (-5).$$

Note that coefficients can be negative.

Expressions like the following are called **polynomials in several variables:**

$$5x^2 y^3 + 17 x^2 y - 2, \qquad 14 a^2 b, \qquad \pi r^2 + 2\pi rh.$$

Example 1   Find the terms and the coefficients of the polynomial

$$5x^3 y - 7xy^2 + 2 + x^4 y.$$

Solution   The terms are

$$5x^3 y, \qquad -7xy^2, \qquad 2, \quad \text{and} \quad x^4 y.$$

The coefficients of the terms are 5, -7, 2, and 1.   ◀

The **degree of a term** is the sum of the exponents of the variables. For example, the degree of the term $-10x^3 y$ is 4. The degree of a nonzero constant term, such as the number -10 by itself, is 0. We can think of -10 as $-10x^0$. The **degree of a nonzero polynomial** is the degree of the term of highest degree.

The polynomial consisting only of the number 0 is a special case. Mathematicians agree that it has *no* degree either as a term or as a polynomial. This is because we can express 0 as $0 = 0x^5 = 0x^{13}$, and so on, using any exponent we wish.

Example 2 In the polynomial $5x^3y - 7xy^2 + 2$, find the degrees of the terms and the degree of the polynomial.

Solution   The degrees of the terms are 4, 3, and 0. The polynomial is of degree 4.   ◀

A polynomial with just one term is called a **monomial.** If there are just two terms, a polynomial is called a **binomial.** If there are just three terms, it is called a **trinomial.** A polynomial in one variable of second degree is also called a **quadratic polynomial.**

DO EXERCISES 1 AND 2. _____

 Addition

If two terms of an expression have the same variables raised to the same powers, then the terms are called **like,** or **similar.** Similar terms can be "combined" using the distributive laws.

Examples

3. $3x^2 - 4y + 2x^2 = 3x^2 + 2x^2 - 4y$ Rearranging using the commutative and the associative laws

$\qquad\qquad\qquad\quad = (3 + 2)x^2 - 4y$ Using a distributive law

$\qquad\qquad\qquad\quad = 5x^2 - 4y$

4. $4x^{1/2}y + 7x^{1/2}y = 11x^{1/2}y$

5. $-2x^2\sqrt{y^3} + 5x^2\sqrt{y^3} = 3x^2\sqrt{y^3}$ ◀

DO EXERCISES 3–5. _____

We can find the sum of two polynomials by writing a plus sign between them and then combining similar terms. Ordinarily, this can be done mentally.

Example 6 Add $-3x^3 + 2x - 4$ and $4x^3 + 3x^2 + 2$.

Solution

$\qquad (-3x^3 + 2x - 4) + (4x^3 + 3x^2 + 2) = x^3 + 3x^2 + 2x - 2$ ◀

DO EXERCISES 6 AND 7. _____

 Opposites

We can find the opposite, or additive inverse, of an expression by placing an opposite sign before the expression. Another equivalent expression is then found by removing the parentheses.

Example 7 Find two equivalent expressions for the opposite of $-3xy^2 + 4x^2y - 5x - 3$.

Solution   One expression is

$\qquad -(-3xy^2 + 4x^2y - 5x - 3).$

Determine the degree of each term and the degree of the polynomial.

1. $x^8 - 7x^6 + 2x^4 - 3x^9 + 2$

2. $8y^4 - 7xy^3 + 6x^2y^3 - 9x^5y - 1$

Combine similar terms.

3. $5x^3y^2 - 2x^2y^3 + 4x^3y^2$

4. $3xy^2 - 4x^2y + 4xy^2 + 2x^2y$

5. $5x^4\sqrt{y} - 2x^4\sqrt{y} + 2$

Add.

6. $(3x^3 + 4x^2 - 7x - 2) +$
    $(-7x^3 - 2x^2 + 3x + \frac{1}{2})$

7. $(5p^2q^4 - 2p^2q^2 - 3q) +$
    $(-6pq^2 + 3p^2q^2 + 5)$

Find two equivalent expressions for the opposite of the expression.

8. $5x^2t^2 - 4xy^2t - 3xt + 6x - 5$

9. $-3x^2y + 5xy - 7x + 4y + 2$

Subtract.

10. $(5xy^4 - 7xy^2 + 4x^2 - 3) - (-3xy^4 + 2xy^2 - 2y + 4)$

11. $(5x^2y - 7x^3y^2 - x^2y^2 + 4y) - (2x^2y + 2x^3y^2 - 5x^2y^3 - 5y)$

Another equivalent expression is

$$3xy^2 - 4x^2y + 5x + 3.$$

◀

Example 8   Find two equivalent expressions for the opposite of $7ab^2 - 6ab - 4b + 8$.

◀

Solution   One expression is

$$-(7ab^2 - 6ab - 4b + 8).$$

Another equivalent expression is

$$-7ab^2 + 6ab + 4b - 8.$$

◀

The preceding examples may bring to mind the following rule.

> To remove parentheses preceded by an opposite sign, change the sign of every term inside the parentheses.

DO EXERCISES 8 AND 9.

▷   **Subtraction**

We know by Theorem 2 that we can subtract by adding the opposite of the number being subtracted. Thus to subtract one polynomial or other algebraic expression from another, we add the opposite of the expression being subtracted. We change the sign of each term of the expression to be subtracted and then add.

Example 9   Subtract:

$$(-9x^5 - x^3 + 2x^2 + 4) - (2x^5 - x^4 + 4x^3 - 3x^2).$$

Solution

$(-9x^5 - x^3 + 2x^2 + 4) - (2x^5 - x^4 + 4x^3 - 3x^2)$

$\quad = (-9x^5 - x^3 + 2x^2 + 4) + [-(2x^5 - x^4 + 4x^3 - 3x^2)]$ Adding the opposite

$\quad = (-9x^5 - x^3 + 2x^2 + 4) + (-2x^5 + x^4 - 4x^3 + 3x^2)$

$\quad = -11x^5 + x^4 - 5x^3 + 5x^2 + 4$

◀

Example 10   Subtract:

$$(4x^2y - 6x^3y^2 + x^2y^2 - 5y) - (4x^2y + x^3y^2 + 3x^2y^3 + 6y).$$

Solution

$(4x^2y - 6x^3y^2 + x^2y^2 - 5y) - (4x^2y + x^3y^2 + 3x^2y^3 + 6y)$

$\quad = -7x^3y^2 - 3x^2y^3 + x^2y^2 - 11y.$

◀

DO EXERCISES 10 AND 11.

● **EXERCISE SET** 1.3

Ⓐ Determine the degree of each term and the degree of the polynomial.

1. $-11x^4 - x^3 + x^2 + 3x - 9$
2. $t^3 - 3t^2 + t + 1$
3. $y^3 + 2y^6 + x^2y^4 - 8$
4. $u^2 + 3v^5 - u^3v^4 - 7$
5. $a^5 + 4a^2b^4 + 6ab + 4a - 3$
6. $8p^6 + 2p^4t^4 - 7p^3t + 5p^2 - 14$

Ⓑ Add.

7. $(5x^2y - 2xy^2 + 3xy - 5) + (-2x^2y - 3xy^2 + 4xy + 7)$
8. $(6x^2y - 3xy^2 + 5xy - 3) + (-4x^2y - 4xy^2 + 3xy + 8)$
9. $(-3pq^2 - 5p^2q + 4pq + 3) + (-7pq^2 + 3pq - 4p + 2q)$
10. $(-5pq^2 - 3p^2q + 6pq + 5) + (-4pq^2 + 5pq - 6p + 4q)$
11. $(2x + 3y + z - 7) + (4x - 2y - z + 8) + (-3x + y - 2z - 4)$
12. $(2x^2 + 12xy - 11) + (6x^2 - 2x + 4) + (-x^2 - y - 2)$
13. $(7x\sqrt{y} - 3y\sqrt{x} + \frac{1}{5}) + (-2x\sqrt{y} - y\sqrt{x} - \frac{3}{5})$
14. $(10x\sqrt{y} - 4y\sqrt{x} + \frac{4}{3}) + (-3x\sqrt{y} - y\sqrt{x} - \frac{1}{3})$

Ⓒ Find two equivalent expressions for the opposite of the expression.

15. $5x^3 - 7x^2 + 3x - 6$ **16.** $-4y^4 + 7y^2 - 2y - 1$

Ⓓ Subtract.

17. $(3x^2 - 2x - x^3 + 2) - (5x^2 - 8x - x^3 + 4)$
18. $(5x^2 + 4xy - 3y^2 + 2) - (9x^2 - 4xy + 2y^2 - 1)$
19. $(4a - 2b - c + 3d) - (-2a + 3b + c - d)$
20. $(5a - 3b - c + 4d) - (-3a + 5b + c - 2d)$
21. $(x^4 - 3x^2 + 4x) - (3x^3 + x^2 - 5x + 3)$
22. $(2x^4 - 5x^2 + 7x) - (5x^3 + 2x^2 - 3x + 5)$
23. $(7x\sqrt{y} - 4y\sqrt{x} + 7.5) - (-2x\sqrt{y} - y\sqrt{x} - 1.6)$
24. $(10x\sqrt{y} - 4y\sqrt{x} + \frac{4}{3}) - (-3x\sqrt{y} + y\sqrt{x} - \frac{1}{3})$

● **SYNTHESIS** _____

Simplify.

25. 🖩 $(0.565p^2q - 2.167pq^2 + 16.02pq - 17.1) + (-1.612p^2q - 0.312pq^2 - 7.141pq - 87.044)$
26. 🖩 $(5003.2xy^{-2} + 3102.4\sqrt{xy} - 5280) - (2143.6xy^{-2} + 6153.8xy - 4141\sqrt{xy} + 4979.12)$

1.4

MULTIPLICATION OF ALGEBRAIC EXPRESSIONS

Ⓐ **Multiplication of Any Two Polynomials**

Multiplication of polynomials is based on the distributive laws. For example,

$$(x - 3)(x + y + 5)$$

$$= (x - 3)x + (x - 3)y + (x - 3)5 \quad \text{Using the distributive law}$$
$$= x \cdot x - 3x + xy - 3y + 5x - 3 \cdot 5$$
$$= x^2 + 2x + xy - 3y - 15.$$

What we have done is to multiply each term of one polynomial by every term of the other and then add the results. We can also do such a multiplication using columns, as follows.

Example 1 Multiply: $(4x^4y - 7x^2y + 3y)(2y - 3x^2y)$.

OBJECTIVE

You should be able to:

Ⓐ Multiply any two polynomials, striving for speed and accuracy. Whenever possible, you should write only the answer. In particular, you should be able to multiply any two binomials, square a binomial, multiply the sum and difference of the same two expressions, and cube a binomial.

Multiply.

1. $(3x^2y - 2xy + 3y)(xy + 2y)$

2. $(p^2q + 2pq + 2q)(2p^2q - pq + q)$

Multiply.

3. $(2xy + 3x)(x^2 - 2)$

4. $(3x - 2y)(5x + 3y)$

5. $(2x + \sqrt{2})(3y - \sqrt{2})$

Solution

$$
\begin{array}{r}
4x^4y - 7x^2y + 3y \\
2y - 3x^2y \\
\hline
-12x^6y^2 + 21x^4y^2 - 9x^2y^2 \\
8x^4y^2 - 14x^2y^2 + 6y^2 \\
\hline
-12x^6y^2 + 29x^4y^2 - 23x^2y^2 + 6y^2 \\
\end{array}
$$

Multiplying by $-3x^2y$
Multiplying by $2y$
Adding ◀

DO EXERCISES 1 AND 2. _____

The following methods allow us to multiply binomials more efficiently.

Products of Two Binomials

We can find a product of two binomials mentally. We multiply the First terms, then the Outside terms, then the Inside terms, then the Last terms (this procedure is sometimes abbreviated **FOIL**), and then add the results. This procedure also works for multiplying algebraic expressions that are not polynomials.

Examples   Multiply.

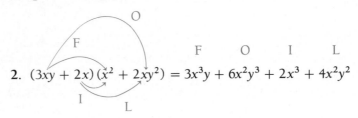

2. $(3xy + 2x)(x^2 + 2xy^2) = 3x^3y + 6x^2y^3 + 2x^3 + 4x^2y^2$

3. $(x + \sqrt{2})(y - \sqrt{2}) = xy - \sqrt{2}x + \sqrt{2}y - 2$
4. $(2x - \sqrt{3})(y + 2) = 2xy + 4x - \sqrt{3}y - 2\sqrt{3}$
5. $(2x + 3y)(x - 4y) = 2x^2 - 5xy - 12y^2$ ◀

DO EXERCISES 3–5. _____

Squares of Binomials

Using FOIL to multiply a binomial $A + B$ by itself, we obtain the following:

$$(A + B)^2 = A^2 + AB + BA + B^2 = A^2 + 2AB + B^2$$

and

$$(A - B)^2 = A^2 - 2AB + B^2.$$

This gives us a way to square a binomial that is faster than FOIL. We square the first term, add twice the product of the terms, and then add the square of the second term.

Examples   Multiply.

6. $(2x + 9y^2)^2 = (2x)^2 + 2(2x)(9y^2) + (9y^2)^2$
$$= 4x^2 + 36xy^2 + 81y^4$$

7. $(3x^2 - 5xy^2)^2$

$\quad = (3x^2)^2 + 2(3x^2)(-5xy^2) + (-5xy^2)^2$

The second term is $-5xy^2$, so twice the product of the terms is $2(3x^2)(-5xy^2)$.

$\quad = 9x^4 - 30x^3y^2 + 25x^2y^4$ ◀

CAUTION! The square of a sum is *not* the sum of the squares; that is,

$\quad (A + B)^2 \neq A^2 + B^2.$

DO EXERCISES 6 AND 7.

Products of Sums and Differences

We can also use FOIL to find the product of a sum and a difference of the same two expressions:

$$(A + B)(A - B) = A^2 - AB + AB - B^2$$
$$= A^2 - B^2.$$

The product of a sum and a difference of the same two terms is the difference of their squares. Thus to find such a product, we square the first term, square the second term, and then write a minus sign between the results:

$$(A + B)(A - B) = A^2 - B^2.$$

Examples Multiply.

8. $(y + 5)(y - 5) = y^2 - 5^2$
$$\qquad\qquad\qquad = y^2 - 25$$

9. $(3x + 2)(3x - 2) = (3x)^2 - 2^2$
$$\qquad\qquad\qquad\quad = 9x^2 - 4$$

10. $(2xy^2 + 3x)(2xy^2 - 3x) = (2xy^2)^2 - (3x)^2$
$$\qquad\qquad\qquad\qquad\qquad = 4x^2y^4 - 9x^2$$

11. $(5x + \sqrt{2})(5x - \sqrt{2}) = (5x)^2 - (\sqrt{2})^2$
$$\qquad\qquad\qquad\qquad\quad = 25x^2 - 2$$

12. $(\boxed{5y + 4} + 3x)(\boxed{5y + 4} - 3x) = (\boxed{5y + 4})^2 - (3x)^2$
$$\qquad\qquad\qquad\qquad\qquad\qquad = 25y^2 + 40y + 16 - 9x^2$$

13. $(3xy^2 + 4y)(-3xy^2 + 4y) = (4y + 3xy^2)(4y - 3xy^2)$
$$\qquad\qquad\qquad\qquad\qquad\quad = (4y)^2 - (3xy^2)^2$$
$$\qquad\qquad\qquad\qquad\qquad\quad = 16y^2 - 9x^2y^4$$ ◀

DO EXERCISES 8–12.

Multiply.

6. $(4x - 5y)^2$

7. $(2y^2 + 6x^2y)^2$

Multiply.

8. $(4x + 7)(4x - 7)$

9. $(5x^2y + 2y)(5x^2y - 2y)$

10. $(4y^2 + \sqrt{3})(4y^2 - \sqrt{3})$

11. $(2x + 3 + 5y)(2x + 3 - 5y)$

12. $(-2x^3y^2 + 5t)(2x^3y^2 + 5t)$

Multiply.

13. $(x + 1)^3$

14. $(x - 1)^3$

15. $(t^2 - 3b)^3$

16. $(2a^3 - 5b^2)^3$

Cubing Binomials

The following multiplication gives another important result:

$$
\begin{aligned}
(A + B)^3 &= (A + B)(A + B)^2 \\
&= (A + B)(A^2 + 2AB + B^2) \\
&= (A + B)A^2 + (A + B)2AB + (A + B)B^2 \\
&= A^3 + A^2B + 2A^2B + 2AB^2 + AB^2 + B^3 \\
&= A^3 + 3A^2B + 3AB^2 + B^3.
\end{aligned}
$$

The result to be remembered is as follows:

$$(A + B)^3 = A^3 + 3A^2B + 3AB^2 + B^3.$$

Examples   Multiply.

14. $(x + 2)^3 = x^3 + 3x^2(2) + 3x(2)^2 + 2^3$
$$= x^3 + 6x^2 + 12x + 8$$

15. $(x - 2)^3 = [x + (-2)]^3$
$$= x^3 + 3x^2(-2) + 3x(-2)^2 + (-2)^3$$
$$= x^3 - 6x^2 + 12x - 8$$

16. $(5m^2 - 4n^3)^3 = (5m^2)^3 + 3(5m^2)^2(-4n^3)$
$$+ 3(5m^2)(-4n^3)^2 + (-4n^3)^3$$
$$= 125m^6 - 300m^4n^3 + 240m^2n^6 - 64n^9$$

Note in Examples 15 and 16 that a separate formula for $(A - B)^3$ need not be memorized. We can think of $(A - B)^3$ as $[A + (-B)]^3$.

DO EXERCISES 13–16.

> ### Summary of Rules for Multiplication of Polynomials
>
> $$(A + B)(C + D) = AC + AD + BC + BD,$$
> $$(A + B)^2 = A^2 + 2AB + B^2,$$
> $$(A - B)^2 = A^2 - 2AB + B^2,$$
> $$(A + B)(A - B) = A^2 - B^2,$$
> $$(A + B)^3 = A^3 + 3A^2B + 3AB^2 + B^3$$

 EXERCISE SET **1.4**

A   Multiply.

1. $(2x^2 + 4x + 16)(3x - 4)$

2. $(3y^2 - 3y + 9)(2y + 3)$

3. $(4a^2b - 2ab + 3b^2)(ab - 2b + 1)$

4. $(2x^2 + y^2 - 2xy)(x^2 - 2y^2 - xy)$

5. $(a - b)(a^2 + ab + b^2)$ **6.** $(t + 1)(t^2 - t + 1)$

7. $(2x + 3y)(2x + y)$ **8.** $(2a - 3b)(2a - b)$

9. $(4x^2 - \frac{1}{2}y)(3x + \frac{1}{4}y)$ **10.** $(2y^3 + \frac{1}{5}x)(3y - \frac{1}{4}x)$

11. $(2p^2q^3 - r^2)(5pq - 2r)$ **12.** $(3y^2 - 2)(3y - x)$

13. $(2x + 3y)^2$ **14.** $(5x + 2y)^2$

15. $(2x^2 - 3y)^2$ **16.** $(4x^2 - 5y)^2$

17. $(2x^3 + 3y^2)^2$ **18.** $(5x^3 + 2y^2)^2$

19. $\left(\frac{1}{2}x^2 - \frac{3}{5}y\right)^2$

20. $\left(\frac{1}{4}x^2 - \frac{2}{3}y\right)^2$

21. $(0.5x + 0.7y^2)^2$

22. $(0.3x + 0.8y^2)^2$

23. $(3x - 2y)(3x + 2y)$

24. $(3x + 5y)(3x - 5y)$

25. $(x^2 + yz)(x^2 - yz)$

26. $(2x^2 + 5xy)(2x^2 - 5xy)$

27. $(3x^2 - \sqrt{2})(3x^2 + \sqrt{2})$

28. $(5x^2 - \sqrt{3})(5x^2 + \sqrt{3})$

29. $(2x + 3y + 4)(2x + 3y - 4)$

30. $(5x + 2y + 3)(5x + 2y - 3)$

31. $(x^2 + 3y + y^2)(x^2 + 3y - y^2)$

32. $(2x^2 + y + y^2)(2x^2 + y - y^2)$

33. $(x + 1)(x - 1)(x^2 + 1)$

34. $(y - 2)(y + 2)(y^2 + 4)$

35. $(2x + y)(2x - y)(4x^2 + y^2)$

36. $(5x + y)(5x - y)(25x^2 + y^2)$

37. ▦ $(0.051x + 0.04y)^2$

38. ▦ $(1.032x - 2.512y)^2$

39. ▦ $(37.86x + 1.42)(65.03x - 27.4)$

40. ▦ $(3.601x - 17.5)(47.105x + 31.23)$

41. $(y + 5)^3$

42. $(t - 7)^3$

43. $(m^2 - 2n)^3$

44. $(3t^2 + 4)^3$

45. $(\sqrt{2}x^2 - y^2)(\sqrt{2}x - 2y)$

46. $(\sqrt{3}y^2 - 2)(\sqrt{3}y - x)$

● SYNTHESIS _____

Multiply. Assume that all exponents are natural numbers.

47. $(a^n + b^n)(a^n - b^n)$

48. $(t^a + 4)(t^a - 7)$

49. $(x^m - t^n)^3$

50. $y^3 z^n (y^{3n} z^3 - 4yz^{2n})$

51. $(x - 1)(x^2 + x + 1)(x^3 + 1)$

52. $(a^n + b^n)^2$

53. $[(2x - 1)^2 - 1]^2$

54. $[(a + b)(a - b)][5 - (a + b)][5 + (a + b)]$

55. $(x^{a-b})^{a+b}$

56. $(t^{m+n})^{m+n} \cdot (t^{m-n})^{m-n}$

57. $(a + b + c)^2$

58. $(a + b + c)^3$

59. $(a + b)^4$

60. $(x - y)(x^4 + x^3 y + x^2 y^2 + xy^3 + y^4)$

61. $(m + t)(m^4 - m^3 t + m^2 t^2 - mt^3 + t^4)$

62. $(a - b)(a^7 + a^6 b + a^5 b^2 + a^4 b^3 + a^3 b^4 + a^2 b^5 + ab^6 + b^7)$

Find the error(s) in each of the following. Explain why each is an error. Then find the correct answer.

63. $(3a + b)^2 = 3a^2 + b^2$

64. $(2x - 3y)(2x - 3y) = 4x^2 - 9y^2$

65. $2x(x + 3) + 4(x^2 - 3) = 2x^2 + 3x + 4x^2 - 3$ (1)
$= 6x^2 + x$ (2)

66. $(2a - 3b)(3a + 2b) = 6a^2 - 6b^2$ (1)
$= a^2 - b^2$ (2)

● CHALLENGE _____

67. Multiply. Assume that n is a natural number. (*Hint:* See Exercises 60 and 62.)

$$(x - y)(x^{n-1} + x^{n-2}y + x^{n-3}y^2 + \cdots + x^3 y^{n-4} + x^2 y^{n-3} + xy^{n-2} + y^{n-1})$$

1.5
FACTORING

◢ Factoring Polynomials

To **factor** a polynomial, we do the reverse of multiplying; that is, we find an equivalent expression that is a product. Factoring is an important algebraic skill.

Terms with Common Factors

When an expression is to be factored, we should always look first for a possible factor that is common to all terms. We then "factor it out" using the distributive laws. We usually look for a constant with the largest absolute value and variables with the largest exponent.

OBJECTIVE

You should be able to:

◢ Determine the kind of factoring to try when an expression is to be factored. Then factor expressions: by removing a common factor, that are differences of squares, that are trinomials, that are trinomial squares, or that are sums or differences of cubes.

Factor.

1. $20x^3y + 12x^2y$

2. $(p + q)(x + 2) + (p + q)(x + y)$

3. $4x^3 + 20x^2 - 3x - 15$

Factor.

4. $x^2 - 16$

5. $25y^4 - 16x^2$

6. $2y^4 - 32x^4$

7. $x^2 - 3$

Example 1   Factor: $4x^2 + 8$.

Solution

$$4x^2 + 8 = 4 \cdot x^2 + 4 \cdot 2 = 4(x^2 + 2)$$

Note that $4x^2$ and 8 are *terms* of the expression. The number 4 is a common factor of each term, so we factor it out. The expression $4 \cdot x^2 + 4 \cdot 2$ is not a correct answer. Although each term is factored, the entire expression has not been factored, that is, expressed as a product. The expression $4(x^2 + 2)$ is an equivalent expression that is a product.   ◀

Examples   Factor.

2. $12x^2y - 20x^3y = 4x^2y \cdot 3 - 4x^2y \cdot 5x = 4x^2y(3 - 5x)$
3. $7x\sqrt{y} + 14x^2\sqrt{y} - 21\sqrt{y} = 7\sqrt{y}(x + 2x^2 - 3)$
4. $(a - b)(x + 5) + (a - b)(x - y^2) = (a - b)[(x + 5) + (x - y^2)]$
$$= (a - b)(2x + 5 - y^2) \quad ◀$$

In some polynomials, pairs of terms have a common factor that can be removed, as in the following examples. This process is called **factoring by grouping,** and uses the distributive laws repeatedly.

Examples   Factor.

5. $y^3 + 3y^2 - 5y - 15 = y^2(y + 3) - 5(y + 3)$
$$= (y + 3)(y^2 - 5)$$
6. $ax^2 + ay + bx^2 + by = a(x^2 + y) + b(x^2 + y)$
$$= (a + b)(x^2 + y) \quad ◀$$

DO EXERCISES 1–3. _____

Differences of Squares

Recall that $(A + B)(A - B) = A^2 - B^2$. We can use this equation in reverse to factor an expression that is a *difference of two squares.*

Examples   Factor.

7. $x^2 - 9 = (x + 3)(x - 3)$
8. $y^2 - 2 = (y + \sqrt{2})(y - \sqrt{2})$
9. $9a^2 - 16x^4 = (3a)^2 - (4x^2)^2 = (3a + 4x^2)(3a - 4x^2)$
10. $9y^4 - 9x^4 = 9(y^4 - x^4)$ Remove the common factor first.
$$= 9(y^2 + x^2)(y^2 - x^2)$$
$$= 9(y^2 + x^2)(y + x)(y - x) \quad ◀$$

DO EXERCISES 4–7. _____

Factoring Trinomials

Some trinomials can be factored into two binomials. To do this, we factor by trial and error and check by multiplying using the **FOIL** equation.

Example 11   Factor: $x^2 + 7x + 12$.

Solution   We look for factors of 12 whose sum is 7. Since the constant term 12 is positive, its factors are either both negative or both positive. We want the sum of the factors to be 7, which is positive, so we consider only the positive factors of 12. By trial, we determine the factors to be 3 and 4 and the factorization to be

$$(x + 4)(x + 3).$$ ◄

Example 12   Factor: $x^4 + 3x^2 - 10$.

Solution   We can think of this polynomial mentally as $u^2 + 3u - 10$, where we have mentally substituted u for x^2. The constant term is negative this time, so any pair of factors must have one positive number and one negative number. The coefficient of the middle term is positive, so we look for pairs of factors for which the positive number has the larger absolute value. By trial, we determine the factors to be 5 and -2 and the factorization to be

$$u^2 + 3u - 10 = (u + 5)(u - 2).$$

Then substituting x^2 for u, we obtain the factorization of the original trinomial:

$$(x^2 + 5)(x^2 - 2).$$ ◄

Example 13   Factor: $3x^2 - 10x - 8$.

Solution

  Method 1.   We look for binomials $ax + b$ and $cx + d$ for which the product of the first terms is $3x^2$. The product of the last terms must be -8. When we multiply the inside terms, then the outside terms, and add, we must have $-10x$. By trial, we determine the factorization to be

$$(3x + 2)(x - 4).$$

  Method 2.   We multiply the leading coefficient 3 and the constant -8: $3(-8) = -24$. Then we try to factor -24 so that the sum of the factors is -10. By trial, we find these factors to be -12 and 2. We then write the middle term $-10x$ as a sum using -12 and 2. That is, we split the middle term as follows:

$$-10x = -12x + 2x.$$

Now we factor by grouping:

$$\begin{aligned}
3x^2 - 10x - 8 &= 3x^2 - 12x + 2x - 8 \\
&= 3x(x - 4) + 2(x - 4) \\
&= (3x + 2)(x - 4).
\end{aligned}$$ ◄

> **CAUTION!**   Keep in mind that any factoring you do can and should be checked by multiplying. It is an easy check and can help you avoid many mistakes.

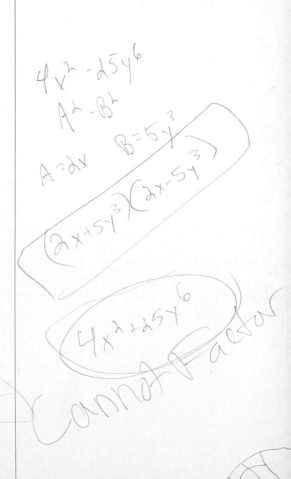

Factor.

8. $x^2 + 6x + 5$

9. $x^2 + 5x - 14$

10. $w^4 - 7w^2 + 10$

11. $3x^2 + 5x + 2$

12. $6x^4y^6 - 9x^2y^3 - 60$

13. $t^2 - t + 5$

Factor.

14. $9y^2 - 30y + 25$

15. $16x^2 + 72xy + 81y^2$

16. $-12x^4y^2 + 60x^2y^5 - 75y^8$

Not all polynomials can be factored into polynomials with integer, rational, or real coefficients. An example is $x^2 - x + 7$. There are no real factors of 7 whose sum is -1. In such a case, we say that the polynomial is "not factorable."

DO EXERCISES 8–13.

Trinomial Squares

Certain trinomials are squares of binomials. You can use trial and error to factor such trinomials, but it is more efficient to make use of the following rules, which reverse the rules for squaring binomials. You should recall that

$$A^2 + 2AB + B^2 = (A + B)^2 \quad \text{and} \quad A^2 - 2AB + B^2 = (A - B)^2.$$

We can use these equations to factor trinomials that are squares. To factor a trinomial, you should check to see if it is a square. For this to be the case, two of the terms must be squares and the other term must be twice the product of the square roots, or the opposite of that product.

Examples Factor.

14. $x^2 - 10x + 25 = (x - 5)^2$

15. $16y^2 + 56y + 49 = (4y + 7)^2$

16. $-4y^2 - 144y^8 + 48y^5$

$= -4y^2(1 + 36y^6 - 12y^3)$ We first removed the common factor.

$= -4y^2(1 - 12y^3 + 36y^6)$

$= -4y^2(1 - 6y^3)^2$ ◀

DO EXERCISES 14–16.

Sums or Differences of Cubes

We can use the following equations to factor a sum or a difference of two cubes:

$$A^3 + B^3 = (A + B)(A^2 - AB + B^2),$$
$$A^3 - B^3 = (A - B)(A^2 + AB + B^2).$$

Check them by multiplying the right-hand sides.

Example 17 Factor: $x^3 - 27$.

Solution   We have $x^3 - 27 = x^3 - 3^3$. In one set of parentheses, we write the cube root of the first expression x^3, then we write a minus sign, and then the cube root of the second expression 27. This gives us $x - 3$.

$$(x - 3)(\qquad\qquad)$$

To get the next factor, we think of $x - 3$ and do the following.

1. Square the first expression: x^2.
2. Multiply the expressions and then change the sign: $3x$.
3. Square the second expression: 9.

$$(x - 3)(x^2 + 3x + 9)$$

Note: We cannot factor $x^2 + 3x + 9$ as a product of polynomials with real coefficients. (It is not a trinomial square nor can it be factored by trial and error.) ◀

DO EXERCISES 17 AND 18.

Example 18   Factor: $125x^3 + y^3$.

Solution   We have $125x^3 + y^3 = (5x)^3 + y^3$. In one set of parentheses, we write the cube root of the first expression, then a plus sign, and then the cube root of the second expression.

$(5x + y) ($ $)$

To get the next factor, we think of $5x + y$ and do the following.

1. Square the first expression: $(5x)^2$ or $25x^2$.

2. Multiply the expressions and then change the sign: $-5xy$.

3. Square the second expression: y^2.

$(5x + y)(25x^2 - 5xy + y^2)$ ◀

DO EXERCISES 19 AND 20.

Example 19   Factor: $16x^7y + 54xy^7$.

Solution   We first look for a common factor.

$$
\begin{aligned}
16x^7y + 54xy^7 &= 2xy(8x^6 + 27y^6) \\
&= 2xy[(2x^2)^3 + (3y^2)^3] \\
&= 2xy(2x^2 + 3y^2)(4x^4 - 6x^2y^2 + 9y^4)
\end{aligned}
$$ ◀

DO EXERCISE 21.

Example 20   Factor: $a^6 - b^6$.

Solution   We can express this polynomial as a difference of squares:

$(a^3)^2 - (b^3)^2$.

We factor as follows:

$(a^3 + b^3)(a^3 - b^3)$.

One factor is a sum of cubes, and the other is a difference of cubes. We factor them:

$(a + b)(a^2 - ab + b^2)(a - b)(a^2 + ab + b^2)$.

The factoring is complete. ◀

In Example 20, had we thought of factoring first as a difference of cubes, we would have had

$$
\begin{aligned}
(a^2)^3 - (b^2)^3 &= (a^2 - b^2)(a^4 + a^2b^2 + b^4) \\
&= (a + b)(a - b)(a^4 + a^2b^2 + b^4).
\end{aligned}
$$

Factor.

17. $x^3 - 8$

18. $64 - t^3$

Factor.

19. $27x^3 + y^3$

20. $8m^3 + 125t^3$

21. Factor: $128y^7 - 250x^6y$.

22. Factor: $p^6 - 64$.

The last factor, $a^4 + a^2b^2 + b^4$, can be factored further as

$$(a^2 - ab + b^2)(a^2 + ab + b^2),$$

but it is in a form that is not easily factored.

DO EXERCISE 22.

Remember the following about factoring:

Trinomial squares:	$A^2 + 2AB + B^2 = (A + B)^2$,
	$A^2 - 2AB + B^2 = (A - B)^2$
Sum of cubes:	$A^3 + B^3 = (A + B)(A^2 - AB + B^2)$
Difference of cubes:	$A^3 - B^3 = (A - B)(A^2 + AB + B^2)$
Difference of squares:	$A^2 - B^2 = (A + B)(A - B)$
Sum of squares:	$A^2 + B^2$ cannot be factored using real-number coefficients.

● EXERCISE SET 1.5

A   Factor.

1. $p^2 + 6p + 8$

2. $w^2 - 7w + 10$

3. $2n^2 + 9n - 56$

4. $3y^2 + 7y - 20$

5. $y^4 - 4y^2 - 21$

6. $m^4 - m^2 - 90$

7. $18a^2b - 15ab^2$

8. $4x^2y + 12xy^2$

9. $a(b - 2) + c(b - 2)$

10. $a(x^2 - 3) - 2(x^2 - 3)$

11. $x^3 + 3x^2 + 6x + 18$

12. $3x^3 + x^2 - 18x - 6$

13. $y^3 - 3y^2 - 4y + 12$

14. $p^3 - 2p^2 - 9p + 18$

15. $9x^2 - 25$

16. $16x^2 - 9$

17. $4xy^4 - 4xz^2$

18. $5xy^4 - 5xz^4$

19. $y^2 - 6y + 9$

20. $x^2 + 8x + 16$

21. $1 - 8x + 16x^2$

22. $1 + 10x + 25x^2$

23. $4x^2 - 5$

24. $16x^2 - 7$

25. $x^2y^2 - 14xy + 49$

26. $x^2y^2 - 16xy + 64$

27. $4ax^2 + 20ax - 56a$

28. $21x^2y + 2xy - 8y$

29. $a^2 + 2ab + b^2 - c^2$

30. $x^2 - 2xy + y^2 - z^2$

31. $x^2 + 2xy + y^2 - a^2 - 2ab - b^2$
[Hint: Factor $x^2 + 2xy + y^2$ and $-1(a^2 + 2ab + b^2)$.]

32. $r^2 + 2rs + s^2 - t^2 + 2tv - v^2$
[Hint: Factor $r^2 + 2rs + s^2$ and $-1(t^2 - 2tv + v^2)$.]

33. $5y^4 - 80x^4$

34. $6y^4 - 96x^4$

35. $x^3 + 8$

36. $y^3 - 64$

37. $3x^3 - \frac{3}{8}$

38. $5y^3 + \frac{5}{27}$

39. $x^3 + 0.001$

40. $y^3 - 0.125$

41. $3z^3 - 24$

42. $4t^3 + 108$

43. $a^6 - t^6$

44. $64m^6 + y^6$

45. $16a^7b + 54ab^7$

46. $24a^2x^4 - 375a^8x$

47. ▨ $x^2 - 17.6$

48. ▨ $x^2 - 8.03$

49. ▨ $37x^2 - 14.5y^2$
(Hint: First remove the common factor 37.)

50. ▨ $1.96x^2 - 17.4y^2$
(Hint: First remove the common factor 1.96.)

51. $(x + 3)^2 - 2(x + 3) - 35$

52. $(y - 4)^2 + 5(y - 4) - 24$

53. $3(a - b)^2 + 10(a - b) - 8$

54. $6(2p + q)^2 - 5(2p + q) - 25$

● SYNTHESIS

Factor.

55. $(x + h)^3 - x^3$

56. $(x + 0.01)^2 - x^2$

57. $y^4 - 84 + 5y^2$

58. $11x^2 + x^4 - 80$

59. $y^2 - \frac{8}{49} + \frac{2}{7}y$

60. $x^2 + \frac{3}{5}x - \frac{4}{25}$

61. $t^2 - 0.27 + 0.6t$

62. $0.4m - 0.05 + m^2$

Factor. Assume that variables in exponents represent natural numbers.

63. $x^{2n} + 5x^n - 24$

64. $4x^{2n} - 4x^n - 3$

65. $x^2 + ax + bx + ab$

66. $bdy^2 + ady + bcy + ac$

67. $\frac{1}{4}t^2 - \frac{2}{5}t + \frac{4}{25}$

68. $\frac{4}{27}r^2 + \frac{5}{9}rs + \frac{1}{12}s^2 - \frac{1}{3}rs$

69. $25y^{2m} - (x^{2n} - 2x^n + 1)$

70. $4x^{4a} + 12x^{2a} + 10x^{2a} + 30$

71. $3x^{3n} - 24y^{3m}$

72. $x^{6a} - t^{3b}$

73. $(y - 1)^4 - (y - 1)^2$

74. $x^6 - 2x^5 + x^4 - x^2 + 2x - 1$

Express in the form $A(x + B)$.

75. $5x - 9$ **76.** $\frac{2}{3}x - 7$

77. **a)** Multiply: $(x^2 - x + 1)(x^3 + x^2 - 1)$.

  b) Factor: $x^5 + x - 1$.

1.6

RATIONAL EXPRESSIONS

A Replacements in Rational Expressions

Expressions like the following are called rational expressions or fractional expressions:

$$\frac{8}{5}, \quad \frac{x^2 - 9}{x - 3}, \quad \frac{3x^2 + 5\sqrt{x} - 2}{x^2 - y^2}, \quad \frac{x - 3}{x^2 - x - 2}.$$

Rational expressions represent division. Certain substitutions are not meaningful in such expressions. Since division by zero is not defined, any number that makes a denominator zero is not a *meaningful* replacement. For example, 3 is not a meaningful replacement in

$$\frac{x^2 - 9}{x - 3}$$

because the denominator $x - 3$ is 0 when x is replaced by 3. All real numbers other than 3 are meaningful replacements. As another example, consider

$$\frac{x - 3}{x^2 - x - 2}.$$

To determine the meaningful replacements, we can first factor the denominator:

$$\frac{x - 3}{x^2 - x - 2} = \frac{x - 3}{(x + 1)(x - 2)}.$$

The factor $x + 1$ is 0 when $x = -1$. The factor $x - 2$ is 0 when $x = 2$. Thus, -1 and 2 are not meaningful replacements. All real numbers except -1 and 2 are meaningful replacements.

DO EXERCISES 1 AND 2.

Multiplication and Division

To multiply two rational expressions, we multiply their numerators and also their denominators. By Theorem 4, when we divide, we multiply by the reciprocal of the divisor.

Example 1 Multiply: $\dfrac{x + 3}{y - 4} \cdot \dfrac{x^3}{y + 5}$.

Solution

$$\frac{x + 3}{y - 4} \cdot \frac{x^3}{y + 5} = \frac{(x + 3)x^3}{(y - 4)(y + 5)}$$

◀

OBJECTIVES

You should be able to:

A Determine meaningful replacements in rational expressions.

B Simplify rational expressions.

C Multiply or divide rational expressions, and simplify.

D Add or subtract rational expressions, and simplify.

E Simplify complex rational expressions.

Determine the meaningful replacements.

1. $\dfrac{x^2 - 25}{x - 5}$

2. $\dfrac{x^3 - xy^2}{x^2 + 7x + 12}$

3. Multiply:

$$\frac{x+y}{2x^2-1}\cdot\frac{x+y}{7x}.$$

Example 2   Divide: $\dfrac{x-2}{x+1}\div\dfrac{x+5}{x-3}$.

Solution

$$\frac{x-2}{x+1}\div\frac{x+5}{x-3}=\frac{x-2}{x+1}\cdot\frac{x-3}{x+5}$$ Multiplying by the reciprocal

$$=\frac{(x-2)(x-3)}{(x+1)(x+5)}$$ Multiplying ◀

In Example 2, we could go on and finish multiplying in the numerator and the denominator, but we choose not to do so because we want to simplify, if possible. It also eases addition, subtraction, and equation solving if we do not carry out the multiplication.

DO EXERCISES 3 AND 4.

4. Divide:

$$\frac{x-2}{x+2}\div\frac{x+2}{x+4}.$$

B   Simplifying

The basis for simplifying rational expressions lies in the fact that certain expressions have a value of 1 for all meaningful replacements. Such expressions have the same numerator and denominator.* Here are some examples:

$$\frac{x-2}{x-2}=1,\qquad\frac{3x^2-4x+2}{3x^2-4x+2}=1,\qquad\frac{4x-5}{4x-5}=1.$$

When we multiply by such an expression, we obtain an *equivalent expression*. This means that the new expression will name the same number as the first for all meaningful replacements. The set of meaningful replacements may not be the same for the two expressions.

5. Multiply

$$\frac{x+2}{x-5}\quad\text{by}\quad\frac{x+3}{x+3}$$

to obtain an equivalent expression. Name the meaningful replacements for the two expressions.

Example 3   Multiply: $\dfrac{y+4}{y-3}\cdot\dfrac{y-2}{y-2}$.

Solution

$$\frac{y+4}{y-3}\cdot\frac{y-2}{y-2}=\frac{(y+4)(y-2)}{(y-3)(y-2)}=\frac{y^2+2y-8}{y^2-5y+6}$$ ◀

The only replacement that is not meaningful in $(y+4)/(y-3)$ is 3. For the expression $(y^2+2y-8)/(y^2-5y+6)$, or $(y^2+2y-8)/[(y-3)(y-2)]$, the replacements that are not meaningful are 3 and 2. The expressions $(y+4)/(y-3)$ and $(y^2+2y-8)/(y^2-5y+6)$ are equivalent. That is, they name the same number for replacements that are meaningful in *both* expressions.

DO EXERCISE 5.

Simplification can be accomplished if we reverse the procedure in the above example; that is, we try to factor the rational expression in such a way that one of the factors is equal to 1 and then we "remove" that factor. In this way, we obtain an equivalent expression that is simpler, or less com-

---

* By Theorem 4, $\dfrac{a}{a}=a\div a=a\left(\dfrac{1}{a}\right)=1$, so $\dfrac{a}{a}=1$.

plicated, than the original.

Example 4   Simplify: $\dfrac{9x^2 + 6xy - 3y^2}{12x^2 - 12y^2}$.

Solution

$$\frac{9x^2 + 6xy - 3y^2}{12x^2 - 12y^2} = \frac{3(x+y)(3x-y)}{3(4)(x+y)(x-y)} \qquad \text{Factoring the numerator and the denominator}$$

$$= \frac{3(x+y)}{3(x+y)} \cdot \frac{3x-y}{4(x-y)} \qquad \text{Factoring the rational expression}$$

$$= \frac{3x-y}{4(x-y)} \qquad \text{Removing a factor of 1}$$

For purposes of later work, we usually do not multiply out the numerator and the denominator.   ◀

Canceling

Canceling is a shortcut that you may have used for removing a factor of 1 when working with rational expressions. With great concern, we mention it as a possibility of speeding up your work here. Canceling may be done for removing factors of 1 only in products. It may *not* be done in sums or when adding expressions together. Our concern is that canceling be done with care and understanding. Example 4 might have been done faster as follows:

$$\frac{9x^2 + 6xy - 3y^2}{12x^2 - 12y^2} = \frac{\cancel{3}\cancel{(x+y)}(3x-y)}{\cancel{3}(4)\cancel{(x+y)}(x-y)} \qquad \text{When a factor of 1 is noted, it is "canceled" as shown.}$$

$$= \frac{3x-y}{4(x-y)}. \qquad \text{Removing a factor of 1: } \frac{3(x+y)}{3(x+y)} = 1$$

CAUTION!   The difficulty with canceling is that it can be applied incorrectly in situations such as the following:

$$\frac{\cancel{2}+3}{\cancel{2}} = 3, \qquad \frac{\cancel{4}+1}{\cancel{4}+2} = \frac{1}{2}, \qquad \frac{1\cancel{5}}{\cancel{5}4} = \frac{1}{4}.$$

$$\quad\text{Wrong!} \qquad\qquad \text{Wrong!} \qquad\qquad \text{Wrong!}$$

In each of these situations, the expressions canceled were *not* factors of 1. Factors are parts of products. For example, in $2 \cdot 3$, 2 and 3 are factors, but in $2 + 3$, 2 and 3 are *not* factors. **If you can't factor, you can't cancel!** If in doubt, don't cancel!

Example 5   Simplify: $\dfrac{x^2 - 1}{2x^2 - x - 1}$.

Solution

$$\frac{x^2 - 1}{2x^2 - x - 1} = \frac{(x-1)(x+1)}{(2x+1)(x-1)} \qquad \text{Factoring the numerator and the denominator}$$

$$= \frac{\cancel{(x-1)}(x+1)}{(2x+1)\cancel{(x-1)}} \qquad \text{Removing a factor of 1: } \frac{x-1}{x-1} = 1$$

$$= \frac{x+1}{2x+1} \qquad\qquad\qquad ◀$$

Simplify. Name the meaningful replacements in both the original and the simplified expressions.

6. $\dfrac{6x^2 + 4x}{2x^2 + 4x}$

7. $\dfrac{y^2 + 3y + 2}{y^2 - 1}$

Multiply or divide and simplify.

8. $\dfrac{x^2 - 2xy + y^2}{x + y} \cdot \dfrac{3x + 3y}{x^2 - y^2}$

9. $\dfrac{a^2 - b^2}{ab} \div \dfrac{a^2 - 2ab + b^2}{2a^2b^2}$

10. $\dfrac{x^3 + y^3}{x^2 - y^2} \div \dfrac{x^2 - xy + y^2}{x^2 - 2xy + y^2}$

In the original expression in Example 5, the meaningful replacements are all real numbers except 1 and $-\frac{1}{2}$. In the simplified expression, all real numbers except $-\frac{1}{2}$ are meaningful replacements. The expressions are equivalent.

DO EXERCISES 6 AND 7.

C Multiplying, Dividing, and Simplifying

We will now multiply or divide and simplify results.

Examples

6. Multiply and simplify.

$$\frac{x + 2}{x - 3} \cdot \frac{x^2 - 4}{x^2 + x - 2} = \frac{(x + 2)(x^2 - 4)}{(x - 3)(x^2 + x - 2)} \quad \text{Multiplying the numerators and the denominators}$$

$$= \frac{(x + 2)(x + 2)(x - 2)}{(x - 3)(x + 2)(x - 1)} \quad \text{Factoring and removing a factor of 1: } \frac{x + 2}{x + 2} = 1$$

$$= \frac{(x + 2)(x - 2)}{(x - 3)(x - 1)} \quad \text{Simplifying}$$

7. Divide and simplify.

$$\frac{a^3 - b^3}{a^2 - b^2} \div \frac{a^2 + ab + b^2}{a^2 + 2ab + b^2}$$

$$= \frac{a^3 - b^3}{a^2 - b^2} \cdot \frac{a^2 + 2ab + b^2}{a^2 + ab + b^2}$$

$$= \frac{(a^3 - b^3)(a^2 + 2ab + b^2)}{(a^2 - b^2)(a^2 + ab + b^2)}$$

$$= \frac{(a - b)(a^2 + ab + b^2)(a + b)(a + b)}{(a - b)(a + b)(a^2 + ab + b^2) \cdot 1} \quad \begin{array}{l}\text{Factoring and removing} \\ \text{a factor of 1:} \\ \frac{(a - b)(a^2 + ab + b^2)(a + b)}{(a - b)(a^2 + ab + b^2)(a + b)} = 1\end{array}$$

$$= a + b \quad \text{Simplifying}$$

DO EXERCISES 8–10.

D Adding and Subtracting

When rational expressions have the same denominator, we can add or subtract them by adding or subtracting the numerators and retaining the common denominator. If the denominators are not the same, we then find equivalent expressions with the same denominator and add.

Example 8 Add: $\dfrac{3x^2 + 4x - 8}{x^2 + y^2} + \dfrac{-5x^2 + 5x + 7}{x^2 + y^2}$.

Solution

$$\frac{3x^2 + 4x - 8}{x^2 + y^2} + \frac{-5x^2 + 5x + 7}{x^2 + y^2} = \frac{-2x^2 + 9x - 1}{x^2 + y^2}$$

In the following example, one denominator is the opposite of the other. We find a common denominator by multiplying by $-1/-1$.

Example 9 Add: $\dfrac{3x^2 + 4}{x - y} + \dfrac{5x^2 - 11}{y - x}$.

Solution

$$\frac{3x^2 + 4}{x - y} + \frac{5x^2 - 11}{y - x} = \frac{3x^2 + 4}{x - y} + \frac{-1}{-1} \cdot \frac{5x^2 - 11}{y - x}$$

We multiply by 1 using $-1/-1$ to convert the second denominator to its opposite.

$$= \frac{3x^2 + 4}{x - y} + \frac{-1(5x^2 - 11)}{-1(y - x)}$$

$$= \frac{3x^2 + 4}{x - y} + \frac{11 - 5x^2}{x - y} \qquad -1(y - x) = -y + x = x - y$$

$$= \frac{-2x^2 + 15}{x - y}$$

DO EXERCISES 11 AND 12.

When denominators are different, but not opposites, we find a common denominator by factoring the denominators. Then we multiply each term by 1 in such a way as to get the common denominator in each expression.

Example 10 Add: $\dfrac{1}{2x} + \dfrac{5x}{x^2 - 1} + \dfrac{3}{x + 1}$.

Solution We first find the **least common multiple (LCM)** of the denominators, also referred to as the **least common denominator (LCD)**. To find the LCD, we first factor each denominator:

$$\left. \begin{array}{l} 2x = 2x, \\ x^2 - 1 = (x + 1)(x - 1), \\ x + 1 = x + 1. \end{array} \right\} \qquad \text{The LCD is } 2x(x + 1)(x - 1).$$

We consider how often each factor occurs in each factorization. We make up a product using each factor the greatest number of times that it occurs in each factorization. We use 2 as a factor once, x as a factor once, $x + 1$ as a factor once even though it occurs in two of the factorizations, and $x - 1$ as a factor once. The LCD is $2x(x + 1)(x - 1)$. Now we multiply each rational expression by 1 in such a way as to get the LCD:

$$\frac{1}{2x} + \frac{5x}{x^2 - 1} + \frac{3}{x + 1}$$

$$= \frac{1}{2x} \cdot \frac{(x + 1)(x - 1)}{(x + 1)(x - 1)} + \frac{5x}{(x + 1)(x - 1)} \cdot \frac{2x}{2x} + \frac{3}{x + 1} \cdot \frac{2x(x - 1)}{2x(x - 1)}$$

$$= \frac{1(x + 1)(x - 1)}{2x(x + 1)(x - 1)} + \frac{5x(2x)}{(x + 1)(x - 1)(2x)} + \frac{3(2x)(x - 1)}{(x + 1)(2x)(x - 1)}$$

$$= \frac{(x + 1)(x - 1) + 10x^2 + 6x(x - 1)}{2x(x + 1)(x - 1)}$$

$$= \frac{17x^2 - 6x - 1}{2x(x + 1)(x - 1)}.$$

Add.

11. $\dfrac{2x^2 + 5x - 9}{x - 5} + \dfrac{x^2 - x + 11}{x - 5}$

12. $\dfrac{3x^2 + 4}{x - 5} + \dfrac{x^2 - 7}{5 - x}$

13. Add:

$$\frac{x^2 - 4xy + 4y^2}{2x^2 - 3xy + y^2} + \frac{x + 4y}{2x - 2y}.$$

14. Subtract:

$$\frac{x}{x^2 + 11x + 30} - \frac{5}{x^2 + 9x + 20}.$$

We usually do not multiply out a numerator or a denominator. Not doing so will be helpful when we solve rational equations.

DO EXERCISE 13.

Example 11   Subtract: $\dfrac{x}{x^2 + 5x + 6} - \dfrac{2}{x^2 + 3x + 2}$.

Solution

$$\frac{x}{x^2 + 5x + 6} - \frac{2}{x^2 + 3x + 2}$$

$$= \frac{x}{(x + 2)(x + 3)} - \frac{2}{(x + 1)(x + 2)}$$

The denominators have been factored in order to determine the LCD. We use each factor the greatest number of times that it occurs in each factorization. The LCD is $(x + 1)(x + 2)(x + 3)$.

$$= \frac{x}{(x + 2)(x + 3)} \cdot \frac{x + 1}{x + 1} - \frac{2}{(x + 1)(x + 2)} \cdot \frac{x + 3}{x + 3}$$

$$= \frac{x(x + 1) - [2(x + 3)]}{(x + 1)(x + 2)(x + 3)}$$

We use the color brackets here to make sure that we subtract the *entire* numerator, not just part of it.

$$= \frac{x^2 + x - [2x + 6]}{(x + 1)(x + 2)(x + 3)}$$

$$= \frac{x^2 + x - 2x - 6}{(x + 1)(x + 2)(x + 3)}$$

$$= \frac{x^2 - x - 6}{(x + 1)(x + 2)(x + 3)}$$

$$= \frac{(x - 3)(x + 2)}{(x + 1)(x + 2)(x + 3)}$$

Factoring and removing a factor of 1: $\frac{x + 2}{x + 2} = 1$

$$= \frac{x - 3}{(x + 1)(x + 3)}$$

DO EXERCISE 14.

> **CAUTION!**   When subtracting one rational expression from another, subtract numerators:
>
> $$\frac{A}{C} - \frac{B}{C} = \frac{A - (B)}{C}.$$
>
> Always be sure to subtract the *entire* numerator B and not just part of it. The use of parentheses or brackets helps. See Example 11.

▶ Complex Rational Expressions

A complex rational expression has a rational expression within its numerator or denominator or both. To simplify such an expression, we can use either of two methods.

To simplify a complex rational expression:

Method 1.   Find the LCM of all the denominators *within* the complex rational expression. Then multiply by 1 using that LCM as the numerator and the denominator of that expression for 1.

Method 2.   First add or subtract, if necessary, to get a single rational expression in both numerator and denominator. Then divide by multiplying by the reciprocal of the denominator.

Example 12   Simplify: $\dfrac{x + \frac{1}{5}}{x - \frac{1}{3}}$.

Solution

Method 1.   The denominators within the complex rational expression are 3 and 5. The LCM is $3 \cdot 5$, or 15. We multiply by 1 using 15/15:

$$\frac{x + \frac{1}{5}}{x - \frac{1}{3}} = \left(\frac{x + \frac{1}{5}}{x - \frac{1}{3}}\right)\frac{15}{15} = \frac{(x + \frac{1}{5})15}{(x - \frac{1}{3})15} = \frac{15x + \frac{1}{5} \cdot 15}{15x - \frac{1}{3} \cdot 15} = \frac{15x + 3}{15x - 5}.$$

Method 2.   We carry out the addition in the numerator and the subtraction in the denominator separately to obtain a single rational expression for both the numerator and the denominator. Then we divide:

$$\frac{x + \frac{1}{5}}{x - \frac{1}{3}} = \frac{x \cdot \frac{5}{5} + \frac{1}{5}}{x \cdot \frac{3}{3} - \frac{1}{3}}$$

$$= \frac{\dfrac{5x + 1}{5}}{\dfrac{3x - 1}{3}} \qquad \text{Now we have a single rational expression for both numerator and denominator.}$$

$$= \frac{5x + 1}{5} \cdot \frac{3}{3x - 1} \qquad \text{Here we divided by multiplying by the reciprocal of the denominator.}$$

$$= \frac{15x + 3}{15x - 5}.$$

◀

Example 13   Simplify: $\dfrac{a^{-3} - b^{-3}}{a^{-1} - b^{-1}}$.

Solution   We first note that

$$\frac{a^{-3} - b^{-3}}{a^{-1} - b^{-1}} = \frac{\dfrac{1}{a^3} - \dfrac{1}{b^3}}{\dfrac{1}{a} - \dfrac{1}{b}}.$$

Method 1.   The denominators within the complex rational expression are a, b, a^3, and b^3. The LCM of these expressions is a^3b^3. We multiply by 1 using a^3b^3/a^3b^3:

$$\frac{a^{-3} - b^{-3}}{a^{-1} - b^{-1}} = \frac{\dfrac{1}{a^3} - \dfrac{1}{b^3}}{\dfrac{1}{a} - \dfrac{1}{b}} = \frac{\dfrac{1}{a^3} - \dfrac{1}{b^3}}{\dfrac{1}{a} - \dfrac{1}{b}} \cdot \frac{a^3b^3}{a^3b^3} = \frac{\left(\dfrac{1}{a^3} - \dfrac{1}{b^3}\right)a^3b^3}{\left(\dfrac{1}{a} - \dfrac{1}{b}\right)a^3b^3}$$

Simplify.

15. $\dfrac{1 + \dfrac{x}{a}}{a - \dfrac{x^2}{a}}$

16. $\dfrac{\dfrac{1}{a} + \dfrac{1}{b}}{\dfrac{1}{a^3} + \dfrac{1}{b^3}}$

Then

$$= \dfrac{\dfrac{1}{a^3}(a^3b^3) - \dfrac{1}{b^3}(a^3b^3)}{\dfrac{1}{a}(a^3b^3) - \dfrac{1}{b}(a^3b^3)} = \dfrac{b^3 - a^3}{a^2b^3 - a^3b^2}$$

$$= \dfrac{(b - a)(b^2 + ab + a^2)}{a^2b^2(b - a)}$$

$$= \dfrac{b^2 + ab + a^2}{a^2b^2}.$$

Method 2. We carry out the subtractions in the numerator and the denominator separately to obtain a single rational expression for both the numerator and the denominator. Then we divide:

$$\dfrac{a^{-3} - b^{-3}}{a^{-1} - b^{-1}} = \dfrac{\dfrac{1}{a^3} - \dfrac{1}{b^3}}{\dfrac{1}{a} - \dfrac{1}{b}} = \dfrac{\dfrac{1}{a^3} \cdot \dfrac{b^3}{b^3} - \dfrac{1}{b^3} \cdot \dfrac{a^3}{a^3}}{\dfrac{1}{a} \cdot \dfrac{b}{b} - \dfrac{1}{b} \cdot \dfrac{a}{a}} = \dfrac{\dfrac{b^3}{a^3b^3} - \dfrac{a^3}{a^3b^3}}{\dfrac{b}{ab} - \dfrac{a}{ab}} = \dfrac{\dfrac{b^3 - a^3}{a^3b^3}}{\dfrac{b - a}{ab}}$$

$$= \dfrac{b^3 - a^3}{a^3b^3} \cdot \dfrac{ab}{b - a} = \dfrac{(b - a)(b^2 + ab + a^2)ab}{ab(a^2b^2)(b - a)}$$

$$= \dfrac{b^2 + ab + a^2}{a^2b^2}.$$

DO EXERCISES 15 AND 16.

● EXERCISE SET 1.6

A Determine the meaningful replacements.

1. $\dfrac{3x - 3}{x(x - 1)}$

2. $\dfrac{(x^2 - 4)(x + 1)}{(x + 2)(x^2 - 1)}$

3. $\dfrac{7x^2 - 28x + 28}{(x^2 - 4)(x^2 + 3x - 10)}$

4. $\dfrac{7x^2 + 11x - 6}{x(x^2 - x - 6)}$

B Simplify. Then determine replacements that are meaningful in the simplified expression.

5. $\dfrac{3x - 3}{x(x - 1)}$

6. $\dfrac{(x^2 - 4)(x + 1)}{(x + 2)(x^2 - 1)}$

7. $\dfrac{7x^2 - 28x + 28}{(x^2 - 4)(x^2 + 3x - 10)}$

8. $\dfrac{7x^2 + 11x - 6}{x(x^2 - x - 6)}$

9. $\dfrac{25x^2y^2}{10xy^2}$

10. $\dfrac{x^2 - 4}{x^2 + 5x + 6}$

11. $\dfrac{x^2 - 3x + 2}{x^2 + x - 2}$

12. $\dfrac{a^2 + 2a + 4}{(a^3 - 8)(a + 5)}$

C Multiply or divide, and simplify.

13. $\dfrac{x^2 - y^2}{(x - y)^2} \cdot \dfrac{1}{x + y}$

14. $\dfrac{r - s}{r + s} \cdot \dfrac{r^2 - s^2}{(r - s)^2}$

15. $\dfrac{x^2 - 2x - 35}{2x^3 - 3x^2} \cdot \dfrac{4x^3 - 9x}{7x - 49}$

16. $\dfrac{x^2 + 2x - 35}{3x^3 - 2x^2} \cdot \dfrac{9x^3 - 4x}{7x + 49}$

17. $\dfrac{a^2 - a - 6}{a^2 - 7a + 12} \cdot \dfrac{a^2 - 2a - 8}{a^2 - 3a - 10}$

18. $\dfrac{a^2 - a - 12}{a^2 - 6a + 8} \cdot \dfrac{a^2 + a - 6}{a^2 - 2a - 24}$

19. $\dfrac{m^2 - n^2}{r + s} \div \dfrac{m - n}{r + s}$

20. $\dfrac{a^2 - b^2}{x - y} \div \dfrac{a + b}{x - y}$

21. $\dfrac{3x + 12}{2x - 8} \div \dfrac{(x + 4)^2}{(x - 4)^2}$

22. $\dfrac{a^2 - a - 2}{a^2 - a - 6} \div \dfrac{a^2 - 2a}{2a + a^2}$

23. $\dfrac{x^2 - y^2}{x^3 - y^3} \cdot \dfrac{x^2 + xy + y^2}{x^2 + 2xy + y^2}$

24. $\dfrac{c^3 + 8}{c^2 - 4} \div \dfrac{c^2 - 2c + 4}{c^2 - 4c + 4}$

25. $\dfrac{(x - y)^2 - z^2}{(x + y)^2 - z^2} \div \dfrac{x - y + z}{x + y - z}$

26. $\dfrac{(a + b)^2 - 9}{(a - b)^2 - 9} \cdot \dfrac{a - b - 3}{a + b + 3}$

D Add or subtract, and simplify.

27. $\dfrac{3}{2a+3} + \dfrac{2a}{2a+3}$

28. $\dfrac{a-3b}{a+b} + \dfrac{a+5b}{a+b}$

29. $\dfrac{y}{y-1} + \dfrac{2}{1-y}$

30. $\dfrac{a}{a-b} + \dfrac{b}{b-a}$

31. $\dfrac{x}{2x-3y} - \dfrac{y}{3y-2x}$

32. $\dfrac{3a}{3a-2b} - \dfrac{2a}{2b-3a}$

33. $\dfrac{3}{x+2} + \dfrac{2}{x^2-4}$

34. $\dfrac{5}{a-3} - \dfrac{2}{a^2-9}$

35. $\dfrac{y}{y^2-y-20} + \dfrac{2}{y+4}$

36. $\dfrac{6}{y^2+6y+9} - \dfrac{5}{y+3}$

37. $\dfrac{3}{x+y} + \dfrac{x-5y}{x^2-y^2}$

38. $\dfrac{a^2+1}{a^2-1} - \dfrac{a-1}{a+1}$

39. $\dfrac{9x+2}{3x^2-2x-8} + \dfrac{7}{3x^2+x-4}$

40. $\dfrac{3y}{y^2-7y+10} - \dfrac{2y}{y^2-8y+15}$

41. $\dfrac{5a}{a-b} + \dfrac{ab}{a^2-b^2} + \dfrac{4b}{a+b}$

42. $\dfrac{6a}{a-b} - \dfrac{3b}{b-a} + \dfrac{5}{a^2-b^2}$

43. $\dfrac{7}{x+2} - \dfrac{x+8}{4-x^2} + \dfrac{3x-2}{4-4x+x^2}$

44. $\dfrac{6}{x+3} - \dfrac{x+4}{9-x^2} + \dfrac{2x-3}{9-6x+x^2}$

45. $\dfrac{1}{x+1} + \dfrac{x}{2-x} + \dfrac{x^2+2}{x^2-x-2}$

46. $\dfrac{x-1}{x-2} - \dfrac{x+1}{x+2} - \dfrac{x-6}{4-x^2}$

E Simplify.

47. $\dfrac{\dfrac{x^2-y^2}{xy}}{\dfrac{x-y}{y}}$

48. $\dfrac{\dfrac{a-b}{b}}{\dfrac{a^2-b^2}{ab}}$

49. $\dfrac{a-a^{-1}}{a+a^{-1}}$

50. $\dfrac{a-\dfrac{a}{b}}{b-\dfrac{b}{a}}$

51. $\dfrac{c+\dfrac{8}{c^2}}{1+\dfrac{2}{c}}$

52. $\dfrac{x^{-1}+y^{-1}}{x^{-3}+y^{-3}}$

53. $\dfrac{x^2+xy+y^2}{\dfrac{x^2}{y}-\dfrac{y^2}{x}}$

54. $\dfrac{\dfrac{a^2}{b}+\dfrac{b^2}{a}}{a^2-ab+b^2}$

55. $\dfrac{\dfrac{x}{y}-\dfrac{y}{x}}{\dfrac{1}{y}+\dfrac{1}{x}}$

56. $\dfrac{\dfrac{a}{b}-\dfrac{b}{a}}{\dfrac{1}{a}-\dfrac{1}{b}}$

57. $\dfrac{x^2y^{-2}-y^2x^{-2}}{xy^{-1}+yx^{-1}}$

58. $\dfrac{a^2b^{-2}-b^2a^{-2}}{ab^{-1}-ba^{-1}}$

59. $\dfrac{\dfrac{a}{1-a}+\dfrac{1+a}{a}}{\dfrac{1-a}{a}+\dfrac{.a}{1+a}}$

60. $\dfrac{\dfrac{1-x}{x}+\dfrac{x}{1+x}}{\dfrac{1+x}{x}+\dfrac{x}{1-x}}$

61. $\dfrac{\dfrac{1}{a^2}+\dfrac{2}{ab}+\dfrac{1}{b^2}}{\dfrac{1}{a^2}-\dfrac{1}{b^2}}$

62. $\dfrac{\dfrac{1}{x^2}-\dfrac{1}{y^2}}{\dfrac{1}{x^2}-\dfrac{2}{xy}+\dfrac{1}{y^2}}$

● **SYNTHESIS** ——————————

Simplify.

63. $\dfrac{(x+h)^2-x^2}{h}$

64. $\dfrac{\dfrac{1}{x+h}-\dfrac{1}{x}}{h}$

65. $\dfrac{(x+h)^3-x^3}{h}$

66. $\dfrac{\dfrac{1}{(x+h)^2}-\dfrac{1}{x^2}}{h}$

67. $\left[\dfrac{\dfrac{x+1}{x-1}+1}{\dfrac{x+1}{x-1}-1}\right]^5$

68. $1+\dfrac{1}{1+\dfrac{1}{1+\dfrac{1}{1+\dfrac{1}{x}}}}$

Find the error(s) in each of the following. Explain why each is an error. Then find the correct answer.

69. $\dfrac{a}{b} \div \left(\dfrac{a}{3}+\dfrac{b}{4}\right) = \dfrac{a}{b} \cdot \left(\dfrac{3}{a}+\dfrac{4}{b}\right)$ (1)

$= \dfrac{a}{b} \cdot \left(\dfrac{3b+4a}{ab}\right)$ (2)

$= \dfrac{a(3b+4a)}{ab^2}$ (3)

$= \dfrac{4a+3b}{b}$ (4)

70. $\dfrac{5x}{2y} + \dfrac{3y}{4x} = \dfrac{5x^2}{4xy} + \dfrac{3y^2}{4xy}$ (1)

$= \dfrac{5x^2+3y^2}{4xy}$ (2)

$= \dfrac{5x+3y^2}{4y}$ (3)

$= \dfrac{5x+3y}{4}$ (4)

Add and simplify.

71. $\dfrac{n(n+1)(n+2)}{2\cdot 3} + \dfrac{(n+1)(n+2)}{2}$

72. $\dfrac{n(n+1)(n+2)(n+3)}{2\cdot 3\cdot 4} + \dfrac{(n+1)(n+2)(n+3)}{2\cdot 3}$

1.7

RADICAL NOTATION

A number a is said to be a **square root** of c if $a^2 = c$. Thus, -3 is a square root of 9 because $(-3)^2 = 9$. Similarly, 3 is also a square root of 9 because $3^2 = 9$. A number a is said to be an **nth root** of c if $a^n = c$. For example, 5 is a third root (called a **cube root**) of 125 because $5^3 = 125$. The number 125 has no other real-number cube root. Any real number has only one real-number cube root.

The symbol $\sqrt{a}$ denotes the nonnegative square root of the number a. The symbol $\sqrt[3]{a}$ denotes the real-number cube root of a, and $\sqrt[n]{a}$ denotes the nth root of a, that is, a number whose nth power is a. The symbol $\sqrt[n]{}$ is called a **radical,** and the symbol under the radical is called the **radicand.** The number n (which is omitted when it is 2) is called the **index.** Examples of nth roots are

$$\sqrt[3]{125} = 5 \quad \text{and} \quad -\sqrt[4]{16} = -2.$$

Odd and Even Roots

Any positive real number has two square roots, one positive and one negative. The same is true for fourth roots, or roots of any even index. The positive root is called the **principal root.** When a radical such as $\sqrt{4}$ or $\sqrt[4]{18}$ is used, it is understood to represent the principal (nonnegative) root. To denote a nonpositive root, we use $-\sqrt{4}$, $-\sqrt[4]{18}$, and so on.

DEFINITION **Principal Root**

A radical expression $\sqrt[n]{a}$, where n is even, represents the *principal* (nonnegative) nth *root* of a. The nonpositive root is denoted $-\sqrt[n]{a}$.

CAUTION! Again, keep in mind that when the index is an even number E, then $\sqrt[E]{x}$ *never* represents a negative number. For example, $\sqrt{9}$ represents 3, and *not* -3, which is represented by $-\sqrt{9}$!

A Meaningful Replacements

Since negative numbers do not have even roots in the system of real numbers, any replacement that makes a radicand negative when the index is even is not meaningful. For example, $\sqrt{-1}$ and $\sqrt[4]{-8.37}$ do not represent real numbers.

Every real number—positive, negative, or zero—has just one real cube root, and the same is true for any odd root. Thus, $\sqrt[n]{a}$, where n is odd, represents the only real nth root of a. In this case, all real numbers are meaningful replacements in the radicand.

Example 1 Determine whether 0 and 3 are meaningful replacements in $\sqrt{5x - 4}$.

Solution   We substitute 0 for x in the radicand $5x - 4$:

$$5(0) - 4 = 0 - 4 = -4.$$

Since the radicand is negative, 0 is not a meaningful replacement. We substitute 3 for x in $5x - 4$:

$$5(3) - 4 = 15 - 4 = 11.$$

Since the radicand is not negative, 3 is a meaningful replacement. ◀

DO EXERCISES 1–4.

B▷ Simplifying Radical Expressions

Consider the expression $\sqrt{(-3)^2}$. This is equivalent to $\sqrt{9}$, which simplifies to 3. Similarly, $\sqrt{3^2} = 3$. This illustrates an important general principle for simplifying radical expressions of even index.

> **THEOREM 10**
>
> For any radicand R, $\sqrt{R^2} = |R|$. Similarly, for any even index n, $\sqrt[n]{R^n} = |R|$.

Examples

2. $\sqrt{x^2} = |x|$
3. $\sqrt{x^2 - 2ax + a^2} = \sqrt{(x-a)^2} = |x - a|$
4. $\sqrt{x^2 y^6} = \sqrt{(xy^3)^2} = |xy^3| = |y^2 xy| = y^2 |xy|$ ◀

If an index is odd, no absolute-value signs are necessary, because there is only one real root and it has the same sign as the radicand.

> **THEOREM 11**
>
> For any radicand R and any odd index n, $\sqrt[n]{R^n} = R$.

Example 5   Simplify: $\sqrt[3]{(15ab)^3}$.

Solution

$$\sqrt[3]{(15ab)^3} = 15ab$$ ◀

Example 6   Simplify: $\sqrt[5]{(-3xy^2)^5}$.

Solution

$$\sqrt[5]{(-3xy^2)^5} = -3xy^2$$ ◀

DO EXERCISES 5–9.

The next property enables us to multiply radicals. We illustrate it with an example.

Determine whether the numbers are meaningful replacements in the given expression.

1. $\sqrt{x - 2}$;   5, −1

2. $\sqrt{x + 3}$;   −7, −3

3. $\sqrt{x^2}$;   18, $-\dfrac{1}{2}$

4. $\sqrt{x^2 + 1}$;   −23, 47.2

Simplify.
5. $\sqrt{(x + 2)^2}$

6. $\sqrt{x^2(y - 2)^2}$

7. $\sqrt[4]{(x + 2)^4}$

8. $\sqrt{x^2 + 8x + 16}$

9. $\sqrt[3]{(-4xy)^3}$

Multiply.

10. $\sqrt{19} \cdot \sqrt{7}$

11. $\sqrt{x + 2y} \cdot \sqrt{x - 2y}$

12. $\sqrt[4]{27} \cdot \sqrt[4]{3}$

Simplify.

13. $\sqrt{300}$

14. $\sqrt{36y^2}$

15. $\sqrt{2x^2 + 4x + 2}$

16. $\sqrt[3]{16}$

17. $\sqrt[3]{(a + b)^4}$

Example 7   Compare $\sqrt{4} \cdot \sqrt{9}$ and $\sqrt{4 \cdot 9}$.

Solution

$$\sqrt{4} \cdot \sqrt{9} = 2 \cdot 3 = 6 \quad \text{and} \quad \sqrt{4 \cdot 9} = \sqrt{36} = 6,$$
$$\text{so} \quad \sqrt{4} \cdot \sqrt{9} = \sqrt{4 \cdot 9} \qquad \blacktriangleleft$$

THEOREM 12

For any nonnegative real numbers a and b and any index n,

$$\sqrt[n]{a} \cdot \sqrt[n]{b} = \sqrt[n]{a \cdot b}.$$

Examples   Multiply.

8. $\sqrt{3} \cdot \sqrt{5} = \sqrt{3 \cdot 5} = \sqrt{15}$

9. $\sqrt{x + 2} \cdot \sqrt{x - 2} = \sqrt{(x + 2)(x - 2)} = \sqrt{x^2 - 4}$

10. $\sqrt[3]{4} \cdot \sqrt[3]{5} = \sqrt[3]{4 \cdot 5} = \sqrt[3]{20}$ $\qquad \blacktriangleleft$

DO EXERCISES 10–12.

Theorem 12 also enables us to simplify radical expressions. The idea is to factor the radicand, obtaining factors that are perfect nth powers.

Examples   Simplify.

11. $\sqrt{50} = \sqrt{25 \cdot 2} = \sqrt{25} \cdot \sqrt{2} = 5\sqrt{2}$

12. $\sqrt{5x^2} = \sqrt{x^2 \cdot 5} = \sqrt{x^2} \cdot \sqrt{5} = |x|\sqrt{5}$

13. $\sqrt[3]{32} = \sqrt[3]{8 \cdot 4} = \sqrt[3]{8} \cdot \sqrt[3]{4} = 2\sqrt[3]{4}$

14. $\sqrt{216x^5y^3} = \sqrt{36 \cdot 6 \cdot x^4 \cdot x \cdot y^2 \cdot y} = |6x^2y|\sqrt{6xy} = 6x^2|y|\sqrt{6xy}$

15. $\sqrt{2x^2 - 4x + 2} = \sqrt{2(x - 1)^2} = |x - 1|\sqrt{2}$ $\qquad \blacktriangleleft$

DO EXERCISES 13–17.

The next property of radicals involves division.

THEOREM 13

For any nonnegative number a and any positive number b, and any index n,

$$\sqrt[n]{\frac{a}{b}} = \frac{\sqrt[n]{a}}{\sqrt[n]{b}}.$$

This property can be used to divide and to simplify radical expressions.

Examples   Simplify.

16. $\sqrt{16x^3y^{-4}} = \sqrt{\frac{16x^3}{y^4}} = \frac{\sqrt{16x^3}}{\sqrt{y^4}} = \frac{\sqrt{16x^2 \cdot x}}{\sqrt{y^4}} = \frac{4|x|\sqrt{x}}{y^2}$

17. $\sqrt[3]{\dfrac{27y^5}{343x^3}} = \dfrac{\sqrt[3]{27y^5}}{\sqrt[3]{343x^3}} = \dfrac{\sqrt[3]{27y^3 \cdot y^2}}{\sqrt[3]{343x^3}} = \dfrac{3y\sqrt[3]{y^2}}{7x}$

18. $\dfrac{18\sqrt{72}}{6\sqrt{6}} = 3\sqrt{\dfrac{72}{6}} = 3\sqrt{12} = 3\sqrt{4 \cdot 3} = 3 \cdot 2\sqrt{3} = 6\sqrt{3}$

19. $\dfrac{\sqrt[3]{32}}{\sqrt[3]{2}} = \sqrt[3]{\dfrac{32}{2}} = \sqrt[3]{16} = \sqrt[3]{8 \cdot 2} = 2\sqrt[3]{2}$ ◀

DO EXERCISES 18–24. _____

Another fundamental principle of radicals involves an exponent under the radical. We illustrate with an example.

Example 20 Compare $\sqrt{3^4}$ and $(\sqrt{3})^4$.

Solution We have

$$\sqrt{3^4} = \sqrt{(3^2)^2} = 3^2 = 9 \quad \text{and} \quad (\sqrt{3})^4 = \sqrt{3} \cdot \sqrt{3} \cdot \sqrt{3} \cdot \sqrt{3}$$
$$= 3 \cdot 3 = 9.$$

Thus, $\sqrt{3^4} = (\sqrt{3})^4$. ◀

The general principle is given in Theorem 14.

THEOREM 14

For any nonnegative number a and any index n and any natural number m,

$$\sqrt[n]{a^m} = (\sqrt[n]{a})^m.$$

Theorem 14 sometimes facilitates radical simplification.

Examples Simplify.

21. $\sqrt[3]{8^5} = (\sqrt[3]{8})^5 = 2^5 = 32$
22. $(\sqrt{2})^6 = \sqrt{2^6} = \sqrt{(2^3)^2} = 2^3 = 8$ ◀

DO EXERCISES 25 AND 26. _____

Various calculations with radicals can be carried out using the properties of radicals and the properties of numbers, such as the distributive property. The following examples illustrate.

Examples Simplify.

23. $3\sqrt{8} - 5\sqrt{2} = 3\sqrt{4 \cdot 2} - 5\sqrt{2}$
$\qquad = 3 \cdot 2\sqrt{2} - 5\sqrt{2}$
$\qquad = 6\sqrt{2} - 5\sqrt{2}$
$\qquad = (6 - 5)\sqrt{2}$ Here we use a distributive law.
$\qquad = \sqrt{2}$

Simplify.

18. $\sqrt{\dfrac{49}{64}}$

19. $\sqrt{\dfrac{25}{y^2}}$

20. $\sqrt[4]{\dfrac{16}{81}}$

21. $\sqrt[3]{\dfrac{7}{125}}$

22. $\dfrac{\sqrt{75}}{\sqrt{3}}$

23. $\dfrac{\sqrt{2x^3}}{\sqrt{50x}}$

24. $\dfrac{\sqrt[3]{24x^3y}}{\sqrt[3]{3y^4}}$

Simplify.

25. $\sqrt[3]{27^{10}}$

26. $(\sqrt{3})^8$

Simplify.

27. $7\sqrt{5} + 3\sqrt{5} - 8\sqrt{20}$

28. $5\sqrt[3]{16y^4} + 7\sqrt[3]{2y}$

29. $(\sqrt{3} - 5\sqrt{2})(2\sqrt{3} + \sqrt{2})$

30. What was the speed of a car that left skid marks 70 ft long?

24. $(4\sqrt{3} + \sqrt{2})(\sqrt{3} - 5\sqrt{2})$
$= 4(\sqrt{3})^2 - 20\sqrt{3}\sqrt{2} + \sqrt{2}\sqrt{3} - 5(\sqrt{2})^2$
$= 4 \cdot 3 - 20\sqrt{6} + \sqrt{6} - 5 \cdot 2$
$= 12 - 19\sqrt{6} - 10$
$= 2 - 19\sqrt{6}$ ◀

DO EXERCISES 27–29.

 Applications

Example 25 *Speed of a skidding car.* How do police determine the speed of a car that has skidded? The formula

$$r = 2\sqrt{5L}$$

can be used to approximate the speed r, in miles per hour, of a car that has left a skid mark of length L, in feet. What was the speed of a car that left skid marks 306 ft long?

Skid marks

Solution We substitute 306 for L in the formula $r = 2\sqrt{5L}$ and use a calculator to approximate r:

$$r = 2\sqrt{5(306)} = 2\sqrt{1530}$$
$$\approx 78.23.$$

Using a calculator, finding the square root, multiplying, and rounding

The speed of the car was about 78.23 mph. ◀

DO EXERCISE 30.

Example 26 By the **Pythagorean theorem,** in right triangles, $c^2 = a^2 + b^2$ and $c = \sqrt{a^2 + b^2}$, where a and b are the lengths of the legs and c is the length of the hypotenuse. A surveyor is trying to measure a certain distance PR across a pond. Poles are placed at points P, Q, and R. The distances that the surveyor was able to measure are shown in the figure. What is the approximate distance from P to R?

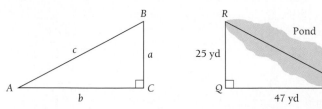

Solution We have the lengths of two of the legs given. We find PR, the hypotenuse, as follows:

$$PR^2 = RQ^2 + PQ^2$$

so

$$PR = \sqrt{RQ^2 + PQ^2}$$
$$= \sqrt{25^2 + 47^2}$$
$$= \sqrt{625 + 2209}$$
$$= \sqrt{2834} \approx 53.2.$$

Thus the distance across the pond is about 53.2 yd. ◀

DO EXERCISE 31.

▶ Rationalizing Denominators or Numerators

Rational expressions are often considered simpler when the denominator is free of radicals. Thus in simplifying, we generally remove the radicals in a denominator. This is called **rationalizing the denominator,** and it can be done by multiplying by 1 in such a way as to obtain a perfect power in the denominator. On occasion, we prefer to rationalize the numerator. This is often the case in calculus. In either case, we can accomplish the rationalization by multiplying by 1, as the following examples show.

Examples   Simplify.

27. $\sqrt{\dfrac{1}{2}} = \sqrt{\dfrac{1}{2} \cdot \dfrac{2}{2}} = \sqrt{\dfrac{2}{4}} = \dfrac{\sqrt{2}}{\sqrt{4}} = \dfrac{\sqrt{2}}{2}$ We multiply by $\frac{2}{2}$ so that we have a perfect square in the denominator.

28. $\sqrt[3]{\dfrac{7}{9}} = \sqrt[3]{\dfrac{7}{9} \cdot \dfrac{3}{3}} = \sqrt[3]{\dfrac{21}{27}} = \dfrac{\sqrt[3]{21}}{\sqrt[3]{27}} = \dfrac{\sqrt[3]{21}}{3}$ We multiply by $\frac{3}{3}$ so that we have a perfect cube in the denominator.

29. $\dfrac{\sqrt{7}}{\sqrt{5}} = \dfrac{\sqrt{7}}{\sqrt{5}} \cdot \dfrac{\sqrt{5}}{\sqrt{5}} = \dfrac{\sqrt{35}}{\sqrt{25}} = \dfrac{\sqrt{35}}{5}$

30. $\dfrac{\sqrt{2a}}{\sqrt{5b}} = \dfrac{\sqrt{2a}}{\sqrt{5b}} \cdot \dfrac{\sqrt{5b}}{\sqrt{5b}} = \dfrac{\sqrt{10ab}}{\sqrt{(5b)^2}}$

$\qquad = \dfrac{\sqrt{10ab}}{|5b|} = \dfrac{\sqrt{10ab}}{5b}$ The absolute-value sign in the denominator is not necessary since $\sqrt{5b}$ would not exist at the outset unless $b > 0$.

31. $\dfrac{\sqrt[3]{54x^3}}{\sqrt[3]{4y^5}} = \sqrt[3]{\dfrac{54x^3}{4y^5} \cdot \dfrac{2y}{2y}}$

$\qquad = \sqrt[3]{\dfrac{27x^3 \cdot 4y}{8y^6}}$

$\qquad = \dfrac{\sqrt[3]{27x^3} \cdot \sqrt[3]{4y}}{\sqrt[3]{8y^6}}$

$\qquad = \dfrac{3x \cdot \sqrt[3]{4y}}{2y^2}$ ◀

DO EXERCISES 32–34.

Pairs of expressions like $c - \sqrt{b}$, $c + \sqrt{b}$ and $\sqrt{a} - \sqrt{b}$, $\sqrt{a} + \sqrt{b}$ are called **conjugates.** The product of a pair of conjugates has no radicals in it. Thus when we wish to rationalize a denominator that has two terms and one or more of them involves a square-root radical, we multiply by 1 using the conjugate of the denominator to write a symbol for 1.

---

31. How long must a wire be to reach from the top of a 14-m telephone pole to a point on the ground 7 m from the foot of the pole?

Simplify.

32. $\sqrt[3]{\dfrac{3}{25}}$

33. $\dfrac{\sqrt{3x}}{\sqrt{7y}}$

34. $\dfrac{\sqrt[3]{48a^3}}{\sqrt[3]{9b^7}}$

Rationalize the denominator. Assume that all letters represent positive numbers.

35. $\dfrac{1}{\sqrt{3} - \sqrt{5}}$

36. $\dfrac{\sqrt{x} - 5}{\sqrt{x} + 2}$

Rationalize the numerator. Assume that all letters represent positive numbers.

37. $\dfrac{\sqrt{a+2} - \sqrt{a}}{2}$

38. $\dfrac{\sqrt{x} - \sqrt{5}}{\sqrt{x} + \sqrt{5}}$

Examples Rationalize the denominator. Assume that all letters represent positive numbers.

32. $\dfrac{1}{\sqrt{2} + \sqrt{3}} = \dfrac{1}{\sqrt{2} + \sqrt{3}} \cdot \dfrac{\sqrt{2} - \sqrt{3}}{\sqrt{2} - \sqrt{3}}$ The number $\sqrt{2} - \sqrt{3}$ is the *conjugate* of $\sqrt{2} + \sqrt{3}$. It is found by changing the middle sign. We use the conjugate to form the symbol for 1.

$$= \dfrac{\sqrt{2} - \sqrt{3}}{(\sqrt{2} + \sqrt{3})(\sqrt{2} - \sqrt{3})}$$

$$= \dfrac{\sqrt{2} - \sqrt{3}}{(\sqrt{2})^2 - (\sqrt{3})^2}$$

$$= \dfrac{\sqrt{2} - \sqrt{3}}{2 - 3} = \dfrac{\sqrt{2} - \sqrt{3}}{-1}$$

$$= \sqrt{3} - \sqrt{2}$$

33. $\dfrac{\sqrt{x} + \sqrt{y}}{\sqrt{x} - \sqrt{y}} = \dfrac{\sqrt{x} + \sqrt{y}}{\sqrt{x} - \sqrt{y}} \cdot \dfrac{\sqrt{x} + \sqrt{y}}{\sqrt{x} + \sqrt{y}}$ The conjugate of $\sqrt{x} - \sqrt{y}$ is $\sqrt{x} + \sqrt{y}$.

$$= \dfrac{(\sqrt{x} + \sqrt{y})^2}{(\sqrt{x})^2 - (\sqrt{y})^2}$$

$$= \dfrac{x + 2\sqrt{xy} + y}{x - y}$$

Examples Rationalize the numerator. Assume that all letters represent positive numbers and that all radicands are positive.

34. $\dfrac{1 - \sqrt{2}}{5} = \dfrac{1 - \sqrt{2}}{5} \cdot \dfrac{1 + \sqrt{2}}{1 + \sqrt{2}}$ The conjugate of $1 - \sqrt{2}$ is $1 + \sqrt{2}$.

$$= \dfrac{(1 - \sqrt{2})(1 + \sqrt{2})}{5(1 + \sqrt{2})}$$

$$= \dfrac{1 - 2}{5(1 + \sqrt{2})}$$

$$= \dfrac{-1}{5 + 5\sqrt{2}}$$

35. $\dfrac{\sqrt{x+h} - \sqrt{x}}{h} = \dfrac{\sqrt{x+h} - \sqrt{x}}{h} \cdot \dfrac{\sqrt{x+h} + \sqrt{x}}{\sqrt{x+h} + \sqrt{x}}$

$$= \dfrac{(x + h) - x}{h(\sqrt{x+h} + \sqrt{x})}$$

$$= \dfrac{h}{h(\sqrt{x+h} + \sqrt{x})}$$

$$= \dfrac{1}{\sqrt{x+h} + \sqrt{x}}$$

The simplification in Example 35 is often needed in calculus.

DO EXERCISES 35–38.

● **EXERCISE SET** **1.7**

A Determine whether the given numbers are meaningful replacements in the expression.

1. $\sqrt{x-3}$; $-2, 5$
2. $\sqrt{2x-5}$; $3, 2$
3. $\sqrt{3-4x}$; $-1, 1$
4. $\sqrt{x^2+3}$; $0, 4.3$
5. $\sqrt{1-x^2}$; $1, 3$
6. $\sqrt{x^2+2x+1}$; $-3, 4$
7. $\sqrt[3]{2x+7}$; $-4, 5$
8. $\sqrt[4]{3-5x}$; $1, 2$

B Simplify.

9. $\sqrt{(-11)^2}$
10. $\sqrt{(-1)^2}$
11. $\sqrt{16x^2}$
12. $\sqrt{36t^2}$
13. $\sqrt{(b+1)^2}$
14. $\sqrt{(2c-3)^2}$
15. $\sqrt[3]{-27x^3}$
16. $\sqrt[3]{-8y^3}$
17. $\sqrt{x^2-4x+4}$
18. $\sqrt{y^2+16y+64}$
19. $\sqrt[5]{32}$
20. $\sqrt[5]{-32}$
21. $\sqrt{180}$
22. $\sqrt{48}$
23. $\sqrt[3]{54}$
24. $\sqrt[3]{135}$
25. $\sqrt{128c^2d^4}$
26. $\sqrt{162c^4d^6}$
27. $\sqrt{3}\cdot\sqrt{6}$
28. $\sqrt{6}\cdot\sqrt{8}$

Simplify, assuming that all letters represent positive numbers and that all radicands are positive. Thus no absolute-value signs will be needed.

29. $\sqrt{2x^3y}\sqrt{12xy}$
30. $\sqrt{3y^4z}\sqrt{20z}$
31. $\sqrt[3]{3x^2y}\sqrt[3]{36x}$
32. $\sqrt[5]{8x^3y^4}\sqrt[5]{4x^4y}$
33. $\sqrt[3]{2(x+4)}\sqrt[3]{4(x+4)^4}$
34. $\sqrt[3]{4(x+1)^2}\sqrt[3]{18(x+1)^2}$

35. $\dfrac{\sqrt{21ab^2}}{\sqrt{3ab}}$
36. $\dfrac{\sqrt{128ab^2}}{\sqrt{16a^2b}}$
37. $\dfrac{\sqrt[3]{40m}}{\sqrt[3]{5m}}$
38. $\dfrac{\sqrt{40xy}}{\sqrt{8x}}$
39. $\dfrac{\sqrt[3]{3x^2}}{\sqrt[3]{24x^5}}$
40. $\dfrac{\sqrt[3]{40xy^3}}{\sqrt[3]{8x}}$
41. $\dfrac{\sqrt{a^2-b^2}}{\sqrt{a-b}}$
42. $\dfrac{\sqrt{x^3-y^3}}{\sqrt{x-y}}$
43. $\sqrt{\dfrac{9a^2}{8b}}$
44. $\sqrt{\dfrac{5b^2}{12a}}$
45. $\sqrt[3]{\dfrac{2x^2y^3}{25z^4}}$
46. $\sqrt[3]{\dfrac{24x^3y}{3y^4}}$
47. $\dfrac{(\sqrt[3]{32x^4y})^2}{(\sqrt[3]{xy})^2}$
48. $\dfrac{(\sqrt[3]{16x^2y})^2}{(\sqrt[3]{xy})^2}$
49. $\dfrac{3\sqrt{a^2b^2}\sqrt{4xy}}{2\sqrt{a^{-1}b^{-2}}\sqrt{9x^{-3}y^{-1}}}$

50. $\dfrac{4\sqrt{xy^2}\sqrt{9ab}}{3\sqrt{x^{-1}y^{-2}}\sqrt{16a^{-5}b^{-1}}}$

51. $9\sqrt{50}+6\sqrt{2}$
52. $11\sqrt{27}-4\sqrt{3}$
53. $8\sqrt{2}-6\sqrt{20}-5\sqrt{8}$
54. $\sqrt{12}-\sqrt{27}+\sqrt{75}$
55. $2\sqrt[3]{8x^2}+5\sqrt[3]{27x^2}-3\sqrt[3]{x^3}$
56. $5a\sqrt{(a+b)^3}-2ab\sqrt{a+b}-3b\sqrt{(a+b)^3}$
57. $3\sqrt{3y^2}-\dfrac{y\sqrt{48}}{\sqrt{2}}+\sqrt{\dfrac{12}{4y^{-2}}}$
58. $\sqrt[3]{x^5}-\dfrac{2\sqrt[3]{x}}{\sqrt[3]{x^{-1}}}+\sqrt[3]{\dfrac{8}{x^{-5}}}$
59. $(\sqrt{3}-\sqrt{2})(\sqrt{3}+\sqrt{2})$
60. $(\sqrt{8}+2\sqrt{5})(\sqrt{8}-2\sqrt{5})$
61. $(1+\sqrt{3})^2$
62. $(\sqrt{2}-5)^2$
63. $(\sqrt{t}-x)^2$
64. $\left(\sqrt{a}+\dfrac{1}{\sqrt{a}}\right)^2$
65. $5\sqrt{7}+\dfrac{35}{\sqrt{7}}$
66. $(\sqrt{a^2b}+3\sqrt{y})(2a\sqrt{b}-\sqrt{y})$
67. $(\sqrt{x+3}-\sqrt{3})(\sqrt{x+3}+\sqrt{3})$
68. $(\sqrt{x+h}-\sqrt{x})(\sqrt{x+h}+\sqrt{x})$

C

69. In Example 25, what was the speed of a car that left skid marks 90 ft long?

70. In Example 25, what was the speed of a car that left skid marks 110 ft long?

71. An airplane is flying at an altitude of 3700 ft. The slanted distance directly to the airport is 14,200 ft. How far horizontally is the airplane from the airport?

72. During the summer heat, a 2-mi bridge expands 2 ft in length. Assuming that the bulge occurs straight up the middle, estimate the height of the bulge. (The answer may surprise you. In reality, bridges are built with expansion joints to control such buckling.)

73. An *equilateral* triangle is shown below.

a) Find an expression for its height h in terms of a.
b) Find an expression for its area A in terms of a.

74. An isosceles right triangle has two sides of length s. Find a formula for the length of the third side.

75. The diagonal of a square has length $8\sqrt{2}$. Find the length of a side of the square.

76. The area of square $PQRS$ is 100 ft² (square feet), and A, B, C, and D are the midpoints of the sides. Find the area of square $ABCD$.

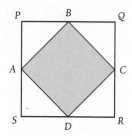

▷   Rationalize the denominator. Assume that all letters represent positive numbers and that all radicands are positive.

77. $\dfrac{6}{3 + \sqrt{5}}$

78. $\dfrac{2}{\sqrt{3} - 1}$

79. $\sqrt[3]{\dfrac{16}{9}}$

80. $\dfrac{\sqrt[3]{3}}{\sqrt[3]{6}}$

81. $\dfrac{4\sqrt{x} - 3\sqrt{xy}}{2\sqrt{x} + 5\sqrt{y}}$

82. $\dfrac{5\sqrt{x} + 2\sqrt{xy}}{3\sqrt{x} - 2\sqrt{y}}$

83. $\dfrac{8\sqrt{2} + 5\sqrt{3}}{5\sqrt{3} - 7\sqrt{2}}$

84. $\dfrac{4\sqrt{6} - 5\sqrt{3}}{2\sqrt{3} + 7\sqrt{6}}$

85. $\dfrac{p - q\sqrt{s}}{q + \sqrt{s}}$

86. $\dfrac{c\sqrt{t} + d}{c - d\sqrt{t}}$

Rationalize the numerator. Assume that all letters represent positive numbers and that all radicands are positive.

87. $\dfrac{\sqrt{2} + \sqrt{5a}}{6}$

88. $\dfrac{\sqrt{3} + \sqrt{5y}}{4}$

89. $\dfrac{\sqrt{x + 1} + 1}{\sqrt{x + 1} - 1}$

90. $\dfrac{\sqrt{x + 4} - 2}{\sqrt{x + 4} + 2}$

91. $\dfrac{\sqrt{a + 3} - \sqrt{3}}{3}$

92. $\dfrac{\sqrt{a + h} - \sqrt{a}}{h}$

93. $\dfrac{\sqrt{x + h} - \sqrt{x}}{h}$

94. $\dfrac{\sqrt{p + 7} - \sqrt{p}}{7}$

95. $\dfrac{\sqrt{x} + \sqrt{y}}{\sqrt{x} - \sqrt{y}}$

96. $\dfrac{\sqrt{a} - \sqrt{b}}{\sqrt{a} + \sqrt{b}}$

● **SYNTHESIS**

Simplify.

97. $\sqrt{1 + x^2} + \dfrac{1}{\sqrt{1 + x^2}}$

98. $\sqrt{1 - x^2} - \dfrac{x^2}{2\sqrt{1 - x^2}}$

99. Show that $\sqrt{a + b}$ and $\sqrt{a} + \sqrt{b}$ are not equivalent for positive real numbers a and b by finding two positive numbers a and b for which $\sqrt{a + b} \neq \sqrt{a} + \sqrt{b}$.

100. Show that $(\sqrt{5 + \sqrt{24}})^2 = (\sqrt{2} + \sqrt{3})^2$.

101. ▨ **Water flow from a faucet.**   As water flows at a velocity v_0 from a faucet with internal diameter d, the width w of the stream at a distance h below the outlet is given by

$$w = d\sqrt{\dfrac{v_0}{\sqrt{v_0^2 + 19.6h}}}.$$

Find the width of a stream 0.1 m below a faucet of internal diameter 0.03 m if the water is flowing at a rate of 0.6 m/sec.

102. ▨ Select the smallest positive number from the following list:

$$10\sqrt{26} - 51, \quad 51 - 10\sqrt{26}, \quad 18 - 5\sqrt{13},$$
$$3\sqrt{11} - 10, \quad 10 - 3\sqrt{11}.$$

103. ▨ **Escape velocity.**   The *escape velocity* V_0 of a projectile is the initial velocity needed in order for it to escape the gravitational pull of the planet. Escape velocity is given in meters per second by

$$V_0 = \sqrt{\dfrac{2GM}{R}},$$

where G is the universal gravitational constant given by $G = 6.672 \times 10^{-11}$ N-m²/kg², M is the mass of the planet, in kilograms, and R is its radius, in meters, where N is newtons in kg-m/sec².

a) The mass of the earth is 5.97×10^{24} kg, and its radius is 6.37×10^6 m. What velocity is necessary for a rocket to escape the gravitational pull of the earth?

b) The mass of Mars is 6.27×10^{23} kg, and its radius is 3.3917×10^6 m. What velocity is necessary for a rocket to escape the gravitational pull of Mars?

1.8

RATIONAL EXPONENTS

We are motivated to define *rational exponents* so that the same rules, or laws, hold for them as for integer exponents. For example, if the laws of exponents are to hold, we must have

$$a^{1/2} \cdot a^{1/2} = a^{1/2+1/2} = a^1 = a.$$

Thus we are led to define $a^{1/2}$ to mean $\sqrt{a}$. Similarly, $a^{1/n}$ would mean $\sqrt[n]{a}$. Again, if the usual laws of exponents are to hold, we must have

$$(a^{1/n})^m = (a^m)^{1/n} = a^{m/n}.$$

Thus we are led to define $a^{m/n}$ to mean $(\sqrt[n]{a})^m$, or $\sqrt[n]{a^m}$.

DEFINITION   **Rational Exponents**

For any natural numbers m and n ($n \neq 1$) and any nonnegative real number a,

$$a^{m/n} = \sqrt[n]{a^m} = (\sqrt[n]{a})^m \quad \text{and} \quad a^{-m/n} = \frac{1}{a^{m/n}}.$$

Note that in this definition we require a to be positive. Thus in manipulations with rational exponents, we assume that all letters represent positive numbers and that all radicands are positive. No absolute-value signs need be used.

Once the definition of rational exponents is made, the question arises whether the usual laws of exponents actually do hold. We will not prove it here, but the answer is that they do. Thus we can simplify or otherwise manipulate expressions containing rational exponents using those laws and the usual arithmetic of rational numbers.

A Converting to Radical Notation

Examples   Convert to radical notation and simplify, if possible.

1. $m^{2/3} = \sqrt[3]{m^2}$

2. $t^{-1/2} = \dfrac{1}{t^{1/2}} = \dfrac{1}{\sqrt{t}}, \quad \text{or} \quad \dfrac{\sqrt{t}}{t}$

3. $64^{5/2} = (64^{1/2})^5$
   $= (\sqrt{64})^5 = 8^5 = 32{,}768$

4. $16^{-3/4} = \dfrac{1}{16^{3/4}}$

   $= \dfrac{1}{(16^{1/4})^3} = \dfrac{1}{(\sqrt[4]{16})^3} = \dfrac{1}{2^3} = \dfrac{1}{8}$ ◀

DO EXERCISES 1–4.

Convert to radical notation and simplify.

1. $n^{3/2}$

2. $y^{-6/7}$

3. $32^{4/5}$

4. $64^{-2/3}$

Convert to exponential notation and simplify.

5. $\sqrt[3]{(5ab)^4}$

6. $\sqrt[4]{16^3}$

7. $\sqrt[6]{a^4}$

8. $\sqrt{\sqrt[3]{4}}$

9. $\sqrt{5^3}\,\sqrt[3]{5}$

Simplify and then write radical notation.

10. $a^{3/4} \cdot a^{1/2}$

11. $(x^{-3})^{2/5}$

12. $(2^{1/4} + 2^{-3/4}) \cdot 2^{1/2}$

 Converting to Exponential Notation

Examples   Convert to exponential notation and simplify.

5. $(\sqrt[4]{7xy})^5 = (7xy)^{5/4}$

6. $\sqrt[3]{8^4} = 8^{4/3} = (8^{1/3})^4 = 2^4 = 16$

7. $\sqrt[6]{x^3} = x^{3/6} = x^{1/2}$,   or $\sqrt{x}$

8. $\sqrt[6]{4} = 4^{1/6} = (2^2)^{1/6} = 2^{2/6} = 2^{1/3}$,   or $\sqrt[3]{2}$

9. $\sqrt[3]{\sqrt{7}} = \sqrt[3]{7^{1/2}} = (7^{1/2})^{1/3} = 7^{1/6}$,   or $\sqrt[6]{7}$

10. $\sqrt{6}\,\sqrt[3]{6} = 6^{1/2} \cdot 6^{1/3} = 6^{1/2 + 1/3} = 6^{5/6}$,   or $\sqrt[6]{6^5}$

DO EXERCISES 5–9. ⎯⎯⎯⎯⎯⎯⎯⎯⎯⎯⎯⎯⎯⎯⎯⎯⎯⎯⎯⎯

 Simplifying Expressions with Rational Exponents

Examples   Simplify and then write radical notation.

11. $x^{5/6} \cdot x^{2/3} = x^{5/6 + 2/3}$ *Adding exponents*

$\qquad\qquad\quad = x^{9/6} = x^{3/2} = \sqrt{x^3}$

$\qquad\qquad\quad = x\sqrt{x}$

12. $(a^5)^{-2/3} = a^{-10/3}$ *Multiplying exponents*

$\qquad\qquad = \dfrac{1}{a^{10/3}} = \dfrac{1}{\sqrt[3]{a^{10}}}$

$\qquad\qquad = \dfrac{1}{\sqrt[3]{a^9 \cdot a}}$

$\qquad\qquad = \dfrac{1}{a^3 \cdot \sqrt[3]{a}}$

13. $(5^{1/3} - 5^{-5/3}) \cdot 5^{1/3} = 5^{1/3} \cdot 5^{1/3} - 5^{-5/3} \cdot 5^{1/3}$ *Using a distributive law*

$\qquad = 5^{2/3} - 5^{-4/3} = \sqrt[3]{5^2} - \dfrac{1}{\sqrt[3]{5^4}}$

$\qquad = \sqrt[3]{25} - \dfrac{1}{5\sqrt[3]{5}} = \sqrt[3]{25} - \dfrac{1}{5\sqrt[3]{5}} \cdot \dfrac{\sqrt[3]{25}}{\sqrt[3]{25}}$

$\qquad = \sqrt[3]{25} - \dfrac{\sqrt[3]{25}}{25} = \dfrac{25}{25} \cdot \sqrt[3]{25} - \dfrac{\sqrt[3]{25}}{25}$

$\qquad = \dfrac{24}{25}\sqrt[3]{25}$

DO EXERCISES 10–12. ⎯⎯⎯⎯⎯⎯⎯⎯⎯⎯⎯⎯⎯⎯⎯⎯⎯⎯⎯⎯

 Writing Single Radicals

In certain expressions containing radicals or rational exponents, it is possible to simplify in such a way that there is a single radical.

Examples Write an expression containing a single radical.

14. $a^{1/2}b^{-1/2}c^{5/6} = a^{3/6}b^{-3/6}c^{5/6}$
$$= (a^3b^{-3}c^5)^{1/6}$$
$$= \sqrt[6]{a^3b^{-3}c^5}$$

15. $\dfrac{a^{1/4}b^{3/8}}{a^{1/2}b^{1/8}} = a^{-1/4}b^{1/4}$
$$= (a^{-1}b)^{1/4} = \sqrt[4]{a^{-1}b}$$
$$= \sqrt[4]{\dfrac{b}{a}}$$

16. $\sqrt[4]{7}\sqrt{3} = 7^{1/4} \cdot 3^{1/2} = 7^{1/4} \cdot 3^{2/4}$
$$= (7 \cdot 3^2)^{1/4}$$
$$= \sqrt[4]{63}$$

17. $\dfrac{\sqrt[4]{(x+2)^3}\sqrt[5]{x+2}}{\sqrt{x+2}} = \dfrac{(x+2)^{3/4}(x+2)^{1/5}}{(x+2)^{1/2}}$
$$= (x+2)^{3/4+1/5-1/2}$$
$$= (x+2)^{9/20}$$
$$= \sqrt[20]{(x+2)^9}$$ ◀

DO EXERCISES 13–15.

Example 18 ***Road pavement signs.*** In a psychological study, pavement signs were found to be most readable by a driver when the letters in the sign are of length L, given by

$$L = \frac{0.000169d^{2.27}}{h},$$

where d is the distance from the car to the lettering and h is the height of the eye above the road. All units are in feet. Find L from a view from the window of a truck for which $h = 7$ ft and $d = 175$ ft.

Solution We substitute 7 for h and 175 for d in the formula for L, and calculate its value using a calculator with a y^x key:

$$L = \frac{0.000169d^{2.27}}{h} = \frac{0.000169(175)^{2.27}}{7} \approx 2.98 \text{ ft.}$$ ◀

DO EXERCISE 16.

 ## Factoring Expressions with Rational Exponents in Calculus

In calculus, it is often important to be able to factor expressions involving rational exponents. Before considering such factoring, let us look again at the kind of factoring we have done in Section 1.5. Consider the expression

$$a^2b^6 + a^4b^3.$$

Write an expression containing a single radical.

13. $\sqrt[3]{5} \cdot \sqrt{2}$

14. $x^{2/3}y^{1/2}z^{5/6}$

15. $\dfrac{\sqrt[4]{(x+y)^3}}{\sqrt{x+y}}$

16. Find the length L of the letters in a road pavement sign when $h = 5$ ft and $d = 205$ ft.

Factor and simplify.

17. $p^8q^{-5} - p^{-6}q^7$

18. $5x^{2/3}y^{-1/4} + 4x^{-1/3}y^{1/2}$

After factoring, we have

$$a^2b^3(b^3 + a^2).$$

We can use the following procedure to factor:

a) We decide which constants and variables are common in the terms. In this case, there are no common constants (other than -1 and 1), but the common variables are powers of a and b.

b) The largest common factor involving a is the smallest exponent of the a-factors. The largest common factor involving b is the smallest exponent of the b-factors. Thus, a^2 and b^3 make up the largest common factor.

c) Determine the factors that remain in each term:

$$a^2b^3 \text{ factored out of } a^2b^6 \text{ leaves } a^{2-2}b^{6-3} = a^0b^3 = b^3;$$
$$a^2b^3 \text{ factored out of } a^4b^3 \text{ leaves } a^{4-2}b^{3-3} = a^2b^0 = a^2.$$

d) Thus, $a^2b^6 + a^4b^3 = a^2b^3 \cdot b^3 + a^2b^3 \cdot a^2 = a^2b^3(b^3 + a^2).$

We just factored an expression involving nonnegative exponents, but the process does not change with negative exponents.

Example 19   Factor: $a^{-3}b^5 + a^4b^{-2}$.

Solution

a) The variables common to both terms involve powers of a and b.

b) a^{-3} has a smaller exponent than a^4.
   b^{-2} has a smaller exponent than b^5.

c) $a^{-3}b^{-2}$ factored out of $a^{-3}b^5$ leaves $a^{-3-(-3)}b^{5-(-2)} = a^0b^7 = b^7$.
   $a^{-3}b^{-2}$ factored out of a^4b^{-2} leaves $a^{4-(-3)}b^{-2-(-2)} = a^7b^0 = a^7$.

d) Thus,

$$a^{-3}b^5 + a^4b^{-2} = a^{-3}b^{-2}(b^7 + a^7) = \frac{b^7 + a^7}{a^3b^2}.$$

◀

Example 20   Factor: $3a^{1/2}b^{-3/4} - a^{-1/2}b^{1/4}$.

Solution

a) The variables common to both terms involve powers of a and b.

b) $a^{-1/2}$ has a smaller exponent than $a^{1/2}$.
   $b^{-3/4}$ has a smaller exponent than $b^{1/4}$.

c) $a^{-1/2}b^{-3/4}$ factored out of $3a^{1/2}b^{-3/4}$ leaves
   $3a^{1/2-(-1/2)}b^{-3/4-(-3/4)} = 3a^1b^0 = 3a.$
   $a^{-1/2}b^{-3/4}$ factored out of $a^{-1/2}b^{1/4}$ leaves
   $a^{-1/2-(-1/2)}b^{1/4-(-3/4)} = a^0b^1 = b.$

d) Thus,

$$3a^{1/2}b^{-3/4} - a^{-1/2}b^{1/4} = a^{-1/2}b^{-3/4}(3a - b)$$
$$= \frac{3a - b}{a^{1/2}b^{3/4}}.$$

◀

DO EXERCISES 17 AND 18.

Example 21 Factor and simplify:

$$\frac{3x^2(2x-1)^{1/2} - x^3(\tfrac{1}{2})(2x-1)^{-1/2}(2)}{[(2x-1)^{1/2}]^2}.$$

Solution

$$\frac{3x^2(2x-1)^{1/2} - x^3(\tfrac{1}{2})(2x-1)^{-1/2}(2)}{[(2x-1)^{1/2}]^2}$$

$$= \frac{3x^2(2x-1)^{1/2} - x^3(2x-1)^{-1/2}}{2x-1} \qquad \text{Simplifying}$$

$$= \frac{x^2(2x-1)^{-1/2}[3(2x-1) - x]}{2x-1} \qquad \text{Factoring the numerator}$$

$$= \frac{x^2(2x-1)^{-1/2}[6x-3-x]}{2x-1}$$

$$= \frac{x^2(5x-3)}{(2x-1)^{1/2}(2x-1)}$$

$$= \frac{x^2(5x-3)}{(2x-1)^{3/2}} \qquad \blacktriangleleft$$

DO EXERCISE 19.

19. Factor and simplify.

$$\frac{3x^2(2x+5)^{1/2} - x^3(\tfrac{1}{2})(2x+5)^{-1/2}(2)}{[(2x+5)^{1/2}]^2}.$$

● EXERCISE SET 1.8

A Convert to radical notation and simplify.

1. $x^{3/4}$ **2.** $y^{2/5}$

3. $16^{3/4}$ **4.** $4^{7/2}$

5. $125^{-1/3}$ **6.** $32^{-4/5}$

7. $a^{5/4}b^{-3/4}$ **8.** $x^{2/5}y^{-1/5}$

B Convert to exponential notation and simplify.

9. $\sqrt[3]{20^2}$ **10.** $\sqrt[5]{17^3}$ **11.** $(\sqrt[4]{13})^5$

12. $(\sqrt[5]{12})^4$ **13.** $\sqrt[3]{\sqrt{11}}$ **14.** $\sqrt[3]{\sqrt[4]{7}}$

15. $\sqrt{5}\,\sqrt[3]{5}$ **16.** $\sqrt[3]{2}\,\sqrt{2}$ **17.** $\sqrt[5]{32^2}$

18. $\sqrt[3]{64^{-2}}$ **19.** $\sqrt[3]{8y^6}$ **20.** $\sqrt[5]{32c^{10}d^{15}}$

21. $\sqrt[3]{a^2+b^2}$ **22.** $\sqrt[4]{a^3-b^3}$ **23.** $\sqrt[3]{27a^3b^9}$

24. $\sqrt[4]{81x^8y^8}$ **25.** $\sqrt[6]{\dfrac{m^{12}n^{24}}{64}}$ **26.** $\sqrt[8]{\dfrac{m^{16}n^{24}}{2^8}}$

C Simplify and then write radical notation, unless it is inappropriate.

27. $(2a^{3/2})(4a^{1/2})$ **28.** $(3a^{5/6})(8a^{2/3})$

29. $\left(\dfrac{x^6}{9b^{-4}}\right)^{-1/2}$ **30.** $\left(\dfrac{x^{2/3}}{4y^{-2}}\right)^{-1/2}$

31. $\dfrac{x^{2/3}y^{5/6}}{x^{-1/3}y^{1/2}}$ **32.** $\dfrac{a^{1/2}b^{5/8}}{a^{1/4}b^{3/8}}$

D Write an expression containing a single radical and simplify.

33. $\sqrt[3]{6}\,\sqrt{2}$ **34.** $\sqrt{2}\,\sqrt[4]{8}$

35. $\sqrt[4]{xy}\,\sqrt[3]{x^2y}$ **36.** $\sqrt[3]{ab^2}\,\sqrt{ab}$

37. $\sqrt[3]{a^4\sqrt{a^3}}$ **38.** $\sqrt{a^3\sqrt[3]{a^2}}$

39. $\dfrac{\sqrt{(a+x)^3}\,\sqrt[3]{(a+x)^2}}{\sqrt[4]{a+x}}$ **40.** $\dfrac{\sqrt[4]{(x+y)^2}\,\sqrt[3]{x+y}}{\sqrt{(x+y)^3}}$

Simplify. Round to three decimal places. (*Note:* Since $x^{1/4} = (x^{1/2})^{1/2}$, you can take a fourth root by taking a square root, and then the square root of the result. Or you can find decimal notation for the exponent, obtaining $x^{0.25}$, and use the power key, x^y. Remember, answers can vary depending on the type and the readout of your calculator.)

41. ▤ $(\sqrt[4]{13})^5$ **42.** ▤ $\sqrt[4]{17^3}$ **43.** ▤ $12.3^{3/2}$

44. ▤ $1.345^{5/2}$ **45.** ▤ $105.6^{3/4}$ **46.** ▤ $7.14^{5/4}$

Using Example 18, find the length L of the letters in the road-pavement sign given the values of h and d.

47. ▤ $h = 4$ ft, $d = 180$ ft   **48.** ▤ $h = 4$ ft, $d = 100$ ft

49. ▤ $h = 4$ ft, $d = 200$ ft   **50.** ▤ $h = 4$ ft, $d = 300$ ft

Task time. In most working situations, the more times a task is performed, the less time it takes to do the task. In a certain situation, industrial psychologists discovered that

$$T = 34x^{-0.41},$$

where T is the number of hours of labor required for the xth unit to be produced.

51. Find T when $x = 1, 6, 8, 10, 32$, and 64.

52. Find T when $x = 2, 3, 5, 9, 11, 19$, and 100.

E Factor and simplify.

53. $a^{-2}b^5 - a^3b^{-5}$

54. $p^8q^{-2} + p^{-5}q^4$

55. $5a^{2/3}b^{-1/2} + 2a^{-1/3}b^{1/2}$

56. $p^{4/5}q^{-2} + 2p^{-1/5}q^2$

57. $x^{-1/3}y^{3/4} - x^{2/3}y^{-1/4}$

58. $4a^{1/2}b^{-3/4} - 6a^{-1/2}b^{1/4}$

59. $(2x - 3)^{-3}(x + 1)^{5/4} + (2x - 3)^{-2}(x + 1)^{1/4}$

60. $2x(5x + 3)^{2/3} + 3x^2(5x + 3)^{-1/3}$

61. $2(x + 1)^{1/2}(3x + 4)^{-3/4} - 10(x + 1)^{-1/2}(3x + 4)^{1/4}$

62. $-4(2x - 5)^{-3}(3x + 1)^{1/3} + 8(2x - 5)^{-2}(3x + 1)^{-2/3}$

63. $3(x^2 + 1)^3 + 3(3x - 5)(x^2 + 1)^2(2x)$

64. $3x^2(x - 1)^4 + x^3(4)(x - 1)^3$

65. $\dfrac{x^3(2x) - (x^2 + 1)(3x^2)}{x^6}$

66. $\frac{1}{2}x^{-1/2}(x^3 - 4) + x^{1/2}(3x^2)$

67. $\dfrac{x^2(x^2 + 1)^{-1/2}(x) - (2x)(x^2 + 1)^{1/2}}{x^4}$

68. $\dfrac{(x - 1)^{1/2} - (x + 1)(\frac{1}{2})(x - 1)^{-1/2}}{x - 1}$

● **SYNTHESIS** _____

Simplify.

69. $(\sqrt{a^{\sqrt{a}}})^{\sqrt{a}}$

70. $(2a^3b^{5/4}c^{1/7})^4 \div (54a^{-2}b^{2/3}c^{6/5})^{-1/3}$

1.9
HANDLING DIMENSION SYMBOLS

We now learn to manipulate dimension symbols. We make calculations, simplifications, and changes of unit. The algebraic skills we have reviewed up to now are quite useful for this. The following table contains abbreviations for some dimension symbols.

Dimension Symbol	Unit
m	meter
cm	centimeter (0.01 m)
km	kilometer (1000 m)
g	gram
cg	centigram (0.01 g)
kg	kilogram (1000 g)
s or sec	second
h or hr	hour
L	liter
mL	milliliter (0.001 L)

Speed

Speed is often determined by measuring a distance and a time and then dividing the distance by the time (this is **average speed**):

$$\text{Speed} = \frac{\text{Distance}}{\text{Time}}.$$

If a distance is measured in kilometers and the time required to travel that distance is measured in hours, the speed will be computed in *kilometers* *per hour* (km/h*). For example, if a car travels 100 km in 2 hr, the average speed is

$$\frac{100 \text{ km}}{2 \text{ hr}}, \quad \text{or} \quad 50 \frac{\text{km}}{\text{hr}}, \quad \text{or} \quad 50 \text{ km/h}.$$

DO EXERCISES 1 AND 2. ⎯⎯⎯⎯⎯⎯⎯⎯⎯⎯⎯⎯⎯⎯⎯⎯⎯

 ## Dimension Symbols

The symbol 100 km/2 hr makes it look as though we are dividing 100 km by 2 hr. It can be argued that we cannot divide 100 km by 2 hr (we can only divide 100 by 2). Nevertheless, it is convenient to treat dimension symbols such as *kilometers, hours, feet, seconds,* and *pounds* as though they were numerals or variables, because correct results can thus be obtained mechanically. Compare, for example,

$$\frac{100x}{2y} = \frac{100}{2} \cdot \frac{x}{y} = 50 \frac{x}{y}$$

with

$$\frac{100 \text{ km}}{2 \text{ hr}} = \frac{100}{2} \cdot \frac{\text{km}}{\text{hr}} = 50 \frac{\text{km}}{\text{hr}}.$$

The analogy holds in other situations, as shown in the following examples.

Example 1 Compare

$$3 \text{ ft} + 2 \text{ ft} = (3 + 2) \text{ ft} = 5 \text{ ft}$$

with

$$3x + 2x = (3 + 2)x = 5x.$$

This looks like a distributive law in use. ◄

DO EXERCISES 3–5. ⎯⎯⎯⎯⎯⎯⎯⎯⎯⎯⎯⎯⎯⎯⎯⎯⎯

Example 2 Compare

$$4 \text{ m} \cdot 3 \text{ m} = 3 \cdot 4 \cdot \text{m} \cdot \text{m} = 12 \text{ m}^2 \text{ (sq m)}$$

with

$$4x \cdot 3x = 4 \cdot 3 \cdot x \cdot x = 12x^2.$$ ◄

4 m

3 m

Example 3 Compare

$$5 \text{ men} \cdot 8 \text{ hr} = 5 \cdot 8 \cdot \text{man-hr} = 40 \text{ man-hr}$$

with

$$5x \cdot 8y = 5 \cdot 8 \cdot x \cdot y = 40xy.$$ ◄

---

* The standard abbreviation for kilometers per hour is km/h and for meters per second is m/s. When the abbreviations for hour and second stand alone, "hr" and "sec" are used instead of "h" and "s."

What is the speed in meters per second?

1. 186,000 m, 10 sec

2. 8 m, 16 sec

Add the measures.

3. 45 ft, 17 ft

4. $\frac{3}{4}$ kg, $\frac{2}{5}$ kg

5. $70 \frac{\text{cm}}{\text{sec}}$, $35 \frac{\text{cm}}{\text{sec}}$

Perform these calculations and simplify, if possible. Do not make any unit changes.

6. $36 \text{ ft} \cdot \dfrac{1 \text{ yd}}{3 \text{ ft}}$

7. $5 \text{ lb} \cdot \dfrac{16 \text{ oz}}{1 \text{ lb}}$

8. $\dfrac{4 \text{ kg}}{5 \text{ ft}} \cdot \dfrac{7 \text{ ft}}{8 \text{ kg}}$

9. $\dfrac{5 \text{ in.} \cdot 9 \text{ lb/hr}}{4 \text{ hr}}$

10. $\dfrac{10 \text{ lb}}{7 \text{ m}} \cdot \dfrac{14 \text{ lb}}{5 \text{ m}}$

Perform the following changes of unit. Use substitution.

11. 34 yd; change to in.

12. 11 mi; change to ft

13. 5 hr; change to sec

In each of the examples above, dimension symbols are treated as though they were variables or numerals, and as though a symbol such as 3 m represents a product 3 times m. A symbol like km/h is treated as though it represents a division of km by hr (*kilometers* by *hours*). Any two measures can be "multiplied" or "divided."

Example 4   Perform this calculation and simplify, if possible. Do not make any unit changes.

$$7 \text{ hr} \cdot \frac{5 \text{ mi}}{8 \text{ hr}}$$

Solution   We have

$$7 \text{ hr} \cdot \frac{5 \text{ mi}}{8 \text{ hr}} = \frac{7 \cdot 5}{8} \cdot \frac{\text{hr}}{\text{hr}} \cdot \text{mi} = 4.375 \text{ mi}.$$

We treated hr/hr as a symbol for 1.   ◄

DO EXERCISES 6–10.

▶ Changes of Unit

Changes of unit can be accomplished by *substitutions*.

Example 5   Change to inches: 25 yd.

Solution

$$
\begin{aligned}
25 \text{ yd} &= 25 \cdot 1 \text{ yd} \\
&= 25 \cdot 3 \text{ ft} && \text{Substituting 3 ft for 1 yd} \\
&= 25 \cdot 3 \cdot 1 \text{ ft} \\
&= 25 \cdot 3 \cdot 12 \text{ in.} && \text{Substituting 12 in. for 1 ft} \\
&= 900 \text{ in.}
\end{aligned}
$$
◄

DO EXERCISES 11–13.

The notion of *multiplying by 1* can also be used to change units.

Example 6   Change to yards: 72 in.

Solution

$$72 \text{ in.} = 72 \text{ in.} \cdot \underset{\uparrow}{\frac{1 \text{ ft}}{12 \text{ in.}}} \cdot \underset{\uparrow}{\frac{1 \text{ yd}}{3 \text{ ft}}}$$

Each of these is equal to 1.

$$
\begin{aligned}
&= \frac{72}{12 \cdot 3} \cdot \frac{\text{in.}}{\text{in.}} \cdot \frac{\text{ft}}{\text{ft}} \cdot \text{yd} \\
&= 2 \text{ yd}
\end{aligned}
$$
◄

In Example 6, we first used the following symbol for 1:

$$\frac{1 \text{ ft}}{12 \text{ in.}} \quad \begin{array}{l} \leftarrow \text{ "ft" in the numerator is the unit we are changing } to. \\ \leftarrow \text{ "in." in the denominator is the unit we are changing } from. \end{array}$$

Then we were converting from ft to yd, so in the second symbol for 1, 1 yd was in the numerator and 3 ft was in the denominator.

DO EXERCISES 14–16.

Example 7   Change $60 \dfrac{\text{km}}{\text{hr}}$ to $\dfrac{\text{m}}{\text{sec}}$.

Solution

$$60 \frac{\text{km}}{\text{hr}} = 60 \frac{\text{km}}{\text{hr}} \cdot \frac{1000 \text{ m}}{1 \text{ km}} \cdot \frac{1 \text{ hr}}{60 \text{ min}} \cdot \frac{1 \text{ min}}{60 \text{ sec}}$$

$$= \frac{60 \cdot 1000}{60 \cdot 60} \cdot \frac{\text{km}}{\text{km}} \cdot \frac{\text{hr}}{\text{hr}} \cdot \frac{\text{min}}{\text{min}} \cdot \frac{\text{m}}{\text{sec}}$$

$$= 16.67 \frac{\text{m}}{\text{sec}}, \quad \text{or } 16.67 \text{ m/s.}$$

Example 8   Change $55 \dfrac{\text{mi}}{\text{hr}}$ to $\dfrac{\text{ft}}{\text{sec}}$.

Solution

$$55 \frac{\text{mi}}{\text{hr}} = 55 \frac{\text{mi}}{\text{hr}} \cdot \frac{5280 \text{ ft}}{1 \text{ mi}} \cdot \frac{1 \text{ hr}}{60 \text{ min}} \cdot \frac{1 \text{ min}}{60 \text{ sec}}$$

$$= \frac{55 \cdot 5280}{60 \cdot 60} \cdot \frac{\text{mi}}{\text{mi}} \cdot \frac{\text{hr}}{\text{hr}} \cdot \frac{\text{min}}{\text{min}} \cdot \frac{\text{ft}}{\text{sec}}$$

$$= 80 \frac{2}{3} \frac{\text{ft}}{\text{sec}}$$

DO EXERCISES 17–20.

Perform the following changes of unit. Use multiplying by 1.

14. 720 in.; change to yd

15. 36,960 m; change to km

16. 360,000 sec; change to hr

Perform the following changes of unit. Use multiplying by 1.

17. $120 \dfrac{\text{mi}}{\text{hr}}$; change to $\dfrac{\text{ft}}{\text{sec}}$

18. 3600 cm²; change to m²

19. $50 \dfrac{\text{kg}}{\text{L}}$; change to $\dfrac{\text{g}}{\text{cm}^3}$
(*Hint:* 1 L = 1000 cm³.)

20. $\dfrac{\$72}{\text{day}}$; change to $\dfrac{\text{¢}}{\text{hr}}$

● EXERCISE SET 1.9

A   Perform the calculations and simplify, if possible. Do not make any unit changes.

1. $36 \text{ ft} \cdot \dfrac{1 \text{ yd}}{3 \text{ ft}}$

2. $6 \text{ lb} \cdot \dfrac{16 \text{ oz}}{1 \text{ lb}}$

3. $6 \text{ kg} \cdot 8 \dfrac{\text{hr}}{\text{kg}}$

4. $9 \dfrac{\text{km}}{\text{hr}} \cdot 3 \text{ hr}$

5. $3 \text{ cm} \cdot \dfrac{2 \text{ g}}{2 \text{ cm}}$

6. $\dfrac{9 \text{ km}}{3 \text{ days}} \cdot 6 \text{ days}$

7. $6 \text{ m} + 2 \text{ m}$

8. $10 \text{ tons} + 6 \text{ tons}$

9. $5 \text{ ft}^3 + 7 \text{ ft}^3$

10. $10 \text{ yd}^3 + 17 \text{ yd}^3$

11. $\dfrac{3 \text{ kg}}{5 \text{ m}} \cdot \dfrac{7 \text{ kg}}{6 \text{ m}}$

12. $3 \text{ acres} \times 60 \dfrac{1}{\text{acre}}$

13. $\dfrac{2000 \text{ lb} \cdot (6 \text{ mi/hr})^2}{100 \text{ ft}}$

14. $\dfrac{7 \text{ m} \cdot 8 \text{ kg/sec}}{4 \text{ sec}}$

15. $\dfrac{6 \text{ cm}^2 \cdot 5 \text{ cm/sec}}{2 \text{ sec}^2/\text{cm}^2 \cdot 2 \dfrac{1}{\text{kg}}}$

16. $\dfrac{320 \text{ lb} \cdot (5 \text{ ft/sec})^2}{2 \cdot 32 \dfrac{\text{ft}}{\text{sec}^2}}$

B   Perform the following changes of unit, using substitution or multiplying by 1.

17. 72 in.; change to ft

18. 17 hr; change to min

19. 2 days; change to sec

20. 360 sec; change to hr

21. $60 \, \dfrac{\text{kg}}{\text{m}}$; change to $\dfrac{\text{g}}{\text{cm}}$

22. $44 \, \dfrac{\text{ft}}{\text{sec}}$; change to $\dfrac{\text{mi}}{\text{hr}}$

23. 216 m²; change to cm²

24. $60 \, \dfrac{\text{lb}}{\text{ft}^3}$; change to $\dfrac{\text{ton}}{\text{yd}^3}$

25. $\dfrac{\$36}{\text{day}}$; change to $\dfrac{\text{¢}}{\text{hr}}$

26. 1440 man-hr; change to man-days

27. $1.73 \, \dfrac{\text{mL}}{\text{sec}}$; change to $\dfrac{\text{L}}{\text{hr}}$

(*Hint:* 1 liter = 1 L = 1000 mL = 1000 milliliters.)

28. $1800 \, \dfrac{\text{g}}{\text{L}}$; change to $\dfrac{\text{cg}}{\text{mL}}$

29. $186{,}000 \, \dfrac{\text{mi}}{\text{sec}}$ (speed of light); change to $\dfrac{\text{mi}}{\text{yr}}$

(Let 365 days = 1 yr.)

30. $1100 \, \dfrac{\text{ft}}{\text{sec}}$ (speed of sound); change to $\dfrac{\text{mi}}{\text{yr}}$

(Let 365 days = 1 yr.)

31. ▤ $89.2 \, \dfrac{\text{ft}}{\text{sec}}$; change to $\dfrac{\text{m}}{\text{min}}$ (1 m ≈ 3.3 ft)

32. ▤ 1013 yd³; change to m³ (1 m ≈ 1.1 yd)

33. ▤ 640 mi²; change to km² (1 km ≈ 0.62 mi)

34. ▤ $312.2 \, \dfrac{\text{kg}}{\text{m}}$; change to $\dfrac{\text{lb}}{\text{ft}}$ (1 kg ≈ 2.2 lb)

35. If a steel rod 2 cm long weighs 5 g, how much does a rod of the same type weigh whose length is 3 cm? 5 m?

36. In Exercise 35, how long is a rod that weighs 4.3 g? 20 cg?

In chemistry, 1 *mole* of a substance is that mass of the substance, in grams, equal to its molecular weight. For example, 1 mole of oxygen is 32 *grams* because its molecular weight is 32, and 1 mole of neon is 20.2 grams.

Convert to grams.

37. 50 moles of oxygen

38. 44 moles of neon

Convert to moles.

39. 303 grams of neon

40. 377.6 grams of oxygen

● **SYNTHESIS** ————————————

41. *Einstein's equation of relativity* states that

$$E = mc^2,$$

where E is the energy emitted from a mass m and c is the speed of light. An atom bomb is exploded. It contains 5000 g of radioactive uranium. The speed of light is 2.9979×10^8 m/sec. Find the amount of energy created from the explosion.

42. Density is mass divided by volume, $D = M/V$. The density of potassium is 0.86 g/mL. What is the mass of 18 cm³ of potassium? (*Hint:* 1 mL = 1 cm³.)

43. 1 nanosecond = 10^{-9} sec. Estimate how far light travels in 1 nanosecond.

——————— 1 ———————

SUMMARY AND REVIEW

Review the objectives found at the beginning of each section. Use the following terms to quiz yourself on your knowledge of each. Then do the review exercises.

● **TERMS TO KNOW**

Real number, p. 1
Natural number, p. 2
Whole number, p. 2
Integer, p. 2
Rational number, p. 2
Irrational number, p. 2
Equivalent expressions, p. 4
Commutative, p. 5
Associative, p. 5
Additive identity, p. 5
Additive inverse, p. 5
Opposite, p. 5
Multiplicative identity, p. 7
Multiplicative inverse, p. 7
Reciprocal, p. 7
Distributive, p. 7, 8

Exponent, p. 11
Scientific notation, p. 14
Order of operations, p. 16
Absolute value, p. 16
Polynomial, p. 20
Degree of a term, p. 20
Degree of a polynomial, p. 20
Binomial, p. 21
Trinomial, p. 21
FOIL, p. 24
Factoring, p. 27
Factoring by grouping, p. 28
Difference of squares, p. 28
Trinomial square, p. 30
Sum or difference of cubes, p. 30

Rational expression, p. 33
Least common multiple (LCM),
    p. 37
Least common denominator (LCD),
    p. 37
Complex rational expression, p. 38
Radical, p. 42
Radicand, p. 42
Index, p. 42
Principal root, p. 42
Pythagorean theorem, p. 46
Rationalizing numerators or
    denominators, p. 47
Conjugates, p. 47
Rational exponent, p. 51

The following review exercises are for practice. Answers are at the back of the book. If you miss an exercise, restudy the section indicated alongside the answer.

Consider the numbers

$$-43.89, 12, -3, -\tfrac{1}{5}, \sqrt{7}, \sqrt[3]{10}, -1, -\tfrac{4}{3}, 7\tfrac{2}{3}, -19, 31, 0.$$

1. Which are integers?
2. Which are natural numbers?
3. Which are rational numbers?
4. Which are real numbers?
5. Which are irrational numbers?
6. Which are whole numbers?

Compute.

7. $15 + (-19)$
8. $-12 + |-4|$
9. $-2.5 + (-2.5)$
10. $22 - (-8)$
11. $\dfrac{18}{-3}$
12. $(-17)(-9)$
13. $-10(20)(-5)(-3)$
14. $-\dfrac{15}{16} + \dfrac{3}{4}$
15. $\dfrac{5}{12} - \left(-\dfrac{7}{8}\right)$
16. $5^3 - [2(4^2 - 3^2 - 6)]^3$
17. $\dfrac{3^4 - (6-7)^4}{2^3 - 2^4}$

Convert to decimal notation.

18. 3.261×10^6
19. 4.1×10^{-4}
20. 2.77 E 8
21. 1.009 E -4

Convert to scientific notation.

22. 0.01432
23. $43,210$

Compute. Write scientific notation for the answer.

24. $\dfrac{2.5 \times 10^{-8}}{3.2 \times 10^{13}}$
25. $(8.4 \times 10^{-17})(6.5 \times 10^{-16})$

Simplify. Write answers using positive exponents where possible.

26. $(7a^2b^4)(-2a^{-4}b^3)$
27. $\dfrac{54x^6y^{-4}z^2}{9x^{-3}y^2z^{-4}}$
28. $\sqrt[4]{81}$
29. $\sqrt[5]{-32}$
30. $\dfrac{b - a^{-1}}{a - b^{-1}}$
31. $\dfrac{\dfrac{x^2}{y} + \dfrac{y^2}{x}}{y^2 - xy + x^2}$
32. $(\sqrt{3} - \sqrt{7})(\sqrt{3} + \sqrt{7})$
33. $(5x^2 - \sqrt{2})^2$
34. $8\sqrt{5} + \dfrac{25}{\sqrt{5}}$
35. $(x + t)(x^2 - xt + t^2)$
36. $(5a + 4b)^3$
37. $(5xy^4 - 7xy^2 + 4x^2 - 3) - (-3xy^4 + 2xy^2 - 2y + 4)$

Factor.

38. $x^3 + 2x^2 - 3x - 6$
39. $12a^3 - 27ab^4$
40. $24x + 144 + x^2$
41. $9x^3 + 35x^2 - 4x$
42. $8x^3 - 1$
43. $27x^6 + 125y^6$
44. $6x^3 + 48$
45. $4x^3 - 4x^2 - 9x + 9$
46. $9x^2 - 30x + 25$
47. $18x^2 - 3x + 6$
48. $9x^2 + 6xy + y^2 + 9x + 3y - 4$

Write an expression containing a single radical.

49. $\sqrt{y^5}\,\sqrt[3]{y^2}$
50. $\dfrac{\sqrt{(a+b)^3}\,\sqrt[3]{a+b}}{\sqrt[6]{(a+b)^7}}$

51. Convert to radical notation: $b^{7/5}$.
52. Convert to exponential notation and simplify:
$$\sqrt[8]{\dfrac{m^{32}n^{16}}{3^8}}.$$
53. Divide and simplify:
$$\dfrac{3x^2 - 12}{x^2 + 4x + 4} \div \dfrac{x-2}{x+2}.$$
54. Subtract and simplify:
$$\dfrac{x}{x^2 + 9x + 20} - \dfrac{4}{x^2 + 7x + 12}.$$
55. Rationalize the numerator:
$$\dfrac{\sqrt{x} - \sqrt{y}}{\sqrt{x} + \sqrt{y}}.$$
56. Rationalize the denominator:
$$\dfrac{\sqrt{x} - \sqrt{y}}{\sqrt{x} + \sqrt{y}}.$$

Factor and simplify.

57. $2x^{1/2}y^{-3/4} - 3x^{-1/2}y^{1/4}$
58. $(x-2)^{-3/4}(3x+5)^{5/2} + (x-2)^{1/4}(3x+5)^{3/2}$

59. How long is a guy wire reaching from the top of a 17-ft pole to a point on the ground 8 ft from the pole?

60. Change $10\,\dfrac{\text{km}}{\text{h}}$ to $\dfrac{\text{m}}{\text{min}}$.

What property is illustrated by the sentence?

61. $t + (-t) = 0$
62. $8(a+b) = 8a + 8b$
63. $-3(ab) = (-3a)b$
64. $tx = xt$

Multiply. Assume that all exponents are integers.

65. $(x^n + 10)(x^n - 4)$
66. $(t^a + t^{-a})^2$
67. $(y^b - z^c)(y^b + z^c)$
68. $(a^n - b^m)^3$

Factor.

69. $y^{2n} + 16y^n + 64$
70. $x^{2t} - 3x^t - 28$

71. $m^{6n} - m^{3n}$

72. Simplify:
$$\frac{\dfrac{2^{n+1}x^{n+1}}{(n+1)^5}}{\dfrac{2^n x^n}{n^5}}.$$

73. Subtract and simplify:
$$\frac{(n-1)(n-2)(n-3)}{2 \cdot 3} - \frac{n(n-1)(n-2)(n-3)}{2 \cdot 3 \cdot 4}.$$

● **THINKING AND WRITING** ──────────

1. Explain the meaning of two expressions being equivalent.

2. Give five examples of real numbers that are not rational and explain why they are not rational.

3. Explain and compare the commutative, associative, and distributive laws.

4. Give as many uses as you can in this chapter of the distributive law.

5. What does n tell you about $\sqrt[n]{x^n}$? Explain.

CHAPTER TEST 1

Consider the numbers
$$-14,\ 23.77,\ -5,\ -\tfrac{4}{7},\ \sqrt{8},\ -\sqrt[3]{11},\ 56\tfrac{1}{4},\ \tfrac{9}{2},\ 0,\ 233.$$

1. Which are whole numbers?

2. Which are irrational numbers?

3. Which are real numbers?

4. Which are rational numbers?

5. Which are natural numbers?

6. Which are integers?

Compute.

7. $-7 + |-7|$

8. $-7.4 + 9.4$

9. $(-6)(-2)$

10. $3 - (-5)$

11. $\dfrac{15}{-5}$

12. $\dfrac{27 \div 3^2 - 3^4 \cdot 3^2}{5(7-16) + 4 \cdot 10}$

Convert to decimal notation.

13. 2.834×10^{-3}

14. 4.7×10^2

15. $4.45\,\text{E}\,9$

16. $4.45\,\text{E}\,{-5}$

Convert to scientific notation.

17. 0.000816

18. 480.57

Compute. Write scientific notation for the answer.

19. $(5.8 \times 10^7)(6.3 \times 10^{-23})$

20. $\dfrac{1.1 \times 10^{-24}}{2.5 \times 10^{-16}}$

Simplify.

21. $(4x^4 y^{-2})(-3x^5 y^{-4})$

22. $\dfrac{24pq^8 r^{-4}}{36p^{-3}q^{-6}r^5}$

23. $\sqrt[3]{-27}$

24. $\sqrt[4]{625}$

25. $(\sqrt{8} - \sqrt{2})(\sqrt{8} + \sqrt{2})$

26. $\dfrac{\dfrac{x}{y^2} + \dfrac{y}{x^2}}{y^2 - yx + x^2}$

27. $(5a^2 - 2b)^2$

28. $(2x^2 y - 3xy + y^2 - 4) - (-6x^2 y + 2xy - y + 8)$

29. $(4y - 3)^3$

30. Write an expression containing a single radical:
$$\frac{\sqrt[4]{(c+d)^3} \cdot \sqrt{c+d}}{\sqrt[5]{(c+d)^4}}.$$

31. Convert to radical notation: $t^{2/7}$.

Factor.

32. $16x^2 + 48x + 36$

33. $t^3 - 343$

34. $m^5 - 9m^3 n^2$

35. $12p^4 + 9p^2 - 30$

36. $64 + 125a^6$

37. $3x + 9y + x^2 + 3xy$

38. $a^8 - 8a^2 b^6$

39. Divide and simplify:
$$\frac{x^2 - 10x + 25}{x^2 + 10x + 25} \div \frac{x^2 - 25}{(x+5)^3}.$$

40. Subtract and simplify:
$$\frac{x}{x^2 - 4x - 12} - \frac{9}{x^2 - 36}.$$

41. Rationalize the denominator:
$$\frac{7 - \sqrt{x}}{7 + \sqrt{x}}.$$

42. Multiply:
$$(x^t + x^{-t})^3.$$

43. Rationalize the numerator:
$$\frac{7 - \sqrt{x}}{7 + \sqrt{x}}.$$

44. Factor:
$$a^{-2/3}b^{8/5} - a^{4/3}b^{3/5}.$$

45. A baseball diamond is actually a square 90 ft on a side. A catcher fields a bunt along the third-base line 12 ft from home plate. How far would the catcher have to throw the ball to first base?

46. Factor and simplify:

$$\frac{3x^2(2x + 3)^{1/2} - x^3(\frac{1}{2})(2x + 3)^{-1/2}(2)}{[(2x + 3)^{1/2}]^2}.$$

47. Change $1200 \dfrac{\text{m}}{\text{min}}$ to $\dfrac{\text{km}}{\text{hr}}$.

● S Y N T H E S I S

48. Factor: $x^{16} - 16$.

49. Simplify:

$$\left| \frac{\dfrac{(-1)^{n+2}(x - 2)^{n+1}}{(n + 1)2^{n+1}}}{\dfrac{(-1)^{n+1}(x - 2)^n}{n \cdot 2^n}} \right|.$$

In this chapter, we continue the review that we began in Chapter 1. In particular, we review basic equation-solving techniques. We also study the system of numbers known as the *complex numbers.* The use of the complex numbers will allow us to find complete solutions to quadratic equations. ● We also begin to consider the payoff that results from the ability to solve equations: *problem solving.* From this introduction to problem solving, we will go on to consider problem solving many times in later chapters. ●

Equations, Inequalities, and Problem Solving

2.1
SOLVING EQUATIONS AND INEQUALITIES

 Solving Equations

> **OBJECTIVES**
>
> You should be able to:
>
> **A** Solve simple equations using the addition and multiplication principles and the principle of zero products.
>
> **B** Solve simple inequalities, and write set-builder notation for a set.
>
> **C** Use inequalities to find meaningful replacements in radical expressions.

DEFINITION	**Solution of an Equation**

A *solution of an equation* is any number that makes the equation true when that number is substituted for the variable.

The number 3 is a solution of $5x = 15$ because $5(3) = 15$ is true. The number -4 is not a solution of $5x = 15$, because $5(-4) = 15$ is false.

DEFINITION	**Solution Set**

The set of all solutions of an equation is called its *solution set.*

When we find all the solutions of an equation (find its solution set), we say that we have *solved* the equation. The number 3 is the only solution of $5x = 15$, so the solution set consists of the number 3 and is denoted $\{3\}$. As another example, consider

$$y^2 - y = 0.$$

65

Solve (find the solution set) by trial and error.

1. $x - 2 = 7$

2. $y^2 + y = 0$

The number 0 is a solution, and so is the number 1. There are no other solutions, so the solution set is $\{0, 1\}$.

DO EXERCISES 1 AND 2.

Equation-Solving Principles

The Addition and Multiplication Principles

Two simple principles allow us to solve many equations. The first of these is as follows.

> **The Addition Principle**
>
> For any real numbers a, b, and c, if an equation $a = b$ is true, then $a + c = b + c$ is true.

This principle and the next are actually very easy theorems. Suppose that $a = b$ is true. Then a and b are the same number. If we add c to this number, the result is $a + c$. It is also $b + c$. Similarly, if we multiply by c, the result is ac. It is also bc. The second principle is similar to the first.

> **The Multiplication Principle**
>
> For any real numbers a, b, and c, if an equation $a = b$ is true, then $ac = bc$ is true.

Note that these principles also cover "subtracting on both sides" and "dividing on both sides," because subtracting c can be accomplished by adding $-c$, and dividing by c can be accomplished by multiplying by $1/c$, provided that $c \neq 0$.

Now let us use the principles to solve some equations.

Example 1 Solve: $3x + 4 = 15$.

Solution

$$3x + 4 + (-4) = 15 + (-4) \qquad \text{Using the addition principle; adding } -4$$
$$3x = 11 \qquad \text{Simplifying}$$
$$\tfrac{1}{3} \cdot 3x = \tfrac{1}{3} \cdot 11 \qquad \text{Using the multiplication principle; multiplying by } \tfrac{1}{3}$$
$$x = \tfrac{11}{3}$$

Check:

$$\begin{array}{c|c} 3x + 4 = 15 \\ \hline 3 \cdot \tfrac{11}{3} + 4 & 15 \qquad \text{Substituting } \tfrac{11}{3} \text{ for } x \\ 11 + 4 & \\ 15 & \text{TRUE} \end{array}$$

The solution is $\tfrac{11}{3}$. The solution set is $\{\tfrac{11}{3}\}$. ◀

Example 2 Solve: $3(7 - 2x) = 14 - 8(x - 1)$.

Solution

$$21 - 6x = 14 - 8x + 8 \qquad \text{Multiplying, using the distributive laws, to remove parentheses}$$

$$21 - 6x = 22 - 8x \qquad \text{Collecting like terms}$$

$$8x - 6x = 22 - 21 \qquad \text{Adding } -21 \text{ and also } 8x$$

$$2x = 1 \qquad \text{Collecting like terms}$$

$$x = \tfrac{1}{2} \qquad \text{Multiplying by } \tfrac{1}{2}$$

Check:

$3(7 - 2x) = 14 - 8(x - 1)$	
$3(7 - 2 \cdot \tfrac{1}{2})$	$14 - 8(\tfrac{1}{2} - 1)$
$3(7 - 1)$	$14 - 8(-\tfrac{1}{2})$
$3 \cdot 6$	$14 + 4$
18	18 TRUE

The solution is $\tfrac{1}{2}$. The solution set is $\{\tfrac{1}{2}\}$. ◀

Example 3 Solve: $x + 3 = x$.

Solution

$$-x + x + 3 = -x + x \qquad \text{Adding } -x$$

$$3 = 0 \qquad \text{Collecting like terms}$$

We get a false equation. No replacement for x will make the equation true. Since there are no solutions, the solution set is the **empty set,** denoted $\varnothing$. ◀

Example 4 Solve: $x + 8 = 8 + x$.

Solution

$$-x + x + 8 = -x + 8 + x \qquad \text{Adding } -x$$

$$8 = 8$$

We get a true equation. Any replacement for x will make the equation true. Thus the solution set is the entire set of real numbers. We also know this by the commutative law of addition. ◀

An equation that is true for all meaningful replacements of the variable is called an **identity.** The equation $x + 8 = 8 + x$ of Example 4 is an identity. The equation $3(7 - 2x) = 14 - 8(x - 1)$ of Example 2 is *not* an identity, because there is only one replacement for which it is true, even though all real numbers are meaningful replacements.

DO EXERCISES 3–7. _____

The Principle of Zero Products

A third principle for solving equations is called the *principle of zero products,* defined as follows.

Solve.

3. $9x - 4 = 8$

4. $-4x + 2 + 5x = 3x - 15$

5. $3(y - 1) - 1 = 2 - 5(y + 5)$

6. $x - 7 = x$

7. $y + 6 = 6 + y$

Solve.

8. $(x - 7)(2x + 3) = 0$

9. $x^2 - x = 20$

10. $x^2 = 5x$

11. $25x^2 + 10x + 1 = 0$

12. $3x^3 - 11x^2 = 4x$

13. $5x^3 + x^2 - 5x - 1 = 0$

> ### The Principle of Zero Products
>
> For any numbers a and b, if $ab = 0$, then $a = 0$ or $b = 0$; and if $a = 0$ or $b = 0$, then $ab = 0$.

We can abbreviate the principle of zero products by saying "$ab = 0$ if and only if $a = 0$ or $b = 0$." To solve an equation using this principle, we must have a 0 on one side of the equation and a product on the other. We then obtain the solutions by setting the factors equal to 0 separately.

Example 5 Solve: $x^2 + x - 12 = 0$.

Solution

$$
\begin{array}{lll}
(x + 4)(x - 3) = 0 & & \text{Factoring} \\
x + 4 = 0 & \text{or} \quad x - 3 = 0 & \text{Using the principle of zero products} \\
x = -4 & \text{or} \qquad x = 3 &
\end{array}
$$

The solutions are -4 and 3. The solution set is $\{-4, 3\}$. ◀

Example 6 Solve: $2x^3 - x^2 = 3x$.

Solution

$$
\begin{array}{lll}
2x^3 - x^2 - 3x = 0 & & \text{Using the addition principle} \\
x(2x - 3)(x + 1) = 0 & & \text{Factoring} \\
x = 0 \quad \text{or} \quad 2x - 3 = 0 \quad \text{or} \quad x + 1 = 0 & & \text{Principle of zero products} \\
x = 0 \quad \text{or} \qquad x = \tfrac{3}{2} \quad \text{or} \qquad x = -1 &
\end{array}
$$

The solution set is $\{0, \tfrac{3}{2}, -1\}$. ◀

Example 7 Solve: $x^3 + 5x^2 - 4x - 20 = 0$.

Solution We first factor by grouping:

$$
\begin{array}{ll}
x^2(x + 5) - 4(x + 5) = 0 & \\
(x^2 - 4)(x + 5) = 0 & \\
(x - 2)(x + 2)(x + 5) = 0 & \\
x - 2 = 0 \quad \text{or} \quad x + 2 = 0 \quad \text{or} \quad x + 5 = 0 & \text{Principle of zero products} \\
x = 2 \quad \text{or} \qquad x = -2 \quad \text{or} \qquad x = -5 &
\end{array}
$$

The solution set is $\{2, -2, -5\}$. ◀

DO EXERCISES 8–13.

Not all equations with four terms can be solved using factoring by grouping. An example is $x^3 - 2x^2 - 2x - 3 = 0$.

▶ B Solving Inequalities

Principles for solving inequalities are similar to those for solving equations. We can add the same number on both sides of an inequality.

The Addition Principle for Inequalities

For any real numbers a, b, and c, if $a < b$ is true, then $a + c < b + c$ is true.

We can also multiply on both sides by the same nonzero number, but if that number is negative, we must reverse the inequality sign. To see this, consider this true inequality:

$-4 < 9.$ True

If we multiply both numbers by 2, we get another true inequality:

$-4(2) < 9(2),$ or $-8 < 18.$ True

If we multiply both numbers by -3, we get a false inequality:

$-4(-3) < 9(-3),$ or $12 < -27.$ False

However, if we now reverse the inequality symbol above, we get a true inequality:

$12 > -27.$ True

| The $<$ symbol has been reversed! |

The Multiplication Principle for Inequalities

1. For any real numbers a, b, and c, if $a < b$ and $c > 0$ are true, then $ac < bc$ is true.
2. For any real numbers a, b, and c, if $a < b$ and $c < 0$ are true, then $ac > bc$ is true.

Similar statements hold when $<$ is replaced by $\leq$.

Example 8 Solve: $3x < 11 - 2x$.

Solution

$$3x + 2x < 11 \qquad \text{Adding } 2x$$
$$5x < 11 \qquad \text{Collecting like terms}$$
$$x < \tfrac{11}{5} \qquad \text{Multiplying by } \tfrac{1}{5}$$

Any number less than $\frac{11}{5}$ is a solution. The solution set is the set of all x such that $x < \frac{11}{5}$. We abbreviate this using *set-builder notation* as follows:

$$\{x \,|\, x < \tfrac{11}{5}\}.$$

We can make a graph of the solution set, as shown here:

A **graph** is a drawing that represents the solution set of an equation.

Solve.

14. $5x > 12 - 3x$

15. $17 - 5y \le 8y - 5$

16. $12x - 6 < 10x + 4$

Write set-builder notation for the set.

17. The set of all x such that $x > \frac{5}{2}$

18. The set of all y such that $y \ge -7$

19. The set of all x such that $x^2 = 5$

Determine the meaningful replacements in the expression.

20. $\sqrt{x - 2}$

21. $\sqrt{x + 3}$

22. $\sqrt{22 - 4x}$

Example 9   Solve: $16 - 7y \ge 10y - 4$.

Solution

$$-16 + 16 - 7y \ge -16 + 10y - 4 \qquad \text{Adding } -16$$
$$-7y \ge 10y - 20 \qquad \text{Simplifying}$$
$$-10y - 7y \ge -10y + 10y - 20 \qquad \text{Adding } -10y$$
$$-17y \ge -20 \qquad \text{Simplifying}$$
$$y \le \tfrac{20}{17} \qquad \text{Multiplying by } -\tfrac{1}{17} \text{ and reversing the inequality sign}$$

Any number less than or equal to $\frac{20}{17}$ is a solution. Thus the solution set is $\{y \mid y \le \frac{20}{17}\}$. ◀

DO EXERCISES 14–19. _____

C   Finding Meaningful Replacements in Radical Expressions

We can solve an inequality to find the *meaningful replacements* in a radical expression.

Example 10   Determine the meaningful replacements in $\sqrt{5x - 4}$.

Solution   The meaningful replacements are those values of x for which $\sqrt{5x - 4}$ is a real number. Such replacements are those that make the radicand nonnegative, that is, numbers x for which

$$5x - 4 \ge 0$$
$$5x \ge 4$$
$$x \ge \tfrac{4}{5}.$$

Thus the meaningful replacements are any numbers x for which $x \ge \frac{4}{5}$. These form the set $\{x \mid x \ge \frac{4}{5}\}$. ◀

DO EXERCISES 20–22. _____

● **EXERCISE SET** **2.1**

A   Solve.

1. $4x + 12 = 60$

2. $2y - 11 = 37$

3. $4 + \frac{1}{2}x = 1$

4. $4.1 - 0.2y = 1.3$

5. $y + 1 = 2y - 7$

6. $5 - 4x = x - 13$

7. $5x - 2 + 3x = 2x + 6 - 4x$

8. $5x - 17 - 2x = 6x - 1 - x$

9. $1.9x - 7.8 + 5.3x = 3.0 + 1.8x$

10. $2.2y - 5 + 4.5y = 1.7y - 20$

11. $7(3x + 6) = 11 - (x + 2)$

12. $4(5y + 3) = 3(2y - 5)$

13. $2x - (5 + 7x) = 4 - [x - (2x + 3)]$

14. $y - (9y - 8) = 5 - 2y - 3(2y - 3) + 29$

15. $(2x - 3)(3x - 2) = 0$

16. $(5x - 2)(2x + 3) = 0$

17. $x(x - 1)(x + 2) = 0$

18. $x(x + 2)(x - 3) = 0$

19. $3x^2 + x - 2 = 0$

20. $10x^2 - 16x + 6 = 0$

21. $(x - 1)(x + 1) = 5(x - 1)$

22. $6(y - 3) = (y - 3)(y - 2)$

23. $x[4(x - 2) - 5(x - 1)] = 2$

24. $14[(x - 4) - \frac{1}{14}(x + 2)] = (x + 2)(x - 4)$

25. $(3x^2 - 7x - 20)(2x - 5) = 0$

26. $(8x + 11)(12x^2 - 5x - 2) = 0$

27. $16x^3 = x$ **28.** $9x^3 = x$

29. $2x^2 = 6x$ **30.** $18x + 9x^2 = 0$

31. $3y^3 - 5y^2 - 2y = 0$ **32.** $3t^3 + 2t = 5t^2$

33. $(2x - 3)(3x + 2)(x - 1) = 0$

34. $(y - 4)(4y + 12)(2y + 1) = 0$

35. $(2 - 4y)(y^2 + 3y) = 0$

36. $(y^2 - 9)(y^2 - 36) = 0$

37. $x + 4 = 8 + x$

38. $x - 7 = 9 + x$

39. $7x^3 + x^2 - 7x - 1 = 0$

40. $3x^3 + x^2 - 12x - 4 = 0$

41. $y^3 + 2y^2 - y - 2 = 0$

42. $t^3 + t^2 - 25t - 25 = 0$

43. $11 + x = x + 11$

44. $0 \cdot x = 0$

B   Solve.

45. $x + 6 < 5x - 6$

46. $3 - x < 4x + 7$

47. $3x - 3 + 2x \geq 1 - 7x - 9$

48. $5y - 5 + y \leq 2 - 6y - 8$

49. $14 - 5y \leq 8y - 8$

50. $8x - 7 < 6x + 3$

51. $-\frac{3}{4}x \geq -\frac{5}{8} + \frac{2}{3}x$

52. $-\frac{5}{6}x \leq \frac{3}{4} + \frac{8}{3}x$

53. $4x(x - 2) < 2(2x - 1)(x - 3)$

54. $(x + 1)(x + 2) > x(x + 1)$

Write set-builder notation for the set.

55. The set of all x such that $x > 2.5$

56. The set of all y such that $y \leq -7$

57. The set of all t such that $2t^2 = 10$

58. The set of all m such that $m^3 + 3 = m^2 - 2$

C   Determine the meaningful replacements.

59. $\sqrt{x - 3}$ **60.** $\sqrt{2x - 5}$

61. $\sqrt{3 - 4x}$ **62.** $\sqrt{x^2 + 3}$

● **SYNTHESIS** _____

Solve.

63. ▤ $2.905x - 3.214 + 6.789x = 3.012 + 1.805x$

64. ▤ $(13.14x + 17.152)(15.15 - 7.616x) = 0$

65. ▤ $3.12x^2 - 6.715x = 0$

66. ▤ $9.25x^2 + 18.03x = 0$

67. ▤ $1.52(6.51x + 7.3) < 11.2 - (7.2x + 13.52)$

68. ▤ $4.73(5.16y + 3.62) \geq 3.005(2.75y - 6.31)$

● **CHALLENGE** _____

Solve.

69. $(x + 1)^3 = (x - 1)^3 + 26$

70. $(x - 2)^3 = x^3 - 2$

71. $(x^2 - x - 20)(x^2 - 25) = 0$

72. $(x^3 - 3x^2 - 16x + 48)(x^3 + x^2 - 9x - 9) = 0$

2.2

RATIONAL EQUATIONS

A **Equivalent Equations**

DEFINITION	**Equivalent Equations**

Equations that have the same solution set are called *equivalent equations*.

Example 1   Determine whether the equations $3x = 6$, $3x + 5 = 11$, and $-12x = -24$ are equivalent.

Solution

    $3x = 6$ $3x + 5 = 11$ $-12x = -24$

    Solution set: $\{2\}$ Solution set: $\{2\}$ Solution set: $\{2\}$

Determine whether the equations of the pair are equivalent.

1. $3x = 7$,
   $-15x = -35$

2. $x = 7$,
   $5x = 35$

3. $x = -5$,
   $x^2 = 25$

4. $x = -5$,
   $x + 1 = -4$

5. $\dfrac{(x - 2)(x + 8)}{x - 2} = x + 8$,

   $x + 8 = x + 8$

6. $3x^2 = 5x$,
   $3x = 5$

7. Explain how you would derive $2x + 5x^2 = 5x^2$ from $2x = 0$.

8. Explain how you would derive $2x = 0$ from $2x + 5x^2 = 5x^2$.

Each equation has the same solution set as the others, $\{2\}$. Thus all three equations are equivalent. ◀

Example 2 Determine whether the equations $3x = 4x$ and $3/x = 4/x$ are equivalent.

Solution

$$3x = 4x \qquad\qquad \frac{3}{x} = \frac{4}{x}$$

Solution set: $\{0\}$ The solution set is $\varnothing$, the empty set (no solution, since division by 0 is not defined).

The empty set $\varnothing$ and the set containing the number 0, $\{0\}$, are not the same set. There are no elements in $\varnothing$. There is one element in $\{0\}$, the number 0. The solution sets are *not* the same, so the equations are *not* equivalent. ◀

Example 3 Determine whether the equations $x = 1$ and $x^2 = x$ are equivalent.

Solution

$$x = 1 \qquad\qquad x^2 = x$$

Solution set: $\{1\}$ Solution set: $\{0, 1\}$

The solution sets are *not* the same, so the equations are *not* equivalent. ◀

DO EXERCISES 1–6.

Let us examine our equation-solving principles from the standpoint of equivalent equations. Consider the following example illustrating the addition principle:

$$x + 5 = 9$$
$$x + 5 + (-5) = 9 + (-5) \qquad \text{Adding } -5 \text{ on both sides}$$
$$x = 4.$$

In this example, the original equation and the last equation have exactly the same solutions. They are equivalent. Whenever the steps in an argument are reversible, the equations are equivalent. When we use the addition principle, an equivalent equation is obtained unless an expression is added having replacements that are not meaningful. To see that, note that we start with an equation $a = b$ and obtain $a + c = b + c$. By adding $-c$, we can always obtain $a = b$ again, so the steps are reversible.

DO EXERCISES 7 AND 8.

> The use of the addition principle produces an equation equivalent to the original, unless an expression is added having replacements that are not meaningful. Checking by substituting is therefore not necessary except to detect errors in solving.

Now let us consider the multiplication principle, which says that if $a = b$ is true, then $ac = bc$ is true. Does the multiplication principle yield equivalent equations? In other words, are the steps reversible? If the number c by which we multiply is not 0, then we can reverse the step by multiplying by $1/c$; but if c is 0, then $1/c$ does not exist, and the step is not reversible.

> The use of the multiplication principle produces an equation equivalent to the original, if we multiply by a nonzero number.

Now let us look at the principle of zero products. According to this principle, if we start with an equation $ab = 0$, we obtain $a = 0$ or $b = 0$. Also, if we start with $a = 0$ or $b = 0$, we obtain $ab = 0$. Thus when we use this principle, that step is reversible, so we have equivalent equations.

> The use of the principle of zero products yields the solutions of the original equation. Checking by substituting is not necessary except to detect errors in carrying out the algebra.

 ## Rational Equations

We now solve equations containing rational expressions. These are called **rational equations.** Finding the solutions of such equations often involves multiplying by expressions with variables, as in the following example. We multiply by the LCM of all the denominators.

Example 4   Solve: $\dfrac{x-3}{x-7} = \dfrac{4}{x-7}$.

Solution   The LCM of the denominators is $x - 7$. We multiply by the LCM:

$$(x-7) \cdot \frac{x-3}{x-7} = (x-7) \cdot \frac{4}{x-7} \qquad \text{Multiplying by } x-7$$

$$x - 3 = 4 \qquad\qquad \text{Simplifying}$$

$$x = 7.$$

The possible solution is 7. We check:

Check:
$$\frac{x-3}{x-7} = \frac{4}{x-7}$$

$$\begin{array}{c|c} \dfrac{7-3}{7-7} & \dfrac{4}{7-7} \\[2mm] \dfrac{4}{0} & \dfrac{4}{0} \end{array}$$

Since division by 0 is undefined, 7 is not a solution. The equation has no solutions. The solution set is $\varnothing$.

9. Solve. Don't forget to check.

$$\frac{x+4}{x+5} = \frac{-1}{x+5}$$

In Example 4, we did not obtain equivalent equations because the multiplication principle gives equivalent equations only when we are multiplying by nonzero numbers.

> When we use the multiplication principle and multiply (or divide) by an expression with a variable, we may not obtain equivalent equations. We must check possible solutions by substituting in the original equation.

DO EXERCISE 9.

Example 5   Solve: $\dfrac{x^2}{x-3} = \dfrac{9}{x-3}$.

Solution   The LCM of the denominators is $x - 3$. We multiply by the LCM:

$$(x-3) \cdot \frac{x^2}{x-3} = (x-3) \cdot \frac{9}{x-3} \qquad \text{Multiplying by } x-3$$

$$x^2 = 9 \qquad \text{Simplifying}$$

$$x^2 - 9 = 0$$

$$(x+3)(x-3) = 0$$

$$x + 3 = 0 \quad \text{or} \quad x - 3 = 0 \qquad \text{Principle of zero products}$$

$$x = -3 \quad \text{or} \qquad x = 3.$$

Solve. Don't forget to check.

10. $\dfrac{y^2}{y+4} = \dfrac{16}{y+4}$

The possible solutions are 3 and -3. We must check, since we have multiplied by an expression with a variable.

For 3:

$$\frac{x^2}{x-3} = \frac{9}{x-3}$$

$$\begin{array}{c|c} \dfrac{3^2}{3-3} & \dfrac{9}{3-3} \\[2ex] \hline \dfrac{9}{0} & \dfrac{9}{0} \end{array} \qquad \text{3 does not check.}$$

For -3:

$$\frac{x^2}{x-3} = \frac{9}{x-3}$$

$$\begin{array}{c|c} \dfrac{(-3)^2}{-3-3} & \dfrac{9}{-3-3} \\[2ex] \hline -\dfrac{9}{6} & -\dfrac{9}{6} \end{array} \qquad \text{-3 checks.}$$

11. $\dfrac{x^2}{x-5} = \dfrac{36}{x-5}$

Thus the solution set is $\{-3\}$.

We can actually determine the nonmeaningful replacements before we start solving by noting when the denominators of the original equation are 0.

DO EXERCISES 10 AND 11.

The general procedure for solving rational equations involves multiplying on both sides by the LCM of all the denominators. This procedure is called **clearing of fractions.**

Example 6   Solve: $\dfrac{14}{x+2} - \dfrac{1}{x-4} = 1$.

Solution   We note at the outset that -2 and 4 are *not* meaningful replacements. We multiply by the LCM of all the denominators: $(x + 2)(x - 4)$.

$$(x + 2)(x - 4) \cdot \left[\frac{14}{x + 2} - \frac{1}{x - 4} \right] = (x + 2)(x - 4) \cdot 1$$

$$(x + 2)(x - 4) \cdot \frac{14}{x + 2} - (x + 2)(x - 4) \cdot \frac{1}{x - 4} = (x + 2)(x - 4) \cdot 1$$

Using the distributive law

$$14(x - 4) - (x + 2) = (x + 2)(x - 4)$$

Simplifying

$$14x - 56 - x - 2 = x^2 - 2x - 8$$
$$13x - 58 = x^2 - 2x - 8$$
$$0 = x^2 - 15x + 50$$
$$0 = (x - 10)(x - 5)$$
$$x - 10 = 0 \quad \text{or} \quad x - 5 = 0$$

Principle of zero products

$$x = 10 \quad \text{or} \quad x = 5$$

The possible solutions are 10 and 5. These check, so the solution set is $\{10, 5\}$. ◀

DO EXERCISE 12.

Example 7 Solve: $\dfrac{12x}{x - 4} - \dfrac{3x^2}{x + 4} = \dfrac{384}{x^2 - 16}$.

Solution We note at the outset that 4 and -4 are *not* meaningful replacements. We first find the LCM by factoring the denominators:

$$\frac{12x}{x - 4} - \frac{3x^2}{x + 4} = \frac{384}{(x - 4)(x + 4)}.$$

The LCM is $(x - 4)(x + 4)$, or $x^2 - 16$. We multiply by the LCM. After clearing of fractions, we factor by grouping:

$$(x - 4)(x + 4) \left[\frac{12x}{x - 4} - \frac{3x^2}{x + 4} \right] = (x - 4)(x + 4) \cdot \frac{384}{x^2 - 16}$$

$$(x - 4)(x + 4) \cdot \frac{12x}{x - 4} - (x - 4)(x + 4) \cdot \frac{3x^2}{x + 4} = (x - 4)(x + 4) \cdot \frac{384}{x^2 - 16}$$

Using the distributive law

$$12x(x + 4) - 3x^2(x - 4) = 384 \qquad \text{Simplifying}$$
$$12x^2 + 48x - 3x^3 + 12x^2 = 384$$
$$-3x^3 + 24x^2 + 48x - 384 = 0$$
$$3x^3 - 24x^2 - 48x + 384 = 0$$

Multiplying by -1 to ease the factoring

$$3(x^3 - 8x^2 - 16x + 128) = 0$$

Factoring out the common factor 3

$$x^3 - 8x^2 - 16x + 128 = 0 \qquad \text{Multiplying by } \tfrac{1}{3}$$
$$x^2(x - 8) - 16(x - 8) = 0 \qquad \text{Factoring by grouping}$$
$$(x^2 - 16)(x - 8) = 0$$
$$(x - 4)(x + 4)(x - 8) = 0$$

12. Solve:

$$\frac{4}{x + 5} + \frac{1}{x - 5} = \frac{1}{x^2 - 25}.$$

13. Solve:

$$\frac{7x - 12}{x - 3} - \frac{x^2}{x + 3} = \frac{54}{x^2 - 9}.$$

Using the principle of zero products gives us

$$x - 4 = 0 \quad \text{or} \quad x + 4 = 0 \quad \text{or} \quad x - 8 = 0$$
$$x = 4 \quad \text{or} \quad x = -4 \quad \text{or} \quad x = 8.$$

The number 8 checks, but the numbers -4 and 4 do not. The solution is 8. The solution set is $\{8\}$. ◀

DO EXERCISE 13.

● EXERCISE SET 2.2

A Determine whether the equations of the pair are equivalent.

1. $3x + 5 = 12$,
$3x = 7$

2. $x^2 = -7x$,
$x = -7$

3. $x = 3$,
$x^2 = 9$

4. $2y + 1 = -3$,
$8y + 4 = -12$

5. $\dfrac{(x - 3)(x + 9)}{(x - 3)} = x + 9$,

$x + 9 = x + 9$

6. $x^2 + x - 20 = 0$,
$x^2 - 25 = 0$

B Solve.

7. $\dfrac{1}{4} + \dfrac{1}{5} = \dfrac{1}{t}$

8. $\dfrac{1}{3} - \dfrac{5}{6} = \dfrac{1}{x}$

9. $\dfrac{3}{x - 8} = \dfrac{x - 5}{x - 8}$

10. $\dfrac{23}{y} = \dfrac{-5}{y}$

11. $\dfrac{x + 2}{4} - \dfrac{x - 1}{5} = 15$

12. $\dfrac{t + 1}{3} - \dfrac{t - 1}{2} = 1$

13. $\dfrac{3x}{x + 2} + \dfrac{6}{x} = \dfrac{12}{x^2 + 2x}$

14. $\dfrac{5x}{x - 4} - \dfrac{20}{x} = \dfrac{80}{x^2 - 4x}$

15. $\dfrac{x + 2}{2} + \dfrac{3x + 1}{5} = \dfrac{x - 2}{4}$

16. $\dfrac{2x - 1}{3} - \dfrac{x - 2}{5} = \dfrac{x}{2}$

17. $\dfrac{1}{2} + \dfrac{2}{x} = \dfrac{1}{3} + \dfrac{3}{x}$

18. $\dfrac{1}{t} + \dfrac{1}{2t} + \dfrac{1}{3t} = 5$

19. $\dfrac{4}{x^2 - 1} - \dfrac{2}{x - 1} = \dfrac{3}{x + 1}$

20. $\dfrac{3y + 5}{y^2 + 5y} + \dfrac{y + 4}{y + 5} = \dfrac{y + 1}{y}$

21. $\dfrac{490}{x^2 - 49} = \dfrac{5x}{x - 7} - \dfrac{35}{x + 7}$

22. $\dfrac{3}{m + 2} + \dfrac{2}{m - 2} = \dfrac{4m - 4}{m^2 - 4}$

23. $\dfrac{4}{x} - \dfrac{4}{x - 6} = \dfrac{24}{6x - x^2}$

24. $\dfrac{24}{4 - x^2} = \dfrac{6}{x + 2} - \dfrac{3x}{x - 2}$

25. $\dfrac{11 - t^2}{3t^2 - 5t + 2} = \dfrac{2t + 3}{3t - 2} + \dfrac{t - 3}{1 - t}$

26. $\dfrac{1}{3y^2 - 10y + 3} = \dfrac{6y}{9y^2 - 1} + \dfrac{2}{1 - 3y}$

27. $\dfrac{7x}{x - 3} - \dfrac{21}{x} + 11 = \dfrac{63}{x^2 - 3x}$

28. $\dfrac{3x}{x - 5} - \dfrac{15}{x + 5} = \dfrac{150}{x^2 - 25}$

29. ▦ $\dfrac{2.315}{y} - \dfrac{12.6}{17.4} = \dfrac{6.71}{7} + 0.763$

30. ▦ $\dfrac{6.034}{x} - 43.17 = \dfrac{0.793}{x} + 18.15$

31. $\dfrac{2x^2}{x - 3} + \dfrac{4x - 6}{x + 3} = \dfrac{108}{x^2 - 9}$

32. $\dfrac{3x^2}{x + 2} + \dfrac{48}{x^2 - 4} = \dfrac{2x + 8}{x - 2}$

33. $\dfrac{24}{x^2 - 2x + 4} = \dfrac{3x}{x + 2} + \dfrac{72}{x^3 + 8}$

34. $\dfrac{90}{x^2 - 3x + 9} - \dfrac{5x}{x + 3} = \dfrac{405}{x^3 + 27}$

35. $\dfrac{5}{x - 1} + \dfrac{9}{x^2 + x + 1} = \dfrac{15}{x^3 - 1}$

36. $\dfrac{7}{x + 2} + \dfrac{5}{x^2 - 2x + 4} = \dfrac{84}{x^3 + 8}$

37. $\dfrac{7}{x - 9} - \dfrac{7}{x} = \dfrac{63}{x^2 - 9x}$

38. $\dfrac{26}{x + 13} - \dfrac{14}{x + 7} = \dfrac{12x}{x^2 + 20x + 91}$

39. $\dfrac{(x - 3)^2}{x - 3} = x - 3$

40. $\dfrac{x^2 + 6x - 16}{x - 2} = x + 8$

41. $\dfrac{x^3 + 8}{x + 2} = x^2 - 2x + 4$

42. $\dfrac{x + 8}{x - 2} = \dfrac{8 + x}{-2 + x}$

● SYNTHESIS

43. Determine whether each equation is equivalent to the one that follows:

$$x^2 - x - 20 = x^2 - 25 \qquad (1)$$
$$(x - 5)(x + 4) = (x - 5)(x + 5) \qquad (2)$$
$$x + 4 = x + 5 \qquad (3)$$
$$4 = 5. \qquad (4)$$

Equations that are true for all meaningful replacements of the variables are called **identities.** (See Section 2.1.) Determine whether each of the following equations is an identity.

44. $\dfrac{x^2 + 6x - 16}{x - 2} = x + 8$ **45.** $x + 4 = 4 + x$

46. $(x - 1)(x^2 + x + 1) = x^3 - 1$

47. $\dfrac{x^3 + 8}{x^2 - 4} = \dfrac{x^2 - 2x + 4}{x - 2}$

48. $(x + 7)^2 = x^2 + 49$

49. $\sqrt{x^2 - 16} = x - 4$

● CHALLENGE

50. Solve:

$$\frac{x + 3}{x + 2} - \frac{x + 4}{x + 3} = \frac{x + 5}{x + 4} - \frac{x + 6}{x + 5}.$$

2.3

FORMULAS AND PROBLEM SOLVING

OBJECTIVES

You should be able to:

A ▸ Solve a formula for a given variable.

B ▸ Use the problem-solving strategy to solve applied problems.

A ▸ Formulas

A formula is a "recipe" for doing a calculation. An example is $A = \pi rs + \pi r^2$, which gives the area A of a cone in terms of the slant height s and the radius of the base r.

Suppose we want to find the slant height s when the area A and the radius r are known. Our knowledge of equations allows us to get s alone on one side or, as we say, "solve the formula for s."

Example 1 Solve $A = \pi rs + \pi r^2$ for s.

Solution We have

$A - \pi r^2 = \pi rs$ Adding $-\pi r^2$

$\dfrac{A - \pi r^2}{\pi r} = s.$ Multiplying by $\dfrac{1}{\pi r}$

An equivalent expression for s is

$$\frac{A}{\pi r} - r.$$

◀

Example 2 Solve

$\dfrac{1}{R} = \dfrac{1}{r_1} + \dfrac{1}{r_2}$

for R. (This is a formula for resistance in parallel from electricity.)

1. Solve

$$C = \frac{5}{9}(F - 32)$$

for F. (This is a formula for Celsius, or Centigrade, temperature in terms of Fahrenheit temperature F.)

2. Solve

$$\frac{1}{R} = \frac{1}{r_1} + \frac{1}{r_2}$$

for r_2.

Solution   We first multiply by the LCM, which is Rr_1r_2:

$$Rr_1r_2 \cdot \frac{1}{R} = Rr_1r_2 \cdot \left(\frac{1}{r_1} + \frac{1}{r_2} \right)$$

$$Rr_1r_2 \cdot \frac{1}{R} = Rr_1r_2 \cdot \frac{1}{r_1} + Rr_1r_2 \cdot \frac{1}{r_2} \qquad \text{Using a distributive law}$$

$$r_1r_2 = Rr_2 + Rr_1 \qquad \text{Simplifying}$$

$$r_1r_2 = (r_2 + r_1)R \qquad \text{Factoring out the common factor } R$$

$$\frac{r_1r_2}{r_2 + r_1} = R. \qquad \text{Multiplying by } \frac{1}{r_2 + r_1} \qquad \blacktriangleleft$$

DO EXERCISES 1 AND 2.

◗ **Problem Solving**

An **applied problem** is a problem in which mathematical techniques are used to answer some question. Problems like this may be posed orally. They may be real-life problems that come about in the course of a conversation, or they can be hatched within the mind of one person. Thus to call them "word problems" or "story problems" is misleading.

There is no one rule that will enable us to solve applied problems, because they are of many different kinds. We can, however, describe an overall, or general, strategy. The idea is to translate the problem situation to mathematical language and then calculate to find a solution.

Five Steps for Problem Solving

1. **Familiarize** yourself with the problem situation. If the problem is presented to you in written words, then, of course, this means to read carefully. Some or all of the following can also be helpful.

   a) Make a drawing, if it makes sense to do so. It is difficult to overemphasize the importance of this!
   b) Make a written list of the known facts and a list of what you wish to find out.
   c) Organize the information in a chart or table.
   d) Assign variables to represent unknown quantities.
   e) Find further information. Look up a formula or consult a reference book or an expert in the field.
   f) Guess or estimate the answer.

2. **Translate** the problem situation to mathematical language or symbolism. For most of the problems you will encounter in algebra, this means to write one or more equations, but sometimes an inequality or some other mathematical symbolism may be appropriate.

3. **Carry out** some kind of mathematical manipulation. This means to use your mathematical knowledge to find a possible solution. In algebra, this usually means to solve an equation or system of equations.

4. **Check** to see whether your possible solution actually fits the problem situation and is thus really a solution of the problem. You may be able to solve an equation, but the solution(s) of the equation may or may not be solution(s) of the original problem.

5. **State** the answer clearly.

Whether you use all or part of the process depends on the difficulty of the problem. You may need only to guess a possible solution and check it in the problem. In this text, you will generally need to do more than that. We may sometimes illustrate all of the steps, and sometimes not. In general, the more difficult the problem, the more steps you will need to use.

Problems stated in textbooks are of necessity somewhat contrived. Problems that you encounter in nonclassroom situations will almost invariably contain insufficient information to obtain a firm, or exact, answer. They may also contain extraneous, useless information.

Let us consider a problem.

Example 3 Assume that a hydrogen atom is a sphere. Find its volume.

Solution

1. **Familiarize.** This problem is typical of those encountered in real life, because there is insufficient information to obtain an answer. To begin to find an answer, we might do the following:

   a) Try to look up the answer in a reference book or simply call a chemist, or
   b) Look up a formula for finding the volume of a sphere. On the basis of the variables in the formula, look up certain information.

   Let us suppose we did the latter. We look up a formula for the volume of a sphere (see the table of geometric formulas at the back of the book) and find it to be

   $$V = \tfrac{4}{3}\pi r^3,$$

   where $V = $ the volume and $r = $ the radius of the sphere. In a reference book, we find that the diameter d of a hydrogen atom is 0.0000000001 m. Now we know that $r = d/2$, so we can indeed complete the problem.

2. **Translate.**

   $$V = \tfrac{4}{3}\pi r^3 \quad \text{and} \quad r = d/2$$

3. **Carry out.** We substitute the value for d and calculate r. Then we compute V.

   $$r = d/2 = 0.0000000001/2 = 5.0 \times 10^{-11},$$
   $$V = \tfrac{4}{3}\pi r^3 = \tfrac{4}{3}(\pi)(5.0 \times 10^{-11})^3$$
   $$\approx 5.236 \times 10^{-31} \qquad \text{⊞ Using a calculator}$$

4. **Check.** In a problem such as this, simply recalculating is a sufficient check, because we are merely substituting into a formula.

5. **State.** The volume of a hydrogen atom is about 5.236×10^{-31} cubic meters (m^3). ◀

In the remainder of this section, we consider applied problems of various types. Although there is no rule for solving applied problems because they can be so different, it does help somewhat to consider a few different types of problems. *The best way to learn to solve applied problems is to solve a lot of them and to use the five steps that we have given you.*

3. An investment is made at 7%, compounded annually. It grows to $2782 at the end of one year. How much was originally invested?

Complete the five steps for problem solving:

1. **Familiarize.**

2. **Translate.**

3. **Carry out.**

4. **Check.**

5. **State.**

In certain simple problem situations, the translation is easy because certain words translate directly to mathematical symbols. Note that the word "is" translates to an equals sign and the word "of" translates to a multiplication sign.

Problems Involving Compound Interest

Example 4   An investment is made at 8%, compounded annually. It grows to $1404 at the end of one year. How much was originally invested?

Solution

1. **Familiarize.**   We first restate the situation as follows:

   The invested amount *plus* the interest is $1404.

   We might guess an answer, say, $1200. We take $1200 plus 8% of $1200, and get $1200 + $96, or $1296. We see that our guess is too small. We could continue our guessing, but we want to use our algebra as a tool. Though we have not found the answer, we are much more familiar with the problem.
      We let $x =$ the amount originally invested. Then we translate.

2. **Translate.**   Since the interest is 8% of the amount invested, we have the following, which translates directly:

$$\underbrace{\text{Invested amount}}_{x} \quad \underbrace{\text{plus}}_{+} \quad \underbrace{\text{8% of invested amount}}_{8\% \cdot x} \quad \underbrace{\text{is}}_{=} \quad \underbrace{\text{\$1404.}}_{\$1404}$$

3. **Carry out.**   We solve the equation:

$$x + 8\%x = 1404$$
$$x + 0.08x = 1404$$
$$(1 + 0.08)x = 1404$$
$$1.08x = 1404$$
$$x = \frac{1404}{1.08} = 1300.$$

4. **Check.**   We check in much the same manner that we made a guess when familiarizing:

   $1300 + 8% of $1300 = $1300 + $104,   or $1404.

5. **State.**   The amount originally invested was $1300.   ◀

DO EXERCISE 3.

Let us now consider an investment over a period longer than one year. If we invest P dollars at an interest rate i, compounded annually, we will have an amount in the account at the end of a year that we will call A_1. Now, $A_1 = P + Pi$, or

$$A_1 = P(1 + i).$$

Going into the second year, we have $P(1 + i)$ dollars. By the end of the second year, we will have A_2 dollars, given by

$$A_2 = A_1 \cdot (1 + i).$$

But $A_1 = P(1 + i)$, so $A_2 = P(1 + i)(1 + i)$, or

$$A_2 = P(1 + i)^2.$$

Similarly, the amount A_3 in the account at the end of three years is given by

$$A_3 = P(1 + i)^3,$$

and so on. In general, the following applies.

THEOREM 1   **Interest Compounded Annually**

If principal P is invested at an interest rate i, compounded annually, in t years it will grow to an amount A given by

$$A = P(1 + i)^t.$$

Example 5   Suppose that $1000 is invested at 9%, compounded annually. What amount will be in the account at the end of 10 years?

Solution   We do not need the entire problem-solving strategy. We have a formula. We arrive at the solution by substituting into the formula. Using the equation $A = P(1 + i)^t$, we get

$$A = 1000(1 + 0.09)^{10} = 1000(1.09)^{10}$$
$$\approx 2367.36. \quad \blacksquare \text{ Using the } y^x \text{ key}$$

The answer is $2367.36.   ◀

DO EXERCISE 4. _____

Interest may be compounded more often than once a year. Suppose it is compounded four times a year, or *quarterly*. The formula derived above can be altered to apply. We consider one fourth of a year to be an *interest period*. The *rate of interest* for such a period is then $i/4$. The number of periods will be four times the number of years. This is shown in the following diagram.

$$A = P(1 + i)^t \text{——} \boxed{\text{The number of times that interest is compounded (interest periods) is } 4t.}$$

$$\boxed{\text{For } \frac{1}{4} \text{ year, the interest rate is } \frac{i}{4}.}$$

$$A = P\left(1 + \frac{i}{4}\right)^{4t}$$

4. Suppose that $1000 is invested at 8.5%, compounded annually. How much will be in the account at the end of 8 years?

5. Suppose that $1000 is invested at 8.5%, compounded semi-annually ($n = 2$). How much will be in the account at the end of 8 years?

Now suppose that the number of interest periods per year is something other than 4, say n. Using the reasoning illustrated above, we obtain a general formula.

THEOREM 2	**Interest Compounded n Times per Year**

If principal P is invested at an interest rate i, compounded n times per year, in t years it will grow to an amount A given by

$$A = P\left(1 + \frac{i}{n}\right)^{nt}.$$

When problems involving compound interest are translated to mathematical language, the preceding formula is almost always used.

Example 6   Suppose that $1000 is invested at 9%, compounded quarterly. How much will be in the account at the end of 10 years?

Solution   In this case, $n = 4$ and $t = 10$. We substitute into the formula:

$$A = P\left(1 + \frac{i}{n}\right)^{nt} = 1000\left(1 + \frac{0.09}{4}\right)^{4 \cdot 10}$$

$$= 1000(1.0225)^{40}$$

$$\approx 2435.19. \qquad \text{Using the } y^x \text{ key}$$

The answer is $2435.19.

DO EXERCISE 5.

Problems Involving Area

Example 7   The radius of a circular swimming pool is 10 ft. A sidewalk of uniform width is constructed around the outside and has an area of 44π ft². How wide is the sidewalk?

Solution

1. **Familiarize.**   First make a drawing:

Let $x =$ the width of the walk. Then, recalling that a formula for the area of a circle is $A = \pi r^2$, we have

Area of pool $= \pi \cdot 10^2 = 100\pi$;

Area of sidewalk plus pool $= \pi \cdot (10 + x)^2$

$$= \pi(100 + 20x + x^2).$$

2. **Translate.** The translation is as follows:

$$\underbrace{\text{Area of sidewalk}}_{\text{plus pool}} - \underbrace{\text{Area of}}_{\text{pool}} = \underbrace{\text{Area of}}_{\text{sidewalk}}$$

$$\pi(100 + 20x + x^2) - 100\pi = 44\pi.$$

3. **Carry out.** We solve the equation:

$$100 + 20x + x^2 - 100 = 44 \qquad \text{Multiplying by } 1/\pi$$
$$x^2 + 20x = 44$$
$$x^2 + 20x - 44 = 0$$
$$(x + 22)(x - 2) = 0$$
$$x = -22 \quad \text{or} \quad x = 2.$$

4. **Check.** We know that -22 ft is not a solution of the original problem because width must be positive. The number 2 checks; that is, when the sidewalk is 2 ft wide, the area of the pool plus the sidewalk is $\pi \cdot 12^2$, or 144π ft^2.

5. **State.** The width of the sidewalk is 2 ft. ◀

DO EXERCISE 6.

Problems Involving Motion

For problems that deal with distance, time, and speed, we almost always need to recall the *distance formula,* or something equivalent to it.

The Distance Formula

Distance = Rate (or speed) times Time or $d = rt$, where d = distance, r = speed, and t = time.

If you memorize the equation $d = rt$, you can easily obtain either of the two equivalent equations, $r = d/t$ or $t = d/r$, as needed.

When translating motion problems to mathematical language, it is often helpful to look for some quantity that is constant in the problem. For example, two cars may travel for the same length of time, or two boats may travel the same distance. Such facts often provide the basis for setting up an equation.

Example 8 A boat travels 246 km downstream in the same time that it takes to travel 180 km upstream. The speed of the current in the stream is 5.5 km/h. Find the speed of the boat in still water.

Solution

1. **Familiarize.** We first make a drawing and lay out the known facts and any other pertinent information. The time to go downstream is the same as the time to go upstream. We call it t. We know that the speed of the current is 5.5 km/h, but we do not know the speed of the boat. Let's call it r.

6. A rectangular garden is 60 ft by 80 ft. Part of the garden is torn up to install a sidewalk of uniform width around the garden. The new area of the garden is one sixth of the old area. How wide is the sidewalk?

Complete the five steps for problem solving:

1. **Familiarize.**

2. **Translate.**

3. **Carry out.**

4. **Check.**

5. **State.**

7. An airplane flies 1062 km with the wind in the same time that it takes to fly 738 km against the wind. The speed of the plane in still air is 200 km/h. Find the speed of the wind.

Complete the five steps for problem solving:

1. Familiarize.

	Distance	Speed	Time
Downstream			
Upstream			

2. Translate.

3. Carry out.

4. Check.

5. State.

Downstream $d_1 = 246$ km t hr   Rate unknown, but equals speed of boat plus speed of current: $r + 5.5$

Upstream $d_2 = 180$ km t hr   Rate unknown, but equals speed of boat minus speed of current: $r - 5.5$

We can also organize the data in a table.

	Distance	Speed	Time
Downstream	246	$r + 5.5$	t
Upstream	180	$r - 5.5$	t

2. Translate. The table suggests that we use the formula $t = d/r$, because we can get two different expressions for t. We get two equations:

$$t = \frac{246}{r + 5.5} \quad \text{and} \quad t = \frac{180}{r - 5.5}.$$

Thus,

$$\frac{246}{r + 5.5} = \frac{180}{r - 5.5}.$$

3. Carry out. Solving for r, we get 35.5 km/h.
4. Check. We leave the check to the student.
5. State. The speed of the boat in still water is 35.5 km/h.   ◀

DO EXERCISE 7. _____

In the following example, although the distance is the same in both directions, the key to the translation lies in an additional piece of given information.

Example 9 The speed of a boat in still water is 10 mph. It travels 24 mi upstream and 24 mi downstream in a total time of 5 hr. What is the speed of the current?

Solution

1. **Familiarize.** We first make a drawing and write out the pertinent information. We know that the speed of the boat is 10 mph, but we do not know the speed of the current. Let's call it c.

Downstream $d_1 = 24$ mi t_1 Rate unknown, but is 10 mph plus speed of current: $10 + c$

Upstream $d_2 - 24$ mi t_2 unknown, but $t_1 + t_2 = 5$ hr   Rate unknown, but is 10 mph minus speed of current: $10 - c$

We can also organize the information in a table.

	Distance	Speed	Time
Downstream	24	$10 + c$	t_1
Upstream	24	$10 - c$	t_2

2. **Translate.** This time, the basis of our translation is the equation involving the times:

$$t_1 + t_2 = 5.$$

We use $t = d/r$ with the information from the table. This leads us to

$$\frac{d_1}{r_1} + \frac{d_2}{r_2} = 5$$

and then

$$\frac{24}{10 + c} + \frac{24}{10 - c} = 5,$$

where c is the speed of the current.
3. **Carry out.** Solving for c, we get $c = -2$ or $c = 2$.
4. **Check.** Since speed cannot be negative in this problem, -2 cannot be a solution. But 2 checks.
5. **State.** The speed of the current is 2 mph. ◀

DO EXERCISES 8 AND 9.

Problems Involving Work

Example 10 Typist A can do a certain job in 3 hr. Typist B can do the same job in 5 hr. How long would it take both, working together, to do the same job?

Solution

1. **Familiarize.** We list the facts:

   A can do the typing job in 3 hr;
   B can do the same typing job in 5 hr.

8. A train leaves a station and travels north at a speed of 75 km/h. Two hours later, a second train leaves on a parallel track traveling north at a speed of 125 km/h. How far from the station will the second train overtake the first train?

9. A train leaves Pittsburgh and travels 300 mi to Philadelphia. On the return trip, it travels 10 mph faster. The total time for the round trip was 11 hr. How fast did the train travel on each part of the trip?

We want to know how long it will take them working together. Let's let t be that number of hours. We first consider two common, *but incorrect*, approaches to the problem.

a) One *incorrect* approach is to simply add the two times:

$$3 \text{ hr} + 5 \text{ hr} = 8 \text{ hr}.$$

This cannot be correct since we know that either A or B can do the typing job alone in less than 8 hr.

b) Another *incorrect* approach is to split up the job so that each does half the job. Then

A does $\frac{1}{2}$ the job in $\frac{1}{2}(3 \text{ hr})$, or 1.5 hr, and

B does $\frac{1}{2}$ the job in $\frac{1}{2}(5 \text{ hr})$, or 2.5 hr.

But this would leave A idle for 1 hr, and they would not be working *together* the entire time. We can at least see that, by working together, A and B can finish the job in a time somewhere between 1.5 hr and 2.5 hr.

Let's consider how much of the job is done in 1 hr, 2 hr, 3 hr, and so on. Since A can do the whole typing job in 3 hr, A can do $\frac{1}{3}$ of it in 1 hr. Since B can do the whole job in 5 hr, B can do $\frac{1}{5}$ of it in 1 hr. Working together, they can do

$$\frac{1}{3} + \frac{1}{5} = \frac{8}{15} \text{ in 1 hr.} \tag{1}$$

In 2 hr, A can do $2(\frac{1}{3})$ of the job and B can do $2(\frac{1}{5})$ of the job. Working together, they can do

$$2\left(\frac{1}{3}\right) + 2\left(\frac{1}{5}\right) = \frac{16}{15}, \quad \text{or } 1\frac{1}{15} \text{ in 2 hr.} \tag{2}$$

But $1\frac{1}{15}$ would represent doing more than 1 job.

We let $t =$ the number of hours required for A and B, working together, to do the job.

2. **Translate.** From equations (1) and (2), we see that the time we want is some number t for which

$$t\left(\frac{1}{3}\right) + t\left(\frac{1}{5}\right) = 1, \quad \text{or} \quad \frac{t}{3} + \frac{t}{5} = 1.$$

3. **Carry out.** We solve the equation:

$$\frac{t}{3} + \frac{t}{5} = 1$$

$$15\left(\frac{t}{3}\right) + 15\left(\frac{t}{5}\right) = 15 \cdot 1 \qquad \text{Multiplying by the LCM, 15}$$

$$5t + 3t = 15$$

$$8t = 15$$

$$t = \frac{15}{8}, \quad \text{or } 1\frac{7}{8} \text{ hr.}$$

4. **Check.** Consider $1\frac{7}{8}$ hr. If A works $1\frac{7}{8}$ hr, then A will do $(\frac{1}{3})(\frac{15}{8})$, or $\frac{5}{8}$ of the job. Then B can do $(\frac{1}{5})(\frac{15}{8})$, or $\frac{3}{8}$ of the job. Together they will do $\frac{5}{8} + \frac{3}{8}$, or 1 complete job.

We have another partial check in noting from the familiarization that the entire job can be done between 1.5 hr, the time it takes A to do half the job, and 3 hr, the time it takes A alone.

5. **State.** It will take $1\frac{7}{8}$ hr for A and B to do the job together. ◄

DO EXERCISE 10.

Example 11 It takes Matt 9 hr longer to build a wall than it takes Chris. If they work together, they can build the wall in 20 hr. How long would it take each, working alone, to build the wall?

Solution Let $t =$ the amount of time it takes Chris working alone. Then $t + 9 =$ the amount of time it takes Matt working alone.

Thus Chris can do $1/t$ of the work in 1 hr and Matt can do $1/(t + 9)$ of it in 1 hr. We know that working together, they can do the job in 20 hr. In 20 hr, Chris does $20(1/t)$ of the job and Matt does $20(1/(t + 9))$ of the job. If we add these fractional parts, we get the entire job, represented by 1. This gives us the following equation:

$$20\left(\frac{1}{t}\right) + 20\left(\frac{1}{t + 9}\right) = 1.$$

Solving the equation, we get $t = -5$ or $t = 36$. Since negative time has no meaning in the problem, -5 is not a solution of the original problem. The number 36 checks in the original problem. Thus it would take Chris 36 hr and Matt 45 hr. ◄

DO EXERCISE 11.

10. A can mow a lawn in 4 hr, whereas B can mow the same lawn in 5 hr. How long will it take them to mow the lawn if they work together?

11. Stacy and Laurie work together and get a certain job completed in 4 hr. It would take Laurie 6 hr longer, working alone, to do the job than it would Stacy. How much time would each need to do the job working alone?

● **EXERCISE SET** **2.3**

Ⓐ Solve the formula for the indicated letter.

1. Solve $P = 2l + 2w$ for w. (Perimeter of a rectangle)

2. Solve $C = 2\pi r$ for r. (Circumference of a circle)

3. Solve $A = \frac{1}{2}bh$ for b. (Area of a triangle)

4. Solve $A = \pi r^2$ for π. (Area of a circle)

5. Solve $d = rt$ for r. (Distance formula)

6. Solve $F = ma$ for a. (Force = Mass times Acceleration)

7. Solve $E = IR$ for I.

8. Solve $F = \frac{km_1 m_2}{d^2}$ for m_2.

9. Solve $\frac{P_1 V_1}{T_1} = \frac{P_2 V_2}{T_2}$ for T_1.

10. Solve $\frac{P_1 V_1}{T_1} = \frac{P_2 V_2}{T_2}$ for V_2.

11. Solve $S = \frac{H}{m(v_1 - v_2)}$ for v_1.

12. Solve $S = \frac{H}{m(v_1 - v_2)}$ for v_2.

13. Solve $\frac{1}{F} = \frac{1}{m} + \frac{1}{p}$ for p.

14. Solve $\frac{1}{F} = \frac{1}{m} + \frac{1}{p}$ for F.

Solve for x.

15. $(x + a)(x - b) = x^2 + 5$

16. $(c + d)x + (c - d)x = c^2$

17. $10(a + x) = 8(a - x)$

18. $4(a + b + x) + 3(a + b - x) = 8a$

Ⓑ **Problem Solving**

19. An investment is made at 8%, compounded annually. It grows to $702 at the end of one year. How much was originally invested?

20. An investment is made at 9%, compounded annually. It grows to $926.50 at the end of one year. How much was originally invested?

21. In triangle ABC, angle B is five times as large as angle A. The measure of angle C is 2° less than that of angle A. Find the measures of the angles. (*Hint:* The sum of the angle measures is 180°.)

22. In triangle ABC, angle B is twice as large as angle A. Angle C measures 20° more than angle A. Find the measures of the angles.

23. The perimeter of a rectangle is 322 m. The length is 25 m more than the width. Find the dimensions.

24. The length of a rectangle is twice the width. The perimeter is 39 m. Find the dimensions.

25. A student's scores on three tests are 87%, 64%, and 78%. What must the student score on the fourth test so that the average will be 80%?

26. A student's scores on three tests are 74%, 55%, and 68%. What must the student score on the fourth test so that the average will be 70%?

27. An open box is made from a 10-cm by 20-cm piece of tin by cutting a square from each corner and folding up the edges. The area of the resulting base is 96 cm². What is the length of the sides of the squares?

28. The frame of a picture is 28 cm by 32 cm outside and is of uniform width. What is the width of the frame if 192 cm² of the picture shows?

29. After a 2% increase, the population of a city is 826,200. What was the former population?

30. After a 3% increase, the population of a city is 741,600. What was the former population?

31. A boat travels 50 km downstream in the same time that it takes to go 30 km upstream. The speed of the stream is 3 km/h. Find the speed of the boat in still water.

32. A boat travels 50 km downstream in the same time that it takes to go 30 km upstream. The speed of the boat in still water is 16 km/h. Find the speed of the stream.

33. The speed of train A is 12 mph slower than the speed of train B. Train A travels 230 mi in the same time that it takes train B to travel 290 mi. Find the speed of each train.

34. The speed of a passenger train is 14 mph faster than the speed of a freight train. The passenger train travels 400 mi in the same time that it takes the freight train to travel 330 mi. Find the speed of each train.

35. An airplane leaves Chicago for Cleveland at a speed of 475 mph. Twenty minutes later, a plane going to Chicago leaves Cleveland, which is 350 mi from Chicago, at a speed of 500 mph. When they meet, how far are they from Cleveland? (*Hint:* It is usually best to make the units consistent. That is, consider 20 min as $\frac{1}{3}$ hr.)

36. A private airplane leaves an airport and flies due east at a speed of 180 km/h. Two hours later, a jet leaves the same airport and flies due east at a speed of 900 km/h. How far from the airport will the jet overtake the private plane?

37. A can do a certain job in 3 hr, B can do the same job in 5 hr, and C can do the same job in 7 hr. How long would the job take with all three working together?

38. Pipe A can fill a tank in 4 hr, pipe B can fill it in 10 hr, and pipe C can fill it in 12 hr. The pipes are connected to the same tank. How long does it take to fill the tank if all three are running together?

39. A can do a certain job, working alone, in 3.15 hr. Working with B, A can do the same job in 2.09 hr. How long would it take B, working alone, to do the job?

40. At a factory, smokestack A pollutes the air 2.13 times as fast as smokestack B. When both stacks operate together, they yield a certain amount of pollution in 16.3 hr. Find the amount of time that it would take each to yield the same amount of pollution if it operated alone.

41. Suppose that $1000 is invested at $8\frac{3}{4}$%. How much is in the account at the end of one year, if interest is compounded (a) annually? (b) semiannually? (c) quarterly? (d) daily (use 365 days per yr)? (e) hourly?

42. Suppose that $1000 is invested at 7.5%. How much is in the account at the end of five years, if interest is compounded (a) annually? (b) semiannually? (c) quarterly? (d) daily (use 365 days per yr)? (e) hourly?

● SYNTHESIS

43. A car is driven 144 mi. If it had gone 4 mph faster, it could have made the trip in $\frac{1}{2}$ hr less time. What was the speed?

44. A car is driven 280 mi. If it had gone 5 mph faster, it could have made the trip in 1 hr less time. What was the speed?

45. A student drove 3 hr on a freeway at a speed of 55 mph and then drove 10 mi in the city at 35 mph. What was the average speed? (*Average speed* is defined as total distance divided by total time.)

46. For the first 100 km of a 200-km trip, a student drove at a speed of 40 km/h. For the second half of the trip, the student drove at a speed of 60 km/h. What was the average speed for the entire trip? (It is not 50 km/h.)

47. A driver drove half the distance of a trip at a speed of 40 mph. At what speed would the driver have to drive for the rest of the distance so that the average speed for the entire trip would be 45 mph and the trip would be completed in 1 hr?

48. At what time after 4:00 will the minute hand and the hour hand of a clock first be in the same position?

49. At what time after 10:30 will the hands of a clock first be perpendicular?

50. Three trucks, A, B, C, working together, can move a load of sand in t hours. When working alone, it takes A 1 extra hour to move the sand; B, 6 extra hours; and C, t extra hours. Find t.

● CHALLENGE

51. An airplane is flying from Los Angeles to Hawaii at a speed of 750 mph with a tailwind of 50 mph. The distance, in statute miles, from Los Angeles to Honolulu is 2574 mi.

Los Angeles

Honolulu

a) Find the point at which it takes the same amount of time to go back to Los Angeles as it does to go on to Honolulu.

b) After traveling 1187 mi, the pilot determines that it is necessary to make an emergency landing. Which would require less time: to continue to Honolulu or to return to Los Angeles?

52. A commuter drives to work at a speed of 45 mph and arrives one minute early. At 40 mph, the commuter would arrive one minute late. How far is it to work?

53. Suppose your father gathers the family together and gives half of all the money in his pocket to your mother. Then he gives one fourth of what is left to your sister, and one third of what is left after that to your brother. He then gives you half of what is left, which happens to be $2. How much was in his pocket at the outset?

54. A student walks into a bakery and says to the owner, "I will buy half of all the pies in the store, plus half a pie." The sale is then made. Another student comes into the store and makes the same statement and purchase. Then so does a third student. The owner then has exactly one pie left. How many pies did the owner have in the store to begin with?

2.4

THE COMPLEX NUMBERS

A Imaginary Numbers

Since negative numbers do not have square roots in the system of real numbers, certain equations such as $x^2 = -1$ have no solutions. A new kind of number, called *imaginary*, was invented so that negative numbers would have square roots and certain equations would have solutions. These numbers were devised, starting with an imaginary unit named i, with the agreement that $i^2 = -1$ and $i = \sqrt{-1}$.

DEFINITION   **The Number i**

The number i is defined such that $i = \sqrt{-1}$ and $i^2 = -1$.

All other imaginary numbers can then be expressed as a product of i and a nonzero real number.

DEFINITION   **Imaginary Number**

An *imaginary number* is a number that can be named bi, where b is some nonzero real number.

In the complex-number system, when $a > 0$, $\sqrt{-a} = \sqrt{-1} \cdot \sqrt{a}$. We can use this property to express certain roots in terms of i.

OBJECTIVES

You should be able to:

A Express imaginary numbers (that is, square roots of negative numbers) in terms of i, and simplify.

B Add, subtract, and multiply complex numbers, expressing the answer as $a + bi$. Also, factor sums of squares.

C Determine whether a complex number is a solution of an equation.

D Use the fact that for equality of complex numbers, the real parts must be the same and the imaginary parts must be the same.

E Find the conjugate of a complex number and divide complex numbers.

F Find the reciprocal of a complex number and express it in the form $a + bi$.

G Solve linear equations with complex-number coefficients.

Express in terms of i.

1. $\sqrt{-6}$

2. $-\sqrt{-10}$

3. $\sqrt{-4}$

4. $-\sqrt{-25}$

5. Simplify: $\sqrt{-5}\sqrt{-2}$.

Simplify.

6. $\dfrac{\sqrt{-22}}{\sqrt{-2}}$

7. $\dfrac{\sqrt{-21}}{\sqrt{3}}$

8. $\sqrt{-16} + \sqrt{-9}$

9. $\sqrt{-25} - \sqrt{-4}$

10. $\sqrt{-17} + \sqrt{-9}$

Examples Express in terms of i.

1. $\sqrt{-7} = \sqrt{-1 \cdot 7} = \sqrt{-1} \cdot \sqrt{7} = i\sqrt{7},$ or $\sqrt{7}i$ ⟵ i is *not* under the radical.
2. $\sqrt{-16} = \sqrt{-1 \cdot 16} = \sqrt{-1} \cdot \sqrt{16} = i \cdot 4 = 4i$
3. $-\sqrt{-13} = -\sqrt{-1 \cdot 13} = -\sqrt{-1} \cdot \sqrt{13} = -i\sqrt{13},$ or $-\sqrt{13}\,i$
4. $-\sqrt{-64} = -\sqrt{-1 \cdot 64} = -\sqrt{-1} \cdot \sqrt{64} = -i \cdot 8 = -8i$
5. $\sqrt{-48} = \sqrt{-1 \cdot 48} = \sqrt{-1} \cdot \sqrt{48}$
$\qquad = i \cdot \sqrt{48} = i \cdot 4\sqrt{3} = 4i\sqrt{3},$ or $4\sqrt{3}\,i$ ◀

It is also to be understood that the imaginary numbers obey the familiar laws of real numbers, such as the commutative and the associative laws.

Example 6 Simplify: $\sqrt{-3}\sqrt{-7}$.

Solution *Important:* We first express the two imaginary numbers in terms of i:

$$\sqrt{-3}\sqrt{-7} = i\sqrt{3} \cdot i\sqrt{7}.$$

Now, rearranging and combining, we have

$$i^2\sqrt{3}\sqrt{7} = -1 \cdot \sqrt{21} = -\sqrt{21}.$$

Had we not expressed the imaginary numbers in terms of i at the outset, we would have obtained $\sqrt{21}$ instead of $-\sqrt{21}$. ◀

> **CAUTION!** All imaginary numbers must be expressed in terms of i before simplifying.

DO EXERCISES 1–5.

Example 7 Simplify: $-\sqrt{20}/\sqrt{-5}$.

Solution

$$\frac{-\sqrt{20}}{\sqrt{-5}} = \frac{-\sqrt{20}}{i\sqrt{5}} = \frac{-\sqrt{20}}{i\sqrt{5}} \cdot \frac{i}{i}$$
$$= \frac{-i\sqrt{20}}{i^2\sqrt{5}} = \frac{-i}{-1}\sqrt{\frac{20}{5}}$$
$$= i\sqrt{4}$$
$$= 2i$$ ◀

Example 8 Simplify: $\sqrt{-9} + \sqrt{-25}$.

Solution

$$\sqrt{-9} + \sqrt{-25} = i\sqrt{9} + i\sqrt{25}$$
$$= 3i + 5i = (3 + 5)i$$
$$= 8i$$ ◀

DO EXERCISES 6–10.

Powers of i

We can simplify powers of i using the fact that $i^2 = -1$ and expressing the given power of i in terms of i^2. Consider the following:

$$i,$$
$$i^2 = -1,$$
$$i^3 = i^2 \cdot i = (-1)i = -i,$$
$$i^4 = (i^2)^2 = (-1)^2 = 1,$$
$$i^5 = i^4 \cdot i = (i^2)^2 \cdot i = (-1)^2 \cdot i = i,$$
$$i^6 = (i^2)^3 = (-1)^3 = -1.$$

Note that the powers of i cycle themselves through the values of i, -1, $-i$, and 1.

Examples   Simplify.

9. $i^{37} = i^{36} \cdot i = (i^2)^{18} \cdot i = (-1)^{18} \cdot i = 1 \cdot i = i$ $i^2 = -1$

10. $i^{58} = (i^2)^{29} = (-1)^{29} = -1$

11. $i^{75} = i^{74} \cdot i = (i^2)^{37} \cdot i = -1 \cdot i = -i$

12. $i^{80} = (i^2)^{40} = 1$ ◀

DO EXERCISES 11–13.

Complex Numbers

The equation $x^2 + 1 = 0$ has no solution in the real-number system, but it has the imaginary solutions i and $-i$ in the complex-number system. There are still rather simple-looking equations that do not have either real or imaginary solutions. For example, $x^2 - 2x + 2 = 0$ does not. If we allow sums of real and imaginary numbers, however, this equation and many others have solutions, as we will show.

In order that more equations will have solutions, we invent a new system of numbers called the **system of complex numbers.*** To form the system of complex numbers, we take the imaginary numbers and the real numbers and all the possible sums of real and imaginary numbers. These are examples of complex numbers:

$$7 - 4i, \quad -\pi + 9i, \quad 37, \quad i\sqrt{6}.$$

DEFINITION	**Complex Number**

A *complex number* is any number that can be named $a + bi$, where a and b are any real numbers. Note that a and b can both be 0.

For the complex number $a + bi$, we say that the **real part** is a and the **imaginary part** is bi.

We also agree that the familiar properties of real numbers (the *field*

---

* You may wonder why we do not invent a system of imaginary numbers. That would not make sense because products of imaginary numbers are not necessarily imaginary. Consider, for example, $i^2 = -1$.

Simplify.

11. i^{25}

12. i^{18}

13. i^{31}

properties) hold for complex numbers.* We list them here.

Commutative Law.   Addition and multiplication are commutative.

Associative Law.   Addition and multiplication are associative.

Distributive Law.   Multiplication is distributive over addition and also over subtraction.

Identities.   The additive identity is $0 + 0i$, or 0. The multiplicative identity is $1 + 0i$, or 1.

Additive Inverse.   Every complex number $a + bi$ has the opposite, or additive inverse, $-a - bi$.

Multiplicative Inverse.   Every nonzero complex number has a reciprocal, or multiplicative inverse.

The complex-number system is an extension of the real-number system. Since $0 + bi = bi$, every imaginary number is a complex number. Similarly, $a + 0i = a$, so every real number is a complex number. The relationships among various real and complex numbers is shown in the following diagram.

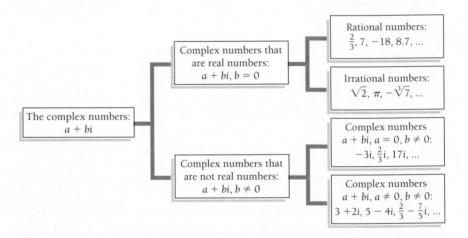

It is important to keep in mind some comparisons between numbers that have real-number roots and those that have complex-number roots that are not real. For example, $\sqrt{-48}$ is a complex number that is not a real number because we are taking the square root of a negative number. *But,* $\sqrt[3]{-125}$ is a real number because we are taking the cube root of a negative number and *any* real number has a cube root that is a real number.

Calculations

Since the field properties hold in the system of complex numbers, calculations are much the same as those for real numbers. The primary difference is that one must remember that $i^2 = -1$.

Example 13   Simplify: $(8 + 6i) + (3 + 2i)$.

---

* In a more rigorous treatment, addition and multiplication are defined for complex numbers. It is then proved that the field properties hold.

Solution

$$(8 + 6i) + (3 + 2i) = 8 + 3 + 6i + 2i$$
$$= 11 + 8i \qquad \blacktriangleleft$$

Example 14 Simplify: $(1 + 2i)(1 + 3i)$.

Solution

$$(1 + 2i)(1 + 3i) = 1 + 3i + 2i + 6i^2$$
$$= 1 - 6 + 3i + 2i \qquad i^2 = -1$$
$$= -5 + 5i \qquad \blacktriangleleft$$

Example 15 Simplify: $(3 + 2i) - (5 - 2i)$.

Solution

$$(3 + 2i) - (5 - 2i) = 3 + 2i - 5 + 2i$$
$$= 3 - 5 + 2i + 2i$$
$$= -2 + 4i \qquad \blacktriangleleft$$

DO EXERCISES 14–19.

In the system of real numbers, a sum of two squares cannot be factored. In the system of complex numbers, a sum of squares is always factorable.

Example 16 Factor: $x^2 + y^2$.

Solution

$$x^2 + y^2 = (x + yi)(x - yi)$$

A check by multiplying will show that this is correct. $\qquad \blacktriangleleft$

DO EXERCISES 20 AND 21.

▶ **C Solutions of Equations**

In the system of complex numbers, a great many equations have solutions. In fact, any equation $P(x) = 0$, where $P(x)$ is a nonconstant polynomial, has a solution.

Example 17 Determine whether $1 + i$ is a solution of $x^2 - 2x + 2 = 0$.

Solution

$$
\begin{array}{c|c}
x^2 - 2x + 2 = 0 & \\
\hline
(1 + i)^2 - 2(1 + i) + 2 & 0 \\
1 + 2i + i^2 - 2 - 2i + 2 & \\
1 + 2i - 1 - 2 - 2i + 2 & \\
& 0 \\
\end{array}
\qquad \text{Substituting}
$$

The number $1 + i$ is a solution. $\qquad \blacktriangleleft$

DO EXERCISE 22.

Simplify.

14. $(8 - i) + (4 + 2i)$

15. $(9 + 2i) - (4 + 3i)$

16. $(2 + 4i)(3 + i)$

17. $(4 + 5i) + (4 - 5i)$

18. $2i(4 + 3i)$

19. $(5 + 6i) - (5 + 3i)$

Factor.

20. $x^2 + 4$

21. $9 + y^2$

22. Determine whether $1 - i$ is a solution of $x^2 - 2x + 2 = 0$.

23. Given that

$3x + 1 + (y + 2)i = 2x + 2yi,$

find x and y.

Find the conjugate.

24. $7 + 2i$

25. $6 - 4i$

26. $-5i$

27. $3i$

28. -3

29. 8

 Equality for Complex Numbers

An equation $a + bi = c + di$ is true if and only if the real parts are the same and the imaginary parts are the same. In other words, we have the following:

$$a + bi = c + di \quad \text{if and only if} \quad a = c \text{ and } b = d.$$

Example 18 Suppose that $3x + yi = 5x + 1 + 2i$. Find x and y.

Solution   We equate the real parts: $3x = 5x + 1$. Solving this equation, we obtain $x = -\frac{1}{2}$. We then equate the imaginary parts: $yi = 2i$. Thus, $y = 2$. ◀

DO EXERCISE 23.

E **Conjugates and Division**

We define the *conjugate* of a complex number as follows.

DEFINITION	**Conjugate**

The *conjugate* of a complex number $a + bi$ is $a - bi$, and the *conjugate* of a complex number $a - bi$ is $a + bi$.

We illustrate:

The conjugate of $3 + 4i$ is $3 - 4i$.

The conjugate of $5 - 7i$ is $5 + 7i$.

The conjugate of $5i$ is $-5i$.

The conjugate of 6 is 6.

DO EXERCISES 24–29.

Division

Fractional notation is useful for division of complex numbers. We also use the notion of conjugates.

Example 19   Divide $4 + 5i$ by $1 + 4i$.

Solution   We write fractional notation and then multiply by 1:

$$\frac{4 + 5i}{1 + 4i} = \frac{4 + 5i}{1 + 4i} \cdot \frac{1 - 4i}{1 - 4i} \qquad \text{Note that } 1 - 4i \text{ is the conjugate of the divisor.}$$

$$= \frac{(4 + 5i)(1 - 4i)}{1^2 - 4^2 i^2}$$

$$= \frac{4 - 11i - 20i^2}{1 + 16} \qquad i^2 = -1$$

$$= \frac{24 - 11i}{17}$$

$$= \frac{24}{17} - \frac{11}{17}i. \qquad ◀$$

The procedure shown in Example 19 allows us always to find a quotient of two numbers and express it in the form $a + bi$. This is true because the product of a number and its conjugate is always a real number, which gives us a real-number denominator:

$$(a - bi)(a + bi) = a^2 - (bi)^2$$
$$= a^2 - b^2 i^2 = a^2 - b^2(-1)$$
$$= a^2 + b^2.$$

DO EXERCISES 30 AND 31.

Reciprocals

We can find the reciprocal, or multiplicative inverse, of a complex number by division. The reciprocal of a complex number $a + bi$ is $1/(a + bi)$.

Example 20   Find the reciprocal of $2 - 3i$ and express it in the form $a + bi$.

Solution

a) The reciprocal of $2 - 3i$ is $1/(2 - 3i)$.
b) We can express it in the form $a + bi$ as follows:

$$\frac{1}{2 - 3i} = \frac{1}{2 - 3i} \cdot \frac{2 + 3i}{2 + 3i}$$
$$= \frac{2 + 3i}{2^2 - 3^2 i^2} = \frac{2 + 3i}{4 + 9}$$
$$= \frac{2}{13} + \frac{3}{13}i. \qquad \blacktriangleleft$$

DO EXERCISE 32.

Linear Equations

Linear equations with complex-number coefficients are solved in the same way as equations with real-number coefficients. The steps used in solving depend on the field properties in each case.

Example 21   Solve: $3ix + 4 - 5i = (1 + i)x + 2i$.

Solution

$$3ix - (1 + i)x = 2i - (4 - 5i) \qquad \text{Adding } -(1 + i)x \text{ and } -(4 - 5i)$$
$$(-1 + 2i)x = -4 + 7i \qquad \text{Simplifying}$$
$$x = \frac{-4 + 7i}{-1 + 2i} \qquad \text{Dividing}$$
$$= \frac{-4 + 7i}{-1 + 2i} \cdot \frac{-1 - 2i}{-1 - 2i} \qquad \text{Simplifying}$$
$$= \frac{18 + i}{5} = \frac{18}{5} + \frac{1}{5}i \qquad \blacktriangleleft$$

DO EXERCISE 33.

Divide.

30. $\dfrac{1 + 3i}{3 + 2i}$

31. $\dfrac{2 + i}{3 - 2i}$

32. Find the reciprocal of $3 + 4i$ and express it in the form $a + bi$.

33. Solve:
$$3 - 4i + 2ix = 3i - (1 - i)x.$$

● **EXERCISE SET** **2.4**

A   Express in terms of i.

1. $\sqrt{-15}$ **2.** $\sqrt{-17}$ **3.** $\sqrt{-81}$

4. $\sqrt{-25}$ **5.** $-\sqrt{-12}$ **6.** $-\sqrt{-20}$

Simplify. Leave answers in terms of i in Exercises 7–10.

7. $\sqrt{-16} + \sqrt{-25}$

8. $\sqrt{-36} - \sqrt{-4}$

9. $\sqrt{-7} - \sqrt{-10}$

10. $\sqrt{-5} + \sqrt{-7}$

11. $\sqrt{-5}\sqrt{-11}$

12. $\sqrt{-7}\sqrt{-8}$

13. $-\sqrt{-4}\sqrt{-5}$

14. $-\sqrt{-9}\sqrt{-7}$

15. $\dfrac{-\sqrt{5}}{\sqrt{-2}}$

16. $\dfrac{\sqrt{-7}}{-\sqrt{5}}$

17. $\dfrac{\sqrt{-9}}{-\sqrt{4}}$

18. $\dfrac{-\sqrt{25}}{\sqrt{-16}}$

19. $\dfrac{-\sqrt{-36}}{\sqrt{-9}}$

20. $\dfrac{\sqrt{-25}}{-\sqrt{-16}}$

Simplify.

21. i^{18} **22.** i^{14} **23.** i^{15}

24. i^{16} **25.** i^{39} **26.** i^{40}

27. i^{46} **28.** i^{72}

B   Simplify.

29. $(2 + 3i) + (4 + 2i)$ **30.** $(5 - 2i) + (6 + 3i)$

31. $(4 + 3i) + (4 - 3i)$ **32.** $(2 + 3i) + (-2 - 3i)$

33. $(8 + 11i) - (6 + 7i)$ **34.** $(9 - 5i) - (4 + 2i)$

35. $2i - (4 + 3i)$ **36.** $3i - (5 + 2i)$

37. $(1 + 2i)(1 + 3i)$ **38.** $(1 + 4i)(1 - 3i)$

39. $(1 + 2i)(1 - 3i)$ **40.** $(2 + 3i)(2 - 3i)$

41. $3i(4 + 2i)$ **42.** $5i(3 - 4i)$

43. $(2 + 3i)^2$ **44.** $(3 - 2i)^2$

Factor.

45. $4x^2 + 25y^2$ **46.** $16a^2 + 49b^2$

C

47. Determine whether $1 + 2i$ is a solution of
$$x^2 - 2x + 5 = 0.$$

48. Determine whether $1 - 2i$ is a solution of
$$x^2 - 2x + 5 = 0.$$

D   Solve for x and y.

49. $4x + 7i = -6 + yi$

50. $5x + (y - 3)i = x - 4 + 6yi$

E   Simplify. Write the answer in the form $a + bi$.

51. $\dfrac{4 + 3i}{1 - i}$ **52.** $\dfrac{2 - 3i}{5 - 4i}$

53. $\dfrac{\sqrt{2} + i}{\sqrt{2} - i}$ **54.** $\dfrac{\sqrt{3} + i}{\sqrt{3} - i}$

55. $\dfrac{3 + 2i}{i}$ **56.** $\dfrac{2 + 3i}{i}$

57. $\dfrac{i}{2 + i}$ **58.** $\dfrac{3}{5 - 11i}$

59. $\dfrac{1 - i}{(1 + i)^2}$ **60.** $\dfrac{1 + i}{(1 - i)^2}$

61. $\dfrac{3 - 4i}{(2 + i)(3 - 2i)}$ **62.** $\dfrac{(4 - i)(5 + i)}{(6 - 5i)(7 - 2i)}$

63. $\dfrac{1 + i}{1 - i} \cdot \dfrac{2 - i}{1 - i}$ **64.** $\dfrac{1 - i}{1 + i} \cdot \dfrac{2 + i}{1 + i}$

65. $\dfrac{3 + 2i}{1 - i} + \dfrac{6 + 2i}{1 - i}$ **66.** $\dfrac{4 - 2i}{1 + i} + \dfrac{2 - 5i}{1 + i}$

F   Find the reciprocal and express it in the form $a + bi$.

67. $4 + 3i$ **68.** $4 - 3i$

69. $5 - 2i$ **70.** $2 + 5i$

71. i **72.** $-i$

73. $-4i$ **74.** $5i$

G   Solve.

75. $(3 + i)x + i = 5i$

76. $(2 + i)x - i = 5 + i$

77. $2ix + 5 - 4i = (2 + 3i)x - 2i$

78. $5ix + 3 + 2i = (3 - 2i)x + 3i$

79. $(1 + 2i)x + 3 - 2i = 4 - 5i + 3ix$

80. $(1 - 2i)x + 2 - 3i = 5 - 4i + 2x$

81. $(5 + i)x + 1 - 3i = (2 - 3i)x + 2 - i$

82. $(5 - i)x + 2 - 3i = (3 - 2i)x + 3 - i$

● **SYNTHESIS** _____

83. Show that the general rule for radicals, in real numbers, $\sqrt{a \cdot b} = \sqrt{a} \cdot \sqrt{b}$, does not hold for complex numbers.

84. Show that the general rule for radicals, in real numbers, $\sqrt{a/b} = \sqrt{a}/\sqrt{b}$, does not hold for complex numbers.

We can use a single letter for a complex number. For example, we could shorten $a + bi$ to z. To denote the conjugate of a number, we use a bar. The conjugate of z is $\bar{z}$. Or, the conjugate of $a + bi$ is $\overline{a + bi}$. Of course, by the definition of conjugates, $\overline{a + bi} = a - bi$ and $\overline{a - bi} = a + bi$. Using this notation, prove the following properties of conjugates.

85. For any complex number z, $z \cdot \bar{z}$ is a real number.

86. For any complex number z, $z + \bar{z}$ is a real number.

87. For any complex numbers z and w, $\overline{z + w} = \bar{z} + \bar{w}$.

88. For any complex numbers z and w, $\overline{z \cdot w} = \bar{z} \cdot \bar{w}$.

89. For any complex number z, $\overline{z^n} = \overline{z}^n$, where n is a natural number.

90. If z is a real number, then $\overline{z} = z$.

91. Using the properties proved in Exercises 85–90, find a polynomial in $\overline{z}$ that is the conjugate of
$$3z^5 - 4z^2 + 3z - 5.$$

92. Solve: $z + 6\overline{z} = 7$.

93. Solve: $5z - 4\overline{z} = 7 + 8i$.

94. Let $z = a + bi$. Find $\frac{1}{2}(z + \overline{z})$.

95. Let $z = a + bi$. Find $\frac{1}{2}(\overline{z} - z)$.

● **CHALLENGE** _____

96. Solve: $z^2 = -2i$. (*Hint:* Let $z = a + bi$.)

97. Solve: $z^2/4 = i$. (*Hint:* Let $z = a + bi$.)

98. Let $z = a + bi$. Find a general expression for $1/z$.

99. Let $z = a + bi$ and $w = c + di$. Find a general expression for w/z.

2.5

QUADRATIC EQUATIONS

A Solving Quadratic Equations

OBJECTIVES

You should be able to:

A Solve quadratic equations by: taking the square root on both sides; completing the square; and using the quadratic formula.

B Use the discriminant to determine the nature of the solutions of a given quadratic equation.

C Write a quadratic equation having specified solutions.

DEFINITION   **Quadratic Equation**

A *quadratic equation* is an equation equivalent to one of the form
$$ax^2 + bx + c = 0,$$
where the numbers a, b, and c are real numbers, called *coefficients*, and $a \neq 0$.

The **standard form of a quadratic equation** is
$$ax^2 + bx + c = 0,$$
where the coefficients a, b, and c are real numbers. The coefficients b and c might be 0, but the **leading coefficient** a cannot be. If it were, the polynomial would not be quadratic, that is, of second degree. You have solved such equations by factoring in Section 2.1. We now consider other methods.

Solving $(x + h)^2 = k$

First consider the equation $x^2 = k$, where k is positive. Since a positive real number has two square roots, the solutions of this equation are $\sqrt{k}$ and $-\sqrt{k}$. We will often abbreviate this by saying that the solutions are $\pm\sqrt{k}$. If $k = 0$, there is just one solution, 0. If k is negative, there are two nonreal complex solutions.

Example 1 Solve: $5x^2 = 15$.

Solution

$$\begin{aligned} x^2 &= 3 &&\text{Multiplying by } \tfrac{1}{5} \\ x = \sqrt{3} \quad &\text{or} \quad x = -\sqrt{3} &&\text{Taking square roots} \end{aligned}$$

These numbers check, so the solutions are $\sqrt{3}$ and $-\sqrt{3}$. The solution set is $\{\sqrt{3}, -\sqrt{3}\}$, or $\{\pm\sqrt{3}\}$. ◀

DO EXERCISES 1–4.

Solve for x.

1. $x^2 = 7$

2. $5x^2 = 0$

3. $3x^2 = \pi$

4. $mx^2 = n$

5. Solve: $2x^2 + 1 = 0$.

Sometimes we get solutions that are not real numbers (nonreal) but are still, of course, complex numbers.

Example 2   Solve: $4x^2 + 9 = 0$.

Solution

$$x^2 = -\frac{9}{4}$$ Adding -9 and multiplying by $\frac{1}{4}$

$$x = \sqrt{-\frac{9}{4}} \quad \text{or} \quad x = -\sqrt{-\frac{9}{4}}$$ Taking square roots

$$x = \frac{3}{2}i \quad \text{or} \quad x = -\frac{3}{2}i$$ Simplifying

The numbers $\frac{3}{2}i$ and $-\frac{3}{2}i$ check, so they are solutions. Thus the solution set is $\{\frac{3}{2}i, -\frac{3}{2}i\}$, or $\{\pm\frac{3}{2}i\}$. ◄

DO EXERCISE 5.

Solve.

6. $(x + 4)^2 = 7$

Example 3   Solve: $(x + 5)^2 = 3$.

Solution
$$x + 5 = \pm\sqrt{3}$$ Taking square roots
$$x = -5 \pm \sqrt{3}$$
The solution set is $\{-5 - \sqrt{3}, -5 + \sqrt{3}\}$, or $\{-5 \pm \sqrt{3}\}$. ◄

DO EXERCISES 6–8.

7. $(x - 5)^2 = 3$

Completing the Square

If the equation is not in the form $(x + h)^2 = k$, we can put it into that form by *completing the square*.

Example 4   Solve by completing the square: $x^2 - 6x - 12 = 0$.

Solution   We consider first the x^2- and x-terms:
$$x^2 - 6x \qquad - 12 = 0.$$

8. $(x + 5)^2 = 4$

We construct a trinomial square. First note that $x^2 - 6x + 9$ is a perfect square: $(x - 3)^2$. So if we add 9 to $x^2 - 6x$, we will have a perfect square. To get an equivalent equation, however, we must subtract 9 as well. This is the same as adding $9 - 9$, or 0. Then we proceed as follows:

$$x^2 - 6x + 9 - 9 - 12 = 0$$
$$(x - 3)^2 - 21 = 0$$
$$(x - 3)^2 = 21$$
$$x - 3 = \pm\sqrt{21}$$
$$x = 3 \pm \sqrt{21}.$$

The solution set is $\{3 + \sqrt{21}, 3 - \sqrt{21}\}$, or $\{3 \pm \sqrt{21}\}$. ◄

The following is the general procedure that we call *completing the square*.

> To solve $ax^2 + bx + c = 0$, by *completing the square*:
>
> 1. If $a \neq 1$, multiply both sides of the equation by $1/a$ to get the equation in the form $x^2 + Bx + C = 0$.
> 2. To complete the square on $x^2 + Bx$, take half the coefficient of x and square it. Then add and subtract that number: $(B/2)^2$.
> 3. Take the square roots and solve for x.

Example 5 Solve by completing the square: $x^2 + 3x - 5 = 0$.

Solution

1. Since $a = 1$, no multiplication by $1/a$ is necessary to get the equation in the proper form: $x^2 + 3x - 5 = 0$.
2. To complete the square on $x^2 + 3x$, we take half the coefficient of x and square it. Then we add and subtract that number: $(\frac{1}{2} \cdot 3)^2$, or $\frac{9}{4}$:

$$x^2 + 3x + \frac{9}{4} - \frac{9}{4} - 5 = 0$$

$$x^2 + 3x + \frac{9}{4} - \frac{29}{4} = 0$$

$$\left(x + \frac{3}{2}\right)^2 - \frac{29}{4} = 0$$

$$\left(x + \frac{3}{2}\right)^2 = \frac{29}{4}.$$

3. We now take the square roots and solve for x:

$$x + \frac{3}{2} = \pm\sqrt{\frac{29}{4}} = \pm\frac{\sqrt{29}}{2}$$

$$x = -\frac{3}{2} \pm \frac{\sqrt{29}}{2} = \frac{-3 \pm \sqrt{29}}{2}.$$

The solutions are

$$\frac{-3 + \sqrt{29}}{2} \quad \text{and} \quad \frac{-3 - \sqrt{29}}{2}, \quad \text{or} \quad \frac{-3 \pm \sqrt{29}}{2}.$$

The solution set is

$$\left\{\frac{-3 \pm \sqrt{29}}{2}\right\}.$$

◀

DO EXERCISES 9–17.

Example 6 Solve by completing the square: $2x^2 - 3x - 1 = 0$.

Solution

1. Since $a \neq 1$, we multiply on both sides by $1/a$, which is $\frac{1}{2}$:

$$x^2 - \tfrac{3}{2}x - \tfrac{1}{2} = 0. \qquad \text{Multiplying by } \tfrac{1}{2}$$

Find the term that completes the square; then fill in the second expression.

9. $x^2 + 4x + \underline{\hspace{1cm}} = ($ $)^2$

10. $x^2 - 6x + \underline{\hspace{1cm}} = ($ $)^2$

11. $x^2 + 5x + \underline{\hspace{1cm}} = ($ $)^2$

12. $x^2 - 7x + \underline{\hspace{1cm}} = ($ $)^2$

13. $x^2 + \frac{3}{4}x + \underline{\hspace{1cm}} = ($ $)^2$

14. $x^2 - x + \underline{\hspace{1cm}} = ($ $)^2$

Solve by completing the square.

15. $x^2 + 4x - 3 = 0$

16. $x^2 - 6x + 8 = 0$

17. $x^2 - 5x + 6 = 0$

Solve by completing the square.

18. $2x^2 + 2x - 3 = 0$

2. To complete the square on $x^2 - \frac{3}{2}x$, we take half the coefficient of x and square it. Then we add and subtract that number: $[\frac{1}{2}(-\frac{3}{2})]^2$, or $\frac{9}{16}$:

$$x^2 - \frac{3}{2}x + \frac{9}{16} - \frac{9}{16} - \frac{1}{2} = 0$$

$$\left(x - \frac{3}{4}\right)^2 - \frac{17}{16} = 0$$

$$\left(x - \frac{3}{4}\right)^2 = \frac{17}{16}.$$

3. We now take the square roots and solve for x:

$$x - \frac{3}{4} = \pm\sqrt{\frac{17}{16}} = \pm\frac{\sqrt{17}}{4}$$

$$x = \frac{3}{4} \pm \frac{\sqrt{17}}{4} = \frac{3 \pm \sqrt{17}}{4}.$$

The solution set is

$$\left\{\frac{3 + \sqrt{17}}{4}, \frac{3 - \sqrt{17}}{4}\right\}, \quad \text{or} \quad \left\{\frac{3 \pm \sqrt{17}}{4}\right\}.$$

DO EXERCISES 18 AND 19.

19. $4x^2 + 3x - 1 = 0$

The Quadratic Formula

We studied completing the square for two reasons. The most important is that it is a useful tool in other places in mathematics. The second reason is that it can be used to derive a general formula for solving quadratic equations, called the **quadratic formula.** We consider the standard form of the quadratic equation $ax^2 + bx + c = 0$, with unspecified coefficients, solving as we have in the preceding examples. We assume that $a > 0$. If $a < 0$, we can first multiply on both sides by -1. Let's solve by completing the square:

$$ax^2 + bx + c = 0$$

$$x^2 + \frac{b}{a}x + \frac{c}{a} = 0. \qquad \text{Multiplying by } \frac{1}{a}$$

Half of b/a is $\frac{1}{2} \cdot b/a$, or $b/2a$. The square is $b^2/4a^2$. Thus we add and subtract $b^2/4a^2$:

$$x^2 + \frac{b}{a}x + \frac{b^2}{4a^2} - \frac{b^2}{4a^2} + \frac{c}{a} = 0$$

$$x^2 + \frac{b}{a}x + \frac{b^2}{4a^2} = \frac{b^2}{4a^2} - \frac{c}{a}$$

$$\left(x + \frac{b}{2a}\right)^2 = \frac{b^2}{4a^2} - \frac{4ac}{4a^2}$$

$$\left(x + \frac{b}{2a}\right)^2 = \frac{b^2 - 4ac}{4a^2}$$

$$x + \frac{b}{2a} = \sqrt{\frac{b^2 - 4ac}{4a^2}} \quad \text{or} \quad x + \frac{b}{2a} = -\sqrt{\frac{b^2 - 4ac}{4a^2}}.$$

Since $a > 0$, then $\sqrt{4a^2} = 2|a| = 2a$, so

$$x + \frac{b}{2a} = \frac{\sqrt{b^2 - 4ac}}{2a} \quad \text{or} \quad x + \frac{b}{2a} = -\frac{\sqrt{b^2 - 4ac}}{2a}.$$

Thus,

$$x = -\frac{b}{2a} + \frac{\sqrt{b^2 - 4ac}}{2a} \quad \text{or} \quad x = -\frac{b}{2a} - \frac{\sqrt{b^2 - 4ac}}{2a}.$$

The solution set is

$$\left\{ \frac{-b \pm \sqrt{b^2 - 4ac}}{2a} \right\}.$$

This gives us the quadratic formula.

THEOREM 3   **The Quadratic Formula**

The solutions of a quadratic equation $ax^2 + bx + c = 0$ are given by

$$x = \frac{-b \pm \sqrt{b^2 - 4ac}}{2a}.$$

When using the quadratic formula, it is helpful to first find the standard form so that the coefficients a, b, and c can be determined.

Example 7 Solve $3x^2 + 2x = 7$. Find exact and approximate solutions. Round to the nearest hundredth.

Solution We first find the standard form and determine a, b, and c:

$$3x^2 + 2x - 7 = 0;$$
$$a = 3, \quad b = 2, \quad c = -7.$$

We then use the quadratic formula:

$$
\begin{aligned}
x &= \frac{-b \pm \sqrt{b^2 - 4ac}}{2a} \\
&= \frac{-2 \pm \sqrt{2^2 - 4 \cdot 3 \cdot (-7)}}{2 \cdot 3} \\
&= \frac{-2 \pm \sqrt{4 + 84}}{6} \quad \text{To prevent careless mistakes, it helps to write out } all \text{ the steps.} \\
&= \frac{-2 \pm \sqrt{88}}{6} = \frac{-2 \pm \sqrt{4 \cdot 22}}{6} \\
&= \frac{-2 \pm 2\sqrt{22}}{6} = \frac{2(-1 \pm \sqrt{22})}{2 \cdot 3} \\
&= \frac{-1 \pm \sqrt{22}}{3}.
\end{aligned}
$$

The solution set is

$$\left\{ \frac{-1 \pm \sqrt{22}}{3} \right\}.$$

20. a) Solve $2x^2 + 7x = 4$ by factoring.

b) Solve the equation in part (a) using the quadratic formula. Compare your answers.

Should such irrational solutions arise in an applied problem, we can find approximations using a calculator:

$$\frac{-1 + \sqrt{22}}{3} \approx 1.23; \qquad \text{Rounding to the nearest hundredth}$$

$$\frac{-1 - \sqrt{22}}{3} \approx -1.90. \qquad \text{Rounding to the nearest hundredth}$$

The set of approximate solutions is $\{1.23, -1.90\}$.   ◀

Example 8 Solve: $x^2 + x + 1 = 0$.

Solution

$$x^2 + x + 1 = 0$$
$$a = 1, \quad b = 1, \quad c = 1;$$
$$x = \frac{-b \pm \sqrt{b^2 - 4ac}}{2a}$$
$$= \frac{-1 \pm \sqrt{1^2 - 4 \cdot 1 \cdot 1}}{2 \cdot 1}$$
$$= \frac{-1 \pm \sqrt{1 - 4}}{2}$$
$$= \frac{-1 \pm \sqrt{-3}}{2}$$
$$= \frac{-1 \pm i\sqrt{3}}{2}$$

◀

Solve using the quadratic formula.

21. $5x^2 - 8x = 3$

The solutions of a quadratic equation can *always* be found using the quadratic formula. Solutions that are irrational are difficult to find by factoring. A general strategy for solving quadratic equations is as follows.

1. Try factoring.
2. If factoring seems difficult, use the quadratic formula. It *always works*!

---

22. $x^2 - x + 2 = 0$

DO EXERCISES 20–22.

▶B The Discriminant

From the quadratic formula, we know that the solutions x_1 and x_2 of a quadratic equation are given by

$$x_1 = \frac{-b + \sqrt{b^2 - 4ac}}{2a} \quad \text{and} \quad x_2 = \frac{-b - \sqrt{b^2 - 4ac}}{2a}.$$

The expression $b^2 - 4ac$ shows the nature of the solutions. This expression is called the **discriminant.** If it is 0, then it doesn't matter whether we choose the plus or the minus sign in the formula. There is just one solution, and we sometimes say that there is one repeated real solution. If the discriminant is positive, there will be two real solutions. If it is negative, we will be taking the square root of a negative number; hence there will be two nonreal solutions, and they will be complex conjugates.

> ### THEOREM 4
>
> For $ax^2 + bx + c = 0$:
>
> $b^2 - 4ac = 0$ $\Rightarrow$ One real-number solution;
>
> $b^2 - 4ac > 0$ $\Rightarrow$ Two different real-number solutions;
>
> $b^2 - 4ac < 0$ $\Rightarrow$ Two different nonreal-number solutions (complex conjugates).

Example 9 Determine the nature of the solutions of $9x^2 - 12x + 4 = 0$.

Solution We have $a = 9$, $b = -12$, and $c = 4$. We compute the discriminant:

$$b^2 - 4ac = (-12)^2 - 4 \cdot 9 \cdot 4$$
$$= 144 - 144 = 0.$$

Since the discriminant is 0, there is just one solution and it is a real number. ◀

Example 10 Determine the nature of the solutions of $x^2 + 5x + 8 = 0$.

Solution We have $a = 1$, $b = 5$, and $c = 8$. We compute the discriminant:

$$b^2 - 4ac = 5^2 - 4 \cdot 1 \cdot 8$$
$$= 25 - 32 = -7.$$

Since the discriminant is negative, there are two nonreal solutions. ◀

Example 11 Determine the nature of the solutions of $x^2 + 5x + 6 = 0$.

Solution We have $a = 1$, $b = 5$, and $c = 6$. We compute the discriminant:

$$b^2 - 4ac = 5^2 - 4 \cdot 1 \cdot 6 = 1.$$

Since the discriminant is positive, there are two solutions and they are real numbers. ◀

DO EXERCISES 23–25.

 ## Writing Equations from Solutions

We know by the principle of zero products that $(x - 2)(x + 3) = 0$ has solutions 2 and -3. If we know the solutions of an equation, we can write the equation.

Example 12 Find a quadratic equation whose solutions are 3 and $-\frac{2}{5}$.

Determine the nature of the solutions by evaluating the discriminant.

23. $x^2 + 5x - 3 = 0$

24. $9x^2 - 6x + 1 = 0$

25. $3x^2 - 2x + 1 = 0$

Find a quadratic equation having the following solutions.

26. -4 and $\dfrac{5}{3}$

27. $-2\sqrt{2}$ and $\sqrt{2}$

28. $5i$ and $-5i$

Solution

$$x = 3 \quad \text{or} \qquad x = -\tfrac{2}{5}$$
$$x - 3 = 0 \quad \text{or} \quad x + \tfrac{2}{5} = 0 \qquad \text{Getting the 0's on one side}$$
$$(x - 3)(x + \tfrac{2}{5}) = 0 \qquad \text{Principle of zero products (multiplying)}$$
$$x^2 + \tfrac{2}{5}x - 3x - 3 \cdot \tfrac{2}{5} = 0 \qquad \text{Using FOIL}$$
$$x^2 - \tfrac{13}{5}x - \tfrac{6}{5} = 0 \qquad \text{Collecting like terms}$$
$$5x^2 - 13x - 6 = 0 \qquad \text{Multiplying by 5} \qquad \blacktriangleleft$$

Example 13   Write a quadratic equation whose solutions are $\sqrt{3}$ and $-2\sqrt{3}$.

Solution

$$x = \sqrt{3} \quad \text{or} \qquad\qquad x = -2\sqrt{3}$$
$$x - \sqrt{3} = 0 \quad \text{or} \quad x + 2\sqrt{3} = 0$$
$$(x - \sqrt{3})(x + 2\sqrt{3}) = 0 \qquad \text{Getting the 0's on one side}$$
$$x^2 + 2\sqrt{3}x - \sqrt{3}x - 2(\sqrt{3})^2 = 0 \qquad \text{Principle of zero products}$$
$$x^2 + \sqrt{3}x - 6 = 0 \qquad\qquad \text{Using FOIL}$$
$$\qquad\qquad\qquad\qquad\qquad \text{Collecting like terms} \quad \blacktriangleleft$$

Example 14   Write a quadratic equation whose solutions are $2i$ and $-2i$.

Solution

$$x = 2i \quad \text{or} \qquad x = -2i$$
$$x - 2i = 0 \quad \text{or} \quad x + 2i = 0 \qquad \text{Getting the 0's on one side}$$
$$(x - 2i)(x + 2i) = 0 \qquad \text{Principle of zero products (multiplying)}$$
$$x^2 + 2ix - 2ix - (2i)^2 = 0 \qquad \text{Using FOIL and simplifying}$$
$$x^2 - 4i^2 = 0$$
$$x^2 + 4 = 0 \qquad\qquad\qquad\qquad \blacktriangleleft$$

DO EXERCISES 26–28.

● **EXERCISE SET** 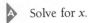 **2.5**

A   Solve for x.

1. $3x^2 = 27$
2. $5x^2 = 80$
3. $x^2 = -1$
4. $x^2 = -4$
5. $4x^2 = 20$
6. $3x^2 = 21$
7. $10x^2 = 0$
8. $9x^2 = 0$
9. $2x^2 - 3 = 0$
10. $3x^2 - 7 = 0$
11. $2x^2 + 14 = 0$
12. $3x^2 + 15 = 0$
13. $ax^2 = b$
14. $\pi x^2 = k$
15. $(x - 7)^2 = 5$
16. $(x + 3)^2 = 2$
17. $\tfrac{4}{9}x^2 - 1 = 0$
18. $\tfrac{16}{25}x^2 - 1 = 0$
19. $(x - h)^2 - 1 = a$
20. $y = a(x - h)^2 + k$

A   Solve by completing the square. (It is important to practice this method because we will need it later.)

21. $x^2 + 6x + 4 = 0$
22. $x^2 - 6x - 4 = 0$

23. $y^2 + 7y - 30 = 0$

24. $y^2 - 7y - 30 = 0$

25. $5x^2 - 4x - 2 = 0$

26. $12y^2 - 14y + 3 = 0$

27. $2x^2 + 7x - 15 = 0$

28. $9x^2 - 30x = -25$

A   Solve using the quadratic formula.

29. $x^2 + 4x = 5$

30. $x^2 = 2x + 15$

31. $2y^2 - 3y - 2 = 0$

32. $5m^2 + 3m - 2 = 0$

33. $3t^2 + 8t + 3 = 0$

34. $3u^2 = 18u - 6$

35. $3 + u^2 = 12u$

36. $40 + 30p + 5p^2 = 0$

37. $x^2 - x + 1 = 0$

38. $x^2 + x + 2 = 0$

39. $x^2 + 13 = 4x$

40. $2x + 1 = -5x^2$

41. $5x^2 = 13x + 17$

42. $x^2 = \frac{5}{3}x + \frac{4}{3}$

43. $0.03 + 0.08v = v^2$

44. $3x^2 - 2x - 5\frac{1}{3} = 0$

45. $\dfrac{1}{x} + \dfrac{1}{x+3} = 7$

46. $\dfrac{1}{x-2} - \dfrac{1}{x+2} = 5$

47. $1 + \dfrac{x+5}{(x+1)^2} - \dfrac{2}{x+1} = 0$

48. $\dfrac{x^2}{x+11} + 1 = 0$

B   Determine the nature of the solutions by evaluating the discriminant.

49. $x^2 - 6x + 9 = 0$

50. $x^2 + 10x + 25 = 0$

51. $x^2 + 7 = 0$

52. $x^2 + 2 = 0$

53. $x^2 - 2 = 0$

54. $x^2 - 5 = 0$

55. $4x^2 - 12x + 9 = 0$

56. $4x^2 + 8x - 5 = 0$

57. $x^2 - 2x + 4 = 0$

58. $x^2 + 3x + 4 = 0$

59. $9t^2 - 3t = 0$

60. $4m^2 + 7m = 0$

61. $y^2 = \frac{1}{2}y + \frac{3}{5}$

62. $y^2 + \frac{9}{4} = 4y$

63. $4x^2 - 4\sqrt{3}x + 3 = 0$

64. $6y^2 - 2\sqrt{3}y - 1 = 0$

C   Write a quadratic equation with the given solutions.

65. $-11, 9$

66. $-4, 4$

67. 7, only solution

68. $-\frac{2}{3}$, only solution

69. $-\frac{2}{5}, \frac{6}{5}$

70. $-\frac{1}{4}, -\frac{1}{2}$

71. $\dfrac{c}{2}, \dfrac{d}{2}$

72. $\dfrac{k}{3}, \dfrac{m}{4}$

73. $\sqrt{2}, 3\sqrt{2}$

74. $-\sqrt{3}, 2\sqrt{3}$

75. $3i, -3i$

76. $4i, -4i$

● **SYNTHESIS** _____

 Solve.

77. $x^2 - 0.75x - 0.5 = 0$

78. $5.33x^2 - 8.23x - 3.24 = 0$

Solve using any method. (In general, try factoring first. Then use the quadratic formula if factoring is not possible.)

79. $x + \dfrac{1}{x} = \dfrac{13}{6}$

80. $\dfrac{3}{x} + \dfrac{x}{3} = \dfrac{5}{2}$

81. $x^2 + x - \sqrt{2} = 0$

82. $x^2 + \sqrt{5}x - \sqrt{3} = 0$

83. $(2t - 3)^2 + 17t = 15$

84. $2y^2 - (y + 2)(y - 3) = 12$

85. $(x + 3)(x - 2) = 2(x + 11)$

86. $9t(t + 2) - 3t(t - 2) = 2(t + 4)(t + 6)$

87. $2x^2 + (x - 4)^2 = 5x(x - 4) + 24$

88. $(c + 2)^2 + (c - 2)(c + 2) = 44 + (c - 2)^2$

89. Prove each of the following.

 a) The sum of the solutions of $ax^2 + bx + c = 0$ is $-b/a$.

 b) The product of the solutions of $ax^2 + bx + c = 0$ is c/a.

For each equation under the given condition, (a) find k and (b) find the other solution.

90. $kx^2 - 17x + 33 = 0$; one solution is 3

91. $kx^2 - 2x + k = 0$; one solution is -3

92. $x^2 - kx + 2 = 0$; one solution is $1 + i$

93. $x^2 - (6 + 3i)x + k = 0$; one solution is 3

94. Find k for which $kx^2 - 4x + (2k - 1) = 0$ and the product of the solutions is 3.

95. Find a quadratic equation for which the sum of the solutions is $\sqrt{3}$ and the product is 8.

96. Write a quadratic equation having the given numbers as solutions.

 a) $\dfrac{2 + \sqrt{3}}{2}, \dfrac{2 - \sqrt{3}}{2}$

 b) $\dfrac{g}{h}, -\dfrac{h}{g}$

 c) $2 - 5i, 2 + 5i$

97. Find h and k, where $3x^2 - hx + 4k = 0$, the sum of the solutions is -12, and the product of the solutions is 20.

● **CHALLENGE** _____

98. One solution of the equation
$$p(q - r)y^2 + q(r - p)y + r(p - q) = 0$$
is 2. Find the other.

99. The sum of the squares of the solutions of
$$x^2 + 2kx - 5 = 0$$
is 26. Find the absolute value of k.

100. Prove that the solutions of $ax^2 + bx + c = 0$ are the reciprocals of the solutions of the equation
$$cx^2 + bx + a = 0, \qquad c \neq 0, \quad a \neq 0.$$

1. Solve $V = \frac{1}{3}\pi r^2 h$ for r.

2. Solve $S = 16t^2 + v_0 t$ for t.

2.6
FORMULAS AND PROBLEM SOLVING

A Formulas

To solve a formula for a certain letter, we use the principles of equation solving that we have developed until we have an equation with the letter alone on one side. In most formulas, the letters represent nonnegative numbers, so we generally need not use absolute values when taking principal square roots.

Example 1 Solve $V = \pi r^2 h$ for r.

Solution Multiplying by $1/\pi h$, we have

$$\frac{V}{\pi h} = r^2$$
$$\sqrt{\frac{V}{\pi h}} = r.$$

DO EXERCISE 1.

Example 2 Solve $A = \pi rs + \pi r^2$ for r.

Solution In standard form, we have $\pi r^2 + \pi rs - A = 0$. Then $a = \pi, b = \pi s$, and $c = -A$, and we use the quadratic formula:

$$r = \frac{-b \pm \sqrt{b^2 - 4ac}}{2a}$$

$$= \frac{-\pi s \pm \sqrt{(\pi s)^2 - 4 \cdot \pi \cdot (-A)}}{2\pi}$$

$$= \frac{-\pi s \pm \sqrt{\pi^2 s^2 + 4\pi A}}{2\pi},$$

or just

$$r = \frac{- + \sqrt{\pi^2 s^2 + 4\pi A}}{2\pi},$$

since the negative square root would result in a negative solution.

DO EXERCISE 2.

Problem Solving

Example 3 *Compound interest.* An investment of $2560 is made at interest rate i, compounded annually. In 2 years, it grows to $3240. What is the interest rate?

Solution We substitute 2560 for P, 3240 for A, and 2 for t in the formula $A = P(1 + i)^t$, and solve for i:

$$A = P(1 + i)^t$$

$$3240 = 2560(1 + i)^2$$

$$\frac{3240}{2560} = (1 + i)^2$$

$$\pm\frac{9}{8} = 1 + i \qquad \text{Taking the square roots}$$

$$-1 + \frac{9}{8} = i \quad \text{or} \quad -1 - \frac{9}{8} = i$$

$$\frac{1}{8} = i \quad \text{or} \quad -\frac{17}{8} = i.$$

Since the interest rate cannot be negative, $i = \frac{1}{8} = 0.125 = 12.5\%$. ◀

DO EXERCISE 3.

Example 4 A ladder 10 ft long leans against a wall. The bottom of the ladder is 6 ft from the wall. The bottom of the ladder is then pulled out 3 ft farther. How much does the top end move down the wall?

Solution

1. **Familiarize.** The first thing to do is to make a drawing and label it with both the known and unknown data. The dashed line in the figure shows the ladder in its original position, with the lower end 6 ft from the wall. Let $d =$ the distance the top of the ladder moves down the wall and $h =$ the vertical height of the ladder before it is moved.

2. **Translate.** We see that there are right triangles in the figure. This is a clue that we may wish to use the Pythagorean theorem. We use that theorem and begin to write equations. From the taller triangle, we get

$$h^2 + 6^2 = 10^2. \tag{1}$$

3. An investment of $2560 is made at interest rate i, compounded annually. In 2 years, it grows to $2890. What is the interest rate?

4. A 13-ft ladder leans against a wall. The bottom of the ladder is 5 ft from the wall. The bottom is then pulled out 4 ft farther. How much does the top end move down the wall?

From the other triangle, we get

$$(h - d)^2 + 9^2 = 10^2. \tag{2}$$

3. **Carry out.**   We solve equation (1) for h and get $h = 8$ ft. Thus we know that $h = 8$ in equation (2), so we have

$$(8 - d)^2 + 9^2 = 10^2,$$

or

$$d^2 - 16d + 45 = 0.$$

Using the quadratic formula, we get

$$d = \frac{16 \pm \sqrt{76}}{2} = \frac{16 \pm 2\sqrt{19}}{2} = 8 \pm \sqrt{19}.$$

4, 5. **Check and state.**   The length $8 + \sqrt{19}$ is not a solution since it exceeds the original length. The number $8 - \sqrt{19} \approx 3.6$ checks and is the solution. Therefore, the top of the ladder moves down 3.6 ft when the bottom is moved out 3 ft.   ◀

DO EXERCISE 4. ——————————————————————

We use the following information in Example 5.

When an object is dropped or thrown downward, the distance, in meters, that it falls in t seconds is given by the following formula:

$$s = 4.9t^2 + v_0 t.$$

In this formula, v_0 is the initial velocity, in meters per second.

The distance, in feet, that the object falls in t seconds is given by

$$s = 16t^2 + v_0 t,$$

where v_0 is the initial velocity in feet per second.

Example 5

a) An object is dropped from the top of the Gateway Arch, which is 195 m high, in St. Louis. How long does it take for the object to reach the ground?

195 m

b) An object is thrown downward from the arch at an initial velocity of 16 m/sec. How long does it take to reach the ground?

c) How far will an object fall in 3 sec if it is thrown downward from the arch at an initial velocity of 16 m/sec?

Solution

a) Since the object was *dropped,* its initial velocity was 0. We substitute 0 for v_0 and 195 for s and then solve for t:

$$s = 4.9t^2 + v_0 t$$
$$195 = 4.9t^2 + 0 \cdot t$$
$$195 = 4.9t^2$$
$$t^2 \approx 39.8$$
$$t \approx \sqrt{39.8} \approx 6.3. \qquad \text{Using the square root key}$$

Thus it takes about 6.3 sec for the object to reach the ground.

b) We substitute 195 for s and 16 for v_0 and solve for t:

$$195 = 4.9t^2 + 16t$$
$$0 = 4.9t^2 + 16t - 195.$$

Using the quadratic formula and a calculator, we obtain

$$t \approx -8.1 \quad \text{or} \quad t \approx 4.9.$$

The negative answer is meaningless in this problem, so the answer is about 4.9 sec.

c) We substitute 16 for v_0 and 3 for t and solve for s:

$$s - 4.9t^2 + v_0 t = 4.9(3)^2 + 16 \cdot 3 = 92.1.$$

Thus the object falls about 92.1 m in 3 sec.   ◀

DO EXERCISE 5.

5. a) An object is dropped from the top of the Statue of Liberty, which is 92 m tall. How long does it take for the object to reach the ground?

b) An object is thrown downward from the statue at an initial velocity of 40 m/sec. How long does it take for the object to reach the ground?

c) How far will an object fall in 1 sec if it is thrown downward from the statue at an initial velocity of 40 m/sec?

● **EXERCISE SET** **2.6**

A   Solve the formula for the given letter. Assume that all letters represent positive numbers.

1. $F = \dfrac{kM_1 M_2}{d^2}$,   for d

2. $E = mc^2$,   for c

3. $S = \dfrac{1}{2} at^2$,   for t

4. $S = 4\pi r^2$,   for r

5. $s = -16t^2 + v_0 t$,   for t

6. $A = 2\pi r^2 + 3\pi rh$,   for r

7. $d = \dfrac{n^2 - 3n}{2}$,   for n

8. $\sqrt{2}t^2 + 3k = \pi t$,   for t

9. $A = P(1 + i)^2$,   for i

10. $A = P\left(1 + \dfrac{i}{2}\right)^2$,   for i

B   **Problem Solving**

What is the interest rate if interest is compounded annually?

11. $6250 grows to $7290 in 2 years

12. $6250 grows to $6760 in 2 years

13. $4000 grows to $4410 in 2 years

14. $4000 grows to $4840 in 2 years

The number of diagonals, d, of a polygon of n sides is given by

$$d = \frac{n^2 - 3n}{2}.$$

15. A polygon has 27 diagonals. How many sides does it have?

16. A polygon has 44 diagonals. How many sides does it have?

17. A ladder 25 ft long leans against a wall. The bottom of the ladder is 7 ft from the wall. The bottom of the ladder is then pulled out 2 ft farther. How much does the top end move down the wall?

18. A 15-ft ladder leans against a wall. The bottom of the ladder is 4 ft from the wall. The bottom is then pulled out 3 ft farther. How much does the top end move down the wall?

19. A ladder 10 ft long leans against a wall. The bottom of the ladder is 6 ft from the wall. How much would the lower end of the ladder have to be pulled away so that the top end would be pulled down the same amount?

20. A ladder 13 ft long leans against a wall. The bottom of the ladder is 5 ft from the wall. How much would the lower end of the ladder have to be pulled away so that the top end would be pulled down the same amount?

21. The area of a triangle is 18 cm². The base is 3 cm longer than the height. Find the height.

22. A baseball diamond is a square 90 ft on a side. How far is it directly from second base to home?

23. Trains A and B leave the same city at right angles at the same time. Train B travels 5 mph faster than train A. After 2 hr, they are 50 mi apart. Find the speed of each train.

24. Trains A and B leave the same city at right angles at the same time. Train A travels 14 km/h faster than train B. After 5 hr, they are 130 km apart. Find the speed of each train.

For Exercises 25 and 26, use the formula $s = 4.9t^2 + v_0 t$.

25. a) An object is dropped 75 m from an airplane. How long does it take for the object to reach the ground?

b) An object is thrown downward from the plane at an initial velocity of 30 m/sec. How long does it take for the object to reach the ground?

c) How far will an object fall in 2 sec if it is thrown downward at an initial velocity of 30 m/sec?

26. a) An object is dropped 500 m from an airplane. How long does it take for the object to reach the ground?

b) An object is thrown downward from the plane at an initial velocity of 30 m/sec. How long does it take for the object to reach the ground?

c) How far will an object fall in 5 sec if it is thrown downward at an initial velocity of 30 m/sec?

27. The diagonal of a square is 1.341 cm longer than a side. Find the length of the side.

28. The hypotenuse of a right triangle is 8.312 cm long. The sum of the lengths of the legs is 10.23 cm. Find the lengths of the legs.

29. A rectangular garden is 60 ft by 80 ft. Part of the garden is torn up to install a sidewalk of uniform width around the garden. The new area of the garden is $\frac{2}{3}$ of the old area. How wide is the sidewalk?

30. The speed of a boat in still water is 20 mph. It travels 24 mi upstream and 24 mi downstream in a total time of 10 hr. What is the speed of the current?

31. An open box is made from a 10-cm by 20-cm piece of tin by cutting a square from each corner and folding up the edges. The area of the resulting base is 90 cm². What is the length of the sides of the squares?

32. The frame of a picture is 28 cm by 32 cm outside and is of uniform width. What is the width of the frame if 200 cm² of the picture shows?

33. During the first part of a trip, a car travels 50 mi at a certain speed. It travels 80 mi on the second part of a trip at a speed of 10 mph slower. The total time for the trip is 2 hr. Find the speed of the car on each part of the trip.

34. A car travels 120 mi at a certain speed. If the speed had been 10 mph faster, the trip could have been made in 2 hr less time. Find the speed.

35. A boat travels 12 mi upstream and 12 mi back. The time required for the round trip is 2 hr. The speed of the stream is 3 mph. Find the speed of the boat in still water.

36. Working together, two people can do a job in 4 hr. Person A takes 5 hr longer, working alone, than person B alone. How long would it take person B to do the job?

The following formula will be helpful in Exercises 37–40:

$T = c \cdot N$

Total cost = (Cost per item) · (Number of items)

Total cost = (Cost per person) · (Number of persons)

37. A group of students share equally in the $140 cost of a boat. At the last minute, 3 students drop out, and this raises the share of each remaining student $15. How many students were in the group at the outset?

38. An investor bought a group of lots for $8400. All but 4 of the lots were sold for $8400. The selling price for each lot was $350 greater than the cost. How many lots were bought?

39. An investor buys some stock for $720. If each share had cost $15 less, 4 more shares could have been bought for the same $720. How many shares of stock were bought?

40. A sorority is going to spend $112 for a party. When 14 new pledges join the sorority, this reduces each student's cost by $4. How much did it cost each student before?

● **SYNTHESIS** _____

Solve for x.

41. $kx^2 + (3 - 2k)x - 6 = 0$

42. $x^2 - 2x + kx + 1 = kx^2$

43. $(m + n)^2 x^2 + (m + n)x = 2$

44. Solve $x^2 - 3xy - 4y^2 = 0$ (a) for x; (b) for y.

45. For interest compounded annually, what is the interest rate when $9826 grows to $13,704 in 3 years?

46. An equilateral triangle is inscribed in a circle whose circumference is 6π. Find the area of the triangle.

47. Two equilateral triangles have sides of respective lengths a_1 and a_2. Find the length of a side a_3 of a third equilateral triangle whose area is the sum of the areas of the two triangles.

48. The sides of triangle A are 25 ft, 25 ft, and 30 ft. The sides of triangle B are 25 ft, 25 ft, and 40 ft. Which triangle has the greatest area?

● CHALLENGE _____

49. A rectangle of 12-cm² area is inscribed in the right triangle ABC, as shown in the figure at the right. What are its dimensions?

50. The world record for free-fall to the earth by a woman without a parachute is 175 ft and is held by Kitty O'Neill. (She fell into a bed of foam padding.) Approximately how long did the fall take?

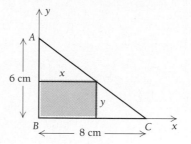

2.7
RADICAL EQUATIONS

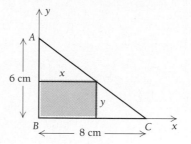

A ## Solving Radical Equations

A **radical equation** is an equation in which variables occur in one or more radicands. An example is the equation $\sqrt{2x-5} - \sqrt{x-3} = 1$. To solve such equations, we need a new principle.

THEOREM 5	The Principle of Powers

For any positive number n, if an equation $a = b$ is true, then $a^n = b^n$ is true.

It is important to check when using the principle of powers. This principle may *not* produce equivalent equations. For example, when we square both sides of an equation, the new equation may have solutions that the first equation does not. Consider, for example, the equation $x = 3$. This equation has just one solution, the number 3. When we square both sides, we get $x^2 = 9$, which has two solutions, 3 and -3. Thus the equations $x = 3$ and $x^2 = 9$ are *not* equivalent.

As another example, consider $\sqrt{x} = -3$. At the outset, we should note that this equation has no real-number solution because, by definition, $\sqrt{x}$ must be a *nonnegative* number. Suppose, though, that we try to solve by squaring both sides. We would get $(\sqrt{x})^2 = (-3)^2$, or $x = 9$. The number 9 does not check.

CAUTION! When using the principle of powers, it is imperative to check possible solutions in the original equation.

Example 1 Solve: $x - \sqrt{x+7} = 5$.

Solve. Don't forget to check!

1. $\sqrt{2x} = -5$

Solution   We must first get the radical alone on one side. That is, we isolate the radical:

$$x - 5 = \sqrt{x + 7} \qquad \text{Adding } -5 \text{ and } \sqrt{x+7} \text{ on both sides to isolate the radical}$$

$$(x - 5)^2 = (\sqrt{x + 7})^2 \qquad \text{Principle of powers; squaring both sides}$$

$$x^2 - 10x + 25 = x + 7$$

$$x^2 - 11x + 18 = 0$$

$$(x - 9)(x - 2) = 0$$

$$x = 9 \quad \text{or} \quad x = 2$$

The possible solutions are 9 and 2. We check, as follows. For 9, we substitute 9 for x on each side of the equation and simplify each side separately:

$$
\begin{array}{c|c}
x - \sqrt{x + 7} = 5 \\
\hline
9 - \sqrt{9 + 7} & 5 \\
9 - \sqrt{16} & \\
9 - 4 & \\
5 & \text{TRUE}
\end{array}
$$

2. $x - \sqrt{x + 5} = 1$

Since the results, 5, are the same, 9 checks. Thus it is a solution.

For 2, we substitute 2 for x on each side and simplify each side separately:

$$
\begin{array}{c|c}
x - \sqrt{x + 7} = 5 \\
\hline
2 - \sqrt{2 + 7} & 5 \\
2 - \sqrt{9} & \\
2 - 3 & \\
-1 & \text{FALSE}
\end{array}
$$

Since the results are not the same, 2 is not a solution. The only solution is 9. Thus the solution set is $\{9\}$. ◀

DO EXERCISES 1 AND 2. _____

3. Solve. Don't forget to check.

$\sqrt[4]{3x - 1} = 2$

Example 2   Solve: $\sqrt[3]{4x^2 + 1} = 5$.

Solution

$$(\sqrt[3]{4x^2 + 1})^3 = 5^3 \qquad \text{Principle of powers; cubing both sides}$$

$$4x^2 + 1 = 125$$

$$4x^2 = 124$$

$$x^2 = 31$$

$$x = \pm\sqrt{31}$$

Both $\sqrt{31}$ and $-\sqrt{31}$ check. The solution set is $\{\pm\sqrt{31}\}$. ◀

DO EXERCISE 3. _____

Equations with Two Radical Terms

A general strategy for solving equations with two square-root radical terms is as follows.

> 1. Isolate one of the radical terms.
> 2. Use the principle of powers.

3. If a radical term remains, perform steps (1) and (2) again.
4. Check possible solutions.

4. Solve: $\sqrt{x} - \sqrt{x - 5} = 1$.

Example 3 Solve: $\sqrt{x - 3} + \sqrt{x + 5} = 4$.

Solution

$$\sqrt{x - 3} = 4 - \sqrt{x + 5}$$ Adding $-\sqrt{x + 5}$, which isolates one of the radical terms

$$(\sqrt{x - 3})^2 = \underbrace{(4 - \sqrt{x + 5})^2}$$ Principle of powers; squaring both sides

Here we are squaring the binomial. We square 4, then we subtract twice the product of 4 and $\sqrt{x + 5}$, and then we add the square of $\sqrt{x + 5}$. Recall that $(A - B)^2 = A^2 - 2AB + B^2$.

$$x - 3 = \overbrace{16 - 8\sqrt{x + 5} + (x + 5)}$$

$$-3 = 21 - 8\sqrt{x + 5}$$ Adding $-x$ and collecting like terms

$$-24 = -8\sqrt{x + 5}$$ Isolating the remaining radical term

$$3 = \sqrt{x + 5}$$

$$3^2 = (\sqrt{x + 5})^2$$ Squaring

$$9 = x + 5$$

$$4 = x$$

The number 4 checks and is the solution. The solution set is $\{4\}$. ◄

> **CAUTION!** A common error in solving equations like
>
> $$\sqrt{x - 3} + \sqrt{x + 5} = 4$$
>
> is to square the left side, obtaining $(x - 3) + (x + 5)$. This is wrong because the square of a sum is *not* the sum of the squares.
> *Example:* $\sqrt{9} + \sqrt{16} = 7$, but $9 + 16 \neq 7^2$.

DO EXERCISE 4.

Example 4 Solve: $\sqrt{2x - 5} = 1 + \sqrt{x - 3}$.

Solution

$$(\sqrt{2x - 5})^2 = (1 + \sqrt{x - 3})^2$$ One radical is already isolated; we square both sides.

$$2x - 5 = 1 + 2\sqrt{x - 3} + (x - 3)$$

$$x - 3 = 2\sqrt{x - 3}$$ Isolating the remaining radical term

$$(x - 3)^2 = (2\sqrt{x - 3})^2$$ Squaring both sides

$$x^2 - 6x + 9 = 4(x - 3)$$

$$x^2 - 6x + 9 = 4x - 12$$

$$x^2 - 10x + 21 = 0$$

$$(x - 7)(x - 3) = 0$$ Factoring

$$x = 7 \quad \text{or} \quad x = 3$$ Using the principle of zero products

5. Solve: $\sqrt{3x+1} = 1 + \sqrt{x+4}$.

The numbers 7 and 3 check and are the solutions. Thus the solution set is $\{3, 7\}$. ◀

DO EXERCISE 5.

Example 5 Solve

$$A = \sqrt{1 + \frac{a^2}{b^2}}$$

for a. Assume that the variables represent positive numbers.

Solution

$$A^2 = 1 + \frac{a^2}{b^2} \qquad \text{Squaring both sides}$$

$$b^2 A^2 = b^2 + a^2 \qquad \text{Multiplying by } b^2$$

$$b^2 A^2 - b^2 = a^2$$

$$\sqrt{b^2 A^2 - b^2} = a$$

$$\sqrt{b^2 (A^2 - 1)} = a$$

$$b\sqrt{A^2 - 1} = a \qquad ◀$$

6. Solve

$$P = \sqrt{\frac{m^2 - 1}{m^2}}$$

for m. Assume that the variables represent positive numbers.

DO EXERCISE 6.

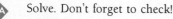

● **EXERCISE SET 2.7**

A   Solve. Don't forget to check!

1. $\sqrt{3x - 4} = 1$
2. $\sqrt[3]{2x + 1} = -5$
3. $\sqrt[4]{x^2 - 1} = 1$
4. $\sqrt{m + 1} - 5 = 8$
5. $\sqrt{y - 1} + 4 = 0$
6. $5 + \sqrt{3x^2 + \pi} = 0$
7. $\sqrt{x - 3} + \sqrt{x + 2} = 5$
8. $\sqrt{x} - \sqrt{x - 5} = 1$
9. $\sqrt{3x - 5} + \sqrt{2x + 3} + 1 = 0$
10. $\sqrt{2m - 3} = \sqrt{m + 7} - 2$
11. $\sqrt[3]{6x + 9} + 8 = 5$
12. $\sqrt[5]{3x + 4} = 2$
13. $\sqrt{6x + 7} = x + 2$
14. $\sqrt{6x + 7} - \sqrt{3x + 3} = 1$
15. $\sqrt{20 - x} = \sqrt{9 - x} + 3$
16. $\sqrt{n + 2} + \sqrt{3n + 4} = 2$
17. $\sqrt{x} - \sqrt{3x - 3} = 1$
18. $\sqrt{2x + 1} - \sqrt{x} = 1$
19. $\sqrt{2y - 5} - \sqrt{y - 3} = 1$
20. $\sqrt{4p - 5} - \sqrt{p - 2} = 3$
21. ▣ $\sqrt{7.35x + 8.051} = 0.345x + 0.067$
22. ▣ $\sqrt{1.213x + 9.333} = 5.343x + 2.312$
23. $x^{1/3} = -2$
24. $t^{1/5} = 2$
25. $t^{1/4} = 3$
26. $m^{1/2} = -7$

27. $8 = \dfrac{1}{\sqrt{x}}$
28. $3 = \dfrac{1}{\sqrt{y}}$
29. $\sqrt[3]{m} = -5$
30. $\sqrt[4]{t} = -5$

For Exercises 31 and 32, assume that the variables represent positive numbers.

31. Solve $T = 2\pi\sqrt{L/g}$ for L; for g.
32. Solve $H = \sqrt{c^2 + d^2}$ for c.

Distance to the horizon. The formula $V = 1.2\sqrt{h}$ can be used to approximate the distance V, in miles, that a person can see to the horizon from a height h, in feet.

33. How far can you see to the horizon through an airplane window at a height of 30,000 ft?

34. How far can a sailor see to the horizon from the top of a 72-ft mast?

35. A person can see 144 mi to the horizon from an airplane window. How high is the airplane?

36. A sailor can see 11 mi to the horizon from the top of a mast. How high is the mast?

● S Y N T H E S I S

Solve.

37. $(x - 5)^{2/3} = 2$

38. $(x - 3)^{2/3} = 2$

39. $\dfrac{x + \sqrt{x + 1}}{x - \sqrt{x + 1}} = \dfrac{5}{11}$

40. $\sqrt{\sqrt{x + 25} - \sqrt{x}} = 5$

41. $\sqrt{x + 2} - \sqrt{x - 2} = \sqrt{2x}$

42. $2\sqrt{x + 3} = \sqrt{x} + \sqrt{x + 8}$

43. $\sqrt[4]{x + 2} = \sqrt{3x + 1}$

44. $\sqrt[3]{2x - 1} = \sqrt[6]{x + 1}$

45. $\dfrac{14}{3 + \sqrt{7 + x}} - \dfrac{\sqrt{7 + x}}{2} = 0$

46. $\dfrac{\sqrt{10 + x}}{4} = \dfrac{6}{2 + \sqrt{10 + x}}$

47. $\sqrt{3x + 1} - \sqrt{2x} = \dfrac{5}{\sqrt{3x + 1}}$

48. $\sqrt{3 + x} + \sqrt{x} = \dfrac{5}{\sqrt{x}}$

49. $\sqrt{15 + \sqrt{2x + 80}} = 5$

50. $\sqrt{x + 5} + 1 = \dfrac{6}{\sqrt{x + 5}}$

2.8

EQUATIONS REDUCIBLE TO QUADRATIC

OBJECTIVES

You should be able to:

A Solve equations that are reducible to quadratic.

B Solve applied problems involving equations reducible to quadratic.

A Solving Equations Reducible to Quadratic

Certain equations that are not really quadratic can be thought of in such a way that they can be solved as quadratic. For example,

$$x \quad + 3\sqrt{x} - 10 = 0$$

$$(\sqrt{x})^2 + 3\sqrt{x} - 10 = 0 \qquad \text{Thinking of } x \text{ as } (\sqrt{x})^2$$

$$u^2 \quad + \quad 3u \quad - 10 = 0. \qquad \text{To make this clearer, write } u \text{ instead of } \sqrt{x}.$$

We can solve the equation $u^2 + 3u - 10 = 0$ by factoring or by using the quadratic formula. After that, we can find x by remembering that $\sqrt{x} = u$. Equations that can be solved in this way are said to be **reducible to quadratic.**

> To solve equations reducible to quadratic, we first make a substitution, solve for the new variable, and then solve for the original variable.

Example 1 Solve: $x + 3\sqrt{x} - 10 = 0$.

Solution Let $u = \sqrt{x}$. Then we solve the equation resulting from substituting u for $\sqrt{x}$:

$$x + 3\sqrt{x} - 10 = 0$$

$$u^2 + 3u - 10 = 0$$

$$(u + 5)(u - 2) = 0$$

$$u = -5 \quad \text{or} \quad u = 2.$$

1. a) Solve: $x + \sqrt{x} - 12 = 0$.

b) Can you think of another procedure for solving this equation? See Section 2.7. Which procedure seemed easier?

We have solved for u. Now we solve for x. We substitute $\sqrt{x}$ for u and solve these equations:

$$\sqrt{x} = -5 \quad \text{or} \quad \sqrt{x} = 2$$
$$\text{(no solution)} \quad \text{or} \quad x = 4.$$

Check:

$$\frac{x + 3\sqrt{x} - 10 = 0}{\begin{array}{c|c} 4 + 3\sqrt{4} - 10 & 0 \\ 4 + 6 - 10 & \\ 0 & \end{array}}$$

The solution is 4. The solution set is $\{4\}$. ◀

DO EXERCISE 1. _____

Example 2 Solve: $x^4 - 6x^2 + 7 = 0$.

Solution Let $u = x^2$. Then we solve the equation resulting from substituting u for x^2. We have

$$x^4 - 6x^2 + 7 = 0$$
$$(x^2)^2 - 6x^2 + 7 = 0$$
$$u^2 - 6u + 7 = 0;$$
$$a = 1, \quad b = -6, \quad c = 7;$$
$$u = \frac{-b \pm \sqrt{b^2 - 4ac}}{2a}$$
$$= \frac{-(-6) \pm \sqrt{(-6)^2 - 4 \cdot 1 \cdot 7}}{2 \cdot 1}$$
$$= \frac{6 \pm \sqrt{8}}{2} = \frac{2 \cdot 3 \pm 2\sqrt{2}}{2 \cdot 1}$$
$$= 3 \pm \sqrt{2}.$$

2. Solve: $2x^4 - 10x^2 + 11 = 0$.

We have solved for u. Now we solve for x. We substitute x^2 for u and solve for x:

$$x^2 = 3 + \sqrt{2} \qquad \text{or} \quad x^2 = 3 - \sqrt{2}$$
$$x = \pm\sqrt{3 + \sqrt{2}} \quad \text{or} \quad x = \pm\sqrt{3 - \sqrt{2}}.$$

Thus the solution set is

$$\{\sqrt{3 + \sqrt{2}}, \ -\sqrt{3 + \sqrt{2}}, \ \sqrt{3 - \sqrt{2}}, \ -\sqrt{3 - \sqrt{2}}\}. ◀$$

DO EXERCISE 2. _____

Example 3 Solve: $(x^2 - x)^2 - 14(x^2 - x) + 24 = 0$.

Solution Let $u = x^2 - x$. Then we solve the equation resulting from substituting u for $x^2 - x$:

$$u^2 - 14u + 24 = 0$$
$$(u - 12)(u - 2) = 0$$
$$u = 12 \quad \text{or} \quad u = 2.$$

We have solved for u. Now we solve for x. We substitute $x^2 - x$ for u and solve:

$$x^2 - x = 12 \quad \text{or} \quad x^2 - x = 2$$
$$x^2 - x - 12 = 0 \quad \text{or} \quad x^2 - x - 2 = 0$$
$$(x - 4)(x + 3) = 0 \quad \text{or} \quad (x - 2)(x + 1) = 0$$
$$x = 4 \quad \text{or} \quad x = -3 \quad \text{or} \quad x = 2 \quad \text{or} \quad x = -1.$$

The solutions are 4, -3, 2, and -1. The solution set is $\{4, -3, 2, -1\}$.

DO EXERCISE 3.

Example 4 Solve: $t^{2/5} - t^{1/5} - 2 = 0$.

Solution Note that $t^{2/5} = (t^{1/5})^2$. Let $u = t^{1/5}$. Then we solve the equation resulting from substituting u for $t^{1/5}$:

$$(t^{1/5})^2 - t^{1/5} - 2 = 0$$
$$u^2 - u - 2 = 0$$
$$(u - 2)(u + 1) = 0$$
$$u = 2 \quad \text{or} \quad u = -1.$$

Now we substitute $t^{1/5}$ for u and solve:

$$t^{1/5} = 2 \quad \text{or} \quad t^{1/5} = -1$$
$$t = 32 \quad \text{or} \quad t = -1. \quad \text{Principle of powers; raising to the fifth power}$$

The solutions are 32 and -1. The solution set is $\{-1, 32\}$.

DO EXERCISE 4.

Example 5 Solve: $x^4 + 3x^2 - 4 = 0$.

Solution Let $u = x^2$. Then we solve the equation resulting from substituting u for x^2:

$$u^2 + 3u - 4 = 0$$
$$(u + 4)(u - 1) = 0$$
$$u = -4 \quad \text{or} \quad u = 1.$$

> **CAUTION!** Remember that you are solving for x, *not* u.

Now we substitute x^2 for u and solve for x:

$$x^2 = -4 \quad \text{or} \quad x^2 = 1$$
$$x = \pm 2i \quad \text{or} \quad x = \pm 1.$$

The solutions are $2i$, $-2i$, 1, and -1. The solution set is $\{2i, -2i, 1, -1\}$.

DO EXERCISE 5.

3. Solve:
$$(x^2 - 1)^2 - (x^2 - 1) - 2 = 0.$$

4. Solve: $t^{2/3} - 3t^{1/3} - 10 = 0$.

5. Solve: $x^4 + 5x^2 - 36 = 0$.

▷ Problem Solving

Example 6   A well problem.   An object is dropped into a well. Two seconds later the sound of the splash is heard at the top. The speed of sound is 1100 ft/sec. How deep is the well?

Solution

1. **Familiarize.**   We first make a drawing and label it with known and unknown information. We can picture the situation as follows. We have a well s feet deep. The time it takes for the object to fall to the bottom of the well can be represented by t_1. The time it takes for the sound to get back to the top of the well can be represented by t_2. The total amount of time is the sum of t_1 and t_2. This gives us the equation

$$t_1 + t_2 = 2. \tag{1}$$

2. **Translate.**   Now can we find any relationship between the two times and the distance s? Often in problem solving you may need to look up related formulas in a physics book, an encyclopedia, or maybe another mathematics book. It turns out that the formula for falling objects, which we considered in Section 2.6, becomes

$$s = 16t^2 + v_0 t,$$

when the distance is given in feet. Since the object is dropped, $v_0 = 0$. The time t_1 that it takes for the object to reach the bottom of the well can be found as follows:

$$s = 16t_1^2, \quad \text{or} \quad t_1 = \frac{\sqrt{s}}{4}. \tag{2}$$

We now have an expression for t_1. What about t_2? To find how long it takes for the sound to get back to the top of the well, we can use the formula $d = rt$, which we considered in Section 2.3. We want to find s, so $d = s$. The speed r is the speed of sound, which is 1100 ft/sec. Then

$$s = 1100 \cdot t_2, \quad \text{or} \quad t_2 = \frac{s}{1100}. \tag{3}$$

We have an expression for t_1 in equation (2) and an expression for t_2 in equation (3) both in terms of s. We substitute these into equation (1) and obtain

$$t_1 + t_2 = 2, \quad \text{or} \quad \frac{\sqrt{s}}{4} + \frac{s}{1100} = 2. \tag{4}$$

3. **Carry out.** We solve equation (4) for s. Multiplying by 1100, we get

$$275\sqrt{s} + s = 2200, \quad \text{or} \quad s + 275\sqrt{s} - 2200 = 0.$$

This equation is reducible to quadratic with $u = \sqrt{s}$. Substituting, we get

$$u^2 + 275u - 2200 = 0.$$

Using the quadratic formula, we can solve for u:

$$u = \frac{-275 + \sqrt{275^2 + 8800}}{2} \qquad \text{We want the positive solution.}$$

$$= \frac{-275 + \sqrt{84{,}425}}{2} \approx 7.78. \qquad \text{Use your calculator and approximate.}$$

Thus, $u \approx 7.78 \approx \sqrt{s}$, so $s \approx 60.5284 \approx 60.5$.

4. **Check.** We check by substituting 60.5 for s back through the related equations.

5. **State.** The well is about 60.5 ft deep. ◀

DO EXERCISE 6.

6. An object is dropped into a well. In 5 sec, the sound of the splash reaches the top of the well. If we assume that the speed of sound is 1100 ft/sec, how deep is the well?

● **EXERCISE SET** **2.8**

A Solve.

1. $x - 10\sqrt{x} + 9 = 0$
2. $2x - 9\sqrt{x} + 4 = 0$
3. $x^4 - 10x^2 + 25 = 0$
4. $x^4 - 3x^2 + 2 = 0$
5. $t^{2/3} + t^{1/3} - 6 = 0$
6. $w^{2/3} - 2w^{1/3} - 8 = 0$
7. $z^{1/2} = z^{1/4} + 2$
8. $6 = m^{1/3} - m^{1/6}$
9. $(x^2 - 6x)^2 - 2(x^2 - 6x) - 35 = 0$
10. $(1 + \sqrt{x})^2 + (1 + \sqrt{x}) - 6 = 0$
11. $(y^2 - 5y)^2 + (y^2 - 5y) - 12 = 0$
12. $(2t^2 + t)^2 - 4(2t^2 + t) + 3 = 0$
13. $w^4 - 4w^2 - 2 = 0$
14. $t^4 - 5t^2 + 5 = 0$
15. $x^{-2} - x^{-1} - 6 = 0$
16. $4x^{-2} - x^{-1} - 5 = 0$
17. $2x^{-2} + x^{-1} = 1$
18. $10 - 9m^{-1} = m^{-2}$
19. $x^4 - 24x^2 - 25 = 0$
20. $x^4 - 5x^2 - 36 = 0$
21. $\left(\dfrac{x^2 - 2}{x}\right)^2 - 7\left(\dfrac{x^2 - 2}{x}\right) - 18 = 0$
22. $\left(\dfrac{x^2 + 1}{x}\right)^2 - 8\left(\dfrac{x^2 + 1}{x}\right) + 15 = 0$
23. $\dfrac{x}{x - 1} - 6\sqrt{\dfrac{x}{x - 1}} - 40 = 0$
$\left(\text{Hint: } u = \sqrt{\dfrac{x}{x - 1}}.\right)$
24. $\dfrac{2x + 1}{x} - 30 = 7\sqrt{\dfrac{2x + 1}{x}}$
25. $5\left(\dfrac{x + 2}{x - 2}\right)^2 = 3\left(\dfrac{x + 2}{x - 2}\right) + 2$
26. $\left(\dfrac{x + 1}{x + 3}\right)^2 + \left(\dfrac{x + 1}{x + 3}\right) - 6 = 0$
27. $\left(\dfrac{x^2 - 1}{x}\right)^2 - \left(\dfrac{x^2 - 1}{x}\right) - 2 = 0$
28. $\left(\dfrac{x + 8}{x - 8}\right)^2 - 6 = 5\left(\dfrac{x + 8}{x - 8}\right)$

B **Problem Solving**

29. A stone is dropped from a cliff. In 3 sec, the sound of the stone striking the ground reaches the top of the cliff. If we assume that the speed of sound is 1100 ft/sec, how high is the cliff?

30. A stone is dropped from a cliff. In 4 sec, the sound of the stone striking the ground reaches the top of the cliff. If we assume that the speed of sound is 1100 ft/sec, how high is the cliff?

31. At the beginning of the year, $2000 is deposited in a bank at a certain interest rate. At the beginning of the next year, $1200 is deposited in another bank at the same interest rate. At the beginning of the third year, there is a total of $3573.80 in both accounts. Interest is compounded annually. What is the interest rate?

32. At the beginning of the year, $3500 is deposited in a bank at a certain interest rate. At the beginning of the next year, $4000 is deposited in another bank at the same interest rate. At the beginning of the third year, there is a total of $8518.35 in both accounts. Interest is compounded annually. What is the interest rate?

● SYNTHESIS

Solve. Check possible solutions by substituting into the original equation.

33. 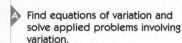 $6.75x = \sqrt{35x} + 5.36$ **34.** $\pi x^4 - \sqrt{99.3} = \pi^2 x^2$

Solve.

35. $9x^{3/2} - 8 = x^3$

36. $\sqrt[3]{2x + 3} = \sqrt[6]{2x + 3}$

37. $\sqrt{x - 3} - \sqrt[4]{x - 3} = 2$

38. $a^3 - 26a^{3/2} - 27 = 0$

39. $x^6 - 28x^3 + 27 = 0$

40. $x^6 + 7x^3 = 8$

41. $(x^2 - 5x - 2)^2 - 5(x^2 - 5x - 2) + 4 = 0$

42. $x^{1/2} + \dfrac{1}{x^{1/2}} = \dfrac{13}{6}$

43. $\left(y + \dfrac{2}{y}\right)^2 + 3y + \dfrac{6}{y} = 4$

44. $x^2 + 3x + 1 - \sqrt{x^2 + 3x + 1} = 8$

● CHALLENGE

45. Solve: $\dfrac{2x + 1}{x} = 3 + 7\sqrt{\dfrac{2x + 1}{x}}$.

46. At the beginning of the year, $2000 is deposited. Six months later, $3000 is deposited in another account at the same interest rate. At the beginning of the next year, there is $956.80 more in the second account than in the first account. If the interest is compounded semiannually, what is the interest rate?

---

1. Find an equation of variation in which y varies directly as x, and $y = 32$ when $x = 0.2$.

2.9
VARIATION

Types of Variation

Direct Variation

There are many situations that yield linear equations like $y = kx$, where k is some positive constant. Note that as x increases, y increases. In such a situation, we say that we have **direct variation,** and k is called the **variation constant.** Usually only positive values of x and y are considered.

> **DEFINITION**   **Variation: Directly Proportional**
>
> If two variables x and y are related as in the equation $y = kx$, where k is a positive constant, we say that "y varies directly as x," or that "y is directly proportional to x."

For example, the circumference C of a circle varies directly as its diameter D: $C = \pi D$. A car moves at a constant speed of 65 mph. The distance d that it travels varies directly as the time t: $d = 65t$.

Example 1   Find an equation of variation in which y varies directly as x, and $y = 5.6$ when $x = 8$.

Solution   We know that $y = kx$, so $5.6 = k \cdot 8$ and $0.7 = k$. Thus the equation of variation is $y = 0.7x$.   ◀

DO EXERCISE 1.

Example 2   *A spring problem.*   *Hooke's law* states that the distance d that an elastic object such as a spring is stretched by placing a certain weight on it varies directly as the weight w of the object. If the distance is 40 cm when the weight is 3 kg, what is the distance stretched when a 2-kg weight is attached?

Spring at rest

d

w

Solution The equation of variation is

$d = kw$.

We first find k using the fact that $d = 40$ cm when $w = 3$ kg:

$$40 = k \cdot 3$$

$$\frac{40}{3} = k.$$

Then the equation of variation becomes

$$d = \frac{40}{3} w.$$

We then substitute 2 kg for w and compute d:

$$d = \frac{40}{3} \cdot 2 = \frac{80}{3} \approx 26.7 \text{ cm}.$$

Thus a 2-kg weight will stretch the spring about 26.7 cm. ◀

DO EXERCISES 2–4. _____

Inverse Variation

There are also situations that yield equations of the type $y = k/x$, where k is a positive constant. Note that as positive values of x increase, y decreases. In such a situation, we say that we have **inverse variation,** and k is called the **variation constant.** In applications, we are generally considering only positive values of x and y.

DEFINITION	**Variation: Inversely Proportional**

If two variables x and y are related as in the equation $y = k/x$, where k is a positive constant, we say that "y varies inversely as x," or that "y is inversely proportional to x."

Example 3 Find an equation of variation in which y varies inversely as x, and $y = 5.6$ when $x = 8$.

Solution We know that $y = k/x$, so $5.6 = k/8$ and $k = 44.8$. Thus the equation of variation is $y = 44.8/x$. ◀

2. Under the conditions of Example 2, what weight would be required to stretch the spring 60 cm?

3. *Ohm's law* states that the voltage V in an electric circuit varies directly as the number of amperes I of electric current in the circuit. If the voltage is 10 volts when the current is 3 amperes, what is the voltage when the current is 15 amperes?

4. The amount of garbage G produced in the United States varies directly as the number of people N who produce the garbage. It is known that 50 tons of garbage is produced by 200 people in 1 year. The population of San Francisco is 705,000. How much garbage is produced by the people of San Francisco in 1 year?

5. Find an equation of variation in which y varies inversely as x, and $y = 32$ when $x = 0.2$.

> **CAUTION!** Keep in mind that an answer to an example like Examples 1 and 3 is an *equation*. The value of k is *not* the answer!

DO EXERCISE 5.

Example 4 *Stocks and gold.* Certain economists theorize that stock prices are inversely proportional to the price of gold. That is, when the price of gold goes up, the prices of stock go down; and when the price of gold goes down, the prices of stock go up. Let us assume that the Dow-Jones Industrial Average, D, an index of the overall price of stock, is inversely proportional to the price of gold, G, in dollars per ounce. One day the Dow-Jones Industrial Average was 3030 and the price of gold was $348 per ounce. What will the Dow-Jones Average be if the price of gold rises to $440 per ounce?

$$D = \frac{k}{G}$$

6. The time t required to drive a fixed distance varies inversely as the speed r. It takes 5 hr at 60 km/h to drive a fixed distance. How long would it take to drive that distance at 40 km/h?

Solution The equation of variation is

$$D = \frac{k}{G}.$$

We first find k using the fact that $D = 3030$ when $G = \$348$:

$$3030 = \frac{k}{348}$$
$$k = 3030(348) = 1{,}054{,}440.$$

Then the equation of variation becomes

$$D = \frac{1{,}054{,}440}{G}.$$

We substitute 440 for G and compute D:

$$D = \frac{1{,}054{,}440}{440} \approx 2396.$$

Thus the Dow-Jones Average will be 2396 when the price of gold is $440 per ounce. ◀

DO EXERCISE 6.

Other Kinds of Variation

There are many other kinds of variation. For example, the area A of a circle varies **directly as the square** of the radius r: $A = \pi r^2$. Also, the weight W of a body varies **inversely as the square** of the distance d from the center of the earth: $W = k/d^2$. As another example, consider the equation for the area of a triangle: $A = \frac{1}{2}bh$. We say that the area varies **jointly as b and h.** The variation constant is $\frac{1}{2}$. Joint variation implies a product of variables. Several kinds of variation can occur together. For example, if

$$y = k \cdot \frac{xz^3}{w^2},$$

then y varies jointly as x and the cube of z and inversely as the square of w.

DO EXERCISES 7–10.

Example 5 *The volume of a tree trunk.* The volume of wood V in a tree trunk varies jointly as the height h and the square of the girth g (girth is distance around). If the volume is 2050 ft³ when the height is 100 ft and the girth is 5 ft, what is the height when the volume is 10,045 ft³ and the girth is 7 ft?

Solution The equation of variation is

$V = khg^2$.

We first find k using the fact that $V = 2050$ ft³ when $h = 100$ ft and $g = 5$ ft:

$2050 = k \cdot 100 \cdot 5^2$

$0.82 = k.$

Then the equation of variation becomes

$V = 0.82hg^2$.

We substitute 10,045 ft³ for V and 7 ft for g and solve for h:

$10{,}045 = 0.82 \cdot h \cdot 7^2$

$h = 250$ ft. ◀

DO EXERCISE 11.

7. Find an equation of variation in which y varies directly as the square of x, and $y = 12$ when $x = 2$.

8. Find an equation of variation in which y varies inversely as the square of x, and $y = \frac{1}{4}$ when $x = 6$.

9. Find an equation of variation in which y varies jointly as x and z, and $y = 42$ when $x = 2$ and $z = 3$.

10. Find an equation of variation in which y varies jointly as x and z and inversely as the square of w, and $y = 105$ when $x = 3$, $z = 20$, and $w = 2$.

11. The distance S that an object falls from some point above the ground varies directly as the square of the time t that it falls. If the object falls 4 ft in 0.5 sec, how long will it take the object to fall 64 ft?

12. a) How much will the 200-lb astronaut weigh when he is 1000 mi from earth?

b) How far is he from earth when his weight is reduced to 50 lb?

Example 6 *The weight of an astronaut.* The weight W of an object varies inversely as the square of the distance d from the object to the center of the earth. At sea level (4000 mi from the center of the earth), an astronaut weighs 200 lb. Find his weight when he is 100 mi above the surface of the earth and the spacecraft is not in motion.

Solution The equation of variation is

$$W = \frac{k}{d^2}.$$

We first find k using the fact that W = 200 lb when d = 4000 mi:

$$200 = \frac{k}{(4000)^2}$$

$$200 = \frac{k}{16{,}000{,}000}$$

$$3{,}200{,}000{,}000 = k.$$

Then the equation of variation becomes

$$W = \frac{3{,}200{,}000{,}000}{d^2}.$$

We add 100 mi to 4000 mi and substitute 4100 for d, and compute W:

$$W = \frac{3{,}200{,}000{,}000}{(4100)^2} = \frac{3{,}200{,}000{,}000}{16{,}810{,}000}$$

$$= 190 \text{ lb} \quad \text{(to the nearest pound)}.$$

Thus the weight of the astronaut 100 mi above the surface of the earth is 190 lb. ◀

DO EXERCISE 12.

● **EXERCISE SET** **2.9**

A Find an equation of variation for the given situation.

1. y varies directly as x, and y = 0.6 when x = 0.4

2. y varies directly as x, and y = 125 when x = 32

3. y varies inversely as x, and y = 125 when x = 32

4. y varies inversely as x, and y = 0.4 when x = 0.8

5. y varies directly as x, and y = 8.6 when x = 1.6

6. y varies inversely as x, and y = 5.4 when x = 3.8

7. y varies inversely as the square of x, and y = 0.15 when x = 0.1

8. y varies jointly as x and z, and y = 56 when x = 7 and z = 10

9. y varies jointly as x and z and inversely as w, and y = $\frac{3}{2}$ when x = 2, z = 3, and w = 4

10. y varies jointly as x and the square of z, and y = 105 when x = 14 and z = 5

11. y varies jointly as x and z and inversely as the square of w, and y = $\frac{12}{5}$ when x = 16, z = 3, and w = 5

12. y varies jointly as x and z and inversely as the product of w and p, and y = $\frac{3}{28}$ when x = 3, z = 10, w = 7, and p = 8

13. Suppose that y varies directly as x and x is doubled. What is the effect on y?

14. Suppose that y varies inversely as x and x is tripled. What is the effect on y?

15. Suppose that y varies inversely as the square of x and x is multiplied by n. What is the effect on y?

16. Suppose that y varies directly as the square of x and x is multiplied by n. What is the effect on y?

17. **Amount of pollution.** The amount of pollution A entering the atmosphere varies directly as the amount of people N living in an area. If 60,000 people cause 42,600 tons of pollutants, how many tons entered the atmosphere in a city with a population of 750,000?

18. **Volume of a gas.** The volume V of a given mass of gas varies directly as the temperature T and inversely as the pressure P. If $V = 231$ in^3 when $T = 420°$ and $P = 20$ lb/in^2, what is the volume when $T = 300°$ and $P = 15$ lb/in^2?

19. **The strength of a beam.** The safe load (the amount it supports without breaking) L of a beam varies jointly as its width w and the square of its height h and inversely as its length l. If the width and the height are doubled at the same time that the length is halved, what is the effect on L?

20. For a chord $\overline{PQ}$ through a fixed point A in a circle, the length of $\overline{PA}$ is inversely proportional to the length of $\overline{AQ}$. If the length of $\overline{PA} = 64$ when the length of $\overline{AQ} = 16$, what is the length of $\overline{AQ}$ when the length of $\overline{PA} = 4$?

21. **A sighting problem.** The distance d that one can see to the horizon varies directly as the square root of the height above sea level. If a person 19.5 m above sea level can see 28.97 km, how high above sea level must one be in order to see 54.32 km?

22. **Electrical resistance.** At a fixed temperature and chemical composition, the resistance of a wire varies directly as the length l and inversely as the square of the diameter d. If the resistance of a certain kind of wire is 0.112 ohm when the diameter is 0.254 cm and the length is 15.24 m, what is the resistance of a wire whose length is 608.7 m and whose diameter is 0.478 cm?

23. **Area of a cube.** The area of a cube varies directly as the square of the length of a side. If a cube has an area of 168.54 m^2 when the length of a side is 5.3 m, what will the area be when the length of a side is 10.2 m?

24. **Intensity of light.** The intensity of light l from a light bulb varies inversely as the square of the distance d from the bulb. Suppose that l is 90 W/m^2 when the distance is 5 m. Find the intensity at a distance of 10 m.

25. **Earned run average.** A pitcher's earned run average A varies directly as the number of earned runs R allowed and inversely as the number of innings pitched I. In a recent year a pitcher had an earned run average of 2.92. He gave up 85 earned runs in 262 innings. How many earned runs would he have given up had he pitched 300 innings? Round to the nearest whole number.

26. **Volume of a gas.** The volume V of a given mass of a gas varies directly as the temperature T and inversely as the pressure P. If $V = 40$ cm^3 when $T = 60°$ and $P = 24$ kg/cm^2, what is the volume when $T = 30°$ and $P = 5$ kg/cm^2?

● SYNTHESIS ─────────────────

27. Show that if p varies directly as q, then q varies directly as p.

28. Show that if u varies inversely as v, then v varies inversely as u, and $1/u$ varies directly as v.

29. The area of a circle varies directly as the square of the length of a diameter. What is the variation constant?

30. P varies directly as the square of t. How does t vary in relationship to P?

● CHALLENGE ─────────────────

31. **The gravity model in sociology.** It has been determined that the average number of telephone calls in a day, N, between two cities is directly proportional to the populations P_1 and P_2 of the cities and inversely proportional to the square of the distance d between the cities. That is,

$$N = \frac{kP_1P_2}{d^2}.$$

This theory is known as the *gravity model* because the equation is similar to Newton's theory of gravity. Use a calculator to find solutions to these problems.

a) The population of Indianapolis is about 741,952, the population of Cincinnati is about 364,040, and the distance between the cities is 174 km. The average number of daily phone calls between the two cities is 11,153. Find the value k and write the equation of variation.

b) The population of Detroit is about 1,027,974, and Detroit is 446 km from Indianapolis. Find the average number of daily phone calls between Detroit and Indianapolis.

c) The average number of daily phone calls between Indianapolis and New York City is 4270 and the population of New York City is about 7,322,564. Find the distance between Indianapolis and New York City.

d) Why is this model not appropriate for adjoining cities such as Minneapolis and St. Paul? (Sociologists say that as communication between two cities increases, the cities tend to merge.)

SUMMARY AND REVIEW 2

Review the objectives found at the beginning of each section. Use the following terms to quiz yourself on your knowledge of each. Then do the review exercises.

● TERMS TO KNOW

Solution, p. 65
Solution set, p. 65
Addition principle, p. 66
Multiplication principle, p. 66
Identity, p. 67
Principle of zero products, p. 68
Meaningful replacements, p. 70
Equivalent equations, p. 71
kational equations, p. 73

Clearing of fractions, p. 74
Compound interest, p. 80
Distance formula, p. 83
The number i, p. 89
Imaginary number, p. 89
Complex number, p. 91
Conjugate, p. 94
Quadratic equation, p. 97
Completing the square, p. 98

Quadratic formula, p. 101
Discriminant, p. 102
Radical equation, p. 111
Reducible to quadratic, p. 115
Direct variation, p. 120
Inverse variation, p. 121
Variation constant, p. 121
Joint variation, p. 123

● REVIEW EXERCISES

Simplify. Leave answers in terms of i.

1. $-\sqrt{-40}$

2. $\sqrt{-12} \cdot \sqrt{-20}$

Simplify.

3. $(2 - 2i)(3 + 4i)$

4. $(3 - 5i) - (2 - i)$

5. $(6 + 2i) + (-4 - 3i)$

6. $\dfrac{2 - 3i}{1 - 3i}$

7. Determine whether $1 - 3i$ is a solution of
$$x^2 - 2x - 10 = 0.$$

8. Find the reciprocal of $6 - 7i$ and express it in the form $a + bi$.

9. Solve for x and y: $4x + 2i = 8 - (2 + y)i$.

Solve.

10. $(6 - i)x + 4 - 8i = 3 - 3i + 2ix$

11. $\dfrac{3x + 2}{7} - \dfrac{5x}{3} = \dfrac{32}{21}$

12. $(p - 3)(3p + 2)(p + 2) = 0$

13. $3x^2 + 2x - 8 = 0$

14. $r^2 - 2r + 10 = 0$

15. $x^3 + 5x^2 - 4x - 20 = 0$

16. $\dfrac{5}{2x + 3} + \dfrac{1}{x - 6} = 0$

17. $y^4 - 3y^2 + 1 = 0$

18. $x = 2\sqrt{x} - 1$

19. $(x^2 - 1)^2 - (x^2 - 1) - 2 = 0$

20. $t^{2/3} - 10 = 3t^{1/3}$

21. $\sqrt{x - 1} - \sqrt{x - 4} = 1$

22. $\sqrt{5x + 1} - 1 = \sqrt{3x}$

23. Solve by completing the square: $y^2 - 6y = 16$. Show your work.

24. $y^2 - 3y = 18$

25. $3[x - 5(4 + 2x)] = 7x - 10(3x - 2)$

26. $(z^2 - 1) + z = 14 - z$

27. $(x - 2)(x + 3) + 4 = 0$

28. $14 - 4y < 22$

29. $(x - 5)(x + 5) \geq (x - 5)(x + 4)$

30. Determine the meaningful replacements in the radical expression $\sqrt{24 - 6x}$.

Determine the nature of the solutions of the equation by evaluating the discriminant.

31. $4y^2 + 5y + 10 = 0$

32. $3x^2 + 2x - 1 = 0$

33. Write a quadratic equation whose solutions are -3 and $\frac{1}{2}$.

34. Find an equation having the solutions $1 - 2i$, $1 + 2i$.

35. Solve $v = \sqrt{2gh}$ for h.

36. Solve $\dfrac{1}{a} + \dfrac{1}{b} = \dfrac{1}{t}$ for t.

Solve.

37. A student scores 73% and 79% on two tests. If the third test counts as though it were two tests, what score must the student make on the third test so the average will be 85%?

38. A can mow a lawn in 4 hr and B can mow it in 2 hr. How long would it take if they worked together?

39. A can mow a lawn in 3 hr. Working together, A and B can mow the lawn in 1 hr. How long would it take B, working alone, to mow the lawn?

40. Two trains leave the same city at right angles. The first train travels at a speed of 60 km/h. In 1 hr, the trains are 100 km apart. How fast is the second train traveling?

41. A boat travels 2 km upstream and 2 km downstream. The total time for both parts of the trip is 1 hr. The speed of the stream is 2 km/h. What is the speed of the boat in still water? Round to the nearest tenth.

42. Find an equation of variation in which y varies inversely as the square of x, and $y = 0.005$ when $x = 10$.

43. Find an equation of variation in which T varies directly as the square of x and inversely as p, and $T = 0.01$ when $x = 6$ and $p = 20$.

44. It is theorized that the dividends paid on utilities stocks are inversely proportional to the prime (interest) rate. Recently, the dividends D on the stock of IBM were $4.84 per share and the prime rate was 8.0%. The prime rate, R, dropped to 7.8%. What dividends were then paid?

45. For a body falling freely from rest, the distance (s ft) that the body falls varies directly as the square of the time (t sec). Given that $s = 64$ when $t = 2$, find a formula for s in terms of t. How long will it take the body to fall 900 ft?

Determine whether the equations of the pair are equivalent.

46. $x = 5$
$x^2 = 25$

47. $x - 7 = \dfrac{x^2 - 49}{x + 7}$
$x - 7 = x - 7$

● **SYNTHESIS**

48. Suppose that $a, b, c,$ and d are nonzero real numbers such that c and d are solutions of
$$x^2 + ax + b = 0$$
and a and b are solutions of
$$x^2 + cx + d = 0.$$
Find $a + b + c + d$.

49. Determine whether
$$x - 7 = \frac{x^2 - 49}{x + 7}$$
is an identity.

50. The area of a circle varies directly as the square of its circumference. Write an equation of variation and find the variation constant.

51. Solve: $\sqrt{\sqrt{\sqrt{x}}} = 2$.

52. Determine whether
$$\frac{x + 1}{x + 2} = \frac{1}{2}$$
is an identity.

53. Solve:
$$2x - 3y = 7 + 7i$$
$$3x + 2y = 4 - 9i.$$

● **THINKING AND WRITING**

1. Explain the difference in meaning between equivalent expressions and equivalent equations.

2. Explain the difference in the following tasks.

 a) Solve: $\dfrac{7x}{2} + \dfrac{1}{x + 1} = 5$.

 b) Add and simplify: $\dfrac{7x}{2} + \dfrac{1}{x + 1}$.

CHAPTER TEST 2

Simplify.

1. i^{83}

2. $-\sqrt{-3}\sqrt{-16}$

3. $(7 - i)(2 + 4i)$

4. $(1 + 6i) - (-3 - 8i)$

5. $\dfrac{5 - 2i}{2 - 3i}$

6. $(1 - i)(1 + i)$

7. Find the reciprocal of $4 + 2i$ and express it in the form $a + bi$.

8. Solve for x and y:
$$3x + 4 - i = 10 + (2 + y)i.$$

Solve.

9. $(2y - 9)(y + 4)(y - 5) = 0$

10. $x - 7\sqrt{x} + 6 = 0$

11. $2t^2 - t - 21 = 0$

12. $4t^2 - 3t - 5 = 0$

13. $\dfrac{8}{3x - 5} = \dfrac{4}{x + 3}$

14. $5x^2 - 4x + 12 = 0$

15. $\sqrt{y - 11} = \sqrt{y + 10} - 3$

16. $\dfrac{2x - 3}{3} = \dfrac{x + 4}{4} - 2$

17. $(a + 3)(a - 7) - 11 = 0$

18. $22 - 8y < 38$

19. $x^3 - 7x^2 - x + 7 = 0$

20. $3x - 4(x + 6) = 2[x - 3(4 - x)]$

21. Grain flows through spout A five times faster than through spout B. When grain flows through both spouts, a grain bin is filled in 4 hr. How many hours would it take to fill the grain bin if grain flows through spout B alone?

22. The speed of train A is 8 mph slower than the speed of train B. Train A travels 144 mi in the same time that it takes train B to travel 240 mi. Find the speed of train A.

23. Solve by completing the square. Show your work.
$$3x^2 - 12x - 6 = 0$$

24. Determine the nature of the solutions of
$$5z^2 + 8z - 4 = 0$$
by evaluating the discriminant.

25. Write a quadratic equation whose solutions are $5i$ and $-5i$.

26. Solve
$$\frac{W_1}{S_1 T_1} = \frac{W_2}{S_2 T_2}$$
for T_2.

27. The hypotenuse of a triangle is 50 ft. One leg is 10 ft longer than the other. What are the lengths of the legs?

28. Find an equation of variation in which y varies jointly as x and w and inversely as the square of z, and $y = 10$ when $x = \frac{4}{3}$, $w = 5$, and $z = 2$.

29. The length l of rectangles of fixed area is inversely proportional to the width w. Suppose that the length is 64 cm when the width is 3 cm. Find the length when the width is 12 cm.

30. Determine whether these equations are equivalent.
$$2x + 3 = -4$$
$$3x + 3 = x - 4$$

31. Determine the meaningful replacements in the radical expression $\sqrt{10 - 5x}$.

● SYNTHESIS _____

Solve.

32. $x = \dfrac{1}{1 + x}$

33. $\sqrt{7x} - \sqrt{3x} = 7 - 3$

In this chapter we are concerned primarily with functions. A *function* is defined to be a special kind of relation, and probably the most important in mathematics. For this reason, we will consider a function from many standpoints, for example, as an input–output machine and as a mapping. We also consider graphs of functions and many of their applications. ● When we draw a picture that represents the solution set of a function, we say that we have *graphed* the function. We will also consider alterations of the graph called *transformations*. ●

Functions, Graphs, and Transformations

3.1

GRAPHS OF EQUATIONS

We see graphs like the following often in newspapers and magazines. Graphs are useful because they allow us to visualize relationships quickly and clearly.

OBJECTIVES

You should be able to:

A Graph ordered pairs and simple relations.

B Determine whether an ordered pair of numbers is a solution of an equation with two variables.

C Graph equations.

D Given the graph of a relation, describe the domain and the range.

WHAT IS SOLID WASTE ?
The U.S. Environmental Protection Agency has found the make-up of solid waste to consist of the materials in this chart.

The following figure relates months of the year and the percentage of birthdays that occur in those months. *Months* are shown horizontally. With

1. Graph the relation

$\{(3, 2), (-5, -2), (-4, 3),$
$(-2, 0), (0, 4)\}.$

each month there is associated a vertical number, or unit. In this case the vertical unit is the *percentage of total births*.

AUGUST IS THE BIRTHDAY MONTH !

A Graphs of Ordered Pairs

We use a number line to graph a single number. We use a plane to graph a pair of numbers. To locate points in a plane, we use two perpendicular number lines, called **axes,** each of which crosses the other at 0. We call this point the **origin.** Usually the horizontal axis is also called the **x-axis,** and the vertical axis is also called the **y-axis.** The arrows show the positive directions. We call such a setup the **Cartesian coordinate system** in honor of the great French mathematician and philosopher René Descartes (1596–1650).

The first member of an ordered pair is called the **first coordinate,** or the **x-coordinate,** or the **abscissa.** The second member is called the **second coordinate,** or the **y-coordinate,** or the **ordinate.** Together these are called the **coordinates of a point.**

The axes divide the plane into four regions called **quadrants,** indicated by the Roman numerals numbered counterclockwise from the upper right (see the figure in Example 1).

A **relation** is any set of ordered pairs. To **graph** a relation, we plot the points that correspond to the ordered pairs in the relation.

Example 1 Graph the relation

$\{(4, 3), (-3, 5), (-4, -2), (3, -4), (0, 0), (-3, 0), (0, 4)\}.$

Solution   The origin O has coordinates $(0, 0)$. For the ordered pair $(-3, 5)$, the x-coordinate tells us to move from the origin 3 units to the left of the y-axis. The y-coordinate tells us to move 5 units up from the x-axis.

DO EXERCISE 1.

TECHNOLOGY CONNECTION

Beginning with this chapter, activities will be included in the margins and in the exercise sets that make use of graphing calculators and/or computer graphing software. We will refer to all such calculators and software simply as **graphers.**

While features vary widely among available graphers, most of our activities use only the most basic options found on all of them. In addition to graphing different types of functions, the only other required feature is the ability to change the **ranges** on the x- and y-axes.

For example, we might examine a graph with the ranges $-10 \leq x \leq 10$, $-10 \leq y \leq 10$.

Then we might examine a portion of the graph by changing the ranges to $2 \leq x \leq 7$, $1 \leq y \leq 6$.

Your grapher must be able to make such changes easily.

We will refer to the portion of the graph that is shown as the **viewing box.** It will be described by giving 4 values: $[L, R] \times [B, T]$. These represent the four corners of the box, as shown below.

In our first example above, the viewing box was $[-10, 10] \times [-10, 10]$. In the second example, the viewing box was $[2, 7] \times [1, 6]$.

As shown in the first two graphs, the equation being graphed will always appear above the upper left corner. Also, the dimensions of the viewing box will appear below the lower right corner using our standard $[L, R] \times [B, T]$ notation.

Many graphers allow individual points to be plotted. (Check your grapher's documentation to see if this feature is available to you. If it is not, then skip the rest of this activity.) Using a viewing box of $[-10, 10] \times [-10, 10]$, plot the following relation:

$$R = \{(3, 4), (-8.5, -4), (2, 0.15), (0, 7), (-8, 8)\}.$$

Note: Depending on the resolution of your screen and whether or not your grapher uses different colors, you may find that points on or near the axes are "hidden." If you can't see $(2, 0.15)$, for example, try changing the values on the ranges for the y-axis and replotting the point (for example, try a viewing box of $[-10, 10] \times [-2, 2]$). Unfortunately, if you can't see $(0, 7)$ in the original viewing box, there is probably nothing you can do to make it visible.

TC 1. Using a $[-10, 10] \times [-10, 10]$ viewing box, plot the points $(9, 3)$, $(-4, -5)$, $(5, -8.4)$, and $(-6.4, 8)$.

TC 2. Select an appropriate viewing box that will clearly show all these points:

$$(42, 50), (-20, 14), (6, -40), (-23, -33).$$

TC 3. Select an appropriate viewing box that will clearly show all these points:

$$(0.2, -0.3), (0.35, 0.45), (-0.05, 0.10).$$

2. Determine whether $(1, 7)$ is a solution of $y = 2x + 5$.

3. Determine whether $(-1, 4)$ is a solution of $y = 2x + 5$.

4. Determine whether $(-2, 5)$ is a solution of $y = x^2$.

5. Determine whether $(4, -5)$ is a solution of $x^3 - y^2 = 39$.

Graphing suggestions

a) Use graph paper.
b) Label axes with symbols for the variables.
c) Use arrows on the axes to indicate positive directions.
d) Scale the axes; that is, mark numbers on the axes.
e) Calculate solutions and list the ordered pairs in a table.
f) Plot solutions, look for patterns, and complete the graph. When finished, label the graph with the equation or relation being graphed.

 Solutions of Equations

If an equation has two variables, its solutions are ordered pairs of numbers. Suppose the variables in the equation are x and y. A **solution** is an ordered pair of numbers such that when the first coordinate is substituted for x and the second coordinate is substituted for y, the result is a true equation. Note that substitutions of numbers for variables are usually made in alphabetical order of the variables.

Example 2   Determine whether the ordered pairs $(-1, -4)$ and $(7, 5)$ are solutions of the equation $y = 3x - 1$.

Solution   We have

$$y = 3x - 1$$

$$
\begin{array}{c|l}
-4 & 3(-1) - 1 \\
 & -3 - 1 \\
 & -4
\end{array}
$$

We substitute -1 for x and -4 for y (alphabetical order of variables).

The equation $-4 = -4$ is true, so $(-1, -4)$ is a solution. Now

$$y = 3x - 1$$

$$
\begin{array}{c|l}
5 & 3 \cdot 7 - 1 \\
 & 21 - 1 \\
 & 20
\end{array}
$$

We substitute.

The equation $5 = 20$ is false, so $(7, 5)$ is not a solution.

DO EXERCISES 2–5.

 Graphs of Equations

The equation $y = 3x - 1$ considered in Example 2 actually has an infinite number of solutions. Rather than attempt to list all the solutions, we will use a graph as a convenient representation of all the solutions. The solutions of an equation in two variables are ordered pairs and thus constitute a relation.

> **DEFINITION**   **The Graph of an Equation**
>
> To *graph* an equation is to make a drawing that represents its solutions.

At the left are some general suggestions for graphing.

Example 3   Graph: $y = 3x - 1$.

Solution   First find some ordered pairs that are solutions. To find an ordered pair, choose *any* number that is a meaningful replacement for x and then determine y. For example, if we choose 2 for x, then $y = 3(2) - 1$, or 5. Thus we have found the solution $(2, 5)$. Continue making choices for x and finding the corresponding values for y. Make some negative choices for x, as well as positive ones, keeping track of the solutions in a table.

The table gives us the ordered pairs $(0, -1)$, $(1, 2)$, $(2, 5)$, and so on. Next, plot these points. If we could calculate enough of them, they would make a solid line. We can draw the line with a ruler and label it $y = 3x - 1$.

x	y	(x, y)
0	−1	(0, −1)
1	2	(1, 2)
2	5	(2, 5)
−1	−4	(−1, −4)

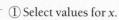
① Select values for x.
② Compute values for y.

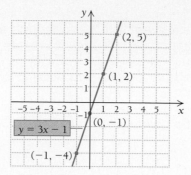

Note that the equation $y = 3x - 1$ has an infinite (unending) set of solutions. The *graph of the equation* is a drawing that represents the relation that is made up of its solutions. Thus the relation consists of all pairs (x, y) such that $y = 3x - 1$ is true, that is, $\{(x, y) | y = 3x - 1\}$. ◀

DO EXERCISE 6.

Example 4   Graph: $y = x^2 - 5$.

Solution   We select numbers for x and find the corresponding values for y. The table gives us the ordered pairs $(0, -5)$, $(-1, -4)$, and so on.

x	y	(x, y)
0	−5	(0, −5)
−1	−4	(−1, −4)
1	−4	(1, −4)
−2	−1	(−2, −1)
2	−1	(2, −1)
−3	4	(−3, 4)
3	4	(3, 4)

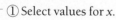
① Select values for x.
② Compute values for y.

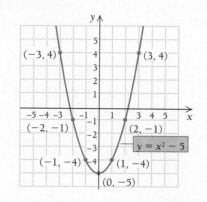

Next we plot these points. We note that as the absolute value of x increases, $x^2 - 5$ also increases. Thus the graph is a curve that rises gradually on either side of the y-axis, as shown above. This graph shows the relation $\{(x, y) | y = x^2 - 5\}$. ◀

DO EXERCISE 7.

Example 5   Graph: $x = y^2 + 1$.

Solution   Since x is expressed in terms of y, we will select numbers for y and then find the corresponding values for x.

x	y	(x, y)
1	0	(1, 0)
2	−1	(2, −1)
2	1	(2, 1)
5	−2	(5, −2)
5	2	(5, 2)

① Select values for y.
② Compute values for x.

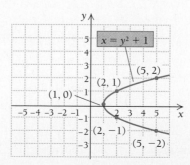

6. Graph: $y = -3x + 1$.

7. Graph: $y = 3 - x^2$.

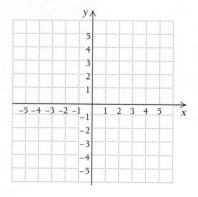

We must remember, however, that x is the first coordinate and y is the second coordinate. The table gives us the ordered pairs $(1, 0)$, $(2, -1)$, $(2, 1)$, and so on. We note that as the absolute value of y gradually increases, the value of x also gradually increases. Thus the graph is a curve that gets farther from the x-axis as it extends farther to the right. This graph shows the relation $\{(x, y)\,|\,x = y^2 + 1\}$. ◄

DO EXERCISE 8 ON THE FOLLOWING PAGE.

You can always use a calculator to find as many values as you wish. This can be helpful when you are uncertain about the shape of a graph.

 TECHNOLOGY CONNECTION

Drawing the graphs of equations on a grapher is usually a very simple matter. There are generally two steps involved: (1) entering the equation to be plotted and (2) selecting the portion of the graph that is to be shown. Although each grapher has its own set of procedures for drawing graphs, almost all of them require that the equation be written in the form $y = \dots$. Therefore, unless it is already in that form, your first step is to solve the equation for y. Then enter the equation into the grapher, using whatever special rules and symbols your grapher requires for math operations. For example, * often represents multiplication, / stands for division, ^ is used for raising to a power, **ABS(x)** is used for the absolute value of x, and so on. Consult your grapher's documentation for specific details and examples.

Some equations require a special "trick" in order to see the entire graph. For example, in Example 5 we were given the equation $x = y^2 + 1$. When this is solved for y, we must take the square root of both sides. When we do, we are left with $y = \pm\sqrt{x - 1}$. These are actually two different equations, $y = +\sqrt{x - 1}$ and $y = -\sqrt{x - 1}$. In order to see the entire graph, we must graph both of these equations together on the same set of axes.

$y = \sqrt{x-1}$
$y = -\sqrt{x-1}$

$[-10, 10] \times [-10, 10]$

Once the equation has been entered, the portion of the graph to be shown must be selected. Most graphers

have a standard viewing box of $[-10, 10] \times [-10, 10]$ that will be used unless you change it. These values are suitable for a large majority of the functions in this book. Other graphers require that you set the dimensions of the viewing box each time you draw a graph. In either case, you are usually free to change them if you want to focus on a small section of the graph or examine a larger portion of the graph.

Many graphers have an automatic Zoom feature that allows you to quickly examine a small region of a graph with a minimum of effort. If such a feature is available on your grapher, learning how to use it will shorten the time it takes to analyze a graph. The method that we will describe here, however, does not require such an automatic feature, and should work on any grapher.

When we wish to zoom in on a particular area, we will describe it in terms of a rectangle that includes that area. The left side of this small rectangle becomes our minimum x-value, the right side is the maximum x-value, the bottom is the minimum y-value, and the top is the maximum y-value. You may want to refer to the first Technology Connection in which the viewing box was described. If you set each of these four values and have the graph redrawn, the small rectangle will now fill the viewing box. If we wish to zoom in further, the process is repeated as often as necessary. (Of course, we can also zoom out, by using minimum and maximum x- and y-values that are *outside* the limits of the current viewing box.)

Graph each of the following equations. First use your grapher's standard viewing box, and then try another viewing box with your own dimensions.

TC 4. $y = x$

TC 5. $y = x + 200$

TC 6. $y = x^2$

TC 7. $y = -4x^2$

TC 8. $y = x^3$

TC 9. $y = x^3 + 2x^2 - 4x - 13$

Example 6   Graph: $xy = 12$.

Solution   We first find numbers that satisfy the equation. To do so, it helps to solve for y (that is, $y = 12/x$) or solve for x (that is, $x = 12/y$). Then we make substitutions.

x	1	-1	2	-2	3	-3	4	-4	6	-6	12	-12
y	12	-12	6	-6	4	-4	3	-3	2	-2	1	-1

We plot these points and connect them. Note that since neither x nor y can be 0, the graph does not cross either axis. As the absolute value of x gets small, the absolute value of y must get large, and vice versa. Thus the graph consists of two curves, shown at the right.

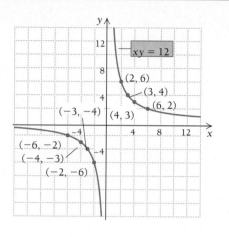

The graph shows the relation $\{(x, y) | xy = 12\}$.

DO EXERCISE 9.

Example 7   Graph: $y = |x|$.

Solution   We find numbers that satisfy the equation. For example, when $x = -3$, $y = |-3| = 3$. When $x = 2$, $y = |2| = 2$, and when $x = 0$, $y = |0| = 0$.

x	0	1	-1	2	-2	3	-3	4	-4
y	0	1	1	2	2	3	3	4	4

We plot these points and connect them. Note that as we get farther from the origin, to the left or right, the absolute value of x increases. Thus the graph is a V-shaped curve that rises to the left and right of the y-axis. It actually consists of parts of two straight lines, as shown at the right.

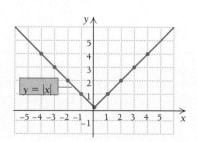

The graph shows the relation $\{(x, y) | y = |x|\}$.

DO EXERCISE 10.

8. Use graph paper. Graph $x = y^2 - 5$. (*Hint:* Select values for y and then find the corresponding values of x. When you plot, be sure to find x (horizontally) first.) Compare it with the graph of Example 4.

9. Graph: $xy = 1$.

10. Graph $x = |y|$. Compare it with the graph of Example 7.

11. List the domain and the range of the relation

$$\{(2, 2), (2, 3), (-4, 5), (-6, 7)\}.$$

On each diagram in Exercises 12 and 13:

a) Shade (on the *x*-axis) the domain. Then describe the domain.

b) Shade (on the *y*-axis) the range. Then describe the range.

12.

13.

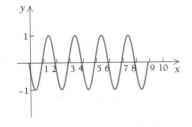

▷ Domains and Ranges

A relation is a set of ordered pairs.

DEFINITION   **Domain and Range**

The *domain* of a relation is the set of all first coordinates, and the *range* is the set of all second coordinates.

Example 8   List the domain and the range of the relation

$$\{(4, 3), (-3, 5), (-4, -2), (3, -4), (0, 0), (-3, 0), (0, 4)\}.$$

Solution

Domain $= \{-4, -3, 0, 3, 4\}$.
Range $= \{-4, -2, 0, 3, 4, 5\}$. ◀

Example 9   For the relation in part (a) of the figure, locate or shade the domain on the *x*-axis and the range on the *y*-axis.

Solution

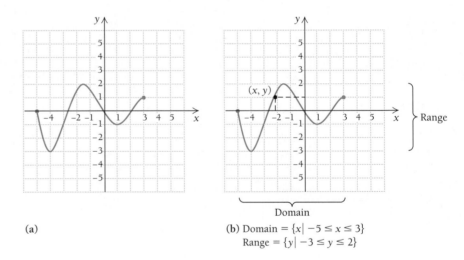

(a)

(b) Domain $= \{x \mid -5 \le x \le 3\}$
Range $= \{y \mid -3 \le y \le 2\}$

From a point (x, y) on the graph, we find *x* by drawing or thinking of a vertical line back to the *x*-axis. To locate the domain, we want all such numbers on the *x*-axis. For this relation, we determine the left and right extremities of the relation, if possible, and draw vertical lines back to the *x*-axis. The domain consists of the numbers where the lines cross the *x*-axis and all numbers in between.

From a point (x, y) on the graph, we find *y* by drawing or thinking of a horizontal line back to the *y*-axis. To locate the range, we want all such numbers on the *y*-axis. For this relation, we determine the upper and lower extremities of the relation, and draw horizontal lines from them to the *y*-axis. The range consists of the numbers where the lines cross the *y*-axis and all numbers in between. The domain and the range are shown in part (b) of the figure. ◀

Example 10 Graph $x = y^2 + 1$. Locate the domain on the x-axis and the range on the y-axis.

Solution The graph is shown in part (a) of the figure. It is the relation that we considered in Example 5. The left extremity of the graph yields the number 1 on the x-axis. There is no right extremity. Thus the domain is the set of all real numbers greater than or equal to 1. The graph extends up and to the right and down and to the right. Since there are no upper or lower extremities, the range consists of the entire set of real numbers. The domain and the range are shown in part (b) of the figure.

(a)

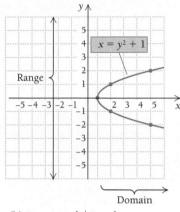

(b) Domain = $\{x \mid 1 \le x\}$
Range = All real numbers

DO EXERCISES 11–14. (EXERCISES 11–13 ARE ON THE PRECEDING PAGE.)

14. Graph $y = x^2$. Then shade and describe the domain on the x-axis and the range on the y-axis.

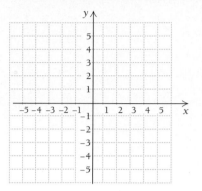

● **EXERCISE SET** **3.1**

Ⓐ Graph the set of ordered pairs.

1. $\{(3, 0), (4, 2), (5, 4), (6, 6)\}$
2. $\{(1, 1), (2, 3), (3, 5), (4, 7)\}$
3. $\{(3, -4), (3, -3), (3, -2), (3, -1), (3, 0)\}$
4. $\{(-2, 1), (-2, 2), (-2, 3), (-2, 4), (-2, 5)\}$
5. $\{(4, 3), (4, 2), (3, 2), (3, 3), (5, 2), (5, 3)\}$
6. $\{(2, -2), (3, -2), (2, -3), (3, -3), (2, -4), (3, -4)\}$
7. $\{(-1, 1), (-2, 1), (-2, 2), (-3, 1), (-3, 2), (-3, 3)\}$
8. $\{(-1, -1), (-1, -2), (-1, -3), (-2, -2),$
$(-2, -3), (-3, -3)\}$

Ⓑ Determine whether the given ordered pairs are solutions of the given equation.

9. $(1, -1), (0, 3);\quad y = 2x - 3$
10. $(2, 5), (-2, -7);\quad y = 3x - 1$
11. $(3, 4), (-3, 5);\quad 3s + t = 4$
12. $(2, 3), (-5, -15);\quad 2p + q = 5$
13. $(0, \frac{3}{5}), (-\frac{1}{2}, -\frac{4}{5});\quad 2a + 5b = 3$
14. $(0, \frac{3}{2}), (\frac{2}{3}, 1);\quad 3f + 4g = 6$
15. $(2, -1), (-0.75, 2.75);\quad r^2 - s^2 = 3$
16. $(2, -4), (4, -5);\quad 5w + 2z^2 = 70$

Ⓒ Graph.

17. $y = x$ **18.** $y = 2x$ **19.** $y = -2x$
20. $y = -\frac{1}{2}x$ **21.** $y = x + 3$ **22.** $y = x - 2$
23. $y = 3x - 2$ **24.** $y = -4x + 1$ **25.** $y = x^2$
26. $y = -x^2$, or $-1 \cdot x^2$ **27.** $y = x^2 + 2$
28. $y = x^2 - 2$ **29.** $x = y^2 + 2$
30. $x = y^2 - 2$ **31.** $y = |x + 1|$ **32.** $y = |x - 1|$
33. $x = |y + 1|$ **34.** $x = |y - 1|$ **35.** $xy = 10$
36. $xy = -18$ **37.** $y = \frac{1}{x}$ **38.** $y = -\frac{2}{x}$
39. $y = \frac{1}{x^2}$ **40.** $y = x^3$ **41.** $y = \sqrt{x}$
42. $y = x^{1/3}$ **43.** $y = 8 - x^2$ **44.** $x = 4 - y^2$

Graph and compare.

45. $y = x^2 + 1$, $y = (-x)^2 + 1$

46. $y = x^2 - 2$, $y = 2 - x^2$

 Graphs of relations follow. In each case, describe the domain and the range.

47.

48.

49. Graph a relation as follows.

    a) Using a compass, draw a circle with radius 2, centered at the point $(4, 3)$.

    b) Shade (on the x-axis) the domain. Describe the domain.

    c) Shade (on the y-axis) the range. Describe the range.

50. Graph a relation as follows.

    a) Draw a triangle with vertices at $(1, 1)$, $(4, 2)$, and $(3, 6)$.

    b) Shade (on the x-axis) the domain. Describe the domain.

    c) Shade (on the y-axis) the range. Describe the range.

Graphs of the following were created in the indicated exercises. Describe the domain and the range of each relation.

51. $y = -x^2$ (Exercise 26)

52. $y = x - 2$ (Exercise 22)

53. $x = |y + 1|$ (Exercise 33)

54. $y = \dfrac{1}{x}$ (Exercise 37)

55. $y = \sqrt{x}$ (Exercise 41)

56. $y = x^{1/3}$ (Exercise 42)

57. $y = 8 - x^2$ (Exercise 43)

58. $x = 4 - y^2$ (Exercise 44)

● CHALLENGE

59. Graph: $|x| + |y| = 1$.

60. Graph the relation $\{(x, y) | |x| \le 1 \ and \ |y| \le 2\}$.

▨ TECHNOLOGY CONNECTION

Use a grapher to graph the equation. In each case, select a viewing box that includes all the major features of the graph.

61. $y = |x| + x$

62. $y = x|x|$

63. $y = |x^2 - 4|$

64. $y = x^{2/3}$

65. $y = x^2|x|$

66. $y = |x^3|$

67. $y = \frac{1}{3}x^3 - x + \frac{2}{3}$

68. $y = -\frac{1}{3}x^3 + \frac{1}{2}x^2 + 2x - 1$

69. $y = 1 + \sqrt{2 - x}$

70. $y = 1 + \dfrac{2}{x + 1}$

OBJECTIVES

You should be able to:

A Given the coordinates of two points, find the distance between them and determine whether three points with given coordinates are vertices of a right triangle.

B Given the coordinates of the endpoints of a segment, find the coordinates of its midpoint.

C Given the center and the radius of a circle, find an equation for the circle. Given an equation of a circle, complete the square if necessary, and find the center and the radius. Given the center and a point through which a circle passes, find an equation of the circle.

3.2
THE DISTANCE FORMULA AND CIRCLES

A The Distance Formula

We develop a formula for finding the *distance between two points* whose coordinates are known. Suppose that the points are on a horizontal line, thus having the same second coordinate. We can find the distance between them by subtracting their first coordinates. This difference may be negative, depending on the order in which we subtract. So to make sure we get a positive number, we take the absolute value of this difference. The distance between two points on a horizontal line (x_1, y) and (x_2, y) is thus $|x_2 - x_1|$. Similarly, the distance between two points on a vertical line (x, y_1) and (x, y_2) is $|y_2 - y_1|$.

Now consider any two points (x_1, y_1) and (x_2, y_2) not on a horizontal or vertical line. These points are vertices of a right triangle, as shown. The other vertex is (x_2, y_1). The legs of this triangle have the lengths $|x_2 - x_1|$ and $|y_2 - y_1|$. Now by the Pythagorean theorem, we obtain a relation between the length of the hypotenuse d and the lengths of the legs:

$$d^2 = |x_2 - x_1|^2 + |y_2 - y_1|^2.$$

We can now dispense with the absolute-value signs because squares of numbers are never negative. Thus we have

$$d^2 = (x_2 - x_1)^2 + (y_2 - y_1)^2.$$

By taking the principal square root, we get the distance formula.

THEOREM 1   **The Distance Formula**

The distance between any two points (x_1, y_1) and (x_2, y_2) is given by

$$d = \sqrt{(x_2 - x_1)^2 + (y_2 - y_1)^2}.$$

The subtraction of the x-coordinates can be done in any order, as can the subtraction of the y-coordinates. Although we derived the distance formula by considering two points not on a horizontal or a vertical line, the formula holds for *any* two points.

Example 1   Find the distance between the points $A(-2, 2)$ and $B(4, -3)$ on the islands in the figure. Find an exact answer and an approximation to three decimal places.

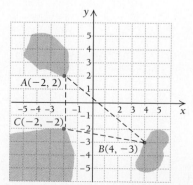

Find the distance between the pair of points. Where appropriate, find an exact answer and an approximation to three decimal places.

1. $(-5, 3)$ and $(2, -7)$

2. $(3, 3)$ and $(-3, -3)$

3. $(9, -5)$ and $(9, 11)$

4. $(0, \pi)$ and $(8, \pi)$

Determine whether the points are vertices of a right triangle.

5. $(11, 1), (6, 6), (2, 2)$

6. $(10, -4), (3, 5), (0, 0)$

Solution   We substitute into the distance formula:

$$d = \sqrt{[4 - (-2)]^2 + (-3 - 2)^2}$$
$$= \sqrt{(6)^2 + (-5)^2}$$
$$= \sqrt{36 + 25}$$
$$= \sqrt{61} \approx 7.810.$$

◀

DO EXERCISES 1–4. _____

We can use the distance formula to determine whether three points are vertices of a right triangle.

Example 2   Determine whether the points $A(-2, 2)$, $B(4, -3)$, and $C(-2, -2)$ of Example 1 are vertices of a right triangle.

Solution   First we find the squares of the distances between the points:

$$d_1^2 = [4 - (-2)]^2 + (-3 - 2)^2$$
$$= (6)^2 + (-5)^2 = 61;$$
$$d_2^2 = [-2 - (-2)]^2 + (-2 - 2)^2$$
$$= (0)^2 + (-4)^2 = 16;$$
$$d_3^2 = (-2 - 4)^2 + [-2 - (-3)]^2$$
$$= (-6)^2 + (1)^2 = 37.$$

Since no sum of any two squares is another, the points are not the vertices of a right triangle.

◀

DO EXERCISES 5 AND 6. _____

 Midpoints of Segments

The distance formula can be used to verify or derive a formula for finding the coordinates of the *midpoint* of a segment when the coordinates of its endpoints are known. We will not derive this formula but simply state it.

THEOREM 2	**The Midpoint Formula**

If the endpoints of a segment are (x_1, y_1) and (x_2, y_2), then the coordinates of the midpoint are

$$\left(\frac{x_1 + x_2}{2}, \frac{y_1 + y_2}{2}\right).$$

Note that we obtain the coordinates of the midpoint by averaging the coordinates of the endpoints. This is an easy way to remember the midpoint formula.

Example 3   Find the midpoint of the segment in Example 2 with endpoints $A(-2, 2)$ and $B(4, -3)$.

Solution Using the midpoint formula, we obtain

$$\left(\frac{-2+4}{2}, \frac{2+(-3)}{2}\right), \quad \text{or} \quad \left(1, -\frac{1}{2}\right).$$ ◀

DO EXERCISES 7 AND 8.

 Circles

A **circle** is a set of points in a plane that are a fixed distance r, called the **radius**, from a fixed point (h, k), called the **center.** If a point (x, y) is on the circle, then by the definition of a circle and the distance formula, it must follow that

$$r = \sqrt{(x - h)^2 + (y - k)^2}, \quad \text{or}$$
$$r^2 = (x - h)^2 + (y - k)^2.$$

> **The Equation of a Circle**
>
> The equation, in standard form, of a circle with center (h, k) and radius r is
> $$(x - h)^2 + (y - k)^2 = r^2.$$

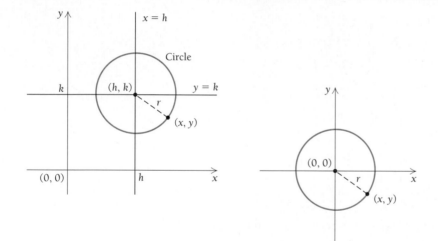

If $h = 0$ and $k = 0$, then the circle is centered at the origin. If a circle centered at the origin has radius $r = 1$, the circle is called a **unit circle.**

Example 4 Find an equation of the circle having center $(4, -5)$ and radius 6.

Solution Using the standard form, we obtain

$$(x - 4)^2 + [y - (-5)]^2 = 6^2,$$

or

$$(x - 4)^2 + (y + 5)^2 = 36.$$ ◀

DO EXERCISES 9 AND 10.

Find the midpoint of the segment having the given endpoints.

7. $(-2, 1)$ and $(5, -6)$

8. $(9, -6)$ and $(9, -4)$

9. Find an equation of the circle centered at the origin, $(0, 0)$, and having radius $2\sqrt{5}$.

10. Find an equation of the circle having center $(-3, 7)$ and radius 5.

11. Find the center and the radius of

$(x + 1)^2 + (y - 3)^2 = 4.$

Then graph the circle.

Example 5   Find the center and the radius of $(x - 2)^2 + (y + 3)^2 = 16$. Then graph the circle.

Solution   We can first write standard form: $(x - 2)^2 + [y - (-3)]^2 = 4^2$. Then we see that the center is $(2, -3)$ and the radius is 4. Now the graph is easy to draw, as shown, using a compass.

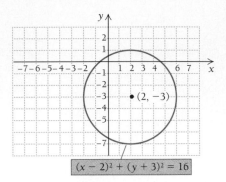

$(x - 2)^2 + (y + 3)^2 = 16$

DO EXERCISE 11.

Completing the square allows us to find the standard form for the equation of a circle.

Example 6   Find the center and the radius of the circle

$x^2 + y^2 + 8x - 2y + 15 = 0.$

Solution   We complete the square twice to get standard form:

$(x^2 + 8x + \quad) + (y^2 - 2y + \quad) = -15.$

We take half the coefficient of the x-term and square it, obtaining 16. We add $16 - 16$ in the first parentheses. Similarly, we add $1 - 1$ in the second parentheses:

$(x^2 + 8x + 16 - 16) + (y^2 - 2y + 1 - 1) = -15.$

Next we do some rearranging and factoring:

12. Find the center and the radius of the circle

$x^2 + y^2 - 14x + 4y - 11 = 0.$

$(x^2 + 8x + 16) + (y^2 - 2y + 1) - 16 - 1 = -15$

$(x + 4)^2 + (y - 1)^2 = 2$

$[x - (-4)]^2 + (y - 1)^2 = (\sqrt{2})^2$ This is standard form.

The center is $(-4, 1)$ and the radius is $\sqrt{2}$.

DO EXERCISE 12.

Example 7   Find an equation of the circle with center $(-2, -3)$ that passes through the point $(1, 1)$.

Solution   Since $(-2, -3)$ is the center, we have

$(x + 2)^2 + (y + 3)^2 = r^2.$

The circle passes through $(1, 1)$. We find r^2 by substituting 1 for x and 1 for y in the above equation:

$$(1 + 2)^2 + (1 + 3)^2 = r^2$$
$$9 + 16 = r^2$$
$$25 = r^2.$$

Thus, $(x + 2)^2 + (y + 3)^2 = 25$ is an equation of the circle. ◀

DO EXERCISE 13.

13. Find an equation of the circle with center $(-1, 4)$ that passes through $(3, -1)$.

●EXERCISE SET 3.2

Ⓐ Find the distance between the pair of points. Where appropriate, find the exact answer and an approximation to three decimal places.

1. $(-3, -2)$ and $(1, 1)$
2. $(5, 9)$ and $(-1, 6)$
3. $(0, -7)$ and $(3, -4)$
4. $(2, 2)$ and $(-2, -2)$
5. $(a, -3)$ and $(2a, 5)$
6. $(5, 2k)$ and $(-3, k)$
7. $(0, 0)$ and (a, b)
8. $(\sqrt{2}, \sqrt{3})$ and $(0, 0)$
9. $(\sqrt{a}, \sqrt{b})$ and $(-\sqrt{a}, \sqrt{b})$
10. $(c - f, c + f)$ and $(c + f, f - c)$
11. ▤ $(7.3482, -3.0991)$ and $(18.9431, -17.9054)$
12. ▤ $(-25.414, 175.31)$ and $(275.34, -95.144)$

Determine whether the points are vertices of a right triangle.
13. $(9, 6)$, $(-1, 2)$, and $(1, -3)$
14. $(-5, -8)$, $(1, 6)$, and $(5, -4)$

Ⓑ Find the midpoints of the segments having the following endpoints.

15. $(-4, 7)$ and $(3, -9)$
16. $(4, 5)$ and $(6, -7)$
17. (a, b) and $(a, -b)$
18. $(-c, d)$ and (c, d)
19. ▤ $(-3.895, 8.1212)$ and $(2.998, -8.6677)$
20. ▤ $(4.1112, 6.9898)$ and $(5.1928, 6.9143)$

Ⓒ Find the center and the radius of the circle. Then graph the circle.

21. $x^2 + y^2 = 36$
22. $x^2 + y^2 = 25$
23. $(x - 3)^2 + y^2 = 3$
24. $(x - 4)^2 + (y - 1)^2 = 2$
25. $(x + 1)^2 + (y + 3)^2 = 4$
26. $(x - 2)^2 + (y + 3)^2 = 1$

Find the center and the radius of the circle.
27. $(x - 8)^2 + (y + 3)^2 = 40$
28. $(x + 5)^2 + (y - 1)^2 = 75$
29. $(x - 3)^2 + y^2 = \frac{1}{25}$
30. $x^2 + (y - 1)^2 = \frac{1}{4}$

31. $x^2 + y^2 + 8x - 6y - 15 = 0$
32. $x^2 + y^2 + 25x + 10y + 12 = 0$
33. $x^2 + y^2 + 6x = 0$
34. $x^2 + y^2 - 4x = 0$
35. $x^2 + y^2 + 8x = 84$
36. $x^2 + y^2 - 75 = 10y$
37. $x^2 + y^2 + 21x + 33y + 17 = 0$
38. $x^2 + y^2 - 7x + 3y - 10 = 0$
39. ▤ $x^2 + y^2 + 8.246x - 6.348y - 74.35 = 0$
40. ▤ $x^2 + y^2 + 25.074x + 10.004y + 12.054 = 0$
41. $9x^2 + 9y^2 = 1$
42. $16x^2 + 16y^2 = 1$

Find an equation of the circle satisfying the given conditions.
43. Center $(0, 0)$, passing through $(-3, 4)$
44. Center $(3, -2)$, passing through $(11, -2)$
45. Center $(-4, 1)$, passing through $(-2, 5)$
46. Center $(-3, -3)$, passing through $(1.8, 2.6)$

● SYNTHESIS

47. Find the point on the x-axis that is equidistant from the points $(1, 3)$ and $(8, 4)$.
48. Find the point on the y-axis that is equidistant from the points $(-2, 0)$ and $(4, 6)$.

Find an equation of a circle satisfying the given conditions.
49. Center $(2, 4)$ and tangent (touching at one point) to the x-axis
50. Center $(-3, -2)$ and tangent to the y-axis
51. The endpoints of a diameter are $(5, -3)$ and $(-3, 7)$.
52. The endpoints of a diameter are $(9, 8)$ and $(-4, -3)$.
53. Center $(-8, 5)$ with a circumference of 10π units
54. Center $(-3, -8)$ with an area of 36π square units
55. Find the center and the radius of the circle that is inscribed in the square with vertices $A(8, 4)$, $B(3, 9)$, $C(8, 14)$, and $D(13, 9)$.

56. ▇ A swimming pool is being constructed in the corner of a lot as shown. In laying out the pool, the contractor wishes to know the distances a_1 and a_2. Find them.

Determine whether each of the following lies on the unit circle $x^2 + y^2 = 1$.

57. $(-1, 0)$

58. $\left(\dfrac{\sqrt{3}}{2}, -\dfrac{1}{2} \right)$

59. ▇ $(0.838670568, -0.544639035)$

60. $\left(\dfrac{\pi}{3}, \dfrac{3}{\pi} \right)$

61. $\left(\dfrac{\sqrt{2}}{2}, \dfrac{\sqrt{2}}{2} \right)$

62. $\left(\dfrac{1}{4}, -\dfrac{3}{4} \right)$

63. $(0, 2)$

64. $\left(\dfrac{1}{2}, -\dfrac{\sqrt{3}}{2} \right)$

● CHALLENGE

65. Consider any right triangle with base b and height h, situated as shown. Show that the midpoint of the hypotenuse P is equidistant from the three vertices of the triangle.

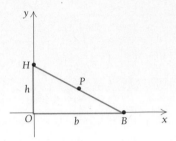

66. Prove that the distance formula holds when two points are on either a vertical line or a horizontal line.

▨ TECHNOLOGY CONNECTION

67. Using a grapher, graph the equations $y = \sqrt{4 - x^2}$ and $y = -\sqrt{4 - x^2}$ using the same set of axes. Describe the result of combining the graphs.

68. Find a way to graph the circle $x^2 + y^2 = 9$ using a grapher.

OBJECTIVES

You should be able to:

A Determine whether a correspondence is a function.

B Determine whether a relation is a function.

C Use a formula to find function values.

D Find the domain of a function given by a formula.

3.3

FUNCTIONS

One of the most important concepts in mathematics is a *function*. It is so important that we will consider it from several viewpoints. If you have trouble with one viewpoint of a function, keep reading, as the next may clarify the concept for you.

A ## Functions as Correspondences

A *function* is a certain rule of correspondence from one set to another. For example:

To each person in a math class	There corresponds	His or her grade;
To each automobile in a state	There corresponds	Its license plate number;
To each triangle	There corresponds	Its area.

In each example, the first set is called the **domain.** The second set is called the **range.** Given a member of the domain, there is *exactly one* member of the range to which it corresponds. Such a rule of correspondence is called a **function.**

DEFINITION	**Function**

A *function* is a rule of correspondence between a first set, called a *domain,* and a second set, called a *range,* such that to any member of the domain, there corresponds *exactly one* member of the range.

Example 1 Which of the following correspondences are functions?

a)

Year (domain)	First-class postage cost, in cents (range)
1978	→ 15
1983	→ 20
1984	↗
1989	→ 25
1991	→ 29

b)

State (domain)	U.S. Senator (range)
Florida	→ Mack
	↗ Graham
Illinois	→ Dixon
	↗ Simon
Arizona	→ DeConcini
	↗ McCain

c)

Number (domain)	Cube (range)
−3	→ −27
−2	→ −8
−1	→ −1
0	→ 0
1	→ 1
2	→ 8
3	→ 27

d)

Number (domain)	Square (range)
−3	
3	→ 9
−2	
2	→ 4
−1	
1	→ 1
0	→ 0

Solution

a) This is a function. Each member in the domain corresponds to exactly one member of the range, even though both 1983 and 1984 correspond to 20. There is one and only one member of the range, 20, that corresponds to 1983. Similarly, there is one and only one member of the range, 20, that corresponds to 1984.

b) This is *not* a function. There are two members of the range, *DeConcini* and *McCain,* that correspond to *Arizona.* There are other instances that show that this is not a function, but one case is all it takes.

c) This is a function.

d) This is a function.   ◀

DO EXERCISES 1–4.

B **Functions as Relations**

Rather than draw arrows to show a rule of correspondence, we can use ordered pairs. For example, instead of

$$1991 \longrightarrow 29$$

in Example 1(a), we write the ordered pair

$$(1991, 29).$$

Determine whether the correspondence is a function.

1.

Year (domain)	Revenue, in millions (range)
1989	→ $35,883
1990	→ 38,828
1991	→ 40,715
1992	→ 48,440

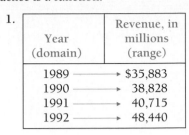

2.

U.S. Senator (domain)	State (range)
DeConcini	→ Arizona
McCain	↗
Mack	→ Florida
Graham	
Dixon	→ Illinois
Simon	↗

3.

Domain	Range
Sigourney	→ David
	↗ Sean
	↗ Fred
Paula	→ Louis
	↗ Reggie
	↗ Bill
Debbie	→ Michael
	↗ John

4.

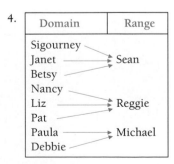

Domain	Range
Sigourney	
Janet	→ Sean
Betsy	
Nancy	
Liz	→ Reggie
Pat	
Paula	→ Michael
Debbie	

5. Which of the following relations are functions?

$A = \{(9, 0), (3, 8), (5, 8), (9, -1)\}$

$B = \{(0, t), (9, e), (-2, q), (-5, b)\}$

$C = \{(-3, 5), (7, -2), (4, -6)\}$

$D = \{(0, 1), (1, 0), (-1, 1)\}$

$E = \{(7, -7), (-7, -7)\}$

Let us write the correspondences in Example 1 as relations.

a) *First-class postage* $= \{(1978, 15), (1983, 20), (1984, 20), (1989, 25), (1991, 29)\}$

b) *U.S. Senator* $= \{(\text{Florida, Mack}), (\text{Florida, Graham}), (\text{Illinois, Dixon}), (\text{Illinois, Simon}), (\text{Arizona, DeConcini}), (\text{Arizona, McCain})\}$

c) *Cube* $= \{(-3, -27), (-2, -8), (-1, -1), (0, 0), (1, 1), (2, 8), (3, 27)\}$

d) *Square* $= \{(-3, 9), (-2, 4), (-1, 1), (0, 0), (1, 1), (2, 4), (3, 9)\}$

In terms of ordered pairs, we know that (b) is not a function because it has at least two ordered pairs, (Arizona, DeConcini) and (Arizona, McCain), with the same first coordinate and different second coordinates. This does not happen with (a), (c), and (d). Thus we know that (a), (c), and (d) are functions, but (b) is not.

DEFINITION   **Function**

A *function* is a relation in which *no* two ordered pairs can have the same first coordinate and different second coordinates.

Example 2   Determine whether the relation

$$A = \{(2, 3), (5, 9), (1, 0), (10, -2)\}$$

is a function.

Solution   The relation A is a function because no two ordered pairs have the same first coordinate and different second coordinates.   ◀

Example 3   Determine whether the relation

$$B = \{(4, 5), (-3, 2), (4, 0), (-1, 9)\}$$

is a function.

Solution   The relation B is *not* a function because the ordered pairs $(4, 5)$ and $(4, 0)$ have the same first coordinate and different second coordinates.   ◀

DO EXERCISE 5.

 Formulas of Functions

Suppose you see a flash of lightning and count the number of seconds until you hear the thunder. If you multiply the number of seconds by $\frac{1}{5}$, you can determine how far away the lightning was. If it takes 10 seconds to hear the thunder, then the lightning was 2 miles away. This sets up a rule of correspondence

$$10 \longrightarrow 2$$

and an ordered pair

$$(10, 2).$$

Suppose that $x =$ the number of seconds until you hear the thunder and $y =$ the distance; then you have $y = \frac{1}{5}x$. The value of y depends on the value of x. We have a formula or "recipe" for determining y given x. Since the value of y depends on the value of x, we call y the **dependent variable** and x the **independent variable.**

Input, x

Output, $f(x)$

To better understand function notation, it helps to think of a **function machine.** We name a function f. Think of putting a member of the domain x, an **input,** into the machine. The machine knows the correspondence and gives you a member of the range $f(x)$, the **output.** Since we have a function, for each input there is one and only one output. We read "$f(x)$" as "f of x," or "f at x." Note that $f(x)$ does *not* mean "f times x."

Most of the functions we will consider in this text will have real-number domains and ranges and can be described by formulas or equations. For example, $f(x) = 3x - 4$ describes the function that takes an input x, multiplies it by 3, and then subtracts 4. To find the output $f(-1)$, we take the input -1, multiply it by 3 to get -3, and then subtract 4 to get -7:

$$f(-1) = 3(-1) - 4$$
$$= -7.$$

Sometimes, instead of writing $f(x) = 3x - 4$, we might write $y = 3x - 4$, where it is understood that the value of y, the *dependent variable,* is calculated after first choosing a value for x, the *independent variable.* To understand why $f(x)$ notation is so useful, consider these two statements, which have the same meaning:

a) If $f(x) = 3x - 4$, then $f(-1) = -7$.
b) If $y = 3x - 4$, then when x is -1, the value of y is -7.

Note that the notation used in (a) is more concise.

Outputs are also called **function values.** In the preceding, $f(-1) = -7$. We say that "the function value at -1 is -7," or "when x is -1, the value of the function is -7." We find function values by making substitutions for the variables. When a substitution is not meaningful, we say that the function value *does not exist.*

Example 4 For the function f given by $f(z) = 2z^2 - z + 3$, find $f(0)$, $f(-7)$, $f(5a)$, and $f(a - 4)$.

Solution One way to find function values when a formula is given is to think of the formula as follows:

$$f(\ ) = 2(\ )^2 - (\ ) + 3.$$

Then whatever goes in the blank on the left between parentheses goes in

6. For the function f given by

$$f(x) = 2x - 3,$$

find $f(-1), f(0), f(5.6)$, and $f(10)$.

7. For the function f given by

$$f(x) = x^2 + 5x - 1,$$

find $f(0), f(1), f(-1), f(2a)$, and $f(a + 1)$.

8. For the function f given by

$$f(x) = \sqrt{x},$$

find $f(16), f(3)$, and $f(-4)$, if possible.

9. For the function G given by $G(x) = 7$, find $G(0), G(-3)$, and $G(\frac{1}{2})$. What is the range?

the blanks on the right between parentheses.

$$f(0) = 2 \cdot (0)^2 - 0 + 3 = 0 - 0 + 3 = 3;$$
$$f(-7) = 2(-7)^2 - (-7) + 3 = 2 \cdot 49 + 7 + 3 = 108;$$
$$f(5a) = 2(5a)^2 - 5a + 3 = 2 \cdot 25a^2 - 5a + 3$$
$$= 50a^2 - 5a + 3;$$
$$f(a - 4) = 2(a - 4)^2 - (a - 4) + 3$$
$$= 2(a^2 - 8a + 16) - a + 4 + 3 = 2a^2 - 17a + 39 \quad \blacktriangleleft$$

Note above that whether we write $f(z) = 2z^2 - z + 3$ or $f(x) = 2x^2 - x + 3$ or $f(\ ) = 2(\ )^2 - (\ ) + 3$, we still have $f(-7) = 108$. Thus our independent variable can be regarded as a *dummy* variable. The letter chosen for the dummy variable is not as important as the algebraic manipulation to which it is subjected.

DO EXERCISES 6 AND 7.

Example 5   For the function f given by $f(x) = 1/x$, find $f(2), f(-\frac{1}{4})$, and $f(0)$, if possible.

Solution

$$f(2) = \frac{1}{2};$$

$$f\left(-\frac{1}{4}\right) = \frac{1}{-\frac{1}{4}} = -4;$$

$f(0)$ is not defined, so $f(0)$ does not exist. $\quad \blacktriangleleft$

DO EXERCISE 8.

TECHNOLOGY CONNECTION

Once you have graphed a function on a grapher, you can investigate some of the points on the graph by using the "Trace" feature that most graphers offer. (Again, check your grapher's documentation for the specific procedures.) Usually, there is a key marked "TRACE" that is pressed to access this feature. Once it is pressed, a cursor (often blinking) appears somewhere on the graph while the x- and y-coordinates of that point are displayed elsewhere on the screen. This cursor can be moved left and right on the function's graph, usually by pressing left and right arrows on the keyboard. The x- and y-coordinates on the screen will change as the cursor moves around the screen.

Graph each of the following functions in a $[-10, 10] \times [-10, 10]$ viewing box. Then use the Trace function to find the coordinates of at least 3 points on the graph.

TC 1. $y = 5x - 3$

TC 2. $y = x^2 - 4x + 3$

TC 3. $y = (x + 4)^2$

Sometimes all the outputs of a function are the same. In such a case, we have what is called a **constant function**.

Example 6   For the constant function $g(x) = 3$, find $g(5), g(-7)$, and $g(24.96)$.

Solution   All the outputs are the same number, 3. Thus,

$$g(5) = 3, \qquad g(-7) = 3, \quad \text{and} \quad g(24.96) = 3.$$

Note that each of the ordered pairs $(5, 3)$, $(-7, 3)$, and $(24.96, 3)$ is in the constant function g. The range contains only one number, 3. $\quad \blacktriangleleft$

DO EXERCISE 9.

Since many students using this text are preparing for

calculus, learning to simplify rational expressions like

$$\frac{f(a + h) - f(a)}{h}$$

is a very important skill.

Example 7   For the function f given by $f(x) = 2x^2 - 3$, construct and simplify the expression

$$\frac{f(a + h) - f(a)}{h}.$$

Solution   We find $f(a + h)$ and $f(a)$:

$$f(a + h) = 2(a + h)^2 - 3; \qquad f(a) = 2a^2 - 3.$$

Then

$$\frac{f(a + h) - f(a)}{h} = \frac{[2(a + h)^2 - 3] - [2a^2 - 3]}{h}$$

$$= \frac{2a^2 + 4ah + 2h^2 - 3 - 2a^2 + 3}{h} \qquad \text{Simplifying}$$

$$= \frac{4ah + 2h^2}{h}$$

$$= 4a + 2h. \qquad \blacktriangleleft$$

DO EXERCISE 10.

Example 8   For the function f given by $f(x) = x^3 + x$, construct and simplify the expression

$$\frac{f(a + h) - f(a)}{h}.$$

Solution   We find $f(a + h)$ and $f(a)$:

$$f(a + h) = (a + h)^3 + (a + h) \qquad \text{Replacing each occurrence of } x \text{ by } a + h$$
$$\qquad\qquad = a^3 + 3a^2h + 3ah^2 + h^3 + a + h;$$
$$f(a) = a^3 + a.$$

Then

$$\frac{f(a + h) - f(a)}{h} = \frac{[a^3 + 3a^2h + 3ah^2 + h^3 + a + h] - [a^3 + a]}{h}$$

$$= \frac{a^3 + 3a^2h + 3ah^2 + h^3 + a + h - a^3 - a}{h}$$

$$= \frac{3a^2h + 3ah^2 + h^3 + h}{h}$$

$$= \frac{h(3a^2 + 3ah + h^2 + 1)}{h}$$

$$= 3a^2 + 3ah + h^2 + 1. \qquad \blacktriangleleft$$

DO EXERCISE 11.

10. For the function f given by $f(x) = 3x^2 + 1$, construct and simplify the expression

$$\frac{f(a + h) - f(a)}{h}.$$

11. For the function f given by $f(x) = x^2 - x$, construct and simplify the expression

$$\frac{f(a + h) - f(a)}{h}.$$

TECHNOLOGY CONNECTION

Once the graph of a function has been drawn, determining the domain and the range becomes a matter of simply examining the graph. The domain of the function is all the x-values that are meaningful replacements, and the range of the function is all the resulting y-values. For example, the function $f(x) = 3 - x^4$ has this graph:

$f(x) = 3 - x^4$

$[-5, 5]$ x $[-10, 10]$

By examining the graph, we see that all values of x are meaningful replacements. Thus the domain of this function is *all real numbers*. It is also clear that the y-values never reach above $+3$. Therefore, the range of this function is *all real numbers less than or equal to 3*. (To verify this, we can increase the range on both axes, effectively "zooming out" to see more and more of the graph.)

As another example, graph the function

$$f(x) = \frac{x}{\sqrt{x^2 - 4}},$$

examining it in several different viewing boxes. One view is shown here:

$f(x) = \dfrac{x}{\sqrt{x^2 - 4}}$

$[-5, 5]$ x $[-4, 4]$

The domain is $-\infty < x < -2$ and $2 < x < \infty$. All values between -2 and $+2$ would make the expression under the radical negative, while the values -2 and $+2$ would make the denominator 0. We also see that the range of this function is $-\infty < y < -1$ and $1 < y < \infty$.

Graph each of the following functions and examine them in different viewing boxes to determine the domain and the range.

TC 4. $f(x) = \dfrac{1}{x^2 - 1}$ **TC 5.** $f(x) = |7 - 4x^2|$

TC 6. $f(x) = \dfrac{1}{\sqrt{5 - x}}$ **TC 7.** $f(x) = \dfrac{x^2}{\sqrt{5 - x}}$

Some comment is in order regarding the notation $f(x)$ and f. When we refer to a function f, we are referring to a set of ordered pairs. The notation $f(x)$ refers to the second coordinate of an ordered pair that has x as its first coordinate. Nevertheless, it is a fact of life in the literature of mathematics to be careless about this notation, referring instead to "the function $f(x)$" and "the function $f(x) = x^2$."

▷ Finding Domains of Functions

When a function f, whose inputs and outputs are real numbers, is given by a formula, the domain is understood to be the set of all real numbers that are meaningful replacements for x. For example, consider the function f given by $f(x) = 1/x$. The number 0 is not a meaningful replacement, because division by 0 is not possible. All other real numbers are meaningful replacements, so the domain of f consists of all nonzero real numbers, $\{x | x \neq 0\}$.

Consider the function g given by $g(x) = \sqrt{x}$. The negative numbers are not meaningful replacements. Thus the domain of g consists of all nonnegative numbers, $\{x | x \geq 0\}$.

Example 9 Find the domain of the function g given by

$$g(x) = \frac{x}{x^2 + 2x - 3}.$$

Solution The formula is meaningful so long as a replacement for x does not make the denominator 0. To find those replacements that do make the denominator 0, we solve $x^2 + 2x - 3 = 0$:

$$x^2 + 2x - 3 = 0$$
$$(x - 1)(x + 3) = 0$$
$$x - 1 = 0 \quad \text{or} \quad x + 3 = 0$$
$$x = 1 \quad \text{or} \quad x = -3.$$

Thus the domain consists of the set of all real numbers except 1 and -3. We can name this set $\{x | x \neq 1 \text{ and } x \neq -3\}$. ◀

Example 10 Find the domain of the function f given by

$$f(x) = \frac{\sqrt{5x - 3}}{x^2 + 1}.$$

Solution Let's first consider the numerator $\sqrt{5x - 3}$. The formula is meaningful so long as replacements make the radicand nonnegative (no negative number has a real square root). Thus to find the domain, we solve the inequality $5x - 3 \geq 0$:

$$5x - 3 \geq 0$$
$$5x \geq 3$$
$$x \geq \tfrac{3}{5}.$$

Now let us consider the denominator. Since x^2 is nonnegative, $x^2 + 1$ is always positive. Thus the denominator is always nonzero, and there are no restrictions on numbers that can be substituted into the denominator. The domain of the function is $\{x \mid x \geq \tfrac{3}{5}\}$. ◀

Example 11   Find the domain of the function t given by $t(x) = x^3 + |x|$.

Solution   There are no restrictions on the numbers that we can substitute into this formula. We can cube any real number, we can take the absolute value of any real number, and we can add the results. Thus the domain is the entire set of real numbers. ◀

DO EXERCISES 12–14.

In Margin Exercises 12–14, functions are given by formulas. Find the domain of each function.

12. $f(x) = \dfrac{x + 1}{3x^2 + 10x + 8}$

13. $g(x) = \dfrac{\sqrt{10x + 25}}{x^2 + 3}$

14. $p(x) = x^3 - 4x^2 + 2x + 8$

● **EXERCISE SET** **3.3**

A   Determine whether the correspondence is a function.

1.

City (domain)	Team (range)
New York	Mets
	Giants
Atlanta	Braves
	Falcons
Houston	Astros
	Oilers
San Diego	Padres
	Chargers

2.

Year (domain)	Population of Brazil, in millions (range)
1950	52
1960	70
1970	93
1980	121
1990	154

3.

Number, x (domain)	$|x|$ (range)
-3	3
3	
-2	2
2	
-1	1
1	
0	0

4.

Number, x (domain)	Square roots (range)
16	-4
	4
9	-3
	3
4	-2
	2
0	0

B   Determine whether the relation is a function.

5. $K = \{(1, 2), (2, 3), (3, 4), (4, 1)\}$

6. $L = \{(-3, 5), (0, 0), (-5, 3), (-3, -5)\}$

7. $T = \{(-4, -4), (-1, -6), (-4, 4), (-6, -1)\}$

8. $V = \{(2, -9), (4, 2), (0, 5), (4, -9)\}$

9. $B = \{(6, -6), (-2, 2), (0, 0), (2, -2), (-6, 6)\}$

10. $F = \{(p, r), (q, r), (a, r), (z, m), (q, t), (q, w)\}$

In Exercises 11–18, functions are given by formulas. Find the indicated function values, and construct and simplify the given expressions.

11. Given $f(x) = 5x^2 + 4x$, find:

a) $f(0)$; **b)** $f(-1)$;
c) $f(3)$; **d)** $f(t)$;
e) $f(t - 1)$; **f)** $\dfrac{f(a + h) - f(a)}{h}$.

12. Given $g(x) = 3x^2 - 2x + 1$, find:

a) $g(0)$; **b)** $g(-1)$;
c) $g(3)$; **d)** $g(t)$;
e) $g(a + h)$; **f)** $\dfrac{g(a + h) - g(a)}{h}$.

13. Given $f(x) = 2|x| + 3x$, find:

a) $f(1)$; **b)** $f(-2)$;
c) $f(-4)$; **d)** $f(2y)$;
e) $f(a + h)$; **f)** $\dfrac{f(a + h) - f(a)}{h}$.

14. Given $g(x) = x^3 - 2x$, find:

a) $g(1)$; **b)** $g(-2)$;
c) $g(-4)$; **d)** $g(3y)$;
e) $g(2 + h)$; **f)** $\dfrac{g(2 + h) - g(2)}{h}$.

15. Given $f(x) = \dfrac{x}{2 - x}$, find:

a) $f(1)$; **b)** $f(2)$;
c) $f(-3)$; **d)** $f(-16)$;
e) $f(x + h)$; **f)** $\dfrac{f(x + h) - f(x)}{h}$.

16. Given $g(x) = \dfrac{x - 4}{x + 3}$, find:

a) $g(5)$; **b)** $g(4)$;
c) $g(-3)$; **d)** $g(-16.25)$;
e) $g(x + h)$; **f)** $\dfrac{g(x + h) - g(x)}{h}$.

17. Given $f(x) = \dfrac{x^2 - x - 2}{2x^2 - 5x - 3}$, find:

a) $f(0)$; **b)** $f(4)$;
c) $f(-1)$; **d)** $f(3)$;
e) $f(2 - h)$; **f)** $f(a + b)$.

18. Given $s(x) = \sqrt{\dfrac{3x - 4}{2x + 5}}$, find:

a) $s(10)$; **b)** $s(2)$;
c) $s(1)$; **d)** $s(-1)$;
e) $s(3 + h)$; **f)** $s(a - b)$.

19. Given $f(x) = x^2$, find
$\dfrac{f(a + h) - f(a)}{h}$.

20. Given $f(x) = x^3$, find
$\dfrac{f(a + h) - f(a)}{h}$.

21. Given $f(x) = x + \sqrt{x^2 - 1}$, find $f(0)$, $f(2)$, and $f(10)$.

22. Given $g(x) = x/\sqrt{1 - x^2}$, find $g(0)$, $g(3)$, and $g(\frac{1}{2})$.

In Exercises 23–30, functions are given by formulas. Find the domain of each function.

23. $f(x) = 7x + 4$ **24.** $f(x) = |3x - 2|$

25. $f(x) = 4 - \dfrac{2}{x}$ **26.** $f(x) = \dfrac{1}{x^4}$

27. $f(x) = \sqrt{7x + 4}$ **28.** $f(x) = \sqrt{x - 3}$

29. $f(x) = \dfrac{1}{x^2 - 4}$ **30.** $f(x) = \dfrac{1}{9 - x^2}$

● **SYNTHESIS**

31. Given the function
$$f = \{(-1, 2), (-3, 4), (5, -6), (7, 9)\},$$
find $f(-1)$, $f(7)$, $f(5)$, and $f(-3)$. What is the domain? What is the range?

32. Given the function
$$g = \{(2, -3), (-4, 5), (6, -7), (8, -7)\},$$
find $g(8)$, $g(6)$, $g(-4)$, and $g(2)$. What is the domain? What is the range?

33. A complex-valued function f is defined as follows:
$$f(z) = z^2 - 4z + i.$$
Find $f(3 + i)$.

34. A complex-valued function g is defined as follows:
$$g(z) = 2z^2 + z - 2i.$$
Find $g(2 - i)$.

For each function, construct and simplify $\dfrac{f(x + h) - f(x)}{h}$.

35. $f(x) = \dfrac{1}{x}$ **36.** $f(x) = \dfrac{1}{x^2}$

37. $f(x) = \sqrt{x}$
(*Hint*: Rationalize the numerator.)

38. Determine whether the relation $\{(x, y) | x^2 + y^2 = 9\}$ is a function.

TECHNOLOGY CONNECTION

Use a grapher to draw the graph of the function. Use the Trace and Zoom features to find $f(-25)$ and $f(3.5)$ for each. Then from the graph, determine the domain.

39. $f(x) = \dfrac{4x^3 + 4}{4x^2 - 5x - 6}$ **40.** $f(x) = \dfrac{2x + 6}{x^3 - 4x}$

41. $f(x) = \dfrac{4x^3 + 4}{x(x + 2)(x - 1)}$ **42.** $f(x) = x^3 - x^2 + x - 2$

43. $f(x) = \dfrac{\sqrt{x + 3}}{x^2 - x - 2}$ **44.** $f(x) = \dfrac{\sqrt{x}}{2x^2 - 3x - 5}$

45. $f(x) = \sqrt{x^2 + 1}$ **46.** $f(x) = \sqrt{\dfrac{x + 1}{x + |x|}}$

3.4
LINES AND LINEAR FUNCTIONS

A   Linear Equations and Graphs

An equation $Ax + By = C$ is called the **standard linear equation** because its graph is a straight line. In this case, A, B, and C are constants, but A and B cannot both be 0. Any equation that is equivalent to one of this form has a graph that is a straight line.

Since two points determine a line, we can graph a linear equation by finding two of its points. Then we draw a line through those points. The easiest points to find are often the intercepts (the points at which the line crosses the axes). A third point should always be used as a check.

You should be able to:

A Graph linear equations.

B Find the slope, if it exists, of the line containing two given points.

C Given the slope and the coordinates of one point on the line, find an equation of the line.

D Given the coordinates of two points, find an equation of the line containing them.

E Given an equation of a line, find its slope, if it exists, and its y-intercept, and graph linear equations using slope and y-intercept.

F Solve problems involving applications of linear functions.

G Given equations of two lines, tell whether the lines are parallel, perpendicular, or neither. Given an equation of a line and the coordinates of a point, find equations of lines parallel or perpendicular to that line and containing the given point.

Example 1   Graph: $4x + 5y = 20$.

Solution   We set $x = 0$ and find that $y = 4$. Thus, $(0, 4)$ is a point of the graph (the y-intercept). We set $y = 0$ and find that $x = 5$. Thus, $(5, 0)$ is a point of the graph (the x-intercept). The graph is shown below.

A third point can be used as a check. We substitute any point, say -2, for x, and solve for y:

$$4(-2) + 5y = 20$$
$$-8 + 5y = 20$$
$$5y = 28$$
$$y = 5\tfrac{3}{5}.$$

This gives us a check point, $(-2, 5\tfrac{3}{5})$, in case we have made a computational error.

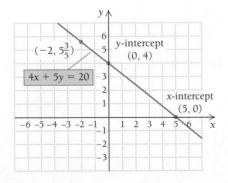

DO EXERCISES 1 AND 2.

If a graph goes through the origin, then it has only one intercept, and other points will be needed for graphing. If an equation has a missing variable ($A = 0$ or $B = 0$), then its graph is parallel to one of the axes. Vertical lines are parallel to the y-axis and are of the form $x = a$. Horizontal lines are parallel to the x-axis and are of the form $y = b$.

Graph.

1. $6x - 4y = 12$

2. $3x + 2y = 6$

Graph.

3. $x = 4$

4. $y = -3$

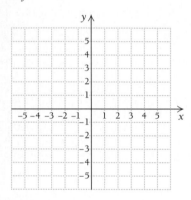

Example 2   Graph: $y = 3$.

Solution   The graph is shown below. All y-coordinates are 3.

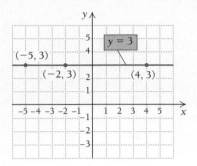

Example 3   Graph: $x = -2$.

Solution   The graph is shown below. All x-coordinates are -2.

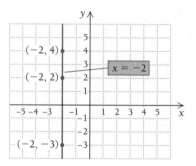

DO EXERCISES 3 AND 4.

TECHNOLOGY CONNECTION

Most graphers can graph any line except a vertical line, which would be written as $x = a$, where a = any constant. In order to draw the graph of a nonvertical line, we first solve for y. Therefore, if we want to graph the line $4x - 7y + 8 = 0$, we first find the equivalent form

$$y = \tfrac{4}{7}x + \tfrac{8}{7}.$$

Since most graphers accept only decimal notation for a number, we express this line in the form $y = 0.5714x + 1.1429$. We can then enter it into the grapher. (The standard $[-10, 10] \times [-10, 10]$ viewing box is fine for this example.)

Use a grapher to draw the graph of each of the following lines.

$y = 0.5714x + 1.1429$

$[-10, 10] \times [-10, 10]$

TC 1. $13x + 12y - 9 = 0$

TC 2. $\dfrac{x}{3} + \dfrac{y}{4} = 1$

TC 3. $12.7x - 3.4y = 23.9$

Slope

The graph of a linear equation may slant upward or downward, or be horizontal or vertical. We need a way to describe the steepness of a line using a number.

Suppose that points P_1 and P_2 with coordinates (x_1, y_1) and (x_2, y_2) are two different points on a line not parallel to an axis. Consider a right triangle, as shown, with legs parallel to the axes. The point P_3 with coordinates (x_2, y_1) is the third vertex of the triangle. As we move from P_1 to P_2, y changes from y_1 to y_2. The change in y is $y_2 - y_1$. Similarly, the change in x is $x_2 - x_1$. The ratio of these changes is called the **slope.**

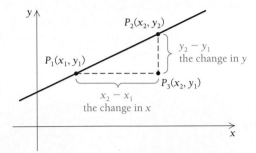

DEFINITION	**Slope of a Line**

The *slope m* of a line containing two points (x_1, y_1) and (x_2, y_2) is defined by

$$m = \frac{y_2 - y_1}{x_2 - x_1}, \quad \text{where } x_2 \neq x_1.$$

Note that when $x_2 = x_1, x_2 - x_1 = 0$, and thus the slope is not defined.

Example 4 Graph the line through the points $(1, 2)$ and $(3, 6)$ and find its slope.

Solution Let us call the slope m and let $(1, 2)$ be (x_1, y_1) and $(3, 6)$ be (x_2, y_2). Applying the definition, we obtain

$$m = \frac{y_2 - y_1}{x_2 - x_1} = \frac{6 - 2}{3 - 1} = \frac{4}{2} = \frac{2}{1} = 2.$$

This means that for every 2 units that the line rises vertically, it runs 1 unit to the right horizontally. Note that we can also use the points in the opposite order, so long as we are consistent. We get the same slope:

$$m = \frac{y_2 - y_1}{x_2 - x_1} = \frac{2 - 6}{1 - 3}$$

$$= \frac{-4}{-2} = 2.$$

Use graph paper. Graph the line through the points and find its slope.

5. $(1, 3)$ and $(2, 5)$

6. $(3, 7)$ and $(5, 3)$

7. $(3, 5)$ and $(2, -1)$

8. $(-1, -1)$ and $(2, -4)$

Find the slope, if it exists, of the line containing the given points.

9. $(4, 6)$ and $(-2, 6)$

10. $(-3, 5)$ and $(-3, 7)$

Extending Example 4, we can also see that it does not matter which two points of a line we choose to determine the slope. No matter what points we use, we get the same number for the slope. For example, if we choose $(2, 4)$ and $(4, 8)$, the slope is

$$\frac{8 - 4}{4 - 2} = \frac{4}{2} = 2.$$

DO EXERCISES 5–8.

If a line slants upward from left to right, it has a positive slope, and the larger the slope is, the steeper the line. If a line slants downward from left to right, the change in x and the change in y are of opposite signs, so the line has a negative slope. The larger the absolute value of the slope, the steeper the line. This is illustrated in the following graphs of lines of the type $y = mx$.

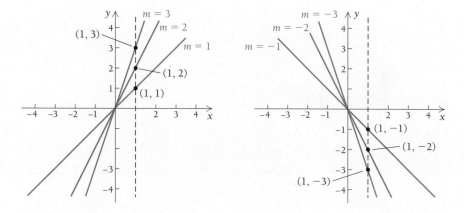

The following summarizes the results for horizontal and vertical lines.

a) If a line is horizontal, the change in y for any two points is 0. Thus a horizontal line has zero slope.
b) If a line is vertical, the change in x for any two points is 0. Thus the slope is not defined, because we cannot divide by zero.

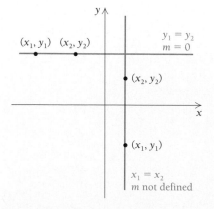

DO EXERCISES 9 AND 10.

Applications of Slope

Slope has many real-world applications. For example, numbers like 2%, 3%, and 10% are often used to represent the **grade** of a road. Such a number is meant to tell how steep a road up a hill or mountain is. For example, a 5% grade means that for every horizontal distance of 100 ft, the road rises 5 ft, if a vehicle is going up; and -5% means that the road is dropping 5 ft for every 100 ft, if the vehicle is going down.

The concept of grade is also relevant in cardiology when a person runs on a treadmill. A physician may change the slope or grade to measure its effects on the patient's heartbeat. Another example occurs in hydrology. When a river flows, the strength or force of the river depends on how much the river falls vertically compared to how much it flows horizontally.

Example 5   A heart patient is running on a treadmill with a 0.6-ft rise and a 5-ft run. A heart arrhythmia occurs. What is the grade of the treadmill?

Solution   The grade is positive for treadmill tests and is the slope of the line on which the patient is jogging. Thus,

$$m = \frac{0.6 \text{ ft}}{5 \text{ ft}} = 12\%.$$

◀

DO EXERCISE 11.

▶ **c** Point–Slope Equations of Lines

Suppose that we have a nonvertical line and that the coordinates of one point P_1 are (x_1, y_1). We think of P_1 as fixed. Suppose, also, that we have a movable point P on the line with coordinates (x, y). Thus the slope is given by

$$\frac{y - y_1}{x - x_1} = m. \qquad (1)$$

11. Find the slope, or pitch, of the right side of this roof.

Find an equation of the line.

12. Containing the point $(-2, \frac{1}{4})$ with slope -3

Note that this is true only when (x, y) is a point different from (x_1, y_1). If we use the multiplication principle, we get

$$y - y_1 = m(x - x_1). \qquad \textit{Point–slope equation} \qquad (2)$$

Equation (2) is called the **point–slope equation** of a line and will be true even if $(x, y) = (x_1, y_1)$. Thus if we know the slope of a line and the coordinates of a point on the line, we can find an equation of the line.

Example 6 Find an equation of the line containing the point $(\frac{1}{2}, -1)$ with slope 5.

Solution If we substitute in $y - y_1 = m(x - x_1)$, we get

$$y - (-1) = 5(x - \tfrac{1}{2}),$$

which simplifies to

$$y + 1 = 5(x - \tfrac{1}{2}),$$

or

$$y = 5x - \tfrac{5}{2} - 1,$$

or

$$y = 5x - \tfrac{7}{2}.$$

13. With y-intercept $(0, -9)$ and slope $\frac{1}{4}$

DO EXERCISES 12–14.

14. With x-intercept $(5, 0)$ and slope $-\frac{1}{2}$

▶ Two-Point Equations of Lines

Suppose that a nonvertical line contains the points $P_1(x_1, y_1)$ and $P_2(x_2, y_2)$. The slope of the line is

$$\frac{y_2 - y_1}{x_2 - x_1}.$$

If we substitute $(y_2 - y_1)/(x_2 - x_1)$ for m in the point–slope equation,

$$y - y_1 = m(x - x_1),$$

we have

$$y - y_1 = \frac{y_2 - y_1}{x_2 - x_1}(x - x_1). \qquad \textit{Two-point equation}$$

This is known as the **two-point equation** of a line. Note that either of the two given points can be called P_1 or P_2 and (x, y) is any point on the line.

Example 7 Find an equation of the line containing the points $(2, 3)$ and $(1, -4)$.

Solution If we take $(2, 3)$ as P_1 and $(1, -4)$ as P_2 and use the two-point

equation, we get

$$y - 3 = \frac{-4-3}{1-2}(x-2),$$

which simplifies to $y = 7x - 11$. ◀

DO EXERCISES 15 AND 16.

 Slope–Intercept Equations of Lines

Suppose that a nonvertical line has slope m and y-intercept $(0, b)$. Sometimes, for brevity, we refer to the number b as the y-intercept. Let us substitute m and $(0, b)$ in the point–slope equation. We get

$$y - y_1 = m(x - x_1)$$
$$y - b = m(x - 0).$$

This simplifies to

$$y = mx + b. \qquad \textit{Slope–intercept equation}$$

This is called the **slope–intercept equation** of a line. The advantage of such an equation is that we can read the slope m and the y-intercept b directly from the equation. This equation is the most useful when graphing.

Example 8   Find the slope and the y-intercept of $y = 5x - \frac{1}{4}$.

Solution

$$y = 5 - \frac{1}{4}$$

Slope: 5 y-intercept: $-\frac{1}{4}$ ◀

Example 9   Find the slope and the y-intercept of $y = 8$.

Solution   We can rewrite this equation as $y = 0x + 8$. We then see that the slope is 0 and the y-intercept is $(0, 8)$, or 8. Thus the graph of the equation is a horizontal line.   ◀

DO EXERCISES 17 AND 18.

To find the slope–intercept equation of a line, when given another equation, we solve for y.

Example 10   Find the slope and the y-intercept of the line having the equation $3x - 6y - 7 = 0$.

Solution   We solve for y, obtaining $y = \frac{1}{2}x - \frac{7}{6}$. Thus the slope is $\frac{1}{2}$ and the y-intercept is $(0, -\frac{7}{6})$, or $-\frac{7}{6}$.   ◀

Find an equation of the line containing the given pair of points.

15. $(1, 4)$ and $(3, -2)$

16. $(3, -6)$ and $(0, 4)$

Find the slope and the y-intercept.

17. $y = -7x + 11$

18. $y = -4$

19. a) Find the slope–intercept equation of the line whose equation is

$$-2x + 3y - 6 = 0.$$

b) Find the slope and the y-intercept of this line.

Graph.

20. $y = \dfrac{3}{5}x + 2$

21. $y = -\dfrac{3}{5}x - 1$

If a line is vertical, its slope is not defined. Thus it has no slope–intercept equation. Such a line does have a simple equation, however. As previously developed, all vertical lines have equations $x = a$, where a is some constant.

DO EXERCISE 19.

We can graph linear equations using the slope and the y-intercept.

Example 11   Graph: $y = -\frac{2}{3}x + 1$.

Solution   First we plot the y-intercept $(0, 1)$. We think of the slope as $-2/3$. Starting at the y-intercept and using the slope, we find another point by moving 2 units down (since the numerator is *negative* and corresponds to the change in y) and 3 units right (since the denominator is *positive* and corresponds to the change in x). We get to a new point, $(3, -1)$. Similarly, we can find other points such as $(6, -3)$ by moving on from $(3, -1)$.

If we think of the slope as $2/-3$, then we start again at the y-intercept, $(0, 1)$, and find another point by moving 2 units up (since the numerator is *positive* and corresponds to the change in y) and 3 units left (since the denominator is *negative* and corresponds to the change in x). We get to another point on the line, $(-3, 3)$.

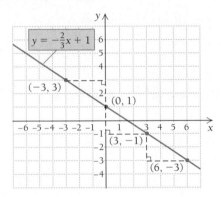

Example 12   Graph: $y = \frac{2}{5}x + 4$.

Solution   First we plot the y-intercept, $(0, 4)$. We then consider the slope, $\frac{2}{5}$. Starting at the y-intercept and using the slope, we find another point by moving 2 units up (since the numerator is *positive* and corresponds to the change in y) and 5 units right (since the denominator is *positive* and corresponds to the change in x). We get to a new point, $(5, 6)$.

By thinking of the slope as $-2/-5$, we can find another point, $(-5, 2)$.

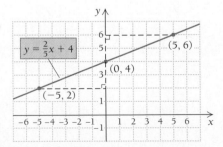

DO EXERCISES 20 AND 21.

F   Linear Functions and Applications

A function f is a **linear function** if it has an equation $f(x) = mx + b$. If the slope m is zero, then the equation simplifies to the constant function $f(x) = b$.
    Let us consider an application of a linear function.

Example 13   *Stopping distance on glare ice.*   The stopping distance D (at some fixed speed) of regular tires on glare ice is a function of the air temperature F, in degrees Fahrenheit. This function D is estimated by

$$D(F) = 2F + 115,$$

where $D(F) =$ the stopping distance, in feet, when the air temperature is F, in degrees Fahrenheit.

a) Find $D(0°)$, $D(-20°)$, $D(10°)$, and $D(32°)$.
b) Graph D.
c) Explain why the domain of the function should be restricted to the interval from $-57.5°$ to $32°$, inclusive.

Solution

a) We find the function values by substitution:

$$\begin{aligned}
D(0°) &= 2(0) + 115 \\
&= 115 \text{ ft;}
\end{aligned}$$

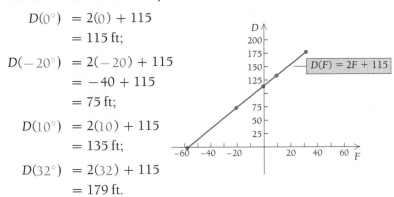

$$\begin{aligned}
D(-20°) &= 2(-20) + 115 \\
&= -40 + 115 \\
&= 75 \text{ ft;}
\end{aligned}$$

$$\begin{aligned}
D(10°) &= 2(10) + 115 \\
&= 135 \text{ ft;}
\end{aligned}$$

$$\begin{aligned}
D(32°) &= 2(32) + 115 \\
&= 179 \text{ ft.}
\end{aligned}$$

b) The graph of the function is shown above.
c) The function has certain restrictions. Stopping distance can only be nonnegative. Thus the function is meaningful only for values of x for which

$$2F + 115 \geq 0.$$

Solving, we get $F \geq -57.5°$. Since ice occurs only when $F \leq 32°$, we have another restriction. The domain is thus the set

$$\{F \mid -57.5° \leq F \leq 32°\}. \qquad \blacktriangleleft$$

DO EXERCISE 22.

G   Parallel and Perpendicular Lines

If two lines are vertical, then they are parallel. That is, they do not intersect. Thus equations such as $x = a_1$ and $x = a_2$ (where a_1 and a_2 are unequal constants) have graphs that are *parallel lines*. Nonvertical lines are parallel if they have the same slope but different y-intercepts. Thus equations such as $y = mx + b_1$ and $y = mx + b_2$, $b_1 \neq b_2$, have graphs that are *parallel lines*.

---

22.   *Pressure at sea depth.*   The pressure P, given by

$$P(d) = 1 + \frac{1}{33}d,$$

gives the pressure, in atmospheres (atm), at a depth d, in feet, in the sea.

a) Find $P(0)$, $P(5)$, $P(10)$, $P(33)$, and $P(200)$.

b) Graph P.

c) Discuss the restrictions on the domain of the function.

In each situation, slopes of the lines are given. Determine whether the lines are perpendicular.

23.

THEOREM 3

Vertical lines are *parallel*. Nonvertical lines are *parallel* if and only if they have the same slope and different *y*-intercepts.

If two equations are equivalent, then they represent the same line. Thus if two lines have the same slope and the same *y*-intercept, then they are not really two different lines. They are the same line.

If one line is vertical and the other is horizontal, such as $x = a$ and $y = b$, then they meet at right angles and are perpendicular. Is there any other way to tell whether two lines are perpendicular?

Consider a line $\overleftrightarrow{AB}$ as shown, with slope p/q. Then think of rotating the figure 90° to get a line perpendicular to $\overleftrightarrow{AB}$. For the new line, the change in *y* and the change in *x* are interchanged, but the change in *x* is now the opposite of what was the change in *y*. Thus the slope of the new line is $-q/p$.

24.

Let us multiply the slopes:

$$\frac{p}{q}\left(-\frac{q}{p}\right) = -1.$$

This is the condition under which lines will be *perpendicular*.

THEOREM 4

Two lines with slopes m_1 and m_2 are *perpendicular* if and only if $m_1 m_2 = -1$.

If one line has slope m_1, the slope m_2 of a line perpendicular to it is $-1/m_1$.

DO EXERCISES 23 AND 24.

Example 14 Determine whether the pair of lines is parallel, perpendicular, or neither.

  a) $y + 2 = 5x$, $5y + x = -15$
  b) $2y + 4x = 8$, $5 + 2x = -y$
  c) $2x + 1 = y$, $y + 3x = 4$

Solution

a) We solve each equation for y:

$$y = 5x - 2, \qquad y = -\frac{1}{5}x - 3.$$

The slopes are 5 and $-\frac{1}{5}$. Their product is -1, so the lines are perpendicular.

b) Solving for y, we get $y = -2x + 4$ and $y = -2x - 5$ and can determine that $m_1 = -2$ and $m_2 = -2$. Since the y-intercepts, 4 and -5, are different, the lines are parallel.

c) Solving for y, we determine that $m_1 = 2$ and $m_2 = -3$, so the lines are neither parallel nor perpendicular. ◀

DO EXERCISES 25–27.

Example 15 Write equations of the lines (a) parallel and (b) perpendicular to the line $4y - x = 20$ and both containing the point $(2, -3)$.

Solution   We first solve for y: $y = \frac{1}{4}x + 5$, so the slope is $\frac{1}{4}$.

a) The line parallel to the given line will have slope $\frac{1}{4}$. Then we use the point–slope equation with slope $\frac{1}{4}$ and containing the point $(2, -3)$:

$$y - y_1 = m(x - x_1)$$

$$y - (-3) = \frac{1}{4}(x - 2)$$

$$y = \frac{1}{4}x - \frac{7}{2}.$$

b) The slope of the perpendicular line is -4. Now we use the point–slope equation to write an equation with slope -4 and containing the point $(2, -3)$:

$$y - y_1 = m(x - x_1)$$

$$y - (-3) = -4(x - 2)$$

$$y = -4x + 5.$$

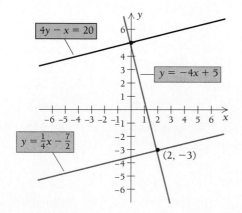

DO EXERCISE 28.

Determine whether the pair of lines is parallel, perpendicular, or neither.

25. $2y - x = 2, \quad y + 2x = 4$

26. $3y = 2x + 15, \quad 2y = 3x + 10$

27. $5y = 3 - 4x, \quad 8x + 10y = 1$

28. Find equations of the lines parallel and perpendicular to the line $4 - y = 2x$ and containing the point $(3, 4)$.

SUMMARY OF TERMINOLOGY ABOUT LINES

Terminology	Mathematical Interpretation
Standard form	$Ax + By = C$
Slope	$m = \dfrac{y_2 - y_1}{x_2 - x_1}$
Point–slope equation	$y - y_1 = m(x - x_1)$
Two-point equation	$y - y_1 = \dfrac{y_2 - y_1}{x_2 - x_1}(x - x_1)$
Slope–intercept equation	$y = mx + b$
Vertical lines	$x = a$
Horizontal lines	$y = b$
Parallel lines	$m_1 = m_2, b_1 \neq b_2$
Perpendicular lines	$m_1 m_2 = -1, \text{ or } m_1 = -\dfrac{1}{m_2}$

● EXERCISE SET 3.4

A Graph.

1. $8x - 3y = 24$ **2.** $5x - 10y = 50$

3. $3x + 12 = 4y$ **4.** $4x - 20 = 5y$

5. $y = -2$ **6.** $2y - 3 = 9$

7. $5x + 2 = 17$ **8.** $19 = 5 - 2x$

B Find the slope of the line containing the given points.

9. $(6, 2)$ and $(-2, 1)$ **10.** $(-2, 1)$ and $(-4, -2)$

11. $(2, -4)$ and $(4, -3)$ **12.** $(5, -3)$ and $(-5, 8)$

13. $(\pi, 5)$ and $(\pi, 4)$ **14.** $(\sqrt{2}, -4)$ and $(\pi, -4)$

Find the road grade and an equation giving the height y as a function of the horizontal distance x.

15.

920.58 m

13,740 m

16.

50 ft

1250 ft

17. A treadmill is 5 ft long and is set at an 8% grade when a heart arrhythmia occurs. How high is the end of the treadmill?

18. A river flows at a slope of 0.12. How many feet does it fall for every 250 ft that it flows horizontally?

C Find the equation of the line.

19. Through $(3, 2)$ with $m = 4$

20. Through $(4, 7)$ with $m = -2$

21. With y-intercept -5 and $m = 2$

22. With y-intercept π and $m = \frac{1}{4}$

23. Through $(-4, 7)$ with $m = -\frac{2}{3}$

24. Through $(-3, -5)$ with $m = \frac{3}{4}$

25. Through $(5, -8)$ with $m = 0$

26. Through $(5, -8)$ with m undefined

D Find the equation of the line.

27. Containing $(1, 4)$ and $(5, 6)$

28. Containing $(-2, 0)$ and $(2, 3)$

29. Containing $(-2, 5)$ and $(-4, -7)$

30. Containing $(\frac{2}{3}, -\frac{4}{5})$ and $(-\frac{1}{2}, 8.2)$

31. Containing $(3, 6)$ and $(-2, 6)$

32. Containing $(-\frac{3}{8}, 0)$ and $(-\frac{3}{8}, \frac{8}{3})$

E Find the slope and the y-intercept of the line.

33. $y = 2x + 3$ **34.** $y = 6 - x$

35. $2y = -6x + 10$ **36.** $-3y = -12x + 9$

37. $3x - 4y = 12$ **38.** $5x + 2y = -7$

39. $3y + 10 = 0$ **40.** $y = 7$

Graph.

41. $y = -\frac{3}{2}x$ **42.** $y = \frac{2}{3}x$

43. $y = -\frac{5}{2}x - 2$

44. $y = -\frac{5}{3}x + 3$

45. $y = \frac{1}{2}x + 1$

46. $y = \frac{1}{3}x - 1$

47. $y = \frac{4}{3} - \frac{1}{3}x$

48. $y = -\frac{1}{4}x - \frac{1}{2}$

 Find the equation of the line.

49. Through $(3.014, -2.563)$ with slope 3.516

50. Through the points $(1.103, 2.443)$ and $(8.114, 11.012)$

51. **Temperature and depth in the earth.**   The function T given by

$$T(d) = 10d + 20$$

can be used to determine the temperature T, in degrees Celsius, at a depth d, in kilometers, inside the earth.

a) Find $T(5\text{ km})$, $T(20\text{ km})$, and $T(1000\text{ km})$.

b) Graph T.

c) The radius of the earth is about 5600 km. Use this fact to determine the domain of the function.

52. **Tail length of a snake.**   It has been found that the total length L and the tail length T, both in millimeters, of females of the snake species *Lampropeltis Polyzona* are related by the linear function

$$T(L) = 0.143L - 1.18.$$

a) Find $T(0\text{ mm})$, $T(50\text{ mm})$, and $T(80\text{ mm})$.

b) Graph T.

c) Determine the domain of the function.

53. **Straight-line depreciation.**   A company buys an office machine for $5200 on January 1 of a given year. The machine is expected to last 8 years, at the end of which time its *trade-in*, or *salvage, value* will be $1100. If the company figures the decline or depreciation in value to be the same each year, then the salvage value V, after t years, $0 \le t \le 8$, is given by the linear function

$$V(t) = \$5200 - \$512.50t.$$

a) Find $V(0)$, $V(1)$, $V(2)$, $V(3)$, and $V(8)$.

b) Graph V.

54. **Estimating heights.**   An anthropologist can use certain linear functions to estimate the height of a male or female, given the length of certain bones. A *humerus* is the bone from the elbow to the shoulder. Let $x =$ the length of the humerus, in centimeters. The height, in centimeters, of a male with a humerus of length x is given by

$$M(x) = 2.97x + 73.6.$$

The height, in centimeters, of a female with a humerus of length x is given by

$$F(x) = 3.14x + 65.0.$$

A 45-cm humerus was uncovered in a ruins.

a) If we assume that it was from a male, how tall was he?

b) If we assume that it was from a female, how tall was she?

c) Discuss the domain of each function.

Humerus

Determine whether the lines are parallel, perpendicular, or neither.

55. $2x - 5y = -3$,
$2x + 5y = 4$

56. $x + 2y = 5$,
$2x + 4y = 8$

57. $y = 4x - 5$,
$4y = 8 - x$

58. $y = -x + 7$,
$y = x + 3$

Find equations of the lines parallel and perpendicular to the given line and containing the given point.

59. $(0, 3)$, $3x - y = 7$

60. $(-4, -5)$, $2x + y = -4$

61. $(-3, -5)$, $5x - 2y = 4$

62. $(3, -2)$, $3x + 4y = 5$

63. $(3, -3)$, $x = -1$

64. $(4, -5)$, $y = -1$

● **SYNTHESIS** _____

Suppose that f is a linear function. Then $f(x) = mx + b$. Find a formula for $f(x)$ given each of the following.

65. $f(3x) = 3f(x)$

66. $f(x + 2) = f(x) + 2$

Suppose that f is a linear function. Then $f(x) = mx + b$. Determine whether each of the following is true or false.

67. $f(c + d) = f(c) + f(d)$

68. $f(cd) = f(c)f(d)$

69. $f(kx) = kf(x)$

70. $f(c - d) = f(c) - f(d)$

71. Determine whether these three points are on a line. [*Hint:* Compare the slopes of $\overline{AB}$ and $\overline{BC}$. ($\overline{AB}$ refers to the segment from A to B.)]

$$A(9, 4), \quad B(-1, 2), \quad C(4, 3)$$

72. Determine whether these three points are on a line. (See the hint for Exercise 71.)

$$A(-1, -1), \quad B(2, 2), \quad C(-3, -4)$$

73. Use graph paper. Plot the points $A(0, 0)$, $B(8, 2)$, $C(11, 6)$, and $D(3, 4)$. Draw $\overline{AB}$, $\overline{BC}$, $\overline{CD}$, and $\overline{DA}$. Find the slopes of these four segments. Compare the slopes of $\overline{AB}$ and $\overline{CD}$. Compare the slopes of $\overline{BC}$ and $\overline{DA}$. (Figure $ABCD$ is a parallelogram and its opposite sides are parallel.)

74. Use graph paper. Plot the points $E(-2, -5)$, $F(2, -2)$, $G(7, -2)$, and $H(3, -5)$. Draw $\overline{EF}$, $\overline{FG}$, $\overline{GH}$, $\overline{HE}$, $\overline{EG}$, and $\overline{FH}$. Compare the slopes of $\overline{EG}$ and $\overline{FH}$. (Figure $EFGH$ is a rhombus and its diagonals are perpendicular.)

75. ***Fahrenheit temperature as a function of Celsius tempera-*
    *ture.***   Fahrenheit temperature F is a linear function of
    Celsius (or Centigrade) temperature C. When C is 0, F
    is 32. When C is 100, F is 212. Use these data to express
    F as a linear function of C.

76. ***Celsius temperature as a function of Fahrenheit temper-*
    *ature.***   Celsius (Centigrade) temperature C is a linear
    function of Fahrenheit temperature F. When F is 32, C is
    0. When F is 212, C is 100. Use these data to express C
    as a linear function of F.

77. Suppose that P is a nonconstant linear function of Q. Show
    that Q is a linear function of P.

78. Suppose that y is directly proportional to x. Show that y
    is a linear function of x.

79. Find equations of the lines containing the point $(4, -2)$
    parallel and perpendicular to the line containing $(-1, 4)$
    and $(2, -3)$.

80. Find equations of the lines containing the point $(-1, 3)$
    parallel and perpendicular to the line containing $(3, -5)$
    and $(-2, -7)$.

81. Find k so that the line containing $(-3, k)$ and $(4, 8)$ is
    parallel to the line containing $(6, 4)$ and $(2, -5)$.

82. Find k so that the line containing $(-3, k)$ and $(4, 8)$ is
    perpendicular to the line containing $(6, 4)$ and $(2, -5)$.

83. Find an equation of the perpendicular bisector of the line
    segment with endpoints $(-1, 3)$ and $(-6, 7)$.

84. Find an equation of the perpendicular bisector of the line
    segment with endpoints $(1, 0)$ and $(-7, -12)$.

● CHALLENGE _____

85. Consider any quadrilateral situated as shown. Show that
    the segments joining the midpoints of the sides, in order
    as shown, form a parallelogram.

86. Prove that ABC is a right angle. Assume that point B is
    on the circle whose radius is a and whose center is at the
    origin. (*Hint:* Use slopes and an equation of the circle.)

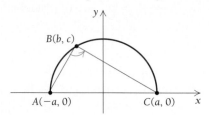

3.5

GRAPHS AND APPLICATIONS OF FUNCTIONS

A Graphs of Functions

Most of the functions that we study in this course and in calculus are given
by formulas. We can graph them in much the same way as we did equations.
We consider various inputs x, and then compute outputs $f(x)$. Then we plot
the ordered pairs $(x, f(x))$, look for patterns, and complete the graph. As
you learn more about functions and graphs, you will refine your skills. For
example, you will know the shape of many graphs from their formulas, and
you will know how to derive one graph from another.

Example 1   Graph the function f given by $f(x) = x^3 - x$. We will often
just say "Graph $f(x) = x^3 - x$."

Solution We compute some function values and organize them in a table:

$f(-2) = (-2)^3 - (-2)$
$\qquad = -8 + 2 = -6;$

$f(-1) = (-1)^3 - (-1)$
$\qquad = -1 + 1 = 0;$

$f\left(-\dfrac{1}{2}\right) = \left(-\dfrac{1}{2}\right)^3 - \left(-\dfrac{1}{2}\right) = \dfrac{3}{8};$

$f(0) = 0^3 - 0 = 0;$

$f\left(\dfrac{1}{2}\right) = \left(\dfrac{1}{2}\right)^3 - \dfrac{1}{2} = -\dfrac{3}{8};$

$f(1) = 1^3 - 1 = 0;$

$f(2) = 2^3 - 2 = 6.$

x	$f(x)$	$(x, f(x))$
-2	-6	$(-2, -6)$
-1	0	$(-1, 0)$
$-\dfrac{1}{2}$	$\dfrac{3}{8}$	$\left(-\dfrac{1}{2}, \dfrac{3}{8}\right)$
0	0	$(0, 0)$
$\dfrac{1}{2}$	$-\dfrac{3}{8}$	$\left(\dfrac{1}{2}, -\dfrac{3}{8}\right)$
1	0	$(1, 0)$
2	6	$(2, 6)$

We plot these points, look for a pattern, and sketch the graph.

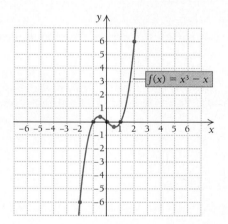

$f(x) = x^3 - x$

DO EXERCISES 1 AND 2.

Suppose a relation has two ordered pairs with the same first coordinate but different second coordinates. The graphs of these two ordered pairs would be points on the same vertical line. This gives a method to test whether a graph of a relation is the graph of a function.

> **The Vertical-Line Test**
>
> If it is possible for a vertical line to intersect a graph more than once, then the graph is not the graph of a function.

We try to find a vertical line that meets the graph more than once. If we do, then the relation is not a function. If we do not, then the relation is a function.

Graph.

1. $f(x) = x - x^3$

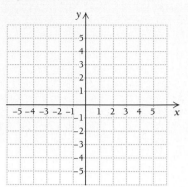

2. $g(x) = -3$
  (a constant function)

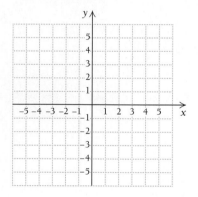

3. Which of the following are graphs of functions?

a)

b)

c)

d)

e)

f)

Example 2 Which of the following are graphs of functions?

a)

b)

c)

d)

e)

f)

In graph (f), the solid dot shows that $(-1, 1)$ belongs to the graph. The open circle shows that $(-1, -2)$ does *not* belong to the graph.

Solution   Graphs (a), (e), and (f) are graphs of functions. In (b) the vertical line crosses the graph in three points and so it is not a function. Also, in (c) and (d), we can find a vertical line that crosses the graph more than once. ◀

DO EXERCISE 3.

B▸ Interval Notation

The set of real numbers corresponds to the set of points on a line.

We need to define a new kind of notation for certain sets of real numbers. This is called *interval notation*. For real numbers a and b such that $a < b$ (a is to the left of b on a number line), we define the **open interval** (a, b) to be the set of numbers between, but not including, a and b. That is,

$(a, b) =$ the set of all numbers x such that $a < x < b$
$= \{x \mid a < x < b\}.$

The graph of (a, b) is shown above. The points a and b are called **endpoints.** The open circles and the parentheses indicate that the endpoints a and b

are not included in the interval. Be careful not to confuse this notation with that of an ordered pair. The context of the writing should make the meaning clear. If not, we might say "the interval $(-2, 3)$." When we mean an ordered pair, we might say "the pair $(-2, 3)$."

DO EXERCISES 4 AND 5.

The **closed interval** $[a, b]$ is the set of numbers between and including a and b. That is,

$[a, b]$ = the set of all numbers x such that $a \le x \le b$
$\quad = \{x | a \le x \le b\}$.

The graph of $[a, b]$ is shown above. The solid circles and the brackets indicate that a and b are included.

There are two kinds of **half-open intervals,** defined as follows:

$(a, b]$ = the set of all numbers x such that $a < x \le b$
$\quad = \{x | a < x \le b\}$.

The open circle and the parenthesis indicate that a is not included. The solid circle and the bracket indicate that b is included. Also,

$[a, b)$ = the set of all numbers x such that $a \le x < b$
$\quad = \{x | a \le x < b\}$.

The solid circle and the bracket indicate that a is included. The open circle and the parenthesis indicate that b is not included.

DO EXERCISES 6 AND 7.

Some intervals are of unlimited extent in one or both directions. In such cases, we use the infinity symbol ∞. For example,

$[a, \infty)$ = the set of all numbers x such that $x \ge a$.

Note that ∞ is not a number.

(a, ∞) = the set of all numbers x such that $x > a$.

$(-\infty, b]$ = the set of all numbers x such that $x \le b$.

$(-\infty, b)$ = the set of all numbers x such that $x < b$.

4. Write interval notation for the graph.
a)

b)

5. Write interval notation for the set.
a) $\{x | -2 < x < 3\}$
b) $\{x | 0 < x < 1\}$
c) $\{x | -\frac{1}{4} < x < \sqrt{2}\}$

6. Write interval notation for the graph.
a)

b)

c)

d)

7. Write interval notation for the set.
a) $\{x | 4 \le x \le 5\frac{1}{2}\}$
b) $\{x | -3 < x \le 0\}$
c) $\{x | -\frac{1}{2} \le x < \frac{1}{2}\}$
d) $\{x | -\pi < x < \pi\}$

8. Write interval notation for the graph.

a)

b)

c)

4.8

d)

9. Write interval notation for the set.

a) $\{x|x \geq 8\}$
b) $\{x|x < -7\}$
c) $\{x|x > 10\}$
d) $\{x|x \leq 0.78\}$

We can name the entire set of real numbers using $(-\infty, \infty)$.

$(-\infty, \infty)$

DO EXERCISES 8 AND 9.

Any point in an interval that is not an endpoint is an **interior point**.

Endpoints

Interior point

Note that all the points in an open interval are interior points.

Interval notation gives us an alternative notation for the solution set of an inequality. For example, we might say that the solution set of $x + 2 < 7$ is the set $\{x|x < 5\}$ or the interval $(-\infty, 5)$.

◤ **Increasing and Decreasing Functions**

If the graph of a function rises from left to right, it is said to be an **increasing function**. If the graph of a function drops from left to right, it is said to be a **decreasing function**. This can be stated more formally.

DEFINITION	**Increasing and Decreasing Functions**

1. A function f is an *increasing* function when for all a and b in the domain of f, if $a < b$, then $f(a) < f(b)$.
2. A function f is a *decreasing* function when for all a and b in the domain of f, if $a < b$, then $f(a) > f(b)$.

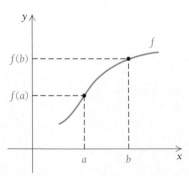

An increasing function
If $a < b$, then $f(a) < f(b)$.

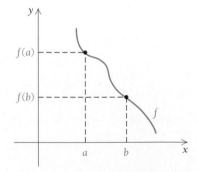

A decreasing function
If $a < b$, then $f(a) > f(b)$.

Examples Determine whether the function is increasing, decreasing, or neither.

3.

f is increasing.

4.

g is decreasing.

5.

h is neither increasing nor decreasing—it is a constant function.

6.

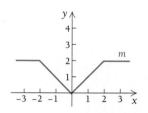

m is neither increasing nor decreasing.

In Example 6, the function m is neither increasing nor decreasing on the entire real line. But it is increasing on the interval $[0, 2]$, decreasing on the interval $[-2, 0]$, and constant on the intervals $(-\infty, -2]$ and $[2, \infty)$.

DO EXERCISES 10 AND 11. _____

 Functions Defined Piecewise

Sometimes functions are defined **piecewise.** That is, the function has different output formulas for different parts of the domain.

Example 7 Graph the function defined as

$$f(x) = \begin{cases} 4, & \text{for } x \le 0, \\ 4 - x^2, & \text{for } 0 < x \le 2, \\ 2x - 6, & \text{for } x > 2. \end{cases}$$

(This means that for any input x less than or equal to 0, the output is 4.)

(This means that for any input x greater than 0 and less than or equal to 2, the output is $4 - x^2$.)

(This means that for any input x greater than 2, the output is $2x - 6$.)

---

10. Determine whether the function is increasing, decreasing, or neither increasing nor decreasing.

a)

b)

c)

d)

11. For the function in Margin Exercise 10(d), find an interval on which the function is (a) increasing and (b) decreasing.

 TECHNOLOGY CONNECTION

Determining where a function is increasing and where it is decreasing is easy when using a grapher. You can first check visually after you have graphed the function. Then activate the Trace feature and move the cursor to the point on the function at the far left of the viewing box. Then move the cursor to the right along the function while watching the value of the y-coordinate. If it is increasing as you move, then you are on an *increasing* portion of the graph. If it is decreasing, then you are on a *decreasing* portion. You must make sure that you watch only the y-coordinate as you move the cursor *to the right.*

For the function, determine the interval(s) on which $f(x)$ is increasing and the interval(s) on which it is decreasing.

TC 1. $f(x) = x^3 - 6x^2 + 9x + 4$

TC 2. $f(x) = x|x^2 - 4|$

TECHNOLOGY CONNECTION

Some graphers have the ability to draw the graphs of piecewise-defined functions. Check your grapher's documentation to see if yours is included in this group. If so, the function is usually entered as follows (we will use the function from Example 7):

$$y = 4(x \le 0) + (4 - x^2)(x > 0)(x \le 2)$$
$$+ (2x - 6)(x > 2).$$

Each individual piece is written like the above. More pieces can be added if necessary. (*Note:* If your grapher provides this feature, then it probably also allows you to choose the mode in which the graph is displayed: CONNECTED, in which adjacent points are connected by segments, or DOT, in which only the points are plotted and they are not connected. For piecewise-defined functions, you should select the DOT mode.)

Graph each of the following functions.

TC 3. $f(x) = \begin{cases} x^3, & x \le -3, \\ 4x, & -3 < x < 2, \\ 2x^2, & x \ge 2 \end{cases}$

TC 4. $f(x) = \begin{cases} \dfrac{1}{x}, & x < 0, \\ |x|, & x \ge 0 \end{cases}$

12. Graph the function defined as

$$f(x) = \begin{cases} x + 3, & \text{for } x \le -2, \\ 1, & \text{for } -2 < x \le 3, \\ x^2 - 10, & \text{for } 3 < x. \end{cases}$$

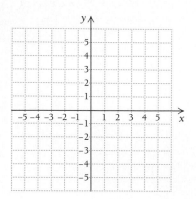

Solution   See the following graph.

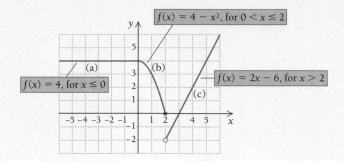

a) We graph $f(x) = 4$ for inputs less than or equal to 0 (that is, $x \le 0$). Note that $f(x) = 4$ *only* for numbers x on the interval $(-\infty, 0]$.

b) We graph $f(x) = 4 - x^2$ for inputs greater than 0 and less than or equal to 2 (that is, $0 < x \le 2$). Note that $f(x) = 4 - x^2$ *only* on the interval $(0, 2]$.

c) We graph $f(x) = 2x - 6$ for inputs greater than 2 (that is, $x > 2$). Note that $f(x) = 2x - 6$ *only* for numbers x on the interval $(2, \infty)$.   ◀

DO EXERCISE 12.

Example 8   Graph the function f defined as

$$f(x) = \begin{cases} \dfrac{x^2 - 4}{x + 2}, & \text{for } x \ne -2, \\ 3, & \text{for } x = -2. \end{cases}$$

Solution   When $x \ne -2$, the denominator of $(x^2 - 4)/(x + 2)$ is nonzero, so we can simplify:

$$\frac{x^2 - 4}{x + 2} = \frac{(x + 2)(x - 2)}{x + 2}$$
$$= x - 2.$$

Thus,

$$f(x) = x - 2 \quad \text{for } x \ne -2.$$

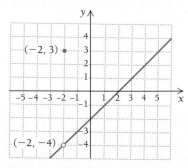

The graph of this part of the function consists of a line with a hole at the point $(-2, -4)$ indicated by the open circle. At $x = -2$, $f(-2) = 3$, so the point $(-2, 3)$ is plotted above $(-2, -4)$.   ◀

DO EXERCISE 13 ON THE FOLLOWING PAGE.

A function with importance in calculus and computer programming is the *greatest integer function* f, given by $f(x) = \text{INT}(x)$ and defined as follows.

DEFINITION   **Greatest Integer Function**

$\text{INT}(x)$ = the greatest integer less than or equal to x.

For example, $\text{INT}(4.5) = 4$, $\text{INT}(-1) = -1$, and $\text{INT}(-3.8) = -4$. In some texts, the notation $[x]$ is used for $\text{INT}(x)$.

Example 9 Graph the greatest integer function.

Solution The greatest integer function can also be defined by a piecewise function with an infinite number of statements:

$$f(x) = \text{INT}(x) = \begin{cases} \vdots \\ -3 & \text{if } -3 \le x < -2 \\ -2 & \text{if } -2 \le x < -1 \\ -1 & \text{if } -1 \le x < 0 \\ 0 & \text{if } 0 \le x < 1 \\ 1 & \text{if } 1 \le x < 2 \\ 2 & \text{if } 2 \le x < 3 \\ 3 & \text{if } 3 \le x < 4 \\ \vdots \end{cases}$$

DO EXERCISE 14.

E Applications of Functions

Many real-life applications involve functions.

Example 10 *Speed of sound in air.* The speed of sound S in air is a function of the temperature T, in degrees Fahrenheit, and is given by

$$S(T) = 1087.7 \sqrt{\frac{5T + 2457}{2457}},$$

where S is in feet per second. Find the speed of sound in air when the temperature is $0°$, $32°$, and $70°$.

13. Graph the function f defined as

$$f(x) = \begin{cases} \dfrac{x^2 - 4}{x - 2} & \text{for } x \ne 2. \\ -1, & \text{for } x = 2. \end{cases}$$

14. a) Find $\text{INT}(4.3)$, $\text{INT}(-2)$, $\text{INT}(1.986734199)$, $\text{INT}(3)$, $\text{INT}(-2.6)$, and $\text{INT}(-0.9)$.

b) Graph $f(x) = \text{INT}(x) + 1$.

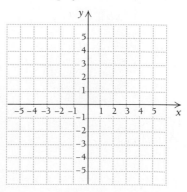

c) Graph $f(x) = \text{INT}(x + 1)$.

TECHNOLOGY CONNECTION

Functions like the one used in Example 10, with noninteger coefficients, are fairly tedious to graph by hand. Graphers, however, can easily create these graphs. Try graphing the function from Example 10 on your grapher, using a viewing box of $[-50, 120] \times [1000, 1250]$. Remember to convert this to the form $y = f(x)$. Note that S is changed to y and T is changed to x.

    Once you have the graph drawn, use the Trace feature to find the values of y for $x = 0, 32$, and 70. Compare these values to those given in the solutions shown in the text. (Due to the limitations of some graphers, you may not be able to find exact values for $x = 0, 32$, and 70. However, you should be able to get reasonably close to these values.)

Solution   We find the function values as follows. A calculator with a square root key would be helpful.

$$S(0) = 1087.7 \sqrt{\frac{5(0) + 2457}{2457}}$$

$$= 1087.7 \text{ ft/sec;}$$

$$S(32) = 1087.7 \sqrt{\frac{5(32) + 2457}{2457}}$$

$$= 1087.7 \sqrt{\frac{160 + 2457}{2457}} \approx 1122.6 \text{ ft/sec;}$$

$$S(70) = 1087.7 \sqrt{\frac{5(70) + 2457}{2457}}$$

$$= 1087.7 \sqrt{\frac{350 + 2457}{2457}} \approx 1162.6 \text{ ft/sec.} \quad \blacktriangleleft$$

---

15. Find the speed of sound in air when the temperature is $-10°$, $55°$, and $90°$.

> **CAUTION!** When using a calculator to do exercises like this, do not stop to round as you make successive calculations. Answers at the back of the text generally will have been found by not stopping to round.

DO EXERCISES 15 AND 16.

16. The volume V of a sphere is a function of the radius r, given by

$$V(r) = \tfrac{4}{3}\pi r^3.$$

The radius of a bowling ball is 4.08 in. Approximate its volume.

F Finding Formulas for Functions

In calculus, many problems are solved that seek to find the *maximum* or *minimum* values of a function. These occur in so-called *maximum–minimum* problems. The critical skill in solving this type of problem is to find a formula for the function. Typically, we are trying to represent one variable in terms of another variable even though there may be many variables involved in the problem. This task involves the *Familiarize* and *Translate* skills of our five steps for problem solving. Let us consider some examples.

Example 11 A music store has 20 ft of dividers to set off a rectangular area for a compact-disc display in one corner of its sales area. The sides against the wall require no partition. Express the area A as a function of the length x of one side of the rectangle, and determine a meaningful domain for the function.

Solution We first draw a picture, letting $x =$ the length of one side and $y =$ the length of the other. Then since the sum of the lengths must be 20 ft, we have

$$x + y = 20, \quad \text{or} \quad y = 20 - x.$$

Then the area is given by

$$A = xy$$
$$= x(20 - x) \qquad \text{Substituting } 20 - x \text{ for } y$$
$$= 20x - x^2.$$

We have thus expressed A as a function of x: $A(x) = 20x - x^2$. This function gives the area A of the rectangular region as a function of the length x of one side. The problem restricts the domain of the function to values of x such that $0 < x < 20$. Otherwise, the lengths would not be positive. ◀

DO EXERCISE 17.

Example 12   Two cars leave the same intersection at right angles to each other. One travels at a speed of 55 mph and the other at 65 mph. Express the distance between the cars as a function of the time t.

Solution   We let $d =$ the distance, in miles, between the cars and $t =$ the time, in hours. Then we make a drawing, as shown here.

The first car travels at 55 mph, so after t hours, it has traveled $55t$ miles. Similarly, in time t, the second car has traveled $65t$ miles. We can then apply the Pythagorean theorem:

$$d^2 = a^2 + b^2$$
$$= (55t)^2 + (65t)^2.$$

Thus,

$$d = \sqrt{(55t)^2 + (65t)^2} = \sqrt{3025t^2 + 4225t^2}$$
$$= \sqrt{7250t^2}$$
$$= \sqrt{7250}\,t, \quad \text{or } 5\sqrt{290}\,t.$$

◀

DO EXERCISE 18.

17. A store has 40 ft of dividers to set off a rectangular area in one corner of its sales area. The sides against the wall require no partition. Express the area A as a function of the length x of one side of the rectangle, and determine a meaningful domain for the function.

18. A hot-air balloon is released and rises straight up from the ground at the rate of 120 ft/min. The balloon is tracked from a rangefinder at point P, which is 400 ft from the release point Q of the balloon. Let $d =$ the distance from the balloon to the rangefinder at point P and $t =$ the time, in minutes, after which the balloon is released. Express d as a function of t.

 EXERCISE SET ## 3.5

Ⓐ Graph.

1. $f(x) = |x| + 2$

2. $f(x) = 2 - |x|$

3. $g(x) = 4 - x^2$

4. $f(x) = 3 - x^2$

5. $f(x) = \dfrac{2}{x}$

6. $f(x) = \dfrac{-3}{x}$

7. $f(x) = x^2 - 3$

8. $g(x) = x^2 - 4$

9. $g(x) = 3$
(constant function)

10. $g(x) = -2$
(constant function)

11. $f(x) = x^3 - 3x$

12. $f(x) = 3x - x^3$

13. $f(x) = |x| + x$

14. $g(x) = |x| - x$

15. $f(x) = \sqrt{x + 3}$

16. $g(x) = \sqrt{x - 5}$

Determine whether the graph is that of a function. An open circle indicates that the point does not belong on the graph.

17.

18.

19.

20.

21.

22.

23.

24.

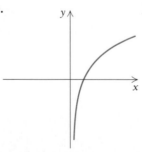

B Write interval notation for the graph.

25.

26.

27.

28.

29.

30.

31.

32.

Write interval notation for the set.

33. $\{x \mid -3 \le x \le 3\}$

34. $\{x \mid -4 < x < 4\}$

35. $\{x \mid -14 \le x < -11\}$

36. $\{x \mid 6 < x \le 20\}$

37. $\{x \mid x \le -4\}$

38. $\{x \mid x > -5\}$

39. $\{x \mid x < 3.8\}$

40. $\{x \mid x \ge \sqrt{3}\}$

◢c

41. Determine whether the function is increasing, decreasing, or neither increasing nor decreasing.

a)

b)

c)

d)

42. Determine whether the function is increasing, decreasing, or neither increasing nor decreasing.

a)

b)

c)

d)

where y = the year and $P(y)$ = the average price, in dollars. Thus, $P(1993)$ is the average price of a movie ticket in 1993. The price is lower than what might be expected due to senior citizen's discounts, children's prices, and special volume discounts.

a) Use the function to predict the average price of a ticket in 1993, 1995, and 2000.

b) When will the average price of a movie ticket be $8.00?

Graph.

43. $f(x) = \begin{cases} 1, & \text{for } x < 0, \\ -1, & \text{for } x \geq 0 \end{cases}$

44. $f(x) = \begin{cases} 2, & \text{for } x \text{ an integer}, \\ -2, & \text{for } x \text{ not an integer} \end{cases}$

45. $f(x) = \begin{cases} 3, & \text{for } x \leq -3, \\ |x|, & \text{for } -3 < x \leq 3, \\ -3, & \text{for } x > 3 \end{cases}$

46. $f(x) = \begin{cases} -2x - 6, & \text{for } x \leq -2, \\ 2 - x^2, & \text{for } -2 < x < 2, \\ 2x - 6, & \text{for } x \geq 2 \end{cases}$

47. $f(x) = \begin{cases} \dfrac{x^2 - 1}{x - 1}, & \text{for } x \neq 1, \\ -2, & \text{for } x = 1 \end{cases}$

48. $f(x) = \begin{cases} \dfrac{x^2 - 9}{x + 3}, & \text{for } x \neq -3, \\ 4, & \text{for } x = -3 \end{cases}$

49. The postage function. Postage rates are as follows: 29 cents for the first ounce plus 23 cents for each additional ounce or fraction thereof. Thus if x is the weight of a letter, in ounces, then $p(x)$ is the cost of mailing the letter, where

$$p(x) = \begin{cases} 29 \text{ cents}, & \text{if } 0 < x \leq 1, \\ 52 \text{ cents}, & \text{if } 1 < x \leq 2, \\ 75 \text{ cents}, & \text{if } 2 < x \leq 3, \end{cases}$$

and so on, up to 12 ounces, after which postal cost also depends on distance. Graph this function for x such that $0 < x \leq 12$.

50. Graph

$$f(x) = \begin{cases} 3 + x, & \text{for } x \leq 0, \\ \sqrt{x}, & \text{for } 0 < x < 4, \\ x^2 - 4x - 1, & \text{for } x \geq 4. \end{cases}$$

Graph.

51. $f(x) = \text{INT}(x - 2)$

52. $f(x) = \text{INT}(x) - 1$

53. $f(x) = \text{INT}(x) + 2$

54. $f(x) = \text{INT}(x - 1)$

55. Average price of a movie ticket. The average price of a movie ticket can be estimated by the function

$$P(y) = 0.1522y - 298.592,$$

56. Number of games in a sports league. The number N of games in a sports league is a function of the number of teams n, where each team plays every other team twice, and is given by

$$N(n) = n(n - 1).$$

a) Find the number of games in a women's softball league of 6 teams, where each team plays every other team twice.

b) Find the number of games in a Big-10 basketball season. There are 10 teams and each team plays every other team twice.

57. Boiling point and elevation. The elevation E, in meters, above sea level at which the boiling point of water is T degrees Celsius is given by

$$E(T) = 1000(100 - T) + 580(100 - T)^2.$$

a) At what elevation is the boiling point $99.5°$?

b) At what elevation is the boiling point $100°$?

58. Territorial area of an animal. The territorial area of an animal is defined to be its defended, or exclusive, region. For example, a lion has a certain region over which it is ruler. It has been shown that the territorial region T, in acres, of predatory animals is a function of body weight w, in pounds, and is given by

$$T(w) = w^{1.31}.$$

Find the territorial area of animals whose body weights are 0.5 lb, 10 lb, 20 lb, 40 lb, 100 lb, and 200 lb.

59. The length L of a rectangle is 4 m longer than the width W.

a) Express the area A of the rectangle as a function of the length L.

b) Express the area A of the rectangle as a function of the width W.

60. The base b of a triangle is 3 less than twice the height h.

a) Express the area A of the triangle as a function of the height h.

b) Express the area A of the triangle as a function of the base b.

61. A rectangle of dimensions x by y has a perimeter of 34 ft. Express the area A as a function of x.

62. A rectangle of dimensions x by y has a perimeter of 48 m. Express the area A as a function of x.

63. A rectangle of dimensions x by y is inscribed in a circle of radius 8 ft. Express the area A as a function of x.

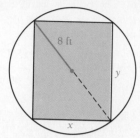

64. A rectangle of dimensions x by y is inscribed in a circle of diameter 20 yd. Express the area A as a function of x.

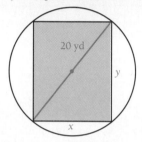

65. From a thin piece of cardboard 10 in. by 10 in., square corners are cut out so that the sides can be folded up to make a box. Express the volume V as a function of the length x of a side of the square base.

66. A store has 30 ft of dividers to set off a rectangular area in one corner of its sales area. The sides against the wall require no partition. Express the area A as a function of the length x of one side of the rectangle.

67. A container firm is designing an open-top rectangular box, with a square base, that will hold 108 in^3 (see the figure). Let x = the length of a side of the base. Express the surface area as a function of x.

68. A rectangular box with a volume of 320 ft^3 is to be constructed with a square base and top (see the figure). The cost per square foot for the bottom is $1.50, for the top is $1, and for the sides is $2.50. Express the cost C as a function of x.

69. *Golf distance finder.* A device used in golf to estimate the distance d, in yards, to a hole measures the size s, in inches, that the 7-ft pin *appears* to be in a viewfinder. Express the distance d as a function of s.

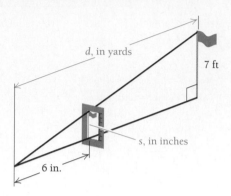

70. An airplane is flying at an altitude of 3700 ft. The slanted distance directly to the airport is d ft. Express the horizontal distance h as a function of d.

71. A gas tank has ends that are hemispheres with radius r and a cylindrical center with the same radius and a height, or length, of 6 ft.

a) Express the volume V as a function of r.
b) Express the surface area S as a function of r.

72. Express the area of an equilateral triangle as a function of the length of a side a.

● SYNTHESIS _____

73. A power line is to be constructed from a power station at point A (see the figure) to an island at point C, which is 1 mi directly out in the water from a point B on the shore. Point B is 4 mi downshore from the power station at A. It costs $5000 per mile to lay the power line under water and $3000 per mile to lay the line under ground. The line comes to the shore at point S downshore from A. Note that S could very well be A or B. Let x = the distance from B to S. Express the cost C of laying the line as a function of x.

74. A right circular cylinder of height h and radius r is inscribed in a right circular cone of height 10 ft and base with a radius of 6 ft.

10 ft

h

r

6 ft

a) Express the height h of the cylinder as a function of r.
b) Express the volume V as a function of r.
c) Express the volume V as a function of h.

Graph.

75. $f(x) = \dfrac{|x|}{x}$

76. $g(x) = \dfrac{x}{|x|}$

77. If $\text{INT}(x) = 4$, what are the possible inputs for x?

78. If $\text{INT}(x) = -3$, what are the possible inputs for x?

79. *Bowling average.* A person's bowling average is defined as follows: *Bowling average* $= \text{INT}(p/n)$, where $p =$ the

total number of pins and $n =$ the total number of games bowled. Find the following bowling averages.

a) 547 pins in 3 games
b) 4621 pins in 27 games

80. ▦ *Decimal place function.* We define a *decimal place function* f_n as follows: Given a number x, find decimal notation for x. Then $f_n(x) =$ the digit in the nth decimal place. For example,

$$f_4(3/5) = f_4(0.600000\ldots) = 0,$$
$$f_4(17/21) = f_4(0.8095238\ldots) = 5,$$
$$f_4(-89/56) = f_4(-1.5892857\ldots) = 2,$$

and so on.

a) Find $f_4(7/8)$, $f_4(17/23)$, $f_4(-49/37)$, and $f_4(\sqrt{11})$.
b) Determine the domain and the range of f_4.
c) Find $f_4(\pi) + f_5(\pi) + f_6(\pi) + f_7(\pi)$.

● **CHALLENGE** _____

81. Graph the equation $\text{INT}(y) = \text{INT}(x)$. Is this the graph of a function?

▧ **TECHNOLOGY CONNECTION** _____

Use a grapher to graph the piecewise-defined function.

82. $f(x) = \begin{cases} 4x^2 - 5x + 3, & x \leq -1, \\ 4.4x^3, & -1 < x < 4, \\ |-3x^2 - 4x|, & x \geq 4 \end{cases}$

83. $f(x) = \begin{cases} x|x - x^2|, & x < 0, \\ \dfrac{x^3}{1 + x}, & x \geq 0 \end{cases}$

84. Use a grapher to plot the function in Exercise 58. Then use that graph to determine, to the nearest tenth of a pound, the body weight of an animal that requires 269 acres of territorial area.

3.6
SYMMETRY

▶ Symmetry with Respect to a Line

In the figure, points P and P_1 are said to be **symmetric with respect to line L.** They are the <u>same</u> distance from L, and the segment $\overline{PP_1}$ is perpendicular to L.

OBJECTIVES

You should be able to:

A Given an equation defining a relation, determine whether the graph is symmetric with respect to a coordinate axis.

B Given an equation defining a relation, determine whether the graph is symmetric with respect to the origin.

C Given the graph of a function or a formula, determine whether the function is even, odd, or neither even nor odd.

DEFINITION   **Points Symmetric with Respect to a Line**

Two points P and P_1 are *symmetric with respect to a line L* if and only if L is the perpendicular bisector of the segment $\overline{PP_1}$. The line L is known as the *line of symmetry*.

We also say that the two points P and P_1 are **reflections** of each other across the line. The line is therefore also known as a **line of reflection.**

Now consider a set of points (geometric figure) as shown in the color curve below. This figure is said to be symmetric with respect to the line L, because if you pick any point Q in the set, you can find another point Q_1 in the set such that Q and Q_1 are symmetric with respect to L.

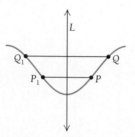

DEFINITION   **Figure Symmetry with Respect to a Line**

A figure, or set of points, is *symmetric with respect to a line L* if and only if for each point Q in the set there exists another point Q_1 in the set for which Q and Q_1 are symmetric with respect to line L.

Think of folding the preceding figure on the line L. The curve would fold over on itself. Also, imagine picking up the preceding figure and flipping it about the line L. Points P and P_1 would be interchanged, as would points Q and Q_1. These, then, are pairs of symmetric points. The entire figure would look exactly like it did before it was flipped. This means that *each* point of the figure is symmetric with *some* point of the figure. Thus the figure is symmetric with respect to the line. A point and its reflection are known as **images** of each other. Thus P_1 is the image of P, for example. The line L is known as an **axis of symmetry.**

Symmetry with Respect to the Axes

There are special and interesting kinds of symmetry in which a coordinate axis is a line of symmetry. The following example shows figures that are symmetric with respect to an axis and a figure that is not.

Example 1   In graph (a), flipping the figure about the y-axis would not change the figure. In graph (b), flipping the graph about the x-axis would not change the figure. In graph (c), flipping about either axis would change the figure.

Symmetric with respect to y-axis (a)

Symmetric with respect to x-axis (b)

Not symmetric with respect to either axis (c)

Let us consider a figure like graph (a), symmetric with respect to the y-axis. For every point of the figure, there is another point that is the same distance across the y-axis. The first coordinates of such a pair of points are opposites of each other.

Example 2 In the relation $y = x^2$, there are points $(2, 4)$ and $(-2, 4)$. The first coordinates, 2 and -2, are opposites of each other, whereas the second coordinates are the same. For every point of the figure (x, y), there is another point $(-x, y)$.

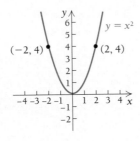

DO EXERCISE 1.

Let us consider a figure symmetric with respect to the x-axis, like graph (b) in Example 1. For every point of such a figure, there is another point that is the same distance across the x-axis. The second coordinates of such a pair of points are opposites of each other.

Example 3 In the relationship $x = y^2$, there are points $(4, 2)$ and $(4, -2)$. The second coordinates, 2 and -2, are opposites of each other, whereas the first coordinates are the same. For every point of the figure (x, y), there is another point $(x, -y)$.

DO EXERCISE 2.

1. a) Plot the point $(3, 2)$. Let the y-axis be a line of symmetry. Plot the point symmetric to $(3, 2)$. What are its coordinates?

b) Plot the point $(-4, -5)$. Let the y-axis be a line of reflection. Plot the image of $(-4, -5)$. What are its coordinates?

2. Let the x-axis be a line of symmetry.

a) Plot the point $(4, 3)$. Plot the point symmetric to it. What are its coordinates?

b) Plot the point $(3, -5)$. Plot its image after reflection across the x-axis. What are its coordinates?

Let us consider a figure that is symmetric with respect to the y-axis, as in Example 2. Suppose it is defined by an equation. If we replace x by $-x$ in this equation, we obtain a new equation equivalent to the original with the same graph. This is true because any number x gives us the same y-value as its opposite, $-x$.

Let us consider a figure that is symmetric with respect to the x-axis, as in Example 3. Suppose it is defined by an equation. If we replace y by $-y$ in the equation, we obtain a new equation equivalent to the original with the same graph. This is true because any number y gives us the same x-value as $-y$. Thus we have a means of testing a relation for symmetry with respect to the axes, when that relation is defined by an equation.

DEFINITION **Symmetry with Respect to the Axes**

When a relation is defined by an equation:

1. If replacing x by $-x$ produces an equivalent equation, then the graph is symmetric with respect to the y-axis.
2. If replacing y by $-y$ produces an equivalent equation, then the graph is symmetric with respect to the x-axis.

Example 4 Test $y = x^2 + 2$ for symmetry with respect to the y-axis.

Solution

a) Replace x by $-x$.

$$y = x^2 + 2 \quad (1)$$

$$y = (-x)^2 + 2$$

b) Simplify, if possible.

$$y = (-x)^2 + 2 = x^2 + 2 \quad (2)$$

c) Is the resulting equation (2) equivalent to the original (1)?

Since the answer is yes, the graph is symmetric with respect to the y-axis.

Example 5 Test $y = x^2 + 2$ for symmetry with respect to the x-axis.

Solution

a) Replace y by $-y$.

$$y = x^2 + 2 \quad (1)$$

$$-y = x^2 + 2$$

b) Simplify, if possible.

The equation is simplified for the most part, although we could multiply on both sides by -1, obtaining

$$y = -x^2 - 2. \quad (2)$$

c) Is the resulting equation (2) equivalent to the original (1)?

This answer is no. To be sure, one might use some trial-and-error reasoning, as follows. Suppose we substitute 0 for x in equation (1). Then

$$y = 0^2 + 2 = 2,$$

so $(0, 2)$ is a solution of equation (1). In order for equations (1) and (2) to be equivalent, $(0, 2)$ must also be a solution of equation (2). We substitute to find out:

$$\frac{y = -x^2 - 2}{2 \,\bigl|\, \begin{matrix} -0^2 - 2 \\ -2 \end{matrix}}$$

Since $(0, 2)$ is not a solution of equation (2), the equations are not equivalent and the graph is not symmetric with respect to the x-axis. ◀

Example 6 Test $a^2 + b^4 - 5 = 0$ for symmetry with respect to the a-axis. (The a-axis is ordinarily the horizontal axis.)

Solution

a) Replace b by $-b$.

$$a^2 + b^4 - 5 = 0 \tag{1}$$

$$a^2 + (-b)^4 - 5 = 0$$

b) Simplify, if possible.

$$a^2 + (-b)^4 - 5 = a^2 + b^4 - 5 = 0 \tag{2}$$

c) Is the resulting equation equivalent to the first?

Since the answer is yes, the graph is symmetric with respect to the a-axis. ◀

DO EXERCISES 3–8.

 Symmetry with Respect to a Point

Two points P and P_1 are **symmetric with respect to a point Q** when they are situated as shown in the following figure. That is, the two points are the same distance from Q, and all three points are on a line.

DEFINITION	**Points Symmetric with Respect to a Point**

Two points P and P_1 are *symmetric with respect to a point Q* if and only if Q (the point of symmetry) is the midpoint of segment $\overline{PP_1}$.

Test for symmetry with respect to the coordinate axes.

3. $y = x^2 - 3$

4. $y^2 = x^3$

5. $x^4 = y^2 + 2$

6. $3y^2 + 4x^2 = 12$

7. $a + 3b = 5$

8. $2p^3 + 4q^3 = 1$

A *set* of points is symmetric with respect to a point if for each point P in the set, there exists another point P_1 in the set such that P and P_1 are symmetric with respect to the point. This is illustrated below. Imagine sticking a pin in this figure at O and then rotating the figure $180°$. Points P and P_1 would be interchanged, as would points Q and Q_1. These are pairs of symmetric points. The entire figure would look exactly as it did before it was rotated. This means that for *each* point P of the figure, there is a point P_1 of the figure such that the points P and P_1 are symmetric with respect to O. Thus the figure is symmetric with respect to the point O.

DEFINITION **Symmetry with Respect to a Point**

A set of points is *symmetric with respect to a point B* if and only if for every point P in the set there exists another point P_1 in the set for which P and P_1 are symmetric with respect to B.

Equivalently, a set of points is symmetric with respect to a point B if when the set is rotated $180°$, the resulting figure coincides with the original.

Symmetry with Respect to the Origin

A special kind of symmetry with respect to a point is symmetry with respect to the origin.

Example 7 In graphs (a) and (b), rotating the figure $180°$ about the origin would not change the figure. In graph (c), such a rotation would change the figure.

Symmetric with respect to origin	Symmetric with respect to origin	Not symmetric with respect to origin
(a)	(b)	(c)

Let us consider figures like (a) and (b) above, symmetric with respect to the origin. For every point of the figure, there is another point that is the same distance across the origin. The first coordinates of such pairs are opposites of each other, and the second coordinates are opposites of each other.

DO EXERCISE 9.

Example 8 The relation $y = x^3$ is symmetric with respect to the origin. In this relation are the points $(2, 8)$ and $(-2, -8)$. The first coordinates are opposites of each other. The second coordinates are opposites of each other. For every point of the figure (x, y), there is another point $(-x, -y)$.

In a relation that is symmetric with respect to the origin, as in Example 8, if we replace x by $-x$ and y by $-y$, we obtain a new equation, but we will get the same figure. This is true because whenever a point (x, y) is in the relation, the point $(-x, -y)$ is also in the relation. This gives us a means of testing a relation for symmetry with respect to the origin when it is defined by an equation.

> **DEFINITION** **Symmetry with Respect to the Origin**
>
> When a relation is defined by an equation, if replacing x by $-x$ and replacing y by $-y$ produces an equivalent equation, then the graph is symmetric with respect to the origin.

Example 9 Test $x^2 = y^2 + 2$ for symmetry with respect to the origin.

Solution

a) Replace x by $-x$ and y by $-y$.

$$x^2 = y^2 + 2$$

$$(-x)^2 = (-y)^2 + 2$$

$x^2 = y^2 + 2$

b) Simplify, if possible.

Since $(-x)^2 = x^2$ and $(-y)^2 = y^2$, then $(-x)^2 = (-y)^2 + 2$ simplifies to

$$x^2 = y^2 + 2.$$

c) Is the resulting equation equivalent to the original?

The answer is yes, so the graph is symmetric with respect to the origin.

9. Draw coordinate axes. Let the origin be a point of symmetry.

a) Plot the point $(3, 2)$. Plot the point symmetric to it. What are its coordinates?

b) Plot the point $(-4, 3)$. Plot the point symmetric to it. What are its coordinates?

c) Plot the point $(-5, -7)$. Plot the point symmetric to it. What are its coordinates?

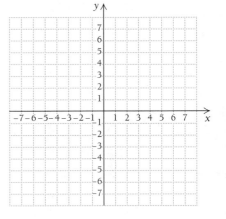

Test for symmetry with respect to the origin.

10. $x^2 + 3y^2 = 4$

11. $x = y$ (*Hint:* After you substitute, multiply on both sides by -1.)

12. $x = -y$

13. $xy = 5$

14. $ab = -5$

15. $u = |v|$

Example 10   Test $2a + 3b = 8$ for symmetry with respect to the origin.

Solution

a) Replace a by $-a$ and b by $-b$.

$$2a \quad + \quad 3b = 8$$
$$2(-a) + 3(-b) = 8$$

b) Simplify, if possible,
We have $2(-a) + 3(-b) = -2a - 3b$, so

$$-(2a + 3b) = 8 \quad \text{and} \quad 2a + 3b = -8.$$

c) Is the resulting equation equivalent to the original?
The equation $2a + 3b = -8$ is *not* equivalent to $2a + 3b = 8$, so the graph is not symmetric with respect to the origin.   ◀

DO EXERCISES 10–15.

▶ Even and Odd Functions

If the graph of a function is symmetric with respect to the y-axis, then it is an **even function.** A function will be symmetric to the y-axis if in its equation we can replace x by $-x$ and obtain an equivalent equation. Thus if we have a function given by $y = f(x)$, then $y = f(-x)$ will give the same function if the function is even. In other words, an even function is one for which $f(x) = f(-x)$ for all x in its domain.

DEFINITION	**Even Function**

A function f is an *even function* if $f(x) = f(-x)$ for all x in the domain of f.

Example 11   Determine whether the function $f(x) = x^2 + 1$ is even.

Solution

a) Find $f(-x)$ and simplify.
$$f(-x) = (-x)^2 + 1$$
$$= x^2 + 1$$

b) Compare $f(x)$ and $f(-x)$.
Since $f(x) = f(-x)$ for all x in the domain, f is an even function.

Note that the graph is symmetric with respect to the y-axis.

◀

Example 12 Determine whether the function $f(x) = x^2 + 8x^3$ is even.

Solution

a) Find $f(-x)$ and simplify.

$$f(-x) = (-x)^2 + 8(-x)^3$$
$$= x^2 - 8x^3$$

b) Compare $f(x)$ and $f(-x)$.

Since $f(x)$ and $f(-x)$ are not the same for all x in the domain, f is not an even function. ◀

DO EXERCISES 16 AND 17.

If the graph of a function is symmetric with respect to the origin, then it is an **odd function.** A function will be symmetric with respect to the origin if in its equation we can replace x by $-x$ and y by $-y$ and obtain an equivalent equation. If we have a function given by $y = f(x)$, then replacing y by $-y$ and x by $-x$ gives us the equation $-y = f(-x)$. If this equation $-y = f(-x)$ is equivalent to $y = f(x)$, then f is an odd function. In other words, an odd function is one for which $f(-x) = -f(x)$ for all x in the domain.

DEFINITION **Odd Function**

A function f is an *odd function* if $f(-x) = -f(x)$ for all x in the domain of f.

Example 13 Determine whether $f(x) = x^3$ is an odd function.

Solution

a) Find $f(-x)$ and $-f(x)$ and simplify.

$$f(-x) = (-x)^3 = -x^3,$$
$$-f(x) = -x^3$$

b) Compare $f(-x)$ and $-f(x)$.

Since $f(-x) = -f(x)$ for all x in the domain, f is odd.

Note that the graph is symmetric with respect to the origin.

16. Determine whether the function is even.

a)   **b)**

c) 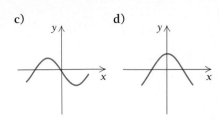 **d)**

17. Determine whether the function is even.

a) $f(x) = x^2 + 3x$
b) $f(x) = |x|$
c) $f(x) = 3x^2 - x^4$
d) $f(x) = 2x^2 + 1$

18. Determine whether the function is even, odd, or neither even nor odd.

a)   **b)**

c) 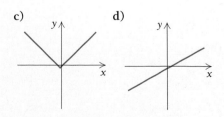 **d)**

19. Determine whether the function is even, odd, or neither even nor odd.

   a) $f(x) = x^3 + 2$
   b) $f(x) = x^4 - x^6$
   c) $f(x) = x^3 + x$
   d) $f(x) = 3x^2 + 3x^5$

   e) $f(x) = x^2 - \dfrac{1}{x}$

Example 14   Determine whether $f(x) = x^2 - 4x^3$ is even, odd, or neither even nor odd.

Solution

   a) Find $f(-x)$ and $-f(x)$ and simplify.
$$f(x) = x^2 - 4x^3,$$
$$f(-x) = (-x)^2 - 4(-x)^3$$
$$= x^2 + 4x^3,$$
$$-f(x) = -x^2 + 4x^3$$

   b) Compare $f(x)$ and $f(-x)$ to determine whether f is even.

   Since $f(x)$ and $f(-x)$ are *not* the same for all x in the domain, f is *not* even.

   c) Compare $f(-x)$ and $-f(x)$ to determine whether f is odd.

   Since $f(-x)$ and $-f(x)$ are *not* the same for all x in the domain, f is *not* odd.

   Thus f is *neither* even nor odd.   ◀

DO EXERCISES 18 AND 19. (EXERCISE 18 IS ON THE PRECEDING PAGE.)

•EXERCISE SET 3.6

A , B   Test for symmetry with respect to the coordinate axes and the origin.

1. $3y = x^2 + 4$
2. $5y = 2x^2 - 3$
3. $y^3 = 2x^2$
4. $3y^3 = 4x^2$
5. $2x^4 + 3 = y^2$
6. $3y^2 = 2x^4 - 5$
7. $2y^2 = 5x^2 + 12$
8. $3x^2 - 2y^2 = 7$
9. $2x - 5 = 3y$
10. $5y = 4x + 5$
11. $3b^3 = 4a^3 + 2$
12. $p^3 - 4q^3 = 12$

B   Test for symmetry with respect to the origin.

13. $3x^2 - 2y^2 = 3$
14. $5y^2 = -7x^2 + 4$
15. $5x - 5y = 0$
16. $3x = 3y$
17. $3x + 3y = 0$
18. $7x = -7y$
19. $3x = \dfrac{5}{y}$
20. $3y = \dfrac{7}{x}$
21. $y = |2x|$
22. $3x = |y|$
23. $3a^2 + 4a = 2b$
24. $5v = 7u^2 - 2u$
25. $3x = 4y$
26. $6x + 7y = 0$
27. $xy = 12$
28. $y = -\dfrac{4}{x}$

C

29. Determine whether the function is even, odd, or neither even nor odd.

   a)

   b)

   c)
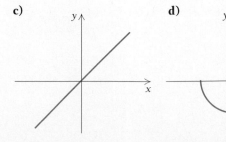

   d)

30. Determine whether the function is even, odd, or neither even nor odd.

a)

b)

c)

d)

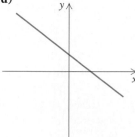

Determine whether the function is even, odd, or neither even nor odd.

31. $f(x) = 2x^2 + 4x$

32. $f(x) = -3x^3 + 2x$

33. $f(x) = 3x^4 - 4x^2$

34. $f(x) = 5x^2 + 2x^4 - 1$

35. $f(x) = 7x^3 + 4x - 2$

36. $f(x) = 4x$

37. $f(x) = |3x|$

38. $f(x) = x^{24}$

39. $f(x) = x^{17}$

40. $f(x) = x + \dfrac{1}{x}$

41. $f(x) = x - |x|$

42. $f(x) = \sqrt{x}$

43. $f(x) = \sqrt[3]{x}$

44. $f(x) = 7$

45. $f(x) = 0$

46. $f(x) = -\dfrac{1}{x}$

Consider this figure for Exercises 47–50.

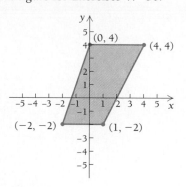

47. Graph the reflection across the x-axis.

48. Graph the reflection across the y-axis.

49. Graph the reflection across the line $y = x$.

50. Graph the figure formed by reflecting each point through the origin.

51. Consider symmetries with respect to the x-axis, the y-axis, and the origin. Prove that symmetry with respect to any two of these implies symmetry with respect to the other.

Use a grapher to determine whether the function is even, odd, or neither even nor odd.

52. $f(x) = \sqrt[3]{x - 2}$

53. $f(x) = x|x^3|$

54. $f(x) = x^2(5 - |x|)$

55. $f(x) = \dfrac{1 - x^4}{x^3 - 1}$

56. $f(x) = \dfrac{x^2 + 1}{x^3 - x}$

57. $f(x) = \sqrt{x^2 + 1}$

3.7

COMBINATIONS OF FUNCTIONS

Next we look at five methods of combining two functions to obtain a new function.

Sums, Differences, Products, and Quotients of Functions

Consider two functions f and g given as follows:

$$f(x) = x^2 - 5 \quad \text{and} \quad g(x) = x + 7.$$

OBJECTIVES

You should be able to:

A Compute function values for the sum, difference, product, and quotient of two functions, and determine their domains.

B Find the composition of two functions f and g, giving formulas for $f \circ g(x)$ and $g \circ f(x)$, find the domains of the composition, and decompose a function as a composition of two functions.

Suppose we have an input 2. This number is in the domain of both functions. That is, it can be substituted into each formula. Let us find the outputs $f(2)$ and $g(2)$:

$$f(2) = 2^2 - 5 = 4 - 5 = -1,$$
$$g(2) = 2 + 7 = 9.$$

We can then add the outputs: $f(2) + g(2) = -1 + 9 = 8$. We start with an input 2 and we end up with exactly one output, 8. Doing this for any input x that is in the domain of *both* functions creates a new function $f + g$, called the **sum** of f and g, described by $f(x) + g(x)$.

Suppose we subtract the function values: $f(2) - g(2) = -1 - 9 = -10$. Doing this for any input x that is in the domain of *both* functions creates a new function $f - g$, called the **difference** of f and g, described by $f(x) - g(x)$.

Suppose we multiply the function values: $f(2) \cdot g(2) = (-1)9 = -9$. Doing this for any input x that is in the domain of *both* functions creates a new function fg, called the **product** of f and g, described by $f(x)g(x)$.

Suppose we divide the function values: $f(2)/g(2) = -1/9 = -\frac{1}{9}$. Doing this for any input x that is in the domain of *both* functions and for which $g(x)$ is not zero creates a new function f/g, called the **quotient** of f and g, described by $f(x)/g(x)$.

DEFINITION   **Sum, Difference, Product, and Quotient of Functions**

From any functions f and g, we can form new functions defined as follows, provided x is in the domain of f and the domain of g.

1. The *sum* of f and g: $(f + g)(x) = f(x) + g(x)$.
2. The *difference* of f and g: $(f - g)(x) = f(x) - g(x)$.
3. The *product* of f and g: $fg(x) = f(x)g(x)$.
4. The *quotient* of f and g: $(f/g)(x) = f(x)/g(x)$, where $g(x) \neq 0$.

Examples   Given f and g described by $f(x) = 8 - x$ and $g(x) = \sqrt{2x + 3}$.

1. Find $(f + g)(5)$ and $(f + g)(-4)$.

$$(f + g)(5) = f(5) + g(5) = (8 - 5) + (\sqrt{2(5) + 3})$$
$$= 3 + \sqrt{13},$$
$$(f + g)(-4) = f(-4) + g(-4) = [8 - (-4)] + \sqrt{2(-4) + 3}$$
$$= 12 + \sqrt{-5}, \quad \text{which does not exist as a real number.}$$

Note that $(f + g)(-4)$ does not exist because -4 is *not* in the domain of g.

2. Find $(f - g)(7)$ and $(f - g)(-10)$.

$$(f - g)(7) = f(7) - g(7) = (8 - 7) - \sqrt{2(7) + 3} = 1 - \sqrt{17}$$

$(f - g)(-10)$ does not exist because -10 is not in the domain of g since it produces a negative radicand.

3. Find $fg(7)$ and $gg(7)$.

$$fg(7) = f(7)g(7) = (8 - 7)(\sqrt{2(7) + 3}) = \sqrt{17},$$
$$gg(7) = g(7)g(7) = [g(7)]^2 = [\sqrt{17}]^2 = 17$$

4. Find $(f/g)(5)$, $(g/f)(5)$, and $(f/g)(-\frac{3}{2})$.

$$(f/g)(5) = \frac{f(5)}{g(5)}$$

$$= \frac{8 - 5}{\sqrt{2(5) + 3}}$$

$$= \frac{3}{\sqrt{13}}, \quad \text{or} \quad \frac{3\sqrt{13}}{13},$$

$$(g/f)(5) = \frac{g(5)}{f(5)} = \frac{\sqrt{13}}{3},$$

$$(f/g)(-\tfrac{3}{2}) = \frac{f(-\frac{3}{2})}{g(-\frac{3}{2})}$$

$$= \frac{8 - (-\frac{3}{2})}{\sqrt{2(-\frac{3}{2}) + 3}} = \frac{\frac{19}{2}}{0}, \quad \text{which does not exist.}$$

5. Find the domain of $f + g$, $f - g$, fg, and f/g.

For the function f above, the domain is the set of all real numbers. For the function g, the domain is the set of all real numbers for which $2x + 3 \geq 0$, which is the set $\{x \mid x \geq -3/2\}$ or the interval $[-3/2, \infty)$. The domain of $f + g$, $f - g$, and fg is therefore the interval $[-3/2, \infty)$. The domain of f/g must exclude any input that makes the denominator 0, namely, $-3/2$. Therefore, the domain of f/g is the interval $(-3/2, \infty)$. ◀

DO EXERCISE 1. _____

It makes sense that if we were to be doing lots of computations with sums, differences, products, and quotients of functions f and g, we might want to find general formulas.

Examples Given f and g described by $f(x) = x^2 - 3$ and $g(x) = x + 4$, find the following.

6. $(f + g)(x) = f(x) + g(x) = (x^2 - 3) + (x + 4) = x^2 + x + 1$

7. $(f - g)(x) = f(x) - g(x) = (x^2 - 3) - (x + 4) = x^2 - x - 7$

8. $fg(x) = f(x)g(x) = (x^2 - 3)(x + 4) = x^3 + 4x^2 - 3x - 12$

9. $gg(x) = [g(x)]^2 = (x + 4)^2 = x^2 + 8x + 16$

10. $(f/g)(x) = \dfrac{f(x)}{g(x)} = \dfrac{x^2 - 3}{x + 4}$, where $x \neq -4$.

11. Find the domain of $f + g$, $f - g$, fg, and f/g.

The domain of f is the set of all real numbers. The domain of g is the set of all real numbers. The domain of $f + g$, $f - g$, and fg is the set of numbers in the intersection of the two domains—that is, the set of numbers in both domains, which is again the set of real numbers. For f/g, we must exclude -4 since $g(-4) = 0$. Thus the domain of f/g is $\{x \mid x \text{ is a real number and } x \neq -4\}$. ◀

If a function is described by a polynomial, such as $f(x) = x^2 - 3$ and $g(x) = x + 4$, we say that it is a **polynomial function.** Note in Examples 6–10 that the sum, difference, and product of polynomial functions

1. Given $f(x) = 1/(x - 2)$ and $g(x) = x^2 - 25$, find each of the following, if possible.

   a) $(f + g)(5)$ and $(f + g)(2)$
   b) $(f - g)(7)$ and $(f - g)(-3)$
   c) $fg(-3)$ and $gg(1)$
   d) $(f/g)(1)$, $(g/f)(5)$, $(f/g)(5)$, and $(f/g)(-5)$
   e) The domains of f, g, $f + g$, $f - g$, fg, and f/g

2. Given $f(x) = x^2 + 3$ and $g(x) = x^2 - 9$, find each of the following.

   a) $(f + g)(x)$
   b) $(f - g)(x)$
   c) $fg(x)$
   d) $(f/g)(x)$
   e) $ff(x)$
   f) The domains of $f + g, f - g, fg,$ and f/g

3. Given

   $$R(x) = 50x - 0.5x^2$$

   and

   $$C(x) = 10x + 3,$$

   find each of the following.

   a) $P(x)$
   b) $R(40), C(40),$ and $P(40)$

are also polynomial functions, but the quotient may not be. The quotient of two polynomial functions is called a **rational function.**

DO EXERCISE 2. _____

Let us now consider an economic application.

Example 12   *Total cost, revenue, and profit.*   In economics, we are frequently concerned with functions defined as follows:

$$\text{Total cost} = C(x) = \text{the total cost of producing } x \text{ units of a}$$
$$\text{product (usually considered in some}$$
$$\text{time period});$$

$$\text{Total revenue} = R(x) = \text{the total revenue from the sale of } x \text{ units}$$
$$\text{of a product;}$$

$$\text{Total profit} = P(x) = \text{the total profit from the production and}$$
$$\text{sale of } x \text{ units of a product}$$
$$P(x) = R(x) - C(x).$$

Given

$$R(x) = 40x - 0.1x^2,$$
$$C(x) = 2x^3 - 12x^2 + 40x + 10,$$

find each of the following.

a) $P(x)$
b) $R(2), C(2),$ and $P(2)$

Solution

a) $P(x) = R(x) - C(x)$
$$= (40x - 0.1x^2) - (2x^3 - 12x^2 + 40x + 10)$$
$$= -2x^3 + 11.9x^2 - 10$$

b) $R(2) = 40(2) - 0.1(2)^2 = 80 - 0.4 = \$79.60$
    (the total revenue from the sale of 2 units);

   $C(2) = 2(2)^3 - 12(2)^2 + 40(2) + 10 = \$58$
    (the total cost of producing 2 units);

   $P(2) = R(2) - C(2) = \$79.60 - \$58 = \$21.60$
    (the total profit from the production and sale of 2 units)

DO EXERCISE 3. _____

B The Composition of Functions

In the real world, functions frequently occur in which some variable depends on the choice of a third variable. For instance, the number of employees hired by a firm may depend on the firm's profits, which may in turn depend on the number of items the firm produces. Functions like this are formed by **composition of functions.**

There is a function g that gives a correspondence between women's shoe sizes in the United States and those in Italy. The function is given by $g(x) = 2(x + 12)$, where x is a shoe size in the United States and $g(x)$

is a shoe size in Italy. For example, a shoe size of 4 in the United States corresponds to a shoe size of $g(4) = 2(4 + 12)$, or 32, in Italy.

$$h(x) = f(g(x))$$

There is also a function that gives a correspondence between women's shoe sizes in Italy and those in Britain. The function is given by $f(x) = \frac{1}{2}x - 14$, where x is a shoe size in Italy and $f(x)$ is the corresponding shoe size in Britain. For example, a shoe size of 32 in Italy corresponds to a shoe size of $f(32) = \frac{1}{2}(32) - 14$, or 2, in Britain.

It seems reasonable to assume that a shoe size of 4 in the United States corresponds to a shoe size of 2 in Britain and that there is a function h that describes this correspondence. Can we find a formula for h? If we look at the tables, we might guess that such a formula is $h(x) = x - 2$, and that is indeed correct. We can obtain the same result using algebra without resorting to the tables, as follows.

A shoe size x in the United States corresponds to a shoe size $g(x)$ in Italy, where

$$g(x) = 2(x + 12).$$

Now $2(x + 12)$ is a shoe size in Italy. If we replace x in $f(x)$ by $2(x + 12)$, we can find the corresponding shoe size in Britain:

$$f(g(x)) = \tfrac{1}{2}[2(x + 12)] - 14 = \tfrac{1}{2}[2x + 24] - 14$$
$$= x + 12 - 14 = x - 2.$$

This gives a formula for h: $h(x) = x - 2$. Thus a shoe size of 4 in the United States corresponds to a shoe size of $h(4) = 4 - 2$, or 2, in Britain. The function h is called the **composition** of f and g and is symbolized $f \circ g$.

DEFINITION   **Composition of Two Functions**

The *composition* of f and g, denoted $f \circ g$, is defined as

$$f \circ g(x) = f(g(x)),$$

for all x in the domain of g for which $g(x)$ is in the domain of f.

We can visualize the composition of functions as follows. To find $f \circ g(x)$, we substitute $g(x)$ for x in $f(x)$.

4. For functions f and g given by $f(x) = x + 5$ and $g(x) = x^2 - 1$, find each of the following.

   a) $f \circ g(-3)$ and $g \circ f(-3)$

   b) $f \circ g(x)$ and $g \circ f(x)$

Inputs, x

$g(x)$

g

f

A composition machine for $f(g(x))$

Outputs, $f(g(x))$
or $f \circ g(x)$

Example 13   Given $f(x) = 3x$ and $g(x) = 1 + x^2$, find each of the following.

a) $f \circ g(5)$ and $g \circ f(5)$ b) $f \circ g(x)$ and $g \circ f(x)$

Solution   Consider each function separately:

$$f(x) = 3x \qquad \text{This function multiplies each input by 3.}$$

and

$$g(x) = 1 + x^2. \qquad \text{This function adds 1 to the square of each input.}$$

a) To find $f \circ g(5)$, we first find $g(5)$ by substituting into the formula for g: We square 5 and add it to 1, to get 26. Then we substitute the result, 26, into the formula for f: We multiply 26 by 3.

$$f \circ g(5) = f(g(5)) = f(1 + 5^2) = f(26)$$
$$= 3(26) = 78$$

To find $g \circ f(5)$, we first find $f(5)$ by substituting into the formula for f: We multiply 5 by 3, to get 15. Then we substitute the result, 15, into the formula for g: We square 15 and add 1.

$$g \circ f(5) = g(f(5)) = g(3 \cdot 5) = g(15)$$
$$= 1 + 15^2$$
$$= 1 + 225 = 226$$

5. For functions f and g given by $f(x) = \sqrt{x + 3}$ and $g(x) = x^2$, find each of the following.

   a) $f \circ g(x)$ and $g \circ f(x)$

   b) The domains of f, g, and $f \circ g$

b) $f \circ g$ first does what g does (adds 1 to the square) and then does what f does (multiplies by 3). We find $f \circ g(x)$ by substituting $g(x)$ for x:

$$f \circ g(x) = f(g(x)) = f(1 + x^2) \qquad \text{Substituting } 1 + x^2 \text{ for } g(x)$$
$$= 3(1 + x^2) = 3 + 3x^2.$$

$g \circ f$ first does what f does (multiplies by 3) and then does what g does (adds 1 to the square). We find $g \circ f(x)$ by substituting $f(x)$ for x:

$$g \circ f(x) = g(f(x)) = g(3x) \qquad \text{Substituting } 3x \text{ for } f(x)$$
$$= 1 + (3x)^2 = 1 + 9x^2. \qquad \blacktriangleleft$$

DO EXERCISE 4.

Note in Example 13 that $f \circ g(5) \neq g \circ f(5)$. In general, $f \circ g(x) \neq g \circ f(x)$.

Example 14   Given $f(x) = \sqrt{1 - x}$ and $g(x) = 2x - 3$, find each of the following.

a) $f \circ g(x)$ and $g \circ f(x)$
b) The domains of $f, g, f \circ g$, and $g \circ f$

Solution

a) $f \circ g(x) = f(g(x)) = f(2x - 3)$
$$= \sqrt{1 - (2x - 3)} = \sqrt{4 - 2x},$$
$g \circ f(x) = g(f(x)) = g(\sqrt{1 - x})$
$$= 2(\sqrt{1 - x}) - 3 = 2\sqrt{1 - x} - 3$$

b) The domain of f is the set of x such that $1 - x \geq 0$, which is $\{x \mid x \leq 1\}$, or the interval $(-\infty, 1]$. The domain of g is the set of all real numbers. To find the domain of $f \circ g$, we must consider that the outputs for g must be acceptable as inputs for f. In other words, if any number $g(x)$ is not in the domain of f, then x is not in the domain of $f \circ g$. The domain of $f \circ g$ is thus the set of all real numbers x such that $4 - 2x \geq 0$, or $x \leq 2$. The domain is $\{x \mid x \leq 2\}$ or the interval $(-\infty, 2]$.

Since the domain of g is the set of all real numbers, any number that is in the domain of f is in the domain of $g \circ f$. The domain of $g \circ f$ is thus the set of real numbers x such that $1 - x \geq 0$, which is $\{x \mid x \leq 1\}$ or the interval $(-\infty, 1]$. ◀

DO EXERCISE 5 ON THE PRECEDING PAGE. _____

Example 15   *Speed of sound in air.*   In Example 10 of Section 3.5, we learned that the speed of sound S in air is a function of the temperature F, in degrees Fahrenheit, and is given by

$$S(F) = 1087.7 \sqrt{\frac{5F + 2457}{2457}},$$

where S is in feet per second. Suppose we wanted a function that would tell us the speed of sound in air when the Celsius temperature is given. The following function can be used to first convert the Celsius temperature C to the corresponding Fahrenheit temperature F:

$$F(C) = \tfrac{9}{5}C + 32.$$

Find a formula for $S \circ F(C)$ that can be used to find the speed of sound in air when the temperature is given in degrees Celsius. Then find $S \circ F(20°)$.

Solution

$$S \circ F(C) = S(F(C)) = 1087.7 \sqrt{\frac{5(\tfrac{9}{5}C + 32) + 2457}{2457}}$$

$$= 1087.7 \sqrt{\frac{9C + 2617}{2457}},$$

$$S \circ F(20°) = 1087.7 \sqrt{\frac{9(20) + 2617}{2457}} \approx 1160.5 \text{ ft/sec.}$$ ◀

DO EXERCISE 6. _____

6. The function K given by
$$K(C) = C + 273$$
converts the Celsius temperature C to degrees in Kelvin units. The function C given by
$$C(F) = \tfrac{5}{9}(F - 32)$$
converts Fahrenheit temperature F to Celsius temperature C.

a) Find $K \circ C(F)$ and explain its meaning and usage.

b) Find the Kelvin temperature when the temperature is $-13°$F.

TECHNOLOGY CONNECTION

The composition of functions can be shown on graphers. For our example, we will use the functions $f(x) = x^2 + 2x + 1$ and $g(x) = x^3$. To see the composite function $f \circ g(x)$, substitute the function $g(x)$ everywhere you see an x in $f(x)$:

$$f \circ g(x) = f(g(x)) = f(x^3)$$
$$= (x^3)^2 + 2(x^3) + 1.$$

It is not necessary to simplify this composite function in order to graph it. Simply enter it into the grapher in the form you see here and it will graph. You can also compare the graph of the composite function with those of the individual functions $f(x)$ and $g(x)$.

The other composite, $g \circ f(x)$, can also be found by a simple substitution:

$$g \circ f(x) = g(f(x)) = g(x^2 + 2x + 1)$$
$$= (x^2 + 2x + 1)^3.$$

Again, the composite function need not be simplified before it is graphed.

TC 1. For the functions $f(x) = 1/(x - 3)$ and $g(x) = x^4 + 2$, graph the two composite functions $f \circ g(x)$ and $g \circ f(x)$.

7. Find $f(x)$ and $g(x)$ such that $h(x) = f \circ g(x)$. Answers may vary, but try to select the most obvious.

a) $h(x) = \sqrt[3]{x^2 + 1}$

b) $h(x) = \dfrac{1}{(x + 5)^4}$

It is important in calculus to be able to recognize how a function can be expressed as a composition. In this way, we are "decomposing" the function.

Example 16 Find $f(x)$ and $g(x)$ such that $h(x) = f \circ g(x)$:

$$h(x) = (2x - 3)^5.$$

Solution   This is $2x - 3$ to the 5th power. Two functions that can be used for the composition are $f(x) = x^5$ and $g(x) = 2x - 3$. We can check by forming the composition:

$$h(x) = f \circ g(x) = f(g(x))$$
$$= f(2x - 3) = (2x - 3)^5.$$

This is the most "obvious" answer to the question. There can be other less obvious answers. For example, if

$$f(x) = (x + 7)^5 \quad \text{and} \quad g(x) = 2x - 10,$$

then

$$h(x) = f \circ g(x) = f(g(x)) = f(2x - 10)$$
$$= [(2x - 10) + 7]^5 = (2x - 3)^5.$$

◀

DO EXERCISE 7.

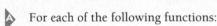
●EXERCISE SET 3.7

A For each of the following functions:

a) Find $(f + g)(x)$, $(f - g)(x)$, $fg(x)$, $ff(x)$, $(f/g)(x)$, $(g/f)(x)$, $f \circ g(x)$, and $g \circ f(x)$.

b) Find the domains of f, g, $f + g$, $f - g$, fg, ff, f/g, g/f, $f \circ g$, and $g \circ f$.

1. $f(x) = x - 3$, $g(x) = x + 4$

2. $f(x) = x^2 - 1$, $g(x) = 2x + 5$

3. $f(x) = x^3$, $g(x) = 2x^2 + 9x - 3$

4. $f(x) = x^2$, $g(x) = \sqrt{x}$

Let $f(x) = x^2 - 4$ and $g(x) = 2x + 5$. Find each of the following.

5. $(f - g)(3)$

6. $(f + g)(-1)$

7. $(f - g)(x)$

8. $(f + g)(x)$

9. $fg(3)$

10. $(f/g)(-1)$

11. $(g/f)(-2)$

12. $(f/g)(-2.5)$

13. $fg(x)$

14. $(f/g)(x)$

15. $(g/f)(x)$

16. $ff(x)$

17. $f \circ g(x)$

18. $f \circ f(x)$

19. $g \circ g(x)$

20. $g \circ f(x)$

21. Given

$$R(x) = 60x - 0.4x^2,$$
$$C(x) = 3x + 13,$$

find each of the following.

a) $P(x)$

b) $R(20)$, $C(20)$, and $P(20)$

22. Given

$$R(x) = 15x,$$
$$C(x) = 0.001x^2 + 1.2x + 60,$$

find each of the following.

a) $P(x)$

b) $R(100)$, $C(100)$, and $P(100)$

B Find $f \circ g(x)$ and $g \circ f(x)$.

23. $f(x) = \frac{4}{5}x$, $g(x) = \frac{5}{4}x$

24. $f(x) = x + 3$, $g(x) = x - 3$

25. $f(x) = 3x - 7$, $g(x) = \dfrac{x + 7}{3}$

26. $f(x) = \frac{2}{3}x - \frac{4}{5}$, $g(x) = 1.5x + 1.2$

27. $f(x) = x^3 - 1$, $g(x) = \sqrt[3]{x + 1}$

28. $f(x) = \sqrt[5]{x + 2}$, $g(x) = x^5 - 2$

29. $f(x) = \sqrt{x + 5}$, $g(x) = x^2 - 5$

30. $f(x) = x^4$, $g(x) = \sqrt[4]{x}$

31. $f(x) = \dfrac{1 - x}{x}$, $g(x) = \dfrac{1}{1 + x}$

32. $f(x) = \dfrac{x^2 - 1}{x^2 + 1}$, $g(x) = \dfrac{3x - 4}{5x - 2}$

33. $f(x) = -6, g(x) = 12$

34. $f(x) = 20, g(x) = 0.2x + 1$

Find $f(x)$ and $g(x)$ such that $h(x) = f \circ g(x)$. Answers may vary, but try to select the most obvious answer.

35. $h(x) = (4 - 3x)^5$

36. $h(x) = \sqrt[3]{x^2 - 8}$

37. $h(x) = \dfrac{1}{(x - 1)^4}$

38. $h(x) = \dfrac{1}{\sqrt{3x + 7}}$

39. $h(x) = \dfrac{x^3 - 1}{x^3 + 1}$

40. $h(x) = |9x^2 - 4|$

41. $h(x) = \left(\dfrac{2 + x^3}{2 - x^3}\right)^6$

42. $h(x) = (\sqrt{x} - 3)^4$

43. $h(x) = \sqrt{\dfrac{x - 5}{x + 2}}$

44. $h(x) = \sqrt{1 + \sqrt{1 + x}}$

45. $h(x) =$
$(x + 3)^5 + (x + 3)^4 + (x + 3)^3 - (x + 3)^2 + 4(x + 3)$

46. $h(x) = 4(x - 1)^{2/3} + 5$

47. An airplane is 300 ft from the control tower at the end of the runway. It takes off at a speed of 250 mph.

    a) Let a be the distance that the plane travels down the runway. Find a formula for a in terms of the time t that the plane travels. That is, find an expression for $a(t)$.

    b) Let P be the distance of the plane from the control tower. Find a formula for P in terms of the distance a. That is, find an expression for $P(a)$.

    c) Find $(P \circ a)(t)$. Explain the meaning of this function.

48. A stone is thrown into a pond. A circular ripple is spreading over the pond in such a way that the radius is increasing at the rate of 3 ft/sec.

    a) Find a function $r(t)$ for the radius in terms of the time t.

    b) Find a function $A(r)$ for the area of the ripple in terms of the radius r.

    c) Find $(A \circ r)(t)$. Explain the meaning of this function.

● **SYNTHESIS** _____

Graph each of the following equations. Then graph $y = (f + g)(x)$ by adding second coordinates.

49. $f(x) = x^2, g(x) = -2x + 3$

50. $f(x) = 2/x, g(x) = x$

51. $f(x) = \sqrt{x}, g(x) = 1 - x^2$

52. $f(x) = 5 - x^2, g(x) = x^2 - 5$

53. Consider $f(x) = 3x + b$ and $g(x) = 2x - 1$. Find b such that $f \circ g(x) = g \circ f(x)$ for all real numbers x.

54. Consider $f(x) = 3x - 4$ and $g(x) = mx + b$. Find m and b such that $f \circ g(x) = g \circ f(x) = x$ for all real numbers x.

55. For $f(x) = 1/(1 - x)$, find $f \circ f(x)$ and $f \circ f \circ f(x)$.

56. Prove that the composition of two even functions is even.

57. Prove that the sum of two even functions is even.

58. Prove that the product of two odd functions is even.

59. Prove that the composition of two odd functions is odd.

60. Prove that the product of an even function and an odd function is odd.

61. Prove that if f and g are increasing functions, then $f \circ g$ is increasing and $f + g$ is increasing.

62. Prove that if f is *any* function, then the function E defined by

$$E(x) = \frac{f(x) + f(-x)}{2}$$

is even.

63. Prove that if f is *any* function, then the function O defined by

$$O(x) = \frac{f(x) - f(-x)}{2}$$

is odd.

64. Consider the functions E and O of Exercises 62 and 63. Prove that

$$f(x) = E(x) + O(x),$$

which shows that every function can be expressed as the sum of an even function and an odd function.

65. Graph: $f(x) = |x| + \text{INT}(x)$.

3.8

TRANSFORMATIONS OF FUNCTIONS

A Transformations

Vertical Translations

Given a function, we can find various ways of altering it to obtain another function. Such an alteration is called a **transformation.** If this alteration consists merely of moving the graph without changing its shape or orientation, the transformation is called a **translation,** or **shift.**

OBJECTIVE
You should be able to:
A Given the graph of a function, graph its transformation under translations, reflections, and shrinkings.

Example 1   Consider functions of the type $f(x) = x^2 + b$. Using the same set of axes, sketch and compare the graphs of

$$f(x) = x^2,$$
$$f(x) = x^2 + 1, \quad \text{and}$$
$$f(x) = x^2 - 3.$$

Solution   The graphs are shown below. Note that the graph of $f(x) = x^2 + 1$ has the same shape as the graph of $f(x) = x^2$, but is moved up a distance of 1 unit. Each function value, or output, is increased by 1 unit. The graph of $f(x) = x^2 - 3$ has the same shape as the graph of $f(x) = x^2$, but is moved down 3 units. Each function value, or output, is decreased by 3 units.

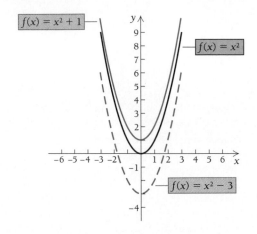

TECHNOLOGY CONNECTION

Select one of the following functions to act as your "base function":

Now add 6 to the base function, forming a new function. For example, if your base function is $f(x) = x^3$, your new function will be $f(x) = x^3 + 6$. Draw the graph of this new function on the same axes as the base function.

Next, instead of adding 6, subtract 4 from the base function. Draw its graph on the same axes with the two that have already been drawn.

Clear the screen. Repeat this process with the other base functions, drawing all three graphs on the same axes in each case. Find a rule that predicts what will happen when you add b to a function $f(x)$, forming $f(x) + b$.

Consider any equation of a function $y = f(x)$. Adding a constant b to produce $y = f(x) + b$ changes each function value by the same amount, b. Thus it produces no change in the shape of the graph, but merely translates or shifts it up if the constant b is positive and down if the constant b is negative.

THEOREM 5   Vertical Translations of Functions

To graph $y = f(x) + b$:

If $b > 0$ $\Rightarrow$ shift up $|b|$ units.

If $b < 0$ $\Rightarrow$ shift down $|b|$ units.

DO EXERCISE 1. _____

Horizontal Translations

We now consider horizontal translations, or shifts.

Example 2   Consider functions of the type $f(x) = (x - d)^2$. Using the same set of axes, sketch and compare the graphs of

$$f(x) = x^2,$$
$$f(x) = (x - 3)^2, \quad \text{and}$$
$$f(x) = (x + 4)^2.$$

Solution   The graphs are shown below. The graph of $f(x) = (x - 3)^2$ is a horizontal translation, or shift, of the graph of $f(x) = x^2$, 3 units to the right. The graph of $f(x) = (x + 4)^2$, or $f(x) = (x - (-4))^2$, is a horizontal translation of the graph of $f(x) = x^2$, 4 units to the left.

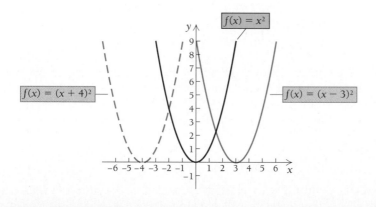

THEOREM 6   Horizontal Translations of Functions

To graph $y = f(x - d)$:

If $d > 0$ $\Rightarrow$ shift to the right $|d|$ units.

If $d < 0$ $\Rightarrow$ shift to the left $|d|$ units.

DO EXERCISE 2. _____

---

1. Use graph paper. Graph

$$f(x) = |x|,$$
$$f(x) - |x| - 2, \quad \text{and}$$
$$f(x) = |x| + 3$$

using the same set of axes.

2. Use graph paper. Graph

$$f(x) = |x|,$$
$$f(x) = |x - 2|, \quad \text{and}$$
$$f(x) = |x + 3|$$

using the same set of axes.

TECHNOLOGY CONNECTION

Begin with the same base function that you started with in the previous Technology Connection and draw its graph on a grapher. Replace x with $x - 3$ in the base function, forming $f(x - 3)$. For example, if the base function is $f(x) = x^3$, the new function will be $f(x) = (x - 3)^3$. Draw this graph using the same axes as the base function. What relationship do you see between the new function and the original function?

Now replace $x - 3$ with $x + 5$ and graph $f(x + 5)$ using the same set of axes with the other two graphs. How does it relate to the base function?

Repeat this process with the other base functions from the initial list, drawing all three graphs using the same axes in each case. Find a rule that predicts what will happen when you substitute $x - d$ for x in any function.

3. Use graph paper. Using the same set of axes, graph

$$f(x) = x^2,$$
$$f(x) = 2x^2, \quad \text{and}$$
$$f(x) = 0.8x^2.$$

Vertical Stretchings and Shrinkings

Let us now consider functions of the type $y = af(x)$, where a is some constant.

Example 3   Using the same set of axes, sketch and compare the graphs of

$$f(x) = |x|, \qquad f(x) = 2|x|, \quad \text{and} \quad f(x) = \tfrac{1}{2}|x|.$$

Solution

x	$2|x|$	$\frac{1}{2}|x|$
-3	6	$\frac{3}{2}$
-2	4	1
0	0	0
2	4	1
3	6	$\frac{3}{2}$

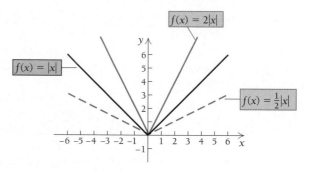

The graph of $f(x) = 2|x|$ looks like that of $f(x) = |x|$, but every function value, or output, is doubled, so the graph is *stretched* in a vertical direction. The graph of $f(x) = \frac{1}{2}|x|$ is flattened, or *shrunk,* in a vertical direction since every output, or function value, is halved.   ◀

Consider any function f given by $y = f(x)$. Multiplying on the right by any constant a greater than 1 to obtain $y = af(x)$ will *stretch* the graph away from the horizontal axis. If the constant a is between 0 and 1, then the graph will be flattened or *shrunk* vertically toward the horizontal axis.

DO EXERCISE 3.

When we multiply by a negative constant, the graph is *reflected,* or flipped, across the x-axis as well as being stretched or shrunk.

Example 4   Using the same set of axes, sketch and compare the graphs of

$$f(x) = x^3 - x, \qquad f(x) = -2(x^3 - x), \quad \text{and} \quad f(x) = -\tfrac{1}{2}(x^3 - x).$$

Solution   The graph of $f(x) = -2(x^3 - x)$ is obtained from the graph of $f(x) = x^3 - x$ by stretching each output by a factor of 2 and reflecting across the x-axis. The graph of $f(x) = -\frac{1}{2}(x^3 - x)$ is obtained by shrinking

TECHNOLOGY CONNECTION

Begin with the same base function that you used in the last two Technology Connections. Multiply by 4, forming $4 \cdot f(x)$. For example, if the base function is $f(x) = x^3$, then the new function is $f(x) = 4x^3$. Draw its graph using the same axes as the base function. What effect does this multiplication have on the graph?

Replace 4 with $\frac{1}{4}$ and graph that function on the same axes. What effect does this multiplication have on the graph?

Repeat this process with the other base functions from the initial list, drawing all three graphs using the same axes in each case. Find a rule that predicts what will happen when a function is multiplied by a number a that is greater than 1. What if a is a fraction between 0 and 1?

each output by a factor of $\frac{1}{2}$ and reflecting across the x-axis.

THEOREM 7	**Vertical Stretchings and Shrinkings of Functions**

Given the graph of $y = f(x)$, the graph of $y = af(x)$ is obtained as follows:

a		*Change in graph*
$|a| > 1$	⇨	Stretched vertically
$|a| < 1$	⇨	Shrunk vertically
$a < 0$	⇨	Reflected across x-axis

Note that multiplying by -1 has the effect of reflecting the graph without stretching or shrinking. That is, the graph of $y = -f(x)$ is a **reflection** of the graph of $y = f(x)$ across the x-axis.

DO EXERCISES 4 AND 5 ON THE FOLLOWING PAGE.

Horizontal Stretchings and Shrinkings

The constant a in the equation $y = af(x)$ has the effect of vertically stretching or shrinking the graph of $y = f(x)$, with a possible reflection across the x-axis if $a < 0$. The constant c in the equation $y = f(cx)$ has the effect of horizontally stretching or shrinking the graph of $y = f(x)$, with a possible reflection across the y-axis if $c < 0$.

THEOREM 8	**Horizontal Stretchings and Shrinkings of Functions**

Given the graph of $y = f(x)$, the graph of $y = f(cx)$ is obtained as follows:

c		*Change in graph*
$|c| > 1$	⇨	Shrunk horizontally
$|c| < 1$	⇨	Stretched horizontally
$c < 0$	⇨	Reflected across y-axis

4. Use graph paper. Using the same set of axes, graph

$$f(x) = -x^2,$$
$$f(x) = -2x^2, \quad \text{and}$$
$$f(x) = -0.8x^2.$$

5. Use graph paper. Using the same set of axes, graph

$$f(x) = x^2 - 4 \quad \text{and}$$
$$f(x) = -(x^2 - 4)$$
$$= 4 - x^2.$$

Below is a graph of $y = f(x)$. No formula for f is given. Sketch a graph of each of the following using graph paper.

6. $y = \dfrac{1}{2}f(x)$

7. $y = -2f(x)$

8. $y = f(2x)$

9. $y = f\left(-\dfrac{1}{2}x\right)$

Example 5   At right is a graph of $y = f(x)$ for some function f. No formula for f is given. Sketch a graph of $y = f(2x)$, $y = f(\frac{1}{2}x)$, and $y = f(-\frac{1}{2}x)$.

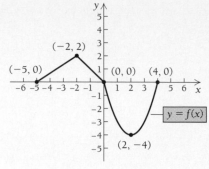

Solution   Since $|2| > 1$, the graph of $y = f(2x)$ is a horizontal shrinking of the graph of $y = f(x)$. We can consider the key points $(-5, 0)$, $(-2, 2)$, $(0, 0)$, $(2, -4)$, and $(4, 0)$. The transformation divides each x-coordinate by 2 to obtain the key points $(-2.5, 0)$, $(-1, 2)$, $(0, 0)$, $(1, -4)$, and $(2, 0)$ of the graph of $y = f(2x)$. The graph is as shown here.

Since $|\frac{1}{2}| < 1$, the graph of $y = f(\frac{1}{2}x)$ is a horizontal stretching of the graph of $y = f(x)$. We can consider the key points $(-5, 0)$, $(-2, 2)$, $(0, 0)$, $(2, -4)$, and $(4, 0)$. The transformation divides each x-coordinate by $1/2$ (which is the same as multiplying by 2) to obtain the key points $(-10, 0)$, $(-4, 2)$, $(0, 0)$, $(4, -4)$, and $(8, 0)$ of the graph of $y = f(\frac{1}{2}x)$. The graph is as follows.

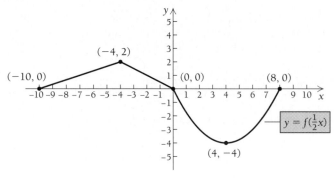

The graph of $y = f(-\frac{1}{2}x)$ can be obtained by reflecting the graph of $y = f(\frac{1}{2}x)$ across the y-axis, as follows.

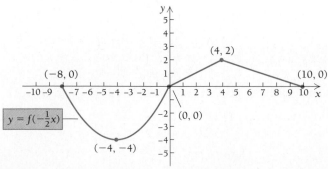

DO EXERCISES 6–9.

Example 6   Use the graph of $y = f(x)$ given in Example 5. Sketch a graph of

$$y = -2f(x - 3) + 1.$$

10. Use the graph of $y = f(x)$ in Margin Exercises 6–9. Using graph paper, graph $y = -3f(x + 1) - 4$.

Solution

 DO EXERCISE 10.

TECHNOLOGY CONNECTION

Begin with the same base function and its graph. On the basis of the rules that you have considered in this section, predict what effect each of the following transformations will have on the graph. Then draw the graph on a grapher to see if your prediction is correct.

1. Start with the base function $f(x)$.
2. Add 3 to x, forming $f(x + 3)$.
3. Multiply the function by 2, forming

$$2 \cdot f(x + 3).$$

4. Subtract 5 from the function, forming

$$2 \cdot f(x + 3) - 5.$$

Next try one of the other base functions in the initial list, and predict what effect each one of these transformations will have on the graph. Check your predictions with a grapher.

1. Start with the base function $f(x)$.
2. Multiply by -2, forming $-2 \cdot f(x)$.
3. Add 6 to the function, forming $-2 \cdot f(x) + 6$.
4. Subtract 5 from x, forming $-2 \cdot f(x - 5) + 6$.

EXERCISE SET **3.8**

A   Sketch the graph by transforming the graph of $f(x) = |x|$.

1. $f(x) = |x| - 3$

2. $f(x) = 2 + |x|$

3. $f(x) = |x - 1|$

4. $f(x) = |x + 2|$

5. $f(x) = -4|x|$

6. $f(x) = 3|x|$

7. $f(x) = \frac{1}{3}|x|$

8. $f(x) = -\frac{1}{4}|x|$

9. $f(x) = |2x|$

10. $f(x) = \left|\dfrac{x}{3}\right|$

11. $f(x) = |x - 2| + 3$

12. $f(x) = 2|x + 1| - 3$

13. $f(x) = -3|x - 2|$

14. $f(x) = \frac{1}{3}|x + 2| + 1$

Here is a graph of $y = f(x)$. No formula will be given for this function. In Exercises 15–34, sketch the graph by transforming this one.

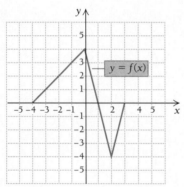

15. $y = 2 + f(x)$

16. $y = f(x) - 1$

17. $y = f(x - 1)$

18. $y = f(x + 2)$

19. $y = -2f(x)$

20. $y = 3f(x)$

21. $y = \frac{1}{3}f(x)$

22. $y = -\frac{1}{2}f(x)$

23. $y = f(2x)$

24. $y = f(3x)$

25. $y = f(-2x)$

26. $y = f(-3x)$

27. $y = f\left(\dfrac{x}{-2}\right)$

28. $y = f(\frac{1}{3}x)$

29. $y = f(x - 2) + 3$

30. $y = -3f(x - 2)$

31. $y = 2 \cdot f(x + 1) - 2$

32. $y = \frac{1}{2}f(x + 2) - 1$

33. $y = -\frac{1}{2}f(x - 3) + 2$

34. $y = -3f(x - 1) + 4$

Here is a graph of $y = f(x)$. No formula will be given for this function. In Exercises 35–38, sketch the graph by transforming this one.

35. $y = -2f(x + 1) - 1$

36. $y = 3f(x + 2) + 1$

37. $y = \frac{5}{2}f(x - 3) - 2$

38. $y = -\frac{2}{3}f(x - 4) + 3$

● **SYNTHESIS** _____

Given the graph of $y = f(x)$ for Exercises 35–38, graph each of the following.

39. $y = -\sqrt{2}f(x + 1.8)$

40. $y = \dfrac{\sqrt{3}}{2} \cdot f(x - 2.5) - 5.3$

A graph of the function $f(x) = x^3 - 3x^2$ is shown below. Exercises 41–44 show graphs of functions transformed from this one. Find a formula for each function.

41.

42.

43.

44.

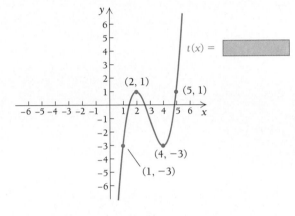

● **CHALLENGE** _____

Given the graph of $y = f(x)$ for Exercises 35–38, graph each of the following.

45. $y = 3f(2x + \frac{1}{2})$

46. $y = -4 \cdot f(5x + 10)$

● **TECHNOLOGY CONNECTION** _____

Use a grapher to graph the original function. Then graph each transformed function.

47. Given $f(x) = x^2|x|$:

    a) Multiply x by 4.7, forming $[f(4.7x)]$.
    b) Multiply $f(x)$ by -2.5, forming $[-2.5f(x)]$.
    c) Subtract 11.5 from $f(x)$, forming $[f(x) - 11.5]$.
    d) Add 9.9 to x, forming $[f(x + 9.9)]$.

48. Given $f(x) = \dfrac{1 - 4x^3}{x^2}$:

    a) Add 34.1 to $f(x)$, forming $[f(x) + 34.1]$.
    b) Divide $f(x)$ by 3, forming $[f(x)/3]$.
    c) Multiply x by 0.35, forming $[f(0.35x)]$.
    d) Subtract 8.1 from x, forming $[f(x - 8.1)]$.

SUMMARY AND REVIEW   3

● **TERMS TO KNOW**

Coordinate, p. 130	Output, p. 147	Greatest integer function, p. 173
Relation, p. 130	Constant function, p. 148	Symmetry with respect to the
Graph, p. 132	Standard linear equation, p. 153	axes, p. 182
Solution, p. 132	Slope, p. 155	Symmetry with respect to the
Domain, p. 136	Point–slope equation, p. 158	origin, p. 185
Range, p. 136	Two-point equation, p. 158	Even function, p. 186
Distance formula, p. 139	Slope–intercept equation, p. 159	Odd function, p. 187
Midpoint formula, p. 140	Linear function, p. 161	Composition of functions, p. 193
Circle, p. 141	Parallel lines, p. 162	Transformation, p. 197
Radius, p. 141	Perpendicular lines, p. 162	Translation, pp. 197, 199
Center, p. 141	Vertical-line test, p. 167	Shift, p. 197
Function, pp. 145, 146	Interval notation, p. 168	Stretching, p. 201
Dependent variable, p. 147	Increasing function, p. 170	Shrinking, p. 201
Independent variable, p. 147	Decreasing function, p. 170	Reflection, p. 201
Input, p. 147		

● REVIEW EXERCISES

Consider the relation $\{(3, 1), (5, 3), (7, 7), (3, 5)\}$ for Exercises 1–3.

1. Determine whether the relation is a function.

2. Find the domain.

3. Find the range.

Graph.

4. $x = |y|$

5. $y = (x + 1)^2$

6. $g(x) = |x| - 2$

7. $f(x) = \sqrt{x}$

8. $f(x) = \sqrt{x - 2}$

9. $f(x) = 2\sqrt{x + 3}$

10. $f(x) = \frac{1}{2}\sqrt{x - 1} + 2$

11. $f(x) = \text{INT}(x)$

12. $f(x) = \text{INT}(x) - 3$

13. $f(x) = \text{INT}(x - 3)$

14. $5y - 2x = 10$

15. Graph $y = -\frac{2}{3}x - 4$ using the slope and the y-intercept.

16. Find the slope and the y-intercept of the line
$$-2x - y = 7.$$

17. Find the slope of the line containing the points $(7, -2)$ and $(1, 4)$.

18. Find an equation of the line through $(-2, -1)$ with $m = 3$.

19. Find an equation of the line containing $(4, 1)$ and $(-2, -1)$.

Consider the following relations for Exercises 20–22.

a) $y = 7$

b) $x^2 + y^2 = 4$

c) $x^3 = y^3 - y$

d) $y^2 = x^2 + 3$

e) $x + y = 3$

f) $x = 3$

g) $y = x^2$

h) $y = x^3$

20. Which are symmetric with respect to the x-axis?

21. Which are symmetric with respect to the y-axis?

22. Which are symmetric with respect to the origin?

23. Given $R(x) = 120x - 0.5x^2$ and $C(x) = 15x + 6$, find $P(x)$.

24. Which of the following are graphs of functions?

a)

b)

Use $f(x) = x^2 - x - 3$ for Exercises 25–27. Find:

25. $f(0)$.

26. $f(-3)$.

27. $\dfrac{f(a + h) - f(a)}{h}$.

Use $g(x) = 2\sqrt{x - 1}$ for Exercises 28–30. Find:

28. $g(1)$.

29. $g(5)$.

30. $g(a + 2)$.

Find the domain.

31. $f(x) = \sqrt{7 - 3x}$

32. $f(x) = \dfrac{1}{x^2 - 6x + 5}$

In Exercises 33 and 34, find each of the following.

a) $(f + g)(x), (f - g)(x), fg(x), (f/g)(x), f \circ g(x)$, and $g \circ f(x)$

b) The domain of $f, g, f + g, f - g, fg, f/g, f \circ g$, and $g \circ f$

33. $f(x) = \dfrac{4}{x^2}$; $g(x) = 3 - 2x$

34. $f(x) = 3x^2 + 4x$; $g(x) = 2x - 1$

Here is a graph of $f(x) = \sqrt{9 - x^2}$. Sketch the graph of each of the following.

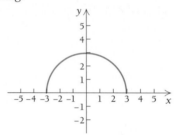

35. $y = 1 + f(x)$

36. $y = \frac{1}{2}f(x)$

37. $y = f(x + 1)$

Use the following for Exercises 38–40.

a)

b)

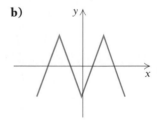

c) $f(x) = 3x^2 - 2$

d) $f(x) = x + 3$

e) $f(x) = 3x^3$

f) $f(x) = x^5 - x^3$

38 Which are even?

39. Which are odd?

40. Which are neither even nor odd?

41. Find the distance between $(3, 7)$ and $(-2, 4)$.

42. Find the midpoint of the segment with endpoints $(3, 7)$ and $(-2, 4)$.

Given the point $(1, -1)$ and the line $2x + 3y = 4$:

43. Find an equation of the line containing the given point and parallel to the given line.

44. Find an equation of the line containing the given point and perpendicular to the given line.

Determine whether the lines are parallel, perpendicular, or neither.

45. $3x - 2y = 8$,
$6x - 4y = 2$

46. $y - 2x = 4$,
$2y - 3x = -7$

47. $y = \frac{3}{2}x + 7$,
$y = -\frac{2}{3}x - 4$

48. Find an equation of the circle with center $(-2, 6)$ and radius $\sqrt{13}$.

49. Find the center and the radius of the circle
$$(x + 1)^2 + (y - 3)^2 = \frac{9}{4}.$$
Then graph the circle.

50. Find the center and the radius of the circle
$$x^2 + y^2 + 4y + 21 = 10x.$$

51. Find an equation of the circle having its center at $(3, 4)$ and passing through the origin.

52. Find an equation of the circle having a diameter with endpoints $(-3, 5)$ and $(7, 3)$.

Use the following graphs of functions for Exercises 53–55.

a)

b)

c)

53. Which are increasing?

54. Which are decreasing?

55. Which are neither increasing nor decreasing?

Write interval notation for the set.

56. $\{x \mid -\pi \le x \le 2\pi\}$

57. $\{x \mid 0 < x \le 1\}$

58. $\{x \mid x < 14\}$

Graph.

59. $f(x) = \begin{cases} x^2 + 2, & \text{for } x < 0, \\ x^3, & \text{for } 0 \le x < 2, \\ -4x + 5, & \text{for } x \ge 2 \end{cases}$

60. $f(x) = \begin{cases} \dfrac{x^2 - 1}{x + 1}, & \text{for } x \ne -1, \\ 3, & \text{for } x = -1 \end{cases}$

61. Two cars leave the same intersection at right angles to each other. One travels at a speed of 50 mph and the other at 55 mph. Express the distance d, in miles, between the cars as a function of time t, in hours.

62. A right circular cylinder with radius r and height 8 is inscribed in a sphere with radius a. Express the volume of the cylinder as a function of a.

63. Find $f(x)$ and $g(x)$ such that $h(x) = f \circ g(x)$.

a) $h(x) = \sqrt{5x + 2}$ **b)** $h(x) = \dfrac{x^3 + 1}{x^3 - 1}$

● **SYNTHESIS** _____

Find the domain.

64. $f(x) = (x - 9x^{-1})^{-1}$ **65.** $f(x) = \dfrac{\sqrt{1 - x}}{x - |x|}$

66. Graph several functions of the type $y = |f(x)|$. Describe a procedure, involving transformations, for graphing such functions.

67. Graph: $|x - y| = 1$.

● **THINKING AND WRITING** _____

1. Explain how a *solution* of the equation $3x - 4 = 7$ differs from a *solution* of the equation $3x - 4y = 7$.

2. Given
$$f(x) = 4x^3 - 2x + 7,$$
find (a) $f(x) + 2$, (b) $f(x + 2)$, and (c) $f(x) + f(2)$. Then discuss how each expression differs from the other.

3. Given the graph of $y = f(x)$, explain and contrast the effect of the constant c on the graphs of $y = cf(x)$ and $y = f(cx)$.

_____ 3 _____

CHAPTER TEST

Consider the relation $\{(2, 7), (-2, -7), (7, -2), (0, 2)\}$ for Questions 1–3.

1. Determine whether the relation is a function.

2. Find the range.

3. Find the domain.

Graph.

4. $f(x) = (x - 2)^2$ **5.** $x = |y + 1|$

6. $f(x) = 2 \text{ INT}(x)$

Consider the following relations for Questions 7 and 8.

a) $2y - \dfrac{5}{x} = 0$ **b)** $x = -5$ **c)** $y = x^2 - 2$

d) $x^3 - x = y^3$ **e)** $x^2 - 1 = y^2$ **f)** $3y = |x|$

7. Which are symmetric with respect to the origin?

8. Which are symmetric with respect to the x-axis?

9. Find the domain of the function f given by

$$\frac{1}{16 - x^2}.$$

Consider the functions f and g given by $f(x) = 2x - 1$ and $g(x) = x^2 + 6$ for Questions 10 and 11.

10. Find $(f + g)(x)$, $(f - g)(x)$, $fg(x)$, $(f/g)(x)$, $f \circ g(x)$, and $g \circ f$.

11. Find the domain of f, g, $f + g$, $f - g$, fg, f/g, $f \circ g$, and $g \circ f$.

12. Which of the following are graphs of functions?

a)   **b)**

Use $g(x) = x^2 - x + 2$ for Questions 13–16. Find:

13. $g(-1)$.   **14.** $g(0)$.

15. $g(a - 1)$.   **16.** $\dfrac{g(a + h) - g(a)}{h}$.

17. Given

$$R(x) = x^2 + 110x + 60$$

and

$$C(x) = 1.1x^2 + 10x + 80,$$

find $P(x)$.

18. Find the slope and the y-intercept of the line $10 - 7y = 3x$.

19. Find an equation of the line through $(-2, 2)$ with $m = 5$.

20. Find an equation of the line containing $(2, -3)$ and $(5, 3)$.

21. Find the distance between $(4, -5)$ and $(6, -2)$.

22. Find the midpoint of the segment with endpoints $(4, -9)$ and $(7, 6)$.

23. Determine whether these lines are parallel, perpendicular, or neither.

$$2x - 3y = -9,$$
$$3x + 2y = 6$$

24. Graph $y = \frac{3}{2}x - 1$ using the slope and the y-intercept.

25. Find an equation of the line containing the given point and parallel to the given line:

$$(2, -4); \quad y = \tfrac{1}{3}x - 6.$$

26. Find an equation of the circle with center $(-2, 4)$ and radius 4.

27. Find the center and the radius of the circle

$$x^2 + y^2 - 2x + 6y + 5 = 0.$$

28. Here is a graph of $y = f(x)$. Sketch the graph of each of the following.

a) $y = f(x - 1)$ **b)** $y = f(2x)$
c) $y = 3 + f(x)$

Use the following for Questions 29–31.

a)   **b)**

c) $f(x) = x^2 - 9$ **d)** $f(x) = x^3 - 2x + 4$
e) $f(x) = |x|$ **f)** $f(x) = x^7 - x^5$

29. Which are even?

30. Which are odd?

31. Which are neither even nor odd?

Write interval notation for the set.

32. $\{x \mid -7 < x < 2\}$ **33.** $\{x \mid x \geq -2\}$

Use the following for Questions 34 and 35.

a)   **b)**

c)

34. Which are decreasing?

35. Which are neither increasing nor decreasing?

36. Graph

$$f(x) = \begin{cases} x^3, & \text{for } x < -2, \\ |x|, & \text{for } -2 \le x < 2, \\ \sqrt{x-1}, & \text{for } x \ge 2. \end{cases}$$

37. A rectangle is inscribed in a semicircle of radius 2 as shown in the figure. The variable $x =$ half the length of the rectangle. Express the area of the rectangle as a function of x.

38. Find $f(x)$ and $g(x)$ such that $h(x) = f \circ g(x)$.

a) $h(x) = \dfrac{1}{\sqrt{7x + 2}}$

b) $h(x) = 4(5x - 1)^2 + 9$

● **SYNTHESIS**

39. Prove that the sum of two odd functions is odd.

A *polynomial function* is a function that can be defined by a polynomial expression. A *rational function* is a function that can be defined as a quotient of two polynomials. We will study both kinds of functions. For polynomials, we will consider the finding of zeros or roots in greater depth. For rational functions, we will consider their graphs. ● We will also discuss absolute value and quadratic and rational inequalities. ●

4

Polynomial and Rational Functions

4.1

QUADRATIC FUNCTIONS

Any function that can be described by a polynomial in one variable, such as

$$f(x) = 5x^7 + 3x^6 - 4x^2 - 5,$$

is called a **polynomial function.** Here is a formal definition.

DEFINITION   **Polynomial Function**

A *polynomial function* f is given by

$$f(x) = a_n x^n + a_{n-1} x^{n-1} + \cdots + a_2 x^2 + a_1 x + a_0,$$

where n is a nonnegative integer and $a_n, a_{n-1}, \ldots, a_2, a_1, a_0$ are real numbers, called the *coefficients* of the polynomial.

The first nonzero coefficient is assumed to be a_n and is called the *leading coefficient.* The *degree* of the polynomial function is n.

A Properties of Quadratic Functions

If a function can be described by a second-degree polynomial, then it is called **quadratic.** The following is a more precise definition.

OBJECTIVES

You should be able to:

A Given a quadratic function, find the vertex of its graph, the line of symmetry, and the maximum or minimum value.

B Graph a quadratic function.

C Find the *x*-intercepts of the graph of a quadratic function.

D Given a situation described by a quadratic function, use that function to make predictions, including finding a maximum or minimum value of the function.

Use graph paper for Margin Exercises 1–4.

1. a) Graph $f(x) = 2x^2$.
   b) Does the graph open up or does it open down?
   c) What is the line of symmetry?
   d) What is the minimum value of the function?
   e) What is the vertex?

2. a) Graph $f(x) = -0.4x^2$.
   b) Does the graph open up or does it open down?
   c) What is the line of symmetry?
   d) What is the maximum value of the function?
   e) What is the vertex?

DEFINITION   **Quadratic Function**

A *quadratic function* is a function that can be described as follows:

$$f(x) = ax^2 + bx + c, \quad \text{where } a \neq 0.$$

In this definition, we insist that $a \neq 0$; otherwise the polynomial would not be of degree two. One or both of the constants b and c can be 0.

Consider $f(x) = x^2$. This is an even function, so the y-axis is a line of symmetry. The graph opens up, as shown below. If we multiply by a constant a to get $f(x) = ax^2$, we obtain a vertical stretching or shrinking, and a reflection if a is negative. Examples of this are shown in the figure. Bear in mind that if a is negative, the graph opens down.

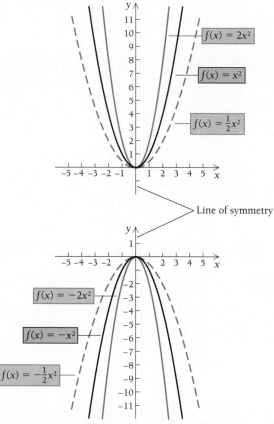

Graphs of quadratic functions. Parabolas.

Graphs of quadratic functions are called **parabolas.** The highest or lowest point at which the graph turns is called the **vertex.** In each parabola shown, the point $(0, 0)$ is the *vertex* and the line $x = 0$, or the y-axis, is the **line of symmetry.**

DO EXERCISES 1 AND 2.

Let us consider $f(x) = a(x - h)^2$. We have replaced x by $x - h$ in ax^2, and will therefore obtain a horizontal translation. The translation will be to the right if h is positive and to the left if h is negative. The examples in the figures below illustrate translations.

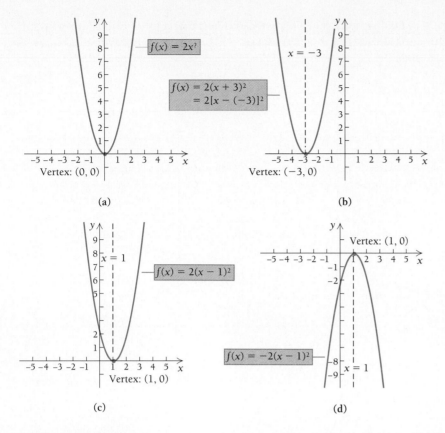

DO EXERCISES 3 AND 4.

Now consider $f(x) = a(x - h)^2 + k$. We have a vertical translation. If k is positive, the translation is up. If k is negative, the translation is down. Consider these examples. Note that the vertex has been moved off the x-axis.

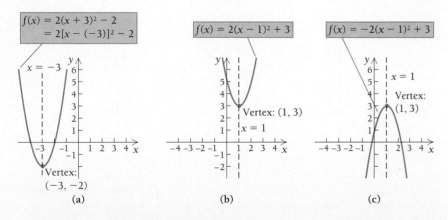

The graph of $f(x) = a(x - h)^2 + k$:

a) opens up if $a > 0$ and down if $a < 0$;
b) has (h, k) as a vertex;
c) has $x = h$ as a line of symmetry;
d) has k as a minimum value (output) if $a > 0$ and has k as a maximum value (output) if $a < 0$.

3. a) Graph $f(x) = 3x^2$.
   b) Use the graph in part (a) to graph $f(x) = 3(x - 2)^2$.
   c) What is the vertex of the graph in part (b)?
   d) What is the line of symmetry of the graph in part (b)?
   e) What is the minimum value?
   f) Does the graph open up or does it open down?
   g) Is the graph of $f(x) = 3(x - 2)^2$ a horizontal translation to the left or to the right?

4. a) Graph $f(x) = -3x^2$.
   b) Use the graph in part (a) to graph
      $$f(x) = -3(x + 2)^2$$
      $$= -3[x - (-2)]^2.$$
   c) What is the vertex of the graph in part (b)?
   d) What is the line of symmetry of the graph in part (b)?
   e) What is the maximum value?
   f) Does the graph open up or does it open down?
   g) Is the graph of $f(x) = -3(x + 2)^2$ a horizontal translation to the left or to the right?

Answer the following questions in Margin Exercises 5–10.

a) What is the vertex?

b) What is the line of symmetry?

c) Is there a maximum? What is it?

d) Is there a minimum? What is it?

Use graph paper for Margin Exercises 5 and 6.

5. Graph: $f(x) = 3(x - 2)^2 + 4$.

6. Graph:

$$f(x) = -3(x + 2)^2 - 1$$
$$= -3[x - (-2)]^2 + (-1).$$

Without graphing, answer the above questions for each function.

7. $f(x) = (x - 5)^2 + \pi$

8. $f(x) = -3(x - 5)^2$

9. $f(x) = 2\left(x + \dfrac{1}{4}\right)^2 - 6$

10. $f(x) = -\dfrac{1}{4}(x + 9)^2 + 3$

11. Use completing the square to put the function

$$f(x) = x^2 - 4x + 7$$

into the form

$$f(x) = a(x - h)^2 + k.$$

Thus without graphing, we can determine a lot of information about a function described by $f(x) = a(x - h)^2 + k$. The following table is an example.

Function	$f(x) = 3(x - \frac{1}{4})^2 - 2$ $= 3(x - \frac{1}{4})^2 + (-2)$	$g(x) = -3(x + 5)^2 + 7$ $= -3[x - (-5)]^2 + 7$
a) What is the vertex?	$(\frac{1}{4}, -2)$	$(-5, 7)$
b) What is the line of symmetry?	$x = \frac{1}{4}$	$x = -5$
c) Is there a maximum? What is it?	No: graph extends up; $3 > 0$.	Yes, 7: graph extends down; $-3 < 0$.
d) Is there a minimum? What is it?	Yes, -2: graph extends up; $3 > 0$.	No: graph extends down; $-3 < 0$.

Note that the vertex (h, k) is used to find the maximum or the minimum. The maximum or minimum is the number k, *not* the ordered pair (h, k).

DO EXERCISES 5–10. _____

Now let us consider a quadratic function $f(x) = ax^2 + bx + c$. Note that it is not in the form $f(x) = a(x - h)^2 + k$. We can put it into that form by *completing the square*.

Example 1 Use completing the square to put the function

$$f(x) = x^2 - 6x + 4$$

into the form $f(x) = a(x - h)^2 + k$.

Solution We consider first the x^2- and x-terms:

$$f(x) = x^2 - 6x \qquad + 4.$$

We construct a trinomial square. To do so, we take half the coefficient of x and square it. The number is $(-6/2)^2$, or 9. We now add that number to complete the square. We also must subtract it to get an equivalent equation. We can think of this simply as adding $9 - 9$, which is 0:

$$f(x) = x^2 - 6x + 9 - 9 + 4$$
$$= x^2 - 6x + 9 - 5.$$

We now factor the trinomial square $x^2 - 6x + 9$ to get

$$f(x) = (x - 3)^2 - 5. \qquad \blacktriangleleft$$

DO EXERCISE 11. _____

If the coefficient of x^2 is not 1, a preliminary step is needed.

Example 2   Use completing the square to put the function

$$f(x) = 2x^2 + 12x - 1$$

into the form $f(x) = a(x - h)^2 + k$.

Solution   Again we consider the x^2- and x-terms, but we begin by factoring out the x^2-coefficient, as follows:

$$f(x) = 2(x^2 + 6x) - 1.$$

Next we proceed as before, *inside* the parentheses. We take half the coefficient of x and square it. That number is $(6/2)^2$, or 9. We add $9 - 9$, but do it *inside* the parentheses:

$$f(x) = 2(x^2 + 6x + 9 - 9) - 1.$$

We now have an extra, unwanted term inside the parentheses, so we use the distributive law and multiplication to get the $(-2 \cdot 9)$ outside, as follows:

$$\begin{aligned} f(x) &= 2(x^2 + 6x + 9) - 2 \cdot 9 - 1 \\ &= 2(x^2 + 6x + 9) - 18 - 1 = 2(x + 3)^2 - 19 \\ &= 2[x - (-3)]^2 - 19. \end{aligned}$$ ◀

DO EXERCISE 12.

In many situations, we want to be able to read off the vertex (h, k) directly from $f(x) = ax^2 + bx + c$ without repeatedly completing the square. We look for a formula to compute h and k. We proceed in a manner quite similar to what we did when we proved the quadratic formula.

THEOREM 1

The vertex of the graph of the quadratic function $f(x) = ax^2 + bx + c$ is

$$\left(-\frac{b}{2a}, -\frac{b^2 - 4ac}{4a} \right).$$

Proof.   We proceed in a manner similar to Example 2. We begin by factoring out the x^2-coefficient, as follows:

$$f(x) = a\left(x^2 + \frac{b}{a}x \right) + c.$$

Next, we proceed *inside* the parentheses. We take half the coefficient of x and square it. That number is $(b/2a)^2$. We add $(b/2a)^2 - (b/2a)^2$, but do it *inside* the parentheses:

$$f(x) = a\left(x^2 + \frac{b}{a}x + \left(\frac{b}{2a}\right)^2 - \left(\frac{b}{2a}\right)^2 \right) + c.$$

We now have an extra, unwanted term inside the parentheses, so we use

12. Use completing the square to put the function

$$f(x) = 3x^2 + 24x + 10$$

into the form

$$f(x) = a(x - h)^2 + k.$$

13. For the function

$$f(x) = 4x^2 - 12x - 5,$$

a) find the vertex and the line of symmetry;

b) determine whether there is a maximum or minimum value and find that value.

the distributive law and multiplication to simplify, as follows:

$$f(x) = a\left(x^2 + \frac{b}{a}x + \left(\frac{b}{2a}\right)^2\right) - a\left(\frac{b}{2a}\right)^2 + c$$

$$= a\left(x^2 + \frac{b}{a}x + \left(\frac{b}{2a}\right)^2\right) - \frac{b^2}{4a} + c$$

$$= a\left(x^2 + \frac{b}{a}x + \left(\frac{b}{2a}\right)^2\right) - \frac{b^2 - 4ac}{4a}$$

$$= a\left(x + \frac{b}{2a}\right)^2 - \frac{b^2 - 4ac}{4a}$$

$$= a\left(x - \left(-\frac{b}{2a}\right)\right)^2 + \left(-\frac{b^2 - 4ac}{4a}\right).$$

The vertex is

$$\left(-\frac{b}{2a}, -\frac{b^2 - 4ac}{4a}\right).$$

Note that the numerator for the second coordinate of the vertex contains the discriminant, used to solve quadratic equations. We can use all of the formula for the vertex, or we can find $-b/2a$ and substitute into the original formula for the function to find the second coordinate, which turns out to be the maximum or minimum value.

Example 3 For the function $f(x) = 4x^2 + 12x + 7$, (a) find the vertex and the line of symmetry and (b) determine whether there is a maximum or minimum value and find that value.

Solution

a) We note that $a = 4$, $b = 12$, and $c = 7$. The first coordinate of the vertex is

$$-\frac{b}{2a} = -\frac{12}{2(4)} = -\frac{3}{2}.$$

We find the second coordinate of the vertex as follows:

$$-\frac{b^2 - 4ac}{4a} = -\frac{12^2 - 4(4)(7)}{4(4)}$$
$$= -2.$$

We can also substitute $-\frac{3}{2}$ into the original function formula:

$$f(-\tfrac{3}{2}) = 4(-\tfrac{3}{2})^2 + 12(-\tfrac{3}{2}) + 7$$
$$= -2.$$

The vertex is $(-\frac{3}{2}, -2)$. The line of symmetry goes through this point, hence is the line $x = -\frac{3}{2}$.

b) Since the coefficient of x^2 is positive, the parabola opens up. Thus we have a minimum value. That value is -2. ◄

DO EXERCISE 13.

Example 4 For the function $f(x) = -2x^2 + 10x - 7$, find the vertex and the maximum or minimum value.

Solution We note that $a = -2, b = 10,$ and $c = -7$. The first coordinate of the vertex is

$$-\frac{b}{2a} = -\frac{10}{2(-2)} = \frac{5}{2}.$$

We find the second coordinate by substituting $\frac{5}{2}$ into the original function formula:

$$f\left(\tfrac{5}{2}\right) = -2\left(\tfrac{5}{2}\right)^2 + 10\left(\tfrac{5}{2}\right) - 7 = \tfrac{11}{2}.$$

The vertex is $\left(\frac{5}{2}, \frac{11}{2}\right)$. The maximum value is $\frac{11}{2}$, since the coefficient of x^2 is negative. ◀

DO EXERCISE 14.

Graphing Quadratic Functions

We know that the graph of any quadratic function is a *parabola*. If the x^2-coefficient is positive, the parabola opens up. If the x^2-coefficient is negative, the graph opens down. We also know how to find the vertex and the line of symmetry. We can compute function values on each side of the vertex. We can also find other values by reflecting across the line of symmetry.

Example 5 Graph: $f(x) = -2x^2 + 10x - 7$.

Solution Since the coefficient of x^2 is negative, the parabola opens down. We find the vertex and the line of symmetry. From Example 4, we know that the vertex is $\left(\frac{5}{2}, \frac{11}{2}\right)$ and that $\frac{11}{2}$ is a maximum value. The line of symmetry is $x = \frac{5}{2}$. We then plot several input–output pairs. We start with $x = 2$, since it is near the vertex and is an integer, making computations easy:

$$f(2) = -2(2)^2 + 10(2) - 7 = 5.$$

This gives us the pair $(2, 5)$. By reflecting this pair across the line of symmetry, we also get the pair $(3, 5)$. We then consider $x = 1$. Since $f(1) = 1$, we get another pair $(1, 1)$ and by reflecting, we get the pair $(4, 1)$. Finding the y-intercept by computing $f(0)$ is easy to do. We get $f(0) = -7$. This gives us another pair $(0, -7)$. We then plot the ordered pairs and complete the graph.

x	$f(x)$	
$\frac{5}{2}$	$\frac{11}{2}$	← Vertex
2	5	
3	5	
1	1	
4	1	
0	−7	←

y-intercept

14. For the function

$$f(x) = -3x^2 - 18x + 7,$$

find the vertex and the maximum or minimum value.

Graph using graph paper.

15. $f(x) = x^2 - 6x + 4$

16. $f(x) = -4x^2 + 12x - 5$

> To graph a quadratic function:
>
> 1. Note whether the x^2-coefficient is positive or negative and thus determine whether the curve opens up or down.
> 2. Find the vertex and the line of symmetry.
> 3. Find several other input–output pairs. Use reflection across the line of symmetry to find other pairs. Compute $f(0)$ to determine the y-intercept.

DO EXERCISES 15 AND 16.

 TECHNOLOGY CONNECTION

Using a grapher to graph quadratic functions is especially useful when the coefficients are complicated, as in the function $f(x) = 3.7185x^2 - 4.3185x + 2.0187$. Although all the techniques for finding the vertex and the line of symmetry would certainly work without a grapher, the formulas would be messy if you had to do them by hand. On a grapher, however, this quadratic function is no easier or harder than any other. By repeatedly zooming in on the vertex, you can determine its coordinates to whatever accuracy you desire. From the coordinates of the vertex, the line of symmetry is easily determined. For the example mentioned above, if we wanted our answers to be accurate to four decimal places, we would find the coordinates of the vertex to be $(0.5807, 0.7649)$, and the line of symmetry to be $x = 0.5807$.

$f(x) = 3.7185x^2 - 4.3185x + 2.0187$

$[-10, 10] \times [-10, 10]$

Use a grapher to graph each of the following functions. Then use the zoom feature to find the vertex and the line of symmetry. (Round to two decimal places.)

TC 1. $f(x) = 2.42x^2 + 15.32x - 4.32$

TC 2. $f(x) = -3.42x^2 - 8.88x + 7.65$

TC 3. $f(x) = 9.11x^2 + 8.71x - 7.82$

x-Intercepts

The points at which a graph crosses the x-axis are called its **x-intercepts.** These are the points at which $f(x) = 0$. Thus the x-values at the intercepts are the solutions of the equation $f(x) = 0$. These are also called **roots,** or **zeros,** of the function.

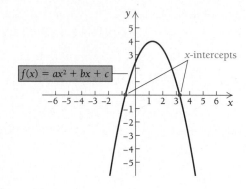

In order to find the x-intercepts of a quadratic function $f(x) = ax^2 + bx + c$, we solve the equation $f(x) = 0$, or

$$ax^2 + bx + c = 0.$$

Example 6   Find the x-intercepts of the graph of $f(x) = x^2 - 2x - 2$.

Solution   To find the x-intercepts, we solve the equation $f(x) = 0$. In this case, we solve

$$x^2 - 2x - 2 = 0.$$

Since the equation is difficult to factor, we use the quadratic formula and get $x = 1 \pm \sqrt{3}$. Thus the x-intercepts are $(1 - \sqrt{3}, 0)$ and $(1 + \sqrt{3}, 0)$. For plotting, we approximate to get $(-0.7, 0)$ and $(2.7, 0)$. We sometimes refer to the x-coordinates as intercepts.   ◄

Note that the x-intercepts could also be used when graphing quadratic functions, but they are not essential.

DO EXERCISE 17.

The **discriminant,** $b^2 - 4ac$, tells us how many real-number solutions the equation $ax^2 + bx + c = 0$ has, so it also indicates how many intercepts there are. Compare.

$y = ax^2 + bx + c$
$b^2 - 4ac > 0$
Two real solutions
Two x-intercepts

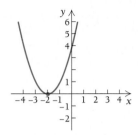

$y = ax^2 + bx + c$
$b^2 - 4ac = 0$
One real solution
One x-intercept

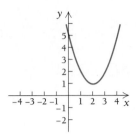

$y = ax^2 + bx + c$
$b^2 - 4ac < 0$
Two nonreal,
  complex solutions
  No x-intercepts

DO EXERCISES 18–20.

▶ Quadratic Functions as Mathematical Models

Example 7 *A projectile problem.* When an object such as a bullet or a ball is shot or thrown upward with an initial velocity v_0, its height is given, approximately, by a quadratic function:

$$s(t) = -4.9t^2 + v_0 t + h.$$

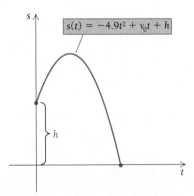

In this function, h is the starting height in meters, s is the actual height (also in meters), and t is the time from projection in seconds.

This model is based on the assumption that there is no air resistance and that the force of gravity pulling the object earthward is constant. Neither of these conditions exists precisely, so this model (as is the case with most mathematical models) gives only approximate results.

A model rocket is fired upward. At the end of the burn, it has an upward velocity of 49 m/sec and is 155 m high. Find (a) its maximum height and when it is attained and (b) when the rocket reaches the ground.

---

17. Find the x-intercepts of
$f(x) = x^2 - 2x - 5$.

TECHNOLOGY CONNECTION

To find the x-intercepts of a function drawn on a grapher, we can use the Zoom feature. We zoom in on the regions in which the graph crosses the x-axis. As an example, we will find the x-intercepts of $f(x) = 1.92x^2 + 4.32x - 3.07$. First, we make sure the viewing box is large enough to see all possible x-intercepts; in this case, the standard $[-10, 10] \times [-10, 10]$ window is fine. When the function is drawn, there seems to be an x-intercept between -3 and -2 and another between 0 and $+1$.

$f(x) = 1.92x^2 + 4.32x - 3.07$

$[-10, 10] \times [-10, 10]$

Next, we zoom in on the negative intercept. By repeating the process several times, we find its value, to two decimal places, to be -2.82. Moving over to the positive intercept, we can use repeated zooming to find its value to be about 0.57.

Use a grapher to find the x-intercept(s), if they exist, of each of the following. (Round to two decimal places.)

TC 4. $f(x) = -0.88x^2 + 4.14x - 3.56$

TC 5. $f(x) = 4.5x^2 - 27.9x + 43.245$

TC 6. $f(x) = 7.41x^2 - 18.39x + 25.11$

Find the x-intercepts, if they exist.

18. $f(x) = x^2 - 2x - 3$

19. $f(x) = x^2 + 8x + 16$

20. $f(x) = -2x^2 - 4x - 3$

21. A ball is thrown upward from the top of a cliff that is 12 m high, at a velocity of 2.8 m/sec. Find:

a) its maximum height and when it is attained;

b) when the ball reaches the ground.

Solution

a) We begin counting time at the end of the burn. Thus, $v_0 = 49$ and $h = 155$. The function is given by

$$s(t) = -4.9t^2 + 49t + 155.$$

We first find the vertex. The first coordinate of the vertex is

$$-\frac{b}{2a} = -\frac{49}{2(-4.9)} = 5.$$

We find the second coordinate by substituting:

$$s(5) = -4.9(5)^2 + 49(5) + 155 = 277.5.$$

The vertex of the graph is the point $(5, 277.5)$. Since the coefficient of t^2, -4.9, is negative, we know that a maximum height is reached and is 277.5 m, attained 5 sec after the end of the burn.

b) To find when the rocket reaches the ground, we set $s(t) = 0$, and solve for t. That is, we find the t-intercept, root, or zero of the function. We do this most conveniently by using the quadratic formula:

$$-4.9t^2 + 49t + 155 = 0$$

$$a = -4.9, \quad b = 49, \quad c = 155$$

$$t = \frac{-b \pm \sqrt{b^2 - 4ac}}{2a}$$

$$= \frac{-49 - \sqrt{49^2 - 4(-4.9)(155)}}{2(-4.9)}$$

$$\approx 12.525. \qquad \text{We use the negative square root so } t \text{ will be positive.}$$

The rocket will reach the ground about 12.525 sec after the end of the burn.

TECHNOLOGY CONNECTION

A grapher makes the following problem easy to solve, even if the numbers are complicated. Assume that a model rocket is 123.5 m high and moving upward with a velocity of 88.7 m/sec when its engine burns out. After how many seconds will it reach its maximum height, and what will that height be? How much time will elapse before it hits the ground?

Substituting the given values into the equation of motion gives the function $s(t) = -4.9t^2 + 88.7t + 123.5$. However, since most graphers require that the function be written as $y = f(x)$, we change t to x and s to y. Thus the function that we enter into the grapher is $f(x) = -4.9x^2 + 88.7x + 123.5$.

When this function is graphed, you will probably need to adjust the viewing box to include the vertex. Since x and y cannot be negative in this problem, we need to see only the first quadrant, where $x > 0$ and $y > 0$. Let's try a viewing box of $[0, 20] \times [0, 600]$.

$f(x) = -4.9x^2 + 88.7x + 123.5$

$[0, 20] \times [0, 600]$

Now, by zooming in on the highest point of the graph, we find that the coordinates of the vertex are $(9.1, 524.9)$. This tells us that the maximum height is 524.9 m, which occurs 9.1 sec after the engine has burned out. Finally, by zooming in on the x-intercept, we find its value to be 19.4. This means that the rocket will strike the ground 19.4 sec after the engine has quit.

Solve.

TC 7. A model rocket is 65.4 m above the ground, moving upward with a velocity of 65.2 m/sec, when its engine burns out. How long does it take to reach its maximum height, and what is that height? How many seconds will elapse before it hits the ground?

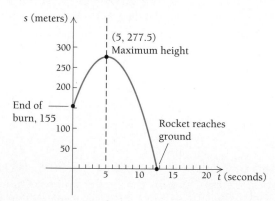

DO EXERCISE 21.

We will fit a quadratic function to a set of data in Chapter 9. For now, let us solve a maximum problem that we began in Example 11 of Section 3.5.

Example 8   *Maximizing area.* A music store has 20 ft of dividers to set off a rectangular area for a compact-disc display in one corner of its sales-room. The sides against the wall require no partition. Find the maximum area that can be enclosed and the dimensions of the maximum area.

Solution

1., 2. ***Familiarize and Translate.*** We first draw a picture. We let x = the length of one side and y = the length of the other. Since the sum of the lengths must be 20 ft, we have

$$x + y = 20 \quad \text{and} \quad y = 20 - x.$$

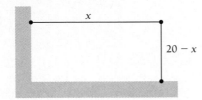

Then the area is given by

$$A = xy$$
$$A = x(20 - x) = 20x - x^2. \quad \text{\small Substituting } 20 - x \text{ for } y$$

We have thus expressed A as a function of x: $A(x) = 20x - x^2$.

3. ***Carry out.*** We want to maximize this function over the interval $(0, 20)$. We first find the vertex of the function. The first coordinate of the vertex is

$$-\frac{b}{2a} = -\frac{20}{2(-1)} = 10.$$

The second coordinate is found by substituting:

$$A(10) = 20(10) - (10)^2 = 200 - 100 = 100.$$

The vertex is thus $(10, 100)$. Since the leading coefficient of $A(x) = 20x - x^2$, -1, is negative, we know that the function has a maximum value. That maximum value is 100 when $x = 10$.

4. ***Check.*** We check our work by going over it another time. We can also check by examining the graph and estimating the maximum value.

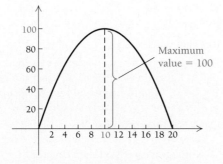

5. ***State.*** The maximum area of 100 ft^2 occurs when $x = 10$ ft. Thus the dimensions of 10 ft by 10 ft will yield the maximum area.

DO EXERCISE 22.

22. ***Maximizing area.*** A store has 40 ft of dividers to set off a rectangular area in one corner of its display room. The sides against the wall require no partition. Find the maximum area and the length and the width that will yield the maximum area.

TECHNOLOGY CONNECTION

Solve.

TC 8. If the music store has 62.8 ft of dividers to use in setting off the corner, what is the maximum area that may be enclosed and what are the dimensions of that area? (Round your answer to the nearest tenth of a foot.)

● **EXERCISE SET 4.1**

Ⓐ   For each of the following functions, (a) find the vertex, (b) find the line of symmetry, and (c) determine whether there is a maximum or minimum function value and find that value.

1. $f(x) = x^2$

2. $f(x) = -5x^2$

3. $f(x) = -2(x - 9)^2$

4. $f(x) = 5(x - 7)^2$

5. $f(x) = 2(x - 1)^2 - 4$

6. $f(x) = -(x + 4)^2 - 3$

For each of the following functions, (a) use completing the square to put each equation into the form $f(x) = a(x - h)^2 + k$, (b) find the vertex, and (c) determine whether there is a maximum or minimum function value and find that value.

7. $f(x) = -x^2 + 2x + 3$

8. $f(x) = -x^2 + 8x - 7$

9. $f(x) = x^2 + 3x$

10. $f(x) = x^2 - 9x$

11. $f(x) = -\frac{3}{4}x^2 + 6x$

12. $f(x) = \frac{3}{2}x^2 + 3x$

13. $f(x) = 3x^2 + x - 4$

14. $f(x) = -2x^2 + x - 1$

Find the maximum or minimum value of the function.

15. $f(x) = -5(x + 2)^2$

16. $f(x) = 0.4(x - 3)^2$

17. $f(x) = 8(x - 1)^2 + 5$

18. $f(x) = -6(x - 4)^2 - 11$

19. $f(x) = -4x^2 + x - 13$

20. $f(x) = \frac{2}{3}x^2 + 0.1x + 3.9$

21. $f(x) = \frac{1}{2}x^2 - \frac{2}{5}x - \frac{67}{100}$

22. $f(x) = -31.8x^2 + 12.3x - 17.2$

23. $g(x) = -\$120,000x^2 + \$430,000x - \$240,000$

24. $g(x) = \sqrt{8}x^2 - \pi x + \sqrt{5}$

Ⓑ   Graph.

25. $f(x) = -x^2 + 2x + 3$

26. $f(x) = x^2 - 3x - 4$

27. $f(x) = x^2 - 8x + 19$

28. $f(x) = -x^2 - 8x - 17$

29. $f(x) = -\frac{1}{2}x^2 - 3x + \frac{1}{2}$

30. $f(x) = 2x^2 - 4x - 2$

31. $f(x) = 3x^2 - 24x + 50$

32. $f(x) = -2x^2 + 2x + 1$

Ⓒ   Find the x-intercepts.

33. $f(x) = -x^2 + 2x + 3$

34. $f(x) = x^2 - 3x - 4$

35. $f(x) = x^2 - 8x + 5$

36. $f(x) = -x^2 - 3x - 3$

37. $f(x) = -5x^2 + 6x - 5$

38. $f(x) = 2x^2 + x - 5$

Ⓓ

39. Of all the numbers whose sum is 50, find the two that have the maximum product.

40. Of all the numbers whose difference is 16, find the two that have the minimum product.

41. The sum of the base and the height of a triangle is 20 cm. Find the dimensions for which the area is a maximum.

42. The sum of the base and the height of a parallelogram is 138 yd. Find the dimensions for which the area is a maximum.

43. A rancher wants to build a rectangular fence next to a river, using 120 yd of fencing. What dimensions of the rectangle will maximize the area? What is the maximum area? Note that the rancher need not fence in the side next to the river.

44. A rancher wants to enclose two congruent rectangular areas near a river, one for sheep and one for cattle. There is 240 yd of fencing available. What is the largest total area that can be enclosed?

45. A carpenter is building a rectangular room with a fixed perimeter of 54 ft. What are the dimensions of the largest room that can be built? What is its area?

46. Of all rectangles that have a perimeter of 34 ft, find the dimensions of the one with the largest area. What is its area?

Business: Maximizing profit.

Total profit = Total revenue − Total cost

$$P(x) = R(x) - C(x)$$

Find the maximum profit and the number of units that must be produced and sold in order to yield the maximum profit.

47. $R(x) = 50x - 0.5x^2, C(x) = 4x + 10$

48. $R(x) = 50x - 0.5x^2, C(x) = 10x + 3$

49. $R(x) = 2x, C(x) = 0.01x^2 + 0.6x + 30$

50. $R(x) = 5x, C(x) = 0.001x^2 + 1.2x + 60$

51. A rocket is fired upward. At the end of the burn, it has an upward velocity of 147 m/sec and is 560 m high. Find (a) its maximum height and when it is attained and (b) when the rocket reaches the ground.

52. A **Norman window** is a rectangle with a semicircle on top. Suppose that the perimeter of a particular Norman window is to be 24 ft. What should its dimensions be in order to allow the maximum amount of light to enter through the window?

A Norman window

● **SYNTHESIS** _____

Graph.

53. $f(x) = |x^2 - 1|$ (*Hint:* Consider two cases,

$$x^2 - 1 \geq 0 \quad \text{and} \quad x^2 - 1 < 0;$$

or graph $y = x^2 - 1$ and then reflect negative values across the x-axis.)

54. $f(x) = |3 - 2x - x^2|$

Find the maximum or minimum value of the function.

55. ▨ $f(x) = 2.31x^2 - 3.135x - 5.89$

56. ▨ $f(x) = -18.8x^2 + 7.92x + 6.18$

57. Find a such that

$$f(x) = ax^2 + 3x - 8$$

has a minimum value at $x = -2$.

58. Find b such that

$$f(x) = -4x^2 + bx + 3$$

has a maximum value of 50.

59. Find c such that

$$f(x) = -0.2x^2 - 3x + c$$

has a maximum value of -225.

60. Find a quadratic function that has $(4, -5)$ as a vertex and contains the point $(-3, 1)$.

61. Find the vertex of the quadratic function

$$f(x) = qx^2 - 2x + q.$$

62. Find the line of symmetry of the graph of

$$f(x) = -3x^2 + px - 7.$$

63. Find c such that

$$f(x) = x^2 - 6x + c$$

has a vertex on the x-axis.

64. Find b such that

$$f(x) = -0.1x^2 + bx - 5$$

has a vertex on the y-axis.

65. A university is trying to determine what price to charge for football tickets. At a price of $6 per ticket, it averages 70,000 people per game. For every increase of $1, it loses 10,000 people from the average number. Every person at the game spends an average of $1.50 on concessions. What price per ticket should be charged in order to maximize revenue? How many people will attend at that price?

66. Suppose that you are the owner of a 30-unit motel. All units are occupied when you charge $20 a day per unit. For every increase of x dollars in the daily rate, there are x units vacant. Each occupied room costs $2 per day to service and maintain. What should you charge per unit in order to maximize profit?

67. An apple farm yields an average of 30 bushels of apples per tree when 20 trees are planted on an acre of ground.

Each time 1 more tree is planted per acre, the yield decreases 1 bushel per tree due to the extra congestion. How many trees should be planted in order to get the highest yield?

68. When a theater owner charges $3 for admission, there is an average attendance of 100 people. For every $0.10 increase in admission, there is a loss of 1 customer from the average. What admission should be charged in order to maximize revenue?

● **CHALLENGE** _____

69. ▨ A 24-in. piece of string is cut into two pieces. One piece is used to form a circle and the other to form a square. How should the string be cut so that the sum of the areas is a minimum?

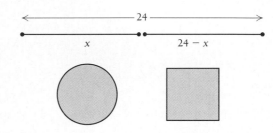

70. Find the dimensions and the area of the largest rectangle that can be inscribed as shown in a right triangle ABC whose sides have lengths 9 cm, 12 cm, and 15 cm.

▨ **TECHNOLOGY CONNECTION** _____

Use a grapher to graph the function. Then find the vertex and the x-intercepts. Round to the nearest tenth.

71. $f(x) = 3.5x^2 - 6.3x - 7.2$

72. $f(x) = -3.7x^2 - 67.34x - 256.4$

73. A model rocket is traveling upward at a velocity of 126.1 m/sec and a height of 109.3 m when its engine burns out. Using a grapher and rounding all values to the nearest tenth, find:

a) the time at which it reaches its maximum height;

b) the maximum height;

c) the time at which it strikes the ground.

OBJECTIVES

You should be able to:

A Find the union and the intersection of sets like $\{1, 2, 3, 4\}$ and $\{3, 4, 5\}$, and graph the union and the intersection of sets like $\{x \mid x > 1\}$ and $\{x \mid x \le 3\}$.

B Solve conjunctions and disjunctions of inequalities.

4.2
SETS, SENTENCES, AND INEQUALITIES

A Intersections and Unions

The **intersection** of two sets consists of those elements common to both sets. Intersection is illustrated in the figure on the left. Note in the figure on the right that the intersection of two sets may be the empty set. The intersection of sets A and B is indicated as $A \cap B$.

The intersection of sets $\{1, 3, 5, 7, 9\}$ and $\{2, 4, 5, 7, 11\}$ is

$$\{1, 3, 5, 7, 9\} \cap \{2, 4, 5, 7, 11\} = \{5, 7\}.$$

Graphs on a number line are set diagrams. They represent the solution set of an equation or inequality. We can find intersections of solution sets using graphs. In the following example, we find the intersection of the set of all x such that 3 is less than x and the set of all x less than or equal to 5. These sets are indicated, and the symbolism is read, as follows:

$\{x \mid 3 < x\}$ "the set of all x such that 3 is less than x."
$\{x \mid x \le 5\}$ "the set of all x such that x is less than or equal to 5."

Example 1   Find the graph of $\{x \mid 3 < x\} \cap \{x \mid x \le 5\}$.

Solution   We first graph the two solution sets separately and then find the intersection.

The open circle at 3 indicates that 3 is not in the solution set. The solid circle at 5 indicates that 5 is in the solution set. The intersection is as follows.

$\{x \mid 3 < x\} \cap \{x \mid x \le 5\}$, or, in interval notation (3, 5]

◀

Example 2   Graph: $\{x \mid -3 \le x\} \cap \{x \mid -1 \le x\}$.

Solution   Again, we graph the solution sets separately and then find the intersection.

$$\{x \mid -3 \le x\}$$

$$\{x \mid -1 \le x\}$$

$$\{x \mid -3 \le x\} \cap \{x \mid -1 \le x\}, \text{ or } [-1, \infty)$$ ◀

DO EXERCISES 1–6.

The **union** of two sets consists of the members that are in one or both of the sets. Union is illustrated in the following figure. The union of sets A and B is indicated as $A \cup B$.

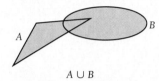

$$A \cup B$$

The union of sets $\{1, 3, 5, 7, 9\}$ and $\{2, 4, 5, 7, 11\}$ is

$$\{1, 3, 5, 7, 9\} \cup \{2, 4, 5, 7, 11\} = \{1, 2, 3, 4, 5, 7, 9, 11\}.$$

In the following examples, we find unions of solution sets. Note that we graph the individual sets separately and then combine them.

Example 3 Graph: $\{x \mid x \ge -1\} \cup \{x \mid x < 2\}$.

Solution

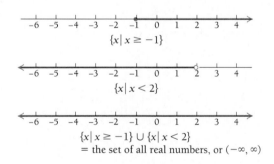

$$\{x \mid x \ge -1\}$$

$$\{x \mid x < 2\}$$

$$\{x \mid x \ge -1\} \cup \{x \mid x < 2\}$$
$$= \text{the set of all real numbers, or } (-\infty, \infty)$$ ◀

Example 4 Graph: $\{x \mid x \le -2\} \cup \{x \mid x > 1\}$.

Solution

$$\{x \mid x \le -2\}$$

$$\{x \mid x > 1\}$$

$$\{x \mid x \le -2\} \cup \{x \mid x > 1\}, \text{ or } (-\infty, -2] \cup (1, \infty)$$ ◀

DO EXERCISES 7–13.

Find the intersection.

1. $\{-3, -4, 2, 3, 4\} \cap$
$\{1, 4, \quad 3, 8, 9, 11\}$

2. $\{2, b, c, d, e\} \cap \{1, 2, d, e, f, g\}$

Graph.

3. $\{x \mid 1 < x\} \cap \{x \mid x \le 3\}$

-6 -5 -4 -3 -2 -1 0 1 2 3 4 5 6

4. $\{x \mid -4 < x\} \cap \{x \mid x < -1\}$

-6 -5 -4 -3 -2 -1 0 1 2 3 4 5 6

5. $\{x \mid 0 \le x\} \cap \{x \mid 2 \le x\}$

-6 -5 -4 -3 -2 -1 0 1 2 3 4 5 6

6. $\{x \mid 1 \le x\} \cap \{x \mid x \le -2\}$

-6 -5 -4 -3 -2 -1 0 1 2 3 4 5 6

Find the union.

7. $\{1, 2\} \cup \{2, 3, 4, 5\}$

8. $\{-3, -4, 2, 3, 4\} \cup$
$\{1, 4, -3, 8, 9, 11\}$

9. $\{2, a, b, c\} \cup \{1, 2, c, d, e\}$

Graph.

10. $\{x \mid x < -3\} \cup \{x \mid x \ge 1\}$

-6 -5 -4 -3 -2 -1 0 1 2 3 4 5 6

11. $\{x \mid x \le -3\} \cup \{x \mid x \le 3\}$

-6 -5 -4 -3 -2 -1 0 1 2 3 4 5 6

12. $\{x \mid x > 0\} \cup \{x \mid x < 1\}$

-6 -5 -4 -3 -2 -1 0 1 2 3 4 5 6

13. $\left\{x \,\middle|\, x > \dfrac{1}{2}\right\} \cup \left\{x \,\middle|\, x = \dfrac{1}{2}\right\}$

-6 -5 -4 -3 -2 -1 0 1 2 3 4 5 6

Abbreviate the conjunction.

14. $4 < x \ and \ x < 8$

15. $-3 \le x \ and \ x < 0$

16. $-5 \le x \ and \ x \le -2$

17. $-1 < x \ and \ x \le -\dfrac{1}{4}$

Rewrite, using the word *and*.

18. $-\dfrac{1}{2} < x < 1$

19. $-\dfrac{17}{3} \le x < -2$

20. $\dfrac{19}{4} \le x \le \dfrac{37}{6}$

 Compound Sentences

When two sentences are joined by the word *and*, a compound sentence is formed. Such a sentence is called a **conjunction.** (The word conjunction used in this way is a logical term; the meaning is not the same as in ordinary grammar.) A conjunction of two sentences is true when both parts are true. Thus the solution set is the intersection of the solution sets of the parts. Consider, for example, the conjunction

$$-2 \le x \ and \ x < 1.$$

Any number that makes this sentence true must make both parts true. We can graph the sentence by graphing the parts and then finding their intersection.

$-2 \le x$
Any number in this set makes the first part true.
(a)

$x < 1$
Any number in this set makes the second part true.
(b)

Any number in this set (the intersection) makes both parts true
($-2 \le x$ and $x < 1$ are both true).
(c)

We often abbreviate certain conjunctions of inequalities. In this case,

$$-2 \le x \ and \ x < 1 \quad \text{is abbreviated} \quad -2 \le x < 1.$$

The latter is read "-2 is less than or equal to x and x is less than 1," or "-2 is less than or equal to x is less than 1." Thus,

$$\{x | -2 \le x\} \cap \{x | x < 1\} = \{x | -2 \le x \ and \ x < 1\}$$
$$= \{x | -2 \le x < 1\}.$$

The word *and* corresponds to set *intersection.*

DO EXERCISES 14–20. _____

Example 5 Solve and graph: $-3 < 2x + 5 < 7$.

Solution

Method 1

$$-3 < 2x + 5 \quad and \quad 2x + 5 < 7 \qquad \text{Rewriting using } and$$
$$-8 < 2x \qquad and \qquad 2x < 2 \qquad \text{Adding } -5$$
$$-4 < x \qquad and \qquad x < 1 \qquad \text{Multiplying by } \tfrac{1}{2}$$

Method 2

$$-3 < 2x + 5 < 7$$
$$-8 < 2x < 2 \qquad \text{Adding } -5$$
$$-4 < x < 1 \qquad \text{Multiplying by } \tfrac{1}{2}$$

The solution set is $\{x|-4 < x\} \cap \{x|x < 1\}$, or $\{x|-4 < x < 1\}$, or $(-4, 1)$. The graph is as follows.

Example 6 Solve and graph: $-4 < \dfrac{5 - 3x}{2} \le 5$.

Solution We have the following:

$-8 < 5 - 3x \le 10$ Multiplying by 2

$-13 < -3x \le 5$ Adding -5

$\dfrac{13}{3} > x \ge -\dfrac{5}{3}$ Multiplying by $-\dfrac{1}{3}$

$-\dfrac{5}{3} \le x < \dfrac{13}{3}$ $x \ge -\dfrac{5}{3}$ means $-\dfrac{5}{3} \le x$, and $\dfrac{13}{3} > x$ means $x < \dfrac{13}{3}$

The solution set is $\{x|-\dfrac{5}{3} \le x < \dfrac{13}{3}\}$, or $[-\dfrac{5}{3}, \dfrac{13}{3})$. The graph is as follows.

DO EXERCISES 21–23.

When two sentences are joined by the word *or*, a compound sentence is formed. Such a sentence is called a **disjunction.** A disjunction of two sentences is true when either part is true. It is also true when both parts are true. The solution set of a disjunction is thus the union of the solution sets of the parts. Consider the disjunction

$x < -2$ *or* $x > \dfrac{1}{4}$.

Any number that makes either or both of the parts true makes the disjunction true. We can graph the sentence by graphing the two parts and then finding their union. The word *or* corresponds to set *union*.

$x < -2$

Any number in this set makes the first part true.

(a)

$x > \dfrac{1}{4}$

Any number in this set makes the second part true.

(b)

$x < -2$ or $x > \dfrac{1}{4}$

$\{x|x < -2\} \cup \{x|x > \dfrac{1}{4}\}$

Any number in this set makes one or both parts true.

(c)

TECHNOLOGY CONNECTION

A grapher can also be used to solve compound inequalities. Using Example 5 as a sample problem, we first graph the three parts of the inequality: $y = -3$, $y = 2x + 5$, and $y = 7$. Make sure your viewing box is large enough to see both intersection points. The standard $[-10, 10] \times [-10, 10]$ box works fine.

$[-10, 10] \times [-10, 10]$

Next, use the Trace feature to determine the coordinates of the point at which $y = -3$ and $y = 2x + 5$ intersect; you should find them to be $(-4, -3)$. Repeating the process for the other intersection, you should find that the coordinates are $(1, 7)$.

From these answers, we have determined that x lies between -4 and $+1$. The final step is deciding whether or not these end values are to be included in the solution. Since the sign used in both cases was $<$, we do *not* include -4 and $+1$ in the solution. The solution set is the interval $(-4, 1)$.

Use a grapher to solve each of the following inequalities.

TC 1. $-5 < x + 3 \le 9$

TC 2. $4 < 2x - 4 < 12$

TC 3. $-5 \le 2 + 2x \le 5$

Solve. Then graph.

21. $-4 < 3x - 2 \le 10$

22. $-1 < \dfrac{5 - 2x}{3} < 1$

23. $\dfrac{2}{3} \le 1 - 2x \le \dfrac{5}{3}$

Solve. Then graph.

24. $x + 4 < -3$ or $x + 4 > 3$

-6 -5 -4 -3 -2 -1 0 1 2 3 4 5 6

25. $2x - 3 \leq -5$ or $2x - 3 > 5$

-6 -5 -4 -3 -2 -1 0 1 2 3 4 5 6

26. $4 - 3x \leq -1$ or $4 - 3x \geq 1$

-6 -5 -4 -3 -2 -1 0 1 2 3 4 5 6

27. $\dfrac{4x + 5}{3} < -2$ or $\dfrac{4x + 5}{3} \geq 2$

-6 -5 -4 -3 -2 -1 0 1 2 3 4 5 6

TECHNOLOGY CONNECTION

This type of problem can be done on a grapher. We will solve Example 7. We begin by graphing three different functions: $y = -7$, $y = 2x - 5$, and $y = 7$.

[–10, 10] x [–10, 10]

We are interested in the coordinates at the intersections of these functions. Using the Trace feature, we can determine the coordinates of the intersections. Specifically, we find that the x-coordinates of the intersections are -1 and $+6$. We want values of x for which the graph of $y = 2x - 5$ lies *below* the graph of $y = -7$ and those for which the graph of $y = 2x - 5$ lies *above* the graph of $y = 7$. These x-values are those to the left of -1 and those to the right of 6. Since the signs were $<$ and $>$, we do *not* include these values in the solution. Therefore, the solution is $(-\infty, -1) \cup (6, +\infty)$.

Use a grapher to solve each of the following inequalities.

TC 4. $x - 8 < -3$ or $x - 8 > 9$

TC 5. $2x + 8 \leq 3$ or $2x + 8 \geq 12$

TC 6. $5(x - 3) \leq 0$ or $5(x - 3) > 20$

CAUTION! There is no compact way to abbreviate disjunctions of inequalities, ordinarily. *Be careful about this!* For example, if you try to abbreviate $-3 < x$ or $x < 4$ as $-3 < x < 4$, you will be *wrong*, because $-3 < x < 4$ is an abbreviation for the conjunction $-3 < x$ *and* $x < 4$. To state this another way, you can write $\{x | -2 \leq x\} \cap \{x | x < 1\}$ as $\{x | -2 \leq x < 1\}$ without the word "and," but you cannot write $\{x | x < -2\} \cup \{x | x > \frac{1}{4}\}$ without the word "or" or without the union symbol $\cup$. Note that $3x \leq 15$ can be written as an abbreviation for $3x < 15$ or $3x = 15$.

Example 7 Solve: $2x - 5 < -7$ or $2x - 5 > 7$. Then graph.

Solution We have

$$2x - 5 < -7 \quad or \quad 2x - 5 > 7$$
$$2x < -2 \quad or \quad 2x > 12 \qquad \text{Adding 5}$$
$$x < -1 \quad or \quad x > 6. \qquad \text{Multiplying by } \tfrac{1}{2}$$

The solution set is $\{x | x < -1 \text{ or } x > 6\}$, which is the union

$$\{x | x < -1\} \cup \{x | x > 6\}, \quad \text{or } (-\infty, -1) \cup (6, \infty).$$

The graph is as follows.

Example 8 Solve: $\dfrac{4 - 3x}{2} < -1$ or $\dfrac{4 - 3x}{2} \geq 1$.

Then graph.

Solution We have

$$\frac{4 - 3x}{2} < -1 \qquad or \qquad \frac{4 - 3x}{2} \geq 1$$
$$4 - 3x < -2 \qquad or \quad 4 - 3x \geq 2$$
$$\qquad\qquad\qquad\qquad\qquad \text{Multiplying by 2}$$
$$4 < -2 + 3x \quad or \qquad 4 \geq 2 + 3x$$
$$\qquad\qquad\qquad\qquad\qquad \text{Adding 3x}$$
$$6 < 3x \qquad or \qquad 2 \geq 3x$$
$$2 < x \qquad or \qquad \frac{2}{3} \geq x.$$

The solution set is

$$\{x | 2 < x \text{ or } \tfrac{2}{3} \geq x\},$$

which is the union

$$\{x | 2 < x\} \cup \{x | \tfrac{2}{3} \geq x\}, \quad \text{or } (2, \infty) \cup (-\infty, \tfrac{2}{3}].$$

The graph is as follows.

DO EXERCISES 24–27.

● **EXERCISE SET** **4.2**

A   Find the union or the intersection.

1. $\{3, 4, 5, 8, 10\} \cap \{1, 2, 3, 4, 5, 6, 7\}$
2. $\{3, 4, 5, 8, 10\} \cup \{1, 2, 3, 4, 5, 6, 7\}$
3. $\{0, 2, 4, 6, 8\} \cup \{4, 6, 9\}$
4. $\{0, 2, 4, 6, 8\} \cap \{4, 6, 9\}$
5. $\{a, b, c\} \cap \{c, d\}$
6. $\{a, b, c\} \cup \{c, d\}$

Graph.

7. $\{x \mid 7 \le x\} \cup \{x \mid x < 9\}$
8. $\{x \mid -\frac{1}{2} \le x\} \cup \{x \mid x < \frac{1}{2}\}$
9. $\{x \mid -\frac{1}{2} \le x\} \cap \{x \mid x < \frac{1}{2}\}$
10. $\{x \mid x > \frac{1}{4}\} \cap \{x \mid 1 \ge x\}$
11. $\{x \mid x < -\pi\} \cup \{x \mid x > \pi\}$
12. $\{x \mid -\pi \le x\} \cap \{x \mid x < \pi\}$
13. $\{x \mid x < -7\} \cup \{x \mid x = -7\}$
14. $\{x \mid x > \frac{1}{2}\} \cup \{x \mid x = \frac{1}{2}\}$
15. $\{x \mid x \ge 5\} \cap \{x \mid x \le -3\}$
16. $\{x \mid x \ge -3\} \cup \{x \mid x \le 0\}$

B   Solve.

17. $-2 \le x + 1 < 4$
18. $-3 < x + 2 \le 5$
19. $5 \le x - 3 \le 7$
20. $-1 < x - 4 < 7$
21. $-3 \le x + 4 \le -3$
22. $-5 < x + 2 < -5$
23. $-2 < 2x + 1 < 5$
24. $-3 \le 5x + 1 \le 3$
25. $-4 \le 6 - 2x < 4$
26. $-3 < 1 - 2x \le 3$
27. $-5 < \frac{1}{2}(3x + 1) \le 7$
28. $\frac{2}{3} \le -\frac{4}{5}(x - 3) < 1$
29. $3x \le -6 \ or \ x - 1 > 0$
30. $2x < 8 \ or \ x + 3 \ge 1$
31. $2x + 3 \le -4 \ or \ 2x + 3 \ge 4$
32. $3x - 1 < -5 \ or \ 3x - 1 > 5$
33. $2x - 20 < -0.8 \ or \ 2x - 20 > 0.8$
34. $5x + 11 \le -4 \ or \ 5x + 11 \ge 4$
35. $x + 14 \le -\frac{1}{4} \ or \ x + 14 \ge \frac{1}{4}$
36. $x - 9 < -\frac{1}{2} \ or \ x - 9 > \frac{1}{2}$

37. ▦ The length of a rectangle is 15.23 cm. What widths will give a perimeter greater than 40.23 cm and less than 137.8 cm?

38. ▦ The height of a triangle is 15 m. What lengths of the base will keep the area less than or equal to 305.4 m² (and, of course, positive)?

39. To get an A in a course, a student's average must be greater than or equal to 90%. It will, of course, be less than or equal to 100%. On the first three tests, a student scored 83%, 87%, and 93%. What scores on the fourth test will produce an A? Is an A possible?

40. In Exercise 39, suppose that the scores on the first three

tests are 75%, 70%, and 83%. What scores on the fourth test will produce an A? Is an A possible?

41. **Temperatures of liquids.** We can use the function $C(F) = \frac{5}{9}(F - 32)$ to convert Fahrenheit temperatures F to Celsius temperatures C. Copper is a liquid at Celsius temperatures C such that $1083° \le C < 2580°$. Find such an inequality for the corresponding Fahrenheit temperatures.

42. **Pressure at sea depth.**   The pressure P, given by
$$P(d) = 1 + \tfrac{1}{33}d,$$
gives the pressure, in atmospheres (atm), at a depth d, in feet, in the sea. Find the depths d for which the pressure P satisfies $2 \le P \le 10$.

43. **Toll road charges.**   The function C, given by
$$C(d) = 0.027d + 0.32,$$
can be used to estimate the cost C, in dollars, of driving a car d miles on the Indiana toll road. Find the distances d if the cost of driving on the toll road is between \$2 and \$4.

44. **Straight-line depreciation.**   A company buys an office machine for \$5200 on January 1 of a given year. The machine is expected to last 8 years, after which its *trade-in*, or *salvage, value* will be \$1100. If the company figures the decline or depreciation in value to be the same each year, then the salvage value V, after t years, $0 \le t \le 8$, is given by the linear function
$$V(t) = \$5200 - \$512.50t.$$
For what time period is the value of the machine between \$2000 and \$4000?

● **SYNTHESIS** _____

Solve.

45. $x \le 3x - 2 \le 2 - x$
46. $2x \le 5 - 7x < 7 + x$
47. $(x + 1)^2 > x(x - 3)$
48. $(x + 4)(x - 5) > (x + 1)(x - 7)$
49. $(x + 1)^2 \le (x + 2)^2 \le (x + 3)^2$
50. $(x - 1)(x + 1) < (x + 1)^2 \le (x - 3)^2$

Find the domain of the function.

51. $f(x) = \dfrac{\sqrt{x + 2}}{\sqrt{x - 2}}$
52. $f(x) = \dfrac{\sqrt{3 - x}}{\sqrt{x + 5}}$

▦▦ **TECHNOLOGY CONNECTION** _____

Use a grapher to solve the inequality.

53. $-9.7 < x - 4.5 \le 4.5$
54. $3.4 \le 3.4x + 1.7 < 15.6$
55. $2.3x - 4 < -4.5 \ or \ 2.3x - 4 \ge 10.3$
56. $9.3(1.8x + 4) \le 1.2 \ or \ 9.3(1.8x + 4) > 17.8$

4.3

EQUATIONS AND INEQUALITIES WITH ABSOLUTE VALUE

 Recall the definition of absolute value:

$$|x| = \begin{cases} x, & \text{if } x \geq 0, \\ -x, & \text{if } x < 0. \end{cases}$$

An informal way of thinking of **absolute value** is that it is the distance from 0 of a number on a number line. For example, $|4|$ is 4 because 4 is 4 units from 0; $|-5|$ is 5 because -5 is 5 units from 0. This idea is helpful in solving *equations* and *inequalities with absolute value*.

Solve and graph.

1. $|x| = 5$

Example 1   Solve: $|x| = 3$.

Solution   We look for all numbers x whose distance from 0 is 3. There are two of them, so there are two solutions, 3 and -3. The solution set is $\{-3, 3\}$. The graph is as follows.

2. $|x| = 3.5$

DO EXERCISES 1 AND 2.

Solve and graph.

3. $|x| < 5$

Example 2   Solve: $|x| < 3$.

Solution   This time, we look for all numbers x whose distance from 0 is less than 3. These are the numbers between -3 and 3. The solution set and its graph are as follows.

$\{x \mid -3 < x < 3\}$, or $(-3, 3)$

4. $|x| \leq 3.5$

DO EXERCISES 3 AND 4.

Solve and graph.

5. $|x| \geq 5$

Example 3   Solve: $|x| \geq 3$.

Solution   This time, we look for all numbers x whose distance from 0 is 3 or greater. The solution set and its graph are as follows.

$\{x \mid x \leq -3 \text{ or } x \geq 3\}$, or $(-\infty, -3] \cup [3, \infty)$

6. $|x| > 3.5$

DO EXERCISES 5 AND 6.

The results of the above examples can be generalized as follows.

For $a > 0$:

	Sentence	Solution Set
A1)	$|x| = a$	$\{-a, a\}$
A2)	$|x| < a$	$\{x \mid -a < x < a\}$, or $(-a, a)$
A3)	$|x| > a$	$\{x \mid x < -a \text{ or } x > a\}$, or $(-\infty, -a) \cup (a, \infty)$

Similar statements hold true for $|x| \leq a$ and $|x| \geq a$.

Example 4   Solve: $|x - 2| = 3$.

Solution   Note that this is a translation of $|x| = 3$, two units to the right. We first graph $|x| = 3$.

This consists of the numbers that are a distance of 3 from 0. We now translate.

This solution set consists of the numbers that are a distance of 3 from 2 (note that 2 is where 0 went in the translation). The solutions of $|x - 2| = 3$ are -1 and 5. The solution set is $\{-1, 5\}$.   ◄

The results of Example 4 can be generalized as follows.

For any x, $|x - a|$ is the distance between a and x.

Example 5

a) $|x - 5|$ is the distance between 5 and x.
b) $|x + 7|$ is the distance between -7 and x
    [because $x + 7 = x - (-7)$].   ◄

Here are some further examples of solving inequalities. In Example 6, we use two methods. The first provides understanding. The second is the most efficient for general solving.

Example 6   Solve: $|x + 2| < 3$.

Solution

Method 1.   An equivalent inequality is $|x - (-2)| < 3$. The solutions are those numbers x whose distance from -2 is less than 3. Thus to find the solutions graphically, we locate -2. Then we locate those numbers that

Solve. Use two methods.

7. $|x + 1| = 4$

8. $|x + 7| < 2$

9. $|x - 3| > 2$

Solve.

10. $|3x - 4| = 7$

11. $\left|5x + \dfrac{1}{2}\right| < 1$

12. $|3x - 4| > 7$

---

are less than 3 units to the left and less than 3 units to the right. Thus the solution set is $\{x \mid -5 < x < 1\}$, or $(-5, 1)$.

Method 2. We use property A2, replacing x by $x + 2$:

$$|x + 2| < 3$$
$$-3 < x + 2 < 3 \qquad \text{Property A2}$$
$$-5 < x < 1. \qquad \text{Adding } -2$$

The solution set is $\{x \mid -5 < x < 1\}$, or $(-5, 1)$. ◀

DO EXERCISES 7–9. _____

The sentences in the next examples are more complicated. They can be solved more efficiently if we use properties A1–A3.

Example 7   Solve: $|2x + 3| = 1$.

Solution   We have

$$2x + 3 = -1 \quad or \quad 2x + 3 = 1 \qquad \text{Property A1}$$
$$2x = -4 \quad or \qquad 2x = -2 \qquad \text{Adding } -3$$
$$x = -2 \quad or \qquad\quad x = -1. \qquad \text{Multiplying by } \tfrac{1}{2}$$

The solution set is $\{-2, -1\}$. ◀

Example 8   Solve: $|2x + 3| \leq 1$.

Solution   We have

$$-1 \leq 2x + 3 \leq 1 \qquad \text{Property A2}$$
$$-4 \leq 2x \leq -2 \qquad \text{Adding } -3$$
$$-2 \leq x \leq -1. \qquad \text{Multiplying by } \tfrac{1}{2}$$

The solution set is $\{x \mid -2 \leq x \leq -1\}$, or $[-2, -1]$. ◀

Example 9   Solve: $|3 - 4x| > 2$.

Solution   We have

$$3 - 4x < -2 \quad or \quad 3 - 4x > 2 \qquad \text{Property A3}$$
$$-4x < -5 \quad or \qquad -4x > -1 \qquad \text{Adding } -3$$
$$x > \tfrac{5}{4} \quad or \qquad\quad x < \tfrac{1}{4} \qquad \text{Multiplying by } -\tfrac{1}{4}$$

Note that the inequality signs must also be reversed in the preceding step. The solution set is $\{x \mid x < \tfrac{1}{4} \ or \ x > \tfrac{5}{4}\}$, or $(-\infty, \tfrac{1}{4}) \cup (\tfrac{5}{4}, \infty)$. ◀

DO EXERCISES 10–12. _____

Example 10   Solve: $|5x - 3| = |x + 4|$.

Solution Consider $|a| = |b|$. This asserts that a and b are the same distance from 0. Thus they are either the same number or opposites of each other. Since $|5x - 3| = |x + 4|$, either $5x - 3 = x + 4$ or $5x - 3 = -(x + 4)$. We now solve each equation separately:

$$5x - 3 = x + 4 \quad \text{or} \quad 5x - 3 = -(x + 4)$$
$$4x = 7 \quad \text{or} \quad 5x - 3 = -x - 4$$
$$x = \tfrac{7}{4} \quad \text{or} \quad 6x = -1$$
$$x = \tfrac{7}{4} \quad \text{or} \quad x = -\tfrac{1}{6}.$$

The solution set is $\left\{ -\tfrac{1}{6}, \tfrac{7}{4} \right\}$. ◀

Example 11 Solve: $|x - 3| = |x + 10|$.

Solution We have

$$x - 3 = x + 10 \quad \text{or} \quad x - 3 = -(x + 10)$$
$$-3 = 10 \quad \text{or} \quad x - 3 = -x - 10$$
$$-3 = 10 \quad \text{or} \quad 2x = -7$$
$$-3 = 10 \quad \text{or} \quad x = -\tfrac{7}{2}.$$

The first equation has no solution. The second equation has solution $-\tfrac{7}{2}$. Thus the solution set is $\varnothing \cup \left\{ -\tfrac{7}{2} \right\}$, or $\left\{ -\tfrac{7}{2} \right\}$. ◀

DO EXERCISES 13 AND 14.

Example 12 Solve: $|2x - 3| \le 4 + 5x$.

Solution We have

$$|2x - 3| \le 4 + 5x$$
$$-(4 + 5x) \le 2x - 3 \le 4 + 5x \qquad \text{Property A2}$$
$$-4 - 5x \le 2x - 3 \le 4 + 5x$$
$$-4 \le 7x - 3 \le 4 + 10x.$$

Note that we have a variable remaining in two parts of the inequality. Any addition, to get the variable alone, must be done to all three parts and will reintroduce the variable on the left side. To deal with this, we write the conjunction, solve each inequality separately, and then find the intersection of the solution sets:

$$-4 \le 7x - 3 \quad \text{and} \quad 7x - 3 \le 4 + 10x$$
$$-1 \le 7x \quad \text{and} \quad -3 \le 4 + 3x$$
$$-\tfrac{1}{7} \le x \quad \text{and} \quad -7 \le 3x$$
$$-\tfrac{1}{7} \le x \quad \text{and} \quad -\tfrac{7}{3} \le x.$$

The solution set of $-\tfrac{1}{7} \le x$ is the interval $\left[-\tfrac{1}{7}, \infty \right)$. The solution set of $-\tfrac{7}{3} \le x$ is $\left[-\tfrac{7}{3}, \infty \right)$. The solution set of the conjunction is the intersection of the intervals:

$$\left[-\tfrac{1}{7}, \infty \right) \cap \left[-\tfrac{7}{3}, \infty \right) = \left[-\tfrac{1}{7}, \infty \right).$$ ◀

DO EXERCISES 15 AND 16.

Solve.

13. $|2x - 3| = |x + 5|$

14. $|x - 5| = |x + 8|$

Solve.

15. $|5 - 6x| \le 10 + 7x$

16. $|5 - 6x| \ge 10 + 7x$

TECHNOLOGY CONNECTION

All the problems in Section 4.3 can be solved using a grapher. We will use $|x + 3|$ and 5 as our two functions, and examine three problems using them:

a) $|x + 3| = 5$; b) $|x + 3| < 5$;
c) $|x + 3| \ge 5$.

In each case, we begin by graphing $y = |x + 3|$ and $y = 5$. (The absolute value function is usually abbreviated ABS on graphers, and parentheses are usually required around the entire expression. Thus $|x + 3|$ becomes ABS$(x + 3)$ on most graphers.)

$[-10, 10] \times [-10, 10]$

In (a) we are interested only in the points at which the graphs intersect. Using the Trace feature, we find the x-coordinates of these intersections to be -8 and 2. Thus the solutions are $x = -8$ and $x = 2$.

In (b) we are looking for the values of x for which the graph of $|x + 3|$ lies *below* ($<$) the graph of $y = 5$. Clearly, that occurs when x is between -8 and 2. Thus the solution is the interval $(-8, 2)$.

In (c) we want to find the values of x for which the graph of $y = |x + 3|$ lies *on or above* ($\ge$) the graph of $y = 5$. The solution in this case is $(-\infty, -8] \cup [2, +\infty)$.

Note: The only drawback to this method is that whenever your answers are nonintegers, you will find the decimal approximations for the values instead of the exact fraction (unless you recognize the fraction from the decimal form). For example, the equation $|6x + 5| = 6$ would give the approximate solutions $x = -1.83$ and $x = 0.17$ instead of the exact solutions $x = -\tfrac{11}{6}$ and $x = \tfrac{1}{6}$.

Use a grapher to solve each of the following.

TC 1. $|x - 6| = 9$ **TC 2.** $|2x - 6| \ge 3$
TC 3. $|x + 1| < 4$

• EXERCISE SET   4.3

A Solve and graph.

1. $|x| = 7$
2. $|x| = 4.5$
3. $|x| < 7$
4. $|x| \leq 4.5$
5. $|x| \geq 4.5$
6. $|x| > 7$

Solve. Use two methods.

7. $|x - 1| = 4$
8. $|x - 7| = 5$
9. $|x + 8| < 9$
10. $|x + 6| \leq 10$
11. $|x + 8| \geq 9$
12. $|x + 6| > 10$
13. $|x - \frac{1}{4}| < \frac{1}{2}$
14. $|x - 0.5| \leq 0.2$

Solve. Use any method.

15. $|3x| = 1$
16. $|5x| = 4$
17. $|3x + 2| = 1$
18. $|7x - 4| = 8$
19. $|3x| < 1$
20. $|5x| \leq 4$
21. $|2x + 3| \leq 9$
22. $|2x + 3| < 13$
23. $|x - 5| > 0.1$
24. $|x - 7| \geq 0.4$
25. $\left| x + \frac{2}{3} \right| \leq \frac{5}{3}$
26. $\left| x + \frac{3}{4} \right| < \frac{1}{4}$
27. $|6 - 4x| \leq 8$
28. $|5 - 2x| > 10$
29. $\left| \frac{2x + 1}{3} \right| > 5$
30. $\left| \frac{3x + 2}{4} \right| \leq 5$
31. $\left| \frac{13}{4} + 2x \right| > \frac{1}{4}$
32. $\left| \frac{5}{6} + 3x \right| < \frac{7}{6}$
33. $\left| \frac{3 - 4x}{2} \right| \leq \frac{3}{4}$
34. $\left| \frac{2x - 1}{3} \right| \geq \frac{5}{6}$
35. $|x| = -3$
36. $|x| < -3$
37. $|2x - 4| < -5$
38. $|3x + 5| < 0$
39. ▣ $|x + 17.217| > 5.0012$
40. ▣ $|x - 2.0245| < 0.1011$
41. ▣ $|-2.1437x + 7.8814| \geq 9.1132$
42. ▣ $|3.0147x - 8.9912| \leq 6.0243$
43. $|2x - 8| = |x + 3|$
44. $|x - 7| = |3x + 4|$
45. $\left| \frac{2x + 3}{6} \right| = \left| \frac{4 - 5x}{8} \right|$
46. $\left| \frac{2}{3} - \frac{5}{6}x \right| = \left| \frac{3}{4} + \frac{3}{5}x \right|$
47. $|x - 2| = x - 2$
48. $|4x - 5| = x + 1$
49. $|7x - 2| = x + 5$
50. $|x - 3| = 3x - 8$
51. $|3x - 4| \leq 2x + 1$
52. $|5x + 8| \leq 4 - 3x$
53. $|3x - 4| \geq 2x + 1$
54. $|5x + 8| \geq 4 - 3x$

• SYNTHESIS

Solve.

55. $|4x - 5| = |x| + 1$
56. $|2x + 3| = |x| + 8$
57. $||x| - 1| = 3$
58. $|x + 2| > x$
59. $|x + 2| \leq |x - 5|$
60. $|3x - 1| > 5x - 2$
61. $|x| + |x - 1| < 10$
62. $|x| - |x - 3| < 7$
63. $|x - 3| + |2x + 5| > 6$
64. $|p - 4| + |p + 4| < 8$

• CHALLENGE

65. Solve: $|x - 3| + |2x + 5| + |3x - 1| = 12$.
66. Solve for x: $|x - x_0| < \delta$.

Prove the following for any real numbers a and b.

67. $-|a| \leq a \leq |a|$
68. $|a + b| \leq |a| + |b|$ (the triangle inequality)
69. Show that if $|a| < \epsilon/2$ and $|b| < \epsilon/2$, then $|a + b| < \epsilon$. (*Note:* ϵ is the Greek letter "epsilon.")
70. **a)** Prove that

$$\left| x - \frac{a + b}{2} \right| < \frac{b - a}{2}$$

is equivalent to $a < x < b$.

Use graphs or the result of part (a) to find an inequality with absolute value for each of the following.

b) $-5 < x < 5$
c) $-6 < x < 6$
d) $-1 < x < 7$
e) $-5 < x < 1$

71. Use absolute value to prove that the number halfway between a and b is $(a + b)/2$.
72. Solve for $f(x)$: $|f(x) - L| < \epsilon$.

▣ TECHNOLOGY CONNECTION

73.–82. Using a grapher, check the solution to each of Exercises 55–64.

4.4

POLYNOMIAL AND RATIONAL INEQUALITIES

A Polynomial Inequalities

Inequalities such as the following are called **quadratic inequalities:**

$$x^2 + 3x - 10 < 0, \qquad 5x^2 - 3x + 2 \geq 0.$$

In each case there is a polynomial of degree 2 on the left. The following are other **polynomial inequalities:**

$$x^3 - x < 0, \qquad x^3 - x \geq 0.$$

If we can easily factor the related polynomials and find the x-intercepts of the graphs, then we can solve the inequalities. Let us consider the graph of the function f given by

$$f(x) = x^3 - x = x(x^2 - 1) = x(x + 1)(x - 1).$$

We graphed this function in Example 1 of Section 3.5. Because we are able to factor the polynomial, we can see that the x-intercepts occur at $x = -1$, $x = 0$, and $x = 1$. These intercepts lead us to natural divisions of the number line on which the function is either positive, negative, or zero.

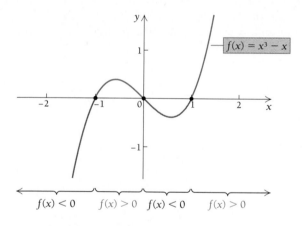

Solution sets of certain inequalities follow:

Inequality	Solution Set
$x^3 - x < 0$	$\{x \mid x < -1\} \cup \{x \mid 0 < x < 1\}$, or $(-\infty, -1) \cup (0, 1)$
$x^3 - x \leq 0$	$\{x \mid x \leq -1\} \cup \{x \mid 0 \leq x \leq 1\}$, or $(-\infty, -1] \cup [0, 1]$
$x^3 - x > 0$	$\{x \mid -1 < x < 0\} \cup \{x \mid x > 1\}$, or $(-1, 0) \cup (1, \infty)$
$x^3 - x \geq 0$	$\{x \mid -1 \leq x \leq 0\} \cup \{x \mid x \geq 1\}$, or $[-1, 0] \cup [1, \infty)$

Note that when an inequality contains $\leq$ or $\geq$, the x-values of the intercepts must be included.

Solve.

1. $x^2 + 2x - 3 < 0$

2. $x^2 + 2x - 3 \le 0$

3. $x^2 + 2x > 3$

4. $x^2 + 2x \ge 3$

The intercepts divide the number line into intervals. If a particular function has a positive output for one number in an interval, it will be positive for all the numbers in the interval. Thus to know whether a function is positive or negative on an interval, we can merely make a test substitution in each interval in order to solve an inequality. This leads us to our method for solving polynomial inequalities.

To solve a polynomial inequality:

1. Get 0 on one side, set the polynomial on the other side equal to 0, factor, and solve to find the intercepts.
2. Use the numbers found in step (1) to divide the number line into intervals.
3. Substitute a number from each interval into the related function. If the value is positive, then the function will be positive for all numbers in the interval. If the value is negative, then the function will be negative for all numbers in the interval.
4. Select the intervals for which the inequality is satisfied and write set-builder notation or interval notation for the solution set. Include the intercepts in the solution sets if the inequality sign is $\le$ or $\ge$.

Example 1   Solve: $x^2 + 3x - 10 < 0$.

Solution   We set the polynomial equal to 0 and solve. The solutions of $x^2 + 3x - 10 = 0$, or $(x + 5)(x - 2) = 0$, are -5 and 2. We locate them on a number line as follows. Note that the numbers divide the number line into three intervals A, B, and C.

We choose a test number in interval A, say -7, and substitute it for x in the function $f(x) = x^2 + 3x - 10$:

$$f(-7) = (-7)^2 + 3(-7) - 10$$
$$= 49 - 21 - 10 = 18.$$

Note that $f(-7) > 0$, so the function will be positive for any number in interval A. Next we try a test number in interval B, say 1, and find $f(1)$:

$$f(1) = 1^2 + 3(1) - 10 = -6.$$

Note that $f(1) < 0$, so the function will be negative for any number in interval B. Next we try a test number in interval C, say 4, and find $f(4)$:

$$f(4) = 4^2 + 3(4) - 10$$
$$= 16 + 12 - 10 = 18.$$

Since $f(4) > 0$, the function will be positive for any number in interval C. We are looking for numbers x for which $x^2 + 3x - 10 < 0$. Thus any number x in interval B is a solution. The solution set is $\{x | -5 < x < 2\}$, or $(-5, 2)$.   ◀

DO EXERCISES 1–4.

Example 2   Solve: $7x(x + 3)(x - 2) \geq 0$.

Solution   The solutions of $f(x) = 7x(x + 3)(x - 2) = 0$ are $-3, 0$, and 2. They divide the real-number line into four intervals as follows.

We try a test number in each interval:

A: Test -5, $f(-5) = 7(-5)(-5 + 3)(-5 - 2) = -490$;

B: Test -2, $f(-2) = 7(-2)(-2 + 3)(-2 - 2) = 56$;

C: Test 1, $f(1) = 7(1)(1 + 3)(1 - 2) = -28$;

D: Test 3, $f(3) = 7(3)(3 + 3)(3 - 2) = 126$.

Function values are positive in intervals B and D. Since the inequality symbol is $\geq$, we need to include the intercepts. Thus the solution set of the inequality is

$$\{x \mid -3 \leq x \leq 0 \text{ or } 2 \leq x\}, \quad \text{or } [-3, 0] \cup [2, \infty).$$

◀

DO EXERCISE 5.

Rational Inequalities

We can adapt the preceding method when an inequality involves rational expressions. We call these **rational inequalities.**

Example 3   Solve: $\dfrac{x - 3}{x + 4} \geq 2$.

Solution   We first write the related equation by changing the $\geq$ symbol to $=$:

$$\frac{x - 3}{x + 4} = 2.$$

Then we solve by multiplying on both sides of the equation by the LCM, which is $x + 4$:

$$(x + 4) \cdot \frac{x - 3}{x + 4} = (x + 4) \cdot 2$$

$$x - 3 = 2x + 8$$

$$-11 = x.$$

In the case of rational inequalities, we also need to determine those replacements that are not meaningful—that is, those that make the denominator 0. We set the denominator equal to 0 and solve:

$$x + 4 = 0$$

$$x = -4.$$

Now we use the numbers -11 and -4 to divide the number line into intervals as follows:

5. Solve: $4x(x + 1)(x - 1) < 0$.

TECHNOLOGY CONNECTION

Using a grapher to solve a polynomial inequality involves 3 steps:

1. Draw the graph, using a viewing box large enough to include all the x-intercepts.
2. Zoom in on each x-intercept to determine its value to whatever accuracy is desired.
3. Examine the graph to determine where it has the desired values.

Using a grapher, solve each of the following inequalities.

TC 1. $-0.88x^2 + 4.14x - 3.56 \leq 0$

TC 2. $3x^3 + 5x^2 - 10x - 12 > 0$

TC 3. $x^4 - 3x^3 + 5x^2 - 9x - 10 < 0$

We try test numbers in each interval to see if each satisfies the original inequality.

A: Test -15, $\dfrac{x-3}{x+4} \geq 2$

$$\dfrac{\dfrac{-15-3}{-15+4}}{} \Bigg| \; 2$$

$$\dfrac{-18}{-11}$$

$$\dfrac{18}{11} \Bigg| \quad \text{FALSE}$$

Since $\frac{18}{11} \geq 2$ is a false inequality, the number -15 is not a solution of the inequality, and the interval A is not part of the solution set.

B: Test -8, $\dfrac{x-3}{x+4} \geq 2$

$$\dfrac{\dfrac{-8-3}{-8+4}}{} \Bigg| \; 2$$

$$\dfrac{-11}{-4}$$

$$\dfrac{11}{4} \Bigg| \quad \text{TRUE}$$

Since $\frac{11}{4} \geq 2$ is a true inequality, the number -8 is a solution of the inequality, and the interval B is part of the solution set.

C: Test 1, $\dfrac{x-3}{x+4} \geq 2$

$$\dfrac{\dfrac{1-3}{1+4}}{} \Bigg| \; 2$$

$$\dfrac{-2}{5}$$

$$-\dfrac{2}{5} \Bigg| \quad \text{FALSE}$$

Since $-\frac{2}{5} \geq 2$ is a false inequality, the number 1 is not a solution of the inequality, and the interval C is not part of the solution set.

The solution set includes the interval B. The number -11 is also included, since the inequality symbol is $\geq$ and -11 is a solution of the related equation. The number -4 is not included, since it is not a meaningful replacement. Thus the solution set of the original inequality is

$$\{x \mid -11 \leq x < -4\}, \quad \text{or} \quad [-11, -4). \qquad \blacktriangleleft$$

TECHNOLOGY CONNECTION

Solving rational inequalities on a grapher is basically the same as solving polynomial inequalities, as discussed in the previous section, except that an extra step is required before using the grapher.

Set the denominator of the function equal to 0 and solve for x. These are the values for which the rational function is not defined. They should appear on the graph as gaps. (On some graphers, vertical lines appear rather than gaps.)

If your grapher offers the choice between plotting only individual dots and plotting the dots and then connecting them with short segments, it is better to select the dots only option when dealing with rational equations or inequalities. Then the "gap" discussed above shows up better.

Using a grapher, solve each of the following inequalities.

TC 4. $\dfrac{x-4}{x} \geq 6$

TC 5. $\dfrac{x^2}{x-9} < -2$

TC 6. $\dfrac{x^2+8}{x^2+x+1} > 8$

To solve a rational inequality:

1. Change the inequality symbol to an equals sign and solve the related equation.
2. Find the replacements that are not meaningful.
3. Use the numbers found in steps (1) and (2) to divide the number line into intervals.
4. Substitute a number from each interval into the inequality. If the number is a solution, then the interval to which it belongs is part of the solution set.
5. Select the intervals for which the inequality is satisfied and write set-builder notation or interval notation for the solution set. If the inequality symbol is ≤ or ≥, then the solutions to step (1) should also be included in the solution set.

Solve.

6. $\dfrac{x+1}{x-2} \geq 3$

7. $\dfrac{x}{x-5} < 2$

DO EXERCISES 6 AND 7.

● EXERCISE SET

A Solve. Use interval notation for the answer, where appropriate.

1. $(x+5)(x-3) > 0$
2. $(x+4)(x-1) > 0$
3. $(x-1)(x+2) \leq 0$
4. $(x+5)(x-3) \leq 0$
5. $x^2 + x - 2 < 0$
6. $x^2 - x - 2 < 0$
7. $x^2 \geq 1$
8. $x^2 < 25$
9. $9 - x^2 \leq 0$
10. $4 - x^2 \geq 0$
11. $x^2 - 2x + 1 \geq 0$
12. $x^2 + 6x + 9 < 0$
13. $x^2 + 8 < 6x$
14. $x^2 - 12 > 4x$
15. $4x^2 + 7x < 15$
16. $4x^2 + 7x \geq 15$
17. $2x^2 + x > 5$
18. $2x^2 + x \leq 2$
19. $3x(x+2)(x-2) < 0$
20. $5x(x+1)(x-1) > 0$
21. $(x+3)(x-2)(x+1) > 0$
22. $(x-1)(x+2)(x-4) < 0$
23. $(x+3)(x+2)(x-1) < 0$
24. $(x-2)(x-3)(x+1) < 0$

B Solve. Use interval notation for the answer, where appropriate.

25. $\dfrac{1}{4-x} < 0$
26. $\dfrac{-4}{2x+5} > 0$
27. $3 < \dfrac{1}{x}$
28. $\dfrac{1}{x} \leq 5$
29. $\dfrac{3x+2}{x-3} > 0$
30. $\dfrac{5-2x}{4x+3} < 0$
31. $\dfrac{x+2}{x} \leq 0$
32. $\dfrac{x}{x-3} \geq 0$

33. $\dfrac{x+1}{2x-3} \geq 1$
34. $\dfrac{x-1}{x-2} \geq 3$
35. $\dfrac{x+1}{x+2} \leq 3$
36. $\dfrac{x+1}{2x-3} \leq 1$
37. $\dfrac{x-6}{x} > 1$
38. $\dfrac{x}{x+3} > -1$
39. $(x+1)(x-2) > (x+3)^2$
40. $(x-4)(x+3) > (x-1)^2$
41. $x^3 - x^2 > 0$
42. $x^3 - 4x > 0$
43. $x + \dfrac{4}{x} > 4$
44. $\dfrac{1}{x^2} \leq \dfrac{1}{x^3}$
45. $\dfrac{1}{x^3} \leq \dfrac{1}{x^2}$
46. $x + \dfrac{1}{x} > 2$
47. $\dfrac{2 + x - x^2}{x^2 + 5x + 6} < 0$
48. $\dfrac{4}{x^2} - 1 > 0$

● **SYNTHESIS**

Solve.

49. $x^4 - 2x^2 \leq 0$
50. $x^4 - 3x^2 > 0$
51. $\left|\dfrac{x+3}{x-4}\right| < 2$
52. $|x^2 - 5| = 5 - x^2$
53. $(7-x)^{-2} < 0$
54. $(1-x)^3 > 0$
55. $\left|1 + \dfrac{1}{x}\right| < 3$
56. $(x+5)^{-2} > 0$
57. $|x|^2 - 4|x| + 4 \geq 9$
58. $2|x|^2 - |x| + 2 \leq 5$
59. $\left|2 - \dfrac{1}{x}\right| \leq 2 + \left|\dfrac{1}{x}\right|$

60. $\dfrac{(x-2)^2(x-3)^3(x+1)}{(x+2)(4-x)} \geq 0$

61. $|x^2 + 3x - 1| < 3$ (*Hint:* Solve
$$-3 < x^2 + 3x - 1 < 3.)$$

62. $|1 + 5x - x^2| \geq 5$ (*Hint:* Solve the disjunction
$$1 + 5x - x^2 \leq -5 \ or \ 1 + 5x - x^2 \geq 5.)$$

63. The base of a triangle is 4 cm greater than the height. Find the possible heights h such that the area of the triangle will be greater than 10 cm^2.

64. The length of a rectangle is 3 m greater than the width. Find the possible widths w such that the area of the rectangle will be greater than 15 m^2.

65. ***Total profit.*** A company determines that its total-profit function is given by
$$P(x) = -3x^2 + 630x - 6000.$$

a) A company makes a profit for those nonnegative values of x for which $P(x) > 0$. Find the values of x for which the company makes a profit.

b) A company loses money for those nonnegative values of x for which $P(x) < 0$. Find the values of x for which the company loses money.

66. ***Height of a thrown object.*** The function
$$s(t) = -16t^2 + 32t + 1920$$
gives the height s of an object thrown from a cliff 1920 ft high, after time t, in seconds.

a) For what times is the height greater than 1920 ft?

b) For what times is the height less than 640 ft?

67. A company has the following total-cost and total-revenue functions to use in producing and selling x units of a certain product:
$$R(x) = 50x - x^2, \qquad C(x) = 5x + 350.$$

a) **Break-even values** are those values of x for which $R(x) = C(x)$. Find the break-even values.

b) Find the values of x that produce a profit. That is, find x such that $R(x) > C(x)$.

c) Find the values of x that result in a loss. That is, find x such that $R(x) < C(x)$.

68. A company has the following total-cost and total-revenue functions to use in producing and selling x units of a certain product:
$$R(x) = 80x - x^2, \qquad C(x) = 10x + 600.$$

a) Find the break-even values.

b) Find the values of x that produce a profit.

c) Find the values of x that result in a loss.

69. Find the numbers k for which the quadratic equation $x^2 + kx + 1 = 0$ has (a) two real-number solutions and (b) no real-number solution.

70. Find the numbers k for which the quadratic equation $2x^2 - kx + 1 = 0$ has (a) two real-number solutions and (b) no real-number solution.

Find the domain of the function.

71. $f(x) = \sqrt{1 - x^2}$

72. $f(x) = \dfrac{1}{\sqrt{1 - x^2}}$

73. $g(x) = \sqrt{x^2 + 2x - 3}$

74. $g(x) = \sqrt{4 - x^2}$

 TECHNOLOGY CONNECTION _____

Using a grapher, graph the function and find solutions of $f(x) = 0$. These numbers are often called **roots,** or **zeros.** Then solve the inequalities $f(x) < 0$ and $f(x) > 0$.

75. $f(x) = x^3 - 2x^2 - 5x + 6$

76. $f(x) = \dfrac{1}{3}x^3 - x + \dfrac{2}{3}$

77. $f(x) = x + \dfrac{1}{x}$

78. $f(x) = x - \sqrt{x}, x \geq 0$

79. $f(x) = x^4 - 4x^3 - x^2 + 16x - 12$

80. $f(x) = \dfrac{x^3 + x^2 - 2x}{x^2 + x - 6}$

OBJECTIVE

You should be able to:

Sketch a graph of a polynomial function.

4.5

GRAPHS OF POLYNOMIAL FUNCTIONS

Graphing Polynomial Functions

Graphs of first-degree linear polynomial functions are lines; graphs of second-degree quadratic functions are parabolas. We now consider polynomials of higher degree.

Example 1 Sketch a graph of each of the functions

$$f(x) = x^2 \quad \text{and} \quad g(x) = x^3$$

using the same set of axes. Find the domain and the range of each function.

Solution Since $f(-x) = (-x)^2 = x^2 = f(x)$, we know that f is an even function, which means that it is symmetric with respect to the y-axis. Thus any time we have a point (a, b) on the graph, we know that $(-a, b)$ is also on the graph. Since $g(-x) = (-x)^3 = -x^3 = -g(x)$, we know that g is an odd function, which means that its graph is symmetric with respect to the origin. Thus any time we have a point (a, b) on the graph, we know that $(-a, -b)$ is also on the graph. We compute some function values and use them and the symmetry ideas to find others. Then we complete the graphs.

x	$f(x) = x^2$	$g(x) = x^3$
-2	4	-8
-1	1	-1
$-\frac{1}{2}$	$\frac{1}{4}$	$-\frac{1}{8}$
0	0	0
$\frac{1}{2}$	$\frac{1}{4}$	$\frac{1}{8}$
1	1	1
2	4	8

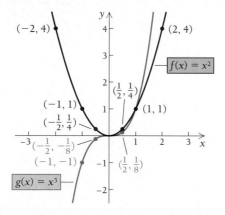

The domain of each function is the set of all real numbers. The range of f is the interval $[0, \infty)$. The range of g is the set of all real numbers, or the interval $(-\infty, \infty)$. ◀

DO EXERCISES 1 AND 2.

Some graphs of $f(x) = ax^n$, where n is *even*, are shown below. Note that $g(x) = x^4$ has the same general shape as $f(x) = x^2$, but the larger the power of n, the closer the graph gets to the x-axis for values of x in the interval $[-1, 1]$. For other values of x, the graphs get steeper. When a is negative, the graphs are reflected across the x-axis. Each function is even, has the y-axis as a line of symmetry, and passes through the origin.

Sketch a graph of the function.

1. $f(x) = (x - 1)^3$

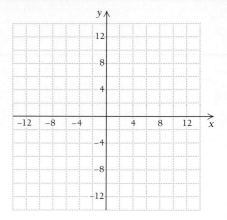

2. $f(x) = -\dfrac{1}{2}x^3 + 2$

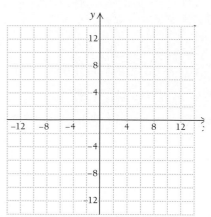

Sketch a graph of the function.

3. $f(x) = x^4$

4. $f(x) = -\dfrac{1}{2}x^5$

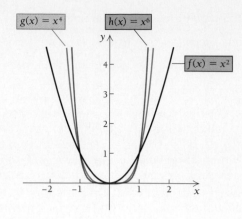

Similarly, the graphs of $f(x) = ax^n$, where n is *odd,* all have the same general shape as the graph of $f(x) = x^3$, as given in Example 1. The larger the power of n, the closer the graph gets to the x-axis for values of x in the interval $[-1, 1]$. For other values of x, the graphs get steeper. When a is negative, the graphs are reflected across the x-axis. Each function is odd, has the origin as a point of symmetry, and passes through the origin.

DO EXERCISES 3 AND 4.

Some functions have graphs that are curves with no breaks or holes in them. Such functions are called **continuous functions.** Loosely speaking, the graph of a continuous function can be drawn without lifting the pencil from the paper. In the following figure, the function f is continuous, but g and h are not.

A continuous function

A discontinuous function

A discontinuous function

The following theorem about polynomials can be helpful in graphing.

THEOREM 2

The domain of any polynomial function is the set of real numbers.
The graph of any polynomial function is a continuous function.

We say that c is a **zero** of f, if $f(c) = 0$. We also call c a **root** of the equation $f(x) = 0$, or simply a **root of the polynomial.** We know by Theorem 2 that every real number can be used as an input and that the graph of any polynomial function can be drawn without lifting the pencil from the paper. So f assumes all real values between $f(a)$ and $f(b)$. Suppose that we have two function values $f(a)$ and $f(b)$ that have opposite signs. Since we have a continuous function, we must be able to draw a curve from $f(a)$

to $f(b)$ without lifting the pencil from the paper. It follows that the line must cross the x-axis somewhere between a and b. Thus, f must have a zero, or root, somewhere between a and b.

| THEOREM 3 | **The Intermediate Value Theorem** |

For any polynomial f with real coefficients, suppose that $f(a) \neq f(b)$ and $a < b$. Also, suppose that the function values have opposite signs. Then f has a real zero between a and b.

Suppose that we have determined all the real-number zeros of a polynomial, which often is no small task. Then between successive zeros, the values of the polynomial are either all positive or all negative. If there were a sign change, then there would be an additional zero. Try drawing a continuous curve between successive zeros. You assume that you have found all the zeros, so the curve cannot cross the x-axis. By using *test values* between successive zeros, we know the sign of the function on the entire interval. We used this idea when solving inequalities in Section 4.4.

Let us now sketch some more complicated graphs. We will be considering polynomials for which it is relatively easy to find the roots.

Example 2   Sketch a graph of the polynomial function f given by

$$f(x) = 2x^3 + x^2 - 8x - 4.$$

Solution   We first try to factor the function in order to find the zeros. In this case, we can use factoring by grouping:

$$\begin{aligned} f(x) &= 2x^3 + x^2 - 8x - 4 \\ &= x^2(2x + 1) - 4(2x + 1) \\ &= (x^2 - 4)(2x + 1) \\ &= (x - 2)(x + 2)(2x + 1). \end{aligned}$$

We see that the solutions of the equation $f(x) = 0$ are -2, -0.5, and 2. These are also the zeros of the polynomial function. The zeros divide the number line into four open intervals: $(-\infty, -2)$, $(-2, -0.5)$, $(-0.5, 2)$, and $(2, \infty)$.

5. Using graph paper, sketch a graph of
$$f(x) = x^3 + 3x^2 - x - 3.$$

6. Using graph paper, sketch a graph of
$$f(x) = x^4 - 10x^2 + 9.$$

7. How many spheres, such as marbles or balls or oranges, are in the pile below? (You might see such a question in connection with a guessing contest in a store.)

If the layers are equilateral triangles and all the spheres are the same size, there is a polynomial function that can be used to determine the count. The number N is given by the polynomial function

$$N(x) = \frac{1}{6}x^3 + \frac{1}{2}x^2 + \frac{1}{3}x,$$

where x is the number of layers and $N(x)$ is the total number of spheres. Thus to win the contest, count the number of layers and substitute the answer into the formula.

a) Find the number of spheres in a pile with 9 layers.

b) Sketch a graph of the function over the interval $[0, \infty)$.

We try test numbers, as follows, in each interval. These also give us function values that can be used for graphing.

A: Test -3, $f(-3) = 2(-3)^3 + (-3)^2 - 8(-3) - 4 = -25$;
B: Test -1, $f(-1) = 2(-1)^3 + (-1)^2 - 8(-1) - 4 = 3$;
C: Test 1, $f(1) = 2(1)^3 + (1)^2 - 8(1) - 4 = -9$;
D: Test 3, $f(3) = 2(3)^3 + (3)^2 - 8(3) - 4 = 35$.

Since $f(-3) = -25$, the function values are all negative and all the points on the graph lie below the x-axis on the interval $(-\infty, -2)$. Since $f(-1) = 3$, the function values are all positive and all the points on the graph lie above the x-axis on the interval $(-2, -0.5)$. Since $f(1) = -9$, the function values are all negative and all the points on the graph lie below the x-axis on the interval $(-0.5, 2)$. Since $f(3) = 35$, the function values are all positive and all the points on the graph lie above the x-axis on the interval $(2, \infty)$. We summarize this information in the following table.

Interval	$(-\infty, -2)$	$(-2, -0.5)$	$(-0.5, 2)$	$(2, \infty)$
Test Value	$f(-3) = -25$	$f(-1) = 3$	$f(1) = -9$	$f(3) = 35$
Sign of $f(x)$	$-$	$+$	$-$	$+$
Location of Graph Points	Below x-axis	Above x-axis	Below x-axis	Above x-axis

Using these results and calculating additional function values, we complete the graph as follows.

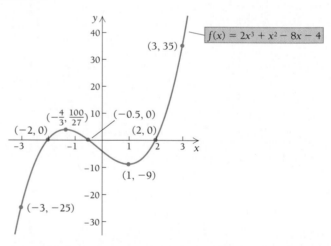

The points $(-\frac{4}{3}, \frac{100}{27})$ and $(1, -9)$ of the preceding graph are called **turning points.** They are points at which the graph changes from increasing to decreasing or from decreasing to increasing. In calculus, you will learn skills that will allow you to find these points easily. The following tip may also help you graph a polynomial function.

THEOREM 4

A polynomial function of degree n has at most $n - 1$ turning points and at most n zeros.

DO EXERCISE 5.

Example 3   Sketch a graph of the polynomial function f given by

$$f(x) = x^4 - 4x^2 + 3.$$

Solution   We first factor the function:

$$
\begin{aligned}
f(x) &= x^4 - 4x^2 + 3 \\
&= (x^2 - 3)(x^2 - 1) \\
&= (x + \sqrt{3})(x - \sqrt{3})(x + 1)(x - 1).
\end{aligned}
$$

The zeros of the function are $-\sqrt{3}$, -1, 1, and $\sqrt{3}$.

These zeros divide the real-number line into five open intervals. We try a test value in each interval and determine the sign of the function for values of x in the interval. The results are summarized in the following table.

Interval	$(-\infty, -\sqrt{3})$	$(-\sqrt{3}, -1)$	$(-1, 1)$	$(1, \sqrt{3})$	$(\sqrt{3}, \infty)$
Test Value	$f(-2) = 3$	$f(-1.5) = -0.9375$	$f(0) = 3$	$f(1.5) = -0.9375$	$f(2) = 3$
Sign of $f(x)$	+	−	+	−	+
Location of Graph Points	Above x-axis	Below x-axis	Above x-axis	Below x-axis	Above x-axis

We use the information listed in the table, calculate some extra function values, if needed, plot points, and sketch the graph, as shown below.

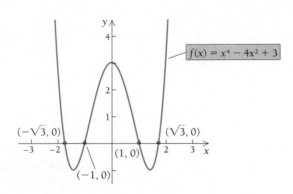

DO EXERCISES 6 AND 7 ON THE PRECEDING PAGE.

   TECHNOLOGY CONNECTION

Obviously, a grapher allows you to easily draw the graph of any polynomial function without going through all the steps described in each example. (Graphers are especially useful in cases where the polynomial does not factor easily, making it difficult or impossible to find the zeros of the function.) If you are permitted to use a grapher to solve these problems, make sure you do not develop too much of a reliance on it, however, to do your thinking for you! You should be sure that you understand the principles described in Theorems 2–4, and be able to apply them to any polynomial function.

Use a grapher to graph each of the following functions.

TC 1. $f(x) = x^3 - 4x - 2$

TC 2. $f(x) = -2x^4 + x^3 - x^2 + 1$

TC 3. $f(x) = -3x^3 + x^2 + x$

● **EXERCISE SET** **4.5**

◤ Sketch a graph of the polynomial function.

1. $f(x) = \frac{1}{3}x^6$

2. $f(x) = -\frac{2}{3}x^3$

3. $f(x) = -0.6x^5$

4. $f(x) = \frac{4}{5}x^4 - 3$

5. $f(x) = (x + 1)^5 - 4$

6. $g(x) = -(x - 2)^3 + 1$

7. $f(x) = \frac{1}{4}(x + 1)^4$

8. $f(x) = -0.7(x + 4)^3$

9. $f(x) = (x + 3)(x - 2)(x + 1)$

10. $f(x) = (x - 1)(x + 2)(x - 4)$

11. $f(x) = 9x^2 - x^4$

12. $f(x) = 4x^2 - x^4$

13. $f(x) = x^4 - x^3$

14. $f(x) = x^4 - 4x^2$

15. $f(x) = x^3 - 4x$

16. $f(x) = 25x - x^3$

17. $f(x) = x^3 + x^2 - 2x$

18. $f(x) = -x^3 - x^2 + 6x$

19. $f(x) = x^4 - 9x^2 + 20$

20. $f(x) = x^4 - 3x^2 + 2$

21. $f(x) = x^3 - 3x^2 - 4x + 12$

22. $f(x) = x^3 - 2x^2 - 9x + 18$

23. $f(x) = -x^4 - 3x^3 - 3x^2$

24. $f(x) = -3x^4 - x^3 + 2x^2$

25. $f(x) = x(x - 2)(x + 1)(x + 3)$

26. $f(x) = x(x - 1)(x - 2)(x - 3)$

27. Beam deflection. A beam rests at two points P and Q and has a concentrated load applied to the center of the beam, as shown in the figure.

Let y denote the deflection of the beam at a distance of x units from the left end of the beam. The deflection depends on the elasticity of the beam, the load, and other physical characteristics. Suppose under certain conditions that y is given by the polynomial function

$$y = \frac{1}{13}x^3 - \frac{1}{14}x.$$

a) Find the deflection of the beam at distances of 1 unit, 2 units, 3 units, 6 units, 8 units, 9 units, and 10 units.

b) Sketch a graph of the function for those values of x for which $y \geq 0$.

28. Medical dosage. The function

$$N(t) = -0.046t^3 + 2.08t + 2$$

gives the bodily concentration in parts per million of a certain dosage of medication after time t, in hours.

a) Find the concentration after 1 hr, 2 hr, 3 hr, 4 hr, 5 hr, 6 hr, and 7 hr.

b) Using a calculator, find several other function values and sketch a graph of the function over the interval $(0, 7)$.

29. Weight threshold. In a medical study done by Alwin Shemesh in 1976, it was found that *threshold weight W,* defined as that weight above which the risk of mortality rises astronomically, is given as a function of height h by

$$W(h) = (h/12.3)^3,$$

where W is in pounds and height h is in inches.

a) What is the threshold weight of a person who is 5 ft 7 in.? 5 ft 10 in.?

b) One of the authors of this book is 6 ft 1 in. tall and weighs 209 lb. Does he need to watch his weight?

30. Windmill power. Under certain conditions, the power P, in watts, generated by a windmill when the speed of the wind is V, in miles per hour, is given by

$$P = 0.015V^3.$$

a) Find the power generated by a wind of speed 8 mph, 10 mph, and 15 mph.

b) How fast would the wind have to blow in order to generate 120 watts of power?

31. Volume. From a thin piece of cardboard 10 in. by 10 in., square corners are cut out so that the sides can be folded up to make a box.

a) Express the volume V as a function of the length x of a side of the square base.

b) Sketch a graph of the function using the entire set of real numbers as the domain.

c) Over what intervals is the function positive?

32. Volume. From a thin piece of cardboard 8 in. by 8 in., square corners are cut out so that the sides can be folded up to make a box.

a) Express the volume V as a function of the length x of a side of the square base.

b) Sketch a graph of the function using the entire set of real numbers as the domain.

c) Over what intervals is the function positive?

● SYNTHESIS _____

33. Which of the functions in Exercises 1–26 is an even function?

34. Which of the functions in Exercises 1–26 is an odd function?

35. Under what conditions is a polynomial function of real variables an even function?

36. Under what conditions is a polynomial function of real variables an odd function?

Using a grapher, graph the function, and approximate its zeros, or roots.

37. $f(x) = x^3 + 4x^2 + x - 6$

38. $f(x) = x^3 + x^2 - 6x + 6$

39. $f(x) = -x^4 + x^3 + 4x^2 - 2x - 4$

40. $f(x) = x^5 - x^4 - 9x^3 + 3x^2 + 18x$

41. Consider the function in Exercise 28.

   a) Graph the function.

   b) Approximate all of its zeros.

   c) After what time will the concentration be 0?

4.6

DIVISION OF POLYNOMIALS

In Section 4.5, we studied graphs of polynomial functions. Most of the graphs we considered in that section had roots, or zeros, that were easy to find because the polynomials were either factored or fairly easy to factor. In general, finding roots, or zeros, is not easy or straightforward. In this section, we will delve more deeply into the problem of finding them. Let us first review some terminology.

   We will now allow the coefficients of a polynomial to be complex numbers. In certain cases, we will restrict the coefficients to be real numbers, rational numbers, or integers, as shown in the following examples.

Polynomial	Type of Coefficient
$5x^3 - 3x^2 + (2 + 4i)x + i$	Complex
$-4x + \sqrt{2}$	Real
-8	Integer
0	Integer

  Zeros, or Roots, of Polynomials

We often refer to polynomials using function notation, such as $P(x)$.

Example 1   Given that $P(x) = x^3 + 2x^2 - 5x - 6$, determine whether 3 is a *zero* of $P(x)$.

Solution   We substitute 3 into the polynomial:

$$P(3) = (3)^3 + 2(3)^2 - 5(3) - 6$$
$$= 24.$$

Since $P(3) \neq 0$, 3 is *not* a zero of the polynomial.   ◀

Example 2   Is -1 a root of $P(x)$ in Example 1?

Solution   We substitute -1 into the polynomial:

$$P(-1) = (-1)^3 + 2(-1)^2 - 5(-1) - 6$$
$$= 0.$$

Since $P(-1) = 0$, we know that -1 is a root of $P(x)$.   ◀

DO EXERCISES 1–3 ON THE FOLLOWING PAGE. _____

  TECHNOLOGY CONNECTION

After performing substitutions of this type, you can use a grapher to verify the results. If a substitution results in a zero for the value of the polynomial, then you should see an x-intercept at that value when you graph the function.

TC 1.   For the function $f(x) = 2.5x^3 - 20x^2 + 2.5x + 105$, use substitution to determine which of the following values are roots of the polynomial: $-5, -2, -1, 2, 3, 7$. Then use a grapher to verify your answer.

1. Determine whether the following numbers are roots of the polynomial $x^2 - 4x - 21$.

   a) 7

   b) 3

   c) -7

2. Determine whether the following numbers are roots of the polynomial $x^4 - 16$.

   a) 2

   b) -2

   c) -1

   d) 0

3. Determine whether the following numbers are roots of the polynomial $x^2 + 1$.

   a) 1

   b) -1

   c) i

   d) $-i$

4. By division, determine whether the following polynomials are factors of the polynomial $x^3 + 2x^2 - 5x - 6$.

   a) $x + 1$

   b) $x - 3$

   c) $x^2 + 3x - 1$

5. By division, determine whether the following polynomials are factors of the polynomial $x^4 - 16$.

   a) $x - 2$

   b) $x^2 + 3x - 1$

B ▶ Determining Factors of Polynomials

When we divide one polynomial by another, we obtain a quotient and a remainder. If the remainder is 0, then the divisor is a **factor** of the dividend.

Example 3 Divide to determine whether $x + 1$ is a factor of $x^3 + 2x^2 - 5x - 6$.

Solution

$$
\begin{array}{r}
x^2 + x - 6 \\
x + 1 \overline{\smash{)}\, x^3 + 2x^2 - 5x - 6} \\
\underline{x^3 + x^2} \\
x^2 - 5x \\
\underline{x^2 + x} \\
-6x - 6 \\
\underline{-6x - 6} \\
0
\end{array}
$$

Since the remainder is 0, we know that $x + 1$ is a factor of $x^3 + 2x^2 - 5x - 6$. ◀

Example 4 Divide to determine whether $x^2 + 3x - 1$ is a factor of $x^4 - 81$.

Solution

$$
\begin{array}{r}
x^2 - 3x + 10 \\
x^2 + 3x - 1 \overline{\smash{)}\, x^4 \qquad\qquad\qquad - 81} \\
\underline{x^4 + 3x^3 - \quad x} \\
-3x^3 + \quad x^2 \\
\underline{-3x^3 - 9x^2 + 3x} \\
10x^2 - 3x - 81 \\
\underline{10x^2 + 30x - 10} \\
-33x - 71
\end{array}
$$

Note that spaces have been left for missing terms.

Since the remainder is not 0, we know that $x^2 + 3x - 1$ is not a factor of $x^4 - 81$. ◀

DO EXERCISES 4 AND 5.

C ▶ Dividends in Terms of Quotients, Divisors, and Remainders

In general, when we divide a polynomial $P(x)$ by a divisor $d(x)$, we obtain some polynomial $Q(x)$ for a quotient and some polynomial $R(x)$ for a remainder. The remainder must either be 0 or have degree less than that of $d(x)$. To check a division, we multiply the quotient by the divisor and add the remainder, to see if we get the dividend. Thus these polynomials are related as follows:

$$P(x) = d(x) \cdot Q(x) + R(x)$$

Dividend Divisor Quotient Remainder

Example 5 If $P(x) = x^3 + 2x^2 - 5x - 6$ and $d(x) = x + 1$, then $Q(x) = x^2 + x - 6$ and $R(x) = 0$, and

$$\underbrace{x^3 + 2x^2 - 5x - 6}_{P(x)} = \underbrace{(x + 1)}_{d(x)} \cdot \underbrace{(x^2 + x - 6)}_{Q(x)} + \underbrace{0}_{R(x).} \blacktriangleleft$$

Example 6 If $P(x) = x^4 - 81$ and $d(x) = x^2 + 3x - 1$, then $Q(x) = x^2 - 3x + 10$ and $R(x) = -33x - 71$, and

$$\underbrace{x^4 - 81}_{P(x)} = \underbrace{(x^2 + 3x - 1)}_{d(x)} \cdot \underbrace{(x^2 - 3x + 10)}_{Q(x)} + \underbrace{(-33x - 71)}_{R(x).} \blacktriangleleft$$

DO EXERCISES 6 AND 7.

▷ **The Remainder Theorem and Synthetic Division**

Margin Exercises 6 and 7 illustrate the following theorem.

THEOREM 5	**The Remainder Theorem**

If a number r is substituted for x in the polynomial $P(x)$, then the result $P(r)$ is the remainder that would be obtained by dividing $P(x)$ by $x - r$. That is, if $P(x) = (x - r) \cdot Q(x) + R$, then $P(r) = R$.

Proof. The equation $P(x) = d(x) \cdot Q(x) + R(x)$, where $d(x) = x - r$, is the basis of this proof. If we divide $P(x)$ by $x - r$, we obtain a quotient $Q(x)$ and a remainder $R(x)$ related as follows:

$$P(x) = (x - r) \cdot Q(x) + R(x).$$

The remainder $R(x)$ must either be 0 or have degree less than $x - r$. Thus, $R(x)$ must be a constant. Let us call this constant R. In the expression above, we get a true sentence whenever we replace x by any number. Let us replace x by r. We get

$$P(r) = (r - r) \cdot Q(r) + R$$
$$= 0 \cdot Q(r) + R$$
$$= R.$$

This tells us that the function value $P(r)$ is the remainder obtained when we divide $P(x)$ by $x - r$.

Theorem 5 motivates us to find a rapid way of dividing by $x - r$, in order to find function values. To streamline division, we can arrange the work so that duplicate and unnecessary writing is avoided. Consider the following.

A.
$$
\begin{array}{r}
4x^2 + 5x\ + 11 \\
x - 2\overline{)\ 4x^3 - 3x^2 +\quad x + 7} \\
\underline{4x^3 - 8x^2\qquad\qquad} \\
5x^2 +\quad x \\
\underline{5x^2 - 10x\qquad} \\
11x + 7 \\
\underline{11x - 22} \\
29
\end{array}
$$

B.
$$
\begin{array}{r}
4\quad 5\quad 11 \\
1 - 2\overline{)\ 4 - 3 +\ 1 + 7} \\
\underline{4 - 8\qquad\qquad} \\
5 + 1 \\
\underline{5 - 10\quad} \\
11 + 7 \\
\underline{11 - 22} \\
29
\end{array}
$$

6. Divide $x^3 + 2x^2 - 5x - 6$ by $x - 3$. Then express the dividend as

$$P(x) = d(x) \cdot Q(x) + R(x).$$

Dividend   Divisor   Quotient   Remainder

7. Consider

$$P(x) = x^3 + 2x^2 - 5x - 6.$$

a) Find $P(3)$.

b) Compare the answer to part (a) with the remainder in Margin Exercise 6.

Use synthetic division to find the quotient and the remainder.

8. $(x^3 + 6x^2 - x - 30) \div (x - 2)$

9. $(x^3 - 2x^2 + 5x - 4) \div (x + 2)$

10. $(y^3 + 1) \div (y + 1)$

11. Let

$P(x) = x^5 - 2x^4 - 7x^3 + x^2 + 20.$

Use synthetic division to find each of the following.

a) $P(10)$

b) $P(-8)$

The division in (B) is the same as that in (A), but we wrote only the coefficients. The color numerals are duplicated, so we look for an arrangement in which they are not duplicated. We can also simplify things by using the opposite of -2 and then adding instead of subtracting. When things are thus "collapsed," we have the algorithm known as **synthetic division.**

C. *Synthetic Division*

We "bring down" the 4. Then we multiply it by the 2 to get 8 and add to get 5. We then multiply 5 by 2 to get 10, add, and so on. The last number, 29, is the remainder. The others, 4, 5, and 11, are the coefficients of the quotient.

We write a 0 in the synthetic division for a missing term in the dividend.

Example 7 Use synthetic division to find the quotient and the remainder:

$$(2x^3 + 7x^2 - 5) \div (x + 3).$$

Solution First we note that $x + 3 = x - (-3)$.

$$\begin{array}{r|rrrr} -3 & 2 & 7 & 0 & -5 \\ & & -6 & -3 & 9 \\ \hline & 2 & 1 & -3 & | \quad 4 \end{array}$$

Note: We must write 0's for missing terms.

The quotient is $2x^2 + x - 3$. The remainder is 4. ◀

DO EXERCISES 8–10. _____

E Function Values for Polynomials

We can now apply synthetic division to find function values for polynomials.

Example 8 Given that $P(x) = 2x^5 - 3x^4 + x^3 - 2x^2 + x - 8$, find $P(10)$.

Solution By Theorem 5, $P(10)$ is the remainder when $P(x)$ is divided by $x - 10$. We use synthetic division to find that remainder.*

$$\begin{array}{r|rrrrrr} 10 & 2 & -3 & 1 & -2 & 1 & -8 \\ & & 20 & 170 & 1710 & 17{,}080 & 170{,}810 \\ \hline & 2 & 17 & 171 & 1708 & 17{,}081 & | \quad 170{,}802 \end{array}$$

Thus, $P(10) = 170{,}802$. ◀

DO EXERCISE 11. _____

_____

* Compare this with the work involved in a direct calculation! 🖩 A calculator is most useful when finding polynomial function values by synthetic division. In Example 8, begin by entering 2. Then multiply by 10, add the result to -3, multiply that result by 10, and so on. The number 10 can be stored and recalled at each step where needed if the calculator has a memory. Since only the last result is needed, no intermediate values need be recorded.

Example 9 Determine whether -4 is a zero, or root, of $P(x)$, where $P(x) = x^3 + 8x^2 + 8x - 32$.

Solution We use synthetic division and Theorem 5 to find $P(-4)$.

$$
\begin{array}{r|rrrr}
-4 & 1 & 8 & 8 & -32 \\
   &   & -4 & -16 & 32 \\
\hline
   & 1 & 4 & -8 & 0
\end{array}
$$

Since $P(-4) = 0$, the number -4 is a root of $P(x)$. ◀

DO EXERCISE 12.

Finding Factors of Polynomials

We now consider the following useful corollary of the remainder theorem.

THEOREM 6	**The Factor Theorem**

For a polynomial $P(x)$, if $P(r) = 0$, then $x - r$ is a factor of $P(x)$.

Proof. If we divide $P(x)$ by $x - r$, we obtain a quotient and a remainder, related as follows:

$$P(x) = (x - r) \cdot Q(x) + P(r).$$

Then if $P(r) = 0$, we have

$$P(x) = (x - r) \cdot Q(x),$$

so $x - r$ is a factor of $P(x)$.

This theorem is very useful in factoring polynomials, and hence in the solving of equations.

Example 10 Let $P(x) = x^3 + 2x^2 - 5x - 6$. Factor $P(x)$ and solve the equation $P(x) = 0$.

Solution We look for linear factors of the form $x - r$. Let us try $x - 1$. We use synthetic division to see whether $P(1) = 0$.

$$
\begin{array}{r|rrrr}
1 & 1 & 2 & -5 & -6 \\
  &   & 1 & 3 & -2 \\
\hline
  & 1 & 3 & -2 & -8
\end{array}
$$

Since $P(1) \neq 0$, we know that $x - 1$ is not a factor of $P(x)$. We try $x + 1$ or $x - (-1)$ in the form $x - r$.

$$
\begin{array}{r|rrrr}
-1 & 1 & 2 & -5 & -6 \\
   &   & -1 & -1 & 6 \\
\hline
   & 1 & 1 & -6 & 0
\end{array}
$$

Since $P(-1) = 0$, we know that $x + 1$ is one factor and the quotient, $x^2 + x - 6$, is another. Thus,

$$P(x) = (x + 1)(x^2 + x - 6).$$

12. Let $P(x) = x^3 + 6x^2 - x - 30$. Using synthetic division, determine whether the given numbers are roots of $P(x)$.

   a) 2

   b) 5

   c) -3

13. Determine whether $x - \frac{1}{2}$ is a factor of $4x^4 + 2x^3 + 8x - 1$.

14. Determine whether $x + 5$ is a factor of $x^3 + 625$.

15. Let

$$P(x) = x^3 + 6x^2 - x - 30.$$

a) Determine whether $x - 2$ is a factor of $P(x)$.

b) Find another factor of $P(x)$.

c) Find a complete factorization of $P(x)$.

d) Solve $P(x) = 0$.

The trinomial is easily factored, so we have

$$P(x) = (x + 1)(x + 3)(x - 2).$$

Our goal is to solve the equation $P(x) = 0$. To do so, we use the principle of zero products. The solutions are -1, -3, and 2. Thus the solution set is $\{-1, -3, 2\}$. ◀

DO EXERCISES 13–15. (EXERCISES 13 AND 14 ARE ON THE PRECEDING PAGE.)

● EXERCISE SET 4.6

1. Determine whether 4, 5, and -2 are roots, or zeros, of

$$P(x) = x^3 - 9x^2 + 14x + 24.$$

2. Determine whether 2, 3, and -1 are roots, or zeros, of

$$P(x) = 2x^3 - 3x^2 + x - 1.$$

B

3. For $P(x)$ in Exercise 1, which of the following are factors of $P(x)$?

  a) $x - 4$ **b)** $x - 5$ **c)** $x + 2$

4. For $P(x)$ in Exercise 2, which of the following are factors of $P(x)$?

  a) $x - 2$ **b)** $x - 3$ **c)** $x + 1$

In each of the following, a polynomial $P(x)$ and a divisor $d(x)$ are given. Find the quotient $Q(x)$ and the remainder $R(x)$ when $P(x)$ is divided by $d(x)$, and express $P(x)$ in the form $d(x) \cdot Q(x) + R(x)$.

5. $P(x) = x^3 + 6x^2 - x - 30$,
   $d(x) = x - 2$

6. $P(x) = 2x^3 - 3x^2 + x - 1$,
   $d(x) = x - 2$

7. $P(x) = x^3 + 6x^2 - x - 30$,
   $d(x) = x - 3$

8. $P(x) = 2x^3 - 3x^2 + x - 1$,
   $d(x) = x - 3$

9. $P(x) = x^3 - 8$,
   $d(x) = x + 2$

10. $P(x) = x^3 + 27$,
   $d(x) = x + 1$

11. $P(x) = x^4 + 9x^2 + 20$,
   $d(x) = x^2 + 4$

12. $P(x) = x^4 + x^2 + 2$,
   $d(x) = x^2 + x + 1$

13. $P(x) = 5x^7 - 3x^4 + 2x^2 - 3$,
   $d(x) = 2x^2 - x + 1$

14. $P(x) = 6x^5 + 4x^4 - 3x^2 + x - 2$,
   $d(x) = 3x^2 + 2x - 1$

D   Use synthetic division to find the quotient and the remainder.

15. $(2x^4 + 7x^3 + x - 12) \div (x + 3)$

16. $(x^3 - 7x^2 + 13x + 3) \div (x - 2)$

17. $(x^3 - 2x^2 - 8) \div (x + 2)$

18. $(x^3 - 3x + 10) \div (x - 2)$

19. $(x^4 - 1) \div (x - 1)$

20. $(x^5 + 32) \div (x + 2)$

21. $(2x^4 + 3x^2 - 1) \div (x - \frac{1}{2})$

22. $(3x^4 - 2x^2 + 2) \div (x - \frac{1}{4})$

23. $(x^4 - y^4) \div (x - y)$

24. $(x^3 + 3ix^2 - 4ix - 2) \div (x + i)$

E   Use synthetic division to find the function values.

25. $P(x) = x^3 - 6x^2 + 11x - 6$; find $P(1)$, $P(-2)$, and $P(3)$.

26. $P(x) = x^3 + 7x^2 - 12x - 3$; find $P(-3)$, $P(-2)$, and $P(1)$.

27. $P(x) = 2x^5 - 3x^4 + 2x^3 - x + 8$; find $P(20)$ and $P(-3)$.

28. $P(x) = x^5 - 10x^4 + 20x^3 - 5x - 100$; find $P(-10)$ and $P(5)$.

29. $P(x) = x^4 - 16$; find $P(2)$, $P(-2)$, and $P(3)$.

30. $P(x) = x^5 + 32$; find $P(2)$, $P(-2)$, and $P(3)$.

F   Using synthetic division, determine whether the numbers are roots of the polynomials.

31. $-3, 2$; $P(x) = 3x^3 + 5x^2 - 6x + 18$

32. $-4, 2$; $P(x) = 3x^3 + 11x^2 - 2x + 8$

33. $-3, \frac{1}{2}$; $P(x) = x^3 - \frac{7}{2}x^2 + x - \frac{3}{2}$

34. $i, -i, -2$; $P(x) = x^3 + 2x^2 + x + 2$

Factor the polynomial $P(x)$. Then solve the equation $P(x) = 0$.

35. $P(x) = x^3 + 4x^2 + x - 6$

36. $P(x) = x^3 + 5x^2 - 2x - 24$

37. $P(x) = x^3 - 6x^2 + 3x + 10$

38. $P(x) = x^3 + 2x^2 - 13x + 10$

39. $P(x) = x^3 - x^2 - 14x + 24$

40. $P(x) = x^3 - 3x^2 - 10x + 24$

41. $P(x) = x^4 - x^3 - 19x^2 + 49x - 30$

42. $P(x) = x^4 + 11x^3 + 41x^2 + 61x + 30$

● SYNTHESIS _____

43. If there are n teams in a sports league and each team plays each other once in a season, we can find the total number of games played by a polynomial function $f(n) = \frac{1}{2}(n^2 - n)$. Find the zeros of the function.

44. ▣ The function

$$N(t) = -0.046t^3 + 2.08t + 2$$

gives the body concentration, in parts per million, of a certain dosage of medication after time t, in hours. Estimate the roots in the interval $[0, 8]$.

45. *Threshold weight.* In a medical study done by Alwin Shemesh in 1976, it was found that **threshold weight** W, defined as that weight above which the risk of mortality rises astronomically, is given as a function of height h by

$$W(h) = \left(\frac{h}{12.3}\right)^3,$$

where W is in pounds and h is in inches. Find the roots of the polynomial on the interval $[0, \infty)$.

46. *Beam deflection.* A beam rests at two points P and Q and has a concentrated load applied to the center of the beam, as shown in the figure. Let y denote the deflection of the beam at a distance of x units from point P to the left of the weight. The deflection depends on the elasticity of the beam, the load, and other physical characteristics.

Suppose under certain conditions that y is given by

$$y = \frac{1}{13}x^3 - \frac{1}{14}x.$$

Find the zeros of the polynomial in the interval $[0, 2]$.

Solve.

47. $\dfrac{6x^2}{x^2 + 11} + \dfrac{60}{x^3 - 7x^2 + 11x - 77} = \dfrac{1}{x - 7}$

48. $\dfrac{2x^2}{x^2 - 1} + \dfrac{4}{x + 3} = \dfrac{32}{x^3 + 3x^2 - x - 3}$

49. $x^3 + 2x^2 - 13x + 10 > 0$

50. $x^4 - x^3 - 19x^2 + 49x - 30 < 0$

51. Find k so that $x + 2$ is a factor of
$$x^3 - kx^2 + 3x + 7k.$$

52. ▣ Given that
$$f(x) = 2.13x^5 - 42.1x^3 + 17.5x^2 + 0.953x - 1.98,$$
find $f(3.21)$ (a) by synthetic division; (b) by substitution.

53. For what values of k will the remainder be the same when $x^2 + kx + 4$ is divided by $x - 1$ or $x + 1$?

54. When $x^2 - 3x + 2k$ is divided by $x + 2$, the remainder is 7. Find the value of k.

● CHALLENGE _____

55. Devise a way to use the method of synthetic division when the divisor is a polynomial such as $bx - r$.

56. Prove that $x - a$ is a factor of $x^n - a^n$, for any natural number n.

4.7

THEOREMS ABOUT ROOTS

The Fundamental Theorem of Algebra

A linear, or first-degree, polynomial $ax + b$ (where $a \neq 0$, of course) has just one root, $-b/a$. It can be shown that any quadratic polynomial with complex numbers for coefficients has at least one, and at most two, complex roots. The following theorem is a generalization. No proof is given.

THEOREM 7	**The Fundamental Theorem of Algebra**

Every polynomial of degree n, $n > 0$, with complex coefficients, has at least one root in the system of complex numbers.

OBJECTIVES

You should be able to:

A Factor polynomials and find their roots and their multiplicities.

B Find a polynomial with specified roots.

C In certain cases, given some of the roots of a polynomial, find such a polynomial and find the rest of its roots.

D Given a polynomial with integer coefficients, find the rational roots and find the other roots, if possible.

E Do the same for polynomials with rational coefficients.

The fundamental theorem of algebra is a very powerful theorem. Note that although it guarantees that a root exists, it does not tell how to find it. Keep in mind that when we speak of "roots," or "zeros," of a polynomial $P(x)$, we are also speaking about "roots" or "solutions" of the polynomial equation $P(x) = 0$. We now develop some theory that can help in finding roots. First, we prove a corollary of the fundamental theorem of algebra.

THEOREM 8

Every polynomial P of degree n, where $n > 0$, having complex coefficients, can be factored into n linear factors (not necessarily unique); that is, $P(x) = a_n(x - r_1)(x - r_2) \cdots (x - r_n)$.

Proof. Let us consider any polynomial of degree n, say $P(x)$. By the fundamental theorem, it has a root r_1. By the factor theorem, $x - r_1$ is a factor of $P(x)$. Thus we know that

$$P(x) = (x - r_1) \cdot Q_1(x),$$

where $Q_1(x)$ is the quotient that would be obtained by dividing $P(x)$ by $x - r_1$. Let the leading coefficient of $P(x)$ be a_n. By considering the actual division process, we see that the leading coefficient of $Q_1(x)$ is also a_n and that the degree of $Q_1(x)$ is $n - 1$. Now if the degree of $Q_1(x)$ is greater than 0, then it has a root r_2, and we have

$$Q_1(x) = (x - r_2) \cdot Q_2(x),$$

where the degree of $Q_2(x)$ is $n - 2$ and the leading coefficient is a_n. Thus we have

$$P(x) = (x - r_1)(x - r_2) \cdot Q_2(x).$$

This process can be continued until a quotient $Q_n(x)$ having degree 0 is obtained. The leading coefficient will be a_n, so $Q_n(x)$ is actually the constant a_n. We now have the following:

$$P(x) = a_n(x - r_1)(x - r_2)(x - r_3) \cdots (x - r_n).$$

This completes the proof. We see that $P(x)$ has been factored with one constant factor a_n and n linear factors having leading coefficient 1.

Finding Roots of Factored Polynomials

When a polynomial is factored into a product of linear factors, it is easy to find the roots by considering the principle of zero products.

Example 1 Find the roots of

$$P(x) = (x - 3)(x + 4)(x + 1)(x - 1).$$

Solution To solve the equation $P(x) = 0$, we use the principle of zero products. The roots are 3, -4, -1, and 1. ◀

Example 2 Find the roots of

$$P(x) = 5(x - 2)(x - 2)(x - 2)(x + 1).$$

Solution   To solve the equation $P(x) = 0$, we use the principle of zero products. The roots are 2 and -1.   ◀

In Example 2, the factor $x - 2$ occurs three times. In a case like this, we sometimes say that the root we obtain, 2, has a **multiplicity** of 3. If we multiply out the right side, we obtain

$$P(x) = 5x^4 - 25x^3 + 30x^2 + 20x - 40.$$

Had we started with this expression, we might have had trouble finding the roots. We will be learning ways to do such factoring. Some polynomials can be factored using techniques we already know, such as factoring by grouping.

Example 3   Find the roots of

$$P(x) = x^3 - 2x^2 - 9x + 18.$$

Solution   We factor by grouping, as follows:

$$\begin{aligned} P(x) &= x^3 - 2x^2 - 9x + 18 \\ &= x^2(x - 2) - 9(x - 2) \\ &= (x^2 - 9)(x - 2) \\ &= (x + 3)(x - 3)(x - 2). \end{aligned}$$

Then by the principle of zero products, the equation $P(x) = 0$ has -3, 3, and 2 as roots.   ◀

Other factoring techniques can be used, as shown in Example 4.

Example 4   Find the roots of

$$P(x) = x^4 + 4x^2 - 45.$$

Solution   We factor as follows:

$$\begin{aligned} P(x) &= x^4 + 4x^2 - 45 \\ &= (x^2 - 5)(x^2 + 9) \\ &= (x - \sqrt{5})(x + \sqrt{5})(x - 3i)(x + 3i). \end{aligned}$$

Then by the principle of zero products, the equation has $\pm\sqrt{5}$ and $\pm 3i$ as roots.   ◀

Theorem 8 and the examples above give us the following.

THEOREM 9

Every polynomial of degree n, where $n > 0$, has at least one root and at most n roots.

Theorem 9 is often stated as follows: "Every polynomial of degree n, where $n > 0$, has *exactly* n roots." This statement is not incompatible with Theorem 9, as it first seems, because to make sense of the statement just

Find the roots of the polynomial and state the multiplicity of each.

1. $P(x) = (x - 5)(x - 5)(x + 6)$

2. $P(x) = 4(x + 7)^2(x - 3)$

3. $P(x) = (x + 2)^3(x^2 - 9)$

4. $P(x) = (x^2 - 7x + 12)^2$

5. $P(x) = 5x^2 - 5$

6. $P(x) = x^4 + x^2 - 12$

7. $P(x) = 2x^3 + x^2 - 8x - 4$

8. Find a polynomial of degree 3 that has -1, 2, and 5 as roots.

9. Find a polynomial of degree 3 that has -1, 2, and $-5i$ as roots.

10. Find a polynomial of degree 5 with -2 as a root of multiplicity 3 and 0 as a root of multiplicity 2.

11. Find a polynomial of degree 4 with 1 as a root of multiplicity 3 and -5 as a root of multiplicity 1.

quoted, one must take multiplicities into account. Theorem 9, as stated here, is simpler and more straightforward.

DO EXERCISES 1–7.

 ### Finding Polynomials with Given Roots

Given several numbers, we can find a polynomial with those numbers as its roots.

Example 5 Find a polynomial of degree 3, having the roots -2, 1, and $3i$.

Solution   By Theorem 6, such a polynomial has factors $x + 2$, $x - 1$, and $x - 3i$, so we have

$$P(x) = a_n(x + 2)(x - 1)(x - 3i).$$

The number a_n can be any nonzero number. The simplest polynomial will be obtained if we let it be 1. If we then multiply the factors, we obtain

$$P(x) = x^3 + (1 - 3i)x^2 + (-2 - 3i)x + 6i. \qquad \blacktriangleleft$$

Example 6 Find a polynomial of degree 5 with -1 as a root of multiplicity 3, 4 as a root of multiplicity 1, and 0 as a root of multiplicity 1.

Solution   Proceeding as in Example 5, letting $a_n = 1$, we obtain

$$(x + 1)^3(x - 4)(x - 0), \quad \text{or} \quad x^5 - x^4 - 9x^3 - 11x^2 - 4x. \qquad \blacktriangleleft$$

DO EXERCISES 8–11.

 ### Roots of Polynomials with Real Coefficients

Consider the quadratic equation $x^2 - 2x + 2 = 0$, with real coefficients. Its roots are $1 + i$ and $1 - i$. Note that they are complex conjugates. This generalizes to any polynomial with real coefficients.

THEOREM 10

If a complex number $a + bi$ is a root of a polynomial $P(x)$ with real coefficients, then its conjugate, $a - bi$, is also a root. (Nonreal roots occur in conjugate pairs.)

For Theorem 10, it is essential that the coefficients be real numbers. We see this in Example 5, where the root $3i$ occurs, but its conjugate does not. This can happen because some of the coefficients of the polynomial are not real.

Rational Coefficients

When a polynomial has rational numbers for coefficients, certain irrational roots also occur in pairs, as described in the following theorem.

THEOREM 11

Suppose that $P(x)$ is a polynomial with rational coefficients. Then if either of the following is a root, so is the other: $a + c\sqrt{b}$, $a - c\sqrt{b}$, a and c rational, b not a square.

Theorem 11 can be used to help in finding roots.

Example 7   Suppose that a polynomial of degree 6 with rational coefficients has $-2 + 5i$, $-2i$, and $1 - \sqrt{3}$ as some of its roots. Find the other roots.

Solution   The other roots are $-2 - 5i$, $2i$, and $1 + \sqrt{3}$. There are no other roots since the degree is 6.   ◀

Example 8   Find a polynomial of lowest degree with rational coefficients that has $1 - \sqrt{2}$ and $1 + 2i$ as some of its roots.

Solution   The polynomial must also have the roots $1 + \sqrt{2}$ and $1 - 2i$. Thus the polynomial is

$$[x - (1 - \sqrt{2})][x - (1 + \sqrt{2})][x - (1 + 2i)][x - (1 - 2i)],$$

or

$$(x^2 - 2x - 1)(x^2 - 2x + 5), \quad \text{or} \quad x^4 - 4x^3 + 8x^2 - 8x - 5. \quad ◀$$

DO EXERCISES 12–14.

Example 9   Let $P(x) = x^4 - 5x^3 + 10x^2 - 20x + 24$. Find the other roots of $P(x)$, given that $2i$ is a root.

Solution   Since $2i$ is a root, we know that $-2i$ is also a root. Thus,

$$P(x) = (x - 2i)(x + 2i) \cdot Q(x)$$

for some $Q(x)$. Since $(x - 2i)(x + 2i) = x^2 + 4$, we know that

$$P(x) = (x^2 + 4) \cdot Q(x).$$

Using division, we find that $Q(x) = x^2 - 5x + 6$, and since we can factor $x^2 - 5x + 6$, we get

$$P(x) = (x^2 + 4)(x - 2)(x - 3).$$

Thus the other roots are $-2i$, 2, and 3.   ◀

DO EXERCISE 15.

▶ **Integer Coefficients and the Rational Roots Theorem**

It is not always easy to find the roots of a polynomial. However, if a polynomial has integer coefficients, there is a procedure that will yield all the rational roots.

---

12. Suppose that a polynomial of degree 5 with rational coefficients has -4, $7 - 2i$, and $3 + \sqrt{5}$ as roots. Find the other roots.

13. Find a polynomial of lowest degree with rational coefficients that has $2 + \sqrt{3}$ and $1 - i$ as some of its roots.

14. Find a polynomial of lowest degree with real coefficients that has $2i$ and 2 as some of its roots.

15. Find the other roots of

$$x^4 + x^3 - x^2 + x - 2,$$

given that i is a root.

THEOREM 12   **The Rational Roots Theorem**

Let

$$P(x) = a_n x^n + a_{n-1} x^{n-1} + \cdots + a_1 x + a_0,$$

where all the coefficients are integers. Consider a rational number denoted by c/d, where c and d are relatively prime (having no common factor besides 1 and -1). If c/d is a root of $P(x)$, then c is a factor of a_0 and d is a factor of a_n.

Proof. Since c/d is a root of $P(x)$, we know that

$$a_n \left(\frac{c}{d}\right)^n + a_{n-1} \left(\frac{c}{d}\right)^{n-1} + \cdots + a_1 \left(\frac{c}{d}\right) + a_0 = 0. \tag{1}$$

We multiply by d^n and get the equation

$$a_n c^n + a_{n-1} c^{n-1} d + \cdots + a_1 c d^{n-1} + a_0 d^n = 0. \tag{2}$$

Then we have

$$a_n c^n = (-a_{n-1} c^{n-1} - \cdots - a_1 c d^{n-2} - a_0 d^{n-1}) d.$$

This shows that d is a factor of $a_n c^n$. Now d has no factor in common with c, other than 1 or -1, because c and d are relatively prime. Thus d has no factor in common with c^n. So d is a factor of a_n.

In a similar way, we can show from Eq. (2) that

$$a_0 d^n = (-a_n c^{n-1} - a_{n-1} c^{n-2} d - \cdots - a_1 d^{n-1}) c.$$

Thus, c is a factor of $a_0 d^n$. Again, c is not a factor of d^n, so it must be a factor of a_0.

Example 10   Let $P(x) = 3x^4 - 11x^3 + 10x - 4$. Find the rational roots of $P(x)$. If possible, find the other roots.

Solution   By the rational roots theorem, if c/d is a root of $P(x)$, then c must be a factor of -4 and d must be a factor of 3. Thus the possibilities for c and d are

$$c: \quad 1, -1, 4, -4, 2, -2; \qquad d: \quad 1, -1, 3, -3.$$

Dividing each c by each d gives us the following possibilities for c/d:

$$\frac{c}{d}: \quad 1, -1, 4, -4, 2, -2, \frac{1}{3}, -\frac{1}{3}, \frac{4}{3}, -\frac{4}{3}, \frac{2}{3}, -\frac{2}{3}.$$

Of these 12 possibilities, we know that at most 4 of them could be roots because $P(x)$ is of degree 4. To find which are roots, we could use substitution, but synthetic division is usually more efficient. It is easier to first consider the integers. Then we consider the fractions, if the integers do not produce all the roots.

We try 1:

$$
\begin{array}{r|rrrrr}
1 & 3 & -11 & 0 & 10 & -4 \\
  &   & 3 & -8 & -8 & 2 \\
\hline
  & 3 & -8 & -8 & 2 & -2.
\end{array}
$$

We try -1:

$$
\begin{array}{r|rrrr}
-1 & 3 & -11 & 0 & 10 & -4 \\
 & & -3 & 14 & -14 & 4 \\
\hline
 & 3 & -14 & 14 & -4 & \;\rule{0.4pt}{8pt}\; 0.
\end{array}
$$

Since $P(1) = -2$, 1 is not a root; but $P(-1) = 0$, so -1 is a root. Using the results of the second synthetic division above, we can express $P(x)$ as

$$P(x) = (x + 1)(3x^3 - 14x^2 + 14x - 4).$$

We now use $3x^3 - 14x^2 + 14x - 4$ and check the other possible roots. We use synthetic division again, to see whether -1 is a double root:

$$
\begin{array}{r|rrr}
-1 & 3 & -14 & 14 & -4 \\
 & & -3 & 17 & -31 \\
\hline
 & 3 & -17 & 31 & \;\rule{0.4pt}{8pt}\; -35.
\end{array}
$$

We see that -1 is not a double root. We leave it to the student to verify that 2, -2, 4, and -4 are not roots. There are no other roots that are integers, so we start checking the fractions. We try $\frac{2}{3}$:

$$
\begin{array}{r|rrr}
\frac{2}{3} & 3 & -14 & 14 & -4 \\
 & & 2 & -8 & 4 \\
\hline
 & 3 & -12 & 6 & \;\rule{0.4pt}{8pt}\; 0.
\end{array}
$$

Since $P(\frac{2}{3}) = 0$, $\frac{2}{3}$ is a root. Again, using the results of the synthetic division, we can express $P(x)$ as

$$P(x) = (x + 1)(x - \tfrac{2}{3})(3x^2 - 12x + 6).$$

Since the factor $3x^2 - 12x + 6$ is quadratic, we can use the quadratic formula to find that the other roots are $2 + \sqrt{2}$ and $2 - \sqrt{2}$. These are irrational numbers. Thus the rational roots are -1 and $\frac{2}{3}$. ◀

DO EXERCISE 16.

Example 11 Let $P(x) = x^3 + 6x^2 + x + 6$. Find the rational roots of $P(x)$. If possible, find the other roots.

Solution By the rational roots theorem, if c/d is a root of $P(x)$, then c must be a factor of 6 and d must be a factor of 1. Thus the possibilities for c and d are

$$c: \quad 1, -1, 2, -2, 3, -3, 6, -6; \qquad d: \quad 1, -1.$$

Then the resulting possibilities for c/d are

$$\frac{c}{d}: \quad 1, -1, 2, -2, 3, -3, 6, -6.$$

Note that these are the same as the possibilities for c. If the leading coefficient is 1, we need check only the factors of the last coefficient as possibilities for rational roots.

There is another aid in eliminating possibilities for rational roots. Note that all coefficients of $P(x)$ are positive. Thus when any positive number is substituted in $P(x)$, we get a positive value, never 0. Therefore, no positive number can be a root. The only possibilities for roots are

$$-1, -2, -3, -6.$$

16. Let
$$P(x) = 2x^4 - 7x^3 - 35x^2 + 13x + 3.$$
If c/d is a rational root of $P(x)$, then:

a) What are the possibilities for c?

b) What are the possibilities for d?

c) What are the possibilities for c/d?

d) Find the rational roots.

e) If possible, find the other roots.

17. Let $P(x) = x^3 + 6x^2 + 10x + 3$. If c/d is a rational root of $P(x)$, then:

a) What are the possibilities for c?

b) What are the possibilities for d?

c) What are the possibilities for c/d?

d) How can you tell without substitution or synthetic division that there are no positive roots?

e) Find the rational roots of $P(x)$.

f) Find the other roots, if they exist.

18. Let $P(x) = x^3 + 5x^2 + 4x + 20$. If c/d is a rational root of $P(x)$, then:

a) What are the possibilities for c?

b) What are the possibilities for d?

c) What are the possibilities for c/d?

d) How can you tell without substitution or synthetic division that there are no positive roots?

e) Find the rational roots of $P(x)$.

f) Find the other roots, if they exist.

19. Let $P(x) = x^4 + x^2 + 2x + 6$.

   a) How do you know at the outset that this polynomial has no positive roots?

   b) Find the rational roots of $P(x)$.

20. a) Find the rational roots of $x^2 + 3x + 3$.

   b) Can you find the other roots of this polynomial? What are they? Why can you find them?

21. Let

$$P(x) = x^4 - \tfrac{1}{6}x^3 - \tfrac{4}{3}x^2 + \tfrac{1}{6}x + \tfrac{1}{3}.$$

   a) Which, if any, of the coefficients are *not* integers?

   b) What is the LCM of the denominators?

   c) Multiply by the LCM.

   d) Find the rational roots of the resulting polynomial.

   e) Are they rational roots of $P(x)$? Why?

We try -6:

$$
\begin{array}{r|rrrr}
-6 & 1 & 6 & 1 & 6 \\
   &   & -6 & 0 & -6 \\
\hline
   & 1 & 0 & 1 & \;|\; 0. \\
\end{array}
$$

Since $P(-6) = 0$, we know that -6 is a root. We could divide again to determine whether -6 is a double root, but since we can now factor $P(x)$ as a product of a linear and a quadratic polynomial, it is preferable to proceed that way. We have

$$P(x) = (x + 6)(x^2 + 1).$$

Now $x^2 + 1$ has the complex roots i and $-i$. Thus the only rational root of $P(x)$ is -6. ◀

THEOREM 13

If all the coefficients of a polynomial are positive real numbers, then there are no positive real roots.

DO EXERCISES 17 AND 18 ON THE PRECEDING PAGE. _____

Example 12   Find the rational roots of $x^4 + 2x^3 + 2x^2 - 4x - 8$.

Solution   Since the leading coefficient is 1, the only possibilities for rational roots are the factors of the last coefficient, -8:

$$1, -1, 2, -2, 4, -4, 8, -8.$$

But, using substitution or synthetic division, we find that none of the possibilities is a root. We leave it to the student to verify this. Thus there are no rational roots. ◀

The polynomial given in Example 12 has no rational roots. We can approximate the irrational roots of such a polynomial by graphing it and determining the x-intercepts (see Section 4.5).

DO EXERCISES 19 AND 20. _____

 Rational Coefficients

Suppose some (or all) of the coefficients of a polynomial are rational, but not integers. After multiplying on both sides by the LCM of the denominators, we can then find the rational roots.

Example 13   Let $P(x) = \tfrac{1}{12}x^3 - \tfrac{1}{12}x^2 - \tfrac{2}{3}x + 1$. Find the rational roots of $P(x)$.

Solution   The LCM of the denominators is 12. When we multiply on both sides by 12, we get

$$12P(x) = x^3 - x^2 - 8x + 12.$$

Note that all coefficients on the right are integers. Since the equation $P(x) = 0$

is equivalent to the equation $12P(x) = 0$, any root of $12P(x)$ is a root of $P(x)$. We leave it to the student to verify that 2 and -3 are the rational roots—in fact, the only roots—of $P(x)$.   ◀

DO EXERCISE 21 ON THE PRECEDING PAGE.

●EXERCISE SET **4.7**

A   Find the roots of the polynomial or polynomial equation, and state the multiplicity of each.

1. $P(x) = (x + 3)^2(x - 1)$

2. $-8(x - 3)^2(x + 4)^3x^4 = 0$

3. $x^3(x - 1)^2(x + 4) = 0$

4. $P(x) = (x^2 - 5x + 6)^2$

5. $P(x) = x^4 - 4x^2 + 3$

6. $P(x) = x^4 - 10x^2 + 9$

7. $P(x) = x^3 + 3x^2 - x - 3$

8. $P(x) = x^3 - x^2 - 2x + 2$

B   Find a polynomial of degree 3 with the given numbers as roots.

9. $-2, 3, 5$

10. $2, i, -i$

11. $-3, 2i, -2i$

12. $1 + 4i, 1 - 4i, -1$

13. $\sqrt{2}, -\sqrt{2}, \sqrt{3}$

14. Find a polynomial equation of degree 4 with -2 as a root of multiplicity 1, 3 as a root of multiplicity 2, and -1 as a root of multiplicity 1.

C   Suppose that a polynomial or polynomial equation of degree 5 with rational coefficients has the given numbers as roots. Find the other roots.

15. $6, -3 + 4i, 4 - \sqrt{5}$

16. $-2, 3, 4, 1 - i$

Find a polynomial of lowest degree with rational coefficients that has the given numbers as some of its roots.

17. $1 + i, 2$

18. $2 - i, -1$

19. $-4i, 5$

20. $2 - \sqrt{3}, 1 + i$

21. $\sqrt{5}, -3i$

22. $-\sqrt{2}, 4i$

Given that the polynomial or polynomial equation has the given root, find the other roots.

23. $x^4 - 5x^3 + 7x^2 - 5x + 6;$ $-i$

24. $x^4 - 16 = 0;$ $2i$

25. $x^3 - 6x^2 + 13x - 20 = 0;$ 4

26. $x^3 - 8;$ 2

D   Use Theorem 12 to list all *possible* rational roots.

27. $x^5 - 3x^2 + 1$

28. $x^7 + 37x^5 - 6x^2 + 12$

29. $15x^6 + 47x^2 + 2$

30. $10x^{25} + 3x^{17} - 35x + 6$

Find the rational roots, if they exist, of the polynomial or equation. If possible, find the other roots. Then write the equation or polynomial in factored form.

31. $x^3 + 3x^2 - 2x - 6$

32. $x^3 - x^2 - 3x + 3 = 0$

33. $x^3 - 3x + 2 = 0$

34. $x^3 - 3x + 4$

35. $x^3 - 5x^2 + 11x - 19$

36. $x^3 - 7x^2 - 2x + 23$

37. $5x^4 - 4x^3 + 19x^2 - 16x - 4 = 0$

38. $3x^4 - 4x^3 + x^2 + 6x - 2$

39. $x^4 - 3x^3 - 20x^2 - 24x - 8$

40. $x^4 + 5x^3 - 27x^2 + 31x - 10$

41. $x^3 - 4x^2 + 2x + 4 = 0$

42. $x^3 - 8x^2 + 17x - 4$

43. $x^3 + 8$

44. $x^3 - 8 = 0$

E

45. $\frac{1}{3}x^3 - \frac{1}{2}x^2 - \frac{1}{6}x + \frac{1}{6}$

46. $\frac{2}{3}x^3 - \frac{1}{2}x^2 + \frac{2}{3}x - \frac{1}{2}$

Find only the rational roots.

47. $x^4 + 32$

48. $x^6 + 8 = 0$

49. $x^3 - x^2 - 4x + 3 = 0$

50. $2x^3 + 3x^2 + 2x + 3$

51. $x^4 + 2x^3 + 2x^2 - 4x - 8 = 0$

52. $x^4 + 6x^3 + 17x^2 + 36x + 66 = 0$

53. $x^5 - 5x^4 + 5x^3 + 15x^2 - 36x + 20$

54. $x^5 - 3x^4 - 3x^3 + 9x^2 - 4x + 12$

● SYNTHESIS

Solve the equation. Use synthetic division, the quadratic formula, the theorems of this section, or whatever else you think might help.

55. $x^3 - 4x^2 + x - 4 = 0$

56. $x^3 - x^2 - 7x + 15 = 0$

57. $x^4 - 2x^3 - 2x - 1 = 0$

58. $x^4 + 9x^2 - 112 = 0$

59. The equation $x^2 + 2ax + b = 0$ has a double root. Find it.

60. Prove that a polynomial with positive coefficients cannot have a positive root.

61. The volume of a cube is 64 cm^3. Find the length of a side. (*Hint:* Solve $x^3 - 64 = 0$.)

62. The volume of a cube is 125 cm^3. Find the length of a side.

63. An open box of volume 48 cm^3 can be made from a piece of tin 10 cm on a side by cutting a square from each corner and folding up the edges. What is the length of a side of the squares?

64. An open box of volume 500 cm^3 can be made from a piece of tin 20 cm on a side by cutting a square from each corner and folding up the edges. What is the length of a side of the squares?

● CHALLENGE

65. Prove that every polynomial of odd degree, with real coefficients, has at least one real root.

66. Prove that every polynomial with real coefficients has a factorization into linear and quadratic factors (with real coefficients).

67. Show that $\sqrt{5}$ is irrational by considering the equation $x^2 - 5 = 0$.

68. Generalize the result of Exercise 67 to find which positive integers have rational square roots.

69. Solve: $x^3 + 7 = 0$.

70. Solve: $x^3 - 5 = 0$.

▱ TECHNOLOGY CONNECTION

71. *Total profit.* The total profit P from the production and sale of x units of a product is given by the polynomial

$$P(x) = x^3 - 31x^2 + 230x - 200.$$

a) Find the break-even values of the function.

b) For what nonnegative values of x does the company make a profit? (See Exercise 67 in Exercise Set 4.4.)

c) For what nonnegative values of x does the company have a loss (negative profit)?

72. The volume of a rectangular crate is $3\frac{3}{8}$ m^3. The length is 3 m longer than the width, and the height is 1 m less than the width. Find the dimensions of the crate.

OBJECTIVES

You should be able to:

A Use Descartes' rule of signs to find information about the number of real roots of a polynomial with real coefficients.

B Find upper and lower bounds of the roots of a polynomial with real coefficients, using synthetic division.

C Approximate zeros, or roots, of polynomial functions.

4.8

FURTHER HELPS IN FINDING ROOTS

A ## Descartes' Rule of Signs

The development of a rule that helps determine the number of positive real roots and the number of negative real roots of a polynomial is credited to Descartes. To use the rule, we must have the polynomial arranged in descending or ascending order, with no zero terms written in, the leading coefficient positive, and the *constant term* not 0. The **constant term** is the term that does not contain a variable. Then we determine the number of *variations of sign,* that is, the number of times, in reading through the polynomial, that successive coefficients are of different sign.

Example 1 Determine the number of variations of sign in the polynomial $P(x) = 2x^5 - 3x^2 + x + 4$.

Solution $\qquad P(x) = 2x^5 - 3x^2 + x + 4$

From positive to negative; a variation

Both positive; no variation

From negative to positive; a variation

The number of variations of sign is 2.

Note the following:

$$P(-x) = 2(-x)^5 - 3(-x)^2 + (-x) + 4$$
$$= -2x^5 - 3x^2 - x + 4.$$

We see that the number of variations of sign in $P(-x)$ is 1.

DO EXERCISES 1 AND 2. _____

We now state Descartes' rule, without proof.

| **THEOREM 14** | **Descartes' Rule of Signs** |

Let $P(x)$ be a polynomial with real coefficients and a nonzero constant term. The number of positive real roots of $P(x)$ is either:

   1. The same as the number of variations of sign in $P(x)$, or
   2. Less than the number of variations of sign in $P(x)$ by a positive even integer.

The number of negative real roots of $P(x)$ is either:

   3. The number of variations of sign in $P(-x)$, or
   4. Less than the number of variations of sign in $P(-x)$ by a positive even integer.

A root of multiplicity m must be counted m times.

Examples In each case, what does Descartes' rule of signs tell you about the number of positive real roots and the number of negative real roots?

2. $P(x) = 2x^5 - 5x^2 - 3x + 6$

The number of variations of sign in $P(x)$ is 2. Therefore, the number of positive real roots is either 2 or less than 2 by 2, 4, 6, and so on. Thus the number of positive real roots is either 2 or 0, since a negative number of roots has no meaning.

$$P(-x) = -2x^5 - 5x^2 + 3x + 6$$

The number of variations of sign in $P(-x)$ is 1. Thus there is exactly 1 negative real root. Since nonreal, complex conjugates occur in pairs, we also know the possible ways in which roots might occur, as summarized in the following table.

Total Number of Roots	5	5
Positives	2	0
Negatives	1	1
Nonreal, Complex	2	4

3. $P(x) = 5x^4 - 3x^3 + 7x^2 - 12x + 4$

There are 4 variations of sign. Thus the number of positive real roots

Determine the number of variations of sign in $P(x)$ and in $P(-x)$.

   1. $P(x) = 3x^5 - 2x^3 - x^2 + x - 2$

   2. $P(x) = 4x^7 + 6x^5 + 2x^3 - 5x^2 + 3$

In each case, what does Descartes' rule of signs tell you about the number of positive real roots and the number of negative real roots?

3. $P(x) = 5x^3 - 4x - 5$

4. $P(x) = 6x^6 - 5x^4 + 3x^3$
$\qquad - 7x^2 + x - 2 = 0$

5. $P(x) = x^5 - x^4 - x^3 + x^2 - x - 1$

6. $P(x) = 3x^3 - 2x^2 + 4x$

is either

$$4 \quad \text{or} \quad 4 - 2 \quad \text{or} \quad 4 - 4.$$

That is, the number of positive real roots is 4, 2, or 0.
$$P(-x) = 5x^4 + 3x^3 + 7x^2 + 12x + 4$$
There are 0 changes in sign, so there are no negative real roots.

4. $P(x) = 6x^6 - 2x^2 - 5x$

As stated, the polynomial does not satisfy the conditions of Descartes' rule of signs because the constant term is 0. But because x is a factor of every term, we know that the polynomial has 0 as a root. We can then factor as follows and get to the roots of the equation:
$$P(x) = x(6x^5 - 2x - 5)$$
Now we analyze $Q(x) = 6x^5 - 2x - 5$ and $Q(-x) = -6x^5 + 2x - 5$. The number of variations of sign in $Q(x)$ is 1. Therefore, there is exactly 1 positive real root. The number of variations of sign in $Q(-x)$ is 2. Thus the number of negative real roots is 2 or 0. The same results apply to $P(x)$.

DO EXERCISES 3–6. _____

B Bounds on Roots

We now look for intervals in which the roots of a polynomial are enclosed. We call b an **upper bound** for the roots of $P(x)$ if any root r is less than or equal to b. We call a a **lower bound** for the roots of $P(x)$ if any root r is greater than or equal to a. Note that upper and lower bounds are not unique because any number greater than b is also an upper bound and any number less than a is also a lower bound.

Upper and lower bounds give us an interval $[a, b]$ that contains the roots of the polynomial. We often look for the largest negative integer that is a lower bound and the smallest positive integer that is an upper bound. We will again use synthetic division, but before doing so note that when the leading coefficient of $P(x)$ is negative, it follows that the leading coefficient of the quotient (the first number in the bottom row in synthetic division) is also negative, regardless of our choice of the divisor. When this situation occurs, we make use of the fact that $P(x) = 0$ and $-P(x) = 0$ are equivalent equations. Thus we can apply the following theorem by just considering polynomials that have positive leading coefficients.

THEOREM 15   **Bounds for Real Roots of Polynomials**

Let $P(x)$ be a polynomial with real coefficients and a positive leading coefficient.

1. Suppose that $P(x)$ is divided by $x - b$, where b is positive and the remainder and all coefficients of the quotient are nonnegative. Then the number b is an upper bound to the roots of the polynomial.

2. Suppose that $P(x)$ is divided by $x - a$, where a is negative and the remainder and the coefficients of the quotient are alternately positive and negative with a 0 being considered either positive or negative. Then the number a is a lower bound to the roots of the polynomial.

Example 5   Determine the smallest positive integer b that is an upper bound and the largest negative integer a that is a lower bound to the roots of

$$P(x) = x^3 + 3x^2 - 9x - 13.$$

Solution   We choose some *positive* number b and divide by $x - b$, using synthetic division. Let's try 2.

$$
\begin{array}{r|rrrr}
2 & 1 & 3 & -9 & -13 \\
  &   & 2 & 10 & 2 \\
\hline
  & 1 & 5 & 1 & -11
\end{array}
$$

Since the remainder is negative, the theorem does not guarantee that 2 is an upper bound. We choose a larger integer, this time 3.

$$
\begin{array}{r|rrrr}
3 & 1 & 3 & -9 & -13 \\
  &   & 3 & 18 & 27 \\
\hline
  & 1 & 6 & 9 & 14
\end{array}
$$

Since there are no negative numbers in the bottom row, 3 is an upper bound. Indeed, 3 is the smallest positive integer that is guaranteed by Theorem 15 to be an upper bound to the roots of $P(x)$.*

Next, we choose some *negative* number a and divide by $x - a$, using synthetic division. Let's try -5.

$$
\begin{array}{r|rrrr}
-5 & 1 & 3 & -9 & -13 \\
   &   & -5 & 10 & -5 \\
\hline
   & 1 & -2 & 1 & -18
\end{array}
$$

Since the numbers in the bottom row alternate from positive to negative, it follows that -5 is a lower bound. We leave it to the student to verify that -4 is not a lower bound.

Interval, $[-5, 3]$, that contains roots

Thus, -5 is the largest negative integer that is a lower bound and 3 is the smallest positive integer that is an upper bound. Look at the graph of $P(x)$ above. The roots are confined to the closed interval $[-5, 3]$. They

---

* The converse of Theorem 15 is not true. That is, if the positive number b is an upper bound to the roots, it is not necessarily true that when we divide by $x - b$, all the coefficients of the quotient and the remainder will be nonnegative. Thus when we divide and find negative numbers in the bottom row, this does not mean that the number b is *not* an upper bound. We simply do not know whether it is or not. For example, while the roots of $P(x) = x^2 - 5x + 6$ are 2 and 3, Theorem 15 would not predict that 3 or 4 is an upper bound to the roots.

In each case, determine, as guaranteed by Theorem 15, the smallest positive integer b that is an upper bound and the largest negative integer a that is a lower bound to the roots.

7. $P(x) = 5x^3 + 18x^2 + 3x - 2$

8. $P(x) = 5x^4 - 18x^2 + 3x - 2$

9. $P(x) = -4x^3 + 5x^2 - x + 2$
   [*Hint:* First find $-P(x)$.]

are approximately -4.4, -1.2, and 2.5. In Example 6, we will consider how to find such approximations.

DO EXERCISES 7–9.

The Method of Bisection: Approximating Solutions

Whenever we find the roots, or zeros, of a function $P(x)$, we have solved the equation $P(x) = 0$. How might a scientist proceed when the need arises to find roots of polynomials? One might use the rational roots theorem to check for rational roots, presuming the coefficients are integers. But even if this is the case, there will be rational roots only about 20% of the time. A more straightforward method of operating is to use a computer software package or a graphing calculator and estimate where the graph crosses the x-axis. Many such tools have "zoom-in" features that allow very good approximation, graphically, by considering the graph over smaller intervals. As in Example 6, we can look near this estimate for function values that differ in sign, which would let us know that there is a root somewhere between the input values.

In fact, most computer software packages also have provisions to go on and estimate the real roots, but let us operate as if that were not the case.

Example 6   Approximate the irrational real roots of the polynomial $P(x) = 2x^3 - x + 2$ to the nearest hundredth.

Solution   We use the rational roots theorem and find that there are no rational roots. If there were, we would factor the polynomial and consider the polynomial that results from factoring out the linear factors involving the rational roots. We then use a computer software package or graphing calculator and graph the polynomial. Below is a computer-generated graph of

$$P(x) = 2x^3 - x + 2.$$

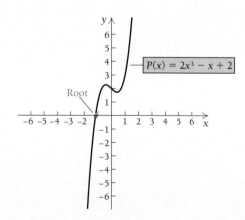

We see from the graph that there seems to be a root between -1 and -2. To be sure, we check function values and find that $P(-1) = 1$ and $P(-2) = -12$. Since the signs differ, there must be a root somewhere between -1 and -2. To find a better approximation, we use computation by hand or a calculator or synthetic division.

x	$P(x)$
-1	1
-1.1	0.44
-1.2	-0.26
-1.15	0.11
-1.17	-0.03
-1.16	0.04

$\left.\begin{array}{l}\end{array}\right\}$ Since the signs of $P(-1)$ and $P(-1.1)$ are the same, there is no root between -1 and -1.1.

$\left.\begin{array}{l}\end{array}\right\}$ Since $P(-1.2) < 0$ and $P(-1.1) > 0$, there is a root somewhere between -1.1 and -1.2. We try -1.15, halfway between.

$\left.\begin{array}{l}\end{array}\right\}$ Since $P(-1.2) < 0$ and $P(-1.15) > 0$, there is a root somewhere between -1.2 and -1.15. We try -1.17.

$\left.\begin{array}{l}\end{array}\right\}$ Since $P(-1.17)$ and $P(-1.15)$ have opposite signs, there is a root somewhere between -1.17 and -1.15. We try -1.16.

$\left.\begin{array}{l}\end{array}\right\}$ Since $P(-1.17) < 0$ and $P(-1.16) > 0$, there is a root somewhere between -1.17 and -1.16.

10. Approximate the irrational roots of $x^3 - 3x^2 + 1$ to the nearest hundredth.

We stop when we have the desired accuracy. In this case, we were asked to approximate to the nearest hundredth. Since -0.03 is closer to zero than 0.04, we will accept -1.17 as a better approximation, to the nearest hundredth. There are no other irrational roots, as evidenced by the graph of $P(x)$. ◀

The method for approximating roots used in Example 6 is known as the **method of bisection.** There are many other methods of approximating used in mathematics. In fact, there is a branch of mathematics known as *numerical analysis,* which deals with approximation in great detail.

DO EXERCISE 10.

● **EXERCISE SET** **4.8**

A   What does Descartes' rule of signs tell you about the number of positive real roots and the number of negative real roots?

1. $P(x) = 3x^5 - 2x^2 + x - 1$
2. $P(x) = 5x^6 - 3x^3 + x^2 - x$
3. $P(x) = 6x^7 + 2x^2 + 5x + 4$
4. $P(x) = -3x^5 - 7x^3 - 4x - 5$
5. $P(x) = 3p^{18} + 2p^4 - 5p^2 + p + 3$
6. $P(x) = 5t^{12} - 7t^4 + 3t^2 + t + 1$
7. $P(x) = 7x^6 + 3x^4 - x - 10$
8. $P(x) = -z^{10} + 8z^7 + z^3 + 6z - 1$
9. $P(x) = -4t^5 - t^3 + 2t^2 + 1$
10. $P(x) = x^6 + 2x^4 - 9x^3 - 4$
11. $P(x) = y^4 + 13y^3 - y + 5$

12. $P(x) = 7y^9 - 6y^6 + 5y^3 + 3y^2 - 2$

B   Determine the smallest positive integer that is guaranteed by Theorem 15 to be an upper bound to the roots. Then find the largest negative integer that is guaranteed to be a lower bound to the roots.

13. $P(x) = 3x^4 - 15x^2 + 2x - 3$
14. $P(x) = 4x^4 - 14x^2 + 4x - 2$
15. $P(x) = 6x^3 - 17x^2 - 3x - 1$
16. $P(x) = 5x^3 - 15x^2 + 5x - 4$
17. $P(x) = -3x^4 + 15x^3 - 2x + 3$
18. $P(x) = -4x^4 + 17x^3 - 3x + 2$
19. $P(x) = 6x^3 + 15x^2 + 3x - 1$
20. $P(x) = 6x^3 + 12x^2 + 5x - 3$

A , B   What does Descartes' rule of signs tell you about the roots of the polynomial? Find upper and lower bounds to the real roots.

21. $P(x) = x^4 - 2x^2 + 12x - 8$

22. $P(x) = x^4 - 6x^2 + 20x - 24$

23. $P(x) = x^4 - 2x^2 - 8$

24. $P(x) = 3x^4 - 5x^2 - 4$

25. $P(x) = x^4 - 9x^2 - 6x + 4$

26. $P(x) = x^4 - 21x^2 + 4x + 6$

27. $P(x) = x^4 + 3x^2 + 2$

28. $P(x) = x^4 + 5x^2 + 6$

C   Approximate the irrational roots to the nearest tenth.

29. $x^3 - 3x - 2 = 0$ **30.** $x^3 - 3x^2 + 3 = 0$

31. $x^3 - 3x - 4 = 0$ **32.** $x^3 - 3x^2 + 5 = 0$

33. $x^4 + x^2 + 1 = 0$ **34.** $x^4 + 2x^2 + 2 = 0$

35. $x^4 - 6x^2 + 8 = 0$ **36.** $x^4 - 4x^2 + 2 = 0$

37. $x^5 + x^4 - x^3 - x^2 - 2x - 2 = 0$

38. $x^5 - 2x^4 - 2x^3 + 4x^2 - 3x + 6 = 0$

39. ▧ The following equation has a solution between 0 and 1. Approximate it to the nearest hundredth.
$$2x^5 + 2x^3 - x^2 - 1 = 0$$

40. ▧ The following equation has a solution between 1 and 2. Approximate it to the nearest hundredth.
$$x^4 - 2x^3 - 3x^2 + 4x + 2 = 0$$

Approximate the irrational roots to the nearest hundredth.

41. ▧ $P(x) = x^3 - 2x^2 - x + 4$

42. ▧ $P(x) = x^3 - 4x^2 + x + 3$

● SYNTHESIS _____

43. Prove that for any positive even integer n, $x^n - 1$ has only two real roots.

44. Prove that for any odd positive integer n, $x^n - 1$ has only one real root.

▧ TECHNOLOGY CONNECTION _____

Use a grapher to graph each of the following functions, approximate its roots, and solve the problems in Exercises 49 and 50.

45. $f(x) = x^3 - 9x^2 + 27x + 50$

46. $f(x) = x^3 - 3x + 1$

47. $f(x) = x^4 + 4x^3 - 36x^2 - 160x + 300$

48. $f(x) = x^6 + 4x^5 - 54x^4 - 160x^3 + 641x^2 + 828x - 1260$

49. See Exercise 46 in Exercise Set 4.6. After what time will the concentration be 0?

50. *Multiple investments.*   At the beginning of a year, $2000 is invested in a bank at a certain interest rate, compounded annually. At the beginning of the next year, $1500 is invested in a second bank at the same interest rate. At the beginning of the third year, $1800 is invested in a third bank at the same interest rate. At the beginning of the fourth year, there is a total of $6213.02 in all three accounts. What is the interest rate?

OBJECTIVE

You should be able to:

A   Sketch a graph of a rational function, being careful to sketch and label all asymptotes.

4.9

RATIONAL FUNCTIONS

A   A **rational function** is a function definable as a quotient of two polynomials. Here are some examples:

$$f(x) = \frac{x^2 + 3x - 5}{x + 4}, \qquad f(x) = \frac{8}{x^2 + 1}, \qquad f(x) = \frac{6x^5 - 7x + 11}{4}.$$

DEFINITION   **Rational Function**

A *rational function* is a function f that can be described by

$$f(x) = \frac{P(x)}{Q(x)},$$

where $P(x)$ and $Q(x)$ are polynomials with no common factor other than 1 and -1, and where $Q(x)$ is not the zero polynomial. The domain of f consists of all inputs x for which $Q(x) \neq 0$.

Polynomial functions are themselves special kinds of rational functions, since $Q(x)$ can be the polynomial 1. Here we are interested in graphing rational functions in which the denominator is not a constant. We begin with the simplest of such functions.

Example 1   Graph: $f(x) = \dfrac{1}{x}$.

Solution   Although we have seen this graph before, we now consider it from the viewpoint of setting up the general graphing of rational functions.

a) The domain of this function consists of all real numbers except 0.
b) The function is odd. Thus the graph is symmetric with respect to the origin (because replacing x with $-x$ and y with $-y$ produces an equivalent equation).
c) Now we find some function values and use the symmetry with respect to the origin to get others.

x	1	2	3	4	5	$\frac{1}{2}$	$\frac{1}{3}$	$\frac{1}{4}$	$\frac{1}{5}$
$f(x)$	1	$\frac{1}{2}$	$\frac{1}{3}$	$\frac{1}{4}$	$\frac{1}{5}$	2	3	4	5

d) We plot these points, using symmetry to find the points in the third quadrant.

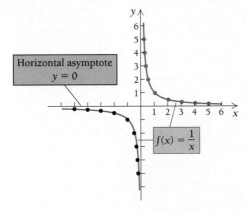

The points indicated by ● are obtained from the table. The points marked ● are obtained by reflection across the origin.   ◄

Asymptotes

Note that the curve in Example 1 does not touch either axis, but comes very close. As x becomes very large positively without bound, the curve comes very near to the x-axis. Similarly, as x becomes very small negatively, the curve also gets closer to the x-axis. In fact, we can find points as close to the x-axis as we please by choosing x large enough or x small enough. You can check this out more thoroughly using your calculator. In other words,

$$f(x) \to 0 \text{ as } x \to \infty \quad \text{and} \quad f(x) \to 0 \text{ as } x \to -\infty.$$

We read "$f(x) \to 0$ as $x \to \infty$" as "$f(x)$ approaches 0 as x gets larger positively without bound." We read "$f(x) \to 0$ as $x \to -\infty$" as "$f(x)$ approaches 0 as x gets smaller negatively without bound."

From these facts, we say that the curve approaches the x-axis *asymptotically* and that the x-axis is a **horizontal asymptote** to the curve.

DEFINITION	**Horizontal Asymptote**

The line $y = b$ is a *horizontal asymptote* if either or both of the following statements are true:

$$f(x) \to b \text{ as } x \to \infty \quad \text{or} \quad f(x) \to b \text{ as } x \to -\infty.$$

The following figures illustrate horizontal asymptotes. Note that the curve gets closer to the line $y = b$ either as $x \to \infty$ or as $x \to -\infty$. Keep in mind that the symbols ∞ and $-\infty$ convey the idea of increasing positively without bound or decreasing negatively without bound.

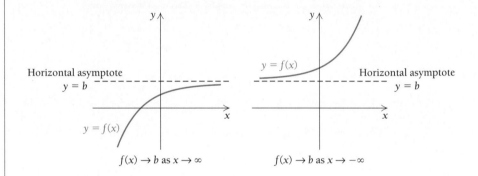

$f(x) \to b$ as $x \to \infty$ $f(x) \to b$ as $x \to -\infty$

Note too in Example 1 that as x-values get closer and closer to 0 from the left, the function values (y-values) decrease negatively without bound. In fact, by selecting x close enough to 0, we can get the function values as small negatively as we wish. Similarly, as we approach 0 from the right, the function values increase positively without bound. You can again check this more thoroughly on a calculator. In other words,

$$f(x) \to -\infty \text{ as } x \to 0^- \quad \text{and} \quad f(x) \to \infty \text{ as } x \to 0^+.$$

We read "$f(x) \to -\infty$ as $x \to 0^-$" as "$f(x)$ gets smaller negatively without bound as x approaches 0 from the left." We read "$f(x) \to \infty$ as $x \to 0^+$" as "$f(x)$ gets larger positively without bound as x approaches 0 from the right." We say that the y-axis is a **vertical asymptote** to this curve.

DEFINITION	**Vertical Asymptote**

The line $x = a$ is a *vertical asymptote* if any of the following limit statements are true:

$$f(x) \to \infty \text{ as } x \to a^- \quad \text{or} \quad f(x) \to -\infty \text{ as } x \to a^- \quad \text{or}$$
$$f(x) \to \infty \text{ as } x \to a^+ \quad \text{or} \quad f(x) \to -\infty \text{ as } x \to a^+.$$

The following figure shows the four ways in which a vertical asymptote can occur.

$f(x) \to \infty$ as $x \to a^-$ $f(x) \to -\infty$ as $x \to a^-$ $f(x) \to \infty$ as $x \to a^+$ $f(x) \to -\infty$ as $x \to a^+$

We now have the following theorem regarding horizontal and vertical asymptotes.

THEOREM 16

The graph of a rational function may or may not cross a horizontal asymptote. The graph of a rational function never crosses a vertical asymptote. If a is a zero of the denominator of the rational function, then the line $x = a$ is a vertical asymptote. (Remember that this theorem applies to rational functions, defined such that the numerator and the denominator have no common factor other than 1 or -1.)

Note that if a is a zero of the denominator, then the function is not defined at $x = a$. Thus it cannot have a function value at a, so the graph cannot cross the line $x = a$, which happens to be a vertical asymptote.

Example 2   Graph: $f(x) = \dfrac{1}{x^2}$.

Solution

a) Note that this function is defined for all values of x except 0. Note too that for all other values of x, $x^2 > 0$, so $1/x^2 > 0$. Therefore, the entire graph is above the x-axis.

b) Since $1/x^2 \to \infty$ as $x \to 0^-$, we know that the line $x = 0$, the y-axis, is a vertical asymptote. We also know this because $1/x^2 \to \infty$ as $x \to 0^+$. As $|x|$ gets very close to 0, $f(x)$ increases positively without bound.

c) Since $1/x^2 \to 0$ as $x \to \infty$ and $1/x^2 \to 0$ as $x \to -\infty$, it follows that the line $y = 0$, the x-axis, is a horizontal asymptote. That is, as $|x|$ gets very large, $f(x)$ approaches 0.

d) This function is even. Therefore, the graph is symmetric with respect to the y-axis (because replacing x with $-x$ produces an equivalent equation).

e) We compute some function values and, using those points and the preceding information, sketch the following graph.

x	1	2	3	4	$\frac{1}{2}$	$\frac{1}{3}$
$f(x)$	1	$\frac{1}{4}$	$\frac{1}{9}$	$\frac{1}{16}$	4	9

Sketch a graph of the rational function. Use graph paper.

1. $f(x) = \dfrac{4}{x}$

2. $f(x) = \dfrac{32}{x^4}$

3. $f(x) = \dfrac{-1}{x^2}$

4. $f(x) = \dfrac{-1}{x^3}$

Sketch a graph of the rational function. Use graph paper.

5. $f(x) = \dfrac{1}{x+5}$

6. $f(x) = \dfrac{3x+1}{x}$

7. $f(x) = 3 - \dfrac{2}{x}$

8. $f(x) = \dfrac{1}{3x+15}$

[*Hint:* $3x + 15 = 3(x + 5)$.]
Compare with Margin Exercise 5.

9. $f(x) = \dfrac{1}{(x-3)^2}$

10. $f(x) = \dfrac{-3x^2 + 1}{x^2}$

(*Hint:* Divide the numerator by the denominator.)

The graph is as follows. Points marked ● are obtained from the table, and points marked ● are obtained by reflection across the *y*-axis.

The following figure shows the four general curves of the graphs of $f(x) = a/x^n$.

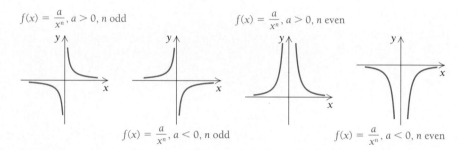

DO EXERCISES 1–4.

Using the idea of transformation (Section 3.8), we can easily graph variations of the preceding curves.

Example 3 Graph: $f(x) = \dfrac{1}{x-2}$.

Solution The graph of $f(x) = 1/(x - 2)$ is a translation of the graph of $f(x) = 1/x$ to the right 2 units. The *x*-axis is a horizontal asymptote to the curve, and the line $x = 2$ is a vertical asymptote.

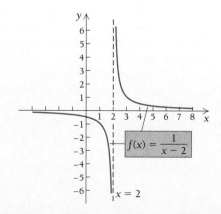

DO EXERCISES 5–10.

Occurrences of Asymptotes

It is important in graphing rational functions to determine where the asymptotes, if any, occur.

> Vertical asymptotes are easy to locate when a denominator is factored. The x-values that make a denominator zero but the numerator nonzero are those of the vertical asymptotes.

Examples Determine the vertical asymptotes.

4. $f(x) = \dfrac{3x - 2}{x(x - 5)(x + 3)}$

The vertical asymptotes are the lines $x = 0$, $x = 5$, and $x = -3$.

5. $f(x) = \dfrac{x - 2}{x^3 - x}$

We factor the denominator:

$$x^3 - x = x(x + 1)(x - 1).$$

The vertical asymptotes are $x = 0$, $x = -1$, and $x = 1$. ◀

DO EXERCISES 11 AND 12.

> Horizontal asymptotes occur when the degree of the numerator is less than or equal to the degree of the denominator.

Let us first consider a function for which the degree of the numerator is less than the degree of the denominator:

$$f(x) = \frac{2x + 3}{x^3 - 2x^2 + 4}.$$

We multiply by $(1/x^3)/(1/x^3)$, to obtain

$$f(x) = \frac{\dfrac{2}{x^2} + \dfrac{3}{x^3}}{1 - \dfrac{2}{x} + \dfrac{4}{x^3}}.$$

Let us now consider what happens to the function values as $|x|$ becomes very large. Each expression with x in its denominator takes on smaller and smaller values, approaching 0. Thus as the numerator approaches 0 and the denominator approaches 1, the entire expression takes on values closer and closer to 0. Therefore,

$$f(x) \to 0 \text{ as } x \to -\infty \quad \text{and} \quad f(x) \to 0 \text{ as } x \to \infty,$$

so the x-axis is a horizontal asymptote. Check this more thoroughly on your calculator.

Next we consider a function for which the numerator and the denomi-

---

Find the vertical asymptotes.

11. $f(x) = \dfrac{x + 3}{x^3 - x^2 - 6x}$

12. $f(x) = \dfrac{x^2 + 5}{2x^3 + x^2 - 8x - 4}$

 TECHNOLOGY CONNECTION

Since rational functions can be some of the most difficult functions to work with, graphers are especially useful in this section. Two notes of caution, however:

1. You may need to examine a function in several different viewing boxes before you see the complete picture. In that case, zoom in to see some of the detail of the graph, and then zoom out to see what happens far out on the positive and negative x-axes.
2. As we discussed in an earlier section, if your grapher offers the option to plot individual points without connecting them, you will probably find it useful to do so when dealing with rational functions that have vertical asymptotes. Otherwise, your grapher may draw a near-vertical line where a "gap" should be.

Use a grapher to draw the graphs of the following rational functions. Examine each graph in different viewing boxes. Find all intercepts and asymptotes.

TC 1. $f(x) = \dfrac{3.5x - 8.6}{x^2 + 2.2x - 19.04}$

TC 2. $f(x) = \dfrac{3x^2 - 4}{2x^2 - 4x + 9}$

TC 3. $f(x) = \dfrac{4x^4 - 5x + 7}{3x^2 + 6}$

13. For which of the following is the x-axis an asymptote?

a) $f(x) = \dfrac{3x^4 - x^2 + 4}{18x^4 - x^3 + 44}$

b) $f(x) = \dfrac{17x^2 + 14x - 5}{x^3 - 2x - 1}$

c) $f(x) = \dfrac{135x^5 - x^2}{x^7}$

Find the horizontal asymptotes.

14. $f(x) = \dfrac{3x^3 + 4x - 9}{6x^3 - 7x^2 + 3}$

15. $f(x) = \dfrac{9x^4 - 7x^2 - 9}{3x^4 + 7x^2 + 9}$

nator have the same degree:

$$f(x) = \frac{3x^2 + 2x - 4}{2x^2 - x + 1} = \frac{3x^2 + 2x - 4}{2x^2 - x + 1} \cdot \frac{\dfrac{1}{x^2}}{\dfrac{1}{x^2}}$$

$$= \frac{3 + \dfrac{2}{x} - \dfrac{4}{x^2}}{2 - \dfrac{1}{x} + \dfrac{1}{x^2}}.$$

As $|x|$ gets very large, the numerator approaches 3 and the denominator approaches 2. Therefore, the function gets very close to $\frac{3}{2}$. Check this more thoroughly on your calculator. Thus,

$$f(x) \to \tfrac{3}{2} \text{ as } x \to \infty \quad \text{and} \quad f(x) \to \tfrac{3}{2} \text{ as } x \to -\infty,$$

so the line $y = \frac{3}{2}$ is a horizontal asymptote.

From this example, we can see that when the degrees of the numerator and the denominator are the same, the asymptote can be determined by dividing the leading coefficients of the two polynomials.

THEOREM 17

Whenever the degree of the numerator is less than the degree of the denominator, the x-axis, or $y = 0$, is a horizontal asymptote.

Whenever the degree of the numerator equals the degree of the denominator, the line $y = a/b$ is a horizontal asymptote, where a is the leading coefficient of the numerator and b is the leading coefficient of the denominator.

Examples Determine the horizontal asymptotes.

6. $f(x) = \dfrac{5x^3 - x^2 + 7}{3x^5 + x - 10}$

The line $y = 0$ is a horizontal asymptote.

7. $f(x) = \dfrac{-7x^4 - 10x^2 + 1}{11x^4 + x - 2}$

The line $y = -\frac{7}{11}$ is a horizontal asymptote. ◀

DO EXERCISES 13–15.

Oblique Asymptotes

There are asymptotes that are neither vertical nor horizontal. Lines that are nonvertical and nonhorizontal can be asymptotes. They are called **oblique asymptotes.** An example is shown in the graph on the following page.

Oblique asymptotes occur when the degree of the numerator is 1 more than the degree of the denominator. One way to find such an asymptote is to divide the numerator by the denominator. Consider

$$f(x) = \frac{2x^2 - 7x - 7}{x - 5}.$$

Find the oblique asymptotes.

16. $f(x) = \dfrac{3x^2 - 7x + 6}{x - 2}$

When we divide the numerator by the denominator, we obtain a quotient of $2x + 3$ and a remainder of 8. Thus,

$$f(x) = (2x + 3) + \frac{8}{x - 5}.$$

Now we can see that when $|x|$ gets very large, $8/(x - 5)$ approaches 0, and the y-values approach $2x + 3$. Thus, for very large and very small values of x, $f(x)$ approaches the line $y = 2x + 3$, and $y = 2x + 3$ is an oblique asymptote.

DEFINITION   **Oblique Asymptote**

The line $y = mx + b$ is an *oblique asymptote* of the rational function $f(x) = P(x)/Q(x)$ if $f(x)$ can be expressed as

$$f(x) = (mx + b) + g(x), \quad \text{where } g(x) \to 0 \text{ as } |x| \to \infty.$$

17. $f(x) = \dfrac{5x^3 + 2x + 1}{x^2 - 4}$

DO EXERCISES 16 AND 17.

The following summarizes the conditions under which asymptotes occur.

Asymptotes of a rational function can occur as follows.

1. *Vertical asymptotes* occur at the x-values that make the denominator zero but the numerator nonzero.
2. The *x-axis is an asymptote* when the degree of the numerator is less than the degree of the denominator.
3. *Horizontal asymptotes other than the x-axis* occur when the numerator and the denominator have the same degree.
4. *Oblique asymptotes* occur when the degree of the numerator is 1 more than the degree of the denominator.

Zeros, or Roots

Zeros, or roots, of a rational function occur when the numerator is 0 but the denominator is not. The zeros of a function occur at points where the

Find the zeros.

18. $f(x) = \dfrac{x(x - 3)(x + 5)}{(x + 2)(x - 4)}$

19. $f(x) = \dfrac{x^3 + 2x^2 - 3x}{x^2 + 5}$

graph crosses the x-axis. Therefore, knowing the zeros helps in graphing. If the numerator can be factored, the zeros are easy to determine.

Example 8 Find the zeros of the rational function

$$f(x) = \frac{x^3 - x^2 - 6x}{x^2 - 3x + 2}.$$

Solution We factor the numerator and the denominator:

$$f(x) = \frac{x(x + 2)(x - 3)}{(x - 1)(x - 2)}.$$

The x-values making the numerator zero are 0, -2, and 3. Since none of these makes the denominator zero, they are the zeros of the function. The graph will cross the x-axis at $(0, 0)$, $(-2, 0)$, and $(3, 0)$. ◀

DO EXERCISES 18 AND 19. _____

The following is an outline of the procedure to be followed in graphing rational functions.

To graph a rational function:

  a) Determine any symmetries.
  b) Determine any horizontal or oblique asymptotes and sketch them.
  c) Factor the numerator and the denominator.
   i) Determine any vertical asymptotes and sketch them.
   ii) Determine the zeros, if possible, and plot them.
  d) Determine where the function values are positive or negative.
  e) Make a table of values and plot them.
  f) Sketch the curve.

Example 9 Graph: $g(x) = \dfrac{1}{x^2 - 3}.$

Solution We follow the outline above.

  a) The function is even, so the graph is symmetric with respect to the y-axis.
  b) Since the degree of the numerator is less than the degree of the denominator, the x-axis is a horizontal asymptote. There are no oblique asymptotes.
  c) We can factor the denominator, but not the numerator:

$$g(x) = \frac{1}{(x - \sqrt{3})(x + \sqrt{3})}.$$

The zeros of the denominator are $\pm\sqrt{3}$. Thus, $x = \sqrt{3}$ and $x = -\sqrt{3}$ are vertical asymptotes of the graph of g. We draw these asymptotes with dashed lines. The function g has no zeros.
  d) The vertical asymptotes divide the x-axis into intervals, as follows. To find where the function is positive or negative, we use the procedure of Section 4.4. We try a test point in each interval.

Interval	$(-\infty, -\sqrt{3})$	$(-\sqrt{3}, \sqrt{3})$	$(\sqrt{3}, \infty)$
Test Value	$g(-4) = \frac{1}{13}$	$g(0) = -\frac{1}{3}$	$g(4) = \frac{1}{13}$
Sign of $g(x)$	$+$	$-$	$+$
Location of Points on Graph	Above x-axis	Below x-axis	Above x-axis

e) and f) We use the preceding information, compute some other function values, and complete the graph of g.

x	$g(x)$, approximately
-5	0.05
-3	0.2
-2	1
-1	-0.5
0	-0.3
1	-0.5
2	1
3	0.2
5	0.05

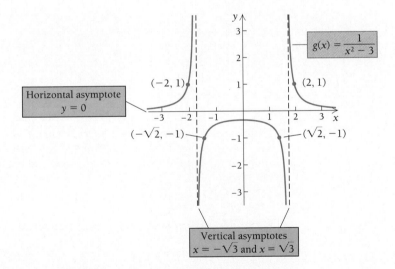

DO EXERCISES 20 AND 21.

It is sometimes written that graphs do not cross their asymptotes. The graph of

$$f(x) = \frac{x^2 - 1}{x^2 + x - 6}$$

shown on the following page crosses the horizontal asymptote $y = 1$. The graph of

$$f(x) = \frac{2x^3}{x^2 + 1}$$

Sketch a graph of the rational function.

20. $f(x) = \dfrac{1}{x^2 - 1}$

21. $f(x) = \dfrac{1}{x^2 + 4}$

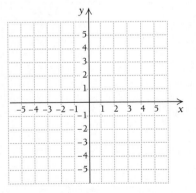

shown below crosses its oblique asymptote $y = 2x$. Thus graphs can cross horizontal or oblique asymptotes. But it is true that graphs of functions do not cross vertical asymptotes.

• EXERCISE SET 4.9

A Sketch a graph of the rational function. Be sure to sketch and label all the asymptotes.

1. $f(x) = \dfrac{1}{x - 3}$

2. $f(x) = \dfrac{1}{x - 5}$

3. $f(x) = \dfrac{-2}{x - 5}$

4. $f(x) = \dfrac{-3}{x - 3}$

5. $f(x) = \dfrac{2x + 1}{x}$

6. $f(x) = \dfrac{3x - 1}{x}$

7. $f(x) = \dfrac{1}{(x - 2)^2}$

8. $f(x) = \dfrac{-2}{(x - 3)^2}$

9. $f(x) = \dfrac{2}{x^2}$

10. $f(x) = \dfrac{1}{3x^2}$

11. $f(x) = \dfrac{1}{x^2 + 3}$

12. $f(x) = \dfrac{-1}{x^2 + 2}$

13. $f(x) = \dfrac{x - 1}{x + 2}$

14. $f(x) = \dfrac{x - 2}{x + 1}$

15. $f(x) = \dfrac{3x}{x^2 + 5x + 4}$

16. $f(x) = \dfrac{x + 3}{2x^2 - 5x - 3}$

17. $f(x) = \dfrac{x^2 - 4}{x - 1}$

18. $f(x) = \dfrac{x^2 - 9}{x + 1}$

19. $f(x) = \dfrac{x^2 + x - 2}{2x^2 + 1}$

20. $f(x) = \dfrac{x^2 - 2x - 3}{3x^2 + 2}$

21. $f(x) = \dfrac{x - 1}{x^2 - 2x - 3}$

22. $f(x) = \dfrac{x + 2}{x^2 + 2x - 15}$

23. $f(x) = \dfrac{x + 2}{(x - 1)^3}$

24. $f(x) = \dfrac{x - 3}{(x + 1)^3}$

25. $f(x) = \dfrac{x^3 + 1}{x}$

26. $f(x) = \dfrac{x^3 - 1}{x}$

27. $f(x) = \dfrac{x^3 + 2x^2 - 15x}{x^2 - 5x - 14}$

28. $f(x) = \dfrac{x^3 + 2x^2 - 3x}{x^2 - 25}$

29. $f(x) = \dfrac{5x^4}{x^4 + 1}$

30. $f(x) = \dfrac{x + 1}{x^2 + x - 6}$

31. $f(x) = \dfrac{x^2 - x - 2}{x + 2}$

32. $f(x) = \dfrac{x^2}{x^2 - x - 2}$

33. ***Distance as a function of time.*** The distance from Kansas City to Indianapolis is 500 miles. A car makes the trip in time t, in hours, at various speeds r, in miles per hour.

a) Express the time t as a rational function of the speed r.

b) Sketch a graph of the function over the interval $(0, \infty)$.

34. ***Electrical resistance.*** The total resistance in a circuit with two resistors in parallel is given by

$$R = \dfrac{R_1 R_2}{R_1 + R_2},$$

where R_1 and R_2 are positive.

a) A circuit has a 10-ohm resistor and a variable resistor in parallel. Express the total resistance of the circuit as a function of the resistance of the variable resistor, r.

b) Graph the function.

c) What are the domain and the range?

d) What number does the resistance of the circuit approach as the resistance of the variable resistor gets larger and larger?

● **S Y N T H E S I S** _____

Sketch a graph of the function.

35. $f(x) = \dfrac{x^3 + 4x^2 + x - 6}{x^2 - x - 2}$

36. $f(x) = \dfrac{2x^3 + x^2 - 8x - 4}{x^3 + x^2 - 9x - 9}$

● **C H A L L E N G E** _____

37. *Time of flight.*   A pilot can make a round-trip flight between two cities of the same latitude in time t_0 flying at
velocity v, if there is no wind. If there is a wind of velocity
w due east or west, the time of flight T is given by

$$T = \frac{v^2 t_0}{v^2 - w^2}.$$

a) Express the time of flight T as a function of the wind
speed w, given that the cities are 300 km apart and
the plane flies between them in still air at a speed of
200 km/h.

b) What is the time of flight if the wind speed is 5 km/h?
10 km/h? 20 km/h?

c) What are the domain and the range of the function?
Use the constraints of the problem in your determination.

d) Graph the function.

38. *Surface area.*   A container firm is designing an open-
top rectangular box, with a square base, that will hold
108 cubic centimeters (cc).

a) Express the surface area S as a function of the length
x of a side of the base.

b) Sketch a graph of the function on the interval $(0, \infty)$.

c) Estimate the minimum surface area and the value of
x that will yield the minimum surface area.

▦ **T E C H N O L O G Y   C O N N E C T I O N** _____

Using a grapher, graph the function.

39. $f(x) = \dfrac{x^4 + 3x^3 + 21x^2 - 50x + 80}{x^4 + 8x^3 - x^2 + 20x - 10}$

40. $f(x) = \dfrac{x^3 + x^2 + x}{x^2 - 1}$

41. $f(x) = \dfrac{x^3}{x^2 - 1}$

42. $f(x) = \dfrac{x^3 + 4}{x}$

SUMMARY AND REVIEW 4

● **T E R M S   T O   K N O W**

Polynomial function, p. 211	Equations with absolute value, p. 230	Factor theorem, p. 251
Quadratic function, p. 212	Inequalities with absolute value, p. 230	Fundamental theorem of algebra, p. 253
Parabola, p. 212	Polynomial inequality, p. 235	Rational roots theorem, p. 258
Vertex, p. 212	Rational inequality, p. 237	Descartes' rule of signs, p. 263
Line of symmetry, p. 212	Continuous function, p. 242	Upper bound on roots, p. 264
x-intercept, p. 218	Zero of a polynomial, p. 242	Lower bound on roots, p. 264
Discriminant, p. 219	Root of a polynomial, p. 242	Rational function, p. 268
Intersection, p. 224	Remainder theorem, p. 249	Horizontal asymptote, p. 270
Union, p. 225	Synthetic division, p. 250	Vertical asymptote, p. 270
Compound sentence, p. 226		Oblique asymptote, p. 275
Conjunction, p. 226		
Disjunction, p. 227		

● **REVIEW EXERCISES**

For the functions in Exercises 1 and 2:

a) use completing the square to put each equation into the form $f(x) = a(x - h)^2 + k$;

b) find the vertex;

c) find the line of symmetry; and

d) determine whether the second coordinate of the vertex is a maximum or minimum, and find the maximum or minimum.

1. $f(x) = 3x^2 + 6x + 1$

2. $f(x) = -2x^2 - 3x + 6$

3. Graph: $f(x) = 3x^2 + 6x + 1$.

4. Find the x-intercepts of $f(x) = -x^2 - x - 1$.

5. Find $\{4, 5, 8, 12, 13\} \cap \{3, 5, 7, 9, 11\}$.

6. Find $\{4, 5, 8, 12, 13\} \cup \{3, 5, 7, 9, 11\}$.

7. Graph $\{x | -2 < x\} \cap \{x | x \leq 3\}$ on a line.

8. Graph $\{x | x < -2\} \cup \{x | x > 2\}$ on a line.

Solve.

9. $-3x + 2 \leq -4$ *and* $x - 1 \leq 3$

10. $|x - 6| < 5$

11. $|-3x - 2| > 2$

12. $|\frac{1}{2}x + 4| \leq 6$

13. $|2x + 5| = 9$

14. $x^2 - 9 < 0$

15. $2x^2 - 3x - 2 > 0$

16. $(1 - x)(x + 4)(x - 2) < 0$

17. $\dfrac{x - 2}{x + 3} < 4$

18. The sum of the length and the width of a rectangle is 40. Find the dimensions for which the area is a maximum.

Sketch a graph of the polynomial function.

19. $f(x) = x^3 + 3x^2 - 2x - 6$

20. $f(x) = x^4 - 3x^3 + 2x^2$

21. Find the remainder when
$$x^4 + 3x^3 + 3x^2 + 3x + 2$$
is divided by $x + 2$.

22. Use synthetic division to find the quotient and the remainder:
$$(2x^4 - 6x^3 + 7x^2 - 5x + 1) \div (x + 2).$$

23. Use synthetic division to find $P(3)$:
$$P(x) = 2x^4 - 3x^3 + x^2 - 3x + 7.$$

24. Factor the polynomial $P(x)$. Then solve the equation $P(x) = 0$.
$$P(x) = x^3 + 7x^2 + 7x - 15$$

25. Determine whether $x + 1$ is a factor of
$$x^3 + 6x^2 + x + 30.$$

26. Find the roots of
$$P(x) = x^2(x^2 + x - 12)^2(x^2 - 16)$$
and state the multiplicity of each.

27. Find a polynomial of degree 3 with roots 0, 1, and 2.

28. Find a polynomial of lowest degree having roots 1 and -1, and having 2 as a root of multiplicity 2 and -3 as a root of multiplicity 3.

29. The equation $x^4 - 81 = 0$ has $3i$ for a root. Find the other roots.

30. A polynomial of degree 4 with rational coefficients has roots $-8 - 7i$ and $10 + \sqrt{5}$. Find the other roots.

31. List all possible rational roots of
$$2x^6 - 12x^4 + 17x^2 + 12.$$

32. Let $P(x) = x^4 - x^3 - 3x^2 - 9x - 108$. Find the rational roots of $P(x)$. If possible, find the other roots.

What does Descartes' rule of signs tell you about the number of positive real roots and the number of negative real roots of the polynomial?

33. $3x^{12} + 3x^4 - 7x^2 + x + 5$

34. $6x^8 - 12x^4 + 5x^2 + x + 2$

35. Find the smallest positive integer that is guaranteed by Theorem 15 to be an upper bound to the roots of
$$2x^4 - 7x^2 + 2x - 1.$$

36. Find the largest negative integer that is guaranteed by Theorem 15 to be a lower bound to the roots of
$$12x^3 + 24x^2 + 10x - 6.$$

37. The equation $x^5 + x^4 - x^3 - x^2 - 2x - 2$ has a root between 1 and 2. Approximate this root to the nearest hundredth.

38. Approximate the irrational roots of
$$P(x) = x^3 - 3x^2 + 3.$$

39. Graph: $f(x) = \dfrac{x^2 + x - 6}{x^2 - x - 20}$.

● **SYNTHESIS** _____

Solve.

40. $4x \leq 5x + 2 < 4 - x$

41. $\left| 1 - \dfrac{1}{x^2} \right| < 3$

42. $(x - 2)^{-3} < 0$

Find the domain of the function.

43. $f(x) = \sqrt{1 - |3x - 2|}$

44. $g(x) = \dfrac{1}{\sqrt{5 - |7x + 2|}}$

45. There are n teams in a sports league. If each team plays each other team once, then the total number of games is given by

$$N = \frac{n(n-1)}{2}.$$

For what values of n is $N \geq 105$?

46. Find a complete factorization of $x^3 - 1$.

47. Find k such that $x + 3$ is a factor of
$$x^3 + kx^2 + kx - 15.$$

48. The equation $x^2 - 8x + c = 0$ has a double root. Find it.

49. When $x^2 - 4x + 3k$ is divided by $x + 5$, the remainder is 33. Find the value of k.

50. Graph: $y = 1 - \dfrac{1}{x^2 + 4}$.

● **THINKING AND WRITING** _____

1. Explain the difference between a polynomial function and a rational function.

2. Explain and contrast the three kinds of asymptotes considered for rational functions.

3. Explain the idea of a root of a function in as many ways as possible.

CHAPTER TEST 4

For the functions in Questions 1 and 2:

a) use completing the square to put each equation into the form $f(x) = a(x - h)^2 + k$;

b) find the vertex; and

c) determine whether there is a maximum or minimum function value and find that value.

1. $f(x) = 5x^2 - 10x + 3$

2. $f(x) = -4x^2 + 3x - 1$

3. Graph: $f(x) = 5x^2 - 10x + 3$.

4. Find the x-intercepts of $f(x) = 3x^2 - x - 1$.

5. Find $\{3, 4, 7, 8, 11, 12\} \cup \{2, 4, 6, 8, 10\}$.

6. Graph: $\{x \mid x \leq \frac{3}{2}\} \cap \{x \mid x > \frac{1}{2}\}$.

Solve.

7. $x - 3 > -8$ and
$-2x + 3 \geq 9$

8. $|-4x + 3| \geq 9$

9. $|x - 3| < 2$

10. $|6x - 3| = 15$

11. $x^2 - 8x + 12 > 0$

12. $8x^2 + 10x - 3 < 0$

13. $\dfrac{x - 4}{2x + 3} < 1$

14. Find two numbers whose sum is -20 and whose product is a maximum.

Sketch a graph of the polynomial function.

15. $f(x) = x^4 - 5x^2$

16. $f(x) = x^4 - 5x^2 + 6$

17. Find the remainder when $3x^4 - 6x^3 + x - 2$ is divided by $x - 3$.

18. Determine whether $x + i$ is a factor of $x^3 + x$.

19. Use synthetic division to find the quotient and the remainder. Show all your work.
$$(5x^3 - x^2 + 4x - 3) \div (x + 1)$$

20. Use synthetic division to find $P(-5)$:
$$P(x) = 2x^4 - 7x^3 + 8x^2 - 10.$$

21. Factor the polynomial $P(x)$. Then solve the equation $P(x) = 0$.
$$P(x) = x^3 - 4x^2 + x + 6$$

22. Find a polynomial of degree 4 with roots 2, -2, $2 + i$, and $2 - i$.

23. Find a polynomial of lowest degree having roots 0 and -1, and having 2 as a root of multiplicity 2 and 1 as a root of multiplicity 3.

24. Find a polynomial of lowest degree with rational coefficients that has $3 - i$ and 2 as two of its roots.

25. List all possible rational roots of
$$3x^5 - 4x^3 - 2x + 6.$$

26. What does Descartes' rule of signs tell you about the number of positive real roots and the number of negative real roots of
$$5x^7 - 2x^6 + x^4 + 2x - 6?$$

27. Find the smallest positive integer that is guaranteed by Theorem 15 to be an upper bound to the roots of
$$3x^4 - 2x^3 - 15x - 10.$$
Then find the largest negative integer that is guaranteed to be a lower bound.

28. Approximate the irrational roots of
$$P(x) = x^4 - x - 2.$$

29. Graph:
$$f(x) = \frac{x - 2}{x^2 - 2x - 15}.$$

● **SYNTHESIS** _____

30. Suppose that an object is thrown upward with an initial velocity of 80 ft/sec from a height of 224 ft. Its height after t seconds is a function s given by

$$s(t) = -16t^2 + 80t + 224.$$

a) Find its maximum height and when the object attains it.

b) Find when the object reaches the ground.

c) Over what interval of time is the height greater than 320 ft?

31. Solve: $\left| 2 + \dfrac{1}{x} \right| < 6$.

32. Find the domain of $f(x) = \sqrt{x^2 + 3x - 10}$.

33. Graph: $y = \left| 2 - \dfrac{1}{3x} \right|$.

34. Solve: $x^4 - 2x^3 + 3x^2 - 2x + 2 = 0$.

In this chapter, we will consider two kinds of functions, closely related. The first kind, called *exponential functions,* are those that have variables for the exponents. Such functions have many applications—for example, exponential functions are used in problems of population growth. •
Imagine putting a function machine in reverse. If it works, it will represent what we call the *inverse* of the original function. Functions that are inverses of each other are thus closely related. The inverses of the exponential functions, called *logarithmic functions,* or *logarithm functions,* are also important in many applications. •

5

Exponential and Logarithmic Functions

5.1

INVERSES OF RELATIONS AND FUNCTIONS

 Inverses of Relations

Consider the relation r given as follows:

$$r = \{(2, 4), (-1, 3), (-2, 0)\}.$$

Suppose we *interchange* the first and second coordinates. The relation we obtain is called the **inverse** of the relation r. It is given as follows:

Inverse of $r = \{(4, 2), (3, -1), (0, -2)\}.$

DO EXERCISE 1 ON THE FOLLOWING PAGE. _____

If a relation is defined by an equation, then an equation of the inverse can be found by interchanging the variables. The solutions of the second equation will be the same as those of the first equation, except that the first and second coordinates will be interchanged in each ordered pair.

OBJECTIVES

You should be able to:

▶ Given a relation that is a set of ordered pairs, find its inverse relation. Given an equation defining a relation, write an equation of the inverse relation.

▶ Given a relation that is a set of ordered pairs, graph the relation and its inverse. Given a graph of a relation, sketch a graph of its inverse, or given an equation defining a relation, graph it and then graph its inverse.

▶ Given a function, determine whether it is one-to-one and thus has an inverse that is a function. If the function is one-to-one, find a formula for the inverse. Graph a function and its inverse using the same set of axes.

▶ Simplify expressions of the type $f \circ f^{-1}(x)$ and $f^{-1} \circ f(x)$.

1. Find the inverse of the relation Q given by

$Q = \{(-1, 4), (2, 5), (0, -3), (5, 1)\}.$

DEFINITION   **Inverse Relation**

When a relation is defined by an equation, interchanging x and y produces an equation of the *inverse relation*.

Example 1   Find an equation of the inverse of the relation $y = x^2 - 5$.

Solution   We interchange x and y and obtain $x = y^2 - 5$. This is an equation of the inverse relation.   ◄

DO EXERCISE 2. _____

2. Write an equation of the inverse of each relation.

a) $y = 3x + 2$

b) $y = x$

c) $x^2 + 3y^2 = 4$

d) $y = 5x^2 + 2$

e) $y^2 = 4x - 5$

f) $xy = 5$

B   Graphs of Inverse Relations

Example 2   Consider the relation r given by

$r = \{(2, 4), (-1, 3), (-2, 0)\}.$

Draw the graph of the relation in black. Find the inverse relation and draw its graph in color.

Solution   The relation r is shown on the graph in black. The inverse of the relation is $\{(4, 2), (3, -1), (0, -2)\}$ and is shown in color.

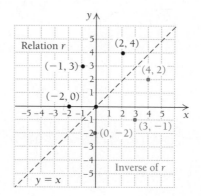

3. Consider the relation P given by

$P = \{(5, 0), (3, -2),$
$\quad (0, -3), (4, -4)\}.$

a) Draw the graph of the relation P in black.

b) Find the inverse of P and draw its graph in color.

DO EXERCISE 3. _____

Compare the relation and its inverse in Example 2. Note that we can obtain the inverse by reflecting each ordered pair across the line $y = x$. Interchanging x and y in an equation to find the inverse has the effect of reflecting each ordered pair in the original relation across the line $y = x$. Thus the graphs of a relation and its inverse are always reflections of each other across the line $y = x$. (This assumes that the same scale is used on both axes.)

Examples In each case, a relation is shown in black. Graph the inverse.

Solution The graph of each inverse is shown in color.

3.

4.

5.

DO EXERCISES 4 AND 5. _____

B ▸ Inverses and One-to-One Functions

Consider the following two functions. They are relations. Let us think of them as correspondences.

Number (domain)	Cube (range)
−3 ⟶	−27
−2 ⟶	−8
−1 ⟶	−1
0 ⟶	0
1 ⟶	1
2 ⟶	8
3 ⟶	27

Year (domain)	First-class Postage Cost, in cents (range)
1978 ⟶	15
1983 ⟶	20
1984 ⟶	
1989 ⟶	25
1991 ⟶	29

Suppose we reverse the arrows. We then obtain what is called the *inverse correspondence,* or *relation.* Now are these new correspondences functions?

Number (domain)	Cube (range)
−3 ⟵	−27
−2 ⟵	−8
−1 ⟵	−1
0 ⟵	0
1 ⟵	1
2 ⟵	8
3 ⟵	27

Year (domain)	First-class Postage Cost, in cents (range)
1978 ⟵	15
1983 ⟵	20
1984 ⟵	
1989 ⟵	25
1991 ⟵	29

We see that the inverse of the first correspondence is a function, but the inverse of the second correspondence is not a function.

Recall that for each input, a function provides exactly one output. However, nothing in our definition of function prevents having the same output

---

4. Graph the inverse of this relation.

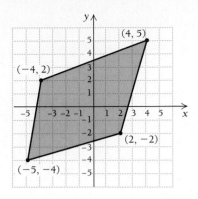

5. Graph the inverse of each relation by reflecting across the line $y = x$.

a)

b)

c) Graph $y = 4 - x^2$. Then by reflecting across the line $y = x$, graph its inverse.

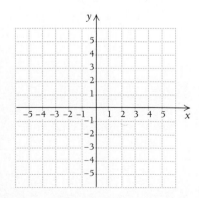

For each function, find its inverse correspondence and determine whether that inverse is a function.

6. WOMEN'S DRESS SIZES

Domain (United States)	Range (France)
6 →	38
8 →	40
10 →	42
12 →	44
14 →	46
16 →	48
18 →	50

7. SPORTS TEAMS

Domain	Range
Lakers	
Dodgers →	Los Angeles
Rams	
Knickerbockers	
Yankees →	New York
Giants	

8. Given the function f described by $f(x) = 5x + 7$, prove that f is one-to-one.

9. Given the function g described by $g(x) = x^2 - 1$, prove that g is not one-to-one.

for two or more different inputs. That is, it is possible with a function for different inputs to correspond to the same output in the range. When this possibility is excluded, the inverse is also a function.

In the cube function, different inputs have different outputs. Thus this function is what is called a **one-to-one function.** In the postage-cost function, the input 1983 has the output 20, and the input 1984 also has the output 20. Thus this is not a one-to-one function.

> **DEFINITION** **One-to-One Function**
>
> A function f is *one-to-one* if different inputs have different outputs. That is, for every a and b in the domain of f, if $a \neq b$, then $f(a) \neq f(b)$.
>
> A function f is *one-to-one* if when the outputs are the same, the inputs are the same. That is, for every a and b in the domain of f, if $f(a) = f(b)$, then $a = b$.
>
> If a function is one-to-one, then its inverse correspondence is a function.

DO EXERCISES 6 AND 7.

Example 6 Given the function f described by $f(x) = 2x - 3$, prove that f is one-to-one.

Solution Assume that $f(a) = f(b)$ for any numbers a and b in the domain of f. Then

$$2a - 3 = 2b - 3$$
$$2a = 2b \qquad \text{Adding 3}$$
$$a = b. \qquad \text{Dividing by 2}$$

Thus, if $f(a) = f(b)$, then $a = b$. This is true for any a and b in the domain of f, so f is one-to-one. ◀

Example 7 Given the function g described by $g(x) = x^2$, prove that g is not one-to-one.

Solution To prove that g is not one-to-one, we need to find two numbers a and b for which $a \neq b$ and $g(a) = g(b)$. Two such numbers are -3 and 3, because $-3 \neq 3$ and $g(-3) = g(3)$. Thus, g is not one-to-one. ◀

DO EXERCISES 8 AND 9.

How can we tell graphically whether a function is one-to-one and thus has an inverse that is a function? The graph below shows a function, in black, and its inverse, in color. To determine whether the inverse is a function, we can apply the vertical-line test to its graph. By reflecting each such vertical line back across the line $y = x$, we obtain an equivalent **horizontal-line test** for the original function.

Function

Inverse

The vertical-line test
shows that the inverse
is not a function.

The horizontal-line test
shows that the function
is not one-to-one.

$f(a) = f(b),$
$a \neq b$

The Horizontal-Line Test

If it is possible for a horizontal line to intersect a graph of a function more than once, then the function is not one-to-one and therefore its inverse is not a function.

Thus a graph is that of a function if no vertical line crosses the graph more than once. A function has an inverse that is also a function if no horizontal line crosses the graph of the original function more than once.

Example 8   Determine whether the function $f(x) = x^2$ is one-to-one and thus has an inverse that is also a function.

Solution   The graph of $f(x) = x^2$ is shown below. Note that there are many horizontal lines that cross the graph more than once. In particular, the line $y = 4$ does so. If you note where the line crosses, the first coordinates are -2 and 2. Although these are different inputs, they have the same output. That is, $-2 \neq 2$, but

$$f(-2) = (-2)^2 = 4 = 2^2 = f(2).$$

Thus the function is not one-to-one and its inverse is not a function.

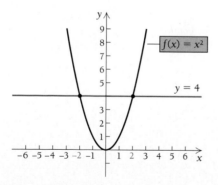

Consider the cube function again. Note that if we begin with the input -3, we get -27 as an output. Then consider the inverse relation. If we start with the input -27, we get -3 as the output. Thus the inverse takes us back to where we started. This does not happen with the postage-cost function. If we start with 1983 as an input, we get 20 as an output. If we go to the inverse relation and start with 20, we have two possibilities for outputs, 1983 and 1984. Thus the postage-cost function is not one-to-one and its inverse is not a function.

Determine whether the function is one-to-one and whether its inverse is also a function.

10. $f(x) = 4 - x$

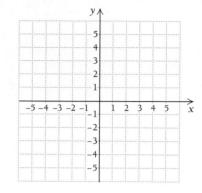

11. $f(x) = x^2 - 1$

12. $f(x) = |x| - 3$

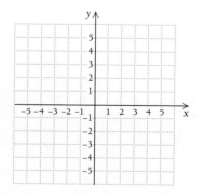

13. Which of the following functions have inverses that are functions?

a)

b)

c)

d)

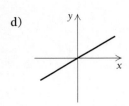

DO EXERCISES 10–13. (EXERCISES 10–12 ARE ON THE PRECEDING PAGE.)

If the inverse of a function f is also a function, it can be named f^{-1}, read "f-inverse."

> **CAUTION!**   The -1 in f^{-1} is *not* an exponent!

Suppose that a function is described by a formula. If it has an inverse that is a function, how do we find a formula for the inverse? If for any equation with two variables such as x and y we interchange the variables, we obtain an equation of the inverse relation. If it is a function, we proceed as follows to find a formula for f^{-1}.

> If a function is one-to-one, a formula for its inverse can be found as follows:
>
> 1. Replace $f(x)$ by y.
> 2. Interchange x and y. (This gives the inverse function.)
> 3. Solve for y.
> 4. Replace y by $f^{-1}(x)$.

Example 9   Given $f(x) = x + 3$:

a) Determine whether the function is one-to-one.
b) If it is one-to-one, find a formula for $f^{-1}(x)$.

Solution

a) The graph of $f(x) = x + 3$ is shown below. The horizontal-line test shows that it is one-to-one. (We could also use an algebraic proof.) Thus its inverse is a function.

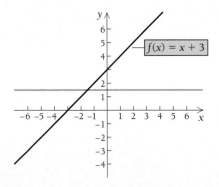

b) 1. Replace $f(x)$ by y: $y = x + 3$.
    2. Interchange x and y: $x = y + 3$.   This gives the inverse function.
    3. Solve for y: $x - 3 = y$.
    4. Replace y by $f^{-1}(x)$: $f^{-1}(x) = x - 3$.   ◀

Example 10   Given $g(x) = 2x - 3$:

a) Determine whether the function is one-to-one.
b) If it is one-to-one, find a formula for $g^{-1}(x)$.

Solution

a) We proved g to be one-to-one algebraically in Example 6.
b) 1. Replace $g(x)$ by y: $y = 2x - 3$.
   2. Interchange x and y: $x = 2y - 3$.
   3. Solve for y: $x + 3 = 2y$

$$\frac{x + 3}{2} = y.$$

   4. Replace y by $g^{-1}(x)$: $g^{-1}(x) = \dfrac{x + 3}{2}.$ ◀

DO EXERCISES 14 AND 15. _____

Let us consider inverses of functions in terms of function machines. Suppose that the function g programmed into a machine has an inverse that is also a function. Suppose then that the function machine has a reverse switch. The machine is programmed to do the inverse, g^{-1}, when the switch is thrown. Inputs then enter at the opposite end and the entire process is reversed.

For example, we know from Example 10 that the inverse of the function $g(x) = 2x - 3$ is

$$g^{-1}(x) = \frac{x + 3}{2}.$$

Consider the input 5. Then

$$g(5) = 2(5) - 3 = 10 - 3 = 7.$$

The output is 7. Now we use 7 for the input in the inverse:

$$g^{-1}(7) = \frac{7 + 3}{2} = \frac{10}{2} = 5.$$

The function g takes 5 to 7. The inverse function g^{-1} takes the number 7 back to 5.

How do the graphs of a function and its inverse compare?

Given each function:

a) Determine whether it is one-to-one.

b) If it is one-to-one, find a formula for the inverse.

14. $f(x) = 3 - x$

15. $g(x) = 3x - 2$

TECHNOLOGY CONNECTION

Only functions that are *one-to-one* have inverses that are also functions. Once a function has been plotted on a grapher, we can use the horizontal-line test to determine whether or not it is also one-to-one.

Use a grapher to plot $f(x)$ in each case. Then determine whether or not it is one-to-one. If it is, find the equation for $f^{-1}(x)$ and graph it on the same axes as the original function.

TC 1. $f(x) = x^2 + 4$
TC 2. $f(x) = x^3 + 4$
TC 3. $f(x) = 2x - 8$

16. Graph

$$g(x) = 3x - 2 \quad \text{and}$$

$$g^{-1}(x) = \frac{x + 2}{3}$$

using the same set of axes.

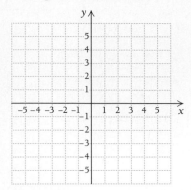

Example 11   Graph $g(x) = 2x - 3$ and $g^{-1}(x) = (x + 3)/2$ using the same set of axes. Then compare.

Solution   The graph of each function is shown below. Note that we can obtain the graph of g^{-1} by reflecting the graph of g across the line $y = x$. That is, if we graph $g(x) = 2x - 3$ and $y = x$ and fold the paper along the line $y = x$, the graph of $g^{-1}(x) = (x + 3)/2$ will be the result of "flipping" the graph of $g(x) = 2x - 3$ across the line.

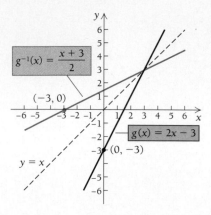

> The graph of f^{-1} is a reflection of the graph of f across the line $y = x$.

DO EXERCISE 16.

Example 12   Given $f(x) = x^3 + 2$:

  a) Determine whether the function is one-to-one.
  b) If it is one-to-one, find a formula for its inverse.
  c) Graph the inverse.

Solution   We solve as follows.

  a) The graph of $f(x) = x^3 + 2$ is shown below. Using the horizontal-line test, we know that it has an inverse.

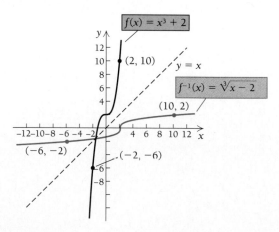

  b) 1. Replace $f(x)$ by y: $y = x^3 + 2.$
     2. Interchange x and y: $x = y^3 + 2.$

3. Solve for y:

$$x - 2 = y^3$$

$$\sqrt[3]{x - 2} = y. \qquad \text{Since a number has only one cube root, we can solve for } y.$$

4. Replace y by $f^{-1}(x)$: $f^{-1}(x) = \sqrt[3]{x - 2}$.

c) We find the graph by flipping the graph of $f(x) = x^3 + 2$ over the line $y = x$, interchanging x- and y-coordinates. The graph can also be found by substituting into $f^{-1}(x) = \sqrt[3]{x - 2}$ to find function values. The graph is shown above using the same set of axes. ◄

DO EXERCISE 17. _____

Now let us consider a situation where the inverse of a function is *not* a function. Consider

$$f(x) = x^2.$$

The graph is shown below. Using the horizontal-line test, we know that it is not one-to-one and does not have an inverse that is a function. Nevertheless, suppose we reflect the graph across the line $y = x$. Using the vertical-line test, we see again that the inverse is not a function.

Suppose we had not known this and had tried to find a formula for the inverse as follows:

$$y = x^2 \qquad \text{Replacing } f(x) \text{ by } y$$

$$x = y^2. \qquad \text{Interchanging } x \text{ and } y$$

We cannot solve for y and get only one value, since most real numbers have two square roots:

$$\pm\sqrt{x} = y.$$

This is not the equation of a function. An input of, say, 4 yields two outputs, -2 and 2.

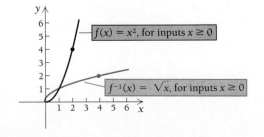

17. Consider $f(x) = x^3 + 1$.

a) Determine whether the function is one-to-one.

b) If it is one-to-one, find a formula for its inverse.

c) Using graph paper, graph the function and its inverse using the same set of axes.

TECHNOLOGY CONNECTION

When you are using a grapher to plot a function and its inverse, there are two steps that will help illustrate the relationship between them more clearly:

1. Select scales on the x- and y-axes so that one unit on each axis has the same length. Consult your grapher's documentation for the steps required to set these "square" axes.

2. Plot the line $y = x$ along with the function and its inverse. If the scales have been set according to suggestion (1) above, the line $y = x$ will form a 45° angle with respect to the x-axis. With all three equations plotted, it becomes clear that the graph of $f^{-1}(x)$ is the reflection of the graph of $f(x)$ across the line $y = x$.

Let's consider $f(x) = \sqrt[3]{x - 2}$ and its inverse.

$f(x) = \sqrt[3]{x - 2}$

$[-9, 9] \times [-6, 6]$

$f(x) = \sqrt[3]{x - 2}$
$f^{-1}(x) = x^3 + 2$

$[-9, 9] \times [-6, 6]$

Square the axes and draw the graph of the line $y = x$. Then draw the graphs of the function and its inverse.

TC 4. $f(x) = 2\sqrt[3]{x}$ **TC 5.** $f(x) = \dfrac{4}{x}$

TC 6. $f(x) = 4x^3 + 2$

18. Given $f(x) = x^2 - 4$, $x \geq 0$, find $f^{-1}(x)$.

In such cases, it is often convenient to consider "part" of the function by restricting the domain of $f(x) = x^2$ to nonnegative numbers. Then its inverse is a function. See the graphs of $f(x) = x^2$, $x \geq 0$, and $f^{-1}(x) = \sqrt{x}$, $x \geq 0$ above.

DO EXERCISE 18.

 Inverse Functions and Composition

19. Let

$$f(x) = \frac{4}{x} - 3.$$

Show that

$$f^{-1}(x) = \frac{4}{x + 3}.$$

THEOREM 1

If a function f is one-to-one and hence has an inverse f^{-1} that is a function, then:

   a) $f^{-1} \circ f(x) = f^{-1}(f(x)) = x$, for each x in the domain of f, and
   b) $f \circ f^{-1}(x) = f(f^{-1}(x)) = x$, for each x in the domain of f^{-1}.

Proof.   Suppose that x is in the domain of f. Then $f(x) = y$, for some number y in the range of f. Thus the ordered pair (x, y) is in f, and by definition of f^{-1}, the pair (y, x) is in f^{-1}. It follows that $f^{-1}(y) = x$. Then by substituting $f(x)$ for y, we get $f^{-1}(f(x)) = x$. A similar proof shows the second part.

   Simply stated, Theorem 1 asserts that if we begin with an input, the composition of a function and its inverse is the input, or the composition of the inverse of a function and the function is the input.

Example 13   Let $f(x) = 5x + 1$. Show that $f^{-1}(x) = (x - 1)/5$.

Solution   We find $f^{-1} \circ f(x)$ and $f \circ f^{-1}(x)$ and check to see that each is x.

   a) $f^{-1} \circ f(x) = f^{-1}(f(x)) = f^{-1}(5x + 1)$

$$= \frac{(5x + 1) - 1}{5} = \frac{5x}{5} = x$$

20. Simplify $f(f^{-1}(1992))$ and $f^{-1}(f(-23{,}456))$.

   b) $f \circ f^{-1}(x) = f(f^{-1}(x)) = f\left(\frac{x - 1}{5}\right)$

$$= 5\left(\frac{x - 1}{5}\right) + 1 = x - 1 + 1 = x$$

Example 14   Simplify $g^{-1}(g(283))$ and $g(g^{-1}(-12{,}045))$.

Solution   Assuming that 283 is in the domain of g, we have

$$g^{-1}(g(283)) = 283.$$

Assuming that $-12{,}045$ is in the domain of g^{-1}, we have

$$g(g^{-1}(-12{,}045)) = -12{,}045.$$

DO EXERCISES 19 AND 20.

EXERCISE SET 5.1

Ⓐ   Find the inverse of the relation.

1. $\{(0, 1), (5, 6), (-2, -4)\}$

2. $\{(-1, 3), (2, 5), (-3, 5), (2, 0)\}$

3. $\{(7, 8), (-2, 8), (3, -4), (8, -8)\}$

4. $\{(-1, -1), (-3, 4)\}$

Write an equation of the inverse relation.

5. $y = 4x - 5$ **6.** $y = 3x + 5$

7. $x^2 - 3y^2 = 3$ **8.** $2x^2 + 5y^2 = 4$

9. $y = 3x^2 + 2$ **10.** $y = 5x^2 - 4$

11. $xy = 7$ **12.** $xy = -5$

Ⓑ   Graph the equation. Then by reflection across the line $y = x$, graph its inverse.

13. $y = x^2 + 1$ **14.** $x = y^2 - 3$

15. $x = |y|$ **16.** $y = |x|$

Ⓒ   Determine whether the function is one-to-one. Use the horizontal-line test or an algebraic procedure.

17. $f(x) = 5x - 8$ **18.** $f(x) = 3 - 2x$

19. $f(x) = x^2 - 7$ **20.** $f(x) = 1 - x^2$

21. $g(x) = 3$ **22.** $g(x) = -0.8$

23. $g(x) = |x|$ **24.** $h(x) = |x| - 2$

25. $f(x) = |x + 1|$ **26.** $f(x) = |x - 3|$

27. $g(x) = \dfrac{-4}{x}$ **28.** $h(x) = \dfrac{1}{x}$

Given each function, (a) determine whether it is one-to-one and (b) if it is one-to-one, find a formula for the inverse.

29. $f(x) = x + 4$ **30.** $f(x) = x + 5$

31. $f(x) = 5 - x$ **32.** $f(x) = 7 - x$

33. $g(x) = x - 3$ **34.** $g(x) = x - 10$

35. $f(x) = 2x$ **36.** $f(x) = 5x$

37. $g(x) = 2x + 5$ **38.** $g(x) = 5x + 8$

39. $h(x) = \dfrac{4}{x + 7}$ **40.** $h(x) = \dfrac{1}{x - 6}$

41. $f(x) = \dfrac{1}{x}$ **42.** $f(x) = -\dfrac{4}{x}$

43. $f(x) = \dfrac{2x + 3}{4}$ **44.** $f(x) = \dfrac{3x - 5}{4}$

45. $g(x) = \dfrac{x + 4}{x - 3}$ **46.** $g(x) = \dfrac{5x - 3}{2x + 1}$

47. $f(x) = x^3 - 1$ **48.** $f(x) = x^3 + 7$

49. $G(x) = (x - 4)^3$ **50.** $G(x) = (x + 5)^3$

51. $f(x) = \sqrt[3]{x}$ **52.** $f(x) = \sqrt[3]{x - 8}$

53. $f(x) = 4x^2 + 3, x > 3$ **54.** $f(x) = 5x^2 - 2, x > 2$

55. $f(x) = \sqrt{x + 1}$ **56.** $g(x) = \sqrt{2x - 3}$

57. Which of the following functions have inverses that are functions?

a)

b)

c)

d)
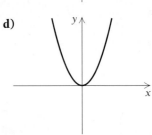

58. Which of the following functions have inverses that are functions?

a)

b)

c)

d)
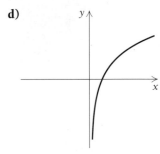

Graph the function and its inverse using the same set of axes.

59. $f(x) = \frac{1}{2}x - 4$ **60.** $g(x) = x + 4$

61. $f(x) = x^3$ **62.** $f(x) = x^3 - 1$

63. $f(x) = \sqrt{x - 3}$ **64.** $f(x) = \sqrt{2x + 5}$

65. $g(x) = \dfrac{1}{x}$ **66.** $f(x) = -\dfrac{2}{x}$

67. $f(x) = 3 - x^2, x \geq 0$ **68.** $f(x) = x^2 - 1, x \leq 0$

Ⓓ

69. Let $f(x) = \frac{7}{8}x$. Show that $f^{-1}(x) = \frac{8}{7}x$.

70. Let $f(x) = (x + 5)/4$. Show that $f^{-1}(x) = 4x - 5$.

71. Let $f(x) = (1 - x)/x$. Show that $f^{-1}(x) = 1/(x + 1)$.

72. Let $f(x) = x^3 - 4$. Show that $f^{-1}(x) = \sqrt[3]{x + 4}$.

73. Let $f(x) = 35x - 173$. Simplify

$$f^{-1}(f(3)) \quad \text{and} \quad f(f^{-1}(-125)).$$

74. Let $g(x) = (15 - 173x)/3$. Simplify

$$g^{-1}(g(5)) \quad \text{and} \quad g(g^{-1}(-12)).$$

75. Let $f(x) = x^3 + 2$. Simplify

$$f^{-1}(f(12{,}053)) \quad \text{and} \quad f(f^{-1}(-17{,}243)).$$

76. Let $g(x) = x^3 - 486$. Simplify

$$g^{-1}(g(489)) \quad \text{and} \quad g(g^{-1}(-17{,}422)).$$

77. *Women's dress sizes in the United States and France.* Sizes of clothing and shoes are not the same numbers in different countries. For example, a size-6 dress in the United States is size 38 in France. A function that will convert dress sizes in the United States to those in France is

$$f(x) = x + 32.$$

a) Find the dress sizes in France that correspond to sizes 8, 10, 14, and 18 in the United States.

b) Determine whether this function has an inverse that is a function. If so, find a formula for the inverse.

c) Use the inverse function to find dress sizes in the United States that correspond to sizes 40, 42, 46, and 50 in France.

78. *Women's dress sizes in the United States and Italy.* A size-6 dress in the United States is size 36 in Italy. A function that will convert dress sizes in the United States to those in Italy is

$$g(x) = 2(x + 12).$$

a) Find the dress sizes in Italy that correspond to sizes 8, 10, 14, and 18 in the United States.

b) Determine whether this function has an inverse that is a function. If so, find a formula for the inverse.

c) Use the inverse function to find dress sizes in the United States that correspond to sizes 40, 44, 52, and 60 in Italy.

● SYNTHESIS _____

79. Does the constant function $f(x) = 5$ have an inverse that is a function? If so, find a formula. If not, explain why.

80. An organization determines that the cost per person of chartering a bus is given by the function

$$C(x) = \frac{100 + 5x}{x},$$

where x = the number of people in the group and $C(x)$ is in dollars. Determine $C^{-1}(x)$ and explain this inverse function.

Determine whether the functions are inverses of each other.

81. $f(x) = \frac{2}{3}, \quad g(x) = \frac{3}{2}$

82. $f(x) = \sqrt[5]{x}, \quad g(x) = x^5$

83. $f(x) = \sqrt[4]{x}, \; x \geq 0, \quad g(x) = x^4$

84. $f(x) = \dfrac{2x - 5}{4x + 7}, \quad g(x) = \dfrac{7x - 4}{5x + 2}$

85. Find three examples of functions that are their own inverses, that is, $f = f^{-1}$.

86. Prove that if a function is increasing, then it is one-to-one and hence has an inverse that is a function.

87. Graph the equation $y = 1/x^2$. Then test for symmetry with respect to the x-axis, the y-axis, and the origin.

88. The following formulas for the conversion between Fahrenheit and Celsius temperatures have been considered several times in the text:

$$C = \tfrac{5}{9}(F - 32), \qquad F = \tfrac{9}{5}C + 32.$$

Discuss the functions from the standpoint of inverses.

● CHALLENGE _____

Graph the equation and its inverse. Then test for symmetry with respect to the x-axis, the y-axis, and the origin.

89. $|x| - |y| = 1$ **90.** $y = \dfrac{|x|}{x}$

▀▄ TECHNOLOGY CONNECTION _____

Use a grapher to plot the graph of the function. If it is one-to-one, find the inverse function and plot it on the same axes, along with the line $y = x$.

91. $y = (x - 1)^{2/5}$ **92.** $y = (3x - 9)^3$

93. $y = \sqrt{2x - 1}$ **94.** $y = \sqrt[4]{90x}$

OBJECTIVES

You should be able to:

A Graph exponential equations and functions.

B Graph exponential equations when x and y have been interchanged.

(continued)

5.2

EXPONENTIAL FUNCTIONS

The following graph shows the years in which the cost of first-class postage was raised. A curve drawn along the graph would approximate the graph of an *exponential function*. We now consider such graphs. We will also find graphs of inverses of exponential functions.

First-Class Postage

Solve problems involving applications of exponential functions and their graphs.

Graphing Exponential Functions

In Section 1.8, we gave meaning to exponential expressions with rational-number exponents such as

$$5^{1/4}, \quad 3^{-3/4}, \quad 7^{2.34}, \quad 8^{1.73}.$$

For example, $5^{1.73}$, or $5^{173/100}$, means to raise 5 to the 173rd power and then take the positive 100th root. We now give meaning to expressions with irrational exponents, such as

$$5^{\sqrt{3}}, \quad 7^{\pi}, \quad 9^{-\sqrt{2}}.$$

Consider $5^{\sqrt{3}}$. Let us think of rational numbers r close to $\sqrt{3}$ and look at 5^r. As r gets closer to $\sqrt{3}$, 5^r gets closer to some real number.

r closes in on $\sqrt{3}$.	5^r closes in on some real number p.
r	5^r
$1 < \sqrt{3} < 2$	$5 = 5^1 < p < 5^2 = 25$
$1.7 < \sqrt{3} < 1.8$	$15.426 = 5^{1.7} < p < 5^{1.8} = 18.119$
$1.73 < \sqrt{3} < 1.74$	$16.189 = 5^{1.73} < p < 5^{1.74} = 16.452$
$1.732 < \sqrt{3} < 1.733$	$16.241 = 5^{1.732} < p < 5^{1.733} = 16.267$

As r closes in on $\sqrt{3}$, 5^r closes in on some real number p. We define $5^{\sqrt{3}}$ to be that number p, given here to seven decimal places:

$$5^{\sqrt{3}} \approx 16.2424508.$$

Any positive irrational exponent can be defined in a similar way. Negative irrational exponents are then defined in the same way as negative integer exponents. Thus the expression a^x has meaning for any real number x. The usual laws of exponents still hold, but we will not prove that here. We now define exponential functions.

DEFINITION   **Exponential Function**

The function $f(x) = a^x$, where x is real, $a > 0$, and $a \neq 1$, is called the *exponential function*, base a.

We restrict the **base** a to being positive to avoid the possibility of taking even roots of negative numbers—for example, the square root of -1, $(-1)^{1/2}$, which is not a real number. We restrict the base from being 1, because $f(x) = 1^x = 1$, and this constant function does not have an inverse. We will see that all other exponential functions, with the base restrictions, do have inverses.

The following are examples of exponential functions:

$$f(x) = 2^x, \qquad f(x) = (\tfrac{1}{2})^x, \qquad f(x) = (0.04)^x.$$

Note that, in contrast to polynomial functions like $f(x) = x^2$ and $f(x) = x^3$, the variable in an exponential function is in the *exponent*. Let us now consider graphs of exponential functions.

Example 1 Graph the exponential function $y = f(x) = 2^x$.

Solution We compute some function values, thinking of y as $f(x)$, and list the results in a table.

$$f(0) = 2^0 = 1,$$
$$f(1) = 2^1 = 2,$$
$$f(2) = 2^2 = 4,$$
$$f(3) = 2^3 = 8,$$
$$f(-1) = 2^{-1} = \frac{1}{2^1} = \frac{1}{2},$$
$$f(-2) = 2^{-2} = \frac{1}{2^2} = \frac{1}{4},$$
$$f(-3) = 2^{-3} = \frac{1}{2^3} = \frac{1}{8}$$

x	y, or $f(x)$
0	1
1	2
2	4
3	8
-1	$\frac{1}{2}$
-2	$\frac{1}{4}$
-3	$\frac{1}{8}$

Next, we plot these points and connect them with a smooth curve. Be sure to plot enough points to determine how steeply the curve rises.

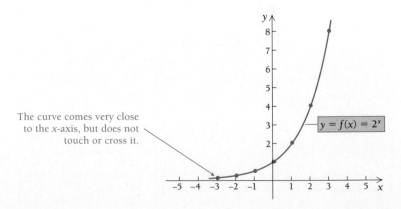

The curve comes very close to the x-axis, but does not touch or cross it.

$y = f(x) = 2^x$

Note that as x increases, the function values increase indefinitely. As x decreases, the function values decrease, getting very close to 0. The x-axis is a *horizontal asymptote*. For this asymptote, the curve comes very close to the x-axis, but never touches it. ◀

DO EXERCISE 1.

Example 2   Graph the exponential function $y = f(x) = (\frac{1}{2})^x$.

Solution   We compute some function values, thinking of y as $f(x)$, and list the results in a table. Before we do so, note that

$$y = f(x) = \left(\frac{1}{2}\right)^x = (2^{-1})^x = 2^{-x}.$$

$f(0) = 2^{-0} = 1,$

$f(1) = 2^{-1} = \dfrac{1}{2^1} = \dfrac{1}{2},$

$f(2) = 2^{-2} = \dfrac{1}{2^2} = \dfrac{1}{4},$

$f(3) = 2^{-3} = \dfrac{1}{2^3} = \dfrac{1}{8},$

$f(-1) = 2^{-(-1)} = 2^1 = 2,$

$f(-2) = 2^{-(-2)} = 2^2 = 4,$

$f(-3) = 2^{-(-3)} = 2^3 = 8$

x	y, or $f(x)$
0	1
1	$\frac{1}{2}$
2	$\frac{1}{4}$
3	$\frac{1}{8}$
-1	2
-2	4
-3	8

We plot these points and draw the curve. Note that this graph is a reflection of the graph in Example 1 across the y-axis.

$y = f(x) = \left(\frac{1}{2}\right)^x$

◀

DO EXERCISE 2 ON THE FOLLOWING PAGE.

The preceding examples illustrate exponential functions with various bases. Let us list some of these characteristics. Keep in mind that the definition of an exponential function, $f(x) = a^x$, requires that the base be positive

1. Graph: $y = f(x) = 3^x$.

   a) Complete this table of solutions.

x	y, or $f(x)$
0	
1	
2	
3	
-1	
-2	
-3	

   b) Plot the points from the table and connect them with a smooth curve.

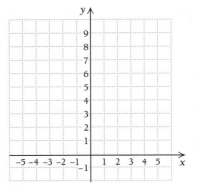

2. Graph: $y = f(x) = \left(\dfrac{1}{3}\right)^x$.

a) Complete this table of solutions.

x	y, or $f(x)$
0	
1	
2	
3	
-1	
-2	
-3	

b) Plot the points from the table and connect them with a smooth curve.

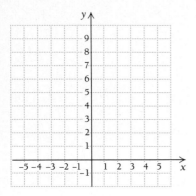

and different from 1. When $a = 1$, the graph is the line $y = 1$, which is not one-to-one and thus does not have an inverse.

THEOREM 2

1. When $a > 1$, the function f given by $f(x) = a^x$ is a continuous, increasing, one-to-one function.
2. When $0 < a < 1$, the function f given by $f(x) = a^x$ is a continuous, decreasing, one-to-one function.

The y-intercept of each exponential function $f(x) = a^x$ is the point $(0, 1)$. The domain of each function is the set of real numbers. The range is the set of all positive real numbers.

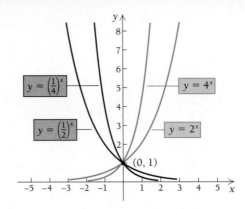

DO EXERCISES 3 AND 4 ON THE FOLLOWING PAGE.

Example 3   Graph: $y = f(x) = 2^{x-2}$.

Solution   We construct a table of values. Then we plot the points and connect them with a smooth curve. Be sure to note that $x - 2$ is the *exponent*.

$$f(0) = 2^{0-2} = 2^{-2} = \frac{1}{2^2} = \frac{1}{4},$$

$$f(1) = 2^{1-2} = 2^{-1} = \frac{1}{2^1} = \frac{1}{2},$$

$$f(2) = 2^{2-2} = 2^0 = 1,$$

$$f(3) = 2^{3-2} = 2^1 = 2,$$

$$f(4) = 2^{4-2} = 2^2 = 4,$$

$$f(-1) = 2^{-1-2} = 2^{-3} = \frac{1}{2^3} = \frac{1}{8},$$

$$f(-2) = 2^{-2-2} = 2^{-4} = \frac{1}{2^4} = \frac{1}{16}$$

x	y, or $f(x)$
0	$\dfrac{1}{4}$
1	$\dfrac{1}{2}$
2	1
3	2
4	4
-1	$\dfrac{1}{8}$
-2	$\dfrac{1}{16}$

The graph is a horizontal translation of the graph of $f(x) = 2^x$, 2 units to the right.

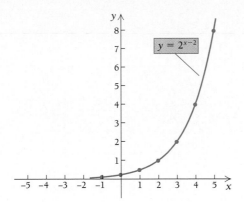

DO EXERCISE 5.

Graphs of Inverses of the Exponential Functions

We have noted that every exponential function with $a > 0$ and $a \neq 1$ is one-to-one and therefore has an inverse that is a function. In the next section, we will consider these inverse functions and give them names. For now, we draw their graphs by interchanging x and y.

Example 4 Graph: $x = 2^y$.

Solution The inverse of $y = 2^x$ is $x = 2^y$. Note that x is alone on one side of the equation. We can find ordered pairs that are solutions more easily by choosing values for y and then computing the x-values.

For $y = 0, x = 2^0 = 1$.

For $y = 1, x = 2^1 = 2$.

For $y = 2, x = 2^2 = 4$.

For $y = 3, x = 2^3 = 8$.

For $y = -1, x = 2^{-1} = \dfrac{1}{2^1} = \dfrac{1}{2}$.

For $y = -2, x = 2^{-2} = \dfrac{1}{2^2} = \dfrac{1}{4}$.

For $y = -3, x = 2^{-3} = \dfrac{1}{2^3} = \dfrac{1}{8}$.

x	y
1	0
2	1
4	2
8	3
$\dfrac{1}{2}$	-1
$\dfrac{1}{4}$	-2
$\dfrac{1}{8}$	-3

(1) Choose values for y.
(2) Compute values for x.

We plot the points and connect them with a smooth curve. Note that the curve does not touch or cross the y-axis. The y-axis is a vertical asymptote.

Graph.

3. $f(x) = 4^x$

4. $f(x) = \left(\dfrac{1}{4}\right)^x$

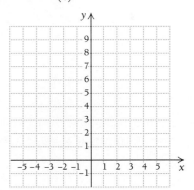

5. Graph: $y = 2^{x+2}$.

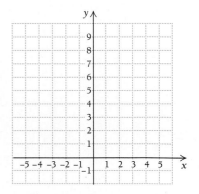

6. Graph: $x = 3^y$.

Note too that this curve looks just like the graph of $y = 2^x$, except that it is reflected across the line $y = x$, as we would expect for an inverse.

DO EXERCISE 6.

▸ Applications of Exponential Functions

Example 5   *Interest compounded annually.*   The amount of money A that a principal P will be worth after t years at interest rate i, compounded annually, is given by the formula

$$A = P(1 + i)^t.$$

Suppose that $100,000 is invested at 8% interest, compounded annually.

a) Find a function for the amount in the account after t years.
b) Find the amount of money in the account at $t = 0$, $t = 4$, $t = 8$, and $t = 10$.
c) Graph the function.

Solution

a) Since $P = \$100,000$ and $i = 8\% = 0.08$, we can substitute these values and form the following function:

$$A(t) = \$100,000(1 + 0.08)^t = \$100,000(1.08)^t.$$

b) To find the function values, we can use a calculator with a power key:

$$A(0) = \$100,000(1.08)^0 = \$100,000,$$
$$A(4) = \$100,000(1.08)^4 \approx \$136,048.90,$$
$$A(8) = \$100,000(1.08)^8 \approx \$185,093.02,$$
$$A(10) = \$100,000(1.08)^{10} \approx \$215,892.50.$$

c) We use the function values computed in part (b) and others if we wish, and draw the graph as follows. Note that the axes are scaled differently because of the large numbers and that t is restricted to non-negative values, because negative time values have no meaning.

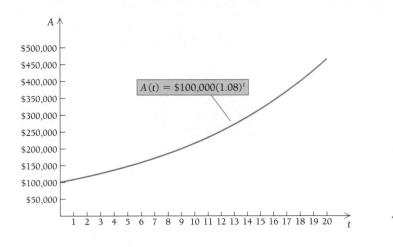

DO EXERCISE 7.

7. Suppose that \$80,000 is invested at 8% interest, compounded annually.

a) Find a function for the amount in the account after t years.

b) Find the amount of money in the account at $t = 0$, $t = 4$, $t = 8$, and $t = 10$.

c) Graph the function.

● EXERCISE SET 5.2

  Graph.

1. $y = f(x) = 2^x$
2. $y = f(x) = 3^x$
3. $y = 5^x$
4. $y = 6^x$
5. $y = 2^{x+1}$
6. $y = 2^{x-1}$
7. $y = 3^{x-2}$
8. $y = 3^{x+2}$
9. $y = 2^x - 3$
10. $y = 2^x + 1$
11. $y = 5^{x+3}$
12. $y = 6^{x-4}$
13. $y = (\frac{1}{2})^x$
14. $y = (\frac{1}{3})^x$
15. $y = (\frac{1}{5})^x$
16. $y = (\frac{1}{4})^x$
17. $y = 2^{2x-1}$
18. $y = 3^{4-x}$
19. $y = 2^{x-1} - 3$
20. $y = 2^{x+3} - 4$

B Graph.

21. $x = 2^y$
22. $x = 6^y$
23. $x = (\frac{1}{2})^y$
24. $x = (\frac{1}{3})^y$
25. $x = 5^y$
26. $x = 3^y$
27. $x = (\frac{2}{3})^y$
28. $x = (\frac{4}{3})^y$

Graph both equations using the same set of axes.

29. $y = 2^x$, $x = 2^y$
30. $y = 3^x$, $x = 3^y$
31. $y = (\frac{1}{2})^x$, $x = (\frac{1}{2})^y$
32. $y = (\frac{1}{4})^x$, $x = (\frac{1}{4})^y$

33. **Compact discs.** The number of compact discs purchased each year is increasing exponentially. The number N purchased, in millions, is given by

$$N(t) = 7.5(6)^{0.5t},$$

where $t = 0$ corresponds to 1985, $t = 1$ corresponds to 1986, and so on, t being the number of years after 1985.

a) Find the number of compact discs sold in 1986, 1987, 1990, 1995, and 2000.
b) Graph the function.

34. **Growth of bacteria Escherichi coli.** The bacteria *Escherichi coli* is commonly found in the human bladder. Suppose that 3000 of the bacteria are present at time $t = 0$. Then t minutes later, the number of bacteria present will be

$$N(t) = 3000(2)^{t/20}.$$

a) How many bacteria will be present after 10 min? 20 min? 30 min? 40 min? 60 min?

b) Graph the function.

35. *Interest compounded annually.* Suppose that $50,000 is invested at 9% interest, compounded annually.

a) Find a function for the amount in the account after t years.

b) Find the amount of money in the account at $t = 0$, $t = 4$, $t = 8$, and $t = 10$.

c) Graph the function.

36. *Recycling aluminum cans.* It is known that $\frac{1}{4}$ of all aluminum cans distributed will be recycled each year. A beverage company distributes 250,000 cans. The number still in use after time t, in years, is given by the function

$$N(t) = 250{,}000 \left(\tfrac{1}{4}\right)^t.$$

a) How many cans are still in use after 0 years? 1 year? 4 years? 10 years?

b) Graph the function.

37. *Salvage value.* An office machine is purchased for $5200. Its value each year is about 75% of the value the preceding year. Its value after t years is given by the exponential function

$$V(t) = \$5200(0.75)^t.$$

a) Find the value of the machine after 0 years, 1 year, 2 years, 5 years, and 10 years.

b) Graph the function.

38. *Turkey consumption.* The amount of turkey consumed by each person in this country is increasing exponentially. Assuming $t = 0$ corresponds to 1937, the amount of turkey, in pounds per person, consumed t years after 1937 is given by the function

$$N(t) = 2.3(3)^{0.033t}.$$

a) How much turkey was consumed per person in 1940? 1950? 1996? 2007?

b) Graph the function.

● **SYNTHESIS** _____

39. Approximate each of the following to six decimal places.

a) 7^3 **b)** $7^{3.1}$ **c)** $7^{3.14}$

d) $7^{3.141}$ **e)** $7^{3.1415}$ **f)** $7^{3.14159}$

Determine which of the two numbers is larger.

40. 7^π or π^7

41. $\pi^{3.2}$ or $\pi^{2.3}$

Graph. You will find a calculator with a power key $\boxed{y^x}$ most helpful.

42. $f(x) = (5.8)^x$ **43.** $f(x) = (2.7)^x$

Graph.

44. $y = \left(\tfrac{1}{2}\right)^x - 1$ **45.** $y = 2^x + 2^{-x}$

46. $f(x) = 2^{|x|}$ **47.** $y = 3^x + 3^{-x}$

48. $y = |2^x - 1|$ **49.** $y = 2^{-(x-1)}$

50. $g(x) = 2^{-|x|}$ **51.** $y = |2^x - 2|$

Graph both equations using the same set of axes.

52. $y = 1^x, \quad x = 1^y$

53. $y = 3^{-(x-1)}, \quad x = 3^{-(y-1)}$

Solve graphically.

54. $3^x \le 1$ **55.** $2^x > 1$

56. *Typing speed.* A person studies typing one semester in college. After he has studied for t hours, his speed, in words per minute, is given by

$$S(t) = 200[1 - (0.86)^t].$$

a) What is the speed of the typist after studying for 10 hr? 20 hr? 40 hr? 100 hr?

b) Graph the function.

● **CHALLENGE** _____

Graph.

57. $y = 2^{-x^2}$ **58.** $y = 3^{-(x+1)^2}$

59. $y = |2^{x^2} - 8|$

▣ **TECHNOLOGY CONNECTION** _____

60. Exactly one million dollars is invested in a certificate of deposit paying 8.85% interest, compounded annually.

a) Find a function that describes the value of the CD t years after the initial investment. Then graph this function on a grapher.

b) Use the Trace and Zoom features to find the number of years required for the investment to reach $2,350,000. Round to the nearest tenth of a year.

c) Find the time it takes for the investment to triple in value, rounded to the nearest tenth of a year.

d) How much is the CD worth, to the nearest $10,000, after 10.5 years?

5.3

LOGARITHMIC FUNCTIONS

The inverse of an exponential function is called a **logarithm function,** or **logarithmic function.** Such functions have many applications to problem solving.

Ⓐ Graphs of Logarithmic Functions

Consider the exponential function $f(x) = 2^x$. Does this function have an inverse that is a function? We see from the graph that this function is one-to-one and does have an inverse f^{-1} that is a function.

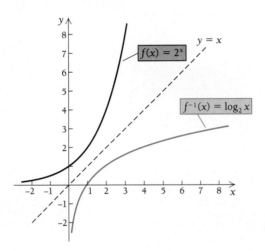

Consider the input 3 and the original function

$$f(x) = 2^x.$$

Then

$$f(3) = 2^3 = 8.$$

This also tells us that $f^{-1}(8) = 3$. If we did not know this, then to find $f^{-1}(8)$, we would be looking for an x such that

$$8 = 2^x.$$

Now we can probably reason that x is 3, but suppose we want to find $f^{-1}(13)$. The input for the inverse function is the number 13. We would be looking for an x such that

$$13 = 2^x.$$

Since $2^3 = 8$ and $2^4 = 16$, it seems reasonable that the number x that we are seeking is somewhere between 3 and 4. We know that the inverse exists, but we do not have a way to name $f^{-1}(13)$ as yet. Mathematicians have invented a name for the *inverse* of $f(x) = 2^x$. It is the *logarithm function, base 2,* denoted

$$f^{-1}(x) = \log_2 x.$$

We read $\log_2 x$ as "the logarithm, base 2, of x." Now $\log_2 8$ is the power to which we raise 2 to get 8. Thus, $\log_2 8 = 3$. Similarly, $\log_2 13$ is the power to which we raise 2 to get 13. We have no simpler way to write this.

For any exponential function $f(x) = a^x$, the inverse is called a **logarithmic function, base a**. The graph of the inverse can, of course, be obtained by reflecting the graph of $y = a^x$ across the line $y = x$, to obtain $x = a^y$, by interchanging x and y. Then $x = a^y$ is equivalent to $y = \log_a x$.

The inverse of $f(x) = a^x$ is given by

$$f^{-1}(x) = \log_a x.$$

We read $\log_a x$ as "the logarithm, base a, of x."

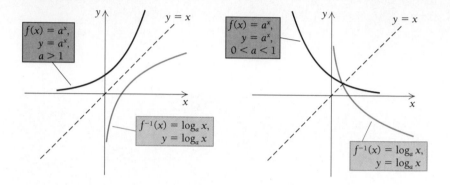

Almost always, we use a number a that is greater than 1 for a logarithmic base.

DEFINITION   **Logarithms**

We define $y = \log_a x$ as that number y such that $x = a^y$, where $x > 0$ and a is a positive constant other than 1.

It is helpful in dealing with logarithmic functions to remember that the logarithm of a number is an *exponent*. It is the exponent y in $x = a^y$. You might also think to yourself, "the logarithm, base a, of a number x is the power to which a must be raised in order to get x."

A logarithm is an exponent.

The following is a comparison of exponential and logarithmic functions.

Exponential Function	Logarithmic Function
$y = a^x$	$x = a^y$
$f(x) = a^x$	$f^{-1}(x) = \log_a x$
$a > 0, a \neq 1$	$a > 0, a \neq 1$
Domain = The set of real numbers	Range = The set of real numbers
Range = The set of positive real numbers	Domain = The set of positive real numbers

Why do we exclude 1 from being a logarithmic base? If we included it, we would be considering $x = 1^y = 1$. The graph of this equation is a vertical line and is not a function, as shown by the vertical-line test. We also exclude it from being a logarithmic base, because we exclude it from being an exponential base.

Example 1 Graph $y = f(x) = \log_5 x$. Determine the domain and the range.

Solution The equation $y = \log_5 x$ is equivalent to $5^y = x$. We can find ordered pairs that are solutions by choosing values for y and computing the x-values.

For $y = 0, x = 5^0 = 1.$
For $y = 1, x = 5^1 = 5.$
For $y = 2, x = 5^2 = 25.$
For $y = 3, x = 5^3 = 125.$

For $y = -1, x = 5^{-1} = \dfrac{1}{5}.$

For $y = -2, x = 5^{-2} = \dfrac{1}{25}.$

x, or 5^y	y
1	0
5	1
25	2
125	3
$\dfrac{1}{5}$	-1
$\dfrac{1}{25}$	-2

(1) Select y.
(2) Compute x.

We plot the ordered pairs and connect them with a smooth curve. The graph of $y = 5^x$ has been shown only for reference.

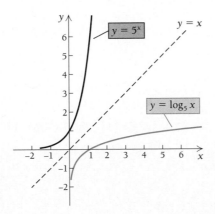

The graph of $f(x) = \log_a x$, for any a, has the x-intercept $(1, 0)$. The domain is the set of positive real numbers. The range is the set of all real numbers.

DO EXERCISE 1.

1. Graph $y = f(x) = \log_3 x$. Determine the domain and the range.

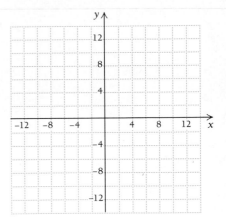

Convert to a logarithmic equation.

2. $6^0 = 1$

3. $10^{-3} = 0.001$

4. $16^{0.25} = 2$

5. $m^T = P$

Convert to an exponential equation.

6. $\log_2 32 = 5$

7. $\log_{10} 1000 = 3$

8. $\log_a Q = 7$

9. $\log_t M = x$

B Converting Between Exponential and Logarithmic Equations

We use the definition of logarithms to convert from exponential to logarithmic equations.

> $y = \log_a x$ is equivalent to $a^y = x$.

Be sure to memorize this relationship! It is probably the most important definition in the chapter. Many times this definition will be a justification for a proof or a procedure that we are considering.

Examples Convert each of the following to a logarithmic equation.

2. $8 = 2^x \rightarrow x = \log_2 8$ — The exponent is the logarithm.
— The base remains the base.

3. $y^{-1} = 4 \rightarrow -1 = \log_y 4$

4. $a^b = c \rightarrow b = \log_a c$

◀

DO EXERCISES 2–5.

We also use the definition of logarithms to convert from logarithmic to exponential equations.

Examples Convert each of the following to an exponential equation.

5. $y = \log_3 5 \rightarrow 3^y = 5$ — The logarithm is the exponent.
— The base does not change.

6. $-2 = \log_a 7 \rightarrow a^{-2} = 7$

7. $a = \log_b d \rightarrow b^a = d$

◀

DO EXERCISES 6–9.

C Solving Certain Logarithmic Equations

Certain equations involving logarithms can be solved by first converting to exponential equations. We will solve more complicated equations later.

Example 8 Solve: $\log_3 x = -2$.

Solution

$3^{-2} = x$ Converting to an exponential equation

$\frac{1}{9} = x$ Computing 3^{-2}

Check: $\log_3 \frac{1}{9}$ is the exponent to which we raise 3 to get $\frac{1}{9}$. Since

$3^{-2} = \frac{1}{9}$, we know that $\frac{1}{9}$ checks and is the solution. The solution set is $\{\frac{1}{9}\}$. ◀

Example 9 Solve: $\log_x 25 = 2$.

Solution

$$x^2 = 25 \qquad \text{Converting to an exponential equation}$$
$$x = 5 \quad \text{or} \quad x = -5$$

Check: $\log_5 25 = 2$ because $5^2 = 25$. Thus, 5 is a solution. Since all logarithmic bases must be positive, $\log_{-5} 25$ is not defined. Therefore, -5 is not a solution. Logarithmic bases must be positive because logarithms are defined in terms of exponential functions, which are defined only for positive bases. ◀

DO EXERCISES 10–12.

Solving an equation like $\log_b a = x$ amounts to finding the logarithm, base b, of the number a. You have done this before in graphing logarithmic functions. To think of finding logarithms as solving equations may help in some cases.

Example 10 Find $\log_{10} 10,000$.

Solution

Method 1. Let $\log_{10} 10,000 = x$. Then

$$10^x = 10,000 \qquad \text{Converting to an exponential equation}$$
$$10^x = 10^4$$
$$x = 4. \qquad \text{The exponents are the same.}$$

Therefore, $\log_{10} 10,000 = 4$.

Method 2. Think of the meaning of $\log_{10} 10,000$. It is the exponent to which you raise 10 to get 10,000. That exponent is 4. Therefore, $\log_{10} 10,000 = 4$. ◀

Example 11 Find $\log_{10} 0.01$.

Solution

Method 1. Let $\log_{10} 0.01 = x$. Then

$$10^x = 0.01$$
$$10^x = \frac{1}{100}$$
$$10^x = 10^{-2}$$
$$x = -2.$$

Therefore, $\log_{10} 0.01 = -2$.

Method 2. We know that $\log_{10} 0.01$ is the exponent to which you raise 10 to get 0.01. Since $0.01 = 1/100 = 10^{-2}$, it follows that the exponent is -2. Therefore, $\log_{10} 0.01 = -2$. ◀

Solve.

10. $\log_{10} x = 4$

11. $\log_x 81 = 4$

12. $\log_2 x = -2$

Find each of the following.

13. $\log_{10} 100{,}000$

14. $\log_{10} 0.0001$

15. $\log_7 1$

Simplify.

16. $\log_7 1$

17. $\log_4 4$

18. $\log_m m$

19. $\log_m 1$

Here are some other examples. Think out mentally how they can be found and compare them.

$$\log_{10} 1000 = 3, \qquad \log_{64} 64 = 1,$$
$$\log_{10} 100 = 2, \qquad \log_8 64 = 2,$$
$$\log_{10} 10 = 1, \qquad \log_4 64 = 3,$$
$$\log_{10} 1 = 0, \qquad \log_2 64 = 6.$$
$$\log_{10} 0.1 = -1,$$
$$\log_{10} 0.01 = -2;$$

Example 12   Find $\log_6 1$.

Solution

Method 1.   Let $\log_6 1 = x$. Then

$$6^x = 1 \qquad \text{Converting to an exponential equation}$$
$$6^x = 6^0 \qquad \text{Renaming 1 as } 6^0$$
$$x = 0.$$

Therefore, $\log_6 1 = 0$.

Method 2.   We know that $\log_6 1$ is the exponent to which 6 is raised to get 1. That exponent is 0. Therefore, $\log_6 1 = 0$. ◄

DO EXERCISES 13–15.

Example 12 illustrates an important property of logarithms.

THEOREM 3

For any base a, where $a > 0$ and $a \neq 1$,

$$\log_a 1 = 0.$$

The logarithm, base a, of 1 is always 0.

The proof follows from the fact that $a^0 = 1$. This is equivalent to the logarithmic equation $\log_a 1 = 0$.

Another property follows similarly. We know that $a^1 = a$ for any real number a. In particular, this holds for any positive number a. This is equivalent to the logarithmic equation $\log_a a = 1$.

THEOREM 4

For any base a, where $a > 0$ and $a \neq 1$,

$$\log_a a = 1.$$

DO EXERCISES 16–19.

EXERCISE SET 5.3

 Graph.

1. $y = \log_7 x$ **2.** $y = \log_6 x$

3. $y = \log_{10} x$ **4.** $y = \log_2 x$

5. $f(x) = \log_4 x$ **6.** $f(x) = \log_5 x$

7. $f(x) = \log_{1/2} x$ **8.** $f(x) = \log_{2.5} x$

9. $f(x) = \log_2 (x + 3)$ **10.** $f(x) = \log_3 (x - 2)$

11. $f(x) = \log_2 x + 3$ **12.** $f(x) = \log_3 x - 1$

Graph each function using the same set of axes.

13. $f(x) = 3^x, \quad f^{-1}(x) = \log_3 x$

14. $f(x) = 4^x, \quad f^{-1}(x) = \log_4 x$

 Convert to a logarithmic equation.

15. $10^3 = 1000$ **16.** $10^2 = 100$

17. $5^{-3} = \frac{1}{125}$ **18.** $4^{-5} = \frac{1}{1024}$

19. $8^{1/3} = 2$ **20.** $16^{1/4} = 2$

21. $10^{0.3010} = 2$ **22.** $10^{0.4771} = 3$

23. $e^3 = t$ **24.** $p^k = 3$

25. $Q^t = x$ **26.** $p^m = V$

27. $e^3 = 20.0855$ **28.** $e^2 = 7.3891$

29. $e^{-1} = 0.3679$ **30.** $e^{-6} = 0.00247$

Convert to an exponential equation.

31. $t = \log_4 7$ **32.** $h = \log_6 29$

33. $\log_2 32 = 5$ **34.** $\log_5 5 = 1$

35. $\log_{10} 0.1 = -1$ **36.** $\log_{10} 0.01 = -2$

37. $\log_{10} 7 = 0.845$ **38.** $\log_{10} 3 = 0.4771$

39. $\log_e 30 = 3.4012$ **40.** $\log_e 10 = 2.3036$

41. $\log_t Q = k$ **42.** $\log_m P = a$

43. $\log_e 0.38 = -0.9676$ **44.** $\log_e 0.906 = -0.0987$

45. $\log_r M = -x$ **46.** $\log_c W = -w$

Solve.

47. $\log_{10} x = 3$ **48.** $\log_2 16 = x$

49. $\log_3 3 = x$ **50.** $\log_5 \frac{1}{25} = x$

51. $\log_3 x = 2$ **52.** $\log_4 x = 3$

53. $\log_x 16 = 2$ **54.** $\log_x 64 = 3$

55. $\log_2 x = -1$ **56.** $\log_3 x = -2$

57. $\log_8 x = \frac{1}{3}$ **58.** $\log_{32} x = \frac{1}{5}$

Find each of the following.

59. $\log_{10} 1000$ **60.** $\log_{10} 10,000,000$

61. $\log_{10} 0.1$ **62.** $\log_{10} 0.001$

63. $\log_{10} 1$ **64.** $\log_{10} 10$

65. $\log_5 625$ **66.** $\log_2 64$

67. $\log_5 \frac{1}{25}$ **68.** $\log_2 \frac{1}{16}$

69. $\log_3 1$ **70.** $\log_8 8$

71. $\log_e 1$ **72.** $\log_e e$

73. $\log_{81} 9$ **74.** $\log_8 2$

75. $\log_e e^5$ **76.** $\log_e e^{-2}$

● **SYNTHESIS**

77. Graph both equations using the same set of axes.

$$y = \left(\tfrac{2}{3}\right)^x, \qquad y = \log_{2/3} x$$

Graph.

78. $y = |\log_2 x|$ **79.** $y = \log_2 |x|$

Solve.

80. $|\log_3 x| = 3$ **81.** $\log_{125} x = \frac{2}{3}$

82. $\log_\pi \pi^4 = x$ **83.** $\log_{\sqrt{5}} x = -3$

84. $\log_b b^{2x^2} = x$ **85.** $\log_4 (3x - 2) = 2$

86. $\log_8 (2x - 3) = -1$ **87.** $\log_x \sqrt[5]{36} = \frac{1}{10}$

88. $\log_{10} (x^2 + 21x) = 2$

Simplify.

89. $\log_{1/4} \frac{1}{64}$ **90.** $\log_{81} 3 \cdot \log_3 81$

91. $\log_{10} (\log_4 (\log_3 81))$ **92.** $\log_2 (\log_2 (\log_4 256))$

93. $\log_{\sqrt{3}} \frac{1}{81}$ **94.** $\log_{1/5} 25$

What is the domain of the function?

95. $f(x) = 3^x$ **96.** $f(x) = \log_{10} x$

97. $f(x) = \log_a x^2$ **98.** $f(x) = \log_4 x^3$

99. $f(x) = \log_{10} (3x - 4)$ **100.** $f(x) = \log_5 |x|$

Solve by graphing.

101. $\log_2 x < 0$ **102.** $\log_2 (x - 3) \geq 4$

OBJECTIVES

You should be able to:

A Express the logarithm of a product as a sum of logarithms, and conversely.

B Express the logarithm of a power as a product, and conversely.

C Express the logarithm of a quotient as a difference of logarithms, and conversely.

D Convert from logarithms of products, quotients, and powers to expressions in terms of individual logarithms, and conversely.

E Simplify expressions of the type $\log_a a^x$ and $a^{\log_a x}$.

5.4

PROPERTIES OF LOGARITHMIC FUNCTIONS

Logarithmic functions are important in many applications and in more advanced mathematics. We now establish some basic properties that are fundamental to the use of the functions. These properties are based on corresponding rules for exponents.

A Logarithms of Products

One property special to logarithmic functions is the following.

PROPERTY 1   **The Product Rule**

For any positive numbers M and N,

$$\log_a M \cdot N = \log_a M + \log_a N.$$

(The logarithm of a product is the sum of the logarithms of the factors. The number a can be any logarithmic base.)

Proof.   Let $\log_a M = x$ and $\log_a N = y$. Converting to exponential equations, we have $a^x = M$ and $a^y = N$. Then we multiply corresponding sides of the latter two equations, to obtain

$$M \cdot N = a^x \cdot a^y$$
$$= a^{x+y}.$$

Converting back to a logarithmic equation, we get

$$\log_a M \cdot N = x + y.$$

Remembering what x and y represent, it follows that

$$\log_a M \cdot N = \log_a M + \log_a N,$$

which was to be shown.

The use of the rule $a^x \cdot a^y = a^{x+y}$, which is the product rule for exponents, gives rise to the name product rule for logarithms. Recall that a logarithm is an exponent. To find the product of exponential expressions with the same base, we add the exponents. To find the logarithm of a product, which is an exponent, we add logarithms of the factors, which is a sum of exponents.

Example 1   Express as a sum of logarithms: $\log_3 (9 \cdot 27)$.

Solution

$$\log_3 (9 \cdot 27) = \log_3 9 + \log_3 27 \qquad \text{By Property 1}$$
$$= 2 + 3 = 5$$

Example 2 Express as a single logarithm: $\log_{10} 0.01 + \log_{10} 3947$.

Solution

$$\log_{10} 0.01 + \log_{10} 3947 = \log_{10} (0.01 \times 3947) \qquad \text{By Property 1}$$
$$= \log_{10} 39.47 \qquad \blacktriangleleft$$

CAUTION! The logarithm of a product is *not* the product of the logarithms—that is,

$$\log_a MN \neq (\log_a M)(\log_a N).$$

DO EXERCISES 1–4.

B **Logarithms of Powers**

The second basic property is as follows.

PROPERTY 2 **The Power Rule**

For any positive number M and any real number p,

$$\log_a M^p = p \cdot \log_a M.$$

(The logarithm of a power of M is the exponent times the logarithm of M. The number a can be any logarithmic base.)

Proof. Let $x = \log_a M$. Then we convert to an exponential equation, to get $a^x = M$. Raising both sides to the pth power, we obtain

$$(a^x)^p = M^p, \quad \text{or} \quad a^{xp} = M^p.$$

Converting back to a logarithmic equation, we get

$$\log_a M^p = xp.$$

But $x = \log_a M$, so

$$\log_a M^p = (\log_a M)p$$
$$= p \cdot \log_a M,$$

which was to be shown.

We call this the power rule for logarithms because of the analogous power rule for exponents.

Examples Express as a product.

3. $\log_a 11^{-3} = -3 \log_a 11$ \qquad By Property 2

4. $\log_a \sqrt[4]{7} = \log_a 7^{1/4}$ \qquad Writing exponential notation

$$= \frac{1}{4} \log_a 7 \qquad \text{By Property 2} \qquad \blacktriangleleft$$

DO EXERCISES 5 AND 6.

Express as a sum of logarithms.

1. $\log_5 (25 \cdot 5)$

2. $\log_b PQ$

Express as a single logarithm.

3. $\log_3 7 + \log_3 5$

4. $\log_a C + \log_a A + \log_a B$
$\quad + \log_a I + \log_a N$

Express as a product.

5. $\log_7 4^5$

6. $\log_a \sqrt{5}$

7. Express as a difference of logarithms:

$$\log_b \frac{P}{Q}.$$

 ## Logarithms of Quotients

Here is the third basic property.

PROPERTY 3 **The Quotient Rule**

For any positive numbers M and N,

$$\log_a \frac{M}{N} = \log_a M - \log_a N.$$

(The logarithm of a quotient is the logarithm of the numerator minus the logarithm of the denominator. The number a can be any logarithmic base.)

Proof. The proof makes use of Property 1 and Property 2:

$$\log_a \frac{M}{N} = \log_a MN^{-1}$$
$$= \log_a M + \log_a N^{-1} \qquad \text{Property 1}$$
$$= \log_a M + (-1)\log_a N \qquad \text{Property 2}$$
$$= \log_a M - \log_a N.$$

Example 5 Express as a difference of logarithms: $\log_t \frac{8}{Q}$.

Solution

$$\log_t \frac{8}{Q} = \log_t 8 - \log_t Q \qquad \text{By Property 3}$$

8. Express as a single logarithm:

$$\log_2 x - \log_2 25.$$

Example 6 Express as a single logarithm: $\log_b 54 - \log_b 27$.

Solution

$$\log_b 54 - \log_b 27 = \log_b \frac{54}{27} = \log_b 2$$

Example 7 Express as a single logarithm: $\log_{10} 10{,}000 - \log_{10} 100$.

Solution

$$\log_{10} 10{,}000 - \log_{10} 100 = \log_{10} \frac{10{,}000}{100} = \log_{10} 100$$

CAUTION! The logarithm of a quotient is *not* the quotient of the logarithms. That is,

$$\log_a \frac{M}{N} \neq \frac{(\log_a M)}{(\log_a N)}.$$

DO EXERCISES 7 AND 8.

▷ **Using the Properties Together**

Examples   Express in terms of logarithms.

8. $\log_a \dfrac{p^3 q^2}{z^4} = \log_a (p^3 q^2) - \log_a z^4$ Using Property 3

$\qquad\qquad = \log_a p^3 + \log_a q^2 - \log_a z^4$ Using Property 1

$\qquad\qquad = 3 \log_a p + 2 \log_a q - 4 \log_a z$ Using Property 2

9. $\log_a \sqrt[4]{\dfrac{ab}{c^3}} = \log_a \left(\dfrac{ab}{c^3}\right)^{1/4}$ Writing exponential notation

$\qquad\qquad = \dfrac{1}{4} \cdot \log_a \dfrac{ab}{c^3}$ Using Property 2

$\qquad\qquad = \dfrac{1}{4}(\log_a ab - \log_a c^3)$ Using Property 3

$\qquad\qquad = \dfrac{1}{4}(\log_a a + \log_a b - 3 \log_a c)$ Using Properties 1 and 2

$\qquad\qquad = \dfrac{1}{4}(1 + \log_a b - 3 \log_a c)$ $\log_a a = 1$

10. $\log_b \dfrac{ab^5}{m^3 n^4}$

$\qquad = \log_b ab^5 - \log_b m^3 n^4$ Using Property 3

$\qquad = (\log_b a + \log_b b^5) - (\log_b m^3 + \log_b n^4)$ Using Property 1

$\qquad = \log_b a + \log_b b^5 - \log_b m^3 - \log_b n^4$ Removing parentheses

$\qquad = \log_b a + 5 \log_b b - 3 \log_b m - 4 \log_b n$ Using Property 2

$\qquad = \log_b a + 5 - 3 \log_b m - 4 \log_b n$ $\log_b b = 1$ ◀

DO EXERCISES 9–11.

Examples   Express as a single logarithm.

11. $\dfrac{1}{2} \log_a x - 7 \log_a y + \log_a z$

$\qquad = \log_a x^{1/2} - \log_a y^7 + \log_a z$ Using Property 2

$\qquad = \log_a \dfrac{\sqrt{x}}{y^7} + \log_a z$ Using Property 3

$\qquad = \log_a \dfrac{z\sqrt{x}}{y^7}$ Using Property 1

Express in terms of logarithms.

9. $\log_a \sqrt{\dfrac{z^3}{xy}}$

10. $\log_a \dfrac{x^2}{y^3 a}$

11. $\log_m \dfrac{a^3 b^4}{m^5 n^9}$

Express as a single logarithm.

12. $5 \log_a x - \log_a y + \frac{1}{4} \log_a z$

13. $\log_a \frac{\sqrt{x}}{b} - \log_a \sqrt{bx}$

12. $\log_a \dfrac{b}{\sqrt{x}} + \log_a \sqrt{bx}$

$$= \log_a b - \log_a \sqrt{x} + \log_a \sqrt{bx} \qquad \text{Using Property 3}$$

$$= \log_a b - \frac{1}{2} \log_a x + \frac{1}{2} \log_a (bx) \qquad \text{Using Property 2}$$

$$= \log_a b - \frac{1}{2} \log_a x + \frac{1}{2}(\log_a b + \log_a x) \qquad \text{Using Property 1}$$

$$= \log_a b - \frac{1}{2} \log_a x + \frac{1}{2} \log_a b + \frac{1}{2} \log_a x$$

$$= \frac{3}{2} \log_a b \qquad \text{Collecting like terms}$$

$$= \log_a b^{3/2} \qquad \text{Using Property 2}$$

Example 12 could also be done as follows:

$$\log_a \frac{b}{\sqrt{x}} + \log_a \sqrt{bx} = \log_a \frac{b}{\sqrt{x}} \sqrt{bx} = \log_a b \sqrt{b} = \log_a b^{3/2}. \qquad \blacktriangleleft$$

DO EXERCISES 12 AND 13.

Examples Given

$$\log_a 2 = 0.301 \quad \text{and}$$
$$\log_a 3 = 0.477,$$

find each of the following.

13. $\log_a 6 \qquad \log_a 6 = \log_a (2 \cdot 3) = \log_a 2 + \log_a 3 \qquad \text{Property 1}$
$$= 0.301 + 0.477$$
$$= 0.778$$

14. $\log_a \frac{2}{3} \qquad \log_a \frac{2}{3} = \log_a 2 - \log_a 3 \qquad \text{Property 3}$
$$= 0.301 - 0.477$$
$$= -0.176$$

15. $\log_a 81 \qquad \log_a 81 = \log_a 3^4 = 4 \log_a 3 \qquad \text{Property 2}$
$$= 4(0.477)$$
$$= 1.908$$

16. $\log_a \frac{1}{3} \qquad \log_a \frac{1}{3} = \log_a 1 - \log_a 3 \qquad \text{Property 3}$
$$= 0 - 0.477$$
$$= -0.477$$

17. $\log_a \sqrt{a} \qquad \log_a \sqrt{a} = \log_a a^{1/2}$
$$= \frac{1}{2} \log_a a \qquad \text{Property 2}$$
$$= \frac{1}{2} \cdot 1$$
$$= \frac{1}{2}$$

18. $\log_a 5 \qquad$ No way to find using these properties and the given information.
$$(\log_a 5 \neq \log_a 2 + \log_a 3)$$

19. $\dfrac{\log_a 3}{\log_a 2}$ $\dfrac{\log_a 3}{\log_a 2} = \dfrac{0.477}{0.301} \approx 1.58$.

We simply divided, not using any of the properties.   ◀

DO EXERCISES 14–21. _____

  Simplifying Expressions $\log_a a^x$ and $a^{\log_a x}$

We have two final properties to consider.

> **PROPERTY 4**
>
> For any base a and any real number x,
>
> $\quad \log_a a^x = x$.
>
> (The logarithm, base a, of a to a power is the power.)

Proof.   The proof involves Property 2 and the fact that $\log_a a = 1$:

$$\log_a a^x = x(\log_a a) \qquad \text{Using Property 2}$$
$$= x \cdot 1 \qquad\qquad \text{Using } \log_a a = 1$$
$$= x.$$

If you forget Property 4, you can apply Property 2 and the fact that $\log_a a = 1$. That is, think of "$\log_a a^x$" as "the power to which we raise a to get a^x." That power is x.

Examples   Simplify.

20. $\log_a a^7 = 7$
21. $\log_{10} 10^{8.6} = 8.6$
22. $\log_e e^{-t} = -t$ ◀

DO EXERCISES 22–24. _____

> **PROPERTY 5**
>
> For any base a and any positive real number x,
>
> $\quad a^{\log_a x} = x$.
>
> (The number a raised to the power $\log_a x$ is x.)

Proof.   The proof follows directly from the definition of logarithms. Let

$$M = \log_a x.$$

Then $a^M = x$ from the definition of logarithms. But $M = \log_a x$, and if we

---

Given

$\quad \log_a 2 = 0.301 \quad$ and
$\quad \log_a 5 = 0.699$,

find each of the following.

14. $\log_a 4$

15. $\log_a 10$

16. $\log_a \dfrac{2}{5}$

17. $\log_a \dfrac{5}{2}$

18. $\log_a \dfrac{1}{5}$

19. $\log_a \sqrt{a^3}$

20. $\log_a 5a$

21. $\log_a 16$

Simplify.

22. $\log_2 2^8$

23. $\log_{10} 10^{4.3}$

24. $\log_e e^{23}$

Simplify.

25. $4^{\log_4 3}$

26. $7^{\log_7 x}$

27. $b^{\log_b 42}$

substitute $\log_a x$ for M, we obtain the desired result:

$$a^{\log_a x} = x.$$

Examples   Simplify.

23. $a^{\log_a 3} = 3$

24. $2^{\log_2 5} = 5$

25. $10^{\log_{10} t} = t$

◀

DO EXERCISES 25–27. _____

Summary of Properties of Logarithms

Property 1: *The Product Rule* $\log_a M \cdot N = \log_a M + \log_a N$

Property 2: *The Power Rule* $\log_a M^p = p \cdot \log_a M$

Property 3: *The Quotient Rule* $\log_a \dfrac{M}{N} = \log_a M - \log_a N$

Property 4: $\log_a a^x = x$

Property 5: $a^{\log_a x} = x$

●**EXERCISE SET** **5.4**

A   Express as a sum of logarithms.

1. $\log_2 (64 \cdot 8)$ **2.** $\log_3 (81 \cdot 27)$

3. $\log_4 (32 \cdot 64)$ **4.** $\log_5 (125 \cdot 25)$

5. $\log_c QP$ **6.** $\log_t 9Y$

Express as a single logarithm.

7. $\log_b 8 + \log_b 90$ **8.** $\log_a 75 + \log_a 2$

9. $\log_c P + \log_c Q$ **10.** $\log_e M + \log_e T$

B   Express as a product.

11. $\log_a x^4$ **12.** $\log_b t^3$

13. $\log_c y^5$ **14.** $\log_{10} y^8$

15. $\log_b Q^{-6}$ **16.** $\log_c K^{-6}$

C   Express as a difference of logarithms.

17. $\log_a \dfrac{76}{13}$ **18.** $\log_t \dfrac{M}{8}$

19. $\log_b \dfrac{5}{4}$ **20.** $\log_a \dfrac{x}{y}$

Express as a single logarithm.

21. $\log_a 18 - \log_a 5$ **22.** $\log_b 54 - \log_b 6$

D   Express in terms of logarithms.

23. $\log_a x^3 y^2 z$ **24.** $\log_a 6xy^5 z^4$

25. $\log_b \dfrac{x^2 y}{b^3}$ **26.** $\log_b \dfrac{p^2 q^5}{m^4 b^9}$

27. $\log_c \sqrt[3]{\dfrac{x^4}{y^3 z^2}}$ **28.** $\log_a \sqrt{\dfrac{x^6}{p^5 q^8}}$

29. $\log_a \sqrt[4]{\dfrac{m^8 n^{12}}{a^3 b^5}}$ **30.** $\log_a \sqrt{\dfrac{a^6 b^8}{a^2 b^5}}$

Express as a single logarithm and simplify if possible.

31. $\frac{2}{5} \log_a x - \frac{1}{3} \log_a y$

32. $\frac{1}{2} \log_a x + 4 \log_a y - 3 \log_a x$

33. $\log_a 2x + 3(\log_a x - \log_a y)$

34. $\log_a x^2 - 2 \log_a \sqrt{x}$

35. $\log_a \dfrac{a}{\sqrt{x}} - \log_a \sqrt{ax}$

36. $\log_a (x^2 - 4) - \log_a (x - 2)$

Given $\log_b 3 = 0.5283$ and $\log_b 5 = 0.7740$, find each of the following.

37. $\log_b 15$ **38.** $\log_b \frac{5}{3}$

39. $\log_b \frac{3}{5}$

40. $\log_b \frac{1}{5}$

41. $\log_b \frac{1}{3}$

42. $\log_b \sqrt{b}$

43. $\log_b \sqrt{b^3}$

44. $\log_b 5b$

45. $\log_b 3b$

46. $\log_b 9$

47. $\log_b 25$

48. $\log_b 75$

Simplify.

49. $\log_t t^{11}$

50. $\log_p p^3$

51. $\log_e e^{|x-4|}$

52. $\log_Q Q^{\sqrt{5}}$

53. $3^{\log_3 4x}$

54. $5^{\log_5 (4x-3)}$

55. $a^{\log_a Q}$

56. $t^{\log_t e^5}$

Solve for x.

57. $a^{\log_a x} = 15$

58. $5^{\log_5 8} = 2x$

59. $\log_e e^x = -7$

60. $\log_a a^x = 2.7$

● SYNTHESIS _____

Determine whether the equation is false.

61. $\dfrac{\log_a M}{\log_a N} = \log_a M - \log_a N$

62. $\dfrac{\log_a M}{\log_a N} = \log_a \dfrac{M}{N}$

63. $\dfrac{\log_a M}{c} = \log_a M^{1/c}$

64. $\log_N (M \cdot N)^x = x \log_N M + x$

65. $\log_a 2x = 2 \log_a x$

66. $\log_a 2x = \log_a 2 + \log_a x$

67. $\log_a (M + N) = \log_a M + \log_a N$

68. $\log_a x^3 = 3 \log_a x$

69. $\log_c a - \log_c b = \log_c \left(\dfrac{a}{b}\right)$

70. $\log_c a - \log_c b = \dfrac{\log_c a}{\log_c b}$

Express as a single logarithm and simplify if possible.

71. $\log_a (x^8 - y^8) - \log_a (x^2 + y^2)$

72. $\log_a (x + y) + \log_a (x^2 - xy + y^2)$

Express as a sum or difference of logarithms.

73. $\log_a \sqrt{4 - x^2}$

74. $\log_a \dfrac{x - y}{\sqrt{x^2 - y^2}}$

75. If $\log_a x = 2$, $\log_a y = 3$, and $\log_a z = 4$, what is

$$\log_a \frac{\sqrt[3]{x^2 z}}{\sqrt[3]{y^2 z^{-2}}}?$$

Solve.

76. $(x - 4) \cdot \log_a a^x = x$

77. $\log_a 3x = \log_a 3 + \log_a x$

78. $\log_\pi \pi^{2x+3} = 4$

79. $3^{\log_3 (8x-4)} = 5$

80. $4^{2\log_4 x} = 7$

81. $8^{2\log_8 x + \log_8 x} = 27$

82. $\log_a x^2 = 2 \log_a x$

83. $\log_b \dfrac{5}{x+2} = \log_b 5 - \log_b (x + 2)$

● CHALLENGE _____

84. If $\log_a x = 2$, what is $\log_a \left(\dfrac{1}{x}\right)$?

85. If $\log_a x = 2$, what is $\log_{1/a} x$?

Prove the following for any base a and any positive number x.

86. $\log_a \left(\dfrac{1}{x}\right) = -\log_a x$

87. $\log_a \left(\dfrac{x + \sqrt{x^2 - 5}}{5}\right) = -\log_a (x - \sqrt{x^2 - 5})$

88. $\log_a \left(\dfrac{1}{x}\right) = \log_{1/a} x$

89. $\log_{a^{1/n}} x = \log_a x^n$

5.5
FINDING LOGARITHMIC FUNCTION VALUES ON A CALCULATOR

You should be able to:

A Find common logarithms on a calculator.

B Find natural logarithms on a calculator.

C Use the change-of-base formula to find logarithms to bases other than 10 or e.

Any positive number other than 1 can be used as the base of a logarithmic function. However, some numbers are easier to use than others, and there are logarithmic bases that fit into certain applications more naturally than others. Base-10 logarithms are called **common logarithms.** They are useful because they are the same base as our *commonly* used decimal system for naming numbers. Before calculators became so widely available, common logarithms were extensively used in calculations. In fact, that is why logarithms were invented.

Using a calculator, find each of the following.

1. log 98,021,544

2. log 0.000617

3. log 53.1

4. log 0.05357

5. Let $f(x) = \log x$.
 a) Find $f(1000)$.
 b) Find $f^{-1}(3)$.
 c) Find $f(0.01)$.
 d) Find $f^{-1}(-2)$.

Another logarithmic base that is used a great deal today is, strangely enough, an irrational number. This number is named e and is about 2.7182818. Logarithms, base e, are called *natural logarithms*. We first consider common logarithms.

A Common Logarithms on a Calculator

Before the invention of calculators, tables were developed in order to find common logarithms. It is faster to use calculators and they are now quite inexpensive to purchase, so here we find common logarithms using calculators.

The abbreviation log, with no base written, is used for logarithms, base 10, or common logarithms. Thus,

$$\log 29 \quad \text{means} \quad \log_{10} 29.$$

On scientific calculators, the key for common logarithms is usually marked $\boxed{\text{LOG}}$. To find the common logarithm of a number, enter that number and press the $\boxed{\text{LOG}}$ key. It is important that you read the instructions for your particular calculator to be sure that you are carrying out the steps correctly.

Example 1 Find log 64,577.

Solution We enter 64,577 and then press the $\boxed{\text{LOG}}$ key. We find that

$$\log 64{,}577 \approx 4.8101. \qquad \text{Rounded to four decimal places} \qquad \blacktriangleleft$$

Keep in mind that 4.8101 is the power to which we raise 10 to get 64,577. That is, $10^{4.8101} \approx 64{,}577$.

Example 2 Find log 0.0000239.

Solution We enter 0.0000239 and then press the $\boxed{\text{LOG}}$ key. We find that

$$\log 0.0000239 \approx -4.6216. \qquad \text{Rounded to four decimal places} \qquad \blacktriangleleft$$

The inverse of a logarithmic function is, of course, an exponential function. That is,

$$\text{if } f(x) = \log_a x, \quad \text{then} \quad f^{-1}(x) = a^x.$$

For example,

$$\text{if } f(100) = \log 100 = 2, \quad \text{then} \quad f^{-1}(2) = 10^2 = 100.$$

Thus to find the inverse you can use the $\boxed{10^x}$ key.

DO EXERCISES 1–5.

B Natural Logarithms on a Calculator

The compound-interest formula, which we considered in Chapter 2, is

$$A = P\left(1 + \frac{i}{n}\right)^{nt},$$

where A is the amount that an initial investment P will be worth after t years at interest rate i, compounded n times per year. Suppose that $1 is an initial investment at 100% interest for 1 year (no bank would pay this). The above formula becomes a function A defined in terms of the number of compounding periods n:

$$A(n) = \left(1 + \frac{1}{n}\right)^n.$$

Let us find some function values. We round to six decimal places. We use a calculator with a power key $\boxed{y^x}$.

n	$A(n) = \left(1 + \dfrac{1}{n}\right)^n$
1 (compounded annually)	\$2.00
2 (compounded semiannually)	\$2.25
3	\$2.370370
4 (compounded quarterly)	\$2.441406
5	\$2.488320
100	\$2.704814
365 (compounded daily)	\$2.714567
8760 (compounded hourly)	\$2.718121

The numbers in this table get closer and closer to a very important number in mathematics, called e. The number e occurs in a great many applications. It may seem like a strange one to use as a logarithmic base, because it is an irrational number. Its decimal representation does not terminate or repeat:

$$e \approx 2.7182818284 \ldots.$$

Logarithms to the base e are called **natural logarithms.**

The abbreviation "ln" is generally used with natural logarithms. Thus

ln 53   means $\log_e 53$.

On scientific calculators, the key for the natural logarithmic function is marked $\boxed{\text{LN}}$, or $\boxed{\ln x}$.

Example 3   Find ln 4568.

Solution   We enter 4568 and then press the $\boxed{\text{LN}}$ key. We find that

ln 4568 ≈ 8.4268.   Rounded to four decimal places   ◀

Example 4   Find ln 2.

Solution

ln 2 ≈ 0.6931   ◀

Example 5   Find ln 0.0005142.

Solution   We enter 0.0005142 and then press the $\boxed{\text{LN}}$ key. We find that

ln 0.0005142 ≈ −7.5729.   ◀

Using a calculator, find each of the following.

6. $\ln 85{,}122$

7. $\ln 0.001127$

8. $\ln 0.39$

9. $\ln 1544.923$

10. Let $f(x) = \ln x$.

a) Find $f(3)$.
b) Find $f^{-1}(1.0986)$.
c) Find $f(0.48)$.
d) Find $f^{-1}(-0.7340)$.

As before, if

$$f(x) = \ln x, \text{ then } f^{-1}(x) = e^x.$$

For example,

$$f(2) = \ln 2 \approx 0.6931, \quad \text{so} \quad f^{-1}(0.6931) = e^{0.6931} \approx 2.$$

Check this on your calculator.

DO EXERCISES 6–10.

Changing Logarithmic Bases

Most calculators give the values of both common logarithms and natural logarithms. To find a logarithm with some other base, we can use the following conversion formula.

THEOREM 5	The Change-of-Base Formula

For any logarithmic bases a and b, and any positive number M,

$$\log_b M = \frac{\log_a M}{\log_a b}.$$

TECHNOLOGY CONNECTION

In Section 5.3, we discussed the inverses of exponential functions. For example, if $y = 3^x$, then the inverse is $x = 3^y$. Unfortunately, very few graphers can plot a function in the form $x = 3^y$. We must solve for y, which introduces the logarithmic function: $y = \log_3 x$.

Using Theorem 5, we can express the function $y = \log_3 x$ in the form

$$y = \frac{\log x}{\log 3}.$$

Now we can use our grapher to draw the inverse of $y = 3^x$.

$y = 3^x$
$y = \log_3 x$

$[-9, 9] \times [-6, 6]$

Draw the graphs of each of the following functions and their inverses. Use a viewing box of $[-10, 10] \times [-10, 10]$.

TC 1. $y = 5^x$

TC 2. $y = 1.2^x$

TC 3. $y = 0.3^x$

Proof. Let $x = \log_b M$. Then, writing an equivalent exponential equation, we have $b^x = M$. Next we take the logarithm base a on both sides. This gives us

$$\log_a b^x = \log_a M.$$

By Property 2,

$$x \log_a b = \log_a M,$$

and solving for x, we obtain

$$x = \frac{\log_a M}{\log_a b}.$$

But $x = \log_b M$, so we have

$$\log_b M = \frac{\log_a M}{\log_a b},$$

which is the change-of-base formula.

Example 6 Find $\log_5 8$ using common logarithms.

Solution Let $a = 10$, $b = 5$, and $M = 8$. Then we substitute into the change-of-base formula:

$$\log_5 8 = \frac{\log_{10} 8}{\log_{10} 5} \quad \text{Substituting}$$

$$\approx \frac{0.9031}{0.6990} \quad \text{When using your calculator, you need not round before dividing.}$$

$$\approx 1.2920.$$

CAUTION! $\dfrac{\log_{10} 8}{\log_{10} 5}$ is *not* $\log_{10} 8 - \log_{10} 5$.

To check, we use a calculator with a power key $\boxed{y^x}$ to verify that

$5^{1.2920} \approx 8$.   ◀

DO EXERCISE 11.

11. Find $\log_6 8$ using common logarithms.

We can also use base e for a conversion.

Example 7   Find $\log_4 31$ using natural logarithms.

Solution   Substituting e for a, 4 for b, and 31 for M, we have

$$\log_4 31 = \frac{\log_e 31}{\log_e 4} \quad \text{Using the change-of-base formula}$$

$$= \frac{\ln 31}{\ln 4}$$

$$\approx \frac{3.4340}{1.3863}$$

$$\approx 2.4771.$$ ◀

12. Find $\log_3 546$ using natural logarithms.

DO EXERCISE 12.

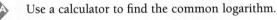

● EXERCISE SET 5.5

Ⓐ   Use a calculator to find the common logarithm.

1. $\log 3$ **2.** $\log 7$

3. $\log 8$ **4.** $\log 13$

5. $\log 2.34$ **6.** $\log 3.07$

7. $\log 65$ **8.** $\log 84$

9. $\log 62.4$ **10.** $\log 10.8$

11. $\log 532$ **12.** $\log 196$

13. $\log 13,400$ **14.** $\log 93,100$

15. $\log 0.57$ **16.** $\log 0.69$

17. $\log 0.052$ **18.** $\log 0.387$

19. $\log 0.009808$ **20.** $\log 0.0005123$

21. $10^{-2.9523}$ **22.** $10^{4.8982}$

23. $\log(-4.923)$ **24.** $\log(-7.891)$

Ⓑ   Find the natural logarithm using a calculator.

25. $\ln 3$ **26.** $\ln 2$ **27.** $\ln 8$

28. $\ln 13$ **29.** $\ln 82$ **30.** $\ln 50$

31. $\ln 8365$ **32.** $\ln 809.3$ **33.** $\ln 0.0059$

34. $\ln 0.00037$ **35.** $e^{1.0312}$ **36.** $e^{-6.3783}$

37. $e^{-12.832}$ **38.** $e^{17.814}$

Ⓒ   Find the logarithm using the change-of-base formula.

39. $\log_4 100$ **40.** $\log_3 20$ **41.** $\log_2 12$

42. $\log_5 40$ **43.** $\log_{100} 0.3$ **44.** $\log_{200} 50$

45. $\log_{0.5} 7$ **46.** $\log_{0.1} 2$ **47.** $\log_3 0.3$

48. $\log_2 0.06$ **49.** $\log_\pi 100$ **50.** $\log_\pi 25$

● SYNTHESIS

Verify each of the following.

51. $\ln x = 2.3026 \log x$

52. $\log x = 0.4343 \ln x$

53. Using function values obtained on a calculator, plot points and draw a precise graph of $y = f(x) = 10^x$.

54. Using function values obtained on a calculator, plot points and draw a precise graph of $y = g(x) = e^x$.

55. Using function values obtained on a calculator and values obtained in Exercise 53, plot points and draw a precise graph of $y = f^{-1}(x) = \log x$.

56. Using function values obtained on a calculator and values obtained in Exercise 54, plot points and draw a precise graph of $y = g^{-1}(x) = \ln x$.

Simplify.

57. $\dfrac{\log_5 8}{\log_5 2}$

58. $\dfrac{\log_3 64}{\log_3 16}$

Solve for x.

59. $\log 872x = 5.3442$

60. $\log 43x^2 = 8.0166$

61. $\log 784 + \log x = \log 2322$

62. $\dfrac{2.34}{\ln x} = \dfrac{57}{4.03}$

Use the change-of-base formula to derive each of the following formulas.

63. $\log e = \dfrac{1}{\ln 10}$

64. $\log M = \dfrac{\ln M}{\ln 10}$

65. $\log_b M = \dfrac{1}{\log_M b}$

66. $\ln M = \dfrac{\log M}{\log e}$

67. $\log_a (\log_a x) = \log_a (\log_b x) - \log_a (\log_b a)$

68. Given $f(x) = (1 + x)^{1/x}$, find $f(1)$, $f(0.5)$, $f(0.2)$, $f(0.1)$, $f(0.01)$, and $f(0.001)$ to six decimal places. This sequence of numbers approaches the number e.

69. Given $f(t) = t^{1/(t-1)}$, find $f(0.5)$, $f(0.9)$, $f(0.99)$, $f(0.999)$, and $f(0.9999)$ to six decimal places. This sequence of numbers approaches the number e.

70. Which is larger, e^π or π^e?

71. Which is larger, $e^{\sqrt{\pi}}$ or $\sqrt{e^\pi}$?

72. In some textbooks and computer applications, $\log x$ is used to represent $\log_e x$. Discuss some methods you might use to discover what the base actually is.

 TECHNOLOGY CONNECTION _____

Using a grapher, graph the function and its inverse.

73. $y = 4.5^x$

74. $y = 6.7^x$

75. $y = 0.15^x$

76. $y = 0.95^x$

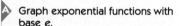

OBJECTIVES

You should be able to:

A Graph exponential functions with base e.

B Graph natural logarithmic functions.

C Sketch graphs of functions in applications.

5.6

GRAPHS OF EXPONENTIAL FUNCTIONS WITH BASE e AND APPLICATIONS

Exponential and logarithmic functions with base e are two of the most valuable functions that we study in mathematics. Because of their importance in many applications, it is helpful to study their graphs.

A Exponential Functions with Base e

Graphs of $f(x) = e^{kx}$ and $f(x) = e^{-kx}$

Example 1 Graph $f(x) = e^x$ and $f(x) = e^{-x}$.

Solution We use a calculator with an $\boxed{e^x}$ key to find approximate values of e^x and e^{-x}. Using these values, we can draw the graphs of the functions.

x	e^x	e^{-x}
0	1	1
1	2.7	0.4
2	7.4	0.1
-1	0.4	2.7
-2	0.1	7.4

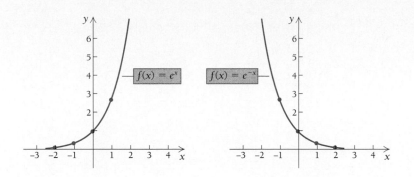

Note that the graph of e^{-x} is a reflection of the graph of e^x across the y-axis. ◀

Example 2   Graph: $f(x) = e^{-0.5x}$.

Solution   We find some solutions with a calculator, plot them, and then draw the graph. For example, $f(2) = e^{-0.5(2)} = e^{-1} \approx 0.4$.

x	$e^{-0.5x}$
0	1
1	0.6
2	0.4
3	0.2
−1	1.6
−2	2.7
−3	4.5

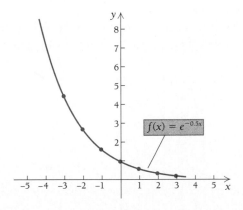

◀

DO EXERCISES 1–3.

Graphs of $f(x) = 1 - e^{-kx}$

Functions of the type $f(x) = 1 - e^{-kx}$ are also important.

Example 3   Graph $f(x) = 1 - e^{-2x}$ for nonnegative values of x.

Graph.

1. $f(x) = e^{2x}$

2. $f(x) = e^{-2x}$

3. $f(x) = e^{0.2x}$

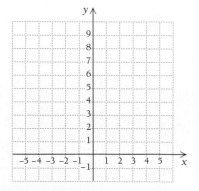

4. Graph $f(x) = 1 - e^{-x}$ for nonnegative values of x.

Solution We obtain these values using a calculator with an $\boxed{e^x}$ key. For example, when $x = 1$,

$$f(1) = 1 - e^{-2(1)} \quad \text{Substituting}$$
$$= 1 - e^{-2}$$
$$\approx 1 - 0.135335 \quad \text{Using a calculator}$$
$$\approx 0.86.$$

x	e^{-2x}	$1 - e^{-2x}$
0	1	0
$\frac{1}{2}$	0.367879	0.63
1	0.135335	0.86
2	0.018316	0.98
3	0.002479	0.998

In general, the graph of $f(x) = 1 - e^{-kx}$, for $k > 0$, increases from 0 and approaches 1 as x gets larger.

DO EXERCISE 4.

B   Natural Logarithmic Functions

Example 4   Graph: $g(x) = \ln x$.

Solution   There are two ways in which we might obtain the graph of $y = g(x) = \ln x$. One is by writing its equivalent equation, $x = e^y$.

We select values of y and use a calculator to find the corresponding values of e^y. We then plot points, remembering that x still is the first coordinate.

x, or e^y	y
0.1	-2
0.4	-1
1	0
2.7	1
7.4	2
20	3

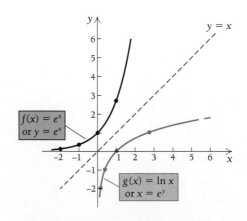

Note that f and g are inverses of each other. That is, the graph of $y = \ln x$ is a reflection across the line $y = x$ of the graph of $y = e^x$.

The second method of graphing $y = \ln x$ is to directly use the $\boxed{\text{ln}}$ key on a calculator. For example, $\ln 2 = 0.6931 \approx 0.7$. ◀

DO EXERCISE 5.

Example 5 Graph: $f(x) = \ln(x + 3)$.

Solution We find some solutions with a calculator, plot them, and then draw the graph. When $x = 2$, $y = \ln(2 + 3) = \ln 5 \approx 1.6$.

x	y, or $\ln(x+3)$
0	1.1
1	1.4
2	1.6
3	1.8
4	1.9
-1	0.7
-2	0
-2.5	-0.7

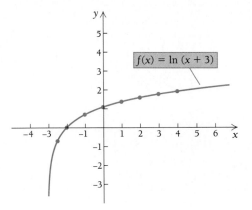

Note that the graph of $y = \ln(x + 3)$ is a horizontal translation (3 units to the left) of the graph of $y = \ln x$. ◀

DO EXERCISE 6.

 Graphs in Applications

Example 6 *Advertising.* A company begins a radio advertising campaign in New York City to market a new video game. The percentage of the target market that buys a product is normally a function of the length of the advertising campaign. The estimated percentage is given by

$$f(t) = 1 - e^{-0.04t},$$

where t = the number of days of the campaign.

a) Find $f(25)$, the percentage of the target market that has bought the product after a 25-day advertising campaign.

b) Sketch a graph of the function.

Solution

a) We evaluate $f(t)$ when $t = 25$:

$$f(25) = 1 - e^{-0.04(25)} \quad \text{Substituting}$$
$$= 1 - e^{-1}$$
$$\approx 1 - 0.367879$$
$$\approx 0.632121$$
$$\approx 63.2\%.$$

5. Graph: $f(x) = 2 \ln x$.

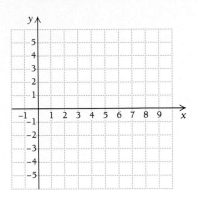

6. Graph: $f(x) = \ln(x - 1)$.

 TECHNOLOGY CONNECTION

Functions involving $\ln x$ are usually simple to graph because most graphers include a $\ln x$ key.

Graph each of the following.

TC 5. $f(x) = \ln(x + 4)$

TC 6. $f(x) = \ln x + 4$

TC 7. $f(x) = 4 \ln x$

TC 8. $f(x) = \dfrac{1}{\ln(x + 4)}$

7. The value of a stock is given by

$$V(t) = \$45(1 - e^{-0.8t}) + \$15,$$

where V is the value of the stock after time t, in months.

a) Find $V(6)$.

b) Find other function values, and sketch a graph of the function.

b) We find other solutions, plot them, and then sketch the graph.

t	$f(t)$
25	63.2%
50	86.5%
75	95.0%
100	98.2%

The function increases from 0 (0%) to 1 (100%). The longer the advertising campaign, the larger the percentage of the market that has bought the product.

DO EXERCISE 7. _____

● **EXERCISE SET** **5.6**

A Graph using a calculator.

1. $f(x) = 2e^x$

2. $f(x) = 0.5e^x$

3. $f(x) = e^{(1/2)x}$

4. $f(x) = e^{-0.3x}$

5. $f(x) = e^{x+1}$

6. $f(x) = e^{-x+1}$

7. $f(x) = e^{2x} + 1$

8. $f(x) = e^x - 2$

9. $f(x) = 1 - e^{-0.01x}$, for nonnegative values of x

10. $f(x) = 1 - e^{-3x}$, for nonnegative values of x

11. $f(x) = 2(1 - e^{-x})$, for nonnegative values of x

12. $f(x) = \frac{1}{2}(1 - e^{-2x})$, for nonnegative values of x

B Graph using a calculator.

13. $f(x) = 4 \ln x$

14. $f(x) = 3 \ln x$

15. $f(x) = \frac{1}{2} \ln x$

16. $f(x) = 0.2 \ln x$

17. $f(x) = \ln(x - 2)$

18. $f(x) = \ln(x + 1)$

19. $f(x) = 2 - \ln x$

20. $f(x) = (\ln x) - 4$

C

21. *Acceptance of a new medicine.* The percentage P of doctors who accept a new medicine is given by

$$P(t) = 1 - e^{-0.2t},$$

where t = time, in months.

a) Find $P(1)$, $P(4)$, $P(6)$, and $P(12)$.

b) Sketch a graph of the function.

22. *Hullian learning model.* A typist learns to type W words per minute after t weeks of practice, where W is given by

$$W(t) = 100(1 - e^{-0.3t}).$$

a) Find $W(1)$, $W(5)$, $W(8)$, and $W(10)$.

b) Sketch a graph of the function.

23. *Growth of a stock.* The value of a stock is given by

$$V(t) = \$58(1 - e^{-1.1t}) + \$20,$$

where V is the value of the stock after t months.

a) Find $V(1)$, $V(2)$, $V(4)$, $V(6)$, and $V(12)$.

b) Sketch a graph of the function.

24. *Advertising.* A toy company begins a television advertising campaign in Houston to market a new product. The television station uses the following function to estimate the percentage of the target market that buys the toy after t days of the campaign:

$$P(t) = 1 - e^{-0.03t}.$$

a) Find $P(10)$, $P(30)$, $P(60)$, and $P(120)$.

b) Sketch a graph of the function.

● SYNTHESIS _____

Graph.

25. $g(x) = e^{|x|}$

26. $f(x) = \ln |x|$

27. $f(x) = |\ln x|$

28. $g(x) = |\ln(x - 1)|$

29. $f(x) = \dfrac{e^x + e^{-x}}{2}$

30. $f(x) = \dfrac{e^x - e^{-x}}{2}$

31. *Spread of a rumor.* In a college with a student population of 800, a group of 6 students spread the rumor, "Men go for women who study calculus, and women go for men who study calculus!" The number of people who have heard the rumor after t minutes is given by

$$N(t) = \frac{4800}{6 + 794e^{-0.4t}}.$$

a) Find $N(3)$, $N(5)$, $N(10)$, and $N(15)$.
b) Sketch the graph of the function.

32. *Spread of an epidemic.*   In a town whose total population is 2000, the disease *Rottenich* creates an epidemic. The initial number of people infected is 10. The number of people infected after time t, in weeks, is given by

$$P(t) = \frac{20,000}{10 + 1990e^{-6t}}.$$

a) Find $P(0.2)$, $P(0.5)$, $P(0.8)$, $P(1)$, $P(1.5)$, and $P(2)$.
b) Sketch the graph of the function.

▨ TECHNOLOGY CONNECTION _____

Use a grapher. For each of the following:

a) Graph the function.
b) Estimate the zeros.
c) Estimate the maximum and the minimum values.

33. $f(x) = x^2 e^{-x}$ **34.** $f(x) = e^{-x^2}$

35. $f(x) = x^2 \ln x$ **36.** $f(x) = \dfrac{\ln x}{x^2}$

37.–42. Use a grapher to check the graphs in Exercises 25–30.

5.7

SOLVING EXPONENTIAL AND LOGARITHMIC EQUATIONS

 Solving Exponential Equations

Equations with variables in exponents, such as $3^x = 20$ and $2^{5x} = 64$, are called **exponential equations.** Sometimes, as is the case with $2^{5x} = 64$, we can write each side as a power of the same number:

$$2^{5x} = 2^6.$$

Then since the exponents are the same, we can set them equal and solve:

$$5x = 6$$
$$x = \tfrac{6}{5}.$$

We use the following property.

THEOREM 6
For any $a > 0$, $a \neq 1$, $a^x = a^y$ is equivalent to $x = y$.

OBJECTIVES

You should be able to:

A Solve exponential equations.

B Solve logarithmic equations.

Proof.   The theorem follows from the fact that $f(x) = a^x$ is a one-to-one function. If $a^x = a^y$ is true, then $f(x) = f(y)$. Then since f is one-to-one (see the definition in Section 5.1), it follows that $x = y$. Conversely, if $x = y$, it follows that $a^x = a^y$, since we are raising a to the same power.

Example 1   Solve: $2^{3x-7} = 32$.

Solution   Note that $32 = 2^5$. Thus we can write each side as a power of

Solve.

1. $5^{2x} = 25$

the same number:

$$2^{3x-7} = 2^5.$$

Since the base is the same, 2, the exponents must be the same. Thus,

$$3x - 7 = 5$$
$$3x = 12$$
$$x = 4.$$

Check:

$$\begin{array}{c|c} 2^{3x-7} = 32 \\ \hline 2^{3(4)-7} & 32 \\ 2^{12-7} \\ 2^5 \\ 32 \end{array}$$

2. $4^{4x-3} = 64$

The solution is 4. The solution set is $\{4\}$. ◀

DO EXERCISES 1 AND 2. ────────────

When it does not seem possible to write each side as a power of the same base, we can take the common or natural logarithm on each side and then use Property 2.

Example 2 Solve: $3^x = 20$.

Solution

3. Solve: $7^x = 20$.

$$\log 3^x = \log 20 \qquad \text{Taking the common logarithm on both sides}$$
$$x \log 3 = \log 20 \qquad \text{Property 2}$$
$$x = \frac{\log 20}{\log 3} \qquad \text{Solving for } x$$
$$\approx \frac{1.3010}{0.4771} \approx 2.7268$$

You can check this answer by finding $3^{2.7268}$ using a $\boxed{y^x}$ key on a calculator. ◀

DO EXERCISE 3. ────────────

If the base is e, we can take the logarithm with e as the base. This will make our work easier.

4. Solve: $e^{0.3t} = 80$.

Example 3 Solve: $e^{0.08t} = 2500$.

Solution We take the natural logarithm on both sides:

$$\ln e^{0.08t} = \ln 2500 \qquad \text{Taking ln on both sides}$$
$$0.08t = \ln 2500 \qquad \text{Here we use Property 4: } \log_a a^x = x.$$
$$t = \frac{\ln 2500}{0.08}$$
$$\approx 97.8. \qquad ◀$$

DO EXERCISE 4. ────────────

Example 4 Solve for x:

$$\frac{e^x + e^{-x}}{2} = t.$$

Solution Note that we are to solve for x. However, we have more than one term with x in the exponent. To get a single expression with x in the exponent, we do the following:

$$e^x + e^{-x} = 2t \qquad \text{Multiplying by 2}$$

$$e^x + \frac{1}{e^x} = 2t \qquad \begin{array}{l}\text{Rewriting with a}\\\text{positive exponent}\end{array}$$

$$e^{2x} + 1 = 2te^x \qquad \begin{array}{l}\text{Multiplying on both}\\\text{sides by } e^x\end{array}$$

$$(e^x)^2 - 2t \cdot e^x + 1 = 0.$$

This equation is reducible to quadratic, with $u = e^x$. The coefficients of the reduced quadratic equation are $a = 1$, $b = -2t$, and $c = 1$. Using the quadratic formula, we obtain

$$e^x = \frac{2t \pm \sqrt{4t^2 - 4}}{2}$$

$$= t \pm \sqrt{t^2 - 1}.$$

We can now take the natural logarithm on both sides:

$$\ln e^x = \ln (t \pm \sqrt{t^2 - 1})$$

$$x = \ln (t \pm \sqrt{t^2 - 1}). \qquad \text{Using Property 4} \quad \blacktriangleleft$$

DO EXERCISE 5. _____

 B **Solving Logarithmic Equations**

Equations containing logarithmic expressions are called **logarithmic equations.** We solved some logarithmic equations in Section 5.3. We did so by converting to an equivalent exponential equation.

Example 5 Solve: $\log_2 x = 4$.

Solution We obtain an equivalent exponential expression:

$$x = 2^4$$

$$x = 16.$$

The solution is 16. The solution set is $\{16\}$.

DO EXERCISE 6. _____

> To solve logarithmic equations, first try to obtain a single logarithmic expression on one side and then write an equivalent exponential equation.

 TECHNOLOGY CONNECTION

Equations of this type can be solved using a grapher. Plot both sides of the equation on the same axes, selecting a viewing box large enough to show all intersections. Then use the Trace and Zoom features to find the solution(s) to whatever accuracy you desire. All answers will be in decimal form instead of fractions or radicals.

As an example, let's solve the exponential equation $e^{5x} - 7 = 2x + 6$. Graphing the functions $y = e^{5x} - 7$ and $y = 2x + 6$ on the same axes, we then look for intersections. (Of course, if the graphs don't intersect anywhere, there are no solutions.)

$$[-10, 5] \times [-10, 10]$$

In this case, the graphs intersect between -7 and -6, and again between 0 and 1. Activating Trace and Zoom, we can find the solution to whatever accuracy we wish. The solutions, accurate to the nearest hundredth, are $x = -6.50$ and $x = 0.53$.

Use a grapher to find the solutions, accurate to the nearest hundredth, to each of the following exponential equations.

TC 1. $e^{7x} = 14$

TC 2. $8e^{0.5x} = 3$

TC 3. $xe^{3x} - 1 = 5$

TC 4. $5e^{5x} + 10 = 3x + 40$

5. Solve $\dfrac{e^x - e^{-x}}{2} = t$ for x.

6. Solve: $\log_2 x = 3$.

7. Solve: $\log_4 (8x - 6) = 3$.

Example 6 Solve: $\log_3 (5x + 7) = 2$.

Solution We already have a single logarithmic expression, so we write an equivalent exponential equation:

$$5x + 7 = 3^2 \qquad \text{Writing an equivalent exponential equation}$$
$$5x + 7 = 9$$
$$5x = 2$$
$$x = \tfrac{2}{5}.$$

Check:

$$\log_3 (5x + 7) = 2$$

$\log_3 (5 \cdot \tfrac{2}{5} + 7)$	2
$\log_3 (2 + 7)$	
$\log_3 9$	
2	

The solution is $\tfrac{2}{5}$. The solution set is $\{\tfrac{2}{5}\}$.

DO EXERCISE 7.

Example 7 Solve: $\log x + \log (x + 3) = 1$.

Solution We have common logarithms here. Writing in the base 10's will help us understand the problem:

$$\log_{10} x + \log_{10} (x + 3) = 1$$
$$\log_{10} [x(x + 3)] = 1 \qquad \text{Using Property 1 to obtain a single logarithm}$$
$$x(x + 3) = 10^1 \qquad \text{Writing an equivalent exponential equation}$$
$$x^2 + 3x = 10$$
$$x^2 + 3x - 10 = 0$$
$$(x - 2)(x + 5) = 0 \qquad \text{Factoring}$$
$$x - 2 = 0 \quad \text{or} \quad x + 5 = 0 \qquad \text{Principle of zero products}$$
$$x = 2 \quad \text{or} \qquad x = -5$$

Check: For 2:

$$\log x + \log (x + 3) = 1$$

$\log 2 + \log (2 + 3)$	1
$\log 2 + \log 5$	
$\log 10$	
1	

For -5:

$$\log x + \log (x + 3) = 1$$

$\log (-5) + \log (-5 + 3)$	1

The number -5 is not a solution because negative numbers do not have logarithms. The solution is 2. The solution set is $\{2\}$.

TECHNOLOGY CONNECTION

Logarithmic equations can also be solved using a grapher by drawing the equations of the functions on either side of the equals sign and looking for intersections.

Let's solve, as an example,

$$\log_3 x + 7 = 4 - \log_5 x,$$

rounding x to 2 decimal places.

We first rewrite the logarithms in the equation, using the change-of-base formula:

$$\frac{\log x}{\log 3} + 7 = 4 - \frac{\log x}{\log 5}.$$

Graphing the left side as

$$y = \frac{\log x}{\log 3} + 7$$

and the right side as

$$y = 4 - \frac{\log x}{\log 5},$$

we find that the graphs intersect at one point. Zooming in on that point, we find its x-coordinate, and therefore the solution, to be $x = 0.14$.

$y = \log_3 x + 7$
$y = 4 - \log_5 x$

$[0, 2] \times [0, 10]$

Use a grapher to solve each of the following equations. Round your answers to the nearest hundredth.

TC 5. $\log_7 x = 3$

TC 6. $\log_8 x + \log_8 (x + 2) = 2$

TC 7. $\log_5 (x + 7) - \log_5 (2x - 3) = 1$

DO EXERCISE 8.

Example 8 Solve: $\log_3 (2x - 1) - \log_3 (x - 4) = 2$.

Solution

$$\log_3 \frac{2x - 1}{x - 4} = 2 \qquad \text{Using Property 3 to obtain a single logarithm}$$

$$\frac{2x - 1}{x - 4} = 3^2 \qquad \text{Writing an equivalent exponential expression}$$

$$\frac{2x - 1}{x - 4} = 9$$

$$2x - 1 = 9(x - 4) \qquad \text{Multiplying by the LCM, } x - 4$$

$$2x - 1 = 9x - 36$$

$$35 = 7x$$

$$5 = x$$

Check:

$$\begin{array}{c|c} \log_3 (2x - 1) - \log_3 (x - 4) = 2 \\ \hline \log_3 (2 \cdot 5 - 1) - \log_3 (5 - 4) & 2 \\ \log_3 9 - \log_3 1 \\ 2 - 0 \\ 2 \end{array}$$

The solution is 5. The solution set is $\{5\}$. ◀

DO EXERCISE 9.

8. Solve: $\log x + \log (x - 3) = 1$.

9. Solve:

$$\log_2 (x + 7) - \log_2 (x - 7) = 3.$$

●**EXERCISE SET** **5.7**

 Solve.

1. $2^x = 32$
2. $4^x = 256$
3. $3^x = 81$
4. $5^x = 625$
5. $2^{2x} = 8$
6. $4^{5x} = 32$
7. $3^{7x} = 27$
8. $5^{3x} = 125$
9. $2^x = 33$
10. $2^x = 20$
11. $2^x = 40$
12. $2^x = 19$
13. $5^{4x-7} = 125$
14. $4^{3x+5} = 16$
15. $3^{x^2+4x} = \frac{1}{27}$
16. $27 = 3^{5x} \cdot 9^{x^2}$
17. $84^x = 70$
18. $28^x = 10$
19. $e^t = 1000$
20. $e^t = 100$
21. $e^{-t} = 0.3$
22. $e^{-t} = 0.04$
23. $e^{-0.03t} = 0.08$
24. $e^{0.09t} = 4$
25. $3^x = 2^{x-1}$
26. $5^{x+2} = 4^{x-1}$
27. $(3.9)^x = 48$
28. $(5.6)^x = 100$
29. $250 - (1.87)^x = 0$
30. $4805 - (21.3)^t = 0$
31. $4^{2x} = 8^{3x-4}$
32. $25^{3x-2} = 625^{2x+7}$
33. $\dfrac{e^x - e^{-x}}{t} = 5$
34. $e^x + e^{-x} = 5$

35. $\dfrac{e^x + e^{-x}}{e^x - e^{-x}} = t$
36. $\dfrac{5^x - 5^{-x}}{5^x + 5^{-x}} = t$

▶ Solve for x.

37. $\log_5 x = 4$
38. $\log_2 x = 2$
39. $\log_5 x = -3$
40. $\log_{25} x = \frac{1}{2}$
41. $\log x = 2$
42. $\log x = 1$
43. $\log x = -3$
44. $\log x = -4$
45. $\ln x = 1$
46. $\ln x = 3$
47. $\ln x = -2$
48. $\ln x = -1$
49. $\log_5 (8 - 7x) = 3$
50. $\log_2 (10 + 3x) = 5$
51. $\log x + \log (x - 9) = 1$
52. $\log x + \log (x + 9) = 1$
53. $\log x - \log (x + 3) = -1$
54. $\log (x + 9) - \log x = 1$
55. $\log_2 (x + 1) + \log_2 (x - 1) = 3$
56. $\log_8 (x + 1) - \log_8 x = 2$
57. $\log_8 (x + 1) - \log_8 x = \log_8 4$
58. $\log (2x + 1) - \log (x - 2) = 1$
59. $\log_4 (x + 3) + \log_4 (x - 3) = 2$

60. $\log_5 (x + 4) + \log_5 (x - 4) = 2$

61. $\log \sqrt[4]{x} = \sqrt{\log x}$

62. $\log \sqrt[3]{x} = \sqrt{\log x}$

63. $\log_5 \sqrt{x^2 + 1} = 1$

64. $\log \sqrt{x} = \sqrt{\log x}$

65. $\log x^2 = (\log x)^2$

66. $(\log_3 x)^2 - \log_3 x^2 = 3$

67. $\log_3 (\log_4 x) = 0$

68. $\log (\log x) = 2$

 SYNTHESIS

Solve.

69. $(\log_a x)^{-1} = \log_a x^{-1}$

70. $|\log_5 x| = 2$

71. $\log_7 \sqrt{x^2 - 9} = 1$

72. $x \log \frac{1}{6} = \log 6$

73. $\log (\log x) = 3$

74. $2^{x^2 + 4x} = \frac{1}{8}$

75. $\log_3 |x| = 2$

76. $\log_5 |x| = 3$

77. $\log x^{\log x} = 4$

78. $\log \sqrt{2x} = \sqrt{\log 2x}$

79. $\log_a a^{x^2} + 5x = 24$

80. $x^{\log x} = \dfrac{x^3}{100}$

81. $x^{\log_{10} x} = \dfrac{x^{-4}}{1000}$

82. $5^{2x} - 9 \cdot 5^x + 14 = 0$

83. $(32^{x-2})(64^{x+1}) = 16^{2x-3}$

84. $49^{x+2} = 5140 + 49^x$

85. $4^{3x} - 4^{3x-1} = 48$

86. $x^{\log x} = 100x$

87. $\dfrac{(e^{3x+1})^2}{e^4} = e^{10x}$

88. $\dfrac{\sqrt{(e^{2x} \cdot e^{-5x})^{-4}}}{e^x \div e^{-x}} = e^7$

Solve for t.

89. $P = P_0 e^{kt}$

90. $P = P_0 e^{-kt}$

91. $T = T_0 + (T_1 - T_0)e^{-kt}$

92. Solve for Q.

$$\log_a Q = \tfrac{1}{3} \log_a y + b$$

 CHALLENGE

93. If $x = (\log_{125} 5)^{\log_5 125}$, what is the value of $\log_3 x$?

94. Given that $2^y = 16^{x-3}$ and $3^{y+2} = 27^x$, find the value of $x + y$.

Solve.

95. $|\log_a x| = \log_a |x|$

96. $|\log_5 x| + 3 \log_5 |x| = 4$

97. $(0.5)^x < \frac{4}{5}$

98. $8x^{0.3} - 8x^{-0.3} = 63$

99. If $2 \log_3 (x - 2y) = \log_3 x + \log_3 y$, find x/y.

100. Given that

$$\log_2 [\log_3 (\log_4 x)] = \log_3 [\log_2 (\log_4 y)]$$
$$= \log_4 [\log_3 (\log_2 z)]$$
$$= 0,$$

find $x + y + z$.

101. Suppose that $a = \log_8 225$ and $b = \log_2 15$. Express a as a function of b.

102. Find the ordered pair (x, y) for which

$$4^{\log_{16} 27} = 2^x 3^y.$$

TECHNOLOGY CONNECTION

103.–122. Using a grapher, check the solutions to Exercises 69–88 graphically.

5.8

APPLICATIONS OF EXPONENTIAL AND LOGARITHMIC FUNCTIONS

A We now consider applications of exponential and logarithmic functions that involve bases other than e.

Example 1 *Interest compounded annually.* The amount A that principal P will be worth after t years at interest rate i, compounded annually, is given by the formula $A = P(1 + i)^t$. Suppose that \$100,000 is invested at 8% interest, compounded annually.

a) Express A as a function of t. (See Example 5 of Section 5.2.)
b) After what amount of time will there be $500,000 in the account?
c) Let $T =$ the amount of time it takes for the $100,000 to double itself. T is called the *doubling time*. Find the doubling time.

Solution

a) We have

$$A(t) = \$100{,}000(1.08)^t.$$

b) We set $A(t) = \$500{,}000$ and solve for t:

$$500{,}000 = 100{,}000(1.08)^t$$

$$\frac{500{,}000}{100{,}000} = (1.08)^t$$

$$5 = (1.08)^t$$

$$\log 5 = \log(1.08)^t \qquad \text{Taking the common logarithm on both sides}$$

$$\log 5 = t \log 1.08 \qquad \text{Property 2}$$

$$\frac{\log 5}{\log 1.08} = t$$

$$20.9 \approx t.$$

It will take about 20.9 years for the $100,000 to grow to $500,000.

c) To find the doubling time, we set $A(t) = \$200{,}000$ and $t = T$ and solve for T:

$$200{,}000 = 100{,}000(1.08)^T$$

$$2 = (1.08)^T$$

$$\log 2 = \log(1.08)^T \qquad \text{Taking the common logarithm on both sides}$$

$$\log 2 = T \log 1.08 \qquad \text{Property 2}$$

$$T = \frac{\log 2}{\log 1.08} \approx 9.0.$$

The doubling time is about 9 years.   ◀

DO EXERCISE 1.

Example 2   *Forgetting.*   Here is a mathematical model from psychology. A group of people take a test and make an average score of A. After a time t, in months, they take an equivalent form of the same test. At that time, the average score is $S(t)$, given by

$$S(t) = A - B \log(t + 1), \quad t \geq 0.$$

The model is appropriate only over the interval $[0, 10^{A/B} - 1]$. Students in a zoology class took a final exam and then took equivalent forms of the exam at monthly intervals thereafter. The average scores, $S(t)$, were found to be given by the function

$$S(t) = 80 - 62 \log(t + 1), \quad t \geq 0.$$

a) What was the average score when they took the test originally? after 1 month? after 9 months? after 1 year?

1. Suppose that $80,000 is invested at 7% interest, compounded annually.

a) Express the amount A as a function of time t, in years.

b) After what amount of time will there be $240,000 in the account?

c) Find the doubling time.

2. *Advertising.* A model for advertising response is given by

$$N(a) = 5000 + 200 \log a, \quad a \geq 1,$$

where $N(a)$ = the number of units sold and a = the amount spent on advertising, in thousands of dollars.

a) How many units were sold after spending $1000 ($a = 1$) on advertising?

b) How many units were sold after spending $7000?

c) Graph the function.

d) How much would have to be spent in order to sell 5140 units?

b) Graph the function for values of t such that $t \geq 0$.

c) After what amount of time will the average score be 20%?

Solution

a) $S(0) = 80 - 62 \log(0 + 1) = 80 - 62(0) = 80\%;$ Original score

$S(1) = 80 - 62 \log(1 + 1) \approx 61\%;$

$S(9) = 80 - 62 \log(9 + 1) = 80 - 62(1) = 18\%;$

$S(12) = 80 - 62 \log(12 + 1) \approx 11\%$

b) Using the values computed in part (a) and any others we want, we can sketch the graph as follows:

c) The time t at which the average score will be 20% is found by setting $S(t) = 20$ and solving for t. Note visually that this is where the line $y = 20$ crosses the graph.

$$20 = 80 - 62 \log(t + 1)$$
$$-60 = -62 \log(t + 1)$$
$$0.96774 \approx \log(t + 1)$$
$$10^{0.96774} \approx t + 1 \qquad \text{Using the definition of logarithms}$$
$$9.3 \approx t + 1$$
$$8.3 \approx t$$

After about 8.3 months, the average score will be 20%. ◀

DO EXERCISE 2.

Example 3   *pH of substances.* In chemistry, the **pH** of a substance is defined as follows:

$$pH = -\log[H^+],$$

where H^+ is the hydrogen ion concentration in moles per liter.

a) The hydrogen ion concentration of a common brand of mouthwash is 6.3×10^{-7} moles per liter. Find the pH.

b) The pH of a common hair rinse is 2.9. Find the hydrogen ion concentration.

Solution

a) To find the pH of the mouthwash, we substitute 6.3×10^{-7} for $[H^+]$ in the formula for pH:

$$\begin{aligned} pH &= -\log[H^+] \\ &= -\log(6.3 \times 10^{-7}) \\ &= -[\log 6.3 + \log 10^{-7}] \\ &= -[\log 6.3 + (-7)] \\ &\approx -[0.7993 + (-7)] \\ &= -[-6.2007] \approx 6.2. \end{aligned}$$

The pH of the mouthwash is about 6.2.

b) To find the hydrogen ion concentration of the hair rinse, we substitute 2.9 for pH in the formula and solve for $[H^+]$:

$$\begin{aligned} 2.9 &= -\log[H^+] \\ -2.9 &= \log[H^+] \\ 10^{-2.9} &= [H^+] \qquad \text{Using the definition of logarithms} \\ 0.0013 &\approx [H^+] \\ 1.3 \times 10^{-3} &\approx [H^+]. \end{aligned}$$

The hydrogen ion concentration of the hair rinse is 1.3×10^{-3}. ◀

DO EXERCISE 3.

Example 4   *Earthquake magnitude.* The magnitude R (measured on the Richter scale) of an earthquake of intensity I is defined as

$$R = \log\frac{I}{I_0},$$

where I_0 is a minimum intensity used for comparison. We can think of I_0 as a threshold intensity that is the weakest earthquake that can be recorded on a seismograph. If one earthquake is 10 times as intense as another, its magnitude on the Richter scale is 1 higher. If one earthquake is 100 times as intense as another, its magnitude on the Richter scale is 2 higher, and so on. Thus an earthquake whose magnitude is 7 on the Richter scale is 10 times as intense as an earthquake whose magnitude is 6. Earthquakes can be interpreted as multiples of the minimum intensity I_0.

A recent devastating earthquake in Armenia had an intensity of $10^{6.9} \cdot I_0$. What was its magnitude on the Richter scale?

Solution   We substitute into the formula:

$$R = \log\frac{10^{6.9}I_0}{I_0} = \log 10^{6.9} = 6.9.$$

The magnitude of the earthquake was 6.9 on the Richter scale. ◀

DO EXERCISE 4.

Example 5   *Loudness of sound.* The sensation of loudness of sound is not proportional to the energy intensity, but rather is a logarithmic function.

3. a) The hydrogen ion concentration of milk is 2.3×10^{-6}. Find the pH.

   b) The pH of vinegar is 5.8. Find the hydrogen ion concentration.

4. The Mexico City earthquake of 1978 had an intensity of $10^{7.85} \cdot I_0$. What was its magnitude on the Richter scale?

5. a) Find the loudness, in decibels, of the sound of an automobile for which the intensity I is $3,100,000 \cdot I_0$.

b) Find the loudness of sound of pain having an intensity $10^{14} \cdot I_0$.

Loudness L, in Bels (after Alexander Graham Bell), of a sound of intensity I is defined to be

$$L = \log \frac{I}{I_0},$$

where I_0 is the minimum intensity detectable by the human ear (such as the tick of a watch at 20 ft under quiet conditions). If a sound is 10 times as intense as another, its loudness is 1 Bel greater. If a sound is 100 times as intense as another, it is louder by 2 Bels, and so on. The Bel is a large unit, so a subunit, a **decibel,** is generally used. For L in decibels, the formula is

$$L = 10 \log \frac{I}{I_0}.$$

a) Find the loudness, in decibels, of the sound in a radio studio for which the intensity I is $199 \cdot I_0$.

b) Find the loudness of a jet aircraft for which the intensity is $10^{12} \cdot I_0$.

Solution   In each case, we substitute into the formula.

a) $L = 10 \log \dfrac{199 I_0}{I_0}$

$\quad = 10 \log 199 \approx 23$ decibels

b) $L = 10 \log \dfrac{10^{12} I_0}{I_0} = 10 \log 10^{12}$

$\quad = 10(12) = 120$ decibels

DO EXERCISE 5.

● **EXERCISE SET** **5.8**

 Solve.

1. Compact discs.   The number of compact discs purchased each year is increasing exponentially. The number N, in millions, purchased is given by

$$N(t) = 7.5(6)^{0.5t},$$

where $t = 0$ corresponds to 1985, $t = 1$ corresponds to 1986, and so on (t being the number of years after 1985).

a) After what amount of time will 1 billion compact discs be sold in a year?

b) What is the doubling time on the sale of compact discs?

2. Growth of bacteria Escherichi coli.   The bacteria *Escherichi coli* is commonly found in the human bladder. Suppose that 3000 of the bacteria are present at time $t = 0$. Then t minutes later, the number of bacteria present will be

$$N(t) = 3000(2)^{t/20}.$$

a) After what amount of time will there be 60,000 bacteria?

b) If the bladder is not emptied and 100,000,000 bacteria accumulate, a bladder infection can occur. What amount of time would have to pass before a possible bladder infection would occur?

c) What is the doubling time?

3. Interest compounded annually.   Suppose that $50,000 is invested at 9% interest, compounded annually. After time t, in years, it grows to an amount A.

a) Express A as a function of t.

b) After what amount of time will there be $450,000 in the account?

c) Find the doubling time.

4. Recycling aluminum cans.   It is known that $\frac{1}{4}$ of all aluminum cans distributed will be recycled each year. A beverage company distributes 250,000 cans. The number still in use after time t, in years, is given by the function

$$N(t) = 250,000 \left(\frac{1}{4}\right)^t.$$

a) After what year will 60,000 cans still be in use?

b) After what amount of time will only 10 cans still be in use?

5. Salvage value. An office machine is purchased for $5200. Its value each year is about 80% of the value the preceding year. Its value after t years is given by the exponential function

$$V(t) = \$5200(0.8)^t.$$

a) After what amount of time will the salvage value be $1200?

b) After what amount of time will the salvage value be half of its original value? This is known as the **half-life.**

6. Turkey consumption. The amount of turkey consumed by each person in this country is increasing exponentially. Assuming $t = 0$ corresponds to 1937, the amount of turkey, in pounds per person, consumed t years after 1937 is given by the function

$$N(t) = 2.3(3)^{0.033t}.$$

a) After what amount of time will each person consume 20 lb of turkey?

b) What is the doubling time of the consumption of turkey?

7. Forgetting. Students in an English class took a final exam. They took equivalent forms of the exam in monthly intervals thereafter. The average score $S(t)$, in percent, after t months was found to be given by

$$S(t) = 68 - 20 \log (t + 1), \quad t \geq 0.$$

a) What was the average score when they initially took the test, $t = 0$?

b) What was the average score after 4 months? after 24 months?

c) Graph the function.

d) After what time t was the average score 50?

8. Forgetting. Students in an accounting class took a final exam. They took equivalent forms of the exam in monthly intervals thereafter. The average score $S(t)$, in percent, after t months was found to be given by

$$S(t) = 78 - 15 \log (t + 1), \quad t \geq 0.$$

a) What was the average score when they initially took the test, $t = 0$?

b) What was the average score after 4 months? after 24 months?

c) Graph the function.

d) After what time t was the average score 30?

9. Advertising. A model for advertising response is given by

$$N(a) = 1000 + 200 \log a, \quad a \geq 1,$$

where $N(a) = $ the number of units sold and $a = $ the amount spent on advertising, in thousands of dollars.

a) How many units were sold after spending $1000 ($a = 1$) on advertising?

b) How many units were sold after spending $5000?

c) Graph the function.

d) How much would have to be spent in order to sell 1276 units?

10. Advertising. A model for advertising response is given by

$$N(a) = 2000 + 500 \log a, \quad a \geq 1,$$

where $N(a) = $ the number of units sold and $a = $ the amount spent on advertising, in thousands of dollars.

a) How many units were sold after spending $1000 ($a = 1$) on advertising?

b) How many units were sold after spending $8000?

c) Graph the function.

d) How much would have to be spent in order to sell 5000 units?

Consider the pH formula for Exercises 11–18.

Find the pH of the substance given the hydrogen ion concentration.

11. Pineapple juice; $[H^+] = 1.6 \times 10^{-4}$

12. A common brand of insect repellent; $[H^+] = 4.0 \times 10^{-8}$

13. Tomato; $[H^+] = 6.3 \times 10^{-5}$

14. Egg; $[H^+] = 1.6 \times 10^{-8}$

Find the hydrogen ion concentration of the substance given the pH.

15. Rainwater; pH $= 5.4$

16. Water; pH $= 7$

17. Wine; pH $= 4.8$

18. Orange juice; pH $= 3.2$

19. The San Francisco earthquake of 1906 had an intensity of $10^{8.25} \cdot I_0$. What was its magnitude on the Richter scale?

20. In 1986, there was an earthquake near Cleveland, Ohio. It had an intensity of $10^5 \cdot I_0$. What was its magnitude on the Richter scale?

21. The Chile earthquake of 1960 had an intensity of $10^{9.6} \cdot I_0$. What was its magnitude on the Richter scale?

22. The Italy earthquake of 1980 had a magnitude of 7.2 on the Richter scale. What was its intensity?

23. Find the loudness, in decibels, of the sound in a library that is 2510 times as intense as the minimum intensity I_0.

24. Find the loudness, in decibels, of the sound of a dishwasher that is 2,500,000 times as intense as the minimum intensity I_0.

25. Find the loudness, in decibels, of conversational speech having an intensity that is 10^6 times as intense as I_0.

26. Find the loudness, in decibels, of the sound of a heavy truck having an intensity 10^9 times as intense as I_0.

● **SYNTHESIS** _____

27. Typing speed. A person is studying typing one semester in college. After t hours, the speed of the typist, in words per minute, is given by

$$S(t) = 200[1 - (0.86)^t].$$

a) When will the typist's speed be 100 words per minute?

b) After the course is completed, the typist's speed is 150 words per minute. How many hours of studying occurred in the course?

28. **Compound interest.**   The amount A that principal P will be worth after t years at interest rate i, compounded n times per year, is given by the formula

$$A = P\left(1 + \frac{i}{n}\right)^{nt}.$$

Suppose that $100,000 is invested at 8% interest, compounded quarterly.

a) Express A as a function of t.

b) How much time will it take for the original investment to grow to $1 million?

c) What is the doubling time?

29. Solve the earthquake-magnitude formula for I.

30. Solve the loudness-of-sound formula for I.

31. Solve the pH formula for $[H^+]$.

32. **Loudness of sound.**   Two sounds have intensities I_1 and I_2, respectively.

a) Show that the difference in the loudness of the sounds can be expressed as

$$L_2 - L_1 = 10 \log \frac{I_2}{I_1}.$$

b) Find a formula for the sum of the loudness of the two sounds.

33. **Apparent magnitude of stars.**   The **apparent magnitude** of a star is the measure of its brightness. In ancient times, the brightest star was of magnitude 1 and those just visible to the naked eye were of magnitude 6. In more recent times, with the advent of improved measuring devices, stars of negative magnitude and of magnitude as great as 25 have been observed. Two stars with *apparent magnitudes* M_1 and M_2 and *apparent brightness* I_1 and I_2 are related by

$$M_2 - M_1 = 2.5 \log \frac{I_1}{I_2}.$$

a) The star Sirius, which is actually made up of more than one star, has an apparent magnitude of -1.5, and is the brightest star known as of this writing. The star closest to the earth is Alpha Centauri and has an apparent magnitude of -0.3. Find the ratio of the intensity of Sirius to the intensity of Alpha Centauri.

b) The star Luyten has an apparent magnitude of 12.6. Find the ratio of the intensity of Sirius to the intensity of Luyten.

c) Two stars have a difference in apparent magnitude of 5. How many times brighter is one over the other?

34. **Absolute magnitude of stars.**   When a star's brightness is compared to that of the sun, we obtain what is called the star's **absolute magnitude.** The **luminosity** of a star is the measure of the total amount of energy radiated. The sun's luminosity is 3.9×10^{26} watts. The absolute magnitude M of a star is, then, a function of its luminosity as given by the function

$$M = 4.75 - 2.5 \log\left(\frac{L}{3.9 \times 10^{26}}\right).$$

a) The star Luyten has an absolute magnitude of 14.9. Find its luminosity.

b) The star Sirius has an absolute magnitude of 1.4. Find its luminosity.

c) Solve the formula for L.

OBJECTIVE

You should be able to:

A Solve problems involving applications of exponential (with base e) and natural logarithmic functions.

5.9
APPLICATIONS OF EXPONENTIAL AND NATURAL LOGARITHMIC FUNCTIONS

A   Exponential (with base e) and natural logarithmic functions are rich in application to many fields such as psychology, business, sociology, and science. We now consider many of these applications. A calculator with logarithmic and power keys would be most helpful for this section.

Example 1 *Walking speed.*   In a study by psychologists Bornstein and Bornstein, it was found that the average walking speed R of a person living in a city of population P, in thousands, is given by the function

$$R(P) = 0.37 \ln P + 0.05,$$

where R is in feet per second.

a) The population of Albuquerque, New Mexico, is 290,000. Find the average walking speed of people living in Albuquerque.

b) Graph the function.

Solution

a) We substitute 290 for P, since P is in thousands:

$$R(290) = 0.37 \ln 290 + 0.05 \qquad \text{Substituting}$$

$$\approx 2.1 \text{ ft/sec.} \qquad \text{Finding the natural logarithm on a calculator}$$

The average walking speed of people living in Albuquerque is 2.1 ft/sec.

b) We find several function values using a calculator and sketch the graph, as follows. Note that the axes must be scaled differently because inputs are very large and outputs are small by comparison.

$$R(P) = 0.37 \ln P + 0.05$$

DO EXERCISE 1.

Population Growth

The equation

$$P(t) = P_0 e^{kt}$$

is an effective model of many kinds of population growth, whether it be a population of people or a population of money. In this equation, P_0 is the number of people at time 0, P is the population after time t, and k is often called the **exponential growth rate.** The graph of such an equation is shown here.

$$P(t) = P_0 e^{kt}$$

Example 2   *Growth of the United States.*   In 1991, the population of the United States was 257 million and the exponential growth rate was 0.9%

1. The population of certain cities is given below. Using the function given in Example 1, find the walking speed of people in each city.

   a) Los Angeles, California: 3,497,000

   b) Key West, Florida: 29,600

   c) Seattle, Washington: 531,000

2. What will the population of the United States be in 1998? in 2020?

per year.

a) Find the exponential growth function.
b) What will the population be in 1996? in 2000?
c) Graph the exponential growth function.

Solution

a) At $t = 0$ (1991), the population was 257 million. We substitute 257 for P_0 and 0.9%, or 0.009, for k to obtain the exponential growth function

$$P(t) = 257e^{0.009t}.$$

b) In 1996, $t = 5$; that is, 5 years have passed. To find the population in 1996, we substitute 5 for t:

$$P(5) = 257e^{0.009(5)} \qquad \text{Substituting 5 for } t$$
$$= 257e^{0.045}$$
$$\approx 269. \qquad \text{Finding } e^{0.045} \text{ using a calculator and multiplying}$$

The population of the United States in 1996 will be about 269 million.
     In 2000, $t = 9$, that is, 9 years have passed. To find the population in 2000, we substitute 9 for t:

$$P(9) = 257e^{0.009(9)} \qquad \text{Substituting 9 for } t$$
$$= 257e^{0.081}$$
$$\approx 279. \qquad \text{Finding } e^{0.081} \text{ using a calculator and multiplying}$$

The population of the United States in 2000 will be about 279 million.
c) We find other function values and sketch the graph as follows.

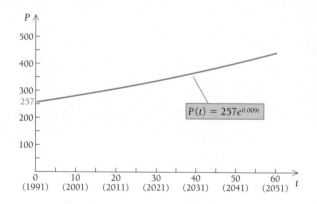

DO EXERCISE 2.

Example 3   ***The cost of a first-class postage stamp.***   The cost of a first-class postage stamp became 3 cents in 1932, and the exponential growth rate of the cost was 3.8% per year. The exponential growth function for the cost is found by substituting 3 for P_0 and 3.8%, or 0.038, for k to obtain the function

$$P(t) = 3e^{0.038t}.$$

a) The cost of first-class postage increased to 29 cents in 1991. Use the function to predict the cost in 1991 and compare.
b) What will the cost of a first-class stamp be in 2000?
c) When will the cost of a first-class postage stamp be $1.00?

Solution

a) At $t = 0$ (1932), the cost of a stamp was 3 cents. Since 1991 is 59 years from 1932, in 1991, we have $t = 59$. To find the cost in 1991, according to the function, we substitute 59 for t:

$$P(59) = 3e^{0.038(59)} \qquad \text{Substituting 59 for } t$$
$$= 3e^{2.242}$$
$$\approx 28.2. \qquad \text{Finding } e^{2.242} \text{ using a calculator and multiplying}$$

The function seems to be a fairly accurate predictor of the cost of first-class postage. The actual amount is 29 cents.

b) In the year 2000, $t = 68$, that is, 68 years have passed. To find the cost in 2000, we substitute 68 for t:

$$P(68) = 3e^{0.038(68)} \qquad \text{Substituting 68 for } t$$
$$= 3e^{2.584}$$
$$\approx 40. \qquad \text{Finding } e^{2.584} \text{ using a calculator and multiplying}$$

The cost of a first-class stamp in the year 2000 will be about 40 cents.

c) To find when the cost will be $1.00, we substitute 100 for $P(t)$ and solve for t:

$$100 = 3e^{0.038t}$$
$$\frac{100}{3} = e^{0.038t}$$
$$\ln \frac{100}{3} = \ln e^{0.038t} \qquad \text{Taking the natural logarithm on both sides}$$
$$3.5066 \approx 0.038t \qquad \text{Using Property 4}$$
$$92 \approx t.$$

In 92 years from 1932, or in 2024, the cost of first-class postage is predicted to be $1.00.   ◀

DO EXERCISE 3.

In order to fit an exponential growth function to a situation, we have to determine P_0 and k from given data. Then we can make predictions.

Example 4   *Heart transplants.*
In 1967, Dr. Christian Barnard, of South Africa, staggered the world by performing the first heart transplant. There was 1 transplant in 1967. In 1987, there were 1418 such transplants. The numbers of transplants through these years are summarized in the table shown here.

Year	Number of Heart Transplants
1967	1
1971	13
1975	23
1980	36
1985	719
1987	1418

Source: The 1987 National Heart Transplant Study and the International Society for Heart Transplantation.

a) Graph the function. Do the table and the graph seem to be of an exponential function?

b) Find an exponential growth function that fits the data.

c) Use the function to predict the number of heart transplants in 1995.

3. a) What will the cost of a first-class stamp be in 1995?

b) When will a first-class stamp cost $2?

4. Predict the number of heart transplants in 2010.

Solution

a) We use the data from the table to sketch the following graph. We draw a curve as close to the data as possible. It is very close to the graph of an exponential function.

If we consider the rapid growth and compare the graph to other exponential functions we have considered, the assumption of exponential growth seems reasonable. A more thorough analysis would require more advanced mathematics.

b) The exponential growth function is

$$N(t) = N_0 e^{kt}.$$

We assume that $t = 0$ corresponds to 1967 and that $N(0) = N_0 e^{k(0)} = N_0 = 1$. Then the growth function is

$$N(t) = e^{kt}.$$

To find k, we can use the fact that at $t = 20$ (1987), the number of transplants was 1418. Then we substitute and solve for k as follows:

$$1418 = e^{k(20)} \qquad \text{Substituting}$$
$$1418 = e^{20k}$$
$$\ln 1418 = \ln e^{20k} \qquad \text{Taking the natural logarithm on both sides}$$
$$\ln 1418 = 20k \qquad \text{Remember, } \log_a a^k = k, \text{ so } \ln e^{20k} = 20k.$$
$$\frac{\ln 1418}{20} = k$$
$$0.363 \approx k.$$

Thus the exponential growth function is

$$N(t) = e^{0.363t}.$$

c) The year 1995 is 28 years from 1967. We let $t = 28$ and find $N(28)$:

$$N(28) = e^{0.363(28)}$$
$$\approx 25{,}952.$$

Thus, according to the exponential growth function, there will be about 25,952 heart transplants in 1995. ◀

DO EXERCISE 4.

Example 5 *Interest compounded continuously.*   Suppose that an amount P_0 is invested in a savings account at interest rate k, compounded continuously. The balance $P(t)$, after t years, is given by the exponential function

$$P(t) = P_0 e^{kt}.$$

a) Suppose that $2000 is invested and grows to $2983.65 in 5 years. What is the interest rate k?
b) Find the exponential growth function.
c) What will the balance be after 10 years?
d) After what amount of time will the $2000 double itself?

Solution

a) At $t = 0$, $P(0) = P_0 = \$2000$. Thus the exponential growth function is

$$P(t) = 2000 e^{kt}.$$

We know that at $t = 5$, $P(5) = \$2983.65$. We substitute and solve for k:

$$2983.65 = 2000 e^{k(5)}$$

$$2983.65 = 2000 e^{5k}$$

$$\frac{2983.65}{2000} = e^{5k} \qquad \text{Dividing both sides by 2000}$$

$$1.491825 = e^{5k}$$

$$\ln 1.491825 = \ln e^{5k} \qquad \text{Taking the natural logarithm on both sides}$$

$$\ln 1.491825 = 5k \qquad \text{Using Property 4}$$

$$0.08 \approx k.$$

The interest rate is about 0.08, or 8%.

b) The exponential growth function is

$$P(t) = 2000 e^{0.08t}.$$

c) The balance after 10 years is

$$P(10) = 2000 e^{0.08(10)} = 2000 e^{0.8} \approx \$4451.08.$$

d) To find the doubling time T, we set $P(T) = \$4000$ and solve for T:

$$4000 = 2000 e^{0.08T}$$

$$2 = e^{0.08T}$$

$$\ln 2 = \ln e^{0.08T}$$

$$\ln 2 = 0.08T$$

$$\frac{\ln 2}{0.08} = T$$

$$8.7 \approx T.$$

Thus the original investment of $2000 will double itself in about 8.7 years.   ◀

DO EXERCISE 5.

We can find a general expression relating the growth rate k and the doubling time T by solving the following equation:

$$2P_0 = P_0 e^{kT} \qquad \text{Substituting } 2P_0 \text{ for } P$$

$$2 = e^{kT} \qquad \text{Multiplying by } 1/P_0$$

$$\ln 2 = \ln e^{kT}$$

$$\ln 2 = kT.$$

5. a) Suppose that $10,000 is invested and grows to $14,049.48 in 4 years. What is the interest rate k, assuming interest is compounded continuously?

b) Find the exponential growth function.

c) What will the balance be after 10 years?

d) After what amount of time will the $10,000 double itself?

6. The exponential growth rate of the Bahamas is about 4.1%. What is the doubling time of the population?

THEOREM 7

The growth rate k and the doubling time T are related by

$$kT = \ln 2 \approx 0.693147,$$

or

$$k = \frac{\ln 2}{T} \approx \frac{0.693147}{T}$$

and

$$T = \frac{\ln 2}{k} \approx \frac{0.693147}{k}.$$

Note that this relationship between k and T does not depend on P_0.

7. The doubling time of the population of Guam is about 13.9 years. What is the exponential growth rate?

Example 6 The population of the world is now doubling every 24.8 years. What is its exponential growth rate?

Solution We have

$$k = \frac{\ln 2}{T} = \frac{\ln 2}{24.8} \approx 2.8\%.$$

The growth rate of the world is about 2.8% per year.

DO EXERCISES 6–8.

8. A financial institution advertises that it will double your money in 6.8 years. What is the interest rate, assuming interest is compounded continuously?

Exponential Decay

The function

$$P(t) = P_0 e^{-kt}$$

is an effective model of the decline, or decay, of a population. An example is the decay of a radioactive substance. Here P_0 is the amount of the substance at time $t = 0$, P is the amount of the substance left after time t, and k is a positive constant that depends on the situation. The constant k is called the **decay rate.**

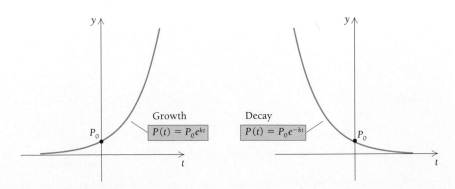

The **half-life** of bismuth is 5 days. This means that half of an amount of bismuth will cease to be radioactive in 5 days. The effect of half-life is shown in the graph below. The exponential function gets close to 0, but never reaches 0, as t gets larger. Thus, in theory, a radioactive substance never completely decays.

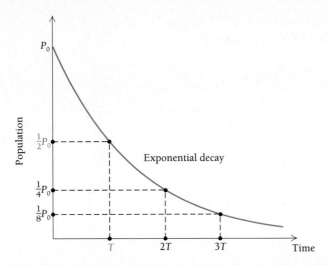

Radioactive decay curve

Example 7   *Carbon dating.* The radioactive element carbon-14 has a half-life of 5750 years. The percentage of carbon-14 present in the remains of living matter can be used to determine their age. Archeologists found that the linen wrapping from one of the Dead Sea Scrolls had lost 22.3% of its carbon-14. How old was the linen wrapping?

Solution We first find k. To do so, we use the concept of half-life. When $t = 5750$ (half-life), P will be half of P_0. We substitute $\frac{1}{2}P_0$ for P and 5750 for t and solve for k. Then

$$\frac{1}{2}P_0 = P_0 e^{-k(5750)},$$

or

$$\frac{1}{2} = e^{-5750k}.$$

We take the natural logarithm on both sides:

$$\ln \frac{1}{2} = \ln e^{-5750k}$$

$$= -5750k.$$

Then

$$k = \frac{\ln 0.5}{-5750}$$

$$\approx 0.00012.$$

Now we have the function

$$P(t) = P_0 e^{-0.00012t}.$$

(*Note:* This equation can be used for any subsequent carbon-dating problem.) If the linen wrapping has lost 22.3% of its carbon-14 from an initial amount P_0, then 77.7%(P_0) is the amount present. To find the age t of the

How can scientists determine that an animal bone has lost 30% of its carbon-14? The assumption is that the percentage of carbon-14 in the atmosphere and in living plants and animals is the same. When a plant or an animal dies, the amount of carbon-14 decays exponentially. The scientist burns the animal bone and uses a Geiger counter to determine the percentage of the smoke that is carbon-14. It is the amount that this varies from the percentage in the atmosphere that tells how much carbon-14 has been lost.

The process of carbon-14 dating was developed by the American chemist Willard E. Libby in 1952. It is known that the radioactivity in a living plant is 16 disintegrations per gram per minute. Since the half-life of carbon-14 is 5750 years, an object with an activity of 8 disintegrations per gram per minute is 5750 years old, one with an activity of 4 disintegrations per gram per minute is 11,500 years old, and so on. Carbon-14 dating can be used to measure the age of objects from 30,000 to 40,000 years old. Beyond such an age, it is too difficult to measure the radioactivity and some other method would have to be used.

Carbon-14 was indeed used to find the age of the Dead Sea Scrolls. It was used recently to refute the authenticity of the Shroud of Turin, presumed to have covered the body of Christ.

9. How old is a skeleton that has lost 80% of its carbon-14?

wrapping, we solve the following equation for t:

$$77.7\%P_0 = P_0 e^{-0.00012t} \qquad \text{Substituting } 77.7\%P_0 \text{ for } P$$
$$0.777 = e^{-0.00012t}$$
$$\ln 0.777 = \ln e^{-0.00012t}$$
$$\ln 0.777 = -0.00012t$$
$$\frac{\ln 0.777}{-0.00012} = t$$
$$t \approx 2103.$$

Thus the linen wrapping on the Dead Sea Scrolls is about 2103 years old. ◀

DO EXERCISE 9.

● EXERCISE SET 5.9

A   Various cities and their populations are given below. Find the walking speed of people in each city. (See Example 1.)

1. New York, New York: 7,900,000

2. Pittsburgh, Pennsylvania: 853,000

3. Rome, New York: 50,400

4. Reno, Nevada: 106,000

5. *Cost of a Hershey bar.*   The cost of a Hershey chocolate bar in 1962 was 5 cents and was increasing at an exponential growth rate of 9.7%.

   a) Find an exponential function describing the growth of the cost of a Hershey bar.

   b) What will a Hershey bar cost in 1990? in 2000?

   c) When will a Hershey bar cost $5?

   d) Graph the function.

   e) What is the doubling time of the cost of a Hershey bar?

6. *World population growth.*   The population of the world passed 5.0 billion in 1987. The exponential growth rate was 2.8% per year.

   a) Find the exponential growth function.

   b) Predict the population of the world in 1996 and in 2000.

   c) When will the world population be 6.0 billion?

   d) Graph the function.

7. *Consumer price index.*   The consumer price index is often in the news. In any one year, the consumer price index is the cost of goods and services that cost $100 in 1967. While the rate of change in the consumer price index varies from year to year, let us assume an average exponential growth rate of 6%.

   a) Find the exponential function for the consumer price index.

   b) Goods and services that cost $100 in 1967 will cost how much in 1995?

   c) Goods and services that cost $100 in 1967 will cost how much in 2000?

   d) Graph the function.

8. *Population growth of rabbits.*   Under ideal conditions, a population increase of rabbits has an exponential growth rate of 11.7% per day. Suppose one starts with a population of 100 rabbits.

   a) Find an exponential function describing the growth of the population of rabbits after t days.

   b) What will the population of rabbits be after 7 days?

   c) Graph the function.

   d) What is the doubling time of the population of rabbits?

9. *Cost of a 60-second commercial during the Super Bowl.* Past data on the cost of a 60-second commercial during the Super Bowl are given in the table below.

Year	Cost of 60-sec TV Commercial During the Super Bowl
1967	$80,000
1970	$200,000
1977	$324,000
1981	$550,000
1983	$800,000
1985	$1,100,000
1988	$1,350,000

   a) Make a graph of the data and analyze the table. Does it appear that we can fit an exponential function to the data?

   b) Find an exponential growth function for the cost of

a Super Bowl commercial. Assume that $C_0 = \$80$ thousand. That is, at $t = 0$ (1967), $C = \$80$ (in thousands). Find k using the data point $C(21) = \$1350$ thousand. That is, in 1988, $1350 thousand, or $1,350,000, was the cost of a 60-second commercial.

c) What will the cost of a 60-second commercial be in 1995?

d) When will the cost of a 60-second commercial be $3,000,000?

e) What is the doubling time for the cost of a 60-second Super Bowl commercial?

10. *Cost of a double-dip ice cream cone.* In 1970, the cost of a double-dip ice cream cone was 52 cents. In 1978, it was 66 cents. Assuming the exponential model:

a) Find the value k ($P_0 = 52$). Write the exponential growth function.

b) Estimate the cost of a cone in 1994.

c) After what period of time will the cost of a cone be twice that of 1970?

d) When will the cost of a cone be $3?

11. *Interest compounded continuously.* Suppose that P_0 is invested in a savings account in which interest is compounded continuously at 9% per year. That is, the balance $P(t)$ after time t, in years, is

$$P(t) = P_0 e^{kt}.$$

a) Find the exponential function for $P(t)$ in terms of P_0 and 0.09.

b) Suppose that $5000 is invested. What is the balance after 1 year? after 2 years?

c) When will an investment of $5000 double itself?

12. *Interest compounded continuously.* Suppose that P_0 is invested in a savings account in which interest is compounded continuously at 10% per year. That is, the balance $P(t)$ after time t, in years, is

$$P(t) = P_0 e^{kt}.$$

a) Find the exponential function for $P(t)$ in terms of P_0 and 0.10.

b) Suppose that $35,000 is invested. What is the balance after 1 year? after 2 years?

c) When will an investment of $35,000 double itself?

13. *Population growth.* The growth rate of the population of Mexico is 3.5% per year (one of the highest in the world). What is the doubling time?

14. *Population growth.* The growth rate of the population of Europe is 1% per year. What is the doubling time?

15. *Annual interest rate.* A bank advertises that it compounds interest continuously and that it will double your money in 7 years. What is its annual interest rate?

16. *Annual interest rate.* A bank advertises that it compounds interest continuously and that it will double your money in 5.4 years. What is its annual interest rate?

17. *Value of a Van Gogh painting.* The Van Gogh painting *Irises* sold for $84,000 in 1947, but was sold again for $53,900,000 in 1987. Assuming that the growth in the value V of the painting was exponential:

a) Find the value k and determine the exponential growth function, assuming $P_0 = 84,000$.

b) Estimate the value of the painting in 2007.

c) What is the doubling time for the value of the painting?

d) After what amount of time will the value of the painting be $1 billion?

Van Gogh's Irises, *a 28-by-32-inch oil on canvas.*

18. *Exponential growth of the value of a baseball card.* The collecting of baseball cards and other memorabilia has become a profitable hobby. The card shown here contains a photograph of Eddie Murray in his rookie season of 1978. The value of that card in 1983 was $7.75. Its value in 1987 was $27.00. The value of the card has increased so much because Murray has turned out to be such an outstanding player. Assume that the value of the card has grown exponentially.

a) Find the value k and determine the exponential growth function, assuming $V_0 = 7.75$.

b) Estimate the value of the card in 1995 and in 2000.

c) What is the doubling time for the value of the card?

d) After what amount of time will the value of the card be $2000?

EDDIE MURRAY

19. **Coal demand.**   The growth rate of the demand for coal in the world is 4% per year. When will the demand be double that of 1990?

20. **Oil demand.**   The growth rate of the demand for oil in the United States is 10% per year. When will the demand be double that of 1990?

21. **Population growth.**   The population of Los Angeles, California, was 2,812,000 in 1970. In 1984, it was 3,097,000. Assuming that growth was exponential:

    a) Find the value of k ($P_0 = 2,812,000$). Write the function.
    b) Estimate the population of Los Angeles in 1996.

22. **Population growth.**   The population of San Antonio, Texas, was 786,000 in 1980. In 1984, it was 843,000. Assuming that growth was exponential:

    a) Find the value of k ($P_0 = 786,000$). Write the function.
    b) Estimate the population of San Antonio in 2000.

23. A mummy discovered in the pyramid Khufu in Egypt has lost 46% of its carbon-14. Determine its age.

24. The statue of Zeus at Olympia in Greece is one of the Seven Wonders of the World. It is made of gold and ivory. The ivory was found to have lost 35% of its carbon-14. Determine the age of the statue.

25. The half-life of polonium is 3 minutes. What is its decay rate?

26. The half-life of lead is 22 years. What is its decay rate?

27. The decay rate of iodine-131 is 9.6% per day. What is its half-life?

28. The decay rate of krypton-85 is 6.3% per year. What is its half-life?

29. **Weight loss.**   The initial weight of a starving animal is W_0. Its weight W after t days is given by

    $$W(t) = W_0 e^{-0.007t}.$$

    a) What percentage of its weight does it lose each day?
    b) What percentage of its initial weight remains after 30 days?

30. **Satellite power.**   The power supply of a satellite is a radioisotope. The power output P, in watts, decreases at a rate proportional to the amount present. P is given by

    $$P(t) = 60e^{-0.006t},$$

    where t = the time in days.

    a) How much power will be available after 365 days?
    b) What is the half-life of the power supply?
    c) The satellite's equipment cannot operate on fewer than 10 watts of power. How long can the satellite stay in operation?
    d) How much power did the satellite have to begin with?

31. **Atmospheric pressure.**   Atmospheric pressure P at altitude a is given by

    $$P = P_0 e^{-0.00005a},$$

    where P_0 = the pressure at sea level. Assume that $P_0 = 14.7$ lb/in² (pounds per square inch).

    a) Find the pressure at an altitude of 2000 ft.
    b) Find the pressure at the top of Mt. Shasta in California, which is 14,162 ft above sea level.
    c) At what altitude is the pressure 1.47 lb/in²?
    d) Blood will boil when atmospheric pressure drops below 0.39 lb/in². At what altitude, in an unpressurized vehicle, will a pilot's blood boil?

32. **Salvage value.**   A business estimates that the salvage value V of a piece of machinery after t years is given by

    $$V(t) = \$46,000e^{-t}.$$

    a) What did the machinery cost initially?
    b) What is the salvage value after 2 years?

● **SYNTHESIS** _____

33. **The Beer–Lambert law.**   A beam of light enters a medium such as water or smog with initial intensity I_0. Its intensity decreases depending on the thickness (or concentration) of the medium. The intensity I at a depth (or concentration) of x units is given by

    $$I = I_0 e^{-\mu x}.$$

    The constant μ (the Greek letter "mu") is called the **coefficient of absorption,** and it varies with the medium. For sea water, $\mu = 1.4$.

    a) What percentage of light intensity I_0 remains at a depth of sea water that is 1 m? 3 m? 5 m? 50 m?
    b) Plant life cannot exist below 10 m. What percentage of I_0 remains at 10 m?

34. **Present value.**   Following the birth of a child, a parent wants to make an initial investment P_0 that will grow to $50,000 for the child's education at age 18. Interest is compounded continuously at 8%. What should the initial investment be? Such an amount is called the **present value** of $50,000 18 years from now.

35. **Velocity of a rocket.**   The theory of rocket flight shows that the velocity of a rocket when its propellant is burned to depletion is expressed by the equation

    $$v = c \ln R,$$

    where v = the velocity gained by the rocket during launch, c = the exhaust velocity of the engine, and R = the mass ratio of the rocket = (Takeoff weight)/(Burnout weight). Solve the formula for R.

36. **Electricity.**   The formula

    $$i = \frac{V}{R}[1 - e^{-(R/L)t}]$$

    occurs in the theory of electricity. Solve for t.

37. In reference to Exercise 31, explain how you might use a barometer, or some other device for measuring atmospheric pressure, to find the height of the Empire State Building.

38. **Newton's Law of Cooling.**   An object whose temperature

differs from that of its surroundings will either cool down or heat up to that of its surroundings. Suppose a body that has a temperature T_1 is placed in surroundings with temperature T_0. The body will either cool or warm to temperature $T(t)$ after time t, in minutes, where

$$T(t) = T_0 + |T_1 - T_0|e^{-kt}.$$

A cup of coffee whose temperature is $105°F$ is placed in a freezer whose temperature is $32°F$. After 5 minutes, its temperature is $70°$. What will its temperature be after 10 minutes?

● CHALLENGE

39. When was the murder committed?   The police discover the body of a math professor. Critical to solving the crime is determining when the murder was committed. The police call the coroner, who arrives at $12:00$ P.M. The coroner immediately takes the temperature of the body and finds it to be $94.6°$. The coroner takes the temperature 1 hour later and finds it to be $93.4°$. The temperature of the room is $70°$. When was the murder committed? (Use Newton's Law of Cooling, Exercise 38.)

SUMMARY AND REVIEW 5

● **TERMS TO KNOW**

Inverse relation, p. 284
One-to-one function, p. 286
Inverse function, p. 286
Horizontal-line test, p. 287

Exponential function, p. 295
Base, p. 296
Logarithmic function, p. 303
Common logarithm, p. 317

Natural logarithm, p. 319
Exponential equation, p. 327
Logarithmic equation, p. 329

● **REVIEW EXERCISES**

1. Find the inverse of the relation H given by
$$H = \{(-4, 5), (2, -3), (1, 7), (8, 8), (5, -4)\}.$$

Write an equation of the inverse.

2. $y = 3x^2 + 2x - 1$ **3.** $y = \sqrt{x + 2}$

4. Which of the following have inverses that are functions?

a)

b)

c)

d)

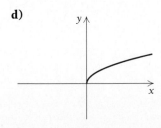

Find a formula for $f^{-1}(x)$.

5. $f(x) = \dfrac{\sqrt{x}}{2} + 2$ **6.** $f(x) = x^3 + 8$

7. Find $f(f^{-1}(a))$:
$$f(x) = x^3 + 2.$$

8. Find $h^{-1}(h(t))$:
$$h(x) = x^{17} + x^{65}.$$

Graph.

9. $y = \log_2(x - 1)$ **10.** $y = \left(\dfrac{1}{2}\right)^x$

11. $f(x) = 3(1 - e^{-x})$, for nonnegative values of x

12. $f(x) = \ln(x - 4)$

13. Find $\log_3 10$ using common logarithms.

14. Find $\log_6 2$ using natural logarithms.

15. Write an exponential equation equivalent to $\log_8 \frac{1}{4} = -\frac{2}{3}$.

16. Write a logarithmic equation equivalent to $7^{2.3} = x$.

17. Write an equivalent expression containing a single logarithm:
$$\frac{1}{2}\log_b a + \frac{3}{2}\log_b c - 4\log_b d.$$

18. Express in terms of logarithms of M and N:
$$\log \sqrt[3]{M^2/N}.$$

Given that $\log_a 2 = 0.301$, $\log_a 3 = 0.477$, and $\log_a 7 = 0.845$, find each of the following.

19. $\log_a 18$

20. $\log_a \frac{7}{2}$

21. $\log_a \frac{1}{4}$

22. $\log_a \sqrt{3}$

Simplify.

23. $\log_{12} 12^{x^2+1}$

24. $\log_8 8^{\sqrt{9}}$

Solve.

25. $\log_x 64 = 3$

26. $\log_{16} 4 = x$

27. $\log_5 125 = x$

28. $3^{1-x} = 9^{2x}$

29. $e^x = 80$

30. $\log x^2 = \log x$

31. $\log(x^2 - 1) - \log(x - 1) = 1$

32. $\log 2 + 2 \log x = \log(5x + 3)$

33. $\log_2(x - 1) + \log_2(x + 1) = 3$

34. How many years will it take an investment of $1000 to double if interest is compounded annually at 13%?

35. What is the loudness, in decibels, of a sound whose intensity is $1000I_0$?

36. The half-life of a radioactive substance is 15 days. How much of a 25-gram sample will remain radioactive after 30 days?

37. *Forgetting.* In an art class, students were tested at the end of the course on a final exam. They were tested again after 6 months. The forgetting formula was determined to be

$$S(t) = 82 - 38 \log(t + 1),$$

where t is the time, in months, after taking the first test.

a) What was the average score when they initially took the test, $t = 0$?

b) What was the average score after 6 months?

c) After what time was the average score 54?

38. *The cost of a prime-rib dinner.* The average cost C of a prime-rib dinner was $4.65 in 1962. In 1986, it was $15.81. Assume that the growth followed the exponential growth function.

a) Find k and write the exponential growth function.

b) How much will a prime-rib dinner cost in 2010?

c) When will the average cost of a prime-rib dinner be $20?

d) What is the doubling time?

39. The population of a city doubled in 18 years. What was the exponential growth rate?

40. How long will it take $7600 to double itself if it is invested at 8.6%, compounded continuously?

41. How old is a skeleton that has lost 27% of its carbon-14?

42. What is the pH of a substance whose hydrogen ion concentration is 3.8×10^{-7} moles per liter?

43. An earthquake has an intensity of $10^8 I_0$. What is its magnitude on the Richter scale?

Find each of the following common and natural logarithms using a calculator.

44. $\log 0.00216$

45. $\log 1{,}342{,}000$

46. $\ln 87{,}380$

47. $\ln 0.00002776$

● **SYNTHESIS** _____

Solve.

48. $|\log_4 x| = 3$

49. $\log x = \ln x$

Graph.

50. $y = |\log_3 x|$

51. $y = |e^x - 4|$

Find the domain.

52. $f(x) = \dfrac{1}{\sqrt{5 \ln x - 6}}$

53. $f(x) = \dfrac{8}{e^{4x} - 10}$

● **THINKING AND WRITING** _____

1. Suppose you were trying to convince a fellow student that $\log_2(x + 3) \neq \log_2 x + \log_2 3$. Give as many explanations as you can.

2. Describe the difference between $f^{-1}(x)$ and $[f(x)]^{-1}$.

3. Describe the difference between $f(x) = 3^x$ and $g(x) = x^3$.

4. Look up data for as many preceding years as you can regarding the number of cases of the disease AIDS that are occurring annually. How might you know whether an exponential function fits the data? If so, use some of the data to find a function that fits and predict the number of AIDS cases in the years 2000 and 2010. Then use other parts of the data to find another function, make predictions for the number of AIDS cases in 2000 and 2010, and compare.

CHAPTER TEST 5

1. Find the inverse of the relation H given by
   $H = \{(3, -5), (6, -3), (8, -3), (-5, 3), (1.3, -2.7)\}$.

2. Write an equation of the inverse of the relation $y = |x|$.

3. Which of the following have inverses that are functions?

   a)   b)

   c)   d)

   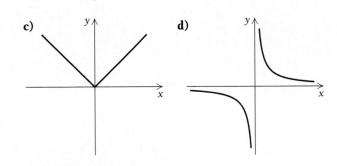

4. Find a formula for $f^{-1}(x)$:
   $$f(x) = \sqrt{x - 6}.$$

5. Find $h(h^{-1}(3))$:
   $$h(x) = \frac{-21x + 3}{20}.$$

6. Graph: $y = \log_3 x$.

7. Graph: $f(x) = e^{x-3}$.

8. Find $\log_5 23$ using natural logarithms.

9. Write an exponential equation equivalent to $\log_{\sqrt{3}} 9 = 4$.

10. Write a logarithmic equation equivalent to $x^5 = 0.03125$.

11. Simplify: $6^{\log_6 3x}$.

12. Write an equivalent expression containing a single logarithm:
    $$3 \log_c x - 4 \log_c y + \tfrac{1}{2} \log_c z.$$

13. Express in terms of logarithms of w and r:
    $$\log \sqrt[4]{wr^3}.$$

Given that $\log_a 2 = 0.301$, $\log_a 5 = 0.699$, and $\log_a 6 = 0.778$, find each of the following.

14. $\log_a 3$

15. $\log_a 50$

16. $\log_a \sqrt[3]{5}$

17. Solve for x: $\log_b b^{2x^2} = x$.

Solve.

18. $\log_4 x = 2$

19. $4^{2x-1} - 3 = 61$

20. $\log_2 x + \log_2 (x - 2) = 3$

21. $e^{-x} = 0.2$

Find using a calculator.

22. $\log 0.005243$

23. $\ln 3.86$

24. $\log 87{,}200$

25. $\ln 0.00000029$

26. $\log (-14.68)$

27. How many years will it take an investment of $5000 to double if interest is compounded annually at 12%?

28. What is the loudness, in decibels, of a sound whose intensity is $50{,}000 I_0$?

29. The population of a city was 80,000 in 1970 and 100,000 in 1980. Estimate the population in 2000.

30. **Walking speed.** The average walking speed R of people living in a city of population P, in thousands, is given by
    $$R = 0.37 \ln P + 0.05,$$
    where R is in feet per second.

    a) The population of Akron, Ohio, is 660,000. Find the average walking speed.

    b) A city's population has an average walking speed of 2.3 ft/sec. Find the population.

31. **Population of Brazil.** The population of Brazil was 52 million in 1959, and the exponential growth rate was 2.8% per year.

    a) Write an exponential function describing the growth of the population of Brazil.

    b) What will the population be in 1998? in 2020?

    c) When will the population be 300 million?

    d) What is the doubling time?

32. The population of a city doubled in 30 years. What was the exponential growth rate?

33. How long will it take an investment to double itself if it is invested at 8.6%, compounded continuously?

34. How old is an animal bone that has lost 38% of its carbon-14?

35. What is the loudness, in decibels, of a sound whose intensity is $230 I_0$?

36. The hydrogen ion concentration of water is 1.8×10^{-9}. What is the pH?

● **SYNTHESIS**

37. True or false:
    $$\log_a (3x^5) = 15 \log_a x.$$

38. Find the domain:
    $$f(x) = \log_3 (\ln x).$$

39. Solve: $5^{\sqrt{x}} = 625$.

In this chapter, we consider an important class of functions called *trigonometric functions*, or *circular functions*. Historically, these functions arose from a study of triangles, hence the name trigonometric. • We will first develop the six basic circular functions from a unit circle (radius = 1), centered at the origin, which allows us to "see" the values of sin *s* and cos *s*, where *s* is a real-number arc length on the circle. Then we will consider trigonometric functions of angles or rotations on any circle centered at the origin, with both radian and degree measure. •

The Trigonometric or Circular Functions

6.1

THE UNIT CIRCLE

We refer to a circle centered at the origin with radius 1 as a **unit circle.** Such a circle has an equation $u^2 + v^2 = 1$. We use *u* and *v* for the variables here because we wish to reserve *x* and *y* for later purposes.

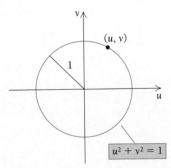

OBJECTIVES

You should be able to:

A Given a real number that is an integral multiple of halves, thirds, fourths, or sixths of π, find on the unit circle the point determined by it. Given a point on the unit circle that is an integral multiple of fourths, sixths, eighths, or twelfths of the distance around the circle, find the real numbers between −2π and 2π that determine that point.

B For the unit circle, state the coordinates of the intercepts and the midpoints of the arcs in each quadrant, and given the coordinates of a point on the unit circle, find its reflection across the *u*-axis, the *v*-axis, and the origin.

To develop the trigonometric, or circular, functions, we use the unit circle. The inputs for the functions that we will define are to be numbers that are associated with points on a unit circle, and the outputs are to be numbers associated with the coordinates of points on the circle.

353

1. How far will a point travel if it goes:

   a) $\frac{1}{4}$ of the way around the unit circle?

   b) $\frac{3}{8}$ of the way around the unit circle?

   c) $\frac{3}{4}$ of the way around the unit circle?

 Distances on the Unit Circle

The circumference of a circle of radius r is $2\pi r$. Thus for the unit circle, where $r = 1$, the circumference is 2π. If a point starts at A and travels counter-clockwise around the circle, it will travel a distance of 2π. If it travels halfway around the circle, it will travel a distance of $\frac{1}{2} \cdot 2\pi$, or π.

Example 1   How far will a point travel if it goes $\frac{1}{12}$ of the way around the unit circle?

Solution   The distance will be $\frac{1}{12}$ of the total distance around the circle, or $\frac{1}{12} \cdot 2\pi$, which is $\pi/6$.

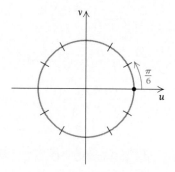

DO EXERCISE 1.

If a point travels $\frac{1}{8}$ of the way around the circle, it will travel a distance of $\frac{1}{8} \cdot 2\pi$, or $\pi/4$. Note that the stopping point is $\frac{1}{4}$ of the way from A to C and $\frac{1}{2}$ of the way from A to B.

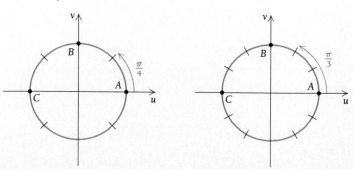

If a point travels $\frac{1}{6}$ of the way around the circle ($\frac{2}{12}$ of the way, as shown), it will travel a distance of $\frac{1}{6} \cdot 2\pi$, or $\pi/3$. Note that the stopping point is $\frac{1}{3}$ of the way from A to C and $\frac{2}{3}$ of the way from A to B.

Example 2 How far will a point move on the unit circle, counterclockwise, going from point A to point Q?

Solution If the point went to C, it would move a distance of π. It goes $\frac{3}{4}$ of that distance to get to Q. Hence it goes a distance of $3\pi/4$ to get to Q.

Example 3 How far will a point move on the unit circle, counterclockwise, going from point A to point G?

Solution If the point went to C, it would move a distance of π. It goes $\frac{5}{6}$ of that distance to G. Hence it goes a distance of $5\pi/6$ to get to G.

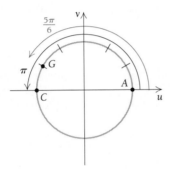

DO EXERCISES 2 AND 3.

A point may travel completely around the circle and then continue. For example, if it goes around once and then continues $\frac{1}{4}$ of the way around, it will have traveled a distance of $2\pi + \frac{1}{4} \cdot 2\pi$, or $5\pi/2$.

Any positive number thus determines a point on the unit circle. For the number 35, for example, we start at A and travel counterclockwise a distance of 35. The point at which we stop is the point "determined" by the number 35.

2. How far will a point move on the unit circle, counterclockwise, going from point A to:

a) point M?

b) point N?

c) point P?

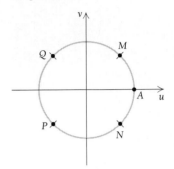

3. How far will a point move on the unit circle, counterclockwise, in going from point A to:

a) point F?

b) point H?

c) point J?

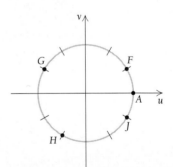

4. On this unit circle, mark the points determined by each of the following.

   a) $\dfrac{\pi}{4}$ b) $\dfrac{7\pi}{4}$

   c) $\dfrac{9\pi}{4}$ d) $\dfrac{13\pi}{4}$

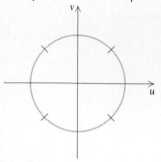

5. On this unit circle, mark the points determined by each of the following.

   a) $\dfrac{\pi}{6}$ b) $\dfrac{7\pi}{6}$

   c) $\dfrac{13\pi}{6}$ d) $\dfrac{25\pi}{6}$

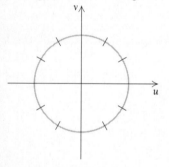

6. On this unit circle, mark the points determined by each of the following.

   a) $-\dfrac{\pi}{2}$ b) $-\dfrac{3\pi}{4}$

   c) $-\dfrac{7\pi}{4}$ d) -2π

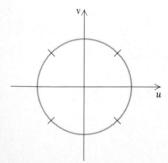

Example 4   On the unit circle, mark the point determined by $5\pi/4$.

Solution   From A to C, the distance is π, or $(\frac{4}{4})\pi$, so we need to go another distance of $\pi/4$. That is, think of $5\pi/4$ as $\pi + \pi/4$.

DO EXERCISES 4 AND 5.

Negative numbers also determine points on the unit circle. For a negative number, we move *clockwise* around the circle. Points for $-\pi/4$ and $-3\pi/2$ are shown. Any negative number thus determines a point on the unit circle. The number 0 determines the point A.

> There is a point on the unit circle for every real number.

Example 5   On the unit circle, mark the point determined by $-7\pi/6$.

Solution   From A to C, the distance is π, or $\frac{6}{6}\pi$, clockwise so we need to go another distance of $\pi/6$, clockwise.

DO EXERCISE 6.

▶ Special Points

The intercepts (points of intersection with the axes) of the unit circle A, B, C, and D have coordinates as shown. Let us find the coordinates of point E, which is the intersection of the circle with the line $u = v$. Note that this point divides the arc in the first quadrant in half. Thus point E is determined by the real number $\pi/4$.

Since for any point (u, v) on the circle, $u^2 + v^2 = 1$, and for point E we have $u = v$, we know that the coordinates of E satisfy $u^2 + u^2 = 1$ (substituting u for v in the equation of the circle). Let us solve this equation:

$$u^2 + u^2 = 1, \quad \text{or } 2u^2 = 1$$

$$u^2 = \frac{1}{2}$$

$$u = \sqrt{\frac{1}{2}}, \quad \text{or } \frac{1}{\sqrt{2}}, \quad \text{or } \frac{\sqrt{2}}{2}.$$

Thus the coordinates of point E are $(\sqrt{2}/2, \sqrt{2}/2)$.

A unit circle is symmetric with respect to the u-axis, the v-axis, and the origin. We can use the coordinates of one point to find coordinates of reflections.

Example 6   The point E on the unit circle, as shown below, has coordinates $(\sqrt{2}/2, \sqrt{2}/2)$. Find the coordinates of points F, G, and H, which are the reflections of E.

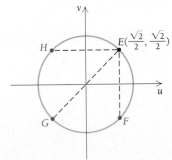

Solution

a) Point F is the reflection of point E across the u-axis. Hence its coordinates are $(\sqrt{2}/2, -\sqrt{2}/2)$.

b) Point G is the reflection of point E across the origin. Hence its coordinates are $(-\sqrt{2}/2, -\sqrt{2}/2)$.

c) Point H is the reflection of point E across the v-axis. Hence its coordinates are $(-\sqrt{2}/2, \sqrt{2}/2)$.

7. The point *M*, as shown, has coordinates ($\sqrt{3}/2$, 1/2). Find the coordinates of the points *N*, *P*, and *R*, which are reflections of *M*.

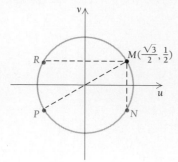

8. The point ($\frac{3}{5}$, $-\frac{4}{5}$) is on the unit circle. Find the coordinates of its reflection across:

a) the *u*-axis;

b) the *v*-axis;

c) the origin.

d) Are these points on the circle? Why?

9. The point ($-\sqrt{35}/6$, $-1/6$) is on the unit circle. Find the coordinates of its reflection across:

a) the *u*-axis;

b) the *v*-axis;

c) the origin.

d) Are these points on the circle? Why?

DO EXERCISE 7.

Example 7 The point (1/2, $\sqrt{3}/2$) is on the unit circle. Find its reflection across (a) the *u*-axis, (b) the *v*-axis, and (c) the origin.

Solution

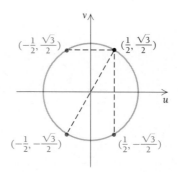

a) The reflection across the *u*-axis is (1/2, $-\sqrt{3}/2$).

b) The reflection across the *v*-axis is ($-1/2$, $\sqrt{3}/2$).

c) The reflection across the origin is ($-1/2$, $-\sqrt{3}/2$).

By symmetry, all these points are on the circle.

DO EXERCISES 8 AND 9.

●**EXERCISE SET** **6.1**

A   For each of Exercises 1–6, sketch a unit circle and mark the points determined by the given real numbers.

1. a) $\dfrac{\pi}{4}$ b) $\dfrac{3\pi}{2}$

   c) $\dfrac{3\pi}{4}$ d) π

   e) $\dfrac{11\pi}{4}$ f) $\dfrac{17\pi}{4}$

2. a) $\dfrac{\pi}{2}$ b) $\dfrac{5\pi}{4}$

   c) 2π d) $\dfrac{9\pi}{4}$

   e) $\dfrac{13\pi}{4}$ f) $\dfrac{23\pi}{4}$

3. a) $\dfrac{\pi}{6}$ b) $\dfrac{2\pi}{3}$

   c) $\dfrac{7\pi}{6}$ d) $\dfrac{10\pi}{6}$

   e) $\dfrac{14\pi}{6}$ f) $\dfrac{23\pi}{4}$

4. a) $\dfrac{\pi}{3}$ b) $\dfrac{5\pi}{6}$

   c) $\dfrac{11\pi}{6}$ d) $\dfrac{13\pi}{6}$

e) $\dfrac{23\pi}{6}$ f) $\dfrac{33\pi}{6}$

5. a) $-\dfrac{\pi}{2}$ **b)** $-\dfrac{3\pi}{4}$

  c) $-\dfrac{5\pi}{6}$ **d)** $-\dfrac{5\pi}{2}$

  e) $-\dfrac{17\pi}{6}$ **f)** $-\dfrac{9\pi}{4}$

6. a) $-\dfrac{3\pi}{2}$ **b)** $-\dfrac{\pi}{3}$

  c) $-\dfrac{7\pi}{6}$ **d)** $-\dfrac{13\pi}{6}$

  e) $-\dfrac{19\pi}{6}$ **f)** $-\dfrac{11\pi}{4}$

Find two real numbers between -2π and 2π that determine each of the points on the unit circle.

7.

8.

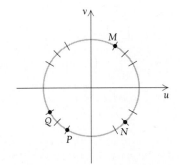

B   The following points are on the unit circle. Find the coordinates of their reflections across (a) the u-axis, (b) the v-axis, and (c) the origin.

9. $\left(-\dfrac{\sqrt{2}}{2}, \dfrac{\sqrt{2}}{2}\right)$ **10.** $\left(-\dfrac{3}{4}, \dfrac{\sqrt{7}}{4}\right)$

11. $\left(\dfrac{2}{3}, \dfrac{\sqrt{5}}{3}\right)$ **12.** $\left(-\dfrac{\sqrt{3}}{2}, -\dfrac{1}{2}\right)$

Find the coordinates of the labeled points on the unit circle.

13.

14.

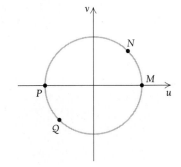

15. The number $\pi/6$ determines a point on the unit circle with coordinates $(\sqrt{3}/2, 1/2)$. What are the coordinates of the point determined by $-\pi/6$?

16. The number $\pi/3$ determines a point on the unit circle with coordinates $(1/2, \sqrt{3}/2)$. What are the coordinates of the point determined by $-\pi/3$?

17. A number α determines a point on the unit circle with coordinates $(3/4, -\sqrt{7}/4)$. What are the coordinates of the point determined by $-\alpha$?

18. A number β determines a point on the unit circle with coordinates $(-2/3, \sqrt{5}/3)$. What are the coordinates of the point determined by $-\beta$?

● **SYNTHESIS** _____

19. For each of the points given in Exercise 7, find a real number between 2π and 4π.

20. For each of the points given in Exercise 8, find a real number between -2π and -4π.

21. A point on the unit circle has u-coordinate $-1/3$. What is its v-coordinate?

22. A point on the unit circle has v-coordinate $-\sqrt{21}/5$. What is its u-coordinate?

6.2

THE SINE AND THE COSINE FUNCTIONS

A Definitions and Function Values

DEFINITION **The Sine Function**

Every real number s determines a point T on a unit circle. To the real number s, we assign the second coordinate of point T as a function value. The function thus defined is called the *sine function*. Function values are denoted by "sine s" or more briefly, "sin s."

DEFINITION **The Cosine Function**

Every real number s determines a point T on a unit circle. To the real number s, we assign the first coordinate of point T as a function value. The function thus defined is called the *cosine function*. Function values are denoted by "cosine s" or more briefly, "cos s."

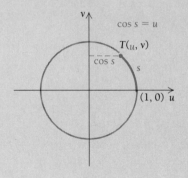

We could use the more general function notation **sin** (s) and **cos** (s), but the parentheses are ordinarily omitted unless the input s has more than one term.

Let us find some function values.

Example 1 Find $\sin 0$, $\cos 0$, $\sin\dfrac{\pi}{2}$, $\cos\dfrac{\pi}{2}$, $\sin\dfrac{\pi}{4}$, $\cos\dfrac{\pi}{4}$, $\sin\left(-\dfrac{\pi}{4}\right)$, and $\cos\left(-\dfrac{\pi}{4}\right)$.

Solution We locate the points determined by 0, $\pi/2$, $\pi/4$, and $-\pi/4$ on a unit circle and label the coordinates.

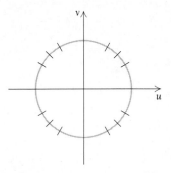

1. Using a unit circle, find each of the following.

   a) $\sin \pi$ b) $\sin \dfrac{3\pi}{2}$

   c) $\sin \dfrac{\pi}{4}$ d) $\sin \left(-\dfrac{3\pi}{4}\right)$

   e) $\sin \dfrac{2\pi}{3}$ f) $\sin \left(-\dfrac{\pi}{6}\right)$

The values of the sine function are the second coordinates. The values of the cosine function are the first coordinates. Thus,

$$\sin 0 = 0, \qquad \sin \frac{\pi}{4} = \frac{\sqrt{2}}{2}, \qquad \sin \frac{\pi}{2} = 1, \qquad \sin \left(-\frac{\pi}{4}\right) = -\frac{\sqrt{2}}{2},$$

$$\cos 0 = 1, \qquad \cos \frac{\pi}{4} = \frac{\sqrt{2}}{2}, \qquad \cos \frac{\pi}{2} = 0, \qquad \cos \left(-\frac{\pi}{4}\right) = \frac{\sqrt{2}}{2}. \quad \blacktriangleleft$$

We will verify in Section 6.5 that the point $\pi/6$ on the unit circle has coordinates $(\sqrt{3}/2, 1/2)$, and the point $\pi/3$ has coordinates $(1/2, \sqrt{3}/2)$. These facts allow us to find other values of the sine function and the cosine function.

2. Using a unit circle, find each of the following.

   a) $\cos \pi$ b) $\cos \dfrac{3\pi}{2}$

   c) $\cos \dfrac{\pi}{4}$ d) $\cos \left(-\dfrac{3\pi}{4}\right)$

   e) $\cos \dfrac{5\pi}{6}$ f) $\cos \left(-\dfrac{\pi}{3}\right)$

Example 2 Find $\sin \dfrac{\pi}{6}$, $\cos \dfrac{\pi}{6}$, $\sin \dfrac{\pi}{3}$, and $\cos \dfrac{\pi}{3}$.

Solution From the coordinates given in the preceding figures, we have

$$\sin \frac{\pi}{6} = \frac{1}{2}, \qquad \cos \frac{\pi}{6} = \frac{\sqrt{3}}{2}, \qquad \sin \frac{\pi}{3} = \frac{\sqrt{3}}{2}, \qquad \cos \frac{\pi}{3} = \frac{1}{2}. \quad \blacktriangleleft$$

DO EXERCISES 1 AND 2.

3. Construct a graph of the sine function. From the unit circle, transfer vertical distances with a compass.

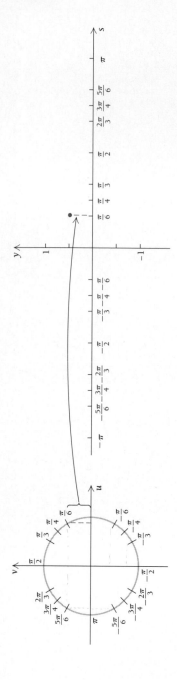

B Properties of the Sine Function

Let us consider making a graph of the sine function. We might do this in two ways. The first is to take the function values we have found in the preceding examples and margin exercises, make a table of values, plot points, look for patterns, and complete the graph. Below is a table of function values.

s	0	$\dfrac{\pi}{6}$	$\dfrac{\pi}{4}$	$\dfrac{\pi}{3}$	$\dfrac{\pi}{2}$	$\dfrac{3\pi}{4}$	π	$\dfrac{5\pi}{4}$	$\dfrac{3\pi}{2}$	2π
$\sin s$	0	$\dfrac{1}{2}$	$\dfrac{\sqrt{2}}{2}$	$\dfrac{\sqrt{3}}{2}$	1	$\dfrac{\sqrt{2}}{2}$	0	$-\dfrac{\sqrt{2}}{2}$	-1	0

s	$-\dfrac{\pi}{6}$	$-\dfrac{\pi}{4}$	$-\dfrac{\pi}{3}$	$-\dfrac{\pi}{2}$	$-\dfrac{3\pi}{4}$	$-\pi$	$-\dfrac{5\pi}{4}$	$-\dfrac{3\pi}{2}$	-2π
$\sin s$	$-\dfrac{1}{2}$	$-\dfrac{\sqrt{2}}{2}$	$-\dfrac{\sqrt{3}}{2}$	-1	$-\dfrac{\sqrt{2}}{2}$	0	$\dfrac{\sqrt{2}}{2}$	1	0

We find decimal approximations of the expressions involving square roots. Then we plot points, look for patterns, and connect them with a curve. The graph is as follows.

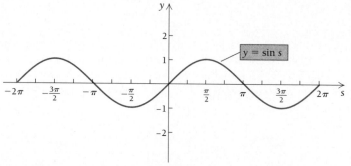

The sine function

The second way to construct a graph is by considering a unit circle. Then from the unit circle, we transfer vertical distances with a compass. You are asked to draw such a graph in Margin Exercise 3.

DO EXERCISES 3–6. (EXERCISES 4–6 ARE ON THE FOLLOWING PAGE.)

Note in the graph of the sine function that function values increase from 0 at 0 to 1 at $\pi/2$, then decrease to 0 at π, decrease further to -1 at $3\pi/2$, and then increase to 0 at 2π. A similar pattern follows for negative inputs.

Certain functions with a repeating pattern are called **periodic.** The sine function is an example. The function values of the sine function repeat themselves every 2π units. In other words, for any s, we have $\sin(s + 2\pi) = \sin s$. To see this another way, think of the part of the graph between 0 and 2π and note that the rest of the graph consists of copies of it. If we translate the graph 2π units to the left or right, the original graph will be obtained. We say that the sine function has a *period* of 2π.

Because their second coordinates are the same, we know that for any number s,

$$\sin(\pi - s) = \sin s.$$

We have proven the following identities concerning the sine function.

> ### Identities for the Sine Function
>
> $$\sin s = \sin(s + 2k\pi), \; k \text{ any integer}$$
> $$\sin(-s) = -\sin s$$
> $$\sin(s \pm \pi) = -\sin s$$
> $$\sin(\pi - s) = \sin s$$

The **amplitude** of a periodic function is defined to be one-half the difference between its maximum and minimum function values. It is always positive. We can see from either the graph or a unit circle that the maximum value of the sine function is 1, whereas the minimum value is -1. Thus,

The amplitude of the sine function $= \frac{1}{2}[1 - (-1)] = 1$.

The following is a summary of the properties of the sine function.

> ### Properties of the Sine Function
>
> 1. The sine function is periodic, with period 2π.
> 2. The sine function is odd. Thus it is symmetric with respect to the origin, and $\sin(-s) = -\sin s$ for all real numbers s.
> 3. It is continuous everywhere.
> 4. The domain is the set of all real numbers. The range is the interval $[-1, 1]$, that is, the set of all real numbers from -1 to 1.
> 5. The amplitude of the sine function is 1.

Properties of the Cosine Function

Let us consider making a graph of the cosine function. There are again two ways in which we can make such a graph. The first is to take the function values that we have found in the preceding examples and margin exercises, make a table of values, plot points, look for patterns, and complete the graph. Below is a table of function values.

s	0	$\dfrac{\pi}{6}$	$\dfrac{\pi}{4}$	$\dfrac{\pi}{3}$	$\dfrac{\pi}{2}$	$\dfrac{3\pi}{4}$	π	$\dfrac{5\pi}{4}$	$\dfrac{3\pi}{2}$	2π
$\cos s$	1	$\dfrac{\sqrt{3}}{2}$	$\dfrac{\sqrt{2}}{2}$	$\dfrac{1}{2}$	0	$-\dfrac{\sqrt{2}}{2}$	-1	$-\dfrac{\sqrt{2}}{2}$	0	1

s	$-\dfrac{\pi}{6}$	$-\dfrac{\pi}{4}$	$-\dfrac{\pi}{3}$	$-\dfrac{\pi}{2}$	$-\dfrac{3\pi}{4}$	$-\pi$	$-\dfrac{5\pi}{4}$	$-\dfrac{3\pi}{2}$	-2π
$\cos s$	$\dfrac{\sqrt{3}}{2}$	$\dfrac{\sqrt{2}}{2}$	$\dfrac{1}{2}$	0	$-\dfrac{\sqrt{2}}{2}$	-1	$-\dfrac{\sqrt{2}}{2}$	0	1

8. Construct a graph of the cosine function. From the unit circle, transfer horizontal distances with a compass.

We find decimal approximations of the expressions involving square roots. Then we plot points, look for patterns, and connect them with a curve. The graph is as follows.

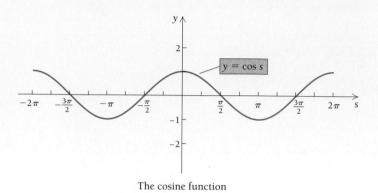

The cosine function

The second way to construct a graph is by considering a unit circle. Then from the unit circle, we transfer horizontal distances with a compass. You are asked to draw such a graph in Margin Exercise 8.

DO EXERCISES 8–11. (EXERCISES 9–11 ARE ON THE FOLLOWING PAGE.)

Note in the graph of the cosine function that function values start at 1 when $s = 0$, and decrease to 0 at $\pi/2$. They decrease further to -1 at π, and then begin to increase, reaching the value 1 at 2π. A similar pattern follows for negative inputs. Moreover, the function is periodic, with period 2π.

We can use the unit circle to verify some properties of the cosine function. Let us verify that the period is 2π. Consider any real number s and the point T that it determines on a unit circle. Now we increase s by 2π. The point for $s + 2\pi$ is again the point T.

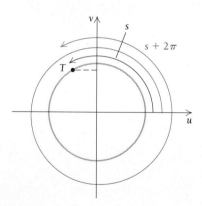

Hence for any real number s,

$$\cos (s + 2\pi) = \cos s.$$

There is no number smaller than 2π for which this is true. Thus the period of the cosine function is 2π.

Identities Involving the Cosine Function

The unit circle can be used to verify that

$$\cos s = \cos (s + 2k\pi)$$

holds for any real number s and any integer k.

Consider any real number s and its opposite, $-s$. These numbers determine points T and T_1 symmetric with respect to the u-axis.

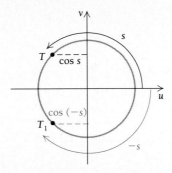

Because T and T_1 have the same first coordinates, we know that for any number s,

$$\cos (-s) = \cos s.$$

We have proved another identity as well as showing that the cosine function is *even*.

Consider any real number s and the point T that it determines on a unit circle. Then $s + \pi$ and $s - \pi$ both determine a point T_1 symmetric to T with respect to the origin.

Thus the cosine of $s + \pi$ and the cosine of $s - \pi$ are the same, and each is the opposite of $\cos s$. This gives us another important identity and property. For any real number s,

$$\cos (s + \pi) = -\cos s$$

and

$$\cos (s - \pi) = -\cos s.$$

Let us consider $\cos (\pi - s)$ as well as any real number s and the point T that it determines on a unit circle. Then $\pi - s$ determines a point T_1 symmetric to T with respect to the v-axis.

9. Does the cosine function appear to be continuous?

10. What is the domain of the cosine function?

11. What is the range of the cosine function?

Because the first coordinates of T and T_1 are opposites, we know that for any number s,

$$\cos(\pi - s) = -\cos s.$$

We have proven the following identities concerning the cosine function.

Identities for the Cosine Function

$$\cos s = \cos(s + 2k\pi), k \text{ any integer}$$
$$\cos(-s) = \cos s$$
$$\cos(s \pm \pi) = -\cos s$$
$$\cos(\pi - s) = -\cos s$$

The following is a summary of the properties of the cosine function.

Properties of the Cosine Function

1. The cosine function is periodic, with period 2π.
2. The cosine function is even. Thus it is symmetric with respect to the y-axis, and $\cos(-s) = \cos s$ for all real numbers s.
3. It is continuous everywhere.
4. The domain is the set of all real numbers. The range is the interval $[-1, 1]$, that is, the set of all real numbers from -1 to 1.
5. The amplitude of the cosine function is 1.

For your understanding, we have used the letter s for arc length and have avoided the letters x and y, which usually represent first and second coordinates. In fact, we can represent the arc length on a unit circle by any variable, such as s, t, x, or θ (the Greek letter "theta"). Each arc length determines a point that can be labeled with an ordered pair. **The first coordinate of that ordered pair is the *cosine* of the arc length, and the second coordinate is the *sine* of the arc length.** This is illustrated in the following figures.

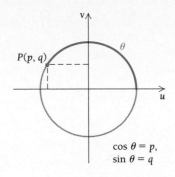

$$\cos x = a,$$
$$\sin x = b$$

$$\cos \theta = p,$$
$$\sin \theta = q$$

The identities we have developed hold no matter what symbols are used for variables. For example, $\cos(-s) = \cos s$, $\cos(-x) = \cos x$, $\cos(-\theta) = \cos\theta$, and $\cos(-y) = \cos y$.

DO EXERCISE 12.

12. Complete each of the following in reference to the figure.

$$\cos y = \underline{\hspace{1.5cm}}$$
$$\sin y = \underline{\hspace{1.5cm}}$$

● **EXERCISE SET** **6.2**

◢ Find the function value.

1. $\sin\dfrac{\pi}{4}$

2. $\sin\pi$

3. $\sin\dfrac{\pi}{6}$

4. $\sin\dfrac{5\pi}{4}$

5. $\sin\dfrac{3\pi}{2}$

6. $\sin\dfrac{\pi}{2}$

7. $\sin\dfrac{\pi}{3}$

8. $\sin\dfrac{3\pi}{4}$

9. $\sin\dfrac{7\pi}{4}$

10. $\sin 2\pi$

11. $\sin\dfrac{5\pi}{6}$

12. $\sin\left(-\dfrac{\pi}{4}\right)$

13. $\sin(-\pi)$

14. $\sin\left(-\dfrac{\pi}{2}\right)$

15. $\sin\left(-\dfrac{3\pi}{4}\right)$

16. $\sin\left(-\dfrac{7\pi}{4}\right)$

17. $\sin\left(-\dfrac{5\pi}{4}\right)$

18. $\sin\left(-\dfrac{3\pi}{2}\right)$

19. $\sin\left(-\dfrac{\pi}{3}\right)$

20. $\sin(-2\pi)$

21. $\sin\left(-\dfrac{2\pi}{3}\right)$

22. $\sin\left(-\dfrac{\pi}{6}\right)$

23. $\cos\dfrac{\pi}{4}$

24. $\cos\pi$

25. $\cos\dfrac{5\pi}{4}$

26. $\cos\dfrac{\pi}{6}$

27. $\cos\dfrac{\pi}{2}$

28. $\cos\dfrac{\pi}{3}$

29. $\cos\dfrac{3\pi}{4}$

30. $\cos\dfrac{7\pi}{4}$

31. $\cos\dfrac{3\pi}{2}$

32. $\cos\dfrac{5\pi}{6}$

33. $\cos\left(-\dfrac{\pi}{4}\right)$

34. $\cos(-\pi)$

35. $\cos 2\pi$

36. $\cos\left(-\dfrac{\pi}{2}\right)$

37. $\cos\left(-\dfrac{3\pi}{4}\right)$

38. $\cos\left(-\dfrac{5\pi}{4}\right)$

39. $\cos\left(-\dfrac{3\pi}{2}\right)$

40. $\cos\left(-\dfrac{5\pi}{3}\right)$

41. $\cos\left(-\dfrac{\pi}{6}\right)$

42. $\cos\left(-\dfrac{7\pi}{4}\right)$

43. $\cos\left(-\dfrac{2\pi}{3}\right)$

44. $\cos(-2\pi)$

◢ ◢ Complete.

45. $\cos(-x) = \underline{\hspace{1cm}}$

46. $\sin(-x) = \underline{\hspace{1cm}}$

47. $\sin(x + \pi) = \underline{\hspace{1cm}}$

48. $\sin(x - \pi) = \underline{\hspace{1cm}}$

49. $\cos(\pi - x) = \underline{\hspace{1cm}}$

50. $\sin(\pi - x) = \underline{\hspace{1cm}}$

51. $\cos(x + 2k\pi) = \underline{\hspace{1cm}}$

52. $\sin(x + 2k\pi) = \underline{\hspace{1cm}}$

53. $\cos(x - \pi) = \underline{\hspace{1cm}}$

54. $\cos(x + \pi) = \underline{\hspace{1cm}}$

55. a) Sketch a graph of $y = \sin x$.
  b) By reflecting the graph in part (a), sketch a graph of $y = \sin(-x)$.
  c) By reflecting the graph in part (a), sketch a graph of $y = -\sin x$.
  d) How do the graphs in parts (b) and (c) compare?

56. a) Sketch a graph of $y = \cos x$.
  b) By reflecting the graph in part (a), sketch a graph of $y = \cos(-x)$.
  c) By reflecting the graph in part (a), sketch a graph of $y = -\cos x$.
  d) How do the graphs in parts (a) and (b) compare?

57. a) Sketch a graph of $y = \sin x$.
  b) By translating, sketch a graph of $y = \sin(x + \pi)$.
  c) By reflecting the graph of part (a), sketch a graph of $y = -\sin x$.
  d) How do the graphs of parts (b) and (c) compare?

58. a) Sketch a graph of $y = \sin x$.
  b) By translating, sketch a graph of $y = \sin(x - \pi)$.

c) By reflecting the graph of part (a), sketch a graph of $y = -\sin x$.

d) How do the graphs of parts (b) and (c) compare?

59. a) Sketch a graph of $y = \cos x$.

b) By translating, sketch a graph of $y = \cos(x + \pi)$.

c) By reflecting the graph of part (a), sketch a graph of $y = -\cos x$.

d) How do the graphs of parts (b) and (c) compare?

60. a) Sketch a graph of $y = \cos x$.

b) By translating, sketch a graph of $y = \cos(x - \pi)$.

c) By reflecting the graph of part (a), sketch a graph of $y = -\cos x$.

d) How do the graphs of parts (b) and (c) compare?

● **SYNTHESIS** _____

61. For which numbers is:

a) $\sin x = 1$?   **b)** $\sin x = -1$?

62. For which numbers is:

a) $\cos x = 1$?   **b)** $\cos x = -1$?

63. Solve for x: $\sin x = 0$.

64. Solve for x: $\cos x = 0$.

65. Find $f \circ g$ and $g \circ f$, where $f(x) = x^2 + 2x$ and $g(x) = \cos x$.

66. Show that the sine function does not have the *linearity* property, that is, it is not true that
$$\sin(x + y) = \sin x + \sin y$$
for all numbers x and y.

67. Graph: $y = 3 \sin x$.

68. Graph: $y = \sin x + \cos x$.

What are the domain, the range, the period, and the amplitude?

69. $y = \sin^2 x$ **70.** $y = |\cos x| + 1$

What is the domain?

71. $f(x) = \sqrt{\cos x}$ **72.** $g(x) = \dfrac{1}{\sin x}$

73. $f(x) = \dfrac{\sin x}{\cos x}$ **74.** $y = \log(\sin x)$

75. Consider $(\sin x)/x$, when x is between 0 and $\pi/2$. What can you say about this function? As x approaches 0, this function approaches a limit. What is it?

76. Solve $\sin x < \cos x$ in the interval $[-\pi, \pi]$.

6.3

The Other Basic Circular Functions

There are six basic circular, or trigonometric, functions. In this section, we define the remaining four. Many other related functions can be obtained from the basic ones by adding them, multiplying them, dividing them, translating them, and so on.

A **Definitions and Function Values**

There are four other circular functions, all defined in terms of the sine and the cosine functions. Their definitions are as follows. We again consider a point T on a unit circle determined by a real number s.

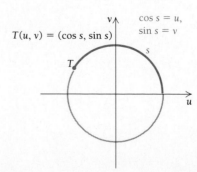

1. Find $\tan \pi$, $\cot \pi$, $\sec \pi$, and $\csc \pi$.

DEFINITIONS

The tangent function: $\tan s = \dfrac{\text{second coordinate}}{\text{first coordinate}} = \dfrac{\sin s}{\cos s} = \dfrac{v}{u}$

The cotangent function: $\cot s = \dfrac{\text{first coordinate}}{\text{second coordinate}} = \dfrac{\cos s}{\sin s} = \dfrac{u}{v}$

The secant function: $\sec s = \dfrac{1}{\text{first coordinate}} = \dfrac{1}{\cos s} = \dfrac{1}{u}$

The cosecant function: $\csc s = \dfrac{1}{\text{second coordinate}} = \dfrac{1}{\sin s} = \dfrac{1}{v}$

Let us find some function values.

Example 1   Find $\tan 0$, $\cot 0$, $\sec 0$, and $\csc 0$.

Solution   We note that 0 determines a point with coordinates $(1, 0)$ on the unit circle.

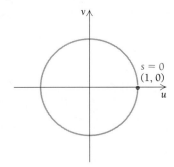

Then

$$\tan 0 = \frac{\text{second coordinate}}{\text{first coordinate}} = \frac{\sin 0}{\cos 0} = \frac{0}{1} = 0;$$

$$\cot 0 = \frac{\text{first coordinate}}{\text{second coordinate}} = \frac{\cos 0}{\sin 0}$$

$$= \frac{1}{0}, \text{ which is not defined;}$$

$$\sec 0 = \frac{1}{\text{first coordinate}} = \frac{1}{\cos 0} = \frac{1}{1} = 1;$$

$$\csc 0 = \frac{1}{\text{second coordinate}} = \frac{1}{\sin 0}$$

$$= \frac{1}{0}, \text{ which is not defined.}$$ ◀

DO EXERCISE 1.

Example 2   Find $\tan \dfrac{\pi}{4}$, $\cot \dfrac{\pi}{4}$, $\sec \dfrac{\pi}{4}$, and $\csc \dfrac{\pi}{4}$.

2. Find $\tan\dfrac{3\pi}{4}$, $\cot\dfrac{3\pi}{4}$, $\sec\dfrac{3\pi}{4}$, and $\csc\dfrac{3\pi}{4}$.

3. Find $\tan\dfrac{\pi}{6}$, $\cot\dfrac{\pi}{6}$, $\sec\dfrac{\pi}{6}$, and $\csc\dfrac{\pi}{6}$.

4. Find $\tan\left(-\dfrac{\pi}{3}\right)$, $\cot\left(-\dfrac{\pi}{3}\right)$, $\sec\left(-\dfrac{\pi}{3}\right)$, and $\csc\left(-\dfrac{\pi}{3}\right)$.

5. Find $\tan\dfrac{\pi}{2}$, $\tan\left(-\dfrac{\pi}{2}\right)$, and $\tan\left(-\dfrac{3\pi}{2}\right)$.

Solution The number $\pi/4$ determines a point with coordinates $(\sqrt{2}/2, \sqrt{2}/2)$ on the unit circle. Then

$$\tan\frac{\pi}{4} = \frac{\sin\dfrac{\pi}{4}}{\cos\dfrac{\pi}{4}} = \frac{\dfrac{\sqrt{2}}{2}}{\dfrac{\sqrt{2}}{2}} = 1;$$

$$\cot\frac{\pi}{4} = \frac{\cos\dfrac{\pi}{4}}{\sin\dfrac{\pi}{4}} = \frac{\dfrac{\sqrt{2}}{2}}{\dfrac{\sqrt{2}}{2}} = 1;$$

$$\sec\frac{\pi}{4} = \frac{1}{\cos\dfrac{\pi}{4}} = \frac{1}{\dfrac{\sqrt{2}}{2}} = \frac{2}{\sqrt{2}} = \frac{2\sqrt{2}}{2} = \sqrt{2};$$

$$\csc\frac{\pi}{4} = \frac{1}{\sin\dfrac{\pi}{4}} = \frac{1}{\dfrac{\sqrt{2}}{2}} = \frac{2}{\sqrt{2}} = \sqrt{2}.$$

DO EXERCISE 2.

Example 3 Find $\tan\dfrac{\pi}{3}$, $\cot\dfrac{\pi}{3}$, $\sec\dfrac{\pi}{3}$, and $\csc\dfrac{\pi}{3}$.

Solution As stated in Section 6.2, the number $\pi/3$ determines a point with coordinates $(1/2, \sqrt{3}/2)$ on the unit circle. Then

$$\tan\frac{\pi}{3} = \frac{\sin\dfrac{\pi}{3}}{\cos\dfrac{\pi}{3}} = \frac{\dfrac{\sqrt{3}}{2}}{\dfrac{1}{2}} = \sqrt{3};$$

$$\cot\frac{\pi}{3} = \frac{\cos\dfrac{\pi}{3}}{\sin\dfrac{\pi}{3}} = \frac{\dfrac{1}{2}}{\dfrac{\sqrt{3}}{2}} = \frac{1}{\sqrt{3}} = \frac{\sqrt{3}}{3};$$

$$\sec\frac{\pi}{3} = \frac{1}{\cos\dfrac{\pi}{3}} = \frac{1}{\dfrac{1}{2}} = 2;$$

$$\csc\frac{\pi}{3} = \frac{1}{\sin\dfrac{\pi}{3}} = \frac{1}{\dfrac{\sqrt{3}}{2}} = \frac{2}{\sqrt{3}} = \frac{2\sqrt{3}}{3}.$$

DO EXERCISES 3–5.

Properties of the Tangent, Cotangent, Secant, and Cosecant Functions

The Tangent Function

Note the following:

$$\tan \frac{\pi}{2} = \frac{\sin \dfrac{\pi}{2}}{\cos \dfrac{\pi}{2}} = \frac{1}{0},$$

which is undefined, since division by 0 is not possible. The tangent function is undefined for any number whose cosine is 0. Thus it is undefined for $(\pi/2) + k\pi$, where k is any integer. In the first quadrant, the function values are positive. In the second quadrant, however, the sine and cosine values have opposite signs, so the tangent values are negative. Tangent values are also negative in the fourth quadrant and positive in the third quadrant.

A graph of the tangent function is shown below. It can be constructed using the values we have found in the preceding examples and margin exercises and by computing other values as needed. Note that the function value is 0 when $x = 0$, and the values increase as x increases toward $\pi/2$. As we approach $\pi/2$, the denominator becomes very small, so the tangent values become very large. In fact, they increase without bound. The dashed vertical line $x = \pi/2$ is called a **vertical asymptote.** The graph gets closer and closer to the asymptote as x gets closer to $\pi/2$, but it never crosses the line. Each of the vertical dashed lines is an asymptote.

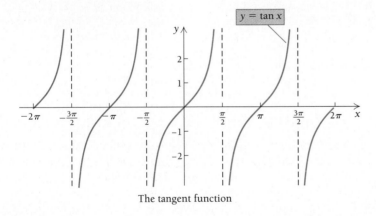

The tangent function

DO EXERCISES 6–11.

The following is a summary of the properties of the tangent function.

Properties of the Tangent Function

1. The tangent function is periodic, with period π.
2. The domain of the tangent function is the set of all real numbers except $(\pi/2) + k\pi$, where k is any integer. Tan x is not defined where $\cos x = 0$.
3. The range of the tangent function is the set of all real numbers.
4. The tangent function is continuous except where it is not defined.

---

6. Find three real numbers other than $\pi/2$ at which the tangent function is undefined.

7. Graph the tangent function.

8. What is the period of the tangent function?

9. What is the domain of the tangent function?

10. What is the range of the tangent function?

11. Does the tangent function appear to be even? Is it odd?

12. Find three real numbers at which the cotangent function is undefined.

13. Graph the cotangent function.

14. What is the period of the cotangent function?

15. What is the domain of the cotangent function?

16. What is the range of the cotangent function?

17. In which quadrants are the function values of the cotangent function positive? negative?

18. Does the cotangent function appear to be even? Is it odd?

The Cotangent Function

The graph of the cotangent function is shown below. The cotangent of a number is found by dividing its cosine by its sine. The cotangent function is undefined for any number whose sine is 0. Thus it is undefined for $k\pi$, where k is any integer.

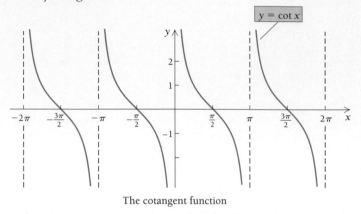

The cotangent function

DO EXERCISES 12–18.

The following is a summary of the properties of the cotangent function.

> **Properties of the Cotangent Function**
>
> 1. The cotangent function is periodic, with period π.
> 2. The domain of the cotangent function is the set of all real numbers except $k\pi$, where k is any integer. Cot x is not defined where $\sin x = 0$.
> 3. The range of the cotangent function is the set of all real numbers.
> 4. The cotangent function is continuous except where it is not defined.

The Secant Function

The secant and cosine functions are reciprocal functions. The graph of the secant function can be constructed by finding the reciprocal of each of the values of the cosine function that we found in Section 6.2. Thus the functions will be positive together and negative together. The secant function is not defined for those numbers x for which $\cos x = 0$. Thus it is undefined for $(\pi/2) + k\pi$, where k is any integer. A graph of the secant function follows. The cosine function is shown for reference by the dashed curve.

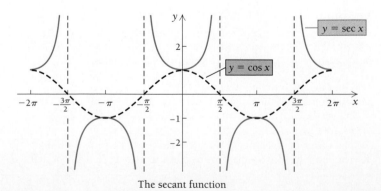

The secant function

The following is a summary of the properties of the secant function.

Properties of the Secant Function

1. The secant function is periodic, with period 2π.
2. The domain of the secant function is the set of all real numbers except $(\pi/2) + k\pi$, where k is any integer. Sec x is not defined where $\cos x = 0$.
3. The range of the secant function consists of all real numbers 1 and greater, in addition to all real numbers -1 and less.
4. The secant function is continuous wherever it is defined.

The Cosecant Function

The cosecant and sine functions are reciprocal functions. The graph of the cosecant function can be constructed by finding the reciprocal of each of the values of the sine function that we found in Section 6.2. Thus the functions will be positive together and negative together. The cosecant function is not defined for those numbers x for which $\sin x = 0$. Thus it is undefined for $k\pi$, where k is any integer. A graph of the cosecant function follows. The sine function is shown for reference by the dashed curve.

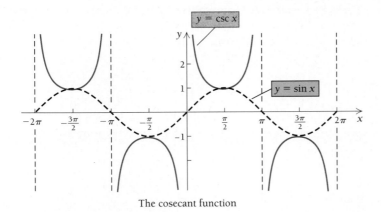

The cosecant function

The following is a summary of the properties of the cosecant function.

Properties of the Cosecant Function

1. The cosecant function is periodic, with period 2π.
2. The domain of the cosecant function is the set of all real numbers except $k\pi$, where k is any integer. Csc x is not defined where $\sin x = 0$.
3. The range of the cosecant function consists of all real numbers 1 and greater, in addition to all real numbers -1 and less.
4. The cosecant function is continuous wherever it is defined.

DO EXERCISES 19–28. _____

Signs of the Functions

Suppose that a real number s determines a point on the unit circle in the first quadrant. Since both coordinates are positive there, the function values

---

19. a) Find three numbers at which the secant function is undefined.

    b) Find three numbers at which the cosecant function is undefined.

20. Graph the secant function.

21. Graph the cosecant function.

22. What is the period of the secant function? the cosecant function?

23. What is the domain of the secant function? the cosecant function?

24. What is the range of the secant function? the cosecant function?

25. In which quadrants are the function values of the secant function positive? negative?

26. In which quadrants are the function values of the cosecant function positive? negative?

27. Does the secant function appear to be even? odd?

28. Does the cosecant function appear to be even? odd?

29. What are the signs of the six circular functions in each of the four quadrants?

for all the circular functions will be positive. In the second quadrant, the first coordinate is negative and the second coordinate is positive. Thus the sine function has positive values and the cosine function has negative values. The secant, being the reciprocal of the cosine, also has negative values, and the cosecant, being the reciprocal of the sine, has positive values. The tangent and the cotangent both have negative values in the second quadrant.

DO EXERCISE 29.

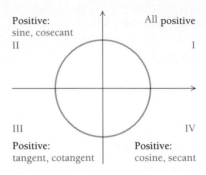

Positive:
sine, cosecant

II

All positive

I

III

IV

Positive:
tangent, cotangent

Positive:
cosine, secant

The diagram shows the quadrants in which the circular function values are positive. It need not be memorized, because the signs can be readily determined by referring to a unit circle.

TECHNOLOGY CONNECTION

A grapher can be used to graph any trigonometric function that can be written in the form $y = f(x)$. Before this can be done, however, there are a few preparations.

First, make sure your grapher is in RADIAN mode. There are two common systems for measuring angles, *radians* and *degrees*. So far, all the work we have done in this chapter has used radian mode; degree mode will be introduced later. All graphers allow you to select a mode. (A third system, *grads*, may also be available on some graphers. We will not use this system.)

Next, the appropriate range values should be set. Some graphers have standard or default settings that are set automatically whenever a trig function is involved. Most of the time, however, you must set the values for the viewing box. The most common values for the x-axis are -2π and $+2\pi$ (or -6.28 and $+6.28$ if your grapher doesn't accept the symbol π). On the y-axis, values slightly larger than the amplitude are generally used so that the entire graph can be viewed.

Finally, if the function includes sec, csc, or cot, it must be rewritten in terms of sin, cos, and tan (these are the only trig keys that are available). For this step, use the following

formulas:

$$\sec x = \frac{1}{\cos x},$$

$$\csc x = \frac{1}{\sin x},$$

$$\cot x = \frac{1}{\tan x} = \frac{\cos x}{\sin x}.$$

Once these preliminaries have been tended to, you are ready to draw the graph. All you need to do is enter the function as you would any other, except that you will use one or more of the trig function keys. In addition, you may need to use parentheses around the part of the trig function that follows *sin* or *cos*. For example, $y = \sin x$ may need to be entered as Y = SIN(X).

Graph each of the following. Use a viewing box of $[-2\pi, 2\pi] \times [-2, 2]$, and make sure your grapher is in radian mode.

TC 1. $y = \cos^2 x \sin x$ **TC 2.** $y = (\cos x + 1)(\tan x)$

TC 3. $y = \csc x + \cot x$ **TC 4.** $y = \sec x \cot x$

The Functions Interrelated

The definitions of the four trigonometric functions introduced in this section give us eight basic identities.

Basic Identities

$$\sin s = \frac{1}{\csc s} \qquad \csc s = \frac{1}{\sin s} \qquad \tan s = \frac{\sin s}{\cos s}$$

$$\cos s = \frac{1}{\sec s} \qquad \sec s = \frac{1}{\cos s} \qquad \cot s = \frac{\cos s}{\sin s}$$

$$\tan s = \frac{1}{\cot s} \qquad \cot s = \frac{1}{\tan s}$$

Some of the above identities have not been proved, but they are straightforward.

Example 4   Prove the identity $\cot s = \dfrac{1}{\tan s}$.

Proof

$$\cot s = \frac{\cos s}{\sin s} = \frac{1}{\dfrac{\sin s}{\cos s}} = \frac{1}{\tan s} \qquad \blacktriangleleft$$

DO EXERCISE 30.

Example 5   Prove the identity $\sin s = \dfrac{1}{\csc s}$.

Proof

$$\sin s = \frac{1}{\dfrac{1}{\sin s}} = \frac{1}{\csc s} \qquad \blacktriangleleft$$

DO EXERCISE 31.

The Pythagorean Identities

Here we consider three other identities that are fundamental to a study of trigonometry. They are called the **Pythagorean identities.** Recall that the equation of a unit circle in a uv-plane is

$$u^2 + v^2 = 1.$$

For any point on a unit circle, the coordinates u and v satisfy this equation. Suppose that a real number s determines a point T on a unit circle, with coordinates (u, v), or $(\cos s, \sin s)$. Then $u = \cos s$ and $v = \sin s$.

30. Prove the identity $\tan s = \dfrac{1}{\cot s}$.

31. Prove the identity $\cos s = \dfrac{1}{\sec s}$.

32. Prove the identity

$$1 + \tan^2 s = \sec^2 s.$$

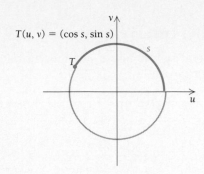

$T(u, v) = (\cos s, \sin s)$

Substituting $\cos s$ for u and $\sin s$ for v in the equation of the unit circle gives us the identity

$$(\cos s)^2 + (\sin s)^2 = 1,$$

which can be expressed as

$$\sin^2 s + \cos^2 s = 1.$$

It is conventional in trigonometry to use the notation $\sin^2 s$ rather than $(\sin s)^2$—there are fewer symbols to write. This identity relates the sine and the cosine of any real number s. It is an important Pythagorean identity. We now develop another. We divide on both sides of the preceding identity by $\sin^2 s$:

$$\frac{\sin^2 s}{\sin^2 s} + \frac{\cos^2 s}{\sin^2 s} = \frac{1}{\sin^2 s}.$$

Simplifying, we get

$$1 + \cot^2 s = \csc^2 s.$$

This equation is true for any replacement of s by a real number for which $\sin^2 s \neq 0$, since we divided by $\sin^2 s$. But the numbers for which $\sin^2 s = 0$ (or $\sin s = 0$) are exactly the ones for which the cotangent and the cosecant functions are undefined. Hence our new equation holds for all numbers s for which $\cot s$ and $\csc s$ are defined, and is thus an identity.

DO EXERCISE 32.

The third Pythagorean identity, obtained by dividing on both sides of the first Pythagorean identity by $\cos^2 s$, is

$$1 + \tan^2 s = \sec^2 s.$$

The Pythagorean Identities

$$\sin^2 s + \cos^2 s = 1$$
$$1 + \cot^2 s = \csc^2 s$$
$$1 + \tan^2 s = \sec^2 s$$

The Pythagorean identities should be memorized. Certain variations of them, although most useful, need not be memorized, because they are so easily obtained from these three.

Example 6   Derive identities that give $\cos^2 x$ and $\cos x$ in terms of $\sin x$.

Solution   We begin with $\sin^2 x + \cos^2 x = 1$, and solve it for $\cos^2 x$. We have

$$\cos^2 x = 1 - \sin^2 x.$$

This gives us $\cos^2 x$ in terms of $\sin x$. Then taking square roots, we have

$$\cos x = \pm\sqrt{1 - \sin^2 x},$$

with the understanding that the sign must be determined by the quadrant in which the point determined by x lies on the unit circle.   ◀

DO EXERCISE 33.

Using identities, we can find the rest of the trigonometric function values from one value if we know the quadrant in which s lies on a unit circle.

Example 7   Given that $\cos s = \frac{1}{5}$ and that s is in the fourth quadrant, find the other function values.

Solution   From Margin Exercise 33, we know that

$$\sin s = \pm\sqrt{1 - \cos^2 s},$$

where the sign of the radical depends on the quadrant in which s lies. Since s is in the fourth quadrant, the sine function is negative. We can then solve for $\sin s$ by substituting $\frac{1}{5}$ for $\cos s$ in the preceding identity:

$$\sin s = -\sqrt{1 - \cos^2 s} = -\sqrt{1 - \left(\frac{1}{5}\right)^2} = -\sqrt{1 - \frac{1}{25}}$$

$$= -\sqrt{\frac{24}{25}} = -\frac{\sqrt{24}}{5} = -\frac{2\sqrt{6}}{5}.$$

Now that we have $\sin s$ and $\cos s$, we can find the other function values:

$$\tan s = \frac{\sin s}{\cos s} = \frac{-\dfrac{2\sqrt{6}}{5}}{\dfrac{1}{5}} = -2\sqrt{6};$$

$$\cot s = \frac{1}{\tan s} = \frac{1}{-2\sqrt{6}} = -\frac{\sqrt{6}}{12};$$

$$\sec s = \frac{1}{\cos s} = \frac{1}{\dfrac{1}{5}} = 5;$$

$$\csc s = \frac{1}{\sin s} = \frac{1}{-\dfrac{2\sqrt{6}}{5}} = -\frac{5}{2\sqrt{6}} = -\frac{5\sqrt{6}}{12}.$$ ◀

DO EXERCISE 34.

33. Derive an identity that gives $\sin^2 x$ and $\sin x$ in terms of $\cos x$.

34. Given that $\sin s = -\frac{3}{8}$ and that s is in the third quadrant, find the other function values.

35. Given that $\tan s = -3$ and that s is in the second quadrant, find the other function values.

Example 8   Given that $\cot s = 2$ and that s is in the third quadrant, find the other function values.

Solution   Solving the identity $1 + \cot^2 s = \csc^2 s$ for $\csc s$, we get

$$\csc s = \pm\sqrt{1 + \cot^2 s},$$

where the sign of the radical depends on the quadrant in which s lies. Since s is in the third quadrant, $\csc s = 1/\sin s$, and the sine function is negative in the third quadrant, it follows that the cosecant function is negative in the third quadrant. Thus,

$$\csc s = -\sqrt{1 + \cot^2 s} = -\sqrt{1 + 2^2} = -\sqrt{5}.$$

From the identity $\sin s = 1/\csc s$, we can find $\sin s$ as follows:

$$\sin s = \frac{1}{\csc s} = \frac{1}{-\sqrt{5}} = -\frac{\sqrt{5}}{5}.$$

From Example 6, we know that

$$\cos s = \pm\sqrt{1 - \sin^2 s}.$$

Since the cosine function is negative in the third quadrant, we can find $\cos s$ as follows:

$$\cos s = -\sqrt{1 - \sin^2 s} = -\sqrt{1 - \left(-\frac{\sqrt{5}}{5}\right)^2}$$

$$= -\sqrt{1 - \frac{5}{25}} = -\sqrt{\frac{4}{5}} = -\frac{2\sqrt{5}}{5}.$$

Finally, we know the secant and the tangent as the reciprocals of the cosine and the cotangent, respectively:

$$\sec s = \frac{1}{\cos s} = \frac{1}{-\frac{2\sqrt{5}}{5}} = -\frac{5}{2\sqrt{5}} = -\frac{\sqrt{5}}{2}, \qquad \tan s = \frac{1}{\cot s} = \frac{1}{2}.$$

◀

DO EXERCISE 35.

● **EXERCISE SET** $\mathbf{6.3}$

A   Find the function values.

1. $\cot\dfrac{\pi}{4}$ 1

2. $\tan\left(-\dfrac{\pi}{4}\right)$

3. $\tan\dfrac{\pi}{6}$

4. $\cot\dfrac{5\pi}{6}$

5. $\sec\dfrac{\pi}{4}$

6. $\csc\dfrac{3\pi}{4}$ $=\sqrt{2}$

7. $\tan\dfrac{3\pi}{2}$

8. $\cot\pi$

9. $\tan\dfrac{2\pi}{3}$

10. $\cot\left(-\dfrac{2\pi}{3}\right)$

11. $\sec\left(-\dfrac{7\pi}{4}\right)$ $\sqrt{2}$

12. $\csc\left(-\dfrac{4\pi}{3}\right)$

13. $\tan\dfrac{5\pi}{6}$

14. $\cot\dfrac{7\pi}{6}$

B

15. Of the six circular functions, which are even?

16. Of the six circular functions, which are odd?

17. Which of the six circular functions have period 2π?

18. Which of the six circular functions have period π?

19. In which quadrants is the tangent function positive? negative?

20. In which quadrants is the cotangent function positive? negative?

21. In which quadrants is the secant function positive? negative?

22. In which quadrants is the cosecant function positive? negative?

23. Complete this table of approximate function values. Do not use trigonometric function keys. Round to five decimal places.

	$\pi/16$	$\pi/8$	$\pi/6$
sin	0.19509	0.38268	
cos	0.98079	0.92388	
tan			
cot			
sec			
csc			

	$\pi/4$	$3\pi/8$	$7\pi/16$
sin		0.92388	0.98079
cos		0.38268	0.19509
tan			
cot			
sec			
csc			

24. Complete this table of approximate function values. Do not use trigonometric function keys. Round to five decimal places.

	$-\pi/16$	$-\pi/8$	$-\pi/6$
sin	-0.19509	-0.38268	-0.50000
cos	0.98079	0.92388	0.86603
tan			
cot			
sec			
csc			

	$-\pi/4$	$-\pi/3$
sin	-0.70711	-0.86603
cos	0.70711	0.50000
tan		
cot		
sec		
csc		

Find the other function values, given each set of conditions.

25. $\cos s = \frac{1}{3}$, s in the first quadrant

26. $\sin s = \frac{2}{3}$, s in the first quadrant

27. $\tan s = 3$, s in the third quadrant

28. $\cot s = 4$, s in the third quadrant

29. $\sec s = -\frac{5}{3}$, s in the second quadrant

30. $\csc s = -\frac{5}{4}$, s in the fourth quadrant

31. $\sin s = -\frac{2}{5}$, s in the third quadrant

32. $\cos s = -\frac{1}{6}$, s in the third quadrant

● **SYNTHESIS** _____

33. Verify the identity $\sec(x - \pi) = -\sec x$ graphically.

34. Verify the identity $\tan(x + \pi) = \tan x$ graphically.

35. Describe how the graphs of the tangent and the cotangent functions are related.

36. Describe how the graphs of the secant and the cosecant functions are related.

37. Which pairs of circular functions have the same zeros? (A *zero* of a function is an input that produces an output of 0.)

38. Describe how the asymptotes of the tangent, cotangent, secant, and cosecant functions are related to the inputs that produce outputs of 0.

Graph.

39. $f(x) = |\tan x|$ **40.** $f(x) = |\sin x|$

41. $g(x) = \sin|x|$ **42.** $f(x) = |\cos x|$

43. One of the motivations for developing trigonometry with a unit circle is that you can actually "see" $\sin \theta$ and $\cos \theta$ on the circle. Note in the figure that $AP = \sin \theta$ and $OA = \cos \theta$. It turns out that you can also "see" the other four trigonometric functions. Prove each of the following.

a) $BD = \tan \theta$ **b)** $OD - \sec \theta$
c) $OE = \csc \theta$ **d)** $CE = \cot \theta$

● **CHALLENGE** _____

Solve graphically.

44. $\sin x > \csc x$ **45.** $\cos x \le \sec x$

OBJECTIVES

You should be able to:

A Given the measure of an angle or rotation in degrees, tell in which quadrant the terminal side lies; find positive and negative angles that are coterminal with a given angle; find the complement and the supplement of a given angle; and convert between angle measure in degrees, minutes, seconds, and decimal measure.

B Convert between radian measure and degree measure.

C Find the length of an arc of a circle, given the measure of its central angle and the length of a radius. Also, find the measure of a central angle of a circle, given the length of its arc and the length of a radius.

D Convert between linear and angular speed.

E Find total distance or total angle, when speed, radius, and time are given.

1. a) Draw $\angle ABC$.

• B

A •

• C

b) Draw $\angle BAC$.

• B

A •

• C

6.4
ANGLES AND ROTATIONS

A ▸ **Angles, Rotations, and Degree Measure**

An *angle* is a familiar figure.

An **angle** is the union of two rays with a common endpoint. Each ray is a **side** of the angle. If B is the common endpoint, it is called the **vertex.** If A is on one ray and C is on the other, and B is the vertex, we can name the angle as $\angle ABC$, where it is understood that the vertex is the middle letter.

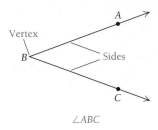

$\angle ABC$

DO EXERCISE 1. _____

In trigonometry, we often think of an angle as a **rotation.** To do so, think of locating a ray along the positive x-axis with its endpoint at the origin. This ray is called the **initial side** of an angle. Though we leave that ray fixed, think of making a copy of it and rotating it. A rotation *counterclockwise* is a **positive rotation,** and a rotation *clockwise* is a **negative rotation.** The ray at the end of the rotation is called the **terminal side** of the angle. The common endpoint is called the *vertex.* An angle so formed is said to be in **standard position.**

It is common to refer to an angle or rotation by Greek letters such as α (alpha), β (beta), γ (gamma), θ (theta), and ϕ (phi). A rotation need not stop after it goes around to the initial side. It can continue for more and more revolutions. If two rotations have the same terminal side, they are said to be **coterminal.** The following are two examples of coterminal angles.

θ and α
coterminal

α and β
coterminal

The measure of an angle or rotation may be given in **degrees.** The Babylonians developed the idea of dividing the circumference of a circle into 360 equal parts. Thus one complete positive revolution or rotation has a measure of 360 *degrees,* which we denote with a degree symbol as $360°$. One half of a revolution has a measure of $180°$, one fourth of a revolution has a measure of $90°$, and so on. We also speak of an angle of measure $60°$, $135°$, $415°$, and so on. Measures of negative rotations can be $-38°$, $-270°$, $-750°$, and so on.

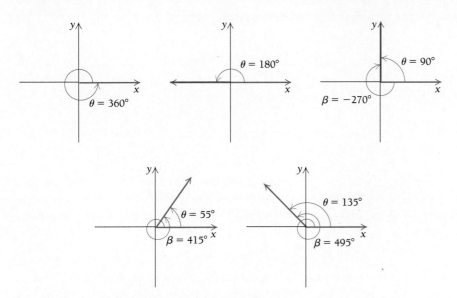

When the measure of an angle is greater than $360°$ or less than $-360°$, the rotating ray has gone through more than one complete revolution. For example, $415°$, shown above, will have the same terminal side as $55°$, since $415° = 360° + 55°$. Thus the terminal side will be in the first quadrant. Note that angles of measure $415°$ and $55°$ are coterminal. We will often speak of an angle by its measure, rather than the more cumbersome terminology "an angle whose measure is ...." Thus we might say "the $45°$ angle," rather than "the angle whose measure is $45°$."

Examples In which quadrant does the terminal side of each angle lie?

1. $53°$

   The terminal side lies in the *first* quadrant.

2. How many degrees are there in:

   a) one revolution?

   b) one half of a revolution?

   c) one fourth of a revolution?

   d) one eighth of a revolution?

   e) one sixth of a revolution?

   f) one twelfth of a revolution?

3. In which quadrant does the terminal side of each angle lie?

   a) $47°$

   b) $212°$

   c) $-43°$

   d) $-145°$

   e) $365°$

   f) $-365°$

   g) $740°$

4. Find two positive angles and two negative angles that are coterminal with a $45°$ angle.

2. $-126°$

The terminal side lies in the *third* quadrant.

3. $460°$

The terminal side lies in the *second* quadrant.

4. $253°$

The terminal side lies in the *third* quadrant.

5. $-373°$

The terminal side lies in the *fourth* quadrant. ◀

DO EXERCISES 2 AND 3. _____

Example 6 Find two positive angles and two negative angles that are coterminal with a $30°$ angle.

Solution Consider a $30°$ angle or rotation. If we have a rotation that goes around another $360°$, we get another angle, coterminal with $30°$. If we go around twice, or $720°$, we get a second angle coterminal with a $30°$ angle. Since

$$30° + 360° = 390° \quad \text{and} \quad 30° + 720° = 750°,$$

we know that $390°$ and $750°$ are two positive angles that are coterminal with a $30°$ angle. Many other answers are possible by adding multiples of $360°$. Similarly, we get the negative coterminal angles by adding $-360°$ and $-720°$:

$$30° + (-360°) = -330° \quad \text{and} \quad 30° + (-720°) = -690°. \quad ◀$$

DO EXERCISE 4. _____

DEFINITION	**Classifications of Angles**
Right angle:	An angle whose measure is $90°$.
Acute angle:	An angle whose measure is between $0°$ and $90°$.
Obtuse angle:	An angle whose measure is between $90°$ and $180°$.
Straight angle:	An angle whose measure is $180°$.

∠DEF is a right angle.   ∠PQR is an acute angle.

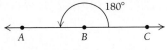

∠XYZ is an obtuse angle.   ∠ABC is a straight angle.

Subunits of degrees are often used. One **minute** is denoted 1′ and is such that

60 minutes = 60′ = 1°.

One **second** is denoted 1″ and is such that

60 seconds = 60″ = 1′.

Then

$34°42′28″$ = 34 degrees, 42 minutes, 28 seconds
= $34° + 42′ + 28″$.

DO EXERCISES 5 AND 6. _____

DEFINITION	Complementary and Supplementary Angles

Two acute angles are *complementary* if their sum is 90°. Two positive angles are *supplementary* if their sum is 180°.

Example 7   Find the complement and the supplement of 56°23′16″.

Solution

$$90° = \begin{array}{r} 89°59′60″ \\ -56°23′16″ \\ \hline 33°36′44″ \end{array} \qquad 180° = \begin{array}{r} 179°59′60″ \\ -56°23′16″ \\ \hline 123°36′44″ \end{array}$$

Thus the complement of 56°23′16″ is 33°36′44″ and the supplement is 123°36′44″.   ◀

DO EXERCISE 7. _____

Example 8   Find the complement and the supplement of 87.46°.

---

5. Classify each of the following angles as right, acute, obtuse, or straight.

a)

b)

c)

d)
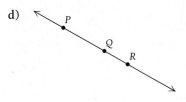

6. Complete.

   a) 1′ = (_____)°
   b) 1″ = (_____)°
   c) 56°23′11″ = (_____)° + (_____)′ + (_____)″

7. Find the complement and the supplement of 85°12′38″.

8. Find the complement and the supplement of 34.89°.

9. Convert 67°13′16″ to degrees and decimal parts of degrees. Round to four decimal places.

10. Convert 37.45° to degrees and minutes.

Solution We have

$$90° - 87.46° = 2.54°,$$
$$180° - 87.46° = 92.54°.$$

Thus the complement of 87.46° is 2.54° and the supplement is 92.54°. ◀

DO EXERCISE 8. _____

Converting a measure like 34°41′52″ to decimal parts of degrees, and vice versa, can be done quite easily with a calculator. Some calculators convert directly. With others we can convert as shown in the following examples.

Example 9 Convert 34°41′52″ to degrees and decimal parts of degrees. Round to two decimal places.

Solution Using $1' = (\frac{1}{60})°$ and $1'' = (\frac{1}{3600})°$, we have

$$34°41′52″ = 34° + 41′ + 52″$$

$$= 34° + \left(\frac{41}{60}\right)° + \left(\frac{52}{3600}\right)°$$

$$\approx 34° + 0.6833 + 0.0144$$

$$\approx 34.70°. \qquad ◀$$

Example 10 Convert 16.35° to degrees and minutes.

Solution We have

$$16.35° = 16° + 0.35 \times 1°.$$

Substituting 60′ for 1° gives us

$$0.35 \times 1° = 0.35 \times 60′ = 21′.$$

Thus,

$$16.35° = 16°21′. \qquad ◀$$

DO EXERCISES 9 AND 10. _____

 Radian Measure

Degree measure is a common unit of angle measure in many everyday applications and in such fields as navigation and surveying. But in many scientific fields and in mathematics and calculus, there is another commonly used unit of measure called the *radian*.

Consider a circle with its center at the origin and radius of length 1 (a *unit circle*). Suppose we measure an arc, moving counterclockwise, of length 1, and mark a point *T* on the circle. If we draw a ray from the origin through *T*, we have formed an angle. The measure of that angle is 1 **radian.** The word radian comes from the word *radius*. Thus measuring 1 radius along the circumference of the circle determines an angle whose measure is 1 *radian*. One radian is about 57.3°.

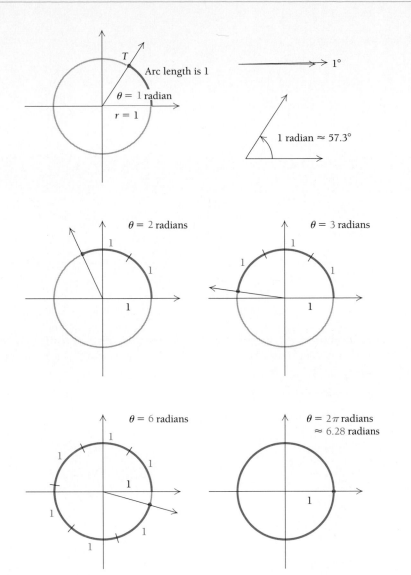

Angles that measure 2 radians and 3 radians are also shown above. When we make a complete (counterclockwise) revolution, the terminal side coincides with the initial side on the positive x-axis. We then have an angle whose measure is 2π radians, or about 6.28 radians, which is the circumference of the circle:

$$2\pi r = 2\pi(1) = 2\pi.$$

Thus a rotation of $360°$ (1 revolution) has a measure of 2π radians. Half of a revolution is a rotation of $180°$, or π radians. A quarter revolution is a rotation of $90°$, or $\pi/2$ radians, and so on.

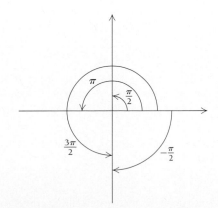

11. In which quadrant does the terminal side of each angle lie?

a) $\dfrac{5\pi}{4}$

b) $\dfrac{17\pi}{8}$

c) $-\dfrac{\pi}{15}$

d) 37.3π

DO EXERCISE 11. _____

To convert between degrees and radians, we first note that

$$360° = 2\pi \text{ radians}.$$

It follows that

$$180° = \pi \text{ radians}.$$

To make conversions, we use the concept of "multiplying by one" (see Section 1.9), noting the following:

$$\frac{\pi \text{ radians}}{180°} = \frac{180°}{\pi \text{ radians}} = 1.$$

When a rotation is given in radians, the word "radians" is optional and is most often omitted. Thus, if no unit is given for a rotation, the rotation is understood to be in radians.

Example 11   Convert $60°$ to radians.

Solution   We have

$$60° = 60° \cdot \frac{\pi \text{ radians}}{180°}$$

$$= \frac{60°}{180°} \cdot \pi \text{ radians}$$

$$= \frac{\pi}{3} \text{ radians,} \quad \text{or} \quad \frac{\pi}{3}.$$

We find that $\pi/3$ radians is about 1.047 radians.   ◀

Example 12   Convert $\dfrac{3\pi}{4}$ radians to degrees.

Solution

$$\frac{3\pi}{4} \text{ radians} = \frac{3\pi}{4} \text{ radians} \cdot \frac{180°}{\pi \text{ radians}}$$

$$= \frac{3\pi}{4\pi} \cdot 180°$$

$$= \frac{3}{4} \cdot 180°$$

$$= 135°$$

◀

Example 13   Convert 1 radian to degrees.

Solution

$$1 \text{ radian} = 1 \text{ radian} \cdot \frac{180°}{\pi \text{ radians}}$$

$$= \frac{180°}{\pi}$$

$$\approx 57.3°$$

DO EXERCISES 12–14.

Arc Length and Central Angles

Radian measure can be determined using a circle other than a unit circle. In the following figure, a unit circle is shown along with another circle. The angle shown is a **central angle** of both circles; hence the arcs that it intercepts have their lengths in the same ratio as the radii of the circles. The radii of the circles are r and 1. The corresponding arc lengths are s and s_1.

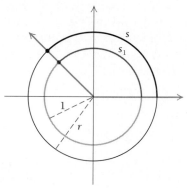

Thus we have the proportion

$$\frac{s}{s_1} = \frac{r}{1}, \quad \text{or} \quad \frac{s}{r} = \frac{s_1}{1}.$$

Now s_1 is the radian measure of the rotation in question. It is more common to use a Greek letter, such as θ, for the measure of an angle or rotation. We commonly use the letter s for arc length. Adopting this convention, we rewrite the proportion above as $s/r = \theta$. In any circle, arc length, central angle, and length of radius are related in this fashion. Or, in general, the following is true.

> **DEFINITION**   **Radian Measure**
>
> The *radian measure* θ of a rotation is the ratio of the distance s traveled by a point at a radius r from the center of rotation, to the length of the radius r:
>
> $$\theta = \frac{s}{r}.$$

12. Convert to radian measure. Leave answers in terms of π.
    a) 225°
    b) 315°
    c) −720°

13. Convert to radian measure. Do not leave answers in terms of π (use 3.14 for π).
    a) $72\frac{1}{2}°$ (a safe working angle for ladders)

    b) 300°
    c) −315°

14. Convert to degree measure.
    a) $\frac{4\pi}{3}$
    b) $\frac{5\pi}{2}$
    c) $-\frac{4\pi}{5}$

15. Find the length of an arc of a circle with 10-cm radius associated with a central angle of measure $11\pi/6$.

Example 14 Find the measure of a rotation in radians where a point 2 m from the center of rotation travels 4 m.

Solution

$$\theta = \frac{s}{r} = \frac{4\,\text{m}}{2\,\text{m}} = 2 \qquad \text{The unit is understood to be radians.} \qquad \blacktriangleleft$$

Example 15 Find the length of an arc of a circle of 5-cm radius associated with a central angle of $\pi/3$ radians.

Solution We have

$$\theta = \frac{s}{r}, \quad \text{or} \quad s = r\theta.$$

Therefore, $s = 5 \cdot \pi/3$ cm, or about 5.23 cm. $\qquad \blacktriangleleft$

16. Find the radian measure of a rotation where a point 2.5 cm from the center of rotation travels 15 cm.

A look at Examples 14 and 15 will show why the word radian is most often omitted. In Example 14, we have the division 4 m/2 m, which simplifies to the number 2, since m/m = 1. From this point of view, it would seem preferable to omit the word radians. In Example 15, had we used the word radians all the way through, our answer would have come out to be 5.23 cm-radians. It is a distance we seek; hence we know that the unit should be centimeters. Thus we must omit the word radians. Since a measure in radians is simply a number, it is usually preferable to omit the word radians.

> **CAUTION!** In using the formula $\theta = s/r$, you must make sure that θ is in radians and that s and r are expressed in the same unit.

17. Find the radian measure of a rotation where a point 24 in. from the center of rotation travels 3 ft.

DO EXERCISES 15–17.

Angular Speed

Linear speed is defined to be distance traveled per unit of time. Similarly, **angular speed** is defined to be amount of rotation per unit of time. For example, we might speak of the angular speed of a wheel as 150 revolutions per minute or the angular speed of the earth as 2π radians per day. The Greek letter ω (omega) is generally used for angular speed. Thus angular speed is defined as

$$\omega = \frac{\theta}{t}.$$

Relating Linear and Angular Speed

For many applications, it is important to know a relationship between *angular speed* and *linear speed*. For example, we might wish to find the linear speed of a point on the earth, knowing its angular speed. Or, we might wish to know the linear speed of an earth satellite, knowing its angular speed. To develop the relationship we seek, we recall the relation between angle and distance: $\theta = s/r$. This is equivalent to

$$s = r\theta.$$

We divide by time, t, to obtain

$$\frac{s}{t} = r\frac{\theta}{t}.$$

$$\downarrow \qquad \downarrow$$
$$v \qquad \omega$$

Now s/t is linear speed v, and θ/t is angular speed ω. Thus we have the relation we seek.

DEFINITION   **Linear Speed**

The *linear speed* v of a point a distance r from the center of rotation is given by

$$v = r\omega,$$

where ω is the *angular speed* in radians per unit time.

In deriving this formula, we used the equation $s = r\theta$, in which the units for s and r must be the same and θ must be in radians. So, for our new formula $v = r\omega$, the units of distance for v and r must be the same, ω must be in radians per unit of time, and the units of time must be the same for v and ω.

Example 16   An earth satellite in circular orbit 1200 km high makes one complete revolution every 90 min. What is its linear speed? Use 6400 km for the length of a radius of the earth.

1200 km

Solution   We will use the formula

$$v = r\omega;$$

thus we will need to know r and ω:

$$r = 6400 \text{ km} + 1200 \text{ km} \qquad \text{Radius of earth plus height of satellite}$$
$$= 7600 \text{ km},$$

$$\omega = \frac{\theta}{t} = \frac{2\pi \text{ radians}}{90 \text{ min}} = \frac{\pi}{45 \text{ min}}. \qquad \text{We have, as usual, omitted the word radians.}$$

Now, using $v = r\omega$, we have

$$v = 7600 \text{ km} \cdot \frac{\pi}{45 \text{ min}} = \frac{7600\pi}{45} \cdot \frac{\text{km}}{\text{min}}$$

$$\approx 531 \frac{\text{km}}{\text{min}}.*$$ ◄

DO EXERCISE 18.

---

* It may be appropriate at this point to review Section 1.9 on handling units.

18. A wheel with 12-cm diameter is rotating at a speed of 10 revolutions per second. What is the velocity of a point on the rim?

19. In using $v = r\omega$, if v is given in centimeters per second, what must be the units for r and ω?

20. In using $v = r\omega$, if ω is given in radians per year and r is in kilometers, what must be the units for v?

21. The old oaken bucket is being raised at a rate of 3 ft/sec. The radius of the drum is 5 in. What is the angular speed of the handle?

5 in.

Example 17 An anchor is being hoisted at a rate of 2 ft/sec, the chain being wound around a capstan with a 1.8-yd diameter. What is the angular speed of the capstan?

Bow of ship

1.8 yd

Capstan

Chain

Anchor

Solution We will use the formula $v = r\omega$ in the form $\omega = v/r$, taking care to use the proper units. Since v is given in feet per second, we need r in feet. Then ω will be in radians per second:

$$r = \frac{1.8}{2}\,\text{yd} \cdot \frac{3\,\text{ft}}{\text{yd}} = 2.7\,\text{ft},$$

$$\omega = \frac{v}{r} = \frac{2}{2.7} \approx 0.741\ \text{radian/sec}. \qquad \blacktriangleleft$$

> **CAUTION!** In applying the formula $v = r\omega$, we must be sure that the distance units for v and r are the same and that ω is in radians per unit of time. The units of time must be the same for v and ω.

DO EXERCISES 19–21.

E Total Distance and Total Angle

The formulas $\theta = s/r$ and $v = r\omega$ can be used in combination to find distances and angles in various situations involving rotational motion.

Example 18 A car is traveling at a speed of 45 mph. Its tires have a 20-in. radius. Find the angle through which a wheel turns in 5 sec.

Solution Recall that $\omega = \theta/t$, or $\theta = \omega t$. Thus we can find θ if we know ω and t. To find ω, we use $v = r\omega$. For convenience, we will first convert 45 mph to ft/sec:

$$v = 45\,\frac{\text{mi}}{\text{hr}} \cdot \frac{1\,\text{hr}}{60\,\text{min}} \cdot \frac{1\,\text{min}}{60\,\text{sec}} \cdot \frac{5280\,\text{ft}}{1\,\text{mi}} = 66\,\frac{\text{ft}}{\text{sec}}.$$

Now $r = 20$ in. We will convert to ft, since v is in ft/sec:

$$r = 20\,\text{in.} \cdot \frac{1\,\text{ft}}{12\,\text{in.}} = \frac{20}{12}\,\text{ft} = \frac{5}{3}\,\text{ft}.$$

Using $v = r\omega$, we have

$$66\,\frac{\text{ft}}{\text{sec}} = \frac{5}{3}\,\text{ft} \cdot \omega, \quad \text{so} \quad \omega = 39.6\,\frac{\text{radians}}{\text{sec}}.$$

Then

$$\theta = \omega t = 39.6 \, \frac{\text{radians}}{\text{sec}} \cdot 5 \text{ sec} = 198 \text{ radians.}$$

◀

DO EXERCISE 22.

22. The diameter of a wheel of a car is 30 in. When the car is traveling at a speed of 60 mph (88 ft/sec), how many revolutions does the wheel make in 1 sec?

● **EXERCISE SET 6.4**

A, B   For the angle of the given measure, state in which quadrant the terminal side lies.

1. 34°
2. 320°
3. −120°
4. 175°
5. $\dfrac{\pi}{3}$
6. $-\dfrac{3\pi}{4}$
7. $\dfrac{11\pi}{4}$
8. $\dfrac{19\pi}{4}$

Find two positive angles and two negative angles that are co-terminal with the given angle.

9. 58°
10. 320°
11. −120°
12. −215°

Find the complement and the supplement.

13. 57°23′
14. 47°38′
15. 73°45′11″
16. 12°03′14″
17. 67.31°
18. 13.68°
19. 11.2344°
20. 85.06312°

Convert to degrees and minutes.

21. 8.6°
22. 47.8°
23. 72.25°
24. 11.75°
25. 46.38°
26. 85.21°
27. 67.84°
28. 38.48°

Convert to degrees, minutes, and seconds.

29. 87.3456°
30. 11.0256°
31. 48.02498°
32. 27.899461°

Convert to degrees and decimal parts of degrees. Round to two decimal places.

33. 9°45′
34. 52°15′
35. 35°50′
36. 64°40′
37. 80°33′
38. 27°19′
39. 3°02′
40. 10°08′

Convert to degrees and decimal parts of degrees. Round to four decimal places.

41. 19°47′23″
42. 49°38′46″
43. 31°57′55″
44. 76°11′34″

B   Convert to radian measure. Leave answers in terms of π.

45. 30°
46. 15°
47. 60°
48. 200°
49. 75°
50. 300°
51. 37.71°
52. 12.73°
53. 214.6°
54. 73.87°

Convert to radian measure. Do not leave answers in terms of π.

55. 120°
56. 240°
57. 320°
58. 75°
59. 200°
60. 300°
61. 117.8°
62. 231.2°
63. 1.354°
64. 327.9°

Convert to degree measure.

65. 1 radian
66. 2 radians
67. 8π
68. −12π
69. $\dfrac{3\pi}{4}$
70. $\dfrac{5\pi}{4}$
71. ▤ 1.303
72. ▤ 2.347
73. ▤ 0.7532π
74. ▤ −1.205π

75. Certain positive angles are marked here in degrees. Find the corresponding radian measures.

76. Certain negative angles are marked here in degrees. Find the corresponding radian measures.

c

77. In a circle with 120-cm radius, an arc 132 cm long subtends a central angle of how many radians? how many degrees, to the nearest degree?

78. In a circle with 200-cm radius, an arc 65 cm long subtends a central angle of how many radians? how many degrees, to the nearest degree?

79. Through how many radians does the minute hand of a clock rotate in 50 min?

80. A wheel on a car has a 14-in. radius. Through what angle (in radians) does the wheel turn while the car travels 1 mi?

81. In a circle with 10-m radius, how long is an arc associated with a central angle of 1.6 radians?

82. In a circle with 5-m radius, how long is an arc associated with a central angle of 2.1 radians?

d

83. A flywheel with 15-cm diameter is rotating at a rate of 7 radians/sec. What is the linear speed of a point on its rim, in centimeters per minute?

84. A wheel with 30-cm radius is rotating at a rate of 3 radians/sec. What is the linear speed of a point on its rim, in meters per minute?

85. A $33\frac{1}{3}$-rpm record has a radius of 15 cm. What is the linear velocity of a point on the rim, in centimeters per second?

86. A 45-rpm record has a radius of 8.7 cm. What is the linear velocity of a point on the rim, in centimeters per second?

87. The earth has a 4000-mi radius and rotates one revolution every 24 hr. What is the linear speed of a point on the equator, in miles per hour?

88. The earth is 93,000,000 miles from the sun and traverses its orbit, which is nearly circular, every 365.25 days. What is the linear velocity of the earth in its orbit, in miles per hour?

89. A wheel has a 32-cm diameter. The speed of a point on its rim is 11 m/s. What is its angular speed?

90. A horse on a merry-go-round is 7 m from the center and travels at a speed of 10 km/h. What is its angular speed?

91. ***Determining the speed of a river.***   A water wheel has a 10-ft radius. To get a good approximation of the speed

of the river, you count the revolutions of the wheel and find that it makes 14 revolutions per minute (rpm). What is the speed of the river?

92. ***Determining the speed of a river.***   A water wheel has a 10-ft radius. To get a good approximation of the speed of the river, you count the revolutions of the wheel and find that it makes 16 rpm. What is the speed of the river?

e

93. The wheels of a bicycle have a 24-in. diameter. When the bicycle is being ridden so that the wheels make 12 rpm, how far will the bike travel in 1 min?

94. The wheels of a car have a 15-in. radius. When the car is being driven so that the wheels make 10 revolutions/second, how far will the car travel in 1 min?

95. A car is traveling at a speed of 30 mph. Its wheels have a 14-in. radius. Find the angle through which a wheel rotates in 10 sec.

96. A car is traveling at a speed of 40 mph. Its wheels have a 15-in. radius. Find the angle through which a wheel rotates in 12 sec.

● **SYNTHESIS**

97. The **grad** is a unit of angle measure similar to a degree. A right angle has a measure of 100 grads. Convert each of the following to grads.
    a) 48°
    b) 153°
    c) $\pi/8$ radian
    d) $5\pi/7$ radians

98. A **mil** is a unit of angle measure. A right angle has a measure of 1600 mils. Convert each of the following to degrees, minutes, and seconds.
    a) 100 mils
    b) 350 mils

99. On the earth, one degree of latitude is how many kilometers? how many miles? (Assume that the radius of the earth is 6400 km, or 4000 mi, approximately.)

100. One minute of latitude on the earth is equivalent to one *nautical mile*. Find the circumference and the radius of the earth in nautical miles.

101. An astronaut on the moon observes the earth, about 240,000 mi away. The diameter of the earth is about 8000 mi. Find the angle α.

102. The circumference of the earth was computed by Eratosthenes (276–195 b.c.). He knew the distance from Aswan to Alexandria to be about 500 mi (see the figure at the top of the following page). From each town he observed the sun at noon, finding the angular difference to be 7°12′. Do Eratosthenes' calculation.

103. Two pulleys, 50 cm and 30 cm in diameter, respectively, are connected by a belt. The larger pulley makes 12 revolutions/minute. Find the angular speed of the smaller pulley, in radians per second.

104. One gear wheel turns another, the teeth being on the rims. The wheels have 40-cm and 50-cm radii, and the smaller wheel rotates at 20 rpm. Find the angular speed of the larger wheel, in radians per second.

105. An airplane engine is idling at 800 rpm. When the throttle is opened, it takes 4.3 sec for the speed to come up to 2500 rpm. What was the angular acceleration:

a) in revolutions per minute per second?
b) in radians per second per second?

106. The linear speed of an airplane, flying at low level over the sea, is 175 knots (nautical miles per hour). It accelerates to 325 knots in 12 sec.

a) What was its linear acceleration in knots per second?
b) What was its angular acceleration in radians per second per second? (*Hint:* One nautical mile is equivalent to one minute of latitude.)

● CHALLENGE _____

107. What is the angle between the hands of a clock at 7:45?

108. At what time between noon and 1:00 P.M. are the hands of a clock perpendicular?

109. The diameter of the earth at the equator has been measured to be 7926.4 statute miles. Use this and the result of Exercise 100 to find the number of statute miles in a nautical mile.

110. To find the distance between two points on the earth when their latitude and longitude are known, we can use a plane right triangle for an excellent approximation if the points are not too far apart. Point A is at latitude 38°27′30″ N, longitude 82°57′15″ W; and point B is at latitude 38°28′45″ N, longitude 82°56′30″ W. Find the distance from A to B in nautical miles (one minute of latitude is one nautical mile).

6.5
TRIGONOMETRIC FUNCTIONS INVOLVING ANGLES OR ROTATIONS

 Function Values of Angles Measured in Degrees

The domains of the circular, or trigonometric, functions are sets of real numbers. We can also consider the domains to be real numbers that represent degree measures. In this context, the number considered is a real-number multiple of a unit called a degree. For example, 47° is really 47 × 1°. If a rotation is in radians, the numbers are the same as those we have already considered. For example, the sine of $\pi/2$ radians is the same as sin $(\pi/2)$. If a rotation is given in degrees, we can merely change to radians to determine

OBJECTIVES

You should be able to:

Ⓐ Find trigonometric function values of angles measured in both degrees and radians. State the definitions of the trigonometric functions, given a point $P(x, y)$ on any circle. Given an ordered pair on the terminal side of an angle θ, or given the equation of the line that is the terminal side, find the six trigonometric function values for the angle.

Ⓑ Determine the signs of the trigonometric function values for a rotation measured in degrees θ in any specified quadrant.

(*continued*)

C State function values (when defined) for any angle whose terminal side lies on an axis.

D State the definitions of the trigonometric function values for any acute angle of a right triangle in terms of the adjacent side, the opposite side, and the hypotenuse. Find the trigonometric function values when the sides of a right triangle are known. Find the function values for any acute angle of 30°, 45°, or 60°.

E Given a function value for an angle and the quadrant in which the terminal side lies, find the other five function values.

Find the following.

1. $\sin 180°$

2. $\tan 0°$

3. $\cos 45°$

4. $\cot 45°$

5. $\sin(-135°)$

the function values. For example,

$$\sin 30° = \sin \frac{\pi}{6}.$$

The diagrams below show unit circles marked in both radians and degrees. You should memorize values in the first quadrant and be able to reason the others.

Examples   Find the following function values.

1. $\sin 90°$

From the diagram, we have $90° = \pi/2$. Hence

$$\sin 90° = \sin \frac{\pi}{2} = 1.$$

2. $\tan 135°$

$$\tan 135° = \tan \frac{3\pi}{4} = -1$$

3. $\cos(-210°)$

$$\cos(-210°) = \cos\left(-\frac{7\pi}{6}\right) = -\frac{\sqrt{3}}{2}$$

DO EXERCISES 1–5.

When we defined the trigonometric functions, we considered a point on a *unit circle*. Now we extend the definition to a point P on *any circle* of radius r. Let us consider a circle of radius r other than a unit circle, as shown below.

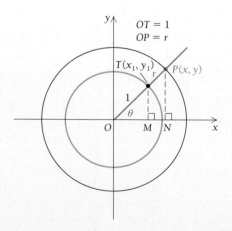

Also shown is a unit circle and a rotation θ. The triangles OMT and ONP are similar, being right triangles with the same acute angles. Thus we know that certain sides have the same ratio. For example,

$$\frac{MT}{OT} = \frac{NP}{OP}.$$

But $MT = y_1$ and $OT = 1$; $NP = y$ and $OP = r$. Substituting, we have

$$\frac{y_1}{1} = \frac{y}{r}.$$

Since T is a point on a unit circle, we know that $y_1 = \sin \theta$. Thus,

$$\boldsymbol{\sin \theta} = \frac{y}{r}.$$

Similarly,

$$\frac{OM}{OT} = \frac{ON}{OP}$$

and

$$\frac{x_1}{1} = \frac{x}{r}.$$

Because T is a point on a unit circle, we know that $x_1 = \cos \theta$. Thus,

$$\boldsymbol{\cos \theta} = \frac{x}{r}.$$

It also follows that

$$\frac{MT}{OM} = \frac{NP}{ON}$$

and

$$\frac{y_1}{x_1} = \frac{y}{x}.$$

Since T is on a unit circle, we know that $y_1/x_1 = \tan \theta$. Thus,

$$\boldsymbol{\tan \theta} = \frac{y}{x}.$$

The point P, which is a point other than the vertex on the terminal side of the angle, can be anywhere on a circle of radius r. Its coordinates may be positive, negative, or zero, depending on the quadrant in which the terminal side lies. The length of the radius, which is also the hypotenuse of what we call a **reference triangle,** is always considered positive. The angle is always measured from the *positive* half of the x-axis. Regardless of the location of point P, we have

$$\boldsymbol{\sin \theta} = \frac{y}{r}, \qquad \boldsymbol{\cos \theta} = \frac{x}{r}, \quad \text{and} \quad \boldsymbol{\tan \theta} = \frac{y}{x}.$$

The following figure shows angles whose terminal sides lie in quadrants II, III, and IV. The reference triangle for the angle θ is triangle ONP, where $\overline{PN}$ is the perpendicular from P to the x-axis.

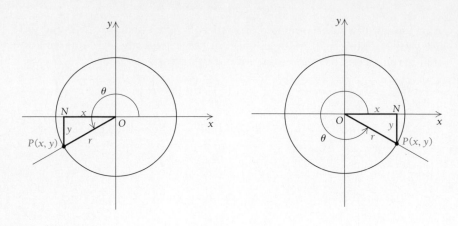

DEFINITION   **Trigonometric Functions: Any Circle**

Suppose that $P(x, y)$ is any point on the terminal side of any angle θ in standard position, and r is the radius, or distance, from the origin to $P(x, y)$. Then the trigonometric functions are defined as follows:

$$\sin \theta = \frac{\text{second coordinate}}{\text{radius}} = \frac{y}{r} \qquad \csc \theta = \frac{\text{radius}}{\text{second coordinate}} = \frac{r}{y}$$

$$\cos \theta = \frac{\text{first coordinate}}{\text{radius}} = \frac{x}{r} \qquad \sec \theta = \frac{\text{radius}}{\text{first coordinate}} = \frac{r}{x}$$

$$\tan \theta = \frac{\text{second coordinate}}{\text{first coordinate}} = \frac{y}{x} \qquad \cot \theta = \frac{\text{first coordinate}}{\text{second coordinate}} = \frac{x}{y}$$

Given any point $P(x, y)$ other than the vertex on the terminal side of an angle θ in standard position, we can find the length r using the distance formula:

$$r = \sqrt{(x-0)^2 + (y-0)^2}$$
$$= \sqrt{x^2 + y^2}.$$

Example 4   Find the six trigonometric function values for the angle shown.

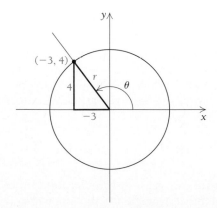

Solution We first determine r, the distance from the origin to the point $(-3, 4)$:

$$r = \sqrt{x^2 + y^2}$$
$$= \sqrt{(-3)^2 + 4^2} \quad \text{Substituting } -3 \text{ for } x \text{ and } 4 \text{ for } y$$
$$= \sqrt{9 + 16}$$
$$= \sqrt{25}, \quad \text{or } 5.$$

Using the definitions of the trigonometric functions, we can now find the function values for θ. We substitute -3 for x, 4 for y, and 5 for r:

$$\sin \theta = \frac{y}{r} = \frac{4}{5} = 0.8, \qquad \csc \theta = \frac{r}{y} = \frac{5}{4} = 1.25,$$

$$\cos \theta = \frac{x}{r} = \frac{-3}{5} = -0.6, \qquad \sec \theta = \frac{r}{x} = \frac{5}{-3} \approx -1.67,$$

$$\tan \theta = \frac{y}{x} = \frac{4}{-3} \approx -1.33, \qquad \cot \theta = \frac{x}{y} = \frac{-3}{4} = -0.75. \qquad \blacktriangleleft$$

Example 5 Find the six trigonometric function values for the angle shown.

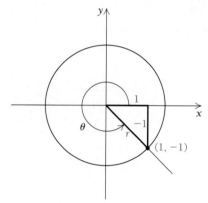

Solution We first determine r, the distance from the origin to the point $(1, -1)$:

$$r = \sqrt{x^2 + y^2}$$
$$= \sqrt{1^2 + (-1)^2} \quad \text{Substituting } 1 \text{ for } x \text{ and } -1 \text{ for } y$$
$$= \sqrt{1 + 1} = \sqrt{2}.$$

Substituting 1 for x, -1 for y, and $\sqrt{2}$ for r, we find that the trigonometric function values of θ are

$$\sin \theta = \frac{y}{r} = \frac{-1}{\sqrt{2}} = -\frac{\sqrt{2}}{2}, \qquad \csc \theta = \frac{r}{y} = \frac{\sqrt{2}}{-1} = -\sqrt{2},$$

$$\cos \theta = \frac{x}{r} = \frac{1}{\sqrt{2}} = \frac{\sqrt{2}}{2}, \qquad \sec \theta = \frac{r}{x} = \frac{\sqrt{2}}{1} = \sqrt{2},$$

$$\tan \theta = \frac{y}{x} = \frac{-1}{1} = -1, \qquad \cot \theta = \frac{x}{y} = \frac{1}{-1} = -1. \qquad \blacktriangleleft$$

DO EXERCISES 6 AND 7.

Find the six trigonometric function values for the angle shown.

6.

7.

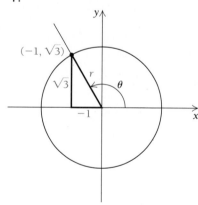

Any point other than the origin on the terminal side of an angle can be used to determine the trigonometric function values of that angle. The function values are the same regardless of which point is used. We illustrate this in the following example.

Example 6   The terminal side of angle θ in standard position lies on the line $x + 2y = 0$. If the terminal side is in quadrant II, find $\sin \theta$, $\cos \theta$, and $\tan \theta$.

Solution   First we draw the graph of $x + 2y = 0$ and determine a second-quadrant solution of the equation.

Using $(-4, 2)$, we then determine r:
$$r = \sqrt{(-4)^2 + 2^2}$$
$$= \sqrt{20} = 2\sqrt{5}.$$

Then using
$$x = -4, \qquad y = 2, \quad \text{and} \quad r = 2\sqrt{5},$$
we find that
$$\sin \theta = \frac{2}{2\sqrt{5}} = \frac{\sqrt{5}}{5}, \qquad \cos \theta = \frac{-4}{2\sqrt{5}} = -\frac{2\sqrt{5}}{5}, \quad \text{and}$$
$$\tan \theta = \frac{2}{-4} = -\frac{1}{2}.$$

Would we get the same function values for θ if we had chosen a different point on the terminal side? The point $(-8, 4)$ is also a solution of the equation that is in the second quadrant and on the terminal side of θ.

We determine r and find the function values:
$$r = \sqrt{(-8)^2 + 4^2} = \sqrt{80} = 4\sqrt{5}.$$

Then

$$\sin \theta = \frac{4}{4\sqrt{5}} = \frac{\sqrt{5}}{5},$$

$$\cos \theta = \frac{-8}{4\sqrt{5}} = -\frac{2\sqrt{5}}{5},$$

$$\tan \theta = \frac{4}{-8} = -\frac{1}{2}.$$

We see that the function values are the same as what we had obtained earlier. ◀

Any point other than the origin on the terminal side of an angle can be used to determine the trigonometric function values.

> The trigonometric function values of θ depend only on the angle, not on the choice of the point on the terminal side.

DO EXERCISE 8.

 Signs of the Functions

Function values of the generalized trigonometric functions can be positive, negative, or zero, depending on where the terminal side of the angle lies. In the first quadrant, all function values are positive because x, y, and r are all positive. In the second quadrant, first coordinates are negative and second coordinates are positive.

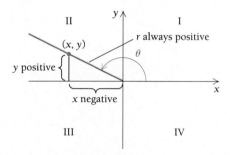

Thus if θ terminates in quadrant II,

$$\sin \theta = \frac{y}{r} \text{ is positive, because } y \text{ and } r \text{ are positive;}$$

$$\cos \theta = \frac{x}{r} \text{ is negative, because } x \text{ is negative and } r \text{ is positive;}$$

$$\tan \theta = \frac{y}{x} \text{ is negative, because } y \text{ is positive and } x \text{ is negative;}$$

and so on.

DO EXERCISE 9.

8. Suppose that θ is an angle whose terminal side lies in quadrant III and has equation $y - 2x = 0$. Find $\sin \theta$, $\cos \theta$, and $\tan \theta$.

9. Complete the following table with the appropriate signs of the trigonometric functions of θ when the terminal side of angle θ lies in each of the four quadrants.

	I	II	III	IV
$\sin \theta$				
$\cos \theta$				
$\tan \theta$				
$\cot \theta$				
$\sec \theta$				
$\csc \theta$				

Determine the signs of the six trigonometric function values for each of the following rotations.

10. $-30°$

11. $160°$

12. $343°$

13. Find the six trigonometric function values for $270°$.

Example 7   Determine the signs of the six trigonometric function values for a rotation of $225°$.

Solution   We have

$$180° < 225° < 270°, \quad \text{so } P(x, y) \text{ is in the third quadrant.}$$

The tangent and the cotangent are positive, and the other four function values are negative.   ◀

DO EXERCISES 10–12.

Terminal Side on an Axis

Now let us suppose that the terminal side of an angle falls on one of the axes. In that case, one of the coordinates is zero. The definitions of the functions still apply, but in some cases, functions will not be defined because a denominator will be 0.

Example 8   Find the trigonometric function values for $180°$.

Solution   Let $(x, 0)$ represent any point, other than the vertex, on the terminal side of a $180°$ angle in standard position. We note that the first coordinate is negative, that the second coordinate is 0, and that x and r have the same absolute value (r being always positive). Thus we have

$$\sin 180° = \frac{y}{r} = \frac{0}{r} = 0,$$

$$\cos 180° = \frac{x}{r} = -1, \qquad \text{Since } |x| = |r|, \text{ but } x \text{ and } r \text{ have opposite signs}$$

$$\tan 180° = \frac{y}{x} = \frac{0}{x} = 0,$$

$$\cot 180° = \frac{x}{y} = \frac{x}{0}, \qquad \text{Thus, } \cot 180° \text{ is undefined.}$$

$$\sec 180° = \frac{r}{x} = -1, \qquad \text{The reciprocal of } \cos 180°$$

$$\csc 180° = \frac{r}{y} = \frac{r}{0}. \qquad \text{Thus, } \csc 180° \text{ is undefined.} \quad ◀$$

DO EXERCISE 13.

Function Values for Acute Angles

We are now ready to consider the trigonometry of right triangles. Historically, this is where trigonometry began. We will develop trigonometric function values of acute angles. These are angles that have a measure greater than $0°$ and less than $90°$. Consider the right $\triangle ONP$ shown at the left below. Note that $\angle O$ has measure θ. Side PN is called the **opposite** side because it is opposite $\angle O$, ON is called the **adjacent** side because it is adjacent to $\angle O$, and side OP is called the **hypotenuse.**

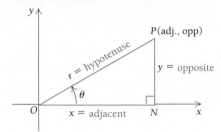

In the figure at the right above, we have moved $\triangle ONP$ onto a coordinate system in such a way that point O is at the origin, with its adjacent side on the x-axis. Then we can express $P(x, y)$ as $P(\text{adjacent}, \text{opposite})$ and the radius is the hypotenuse. That is, $x = \text{adjacent}$, $y = \text{opposite}$, and $r = \text{hypotenuse}$.

This leads us to the following definition of trigonometric function values of acute angles θ.

DEFINITION   **Trigonometric Function Values of an Acute Angle θ**

$$\sin \theta = \frac{\text{opposite}}{\text{hypotenuse}} \qquad \csc \theta = \frac{\text{hypotenuse}}{\text{opposite}}$$

$$\cos \theta = \frac{\text{adjacent}}{\text{hypotenuse}} \qquad \sec \theta = \frac{\text{hypotenuse}}{\text{adjacent}}$$

$$\tan \theta = \frac{\text{opposite}}{\text{adjacent}} \qquad \cot \theta = \frac{\text{adjacent}}{\text{opposite}}$$

Example 9   In this triangle, find each of the following:

a) The trigonometric function values for θ.

b) The trigonometric function values for ϕ(the Greek letter "phi").

Solution

$$\sin \theta = \frac{\text{side opposite } \theta}{\text{hypotenuse}} = \frac{3}{5} \qquad \sin \phi = \frac{\text{side opposite } \phi}{\text{hypotenuse}} = \frac{4}{5}$$

$$\cos \theta = \frac{\text{side adjacent } \theta}{\text{hypotenuse}} = \frac{4}{5} \qquad \cos \phi = \frac{\text{side adjacent } \phi}{\text{hypotenuse}} = \frac{3}{5}$$

$$\tan \theta = \frac{\text{side opposite } \theta}{\text{side adjacent } \theta} = \frac{3}{4} \qquad \tan \phi = \frac{\text{side opposite } \phi}{\text{side adjacent } \phi} = \frac{4}{3}$$

$$\cot \theta = \frac{\text{side adjacent } \theta}{\text{side opposite } \theta} = \frac{4}{3} \qquad \cot \phi = \frac{\text{side adjacent } \phi}{\text{side opposite } \phi} = \frac{3}{4}$$

$$\sec \theta = \frac{\text{hypotenuse}}{\text{side adjacent } \theta} = \frac{5}{4} \qquad \sec \phi = \frac{\text{hypotenuse}}{\text{side adjacent } \phi} = \frac{5}{3}$$

$$\csc \theta = \frac{\text{hypotenuse}}{\text{side opposite } \theta} = \frac{5}{3} \qquad \csc \phi = \frac{\text{hypotenuse}}{\text{side opposite } \phi} = \frac{5}{4}$$

14. Find the following.

 a) $\sin \theta$, $\cos \theta$, $\tan \theta$, $\cot \theta$, $\sec \theta$, $\csc \theta$

 b) $\sin \phi$, $\cos \phi$, $\tan \phi$, $\cot \phi$, $\sec \phi$, $\csc \phi$

DO EXERCISE 14.

In Example 9, we note that the value of $\cot \theta$, $\frac{4}{3}$, is the reciprocal of $\frac{3}{4}$, the value of $\tan \theta$. Likewise, we see the same reciprocal relationship between the values of $\sec \theta$ and $\cos \theta$ and between the values of $\csc \theta$ and $\sin \theta$. For any angle, the cotangent, secant, and cosecant function values are the respective reciprocals of the tangent, cosine, and sine function values.

$$\cot \boldsymbol{\theta} = \frac{1}{\tan \boldsymbol{\theta}}, \qquad \sec \boldsymbol{\theta} = \frac{1}{\cos \boldsymbol{\theta}}, \qquad \csc \boldsymbol{\theta} = \frac{1}{\sin \boldsymbol{\theta}}$$

Function Values for Some Special Angles

We can determine the function values for certain angles using our knowledge of geometry. First, recall the Pythagorean theorem. It says that in any right triangle, $a^2 + b^2 = c^2$, where c is the length of the hypotenuse.

A right triangle with a $45°$ angle actually has two $45°$ angles. Thus the triangle is isosceles, and the legs are the same length. Let us consider such a triangle whose legs have length 1. Then its hypotenuse has length c:

$$1^2 + 1^2 = c^2, \quad \text{or} \quad c^2 = 2, \quad \text{or} \quad c = \sqrt{2}.$$

Such a triangle is shown below. From this diagram, we can easily determine the trigonometric function values for $45°$, or $\pi/4$.

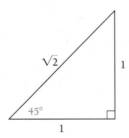

$$\sin 45° = \frac{\text{opposite}}{\text{hypotenuse}} = \frac{1}{\sqrt{2}} = \frac{\sqrt{2}}{2} \approx 0.707$$

$$\cos 45° = \frac{\text{adjacent}}{\text{hypotenuse}} = \frac{1}{\sqrt{2}} = \frac{\sqrt{2}}{2} \approx 0.707$$

$$\tan 45° = \frac{\text{opposite}}{\text{adjacent}} = \frac{1}{1} = 1$$

$$\cot 45° = \frac{\text{adjacent}}{\text{opposite}} = \frac{1}{1} = 1$$

$$\sec 45° = \frac{\text{hypotenuse}}{\text{adjacent}} = \frac{\sqrt{2}}{1} = \sqrt{2} \approx 1.414$$

$$\csc 45° = \frac{\text{hypotenuse}}{\text{opposite}} = \frac{\sqrt{2}}{1} = \sqrt{2} \approx 1.414$$

In a similar way, we can determine function values for 30° and 60°, or $\pi/6$ and $\pi/3$. A right triangle with 30° and 60° acute angles is half of an equilateral triangle, as shown in the following diagram. Thus if we choose an equilateral triangle whose sides have length 2 and take half of it, we obtain a right triangle that has a hypotenuse of length 2 and a leg of length 1. The other leg has length a, which can be found using the Pythagorean theorem as follows:

$$a^2 + 1^2 = 2^2$$
$$a^2 = 3$$
$$a = \sqrt{3}.$$

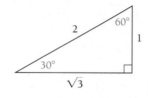

We can now determine function values for 30° and 60°.

$$\sin 30° = \frac{1}{2} \qquad \sin 60° = \frac{\sqrt{3}}{2}$$

$$\cos 30° = \frac{\sqrt{3}}{2} \qquad \cos 60° = \frac{1}{2}$$

$$\tan 30° = \frac{1}{\sqrt{3}} = \frac{\sqrt{3}}{3} \qquad \tan 60° = \frac{\sqrt{3}}{1} = \sqrt{3}$$

$$\cot 30° = \frac{\sqrt{3}}{1} = \sqrt{3} \qquad \cot 60° = \frac{1}{\sqrt{3}} = \frac{\sqrt{3}}{3}$$

$$\sec 30° = \frac{2}{\sqrt{3}} = \frac{2\sqrt{3}}{3} \qquad \sec 60° = \frac{2}{1} = 2$$

$$\csc 30° = \frac{2}{1} = 2 \qquad \csc 60° = \frac{2}{\sqrt{3}} = \frac{2\sqrt{3}}{3}$$

Since we will often use the function values for 45°, 30°, and 60°, they should be memorized. It is sufficient to learn only those for the sine, cosine, and tangent, since the others are their reciprocals.

$$\sin 45° = \frac{\sqrt{2}}{2} \qquad \cos 45° = \frac{\sqrt{2}}{2} \qquad \tan 45° = 1$$

$$\sin 30° = \frac{1}{2} \qquad \cos 30° = \frac{\sqrt{3}}{2} \qquad \tan 30° = \frac{\sqrt{3}}{3}$$

$$\sin 60° = \frac{\sqrt{3}}{2} \qquad \cos 60° = \frac{1}{2} \qquad \tan 60° = \sqrt{3}$$

15. A baseball diamond is really a square 90 ft on a side. If a line is drawn from third base to first base, then a right triangle *QPH* is formed, where ∠*QPH* is 45°. Use the sine function to find the length *PQ* from third base to first base.

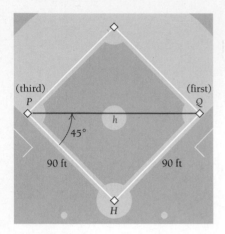

We can now use what we have learned about trigonometric functions of special angles to solve problems. We will consider such applications in much greater detail in Section 8.1.

Example 10 The rays of the sun over the top of a building cast a 40-ft shadow and form an angle of 60° with the ground. Find the height *h* of the building.

Solution We will typically refer to an angle *A* and its measure by the same notation. Thus, *A* = 60°. We know the side adjacent to ∠*A* and also the measure of ∠*A*. Since we want to determine the length of the opposite side, we can use the tangent ratio or the cotangent ratio. Here we use the tangent:

$$\tan 60° = \frac{\text{opposite}}{\text{adjacent}} = \frac{h}{40}$$

$$\sqrt{3} = \frac{h}{40} \qquad \text{Substituting}$$

$$40\sqrt{3} = h$$

$$69.3 \text{ ft} \approx h.$$

The height of the building is about 69.3 ft.

DO EXERCISE 15.

E The Six Trigonometric Functions Related

The six basic circular, or trigonometric, functions are related in a number of interesting ways. When we know one of the function values for an angle, we can find the other five if we know the quadrant in which the terminal side lies. In Section 6.3, we used identities. In this section, we will use diagrams. The idea is to sketch a reference triangle in the appropriate quadrant, use the Pythagorean theorem as needed to find the lengths of its sides, and then read off the appropriate ratios.

Example 11 Given that $\tan \theta = -\frac{2}{3}$ and that θ is in the second quadrant, find the other function values.

Solution We first sketch a second-quadrant triangle. Since $\tan \theta = -\frac{2}{3}$, we make the legs of lengths 2 and 3. The hypotenuse must then have length $\sqrt{13}$. Now we can read off the appropriate ratios:

$$\sin \theta = \frac{2}{\sqrt{13}}, \qquad \csc \theta = \frac{\sqrt{13}}{2},$$

$$\cos \theta = -\frac{3}{\sqrt{13}}, \qquad \sec \theta = -\frac{\sqrt{13}}{3},$$

$$\tan \theta = -\frac{2}{3}, \qquad \cot \theta = -\frac{3}{2}.$$

Example 12   Given that $\cos \theta = -\frac{3}{4}$ and θ is in the third quadrant, find $\sin \theta$ and $\tan \theta$.

Solution   We first sketch a third-quadrant triangle. Since $\cos \theta = -\frac{3}{4}$, we make the hypotenuse 4 and the leg on the x-axis 3. The other leg must have length $\sqrt{7}$. Now we can read off the appropriate ratios:

$$\sin \theta = -\frac{\sqrt{7}}{4},$$

$$\tan \theta = \frac{\sqrt{7}}{3}.$$

DO EXERCISES 16 AND 17.

16. Given that $\cot \theta = -3$ and the terminal side is in quadrant II, find the other function values.

17. Given that $\cos \theta = \frac{3}{4}$ and the terminal side is in quadrant IV, find the other function values.

●EXERCISE SET 6.5

A   Find each of the following, if they exist.

1. $\cos 180°$
2. $\cot 0°$
3. $\sin 45°$
4. $\tan 45°$
5. $\cos (-135°)$
6. $\sin 150°$
7. $\cot (-60°)$
8. $\tan (-120°)$

Find the six trigonometric function values for the angle θ.

9.

10.

11.

12.

The terminal side of angle θ in standard position lies on the given line in the given quadrant. Find $\sin \theta$, $\cos \theta$, and $\tan \theta$.

13. $2x + 3y = 0$; quadrant IV
14. $4x + y = 0$; quadrant II
15. $5x - 4y = 0$; quadrant I
16. $y = 0.8x$; quadrant III

B   Find the signs of the six trigonometric function values for the given rotation.

17. $319°$
18. $-57°$
19. $-620°$
20. $194°$
21. $-215°$
22. $290°$
23. $91°$
24. $-272°$

C   Find the trigonometric function values for the given angle.

25. $90°$
26. $360°$
27. $-180°$
28. $-270°$

D   Find the six trigonometric function values for the specified angle.

29.

30.

31.

32.

33.

34.

35. Complete the following table with exact function values.

θ	sine	cosine	tangent
45°			
30°			
60°			

θ	cotangent	secant	cosecant
45°			
30°			
60°			

36. Complete the following table with exact function values.

θ	sine	cosine	tangent
−45°			
−30°			
−60°			

θ	cotangent	secant	cosecant
−45°			
−30°			
−60°			

37. Find the distance a across the river.

38. Find the length L from point A to the top of the pole.

A function value and a quadrant are specified. Find the other five function values.

39. $\sin \theta = -\frac{1}{3}$; III

40. $\sin \theta = -\frac{1}{5}$; IV

41. $\cos \theta = \frac{3}{5}$; IV

42. $\cos \theta = -\frac{4}{5}$; II

43. $\cot \theta = -2$; IV

44. $\tan \theta = 5$; III

45. $\sin \theta = \frac{1}{3}$; II

46. $\cos \theta = \frac{4}{5}$; IV

Assume that θ is an acute angle. A function value is specified. Find the other five function values.

47. $\sin \theta = \frac{24}{25}$

48. $\cos \theta = 0.7$

49. $\tan \phi = 2$

50. $\sec \phi = \sqrt{17}$

● **SYNTHESIS**

51. This diagram shows a piston of a steam engine, driving a drive wheel of a locomotive. The radius of the drive wheel (from the center of the wheel to the pin P) is R, and the length of the rod is L. Suppose that the drive wheel is rotating at a speed of ω radians per second (so that $\theta = \omega t$). Show that the distance of pin Q from 0, the center of the wheel, is a function of time given by

$$x = \sqrt{L^2 - R^2 \sin^2 \omega t} + R \cos \omega t.$$

6.6
FINDING TRIGONOMETRIC
FUNCTION VALUES

OBJECTIVES

You should be able to:

A Find the six trigonometric function values for angles that are multiples of 30° ($\pi/6$), 45° ($\pi/4$), and 60° ($\pi/3$).

B Using a calculator, find the six trigonometric function values for any angle, measured in either radians or degrees. Given a trigonometric function value of θ, find θ given an interval in which θ lies, providing the answer in either radians or degrees.

A Function Values for Special Angles

We are now able to determine trigonometric function values for many other angles, whether they are measured in radians or degrees. If the terminal side of an angle falls on one of the axes, the function values are 0, 1, -1, or undefined. Thus we can determine function values for any multiple of 90°, or $\pi/2$. We can also determine the function values for any angle whose terminal side makes a 30° ($\pi/6$), 45° ($\pi/4$), or 60° ($\pi/3$) angle with the x-axis.

Consider, for example, an angle of 150°. The terminal side makes a 30° angle with the x-axis, since $180° - 150° = 30°$. As the figure below shows, $\triangle ONR$ is congruent to $\triangle ON'R'$; the ratios of the sides of the two triangles are the same. Thus the trigonometric function values are the same except perhaps for sign. We could determine the function values directly from $\triangle ONR$, but this is not necessary. If we remember that in quadrant II the sine is positive and the cosine and the tangent are negative, we can simply use the values of 30°, prefixing the appropriate sign. The $\triangle ONR$ is called a **reference triangle** and its acute angle, $\angle NOR$, is called a **reference angle.** The reference angle for a rotation, or angle, is the acute angle formed by the terminal side and the x-axis.

If we considered the same angle measured in radians, $5\pi/6$, then we would think of an arc length of $s = 5\pi/6$ on a unit circle. The number s measured on a unit circle determines a point T. Then we would consider a **reference number,** which is the shortest positive arc length s' between the point T and the x-axis. In the case of $5\pi/6$, the reference number is $\pi/6$, since $\pi - 5\pi/6 = \pi/6$.

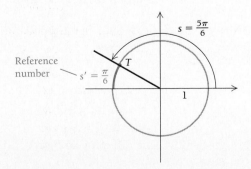

1. Find the trigonometric function values for 120°.

Using Reference Angles to Find Function Values

To find the function values of an angle **measured in degrees,** we find them for the reference angle and prefix the appropriate sign. A reference angle formed by the terminal side and the *x*-axis is always positive and acute.

To find the function values of an angle **measured in radians,** we find them for the reference number and prefix the appropriate sign. A reference number, being the shortest positive arc length between the terminal side and the *x*-axis, is always positive.

Example 1   Find the trigonometric function values for 225°.

Solution   We draw a figure showing the terminal side of a 225° angle. The reference angle is 225° − 180°, or 45°.

We recall from Section 6.5 that sin 45° = $\sqrt{2}/2$, cos 45° = $\sqrt{2}/2$, and tan 45° = 1. We also note that in the third quadrant, the sine and the cosine are negative and the tangent is positive. We can easily determine the other three function values—the cosecant, the secant, and the cotangent— by remembering that they are the respective reciprocals of the sine, cosine, and tangent function values. Thus we have

$$\sin 225° = -\frac{\sqrt{2}}{2}, \qquad \csc 225° = -\frac{2}{\sqrt{2}}, \quad \text{or } -\sqrt{2},$$

$$\cos 225° = -\frac{\sqrt{2}}{2}, \qquad \sec 225° = -\frac{2}{\sqrt{2}}, \quad \text{or } -\sqrt{2},$$

$$\tan 225° = 1, \qquad \cot 225° = 1.$$

◀

DO EXERCISE 1. _____

Example 2   Find the trigonometric function values for 660°.

Solution   We find the multiple of 180° nearest 660°:

$$180° \times 2 = 360°$$
$$180° \times 3 = 540°$$

and   ⟵—— 660°

$$180° \times 4 = 720°.$$

The nearest multiple is 720°. The difference between 720° and 660° is 60°. This gives us the reference angle.

180° or 540°   360° 720° x

660°

Reference angle

60°

660°

We recall that $\sin 60° = \sqrt{3}/2$, $\cos 60° = 1/2$, and $\tan 60° = \sqrt{3}$. In the fourth quadrant, the cosine is positive and the sine and tangent are negative. Thus we have

$$\sin 660° = -\frac{\sqrt{3}}{2}, \qquad \csc 660° = -\frac{2}{\sqrt{3}}, \quad \text{or } -\frac{2\sqrt{3}}{3},$$

$$\cos 660° = \frac{1}{2}, \qquad \sec 660° = \frac{2}{1} = 2,$$

$$\tan 660° = -\sqrt{3}, \qquad \cot 660° = -\frac{1}{\sqrt{3}}, \quad \text{or } -\frac{\sqrt{3}}{3}. \qquad \blacktriangleleft$$

DO EXERCISE 2. _____

We can use the same procedure for negative rotations.

Example 3   Find the sine, the cosine, and the tangent of $-1050°$.

Solution   We find the multiple of $-180°$ nearest $-1050°$:

$$-180° \times 5 = -900° \quad \text{and} \quad -180° \times 6 = -1080°.$$

The nearest multiple is $-1080°$. The difference between $-1050°$ and $-1080°$ is $30°$. This gives us the reference angle.

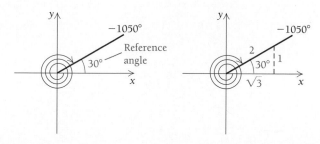

$-1050°$

Reference angle

30°

$-1050°$

30°

All trigonometric function values are positive in the first quadrant, and $\sin 30° = 1/2$, $\cos 30° = \sqrt{3}/2$, and $\tan 30° = \sqrt{3}/3$. Thus we have

$$\sin(-1050°) = \frac{1}{2}, \quad \cos(-1050°) = \frac{\sqrt{3}}{2}, \text{ and } \tan(-1050°) = \frac{\sqrt{3}}{3}. \qquad \blacktriangleleft$$

DO EXERCISE 3. _____

Example 4   Find the trigonometric function values of $\dfrac{7\pi}{6}$.

2. Find the trigonometric function values for 570°.

3. Find the trigonometric function values for $-945°$.

4. Find the trigonometric function values for $\dfrac{17\pi}{6}$.

Solution   Since we have radian measure, we consider a unit circle. Because

$$\frac{7\pi}{6} = \frac{6\pi}{6} + \frac{\pi}{6} = \pi + \frac{\pi}{6},$$

we know that $7\pi/6$ is in the third quadrant and that the reference number is $\pi/6$. Recall that $\sin(\pi/6) = 1/2$, $\cos(\pi/6) = \sqrt{3}/2$, and $\tan(\pi/6) = \sqrt{3}/3$. In the third quadrant, the sine and the cosine are negative and the tangent is positive. Thus

$$\sin\frac{7\pi}{6} = -\frac{1}{2}, \qquad \csc\frac{7\pi}{6} = -2,$$

$$\cos\frac{7\pi}{6} = -\frac{\sqrt{3}}{2}, \qquad \sec\frac{7\pi}{6} = -\frac{2\sqrt{3}}{3},$$

$$\tan\frac{7\pi}{6} = \frac{\sqrt{3}}{3}, \qquad \cot\frac{7\pi}{6} = \sqrt{3}.$$

◀

DO EXERCISE 4.

Example 5   Find the trigonometric function values of $-\dfrac{13\pi}{3}$.

Solution   Since we have radian measure, we consider a unit circle. Because

$$-\frac{13\pi}{3} = -\frac{12\pi}{3} - \frac{\pi}{3}$$

$$= -\left(4\pi + \frac{\pi}{3}\right),$$

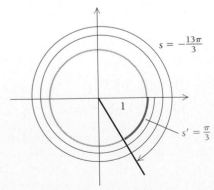

we know that $-13\pi/3$ is in the fourth quadrant and that the reference number is $\pi/3$. Note that $\sin(\pi/3) = \sqrt{3}/2$, $\cos(\pi/3) = 1/2$, and $\tan(\pi/3) = \sqrt{3}$. In the fourth quadrant, the sine and the tangent are negative and the

cosine is positive. Thus

$$\sin\left(-\frac{13\pi}{3}\right) = -\frac{\sqrt{3}}{2}, \qquad \csc\left(-\frac{13\pi}{3}\right) = -\frac{2\sqrt{3}}{3},$$

$$\cos\left(-\frac{13\pi}{3}\right) = \frac{1}{2}, \qquad \sec\left(-\frac{13\pi}{3}\right) = 2,$$

$$\tan\left(-\frac{13\pi}{3}\right) = -\sqrt{3}, \qquad \cot\left(-\frac{13\pi}{3}\right) = -\frac{\sqrt{3}}{3}. \qquad \blacktriangleleft$$

DO EXERCISES 5 AND 6.

 ## Finding Trigonometric Function Values with a Calculator

How do we find function values when the angle or arc length is not one of the special values 0, $\pi/6$, $\pi/4$, $\pi/3$, and $\pi/2$; or 0°, 30°, 45°, 60°, and 90°? Before the advent of scientific calculators, tables were used to find such values. These were prepared by mathematicians using advanced procedures for approximation. Table 3, at the back of the book, is such a table. It contains function values for degrees in the interval [0°, 90°] and for radians in the interval [0, $\pi/2$].

With the advent of the scientific calculator, the need for tables has diminished. The advantage of using a calculator is that no reference angle or reference number need be considered. The main concern is whether the calculator is operating in degree mode or in radian mode. One quick check to keep in mind is that

$$\sin 90° = 1, \quad \text{or} \quad \sin\frac{\pi}{2} = 1.$$

If you try sin 90 and get 0.893996664, you know that you are in radian mode. But if you try sin 90 and get 1, you know that you are in degree mode.

When finding values on a calculator, remember that different calculators round to different decimal places, so there may be variance in answers found on your calculator. Also keep in mind the difference between an exact answer and an approximation. For example,

$$\sin 60° = \frac{\sqrt{3}}{2}. \qquad \textbf{This is exact!}$$

But on a calculator, you might get an answer like

$$\sin 60° \approx 0.866025404. \qquad \textbf{This is approximate!}$$

Scientific calculators usually provide values of the sine, the cosine, and the tangent functions only. You can find values of the cotangent, the secant, and the cosecant by taking reciprocals of the tangent, the cosine, and the sine function, respectively. Since calculators are different, we suggest that you carefully read the directions for the calculator you are using.

Example 6 Find tan 81°.

5. Find the trigonometric function values for $-\frac{22\pi}{3}$.

6. Find the trigonometric function values for $\frac{29\pi}{4}$.

Find each of the following to six decimal places.

7. $\sin 229°$

8. $\tan 22.6°$

9. $\cos(-1007°)$

10. $\csc(143.567°)$

11. $\cot(-789.23°)$

12. $\sec(209.056778°)$

In each of the following cases, a wrong answer was found using a calculator. Explain the incorrect procedure used.

13. $\sin\dfrac{21\pi}{19} = 0.060565745$

14. $\tan(-405) = -1$

15. $\cos\dfrac{16\pi}{17} = 0.99866872$

Find each of the following to four decimal places.

16. $\cos 61°54'$

17. $\tan 347°02'38''$

18. $\sec 55°13'43''$

Solution We check to be sure that the calculator is in degree mode. Then we enter 81 and press the TAN key. We find that

$$\tan 81° \approx 6.3138.$$

We rounded the answer to four decimal places. ◀

Example 7 Find $\cos 715.8234°$.

Solution We enter 715.8234 and press the COS key. We need not consider a reference angle. We find that

$$\cos 715.8234° \approx 0.997344305.$$

This time we rounded to nine decimal places. ◀

Example 8 Find $\sin(-8.0113°)$. Round to four decimal places.

Solution We enter -8.0113 and press the SIN key. Again, we do not need to consider a reference angle. We find that

$$\sin(-8.0113°) \approx -0.1394.$$ ◀

Example 9 Find $\sec 43°$. Round to nine decimal places.

Solution The secant function value can be found by taking the reciprocal of the cosine function value. We enter 43 and press the COS key:

$$\cos 43° \approx 0.731353702.$$

We then press the reciprocal key 1/x, or divide 1 by 0.731353702. You need not round before taking the reciprocal. We find that

$$\sec 43° = \frac{1}{\cos 43°} \approx 1.367327461.$$ ◀

DO EXERCISES 7–15.

When the angle measure is given in degrees, minutes, and seconds, we first convert to degrees and decimal parts of degrees.

Example 10 Find $\sin 38°25'57''$. Round to four decimal places.

Solution We first convert to degrees and decimal parts of degrees. (See Section 6.4.) We get

$$38°25'57'' = 38.4325°.$$

We then press the SIN key and get

$$\sin 38°25'57'' \approx 0.6216.$$ ◀

DO EXERCISES 16–18.

Using a calculator, we can find function values of real numbers, or radians, directly without changing to degrees. Most scientific calculators have both degree and radian modes. When directly finding function values of real numbers or radian measures, we must set the calculator in radian mode.

Example 11   Find $\cos \dfrac{2\pi}{5}$. Round to four decimal places.

Solution   Using the π key, we get

$$\frac{2\pi}{5} \approx 1.256637062.$$

Again, you need not round off before continuing. Then remembering to use the radian mode, we press the $\boxed{\text{COS}}$ key. We find that

$$\cos \frac{2\pi}{5} \approx 0.3090. \qquad \blacktriangleleft$$

Example 12   Find $\tan \left(-\dfrac{\pi}{7} \right)$. Round to six decimal places.

Solution

$$\tan \left(-\frac{\pi}{7} \right) \approx \tan \left(-0.448798951 \right) \approx -0.481575 \qquad \blacktriangleleft$$

Example 13   Find $\sin 13{,}523.86$. Round to six decimal places.

Solution   We enter $13{,}523.86$ and press the $\boxed{\text{SIN}}$ key. The calculator must be in radian mode. We find that

$$\sin 13{,}523.86 \approx 0.641440. \qquad \blacktriangleleft$$

Example 14   Find $\csc 429$. Round to six decimal places.

Solution   The cosecant function value can be found by taking the reciprocal of the sine function value. Remembering to use radian mode, we enter 429, press the $\boxed{\text{SIN}}$ key, and then press the reciprocal key. We find that

$$\csc 429 = \frac{1}{\sin 429} \approx \frac{1}{0.985141084} \approx 1.015083. \qquad \blacktriangleleft$$

DO EXERCISES 19–24.

Sometimes we need to work backward and find an angle that has a specified function value. There are many such values unless restrictions to a quadrant or an interval are specified. For example, since

$$\sin 30° = \sin 150° = \sin 390° = \sin \left(-210° \right) = \tfrac{1}{2},$$

we see that there are many solutions, in degrees, to the equation

$$\sin \theta = \tfrac{1}{2}.$$

Find the function value. Round to five decimal places.

19. $\cos \dfrac{11\pi}{13}$

20. $\tan \dfrac{\pi}{9}$

21. $\sin 21$

22. $\cot 1026$

23. $\cos \left(-\dfrac{3\pi}{11} \right)$

24. $\sec \left(-90{,}234.567 \right)$

Find the acute angle θ, to the nearest hundredth of a degree, for each of the following.

25. $\tan \theta = 0.4621$

26. $\sin \theta = 0.7660$

27. $\cos \theta = 0.7009$

Find θ in the interval indicated.

28. $\sin \theta = -0.2363$, $[-90°, 90°]$

29. $\cos \theta = -0.5712$, $[0, \pi]$

30. $\cos \theta = -0.5712$, $[0°, 180°]$

31. $\tan \theta = 124.84$, $(-\pi/2, \pi/2)$

In fact, there are infinitely many solutions. Similarly, there are infinitely many solutions in radians.

We can use a calculator in reverse—for example, to find an angle when its sine is given—but the answers are restricted to certain intervals. That is, if we ask a calculator to find the angle θ for which

$$\sin \theta = \tfrac{1}{2},$$

the calculator will give us the answer in degrees only in the interval $[-90°, 90°]$, or in radians only in the interval $[-\pi/2, \pi/2]$.

To use a calculator in reverse—for example, to find an angle whose sine is given—we first enter the sine value. Then we press the key marked either SIN^{-1} (on some calculators) or ARCSIN (on other calculators). On still others, two keys must be pressed in sequence: ARC and then SIN, INV and then SIN, or f^{-1} and then SIN. Be sure to read the instructions for your calculator.

Example 15   Find the acute angle θ, to the nearest hundredth of a degree, for which $\cos \theta = 0.5437$.

Solution   We check to be certain that the calculator is in degree mode. We enter 0.5437 and press the $\boxed{\text{COS}^{-1}}$ key. We find that

$$\theta = \cos^{-1} 0.5437 \approx 57.06°.$$

Thus, $\theta \approx 57.06°$.   ◄

DO EXERCISES 25–27.

Inverse functions for the sine, the cosine, and the tangent functions are defined and developed in Section 7.6. The restrictions are given as follows.

The Inverse Function Key

1. SIN^{-1} gives values of θ in degrees in the interval $[-90°, 90°]$ and in radians in the interval $[-\pi/2, \pi/2]$.
2. COS^{-1} gives values of θ in degrees in the interval $[0°, 180°]$ and in radians in the interval $[0, \pi]$.
3. TAN^{-1} gives values of θ in degrees in the interval $(-90°, 90°)$ and in radians in the interval $(-\pi/2, \pi/2)$.

Examples   Find the value of θ satisfying the given condition on the specified interval.

16. $\sin \theta = -0.8789$, θ in $[-90°, 90°]$
Using the $\boxed{\text{SIN}^{-1}}$ key in degree mode, we find that $\theta \approx -61.51°$.

17. $\tan \theta = -6.2051$, θ in $(-\pi/2, \pi/2)$
Using the $\boxed{\text{TAN}^{-1}}$ key in radian mode, we find that $\theta \approx -1.4110$.   ◄

DO EXERCISES 28–31.

To find function values in other intervals, we must consider the reference angle.

Example 18   Given that sin θ = 0.2812, find θ between 90° and 180°.

Solution   We first find the reference angle. We enter 0.2812 and press the [SIN⁻¹] key. The reference angle is approximately 16.33°.

We find the angle θ by subtracting 16.33° from 180°:

$$180° - 16.33° = 163.67°.$$

Thus, θ ≈ 163.67°.

Example 19   Given that tan θ = −6.2051, find θ between 270° and 360°.

Solution   We find the reference angle, ignoring the fact that tan θ is negative. We enter 6.2051 and press the [TAN⁻¹] key. The reference angle is approximately 80.85°.

We find the angle θ by subtracting 80.85° from 360°:

$$360° - 80.85° = 279.15°.$$

Thus, θ ≈ 279.15°.

DO EXERCISES 32–35.

Find θ in the interval indicated.

32. cos θ = 0.5712, (270°, 360°)

33. tan θ = −2.4778, (90°, 180°)

34. sin θ = −0.2363, (180°, 270°)

35. sec θ = −1.2020, (180°, 270°)

 EXERCISE SET 6.6

Find each of the following.

1. sec 315°
2. csc 315°
3. sin 150°
4. cos 150°
5. cot 570°
6. tan 570°
7. csc 270°
8. sin (−450°)
9. cot (−225°)
10. sec (−225°)
11. sin 1050°
12. cos 675°
13. tan (−135°)
14. sin (−135°)
15. sec 1125°
16. csc 1125°

17. sin 495°
18. cot 330°
19. cos 5220°
20. sin (−7560°)
21. $\sin \frac{13\pi}{4}$
22. $\cos \frac{19\pi}{4}$
23. $\tan \frac{11\pi}{3}$
24. $\sec \frac{11\pi}{3}$

Find the six trigonometric function values for each of the following.

25. $\frac{23\pi}{6}$
26. $-\frac{29\pi}{3}$
27. $-\frac{19\pi}{3}$

28. $\dfrac{37\pi}{6}$ **29.** $\dfrac{31\pi}{3}$ **30.** $-\dfrac{25\pi}{6}$

31. $-750°$ **32.** $1590°$

B   Use a calculator to find each of the following function values.

33. $\cos 18°$ **34.** $\sin 37°$

35. $\tan 2.6°$ **36.** $\cos 34.8°$

37. $\sin 62°20'$ **38.** $\tan 55°30'$

39. $\cos 15°35'$ **40.** $\sin 40°55'$

41. $\csc 29°11'36''$ **42.** $\cot 54°44'53''$

43. $\sec 10°31'42''$ **44.** $\csc 45°15'23''$

45. $\sin 561.2344°$ **46.** $\tan(-234.567°)$

47. $\cot(-900.23°)$ **48.** $\sec(0.2045°)$

49. $\tan 295°14'$ **50.** $\cos 230°53'$

51. $\sec 146.9°$ **52.** $\sin 98.4°$

53. $\sin 756°25'$ **54.** $\cot 820°40'$

55. $\cos(-1000.85°)$ **56.** $\tan(-1086.15°)$

57. $\sin 37$ **58.** $\cos 810$

59. $\cos(-10)$ **60.** $\sin(-4)$

61. $\tan 5\pi$ **62.** $\cot 28\pi$

63. $\cot 1000$ **64.** $\tan 13$

65. $\sec\left(-\dfrac{\pi}{5}\right)$ **66.** $\cot\left(-\dfrac{7\pi}{5}\right)$

67. $\sec\dfrac{10\pi}{7}$ **68.** $\cot\dfrac{2\pi}{7}$

69. $\cos(-13\pi)$ **70.** $\sin(-10\pi)$

71. $\tan 2.5$ **72.** $\cot 5.2$

Use a calculator to find the acute angle θ that is a solution of the equation. Give the answer in degrees and decimal parts of degrees, as well as in radians.

73. $\sin \theta = 0.5125$ **74.** $\cos \theta = 0.8241$

75. $\cos \theta = 0.6512$ **76.** $\sin \theta = 0.8220$

77. $\tan \theta = 7.425$ **78.** $\tan \theta = 3.163$

79. $\csc \theta = 6.277$ **80.** $\sec \theta = 1.175$

Find θ in the interval indicated.

81. $\sin \theta = -0.9956$, $(270°, 360°)$

82. $\sin \theta = 0.4313$, $(90°, 180°)$

83. $\cos \theta = -0.9388$, $(180°, 270°)$

84. $\cos \theta = -0.0990$, $(90°, 180°)$

85. $\tan \theta = 0.2460$, $(180°, 270°)$

86. $\tan \theta = -3.0541$, $(270°, 360°)$

87. $\sec \theta = -1.0485$, $(90°, 180°)$

88. $\csc \theta = 1.0480$, $(0°, 90°)$

89. $\sin \theta = 0.5766$, $(\pi/2, \pi)$

90. $\sin \theta = -0.8873$, $(3\pi/2, 2\pi)$

91. $\cos \theta = -0.211499$, $(\pi/2, \pi)$

92. $\cos \theta = -0.9388$, $(\pi, 3\pi/2)$

93. $\tan \theta = -19.0541$, $(3\pi/2, 2\pi)$

94. $\tan \theta = 32.246$, $(\pi, 3\pi/2)$

95. $\csc \theta = 11.21048$, $(0, \pi/2)$

96. $\sec \theta = -4.23785$, $(\pi/2, \pi)$

● **SYNTHESIS** _____

97. The valve cap on a bicycle wheel is 24.5 in. from the center of the wheel. From the position shown, the wheel starts rolling. After the wheel has turned 390°, how far above the ground is the valve cap? Assume that the outer radius of the tire is 26 in.

98. The seats of a ferris wheel are 35 ft from the center of the wheel. When you board the wheel, you are 5 ft above the ground. After you have rotated through an angle of 765°, how far above the ground are you?

99. How big must an angle be in order that it first differs from its tangent in the fourth decimal place?

100. How big must an angle be in order that it first differs from its tangent in the third decimal place?

101. The formula

$$\sin x = x - \frac{x^3}{6} + \frac{x^5}{120}$$

gives an approximation for sine values when x is in radians. Calculate $\sin 0.5$ and compare your answer with the value given by a calculator.

102. Using values from a calculator, graph the tangent function between $-90°$ and $90°$. What is the domain of the tangent function?

103. Use the fact that $\sin \theta \approx \theta$, in radians, when θ is small to calculate the diameter of the sun. The sun is 93 million miles from the earth and the angle it subtends at the earth's surface is about $31'59''$.

104. *Snell's law.*   When light passes from one substance to another, rays are bent, depending on the speed of light in those substances. For example, light traveling in air is bent as it enters water. Angle *i* is called the **angle of incidence,** and angle *r* is called the **angle of refraction.** Snell's law states that $\sin i / \sin r$ is constant. The constant is called the **index of refraction.** The index of refraction of a certain crystal is 1.52. Light strikes it, making an angle of incidence of $27°$. What is the angle of refraction?

● CHALLENGE _____

105. *Projectile motion.*   An object is thrown into the air at a velocity V, in meters per second, at an angle θ. After *t* seconds, the object is at a horizontal distance *x* and a vertical distance *y* from its release point, where

$$x = (V \cos \theta)t$$

and

$$y = -\tfrac{1}{2}gt^2 + (V \sin \theta)t + h,$$

and where h = the height at which the shot is released and g = the acceleration due to gravity (air resistance is neglected).

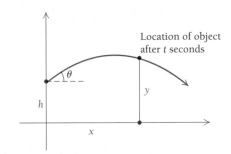

Location of object after *t* seconds

a) At what time will the object hit the ground?

b) At what horizontal distance will the object hit the ground?

c) In August 1975, Marianne Adam of East Germany put a shot 21.6 m. Assume that she was 1.6 m tall and that $\theta = 44°$ and $g = 9.8$ m/sec². What was the velocity of the shot when she released it?

d) All other things being equal, what advantage does a tall shotputter have over a shorter one?

e) When the units are given in feet and seconds, $g = 32$ ft/sec². A softball player hits a softball at a height of 4.5 ft and an angle of $20°$. The ball travels for a home run just over an 8-ft fence that is 300 ft from home plate. How long does it take the ball to get to the fence? What was its velocity when hit?

6.7
GRAPHS OF TRANSFORMED SINE AND COSINE FUNCTIONS

A, B **Variations of Basic Graphs**

In Sections 6.2 and 6.3, we graphed the six circular, or trigonometric, functions. It might be helpful to review those graphs and have their shapes well in mind because in this section, we will consider variations of these graphs. In particular, we are interested in graphs in the form

$$y = A \sin(Cx - D) + B$$

and

$$y = A \cos(Cx - D) + B,$$

1. Sketch a graph of $y = \cos x - 2$.

where A, B, C, and D are constants, some of which may be 0. These constants have the effect of translating, stretching, reflecting, or shrinking the basic graphs. (It might be helpful also to review Section 3.8.) Let us first examine the effect of each constant individually. Then we will consider the combined effect of more than one of the constants.

OPTIONAL

If you have a grapher, you can turn to the Technology Connections on pp. 424–425 to quickly observe the effect of the constants B, A, C, and D on the graphs of

$$y = A \sin (Cx - D) + B$$

and

$$y = A \cos (Cx - D) + B.$$

This discovery approach will enhance your understanding of Examples 1–8.

We first consider the effect of the constant B. (Also see the Technology Connection at the top of p. 424.)

Example 1 Sketch a graph of $y = \sin x + 3$.

Solution The graph of $y = \sin x + 3$ is a vertical translation of the graph of $y = \sin x$ up 3 units. One way to sketch the graph is to first consider $y = \sin x$ on an interval of length 2π, say, $[0, 2\pi]$. The zeros of the function and the maximum and minimum values can be considered key points. These are

$$(0, 0), \quad \left(\frac{\pi}{2}, 1\right), \quad (\pi, 0), \quad \left(\frac{3\pi}{2}, -1\right), \quad (2\pi, 0).$$

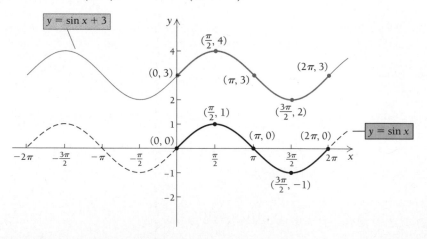

These key points are transformed 3 units up to obtain the key points of the graph of $y = \sin x + 3$. These are

$$(0, 3), \quad \left(\frac{\pi}{2}, 4\right), \quad (\pi, 3), \quad \left(\frac{3\pi}{2}, 2\right), \quad (2\pi, 3).$$

The graph of $y = \sin x + 3$ can be sketched over the interval $[0, 2\pi]$ and extended to obtain the rest of the graph by repeating the graph over intervals of length 2π. ◀

DO EXERCISE 1.

Next we consider the effect of the constant A. (Also see the Technology Connection at the bottom of p. 429).

Example 2 Sketch a graph of $y = 2 \sin x$. What is the amplitude?

Solution   The constant 2 in $y = 2 \sin x$ has the effect of stretching the graph of $y = \sin x$ vertically by a factor of 2 units. The function values of $y = \sin x$ are such that $-1 \le \sin x \le 1$, and $|\sin x| \le 1$. The function values of $y = 2 \sin x$ are such that $-2 \le 2 \sin x \le 2$, or $|2 \sin x| \le 2$. The maximum value of $y = 2 \sin x$ is 2, and the minimum value is -2. Thus the amplitude is $\frac{1}{2}[2 - (-2)]$, or 2.

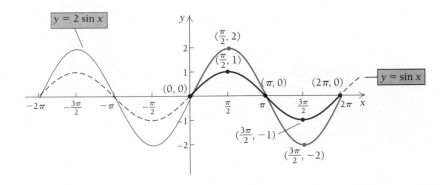

We draw the graph of $y = \sin x$ and consider its key points

$$(0, 0), \quad \left(\frac{\pi}{2}, 1\right), \quad (\pi, 0), \quad \left(\frac{3\pi}{2}, -1\right), \quad (2\pi, 0)$$

over the interval $[0, 2\pi]$.

We then multiply the second coordinates by 2 to obtain the key points of $y = 2 \sin x$. These are

$$(0, 0), \quad \left(\frac{\pi}{2}, 2\right), \quad (\pi, 0), \quad \left(\frac{3\pi}{2}, -2\right), \quad (2\pi, 0).$$

We plot these points and sketch the graph over the interval $[0, 2\pi]$. Then we repeat this part of the graph over other intervals of length 2π. ◀

DO EXERCISE 2.

If the absolute value of A is greater than 1, then there will be a vertical stretching. If the absolute value of A is less than 1, then there will be a vertical shrinking.

DEFINITION	**Amplitude**

The *amplitude* of $y = A \sin (Cx - D) + B$ is $|A|$.

---

2. Sketch a graph of $y = \frac{2}{3} \cos x$. What is the amplitude?

3. Sketch a graph of $y = -2 \cos x$. What is the amplitude?

If the constant A in $y = A \sin x$ is negative, there will also be a reflection across the x-axis.

Example 3 Sketch a graph of $y = -\frac{1}{2} \sin x$.

Solution The amplitude of the graph is $|-\frac{1}{2}|$, or $\frac{1}{2}$. The graph of $y = -\frac{1}{2} \sin x$ is a vertical shrinking and a reflection of the graph of $y = \sin x$. In graphing, the key points of $y = \sin x$,

$$(0, 0), \quad \left(\frac{\pi}{2}, 1\right), \quad (\pi, 0), \quad \left(\frac{3\pi}{2}, -1\right), \quad (2\pi, 0),$$

are transformed to

$$(0, 0), \quad \left(\frac{\pi}{2}, -\frac{1}{2}\right), \quad (\pi, 0), \quad \left(\frac{3\pi}{2}, \frac{1}{2}\right), \quad (2\pi, 0).$$

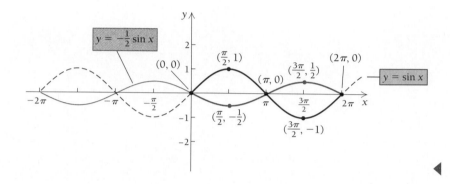

DO EXERCISE 3.

Now we consider the effect of the constant C in $y = A \sin (Cx - D) + B$. (Also see the Technology Connection at the top of p. 425.)

Example 4 Sketch a graph of $y = \sin 2x$. What is the period?

Solution Each of the graphs in Examples 1–3, as well as the graph of $y = \sin x$, has period 2π. The constant C has the effect of changing the period. Recall from Section 3.8 that the graph of $y = f(2x)$ is obtained from the graph of $y = f(x)$ by shrinking the graph horizontally. The new graph is obtained by dividing the first coordinate of each ordered-pair solution of $y = f(x)$ by 2. The key points of $y = \sin x$ are

$$(0, 0), \quad \left(\frac{\pi}{2}, 1\right), \quad (\pi, 0), \quad \left(\frac{3\pi}{2}, -1\right), \quad (2\pi, 0).$$

These are transformed to the key points of $y = \sin 2x$, which are

$$(0, 0), \quad \left(\frac{\pi}{4}, 1\right), \quad \left(\frac{\pi}{2}, 0\right), \quad \left(\frac{3\pi}{4}, -1\right), \quad (\pi, 0).$$

We plot these key points and sketch in the graph over the shortened interval $[0, \pi]$, which is of length π. Then we repeat the graph over other intervals of length π.

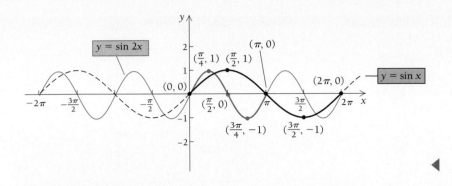

DEFINITION   **Period**

The *period* of the graph of $y = A \sin (Cx - D) + B$ is $\left| \dfrac{2\pi}{C} \right|$.

DO EXERCISES 4 AND 5.

Now we examine the effect of the constant D in $y = A \sin (Cx - D) + B$. (Also see the Technology Connection in the middle of p. 425.)

Example 5   Sketch a graph of $y = \sin \left(x - \dfrac{\pi}{2} \right)$.

Solution   The graph of $y = \sin (x - \pi/2)$ is a translation of the graph of $y = \sin x$ to the right $\pi/2$ units. The key points of $y = \sin x$,

$$(0, 0), \quad \left(\frac{\pi}{2}, 1 \right), \quad (\pi, 0), \quad \left(\frac{3\pi}{2}, -1 \right), \quad (2\pi, 0),$$

are transformed by adding $\pi/2$ to each of the first coordinates to obtain the following key points of $y = \sin (x - \pi/2)$:

$$\left(\frac{\pi}{2}, 0 \right), \quad (\pi, 1), \quad \left(\frac{3\pi}{2}, 0 \right), \quad (2\pi, -1), \quad \left(\frac{5\pi}{2}, 0 \right).$$

We plot these key points and sketch the curve over the interval $[\pi/2, 5\pi/2]$. Then we repeat the graph over other intervals of length 2π.

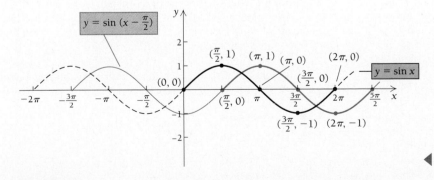

DO EXERCISE 6.

---

4. Sketch a graph of $y = \cos \frac{1}{2}x$. What is the period?

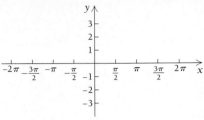

5. Sketch a graph of $y = \sin (-2x)$. What is the period?

6. Sketch a graph of

$$y = \cos \left(x + \frac{\pi}{2} \right).$$

TECHNOLOGY CONNECTION

Just as we did in Section 3.8, we can use a grapher to determine the effect of each of the constants in equations of the form

$$f(x) = A \sin (Cx - D) + B$$

or

$$f(x) = A \cos (Cx - D) + B.$$

The effect of each constant will be investigated first individually, and then in combination.

Again, as we did earlier, we will use simple *base functions* in our investigations. These base functions are shown here:

$f(x) = \sin x$

$[-2\pi, 2\pi] \times [-2, 2]$

$f(x) = \cos x$

$[-2\pi, 2\pi] \times [-2, 2]$

Start by graphing one of the base functions. Use radian mode and a $[-2\pi, 2\pi] \times [-5, 5]$ viewing box. Next, add 3 to the function, forming the new function $f(x) + 3$. (Thus if your base function is $\sin x$, your new function is $f(x) = \sin x + 3$.) Finally, subtract 2 from the base function, forming $f(x) - 2$.

Try repeating this process with the other base function. What is the rule that describes the difference between the graph of $f(x)$ and the graph of $f(x) + B$?

Predict what each of the following graphs will look like. Then check your prediction using a grapher.

TC 1. $f(x) = \sin x + 4$ **TC 2.** $f(x) = \cos x - 2$

TC 3. $f(x) = \cos x + 3.5$ **TC 4.** $f(x) = \sin x - 0.6$

TECHNOLOGY CONNECTION

Beginning with the same base function that you used in the last investigation, draw its graph on your grapher. Next, multiply the function by 3 and draw the graph of this new function, $3 \cdot f(x)$. How does it compare to the graph of $f(x)$?

Now replace the 3 with $\frac{1}{4}$, and draw the graph of $\frac{1}{4} \cdot f(x)$. How does it compare with the graph of $f(x)$? Repeat this three-part process with the other base function.

What rule predicts the shape of the graph that results from multiplying a function by a constant A that is greater than 1? What about the multiplication of a function by a constant A that is between 0 and 1?

Beginning with the graph of your original base function, try multiplying the function by the negative constant -3.

How does its graph compare with the graph of $f(x)$? What about multiplying by the negative constant $-\frac{1}{4}$?

After you have tried multiplying the other base function by negative constants, state a rule that predicts the result of multiplying a function by a negative constant A.

Predict the shape of each of the following graphs. Then use your grapher to check your answer.

TC 5. $f(x) = 5 \sin x$

TC 6. $f(x) = -5 \sin x$

TC 7. $f(x) = \frac{1}{2} \cos x$

TC 8. $f(x) = -\frac{1}{2} \cos x$

TECHNOLOGY CONNECTION

Beginning with the graph of the base function you've been using, graph the function that results when the variable x is multiplied by 3, forming $f(3x)$. (If your base function is $\sin x$, the new function will be $\sin 3x$.) Note the change in the graph. Now replace the 3 with $\frac{1}{2}$, forming $f(\frac{1}{2}x)$. How is its graph different from the graph of the base function? Now try these steps with the other base function.

Start with your base function again, but this time multiply by *negative* constants. Graph $f(-3x)$ and $f(-\frac{1}{2}x)$. What effect does this negative sign have?

What is the general rule that describes the effect of the constant C in a function $f(Cx)$? (Think of both positive *and* negative values of C.)

Predict the shape of each of the following graphs. Then use your grapher to check your prediction. Use radian mode, with a viewing box of $[-2\pi, 2\pi] \times [-2, 2]$.

TC 9. $f(x) = \sin 4x$
TC 10. $f(x) = \sin(-4x)$
TC 11. $f(x) = \cos(-\frac{1}{4}x)$
TC 12. $f(x) = \cos 2.5x$

TECHNOLOGY CONNECTION

Once again, start with the graph of your favorite base function and your grapher in radian mode. This time, subtract $\pi/2$ from x, forming $f(x - \pi/2)$. How does this graph compare with the graph of the base function? Replace the $\pi/2$ with $-\pi/3$, and draw the graph of $f(x + \pi/3)$.

If your function is in degrees, the values of the constants added to, or subtracted from, x will probably be significantly different. For example, instead of the $\pi/2$ and $-\pi/3$ used above, you would use $90°$ and $-60°$, respectively. Change your grapher to degree mode and then graph the same base function you used above. However, this time, the second function will be $f(x - 90°)$ and the third function will be $f(x + 60°)$. The resulting graphs should be identical to the graphs you found at the start of this investigation.

Try this with the other base function, in both radian and degree modes. Can you state a rule that predicts the effect of subtracting (or adding) a constant D from x in a trigonometric function? That is, how is the graph of $f(x - D)$ different from the graph of $f(x)$?

For each of the following functions, predict the shape of the graph. If the constant includes a π, you can assume that all angles are measured in radians; otherwise, assume that the angles are in degrees.

TC 13. $f(x) = \cos(x - \pi/4)$
TC 14. $f(x) = \cos(x + 135°)$
TC 15. $f(x) = \sin(x + 3\pi/2)$
TC 16. $f(x) = \sin(x - 45°)$

TECHNOLOGY CONNECTION

Now, we can put everything together, predicting the shape of functions such as

$$f(x) = A \sin(Cx - D) + B$$

or

$$f(x) = A \cos(Cx - D) + B.$$

In the last four investigations, we have examined the effect of each of the constants A through D individually. By breaking the function down step by step, you will be able to predict the shape of functions that include any combination of these constants. If you have a grapher available, you can check each step of the solution in Examples 6–8 and Margin Exercises 7–9. (You should be able to go through each of these steps *without* a grapher. Use the grapher only to check your work or to deal with cases in which the constants are "difficult" to work with.)

For each of the following functions, predict the shape of its graph by following the steps illustrated in Examples 6–8. Write a description of the effect of each transformation *before* you graph it and then use your grapher to check your prediction.

TC 17. $f(x) = 12.5 \cos(5x - 2.5\pi) - 3$
TC 18. $f(x) = -20 \sin(-3x - 2\pi) + 1.5$
TC 19. $f(x) = 0.25 \cos(4x + \pi/3) + 3.5$

Now we consider combined transformations of graphs. (Also see the Technology Connection at the bottom of p. 425.)

Example 6   Sketch a graph of $y = \sin(2x - \pi)$. What is the period?

Solution   The graph of

$$y = \sin(2x - \pi)$$

is the same as the graph of

$$y = \sin[2(x - \pi/2)].$$

The $\pi/2$ translates the graph of $y = \sin 2x$ to the right $\pi/2$ units. (See Example 5.) The 2 factored out of the parentheses shrinks the period by half, making the period $|2\pi/2|$, or π.

Thus to form this graph, we first graph $y = \sin 2x$, as in Example 4. The key points of $y = \sin 2x$ are

$$(0,0), \quad \left(\frac{\pi}{4}, 1\right), \quad \left(\frac{\pi}{2}, 0\right), \quad \left(\frac{3\pi}{4}, -1\right), \quad (\pi, 0).$$

We use them to obtain the key points of $y = \sin[2(x - \pi/2)]$, which are obtained, in a manner similar to Example 5, by adding $\pi/2$ to each first coordinate:

$$\left(\frac{\pi}{2}, 0\right), \quad \left(\frac{3\pi}{4}, 1\right), \quad (\pi, 0), \quad \left(\frac{5\pi}{4}, -1\right), \quad \left(\frac{3\pi}{2}, 0\right).$$

We draw one period of the graph over the interval $[\pi/2, 3\pi/2]$, which has length π. Then we repeat the graph over other intervals of length π.

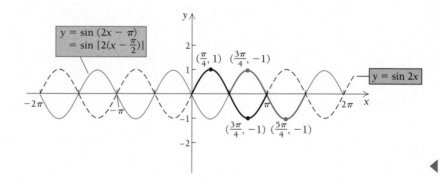

DEFINITION   **Phase Shift**

For the graph of

$$y = A\sin(Cx - D) + B = A\sin\left[C\left(x - \frac{D}{C}\right)\right] + B,$$

the quantity D/C, called the *phase shift*, translates the graph to the right if D/C is positive and to the left if D/C is negative. Be sure to do the horizontal stretching or shrinking based on the constant C *before* the translation based on the phase shift D/C.

In Example 6, the phase shift is $\pi/2$. This means that the graph of $y = \sin 2x$ has been shifted $\pi/2$ units to the right.

DO EXERCISE 7. _____

Let us summarize the effects of the constants. We carry out the procedures in the *order* listed.

> **(1)** Stretches the graph horizontally if $|C| < 1$. Shrinks the graph horizontally if $|C| > 1$. If $C < 0$, the graph is also reflected across the y-axis. The *period* is $|2\pi/C|$.

> **(2)** Stretches the graph vertically if $|A| > 1$. Shrinks the graph vertically if $|A| < 1$. If $A < 0$, the graph is also reflected across the x-axis. The *amplitude* of the graph is $|A|$.

$$y = A \sin (Cx - D) + B = A \sin \left[C\left(x - \frac{D}{C}\right)\right] + B$$

> **(3)** Translates the graph $|D/C|$ units to the right if $D/C > 0$ or $|D/C|$ units to the left if $D/C < 0$. The *phase shift* is D/C.

> **(4)** Translates the graph $|B|$ units up if $B > 0$ or $|B|$ units down if $B < 0$.

Similar statements hold for

$$y = A \cos (Cx - D) + B = A \cos \left[C\left(x - \frac{D}{C}\right)\right] + B.$$

Example 7 Sketch a graph of $y = 3 \sin (2x + \pi/2)$. Find the amplitude, the period, and the phase shift.

Solution We first note that

$$y = 3 \sin \left(2x + \frac{\pi}{2}\right) = 3 \sin \left[2\left(x - \left(-\frac{\pi}{4}\right)\right)\right].$$

7. Sketch a graph of

$$y = \cos (2x + \pi).$$

What is the period? the phase shift?

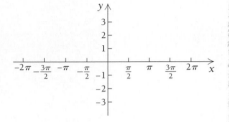

8. Sketch a graph of

$$y = 3 \cos\left(2x - \frac{\pi}{2}\right).$$

What is the amplitude? the period? the phase shift?

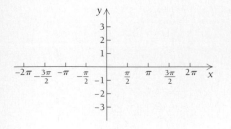

Then

Amplitude $= |A| = |3| = 3,$

Period $= \left|\dfrac{2\pi}{C}\right| = \left|\dfrac{2\pi}{2}\right| = \pi,$

Phase shift $= \dfrac{D}{C} = -\dfrac{\pi}{4}.$

To create the final graph, we sketch graphs of each of the following equations in sequence:

1. $y = \sin 2x,$

2. $y = 3 \sin 2x,$

3. $y = 3 \sin\left[2\left(x - \left(-\dfrac{\pi}{4}\right)\right)\right].$

Start with $y = \sin x.$

1. Then graph $y = \sin 2x.$ (See Example 4.)

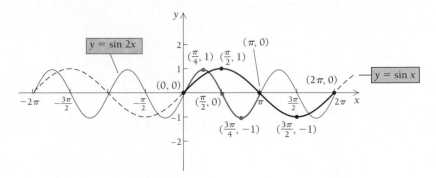

2. Next graph $y = 3 \sin 2x$ by stretching the graph vertically by a factor of 3.

3. Then graph $y = 3 \sin[2(x - (-\pi/4))]$ by translating the graph of $y = 3 \sin 2x$ to the left $\pi/4$ units.

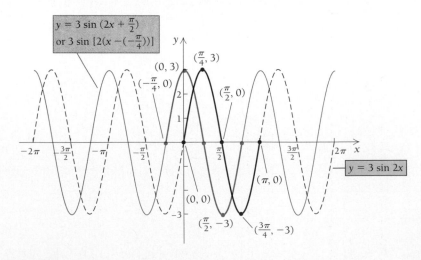

DO EXERCISE 8.

Example 8   Sketch a graph of $y = 3\cos(2\pi x) - 1$. Find the amplitude, the period, and the phase shift.

Solution   First we note the following:

$$\text{Amplitude} = |A| = |3| = 3,$$

$$\text{Period} = \left|\frac{2\pi}{C}\right| = \left|\frac{2\pi}{2\pi}\right| = |1| = 1,$$

$$\text{Phase shift} = \frac{D}{C} = \frac{0}{2\pi} = 0.$$

There is no phase shift in this problem since the constant $D = 0$. Thus there is no horizontal translation.

To create the final graph, we graph each of the following equations in sequence:

1. $y = \cos(2\pi x)$,
2. $y = 3\cos(2\pi x)$,
3. $y = 3\cos(2\pi x) - 1$.

Start with $y = \cos x$.

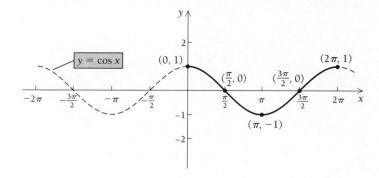

1. Then graph $y = \cos(2\pi x)$. The period is 1 and the graph is shrunk horizontally.

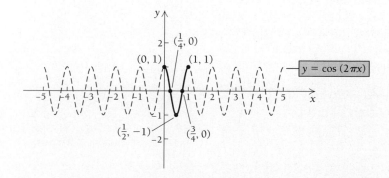

2. Next graph $y = 3\cos(2\pi x)$. The amplitude is 3. The graph is stretched vertically by a factor of 3.

9. Sketch a graph of

$$y = -2\sin(\pi x) + 2.$$

What is the amplitude? the period? the phase shift?

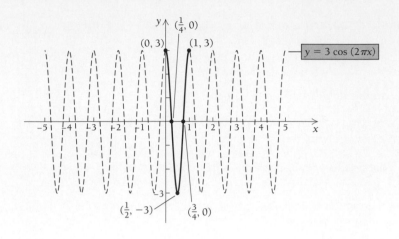

3. Then graph $y = 3\cos(2\pi x) - 1$. The graph is translated down 1 unit.

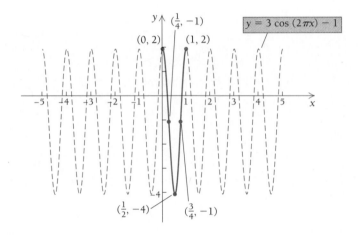

DO EXERCISE 9.

The oscilloscope shown here is an electronic device that draws graphs like those in the preceding examples. These graphs are often called **sinusoidal.** By manipulating the controls, we can change such things as the amplitude, the period, and the phase shift. The oscilloscope has many applications, and the trigonometric functions play a major role in many of them.

An oscilloscope display, showing a sinusoidal (sine-shaped) curve.

▶ Graphs of Sums: Addition of Ordinates

A function that is a sum of two functions can often be graphed by a method called **addition of ordinates.** We graph the two functions separately and then add the second coordinates (called **ordinates**) graphically. A compass may be helpful.

Example 9   Graph: $y = 2 \sin x + \sin 2x$.

Solution   We graph $y = 2 \sin x$ and $y = \sin 2x$ using the same set of axes.

Now we graphically add some ordinates to obtain points on the graph we seek. At $x = \pi/4$, we transfer the distance h, which is the value of $\sin 2x$, up to add it to the value of $2 \sin x$. Point P_1 is on the graph we seek. At $x = -\pi/4$, we do a similar thing, but the distance m is negative. Point P_2 is on our graph. At $x = -5\pi/4$, the distance n is negative, so we in effect subtract it from the value of $2 \sin x$. Point P_3 is on the graph we seek. We continue to plot points in this fashion and then connect them to get the desired graph, shown here.

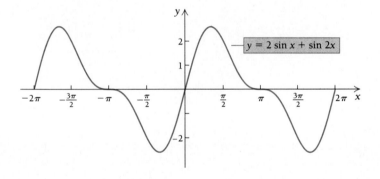

DO EXERCISE 10. _____

A "sawtooth" function, such as the one shown on this oscilloscope, has numerous applications—for example, in television circuits. This sawtooth function can be approximated extremely well by adding, electronically, several sine and cosine functions. The graph in Example 9 could be a first step in such an approximation.

An oscilloscope displaying a sawtooth function. The function has been synthesized using sines and cosines.

10. By addition of ordinates, sketch a graph of $y = \cos x + \sin 2x$.

TECHNOLOGY CONNECTION

A grapher can easily graph any functions shown in this section. It is especially useful when a function is made up of more than one trigonometric function. The only step is to select the appropriate viewing box.

Use a grapher to graph each of the following functions.

TC 20. $f(x) = 5 \cos x + \cos 5x$

TC 21. $f(x) = \cos 2x - \sin \frac{1}{2}x$

TC 22. $f(x) = 4 \cos x + \sin 3x$

● **EXERCISE SET** **6.7**

A , B   Use graph paper to sketch the graph of the function. Determine the amplitude, the period, and the phase shift.

1. $y = \sin x + 2$

2. $y = \cos x - 3$

3. $y = \dfrac{1}{2} \sin x$

4. $y = \dfrac{1}{3} \cos x$

5. $y = 2 \cos x$

6. $y = 3 \cos x$

7. $y = -\dfrac{1}{2} \cos x$

8. $y = -2 \sin x$

9. $y = \cos 2x$

10. $y = \cos 3x$

11. $y = \cos(-2x)$

12. $y = \cos(-3x)$

13. $y = \sin \dfrac{1}{2} x$

14. $y = \cos \dfrac{1}{3} x$

15. $y = \sin\left(-\dfrac{1}{2} x\right)$

16. $y = \cos\left(-\dfrac{1}{3} x\right)$

17. $y = \cos(2x - \pi)$

18. $y = \sin(2x + \pi)$

19. $y = 2 \cos\left(\dfrac{1}{2} x - \dfrac{\pi}{2}\right)$

20. $y = 4 \sin\left(\dfrac{1}{4} x + \dfrac{\pi}{8}\right)$

21. $y = -3 \cos(4x - \pi)$

22. $y = -3 \sin\left(2x + \dfrac{\pi}{2}\right)$

23. $y = 2 + 3 \cos(\pi x - 3)$

24. $y = 5 - 2 \cos\left(\dfrac{\pi}{2} x + \dfrac{\pi}{2}\right)$

B   Determine the amplitude, the period, and the phase shift.

25. $y = 3 \cos\left(3x - \dfrac{\pi}{2}\right)$

26. $y = 4 \sin\left(4x - \dfrac{\pi}{3}\right)$

27. $y = -5 \cos\left(4x + \dfrac{\pi}{3}\right)$

28. $y = -4 \sin\left(5x + \dfrac{\pi}{2}\right)$

29. $y = \dfrac{1}{2} \sin(2\pi x + \pi)$

30. $y = -\dfrac{1}{4} \cos(\pi x - 4)$

C   Graph.

31. $y = 2 \cos x + \cos 2x$

32. $y = 3 \cos x + \cos 3x$

33. $y = \sin x + \cos 2x$

34. $y = 2 \sin x + \cos 2x$

35. $y = \sin x - \cos x$

36. $y = 3 \cos x - \sin x$

37. $y = 3 \cos x + \sin 2x$

38. $y = 3 \sin x - \cos 2x$

● SYNTHESIS _____

39. *Temperature during an illness.* The temperature T of a patient during a 12-day illness is given by

$$T(t) = 101.6° + 3° \sin\left(\dfrac{\pi}{8} t\right).$$

a) Sketch a graph of the function over the interval $[0, 12]$.
b) What are the maximum and the minimum temperatures during the illness?

40. *Periodic sales.* A company in a northern climate has sales of skis as given by

$$S(t) = 10\left(1 - \cos\dfrac{\pi}{6} t\right),$$

where t = time, in months ($t = 0$ corresponds to July 1), and $S(t)$ is in thousands of dollars.

a) Sketch a graph of the function over a 12-month interval $[0, 12]$.
b) What is the period of the function?
c) What is the minimum amount of sales and when does it occur?
d) What is the maximum amount of sales and when does it occur?

41. *Satellite location.* A satellite circles the earth in such a way that it is y miles from the equator (north or south, height not considered) t minutes after its launch, where

$$y = 3000\left[\cos\dfrac{\pi}{45}(t - 10)\right].$$

a) Sketch a graph of the function.
b) What are the amplitude, the period, and the phase shift?

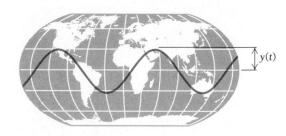

42. *Water wave.* The cross-section of a water wave is given by

$$y = 3 \sin\left(\dfrac{\pi}{4} x + \dfrac{\pi}{4}\right),$$

where y is the vertical height of the water wave and x is the distance from the origin to the wave.

a) Sketch a graph of the function.
b) What are the amplitude, the period, and the phase shift?

43. Find a sine function that has amplitude 8.6, period 5, and phase shift -11.

44. Find a cosine function that has amplitude 6.7, period π, and phase shift $\frac{4}{7}$.

45. Find a cosine function that has amplitude 16, period $\pi/3$, and phase shift $-2/\pi$.

46. Find a sine function that has amplitude 34, period $\pi/5$, and phase shift $-8/\pi$.

Graph.

47. $y = |4 \sin x|$

48. $y = |\cos 3x|$

● **CHALLENGE** _____

Graph using addition of ordinates. You might want to use a calculator to find some function values to plot.

49. $y = x + \sin x$ **50.** $y = \cos x - x$

⊡ **TECHNOLOGY CONNECTION** _____

Use a grapher to graph the function.

51. $y = \cos 2x + 2x$ **52.** $y = \cos 3x + \sin 3x$

53. $y = 4 \cos 2x - 2 \sin x$ **54.** $y = 7.5 \cos x + \sin 2x$

6.8
TRANSFORMED GRAPHS OF OTHER TRIGONOMETRIC FUNCTIONS

OBJECTIVES

You should be able to:

Ⓐ Sketch graphs that are transformations of the tangent, cotangent, secant, and cosecant functions.

Ⓑ Sketch graphs of functions found by multiplying trigonometric functions by other functions.

Ⓐ **Transformations of Tangent, Cotangent, Secant, and Cosecant**

The transformational techniques that we learned in Section 6.7 for graphing the sine and cosine can also be applied to the other trigonometric functions.

Example 1   Sketch a graph of $y = \dfrac{1}{2} \tan \left(x - \dfrac{\pi}{4} \right)$.

Solution   The graph of $y = \tan x$ is shown below. Recall that it has period π and is defined for all real numbers except $\pi/2 + k\pi$.

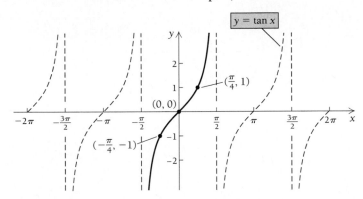

Let us consider the following ordered pairs as key points in the interval $(-\pi/2, \pi/2)$. Note that the function is not defined at the endpoints, so we indicate the following as key points,

$$\left(-\frac{\pi}{2}, \text{undefined} \right), \quad \left(-\frac{\pi}{4}, -1 \right), \quad (0, 0), \quad \left(\frac{\pi}{4}, 1 \right), \quad \left(\frac{\pi}{2}, \text{undefined} \right),$$

keeping in mind that "undefined" means that the first coordinate is not in the domain.

There is no amplitude since the function has no maximum or minimum value. Nevertheless, the constant $\frac{1}{2}$ shrinks each function value of $y = \tan x$ by a factor of $\frac{1}{2}$. Thus the graph of $y = \frac{1}{2} \tan x$ is obtained from the graph of $y = \tan x$ by a vertical shrinking. The key points are transformed by multiplying each second coordinate by $\frac{1}{2}$:

$$\left(-\frac{\pi}{2}, \text{undefined} \right), \quad \left(-\frac{\pi}{4}, -\frac{1}{2} \right), \quad (0, 0), \quad \left(\frac{\pi}{4}, \frac{1}{2} \right), \quad \left(\frac{\pi}{2}, \text{undefined} \right).$$

1. Sketch a graph of

$$y = 2 \cot\left(x + \frac{\pi}{4}\right).$$

We plot these key points and sketch the curve over the interval $(-\pi/2, \pi/2)$. We can use a calculator to find other function values, if desired. Then we repeat the graph over other intervals of length π.

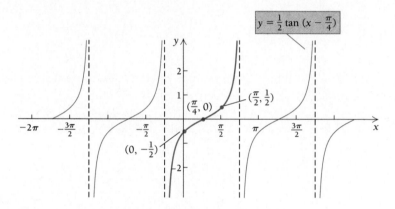

Next, we obtain the graph of $y = \frac{1}{2} \tan(x - \pi/4)$ by translating the graph of $y = \frac{1}{2} \tan x$ to the right $\pi/4$ units.

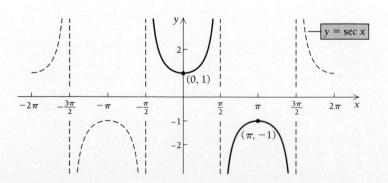

DO EXERCISE 1.

Example 2   Sketch a graph of $y = 3 \sec\left(x + \frac{\pi}{2}\right)$.

Solution   The graph of $y = \sec x$ is shown below. Recall that it has period 2π and is defined for all real numbers except $\pi/2 + k\pi$.

Let us consider the following ordered pairs as key points in the interval $(-\pi/2, 3\pi/2)$. Note that the function is not defined at the endpoints or at $\pi/2$, so we indicate the following as key points,

$$\left(-\frac{\pi}{2}, \text{undefined}\right), \quad (0, 1), \quad \left(\frac{\pi}{2}, \text{undefined}\right),$$

$$(\pi, -1), \quad \left(\frac{3\pi}{2}, \text{undefined}\right),$$

keeping in mind that "undefined" means that the first coordinate is not in the domain.

There is no amplitude since the function has no maximum or minimum value. Nevertheless, the constant 3 stretches each function value of $y = \sec x$ by a factor of 3. Thus the graph of $y = 3 \sec x$ is obtained from the graph of $y = \sec x$ by a vertical stretching. The key points are transformed by multiplying each second coordinate by 3:

$$\left(-\frac{\pi}{2}, \text{undefined}\right), \quad (0, 3), \quad \left(\frac{\pi}{2}, \text{undefined}\right),$$

$$(\pi, -3), \quad \left(\frac{3\pi}{2}, \text{undefined}\right).$$

 TECHNOLOGY CONNECTION

All the work that we did in the last section concerning the graphs of functions of the form

$$f(x) = A \sin (Cx - D) + B$$

or

$$f(x) = A \cos (Cx - D) + B$$

also applies to similar functions involving tan, sec, csc, and cot. In order to use what we learned, we need to review what the base functions look like.

The effects of the constants A, B, C, and D are exactly the same on these four base functions as they are on the functions $\sin x$ and $\cos x$. Therefore, you can use the same steps used in the analysis of $\sin x$ and $\cos x$ functions to predict the shapes of these graphs. Finally, you can use your grapher to verify your predictions or to handle functions that include difficult values.

Predict the shapes of each of the following functions, using the steps learned in Section 6.7. Then check your predictions using your grapher. (These functions are in radians.)

TC 1. $f(x) = 5 \tan (2x - 3\pi) + 6$

TC 2. $f(x) = -4 \csc (-x - \pi/2) + 3.5$

TC 3. $f(x) = 2.5 \sec (-3x + 2\pi) - 4.5$

TC 4. $f(x) = 0.4 \cot (8x + 2\pi) - 1.2$

$f(x) = \tan x$

$[-2\pi, 2\pi] \times [-3, 3]$

$f(x) = \csc x$

$[-2\pi, 2\pi] \times [-3, 3]$

$f(x) = \sec x$

$[-2\pi, 2\pi] \times [-3, 3]$

$f(x) = \cot x$

$[-2\pi, 2\pi] \times [-3, 3]$

2. Sketch a graph of

$$y = -\frac{2}{3} \csc \left(x - \frac{\pi}{2} \right).$$

We plot these key points and sketch the curve over the interval $(-\pi/2, 3\pi/2)$. We can use a calculator to find other function values, if desired. Then we repeat the graph over other intervals of length 2π.

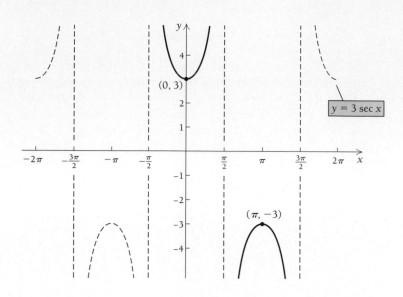

Next, we obtain the graph of $y = 3 \sec [x - (-\pi/2)]$ by translating the graph of $y = 3 \sec x$ to the left $\pi/2$ units.

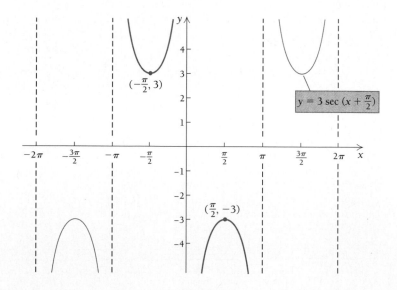

DO EXERCISE 2.

B ▶ Multiplication of Ordinates

Suppose that a weight is attached to a spring and the spring is stretched and put into motion. The weight oscillates up and down. If we could assume falsely that the weight will bob up and down forever, then its height h after time t in seconds might be approximated by a function like

$$h(t) = 5 + 2\sin(6\pi t).$$

Over a short time period, this might be a valid model, but experience tells us that eventually the spring will come to rest. A more appropriate model is provided by the following example, which illustrates **damped oscillation.**

Example 3   Sketch a graph of $f(x) = e^{-x/2}\sin x$.

Solution   The function f is the product of two functions g and h, where

$$g(x) = e^{-x/2} \quad \text{and} \quad h(x) = \sin x.$$

Thus, to find function values, we can **multiply ordinates.** Let us do more analysis before graphing. Note that for any real number x,

$$-1 \le \sin x \le 1.$$

Recall from Chapter 5 that all values of the exponential function are positive. Thus we can multiply by $e^{-x/2}$ and obtain the inequality

$$-e^{-x/2} \le e^{-x/2}\sin x \le e^{-x/2}.$$

The direction of the inequality symbols does not change since $e^{-x/2} > 0$. This also tells us that the original function crosses the x-axis only at values for which $\sin x = 0$. These are the numbers $k\pi$, for any integer k.

The inequality tells us that the function f is constrained between the graphs of $y = -e^{-x/2}$ and $y = e^{-x/2}$. We start by graphing these functions using dashed lines. Since we also know that $f(x) = 0$ when $x = k\pi$, k an integer, we mark these points on the graph. Then we use a calculator and compute other function values. The graph is as follows.

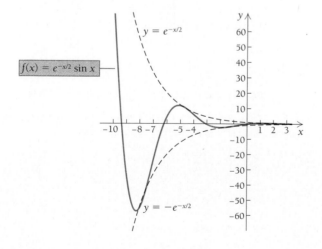

DO EXERCISE 3.

3. Sketch a graph of

$$f(x) = \frac{1}{2}x^2\sin x.$$

TECHNOLOGY CONNECTION

The following functions are very simple to graph using a grapher. In fact, the constants involved can be any real number—they are all equally easy. (You may need to determine how your grapher handles exponential functions of the form e^{ax}.)

Use a grapher to draw the graphs of each of these functions.

TC 5. $f(x) = e^{-0.6x}\cos x$

TC 6. $f(x) = 0.2x^2\sin x$

TC 7. $f(x) = 4^{-x}\cos x$

●EXERCISE SET 6.8

A   Sketch a graph of each of the following.

1. $y = -\tan x$ **2.** $y = -\cot x$

3. $y = -\csc x$ **4.** $y = -\sec x$

5. $y = \sec(-x)$ **6.** $y = \csc(-x)$

7. $y = \cot(-x)$ **8.** $y = \tan(-x)$

9. $y = -2 + \cot x$ **10.** $y = -3 + \sec x$

11. $y = -\dfrac{3}{2}\csc x$ **12.** $y = 0.25\cot x + 3$

13. $y = \cot 2x$ **14.** $y = 2\tan\dfrac{1}{2}x$

15. $y = \sec\left(-\dfrac{1}{2}x\right)$ **16.** $y = \csc\left(-\dfrac{1}{3}x\right)$

17. $y = 2\sec(x - \pi)$ **18.** $y = 3\csc(x + \pi)$

19. $y = 4\tan\left(\dfrac{1}{4}x + \dfrac{\pi}{8}\right)$ **20.** $y = 2\sec\left(\dfrac{1}{2}x - \dfrac{\pi}{2}\right)$

21. $y = -3\cot\left(2x + \dfrac{\pi}{2}\right)$ **22.** $y = -3\tan(4x - \pi)$

23. $y = 4\sec(2x - \pi)$ **24.** $y = 2\csc\left(\dfrac{1}{2}x - \dfrac{3\pi}{4}\right)$

B   Sketch a graph of each of the following.

25. $f(x) = e^{-x/2}\cos x$

26. $f(x) = e^{-0.4x}\sin x$

27. $f(x) = 0.6x^2\cos x$

28. $f(x) = e^{-x/4}\sin x$

29. $f(x) = x\sin x$

30. $f(x) = |x|\cos x$

31. $f(x) = 2^{-x}\sin x$

32. $f(x) = 2^{-x}\cos x$

●   **SYNTHESIS** _____

Sketch a graph of each of the following functions.

33. $f(x) = |\tan x|$

34. $f(x) = |\csc x|$

35. $f(x) = (\ln x)(\sin x)$

36. $f(x) = \sin^2 x$

37. $f(x) = \sec^2 x$

38. $f(x) = (\tan x)(\csc x)$

39. *Damped oscillations.* Suppose that the motion of a spring is given by
$$d(t) = 6e^{-0.8t}\cos(6\pi t) + 4,$$
where d is the distance, in inches, of a weight from the point at which the spring is attached to a ceiling, after t seconds.

a) Sketch the graph over the interval $[0, 10]$.

b) How far do you think the spring is from the ceiling when the spring stops bobbing?

40. *Temperature during an illness.* A patient's temperature during an illness is given by
$$T(t) = 98.6° + 2°e^{-0.6t}\sin t,$$
where T is the temperature after time t, in hours.

a) Sketch the graph over the interval $[0, 12]$.

b) What is the patient's normal temperature after the illness?

41. *Rotating beacon.* A police car is parked 10 ft from a wall. On top of the car is a beacon rotating in such a way that the light is at a distance $d(t)$ from point Q after t seconds, where
$$d(t) = 10\tan(2\pi t).$$

When d is positive, the light is pointing north of Q, and when d is negative, the light is pointing south of Q.

a) Sketch a graph of the function over the interval $[0, 2]$.

b) Explain the meaning of the values of t for which the function is undefined.

●   **TECHNOLOGY CONNECTION** _____

Use a grapher to graph each of the following over the given interval and approximate the zeros.

42. $f(x) = \dfrac{\sin x}{x}; \; [-12, 12]$

43. $f(x) = \dfrac{\sin^2 x}{x}; \; [-4, 4]$

44. $f(x) = \ln|\sec x + \tan x|; \; [-12, 12]$

45. $f(x) = \dfrac{\cos x - 1}{x}; \; [-12, 12]$

46. $f(x) = x^3\sin x; \; [-5, 5]$

47. $f(x) = 3\cos^2 x\sin x + 1; \; [-2, 2]$

SUMMARY AND REVIEW

● **REVIEW EXERCISES**

1. The point $(\frac{3}{5}, -\frac{4}{5})$ is on a unit circle. Find the coordinates of its reflections across the u-axis, the v-axis, and the origin.

2. On a unit circle, mark and label the points determined by $7\pi/6$, $-3\pi/4$, $-\pi/3$, and $9\pi/4$.

3. In which quadrants are the signs of the sine and the tangent the same?

4. Complete the following table. Give exact values. Do not use a calculator.

	$\pi/6$	$\pi/4$	$\pi/2$	$2\pi/3$	$-\pi$	$-5\pi/4$
$\sin x$						
$\cos x$						
$\cot x$						
$\csc x$						

For angles of the following measures, state in which quadrant the terminal side lies, convert to radian measure in terms of π, and convert to radian measure not in terms of π.

5. $87°$ **6.** $145°$ **7.** $-30°$

Convert to degree measure.

8. $\dfrac{3\pi}{2}$ **9.** 3

10. Find two positive angles and two negative angles coterminal with a $65°$ angle.

11. Find the length of an arc of a circle, given a central angle of $\pi/4$ and a radius of 7 cm.

12. An arc 18 m long on a circle of radius 8 m subtends an angle of how many radians? how many degrees, to the nearest degree?

13. A phonograph record revolves at 45 rpm. What is the linear velocity, in centimeters per minute, of a point 4 cm from the center?

14. An automobile wheel has a diameter of 26 in. If the car travels at a speed of 30 mph, what is the angular velocity, in radians per hour, of a point on the edge of the wheel?

15. Find the six trigonometric function values for the angle θ as shown.

16. Find the six trigonometric ratios for the specified angle.

Find the following if they exist. Do not use a calculator.

17. $\sin 495°$

18. $\tan(-315°)$

19. $\cot 210°$

20. $\cos 150°$

21. $\sec(-270°)$

22. $\csc 600°$

23. Given that $\tan\theta = 2/\sqrt{5}$ and the terminal side is in quadrant III, find the other five function values.

24. Convert $22.20°$ to degrees and minutes.

25. Convert $47°33'27''$ to degrees and decimal parts of degrees.

Use a calculator to find each of the following.

26. $\sin(-190°40')$

27. $\cos 0.1134$

28. $\sec 83°18'24''$

Use a calculator to find θ in degrees in the interval indicated.

29. $\tan\theta = 0.1228$, $(180°, 270°)$

30. $\cos\theta = -0.2672$, $(90°, 180°)$

Use a calculator to find the acute angle θ in radians that is a solution of the equation.

31. $\sin\theta = 0.2756$

32. $\tan\theta = 6.197$

Use a calculator to find the function value.

33. $\sin 24$

34. $\csc\left(-\dfrac{\pi}{5}\right)$

35. $\tan 5.2256$

36. $\cot 16\pi$

37. $\sec 14.3$

38. $\cos\left(-\dfrac{3\pi}{7}\right)$

39. Sketch a graph of $y = \sin x$.

40. Sketch a graph of $y = \cot x$.

41. Sketch a graph of $y = \csc x$.

42. What are the period and the range of the sine function?

43. What are the amplitude and the domain of the sine function?

44. What are the period and the range of the cosecant function?

45. Sketch a graph of $y = \sin\left(x + \dfrac{\pi}{2}\right)$.

46. Sketch a graph of $y = 3 + \cos\left(x - \dfrac{\pi}{4}\right)$.

47. What is the phase shift of the function in Exercise 46?

48. What is the period of the function in Exercise 46?

49. Sketch a graph of $y = 3\cos x + \sin x$ for values of x between 0 and 2π.

Sketch the graph.

50. $y = 2\tan\left(x - \dfrac{\pi}{2}\right)$

51. $f(x) = e^{-0.7x}\cos x$

● **SYNTHESIS** _____

52. Does $5\sin x = 7$ have a solution for x? Why or why not?

53. For what values of x in $(0, \pi/2]$ is $\sin x < x$ true?

54. Graph $y = 3\sin(x/2)$ and determine the domain, the range, and the period.

55. Find the domain of $y = \log(\cos x)$.

56. Given that $\sin x = 0.6144$ and the terminal side is in quadrant II, find the other basic circular function values.

● **THINKING AND WRITING** _____

1. Compare the notions of radian and degree.

2. Describe the shape of the graph of the cosine function. How many maximum values are there of the cosine function? Where do they occur?

3. A graphing calculator has many advantages. Explain a disadvantage of such a calculator when graphing a function like

$$f(x) = \frac{\sin x}{x}.$$

CHAPTER TEST

1. The point $\left(-\frac{4}{5}, -\frac{2}{5}\right)$ is on a unit circle. Find the coordinates of its reflection across the origin, the u-axis, and the v-axis.

2. On a unit circle, mark and label the points determined by $5\pi/6$, $-7\pi/2$, and $13\pi/4$.

3. Sketch a graph of $y = \cos x$.

4. What is the range of the cosine function?

5. What is the period of the cosine function?

6. What is the domain of the sine function?

7. What is the amplitude of the sine function?

8. Complete the following table.

	$\pi/2$	$-4\pi/2$	$7\pi/2$
$\sin x$			
$\cos x$			

9. Sketch a graph of the secant function.

10. What is the period of the secant function?

11. What is the domain of the secant function?

12. In which quadrants are the signs of the sine and the secant different?

For angles of the following measures, state in which quadrant the terminal side lies, convert to radian measure in terms of π, and convert to radian measure not in terms of π.

13. $-150°$ **14.** $117°$ **15.** $225°$

Convert to degree measure.

16. $\dfrac{7\pi}{4}$ radians   **17.** 2 radians

18. Find the length of an arc of a circle, given a central angle of $\pi/3$ and a radius of 10 cm.

19. An arc 8 in. long on a circle of radius 16 in. subtends an angle of how many radians? how many degrees, to the nearest degree?

20. A rock on a 4-ft string is rotated at 80 rpm. What is its linear speed, in feet per minute?

21. A pulley belt runs, uncrossed, around two pulleys of radii 10 cm and 4 cm, respectively. A point on the belt travels at a rate of 50 m/sec. Find the angular speed, in radians per second, of the smaller pulley.

22. Find the six trigonometric function values for the angle θ shown.

Find the following exactly. Do *not* use a calculator.

23. $\tan 495°$ **24.** $\sin(-150°)$

25. $\cos 300°$

26. Given that $\sin \theta = \frac{1}{4}$ and the terminal side is in quadrant II, find the other five function values.

27. Convert $19.25°$ to degrees and minutes.

28. Convert $72°48'$ to degrees and decimal parts of degrees.

Use a calculator to find each of the following.

29. $\cos(-213°48'13'')$ **30.** $\tan 1.228$

31. $\sin(-34\pi/19)$ **32.** $\sec 38°17'$

33. $\csc 3.4128$ **34.** $\cot(3472°)$

Use a calculator to find the acute angle θ that is a solution of the equation. Give the answer in degrees and decimal parts of degrees, as well as in radians.

35. $\sin \theta = 0.8714$ **36.** $\cot \theta = 8$

Use a calculator to find θ in the interval indicated.

37. $\sin \theta = 0.4305$, $(90°, 180°)$

38. $\tan \theta = -11.412$, $(-\pi/2, \pi/2)$

39. Sketch a graph of $y = 2\cos(x - \pi/4)$.

40. What is the period of the function in Question 39?

41. What is the phase shift of the function in Question 39?

42. Sketch the graph of $y = 2\cos x - \sin x$ for values between 0 and 2π.

● **SYNTHESIS** _____

43. Given that $\cos(\pi/32) = 0.99518$, find each of the following.

   a) $\sin(15\pi/32)$ **b)** $\sin(17\pi/32)$

44. What is the angle between the hands of a clock at 9:25?

There are a number of relationships among the trigonometric functions, given by identities, that are important in algebraic and trigonometric manipulation. A large part of this chapter is devoted to those identities. ● Among the manipulations in which the identities are used is the solving of *trigonometric equations*. ● We have not yet considered the inverses of the trigonometric functions beyond the use of calculators in reverse. We now look at those inverses. The chapter is an important one, both for problem solving and as a foundation for further work in mathematics. ●

Trigonometric Identities, Inverse Functions, and Equations

7.1

TRIGONOMETRIC MANIPULATIONS AND IDENTITIES

Trigonometric expressions such as $\sin 2x$ or $\tan(x - \pi)$ represent real numbers, in the same way that purely algebraic expressions do. Thus the laws of real numbers hold whether we are working with algebraic expressions or trigonometric expressions, or combinations thereof. We can factor, simplify, and manipulate trigonometric expressions in the same way that we manipulate strictly algebraic expressions.

The laws of real numbers can be combined with the fundamental identities considered in Chapter 6 to further enhance our ability to manipulate trigonometric expressions. The following is a list of these fundamental identities.

OBJECTIVES

You should be able to:

A Multiply and factor expressions containing trigonometric expressions.

B Simplify and manipulate expressions containing trigonometric expressions.

C Simplify and manipulate radical expressions containing trigonometric expressions, including rationalizing numerators or denominators.

1. Multiply and simplify:

$$\sin x \,(\cot x + \csc x).$$

Fundamental Identities

$$\sin x = \frac{1}{\csc x} \qquad \csc x = \frac{1}{\sin x} \qquad \tan x = \frac{\sin x}{\cos x}$$

$$\cos x = \frac{1}{\sec x} \qquad \sec x = \frac{1}{\cos x} \qquad \cot x = \frac{\cos x}{\sin x}$$

$$\tan x = \frac{1}{\cot x} \qquad \cot x = \frac{1}{\tan x}$$

$$\sin^2 x + \cos^2 x = 1 \qquad 1 + \tan^2 x = \sec^2 x \qquad 1 + \cot^2 x = \csc^2 x$$

$$\sin(-x) = -\sin x \qquad \csc(-x) = -\csc x$$

$$\cos(-x) = \cos x \qquad \sec(-x) = \sec x$$

$$\tan(-x) = -\tan x \qquad \cot(-x) = -\cot x$$

With these identities in mind, let us now consider an example.

◢ Multiplying and Factoring

Example 1　Multiply and simplify: $\cos y \,(\tan y - \sec y)$.

Solution

$$\cos y \,(\tan y - \sec y)$$

$$= \cos y \tan y - \cos y \sec y \qquad \text{Multiplying}$$

$$= \cos y \,\frac{\sin y}{\cos y} - \cos y \,\frac{1}{\cos y} \qquad \text{Recalling the identities } \tan x = \frac{\sin x}{\cos x} \text{ and}$$

$$\sec x = \frac{1}{\cos x} \text{ and substituting}$$

$$= \sin y - 1 \qquad \text{Simplifying} \qquad ◀$$

Recall that an **identity** is an equation that is true for all meaningful replacements of the variables. Thus we have proved in Example 1 that

$$\cos y \,(\tan y - \sec y) = \sin y - 1$$

is an identity. There is no general rule for doing a simplification that creates an identity as in Example 1 and no specific result, but it is often helpful to put everything in terms of sines and cosines. For the most part, the main goal of this section is to develop your skill by the practice you receive in manipulating trigonometric expressions. This kind of simplifying is very important for the rest of the trigonometry that you will study in this chapter and in other courses that you will take in mathematics and science.

2. Factor and simplify:

$$\sin^3 x + \sin x \cos^2 x.$$

Example 2　Factor and simplify: $\sin^2 x \cos^2 x + \cos^4 x$.

Solution

$$\sin^2 x \cos^2 x + \cos^4 x = \cos^2 x \,(\sin^2 x + \cos^2 x) \qquad \text{Factoring}$$

$$= \cos^2 x \qquad \text{Using } \sin^2 x + \cos^2 x = 1 \qquad ◀$$

DO EXERCISES 1 AND 2.

▶ Simplifying

Example 3 Simplify:

$$\frac{\cot{(-\theta)}}{\csc{(-\theta)}}.$$

Solution

$$\frac{\cot{(-\theta)}}{\csc{(-\theta)}} = \frac{\dfrac{\cos{(-\theta)}}{\sin{(-\theta)}}}{\dfrac{1}{\sin{(-\theta)}}}$$

$$= \frac{\cos{(-\theta)}}{\sin{(-\theta)}} \cdot \sin{(-\theta)}$$

$$= \cos{(-\theta)}$$

$$= \cos\theta$$ ◀

Example 4 Simplify:

$$\frac{\sin x - \sin x \cos x}{\sin x + \sin x \tan x}.$$

Solution

$$\frac{\sin x - \sin x \cos x}{\sin x + \sin x \tan x} = \frac{\sin x\,(1 - \cos x)}{\sin x\,(1 + \tan x)} \qquad \text{Factoring}$$

$$= \frac{1 - \cos x}{1 + \tan x} \qquad \text{Simplifying}$$ ◀

Example 5 Subtract and simplify:

$$\frac{2}{\sin x - \cos x} - \frac{3}{\sin x + \cos x}.$$

Solution

$$\frac{2}{\sin x - \cos x} - \frac{3}{\sin x + \cos x}$$

$$= \frac{2}{\sin x - \cos x} \cdot \frac{\sin x + \cos x}{\sin x + \cos x} - \frac{3}{\sin x + \cos x} \cdot \frac{\sin x - \cos x}{\sin x - \cos x}$$

$$= \frac{2(\sin x + \cos x) - 3(\sin x - \cos x)}{\sin^2 x - \cos^2 x}$$

$$= \frac{-\sin x + 5 \cos x}{\sin^2 x - \cos^2 x}$$ ◀

TECHNOLOGY CONNECTION

A grapher can be used to provide a partial check of any trigonometric identity. First, graph the expression on the left side of the equals sign, using an appropriate domain. A viewing box of $[-2\pi, 2\pi] \times [-5, 5]$ will usually give satisfactory results, so long as the grapher is in RADIAN mode. Then graph the expression on the right side using the same axes. If the two graphs are indistinguishable, then you have a partial verification that the equation is an identity. Of course, you can never see the entire graph, so there can always be some doubt. If the two graphs are different, then the equation is not an identity.

It may be necessary to make some slight modifications in the equation, due to the limitations of some graphers. If the equation contains $\sec x$, $\csc x$, or $\cot x$, you may need to replace these terms with $1/\cos x$, $1/\sin x$, or $1/\tan x$, respectively.

For example, consider the identity in Example 1:

$$\cos y\,(\tan y - \sec y) = \sin y - 1.$$

Since this equation contains $\sec y$, it will usually be necessary to replace that term with $1/\cos y$ before graphing it. Thus the new equation is

$$\cos y\left(\tan y - \frac{1}{\cos y}\right) = \sin y - 1.$$

Now we graph each side. Using the viewing box of $[-2\pi, 2\pi] \times [-5, 5]$ that is suggested above, we first draw the graph of the expression on the left side. Of course, we first change the y's to x's, since most graphers require the form $y = f(x)$. After we enter

$$y = \cos x\left(\tan x - \frac{1}{\cos x}\right),$$

we graph it, with the result shown here:

$$y = \cos x\left(\tan x - \frac{1}{\cos x}\right)$$

$[-2\pi, 2\pi] \times [-5, 5]$

Then we graph the right side (after changing it to $y = \sin x - 1$). The result is a graph identical to the one shown above. When the two graphs are drawn on the same axes, it is impossible to distinguish between them.

To illustrate a case in which an equation is shown not to be an identity, we can use the equation

$$\sin^2 x \cos^2 x - \cos^4 x = \cos^2 x.$$

Since all the terms are either $\cos x$ or $\sin x$, no modifications are needed before graphing. The graph of the expression on the left side, $y = \sin^2 x \cos^2 x - \cos^4 x$, is shown here:

$$y = \sin^2 x \cos^2 x - \cos^4 x$$

$[-2\pi, 2\pi] \times [-5, 5]$

The expression on the right side, $y = \cos^2 x$, produces this graph:

$$y = \cos^2 x$$

$[-2\pi, 2\pi] \times [-5, 5]$

Clearly, the graphs are different. When graphed on the same set of axes, it is easy to tell them apart. Thus the expression $\sin^2 x \cos^2 x - \cos^4 x = \cos^2 x$ is *not* an identity.

The main limitation to this technique is that it can give only a partial check whether or not a given equation is an identity.

TC 1. Use a grapher to determine which of the following are identities.

A. $\sin x \cot x = \cos x$

B. $\dfrac{\sin^2 x}{\cos x} + \cos x = \sec x$

C. $\dfrac{1 - \sin x}{\cos x} = \dfrac{\cos x}{1 + \sin x}$

D. $(\sec x - \cos x) \csc^2 x = \cot x$

TC 2. Use a grapher to determine which of the following expressions can complete the identity

$$\cos x\,(1 + \sec x) = \dots.$$

A. $1 + \cos x$

B. $\dfrac{1 + \cos x - \sin^2 x}{\cos x}$

C. $\cos x\,(\sec x + 1)$

D. $\sin x - \csc x$

E. $1 + \sin x \cot x$

Example 6 Add and simplify: $\dfrac{2}{\tan^3 x - 2\tan^2 x} + \dfrac{2}{\tan x - 2}$.

Solution

$$\dfrac{2}{\tan^3 x - 2\tan^2 x} + \dfrac{2}{\tan x - 2}$$

$$= \dfrac{2}{\tan^2 x(\tan x - 2)} + \dfrac{2}{\tan x - 2}\cdot\dfrac{\tan^2 x}{\tan^2 x}$$

$$= \dfrac{2 + 2\tan^2 x}{\tan^2 x(\tan x - 2)} = \dfrac{2(1 + \tan^2 x)}{\tan^2 x(\tan x - 2)}$$

$$= \dfrac{2\sec^2 x}{\tan^2 x(\tan x - 2)} \qquad \text{Recalling that } 1 + \tan^2 x = \sec^2 x$$

$$= \dfrac{2\cdot\dfrac{1}{\cos^2 x}}{\dfrac{\sin^2 x}{\cos^2 x}(\tan x - 2)}$$

$$= \dfrac{2}{\sin^2 x(\tan x - 2)} \qquad \blacktriangleleft$$

DO EXERCISES 3–6.

 ## Radical Expressions

When radicals occur, the use of absolute value is sometimes necessary, but it can be difficult to determine when to use it. In the examples and exercises that follow, we will assume that all radicands are nonnegative.

Example 7 Multiply and simplify: $\sqrt{\sin^3 x \cos x}\cdot\sqrt{\cos x}$.

Solution

$$\sqrt{\sin^3 x \cos x}\cdot\sqrt{\cos x} = \sqrt{\sin^3 x \cos^2 x}$$

$$= \sqrt{\sin^2 x \cos^2 x \sin x}$$

$$= \sin x \cos x\sqrt{\sin x} \qquad \blacktriangleleft$$

Example 8 Rationalize the denominator: $\sqrt{\dfrac{2}{\tan x}}$.

Solution

$$\sqrt{\dfrac{2}{\tan x}} = \sqrt{\dfrac{2}{\tan x}\cdot\dfrac{\tan x}{\tan x}} = \sqrt{\dfrac{2\tan x}{\tan^2 x}} = \dfrac{\sqrt{2\tan x}}{\tan x} \qquad \blacktriangleleft$$

DO EXERCISES 7 AND 8.

Often in calculus, a substitution such as the following is a useful manipulation.

Simplify.

3. $\dfrac{\tan(-\theta)}{\sin(-\theta)}$

4. $\dfrac{\cos x + \sin x \cos x}{\cos x - \cos x \cot x}$

Add and simplify.

5. $\dfrac{2}{\sin x - \cos x} + \dfrac{3}{\sin x + \cos x}$

6. $\dfrac{2}{\cot^3 x - 2\cot^2 x} + \dfrac{2}{\cot x - 2}$

7. Multiply and simplify:
$$\sqrt{\tan x \sin^2 x}\cdot\sqrt{\tan x \sin x}.$$

8. Rationalize the numerator:
$$\sqrt{\dfrac{\cos x}{3}}.$$

9. Use the substitution $x = 2 \sin \theta$ to express $\sqrt{4 - x^2}$ as a trigonometric function without radicals. Assume that $-\pi/2 \leq \theta \leq \pi/2$. Then find $\tan \theta$ and $\cos \theta$.

Example 9 Use the substitution $x = a \tan \theta$ to express $\sqrt{a^2 + x^2}$ as a trigonometric function without radicals. Assume that $a > 0$ and that $-\pi/2 < \theta < \pi/2$. Then find $\sin \theta$ and $\cos \theta$.

Solution Substituting $a \tan \theta$ for x, we get

$$\sqrt{a^2 + x^2} = \sqrt{a^2 + (a \tan \theta)^2} = \sqrt{a^2 + a^2 \tan^2 \theta}$$
$$= \sqrt{a^2(1 + \tan^2 \theta)} = \sqrt{a^2 \sec^2 \theta}$$
$$= \sqrt{a^2} \sqrt{\sec^2 \theta} = |a| \cdot |\sec \theta| = a \sec \theta.$$

The latter equality follows from the fact that $a > 0$ and that $\sec \theta > 0$ for values of θ for which $-\pi/2 < \theta < \pi/2$.

We can express the preceding result as

$$\sec \theta = \frac{\sqrt{a^2 + x^2}}{a}.$$

In terms of right triangles, we know that $\sec \theta$ is hypotenuse/adjacent, where θ is one of the acute angles. Now interpreting the hypotenuse to be $\sqrt{a^2 + x^2}$ and the adjacent side to be a, we can find the length of the side opposite θ to be x using the Pythagorean theorem.

Then from the right triangle we see that

$$\sin \theta = \frac{x}{\sqrt{a^2 + x^2}} \quad \text{and} \quad \cos \theta = \frac{a}{\sqrt{a^2 + x^2}}.$$

DO EXERCISE 9.

● EXERCISE SET 7.1

 Multiply and simplify.

1. $(\sin x - \cos x)(\sin x + \cos x)$

2. $(\tan y - \cot y)(\tan y + \cot y)$

3. $\tan x (\cos x - \csc x)$

4. $\cot x (\sin x + \sec x)$

5. $\cos y \sin y (\sec y + \csc y)$

6. $\tan y \sin y (\cot y - \csc y)$

7. $(\sin x + \cos x)(\csc x - \sec x)$

8. $(\sin x + \cos x)(\sec x + \csc x)$

9. $(\sin y - \cos y)^2$

10. $(\sin y + \cos y)^2$

11. $(1 + \tan x)^2$

12. $(1 + \cot x)^2$

13. $(\sin y - \csc y)^2$

14. $(\cos y + \sec y)^2$

15. $(\cos x - \sec x)(\cos^2 x + \sec^2 x + 1)$

16. $(\sin x + \csc x)(\sin^2 x + \csc^2 x - 1)$

17. $(\cot x - \tan x)(\cot^2 x + 1 + \tan^2 x)$

18. $(\cot y + \tan y)(\cot^2 y - 1 + \tan^2 y)$

19. $(1 - \sin x)(1 + \sin x)$

20. $(1 + \cos x)(1 - \cos x)$

Factor and simplify.

21. $\sin x \cos x + \cos^2 x$

22. $\sec x \csc x - \csc^2 x$

23. $\sin^2 \theta - \cos^2 \theta$

24. $\tan^2 \theta - \cot^2 \theta$

25. $\tan x + \sin (\pi - x)$

26. $\cot x - \cos (\pi - x)$

27. $\sin^4 x - \cos^4 x$

28. $\tan^4 x - \sec^4 x$

29. $3 \cot^2 y + 6 \cot y + 3$

30. $4 \sin^2 y + 8 \sin y + 4$

31. $\csc^4 x + 4 \csc^2 x - 5$

32. $-8 + \tan^4 x - 2 \tan^2 x$

33. $\sin^3 y + 27$

34. $1 - 125 \tan^3 y$

35. $\sin^3 v - \csc^3 v$

36. $\cos^3 u - \sec^3 u$

B   Simplify. If you have a grapher, verify your results.

37. $\dfrac{\sin^2 x \cos x}{\cos^2 x \sin x}$

38. $\dfrac{\cos^2 x \sin x}{\sin^2 x \cos x}$

39. $\dfrac{4 \sin x \cos^3 x}{18 \sin^2 x \cos x}$

40. $\dfrac{30 \sin^3 x \cos x}{6 \cos^2 x \sin x}$

41. $\dfrac{\cos^2 x - 2 \cos x + 1}{\cos x - 1}$

42. $\dfrac{\sin^2 x + 2 \sin x + 1}{\sin x + 1}$

43. $\dfrac{\cos^2 \alpha - 1}{\cos \alpha - 1}$

44. $\dfrac{\sin^2 \alpha - 1}{\sin \alpha + 1}$

45. $\dfrac{4 \tan x \sec x + 2 \sec x}{6 \sin x \sec x + 2 \sec x}$

46. $\dfrac{6 \tan x \sin x - 3 \sin x}{9 \sin^2 x + 3 \sin x}$

47. $\dfrac{\csc(-\theta)}{\cot(-\theta)}$

48. $\tan(-\beta) + \cot(-\beta)$

49. $\dfrac{\sin^4 x - \cos^4 x}{\sin^2 x - \cos^2 x}$

50. $\dfrac{\sec^4 x - \tan^4 x}{\sec^2 x + \tan^2 x}$

51. $\dfrac{2 \sin^2 x}{\cos^3 x} \cdot \left(\dfrac{\cos x}{2 \sin x}\right)^2$

52. $\dfrac{4 \cos^3 x}{\sin^2 x} \cdot \left(\dfrac{\sin x}{4 \cos x}\right)^2$

53. $\dfrac{3 \sin x}{\cos^2 x} \cdot \dfrac{\cos^2 x + \cos x \sin x}{\cos^2 x - \sin^2 x}$

54. $\dfrac{5 \cos x}{\sin^2 x} \cdot \dfrac{\sin^2 x - \sin x \cos x}{\sin^2 x - \cos^2 x}$

55. $\dfrac{\tan^2 \gamma}{\sec \gamma} \div \dfrac{3 \tan^3 \gamma}{\sec \gamma}$

56. $\dfrac{\cot^3 \phi}{\csc \phi} \div \dfrac{4 \cot^2 \phi}{\csc \phi}$

57. $\dfrac{1}{\sin^2 y - \cot^2 y} - \dfrac{2}{\cos y + \sin y}$

58. $\dfrac{3}{\cos y - \sin y} - \dfrac{2}{\sin^2 y - \cos^2 y}$

59. $\left(\dfrac{\sin x}{\cos x}\right)^2 - \dfrac{1}{\cos^2 x}$

60. $\left(\dfrac{\cot x}{\csc x}\right)^2 + \dfrac{1}{\csc^2 x}$

61. $\dfrac{\sin^2 x - 9}{2 \cos x + 1} \cdot \dfrac{10 \cos x + 5}{3 \sin x + 9}$

62. $\dfrac{9 \cos^2 x - 25}{2 \cos x - 2} \cdot \dfrac{\cos^2 x - 1}{6 \cos x - 10}$

C   Simplify. Assume that expressions in radicands are non-negative.

63. $\sqrt{\sin^2 x \cos x} \cdot \sqrt{\cos x}$

64. $\sqrt{\cos^2 x \sin x} \cdot \sqrt{\sin x}$

65. $\sqrt{\sin^3 y} + \sqrt{\sin y \cos^2 y}$

66. $\sqrt{\cos y \sin^2 y} - \sqrt{\cos^3 y}$

67. $\sqrt{\sin^2 x + 2 \cos x \sin x + \cos^2 x}$

68. $\sqrt{\tan^2 x - 2 \tan x \sin x + \sin^2 x}$

69. $(1 - \sqrt{\sin y})(\sqrt{\sin y} + 1)$

70. $(2 - \sqrt{\tan y})(\sqrt{\tan y} + 2)$

71. $\sqrt{\sin x}(\sqrt{2 \sin x} + \sqrt{\sin x \cos x})$

72. $\sqrt{\cos x}(\sqrt{2 \cos x} + \sqrt{\sin x \cos x})$

Rationalize the denominator.

73. $\sqrt{\dfrac{\sin x}{\cos x}}$

74. $\sqrt{\dfrac{\cos x}{\sin x}}$

75. $\sqrt{\dfrac{\sin x}{\cot x}}$

76. $\sqrt{\dfrac{\cos x}{\tan x}}$

77. $\sqrt{\dfrac{\cos^2 x}{2 \sin^2 x}}$

78. $\sqrt{\dfrac{\sin^2 x}{3 \cos^2 x}}$

79. $\sqrt{\dfrac{1 + \sin x}{1 - \sin x}}$

80. $\sqrt{\dfrac{1 - \cos x}{1 + \cos x}}$

Rationalize the numerator.

81. $\sqrt{\dfrac{\sin x}{\cos x}}$

82. $\sqrt{\dfrac{\cos x}{\sin x}}$

83. $\sqrt{\dfrac{\sin x}{\cot x}}$

84. $\sqrt{\dfrac{\cos x}{\tan x}}$

85. $\sqrt{\dfrac{\cos^2 x}{2 \sin^2 x}}$

86. $\sqrt{\dfrac{\sin^2 x}{3 \cos^2 x}}$

87. $\sqrt{\dfrac{1 + \sin x}{1 - \sin x}}$

88. $\sqrt{\dfrac{1 - \cos x}{1 + \cos x}}$

Use the given substitution to express the given radical expression as a trigonometric function without radicals. Assume that $a > 0$ and $0 < \theta < \pi/2$. Then find expressions for the indicated trigonometric functions.

89. Let $x = a \sin \theta$ in $\sqrt{a^2 - x^2}$. Then find $\cos \theta$ and $\tan \theta$.

90. Let $x = 2 \tan \theta$ in $\sqrt{4 + x^2}$. Then find $\sin \theta$ and $\cos \theta$.

91. Let $x = 3 \sec \theta$ in $\sqrt{x^2 - 9}$. Then find $\sin \theta$ and $\cos \theta$.

92. Let $x = a \sec \theta$ in $\sqrt{x^2 - a^2}$. Then find $\sin \theta$ and $\cos \theta$.

Use the given substitution to express the given radical expression as a trigonometric function without radicals. Assume that $0 < \theta < \pi/2$.

93. Let $x = \sin \theta$ in $\dfrac{x^2}{\sqrt{1 - x^2}}$.

94. Let $x = 4 \sec \theta$ in $\dfrac{\sqrt{x^2 - 16}}{x^2}$.

●   **SYNTHESIS** _____

Show that each of the following is *not* an identity by finding a replacement or replacements for which each side of the equation does not name the same number.

95. $(\sin x + \cos x)^2 = \sin^2 x + \cos^2 x$

96. $\sqrt{\sin^2 \theta} = \sin \theta$

97. $\dfrac{\sin x}{\sin y} = \dfrac{x}{y}$

98. $\cos x = \sqrt{1 - \sin^2 x}$

99. $\cos(\alpha + \beta) = \cos\alpha + \cos\beta$

100. $\sin(-\theta) = \sin\theta$

101. $\cos(2\theta) = 2\cos\theta$

102. $\tan^2\theta + \cot^2\theta = 1$

103. $\ln(\sin\theta) = \sin(\ln\theta)$

OBJECTIVES

You should be able to:

A Find function values using the sum and difference identities.

B Simplify expressions using the sum and difference identities.

7.2

SUM AND DIFFERENCE IDENTITIES

 ## Sum and Difference Identities

Cosine Sum and Difference Identities

We now develop some important identities involving sums or differences of two numbers (or angles), beginning with an identity for the cosine of the difference of two numbers. We use the Greek letters α (alpha) and β (beta) for these numbers.

Let us consider a real number α in the interval $[\pi/2, \pi]$ and a real number β in the interval $[0, \pi/2]$. These determine points A and B on the unit circle as shown. The arc length s is $\alpha - \beta$, and we know that $0 \le s \le \pi$. Recall that the coordinates of A are $(\cos\alpha, \sin\alpha)$, and the coordinates of B are $(\cos\beta, \sin\beta)$.

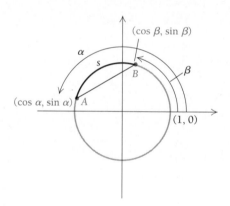

Using the distance formula, we can write an expression for the square of the distance AB:

$$AB^2 = (\cos\alpha - \cos\beta)^2 + (\sin\alpha - \sin\beta)^2.$$

This can be simplified as follows:

$$\begin{aligned} AB^2 &= \cos^2\alpha - 2\cos\alpha\cos\beta + \cos^2\beta + \sin^2\alpha \\ &\quad - 2\sin\alpha\sin\beta + \sin^2\beta \\ &= (\sin^2\alpha + \cos^2\alpha) + (\sin^2\beta + \cos^2\beta) \\ &\quad - 2(\cos\alpha\cos\beta + \sin\alpha\sin\beta) \\ &= 2 - 2(\cos\alpha\cos\beta + \sin\alpha\sin\beta). \end{aligned}$$

Now let us imagine rotating the circle above so that point B is at $(1, 0)$.

Although the coordinates of point A are now $(\cos s, \sin s)$, the distance AB has not changed.

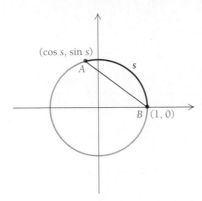

Again we use the distance formula to write an expression for the square of AB:

$$AB^2 = (\cos s - 1)^2 + (\sin s - 0)^2.$$

This simplifies as follows:

$$\begin{aligned}
AB^2 &= \cos^2 s - 2\cos s + 1 + \sin^2 s \\
&= (\sin^2 s + \cos^2 s) + 1 - 2\cos s \\
&= 2 - 2\cos s.
\end{aligned}$$

Equating our two expressions for AB^2, we obtain

$$2 - 2(\cos \alpha \cos \beta + \sin \alpha \sin \beta) = 2 - 2\cos s.$$

Solving this equation for $\cos s$ gives

$$\cos s = \cos \alpha \cos \beta + \sin \alpha \sin \beta. \tag{1}$$

But $s = \alpha - \beta$, so we have the equation

$$\cos(\alpha - \beta) = \cos \alpha \cos \beta + \sin \alpha \sin \beta. \tag{2}$$

This formula holds for any numbers α and β for which $\alpha - \beta$ is the length of the shortest arc from A to B—in other words, when $0 \leq \alpha - \beta \leq \pi$.

Example 1 Find $\cos\left(\dfrac{3\pi}{4} - \dfrac{\pi}{3}\right)$.

Solution

$$\begin{aligned}
\cos\left(\frac{3\pi}{4} - \frac{\pi}{3}\right) &= \cos\frac{3\pi}{4}\cos\frac{\pi}{3} + \sin\frac{3\pi}{4}\sin\frac{\pi}{3} \\
&= -\frac{\sqrt{2}}{2}\cdot\frac{1}{2} + \frac{\sqrt{2}}{2}\cdot\frac{\sqrt{3}}{2} \\
&= \frac{\sqrt{2}}{4}(\sqrt{3} - 1), \quad\text{or}\quad \frac{\sqrt{6} - \sqrt{2}}{4} \qquad \blacktriangleleft
\end{aligned}$$

Example 2 Find $\cos 15°$.

1. Find $\cos\left(\dfrac{\pi}{2} - \dfrac{\pi}{6}\right)$.

Solution

$$\cos 15° = \cos\left(45° - 30°\right)$$
$$= \cos 45° \cos 30° + \sin 45° \sin 30°$$
$$= \frac{\sqrt{2}}{2} \cdot \frac{\sqrt{3}}{2} + \frac{\sqrt{2}}{2} \cdot \frac{1}{2}$$
$$= \frac{\sqrt{2}}{4}\left(\sqrt{3} + 1\right), \quad \text{or} \quad \frac{\sqrt{6} + \sqrt{2}}{4} \qquad \blacktriangleleft$$

DO EXERCISES 1 AND 2. _____

Formula (1) above holds when s is the length of the shortest arc from A to B. Given any real numbers α and β, the length of the shortest arc from A to B is not always $\alpha - \beta$. In fact, it could be $\beta - \alpha$. However, since $\cos(-x) = \cos x$, we know that $\cos(\beta - \alpha) = \cos(\alpha - \beta)$. Thus, $\cos s$ is always equal to $\cos(\alpha - \beta)$.

Formula (2) holds for all real numbers α and β. That formula is thus the identity we sought.

2. Find $\cos 105°$ as $\cos\left(150° - 45°\right)$.

$$\cos(\alpha - \beta) = \cos\alpha\cos\beta + \sin\alpha\sin\beta$$

The cosine sum formula follows easily from the one we have just derived. Let us consider $\cos(\alpha + \beta)$. This is equal to $\cos[\alpha - (-\beta)]$, and by the identity above, we have

$$\cos(\alpha + \beta) = \cos[\alpha - (-\beta)] = \cos\alpha\cos(-\beta) + \sin\alpha\sin(-\beta).$$

But $\cos(-\beta) = \cos\beta$ and $\sin(-\beta) = -\sin\beta$, so the identity we seek is the following:

$$\cos(\alpha + \beta) = \cos\alpha\cos\beta - \sin\alpha\sin\beta.$$

DO EXERCISE 3. _____

Sine Sum and Difference Identities

To develop an identity for the sine of a sum, we first need two intermediary identities. Let us consider $\cos(\pi/2 - \theta)$. We can use the identity for the cosine of a difference to simplify as follows:

3. Find $\cos\left(\dfrac{2\pi}{3} + \dfrac{\pi}{4}\right)$.

$$\cos\left(\frac{\pi}{2} - \theta\right) = \cos\frac{\pi}{2}\cos\theta + \sin\frac{\pi}{2}\sin\theta$$
$$= 0 \cdot \cos\theta + 1 \cdot \sin\theta = \sin\theta.$$

Thus we have developed the identity $\cos(\pi/2 - \theta) = \sin\theta$, which we can rewrite as

$$\sin\theta = \cos\left(\frac{\pi}{2} - \theta\right). \qquad (1)$$

This identity holds for any real number θ. From it we can obtain a similar identity for the sine function. Let α be any real number. Then replace θ in $\sin\theta = \cos(\pi/2 - \theta)$ by $\pi/2 - \alpha$. This gives us

$$\sin\left(\frac{\pi}{2} - \alpha\right) = \cos\left[\frac{\pi}{2} - \left(\frac{\pi}{2} - \alpha\right)\right] = \cos\alpha,$$

which yields the identity

$$\cos \alpha = \sin \left(\frac{\pi}{2} - \alpha \right). \qquad (2)$$

This identity holds for any real number α.

Using identities (1) and (2) and the identity for the cosine of a difference, we now obtain an identity for the sine of a sum. We start with identity (1) and substitute $\alpha + \beta$ for θ, obtaining

$$\sin (\alpha + \beta)$$

$$= \cos \left[\frac{\pi}{2} - (\alpha + \beta) \right]$$

$$= \cos \left[\left(\frac{\pi}{2} - \alpha \right) - \beta \right] \qquad \text{Rewriting}$$

$$= \cos \left(\frac{\pi}{2} - \alpha \right) \cos \beta + \sin \left(\frac{\pi}{2} - \alpha \right) \sin \beta \qquad \begin{array}{l}\text{Using the identity for the}\\ \text{cosine of a difference}\end{array}$$

$$= \sin \alpha \cos \beta + \cos \alpha \sin \beta. \qquad \begin{array}{l}\text{Using identities (1)}\\ \text{and (2)}\end{array}$$

Thus the identity we seek is

$$\sin (\boldsymbol{\alpha} + \boldsymbol{\beta}) = \sin \boldsymbol{\alpha} \cos \boldsymbol{\beta} + \cos \boldsymbol{\alpha} \sin \boldsymbol{\beta}.$$

DO EXERCISES 4 AND 5.

To find a formula for the sine of a difference, we can use the identity just derived, substituting $-\beta$ for β. This gives us

$$\sin (\boldsymbol{\alpha} - \boldsymbol{\beta}) = \sin \boldsymbol{\alpha} \cos \boldsymbol{\beta} - \cos \boldsymbol{\alpha} \sin \boldsymbol{\beta}.$$

Tangent Sum and Difference Identities

A formula for the tangent of a sum can be derived as follows, using identities already established:

$$\tan (\alpha + \beta) = \frac{\sin (\alpha + \beta)}{\cos (\alpha + \beta)}$$

$$= \frac{\sin \alpha \cos \beta + \cos \alpha \sin \beta}{\cos \alpha \cos \beta - \sin \alpha \sin \beta} \cdot \frac{\dfrac{1}{\cos \alpha \cos \beta}}{\dfrac{1}{\cos \alpha \cos \beta}}$$

$$= \frac{\dfrac{\sin \alpha \cos \beta}{\cos \alpha \cos \beta} + \dfrac{\cos \alpha \sin \beta}{\cos \alpha \cos \beta}}{\dfrac{\cos \alpha \cos \beta}{\cos \alpha \cos \beta} - \dfrac{\sin \alpha \sin \beta}{\cos \alpha \cos \beta}}$$

$$= \frac{\dfrac{\sin \alpha}{\cos \alpha} + \dfrac{\sin \beta}{\cos \beta}}{1 - \dfrac{\sin \alpha \sin \beta}{\cos \alpha \cos \beta}}$$

$$= \frac{\tan \alpha + \tan \beta}{1 - \tan \alpha \tan \beta}.$$

4. Find $\sin \left(\dfrac{\pi}{4} + \dfrac{\pi}{3} \right)$.

5. Simplify:

$\sin \alpha \cos (-\beta) + \cos \alpha \sin (-\beta)$.

6. Derive the formula for $\tan(\alpha - \beta)$.

Similarly, a formula for the tangent of a difference can be established. A summary of the **sum and difference identities** follows.* These should be memorized.

Sum and Difference Identities

$$\sin(\alpha \pm \beta) = \sin\alpha\cos\beta \pm \cos\alpha\sin\beta$$
$$\cos(\alpha \pm \beta) = \cos\alpha\cos\beta \mp \sin\alpha\sin\beta$$
$$\tan(\alpha \pm \beta) = \frac{\tan\alpha \pm \tan\beta}{1 \mp \tan\alpha\tan\beta}$$

DO EXERCISE 6. _____

7. Find $\tan 75°$ as $\tan(45° + 30°)$.

Example 3 Find $\tan 15°$.

Solution

$$\tan 15° = \tan(45° - 30°)$$
$$= \frac{\tan 45° - \tan 30°}{1 + \tan 45° \tan 30°}$$
$$= \frac{1 - \sqrt{3}/3}{1 + \sqrt{3}/3}$$
$$= \frac{3 - \sqrt{3}}{3 + \sqrt{3}}$$

◀

Simplify.

8. $\sin\dfrac{\pi}{3}\sin\left(-\dfrac{\pi}{4}\right) + \cos\left(-\dfrac{\pi}{4}\right)\cos\dfrac{\pi}{3}$

DO EXERCISE 7. _____

B **Simplification**

Example 4 Simplify: $\sin\left(-\dfrac{5\pi}{2}\right)\sin\dfrac{\pi}{2} + \cos\dfrac{\pi}{2}\cos\left(-\dfrac{5\pi}{2}\right)$.

Solution The expression is equal to

$$\cos\frac{\pi}{2}\cos\left(-\frac{5\pi}{2}\right) + \sin\frac{\pi}{2}\sin\left(-\frac{5\pi}{2}\right).$$

By the fourth identity in the list above, we can simplify to

9. $\cos 37° \cos 12° + \sin 12° \sin 37°$

$$\cos\left[\frac{\pi}{2} - \left(-\frac{5\pi}{2}\right)\right], \quad \text{or} \quad \cos\frac{6\pi}{2}, \quad \text{or} \quad \cos 3\pi,$$

which is -1.

◀

DO EXERCISES 8 AND 9. _____

Example 5 Simplify: $\sin\dfrac{\pi}{3}\cos\pi + \sin\pi\cos\dfrac{\pi}{3}$.

---

* There are six identities here, half of them obtained by using the signs shown in color.

Solution   The expression is equal to

$$\sin\frac{\pi}{3}\cos\pi + \cos\frac{\pi}{3}\sin\pi.$$

Thus by the first identity above, we can simplify to

$$\sin\left(\frac{\pi}{3}+\pi\right), \quad\text{or}\quad \sin\frac{4\pi}{3}, \quad\text{or}\quad -\frac{\sqrt{3}}{2}. \quad◀$$

DO EXERCISE 10.

Example 6   Assume that $\sin u = \frac{2}{3}$ and $\sin v = \frac{1}{3}$ and that u and v are between 0 and $\pi/2$. Then evaluate $\sin(u+v)$.

Solution   Using the identity for the sine of a sum, we have

$$\sin(u+v) = \sin u \cos v + \cos u \sin v$$
$$= \tfrac{2}{3}\cos v + \tfrac{1}{3}\cos u.$$

To finish, we need to know the values of $\cos v$ and $\cos u$. We first find $\cos v$. We know that $\sin^2 v + \cos^2 v = 1$. Solving for $\cos v$, we get

$$\cos v = \sqrt{1 - \sin^2 v}.$$

We know that the radical is positive because v is between 0 and $\pi/2$. Then substituting $\frac{1}{3}$ for $\sin v$, we get

$$\cos v = \sqrt{1-\left(\frac{1}{3}\right)^2} = \sqrt{1-\frac{1}{9}} = \sqrt{\frac{8}{9}} = \frac{\sqrt{8}}{3} = \frac{2\sqrt{2}}{3}.$$

Similarly,

$$\cos u = \sqrt{1-\sin^2 u} = \sqrt{1-\left(\frac{2}{3}\right)^2} = \sqrt{1-\frac{4}{9}} = \sqrt{\frac{5}{9}} = \frac{\sqrt{5}}{3}.$$

Substituting these values gives us

$$\sin(u+v) = \frac{2}{3}\cdot\frac{2\sqrt{2}}{3} + \frac{1}{3}\cdot\frac{\sqrt{5}}{3} = 4\frac{\sqrt{2}}{9} + \frac{\sqrt{5}}{9} = \frac{4\sqrt{2}+\sqrt{5}}{9}. \quad◀$$

DO EXERCISE 11.

10. Simplify:

$$\sin\frac{\pi}{2}\cos\frac{\pi}{3} - \sin\frac{\pi}{3}\cos\frac{\pi}{2}.$$

11. Assume that $\cos\alpha = \frac{3}{4}$ and $\cos\beta = \frac{1}{4}$ and that α and β are between 0 and $\pi/2$. Then evaluate $\cos(\alpha-\beta)$.

EXERCISE SET 7.2

A   Use the sum and difference identities to evaluate.

1. $\sin 75°$
(*Hint:* $75° = 45° + 30°$.)

2. $\cos 75°$

3. $\sin 15°$
(*Hint:* $15° = 45° - 30°$.)

4. $\cos 15°$

5. $\sin 105°$

6. $\cos 105°$

7. $\tan 75°$

8. $\cot 15°$

B   Simplify or evaluate.

9. $\sin 37° \cos 22° + \cos 37° \sin 22°$

10. $\cos 37° \cos 22° - \sin 22° \sin 37°$

11. $\dfrac{\tan 20° + \tan 32°}{1 - \tan 20° \tan 32°}$

12. $\dfrac{\tan 35° - \tan 12°}{1 + \tan 35° \tan 12°}$

Assume that $\sin u = \frac{3}{5}$ and $\sin v = \frac{4}{5}$ and that u and v are between 0 and $\pi/2$. Then evaluate.

13. $\sin (u + v)$ **14.** $\sin (u - v)$

15. $\cos (u + v)$ **16.** $\cos (u - v)$

17. $\tan (u + v)$ **18.** $\tan (u - v)$

■ Assume that $\sin \theta = 0.6249$ and $\cos \phi = 0.1102$ and that θ and ϕ are both first-quadrant angles. Evaluate.

19. $\sin (\theta + \phi)$ **20.** $\cos (\theta + \phi)$

Simplify.

21. $\sin (\alpha + \beta) + \sin (\alpha - \beta)$

22. $\sin (\alpha + \beta) - \sin (\alpha - \beta)$

23. $\cos (\alpha + \beta) + \cos (\alpha - \beta)$

24. $\cos (\alpha + \beta) - \cos (\alpha - \beta)$

25. $\cos (u + v) \cos v + \sin (u + v) \sin v$

26. $\sin (u - v) \cos v + \cos (u - v) \sin v$

● **SYNTHESIS** _____

Angles between lines. One of the identities just developed gives us an easy way to find an angle formed by two lines. Consider two lines with equations: $l_1: y = m_1 x + b_1$ and $l_2: y = m_2 x + b_2$.

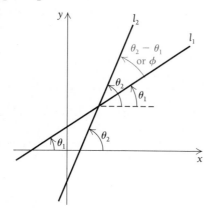

The slopes m_1 and m_2 are the tangents of the angles θ_1 and θ_2 that the lines form with the positive direction of the x-axis. Thus we have $m_1 = \tan \theta_1$ and $m_2 = \tan \theta_2$. To find the measure of the smallest angle—in this case, $\theta_2 - \theta_1$, or ϕ (phi)—through which l_1 can be rotated in the positive direction in order to get to l_2, we proceed as follows:

$$\tan \phi = \tan (\theta_2 - \theta_1) = \frac{\tan \theta_2 - \tan \theta_1}{1 + \tan \theta_2 \tan \theta_1}$$

$$= \frac{m_2 - m_1}{1 + m_2 m_1}.$$

This formula also holds when the lines are taken in the reverse order. When ϕ is an acute angle, $\tan \phi$ will be positive. When ϕ is obtuse, $\tan \phi$ will be negative.

Find the angle from l_1 to l_2.

27. $l_1: 2x = 3 - 2y,$
$l_2: x + y = 5$

28. $l_1: 3y = \sqrt{3}x + 3,$
$l_2: y = \sqrt{3}x + 2$

29. $l_1: y = 3,$
$l_2: x + y = 5$

30. $l_1: 2x + y - 4 = 0,$
$l_2: y - 2x + 5 = 0$

31. In a circus, a guy wire A is attached to the top of a 30-ft pole. Wire B is used for performers to walk up to the tight wire, 10 ft above the ground. Find the angle ϕ between the wires if they are attached to the ground 40 ft from the pole.

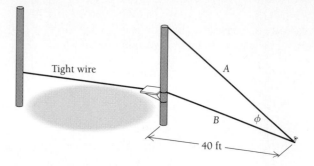

32. Given that $f(x) = \sin x$, show that
$$\frac{f(x + h) - f(x)}{h} = \sin x \left(\frac{\cos h - 1}{h} \right) + \cos x \left(\frac{\sin h}{h} \right).$$

33. Given that $f(x) = \cos x$, show that
$$\frac{f(x + h) - f(x)}{h} = \cos x \left(\frac{\cos h - 1}{h} \right) - \sin x \left(\frac{\sin h}{h} \right).$$

34. Given that $f(x) = \tan x$, show that
$$\frac{f(x + h) - f(x)}{h} = \frac{\sec^2 x}{1 - \tan x \tan h} \left(\frac{\tan h}{h} \right).$$

Find the slope of line l_1, where m_2 is the slope of line l_2 and ϕ is the smallest positive angle from l_1 to l_2.

35. $m_2 = \frac{4}{3}, \phi = 45°$

36. $m_2 = \frac{2}{3}, \phi = 30°$

37. Line l_1 contains the points $(-2, 4)$ and $(5, -1)$. Find the slope of line l_2 such that the angle from l_1 to l_2 is $45°$.

38. Line l_1 contains the points $(-3, 7)$ and $(-3, -2)$. Line l_2 contains $(0, -4)$ and $(2, 6)$. Find the smallest positive angle from l_1 to l_2.

39. Find an identity for $\sin 2\theta$. (*Hint:* $2\theta = \theta + \theta$.)

40. Find an identity for $\cos 2\theta$. (*Hint:* $2\theta = \theta + \theta$.)

41. Derive an identity for $\cot (\alpha + \beta)$ in terms of $\cot \alpha$ and $\cot \beta$.

42. Derive an identity for $\cot (\alpha - \beta)$ in terms of $\cot \alpha$ and $\cot \beta$.

Derive each of the following identities.

43. $\sin \left(x + \dfrac{3\pi}{2} \right) = -\cos x$

44. $\cos\left(x - \dfrac{3\pi}{2}\right) = -\sin x$

45. $\cos\left(x + \dfrac{3\pi}{2}\right) = \sin x$

46. $\sin\left(x - \dfrac{3\pi}{2}\right) = \cos x$

47. $\tan\left(x + \dfrac{\pi}{4}\right) = \dfrac{1 + \tan x}{1 - \tan x}$

48. $\tan\left(x - \dfrac{\pi}{4}\right) = \dfrac{\tan x - 1}{\tan x + 1}$

49. $\dfrac{\sin(\alpha + \beta)}{\cos(\alpha - \beta)} = \dfrac{\tan\alpha + \tan\beta}{1 + \tan\alpha\tan\beta}$

50. $\dfrac{\sin(\alpha + \beta)}{\sin(\alpha - \beta)} = \dfrac{\tan\alpha + \tan\beta}{\tan\alpha - \tan\beta}$

51. $\sin(\alpha + \beta) + \sin(\alpha - \beta) = 2\sin\alpha\cos\beta$

52. $\cos(\alpha + \beta) \cdot \cos(\alpha - \beta) = \cos^2\alpha - \sin^2\beta$

● **CHALLENGE** _____

53. Find an identity for $\cos(\alpha + \beta)$ involving only cosines.

54. Find an identity for $\sin(\alpha + \beta + \gamma)$.

7.3
COFUNCTION AND RELATED IDENTITIES

 Cofunction Identities

In Section 7.2, we developed the following two identities:

$$\cos\left(\frac{\pi}{2} - \theta\right) = \sin\theta \quad \text{and} \quad \sin\left(\frac{\pi}{2} - \theta\right) = \cos\theta.$$

We can now use these to prove another related identity:

$$\tan\left(\frac{\pi}{2} - \theta\right) = \frac{\sin\left(\dfrac{\pi}{2} - \theta\right)}{\cos\left(\dfrac{\pi}{2} - \theta\right)} = \frac{\cos\theta}{\sin\theta} = \cot\theta.$$

DO EXERCISES 1–3. _____

Each of these identities yields a conversion to a cofunction. For this reason, we call them **cofunction identities.** They are summarized below.

> ## Cofunction Identities
>
> $$\sin\left(\frac{\pi}{2} - \theta\right) = \cos\theta \qquad \cos\left(\frac{\pi}{2} - \theta\right) = \sin\theta$$
>
> $$\tan\left(\frac{\pi}{2} - \theta\right) = \cot\theta \qquad \cot\left(\frac{\pi}{2} - \theta\right) = \tan\theta$$
>
> $$\sec\left(\frac{\pi}{2} - \theta\right) = \csc\theta \qquad \csc\left(\frac{\pi}{2} - \theta\right) = \sec\theta$$

These identities hold for all real numbers, and thus, for all degree measures, but if we restrict θ to values such that $0° < \theta < 90°$, then we have a special application to the acute angles of a right triangle.

OBJECTIVE

You should be able to:

A State the cofunction identities for the sine and the cosine, and derive cofunction identities for the other functions. Use the cofunction identities and a given function value for an acute angle to find the function value for its complement.

Prove each of the following identities.

1. $\cot\left(\dfrac{\pi}{2} - \theta\right) = \tan\theta$

2. $\sec\left(\dfrac{\pi}{2} - \theta\right) = \csc\theta$

3. $\csc\left(\dfrac{\pi}{2} - \theta\right) = \sec\theta$

4. Given that

$$\sin 75° = 0.9659,$$
$$\cos 75° = 0.2588,$$
$$\tan 75° = 3.732,$$
$$\cot 75° = 0.2679,$$
$$\sec 75° = 3.864,$$
$$\csc 75° = 1.035,$$

find the six function values for the complement of 75°.

In a right triangle, the acute angles are complementary, since the sum of all three angle measures is $180°$ and the right angle accounts for $90°$ of this total. Thus if one acute angle of a right triangle is θ, the other is $90° - \theta$, or $\pi/2 - \theta$. Note that the sine of $\angle A$ is also the cosine of $\angle B$, its complement:

Similarly, the tangent of $\angle A$ is the cotangent of its complement, and the secant of $\angle A$ is the cosecant of its complement.

These pairs of functions are called **co**functions. The name **cosine** originally meant the sine of the complement, **cotangent** meant the tangent of the complement, and **cosecant** meant the secant of the complement.

Example 1   Given that

$$\sin 18° = 0.3090, \qquad \cos 18° = 0.9511,$$
$$\tan 18° = 0.3249, \qquad \cot 18° = 3.078,$$
$$\sec 18° = 1.051, \qquad \csc 18° = 3.236,$$

find the six function values for 72°.

Solution   Since 72° and 18° are complements, we have $\sin 72° = \cos 18°$, and so on, and the function values are

$$\sin 72° = 0.9511, \qquad \cos 72° = 0.3090,$$
$$\tan 72° = 3.078, \qquad \cot 72° = 0.3249,$$
$$\sec 72° = 3.236, \qquad \csc 72° = 1.051.$$

◀

DO EXERCISE 4.

Other Cofunction Identities

Suppose we want to derive an identity for

$$\sin\left(\theta + \frac{\pi}{2}\right).$$

We have at least two ways to prove this identity. One way to do so is to use the identity for the sine of a sum developed in Section 7.2:

$$\sin\left(\theta + \frac{\pi}{2}\right) = \sin\theta\cos\frac{\pi}{2} + \cos\theta\sin\frac{\pi}{2}$$
$$= \sin\theta \cdot 0 + \cos\theta \cdot 1$$
$$= \cos\theta.$$

Another proof of this identity can be found by considering the left side in terms of x in the form of a function, as follows:

$$g(x) = \sin\left(x + \frac{\pi}{2}\right) = \sin\left[x - \left(-\frac{\pi}{2}\right)\right].$$

The graph of g is found by translating the graph of $f(x) = \sin x$ to the left $\pi/2$ units. The translation is also a graph of the cosine function. Thus we obtain the identity

$$\sin\left(x + \frac{\pi}{2}\right) = \cos x.$$

DO EXERCISE 5. _____

In Margin Exercise 5, you gave a graphical proof of the following identity:

$$\cos\left(x - \frac{\pi}{2}\right) = \sin x.$$

We can also give a proof using the identity $\cos(-x) = \cos x$ and a cofunction identity:

$$\cos\left(x - \frac{\pi}{2}\right) = \cos\left[-\left(\frac{\pi}{2} - x\right)\right] = \cos\left(\frac{\pi}{2} - x\right) = \sin x.$$

Yet another proof can be given using the formula for the cosine of a difference.

DO EXERCISE 6. _____

Consider the following graph. The graph of $f(x) = \sin x$ has been translated $\pi/2$ units to the right, to obtain the graph of $g(x) = \sin(x - \pi/2)$. The latter is a reflection of the cosine function across the x-axis. In other words, it is a graph of $g(x) = -\cos x$. We thus obtain the following identity:

$$\sin\left(x - \frac{\pi}{2}\right) = -\cos x.$$

We have at least two other ways to prove this identity. One way to do so

5. a) Graph $y = \cos x$.
   b) Translate to obtain a graph of
      $$y = \cos\left(x - \frac{\pi}{2}\right).$$
   c) Graph $y = \sin x$.
   d) How do the graphs of parts (b) and (c) compare?
   e) Write the identity thus established.

6. Prove the identity
   $$\cos\left(x - \frac{\pi}{2}\right) = \sin x$$
   using the formula for the cosine of a difference.

7. Prove the identity

$$\sin\left(x - \frac{\pi}{2}\right) = -\cos x$$

using the formula for the sine of a difference.

8. Prove the identity

$$\cos\left(x + \frac{\pi}{2}\right) = -\sin x$$

a) by graphing. Explain your proof.

b) by using the formula for the cosine of a sum.

9. Find an identity for $\cot\left(x + \frac{\pi}{2}\right)$.

---

is to use the identity $\sin(-x) = -\sin x$ and a cofunction identity:

$$\sin\left(x - \frac{\pi}{2}\right) = \sin\left[-\left(\frac{\pi}{2} - x\right)\right] = -\sin\left(\frac{\pi}{2} - x\right) = -\cos x.$$

Yet another proof can be given using the formula for the sine of a difference.

DO EXERCISE 7. _____

By using the formula for the cosine of a sum or a graphical method, we can prove the following identity:

$$\cos\left(x + \frac{\pi}{2}\right) = -\sin x.$$

DO EXERCISE 8. _____

We now have established four more cofunction identities. We give those involving sine and cosine in the following list. Two of them are obtained by taking the top signs, the other two by taking the bottom signs.

Cofunction Identities for Sine and Cosine

$$\sin\left(x \pm \frac{\pi}{2}\right) = \pm\cos x, \qquad \cos\left(x \pm \frac{\pi}{2}\right) = \mp\sin x$$

A good memory device is to think of the "s" in "sine" as "same," meaning that the signs stay the same. For cosine, think of "c" for "change," meaning that the signs change. Cofunction identities for the other functions can be obtained from these easily, using the definitions of the other basic circular functions.

Example 2   Find an identity for $\tan\left(x + \frac{\pi}{2}\right)$.

Solution   By definition of the tangent function,

$$\tan\left(x + \frac{\pi}{2}\right) = \frac{\sin\left(x + \frac{\pi}{2}\right)}{\cos\left(x + \frac{\pi}{2}\right)}.$$

Using a cofunction identity, we obtain

$$\frac{\sin\left(x + \frac{\pi}{2}\right)}{\cos\left(x + \frac{\pi}{2}\right)} = \frac{\cos x}{-\sin x} = -\frac{\cos x}{\sin x} = -\cot x.$$

Thus the identity we seek is

$$\tan\left(x + \frac{\pi}{2}\right) = -\cot x.$$

DO EXERCISE 9. _____

Example 3   Find an identity for $\sec\left(x - \dfrac{\pi}{2}\right)$.

Solution   By definition of the secant function,

$$\sec\left(x - \frac{\pi}{2}\right) = \frac{1}{\cos\left(x - \dfrac{\pi}{2}\right)}.$$

Using a cofunction identity, we obtain

$$\frac{1}{\cos\left(x - \dfrac{\pi}{2}\right)} = \frac{1}{\sin x} = \csc x.$$

Thus the identity we seek is

$$\sec\left(x - \frac{\pi}{2}\right) = \csc x. \qquad \blacktriangleleft$$

DO EXERCISE 10.

Expressions such as

$$\sin\left(x \pm n \cdot \frac{\pi}{2}\right), \qquad \cos\left(x \pm n \cdot \frac{\pi}{2}\right), \quad \text{and} \quad \tan\left(x \pm n \cdot \frac{\pi}{2}\right),$$

where n is an integer, can also be changed to expressions involving only $\sin x$, $\cos x$, $\tan x$, or $\cot x$ using sum and difference formulas for sine and cosine.

Example 4   Find an identity for $\cos\left(x + \dfrac{3\pi}{2}\right)$.

Solution   Using the cosine sum formula, we obtain

$$\cos\left(x + \frac{3\pi}{2}\right) = \cos x \cos\frac{3\pi}{2} - \sin x \sin\frac{3\pi}{2}$$

$$= (\cos x)(0) - (\sin x)(-1)$$

$$= \sin x. \qquad \blacktriangleleft$$

Example 5   Find an identity for $\tan\left(\theta - 270°\right)$.

Solution

$$\tan\left(\theta - 270°\right) = \frac{\sin\left(\theta - 270°\right)}{\cos\left(\theta - 270°\right)} \qquad \text{Using } \tan\theta = \frac{\sin\theta}{\cos\theta}$$

$$= \frac{\sin\theta\cos 270° - \cos\theta\sin 270°}{\cos\theta\cos 270° + \sin\theta\sin 270°}$$

$$= \frac{-\cos\theta\cdot(-1)}{\sin\theta\cdot(-1)} = -\frac{\cos\theta}{\sin\theta}$$

$$= -\cot\theta \qquad \blacktriangleleft$$

DO EXERCISES 11 AND 12.

---

10. Find an identity for $\csc\left(x - \dfrac{\pi}{2}\right)$.

Find an identity for each of the following.

11. $\sin\left(x + \pi\right)$

12. $\tan\left(\theta - 450°\right)$

TECHNOLOGY CONNECTION

A grapher can be used to check identities involving cofunctions. The only drawback is that you must already know both sides of what you *think* is an identity. When using cofunctions, make sure the grapher is in the proper mode (RADIAN or DEGREE). You will be able to tell which mode is required by inspecting the angle that is added or subtracted in the cofunctions. If a term includes an angle involving π, then it is assumed that angles are being measured in RADIANS. If no π appears, then the DEGREE system is being used (usually the degree symbol ° will also be used).

Use a grapher to determine whether each of the following equations is an identity.

TC 1. $\tan\left(x - \pi/2\right) = \cot x$

TC 2. $\cos\left(x - 450°\right) = \sin x$

TC 3. $\cos\left(900° - x\right) - \cos x$
$$= -2\cos x$$

TC 4. $\csc\left(3\pi/2 + x\right) + \sec\left(\pi - x\right)$
$$= 2\csc\left(3\pi/2 - x\right)$$

●**EXERCISE SET** **7.3**

A

1. Given that

$$\sin 65° = 0.9063, \qquad \cos 65° = 0.4226,$$
$$\tan 65° = 2.145, \qquad \cot 65° = 0.4663,$$
$$\sec 65° = 2.366, \qquad \csc 65° = 1.103,$$

find the six function values for 25°.

2. Given that

$$\sin 32° = 0.5299, \qquad \cos 32° = 0.8480,$$
$$\tan 32° = 0.6249, \qquad \cot 32° = 1.600,$$
$$\sec 32° = 1.179, \qquad \csc 32° = 1.887,$$

find the six function values for 58°.

3. Given that

$$\sin 22.5° = \frac{\sqrt{2 - \sqrt{2}}}{2}$$

and

$$\cos 22.5° = \frac{\sqrt{2 + \sqrt{2}}}{2},$$

find each of the following.

a) The other four function values for 22.5°
b) The six function values for 67.5°

4. Given that

$$\sin \frac{\pi}{12} = \frac{\sqrt{2 - \sqrt{3}}}{2}$$

and

$$\cos \frac{\pi}{12} = \frac{\sqrt{2 + \sqrt{3}}}{2},$$

find each of the following.

a) The other four function values for $\pi/12$
b) The six function values for $5\pi/12$

Find and prove an identity for each of the following.

5. $\tan\left(x - \dfrac{\pi}{2}\right)$ 6. $\cot\left(x - \dfrac{\pi}{2}\right)$

7. $\csc\left(x + \dfrac{\pi}{2}\right)$ 8. $\sec\left(x + \dfrac{\pi}{2}\right)$

9. $\sin(x - \pi)$ 10. $\cos(\theta - 270°)$

Find an identity for each of the following.

11. $\tan(\theta + 270°)$ 12. $\sin(\theta - 270°)$
13. $\sin(\theta + 450°)$ 14. $\cos(\theta - 450°)$
15. $\sec(\pi - \theta)$ 16. $\csc(\pi + \theta)$
17. $\tan(x - 4\pi)$ 18. $\sec(x - 10\pi)$
19. $\cos\left(\dfrac{9\pi}{2} + x\right)$ 20. $\cot\left(\dfrac{9\pi}{2} - x\right)$

21. $\csc(540° - \theta)$ 22. $\tan(540° + \theta)$

23. $\cot\left(x - \dfrac{3\pi}{2}\right)$ 24. $\sin\left(x + \dfrac{11\pi}{2}\right)$

● **SYNTHESIS**

25. Given that $\sin \theta = \frac{1}{3}$ and that the terminal side is in quadrant II.

a) Find the other function values for θ.
b) Find the six function values for $\pi/2 - \theta$.
c) Find the six function values for $\pi + \theta$.
d) Find the six function values for $\pi - \theta$.
e) Find the six function values for $2\pi - \theta$.

26. Given that $\cos \theta = \frac{4}{5}$ and that the terminal side is in quadrant IV.

a) Find the other function values for θ.
b) Find the six function values for $\pi/2 + \theta$.
c) Find the six function values for $\pi + \theta$.
d) Find the six function values for $\pi - \theta$.
e) Find the six function values for $2\pi - \theta$.

27. Given that $\sin 27° = 0.45399$.

a) Find the other function values for 27°.
b) Find the six function values for 63°.

28. Given that $\cot 54° = 0.72654$.

a) Find the other function values for 54°.
b) Find the six function values for 36°.

29. Given that $\cos 38° = 0.78801$, find the six function values for 128°.

30. Given that $\tan 73° = 3.27085$, find the six function values for 343°.

Simplify. If you have a grapher, verify your results.

31. $\sin\left(\dfrac{\pi}{2} - x\right)[\sec x - \cos x]$

32. $\cos\left(\dfrac{\pi}{2} - x\right)[\csc x - \sin x]$

33. $\sin x \cos y - \cos\left(x + \dfrac{\pi}{2}\right)\tan y$

34. $\cos\left(x - \dfrac{\pi}{2}\right)\tan y + \sin x \cot y$

35. $\cos(\pi - x) + \cot x \sin\left(x - \dfrac{\pi}{2}\right)$

36. $\sin(\pi - x) - \tan x \cos\left(\dfrac{\pi}{2} - x\right)$

37. $\dfrac{\cos^2 x - 1}{\sin\left(\dfrac{\pi}{2} - x\right) - 1}$

38. $\dfrac{\sin^2 x - 1}{\cos\left(\dfrac{\pi}{2} - x\right) + 1}$

39. $\dfrac{\sin x - \cos\left(x - \dfrac{\pi}{2}\right)\cos x}{-\sin x - \cos\left(x - \dfrac{\pi}{2}\right)\tan x}$

40. $\dfrac{\cos x - \sin\left(\dfrac{\pi}{2} - x\right)\sin x}{\cos x - \cos(\pi - x)\tan x}$

41. $\dfrac{\cos^2 x + 2\sin\left(x - \dfrac{\pi}{2}\right) + 1}{\sin\left(\dfrac{\pi}{2} - x\right) - 1}$

42. $\dfrac{\sin^2 x - 2\cos\left(x - \dfrac{\pi}{2}\right) + 1}{\cos\left(\dfrac{\pi}{2} - x\right) - 1}$

43. $\dfrac{\sin^2 y \cos\left(y + \dfrac{\pi}{2}\right)}{\cos^2 y \cos\left(\dfrac{\pi}{2} - y\right)}$

44. $\dfrac{\cos^2 y \sin\left(y + \dfrac{\pi}{2}\right)}{\sin^2 y \sin\left(\dfrac{\pi}{2} - y\right)}$

7.4
SOME IMPORTANT IDENTITIES

OBJECTIVES

You should be able to:

A Use the double-angle identities to find function values of twice an angle when one function value is known for that angle.

B Use the half-angle identities to find the function values of half an angle when one function value is known for that angle.

C Simplify certain trigonometric expressions using the half-angle and the double-angle formulas. Also, derive some identities by simplifying.

Two important classes of trigonometric identities are known as the *double-angle identities* and the *half-angle identities*.

A   Double-Angle Identities

To develop these identities, we will use the sum formulas from the preceding section. We first develop a formula for $\sin 2\theta$. Recall that

$$\sin(\alpha + \beta) = \sin\alpha\cos\beta + \cos\alpha\sin\beta.$$

We will consider a number θ and substitute it for both α and β in this identity. We obtain

$$\begin{aligned}\sin(\theta + \theta) &= \sin 2\theta \\ &= \sin\theta\cos\theta + \cos\theta\sin\theta \\ &= 2\sin\theta\cos\theta.\end{aligned}$$

Our first double-angle identity is thus

$$\sin 2\theta = 2\sin\theta\cos\theta.$$

1. Given that $\sin\theta = \tfrac{3}{5}$ and θ is in the first quadrant, what is $\sin 2\theta$?

Example 1   If $\sin\theta = \tfrac{3}{8}$ and θ is in the first quadrant, what is $\sin 2\theta$?

Solution   From the figure, we see that $\cos\theta = \sqrt{55}/8$. Thus,

$$\sin 2\theta = 2\sin\theta\cos\theta = 2\cdot\frac{3}{8}\cdot\frac{\sqrt{55}}{8} = \frac{3\sqrt{55}}{32}.$$

◀

DO EXERCISE 1.

2. Given that $\cos \theta = -\frac{5}{13}$ and θ is in the third quadrant, find $\sin 2\theta$, $\cos 2\theta$, and $\tan 2\theta$. Also, determine the quadrant in which 2θ lies.

Double-angle identities for the cosine and tangent functions can be derived in much the same way as the identity above:

$$\cos (\alpha + \beta) = \cos \alpha \cos \beta - \sin \alpha \sin \beta$$
$$\cos 2\theta = \cos (\theta + \theta) = \cos \theta \cos \theta - \sin \theta \sin \theta$$
$$= \cos^2 \theta - \sin^2 \theta$$
$$\boldsymbol{\cos 2\theta = \cos^2 \theta - \sin^2 \theta.}$$

Now we derive an identity for $\tan 2\theta$:

$$\tan (\alpha + \beta) = \frac{\tan \alpha + \tan \beta}{1 - \tan \alpha \tan \beta}$$
$$\tan 2\theta = \tan (\theta + \theta) = \frac{\tan \theta + \tan \theta}{1 - \tan \theta \tan \theta} = \frac{2 \tan \theta}{1 - \tan^2 \theta}$$
$$\boldsymbol{\tan 2\theta = \frac{2 \tan \theta}{1 - \tan^2 \theta}.}$$

Example 2 Given that $\tan \theta = -\frac{3}{4}$ and θ is in the second quadrant, find $\sin 2\theta$, $\cos 2\theta$, $\tan 2\theta$, and the quadrant in which 2θ lies.

Solution   By drawing a diagram as shown, we find that

$$\sin \theta = \tfrac{3}{5} \quad \text{and} \quad \cos \theta = -\tfrac{4}{5}.$$

Now,

$$\sin 2\theta = 2 \sin \theta \cos \theta$$
$$= 2 \cdot \frac{3}{5} \cdot \left(-\frac{4}{5}\right) = -\frac{24}{25};$$
$$\cos 2\theta = \cos^2 \theta - \sin^2 \theta = \left(-\frac{4}{5}\right)^2 - \left(\frac{3}{5}\right)^2$$
$$= \frac{16}{25} - \frac{9}{25} = \frac{7}{25};$$
$$\tan 2\theta = \frac{2 \tan \theta}{1 - \tan^2 \theta} = \frac{2 \cdot \left(-\frac{3}{4}\right)}{1 - \left(-\frac{3}{4}\right)^2}$$
$$= \frac{-\frac{3}{2}}{1 - \frac{9}{16}} = -\frac{24}{7}.$$

Since $\sin 2\theta$ is negative and $\cos 2\theta$ is positive, we know that 2θ is in quadrant IV. Note that $\tan 2\theta$ could have been found more easily in this case by simply dividing the values of $\sin 2\theta$ and $\cos 2\theta$. ◀

DO EXERCISE 2. _____

Two other useful identities for $\cos 2\theta$ can be derived easily, as follows:

$$\cos 2\theta = \cos^2 \theta - \sin^2 \theta$$
$$= (1 - \sin^2 \theta) - \sin^2 \theta \qquad \text{Using } \sin^2 \theta + \cos^2 \theta = 1$$
$$= 1 - 2 \sin^2 \theta.$$

Similarly,

$$\cos 2\theta = \cos^2 \theta - \sin^2 \theta$$
$$= \cos^2 \theta - (1 - \cos^2 \theta) = 2 \cos^2 \theta - 1.$$

Solving these two identities for $\sin^2 \theta$ and $\cos^2 \theta$, respectively, we obtain two more identities that are often useful:

$$\sin^2 \theta = \frac{1 - \cos 2\theta}{2},$$

$$\cos^2 \theta = \frac{1 + \cos 2\theta}{2}.$$

Using division and the last two identities, we can easily deduce the following identity, which is also often useful:

$$\tan^2 \theta = \frac{1 - \cos 2\theta}{1 + \cos 2\theta}.$$

The following list of **double-angle identities** should be memorized.

Double-Angle Identities

$$\sin 2\theta = 2 \sin \theta \cos \theta, \qquad \sin^2 \theta = \frac{1 - \cos 2\theta}{2},$$

$$\cos 2\theta = \cos^2 \theta - \sin^2 \theta \qquad \cos^2 \theta = \frac{1 + \cos 2\theta}{2},$$
$$= 1 - 2 \sin^2 \theta$$
$$= 2 \cos^2 \theta - 1,$$

$$\tan 2\theta = \frac{2 \tan \theta}{1 - \tan^2 \theta}, \qquad \tan^2 \theta = \frac{1 - \cos 2\theta}{1 + \cos 2\theta}$$

From the basic identities (in the lists to be memorized), others can be obtained.

Example 3 Find a formula for $\sin 3\theta$ in terms of function values of θ.

Solution
$$\sin 3\theta = \sin (2\theta + \theta)$$
$$= \sin 2\theta \cos \theta + \cos 2\theta \sin \theta$$
$$= (2 \sin \theta \cos \theta) \cos \theta + (2 \cos^2 \theta - 1) \sin \theta$$
$$= 2 \sin \theta \cos^2 \theta + 2 \sin \theta \cos^2 \theta - \sin \theta$$
$$= 4 \sin \theta \cos^2 \theta - \sin \theta \qquad \blacktriangleleft$$

Example 4 Find a formula for $\cos^3 x$ in terms of function values of x or $2x$, raised only to the first power.

Solution
$$\cos^3 x = \cos^2 x \cos x$$
$$= \frac{1 + \cos 2x}{2} \cos x \qquad \blacktriangleleft$$

DO EXERCISES 3 AND 4.

3. Find a formula for $\cos 3\theta$ in terms of function values of θ.

4. Find a formula for $\sin^3 x$ in terms of function values of x or $2x$, raised only to the first power.

B ▸ Half-Angle Identities

To develop these identities, we use three of the previously formed ones. As shown below, we take square roots and replace θ by $\phi/2$:

$$\sin^2 \theta = \frac{1 - \cos 2\theta}{2} \longrightarrow \left|\sin \frac{\phi}{2}\right| = \sqrt{\frac{1 - \cos \phi}{2}},$$

$$\cos^2 \theta = \frac{1 + \cos 2\theta}{2} \longrightarrow \left|\cos \frac{\phi}{2}\right| = \sqrt{\frac{1 + \cos \phi}{2}},$$

$$\tan^2 \theta = \frac{1 - \cos 2\theta}{1 + \cos 2\theta} \longrightarrow \left|\tan \frac{\phi}{2}\right| = \sqrt{\frac{1 - \cos \phi}{1 + \cos \phi}}.$$

The half-angle formulas are those on the right above. We can eliminate the absolute-value signs by introducing $\pm$ signs, with the understanding that our use of $+$ or $-$ depends on the quadrant in which the angle $\phi/2$ lies. We thus obtain these formulas in the following form.

Half-Angle Identities

$$\sin \frac{\phi}{2} = \pm \sqrt{\frac{1 - \cos \phi}{2}}$$

$$\cos \frac{\phi}{2} = \pm \sqrt{\frac{1 + \cos \phi}{2}}$$

$$\tan \frac{\phi}{2} = \pm \sqrt{\frac{1 - \cos \phi}{1 + \cos \phi}}$$

These formulas should be memorized.

There are two other formulas for $\tan (\phi/2)$ that are often useful. They can be obtained as follows:

$$\left|\tan \frac{\phi}{2}\right| = \sqrt{\frac{1 - \cos \phi}{1 + \cos \phi}} = \sqrt{\frac{1 - \cos \phi}{1 + \cos \phi} \cdot \frac{1 + \cos \phi}{1 + \cos \phi}}$$

$$= \sqrt{\frac{1 - \cos^2 \phi}{(1 + \cos \phi)^2}} = \sqrt{\frac{\sin^2 \phi}{(1 + \cos \phi)^2}}$$

$$= \frac{|\sin \phi|}{|1 + \cos \phi|}.$$

Now $1 + \cos \phi$ cannot be negative because $\cos \phi$ is never less than -1. Thus the absolute-value signs are not necessary in the denominator. As the following graph shows, $\tan (\phi/2)$ and $\sin \phi$ have the same sign for all ϕ for which $\tan (\phi/2)$ is defined.

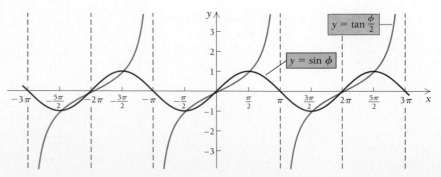

Thus we can dispense with the other absolute-value signs and obtain the formula we seek. A second formula can be obtained in a similar way.

$$\tan\frac{\phi}{2} = \frac{\sin\phi}{1 + \cos\phi}, \qquad \tan\frac{\phi}{2} = \frac{1 - \cos\phi}{\sin\phi}$$

These formulas have the advantage that they give the sign of $\tan(\phi/2)$ directly.

Example 5   Find $\sin 15°$.

Solution

$$\sin 15° = \sin\frac{30°}{2} = \pm\sqrt{\frac{1 - \cos 30°}{2}}$$

$$= \pm\sqrt{\frac{1 - (\sqrt{3}/2)}{2}} = \pm\sqrt{\frac{2 - \sqrt{3}}{4}}$$

$$= \frac{\sqrt{2 - \sqrt{3}}}{2}$$

The expression is positive, because $15°$ is in the first quadrant.   ◀

Example 6   Find $\tan\dfrac{\pi}{8}$.

Solution

$$\tan\frac{\pi}{8} = \tan\frac{\frac{\pi}{4}}{2} = \frac{\sin\frac{\pi}{4}}{1 + \cos\frac{\pi}{4}} = \frac{\frac{\sqrt{2}}{2}}{1 + \frac{\sqrt{2}}{2}}$$

$$= \frac{\sqrt{2}}{2 + \sqrt{2}}$$

$$= \sqrt{2} - 1 \qquad \text{\small Rationalizing the denominator}$$ ◀

DO EXERCISES 5 AND 6.

 Simplification

Many simplifications of trigonometric expressions are possible through the use of the identities that we have developed.

Example 7   Simplify:

$$\frac{1 - \cos 2x}{4\sin x\cos x}.$$

Solution   We search the list of identities, looking for some substitution that will simplify the expression. In this case, one might note that the

5. Find $\cos 15°$.

6. Find $\tan\dfrac{\pi}{12}$.

7. Simplify:

$$\frac{2(\tan x + \tan^3 x)}{1 - \tan^4 x}.$$

8. By simplifying, derive an identity:

$$\frac{\frac{1}{2}(\cos^2 x - \sin^2 x)}{\sin x\cos x}.$$

TECHNOLOGY CONNECTION

A grapher can also verify double-angle and half-angle identities.

TC 1. Use a grapher to determine which of the following expressions completes the identity

$$\frac{\cos 2x}{\cos x - \sin x} = \ldots$$

A. $1 + \cos x$

B. $\cos x + \sin x$

C. $\cos x (\sec x + 1)$

D. $\sin x (\cot x + 1)$

TC 2. Use a grapher to determine which of the following expressions completes the identity

$$2\cos^2 \frac{x}{2} = \ldots$$

A. $\sin x (\csc x + \cot x)$

B. $\sin x - 2 \cos x$

C. $2 (\cos^2 x - \sin^2 x)$

D. $1 + \cos x$

---

denominator is $2(2 \sin x \cos x)$ and thus find a simplification, since $2 \sin x \cos x = \sin 2x$:

$$\frac{1 - \cos 2x}{4 \sin x \cos x} = \frac{1 - \cos 2x}{2(2 \sin x \cos x)} = \frac{1 - \cos 2x}{2 \sin 2x}.$$

Recalling that $(1 - \cos \phi)/\sin \phi = \tan (\phi/2)$ and letting $\phi = 2x$, we have

$$\frac{1 - \cos 2x}{\sin 2x} = \tan x.$$

Multiplying by $\frac{1}{2}$, we obtain

$$\frac{1 - \cos 2x}{2 \sin 2x} = \frac{1}{2} \tan x. \qquad \blacktriangleleft$$

We can derive an identity by simplifying an expression, as in the following example.

Example 8 By simplifying, derive an identity:

$$\frac{\sin x \cos x}{\frac{1}{2} \cos 2x}.$$

Solution We can obtain $2 \sin x \cos x$ in the numerator by multiplying the expression by $\frac{2}{2}$:

$$\frac{2 \sin x \cos x}{2 \cdot \frac{1}{2} \cos 2x} = \frac{\sin 2x}{\cos 2x} \qquad \text{Using } \sin 2x = 2 \sin x \cos x$$

$$= \tan 2x.$$

We have thus derived the identity

$$\frac{\sin x \cos x}{\frac{1}{2} \cos 2x} = \tan 2x. \qquad \blacktriangleleft$$

DO EXERCISES 7 AND 8 ON THE PRECEDING PAGE.

● EXERCISE SET 7.4

A   Find $\sin 2\theta$, $\cos 2\theta$, $\tan 2\theta$, and the quadrant in which 2θ lies.

1. $\sin \theta = \frac{4}{5}$; θ in quadrant I

2. $\sin \theta = \frac{5}{13}$; θ in quadrant I

3. $\cos \theta = -\frac{4}{5}$; θ in quadrant III

4. $\cos \theta = -\frac{3}{5}$; θ in quadrant III

5. $\tan \theta = \frac{4}{3}$; θ in quadrant III

6. $\tan \theta = \frac{3}{4}$; θ in quadrant III

7. Find a formula for $\sin 4\theta$ in terms of function values of θ.

8. Find a formula for $\cos 4\theta$ in terms of function values of θ.

9. Find a formula for $\sin^4 \theta$ in terms of function values of θ or 2θ or 4θ, raised only to the first power.

10. Find a formula for $\cos^4 \theta$ in terms of function values of θ or 2θ or 4θ, raised only to the first power.

B   Find each of the following without using a calculator.

11. $\sin 75°$

12. $\cos 75°$

(*Hint:* $75 = 150/2$.)

13. $\tan 75°$

14. $\tan 67.5°$

(*Hint:* $67.5 = 135/2$.)

15. $\sin \dfrac{5\pi}{8}$

16. $\cos \dfrac{5\pi}{8}$

17. $\sin 112.5°$

18. $\cos 22.5°$

19. $\cos \dfrac{\pi}{8}$

20. $\tan 15°$

21. $\sin 22.5°$

22. $\tan 112.5°$

Ⓐ, Ⓑ 🖩 Given that $\sin \theta = 0.3416$ and θ is in the first quadrant, find each of the following.

23. $\sin 2\theta$

24. $\cos 2\theta$

25. $\sin 4\theta$

26. $\cos 4\theta$

27. $\sin \dfrac{\theta}{2}$

28. $\cos \dfrac{\theta}{2}$

Ⓒ Simplify.

29. $\dfrac{\sin 2x}{2 \sin x}$

30. $\dfrac{\sin 2x}{2 \cos x}$

31. $1 - 2\sin^2 \dfrac{x}{2}$

32. $2\cos^2 \dfrac{x}{2} - 1$

33. $2 \sin \dfrac{x}{2} \cos \dfrac{x}{2}$

34. $2 \sin 2x \cos 2x$

35. $\cos^2 \dfrac{x}{2} - \sin^2 \dfrac{x}{2}$

36. $\cos^4 x - \sin^4 x$

37. $(\sin x + \cos x)^2 - \sin 2x$

38. $(\sin x - \cos x)^2 + \sin 2x$

39. $2 \sin^2 \dfrac{x}{2} + \cos x$

40. $2 \cos^2 \dfrac{x}{2} - \cos x$

41. $(-4 \cos x \sin x + 2 \cos 2x)^2$
$$+ (2 \cos 2x + 4 \sin x \cos x)^2$$

42. $(-4 \cos 2x + 8 \cos x \sin x)^2$
$$+ (8 \sin x \cos x + 4 \cos 2x)^2$$

43. $2 \sin x \cos^3 x + 2 \sin^3 x \cos x$

44. $2 \sin x \cos^3 x - 2 \sin^3 x \cos x$

By simplifying, derive an identity. If you have a grapher, verify your results.

45. $(\sin x + \cos x)^2$

46. $\cos^4 x - \sin^4 x$

47. $\dfrac{2 \cot x}{\cot^2 x - 1}$

48. $\dfrac{2 - \sec^2 x}{\sec^2 x}$

49. $2 \sin^2 2x + \cos 4x$

50. $\dfrac{1 + \sin 2x + \cos 2x}{1 + \sin 2x - \cos 2x}$

● SYNTHESIS _____

Find $\sin \theta$, $\cos \theta$, and $\tan \theta$ under the given conditions.

51. $\sin 2\theta = \dfrac{1}{5}, \dfrac{\pi}{2} \le 2\theta \le \pi$

52. $\cos 2\theta = \dfrac{7}{12}, \dfrac{3\pi}{2} \le 2\theta \le 2\pi$

53. $\tan \dfrac{\theta}{2} = -\dfrac{1}{4}, \dfrac{3\pi}{2} \le \theta \le 2\pi$

54. $\tan \dfrac{\theta}{2} = -\dfrac{5}{3}, \pi < \theta \le \dfrac{3\pi}{2}$

55. *Nautical mile.*   *Latitude* is used to measure north–south location on the earth between the equator and the poles. For example, the north end of Chicago has latitude 42°N. (See the figure.) In Great Britain, the **nautical mile** is defined as the length of a minute of arc of the earth's radius. Since the earth is flattened at the poles, a British nautical mile varies with latitude. In fact, it is given, in feet, by the function

$$N(\phi) = 6066 - 31 \cos 2\phi,$$

where ϕ is the latitude in degrees.

a) What is the length of a British nautical mile at the north end of Chicago?

b) What is the length of a British nautical mile at the North Pole?

c) Express $N(\phi)$ in terms of $\cos \phi$ only. That is, eliminate the double angle.

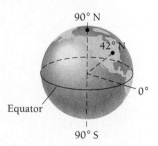

56. *Acceleration due to gravity.*   The acceleration due to gravity is often denoted by g in a formula such as $S = \frac{1}{2}gt^2$, where S is the distance that an object falls in time t. It has to do with the physics of motion near the earth's surface and is usually considered constant. In fact, however, g is not constant, but varies slightly with latitude. If ϕ stands for latitude, in degrees, g is given with good approximation by the formula

$$g = 9.78049(1 + 0.005288 \sin^2 \phi - 0.000006 \sin^2 2\phi),$$

where g is measured in meters per second per second at sea level.

a) Chicago has latitude 42°N. Find g.

b) Philadelphia has latitude 40°N. Find g.

c) Express g in terms of $\sin \phi$ only. That is, eliminate the double angle.

d) Where on earth is g greatest? least?

● CHALLENGE _____

57. Graph: $f(x) = \cos^2 x - \sin^2 x$.

58. Graph: $f(x) = |\sin x \cos x|$.

OBJECTIVES

You should be able to:

 Prove identities using other identities.

Prove the sum–product identities and use them to prove other identities.

7.5
PROVING TRIGONOMETRIC IDENTITIES

Basic Identities

We have proved a great many identities in the last chapter and in this chapter. Here we consider the proving of identities more intensely. You will be amazed how many trigonometric identities we can prove, and how the proofs of some identities enable us to prove others. The proofs we considered in Sections 7.1–7.4 are good examples.

A first step in learning to prove identities is to have at hand as many identities as possible that you have already learned. The following list contains many such identities. Although it is not necessary to learn them all, the more you learn, the better your identity-proving skill will be.

Basic Identities

$$\sin x = \frac{1}{\csc x} \qquad \csc x = \frac{1}{\sin x} \qquad \tan x = \frac{\sin x}{\cos x}$$

$$\cos x = \frac{1}{\sec x} \qquad \sec x = \frac{1}{\cos x} \qquad \cot x = \frac{\cos x}{\sin x}$$

$$\tan x = \frac{1}{\cot x} \qquad \cot x = \frac{1}{\tan x}$$

$$\sin(-x) = -\sin x \qquad \cos(-x) = \cos x \qquad \tan(-x) = -\tan x$$

$$\csc(-x) = -\csc x \qquad \sec(-x) = \sec x \qquad \cot(-x) = -\cot x$$

Pythagorean Identities

$$\sin^2 x + \cos^2 x = 1$$

$$1 + \tan^2 x = \sec^2 x$$

$$1 + \cot^2 x = \csc^2 x$$

Cofunction Identities

$$\sin\left(x \pm \frac{\pi}{2}\right) = \pm\cos x \qquad \sin\left(\frac{\pi}{2} \pm x\right) = \cos x$$

$$\cos\left(x \pm \frac{\pi}{2}\right) = \mp\sin x \qquad \cos\left(\frac{\pi}{2} \pm x\right) = \mp\sin x$$

Sum and Difference Identities

$$\sin(\alpha \pm \beta) = \sin\alpha\cos\beta \pm \cos\alpha\sin\beta$$

$$\cos(\alpha \pm \beta) = \cos\alpha\cos\beta \mp \sin\alpha\sin\beta$$

$$\tan(\alpha \pm \beta) = \frac{\tan\alpha \pm \tan\beta}{1 \mp \tan\alpha\tan\beta}$$

Double-Angle Identities

$$\sin 2x = 2 \sin x \cos x$$

$$\cos 2x = \cos^2 x - \sin^2 x$$

$$= 1 - 2 \sin^2 x$$

$$= 2 \cos^2 x - 1$$

$$\tan 2x = \frac{2 \tan x}{1 - \tan^2 x}$$

$$\sin^2 x = \frac{1 - \cos 2x}{2}$$

$$\cos^2 x = \frac{1 + \cos 2x}{2}$$

$$\tan^2 x = \frac{1 - \cos 2x}{1 + \cos 2x}$$

Half-Angle Identities

$$\sin \frac{x}{2} = \pm \sqrt{\frac{1 - \cos x}{2}}$$

$$\cos \frac{x}{2} = \pm \sqrt{\frac{1 + \cos x}{2}}$$

$$\tan \frac{x}{2} = \pm \sqrt{\frac{1 - \cos x}{1 + \cos x}}$$

$$= \frac{\sin x}{1 + \cos x}$$

$$= \frac{1 - \cos x}{\sin x}$$

Proving Identities

The Logic of Proving Identities

We now outline two methods for proving identities.

Method 1.   Start with either the left or the right side of an identity and deduce the other side. For example, suppose you are trying to prove that the equation $P = Q$ is an identity. You might try to produce a string of statements like the following, which start at P and end with Q:

$$P = S_1$$

$$= S_2$$

$$\vdots$$

$$= Q.$$

Method 2.   Work with each side separately until you deduce the same expression. For example, suppose you are trying to prove that $P = Q$ is an identity. You might be able to produce two strings of statements like the following, each ending with the same statement S.

$$P = S_1 \qquad Q = P_1$$

$$= S_2 \qquad\quad = P_2$$

$$\vdots \qquad\qquad \vdots$$

$$= S. \qquad\quad = S.$$

The number of steps in each string might be different, but in each case the result is S.

Prove each of the following identities.

1. $\dfrac{\csc t - 1}{t \csc t} = \dfrac{1 - \sin t}{t}$

2. $\sec \theta \csc \theta = \tan \theta + \cot \theta$

Prove each of the following identities.

3. $(\sin \theta - \cos \theta)^2 = 1 - \sin 2\theta$

4. $(\sec u - \tan u)(1 + \sin u) = \cos u$

Examples 1–3 illustrate Method 1, and Examples 4 and 5 illustrate Method 2.

Example 1 Prove the identity

$$\frac{\sec t - 1}{t \sec t} = \frac{1 - \cos t}{t}.$$

Proof. We use Method 1, starting with the left side and deducing the right side. We note that the left side involves $\sec t$, whereas the right side involves $\cos t$, so it might be wise to make use of an identity that involves these two expressions. That basic identity is $\sec t = 1/\cos t$.

$$\frac{\sec t - 1}{t \sec t} = \frac{\dfrac{1}{\cos t} - 1}{t \dfrac{1}{\cos t}} \qquad \text{Substituting } \frac{1}{\cos t} \text{ for } \sec t$$

$$= \frac{\left(\dfrac{1}{\cos t} - 1\right)}{\dfrac{t}{\cos t}} \cdot \frac{\cos t}{\cos t} \qquad \begin{array}{l}\text{Multiplying by 1 in order}\\\text{to simplify the complex}\\\text{rational expression}\end{array}$$

$$= \frac{1 - \cos t}{t} \qquad \text{Multiplying and simplifying}$$

We started with the left side and deduced the right side, so the proof is complete. ◄

DO EXERCISES 1 AND 2.

Example 2 Prove the identity $1 + \sin 2\theta = (\sin \theta + \cos \theta)^2$.

Proof. We again use Method 1. This time we start with the right side and deduce the left side.

$(\sin \theta + \cos \theta)^2$

$\qquad = \sin^2 \theta + 2 \sin \theta \cos \theta + \cos^2 \theta \qquad$ Squaring

$\qquad = 1 + 2 \sin \theta \cos \theta \qquad \begin{array}{l}\text{Recalling the identity}\\ \sin^2 \theta + \cos^2 \theta = 1\\\text{and substituting}\end{array}$

$\qquad = 1 + \sin 2\theta \qquad \begin{array}{l}\text{Recalling the identity}\\ \sin 2\theta = 2 \sin \theta \cos \theta\\\text{and substituting}\end{array}$

We started with the right side and deduced the left side, so the proof is complete. ◄

DO EXERCISES 3 AND 4.

Example 3 Prove the identity

$$\frac{\sin \theta - \cos \theta}{\sin \theta + \cos \theta} = -\frac{\cos 2\theta}{1 + \sin 2\theta}.$$

Proof. We start with the left side and deduce the right side. In the first step, we multiply by 1, where the symbol for 1 is formed from the conjugate of the numerator of the original expression.

$$\frac{\sin\theta - \cos\theta}{\sin\theta + \cos\theta} = \frac{\sin\theta - \cos\theta}{\sin\theta + \cos\theta} \cdot \frac{\sin\theta + \cos\theta}{\sin\theta + \cos\theta} \qquad \text{Multiplying by 1}$$

$$= \frac{\sin^2\theta - \cos^2\theta}{(\sin\theta + \cos\theta)^2}$$

$$= \frac{\sin^2\theta - \cos^2\theta}{1 + \sin 2\theta} \qquad \text{Substituting, using the identity of Example 2}$$

$$= \frac{-(\cos^2\theta - \sin^2\theta)}{1 + \sin 2\theta}$$

$$= -\frac{\cos 2\theta}{1 + \sin 2\theta} \qquad \text{Recalling the identity } \cos^2\theta - \sin^2\theta = \cos 2\theta \text{ and substituting}$$

We started with the left side and deduced the right side, so the proof is complete.   ◀

DO EXERCISE 5. _____

Example 4   Prove the identity $\tan^2 x - \sin^2 x = \sin^2 x \tan^2 x$.

Proof.   For this proof, we are going to work with each side separately using Method 2. We try to deduce the same expression. In practice, when carrying out this method of proof, you might work on one side for awhile, then work on the other side separately, and then go back to the other side. That is, you bounce back and forth until you arrive at the same expression. Let us start with the left side:

$$\tan^2 x - \sin^2 x = \frac{\sin^2 x}{\cos^2 x} - \sin^2 x \qquad \text{Recalling the identity } \tan x = \frac{\sin x}{\cos x} \text{ and substituting}$$

$$= \frac{\sin^2 x}{\cos^2 x} - \sin^2 x \cdot \frac{\cos^2 x}{\cos^2 x} \qquad \text{Multiplying by 1 in order to subtract}$$

$$= \frac{\sin^2 x - \sin^2 x \cos^2 x}{\cos^2 x} \qquad \text{Carrying out the subtraction}$$

$$= \frac{\sin^2 x (1 - \cos^2 x)}{\cos^2 x} \qquad \text{Factoring}$$

$$= \frac{\sin^2 x \sin^2 x}{\cos^2 x} \qquad \text{Recalling the identity } \sin^2 x + \cos^2 x = 1 \text{ or } 1 - \cos^2 x = \sin^2 x \text{ and substituting}$$

$$= \frac{\sin^4 x}{\cos^2 x}.$$

At this point, we stop and work with the right side of the original identity and try to end with the same expression that we ended with on the left side:

$$\sin^2 x \tan^2 x = \sin^2 x \frac{\sin^2 x}{\cos^2 x} \qquad \text{Recalling the identity } \tan x = \frac{\sin x}{\cos x} \text{ and substituting}$$

$$= \frac{\sin^4 x}{\cos^2 x}.$$

5. Prove the identity

$$\frac{\sin t}{1 + \cos t} = \frac{1 - \cos t}{\sin t}.$$

6. Prove the identity

$$\cot^2 x - \cos^2 x = \cos^2 x \cot^2 x.$$

From each side we have deduced the same expression, so the proof is complete.   ◀

DO EXERCISE 6. _____

Example 5   Prove the identity

$$\frac{\sin 2\theta}{\sin \theta} - \frac{\cos 2\theta}{\cos \theta} = \sec \theta.$$

Proof.   We are again using Method 2, beginning with the left side:

$$\frac{\sin 2\theta}{\sin \theta} - \frac{\cos 2\theta}{\cos \theta}$$

$$= \frac{2 \sin \theta \cos \theta}{\sin \theta} - \frac{\cos^2 \theta - \sin^2 \theta}{\cos \theta}$$

Using the identities $\sin 2\theta = 2 \sin \theta \cos \theta$ and $\cos 2\theta = \cos^2 \theta - \sin^2 \theta$ and substituting

$$= 2 \cos \theta - \frac{\cos^2 \theta - \sin^2 \theta}{\cos \theta}$$

Simplifying

$$= \frac{2 \cos^2 \theta}{\cos \theta} - \frac{\cos^2 \theta - \sin^2 \theta}{\cos \theta}$$

Multiplying $2 \cos \theta$ by 1, or $\cos \theta / \cos \theta$

$$= \frac{2 \cos^2 \theta - \cos^2 \theta + \sin^2 \theta}{\cos \theta}$$

Carrying out the subtraction

$$= \frac{\cos^2 \theta + \sin^2 \theta}{\cos \theta}$$

$$= \frac{1}{\cos \theta}.$$

Recalling the identity $\sin^2 \theta + \cos^2 \theta = 1$

7. Prove the identity

$$\frac{\sin 2\theta + \sin \theta}{\cos 2\theta + \cos \theta + 1} = \tan \theta.$$

At this point, we stop and work with the right side of the original identity and try to end with the same expression:

$$\sec \theta = \frac{1}{\cos \theta}.$$ Recalling a basic identity

From each side we have deduced the same expression, so the proof is complete.   ◀

DO EXERCISE 7. _____

Hints for Providing Identities

1. Use Methods 1 or 2 previously outlined.
2. Work with the more complex side first.
3. Carry out the algebraic manipulations, such as adding, subtracting, multiplying, or factoring.
4. Multiplying by 1 can often be helpful when rational expressions are involved.
5. Converting all expressions to sines and cosines is often helpful.
6. Try something! Put your pencil to work and get involved. You will be amazed at how often this leads to success.

Sum–Product Identities

On occasion, it is convenient to convert a product of trigonometric expressions to a sum, or the reverse. The following identities are useful in this connection. Proofs are left as exercises.

Sum–Product Identities

$$\sin u \cdot \cos v = \frac{1}{2}[\sin(u+v) + \sin(u-v)]$$

$$\cos u \cdot \sin v = \frac{1}{2}[\sin(u+v) - \sin(u-v)]$$

$$\cos u \cdot \cos v = \frac{1}{2}[\cos(u-v) + \cos(u+v)]$$

$$\sin u \cdot \sin v = \frac{1}{2}[\cos(u-v) - \cos(u+v)]$$

$$\sin x + \sin y = 2\sin\frac{x+y}{2}\cos\frac{x-y}{2}$$

$$\sin x - \sin y = 2\cos\frac{x+y}{2}\sin\frac{x-y}{2}$$

$$\cos y + \cos x = 2\cos\frac{x+y}{2}\cos\frac{x-y}{2}$$

$$\cos y - \cos x = 2\sin\frac{x+y}{2}\sin\frac{x-y}{2}$$

DO EXERCISE 8.

Example 6 Find an identity for $\cos\theta + \cos 5\theta$.

Solution We will use the identity

$$\cos y + \cos x = 2\cos\frac{x+y}{2}\cos\frac{x-y}{2}.$$

Here $x = 5\theta$ and $y = \theta$. Thus,

$$\cos\theta + \cos 5\theta = 2\cos\frac{5\theta+\theta}{2}\cos\frac{5\theta-\theta}{2}$$

$$= 2\cos 3\theta \cos 2\theta. \quad \blacktriangleleft$$

DO EXERCISE 9.

---

8. Prove the first identity in the list of sum–product identities. (*Hint:* Start with the right side and use the sine of a sum and the sine of a difference.)

9. Find an identity for

$$2\sin 7\theta \cos 5\theta.$$

 TECHNOLOGY CONNECTION

Almost any identity can be checked on a grapher. The exception is an identity that involves two different angles, such as u and v, or x and y, in the sum–product identities. Because we must enter each side of the identity as a function of the single variable x, there is no way to graph any function that has two variables.

Match each expression in column A with the expression from column B that would result in an identity.

COLUMN A	COLUMN B
TC 1. $1 - \dfrac{\sin^2 x}{1 + \cos x}$	A. $\sin x$
TC 2. $\dfrac{\sin x \sec x}{1 + \tan^2 x}$	B. $\sin x \cos x + 1$
TC 3. $\frac{1}{2}\sec^2 x \cos x \sin 2x$	C. $\cos x$
TC 4. $\dfrac{(\sin x \cos x + 1)(\cos^2 x \tan x - 1)}{\cos^2 x \tan x - 1}$	D. $\dfrac{\cos x}{\csc x}$

● EXERCISE SET 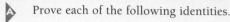 **7.5**

Prove each of the following identities.

1. $\csc x - \cos x \cot x = \sin x$

2. $\sec x - \sin x \tan x = \cos x$

3. $\dfrac{1 + \cos \theta}{\sin \theta} + \dfrac{\sin \theta}{\cos \theta} = \dfrac{\cos \theta + 1}{\sin \theta \cos \theta}$

4. $\dfrac{1}{\sin \theta \cos \theta} - \dfrac{\cos \theta}{\sin \theta} = \dfrac{\sin \theta \cos \theta}{1 - \sin^2 \theta}$

5. $\dfrac{1 - \sin x}{\cos x} = \dfrac{\cos x}{1 + \sin x}$

6. $\dfrac{1 - \cos x}{\sin x} = \dfrac{\sin x}{1 + \cos x}$

7. $\dfrac{1 + \tan \theta}{1 + \cot \theta} = \dfrac{\sec \theta}{\csc \theta}$

8. $\dfrac{\cot \theta - 1}{1 - \tan \theta} = \dfrac{\csc \theta}{\sec \theta}$

9. $\dfrac{\sin x + \cos x}{\sec x + \csc x} = \dfrac{\sin x}{\sec x}$

10. $\dfrac{\sin x - \cos x}{\sec x - \csc x} = \dfrac{\cos x}{\csc x}$

11. $\dfrac{1 + \tan \theta}{1 - \tan \theta} + \dfrac{1 + \cot \theta}{1 - \cot \theta} = 0$

12. $\dfrac{\cos^2 \theta + \cot \theta}{\cos^2 \theta - \cot \theta} = \dfrac{\cos^2 \theta \tan \theta + 1}{\cos^2 \theta \tan \theta - 1}$

13. $\dfrac{1 + \cos 2\theta}{\sin 2\theta} = \cot \theta$

14. $\dfrac{2 \tan \theta}{1 + \tan^2 \theta} = \sin 2\theta$

15. $\sec 2\theta = \dfrac{\sec^2 \theta}{2 - \sec^2 \theta}$

16. $\cot 2\theta = \dfrac{\cot^2 \theta - 1}{2 \cot \theta}$

17. $\dfrac{\sin (\alpha + \beta)}{\cos \alpha \cos \beta} = \tan \alpha + \tan \beta$

18. $\dfrac{\cos (\alpha - \beta)}{\cos \alpha \sin \beta} = \tan \alpha + \cot \beta$

19. $1 - \cos 5\theta \cos 3\theta - \sin 5\theta \sin 3\theta = 2 \sin^2 \theta$

20. $2 \sin \theta \cos^3 \theta + 2 \sin^3 \theta \cos \theta = \sin 2\theta$

21. $\dfrac{\tan \theta + \sin \theta}{2 \tan \theta} = \cos^2 \dfrac{\theta}{2}$

22. $\dfrac{\tan \theta - \sin \theta}{2 \tan \theta} = \sin^2 \dfrac{\theta}{2}$

23. $\cos^4 x - \sin^4 x = \cos 2x$

24. $\dfrac{\cos^4 x - \sin^4 x}{1 - \tan^4 x} = \cos^4 x$

25. $\dfrac{\tan 3\theta - \tan \theta}{1 + \tan 3\theta \tan \theta} = \dfrac{2 \tan \theta}{1 - \tan^2 \theta}$

26. $\left(\dfrac{1 + \tan \theta}{1 - \tan \theta} \right)^2 = \dfrac{1 + \sin 2\theta}{1 - \sin 2\theta}$

27. $\dfrac{\cos^3 x - \sin^3 x}{\cos x - \sin x} = \dfrac{2 + \sin 2x}{2}$

28. $\dfrac{\sin^3 t + \cos^3 t}{\sin t + \cos t} = \dfrac{2 - \sin 2t}{2}$

29. $\sin (\alpha + \beta) \sin (\alpha - \beta) = \sin^2 \alpha - \sin^2 \beta$

30. $\cos (\alpha + \beta) \cos (\alpha - \beta) = \cos^2 \alpha - \sin^2 \beta$

31. $\cos (\alpha + \beta) + \cos (\alpha - \beta) = 2 \cos \alpha \cos \beta$

32. $\sin (\alpha + \beta) + \sin (\alpha - \beta) = 2 \sin \alpha \cos \beta$

33. $\sin^2 x - \cos^2 x = 1 - 2 \cos^2 x$

34. $\cos^2 x (1 - \sec^2 x) = -\sin^2 x$

35. $\sin^2 \theta = \cos^2 \theta (\sec^2 \theta - 1)$

36. $\tan \theta (\tan \theta + \cot \theta) = \sec^2 \theta$

37. $\dfrac{\tan x}{\sec x - \cos x} = \dfrac{\sec x}{\tan x}$

38. $\dfrac{\tan x + \sin x}{1 + \sec x} = \sin x$

39. $\dfrac{\cos \theta + \sin \theta}{\cos \theta} = 1 + \tan \theta$

40. $\dfrac{1 + \tan^2 \theta}{\csc^2 \theta} = \tan^2 \theta$

41. $\dfrac{\tan^2 x}{1 + \tan^2 x} = \sin^2 x$

42. $\dfrac{1 + \cos^2 x}{\sin^2 x} = 2 \csc^2 x - 1$

43. $\dfrac{\cot x + \tan x}{\csc x} = \sec x$

44. $\dfrac{\csc x - \sin x}{\cot x} = \cos x$

45. $\dfrac{\tan \theta + \cot \theta}{\csc \theta} = \sec \theta$

46. $\dfrac{\csc \theta - \sin \theta}{\cos^2 \theta} = \csc \theta$

47. $\dfrac{\sin x}{1 + \cos x} + \dfrac{1 + \cos x}{\sin x} = 2 \csc x$

48. $\dfrac{1 + \sin x}{1 - \sin x} + \dfrac{\sin x - 1}{1 + \sin x} = 4 \sec x \tan x$

49. $\cos \theta (1 + \csc \theta) - \cot \theta = \cos \theta$

50. $\cos^2 \theta \cot^2 \theta = \cot^2 \theta - \cos^2 \theta$

51. $\dfrac{\tan x + \cot x}{\sec x + \csc x} = \dfrac{1}{\cos x + \sin x}$

52. $\dfrac{\cos^2 x - 1}{1 - \sec^2 x} = \dfrac{1}{\tan^2 x + 1}$

53. $(\sec\theta - \tan\theta)(1 + \csc\theta) = \cot\theta$

54. $\tan\theta - \cot\theta = (\sec\theta - \csc\theta)(\sin\theta + \cos\theta)$

55. $(\sec x + \tan x)(1 - \sin x) = \cos x$

56. $\csc x - \cot x = \dfrac{1}{\csc x + \cot x}$

57. $\cos^2\theta - \sin^2\theta = \cos^4\theta - \sin^4\theta$

58. $2\sin^2\theta\cos^2\theta + \cos^4\theta = 1 - \sin^4\theta$

59. $\dfrac{\cos\theta}{1 - \cos\theta} = \dfrac{1 + \sec\theta}{\tan^2\theta}$

60. $\dfrac{\cot\theta}{\csc\theta - 1} = \dfrac{\csc\theta + 1}{\cot\theta}$

61. $\dfrac{\sin x - \cos x}{\cos^2 x} = \dfrac{\tan^2 x - 1}{\sin x + \cos x}$

62. $\dfrac{1 + \sin x}{1 - \sin x} = (\sec x + \tan x)^2$

63. $1 + \sec^4\theta = \tan^4\theta + 2\sec^2\theta$

64. $\sec^4\theta - \tan^2\theta = \tan^4\theta + \sec^2\theta$

65. $\dfrac{1 + \tan x}{1 - \tan x} = \dfrac{\cot x + 1}{\cot x - 1}$

66. $1 + \sec x = \csc x\,(\sin x + \tan x)$

67. $\dfrac{\sin\theta\tan\theta}{\sin\theta + \tan\theta} = \dfrac{\tan\theta - \sin\theta}{\tan\theta\sin\theta}$

68. $\dfrac{\sin^3\theta - \cos^3\theta}{\sin\theta - \cos\theta} = \sin\theta\cos\theta + 1$

69. $\dfrac{\cos x + 1}{\sin x} + \dfrac{\sin x}{\cos x + 1} = \dfrac{2}{\sin x}$

70. $2\sec^2 x + \dfrac{\csc x}{1 - \csc x} = \dfrac{\csc x}{\csc x + 1}$

71. $\sec\theta\csc\theta + \dfrac{\cot\theta}{\tan\theta - 1} = \dfrac{\tan\theta}{1 - \cot\theta} - 1$

72. $\cos\theta + \dfrac{\sin\theta}{\cot\theta - 1} = \dfrac{\cos\theta}{1 - \tan\theta} - \sin\theta$

73. $\cot x + \csc x = \dfrac{\sin x}{1 - \cos x}$

74. $\dfrac{\cos x + \cot x}{1 + \csc x} = \cos x$

75. $\tan\theta + \cot\theta = \dfrac{1}{\cot\theta\sin^2\theta}$

76. $\sec^2\theta - \csc^2\theta = \dfrac{\tan\theta - \cot\theta}{\sin\theta\cos\theta}$

77. $2\cos^2 x - 1 = \cos^4 x - \sin^4 x$

78. $\sec^4 x - 4\tan^2 x = (1 - \tan^2 x)^2$

79. $(\cos x + \sin x)(1 - \sin x\cos x) = \cos^3 x + \sin^3 x$

80. $\dfrac{\tan^2 x + \sec^2 x}{\sec^4 x} = 1 - \sin^4 x$

81. Prove the second, third, and fourth of the sum–product identities using the sum and difference formulas for sine and cosine.

82. Prove the last four of the sum–product identities. (*Hint:* Use the first four sum–product identities.)

Use the sum–product identities to find identities for each of the following.

83. $\sin 3\theta - \sin 5\theta$ **84.** $\sin 7x - \sin 4x$

85. $\sin 8\theta + \sin 5\theta$ **86.** $\cos\theta - \cos 7\theta$

87. $\sin 7u\sin 5u$ **88.** $2\sin 7\theta\cos 3\theta$

89. $7\cos\theta\sin 7\theta$ **90.** $\cos 2t\sin t$

91. $\cos 55°\sin 25°$ **92.** $7\cos 5\theta\cos 7\theta$

Use the sum–product identities to prove each of the following.

93. $\sin 4\theta + \sin 6\theta = \cot\theta\,(\cos 4\theta - \cos 6\theta)$

94. $\tan 2x\,(\cos x + \cos 3x) = \sin x + \sin 3x$

95. $\cot 4x(\sin x + \sin 4x + \sin 7x)$
$$= \cos x + \cos 4x + \cos 7x$$

96. $\tan\dfrac{x + y}{2} = \dfrac{\sin x + \sin y}{\cos x + \cos y}$

97. $\cot\dfrac{x + y}{2} = \dfrac{\sin y - \sin x}{\cos x - \cos y}$

98. $\tan\dfrac{\theta + \phi}{2}\tan\dfrac{\phi - \theta}{2} = \dfrac{\cos\theta - \cos\phi}{\cos\theta + \cos\phi}$

99. $\tan\dfrac{\theta + \phi}{2}(\sin\theta - \sin\phi)$
$$= \tan\dfrac{\theta - \phi}{2}(\sin\theta + \sin\phi)$$

100. $\sin 2\theta + \sin 4\theta + \sin 6\theta = 4\cos\theta\cos 2\theta\sin 3\theta$

●   **SYNTHESIS** _____

Prove each of the following identities. If you have a grapher, verify your results.

101. $\ln|\sec x| = -\ln|\cos x|$

102. $\ln|\tan x| = -\ln|\cot x|$

103. $\ln|\tan x| = \ln|\sin x| - \ln|\cos x|$

104. $\ln|\csc x| = -\ln|\sin x|$

105. $\ln e^{\sin t} = \sin t$

106. $e^{\ln|\cos t|} = |\cos t|$

107. $\ln|\csc\theta - \cot\theta| = -\ln|\csc\theta + \cot\theta|$

108. $\ln|\sec\theta + \tan\theta| = -\ln|\sec\theta - \tan\theta|$

109. Show that $\log(\cos x - \sin x) + \log(\cos x + \sin x) = \log\cos 2x$.

110. The following equation occurs in the study of mechanics:

$$\sin \theta = \frac{I_1 \cos \phi}{\sqrt{(I_1 \cos \phi)^2 + (I_2 \sin \phi)^2}}.$$

It can happen that $I_1 = I_2$. Assuming that this happens, simplify the equation.

111. In the theory of alternating current, the following equation occurs:

$$R = \frac{1}{\omega C(\tan \theta + \tan \phi)}.$$

Show that this equation is equivalent to

$$R = \frac{\cos \theta \cos \phi}{\omega C \sin (\theta + \phi)}.$$

● **CHALLENGE**

112. In electrical theory, the following equations occur:

$$E_1 = \sqrt{2}E_t \cos \left(\theta + \frac{\pi}{P} \right)$$

and

$$E_2 = \sqrt{2}E_t \cos \left(\theta - \frac{\pi}{P} \right).$$

Assuming that these equations hold, show that

$$\frac{E_1 + E_2}{2} = \sqrt{2}E_t \cos \theta \cos \frac{\pi}{P}$$

and

$$\frac{E_1 - E_2}{2} = -\sqrt{2}E_t \sin \theta \sin \frac{\pi}{P}.$$

OBJECTIVE

You should be able to:

▲ Find values of the inverse trigonometric functions.

7.6

INVERSES OF THE TRIGONOMETRIC FUNCTIONS

In this section, we develop inverse trigonometric functions. It may be helpful for you to review the material on inverse functions in Section 5.1.

The graphs of the sine, cosine, tangent, and cotangent functions follow. Do these functions have inverses that are functions? They do if these functions are one-to-one, which means that they pass the horizontal-line test.

The sine function

The cosine function

The tangent function

The cotangent function

Note that each function has a horizontal line that crosses the graph more than once. Therefore, none of them has an inverse that is a function.

Recall that to obtain the inverse of any relation, we interchange the first and second members of each ordered pair in the relation. If a relation is defined in terms of, say, x and y, interchanging x and y produces an equation of the inverse relation. The graphs of an equation and its inverse are reflections of each other across the line $y = x$. Let us examine the inverses of each of the four trigonometric functions graphed above. The graphs are as follows.

We can check again to see whether these are graphs of functions by using the vertical-line test. In each case, there is a vertical line that crosses the graph more than once, so each fails to be a function.

Let us consider specifically the graph of the inverse of $y = \sin x$, which is $x = \sin y$. Consider the input $x = 1/2$. On the graph of the inverse of the sine function, we draw a vertical line at $x = 1/2$, as shown.

It intersects the graph at points whose y-value is such that $1/2 = \sin y$. Some numbers whose sine is $1/2$ are $\pi/6$, $5\pi/6$, $-7\pi/6$, and so on. From the graph, we see that $\pi/6$ plus any multiple of 2π is such a number. Also, $5\pi/6$ plus any multiple of 2π is such a number. The complete set of values is given by $\pi/6 + 2k\pi$, k an integer, and $5\pi/6 + 2k\pi$, k an integer. Indeed, the vertical line crosses the curve at many points, verifying that the inverse of the sine function is not a function.

A   Restricting Ranges to Define Inverse Functions

Recall that a function like $f(x) = x^2$ does not have an inverse that is a function, but by restricting the domain of f to nonnegative numbers, we have a new squaring function, $f(x) = x^2$, $x \geq 0$, that has an inverse, $f^{-1}(x) = \sqrt{x}$, $x \geq 0$. This is equivalent to restricting the range of the inverse relation to exclude ordered pairs that contain negative numbers.

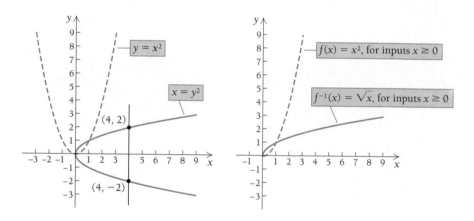

In a similar manner, we can define new trigonometric functions whose inverses are functions. We do this by restricting the ranges of the inverses. This can be done in many ways, but we choose restrictions that are fairly standard in mathematics. For the inverse sine function, we choose a range close to the origin that allows all inputs in the interval $[-1, 1]$ to have function values in the interval $[-\pi/2, \pi/2]$. For the inverse cosine function, we choose a range close to the origin that allows all inputs in the interval $[-1, 1]$ to have function values in the interval $[0, \pi]$. For the inverse tangent function, the domain is the set of all real numbers and the range is the interval $(-\pi/2, \pi/2)$. For the inverse cotangent function, the domain is the set of all real numbers and the range is the interval $(0, \pi)$.

We denote the inverse trigonometric functions as follows:

DEFINITION   **The Inverse Sine Function**

$y = \sin^{-1} x$,   or $y = \arcsin x$,   where $x = \sin y$.

The domain $= [-1, 1]$ and the range $= [-\pi/2, \pi/2]$.

The notation $\arcsin x$ arises because it is the length of an arc on the unit circle for which the sine is x. The notation $\sin^{-1} x$ is *not* exponential notation. It does not mean $1/\sin x$! Either of the two kinds of notation above can be read "the inverse sine of x" or "the arc sine of x" or "the number (or angle) whose sine is x." Notation is chosen similarly for the inverse of the other trigonometric functions: $\cos^{-1} x$, or arccos x, $\tan^{-1} x$, or arctan x, and so on.

The following diagrams show the restricted ranges for the inverse trigonometric functions on a unit circle. These ranges should be memorized. The missing endpoints indicate inputs that are not in the domain of the original function.

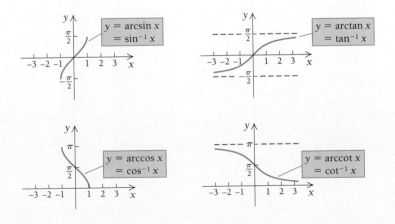

The graphs of the inverse trigonometric functions are as follows.

DO EXERCISES 1–4. (EXERCISE 4 IS ON THE FOLLOWING PAGE.)

In Margin Exercises 1–4, the graph shown is an inverse of a trigonometric function. In each case, shade the restricted range so the inverse is a function.

1.

2.

3.

4.

Find each of the following.

5. $\arcsin \dfrac{\sqrt{3}}{2}$

6. $\cos^{-1}\left(-\dfrac{\sqrt{2}}{2}\right)$

7. $\operatorname{arccot}(-1)$

8. $\tan^{-1}(-1)$

Now let us find some function values.

Example 1   Find $\sin^{-1}\frac{1}{2}$.

Solution   In the restricted range, as shown in the figure, the only number whose sine is $\frac{1}{2}$ is $\pi/6$. Therefore, $\sin^{-1}\frac{1}{2} = \pi/6$. In degrees, we have $\sin^{-1}\frac{1}{2} = 30°$.

arcsine

Example 2   Find $\arcsin \dfrac{\sqrt{2}}{2}$.

Solution   In the restricted range, as shown in the figure, the only number whose sine is $\sqrt{2}/2$ is $\pi/4$. Thus, $\arcsin(\sqrt{2}/2) = \pi/4$, or $45°$.

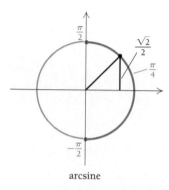

arcsine

Example 3   Find $\cos^{-1}(-\frac{1}{2})$.

Solution   The only number whose cosine is $-\frac{1}{2}$ in the restricted range $[0, \pi]$ is $2\pi/3$. Thus, $\cos^{-1}(-\frac{1}{2}) = 2\pi/3$, or $120°$.

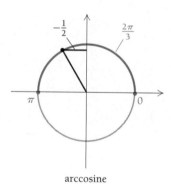

arccosine

Example 4   Find $\tan^{-1} 1$.

Solution   The only number in the restricted range $(-\pi/2, \pi/2)$ whose tangent is 1 is $\pi/4$, or $45°$.

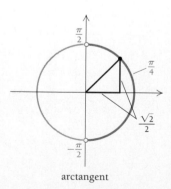

arctangent

DO EXERCISES 5–8.

We can also use a calculator to find inverse trigonometric function values. Some calculators give inverse function values in both radians and degrees. Some give values only in degrees. The key strokes involved in finding inverse function values vary with the calculator. Be sure to read the instructions for the calculator you are using, or try a simple value such as $\sin^{-1} 1$. If you get 90, you know that the values are in degrees. If you get 1.570796327, which is about $\pi/2$, you know the values are in radians.

Example 5   Find $\cos^{-1}(-0.925678)$ in degrees, using a calculator.

Solution   Using the inverse trigonometric keys on a calculator, we find that

$$\cos^{-1}(-0.925678) \approx 157.77°.$$
◀

DO EXERCISES 9–12.

The following is a summary of the domains and the ranges of the trigonometric functions together with a summary of the domains and the ranges of the inverse trigonometric functions. For completeness, we have included the arcsecant and the arccosecant, though there is a lack of uniformity on their definitions in the mathematical literature.

Find each of the following using a calculator.

9. $\cos^{-1} 0.63254$

10. $\arcsin(-0.10203)$

11. $\arctan 1.34568$

12. $\tan^{-1}(-22.467)$

Function	Domain	Range	Inverse Function	Domain	Range
sin	All reals, $(-\infty, \infty)$	$[-1, 1]$	$\sin^{-1}$	$[-1, 1]$	$\left[-\dfrac{\pi}{2}, \dfrac{\pi}{2}\right]$
cos	All reals, $(-\infty, \infty)$	$[-1, 1]$	$\cos^{-1}$	$[-1, 1]$	$[0, \pi]$
tan	All reals except $k\pi/2$, k odd	All reals, $(-\infty, \infty)$	$\tan^{-1}$	All reals, $(-\infty, \infty)$	$\left(-\dfrac{\pi}{2}, \dfrac{\pi}{2}\right)$
cot	All reals except $k\pi$	All reals, $(-\infty, \infty)$	$\cot^{-1}$	All reals, $(-\infty, \infty)$	$(0, \pi)$
sec	All reals except $k\pi/2$, k odd	$(-\infty, -1] \cup [1, \infty)$	$\sec^{-1}$	$(-\infty, -1] \cup [1, \infty)$	$\left[0, \dfrac{\pi}{2}\right) \cup \left[\pi, \dfrac{3\pi}{2}\right)$
csc	All reals except $k\pi$	$(-\infty, -1] \cup [1, \infty)$	$\csc^{-1}$	$(-\infty, -1] \cup [1, \infty)$	$\left(0, \dfrac{\pi}{2}\right] \cup \left(-\pi, -\dfrac{\pi}{2}\right]$

● **EXERCISE SET** **7.6**

A   Find each of the following without using a calculator.

1. $\arcsin \dfrac{\sqrt{2}}{2}$

2. $\arcsin \dfrac{\sqrt{3}}{2}$

3. $\cos^{-1} \dfrac{\sqrt{2}}{2}$

4. $\cos^{-1} \dfrac{\sqrt{3}}{2}$

5. $\sin^{-1}\left(-\dfrac{\sqrt{2}}{2}\right)$

6. $\sin^{-1}\left(-\dfrac{\sqrt{3}}{2}\right)$

7. $\arccos\left(-\dfrac{\sqrt{2}}{2}\right)$

8. $\arccos\left(-\dfrac{\sqrt{3}}{2}\right)$

9. $\arctan \sqrt{3}$

10. $\arctan \dfrac{\sqrt{3}}{3}$

11. $\cot^{-1} 1$

12. $\cot^{-1} \sqrt{3}$

13. $\arctan\left(-\dfrac{\sqrt{3}}{3}\right)$

14. $\arctan(-\sqrt{3})$

15. $\operatorname{arccot}(-1)$

16. $\operatorname{arccot}(-\sqrt{3})$

17. $\operatorname{arcsec} 1$

18. $\operatorname{arcsec} 2$

19. $\csc^{-1} 1$

20. $\csc^{-1} 2$

21. $\arcsin\left(-\dfrac{\sqrt{2}}{2}\right)$

22. $\arcsin \dfrac{1}{2}$

23. $\cos^{-1} \dfrac{1}{2}$

24. $\arccos \dfrac{\sqrt{2}}{2}$

25. $\arcsin\left(-\dfrac{\sqrt{3}}{2}\right)$

26. $\sin^{-1}\left(-\dfrac{1}{2}\right)$

27. $\cos^{-1}\left(-\dfrac{1}{2}\right)$

28. $\cos^{-1} 0$

29. $\tan^{-1} 1$

30. $\tan^{-1} 0$

31. $\operatorname{arccot}\left(-\dfrac{\sqrt{3}}{3}\right)$

32. $\operatorname{arccot}(-\sqrt{3})$

Use a calculator to find each of the following in degrees.

33. $\arcsin 0.3907$

34. $\arcsin 0.9613$

35. $\sin^{-1}(-0.619867)$

36. $\sin^{-1}(-0.867314)$

37. $\arccos 0.7990$

38. $\arccos 0.9265$

39. $\cos^{-1}(-0.981028)$

40. $\cos^{-1}(-0.271568)$

41. $\tan^{-1} 0.3673$

42. $\tan^{-1} 1.091$

43. $\cot^{-1} 1.265$

44. $\cot^{-1} 0.4770$

45. $\sec^{-1} 1.1677$

46. $\sec^{-1}(-1.4402)$

47. $\operatorname{arccsc}(-6.2774)$

48. $\operatorname{arccsc} 1.11123$

49. $\arcsin 0.2334$

50. $\arcsin 0.4514$

51. $\sin^{-1}(-0.6361)$

52. $\sin^{-1}(-0.8192)$

53. $\arcsin(-0.8886)$

54. $\arccos(-0.2935)$

55. $\tan^{-1}(-0.4087)$

56. $\tan^{-1}(-0.2410)$

57. $\cot^{-1}(-5.936)$

58. $\cot^{-1}(-1.319)$

59. $\cot^{-1}(-23)$

60. $\tan^{-1}(158)$

● **SYNTHESIS** _____

61. Graph the function $y = \sec^{-1} x$.

62. Graph the function $y = \csc^{-1} x$.

Show that each of the following is *not* an identity.

63. $\sin^{-1} x = (\sin x)^{-1}$

64. $\cos^{-1} x = (\cos x)^{-1}$

65. $\tan^{-1} x = (\tan x)^{-1}$

66. $\cot^{-1} x = (\cot x)^{-1}$

67. $\tan^{-1} x = \dfrac{\sin^{-1} x}{\cos^{-1} x}$

68. $(\cos^{-1} x)^2 + (\sin^{-1} x)^2 = 1$

69. A guy wire is attached to the top of a 50-ft pole and stretched to a point that is b ft from the bottom of the pole. Show that the angle of inclination θ of the wire to the top is given by

$$\theta = \tan^{-1}\left(\dfrac{50}{b}\right).$$

70. An airplane at an altitude of 2000 ft is flying toward an island. The straight-line distance from the airplane to the island is h ft. Show that θ, which is called the *angle of depression*, is given by

$$\theta = \sin^{-1}\left(\dfrac{2000}{h}\right).$$

71. a) Use your calculator to approximate the following expression:

$$16\tan^{-1}\dfrac{1}{5} - 4\tan^{-1}\dfrac{1}{239}.$$

b) What number does this expression seem to approximate?

7.7
COMPOSITION OF TRIGONOMETRIC FUNCTIONS AND THEIR INVERSES

A Immediate Simplification

Various compositions of trigonometric functions and their inverses often arise in practice. For example, we might want to try to simplify expressions such as

$$\sin\left(\sin^{-1} x\right) \quad \text{or} \quad \sin\left(\text{arccot}\,\frac{x}{2}\right).$$

In the expression on the left, we are finding "the sine of a number whose sine is x." Recall from Section 5.1 that if a function f has an inverse that is also a function, then

$$f(f^{-1}(x)) = x, \quad \text{for all } x \text{ in the domain of } f^{-1},$$

and

$$f^{-1}(f(x)) = x, \quad \text{for all } x \text{ in the domain of } f.$$

Thus, if $f(x) = \sin x$ and $f^{-1}(x) = \sin^{-1} x$, then $\sin\left(\sin^{-1} x\right) = x$, if x is in the domain of $\sin^{-1}$, which is any number in the interval $[-1, 1]$. Similar results hold for the other trigonometric functions. These yield the following theorem.

THEOREM 1

    a) $\sin\left(\sin^{-1} x\right) = x$, for any x in the domain of $\sin^{-1}$.
    b) $\cos\left(\cos^{-1} x\right) = x$, for any x in the domain of $\cos^{-1}$.
    c) $\tan\left(\tan^{-1} x\right) = x$, for any x in the domain of $\tan^{-1}$.
    d) $\cot\left(\cot^{-1} x\right) = x$, for any x in the domain of $\cot^{-1}$.
    e) $\sec\left(\sec^{-1} x\right) = x$, for any x in the domain of $\sec^{-1}$.
    f) $\csc\left(\csc^{-1} x\right) = x$, for any x in the domain of $\csc^{-1}$.

Example 1 Find $\cos\left(\cos^{-1}\dfrac{\sqrt{3}}{2}\right)$.

Solution Since $\sqrt{3}/2$ is in the interval $[-1, 1]$, it follows that

$$\cos\left(\cos^{-1}\frac{\sqrt{3}}{2}\right) = \frac{\sqrt{3}}{2}.$$

◀

Example 2 Find $\sin\left(\sin^{-1} 12.3\right)$.

Solution Since 12.3 is not in the interval $[-1, 1]$, we cannot evaluate this expression. Thinking one step at a time, we know that there is no number whose sine is 12.3. Since we cannot find $\sin^{-1} 12.3$, we cannot evaluate the expression.

◀

DO EXERCISES 1–4.

Find each of the following.

1. $\sin\left(\sin^{-1}\dfrac{1}{2}\right)$

2. $\tan\left(\tan^{-1} 1\right)$

3. $\cos\left(\arccos\dfrac{\sqrt{2}}{2}\right)$

4. $\sin\left[\sin^{-1}(-5.7)\right]$

Find each of the following.

5. $\cos^{-1}\left(\cos\dfrac{2\pi}{3}\right)$

6. $\arctan\left(\tan\dfrac{3\pi}{4}\right)$

7. $\sin^{-1}\left(\sin\dfrac{\pi}{6}\right)$

8. $\arccos\left[\cos\left(-\dfrac{2\pi}{3}\right)\right]$

Find each of the following.

9. $\sin(\arctan 1)$

10. $\cos\left(\arcsin\dfrac{1}{2}\right)$

11. $\cos^{-1}\left(\sin\dfrac{\pi}{6}\right)$

12. $\sin^{-1}\left(\tan\dfrac{\pi}{4}\right)$

13. $\cos(\sin^{-1} 0)$

Now let us consider an expression like $\sin^{-1}(\sin x)$. We might also suspect that this is equal to x for any x, but this is not true unless x is in the range of the $\sin^{-1}$ function. Note that in order to define $\sin^{-1}$, we had to restrict the domain of the sine function. In so doing, we restricted the range of the inverse sine function.

Example 3   Find $\sin^{-1}\left(\sin\dfrac{3\pi}{4}\right)$.

Solution   We first find $\sin(3\pi/4)$. It is $\sqrt{2}/2$. Next we find $\sin^{-1}(\sqrt{2}/2)$. It is $\pi/4$. So

$$\sin^{-1}\left(\sin\frac{3\pi}{4}\right)=\frac{\pi}{4}.$$ ◀

DO EXERCISES 5–8.

For any x in the range of the arcsine function, $[-\pi/2, \pi/2]$, we do have $\arcsin(\sin x)=x$, or $\sin^{-1}(\sin x)=x$. Similar conditions hold for the other functions.

THEOREM 2

The following are true for any x in the range of the inverse function.

a) $\sin^{-1}(\sin x)=x$, for any x in the range of $\sin^{-1}$.
b) $\cos^{-1}(\cos x)=x$, for any x in the range of $\cos^{-1}$.
c) $\tan^{-1}(\tan x)=x$, for any x in the range of $\tan^{-1}$.
d) $\cot^{-1}(\cot x)=x$, for any x in the range of $\cot^{-1}$.
e) $\sec^{-1}(\sec x)=x$, for any x in the range of $\sec^{-1}$.
f) $\csc^{-1}(\csc x)=x$, for any x in the range of $\csc^{-1}$.

B   Simplifying Other Compositions

Now we find some other function compositions.

Example 4   Find $\sin[\arctan(-1)]$.

Solution   We first find $\arctan(-1)$. It is $-\pi/4$. Now we find the sine of this, which is $\sin(-\pi/4)=-\sqrt{2}/2$, so $\sin[\arctan(-1)]=-\sqrt{2}/2$.   ◀

Example 5   Find $\cos^{-1}\left(\sin\dfrac{\pi}{2}\right)$.

Solution   We first find $\sin(\pi/2)$. It is 1. Now we find $\cos^{-1} 1$. It is 0, so $\cos^{-1}[\sin(\pi/2)]=0$.   ◀

DO EXERCISES 9–13.

Now let us consider

$\cos(\arcsin\frac{3}{5})$.

Without using a calculator, we cannot find arcsin $\frac{3}{5}$. However, we can still evaluate the entire expression without using a calculator. We are looking for an angle θ such that arcsin $\frac{3}{5} = \theta$, or $\sin \theta = \frac{3}{5}$.

We sketch a triangle, as shown above. The angle θ in this triangle is an angle whose sine is $\frac{3}{5}$ (it is arcsin $\frac{3}{5}$). We wish to find the cosine of this angle. Since the triangle is a right triangle, we can find the length of the base, b. It is 4. Thus we know that $\cos \theta = b/5$, or $\frac{4}{5}$. Therefore,

$$\cos\left(\arcsin \tfrac{3}{5}\right) = \tfrac{4}{5}.$$

Example 6 Find $\sin\left(\operatorname{arccot}\dfrac{x}{2}\right)$.

Solution   We draw a right triangle whose legs have lengths x and 2, so that $\cot \theta = x/2$. If x is negative, we get the triangle in standard position shown on the left below. If x is positive, we get the triangle shown on the right.

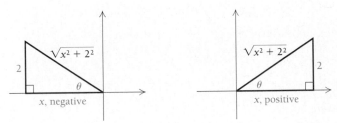

We find the length of the hypotenuse and then read off the sine ratio. In either case, we get

$$\sin\left(\operatorname{arccot}\frac{x}{2}\right) = \frac{2}{\sqrt{x^2 + 2^2}}, \quad \text{or} \quad \frac{2}{\sqrt{x^2 + 4}}. \qquad \blacktriangleleft$$

Example 7 Find $\cos\left(\arctan p\right)$.

Solution   We draw two right triangles, whose legs have lengths p and 1, so that $\tan \theta = p/1$.

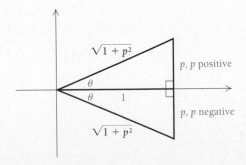

Find each of the following.

14. $\cos\left(\arctan\dfrac{b}{3}\right)$

15. $\sin\left(\arctan 2\right)$

$\left(Hint:\ 2=\dfrac{2}{1}.\right)$

16. $\tan\left(\arcsin t\right)$

Evaluate.

17. $\cos\left(\sin^{-1}\dfrac{\sqrt{3}}{2}-\cos^{-1}\dfrac{1}{2}\right)$

18. $\tan\left(\dfrac{1}{2}\arcsin\dfrac{3}{5}\right)$

(*Hint:* Let $\arcsin\frac{3}{5}=u$ and use a half-angle formula.)

We find the length of the other side and then read off the cosine ratio. In either case, we get

$$\cos\left(\arctan p\right)=\frac{1}{\sqrt{1+p^2}}.$$

◀

The idea in Examples 6 and 7 is to first sketch a triangle, or triangles, two of whose sides have the appropriate ratio. Then use the Pythagorean theorem to find the length of the third side and read off the desired ratio.

DO EXERCISES 14–16.

In some cases, we use certain identities to evaluate expressions.

Example 8   Evaluate $\sin\left(\sin^{-1}\frac{1}{2}+\cos^{-1}\frac{4}{5}\right)$.

Solution   We simplify this expression by letting

$$u=\sin^{-1}\tfrac{1}{2}\quad\text{and}\quad v=\cos^{-1}\tfrac{4}{5}.$$

Then we have $\sin\left(u+v\right)$. By a sum formula, this is equivalent to

$$\sin u\cos v+\cos u\sin v.$$

Now we make the reverse substitutions and obtain

$$\sin\left(\sin^{-1}\tfrac{1}{2}\right)\cdot\cos\left(\cos^{-1}\tfrac{4}{5}\right)+\cos\left(\sin^{-1}\tfrac{1}{2}\right)\cdot\sin\left(\cos^{-1}\tfrac{4}{5}\right).$$

This immediately simplifies to

$$\tfrac{1}{2}\cdot\tfrac{4}{5}+\cos\left(\sin^{-1}\tfrac{1}{2}\right)\cdot\sin\left(\cos^{-1}\tfrac{4}{5}\right).$$

Now $\cos\left(\sin^{-1}\frac{1}{2}\right)$ simplifies to $\sqrt{3}/2$. To find $\sin\left(\cos^{-1}\frac{4}{5}\right)$, we need a triangle. We set up the triangle so that $\cos^{-1}\frac{4}{5}=\theta$, or $\cos\theta=\frac{4}{5}$. Then we find $\sin\theta$.

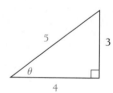

We see that $\sin\left(\cos^{-1}\frac{4}{5}\right)=\frac{3}{5}$. Our expression is now simplified to

$$\frac{1}{2}\cdot\frac{4}{5}+\frac{\sqrt{3}}{2}\cdot\frac{3}{5},$$

or

$$\frac{4+3\sqrt{3}}{10}.$$

Thus,

$$\sin\left(\sin^{-1}\frac{1}{2}+\cos^{-1}\frac{4}{5}\right)=\frac{4+3\sqrt{3}}{10}.$$

◀

DO EXERCISES 17 AND 18.

EXERCISE SET 7.7

▶ Evaluate or simplify.

1. $\sin(\arcsin 0.3)$

2. $\cos(\arccos 0.2)$

3. $\tan[\tan^{-1}(-4.2)]$

4. $\cot[\cot^{-1}(-1.5)]$

5. $\arcsin\left(\sin\dfrac{2\pi}{3}\right)$

6. $\arccos\left(\cos\dfrac{3\pi}{2}\right)$

7. $\sin^{-1}\left[\sin\left(-\dfrac{3\pi}{4}\right)\right]$

8. $\cos^{-1}\left[\cos\left(-\dfrac{\pi}{4}\right)\right]$

9. $\sin^{-1}\left(\sin\dfrac{\pi}{5}\right)$

10. $\cos^{-1}\left(\cos\dfrac{\pi}{7}\right)$

11. $\tan^{-1}\left(\tan\dfrac{2\pi}{3}\right)$

12. $\cot^{-1}\left(\cot\dfrac{2\pi}{3}\right)$

▶ Evaluate or simplify.

13. $\sin(\arctan\sqrt{3})$

14. $\sin\left(\arctan\dfrac{\sqrt{3}}{3}\right)$

15. $\cos\left(\arcsin\dfrac{\sqrt{3}}{2}\right)$

16. $\cos\left(\arcsin\dfrac{\sqrt{2}}{2}\right)$

17. $\tan\left(\cos^{-1}\dfrac{\sqrt{2}}{2}\right)$

18. $\tan\left(\cos^{-1}\dfrac{\sqrt{3}}{2}\right)$

19. $\cos^{-1}\left(\sin\dfrac{\pi}{3}\right)$

20. $\cos^{-1}(\sin\pi)$

21. $\arcsin\left(\cos\dfrac{\pi}{6}\right)$

22. $\arcsin\left(\cos\dfrac{\pi}{4}\right)$

23. $\sin^{-1}\left(\tan\dfrac{\pi}{4}\right)$

24. $\sin^{-1}\left[\tan\left(-\dfrac{\pi}{4}\right)\right]$

25. $\sin\left(\arctan\dfrac{x}{2}\right)$

26. $\sin\left(\arctan\dfrac{a}{3}\right)$

27. $\tan\left(\cos^{-1}\dfrac{3}{x}\right)$

28. $\cot\left(\sin^{-1}\dfrac{5}{y}\right)$

29. $\cot\left(\sin^{-1}\dfrac{a}{b}\right)$

30. $\tan\left(\cos^{-1}\dfrac{p}{q}\right)$

31. $\cos\left(\tan^{-1}\dfrac{\sqrt{2}}{3}\right)$

32. $\cos\left(\tan^{-1}\dfrac{\sqrt{3}}{4}\right)$

33. $\tan(\arcsin 0.1)$

34. $\tan(\arcsin 0.2)$

35. $\cot[\cos^{-1}(-0.2)]$

36. $\cot[\cos^{-1}(-0.3)]$

37. $\sin(\operatorname{arccot} y)$

38. $\cos(\operatorname{arccot} x)$

39. $\cos(\arctan t)$

40. $\sin(\arctan t)$

41. $\cot(\sin^{-1} y)$

42. $\tan(\cos^{-1} y)$

43. $\sin(\cos^{-1} x)$

44. $\cos(\sin^{-1} x)$

45. $\tan\left(\arcsin\dfrac{p}{\sqrt{p^2+9}}\right)$

46. $\csc\left(\tan^{-1}\dfrac{\sqrt{25-p^2}}{p}\right)$

47. $\tan\left(\dfrac{1}{2}\arcsin\dfrac{4}{5}\right)$

48. $\tan\left(\dfrac{1}{2}\arcsin\dfrac{1}{2}\right)$

49. $\cos\left(\dfrac{1}{2}\arcsin\dfrac{1}{2}\right)$

50. $\cos\left(\dfrac{1}{2}\arcsin\dfrac{\sqrt{3}}{2}\right)$

51. $\sin\left(2\cos^{-1}\dfrac{3}{5}\right)$

52. $\sin\left(2\cos^{-1}\dfrac{1}{2}\right)$

53. $\cos\left(2\sin^{-1}\dfrac{5}{13}\right)$

54. $\cos\left(2\cos^{-1}\dfrac{4}{5}\right)$

55. $\sin\left(\sin^{-1}\dfrac{1}{2}+\cos^{-1}\dfrac{3}{5}\right)$

56. $\sin\left(\sin^{-1}\dfrac{1}{2}-\cos^{-1}\dfrac{4}{5}\right)$

57. $\cos\left(\sin^{-1}\dfrac{\sqrt{2}}{2}+\cos^{-1}\dfrac{3}{5}\right)$

58. $\cos\left(\sin^{-1}\dfrac{4}{5}-\cos^{-1}\dfrac{1}{2}\right)$

59. $\sin(\sin^{-1} x+\cos^{-1} y)$

60. $\sin(\sin^{-1} x-\cos^{-1} y)$

61. $\cos(\sin^{-1} x+\cos^{-1} y)$

62. $\cos(\sin^{-1} x-\cos^{-1} y)$

63. ▪ $\sin(\sin^{-1} 0.6032+\cos^{-1} 0.4621)$

64. ▪ $\cos(\sin^{-1} 0.7325-\cos^{-1} 0.4838)$

● **SYNTHESIS** _____

65. Suppose that $\theta=\sin^{-1} x$. Find expressions in terms of x for $\sin\theta$, $\cos\theta$, $\tan\theta$, $\cot\theta$, $\sec\theta$, and $\csc\theta$.

66. Suppose that $\theta=\arccos x$. Find expressions in terms of x for $\sin\theta$, $\cos\theta$, $\tan\theta$, $\cot\theta$, $\sec\theta$, and $\csc\theta$.

67. Suppose that $\theta=\tan^{-1} x$. Find expressions in terms of x for $\sin\theta$, $\cos\theta$, $\tan\theta$, $\cot\theta$, $\sec\theta$, and $\csc\theta$.

68. Suppose that $\theta=\operatorname{arccot} x$. Find expressions in terms of x for $\sin\theta$, $\cos\theta$, $\tan\theta$, $\cot\theta$, $\sec\theta$, and $\csc\theta$.

Prove each of the following identities.

69. $\sin^{-1} x+\cos^{-1} x=\dfrac{\pi}{2}$

70. $\tan^{-1} x+\cot^{-1} x=\dfrac{\pi}{2}$

71. $\sin^{-1} x=\tan^{-1}\dfrac{x}{\sqrt{1-x^2}}$

72. $\tan^{-1} x=\sin^{-1}\dfrac{x}{\sqrt{x^2+1}}$

73. For $x\geq 0$,

$$\arcsin x=\arccos\sqrt{1-x^2}.$$

74. For $x > 0$,

$$\arccos x = \arctan \frac{\sqrt{1 - x^2}}{x}.$$

75. An observer's eye is at a point A, looking at a mural of height h, with the bottom of the mural y feet above the eye. The eye is x feet from the wall. Write an expression for θ in terms of x, y, and h.

76. ▇ Evaluate the expression found in Exercise 75 when $x = 20$ ft, $y = 7$ ft, and $h = 25$ ft.

OBJECTIVES

You should be able to:

A Solve simple trigonometric equations not requiring the use of calculators, finding all solutions, or all solutions in $[0, 2\pi)$.

B Solve simple trigonometric equations requiring the use of calculators, finding all solutions, or all solutions in $[0°, 360°)$.

C Solve trigonometric equations by factoring or using the quadratic formula.

1. Solve. Give answers in both degrees and radians.

$$2 \cos x = 1$$

7.8

TRIGONOMETRIC EQUATIONS

A ▶ **Simple Equations**

When an equation contains a trigonometric expression with a variable, such as $\sin x$, it is called a **trigonometric equation.** We have worked with many trigonometric equations that are identities. Now we consider equations that may not be identities. To solve such an equation, we find all replacements for the variable that make the equation true.

Example 1   Solve: $2 \sin x = 1$.

Solution   We first solve for $\sin x$: $\sin x = \frac{1}{2}$. Now we note that the solutions are those numbers having a sine of $\frac{1}{2}$. We look for them. The unit circle is helpful.

There are just two points on it for which the sine is $\frac{1}{2}$, as shown. They are the points for $\pi/6$ and $5\pi/6$. These numbers, plus any multiple of 2π, are the solutions

$$\frac{\pi}{6} + 2k\pi \quad \text{or} \quad \frac{5\pi}{6} + 2k\pi,$$

where k is any integer. In degrees, the solutions are

$$30° + k \cdot 360° \quad \text{or} \quad 150° + k \cdot 360°,$$

where k is any integer.   ◀

Note that what we did in Example 1 was comparable to solving $x = \sin^{-1} \frac{1}{2}$, only now we get *all* the numbers whose sines are $\frac{1}{2}$, not just $\pi/6$.

DO EXERCISE 1.

Example 2   Solve: $4 \cos^2 x = 1$.

Solution We first solve for $\cos x$:

$$\cos^2 x = \tfrac{1}{4}$$
$$\cos x = \pm \tfrac{1}{2}.$$

Now we use the unit circle to find those numbers having a cosine of $\tfrac{1}{2}$ or $-\tfrac{1}{2}$. The solutions are $\pi/3$, $2\pi/3$, $4\pi/3$, $5\pi/3$, plus any multiple of 2π. In degrees, the solutions are $60°$, $120°$, $240°$, $300°$, plus any multiple of $360°$.

DO EXERCISE 2.

   In most practical applications, it is usually sufficient to find just the solutions from 0 to 2π or from $0°$ to $360°$. We then remember that any multiple of 2π, or $360°$, can be added to obtain the rest of the solutions.
   The following example illustrates that when we look for solutions in the interval $[0, 2\pi)$, or $[0°, 360°)$, we must be cautious.

Example 3 Solve $2 \sin 2x = -1$ in the interval $[0, 2\pi)$.

Solution We first solve for $\sin 2x$:

$$2 \sin 2x = -1$$
$$\sin 2x = -\tfrac{1}{2}.$$

We are looking for solutions x to the equation for which

$$0 \le x < 2\pi.$$

Multiplying by 2, we get

$$0 \le 2x < 4\pi,$$

which is the interval we consider to solve $\sin 2x = -\tfrac{1}{2}$.

   Using the unit circle, we find points $2x$ for which $\sin 2x = -\tfrac{1}{2}$ and $0 \le 2x < 4\pi$. These values of $2x$ are $7\pi/6$, $11\pi/6$, $19\pi/6$, and $23\pi/6$. Thus

---

2. Find all solutions in $[0, 2\pi)$. Give answers in both degrees and radians.

$$4 \sin^2 x = 1$$

3. Find all solutions in $[0, 2\pi)$. Leave answers in terms of π.

$$2 \cos 2x = 1$$

the desired values of x in $[0, 2\pi)$ are half of these. Therefore,

$$x = \frac{7\pi}{12}, \frac{11\pi}{12}, \frac{19\pi}{12}, \frac{23\pi}{12}.$$

◀

DO EXERCISE 3.

 Using a Calculator

In solving some trigonometric equations, it is necessary to use a calculator. Answers can then be found in radians or degrees, depending on how the calculator is set. We will usually find answers in degrees.

Example 4   Solve $2 + \sin x = 2.5263$ in $[0, 360°)$.

Solution   We first solve for $\sin x$:

$$2 + \sin x = 2.5263$$
$$\sin x = 0.5263.$$

From a calculator, we find the reference angle, $x \approx 31.76°$. Since $\sin x$ is positive, the solutions are to be found in the first and second quadrants. The solutions are $31.76°$ and $148.24°$.

Solve in $[0°, 360°)$.

4. $2 + \cos x = 2.7660$

Example 5   Solve $\sin x - 1 = -1.6381$ in $[0°, 360°)$.

Solution   We first solve for $\sin x$:

$$\sin x - 1 = -1.6381$$
$$\sin x = -0.6381.$$

5. $\cos x - 1 = -1.7660$

Using a calculator, we find that the reference angle is $39.65°$. Since $\sin x$ is negative, the solutions are to be found in the third and fourth quadrants. The solutions are $219.65°$ and $320.35°$.

DO EXERCISES 4 AND 5.

Using Algebraic Techniques

In solving trigonometric equations, we can expect to apply some algebra before concerning ourselves with the trigonometric part. In the next examples, we begin by factoring.

Example 6 Solve $8 \cos^2 \theta - 2 \cos \theta = 1$ in $[0°, 360°)$.

Solution Since we will be using the principle of zero products, we first obtain a 0 on one side of the equation:

$$8 \cos^2 \theta - 2 \cos \theta - 1 = 0$$

$$(4 \cos \theta + 1)(2 \cos \theta - 1) = 0 \qquad \text{Factoring}$$

$$4 \cos \theta + 1 = 0 \qquad \text{or} \quad 2 \cos \theta - 1 = 0 \qquad \substack{\text{Principle of} \\ \text{zero products}}$$

$$\cos \theta = -\tfrac{1}{4} = -0.25 \qquad \text{or} \qquad \cos \theta = \tfrac{1}{2}$$

$$\theta = 104.48°, 255.52° \quad \text{or} \qquad \theta = 60°, 300°.$$

The solutions in $[0°, 360°)$ are $104.48°$, $255.52°$, $60°$, and $300°$. ◀

It may be helpful in accomplishing the algebraic part of solving trigonometric equations to make substitutions as in Chapter 2. If we use such an approach, the algebraic part of Example 6 would look like this:

$$8 \cos^2 \theta - 2 \cos \theta = 1.$$

We would then let $u = \cos \theta$:

$$8u^2 - 2u = 1$$

$$(4u + 1)(2u - 1) = 0$$

$$4u + 1 = 0 \quad \text{or} \quad 2u - 1 = 0$$

$$u = -\tfrac{1}{4} \quad \text{or} \qquad u = \tfrac{1}{2}.$$

We then substitute $\cos \theta$ for u and solve for θ.

Example 7 Solve $2 \sin^2 \phi + \sin \phi = 0$ in $[0, 2\pi)$.

Solution

$$2 \sin^2 \phi + \sin \phi = 0$$

$$\sin \phi (2 \sin \phi + 1) = 0 \qquad \text{Factoring}$$

$$\sin \phi = 0 \qquad \text{or} \quad 2 \sin \phi + 1 = 0 \qquad \text{Principle of zero products}$$

$$\sin \phi = 0 \qquad \text{or} \qquad \sin \phi = -\frac{1}{2}$$

$$\phi = 0, \pi \quad \text{or} \qquad \phi = \frac{7\pi}{6}, \frac{11\pi}{6}$$

The solutions in $[0, 2\pi)$ are 0, π, $7\pi/6$, and $11\pi/6$. ◀

DO EXERCISES 6 AND 7. _____

If a trigonometric equation is quadratic but difficult or impossible to factor, we use the quadratic formula.

Solve in $[0°, 360°)$.

6. $8 \cos^2 \theta + 2 \cos \theta = 1$

7. $2 \cos^2 \phi + \cos \phi = 0$

8. Solve in $[0°, 360°)$ using the quadratic formula:

$$10 \cos^2 x - 10 \cos x - 7 = 0.$$

Example 8 Solve $10 \sin^2 x - 12 \sin x - 7 = 0$ in $[0°, 360°)$.

Solution It may help to make the substitution $u = \sin x$, in order to obtain the equation $10u^2 - 12u - 7 = 0$. We then use the quadratic formula to find u, or $\sin x$:

$$\sin x = \frac{12 \pm \sqrt{144 + 280}}{20} \quad \text{Using the quadratic formula}$$

$$= \frac{12 \pm \sqrt{424}}{20}$$

$$= \frac{12 \pm 2\sqrt{106}}{20} = \frac{6 \pm \sqrt{106}}{10}$$

$$\approx \frac{6 \pm 10.296}{10}$$

$$\sin x \approx 1.6296 \quad \text{or} \quad \sin x \approx -0.4296.$$

Since sines are never greater than 1, the first of the equations has no solution. Using the other equation, we find the reference angle to be $25.44°$. Since $\sin x$ is negative, the solutions are to be found in the third and fourth quadrants. Thus the solutions are $205.44°$ and $334.56°$. ◀

DO EXERCISE 8.

 TECHNOLOGY CONNECTION

We have used a grapher to find the solutions of equations by zooming in on their points of intersection. We now solve equations that involve trigonometric functions. For example, if we wish to find the solutions for the equation $5 \cos^2 x = 2$, we graph the left side as one function and the right side as another function and look for point(s) of intersection. The first function is $y = 5 \cos^2 x$, and the second function is $y = 2$. These functions are then graphed together on the same set of axes. If we want to find x in radian measure, the viewing box in RADIAN mode might be $[-\pi, \pi] \times [-5, 5]$. If we want to find x in degrees, the same viewing box in DEGREE mode would be $[-180, 180] \times [-5, 5]$. Either way, the result would be the following:

$y = 5 \cos^2 x$
$y = 2$

$[-\pi, \pi] \times [-5, 5]$

By zooming in on the points of intersection, we find that their coordinates (rounded to two decimal places) are $(-2.26, 2.00)$, $(-0.89, 2.00)$, $(0.89, 2.00)$, and $(2.26, 2.00)$, in RADIAN mode. If we were using DEGREE mode, the same four points would have the coordinates $(-129.23°, 2.00)$, $(-50.77°, 2.00)$, $(50.77°, 2.00)$, and $(129.23°, 2.00)$.

Use a grapher to find the solutions of each of the following trigonometric equations on the specified domain. Use the appropriate mode (RADIAN or DEGREE) to match each domain.

TC 1. $3 \sin^2 x - 4 \sin x = 5 \cos x; \ [-180°, 180°]$

TC 2. $5 \sin x \cos x = 2 + \cos^2 x; \ [0, \pi/2]$

TC 3. $10 \cos^2 x \sin^2 x = 4 + \sin x; \ [-90°, 90°]$

TC 4. $(\cos x + 2)(\sin x - 3) = 6 \cos x; \ [0, 2\pi)$

● **EXERCISE SET** **7.8**

A, B   Solve, finding all solutions.

1. $\sin x = \dfrac{\sqrt{3}}{2}$ 　　　**2.** $\cos x = \dfrac{\sqrt{3}}{2}$

3. $\cos x = \dfrac{1}{\sqrt{2}}$ 　　　**4.** $\tan x = \sqrt{3}$

5. $\sin x = 0.3448$ 　　　**6.** $\cos x = 0.6406$

Solve, finding all solutions in $[0, 2\pi)$ or $[0°, 360°)$.

7. $\cos x = -0.5495$ 　　　**8.** $\sin x = -0.4279$

9. $2 \sin x + \sqrt{3} = 0$ 　　　**10.** $\sqrt{3} \tan x + 1 = 0$

11. $2 \tan x + 3 = 0$ 　　　**12.** $4 \sin x - 1 = 0$

C   Solve, finding all solutions in $[0, 2\pi)$ or $[0°, 360°)$. If you have a grapher, verify your results.

13. $4 \sin^2 x - 1 = 0$ 　　　**14.** $2 \cos^2 x = 1$

15. $\cot^2 x - 3 = 0$ 　　　**16.** $\csc^2 x - 4 = 0$

17. $2 \sin^2 x + \sin x = 1$

18. $2 \cos^2 x + 3 \cos x = -1$

19. $\cos^2 x + 2 \cos x = 3$

20. $2 \sin^2 x - \sin x = 3$

21. $4 \sin^3 x - \sin x = 0$

22. $2 \cos^2 x - \sqrt{3} \cos x = 0$

23. $2 \sin^2 \theta + 7 \sin \theta = 4$

24. $2 \sin^2 \theta - 5 \sin \theta + 2 = 0$

25. $6 \cos^2 \phi + 5 \cos \phi + 1 = 0$

26. $2 \sin^2 \phi + \sin \phi - 1 = 0$

27. $2 \sin t \cos t + 2 \sin t - \cos t - 1 = 0$

28. $2 \sin t \tan t + \tan t - 2 \sin t - 1 = 0$

29. $\cos 2x \sin x + \sin x = 0$

30. $\sin 2x \cos x - \cos x = 0$

31. $5 \sin^2 x - 8 \sin x = 3$

32. $\cos^2 x + 6 \cos x + 4 = 0$

33. $2 \tan^2 x = 3 \tan x + 7$

34. $3 \sin^2 x = 3 \sin x + 2$

35. $7 = \cot^2 x + 4 \cot x$

36. $3 \tan^2 x + 2 \tan x = 7$

● **SYNTHESIS**

Solve, restricting solutions to $[0, 2\pi)$ or $[0°, 360°)$. If you have a grapher, verify your results.

37. $|\sin x| = \dfrac{\sqrt{3}}{2}$ 　　　**38.** $|\cos x| = \dfrac{1}{2}$

39. $\sqrt{\tan x} = \sqrt[4]{3}$

40. $12 \sin x - 7\sqrt{\sin x} + 1 = 0$

41. $16 \cos^4 x - 16 \cos^2 x + 3 = 0$

42. $\ln (\cos x) = 0$

43. $e^{\sin x} = 1$

44. $\sin (\ln x) = -1$

45. $e^{\ln(\sin x)} = 1$

46. A guy wire is attached to the top of a 50-ft pole and stretched to a point that is b ft from the bottom of the pole.

a) Suppose that the angle of inclination of the wire to the top is θ. Show that

$$\tan \theta = \frac{50}{b}.$$

b) For what angle of inclination θ does $b = 40$ ft?

47. An airplane is flying at an altitude of 2000 ft toward an island. The straight-line distance from the airplane to the island is h ft.

a) Suppose that the angle of depression is θ. Show that

$$\sin \theta = \frac{2000}{h}.$$

b) For what angle of depression θ does $h = 3000$ ft?

48. Solve graphically: $\sin x = \tan (x/2)$.

7.9

IDENTITIES IN SOLVING TRIGONOMETRIC EQUATIONS

A   When a trigonometric equation involves more than one function, we can use identities to put it in terms of a single function. This can usually be done in several ways. In Example 1, we illustrate by solving the same equation using two different methods.

OBJECTIVE

You should be able to:

A   Solve trigonometric equations requiring the use of identities.

1. Solve $\sin x - \cos x = 1$, as in Example 1 (method 1).

Example 1

Method 1.   Solve the equation $\sin x + \cos x = 1$.

Solution   To express $\cos x$ in terms of $\sin x$, we use the identity

$$\sin^2 x + \cos^2 x = 1.$$

From this, we obtain $\cos x = \pm\sqrt{1 - \sin^2 x}$. Now we substitute in the original equation:

$$\sin x \pm \sqrt{1 - \sin^2 x} = 1.$$

This is a radical equation. We get the radical alone on one side and then square both sides and simplify:

$$\pm\sqrt{1 - \sin^2 x} = 1 - \sin x$$
$$(\pm\sqrt{1 - \sin^2 x})^2 = (1 - \sin x)^2$$
$$1 - \sin^2 x = 1 - 2\sin x + \sin^2 x$$
$$-2\sin^2 x + 2\sin x = 0.$$

Factoring and using the principle of zero products gives us

$$-2\sin x\,(\sin x - 1) = 0$$
$$-2\sin x = 0 \quad \text{or} \quad \sin x - 1 = 0$$
$$\sin x = 0 \quad \text{or} \quad \sin x = 1.$$

The values of x in $[0, 2\pi)$ satisfying these are

$$x = 0, \quad x = \pi, \quad \text{or} \quad x = \frac{\pi}{2}.$$

Now we check these in the original equation. We find that π does not check but the other values do. Thus the solutions are 0 and $\pi/2$.   ◀

> **CAUTION!**   It is important to check when solving trigonometric equations. Values are often obtained that are not solutions of the original equation.

DO EXERCISE 1. _____

Example 1

Method 2.   Solve $\sin x + \cos x = 1$.

Solution   This time we will square both sides, obtaining

$$\sin^2 x + 2\sin x \cos x + \cos^2 x = 1.$$

Since $\sin^2 x + \cos^2 x = 1$, we can simplify to

$$2\sin x \cos x = 0.$$

Now we use the identity $2\sin x \cos x = \sin 2x$, and obtain

$$\sin 2x = 0.$$

We are looking for solutions x to the equation for which

$$0 \leq x < 2\pi.$$

Multiplying by 2, we get

$$0 \le 2x < 4\pi,$$

the interval we consider to solve $\sin 2x = 0$. These values of $2x$ are 0, π, 2π, and 3π. Thus the desired values of x in $[0, 2\pi)$ satisfying this equation are

$$x = 0, \qquad x = \frac{\pi}{2}, \qquad x = \pi, \quad \text{or} \quad x = \frac{3\pi}{2}.$$

Now we check these in the original equation. We find that π and $3\pi/2$ do not check but the other values do. Thus the solutions are 0 and $\pi/2$. ◀

DO EXERCISE 2. _____

Example 2 Solve: $2\cos^2 x \tan x - \tan x = 0$.

Solution First, we factor:

$$2\cos^2 x \tan x - \tan x = 0$$
$$\tan x (2\cos^2 x - 1) = 0.$$

Now we use the identity $2\cos^2 x - 1 = \cos 2x$:

$$\tan x \cos 2x = 0$$

$$\tan x = 0 \qquad \text{or} \quad \cos 2x = 0 \qquad \text{Principle of zero products}$$

$$x = 0, \pi \quad \text{or} \qquad 2x = \frac{\pi}{2}, \frac{3\pi}{2}, \frac{5\pi}{2}, \frac{7\pi}{2}$$

$$x = 0, \pi \quad \text{or} \qquad x = \frac{\pi}{4}, \frac{3\pi}{4}, \frac{5\pi}{4}, \frac{7\pi}{4}.$$

All values check. The solutions in $[0, 2\pi)$ are 0, π, $\pi/4$, $3\pi/4$, $5\pi/4$, and $7\pi/4$. ◀

DO EXERCISE 3. _____

Example 3 Solve: $\cos 2x + \sin x = 1$.

Solution We first use the identity $\cos 2x = 1 - 2\sin^2 x$, to get

$$1 - 2\sin^2 x + \sin x = 1$$
$$-2\sin^2 x + \sin x = 0$$
$$\sin x (1 - 2\sin x) = 0 \qquad \text{Factoring}$$

$$\sin x = 0 \qquad \text{or} \quad 1 - 2\sin x = 0 \qquad \text{Principle of zero products}$$

$$\sin x = 0 \qquad \text{or} \qquad \sin x = \frac{1}{2}$$

$$x = 0, \pi \quad \text{or} \qquad x = \frac{\pi}{6}, \frac{5\pi}{6}.$$

All values check. The solutions in $[0, 2\pi)$ are 0, π, $\pi/6$, and $5\pi/6$. ◀

DO EXERCISE 4. _____

Example 4 Solve: $\sin 5\theta \cos 2\theta - \cos 5\theta \sin 2\theta = \sqrt{2}/2$.

2. Solve $\sin x - \cos x = 1$, as in Example 1 (method 2).

3. Solve: $\tan^2 x \cos x - \cos x = 0$.

4. Solve: $\sin 2x + \cos x = 0$.

5. Solve:

$\sin 3\theta \cos \theta - \cos 3\theta \sin \theta = \frac{1}{2}$.

Solution   We use the identity

$$\sin(\alpha - \beta) = \sin \alpha \cos \beta - \cos \alpha \sin \beta$$

to get $\sin 3\theta = \sqrt{2}/2$. We are looking for solutions to the equation for which

$$0 \le \theta < 2\pi.$$

Multiplying by 3, we get

$$0 \le 3\theta < 6\pi,$$

which is the interval we consider to solve $\sin 3\theta = \sqrt{2}/2$. These values of 3θ are

$$3\theta = \frac{\pi}{4}, \frac{3\pi}{4}, \frac{9\pi}{4}, \frac{11\pi}{4}, \frac{17\pi}{4}, \frac{19\pi}{4}.$$

Then

$$\theta = \frac{\pi}{12}, \frac{\pi}{4}, \frac{3\pi}{4}, \frac{11\pi}{12}, \frac{17\pi}{12}, \frac{19\pi}{12}.$$

Let us check $\pi/12$:

$$\sin \frac{5\pi}{12} \cos \frac{\pi}{6} - \cos \frac{5\pi}{12} \sin \frac{\pi}{6}.$$

We again use the difference identity to obtain $\sin(5\pi/12 - \pi/6)$. This is equal to $\sin(3\pi/12)$, or $\sin(\pi/4)$, which is equal to $\sqrt{2}/2$. Thus $\pi/12$ checks. All of the above answers check.   ◀

DO EXERCISE 5.

Example 5   Solve: $\tan^2 x + \sec x - 1 = 0$.

Solution   We use the identity $1 + \tan^2 x = \sec^2 x$. Substituting, we then get

$$\sec^2 x - 1 + \sec x - 1 = 0,$$

or

$$\sec^2 x + \sec x - 2 = 0$$
$$(\sec x + 2)(\sec x - 1) = 0 \qquad \text{Factoring}$$
$$\sec x = -2 \qquad \text{or} \quad \sec x = 1 \qquad \text{Principle of zero products}$$
$$x = \frac{2\pi}{3}, \frac{4\pi}{3} \qquad \text{or} \qquad x = 0.$$

All of these values check. The solutions in $[0, 2\pi)$ are 0, $2\pi/3$, and $4\pi/3$.   ◀

Example 6   Solve: $\sin x = \sin(2x - \pi)$.

Solution   We can use either the identity $\sin(y - \pi) = -\sin y$ or the identity $\sin(\alpha - \beta) = \sin \alpha \cos \beta - \cos \alpha \sin \beta$ with the right-hand side. We get

$$\sin x = -\sin 2x.$$

Next, we use the identity $\sin 2x = 2 \sin x \cos x$, and obtain

$$\sin x = -2 \sin x \cos x,$$

or

$$2 \sin x \cos x + \sin x = 0$$

$$\sin x (2 \cos x + 1) = 0 \qquad \text{Factoring}$$

$$\sin x = 0 \quad \text{ or } \quad \cos x = -\frac{1}{2} \qquad \text{Principle of zero products}$$

$$x = 0, \pi \quad \text{ or } \qquad x = \frac{2\pi}{3}, \frac{4\pi}{3}.$$

All of these values check, so the solutions in $[0, 2\pi)$ are $0, \pi, 2\pi/3$, and $4\pi/3$. ◀

DO EXERCISE 6.

6. Solve: $\cos x = \cos (\pi - 2x)$.

● **EXERCISE SET** **7.9**

Ⓐ   Find all solutions of the following equations in $[0, 2\pi)$.

1. $\tan x \sin x - \tan x = 0$

2. $2 \sin x \cos x + \sin x = 0$

3. $2 \sec x \tan x + 2 \sec x + \tan x + 1 = 0$

4. $2 \csc x \cos x - 4 \cos x - \csc x + 2 = 0$

5. $\sin 2x - \cos x = 0$

6. $\cos 2x - \sin x = 1$

7. $\sin 2x \sin x - \cos x = 0$

8. $\sin 2x \cos x - \sin x = 0$

9. $\sin 2x + 2 \sin x \cos x = 0$

10. $\cos 2x \sin x + \sin x = 0$

11. $\cos 2x \cos x + \sin 2x \sin x = 1$

12. $\sin 2x \sin x - \cos 2x \cos x = -\cos x$

13. $\sin 4x - 2 \sin 2x = 0$

14. $\sin 4x + 2 \sin 2x = 0$

15. $\sin 2x + 2 \sin x - \cos x - 1 = 0$

16. $\sin 2x + \sin x + 2 \cos x + 1 = 0$

17. $\sec^2 x = 4 \tan^2 x$

18. $\sec^2 x - 2 \tan^2 x = 0$

19. $\sec^2 x + 3 \tan x - 11 = 0$

20. $\tan^2 x + 4 = 2 \sec^2 x + \tan x$

21. $\cot x = \tan (2x - 3\pi)$

22. $\tan x = \cot (2x + \pi)$

23. $\cos (\pi - x) + \sin \left(x - \frac{\pi}{2} \right) = 1$

24. $\sin (\pi - x) + \cos \left(\frac{\pi}{2} - x \right) = 1$

25. $\dfrac{\cos^2 x - 1}{\sin \left(\dfrac{\pi}{2} - x \right) - 1} = \dfrac{\sqrt{2}}{2} + 1$

26. $\dfrac{\sin^2 x - 1}{\cos \left(\dfrac{\pi}{2} - x \right) + 1} = \dfrac{\sqrt{2}}{2} - 1$

27. $2 \cos x + 2 \sin x = \sqrt{6}$

28. $2 \cos x + 2 \sin x = \sqrt{2}$

29. $\sqrt{3} \cos x - \sin x = 1$

30. $\sqrt{2} \cos x - \sqrt{2} \sin x = 2$

31. $\sec^2 x + 2 \tan x = 6$

32. $6 \tan^2 x = 5 \tan x + \sec^2 x$

33. $3 \cos 2x + \sin x = 1$

34. $5 \cos 2x + \sin x = 4$

● **SYNTHESIS**

35. *Temperature during an illness.* The temperature T of a patient during a 12-day illness is given by

$$T(t) = 101.6° + 3° \sin \left(\frac{\pi}{8} t \right).$$

Find the times t during the illness at which the patient's temperature was $103°$.

36. *Satellite location.* A satellite circles the earth in such a manner that it is y miles from the equator (north or south, height from the surface not considered) t minutes after its launch, where

$$y = 5000\left[\cos\frac{\pi}{45}(t - 10)\right].$$

At what times t in the interval $[0, 240]$, the first 4 hr, is the satellite 3000 mi north of the equator?

37. Nautical mile. (See Exercise 55 in Exercise Set 7.4.) In Great Britain, the **nautical mile** is defined as the length of a minute of arc of the earth's radius. Since the earth is flattened at the poles, a British nautical mile varies with latitude. In fact, it is given, in feet, by the function

$$N(\phi) = 6066 - 31\cos 2\phi,$$

where ϕ is the latitude in degrees. At what latitude north is the length of a British nautical mile found to be 6040 ft?

38. Acceleration due to gravity. (See Exercise 56 in Exercise Set 7.4.) The acceleration due to gravity is often denoted by g in a formula such as $S = \frac{1}{2}gt^2$, where S is the distance that an object falls in time t. It has to do with the

physics of motion near the earth's surface and is usually considered constant. In fact, however, g is not constant, but varies slightly with latitude. If ϕ stands for latitude, in degrees, g is given with good approximation by the formula

$$g = 9.78049(1 + 0.005288\sin^2\phi - 0.000006\sin^2 2\phi),$$

where g is measured in meters per second per second at sea level. At what latitude north does $g = 9.8$?

Solve.

39. $\arccos x = \arccos\frac{3}{5} - \arcsin\frac{4}{5}$

40. $\sin^{-1} x = \tan^{-1}\frac{1}{3} + \tan^{-1}\frac{1}{2}$

Solve graphically.

41. $\sin x - \cos x = \cot x$

42. $x\sin x = 1$

43. Suppose that $\sin x = 5\cos x$. Find $\sin x\cos x$.

SUMMARY AND REVIEW 7

● TERMS TO KNOW

Identity, p. 444
Sum and difference identities, p. 454
Cofunction identities, p. 457
Double-angle identities, p. 465
Half-angle identities, p. 466

Basic identities, p. 470
Pythagorean identities, p. 470
Sum–product identities, p. 475
Inverse trigonometric functions, p. 478

Arcsine function, p. 480
Arccosine function, p. 481
Arctangent function, p. 481
Trigonometric equations, p. 490

● REVIEW EXERCISES

1. Find an identity for $\cot(x - \pi)$.

Complete these Pythagorean identities.

2. $\sin^2 x + \cos^2 x = $ _____

3. $1 + \cot^2 x = $ _____

Complete these cofunction identities.

4. $\cos\left(x + \frac{\pi}{2}\right) = $ _____

5. $\cos\left(\frac{\pi}{2} - x\right) = $ _____

6. $\sin\left(x - \frac{\pi}{2}\right) = $ _____

7. Express $\tan x$ in terms of $\sec x$.

Simplify.

8. $\cos x(\tan x + \cot x)$

9. $\dfrac{\csc x(\sin^2 x + \cos^2 x\tan x)}{\sin x + \cos x}$

10. Rationalize the denominator: $\sqrt{\dfrac{\tan x}{\sec x}}$.

11. Rationalize the numerator: $\sqrt{\dfrac{\tan x}{\sec x}}$.

12. Given that $\sin\theta = 0.6820$, $\cos\theta = 0.7314$, $\tan\theta = 0.9325$, $\cot\theta = 1.0724$, $\sec\theta = 1.3673$, and $\csc\theta = 1.4663$, find the six function values for $90° - \theta$.

Use the sum and difference formulas to write equivalent expressions. You need not simplify.

13. $\cos\left(x + \dfrac{3\pi}{2}\right)$

14. $\tan(45° - 30°)$

15. Simplify: $\cos 27°\cos 16° + \sin 27°\sin 16°$.

16. Find $\cos 165°$ exactly.

17. Given that $\tan \alpha = \sqrt{3}$ and $\sin \beta = \sqrt{2}/2$ and that α and β are between 0 and $\pi/2$, evaluate $\tan(\alpha - \beta)$ exactly.

18. Find $\tan 2\theta$, $\cos 2\theta$, and $\sin 2\theta$ and the quadrant in which 2θ lies, where $\cos \theta = -\frac{3}{5}$ and θ is in quadrant III.

19. Find $\sin(\pi/8)$ without using a table or a calculator.

20. Simplify: $\dfrac{\sin 2\theta}{\sin^2 \theta}$.

Prove each of the following identities.

21. $\tan 2\theta = \dfrac{2 \tan \theta}{1 - \tan^2 \theta}$ **22.** $\dfrac{\sec x - \cos x}{\tan x} = \sin x$

23. Find, in radians, $\sin^{-1} \frac{1}{2}$.

24. Use a calculator to find, in degrees, $\text{arccot } 0.1584$.

Find each of the following without the use of a table or a calculator.

25. $\sin^{-1}\left(-\dfrac{\sqrt{2}}{2}\right)$ **26.** $\text{arccot}\left(-\sqrt{3}\right)$

Evaluate or simplify.

27. $\tan\left(\arctan \dfrac{7}{8}\right)$ **28.** $\cos^{-1}\left[\cos\left(-\dfrac{\pi}{3}\right)\right]$

29. $\cos^{-1}\left(\sin \dfrac{2\pi}{3}\right)$ **30.** $\sin\left(\arctan \dfrac{b}{5}\right)$

Solve, finding all solutions in $[0, 2\pi)$.

31. $\sin^2 x - 7 \sin x = 0$

32. $2 \cos^2 x - 5 \cos x + 2 = 0$

33. $\csc^2 x - 2 \cot^2 x = 0$

34. $2 \cot^2 x = 3 \cot x + 1$

Solve, finding all solutions in $[0, 2\pi)$.

35. $|\sin x| = \frac{1}{2}$

36. $\cos(\ln 2x) = -1$

Graph.

37. $y + 1 = 2 \cos^2 x$

38. $f(x) = 2 \sin^{-1}\left(x + \dfrac{\pi}{2}\right)$

● **THINKING AND WRITING** _____

1. Prove the identity $2 \cos^2 x - 1 = \cos^4 x - \sin^4 x$ in three ways:

a) Start with the left side and deduce the right (Method 1).

b) Start with the right side and deduce the left (Method 1).

c) Work with each side separately until you deduce the same expression (Method 2).

Then determine the most efficient method and explain why.

2. Why are the ranges of the inverse trigonometric functions restricted?

3. Explain why $\tan(x + 450°)$ cannot be simplified using the tangent sum formula, but can be simplified using the sine and cosine sum formulas.

CHAPTER TEST 7

Use the sum and difference formulas to write equivalent expressions. You need not simplify.

1. $\cos(\pi - x)$ **2.** $\tan(83° + 15°)$

3. Simplify: $\sin 40° \cos 5° - \cos 40° \sin 5°$.

4. Find $\cos 105°$ exactly.

5. Given that $\cos \alpha = \sqrt{2}/2$ and $\sin \beta = 1/2$ and that α and β are between 0 and $\pi/2$, evaluate $\tan(\alpha - \beta)$ exactly.

6. Find an identity for $\sec(x - 450°)$.

Complete these identities.

7. $\sin^2 x + \cos^2 x =$ _____

8. $1 + \cot^2 x =$ _____

9. $\sin\left(x + \dfrac{\pi}{2}\right) =$ _____

10. $\cos\left(\dfrac{\pi}{2} - x\right) =$ _____

11. $\cos\left(x - \dfrac{\pi}{2}\right) =$ _____

12. Express $\csc x$ in terms of $\cot x$.

Simplify.

13. $\dfrac{\sqrt{\sec^2 x - 1}}{\sin x}$ **14.** $\dfrac{\tan^2 x \csc^2 x - 1}{\csc x \tan^2 x \sin x}$

15. Rationalize the denominator: $\sqrt{\dfrac{\sec x}{\csc x}}$.

16. Find $\sin 2\theta$, $\cos 2\theta$, and $\tan 2\theta$ and the quadrant in which 2θ lies, where $\sin \theta = \frac{12}{13}$ and θ is in quadrant II.

17. Find $\tan(5\pi/12)$ exactly.

18. Simplify: $4 \sin 2x \cos 2x$.

19. Prove the identity

$$\frac{1 - \cos 2\theta}{\sin 2\theta} = \tan \theta.$$

20. Find, in radians, $\arcsin(-\sqrt{2}/2)$.

21. Use a calculator to find, in degrees, $\tan^{-1} 0.931304$.

22. Find $\cos^{-1}(\sqrt{3}/2)$.

23. Find $\operatorname{arccot}(-1)$.

24. Find $\sin(\tan^{-1}\sqrt{3})$.

25. Find $\arccos\left(\cos \dfrac{\pi}{4}\right)$.

Find all solutions of the following equations in $[0, 2\pi)$.

26. $2 \sin^2 x - 5 \sin x = 3$

27. $2 \cos^3 x - \cos x = 0$

28. $\tan^2 x + 3 \tan x = 5$

29. Given that $\sin \theta = 0.4540$, $\cos \theta = 0.8910$, $\tan \theta = 0.5095$, $\cot \theta = 1.963$, $\sec \theta = 1.122$, and $\csc \theta = 2.203$, find the six function values for $90° - \theta$.

● **SYNTHESIS** _____

30. Solve $\ln(\sin x) = 0$. Restrict the solutions to $[0, 2\pi)$.

8

Triangle trigonometry is important in many practical applications such as surveying and navigation. In this chapter, we first solve right triangles and related problems. We also use trigonometry to solve triangles that are not right (*oblique triangles*). • The idea of a *vector* is related to the study of triangles. A vector is a quantity that has a direction. Vectors have many practical applications in the physical sciences. • We conclude the chapter with a continuation of the study of complex numbers begun in Chapter 2. Complex numbers have applications in both electricity and engineering. •

Triangles, Vectors, and Applications

8.1

APPLICATIONS OF RIGHT TRIANGLES

OBJECTIVES

You should be able to:

A Solve right triangles.

B Solve problems involving the use of right triangles and the trigonometric functions.

As we look at the many applications of trigonometry, we begin with a study of right triangles. The basic circular, or trigonometric, functions can be considered to be ratios of sides of right triangles, placed in standard position, as developed in Chapter 6. For any right triangle, whether or not it is in standard position, the ratios of sides are given by trigonometric functions.

It is customary to label the right angle C. The side opposite C, the hypotenuse, is labeled c. The acute angles are labeled A and B, and the lengths of the sides opposite them are labeled a and b, respectively. We refer to this as **standard lettering.**

Standard position

$\sin \theta = \dfrac{y}{r}$

Standard lettering

$\sin A = \dfrac{a}{c}$

503

1. For the right triangle, using standard lettering, express each of the following using a, b, and c.

  a) $\sin B$

  b) $\cos B$

  c) $\tan B$

  d) $\csc B$

  e) $\sec B$

  f) $\cot B$

We know the following.

$$\sin A = \frac{\text{opposite}}{\text{hypotenuse}} = \frac{a}{c} \qquad \csc A = \frac{\text{hypotenuse}}{\text{opposite}} = \frac{c}{a}$$

$$\cos A = \frac{\text{adjacent}}{\text{hypotenuse}} = \frac{b}{c} \qquad \sec A = \frac{\text{hypotenuse}}{\text{adjacent}} = \frac{c}{b}$$

$$\tan A = \frac{\text{opposite}}{\text{adjacent}} = \frac{a}{b} \qquad \cot A = \frac{\text{adjacent}}{\text{opposite}} = \frac{b}{a}$$

DO EXERCISE 1. _____

A   Solving Right Triangles

Now that we know how to find function values for any acute angle, we can begin to solve right triangles. Triangles that are not right triangles will be studied later. To **solve** a triangle means to find the lengths of all its sides and the measures of all its angles, provided they are not already known.

Example 1   Find $m\angle B$.

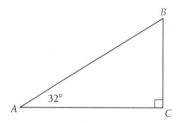

Solution   Since the angle measures of any triangle add up to $180°$ and the right angle measures $90°$, the acute angles must add up to $90°$:

$$m\angle A + m\angle B = 90°.$$

Thus,

$$m\angle B = 90° - 32° = 58°.$$

Example 2   In this right triangle, find a and b.

Solution   By the definition of the sine ratio, we know that

$$\sin 36.75° = \frac{a}{28.2}.$$

Solving for a, we have

$$a = 28.2 \sin 36.75°$$
$$\approx 16.9.$$

To find b, we use the cosine ratio:

$$\cos 36.75° = \frac{b}{28.2}.$$

We solve for b:

$$b = 28.2 \cos 36.75° \approx 22.6.$$

DO EXERCISES 2 AND 3.

Henceforth, for simplicity, we will use an equals sign, $=$, even when an approximation symbol, $\approx$, might be appropriate.

Example 3 In this right triangle, find A and B.* Then find a. That is, *solve* the triangle.

Solution We know the side adjacent to A and also the hypotenuse. That suggests the use of the cosine ratio:

$$\cos A = \frac{12.3}{27.9}$$

$$= 0.4409. \qquad \text{Dividing}$$

We now use a calculator to find the inverse cosine of 0.4409, which is A. We find that

$$A = 63.84°.$$

Now, since the measures of A and B must total $90°$, we can find B by subtracting:

$$B = 90° - 63.84°$$

$$= 26.16°.$$

To find a, we have a choice of how to begin. We could use $\cos B$, or we could use $\sin A$. We could also use the tangent or cotangent ratios for either A or B. Let's use $\tan A$:

$$\tan A = \frac{a}{12.3}, \quad \text{or}$$

$$\tan 63.84° = \frac{a}{12.3}.$$

Then

$$a = 12.3 \tan 63.84° \qquad \text{Solving for } a$$

$$= 25.0. \qquad \text{Using a calculator}$$

DO EXERCISES 4 AND 5.

---

* We will often shorten our writing or speaking, saying that we find B, rather than $m\angle B$.

2. Find $m\angle A$.

3. In this right triangle, find a and b.

4. Solve this triangle.

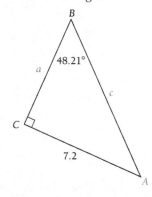

5. Solve this right triangle. Standard lettering has been used.

$$A = 37.8°, \qquad b = 8.62$$

6. An observer stands 120 m from a tree and finds that the line of sight to the top of the tree is 32.3° above the horizontal.

a) Find the height of the tree above eye level.

b) Find the distance from the observer to the top of the tree.

B ▶ Applications of Right Triangles

Right triangles have many applications. To solve a problem, we locate a right triangle and then solve, or at least partially solve, it.

Example 4   An observer stands on level ground, 200 m from the base of a television tower, and looks up at an angle of 26.5° to see the top of the tower.

a) How high is the tower above the observer's eye level?

b) How far is it from the observer's eye to the top of the tower?

Solution   We draw a diagram and see that a right triangle is formed.

a) We use one of the trigonometric functions. The tangent function is most convenient. From the definition of the tangent function, we have

$$\tan 26.5° = \frac{\text{opposite}}{\text{adjacent}} = \frac{h}{200}.$$

Solving for h gives us

$$h = 200 \tan 26.5° = 100.$$

The height of the tower is about 100 m.

b) To find the distance d from the observer's eye to the top of the tower, we use a different trigonometric function, one that involves the hypotenuse. The cosine function gives us

$$\cos 26.5° = \frac{\text{adjacent}}{\text{hypotenuse}} = \frac{200}{d}.$$

Then solving for d, we get

$$d = \frac{200}{\cos 26.5°} = 223.$$

The distance from the observer's eye to the top of the tower is about 223 m.   ◀

DO EXERCISE 6. ──────────────

Example 5   *Safe angle for ladders.*   Suppose that a ladder has length L. It has been determined that the ladder is at its safest position on a wall when it is pulled out a distance D from the wall, where $L = 4D$. What angle does this safest position determine with the ground?

Solution   We draw a diagram and then use the most convenient trigonometric function. From the definition of the cosine function, we have

$$\cos \theta = \frac{\text{adjacent}}{\text{hypotenuse}} = \frac{D}{L} = \frac{D}{4D} = \frac{1}{4}.$$

We find that θ is about 75.52°. Thus the ladder is at its safest position when it makes an angle of about 75.52° with the ground.

DO EXERCISE 7. _____

Many applications with right triangles involve an *angle of elevation* or an *angle of depression*.

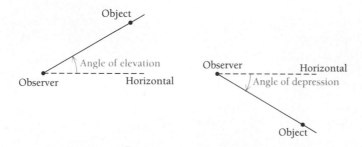

The angle between the horizontal and a line of sight *above* the horizontal is called an **angle of elevation.** The angle between the horizontal and a line of sight *below* the horizontal is called an **angle of depression.** For example, suppose that you were looking straight ahead and then you moved your eyes upward toward an oncoming airplane. The angle your eyes pass through is an angle of elevation. If the pilot of the plane is looking forward and then moves his eyes down toward you, his eyes pass through an angle of depression.

Example 6   *Finding a cloud height.*   A device for measuring cloud height at night consists of a vertical beam of light, which makes a spot on the clouds. The spot is viewed from a point 135 ft away. With a surveying device, the angle of elevation is found to be 67°40′. Find the height of the clouds.

Solution   From the figure, we have

$$\frac{h}{135} = \tan 67°40′$$

$$h = 135 \times \tan 67°40′$$
$$= 329 \text{ ft.}$$

7. A guy wire is 13.6 m long and is fastened from the ground to a pole 6.5 m above the ground. What angle does the wire make with the ground?

8. The length of a guy wire to a pole is 37.7 ft. It makes an angle of 71°20′ with the ground, which is horizontal. How high above the ground is the guy wire attached to the pole?

The following is a general procedure for solving problems involving triangles.

> To solve a triangle problem:
>
> 1. Draw a sketch of the problem situation.
> 2. Look for triangles and sketch them in.
> 3. Mark the known and unknown sides and angles.
> 4. Express the desired side or angle in terms of known trigonometric ratios. Then solve.

DO EXERCISE 8. _____

Example 7 Surveying.   Horizontal distances must often be measured even though terrain is not level. One way of doing so is as follows. Distance down a slope is measured with a surveyor's tape, and the distance d is measured by making a level sighting from A to a pole held vertically at B, or the angle α is measured by an instrument placed at A. Suppose that a slope distance L is measured to be 121.3 ft and the angle α is measured to be 3°25′. Find the horizontal distance H.

Solution   From the figure, we see that $H/L = \cos \alpha$. Thus, $H = L \cos \alpha$, and in this case,

$$H = 121.3 \times \cos 3°25′ = 121.1 \text{ ft.}$$ ◀

DO EXERCISE 9. _____

9. A downslope distance is measured to be 241.3 ft, and the angle of depression α is measured to be 5°15′. Find the horizontal distance.

There are two applications of trigonometry that involve the idea of **direction,** or **bearing.** The first application is to aerial navigation and the second is to sea navigation, surveying, and other fields.

Bearing: First Type   In aerial navigation, directions are given in degrees clockwise from north. Thus, east is 90°, south is 180°, and west is 270°. Several aerial directions or bearings are given below.

Example 8   An airplane leaves an airport and travels for 100 mi in a direction of 300°. How far north of the airport is the plane then? How far west?

Solution   The direction of flight is shown in the figure. In the triangle, d_1 is the northerly distance and d_2 is the westerly distance. Then

$$\frac{d_1}{100} = \sin 30° \quad \text{and} \quad \frac{d_2}{100} = \cos 30°.$$

Thus, to the nearest mile, we have

$$d_1 = 100 \sin 30° = 50 \text{ mi};$$
$$d_2 = 100 \cos 30° = 87 \text{ mi}.$$

The plane is 50 mi north and 87 mi west of the airport.   ◀

DO EXERCISE 10. _____

Bearing: Second Type   The second method of giving direction, or bearing, involves reference to a north–south line using an acute angle. For example, N 43° W means 43° west of north and S 30° E means 30° east of south. Several bearings of this type are shown below.

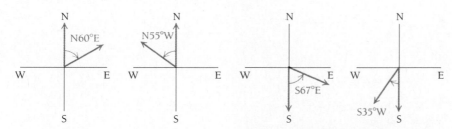

Example 9   A forest ranger at point A sights a fire directly south. A second ranger at point B, 7.5 mi east, sights the same fire at a bearing of S 27°23′ W. How far from A is the fire?

$$\angle B = 90° - 27°23'$$
$$= 89°60' - 27°23'$$
$$= 62°37'$$

10. An airplane flies 150 km from an airport in a direction of 115°. How far east of the airport is the plane then? How far south?

11. Directly east of a lookout station, there is a small forest fire. The bearing of this fire from a station 12.5 km south of the first is N 57°19′ E. How far is the fire from the southerly lookout station?

Solution   From the figure, we see that the desired distance d is part of a right triangle, as shown. Thus we have

$$\frac{d}{7.5} = \tan 62°37′$$

$$d = 7.5 \tan 62°37′ = 14.5 \text{ mi.}$$

The forest ranger at A is 14.5 mi from the fire.   ◀

DO EXERCISE 11. _____

Example 10   From an observation tower, two markers are viewed on the ground. The markers and the base of the tower are on a line, and the observer's eye is 65.3 ft above the ground. The angles of depression to the markers are 53.2° and 27.8°. How far is it from one marker to the other?

Solution   We first make a sketch, looking for right triangles to sketch in. Then we mark the known information on the sketch. The distance we seek is d, which is $d_1 - d_2$.

From the right angles in the figure, we have

$$\frac{d_1}{65.3} = \cot\theta_1 \quad \text{and} \quad \frac{d_2}{65.3} = \cot\theta_2,$$

or

$$d_1 = 65.3 \cot\theta_1 \quad \text{and} \quad d_2 = 65.3 \cot\theta_2.$$

When parallel lines are cut by a transversal, alternate interior angles are equal. Thus,

$$\theta_1 = 27.8° \quad \text{and} \quad \theta_2 = 53.2°.$$

Then

$$d_1 = 65.3 \cot 27.8° \quad \text{and} \quad d_2 = 65.3 \cot 53.2°.$$

We could calculate these and subtract, but the use of a calculator is more efficient if we leave all calculations until the end:

$$d = d_1 - d_2 = 65.3 \cot 27.8° - 65.3 \cot 53.2°$$
$$= 65.3(\cot 27.8° - \cot 53.2°)$$
$$= 75.0 \text{ ft.}$$

The distance between the markers is 75.0 ft.   ◀

DO EXERCISE 12. _____

12. From an airplane flying 7500 ft above level ground, one can see two towns directly to the east. The angles of depression to the towns are 5.3° and 77.5°. How far apart are the towns, to the nearest mile?

● EXERCISE SET 8.1

A Solve each of the following right triangles. (Standard lettering has been used.)

1.

2.

3.

4.

5.

6.

7. $A = 36°12', a = 27.2$ **8.** $A = 87°43', a = 9.73$

9. $B = 12.65°, b = 98.1$ **10.** $B = 69.88°, b = 127$

11. $A = 17.4°, b = 13.6$ **12.** $A = 78.7°, b = 1340$

13. $B = 23°12', a = 350$ **14.** $B = 69°22', a = 240$

15. $A = 47.58°, c = 48.3$ **16.** $A = 88.92°, c = 3950$

17. $B = 82.3°, c = 0.982$ **18.** $B = 56.5°, c = 0.0447$

19. $a = 12.5, b = 18.5$ **20.** $a = 10.2, b = 20.4$

21. $a = 16.0, c = 20.0$ **22.** $a = 15.0, c = 45.0$

23. $b = 1.86, c = 4.02$ **24.** $b = 100, c = 450$

B Solve.

25. A guy wire to a pole makes an angle of 73.2° with the level ground and is 14.5 ft from the pole at the ground. How far above the ground is the wire attached to the pole?

26. A guy wire to a pole makes an angle of 74.4° with the level ground and is attached to the pole 34.2 ft above the ground. How far from the base of the pole is the wire attached to the ground?

27. A kite string makes an angle of 31.7° with the (level) ground when 455 ft of string is out. How high is the kite?

28. A kite string makes an angle of 41.3° with the (level) ground when the kite is 114 ft high. How long is the string?

29. A road rises 3 m per 100 horizontal m. What angle does it make with the horizontal?

30. A kite is 120 ft high when 670 ft of string is out. What angle does the kite make with the ground?

31. What is the angle of elevation of the sun when a 6-ft man casts a 10.3-ft shadow?

32. What is the angle of elevation of the sun when a 35-ft mast casts a 20-ft shadow?

33. From a balloon 2500 ft high, a command post is seen with an angle of depression of 7.7°. How far is it from a point on the ground below the balloon to the command post, to the nearest tenth of a mile?

34. From a lighthouse 55 ft above sea level, the angle of depression to a small boat is 11.3°. How far from the foot of the lighthouse is the boat?

35. An observer at a command post sights a balloon that is 2500 ft high at an angle of elevation of 8°20'. How far is it from the command post to a point directly under the balloon?

36. An observer sights the top of a building 173 ft higher than the eye, at an angle of elevation of 27°50'. How far is it from the observer to the building?

37. Ship A is due west of a lighthouse. Ship B is 12 km south of ship A. From ship B, the bearing to the lighthouse is N 63°20'   How far is ship A from the lighthouse?

38. Lookout station A is 15 km west of station B. The bearing from A to a fire directly south of B is S 37°50' E. How far is the fire from B?

39. A regular pentagon has sides 30.5 cm long. Find the radius of the circumscribed circle.

40. A regular pentagon has sides 42.8 cm long. Find the radius of the inscribed circle.

41. A regular hexagon has a perimeter of 50 cm and is inscribed in a circle. Find the radius of the circle.

42. A regular octagon is inscribed in a circle of radius 15.8 cm. Find the perimeter of the octagon.

43. A vertical antenna is mounted on top of a 50-ft pole. From a point on the level ground 75 ft from the base of the pole, the antenna subtends an angle of 10.5°. Find the length of the antenna.

44. An observer on a ladder looks at a building 100 ft away, noting that the angle of elevation of the top of the building is 18°40′ and the angle of depression of the bottom of the building is 6°20′. How tall is the building?

45. From a balloon 2 km high, the angles of depression to two towns, in line with the balloon, are 81.2° and 13.5°. How far apart are the towns?

46. From a balloon 1000 m high, the angles of depression to two artillery posts, in line with the balloon, are 11.8° and 84.1°. How far apart are the artillery posts?

47. A weather balloon is directly west of two observing stations that are 10 km apart. The angles of elevation of the balloon from the two stations are 17°50′ and 78°10′. How high is the balloon, to the nearest tenth of a kilometer?

48. From two points south of a hill on level ground and 1000 ft apart, the angles of elevation of the hill are 12°20′ and 82°40′. How high is the hill?

49. An airplane travels at 120 km/h for 2 hr in a direction of 243° from Chicago. At the end of this time, how far south of Chicago is the plane?

50. An airplane travels at 150 km/h for 2 hr in a direction of 138° from Omaha. At the end of this time, how far east of Omaha is the plane?

● SYNTHESIS _____

51. Find h, to the nearest tenth.

52. Find a, to the nearest tenth.

53. Show that the area of a right triangle is

$$\frac{1}{4}c^2 \sin 2A.$$

54. Show that the area of a right triangle is

$$\frac{1}{2}bc \sin A.$$

55. Find a formula for the distance to the horizon as a function of the height of the observer above the earth. Calculate the distance, to the nearest tenth of a mile, to the horizon from an airplane that is flying at an altitude of 1000 ft.

56. In finding horizontal distance from slope distance (see Example 7), we have $H = L - C$, where C is a correction. Show that a good approximation to C is $d^2/2L$.

57. *Carpentry.*   A carpenter is contructing picnic pavilions in parks, as shown in the figure. The rafter ends are to be sawed in such a way that they will be vertical when in place. The front wall is 8 ft high, the back wall is $6\frac{1}{2}$ ft high, and the distance between walls is 8 ft. At what angle should the rafters be cut?

58. *Carpentry.*   A V-gauge is used to find diameters of pipes. The advantage of such a device is that it is rugged, it is accurate, and it has no moving parts to break down. In the figure at the top of the following page, the measure of angle AVB is 54°. A pipe is placed in the V-shaped slot and the distance VP is used to predict the diameter.

a) Suppose that the diameter of a pipe is 2 cm. What is the distance VP?
b) Suppose that the distance VP is 3.93 cm. What is the diameter of the pipe?
c) Find a formula for d in terms of VP.
d) Find a formula for VP in terms of d.

The line VP is calibrated by listing as its units the corresponding diameters. This, in effect, establishes a function between VP and d.

● CHALLENGE _____

59. *Sound of an airplane.*   It is a common experience to hear the sound of a low-flying airplane and look at the wrong place in the sky to see the plane. Suppose that a plane is traveling directly at you at a speed of 200 mph and an altitude of 3000 ft, and you hear the sound at what seems to be an angle of inclination of 20°. At what angle θ should you actually look in order to see the plane? Consider the speed of sound to be 1100 ft/sec.

60. *Measuring the radius of the earth.*   One way to measure the radius of the earth is to climb to the top of a mountain whose height above sea level is known and measure the angle between a vertical line to the center of the earth from the top of the mountain and a line drawn from the top of the mountain to the horizon, as shown in the figure. The height of Mt. Shasta in California is 14,162 ft. From the top of Mt. Shasta, one can see the horizon on the Pacific Ocean. The angle formed between a line to the horizon and the vertical is found to be 87°53′. Use this information to estimate the radius of the earth, in miles.

8.2
THE LAW OF SINES

The trigonometric functions can also be used to solve triangles that are not right triangles. Such triangles are called **oblique.** Any triangle, right or oblique, can be solved if at least one side and any other two measures are known. The four possible situations are as follows:

1. Two angles and any side of a triangle are known (AAS and ASA).
2. Two sides of a triangle and an angle opposite one of them are known (SSA). In this case, there may be no solution, one solution, or two solutions. The latter is known as the *ambiguous case.*
3. Two sides of a triangle and the included angle are known (SAS).
4. All three sides of the triangle are known (SSS).

In order to solve oblique triangles, we need to derive some properties. One is the *law of sines* and the other is the *law of cosines.* The law of sines applies to the first two situations listed above. The law of cosines, which we will develop in Section 8.3, applies to the last two situations.

OBJECTIVES

You should be able to:

A Use the law of sines to solve any triangle given a side and two angles.

B Use the law of sines to solve any triangle given two sides and an angle opposite one of them, finding two solutions when they exist, and recognizing when a solution does not exist.

C Find the area of any triangle given the lengths of two sides and the measure of the included angle.

The Law of Sines

We consider any oblique triangle. It may or may not have an obtuse angle. Although we look at both cases, the derivations are essentially the same. In the following figures, h is the same length.

The triangles are lettered in the standard way, with angles A, B, and C and the sides opposite them a, b, and c, respectively. We have drawn an altitude from vertex C. It has length h. In either triangle we now have, from triangle ADC,

$$\frac{h}{b} = \sin A, \quad \text{or}$$

$$h = b \sin A.$$

From triangle DBC on the left above, we have

$$\frac{h}{a} = \sin B, \quad \text{or}$$

$$h = a \sin B.$$

On the right above, we have

$$\frac{h}{a} = \sin(180° - B) = \sin B, \quad \text{or}$$

$$h = a \sin B.$$

So in either triangle we now have

$$h = a \sin B \quad \text{and} \quad h = b \sin A.$$

It follows then that

$$a \sin B = b \sin A.$$

We divide by $\sin A \sin B$ to obtain

$$\frac{a}{\sin A} = \frac{b}{\sin B}.$$

There is no danger of dividing by 0 here because we are dealing with triangles, whose angles are never 0° or 180°.

If we were to consider an altitude from vertex A in the triangles shown above, the same argument would give us

$$\frac{b}{\sin B} = \frac{c}{\sin C}$$

We combine these results to obtain the law of sines, which holds for right triangles as well as oblique triangles.

> ### THEOREM 1
>
> **The Law of Sines**
>
> In any triangle ABC,
>
> $$\frac{a}{\sin A} = \frac{b}{\sin B} = \frac{c}{\sin C}.$$
>
> (The sides are proportional to the sines of the opposite angles.)

◢ Solving Triangles (AAS and ASA)

When two angles and a side of any triangle are known, the law of sines can be used to solve the triangle.

Example 1   In triangle ABC, $a = 4.56$, $A = 43°$, and $C = 57°$. Solve the triangle.

Solution   We first draw a sketch.

We then find B, as follows:

$$B = 180° - (43° + 57°) = 80°.$$

We can now find the other two sides, using the law of sines:

$$\frac{c}{\sin C} = \frac{a}{\sin A}$$

$$\frac{c}{\sin 57°} = \frac{4.56}{\sin 43°}$$

$$c = \frac{4.56 \sin 57°}{\sin 43°} \qquad \text{Solving for } c$$

$$c = 5.61;$$

$$\frac{b}{\sin B} = \frac{a}{\sin A}$$

$$\frac{b}{\sin 80°} = \frac{4.56}{\sin 43°}$$

$$b = \frac{4.56 \sin 80°}{\sin 43°} \qquad \text{Solving for } b$$

$$b = 6.58.$$

We have now found the unknown parts of the triangle: $B = 80°$, $c = 5.61$, and $b = 6.58$.   ◀

1. Solve this triangle.

DO EXERCISE 1.

B Solving Triangles (SSA)

When two sides of a triangle and an angle opposite one of them are known, the law of sines can be used to solve the triangle. However, there may be no solution, one solution, or two solutions.

Suppose that a, b, and A are given. Then the various possibilities are as shown in the five cases below. Note that $a < b$ in cases I, II, and III, $a = b$ in case IV, and $a > b$ in case V. Since there are two solutions in case III, this possibility is referred to as the **ambiguous case.**

Case I: No solution

$a < b$, side a is too short to reach the base.

Case II: One solution

$a < b$, side a just reaches the base and is perpendicular to it.

Case III: Two solutions (Ambiguous case)

$a < b$, an arc of radius a meets the base at two points.

Case IV: One solution

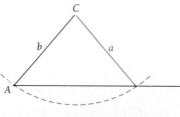

$a = b$, an arc of radius a meets the base at just one point, other than A.

Case V: One solution

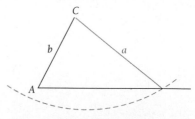

$a > b$, an arc of radius a meets the base at just one point.

The following examples correspond to cases I, III, and V just described.

Example 2 Case I: No solution.   In triangle ABC, $a = 15$, $b = 25$, and $A = 47°$. Solve the triangle.

Solution   We look for B:

$$\frac{a}{\sin A} = \frac{b}{\sin B}$$

$$\frac{15}{\sin 47°} = \frac{25}{\sin B}$$

$$\sin B = \frac{25 \sin 47°}{15} \qquad \text{Solving for } \sin B$$

$$\sin B = 1.219.$$

Since there is no angle with a sine greater than 1, there is *no* solution.   ◀

DO EXERCISE 2. _____

Example 3   *Case III: Two solutions.*   In triangle ABC, $a = 20$, $b = 15$, and $B = 30°$. Solve the triangle.

Solution   We look for A:

$$\frac{a}{\sin A} = \frac{b}{\sin B}$$

$$\frac{20}{\sin A} = \frac{15}{\sin 30°}$$

$$\sin A = \frac{20 \sin 30°}{15}$$

$$\sin A = 0.6667.$$

There are two angles less than $180°$ with a sine of 0.6667. They are $42°$ and $138°$, to the nearest degree. This gives us two possible solutions.

Possible Solution 1.   If $A = 42°$, then

$$C = 180° - (30° + 42°) = 108°.$$

We now find c:

$$\frac{c}{\sin C} = \frac{b}{\sin B}$$

$$\frac{c}{\sin 108°} = \frac{15}{\sin 30°}$$

$$c = \frac{15 \sin 108°}{\sin 30°} \qquad \text{Solving for } c$$

$$c = 29.$$

These parts make a triangle, as shown; thus we have a solution.

Possible Solution 2.   If $A = 138°$, then

$$C = 180° - (30° + 138°) = 12°.$$

2. Solve this triangle.

$$a = 40, \quad b = 12, \quad B = 57°$$

3. In triangle ABC, $a = 25$, $b = 20$, and $B = 33°$. Solve the triangle.

We now find c:

$$\frac{c}{\sin C} = \frac{b}{\sin B}$$

$$\frac{c}{\sin 12°} = \frac{15}{\sin 30°}$$

$$c = \frac{15 \sin 12°}{\sin 30°} \qquad \text{Solving for } c$$

$$c = 6.$$

These parts make a triangle; thus we have a second solution.

DO EXERCISE 3.

Example 4 *Case V: One solution.* In triangle ABC, $a = 25$, $b = 10$, and $A = 42°$. Solve the triangle.

Solution We look for B:

$$\frac{b}{\sin B} = \frac{a}{\sin A}$$

$$\frac{10}{\sin B} = \frac{25}{\sin 42°}$$

$$\sin B = \frac{10 \sin 42°}{25} \qquad \text{Solving for } \sin B$$

$$\sin B = 0.2677.$$

4. In triangle ABC, $b = 20$, $c = 10$, and $B = 38°$. Solve the triangle.

Then $B = 16°$ or $B = 164°$, to the nearest degree. Since $a > b$, we know that there is only one solution. If we had not noticed this, we could tell it now. An angle of $164°$ cannot be an angle of this triangle because it already has an angle of $42°$ and these two angles would total more than $180°$. Thus $16°$ is the only possibility, and

$$C = 180° - (42° + 16°) = 122°.$$

We now find c:

$$\frac{c}{\sin C} = \frac{a}{\sin A}$$

$$\frac{c}{\sin 122°} = \frac{25}{\sin 42°}$$

$$c = \frac{25 \sin 122°}{\sin 42°} \qquad \text{Solving for } c$$

$$c = 32.$$

DO EXERCISE 4.

The Area of a Triangle

The familiar formula for the area of a triangle, $A = \frac{1}{2}bh$, can be used only when h is known. The method used to derive the law of sines can be used to derive an area formula that does not involve the height.

Consider a general triangle ABC with area K.

 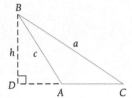

In each triangle ADB,

$$\frac{h}{c} = \sin A, \quad \text{or} \quad h = c \sin A.$$

Substituting into the formula $K = \frac{1}{2}bh$, we get

$$K = \frac{1}{2}b \cdot$$

Any pair of sides and the included angle could have been used. Thus we also have

$$K = \frac{1}{2}ab \sin C \quad \text{and} \quad K = \frac{1}{2}ac \sin B.$$

THEOREM 2   Area of a Triangle

The area K of any triangle ABC is one-half the product of the lengths of two sides and the sine of the included angle:

$$K = \frac{1}{2}bc \sin A = \frac{1}{2}ab \sin C = \frac{1}{2}ac \sin B.$$

Example 5   Find the area of triangle ABC if $b = 10.1$ cm, $c = 6.8$ cm, and $A = 62.1°$.

Solution

$$\begin{aligned}
K &= \tfrac{1}{2}bc \sin A &&\text{Theorem 2} \\
&= \tfrac{1}{2}(10.1)(6.8) \sin 62.1° &&\text{Substituting} \\
&= 30.3 \text{ cm}^2
\end{aligned}$$

◀

DO EXERCISE 5.

5. Find the area of triangle ABC if $a = 9.2$ in., $b = 8.0$ in., and $C = 37.6°$.

 EXERCISE SET 8.2

A   Solve the triangle, if possible.

1. $A = 133°, B = 30°, b = 18$

2. $B = 120°, C = 30°, a = 16$

3. $B = 38°, C = 21°, b = 24$

4. $A = 131°, C = 23°, b = 10$

5. $A = 68.5°$, $C = 42.7°$, $c = 23.5$ cm

6. $B = 118.3°$, $C = 45.6°$, $b = 42.1$ ft

7. $c = 3$ mi, $B = 37.48°$, $C = 32.16°$

8. $a = 200$ m, $A = 32.76°$, $C = 21.97°$

9. $c = 6019$ km, $A = 16°56'$, $B = 59°23'$

10. $A = 120°18'$, $C = 27°46'$, $a = 16,435$ mi

11. $A = 129°32'$, $C = 18°28'$, $b = 1204$ in.

12. $B = 37°51'$, $C = 19°47'$, $b = 240.123$ mm

Solve the triangle, if possible.

13. $A = 36.5°$, $a = 24$, $b = 34$

14. $C = 43.2°$, $c = 28$, $b = 27$

15. $A = 116.3°$, $a = 17.2$, $c = 13.5$

16. $A = 47.8°$, $a = 28.3$, $b = 18.2$

17. $C = 61°10'$, $c = 30.3$, $b = 24.2$

18. $B = 58°40'$, $a = 25.1$, $b = 32.6$

19. $a = 2345$ mi, $b = 2345$ mi, $A = 124.67°$

20. $b = 56.78$ yd, $c = 56.78$ yd, $C = 83.78°$

21. $a = 20.01$ cm, $b = 10.07$ cm, $A = 30.3°$

22. $A = 115°$, $c = 45.6$ yd, $a = 23.8$ yd

23. $a = 4000$ m, $b = 8000$ m, $A = 32°52'$

24. $a = 2$ cm, $b = 6$ cm, $A = 30°$

25. $b = 4.157$ km, $c = 3.446$ km, $C = 51°48'$

26. $b = 43.67$ mm, $c = 38.86$ mm, $C = 48.76°$

27. $A = 89°$, $a = 15.6$ in., $b = 18.4$ in.

28. $B = 36°$, $b = 8.0$ ft, $c = 10.0$ ft

29. $A = 41°50'$, $a = 90.0$ mi, $c = 110.0$ mi

30. $C = 46°30'$, $a = 56.2$ m, $c = 22.1$ m

Find the area of the triangle.

31. $B = 42°$, $a = 7.2$ ft, $c = 3.4$ ft

32. $A = 17°12'$, $b = 10$ in., $c = 13$ in.

33. $C = 82°54'$, $a = 4$ yd, $b = 6$ yd

34. $C = 75.16°$, $a = 1.5$ m, $b = 2.1$ m

35. $B = 135.2°$, $a = 46.12$ ft, $c = 36.74$ ft

36. $A = 113°$, $b = 18.2$ cm, $c = 23.7$ cm

Solve. Keep in mind the two types of bearing considered in Section 8.1.

37. Points A and B are on opposite sides of a lunar crater. Point C is 50 m from A. The measure of $\angle BAC$ is determined to be 112° and the measure of $\angle ACB$ is determined to be 42°. What is the width of the crater?

38. A guy wire to a pole makes a 71° angle with level ground. At a point 25 ft farther from the pole than the guy wire, the angle of elevation of the top of the pole is 37°. How long is the guy wire?

39. A pole leans away from the sun at an angle of 7° to the vertical (see the figure). When the angle of elevation of the sun is 51°, the pole casts a shadow 47 ft long on level ground. How long is the pole?

40. A vertical pole stands by a road that is inclined 10° to the horizontal. When the angle of elevation of the sun is 23°, the pole casts a shadow 38 ft long directly downhill along the road. How long is the pole?

41. A reconnaissance plane leaves its airport on the east coast of the United States and flies in a direction of 085°. Because of bad weather, it returns to another airport 230 km to the north of its home base. For the return trip, it flies in a direction of 283°. What was the total distance that it flew?

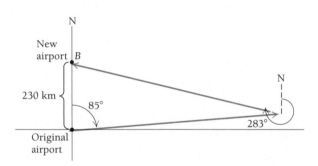

42. Lookout station B is 10.2 km east of station A. The bearing of a fire from A is S 10°40' W. The bearing of the fire from B is S 31°20' W. How far is the fire from A? from B?

43. A ranger in fire tower A spots a fire at a direction of 295°. A ranger in fire tower B, located 45 mi at a direction of 045° from tower A, spots the same fire at a direction of 255°. How far from tower A is the fire? from tower B?

44. An airplane leaves airport A and flies 200 km. At this time its direction from airport B, 250 km to the west of A, is 120°. How far is the airplane from B?

45. A boat leaves a lighthouse A and sails 5.1 km. At this time it is sighted from lighthouse B, 7.2 km west of A. The bearing of the boat from B is N 65°10' E. How far is the boat from B?

46. Mackinac Island is located 35 mi N 65°20' W of Cheboygan, Michigan, where the Coast Guard cutter Mackinaw is stationed. A freighter in distress radios the Coast Guard cutter for help. It radios their position as N 25°40' E of Mackinac Island and N 10°10' W of Cheboygan. How far is the freighter from Cheboygan?

47. Prove each of the following formulas for a general triangle *ABC*.

a) $\dfrac{a+b}{c} = \dfrac{\cos \frac{1}{2}(A-B)}{\sin \frac{1}{2}C}$

b) $\dfrac{a-b}{c} = \dfrac{\sin \frac{1}{2}(A-B)}{\cos \frac{1}{2}C}$

48. Prove the following area formulas for a general triangle *ABC* with area represented by *K*.

$$K = \dfrac{a^2 \sin B \sin C}{2 \sin A} = \dfrac{c^2 \sin A \sin B}{2 \sin C}$$
$$= \dfrac{b^2 \sin C \sin A}{2 \sin B}$$

49. Prove that the area of a parallelogram is the product of two sides and the sine of the included angle.

50. Prove that the area of a quadrilateral is one-half the product of the lengths of its diagonals and the sine of the angle between the diagonals.

51. Consider the following triangle. Prove that

$$\frac{\sin \alpha}{z} + \frac{\sin \beta}{x} = \frac{\sin (\alpha + \beta)}{y}.$$

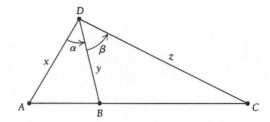

52. When two objects, such as ships, airplanes, or runners, move in straight-line paths, if the distance between them is decreasing and if the bearing from one of them to the other is constant, they will collide. ("Constant bearing means collision," as mariners put it.) Prove that this statement is true.

8.3

THE LAW OF COSINES

The law of sines is used to solve triangles given a side and two angles (AAS and ASA) or given two sides and an angle opposite one of them (SSA). A second law, called the *law of cosines,* is needed to solve triangles given two sides and the included angle (SAS) or given three sides (SSS).

The Law of Cosines

To derive this property, we consider any triangle *ABC* placed on a coordinate system. We position the origin at one of the vertices, say *C*, and the positive half of the *x*-axis along one of the sides, say *CB*. Let (x, y) be the coordinates of vertex *A*. Point *B* has coordinates $(a, 0)$, and point *C* has coordinates $(0, 0)$.

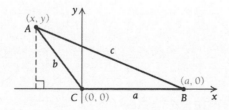

OBJECTIVES

You should be able to:

A Use the law of cosines, with the law of sines, to solve any triangle given two sides and the included angle.

B Use the law of cosines to solve any triangle given three sides.

C Determine whether the law of sines or the law of cosines should be applied to solve a triangle, and then solve the triangle.

Then

$$\cos C = \frac{x}{b}, \quad \text{so} \quad x = b \cos C,$$

and

$$\sin C = \frac{y}{b}, \quad \text{so} \quad y = b \sin C.$$

Thus,

$$(x, y) = (b \cos C, b \sin C),$$

and point A has coordinates $(b \cos C, b \sin C)$.

Next, we use the distance formula to determine c^2:

$$c^2 = (x - a)^2 + (y - 0)^2,$$

or

$$c^2 = (b \cos C - a)^2 + (b \sin C - 0)^2.$$

Now we multiply and simplify:

$$c^2 = b^2 \cos^2 C - 2ab \cos C + a^2 + b^2 \sin^2 C$$
$$= a^2 + b^2 (\sin^2 C + \cos^2 C) - 2ab \cos C$$
$$= a^2 + b^2 - 2ab \cos C.$$

Had we placed the origin at one of the other vertices, we would have obtained

$$a^2 = b^2 + c^2 - 2bc \cos A$$

or

$$b^2 = a^2 + c^2 - 2ac \cos B.$$

These results can be summarized as follows.

THEOREM 3   **The Law of Cosines**

In any triangle ABC,

$$a^2 = b^2 + c^2 - 2bc \cos A,$$
$$b^2 = a^2 + c^2 - 2ac \cos B,$$

and

$$c^2 = a^2 + b^2 - 2ab \cos C.$$

(In any triangle, the square of a side is the sum of the squares of the other two sides, minus twice the product of those sides and the cosine of the included angle.)

CAUTION!   Should the law of sines be applied after the law of cosines, one should be alert for the possibility of the ambiguous case.

 ## Solving Triangles (SAS)

When two sides of a triangle and the included angle are known, we can use the law of cosines to find the third side. The law of cosines or the law of sines can then be used to finish solving the triangle.

Example 1   In triangle ABC, $a = 24$, $c = 32$, and $B = 115°$. Solve the triangle.

Solution   We first sketch the triangle and label it.

Next we find the third side. From the law of cosines,

$b^2 = a^2 + c^2 - 2ac \cos B$

$b^2 = 24^2 + 32^2 - 2 \cdot 24 \cdot 32(\cos 115°)$

$b^2 = 576 + 1024 - 1536 \cos 115°$

$b^2 = 2249$

$b = 47.$

We now have $a = 24$, $b = 47$, and $c = 32$. We need to find the other two angle measures. At this point, we can find them in two ways. One way uses the law of sines. The ambiguous case may arise, however, and we would have to be alert to this possibility. The advantage of using the law of cosines again is that if we solve for the cosine and its value is negative, then we know that the angle is obtuse. If the value of the cosine is positive, then the angle is acute. Thus we use the law of cosines to find a second angle.

Let us find angle A. We select the formula from the law of cosines that contains $\cos A$ and substitute:

$a^2 = b^2 + c^2 - 2bc \cos A$

$24^2 = 47^2 + 32^2 - 2(47)(32) \cos A$

$576 = 2209 + 1024 - 3008 \cos A$

$-2657 = -3008 \cos A$

$\cos A = 0.8833.$

Then $A = 28°$. The third angle is now easy to find:

$C = 180° - (115° + 28°) = 37°.$

Answers may vary depending on the order in which they are found. Had we found the measure of angle C first in Example 1, the angle measures would have been $C = 39°$ and $A = 26°$. The answers at the back of the text were generated by considering alphabetical order. Variances in rounding also changes the answers. Had we used 47.4 for b in Example 1, the angle measures would have been $A = 27°$ and $C = 38°$.   ◀

DO EXERCISE 1.

 Solving Triangles (SSS)

When all three sides of a triangle are known, the law of cosines can be used to solve the triangle.

1. In triangle ABC, $b = 18$, $c = 28$, and $A = 122°$. Solve the triangle.

2. In triangle ABC, $a = 25$, $b = 10$, and $c = 20$. Solve the triangle.

Example 2   In triangle ABC, $a = 18$, $b = 25$, and $c = 12$. Solve the triangle.

Solution   We sketch the triangle.

Let us find angle B. We select the formula from the law of cosines that contains $\cos B$; in other words, $b^2 = a^2 + c^2 - 2ac \cos B$. We substitute and solve for $\cos B$:

$$b^2 = a^2 + c^2 - 2ac \cos B$$
$$25^2 = 18^2 + 12^2 - 2(18)(12) \cos B$$
$$625 = 324 + 144 - 432 \cos B$$
$$157 = -432 \cos B$$
$$\cos B = -0.3634.$$

Then $B = 111.31°$. Similarly, we find angle A*:

$$a^2 = b^2 + c^2 - 2bc \cos A$$
$$18^2 = 25^2 + 12^2 - 2(25)(12) \cos A$$
$$324 = 625 + 144 - 600 \cos A$$
$$-445 = -600 \cos A$$
$$\cos A = 0.7417.$$

Thus, $A = 42.12°$. Then $C = 180° - (111.31° + 42.12°) = 26.57°$.

DO EXERCISE 2. _____

▶ **Choosing the Appropriate Law**

The following summarizes the situations in which to use the law of sines and the law of cosines.

> To solve an oblique triangle:
>
> A) Use the *law of sines* when:
>
>   1. Two angles and any side of a triangle are known (AAS and ASA).
>   2. Two sides of a triangle and an angle opposite one of them are known (SSA). Keep in mind that there may be no solution, one solution, or two solutions. The latter is known as the *ambiguous case*.
>
> B) Use the *law of cosines* when:
>
>   3. Two sides of a triangle and the included angle are known (SAS).
>   4. All three sides of the triangle are known (SSS).

_____

* The law of sines could be used at this point, but be alert for the ambiguous case.

Let us practice determining which law to use.

Example 3 In triangle *ABC*, *a* = 20, *b* = 15, and *B* = 30°. Determine which law applies. Then solve the triangle.

Solution We are given two sides and an angle opposite one of them. In this case, the law of sines applies. Thus there may be no solution, one solution, or two solutions (the ambiguous case). This is indeed the ambiguous case. The triangle is solved in Example 3 of Section 8.2. ◀

Example 4 In triangle *ABC*, *a* = 24, *c* = 32, and *B* = 115°. Determine which law applies. Then solve the triangle.

Solution We are given two sides and the included angle. In this case, the law of cosines applies. The triangle is solved in Example 1 of this section. ◀

DO EXERCISES 3–8.

Determine which law applies. Then solve the triangle.

3. *a* = 50, *b* = 20, *c* = 40

4. *b* = 40, *c* = 20, *B* = 38°

5. *b* = 54, *c* = 84, *A* = 122.4°

6. *a* = 75, *b* = 60, *B* = 33°

7. *A* = 43°35′, *a* = 4.5, *b* = 6.0

8. *b* = 6, *B* = 57.1°, *a* = 20

● **EXERCISE SET** **8.3**

A   Solve the triangle, if possible.

1. *A* = 30°, *b* = 12, *c* = 24
2. *C* = 60°, *a* = 15, *b* = 12
3. *B* = 133°, *a* = 12, *c* = 15
4. *A* = 116°, *b* = 31, *c* = 25
5. *B* = 72°40′, *c* = 16 m, *a* = 78 m
6. *A* = 24°30′, *b* = 68 ft, *c* = 14 ft
7. *C* = 22.28°, *a* = 25.4 cm, *b* = 73.8 cm
8. *B* = 72.66°, *a* = 23.78 km, *c* = 25.74 km
9. *A* = 96°13′, *b* = 15.8 yd, *c* = 18.4 yd
10. *C* = 28°43′, *a* = 6 mm, *b* = 9 mm
11. *a* = 60.12 mi, *b* = 40.23 mi, *C* = 48.7°
12. *b* = 10.2 in., *c* = 17.3 in., *A* = 53.456°

B   Solve the triangle, if possible.

13. *a* = 12, *b* = 14, *c* = 20
14. *a* = 22, *b* = 22, *c* = 35
15. *a* = 16 m, *b* = 20 m, *c* = 32 m
16. *a* = 2.2 cm, *b* = 4.1 cm, *c* = 2.4 cm
17. *a* = 2 ft, *b* = 3 ft, *c* = 8 ft
18. *a* = 17 yd, *b* = 15.4 yd, *c* = 1.5 yd
19. *a* = 11.2 cm, *b* = 5.4 cm, *c* = 7 cm
20. *a* = 26.12 km, *b* = 21.34 km, *c* = 19.25 km

C   Determine which law applies. Then solve the triangle.

21. *A* = 70°, *B* = 12°, *b* = 21.4
22. *a* = 15, *c* = 7, *B* = 62°
23. *a* = 3.3, *b* = 2.7, *c* = 2.8
24. *a* = 1.5, *b* = 2.5, *A* = 58°
25. *a* = 60, *b* = 40, *C* = 47°
26. *a* = 3.6, *b* = 6.2, *c* = 4.1
27. *B* = 110°30′, *C* = 8°10′, *c* = 0.912
28. *B* = 52°, *C* = 15°, *b* = 60.4

A, B

29. A ship leaves a harbor and sails 15 nautical mi east. It then sails 18 nautical mi in a direction of S 27° E. How far is it, then, from the harbor and in what direction?

30. An airplane leaves an airport and flies west 147 km. It then flies 200 km in a direction of 220°. How far is it, then, from the airport and in what direction?

31. Two ships leave harbor at the same time. The first sails N 15° W at 25 knots (a knot is one nautical mile per hour). The second sails N 32° E at 20 knots. After 2 hr, how far apart are the ships?

32. Two airplanes leave an airport at the same time. The first flies 150 km/h in a direction of 320°. The second flies 200 km/h in a direction of 200°. After 3 hr, how far apart are the planes?

33. A hill is inclined 5° to the horizontal. A 45-ft pole stands at the top of the hill. How long a rope will it take to reach from the top of the pole to a point 35 ft downhill from the base of the pole?

34. A hill is inclined 15° to the horizontal. A 40-ft pole stands at the top of the hill. How long a rope will it take to reach from the top of the pole to a point 68 ft downhill from the base of the pole?

35. A piece of wire 5.5 m long is bent into a triangular shape. One side is 1.5 m long and another is 2 m long. Find the angles of the triangle.

36. A triangular lot has sides of 120 ft, 150 ft, and 100 ft. Find the angles of the lot.

37. A slow-pitch softball diamond is a square 65 ft on a side. The pitcher's mound is 46 ft from home. How far is it from the pitcher's mound to first base?

38. A baseball diamond is a square 90 ft on a side. The pitcher's mound is 60.5 ft from home. How far does the pitcher have to run to cover first base?

39. The longer base of an isosceles trapezoid measures 14 ft. The nonparallel sides measure 10 ft, and the base angles measure 80°.

   a) Find the length of a diagonal.
   b) Find the area.

40. An isosceles triangle has a vertex angle of 38° and this angle is included by two sides, each measuring 20 ft. Find the area of the triangle.

41. A field in the shape of a parallelogram has sides that measure 50 yd and 70 yd. One angle of the field measures 78°. Find the area of the field.

42. An aircraft takes off to fly a 180-mi trip. After flying 75 mi, it is 10 mi off course. How much should the heading be corrected to then fly straight to the destination, assuming no wind correction?

43. A fence along one side of a triangular lot is 28 m long and makes angles of 25° and 58° with the sides. How long are the other two sides?

44. A guy wire to a pole makes a 63°40′ angle with level ground. At a point 8.2 m farther from the pole than the guy wire, the angle of elevation of the top of the pole is 42°30′. How long is the guy wire?

45. A triangular swimming pool has sides of length 40 ft, 35.3 ft, and 26.6 ft. Find the angles of the triangle.

46. A triangular swimming pool measures 44 ft on one side and 32.8 ft on another side. These sides form an angle that measures 40.8°. How long is the other side?

47. A pole leans away from the sun at an angle of 8° to the vertical. When the angle of elevation of the sun is 53°, the pole casts a shadow 44 ft long on level ground. How long is the pole?

48. A hill is inclined 6° to the horizontal. A 42-ft pole stands at the top of the hill. How long a wire will it take to reach from the top of the pole to a point 35 ft downhill from the base of the pole?

49. Three circles are arranged as shown in the figure below. Find the length *PQ*.

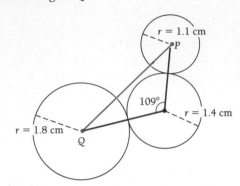

50. Three coins are arranged as shown in the figure below. Find the angle *ϕ*.

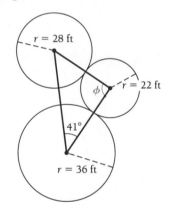

● SYNTHESIS _____

51. *Heron's formula.* If *a*, *b*, and *c* are the lengths of the sides of a triangle, then the area *K* of the triangle is given by

$$K = \sqrt{s(s-a)(s-b)(s-c)},$$

where $s = \frac{1}{2}(a + b + c)$. The number *s* is called the **semiperimeter.** Prove Heron's formula. (*Hint:* Use the area formula $K = \frac{1}{2}bc \sin A$ developed in Section 8.2.)

52. Use Heron's formula to find the area of the triangular swimming pool described in Exercise 45.

53. The measures of two sides of a parallelogram are 50 and 60 in., while one diagonal is 90 in. How long is the other diagonal?

54. The Port Huron-to-Mackinac sailboat race covers a distance of 259 mi from Port Huron, Michigan, to Mackinac Island. During the race, the course is measured at a heading of 035° from Port Huron to Cove Island for a distance of 181 mi. From Cove Island, the boats head 295° to Mackinac Island. With a north wind, one boat started on a course of 320° and sailed for 85 mi. The navigator then determined that the boat could "tack" (change heading) and sail directly for Cove Island. How far is the boat from Cove Island, and what must its new heading be in order to reach Cove Island?

55. From the top of a hill 20 ft above the surface of a lake, a tree is sighted across the lake. The angle of elevation is 11°. From the same position, the top of the tree is seen by reflection in the lake. The angle of depression is 14°. What is the height of the tree? (*Hint:* The angle of incidence, θ, equals the angle of reflection, α.)

56. A bridge is being built across a canyon. The length of the bridge is 5045 ft. From the deepest point in the canyon, the angles of elevation of the ends of the bridge are 78° and 72°. How deep is the canyon?

57. Find a formula for the area of an isosceles triangle in terms of the congruent sides and their included angle. Under what conditions will the area of a triangle with fixed congruent sides be maximum?

58. *Surveying.* In surveying, a series of bearings and distances is called a **transverse.** Measurements are taken as shown in the following figure.

a) Compute the bearing *BC*.
b) Compute the distance and bearing *AC*.

● CHALLENGE _____

59. Show that in any triangle *ABC*,
$$a^2 + b^2 + c^2 = 2(bc \cos A + ac \cos B + ab \cos C).$$

60. Show that in any triangle *ABC*,
$$\frac{\cos A}{a} + \frac{\cos B}{b} + \frac{\cos C}{c} = \frac{a^2 + b^2 + c^2}{2abc}.$$

61. A reconnaissance plane patrolling at 5000 ft sights a submarine at bearing 35° at an angle of depression of 25°. A carrier is at bearing 105° and at an angle of depression of 60°. How far is the submarine from the carrier?

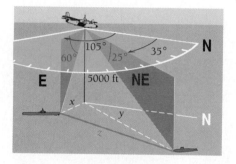

62. Two of NASA's tracking stations are located near the equator: One is in Ethiopia, at 40° east longitude; another is near Quito, Ecuador, at 78° west longitude. Assume that both stations, represented by *E* and *Q* in the figure below, are on the equator and that the radius of the earth is 6380 km. A satellite in orbit over the equator is observed at the same instant from both tracking stations. The angles of elevation above the horizon are 5° from Quito and 10° from Ethiopia. Find the distance of the satellite from earth at the instant of observation.*

_____

* *Source: Space Mathematics,* National Aeronautics and Space Administration, pp. 104–106.

8.4

VECTORS AND APPLICATIONS

In many applications, there occur certain quantities in which a direction is specified. Any such quantity having a *direction* and a *magnitude* is called a **vector quantity,** or **vector.** Here are some examples of vector quantities.

Displacement   An object moves a certain distance in a certain direction.

A train travels 100 mi to the northeast.

A person takes 5 steps to the west.

A batter hits a ball 100 m along the left-field foul line.

Velocity   An object travels at a certain speed in a certain direction.

The wind is blowing 15 mph from the northwest.

An airplane is traveling 450 km/h in a direction of 243°.

Force   A push or pull is exerted on an object in a certain direction.

A force of 15 newtons (a newton is a unit of force used in physics, abbreviated N) is exerted downward on the handle of a jack.

A 25-lb upward force is required to lift a box.

A wagon is being pulled up a 30° incline, requiring an effort of 200 lb.

We represent vectors graphically by directed line segments, or arrows. The length is chosen, according to some scale, to represent the **magnitude of the vector,** and the direction of the arrow represents the **direction of the vector.** For example, if we let 1 cm represent 5 km/h, then a 15-km/h wind from the northwest would be represented by an arrow 3 cm long, as shown below.

A   Vector Addition

To "add" vectors, we find a single vector that would have the same effect as the vectors combined. Suppose a person takes 4 steps east and then 3 steps north. He will then be 5 steps from the starting point in the direction shown. The **sum** of the two vectors is the vector 5 steps in magnitude and in the direction shown. The sum is also called the **resultant** of the two vectors.

In general, if we have two vectors **a** and **b**, we can add them as in the above example.* That is, we place the tail of one arrow at the head of the other and then find the vector that forms the third side of a triangle. Or, we can place the tails of the arrows together, complete a parallelogram, and find the diagonal of the parallelogram.

Example 1 Forces of 15 newtons and 25 newtons act on an object at right angles to each other. Find their sum, or resultant, giving the angle that it makes with the larger force.

Solution

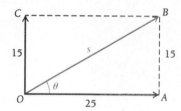

We make a drawing, this time a rectangle, using s for the length of vector **OB**. Since OAB is a right triangle, we have

$$\tan \theta = \frac{15}{25} = 0.6.$$

Thus θ, the angle that the resultant makes with the larger force, is $31°$ to the nearest degree. Now, using the Pythagorean theorem, we have

$$s^2 = 15^2 + 25^2.$$

Thus, $s = \sqrt{15^2 + 25^2} = 29.2$, to the nearest tenth. The resultant **OB** has a magnitude of 29.2 N and makes an angle of $31°$ with the larger force. ◀

DO EXERCISE 1.

Vector subtraction, $\mathbf{a} - \mathbf{b}$, is defined as $\mathbf{a} + (-\mathbf{b})$, where $-\mathbf{b}$ is the vector found from **b** by heading in the opposite direction. We can show $\mathbf{a} - \mathbf{b}$ as follows, though we will not give much attention to vector subtraction.

---

* It is general practice to denote vectors using boldface type; sometimes in handwriting, we write a half arrow over the letters, $\vec{a}$.

1. Two forces of 5.2 N and 14.6 N act at right angles to each other. Find the resultant, specifying, to the nearest degree, the angle that it makes with the larger force.

2. An airplane heads in a direction of 90° at a 140-km/h airspeed. The wind is 50 km/h from 210°. Find the direction and the speed of the airplane over the ground.

 Applications

Example 2 An airplane travels on a bearing of 100° at a 180-km/h airspeed while a wind is blowing 45 km/h from 220°. Find the speed of the airplane over the ground and the direction of its track over the ground.

Solution We first make a drawing. The wind is represented by **OC** and the velocity vector of the airplane by **OA**. The resultant velocity is **v**, the sum of the two vectors. We denote the length of **v** by |**v**|.

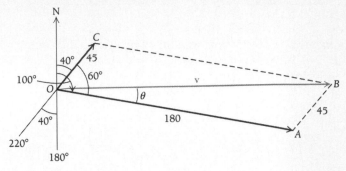

The measure of ∠*COA* is 60°, so ∠*CBA* = 60°. Now since the sum of all the angles of the parallelogram is 360° and ∠*OCB* and ∠*OAB* have the same measure, each must be 120°. By the law of cosines in △*OAB*, we have

$$|\mathbf{v}|^2 = 45^2 + 180^2 - 2 \cdot 45 \cdot 180 \cos 120°$$
$$= 42{,}525.$$

Thus, |**v**| is 206 km/h. By the law of sines in the same triangle,

$$\frac{45}{\sin \theta} = \frac{206}{\sin 120°},$$

or

$$\sin \theta = \frac{45 \sin 120°}{206}$$
$$= 0.1892.$$

Thus, $\theta = 11°$, to the nearest degree. The ground speed of the airplane is 206 km/h, and its track is in the direction of 100° − 11°, or 89°. ◀

DO EXERCISE 2.

 Components of Vectors

Given a vector, it is often convenient to reverse the addition procedure, that is, to find two vectors whose sum is the given vector. Usually the two vectors we seek will be perpendicular. They are called **components** of the given vector, and the process of finding them is called **resolving** a vector into its components.

Example 3 A certain vector **a** has a magnitude of 130 and is inclined 40° with the horizontal. Resolve the vector into horizontal and vertical components.

Solution   We first make a drawing showing horizontal and vertical vectors whose sum is the given vector **a**. From △*OAB*, we see that

$$|\mathbf{h}| = 130 \cos 40° = 100$$

and

$$|\mathbf{v}| = 130 \sin 40° = 84.$$

These are the components we seek.   ◀

Example 4   An airplane is flying at 200 km/h in a direction of 305°. Find the westerly and northerly components of its velocity.

Solution   We first make a drawing showing westerly and northerly vectors whose sum is the given velocity. From △*OAB*, we see that

$$|\mathbf{n}| = 200 \cos 55° = 115 \text{ km/h}$$

and

$$|\mathbf{w}| = 200 \sin 55° = 164 \text{ km/h}.$$

◀

DO EXERCISE 3.

▷ Adding Vectors Using Components

Given a vector, we can resolve it into components. Given components of a vector, we can find the vector. We simply find the vector sum of the components.

Example 5   A vector **v** has a westerly component of 12 and a southerly component of 16. Find the vector.

Solution   We first make a drawing showing westerly and southerly components of vector **v**. From the drawing, we see that

$$\tan \theta = \tfrac{12}{16} = 0.75$$
$$\theta = 37°;$$

$$|\mathbf{v}| = \sqrt{12^2 + 16^2}$$
$$= 20.$$

Thus the vector has a magnitude of 20 and a direction of S 37° W.   ◀

DO EXERCISE 4.

When we have two vectors resolved into components, we can add the two vectors by adding the components.

3. A vector of magnitude 100 points southeast. Resolve the vector into easterly and southerly components.

4. A wind has an easterly component (*from* the east) of 10 km/h and a southerly component (*from* the south) of 16 km/h. Find the magnitude and the direction of the wind.

5. One vector has components of 7 up and 9 to the left. A second vector has components of 3 down and 12 to the left. Find the sum, expressing it:

a) in terms of components;

b) giving magnitude and direction.

Example 6   A vector **v** has a westerly component of 3 and a northerly component of 5. A second vector **w** has an easterly component of 8 and a northerly component of 4. Find the components of the sum. Find the sum.

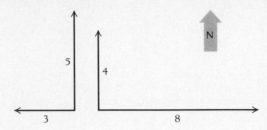

Solution   Adding the east–west components, we obtain 5 east. Adding the north–south components, we obtain 9 north. The components of **v** + **w** are 5 east and 9 north. To find **v** + **w**, we find both the direction and the magnitude:

$$\tan \theta = \tfrac{9}{5} = 1.8$$
$$\theta = 61°;$$

$$|\mathbf{v} + \mathbf{w}| = \sqrt{9^2 + 5^2}$$
$$= 10.3.$$

The sum $|\mathbf{v} + \mathbf{w}|$ has a magnitude of 10.3 and a direction of N 29° E.   ◀

DO EXERCISE 5. ────────────────────────────

Analytic Representation of Vectors

If we place a vector on a coordinate system so that the tail of the vector is at the origin, we say that the vector is in **standard position.** Then if we know the coordinates of the other end of the vector, we know the vector. The coordinates will be the x-component and the y-component of the vector. Thus we can consider an ordered pair (a, b) to be a vector. To emphasize that we are thinking of a vector, we usually write $\langle a, b \rangle$. When vectors are given in this form, it is easy to add them. We simply add the respective components.

Example 7   Find **u** + **v**, where $\mathbf{u} = \langle 3, -7 \rangle$ and $\mathbf{v} = \langle 4, 2 \rangle$.

Solution The sum is found by adding the *x*-components and then adding the *y*-components:

$$\mathbf{u} + \mathbf{v} = \langle 3, -7 \rangle + \langle 4, 2 \rangle = \langle 7, -5 \rangle.$$

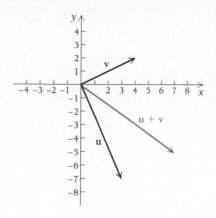

Add the vectors, giving answers as ordered pairs.

6. $\langle 7, 2 \rangle$ and $\langle 5, -10 \rangle$

DO EXERCISES 6 AND 7.

7. $\langle -2, 7 \rangle$ and $\langle 14, -3 \rangle$

 Polar and Rectangular Notation

Vectors can also be specified by giving their length and direction. When this is done, we say that we have **polar notation** for the vector. An example of polar notation is $(15, 260°)$.

The angle is measured from the positive half of the *x*-axis counterclockwise. In Section 8.5, we consider a coordinate system (called a **polar coordinate system**) in which polar notation is very natural.

8. Find polar notation for the vector $\langle -8, 3 \rangle$.

Example 8 Find polar notation for the vector **v**, where $\mathbf{v} = \langle 7, -2 \rangle$.

Solution We see the reference angle ϕ in the figure:

$$\tan \phi = \frac{-2}{7} = -0.2857$$

$$\phi = -16°;$$

$$|\mathbf{v}| = \sqrt{(-2)^2 + 7^2}$$

$$= 7.3.$$

Thus polar notation for **v** is

$$(7.3, -16°), \quad \text{or} \quad (7.3, 344°).$$

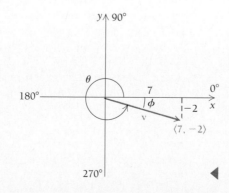

DO EXERCISE 8.

9. Find rectangular notation for the vector $(15, 341°)$.

Given polar notation for a vector, we can also find the ordered-pair notation (**rectangular notation**).

Example 9 Find rectangular notation for the vector **w**, where **w** = $(14, 155°)$.

Solution From the figure, we have

$$x = 14 \cos 155° = -12.7,$$
$$y = 14 \sin 155° = 5.9.$$

Thus the rectangular notation for **w** is $\langle -12.7, 5.9 \rangle$.

DO EXERCISE 9. _____

▶ Properties of Vectors and Vector Algebra

With vectors represented in rectangular notation $\langle a, b \rangle$, it is easy to determine many of their properties. We consider the set of all ordered pairs of real numbers $\langle a, b \rangle$. This is the set of all vectors on a plane. Consider the sum of any two vectors:

$$\mathbf{u} = \langle a, b \rangle, \qquad\qquad \mathbf{v} = \langle c, d \rangle,$$
$$\mathbf{u} + \mathbf{v} = \langle a + c, b + d \rangle, \qquad \mathbf{v} + \mathbf{u} = \langle c + a, d + b \rangle.$$

We have just shown that addition of vectors is commutative. Other properties are also not difficult to show. The system of vectors is not a field, but it does have some of the familiar field properties discussed in Section 1.1.

Properties of Vectors

Commutative law. Addition of vectors is commutative.

Associative law. Addition of vectors is associative.

Identity. There is an additive identity, the vector $\langle 0, 0 \rangle$.

Inverses. Every vector **v**, given by $\mathbf{v} = \langle a, b \rangle$, has an opposite $-\mathbf{v}$, given by $-\mathbf{v} = \langle -a, -b \rangle$.

Subtraction. We define $\mathbf{u} - \mathbf{v}$ to be the vector that when added to **v** gives **u**. It follows that if $\mathbf{u} = \langle a, b \rangle$ and $\mathbf{v} = \langle c, d \rangle$, then $\mathbf{u} - \mathbf{v} = \langle a - c, b - d \rangle$. Thus subtraction is always possible. Also, $\mathbf{u} - \mathbf{v} = \mathbf{u} + (-\mathbf{v})$.

Length, or absolute value. If $\mathbf{u} = \langle a, b \rangle$, then it follows that $|\mathbf{u}| = \sqrt{a^2 + b^2}$.

A vector can be multiplied by a real number. A real number in this context is called a **scalar.** For example, 2**v** is a vector in the same direction as **v** but twice as long. Analytically, we define scalar multiplication as follows.

DEFINITION   **Scalar Multiplication**

If r is any real number and $\mathbf{v} = \langle a, b \rangle$, then $r\mathbf{v} = \langle ar, br \rangle$.

Example 10 Do the following calculations, where $\mathbf{u} = \langle 4, 3 \rangle$ and $\mathbf{v} = \langle -5, 8 \rangle$: (a) $\mathbf{u} + \mathbf{v}$; (b) $\mathbf{u} - \mathbf{v}$; (c) $3\mathbf{u} - 4\mathbf{v}$; (d) $|3\mathbf{u} - 4\mathbf{v}|$.

Solution

a) $\mathbf{u} + \mathbf{v} = \langle 4, 3 \rangle + \langle -5, 8 \rangle = \langle -1, 11 \rangle$

b) $\mathbf{u} - \mathbf{v} = \mathbf{u} + (-\mathbf{v}) = \langle 4, 3 \rangle + \langle 5, -8 \rangle = \langle 9, -5 \rangle$

c) $3\mathbf{u} - 4\mathbf{v} = 3\langle 4, 3 \rangle - 4\langle -5, 8 \rangle = \langle 12, 9 \rangle - \langle -20, 32 \rangle$
$= \langle 32, -23 \rangle$

d) $|3\mathbf{u} - 4\mathbf{v}| = \sqrt{32^2 + (-23)^2} = \sqrt{1553} = 39.4$

DO EXERCISE 10.

10. For $\mathbf{u} = \langle -3, 5 \rangle$ and $\mathbf{v} = \langle 4, 2 \rangle$, find each of the following.

a) $\mathbf{u} + \mathbf{v}$

b) $\mathbf{v} - \mathbf{u}$

c) $5\mathbf{u} - 2\mathbf{v}$

d) $|2\mathbf{u} + 3\mathbf{v}|$

● EXERCISE SET 8.4

 Magnitudes of vectors **a** and **b** and the angle between the vectors θ are given. Find the resultant, to the nearest tenth, giving the direction by specifying to the nearest degree the angle that it makes with the vector **a**.

1. $|\mathbf{a}| = 45$, $|\mathbf{b}| = 35$, $\theta = 90°$

2. $|\mathbf{a}| = 54$, $|\mathbf{b}| = 43$, $\theta = 90°$

3. $|\mathbf{a}| = 10$, $|\mathbf{b}| = 12$, $\theta = 67°$

4. $|\mathbf{a}| = 25$, $|\mathbf{b}| = 30$, $\theta = 75°$

5. $|\mathbf{a}| = 20$, $|\mathbf{b}| = 20$, $\theta = 117°$

6. $|\mathbf{a}| = 30$, $|\mathbf{b}| = 30$, $\theta = 123°$

7. $|\mathbf{a}| = 23$, $|\mathbf{b}| = 47$, $\theta = 27°$

8. $|\mathbf{a}| = 32$, $|\mathbf{b}| = 74$, $\theta = 72°$

B

9. Two forces of 5.8 N and 12.3 N act on an object at right angles. Find the magnitude of the resultant and the angle that it makes with the smaller force.

10. Two forces of 32 N and 45 N act on an object at right angles. Find the magnitude of the resultant and the angle that it makes with the smaller force.

11. Forces of 420 N and 300 N act on an object. The angle between the forces is 50°. Find the resultant, giving the angle that it makes with the larger force.

12. Forces of 410 N and 600 N act on an object. The angle between the forces is 47°. Find the resultant, giving the angle that it makes with the smaller force.

13. A balloon is rising 12 ft/sec while a wind is blowing 18 ft/sec. Find the speed of the balloon and the angle that it makes with the horizontal.

14. A balloon is rising 10 ft/sec while a wind is blowing 5 ft/sec. Find the speed of the balloon and the angle that it makes with the horizontal.

15. A boat heads 35°, propelled by a force of 750 lb. A wind from 320° exerts a force of 150 lb on the boat. How large is the resultant force, and in what direction is the boat moving?

16. A boat heads 220°, propelled by a force of 650 lb. A wind from 080° exerts a force of 100 lb on the boat. How large is the resultant force, and in what direction is the boat moving?

17. A ship sails N 80° E for 120 nautical mi, then S 20° W for 200 nautical mi. How far is it then from the starting point, and in what direction?

18. An airplane flies 032° for 210 km, then 280° for 170 km. How far is it, then, from the starting point, and in what direction?

19. A motorboat has a speed of 15 km/h. It crosses a river whose current has a speed of 3 km/h. In order to cross the river at right angles, in what direction should the boat be pointed?

20. An airplane has an airspeed of 150 km/h. It is to make a flight in a direction of 080° while there is a 25-km/h wind from 350°. What should the airplane's heading be?

A   C

21. A vector **u** with magnitude 150 is inclined to the right and upward 52° from the horizontal. Find the horizontal and vertical components of **u**.

22. A vector **u** with magnitude 170 is inclined to the right and downward 63° from the horizontal. Find the horizontal and vertical components of **u**.

23. An airplane is flying 220° at 250 km/h. Find the magnitude of the southerly and westerly components of its velocity **v**.

24. A wind is blowing from 310° at 25 mph. Find the magnitude of the southerly and easterly components of the wind velocity **v**.

B   D

25. A force **f** has a westerly component of 25 N and a southerly component of 35 N. Find the magnitude and the direction of **f**.

26. A force **f** has a component of 65 N upward and a component of 90 N to the left. Find the magnitude and the direction of **f**.

27. Vector **u** has a westerly component of 15 and a northerly component of 22. Vector **v** has a northerly component of 6 and an easterly component of 8. Find:

a) the components of **u** + **v**;
b) the magnitude and the direction of **u** + **v**.

28. Vector **u** has a component of 18 up and a component of 12 to the left. Vector **v** has a component of 5 down and a component of 35 to the right. Find:

a) the components of **u** + **v**;
b) the magnitude and the direction of **u** + **v**.

   In Exercises 29–42, rectangular notation for vectors is given.

Add, writing rectangular notation for the answer.

29. $\langle 3, 7 \rangle + \langle 2, 9 \rangle$

30. $\langle 5, -7 \rangle + \langle -3, 2 \rangle$

31. $\langle 17, 7.6 \rangle + \langle -12.2, 6.1 \rangle$

32. $\langle -15.2, 37.1 \rangle + \langle 7.9, -17.8 \rangle$

33. $\langle -650, -750 \rangle + \langle -12, 324 \rangle$

34. $\langle -354, -973 \rangle + \langle -75, 256 \rangle$

C   E

35. $\langle 3, 4 \rangle$ **36.** $\langle 4, 3 \rangle$

37. $\langle 10, -15 \rangle$ **38.** $\langle 17, -10 \rangle$

39. $\langle -3, -4 \rangle$ **40.** $\langle -4, -3 \rangle$

41. $\langle -10, 15 \rangle$ **42.** $\langle -17, 10 \rangle$

Find rectangular notation.

43. $(4, 30°)$ **44.** $(8, 60°)$

45. $(10, 235°)$ **46.** $(15, 210°)$

47. $(20, 330°)$ **48.** $(20, 200°)$

49. $(100, -45°)$ **50.** $(150, -60°)$

E   F   Do the calculations for the following vectors:
$\mathbf{u} = \langle 3, 4 \rangle$, $\mathbf{v} = \langle 5, 12 \rangle$, and $\mathbf{w} = \langle -6, 8 \rangle$.

51. $3\mathbf{u} + 2\mathbf{v}$ **52.** $3\mathbf{u} - 2\mathbf{v}$

53. $(\mathbf{u} + \mathbf{v}) - \mathbf{w}$ **54.** $\mathbf{u} - (\mathbf{v} + \mathbf{w})$

55. $|\mathbf{u}| + |\mathbf{v}|$ **56.** $|\mathbf{u}| - |\mathbf{v}|$

57. $|\mathbf{u} + \mathbf{v}|$ **58.** $|\mathbf{u} - \mathbf{v}|$

59. $2|\mathbf{u} + \mathbf{v}|$ **60.** $2|\mathbf{u}| + 2|\mathbf{v}|$

●   **SYNTHESIS** _____

61. If **PQ** is any vector, what is **PQ** + **QP**?

62. The **inner product** of vectors **u** · **v** is a scalar, defined as follows: $\mathbf{u} \cdot \mathbf{v} = |\mathbf{u}| \, |\mathbf{v}| \cos \theta$, where θ is the angle between the vectors. Show that vectors **u** and **v** are perpendicular if and only if $\mathbf{u} \cdot \mathbf{v} = 0$.

●   **CHALLENGE** _____

63. Show that scalar multiplication is distributive over vector addition.

64. Let $\mathbf{u} = \langle 3, 4 \rangle$. Find a vector that has the same direction as **u** but length 1.

65. Prove that for any vectors **u** and **v**, $|\mathbf{u} + \mathbf{v}| \leq |\mathbf{u}| + |\mathbf{v}|$.

OBJECTIVES

You should be able to:

A Plot points given their polar coordinates, and determine polar coordinates of points on a graph.

B Convert from rectangular to polar equations and from polar to rectangular equations.

C Graph polar equations.

8.5
POLAR COORDINATES

A   In graphing, we locate a point with an ordered pair of numbers (a, b). We can consider any such ordered pair to be a vector. From our work with vectors, we know that we could also locate a point with a vector, given a length and a direction. When we use rectangular notation to locate points, we describe the coordinate system as **rectangular.*** When we use polar notation, we describe the coordinate system as **polar.** As this diagram shows, any point has **rectangular coordinates** (x, y) and **polar coordinates** (r, θ). On a polar graph, the origin is called the **pole** and the positive half of the x-axis is called the **polar axis.**

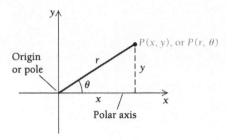

To plot points on a polar graph:

1. Locate θ.
2. Move a distance r from the pole in that direction.
3. If r is negative, move in the opposite direction.

Polar graph paper, shown below, facilitates plotting. Point B illustrates that θ may be in radians. Point E illustrates that the polar coordinates of a point are not unique.

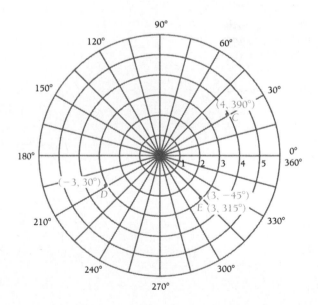

DO EXERCISES 1–4 ON THE FOLLOWING PAGE.

---

* Also called *Cartesian,* after the French mathematician René Descartes.

1. Plot each of the following points.

   a) $(3, 60°)$ b) $(0, 10°)$

   c) $(-5, 120°)$ d) $(5, -60°)$

   e) $(2, 3\pi/2)$

2. Find polar coordinates of each of these points. Give two answers for each.

3. Find the polar coordinates of each of these points. (See Example 8 of Section 8.4.)

   a) $(3, 3)$ b) $(0, -4)$

   c) $(-3, 3\sqrt{3})$ d) $(2\sqrt{3}, -2)$

4. Find the rectangular coordinates of each of these points. (See Example 9 of Section 8.4.)

   a) $(5, 30°)$

   b) $(10, \pi/3)$

   c) $(-5, 45°)$

   d) $(-8, -5\pi/6)$

Polar and Rectangular Equations

Some curves have simpler equations in polar coordinates than in rectangular coordinates. For others, the reverse is true. Applying the definitions of sine and cosine,

$$\frac{x}{r} = \cos\theta \quad \text{and} \quad \frac{y}{r} = \sin\theta,$$

we obtain the following relationships between the two kinds of coordinates. These allow us to convert an equation from rectangular to polar coordinates:

$$x = r\cos\theta, \qquad y = r\sin\theta.$$

Example 1   Convert to a polar equation: $x^2 + y^2 = 25$.

Solution

$$x^2 + y^2 = 25$$
$$(r\cos\theta)^2 + (r\sin\theta)^2 = 25 \qquad \text{Substituting for } x \text{ and } y$$
$$r^2\cos^2\theta + r^2\sin^2\theta = 25$$
$$r^2(\cos^2\theta + \sin^2\theta) = 25$$
$$r^2 = 25 \qquad \blacktriangleleft$$

Example 1 illustrates that the polar equation of a circle centered at the origin is much simpler than the rectangular equation.

Example 2   Convert to a polar equation: $2x - y = 5$.

Solution

$$2x - y = 5$$
$$2(r\cos\theta) - (r\sin\theta) = 5$$
$$r(2\cos\theta - \sin\theta) = 5 \qquad \blacktriangleleft$$

DO EXERCISES 5 AND 6 ON THE FOLLOWING PAGE.

The relationships that we need to convert from polar equations to rectangular equations can be determined easily from the triangle. They are as follows:

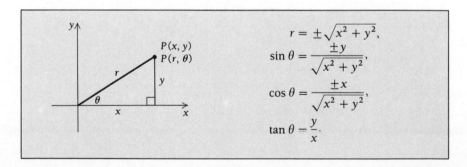

To convert to rectangular notation, we substitute as needed from either of the two lists, choosing the $+$ or the $-$ signs as appropriate in each case.

Example 3   Convert to a rectangular (Cartesian) equation: $r = 4$.

Solution

$$r = 4$$
$$+\sqrt{x^2 + y^2} = 4 \qquad \text{Substituting for } r$$
$$x^2 + y^2 = 16 \qquad \text{Squaring} \qquad \blacktriangleleft$$

In squaring, as we did in Example 3, we must be careful not to introduce solutions of the equation that are not already present. We did not, because the graph is a circle of radius 4 centered at the origin, in both cases.

Example 4   Convert to a rectangular equation: $r \cos \theta = 6$.

Solution   From our first list, we know that $x = r \cos \theta$, so we have $x = 6$.

$\blacktriangleleft$

Example 5   Convert to a rectangular equation: $r = 2 \cos \theta + 3 \sin \theta$.

Solution   We first multiply on both sides by r:

$$r^2 = 2r \cos \theta + 3r \sin \theta$$
$$x^2 + y^2 = 2x + 3y. \qquad \text{Substituting} \qquad \blacktriangleleft$$

DO EXERCISES 7–9.

 Graphing Polar Equations

To graph a polar equation, we usually make a table of values, choosing values of θ and calculating corresponding values of r. We plot the points and then complete the graph, as in the rectangular case. A difference occurs in the case of a polar equation, because as θ increases sufficiently, points may, in some cases, begin to repeat and the curve will be traced again and again. If such a point is reached, the curve is complete.

Example 6   Graph: $r = \cos \theta$.

Solution   We first make a table of values.

θ	0°	30°	45°	60°	90°	120°	135°	150°	180°
r	1	0.866	0.707	0.5	0	−0.5	−0.707	−0.866	−1

Convert to a polar equation.

5. $2x + 5y = 9$

6. $x^2 + y^2 + 8x = 0$

Convert to a Cartesian equation.

7. $r = 7$

8. $r \sin \theta = 5$

9. $r - 3 \cos \theta = 5 \sin \theta$

10. Graph: $r = 1 - \sin \theta$.

We plot these points and note that the last point is the same as the first.

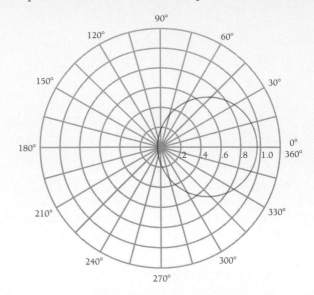

We try another point, $(-0.866, 210°)$, and find that it has already been plotted. Since the repeating has started, we plot no more points, but draw the graph, which is a circle as shown. ◀

DO EXERCISE 10.

Example 7 Graph: $r = 4 \sin 3\theta$.

 TECHNOLOGY CONNECTION

Some graphers allow the graphing of polar equations. Check the documentation for your grapher. Usually the equation must first be written in the form $r = f(\theta)$. When drawing such a graph, it is necessary to decide on not only the best viewing box but also the range of values for θ. Typically, we begin with a range of 0 to 2π for θ for radians, or 0 to 360 for degrees. If the graph does not appear to be complete, these limits are extended, usually in multiples of 2π radians or 360°. (The letter T may be used instead of θ, if the grapher can't use Greek letters.) Because most polar graphs are curved, it is important that you choose a viewing box in which there is no distortion in either the horizontal or vertical direction.

For example, to graph the polar coordinate equation $r - 7 \cos \theta = 7$, the first step is to solve for r, rewriting the equation in the form $r = 7 + 7 \cos \theta$. If we let the values of θ increase from 0 to 360° in steps of 1° and use a viewing box of $[-15, 15] \times [-10, 10]$, the following graph will be drawn.

$r = 7 + 7 \cos \theta$

$[-15, 15] \times [-10, 10]$

Use a grapher to draw the graphs of each of the following polar coordinate equations. Use a viewing box of $[-15, 15] \times [-10, 10]$.

TC 1. $r = 8 \sin 4\theta$

TC 2. $r = 10 \tan \theta \sin \theta$

TC 3. $r = 4 \sin 2\theta - 8$

TC 4. $r = 7 \cos 3\theta + 4 \sin 2\theta$

Solution   We first make a table of values.

θ	0°	15°	30°	45°	60°	75°	90°
r	0	2.83	4	2.83	0	−2.83	−4

θ	105°	120°	135°	150°	165°	180°
r	−2.83	0	2.83	4	2.83	0

11. Graph: $r = -2 \sin 3\theta$.

We plot these points and sketch the curve. We then try another point, $(-2.83, 195°)$, and find that it has already been plotted. We plot no more points since the repeating has begun.

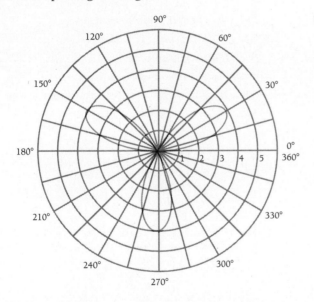

This curve is often referred to as a **three-leafed rose.**

DO EXERCISE 11.

●**EXERCISE SET** **8.5**

A   Using polar coordinates, graph each of the following points.

1. $(4, 30°)$ **2.** $(5, 45°)$

3. $(0, 37°)$ **4.** $(0, 48°)$

5. $(-6, 150°)$ **6.** $(-5, 135°)$

7. $(-8, 210°)$ **8.** $(-5, 270°)$

9. $(3, -30°)$ **10.** $(6, -45°)$

11. $(7, -315°)$ **12.** $(4, -270°)$

13. $(-3, -30°)$ **14.** $(-6, -45°)$

15. $(-3.2, 27°)$ **16.** $(-6.8, 34°)$

17. $\left(6, \dfrac{\pi}{4}\right)$ **18.** $\left(5, \dfrac{\pi}{6}\right)$

19. $\left(4, \dfrac{3\pi}{2}\right)$ **20.** $\left(3, \dfrac{3\pi}{4}\right)$

21. $\left(-6, \dfrac{\pi}{4}\right)$ **22.** $\left(-5, \dfrac{\pi}{6}\right)$

23. $\left(-4, -\dfrac{3\pi}{2}\right)$ **24.** $\left(-3, -\dfrac{3\pi}{4}\right)$

B   Find the polar coordinates of the point.

25. $(4, 4)$ **26.** $(5, 5)$

27. $(0, 5)$ **28.** $(0, -3)$

29. $(4, 0)$ **30.** $(-5, 0)$

31. $(3, 3\sqrt{3})$ **32.** $(-3, -3\sqrt{3})$

33. $(\sqrt{3}, 1)$ **34.** $(-\sqrt{3}, 1)$

35. $(3\sqrt{3}, 3)$ **36.** $(4\sqrt{3}, -4)$

Find the Cartesian coordinates of the point.

37. $(4, 45°)$ **38.** $(5, 60°)$

39. $(0, 23°)$ **40.** $(0, -34°)$

41. $(-3, 45°)$ **42.** $(-5, 30°)$

43. $(6, -60°)$ **44.** $(3, -120°)$

45. $\left(10, \dfrac{\pi}{6}\right)$ **46.** $\left(12, \dfrac{3\pi}{4}\right)$

47. $\left(-5, \dfrac{5\pi}{6}\right)$ **48.** $\left(-6, \dfrac{3\pi}{4}\right)$

Convert to a polar equation.

49. $3x + 4y = 5$ **50.** $5x + 3y = 4$

51. $x = 5$ **52.** $y = 4$

53. $x^2 + y^2 = 36$ **54.** $x^2 + y^2 = 16$

55. $x^2 - 4y^2 = 4$ **56.** $x^2 - 5y^2 = 5$

Convert to a rectangular equation.

57. $r = 5$ **58.** $r = 8$

59. $\theta = \dfrac{\pi}{4}$ **60.** $\theta = \dfrac{3\pi}{4}$

61. $r \sin \theta = 2$ **62.** $r \cos \theta = 5$

63. $r = 4 \cos \theta$ **64.** $r = -3 \sin \theta$

65. $r - r \sin \theta = 2$

66. $r + r \cos \theta = 3$

67. $r - 2 \cos \theta = 3 \sin \theta$

68. $r + 5 \sin \theta = 7 \cos \theta$

C   Graph.

69. $r = 4 \cos \theta$

70. $r = 4 \sin \theta$

71. $r = 1 - \cos \theta$ (Cardioid)

72. $r = 1 + \sin \theta$ (Cardioid)

73. $r = \sin 2\theta$ (Four-leafed rose)

74. $r = 3 \cos 2\theta$ (Four-leafed rose)

75. $r = 2 \cos 3\theta$ (Three-leafed rose)

76. $r = \sin 3\theta$ (Three-leafed rose)

77. $r \cos \theta = 4$

78. $r \sin \theta = 6$

79. $r = \dfrac{5}{1 + \cos \theta}$ (Limaçon)

80. $r = \dfrac{3}{1 + \sin \theta}$ (Limaçon)

81. $r = \theta$ (Spiral of Archimedes)

82. $r = 3\theta$ (Spiral of Archimedes)

83. $r^2 = \sin 2\theta$ (Lemniscate)

84. $r^2 = 4 \cos 2\theta$ (Lemniscate)

85. $r = e^{\theta/10}$ or $\ln r = \theta/10$ (Logarithmic spiral)

86. $r = 10^{2\theta}$ or $\log r = 2\theta$ (Logarithmic spiral)

●   CHALLENGE _____

87. Convert to a rectangular equation:

$$r = \sec^2 \dfrac{\theta}{2}.$$

88. The center of a regular hexagon is at the origin, and one vertex is the point $(4, 0°)$. Find the coordinates of the other vertices.

▣   TECHNOLOGY CONNECTION _____

Use a grapher to graph each of the following.

89. $r = \sin \theta \tan \theta$ (Cissoid)

90. $r = \cos 2\theta \sec \theta$ (Strophoid)

91. $r = 2 \cos 2\theta - 1$ (Bow tie)

92. $r = \cos 2\theta - 2$ (Peanut)

93. $r = \frac{1}{4} \tan^2 \theta \sec \theta$ (Semicubical parabola)

94. $r = \sin 2\theta + \cos \theta$ (Twisted sister)

8.6
COMPLEX NUMBERS

Graphical Representation and Polar Notation

◢ Graphs

The real numbers can be graphed on a line. Complex numbers are graphed on a plane. We graph a complex number $a + bi$ in the same way that we graph an ordered pair of real numbers (a, b). However, in place of an x-axis, we have a **real axis,** and in place of a y-axis, we have an **imaginary axis.**

Examples Graph each of the following.

1. $3 + 2i$
2. $-4 + 5i$
3. $-5 - 4i$

Horizontal distances correspond to the real part of a number. Vertical distances correspond to the imaginary part. Graphs of the numbers are shown above as vectors. The horizontal component of the vector for $a + bi$ is a and the vertical component is b. Vectors are sometimes used to study complex numbers.

DO EXERCISE 1.

Adding complex numbers is like adding vectors using components. For example, to add $3 + 2i$ and $5 + 4i$, we add the real parts and the imaginary parts to obtain $8 + 6i$. Graphically, then, the sum of two complex numbers looks like a vector sum. It is the diagonal of a parallelogram.

Example 4 Show graphically $2 + 2i$ and $3 - i$. Show also their sum.

Solution

1. Graph each of the following complex numbers.
   a) $5 - 3i$
   b) $-3 + 4i$
   c) $-5 - 2i$
   d) $5 + 5i$

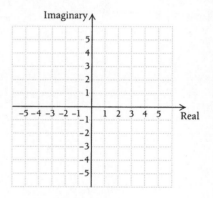

2. Graph each pair of complex numbers and then graph their sum.

a) $2 + 3i$, $1 - 5i$

b) $-5 + 2i$, $-1 - 4i$

DO EXERCISE 2.

 ## Polar Notation for Complex Numbers

We can locate endpoints of arrows as shown in the preceding examples or by giving the length of the arrow and the angle that it makes with the positive half of the real axis, as we did in Section 8.5. We now develop **polar notation** for complex numbers using this idea. From the figure below, we see that the length of the vector is $\sqrt{a^2 + b^2}$. Note that this quantity is a real number. It is called the **absolute value** of $a + bi$.

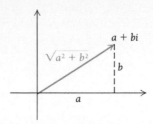

DEFINITION   **Absolute Value of a Complex Number**

The *absolute value of a complex number* $a + bi$ is denoted $|a + bi|$ and is defined to be $\sqrt{a^2 + b^2}$.

Example 5   Find $|3 + 4i|$.

Solution

$$|3 + 4i| = \sqrt{3^2 + 4^2} = \sqrt{9 + 16} = 5$$ ◄

DO EXERCISE 3.

3. Find the absolute value.

a) $|4 - 3i|$

b) $|-12 - 5i|$

Now let us consider any complex number $a + bi$. Suppose that its absolute value is r. Let us also suppose that the angle that the vector makes with the real axis is θ. As this figure shows, we have $a/r = \cos\theta$ and $b/r = \sin\theta$, so

$$a = r\cos\theta \quad \text{and} \quad b = r\sin\theta.$$

Thus,

$$a + bi = r\cos\theta + (r\sin\theta)i \qquad \text{Substituting}$$
$$= r(\cos\theta + i\sin\theta).$$

This is polar notation for $a + bi$. The angle θ is called the **argument.**

Polar notation for complex numbers is also called **trigonometric notation**, and is often abbreviated $r \operatorname{cis} \theta$.

Polar Notation for Complex Numbers

$a + bi = r(\cos \theta + i \sin \theta) = r \operatorname{cis} \theta$

(Polar notation for complex numbers is also called trigonometric notation.)

▶ B, ▶ C   Change of Notation

To change from polar notation to **binomial**, or **rectangular**, notation $a + bi$, we proceed as in Section 8.5, recalling that $a = r \cos \theta$ and $b = r \sin \theta$.

Example 6   Write binomial notation for $2(\cos 120° + i \sin 120°)$.

Solution   Rewriting, we have

$$2(\cos 120° + i \sin 120°) = 2 \cos 120° + (2 \sin 120°)i.$$

Thus,

$$a = 2 \cos 120° = -1,$$
$$b = 2 \sin 120° = \sqrt{3},$$

and

$$2(\cos 120° + i \sin 120°) = -1 + i\sqrt{3}.$$

Example 7   Write binomial notation for $\sqrt{8} \operatorname{cis} \dfrac{7\pi}{4}$.

Solution

$$\sqrt{8} \operatorname{cis} \frac{7\pi}{4} = \sqrt{8}\left(\cos \frac{7\pi}{4} + i \sin \frac{7\pi}{4}\right),$$

$$a = \sqrt{8} \cos \frac{7\pi}{4} = \sqrt{8} \cdot \frac{1}{\sqrt{2}} = 2,$$

$$b = \sqrt{8} \sin \frac{7\pi}{4} = \sqrt{8} \cdot \frac{-1}{\sqrt{2}} = -2.$$

Thus,

$$\sqrt{8} \operatorname{cis} \frac{7\pi}{4} = 2 - 2i. \qquad ◀$$

DO EXERCISES 4 AND 5.

To change from binomial notation to polar notation, we remember that $r = \sqrt{a^2 + b^2}$ and that θ is an angle for which $\sin \theta = b/r$ and $\cos \theta = a/r$.

Example 8   Find polar notation for $1 + i$.

---

4. Write rectangular notation for
$$\sqrt{2}(\cos 315° + i \sin 315°).$$

5. Write binomial notation for
$$2 \operatorname{cis}\left(-\frac{\pi}{6}\right).$$

Write polar notation.

6. $1 - i$

7. $-3\sqrt{2} - 3\sqrt{2}i$

Solution   We note that $a = 1$ and $b = 1$. Then

$$r = \sqrt{1^2 + 1^2} = \sqrt{2},$$

$$\sin \theta = \frac{1}{\sqrt{2}}, \quad \text{or} \quad \frac{\sqrt{2}}{2},$$

and

$$\cos \theta = \frac{1}{\sqrt{2}}, \quad \text{or} \quad \frac{\sqrt{2}}{2}.$$

Thus, $\theta = \pi/4$, or $45°$, and we have

$$1 + i = \sqrt{2} \operatorname{cis} \frac{\pi}{4}, \quad \text{or} \quad 1 + i = \sqrt{2} \operatorname{cis} 45°. \qquad \blacktriangleleft$$

Example 9   Find polar notation for $\sqrt{3} - i$.

Solution   We note that $a = \sqrt{3}$ and $b = -1$. Then

$$r = \sqrt{(\sqrt{3})^2 + (-1)^2} = 2,$$

$$\sin \theta = -\frac{1}{2}, \quad \text{and} \quad \cos \theta = \frac{\sqrt{3}}{2}.$$

Thus, $\theta = 11\pi/6$, or $330°$, and we have

$$\sqrt{3} - i = 2 \operatorname{cis} \frac{11\pi}{6}, \quad \text{or} \quad 2 \operatorname{cis} 330°. \qquad \blacktriangleleft$$

In changing to polar notation, note that there are many angles satisfying the given conditions. We ordinarily choose the smallest positive angle.

DO EXERCISES 6 AND 7.

▶ Multiplication and Division with Polar Notation

Multiplication of complex numbers is somewhat easier to do with polar notation than with rectangular notation. We simply multiply the absolute values and add the arguments. Let us state this more formally and prove it.

THEOREM 4

For any complex numbers $r_1 \operatorname{cis} \theta_1$ and $r_2 \operatorname{cis} \theta_2$,

$$(r_1 \operatorname{cis} \theta_1)(r_2 \operatorname{cis} \theta_2) = r_1 \cdot r_2 \operatorname{cis}(\theta_1 + \theta_2).$$

Proof.   Let us first multiply $a_1 + b_1 i$ by $a_2 + b_2 i$:

$$(a_1 + b_1 i)(a_2 + b_2 i) = (a_1 a_2 - b_1 b_2) + (a_2 b_1 + a_1 b_2)i.$$

Recall that

$$a_1 = r_1 \cos \theta_1, \qquad b_1 = r_1 \sin \theta_1,$$

and

$$a_2 = r_2 \cos \theta_2, \qquad b_2 = r_2 \sin \theta_2.$$

We substitute these in the product above, to obtain

$$r_1(\cos\theta_1 + i\sin\theta_1) \cdot r_2(\cos\theta_2 + i\sin\theta_2)$$
$$= (r_1 r_2 \cos\theta_1 \cos\theta_2 - r_1 r_2 \sin\theta_1 \sin\theta_2)$$
$$+ (r_1 r_2 \sin\theta_1 \cos\theta_2 + r_1 r_2 \cos\theta_1 \sin\theta_2)i.$$

This simplifies to

$$r_1 r_2(\cos\theta_1 \cos\theta_2 - \sin\theta_1 \sin\theta_2) + r_1 r_2(\sin\theta_1 \cos\theta_2 + \cos\theta_1 \sin\theta_2)i.$$

Now, using identities for sums of angles, we simplify, obtaining

$$r_1 r_2 \cos(\theta_1 + \theta_2) + r_1 r_2 \sin(\theta_1 + \theta_2)i,$$

or

$$r_1 r_2 \operatorname{cis}(\theta_1 + \theta_2),$$

which was to be shown.

To divide complex numbers, we do the reverse of the above. We state that fact, but omit the proof.

THEOREM 5

For any complex numbers $r_1 \operatorname{cis}\theta_1$ and $r_2 \operatorname{cis}\theta_2$ $(r_2 \neq 0)$,

$$\frac{r_1 \operatorname{cis}\theta_1}{r_2 \operatorname{cis}\theta_2} = \frac{r_1}{r_2}\operatorname{cis}(\theta_1 - \theta_2).$$

Example 10   Find the product of $3 \operatorname{cis} 40°$ and $7 \operatorname{cis} 20°$.

Solution

$$3 \operatorname{cis} 40° \cdot 7 \operatorname{cis} 20° = 3 \cdot 7 \operatorname{cis}(40° + 20°) \qquad \text{Using Theorem 4}$$
$$= 21 \operatorname{cis} 60° \qquad \blacktriangleleft$$

Example 11   Find the product of $2 \operatorname{cis}\pi$ and $3 \operatorname{cis}\left(-\dfrac{\pi}{2}\right)$.

Solution

$$2 \operatorname{cis}\pi \cdot 3 \operatorname{cis}\left(-\frac{\pi}{2}\right) = 2 \cdot 3 \operatorname{cis}\left(\pi - \frac{\pi}{2}\right)$$
$$= 6 \operatorname{cis}\frac{\pi}{2} \qquad \blacktriangleleft$$

Example 12   Convert to polar notation and multiply: $(1 + i)(\sqrt{3} - i)$.

Solution   We first find polar notation (see Examples 8 and 9):

$$1 + i = \sqrt{2} \operatorname{cis} 45°, \qquad \sqrt{3} - i = 2 \operatorname{cis} 330°.$$

We then multiply, using Theorem 4:

$$(\sqrt{2} \operatorname{cis} 45°)(2 \operatorname{cis} 330°) = 2 \cdot \sqrt{2} \operatorname{cis} 375°, \quad \text{or } 2\sqrt{2} \operatorname{cis} 15°. \qquad \blacktriangleleft$$

Multiply.

8. $5 \operatorname{cis} 25° \cdot 4 \operatorname{cis} 30°$

9. $8 \operatorname{cis} \pi \cdot \dfrac{1}{2} \operatorname{cis} \dfrac{\pi}{4}$

10. Convert to polar notation and multiply:

$$(1 + i)(2 + 2i).$$

11. Divide:

$$10 \operatorname{cis} \frac{\pi}{2} \div 5 \operatorname{cis} \frac{\pi}{4}.$$

12. Convert to polar notation and divide:

$$\frac{\sqrt{3} - i}{1 + i}.$$

Example 13 Divide $2 \operatorname{cis} \pi$ by $4 \operatorname{cis} \dfrac{\pi}{2}$.

Solution

$$\frac{2 \operatorname{cis} \pi}{4 \operatorname{cis} \dfrac{\pi}{2}} = \frac{2}{4} \operatorname{cis} \left(\pi - \frac{\pi}{2} \right) \qquad \text{Using Theorem 5}$$

$$= \frac{1}{2} \operatorname{cis} \frac{\pi}{2}$$

◀

Example 14 Convert to polar notation and divide: $(1 + i)/(1 - i)$.

Solution We first convert to polar notation:

$$1 + i = \sqrt{2} \operatorname{cis} 45°, \qquad \text{See Example 8.}$$
$$1 - i = \sqrt{2} \operatorname{cis} 315°.$$

We now divide, using Theorem 5:

$$\frac{1 + i}{1 - i} = \frac{\sqrt{2} \operatorname{cis} 45°}{\sqrt{2} \operatorname{cis} 315°} = 1 \cdot \operatorname{cis} (45° - 315°)$$

$$= 1 \cdot \operatorname{cis} (-270°), \quad \text{or } \operatorname{cis} 90°.$$

◀

DO EXERCISES 8–12.

● EXERCISE SET 8.6

A Graph each pair of complex numbers and then graph their sum.

1. $3 + 2i, \quad 2 - 5i$
2. $4 + 3i, \quad 3 - 4i$
3. $-5 + 3i, \quad -2 - 3i$
4. $-4 + 2i, \quad -3 - 4i$
5. $2 - 3i, \quad -5 + 4i$
6. $3 - 2i, \quad -5 + 5i$
7. $-2 - 5i, \quad 5 + 3i$
8. $-3 - 4i, \quad 6 + 3i$

  Find rectangular notation.

9. $3(\cos 30° + i \sin 30°)$
10. $6(\cos 150° + i \sin 150°)$
11. $10 \operatorname{cis} 270°$
12. $5 \operatorname{cis} (-60°)$
13. $\sqrt{8} \left(\cos \dfrac{\pi}{4} + i \sin \dfrac{\pi}{4} \right)$
14. $5 \left(\cos \dfrac{\pi}{3} + i \sin \dfrac{\pi}{3} \right)$
15. $\sqrt{8} \operatorname{cis} \dfrac{5\pi}{4}$
16. $\sqrt{8} \operatorname{cis} \left(-\dfrac{\pi}{4} \right)$

C Find polar notation.

17. $1 - i$
18. $\sqrt{3} + i$
19. $10\sqrt{3} - 10i$
20. $-10\sqrt{3} + 10i$
21. -5
22. $-5i$

D Convert to polar notation and then multiply or divide.

23. $(1 - i)(2 + 2i)$
24. $(1 + i\sqrt{3})(1 + i)$
25. $(2\sqrt{3} + 2i)(2i)$
26. $(3\sqrt{3} - 3i)(2i)$
27. $\dfrac{1 - i}{1 + i}$
28. $\dfrac{1 - i}{\sqrt{3} - i}$
29. $\dfrac{2\sqrt{3} - 2i}{1 + \sqrt{3}i}$
30. $\dfrac{3 - 3\sqrt{3}i}{\sqrt{3} - i}$

● SYNTHESIS

31. Show that for any complex number z,

$$|z| = |-z|.$$

   (*Hint*: Let $z = a + bi$.)

32. Show that for any complex number z,

$$|z| = |\bar{z}|.$$

   (*Hint*: Let $z = a + bi$.)

33. Show that for any complex number z,

$$|z\bar{z}| = |z^2|.$$

34. Show that for any complex number z,

$$|z^2| = |z|^2.$$

35. Show that for any complex numbers z and w,
$$|z \cdot w| = |z| \cdot |w|.$$
(*Hint:* Let $z = r_1 \operatorname{cis} \theta_1$ and $w = r_2 \operatorname{cis} \theta_2$.)

36. Show that for any complex number z and any nonzero, complex number w,
$$\left| \frac{z}{w} \right| = \frac{|z|}{|w|}. \qquad \text{(Use the hint for Exercise 35.)}$$

37. On a complex plane, graph $|z| = 1$.

38. On a complex plane, graph $z + \bar{z} = 3$.

● CHALLENGE

39. Find polar notation for $(\cos \theta + i \sin \theta)^{-1}$.

40. Compute $(5 \operatorname{cis} 20°)^3$.

8.7

COMPLEX NUMBERS: DeMOIVRE'S THEOREM AND *n*th ROOTS OF COMPLEX NUMBERS

OBJECTIVES

You should be able to:

 Use DeMoivre's theorem to raise complex numbers to powers.

Find the *n*th roots of a complex number.

A   Powers of Complex Numbers

An important theorem about powers and roots of complex numbers is named for the French mathematician Abraham DeMoivre (1667–1754). Let us consider a number $r \operatorname{cis} \theta$ and its square:

$$(r \operatorname{cis} \theta)^2 = (r \operatorname{cis} \theta)(r \operatorname{cis} \theta) = r \cdot r \operatorname{cis} (\theta + \theta)$$
$$= r^2 \operatorname{cis} 2\theta.$$

Similarly, we see that

$$(r \operatorname{cis} \theta)^3 = r \cdot r \cdot r \operatorname{cis} (\theta + \theta + \theta)$$
$$= r^3 \operatorname{cis} 3\theta.$$

DeMoivre's theorem is the generalization of these results.

THEOREM 6   **DeMoivre's Theorem**

For any complex number $r \operatorname{cis} \theta$ and any natural number n,

$$(r \operatorname{cis} \theta)^n = r^n \operatorname{cis} n\theta,$$

or

$$[r(\cos \theta + i \sin \theta)]^n = r^n[\cos n\theta + i \sin n\theta].$$

Example 1   Find $(1 + i)^9$.

Solution   We first find polar notation: $1 + i = \sqrt{2} \operatorname{cis} 45°$. Then

$$(1 + i)^9 = (\sqrt{2} \operatorname{cis} 45°)^9 = \sqrt{2}^9 \operatorname{cis} (9 \cdot 45°)$$
$$= 2^{9/2} \operatorname{cis} 405°$$
$$= 16\sqrt{2} \operatorname{cis} 45° \qquad \text{405° has the same terminal side as 45°}$$
$$= 16\sqrt{2}(\cos 45° + i \sin 45°)$$
$$= 16\sqrt{2}\left(\frac{\sqrt{2}}{2} + i \frac{\sqrt{2}}{2} \right)$$
$$= 16 + 16i. \qquad ◀$$

1. Find $(1 - i)^{10}$.

2. Find $(\sqrt{3} + i)^4$.

Example 2 Find $(\sqrt{3} - i)^{10}$.

Solution We first find polar notation: $\sqrt{3} - i = 2 \operatorname{cis} 330°$. Then

$$
\begin{aligned}
(\sqrt{3} - i)^{10} &= (2 \operatorname{cis} 330°)^{10} \\
&= 2^{10} \operatorname{cis} (10 \cdot 330°) \\
&= 1024 \operatorname{cis} 3300° \\
&= 1024 \operatorname{cis} 60° \\
&= 1024[\cos 60° + i \sin 60°] \\
&= 1024\left[\frac{1}{2} + i \frac{\sqrt{3}}{2}\right] \\
&= 512 + 512\sqrt{3}i.
\end{aligned}
$$

◀

DO EXERCISES 1 AND 2.

 Roots of Complex Numbers

As we will see, every nonzero complex number has two square roots. A number has three cube roots, four fourth roots, and so on. In general, a nonzero complex number has n different nth roots. They can be found by the formula that we now state and prove.

THEOREM 7

The nth roots of a complex number $r \operatorname{cis} \theta$, $r \neq 0$, are given by

$$
r^{1/n} \operatorname{cis} \left(\frac{\theta}{n} + k \cdot \frac{360°}{n}\right),
$$

where $k = 0, 1, 2, \ldots, n - 1$.

Using DeMoivre's theorem, we show that this formula gives us n different roots. We take the expression for the nth roots and raise it to the nth power, to show that we get $r \operatorname{cis} \theta$:

$$
\begin{aligned}
\left[r^{1/n} \operatorname{cis} \left(\frac{\theta}{n} + k \cdot \frac{360°}{n}\right)\right]^n &= (r^{1/n})^n \operatorname{cis} \left(\frac{\theta}{n} \cdot n + k \cdot n \cdot \frac{360°}{n}\right) \\
&= r \operatorname{cis} (\theta + k \cdot 360°) \\
&= r \operatorname{cis} \theta.
\end{aligned}
$$

Thus we know that the formula gives us nth roots for any natural number k. Next, we show that there are at least n different roots. To see this, consider substituting 0, 1, 2, and so on, for k. When $k = n$, the cycle begins to repeat, but from 0 to $n - 1$, the angles obtained and their sines and cosines are all different. There cannot be more than n different nth roots. That fact follows from the **fundamental theorem of algebra,** considered in the theory of polynomial functions.

Example 3 Find the square roots of $2 + 2\sqrt{3}i$.

Solution   We first find polar notation: $2 + 2\sqrt{3}i = 4 \text{ cis } 60°$. Then using Theorem 7 gives us

$$(4 \text{ cis } 60°)^{1/2} = 4^{1/2} \text{ cis}\left(\frac{60°}{2} + k \cdot \frac{360°}{2}\right), \quad k = 0, 1$$

$$= 2 \text{ cis}\left(30° + k \cdot \frac{360°}{2}\right), \quad k = 0, 1.$$

Thus the roots are $2 \text{ cis } 30°$ and $2 \text{ cis } 210°$, or

$$\sqrt{3} + i \quad \text{and} \quad -\sqrt{3} - i.$$ ◀

DO EXERCISES 3 AND 4.

In Example 3, it should be noted that the two square roots of the number are opposites of each other. The same is true of the square roots of any complex number. To see this, let us find the square roots of any complex number $r \text{ cis } \theta$:

$$(r \text{ cis } \theta)^{1/2} = r^{1/2} \text{ cis}\left(\frac{\theta}{2} + k \cdot \frac{360°}{2}\right), \quad k = 0, 1$$

$$= r^{1/2} \text{ cis }\frac{\theta}{2} \quad \text{or} \quad r^{1/2} \text{ cis}\left(\frac{\theta}{2} + 180°\right).$$

Now let us look at the two numbers on a graph. They lie on a line, so if one number has binomial notation $a + bi$, the other has binomial notation $-a - bi$. Thus their sum is 0 and they are opposites of each other.

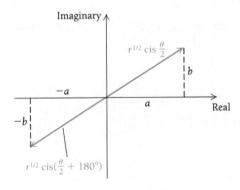

Example 4   Find the cube roots of 1. Then locate them on a graph.

Solution   We have

$$1 = 1 \text{ cis } 0° \qquad \text{Polar notation}$$

$$(1 \text{ cis } 0°)^{1/3} = 1^{1/3} \text{ cis}\left(\frac{0°}{3} + k \cdot \frac{360°}{3}\right), \quad k = 0, 1, 2.$$

The roots are $1 \text{ cis } 0°$, $1 \text{ cis } 120°$, and $1 \text{ cis } 240°$, or

$$1, \qquad -\frac{1}{2} + \frac{\sqrt{3}}{2}i, \quad \text{and} \quad -\frac{1}{2} - \frac{\sqrt{3}}{2}i.$$

Find the square roots.

3. $2i$

4. $10i$

5. Find and graph the cube roots of -1.

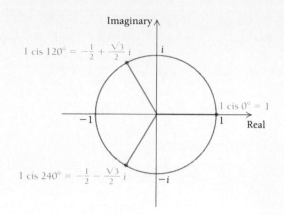

Note in Example 4 that the graphs of the cube roots lie equally spaced about a circle. The same is true of the nth roots of any complex number.

DO EXERCISE 5.

 TECHNOLOGY CONNECTION

If your grapher has a PARAMETRIC mode, you can use it to graphically find all the roots of a number. Parametric equations define x and y in terms of a third variable, usually t. For example, a parametric function is defined by the pair of equations $x = t^2$ and $y = 3t$. To begin the graph of this function, we select a value of t, substitute that value into the two equations to find the corresponding values for x and y, and then plot the point (x, y). This is repeated with many different values of t, producing many ordered pairs (x, y) that are then plotted. By drawing a smooth curve through these points, we obtain a graph of the equation.

To find the fifth roots of the number 8, we do the following:

- Set the mode to DEGREE and PARAMETRIC.
- Set the range for t:

$$t_{\min} = 0, \quad t_{\max} = 360,$$
$$t_{\text{step}} = 72 \quad \text{(this is because } 360/5 = 72\text{)}.$$

- Set the viewing rectangle:

$$[-4, 4] \times [-3, 3].$$

- Set the parametric equations as follows:

$$x_t = (8 \wedge (1/5)) \cos t,$$
$$y_t = (8 \wedge (1/5)) \sin t.$$

- After the graph has been drawn, use the TRACE feature to see the fifth roots of 8.

$X_T = 8^{1/5} \cos T$
$Y_T = 8^{1/5} \sin T$

$[-4, 4] \times [-3, 3]$

Use a grapher to find each of the following graphically.

TC 1. The fourth roots of 5   **TC 2.** The cube roots of -10
TC 3. The 10th roots of 1   **TC 4.** The 15th roots of 8

● EXERCISE SET 8.7

A   Raise the number to the power and write polar notation for the answer.

1. $\left(2 \operatorname{cis} \dfrac{\pi}{3}\right)^3$

2. $\left(3 \operatorname{cis} \dfrac{\pi}{2}\right)^4$

3. $\left(2 \operatorname{cis} \dfrac{\pi}{6}\right)^6$

4. $\left(2 \operatorname{cis} \dfrac{\pi}{5}\right)^5$

5. $(1 + i)^6$

6. $(1 - i)^6$

Raise the number to the power and write rectangular notation for the answer.

7. $(2 \operatorname{cis} 240°)^4$

8. $(2 \operatorname{cis} 120°)^4$

9. $(1 + \sqrt{3}i)^4$

10. $(-\sqrt{3} + i)^6$

11. $\left(\dfrac{1}{\sqrt{2}} + \dfrac{1}{\sqrt{2}}i\right)^{10}$

12. $\left(\dfrac{1}{\sqrt{2}} - \dfrac{1}{\sqrt{2}}i\right)^{12}$

13. $\left(\dfrac{\sqrt{3}}{2} + \dfrac{1}{2}i\right)^{12}$

14. $\left(\dfrac{\sqrt{3}}{2} + \dfrac{1}{2}i\right)^{14}$

B   Solve, that is, find the square roots or the cube roots.

15. $x^2 = i$

16. $x^2 = -i$

17. $x^2 = 2\sqrt{2} - 2\sqrt{2}i$

18. $x^2 = 2\sqrt{2} + 2\sqrt{2}i$

19. $x^2 = -1 + \sqrt{3}i$

20. $x^2 = -\sqrt{3} - i$

21. $x^3 = i$

22. ▪ $x^3 = 68.4321$

23. Find and graph the fourth roots of 16.

24. Find and graph the fourth roots of i.

25. Find and graph the fifth roots of -1.

26. Find and graph the sixth roots of 1.

27. Find the fourth roots of $-8 + i8\sqrt{3}$.

28. Find the cube roots of $-64i$.

29. Find the cube roots of $2\sqrt{3} - 2i$.

30. Find the cube roots of $1 - i\sqrt{3}$.

Find all the complex solutions of the equation.

31. $x^3 = 1$

32. $x^5 - 1 = 0$

33. $x^5 + 1 = 0$

34. $x^4 + i = 0$

35. $x^5 + \sqrt{3} + i = 0$

36. $x^3 + (4 - 4\sqrt{3}i) = 0$

37. $x^4 + 81 = 0$

38. $x^6 + 1 = 0$

● SYNTHESIS _____

Solve. Evaluate square roots.

39. $x^2 + (1 - i)x + i = 0$

40. $3x^2 + (1 + 2i)x + 1 - i = 0$

▨ TECHNOLOGY CONNECTION _____

Use a grapher to find each of the following.

41. The fourth roots of 12

42. The sixth roots of -1

43. The tenth roots of 8

44. The ninth roots of -4

SUMMARY AND REVIEW   8

● TERMS TO KNOW

Right triangle, p. 503
Angle of elevation, p. 507
Angle of depression, p. 507
Bearing, p. 508
Oblique triangle, p. 513
Law of sines, p. 515
Ambiguous case, p. 516
Law of cosines, p. 522
Vector, p. 528

Magnitude of a vector, p. 528
Direction of a vector, p. 528
Vector addition, p. 528
Resultant, p. 528
Components of vectors, p. 530
Polar coordinates, p. 537
Rectangular coordinates, p. 537
Polar equation, p. 538
Rectangular equation, p. 538

Absolute value of a complex
   number, p. 544
Polar notation for complex
   numbers, p. 545
Rectangular notation for complex
   numbers, p. 545
DeMoivre's theorem, p. 549
Roots of complex numbers, p. 550

● **REVIEW EXERCISES**

Solve each of the following right triangles. Standard lettering has been used.

1. $a = 7.3$, $c = 8.6$

2. $a = 30.5$, $B = 51.17°$

3. One leg of a right triangle bears east. The hypotenuse is 734 cm long and bears N 57°23′ E. Find the perimeter of the triangle.

4. An observer's eye is 6 ft above the floor. A mural is being viewed. The bottom of the mural is at floor level. The observer looks downward 13° to see the bottom and upward 17° to see the top. How tall is the mural?

5. Solve triangle ABC, where $a = 25$, $b = 20$, and $B = 35°$.

6. Solve triangle ABC, where $B = 118.3°$, $C = 27.5°$, and $b = 0.974$.

7. In triangle ABC, $a = 3.7$, $c = 4.9$, and $B = 135°13′$. Find b.

8. In an isosceles triangle, the base angles each measure 52.3° and the base is 513 cm long. Find the lengths of the other two sides.

9. A triangular flower garden has sides of lengths 11 ft, 9 ft, and 6 ft. Find the angles of the garden, to the nearest tenth of a degree.

10. A parallelogram has sides of lengths 3.21 cm and 7.85 cm. One of its angles measures 147°. Find the area of the parallelogram.

11. A car is moving north at 45 mph. A ball is thrown from the car at 37 mph in a direction of N 45° E. Find the speed and the direction of the ball over the ground.

12. Find the area of triangle ABC if $b = 9.8$ m, $c = 7.3$ m, and $A = 67.3°$.

13. Vector $\mathbf{u}$ has components of 12 lb up and 18 lb to the right. Vector $\mathbf{v}$ has components of 35 lb down and 35 lb to the left. Find the components of $\mathbf{u} + \mathbf{v}$ and the magnitude and the direction of $\mathbf{u} + \mathbf{v}$.

14. Find rectangular notation for the vector $(20, -200°)$.

15. Find polar notation for the vector $\langle -2, 3 \rangle$.

16. Do the following calculations for $\mathbf{u} = \langle 4, 3 \rangle$ and $\mathbf{v} = \langle -3, 4 \rangle$.

   a) $4\mathbf{u} - 3\mathbf{v}$
   b) $|\mathbf{u} + \mathbf{v}|$

Find the polar coordinates.

17. $(0, -4)$

18. $(\sqrt{3}, -1)$

Find the Cartesian coordinates.

19. $(-4, 60°)$

20. $\left(4, \dfrac{2\pi}{3}\right)$

21. Convert to a polar equation:
$$x^2 + y^2 + 2x - 3y = 0.$$

22. Convert to a rectangular equation:
$$r = -4\sin\theta.$$

Graph.

23. $r = \cos 2\theta$

24. $r = 2\theta$

25. $r = 3(1 - \sin\theta)$

26. Graph the pair of complex numbers $-3 - 2i$ and $4 + 7i$ and their sum.

27. Find rectangular notation for $2(\cos 135° + i\sin 135°)$.

28. Find polar notation for $1 + i$.

29. Find the product of $7\,\mathrm{cis}\,18°$ and $10\,\mathrm{cis}\,32°$.

30. Convert to polar notation and then divide:
$$\frac{1 + i}{\sqrt{3} + i}.$$

31. Find $(1 - i)^5$ and write polar notation for the answer.

32. Find $(3\,\mathrm{cis}\,120°)^4$ and write rectangular notation for the answer.

33. Find the cube roots of $1 + i$.

34. Solve $x^3 - 27 = 0$.

35. Find the square roots of $4i$.

● **SYNTHESIS** _____

36. A parallelogram has sides of lengths 3.42 and 6.97. Its area is 18.4. Find the sizes of its angles.

37. Let $\mathbf{u} = \langle 12, 5 \rangle$. Find a vector that has the same direction as $\mathbf{u}$ but length 3.

38. Convert to a rectangular equation:
$$r = \csc^2\frac{\theta}{2}.$$

● **THINKING AND WRITING** _____

1. Summarize how you can algebraically tell when solving triangles whether there is no solution, one solution, or two solutions.

2. Explain why the law of sines cannot be used to find the first angle when solving a triangle given three sides.

3. We considered five cases of solving triangles given two sides and an angle opposite one of them. Discuss the relationship between side a and the height h in each.

CHAPTER TEST 8

Solve each of the following right triangles. Standard lettering has been used.

1. $A = 22.4°, b = 18.4$

2. $a = 11, c = 28$

3. A building 70 m high casts a shadow 100 m long. What is the angle of elevation of the sun?

4. Solve triangle ABC, where $B = 46°$, $C = 19°$, and $b = 70.2$.

5. In triangle ABC, $b = 11$, $c = 13$, and $A = 58°$. Find a.

6. A triangular flower bed measures 8 ft on one side, and that side makes angles of $38°40'$ and $50°$ with the other sides. How long are the other two sides?

7. Find the area of triangle ABC if $a = 3.1$ cm, $c = 2.7$ cm, and $B = 52.5°$.

8. Two forces of 20 N and 11 N act on an object at right angles. Find the magnitude of the resultant and the angle that it makes with the smaller force.

9. Find rectangular notation for the vector $(25, 150°)$.

10. Find polar notation for the vector $\langle -10, 6 \rangle$.

11. Do the following calculations for $\mathbf{u} = \langle 2, -6 \rangle$ and $\mathbf{v} = \langle 3, 8 \rangle$.

    a) $\mathbf{u} - \mathbf{v}$

    b) $|2\mathbf{u} + \mathbf{v}|$

12. Vector $\mathbf{u}$ has an easterly component of 10 and a northerly component of 15. Vector $\mathbf{v}$ has a southerly component of 8 and a westerly component of 15. Find:

    a) the components of $\mathbf{u} + \mathbf{v}$;

    b) the magnitude and the direction of $\mathbf{u} + \mathbf{v}$.

13. Find Cartesian coordinates of the point $(-3, 30°)$.

14. Convert $r = 25$ to a rectangular equation.

15. Solve $x^2 = 8i$.

16. Graph the pair of complex numbers $2 - 7i$ and $-4 + 2i$ and their sum.

17. Find rectangular notation for $3 \operatorname{cis}(-30°)$.

18. Find polar notation for $-1 + \sqrt{3}i$.

19. Find the product of $2 \operatorname{cis} 50°$ and $5 \operatorname{cis} 40°$.

20. Divide $3 \operatorname{cis} 120°$ by $6 \operatorname{cis} 40°$.

21. Find $(1 + i)^6$. Write polar notation for the answer.

22. Find the cube roots of -8.

Graph.

23. $r = 4 \cos 3\theta$

24. $r = \dfrac{2}{1 + \sin \theta}$

● **SYNTHESIS** _____

25. Let $\mathbf{u} = (-2, 7)$. Find a vector that has the same direction as $\mathbf{u}$ but has length 2.

In this chapter, we study *systems of equations* and how to solve them using graphing, substitution, and elimination. One of the great advantages of using a system of equations is that many problem situations then become easier to translate to mathematical language. ● Systems of equations have extensive application to many fields such as psychology, sociology, business, education, engineering, and science. Systems of inequalities are also useful in a branch of mathematics called *linear programming*. We include a brief introduction to linear programming as well as a study of *matrices*, which can also be used to solve systems of equations. ●

Systems and Matrices

9.1

SYSTEMS OF EQUATIONS
IN TWO VARIABLES

A Identifying Solutions

A **system of equations** is a *conjunction* of equations formed by joining equations with the word *and*. Here is an example:

$$x + y = 11 \quad and \quad 3x - y = 5.$$

A **solution** of an equation with two variables, such as $x + y = 11$, is an ordered pair. Some pairs in the solution set of $x + y = 11$ are

$$(5, 6), \quad (12, -1), \quad (4, 7), \quad (8, 3).$$

Some pairs in the solution set of $3x - y = 5$ are

$$(0, -5), \quad (4, 7), \quad (-2, -11), \quad (9, 22).$$

The **solution set** of the system

$$x + y = 11 \quad and \quad 3x - y = 5$$

consists of all pairs that make *both* equations true. That is, it is the *intersection* of the solution sets. Note that $(4, 7)$ is a solution of the system of equations above—in fact, it is the only solution.

DO EXERCISES 1 AND 2 ON THE FOLLOWING PAGE.

1. Determine whether $(-3, 2)$ is a solution of the conjunction

$$2x - y = -8 \quad and$$
$$3x + 4y = -1.$$

2. Determine whether $(0, \frac{1}{4})$ is a solution of the conjunction

$$5x + 12y = 13 \quad and$$
$$\sqrt{2}x + 9y = 10.$$

3. Solve graphically:

$$y - x = 1,$$
$$y + x = 3.$$

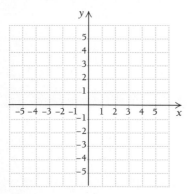

B Solving Systems of Equations Graphically

Recall that the **graph** of an equation is a drawing that represents its solution set. If the graph of an equation is a line, then every point on that line corresponds to an ordered pair that is a solution of the equation. If we graph a *system* of two linear equations, the point at which the lines intersect will be a solution of *both* equations.

In general, we drop the word *and* and write one equation under the other. Consider the system

$$x + y = 11,$$
$$3x - y = 5.$$

The following graphs show the solution set of each equation. Their intersection is the single ordered pair $(4, 7)$, so $(4, 7)$ is a solution of *both* equations.

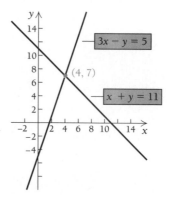

DO EXERCISE 3.

Graphing helps us to picture the solutions of a system of equations, but solving by graphing, though useful in many applied situations, is not always fast or accurate in cases where solutions are not integers.

TECHNOLOGY CONNECTION

Your grapher can be used to solve a system of equations, especially if it has trace and zoom-in features. Begin by graphing both equations in the system. Then activate the trace feature so that the coordinates of any points of interest are displayed. Next, zoom in as many times as necessary to determine the coordinates to the desired accuracy.

Using the grapher is especially useful for systems in which the equations have noninteger coefficients, such as

$$3.41236x + 6.23143y - 8.12399 = 0,$$
$$9.77654x - 7.66235y + 3.11143 = 0.$$

This technique with the grapher can be used to check possible solutions of *any* system of equations.

For each of the systems of equations given here, use a grapher to determine the solution. Round each coordinate to three decimal places.

TC 1. $2.397x + 4.432y = 9.328,$
$\quad\quad 8.122x - 9.332y = -0.883$

TC 2. $0.2y = -0.007x - 5.018,$
$\quad\quad 0.035x + y = -25.09$

TC 3. $y = -7.834x + 3.876,$
$\quad\quad y = 2.219x - 12.345$

TC 4. $2.438y - 13.426x = 11.318,$
$\quad\quad 7.413 + 26.852x = 4.876y$

Given the graphs of two lines, the following can happen:

a) The lines have no point in common—they are parallel. The system has no solution. (See (a) below.)

b) The lines have exactly one point in common. The system has exactly one solution. (See (b) below.)

c) The lines are the same—they have infinitely many points in common. The system has infinitely many solutions. (See (c) below.)

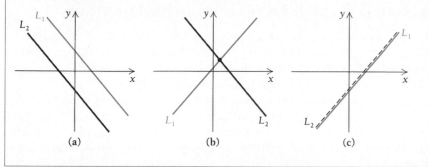

    (a)   (b)   (c)

DO EXERCISES 4–6.

 ## The Substitution Method

We now consider the first of two algebraic methods for solving systems of linear equations: the substitution method. The substitution method is used most often when a variable is alone on one side of an equation or when it is easy to solve for a variable. To use this method, we solve one equation for one of the variables. Then we substitute in the other equation and solve.

Example 1 Solve the system

$$x + y = 11, \tag{1}$$
$$3x - y = 5. \tag{2}$$

Solution First we solve equation (1) for y. (We could just as well solve for x.)

$$y = 11 - x$$

Then we substitute $11 - x$ for y in equation (2). This gives an equation in one variable, which we know how to solve from earlier work:

$$3x - (11 - x) = 5$$
$$x = 4.$$

Now we substitute 4 for x in either equation (1) or (2) and solve for y. Let us use equation (1):

$$4 + y = 11$$
$$y = 7.$$

The solution is $(4, 7)$. We list the coordinates of the solution in alphabetical order, 4 for x and 7 for y.

Solve the system graphically.

4. $2x - y = 1,$
     $-6x + 3y = -3$

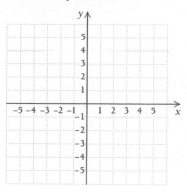

5. $x - 4y = -4,$
    $-x + 4y = 8$

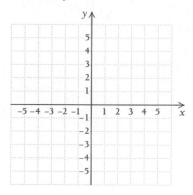

6. $y + x = 1,$
    $y - x = 5$

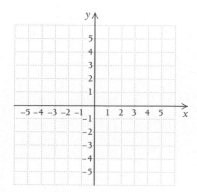

Solve using the substitution method.

7. $2x + y = 6,$
   $3x + 4y = 4$

8. $8x - 3y = -31,$
   $2x + 6y = 26$

Check:

$$\begin{array}{c|c} x + y = 11 \\ \hline 4 + 7 & 11 \\ 11 & \end{array} \qquad \begin{array}{c|c} 3x - y = 5 \\ \hline 3 \cdot 4 - 7 & 5 \\ 12 - 7 & \\ 5 & \end{array}$$

Note that the solution should be checked in *both* equations.

DO EXERCISES 7 AND 8.

Gauss–Jordan Elimination with Equations

The next algebraic method to consider for solving systems of equations is called **Gauss–Jordan elimination,** or simply, **elimination.** It is an adaptation of methods developed by two German mathematicians, Carl Friedrich Gauss (1777–1855) and Wilhelm Jordan (1842–1899).

The **elimination method** makes use of the addition and multiplication principles for solving equations and is based on algebraic methods that you have probably learned in your earlier mathematics courses. Gauss–Jordan elimination is a special **algorithm,** or step-by-step procedure, that can be programmed on a computer. Our goal is to perform special operations on a system that will result in an "equivalent system" for which the solution is obvious. Two systems are **equivalent** if they have exactly the same solutions.

Suppose we want to solve the system

$$3x - 4y = -1,$$
$$-3x + 2y = 0.$$

Our goal is to carry out certain procedures in a special way to obtain an equivalent system of the type

$$Ax + By = C, \tag{1}$$
$$Dy = E. \tag{2}$$

When we obtain such a system, we can easily solve for the variables by dividing on each side of equation (2) by the coefficient of y. After we have solved for y, we substitute into equation (1) to find x.

Now considering the system given above, we add the left-hand sides, obtaining $-2y$, and then add the right-hand sides, obtaining -1. When we do this, we often loosely say that we are "adding the two equations." In this way, we eliminate the x-term in the second equation to obtain a system of equations equivalent (having the same solutions) to the original:

$$\begin{array}{ll} 3x - 4y = -1 & (1) \\ -2y = -1. & (2) \end{array} \quad \Longleftarrow \quad \begin{array}{r} 3x - 4y = -1 \\ \underline{-3x + 2y = 0} \\ -2y = -1 \quad \text{Adding} \end{array}$$

The system is now in the form

$$Ax + By = C,$$
$$Dy = E.$$

We solve for y and then substitute into the first equation to find x:

$$\begin{array}{ll} -2y = -1 & 3x - 4(\tfrac{1}{2}) = -1 \\ y = \tfrac{1}{2}; & 3x - 2 = -1 \\ & 3x = 1 \\ & x = \tfrac{1}{3}. \end{array}$$

This kind of substitution is often called *back substitution.*

We now know that the solution of the original system is $(\frac{1}{3}, \frac{1}{2})$, because we know the solution of this last system and that this system is equivalent to the original. We know that the last system is equivalent to the original, because the only computations we did consisted of using the addition principle, adding only expressions for which all replacements were meaningful, and multiplying by a nonzero constant. Each computation produces results equivalent to the original equations. For this reason, we need to check results only to detect errors in computation.

Example 2   Solve:

$$5x + 3y = 7,$$
$$3x - 5y = -23.$$

Solution   We want to eliminate x from the second equation. Thus we first multiply the second equation by 5 to make the x-coefficient a multiple of 5:

$$5x + 3y = 7,$$
$$15x - 25y = -115.$$

Now we multiply the first equation by -3 and add it to the second equation. This eliminates the x-term:

$$5x + 3y = 7,$$
$$ -34y = -136.$$

$$\left[\begin{array}{l} -15x - 9y = -21 \quad \text{Multiplying by } -3 \\ \underline{15x - 25y = -115} \\ -34y = -136 \quad \text{Adding} \end{array}\right.$$

Next, we solve the second equation for y. Then we substitute the result into the first equation to find x:

$$-34y = -136 \qquad 5x + 3(4) = 7$$
$$y = 4; \qquad\qquad 5x + 12 = 7$$
$$\qquad\qquad\qquad\qquad 5x = -5$$
$$\qquad\qquad\qquad\qquad x = -1.$$

The solution is $(-1, 4)$.   ◀

Some preliminary work may make the steps of the elimination method simpler. First, we write the equations in the form $Ax + By = C$. We can also interchange two equations before beginning. For example, if we have the system

$$5x + y = -2,$$
$$x + 7y = 3,$$

we may prefer to write the second equation first:

$$x + 7y = 3,$$
$$5x + y = -2.$$

This accomplishes two things. One is that the equation whose x-coefficient is 1 is listed first. When we have found y later, this will make solving for x easier. The other is that this makes the x-coefficient in the second equation a multiple of the first.

Another step that might be done is to multiply one or more equations by a power of 10 before beginning in order to eliminate decimal points. For example, if we have the system

$$-0.3x + 0.5y = 0.3,$$
$$0.01x - 0.4y = 1.2,$$

Solve.

9. $4x + 3y = -6,$
    $-4x + 2y = 16$

10. $9x - 2y = -4,$
    $3x + 4y = 1$

11. $0.2x + 0.3y = 0.1,$
    $0.3x - 0.1y = 0.7$

we can multiply the first equation by 10 and the second by 100 to clear of decimals, transforming the system to

$$-3x + \quad 5y = 3,$$
$$x - 40y = 120.$$

We might also multiply in order to clear equations of fractions.

We transform the original system to an equivalent system of equations using any of the following operations, or transformations.

THEOREM 1   **Transformations Producing Equivalent Systems**

Each of the following will produce an equivalent system of equations:

a) Interchanging any two equations.
b) Multiplying each number or term of an equation by the same nonzero number.
c) Multiplying each number or term of one equation by the same nonzero number and adding the result to another equation.

Example 3 Solve:

$$5x + \quad y = -2,$$
$$x + 7y = 3.$$

Solution   We first interchange the equations so that the x-coefficient of the second equation will be a multiple of the first:

$$x + 7y = 3,$$
$$5x + \quad y = -2.$$

Next, we multiply the first equation by -5 and add the result to the second equation. This eliminates the x-term in the new second equation:

$$x + 7y = 3,$$
$$-34y = -17. \quad \longleftarrow \quad \begin{array}{r} -5x - 35y = -15 \\ 5x + \quad y = \quad -2 \\ \hline -34y = -17 \quad \text{Adding} \end{array}$$

Now we solve the second equation for y. Then we substitute the result in the first equation to find x:

$$-34y = -17 \qquad x + 7(\tfrac{1}{2}) = 3$$
$$y = \tfrac{1}{2}; \qquad\qquad x + \tfrac{7}{2} = 3$$
$$x = 3 - \tfrac{7}{2}, \quad \text{or } -\tfrac{1}{2}.$$

The solution is $(-\tfrac{1}{2}, \tfrac{1}{2})$.   ◀

DO EXERCISES 9–11.

 Problem Solving

Recall that in the five-step problem-solving process, the most difficult and time-consuming part is translating a problem situation to mathematical language. Sometimes translating to an equation takes considerable thought and effort. In many cases, this task becomes easier if we translate to more than one equation in more than one variable.

Example 4   An airplane flies the 3000-mi distance from Los Angeles to New York, with a tail wind, in 5 hr. On the return trip, against the wind, it makes the trip in 6 hr. Find the speed of the plane and the speed of the wind.

Solution

1. **Familiarize.**   We first make a drawing, letting $p =$ the speed of the plane in still air and $w =$ the speed of the wind.

Recall from Chapter 2 that distance, speed, and time are related by the motion formula, $d = rt$. We often need to use other formulas that we can derive from this one, namely, $r = d/t$ and $t = d/r$.

The distances are the same. When the plane flies east with the wind, its speed is $p + w$. When it flies west against the wind, its speed is $p - w$. We list the information in a table, the columns of the table coming from the formula $d = rt$.

	Distance	Speed	Time	
East (with the wind)	3000	$p + w$	5	$\rightarrow 3000 = (p + w)5$
West (against the wind)	3000	$p - w$	6	$\rightarrow 3000 = (p - w)6$

2. **Translate.**   Using $d = rt$ in each row of the table, we get an equation. Thus we have a system of equations:

$$3000 = (p + w)5 = 5p + 5w,$$
$$3000 = (p - w)6 = 6p - 6w.$$

3. **Carry out.**   We solve the system

$$5p + 5w = 3000,$$
$$6p - 6w = 3000.$$

There is a common factor in each equation. We multiply by $\frac{1}{5}$ in the first equation and $\frac{1}{6}$ in the second equation to eliminate the common factors. Then we have

$$p + w = 600,$$
$$p - w = 500.$$

Next, we multiply the first equation by -1 and add it to the second equation. This eliminates the p-term:

$$p + w = 600,$$
$$-2w = -100.$$

12. The sum of two numbers is 10. The difference is 1. Find the numbers.

Then we solve the second equation for w and substitute into the first equation to find p:

$$-2w = -100 \qquad p + 50 = 600$$
$$w = 50; \qquad p = 550.$$

4. **Check.**   We leave the check to the student.

5. **State.**   The solution of the system of equations is $(550, 50)$. That is, the speed of the plane is 550 mph and the speed of the wind is 50 mph.

◀

DO EXERCISES 12–14.

13. The sum of two numbers is 1. The difference is 10. Find the numbers.

Example 5   Wine A is 5% alcohol and wine B is 15% alcohol. How many liters of each should be mixed in order to get a 10-L mixture that is 12% alcohol?

Solution

1. **Familiarize.**   We organize the information in a table, letting $x =$ the number of liters of A and $y =$ the number of liters of B.

	Wine A	Wine B	Mixture	
Amount of solution	x	y	10 liters	→ $x + y = 10$
Percent of alcohol	5%	15%	12%	
Amount of alcohol in solution	5% x	15% y	0.12×10, 1.2 liters	→ 5%x + 15%y = 1.2

14. A boat travels 24 km upstream in 3 hr. It travels the 24 km downstream in 2 hr. Find the speed of the boat and the speed of the stream.

To get the amount of alcohol, we multiply by the percentages:

(Amount of solution) × (Percent of alcohol)
$\qquad\qquad$ = (Amount of alcohol in solution).

2. **Translate.**   If we add x and y in the first row, we get 10, and this gives us one equation:

$$x + y = 10.$$

If we add the amounts of alcohol in the third row, we get 1.2, and this gives us another equation:

$$5\%x + 15\%y = 1.2.$$

After changing percents to decimals and clearing, we have this system:

$$x + \quad y = 10,$$
$$5x + 15y = 120.$$

3. **Carry out.**   We then solve the system. (We leave this to the student.) The solution is $(3, 7)$. That is, 3 L of wine A and 7 L of wine B are possibilities for a solution to the original problem.

4. Check. We add the amounts of wine: $3\,L + 7\,L = 10\,L$. Thus the amount of wine checks. Next we check the amount of alcohol:

$$5\%(3) + 15\%(7) = 0.15 + 1.05, \quad \text{or } 1.2\,L.$$

Thus the amount of alcohol checks.

5. State. The solution of the problem is 3 L of wine A and 7 L of wine B. ◀

DO EXERCISE 15.

15. Solution A is 25% acid and solution B is 65% acid. How many liters of each should be mixed in order to get 8 L of a solution that is 40% acid?

● EXERCISE SET **9.1**

1. Determine whether $(\frac{1}{2}, 1)$ is a solution of the system
$$3x + y = \tfrac{5}{2},$$
$$2x - y = \tfrac{1}{4}.$$

2. Determine whether $(-2, \frac{1}{4})$ is a solution of the system
$$x + 4y = -1,$$
$$2x + 8y = -2.$$

Solve graphically.

3. $x + y = 2,$
$3x + y = 0$

4. $x + y = 1,$
$3x + y = 7$

5. $y + 1 = 2x,$
$y - 1 = 2x$

6. $y + 1 = 2x,$
$3y = 6x - 3$

Solve using the substitution method.

7. $x - 5y = 4,$
$y = 7 - 2x$

8. $3x - y = 5,$
$x + y = \tfrac{1}{2}$

Solve using the elimination method.

9. $x - 3y = 2,$
$6x + 5y = -34$

10. $x + 3y = 0,$
$20x - 15y = 75$

11. $0.3x + 0.2y = -0.9,$
$0.2x - 0.3y = -0.6$

12. $0.2x - 0.3y = 0.3,$
$0.4x + 0.6y = -0.2$

13. $\tfrac{1}{5}x + \tfrac{1}{2}y = 6,$
$\tfrac{3}{5}x - \tfrac{1}{2}y = 2$

14. $\tfrac{2}{3}x + \tfrac{3}{5}y = -17,$
$\tfrac{1}{2}x - \tfrac{1}{3}y = -1$

15. $2a = 5 - 3b,$
$4a = 11 - 7b$

16. $7(a - b) = 14,$
$2a = b + 5$

Problem Solving

17. Find two numbers whose sum is -10 and whose difference is 1.

18. Find two numbers whose sum is -1 and whose difference is 10.

19. A boat travels 46 km downstream in 2 hr. It travels 51 km upstream in 3 hr. Find the speed of the boat and the speed of the stream.

20. An airplane travels 3000 km with a tail wind in 3 hr. It travels 3000 km with a head wind in 4 hr. Find the speed of the plane and the speed of the wind.

21. Antifreeze A is 18% alcohol and antifreeze B is 10% alcohol. How many liters of each should be mixed in order to get 20 L of a mixture that is 15% alcohol?

22. Beer A is 6% alcohol and beer B is 2% alcohol. How many liters of each should be mixed in order to get 50 L of a mixture that is 3.2% alcohol?

23. Two cars leave town traveling in opposite directions. One travels at a speed of 80 km/h and the other at 96 km/h. In how many hours will they be 528 km apart?

24. A train leaves a station and travels north at a speed of 75 km/h. Two hours later, a second train leaves on a parallel track and travels north at 125 km/h. How far from the station will they meet?

25. Two planes travel toward each other from cities that are 780 km apart at speeds of 190 and 200 km/h. They started at the same time. In how many hours will they meet?

26. Two motorcycles travel toward each other from Chicago and Indianapolis, which are about 350 km apart, at speeds of 110 and 90 km/h. They start at the same time. In how many hours will they be at the same location?

27. One week, a business sold 40 scarves. White ones cost $4.95 and printed ones cost $7.95. In all, $282 worth of scarves were sold. How many of each kind were sold?

28. One day, a store sold 30 sweatshirts. White ones cost $9.95 and yellow ones cost $10.50. In all, $310.60 worth of sweatshirts were sold. How many of each color were sold?

29. Paula is 12 years older than her brother Bob. Four years from now, Bob will be $\tfrac{2}{3}$ as old as Paula. How old are they now?

30. Carlos is 8 years older than his sister Maria. Four years ago, Maria was $\tfrac{2}{3}$ as old as Carlos. How old are they now?

31. The perimeter of a lot is 190 m. The width is one-fourth the length. Find the dimensions.

32. The perimeter of a rectangular field is 628 m. The width of the field is 6 m less than the length. Find the dimensions.

33. The perimeter of a rectangle is 384 m. The length is 82 m greater than the width. Find the length and the width.

34. The perimeter of a rectangle is 86 cm. The length is 19 cm greater than the width. Find the area.

35. Two investments are made that total $15,000. For a certain year, these investments yield $1432 in simple interest. Part of the $15,000 is invested at 9% and part at 10%. Find the amount invested at each rate.

36. For a certain year, $3900 is received in interest from two investments. A certain amount is invested at 5%, and $10,000 more than this is invested at 6%. Find the amount invested at each rate.

37. A collection of 34 coins consists of dimes and nickels. The total value is $1.90. How many dimes and how many nickels are there?

38. A collection of 43 coins consists of dimes and quarters. The total value is $7.60. How many dimes and how many quarters are there?

39. A tobacco dealer has two kinds of tobacco. One is worth $4.05 per pound and the other is worth $2.70 per pound. The dealer wants to blend the two tobaccos to get a 15-lb mixture worth $3.15 per pound. How much of each kind of tobacco should be used?

40. A grocer mixes candy worth $0.80 per pound with nuts worth $0.70 per pound to get a 20-lb mixture worth $0.77 per pound. How many pounds of candy and how many pounds of nuts are used?

● **SYNTHESIS** _____

Solve.

41. $\dfrac{x+y}{4} - \dfrac{x-y}{3} = 1,$

$\dfrac{x-y}{2} + \dfrac{x+y}{4} = -9$

42. $\dfrac{x+y}{2} - \dfrac{y-x}{3} = 0,$

$\dfrac{x+y}{3} - \dfrac{x+y}{4} = 0$

Problem Solving

43. Nancy jogs and walks to the university each day. She averages 4 km/h walking and 8 km/h jogging. The distance from home to the university is 6 km and she makes the trip in 1 hr. How far does she jog in a trip?

44. James and Joan are mathematics professors. They have a total of 46 years of teaching. Two years ago, James had taught 2.5 times as many years as Joan. How long has each taught?

45. A limited edition of a book published by a historical society was offered for sale to its membership. The cost was one book for $12 or two books for $20. The society sold 880 books, and the total amount of money taken in was $9840. How many members ordered two books?

46. The ten's digit of a two-digit positive number is 2 more than three times the unit's digit. If the digits are interchanged, the new number is 13 less than half the given number. Find the given integer. (*Hint:* Let x = the ten's-place digit and y = the unit's-place digit; then $10x + y$ is the number.)

47. The numerator of a fraction is 12 more than the denominator. The sum of the numerator and the denominator is 5 more than three times the denominator. What is the reciprocal of the fraction?

48. The measure of one of two supplementary angles is 8° more than three times the measure of the other. Find the measure of the larger of the two angles.

49. A train leaves Union Station for Central Station, 216 km away, at 9 A.M. One hour later, a train leaves Central Station for Union Station. They meet at noon. If the second train had started at 9 A.M. and the first train at 10:30 A.M., they would still have met at noon. Find the speed of each train.

50. An automobile radiator contains 16 L of antifreeze and water. This mixture is 30% antifreeze. How much of this mixture should be drained and replaced with pure antifreeze so that there will be 50% antifreeze?

51. A stablehand agreed to work for one year. At the end of that time, he was to receive $240 and one horse. After 7 months, the boy quit the job, but still received the horse and $100. What is the value of the horse?

52. You are in line at a ticket window. There are two more people ahead of you in line than there are behind you. In the entire line, there are three times as many people as there are behind you. How many people are ahead of you in the line?

53. Phil and Phyllis are siblings. Phyllis has twice as many brothers as she has sisters. Phil has the same number of brothers and sisters. How many girls and how many boys are there in the family?

54. An automobile gets 18 miles per gallon (mpg) in city driving and 24 mpg in highway driving. The car is driven 465 mi on a full tank of 23 gal of gasoline. How many miles were driven in the city and how many were driven on the highway?

55. Two solutions of the equation $y = mx + b$ are $(-2, 3)$ and $(4, -5)$. Find m and b.

56. Two solutions of the equation $Ax + By = 1$ are $(3, -1)$ and $(-4, -2)$. Find A and B.

Each of the following is a system of equations that is *not* linear. But each is *linear in form*, in that an appropriate substitution, say u for $1/x$ and v for $1/y$, yields a linear system. Solve for the new variable and then solve for the original variable.

57. $\dfrac{1}{x} - \dfrac{3}{y} = 2,$

$\dfrac{6}{x} + \dfrac{5}{y} = -34$

58. $2\sqrt[3]{x} + \sqrt{y} = 0,$

$5\sqrt[3]{x} + 2\sqrt{y} = -5$

59. $3|x| + 5|y| = 30,$

$5|x| + 3|y| = 34$

60. $15x^2 + 2y^3 = 6,$

$25x^2 - 2y^3 = -6$

● CHALLENGE _____

61. A student, out hiking for the weekend, is standing on a railroad bridge, as shown in the figure below. A train is approaching from the direction shown by the arrow. If the student runs at a speed of 10 mph toward the train, she will reach point P on the bridge at the same moment that the train does. If she runs to point Q at the other end of the bridge at a speed of 10 mph, she will reach point Q also at the same moment that the train does. How fast, in miles per hour, is the train traveling?

⚂ TECHNOLOGY CONNECTION _____

Use a grapher to solve each system of linear equations, rounding to three decimal places.

62. $2.341x + 5.122y = 12.331,$
$4.522x - 0.912y = -3.119$

63. $43.234x + 64.447y = 123.664,$
$-12.384x - 2.543y = 45.297$

64. $0.944x - 0.054y = 2.312,$
$1.118x + 0.854y = 0.483$

65. $23.423x + 99.321y = 45.285,$
$21.866x - 98.432y = -54.336$

9.2

SYSTEMS OF EQUATIONS IN THREE OR MORE VARIABLES

OBJECTIVES

You should be able to:

Ⓐ Determine whether an ordered triple is a solution of a system of equations in three variables.

Ⓑ Solve systems of linear equations in three or more variables.

Ⓒ Solve applied problems by translating them to a system of three linear equations and solving the system.

Ⓓ Fit a quadratic function to data when three data points are given.

Ⓐ Identifying Solutions

A **linear equation in three variables** is an equation equivalent to one of the type $Ax + By + Cz = D$. We now solve systems of these equations.

A **solution** of a system of three equations in three variables is an ordered triple that makes all three equations true.

Example 1 Determine whether $(2, -1, 0)$ is a solution of the system

$4x + 2y + 5z = 6,$
$2x - y + z = 5,$
$x + 2y - z = 2.$

Solution We substitute $(2, -1, 0)$ into each of the three equations:

$$\begin{array}{c|c} 4x + 2y + 5z = 6 \\ \hline 4(2) + 2(-1) + 5(0) & 6 \\ 8 - 2 + 0 & \\ 6 & \end{array} \qquad \begin{array}{c|c} 2x - y + z = 5 \\ \hline 2(2) - (-1) + 0 & 5 \\ 4 + 1 + 0 & \\ 5 & \end{array}$$

$$\begin{array}{c|c} x + 2y - z = 2 \\ \hline 2 + 2(-1) - 0 & 2 \\ 2 - 2 - 0 & \\ 0 & \end{array}$$

Since $(2, -1, 0)$ is a solution of two of the equations but not *all* of the equations, it is not a solution of the system. ◀

1. Consider the following system:

$$4x - y + z = 6,$$
$$2x + y + 2z = 3,$$
$$3x - 2y + z = 3.$$

a) Determine whether $(3, 0, \frac{1}{4})$ is a solution.

b) Determine whether $(2, 1, -1)$ is a solution.

DO EXERCISE 1.

B Solving Systems of Equations in Three or More Variables

Graphical methods of solving linear equations in three variables are unsatisfactory, because a three-dimensional coordinate system is required. The substitution method becomes cumbersome for most systems of more than two equations. Therefore, we will use the elimination method and the transformations of Theorem 1. The method is essentially the same as the method used for two equations in two variables.

Our goal is to transform the original system to an equivalent one of the form

$$Ax + By + Cz = D,$$
$$Ey + Fz = G,$$
$$Hz = K.$$

Then we solve the third equation for z and back-substitute to find the other variables.

Example 2   Solve

$$2x - 4y + 6z = 22, \qquad \text{(P1)}$$
$$4x + 2y - 3z = 4, \qquad \text{(P2)}$$
$$3x + 3y - z = 4, \qquad \text{(P3)}$$

where (P1), (P2), and (P3) indicate the equation that is in the first, second, and third position, respectively. We will maintain this positional order throughout the solution, and refer to the equations by their positional number.

Solution   We begin by multiplying (P3) by 2, to make each x-coefficient a multiple of the first. By proceeding in this manner, we avoid fractions. Although the method will work when fractions are allowed, it is more difficult. Then we have the following system:

$$2x - 4y + 6z = 22, \qquad \text{(P1)}$$
$$4x + 2y - 3z = 4, \qquad \text{(P2)}$$
$$6x + 6y - 2z = 8. \qquad \text{(P3)}$$

Next, we multiply (P1) by -2 and add it to (P2). We also multiply (P1) by -3 and add it to (P3). This gives us the following:

$$2x - 4y + 6z = 22, \qquad \text{(P1)}$$
$$10y - 15z = -40, \qquad \text{(P2)}$$
$$18y - 20z = -58. \qquad \text{(P3)}$$

Now we multiply (P3) by -5 to make the y-coefficient a multiple of the y-coefficient in (P2):

$$2x - 4y + 6z = 22, \qquad \text{(P1)}$$
$$10y - 15z = -40, \qquad \text{(P2)}$$
$$-90y + 100z = 290. \qquad \text{(P3)}$$

Next, we multiply (P2) by 9 and add it to (P3):

$$2x - 4y + 6z = 22, \qquad \text{(P1)}$$
$$10y - 15z = -40, \qquad \text{(P2)}$$
$$-35z = -70. \qquad \text{(P3)}$$

Now we solve (P3) for z:

$$-35z = -70$$
$$z = 2.$$

Next, we back-substitute 2 for z in (P2) and solve for y:

$$10y - 15(2) = -40$$
$$10y - 30 = -40$$
$$10y = -10$$
$$y = -1.$$

Finally, we back-substitute -1 for y and 2 for z in (P1) and solve for x:

$$2x - 4(-1) + 6(2) = 22$$
$$2x + 4 + 12 = 22$$
$$2x + 16 = 22$$
$$2x = 6$$
$$x = 3.$$

The solution is $(3, -1, 2)$. To be sure that computational errors have not been made, we check by substituting 3 for x, -1 for y, and 2 for z in all three original equations. If all are true, then the triple is a solution. ◀

DO EXERCISE 2. ———————————

Although the solution of a system of three linear equations in three variables is difficult to find graphically, it is of interest to "see" what a solution might be. The graph of a linear equation in three variables is a plane. Thus the solution set of such a system is the intersection of three planes. Some possibilities are shown in the following figures.

One solution: planes intersecting in exactly one point.

No solution: three planes; each intersects another; at no point do all intersect.

No solution: parallel planes

▶ Problem Solving

Systems of equations arise very often in the use of statistics in such fields as the social sciences. They also occur in problems of business, science, and engineering.

2. Solve the system using exactly the procedure given in the text.

$$x + 2y - z = 5,$$
$$2x - 4y + z = 0,$$
$$3x + 2y + 2z = 3$$

3. In a suitcase factory, there are three machines A, B, and C. When all three are running, they produce 222 suitcases per day. If A and B work but C does not, they produce 159 suitcases per day. If B and C work but A does not, they produce 147 suitcases per day. What is the daily production of each machine?

Example 3 In a triangle, the largest angle is 70° greater than the smallest angle. The largest angle is twice as large as the remaining angle. Find the measure of each angle.

Solution

1. **Familiarize.** The first thing to do with a problem like this is to make a drawing, or a sketch.

We don't know the size of any angle, so we have used A, B, and C for the measures of the angles. A geometric fact will be needed here—the fact that the measures of the angles of a triangle add up to 180°.

2. **Translate.** The geometric fact about triangles gives us one equation:

$A + B + C = 180$.

There are two statements in the problem that we can translate almost directly.

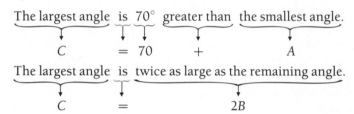

We now have a system of three equations:

$$A + B + C = 180, \qquad A + B + C = 180,$$
$$A + 70 = C, \qquad \text{or} \quad A \quad - C = -70,$$
$$2B = C; \qquad 2B - C = 0.$$

3. **Carry out.** We solve the system. The details are left to the student, but the solution is $(30, 50, 100)$.

4. **Check.** The sum of the numbers is 180, so that checks. The largest angle measures 100° and the smallest measures 30°. The largest angle is thus 70° greater than the smallest. The remaining angle measures 50°. The largest angle measures 100°, so it is twice as large. We do have an answer to the problem.

5. **State.** The measures of the angles of the triangle are 30°, 50°, and 100°.

DO EXERCISE 3. ————————————————————————

▷ **Mathematical Models and Problem Solving**

In a situation in which a quadratic function will serve as a mathematical model, we may wish to find an equation, or formula, for the function. For a linear model, we can find an equation if we know two data points. For a quadratic function, we need three data points.

Example 4 *Curve Fitting.*   In a certain situation, it is believed that a quadratic function will be a good model. Find an equation of the function, given the data points $(1, -4)$, $(-1, -6)$, and $(2, -9)$.

4. *Curve Fitting.* Find a quadratic function that fits the data points $(1, 0)$, $(-1, 4)$, and $(2, 1)$.

Solution   We want to find a quadratic function

$$f(x) = ax^2 + bx + c$$

containing the three given points, that is, a function for which the equation will be true when we substitute any of the ordered pairs of numbers into it. When we substitute, we get

for $(1, -4)$: $-4 = a \cdot 1^2 + b \cdot 1 + c$;
for $(-1, -6)$: $-6 = a(-1)^2 + b(-1) + c$;
for $(2, -9)$: $-9 = a \cdot 2^2 + b \cdot 2 + c$.

We now have a system of equations in the three unknowns a, b, and c:

$$a + b + c = -4,$$
$$a - b + c = -6,$$
$$4a + 2b + c = -9.$$

We solve this system of equations, obtaining $(-2, 1, -3)$. Thus the function we are looking for is

$$f(x) = -2x^2 + x - 3.$$ ◀

DO EXERCISE 4.

Example 5 *The Cost of Operating an Automobile at Various Speeds.* Under certain conditions, it is found that the cost of operating an automobile as a function of speed is approximated by a quadratic function. Use the data shown below to find an equation of the function. Then use the equation to determine the cost of operating the automobile at 60 mph and at 80 mph.

Speed, in Miles per Hour	Operating Cost per Mile, in Cents
10	22
20	20
50	20

Solution   Letting $x =$ the speed and $f(x) =$ the cost, we use the three data points to obtain a, b, and c in the equation $f(x) = ax^2 + bx + c$:

$$22 = 100a + 10b + c,$$
$$20 = 400a + 20b + c, \quad \text{Substituting}$$
$$20 = 2500a + 50b + c.$$

We solve this system of equations, obtaining $(0.005, -0.35, 25)$. Thus,

$$f(x) = 0.005x^2 - 0.35x + 25.$$

To find the cost of operating at 60 mph, we find $f(60)$:

$$f(60) = 0.005(60)^2 - 0.35(60) + 25 = 22¢.$$

5. The following table has values that will fit a quadratic function. Find the average number of accidents as a function of age. Then use the model to calculate the average number of accidents in which 16-year-olds are involved daily.

Age of Driver	Average Number of Accidents per Day
20	400
40	150
60	400

We also find $f(80)$:

$$f(80) = 0.005(80)^2 - 0.35(80) + 25 = 29¢.$$

A graph of the cost function of Example 5 is as follows.

It should be noted that this cost function can give approximate results only within a certain interval. For example, $f(0) = 25$, meaning that it costs 25 cents per mile to stand still. This, of course, is absurd in the sense of mileage, although one does incur costs in owning a car whether one drives it or not. ◄

DO EXERCISE 5. _____

EXERCISE SET 9.2

A Consider the system

$2x + 3y - 5z = 1,$
$6x - 6y + 10z = 3,$
$4x - 9y + 5z = 0.$

1. Determine whether $(-1, 1, 0)$ is a solution of the system.
2. Determine whether $(\frac{1}{2}, \frac{1}{3}, \frac{1}{5})$ is a solution of the system.

B Solve.

3. $x + y + z = 2,$
$6x - 4y + 5z = 31,$
$5x + 2y + 2z = 13$

4. $x + 6y + 3z = 4,$
$2x + y + 2z = 3,$
$3x - 2y + z = 0$

5. $x - y + 2z = -3,$
$x + 2y + 3z = 4,$
$2x + y + z = -3$

6. $x + y + z = 6,$
$2x - y - z = -3,$
$x - 2y + 3z = 6$

7. $4a + 9b = 8,$
$8a + 6c = -1,$
$6b + 6c = -1$

8. $3p + 2r = 11,$
$q - 7r = 4,$
$p - 6q = 1$

9. $w + x + y + z = 2,$
$w + 2x + 2y + 4z = 1,$
$-w + x - y - z = -6,$
$-w + 3x + y - z = -2$

10. $w + x - y + z = 0,$
$-w + 2x + 2y + z = 5,$
$-w + 3x + y - z = -4,$
$-2w + x + y - 3z = -7$

C **Problem Solving**

11. The sum of three numbers is 26. Twice the first minus the second is 2 less than the third. The third is the second minus three times the first. Find the numbers.

12. The sum of three numbers is 5. The first number minus the second plus the third is 1. The first minus the third is 3 more than the second. Find the numbers.

13. In triangle ABC, the measure of angle B is three times the measure of angle A. The measure of angle C is $30°$ greater than the measure of angle A. Find the angle measures.

14. In triangle ABC, the measure of angle B is $2°$ more than three times the measure of angle A. The measure of angle C is $8°$ more than the measure of angle A. Find the angle measures.

15. A farmer picked strawberries on three days. She picked a total of 87 quarts. On Tuesday, she picked 15 quarts more than on Monday. On Wednesday, she picked 3 quarts fewer than on Tuesday. How many quarts did she pick each day?

16. Gina sells magazines part time. On Thursday, Friday, and Saturday, she sold $66 worth. On Thursday, she sold $3 more than on Friday. On Saturday, she sold $6 more than on Thursday. How much did she take in each day?

17. Sawmills A, B, and C can produce 7400 board-feet of lumber per day. Mills A and B together can produce 4700 board-feet, while B and C together can produce 5200 board-feet. How many board-feet can each mill produce by itself?

18. In a factory there are three polishing machines, A, B, and C. When all three of them are working, 5700 lenses can be polished in one week. When only A and B are working, 3400 lenses can be polished in one week. When only B and C are working, 4200 lenses can be polished in one week. How many lenses can be polished in a week by each machine?

19. Three welders, A, B, and C, can weld 37 linear feet per hour when working together. If A and B together can weld 22 linear feet per hour, and A and C together can weld 25 linear feet per hour, how many linear feet per hour can each weld alone?

20. When three pumps, A, B, and C, are running together, they can pump 3700 gallons per hour. When only A and B are running, 2200 gallons per hour can be pumped. When only A and C are running, 2400 gallons per hour can be pumped. What is the pumping capacity of each pump?

21. On an 18-hole golf course, there are par-3 holes, par-4 holes, and par-5 holes. A golfer who shoots par on every hole has a total of 72. The sum of the number of par-3 holes and the number of par-5 holes is 8. How many of each type of hole are there on the golf course?

22. On an 18-hole golf course, there are par-3 holes, par-4 holes, and par-5 holes, A golfer who shoots par on every hole has a total of 70. There are twice as many par-4 holes as there are par-5 holes. How many of each type of hole are there on the golf course?

23. A person receives $212 per year in simple interest from three investments totaling $2500. Part is invested at 7%, part at 8%, and part at 9%. There is $1100 more invested at 9% than at 8%. Find the amount invested at each rate.

24. A person receives $341 per year in simple interest from three investments totaling $3500. Part is invested at 8%, part at 9%, and part at 10%. There is $2600 more invested at 10% than at 9%. Find the amount invested at each rate.

Solve.

25. *Curve fitting.*   Find numbers a, b, and c such that a quadratic function $ax^2 + bx + c$ fits the data points $(1, 4)$, $(-1, -2)$, and $(2, 13)$. Write the equation for the function.

26. *Curve fitting.*   Find numbers a, b, and c such that a quadratic function $ax^2 + bx + c$ fits the data points $(1, 4)$, $(-1, 6)$, and $(-2, 16)$. Write the equation for the function.

27. *Predicting earnings.*   A business earns $38 in the first week, $66 in the second week, and $86 in the third week. The manager graphs the points $(1, 38)$, $(2, 66)$, and $(3, 86)$ and finds that a quadratic function might fit the data.

  a) Find a quadratic function that fits the data.
  b) Using the model, predict the earnings for the fourth week.

28. *Predicting earnings.*   A business earns $1000 in its first month, $2000 in the second month, and $8000 in the third month. The manager plots the points $(1, 1000)$, $(2, 2000)$, and $(3, 8000)$ and finds that a quadratic function might fit the data.

  a) Find a quadratic function that fits the data.
  b) Using the model, predict the earnings for the fourth month.

29. *Death rate as a function of sleep.*   (This problem is based on a study by Dr. Harold J. Morowitz.)

Average Number of Hours of Sleep, x	Death Rate per Year per 100,000 Males, y
5	1121
7	626
9	967

  a) Use the given data points to find a quadratic function $f(x) = ax^2 + bx + c$ that fits the data.
  b) Use the model to find the death rate of males who sleep 4 hr, 6 hr, and 10 hr.

30. *Counter reading on a VCR.*   A person buys a video cassette recorder on which there is a revolution counter. There is also a booklet with a table that relates the time and the counter reading for which the tape has run.

Counter Reading	Time of Tape, in Hours
000	0
300	1
500	2

  a) Find a quadratic function that fits the data.
  b) Use the quadratic function to find what amount of time a tape has run if the counter reading is 650.
  c) Use the function to find the counter reading after the tape has run for $1\frac{1}{2}$ hr.

Hint for Exercises 31 and 32: Let u represent $1/x$, v represent $1/y$, and w represent $1/z$. First solve for u, v, and w.

31. $\dfrac{2}{x} - \dfrac{1}{y} - \dfrac{3}{z} = -1$,

$\dfrac{2}{x} - \dfrac{1}{y} + \dfrac{1}{z} = -9$,

$\dfrac{1}{x} + \dfrac{2}{y} - \dfrac{4}{z} = 17$

32. $\dfrac{2}{x} + \dfrac{2}{y} - \dfrac{3}{z} = 3$,

$\dfrac{1}{x} - \dfrac{2}{y} - \dfrac{3}{z} = 9$,

$\dfrac{7}{x} - \dfrac{2}{y} + \dfrac{9}{z} = -39$

33. When A, B, and C work together, they can do a job in 2 hr. When B and C work together, they can do the job in 4 hr. When A and B work together, they can do the job in $\frac{12}{5}$ hr. How long would it take each, working alone, to do the job?

34. Pipes A, B, and C are connected to the same tank. When all three pipes are running, they can fill the tank in 3 hr. When pipes A and C are running, they can fill the tank in 4 hr. When pipes A and B are running, they can fill the tank in 8 hr. How long would it take each, running alone, to fill the tank?

35. Find the sum of the angle measures at the tips of the star.

36. Find the year in which the first U.S. transcontinental railroad was completed. The following are some facts about the number. The sum of the digits in the year is 24. The unit's digit is 1 more than the hundred's digit. Both the ten's and the unit's digits are multiples of three.

In each case, three solutions of an equation are given. Find the constants and write the equation. Use a system of equations.

37. $Ax + By + Cz = 12$; $(1, \frac{3}{4}, 3)$, $(\frac{4}{3}, 1, 2)$, and $(2, 1, 1)$

38. $y = B - Mx - Nz$; $(1, 1, 2)$, $(3, 2, -6)$, and $(\frac{3}{2}, 1, 1)$

39. A theater had 100 people in attendance. The audience consisted of adults, students, and children. The ticket prices were $10 for adults, $3 for students, and 50 cents for children. The total amount of money taken in was $100. How many adults, students, and children were in attendance? Does there seem to be some information missing? Do some careful reasoning.

40. Art, Bob, Carl, Denny, and Emmett are on the same bowling team. They are all being truthful in the following comments regarding the last game they bowled.

Art:	My score was a prime number. Emmett finished third.
Bob:	None of us bowled a score over 200.
Carl:	Art beat me by exactly 23 pins. Denny's score was divisible by 10.
Denny:	The sum of our five scores was exactly 885 pins. Bob's score was divisible by 8.
Emmett:	Art beat Bob by fewer than 10 pins. Denny beat Bob by exactly 14 pins.

Determine the score of each bowler in the game.

You should be able to:

A Solve systems of equations using the elimination method for special cases where systems may have no solution or infinitely many solutions. Then classify systems of equations as consistent or inconsistent, dependent or independent.

9.3

SPECIAL CASES OF SYSTEMS

A In Sections 9.1 and 9.2, each system had *exactly* one solution. Here we consider special cases where systems have no solution or infinitely many solutions.

Consistent and Inconsistent Systems

DEFINITION **Consistent and Inconsistent Systems**

A system of equations is *consistent* if and only if it has a solution.
A system of equations is *inconsistent* if and only if it has no solution.

Let us consider a system that does not have a solution and see what happens when we apply the elimination method.

Example 1 Solve using the elimination method. Classify the system as consistent or inconsistent.

$$x - 3y = 1,$$
$$-2x + 6y = 5$$

Solution Let us first look at what happens graphically. We graph each equation and find where they intersect. It turns out that the lines are parallel and have no point of intersection.

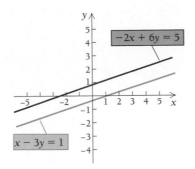

The solution set of the first equation is given by

$$S_1 = \{(x, y) | x - 3y = 1\}.$$

The solution set of the second equation is given by

$$S_2 = \{(x, y) | -2x + 6y = 5\}.$$

Now the solution set S of the system has no ordered pairs in it and is given by

$$S = S_1 \cap S_2 = \varnothing.$$

We see graphically that the system has no solution.

Let us see what happens if we apply the elimination method. We multiply the first equation by 2 and add the result to the second equation. This gives us

$$x - 3y = 1,$$
$$0 = 7.$$

The second equation says that $0 \cdot x + 0 \cdot y = 7$. There are no numbers x and y for which this is true.

Whenever we obtain a statement such as $0 = 7$, which is obviously false, we know that the system we are trying to solve has no solutions. It is *inconsistent*. The solution set is $\varnothing$. ◀

DO EXERCISES 1 AND 2.

Dependent and Independent Systems

> **DEFINITION** **Dependent and Independent Systems**
>
> A system of linear equations is *dependent* if and only if removing one or more equations from the system results in a system that is equivalent to the original system. That is, if there exists a system of fewer equations with the same solutions, then the original system is dependent. Otherwise, the system is *independent*.

Solve using the elimination method. Classify the system as consistent or inconsistent.

1. $4x - 2y = 2,$
   $2x - y = -8$

2. $4x - 2y = 2,$
   $2x + y = -3$

Example 2   Solve the system using the elimination method. Classify it as consistent or inconsistent, dependent or independent.

$$2x + 3y = 6,$$
$$4x + 6y = 12$$

Solution   Let us look at what happens graphically. We graph each equation.

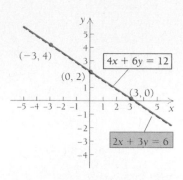

The solution set of the first equation is given by

$$S_1 = \{(x, y) | 2x + 3y = 6\}.$$

Some of the ordered pairs in S_1 are

$$(0, 2), \quad (3, 0), \quad (6, -2), \quad \text{and} \quad (-3, 4).$$

The solution set of the second equation is given by

$$S_2 = \{(x, y) | 4x + 6y = 12\}.$$

Some ordered pairs in S_2 are

$$(0, 2), \quad (3, 0), \quad (6, -2), \quad \text{and} \quad (-3, 4).$$

Indeed, the solution sets are the same:

$$S_1 = S_2.$$

Thus the solution sets of each equation are the same. If we remove one of the equations from the system, we still get the same solution set. That is,

$$\begin{aligned} 2x + 3y &= 6, \\ 4x + 6y &= 12 \end{aligned} \quad \text{is equivalent to} \quad 2x + 3y = 6.$$

Thus the system is *dependent*. It is also *consistent*.

What happens when we apply the elimination method? We multiply the first equation by -2 and add. This gives us

$$2x + 3y = 6$$
$$0 = 0.$$

The equation $0 = 0$ is equivalent to $0x + 0y = 0$, which is true for any values of x and y. Thus it is true for any pair of numbers x and y that constitute a solution of the system. Therefore, the equation $0 = 0$ contributes nothing to the system and can be ignored. We then analyze the rest of the equations to see if the system they form has a solution. In this case, we know that the equation $2x + 3y = 6$ has infinitely many solutions, so the system is *dependent* and *consistent*.

Example 3   Solve using the elimination method. Classify the system as consistent or inconsistent, dependent or independent.

$$x - 3y = 1,$$
$$x + y = 3,$$
$$5x - 7y = 9$$

Solution   Let us look at what happens graphically.

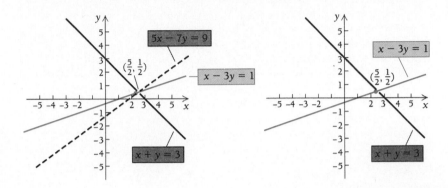

The solution set of the original system is $\{(\frac{5}{2}, \frac{1}{2})\}$. This is where the three graphs intersect. If we ignore the equation $5x - 7y = 9$, we obtain the graphs shown on the right. These still intersect at exactly one point, $(\frac{5}{2}, \frac{1}{2})$. Thus,

$$\begin{array}{l} x - 3y = 1, \\ x + y = 3, \\ 5x - 7y = 9 \end{array} \quad \text{is equivalent to} \quad \begin{array}{l} x - 3y = 1, \\ x + y = 3, \end{array}$$

which means that the system is *dependent.*

Suppose we apply the elimination method to the original system. We start with

$$x - 3y = 1,$$
$$x + y = 3,$$
$$5x - 7y = 9.$$

We multiply the first equation by -1 and add the result to the second equation. We also multiply the first equation by -5 and add the result to the third equation:

$$x - 3y = 1,$$
$$4y = 2,$$
$$8y = 4.$$

Then we multiply the second equation by -2 and add the result to the third equation:

$$x - 3y = 1,$$
$$4y = 2,$$
$$0 = 0. \quad \text{True for all } x \text{ and } y$$

We obtain the equation $0 = 0$, true for all values of x and y. This tells us that the system is *dependent.* To determine consistency, we consider the first

Solve using the elimination method. Classify the system as consistent or inconsistent, dependent or independent.

3. $4x - 2y = 6,$
    $-2x + y = -3$

4. $2x - y = 1,$
    $x + y = 5,$
    $x - 2y = -4$

Solve. Describe the solutions by expressing one variable in terms of the other. Give three of the ordered pairs in the solution set.

5. $-6x + 4y = 10,$
    $3x - 2y = -5$

6. $2x + 5y = 1,$
    $4x + 10y = 2$

two equations and find that the solution is $x = \frac{5}{2}$ and $y = \frac{1}{2}$. Thus the system is *consistent*. ◀

DO EXERCISES 3 AND 4.

In solving a system, how do we know that it is dependent?

> If, at some stage, we find that two of the equations are identical, then we know that the system is *dependent*. If we obtain an obviously true statement, such as $0 = 0$, then we know that the system is *dependent*. We cannot know whether such a system is consistent or inconsistent without further analysis.

A dependent and consistent system may have an infinite number of solutions. In such a case for systems of two variables, we can describe the solutions by expressing one variable in terms of the other.

Example 4 Solve:

$$2x + 3y = 1,$$
$$4x + 6y = 2.$$

Solution We multiply the first equation by -2 and add. This gives us

$$2x + 3y = 1,$$
$$0 = 0.$$

Now we know that the system is dependent. The last equation contributes nothing, so we consider only the first one. Let us solve for x. We obtain

$$x = \frac{1 - 3y}{2}.$$

We can now describe the ordered pairs in the solution set, in terms of y only, as follows:

$$\left(\frac{1 - 3y}{2}, y \right).$$

Any value that we choose for y then gives us a value for x, and thus an ordered pair in the solution set. Some of these solutions are

$$(-4, 3), \qquad \left(-\frac{5}{2}, 2 \right), \qquad \left(\frac{7}{2}, -2 \right). \qquad ◀$$

DO EXERCISES 5 AND 6.

When a system of three or more equations is dependent and consistent, we can describe its solutions by expressing one or more of the variables in terms of the others.

Example 5   Solve:

$$x + 2y + 3z = 4, \qquad\qquad\qquad\text{(P1)}$$
$$2x - y + z = 3, \qquad\qquad\qquad\text{(P2)}$$
$$3x + y + 4z = 7, \qquad\qquad\qquad\text{(P3)}$$

where (P1), (P2), and (P3) indicate the equations in the first, second, and third positions, respectively.

Solution   Each x-coefficient is a multiple of the first. Thus we can begin solving by multiplying (P1) by -2 and adding it to (P2). We also multiply (P1) by -3 and add it to (P3):

$$x + 2y + 3z = 4, \qquad\qquad\qquad\text{(P1)}$$
$$-5y - 5z = -5, \qquad\qquad\qquad\text{(P2)}$$
$$-5y - 5z = -5. \qquad\qquad\qquad\text{(P3)}$$

Now (P2) and (P3) are identical. We no longer have a system of three equations, but a system of two. Thus we know that the system is dependent. If we were to multiply (P2) by -1 and add it to (P3), we would obtain $0 = 0$. We proceed by multiplying (P2) by $-\frac{1}{5}$ since the coefficients have the common factor -5:

$$x + 2y + 3z = 4, \qquad\qquad\qquad\text{(P1)}$$
$$y + z = 1. \qquad\qquad\qquad\text{(P2)}$$

We will express two of the variables in terms of the other one. Let us choose z. Then we solve (P2) for y:

$$y = 1 - z.$$

We substitute this value of y in (P1), obtaining

$$x + 2(1 - z) + 3z = 4.$$

Solving for x, we get

$$x = 2 - z.$$

The solutions, then, are all of the form

$$(2 - z, \ 1 - z, \ z).$$

We obtain the solutions by choosing various values of z. If we let $z = 0$, we obtain the triple $(2, 1, 0)$. If we let $z = 2$, we obtain $(0, -1, 2)$, and so on. ◀

DO EXERCISE 7. ────────────────────────────

Homogeneous Equations

When all the terms of a polynomial have the same degree, we say that the polynomial is **homogeneous.** Here are some examples:

$$3x^2 + 5y^2, \qquad 4x + 5y - 2z, \qquad 17x^3 - 4y^3 + 57z^3.$$

An equation formed by a homogeneous polynomial set equal to 0 is called a **homogeneous equation.** Let us now consider a system of homogeneous linear equations.

7. Solve, giving the general form of the solutions. Then list three of the solutions.

$$2x + y + z = 9,$$
$$x + 2y - 4z = -3,$$
$$x + y - z = 2$$

Solve. If there is more than one solution, list three of them.

8. $x + y - z = 0,$
$x + 2y - 4z = 0,$
$2x + y + z = 0$

9. $x + y - z = 0,$
$x - y - z = 0,$
$x + y + 2z = 0$

Example 6 Solve:

$$4x - 3y + z = 0,$$
$$2x \quad\quad - 3z = 0,$$
$$-8x + 6y - 2z = 0.$$

Solution Any homogeneous system like this always has a solution (it can never be inconsistent), because $(0, 0, 0)$ is a solution. This is called the **trivial solution.** There may or may not be other solutions. To find out, we proceed as in the case of nonhomogeneous equations.

First, we interchange the first two equations so that all the x-coefficients are multiples of the first:

$$2x \quad\quad - 3z = 0, \tag{P1}$$
$$4x - 3y + z = 0, \tag{P2}$$
$$-8x + 6y - 2z = 0. \tag{P3}$$

Now we multiply (P1) by -2 and add it to (P2). We also multiply (P1) by 4 and add it to (P3):

$$2x \quad\quad - 3z = 0, \tag{P1}$$
$$-3y + 7z = 0, \tag{P2}$$
$$6y - 14z = 0. \tag{P3}$$

Next, we multiply (P2) by 2 and add it to (P3):

$$2x \quad\quad - 3z = 0, \tag{P1}$$
$$-3y + 7z = 0, \tag{P2}$$
$$0 = 0. \tag{P3}$$

This system is dependent. It has an infinite set of solutions. Next, we solve (P2) for y:

$$y = \tfrac{7}{3}z.$$

Typically, we would now substitute $\tfrac{7}{3}z$ for y in (P1), but since the y-term is missing, we need only solve for x:

$$x = \tfrac{3}{2}z.$$

We can now describe the members of the solution set as follows:

$$(\tfrac{3}{2}z, \tfrac{7}{3}z, z).$$

Some of the ordered pairs in the solution set are

$$(\tfrac{3}{2}, \tfrac{7}{3}, 1), \quad (3, \tfrac{14}{3}, 2), \quad (-\tfrac{3}{2}, -\tfrac{7}{3}, -1). \quad \blacktriangleleft$$

DO EXERCISES 8 AND 9.

• EXERCISE SET 9.3

Ⓐ   Solve. If a system has more than one solution, list three of them.

1. $9x - 3y = 15,$
$6x - 2y = 10$

2. $2s - 3t = 9,$
$4s - 6t = 9$

3. $5c + 2d = 24,$
$30c + 12d = 10$

4. $3x + 2y = 18,$
$9x + 6y = 5$

5. $3x + 2y = 5,$
$4y = 10 - 6x$

6. $5x + 2 = 7y,$
$-14y + 4 = -10x$

7. $12y - 8x = 6,$
$4x + 3 = 6y$

8. $16x - 12y = 10,$
$6y + 5 = 8x$

9. $x + 2y - z = -8,$
$2x - y + z = 4,$
$8x + y + z = 2$

10. $x + 2y - z = 4,$
$4x - 3y + z = 8,$
$5x - y = 12$

11. $2x + y - 3z = 1,$
$x - 4y + z = 6,$
$4x + 16y + 4z = 24$

12. $4x + 12y + 16z = 4,$
$3x + 4y + 5z = 3,$
$x + 8y + 11z = 1$

13. $2x + y - 3z = 0,$
$x - 4y + z = 0,$
$4x - 16y + 4z = 0$

14. $4x + 12y + 16z = 0,$
$3x + 4y + 5z = 0,$
$x + 8y + 11z = 0$

15. $x + y - z = -3,$
$x + 2y + 2z = -1$

16. $x + y + 13z = 0,$
$x - y - 6z = 0$

17. $2x + y + z = 0,$
$x + y - z = 0,$
$x + 2y + 2z = 0$

18. $5x + 4y + z = 0,$
$10x + 8y - z = 0,$
$x - y - z = 0$

19. Classify each of the systems in the odd-numbered exercises 1–17 as consistent or inconsistent, dependent or independent.

20. Classify each of the systems in the even-numbered exercises 2–18 as consistent or inconsistent, dependent or independent.

● SYNTHESIS _____

Solve.

21. $4.026x - 1.448y = 18.32,$
$0.724y = -9.16 + 2.013x$

22. $0.0284y = 1.052 - 8.114x,$
$0.0142y + 4.057x = 0.526$

23. a) Solve:
$$w + x + y + z = 4,$$
$$w + x + y + z = 3,$$
$$w + x + y + z = 3.$$
b) Classify the system as consistent or inconsistent.
c) Classify the system as dependent or independent.

24. a) Solve:
$$w - 8x + 3y + 2z = 0,$$
$$-w + 5x + y - z = 0,$$
$$-w + 2x + 5y = 0,$$
$$3x - 4y - z = 0.$$
b) Classify the system as consistent or inconsistent.
c) Classify the system as dependent or independent.

● CHALLENGE _____

Determine the constant k such that each system is dependent.

25. $6x - 9y = -3,$
$-4x + 6y = k$

26. $8x - 16y = 20,$
$10x - 20y = k$

27. On an 18-hole golf course, there are par-3 holes, par-4 holes, and par-5 holes. A golfer who shoots par on every hole has a total of 72. There are the same number of par-3 holes as there are par-5 holes. There is at least one of each type of hole. How many of each type of hole are there on the golf course?

28. A two-digit number is such that the number is equal to four times the sum of the digits. Find the number or numbers.

9.4
MATRICES

A Solving Systems Using Matrices

In solving systems of equations, we perform computations with the constants. The variables play no important role in the process. We can simplify writing the systems by omitting the variables. For example, the system

$$3x + 4y = 5,$$
$$x - 2y = 1$$

simplifies to

$$\begin{array}{rrr} 3 & 4 & 5 \\ 1 & -2 & 1 \end{array}$$

if we leave off the variables and omit the operation and equals signs.

OBJECTIVES

You should be able to:

A Solve systems of linear equations using matrices.

B Add, subtract, and multiply matrices when possible.

C Write a matrix equation equivalent to a system of equations.

In the example above, we have written a rectangular array of numbers. Such an array is called a **matrix** (plural, **matrices**). We ordinarily write brackets around matrices, although this is not necessary when calculating with matrices. The following are matrices:

$$\begin{bmatrix} 4 & 1 & 3 & 5 \\ 1 & 0 & 1 & 2 \\ 6 & 3 & -2 & 0 \end{bmatrix}, \quad \begin{bmatrix} 6 & 2 & 1 & 4 & 7 \\ 1 & 2 & 1 & 3 & 1 \\ 4 & 0 & -2 & 0 & -3 \end{bmatrix}, \quad \begin{bmatrix} 1 & 2 \\ 145 & 0 \\ -7 & 9 \\ 8 & 1 \\ 0 & 0 \end{bmatrix}.$$

The **rows** of a matrix are horizontal, and the **columns** are vertical.

$$\begin{bmatrix} 5 & -2 & 2 \\ 1 & 0 & 1 \\ 0 & 1 & 2 \end{bmatrix} \begin{array}{l} \longleftarrow \text{row 1} \\ \longleftarrow \text{row 2} \\ \longleftarrow \text{row 3} \end{array}$$

column 1   column 2   column 3

Let us now use matrices to solve systems of linear equations.

Example 1   Solve:

$$2x - y + 4z = -3,$$
$$x \qquad - 4z = 5,$$
$$6x - y + 2z = 10.$$

Solution   We first write a matrix, using the constants. Note that where there are missing terms, we must write 0's:

$$\begin{bmatrix} 2 & -1 & 4 & -3 \\ 1 & 0 & -4 & 5 \\ 6 & -1 & 2 & 10 \end{bmatrix}.$$

We do exactly the same calculations using the matrix that we would do if we wrote the entire equations. The first step, if possible, is to interchange the rows so that each number in the first column below the first number is a multiple of that number. We do this by interchanging rows 1 and 2. This corresponds to interchanging the first two equations. The symbolism $R_1' = R_2$ and $R_2' = R_1$ means that the new row 1 is the old row 2 and the new row 2 is the old row 1.

$$\begin{bmatrix} 1 & 0 & -4 & 5 \\ 2 & -1 & 4 & -3 \\ 6 & -1 & 2 & 10 \end{bmatrix} \begin{array}{l} \longrightarrow \ R_1' = R_2 \\ \longrightarrow \ R_2' = R_1 \end{array}$$

Next, we multiply the first row by -2 and add it to the second row. The symbolism $R_2' = -2R_1 + R_2$ means that the new row 2 is -2 times the old row 1 added to the old row 2.

$$\begin{bmatrix} 1 & 0 & -4 & 5 \\ 0 & -1 & 12 & -13 \\ 6 & -1 & 2 & 10 \end{bmatrix} \longrightarrow R_2' = -2R_1 + R_2$$

Now we multiply the first row by -6 and add it to the third row:

$$\begin{bmatrix} 1 & 0 & -4 & 5 \\ 0 & -1 & 12 & -13 \\ 0 & -1 & 26 & -20 \end{bmatrix}. \qquad R'_3 = -6R_1 + R_3$$

Next we multiply row 2 by -1 and add it to the third row:

$$\begin{bmatrix} 1 & 0 & -4 & 5 \\ 0 & -1 & 12 & -13 \\ 0 & 0 & 14 & -7 \end{bmatrix}. \qquad R'_3 = -1 \cdot R_2 + R_3$$

If we now put the variables back, we have

$$x \quad - \quad 4z = 5, \qquad \text{(P1)}$$
$$-y + 12z = -13, \qquad \text{(P2)}$$
$$14z = -7. \qquad \text{(P3)}$$

Now we proceed as before. We solve (P3) for z and get $z = -\frac{1}{2}$. Next we back-substitute $-\frac{1}{2}$ for z in (P2) and solve for y: $-y + 12(-\frac{1}{2}) = -13$, so $y = 7$. Since there is no y-term in (P1), we need only substitute $-\frac{1}{2}$ for z in (P1) and solve for x: $x - 4(-\frac{1}{2}) = 5$, so $x = 3$. The solution is $(3, 7, -\frac{1}{2})$. ◀

Note in the preceding that our goal was to get the matrix in the form

$$\begin{bmatrix} a & b & c & d \\ 0 & e & f & g \\ 0 & 0 & h & k \end{bmatrix},$$

where there are just 0's below the **main diagonal**, formed by a, e, and h. Then we put the variables back and complete the solution.

All the operations used in the preceding example correspond to operations with the equations and they produce equivalent systems of equations. We call the matrices **row-equivalent** and the operations that produce them **row-equivalent operations.** We then have a theorem similar to Theorem 1.

THEOREM 2

Each of the following row-equivalent operations produces an equivalent matrix:

a) Interchanging any two rows of a matrix.
b) Multiplying each element of a row by the same nonzero number.
c) Multiplying each element of a row by the same nonzero number and adding the result to another row.

The best overall method for solving systems of equations is by row-equivalent matrices; even computers are programmed to use them.

DO EXERCISES 1 AND 2.

B ▸ Operations on Matrices

We just used matrices to solve systems of equations. Next we study matrices as stand-alone entities, which can be added, subtracted, and multiplied.

Solve using matrices.

1. $5x - 2y = -3,$
   $2x + 5y = -24$

2. $x - 2y + 3z = 4,$
   $2x - y + z = -1,$
   $4x + y + z = 1$

Find the dimensions of the matrix.

3. $\begin{bmatrix} -3 & 5 \\ 4 & \frac{1}{4} \\ -\pi & 0 \end{bmatrix}$

4. $\begin{bmatrix} -3 & 0 \\ 0 & 3 \end{bmatrix}$

5. $\begin{bmatrix} 1 & 2 & 3 \\ 0 & 1 & 8 \\ 0 & 0 & 1 \end{bmatrix}$

6. $[\pi \quad \sqrt{2}]$

7. $\begin{bmatrix} -5 \\ \pi \end{bmatrix}$

8. $[-3]$

9. Which of the above are square matrices?

10. Let

$$A = \begin{bmatrix} 4 & -1 \\ 6 & -3 \end{bmatrix}$$

and

$$B = \begin{bmatrix} -6 & -5 \\ 7 & 3 \end{bmatrix}.$$

a) Find $A + B$.
b) Find $B + A$.

11. Add:

$$\begin{bmatrix} -3 & -4 & -5 \\ 0 & 1 & -1 \end{bmatrix} + \begin{bmatrix} 4 & 5 & -5 \\ 2 & 3 & -2 \end{bmatrix}.$$

Dimensions of a Matrix

A matrix of m rows and n columns is called a matrix with **dimensions** $m \times n$ (read "m by n").

Example 2 Find the dimensions of each matrix.

$$\underbrace{\begin{bmatrix} 2 & -3 & 4 \\ -1 & \frac{1}{2} & \pi \end{bmatrix}}_{2 \times 3 \text{ matrix}}, \qquad \underbrace{\begin{bmatrix} -3 & 8 & 9 \\ \pi & -2 & 5 \\ -6 & 7 & 8 \end{bmatrix}}_{3 \times 3 \text{ matrix}},$$

$$\underbrace{[-3 \quad 4]}_{1 \times 2 \text{ matrix}}, \qquad \underbrace{\begin{bmatrix} 10 \\ -7 \end{bmatrix}}_{2 \times 1 \text{ matrix}}. \qquad \blacktriangleleft$$

Square matrices have the same number of rows as columns. The 3×3 matrix is a square matrix.

DO EXERCISES 3–9.

Matrix Addition

To add matrices, we add the corresponding elements or members. For this to be possible, the matrices must have the same dimensions.

Examples Add.

3. $\begin{bmatrix} -5 & 0 \\ 4 & \frac{1}{2} \end{bmatrix} + \begin{bmatrix} 6 & -3 \\ 2 & 3 \end{bmatrix} = \begin{bmatrix} -5+6 & 0-3 \\ 4+2 & \frac{1}{2}+3 \end{bmatrix}$ The matrices have the same dimensions.

$$= \begin{bmatrix} 1 & -3 \\ 6 & 3\frac{1}{2} \end{bmatrix}$$

4. $\begin{bmatrix} 1 & 3 \\ -1 & 5 \\ 6 & 0 \end{bmatrix} + \begin{bmatrix} -1 & -2 \\ 1 & -2 \\ -3 & 1 \end{bmatrix} = \begin{bmatrix} 0 & 1 \\ 0 & 3 \\ 3 & 1 \end{bmatrix}$ The matrices have the same dimensions. $\blacktriangleleft$

Addition of matrices is both commutative and associative.

DO EXERCISES 10 AND 11.

Zero Matrices

A matrix having zeros for all of its members is called a **zero matrix** and is often denoted by O. When a zero matrix is added to another matrix of the same dimensions, the original matrix is obtained. Thus a zero matrix is an **additive identity.**

Example 5 Add.

$$\begin{bmatrix} 2 & -1 & 3 \\ 1 & 0 & -1 \end{bmatrix} + \begin{bmatrix} 0 & 0 & 0 \\ 0 & 0 & 0 \end{bmatrix} = \begin{bmatrix} 2 & -1 & 3 \\ 1 & 0 & -1 \end{bmatrix} \qquad \blacktriangleleft$$

Opposites and Subtraction

To subtract matrices, we subtract the corresponding members. Of course, the matrices must have the same dimensions for this to be possible.

Example 6 Subtract.

$$\begin{bmatrix} 1 & 2 \\ -2 & 0 \\ -3 & -1 \end{bmatrix} - \begin{bmatrix} 1 & -1 \\ 1 & 3 \\ 2 & 3 \end{bmatrix} = \begin{bmatrix} 0 & 3 \\ -3 & -3 \\ -5 & -4 \end{bmatrix}$$ ◀

DO EXERCISES 13 AND 14.

The opposite, or additive inverse, of a matrix can be obtained by replacing each member by its opposite. Of course, when two matrices that are opposites, or additive inverses, of each other are added, a zero matrix is obtained.

Example 7 Add.

$$\begin{bmatrix} 1 & 0 & 2 \\ 3 & -1 & 5 \end{bmatrix} + \begin{bmatrix} -1 & 0 & -2 \\ -3 & 1 & -5 \end{bmatrix} = \begin{bmatrix} 0 & 0 & 0 \\ 0 & 0 & 0 \end{bmatrix}$$
$$\quad A \qquad + \qquad (-A) \qquad = \qquad O$$ ◀

DO EXERCISES 15 AND 16.

With numbers, we can subtract by adding an opposite. This is also true of matrices. If we denote matrices by **A** and **B** and an opposite by $-\mathbf{B}$, this fact can be stated as follows:

$$\mathbf{A} - \mathbf{B} = \mathbf{A} + (-\mathbf{B}).$$

Example 8

$$\begin{bmatrix} 3 & -1 \\ -2 & 4 \end{bmatrix} - \begin{bmatrix} 2 & 1 \\ 3 & -2 \end{bmatrix} = \begin{bmatrix} 1 & -2 \\ -5 & 6 \end{bmatrix}$$
$$\quad A \qquad - \qquad B$$

$$\begin{bmatrix} 3 & -1 \\ -2 & 4 \end{bmatrix} + \begin{bmatrix} -2 & -1 \\ -3 & 2 \end{bmatrix} = \begin{bmatrix} 1 & -2 \\ -5 & 6 \end{bmatrix}$$
$$\quad A \qquad + \quad (-B)$$ ◀

DO EXERCISE 17.

Multiplying Matrices and Numbers

We define the product of a matrix and a number to obtain what is called a **scalar product**.

12. Let

$$A = \begin{bmatrix} 4 & -3 \\ 5 & 8 \end{bmatrix}$$

and

$$O = \begin{bmatrix} 0 & 0 \\ 0 & 0 \end{bmatrix}.$$

a) Find $A + O$.
b) Find $O + A$.

Subtract.

13. $\begin{bmatrix} 1 & 3 & -2 \\ 4 & 0 & 5 \end{bmatrix} - \begin{bmatrix} 2 & -1 & 5 \\ 6 & 4 & -3 \end{bmatrix}$

14. $\begin{bmatrix} 1 & 2 \\ 4 & 1 \\ -5 & 4 \end{bmatrix} - \begin{bmatrix} 7 & -4 \\ 3 & 5 \\ 2 & -1 \end{bmatrix}$

15. Find the opposite:

$$\begin{bmatrix} 2 & -1 & 5 \\ 6 & 4 & -3 \end{bmatrix}.$$

16. Add:

$$\begin{bmatrix} 2 & -1 & 5 \\ 6 & 4 & -3 \end{bmatrix} + \begin{bmatrix} -2 & 1 & -5 \\ -6 & -4 & 3 \end{bmatrix}.$$

17. Add. Compare with Exercise 13.

$$\begin{bmatrix} 1 & 3 & -2 \\ 4 & 0 & 5 \end{bmatrix} + \begin{bmatrix} -2 & 1 & -5 \\ -6 & -4 & 3 \end{bmatrix}$$

Compute the product.

18. $5\begin{bmatrix} 1 & -2 & x \\ 4 & y & 1 \\ 0 & -5 & x^2 \end{bmatrix}$

19. $t\begin{bmatrix} 1 & -1 & 4 & x \\ y & 3 & -2 & y \\ 1 & 4 & -5 & y \end{bmatrix}$

DEFINITION Scalar Product

The *scalar product* of a number k and a matrix A is the matrix, denoted kA, obtained by multiplying each number in A by the number k. The number k is called a *scalar*.

Example 9 Let

$$A = \begin{bmatrix} -3 & 0 \\ 4 & 5 \end{bmatrix}.$$

Find 3A and (-1)A.

Solution

$$3A = 3\begin{bmatrix} -3 & 0 \\ 4 & 5 \end{bmatrix} = \begin{bmatrix} -9 & 0 \\ 12 & 15 \end{bmatrix},$$

$$(-1)A = -1\begin{bmatrix} -3 & 0 \\ 4 & 5 \end{bmatrix}$$

$$= \begin{bmatrix} 3 & 0 \\ -4 & -5 \end{bmatrix}$$

◀

DO EXERCISES 18 AND 19.

Products of Matrices

We do not multiply two matrices by multiplying their corresponding members. The definition of matrix products comes from a need to convert a system of linear equations to a product of matrices.

Let us begin by considering one equation,

$$3x + 2y - 2z = 4.$$

We will write the coefficients on the left side in a 1×3 matrix (a **row matrix**) and the variables in a 3×1 matrix (a **column matrix**). The 4 on the right is written in a 1×1 matrix:

$$[3 \quad 2 \quad -2]\begin{bmatrix} x \\ y \\ z \end{bmatrix} = [4].$$

We can return to our original equation by multiplying the members of the row matrix by those of the column matrix, and adding:

$$[3 \quad 2 \quad -2]\begin{bmatrix} x \\ y \\ z \end{bmatrix} = [3x + 2y - 2z].$$

We define multiplication accordingly. In this special case, we have a *row matrix* A and a *column matrix* B. Their product AB is a 1×1 matrix, having the single member 4 (also called $3x + 2y - 3z$).

Example 10 Find the product of these matrices.

$$[3 \quad 2 \quad -1] \begin{bmatrix} 1 \\ -2 \\ 3 \end{bmatrix} = [3 \cdot 1 + 2(-2) + (-1) \cdot 3] = [-4] \quad \blacktriangleleft$$

DO EXERCISE 20.

20. Multiply:

$$[4 \quad -2 \quad 3] \begin{bmatrix} 2 \\ 3 \\ -5 \end{bmatrix}.$$

Let us continue by considering a system of equations:

$$3x + 2y - 2z = 4,$$
$$2x - y + 5z = 3,$$
$$-x + y + 4z = 7.$$

Consider the following matrices:

$$\begin{bmatrix} 3 & 2 & -2 \\ 2 & -1 & 5 \\ -1 & 1 & 4 \end{bmatrix} \begin{bmatrix} x \\ y \\ z \end{bmatrix} \begin{bmatrix} 4 \\ 3 \\ 7 \end{bmatrix}$$
$$\quad\; A \qquad\quad X \quad B$$

We call **A** the **coefficient matrix.** If we multiply the first row of **A** by the (only) column of **X**, as we did above, we get $3x + 2y - 2z$. If we multiply the second row of **A** by the column in **X**, in the same way, we get the following:

$$[2 \quad -1 \quad 5] \begin{bmatrix} x \\ y \\ z \end{bmatrix} = 2x - y + 5z.$$

Note that the first members are multiplied, the second members are multiplied, the third members are multiplied, and the results are added, to get the single number $2x - y + 5z$. What do we get when we multiply the third row of **A** by the column in **X**?

$$[-1 \quad 1 \quad 4] \begin{bmatrix} x \\ y \\ z \end{bmatrix} = -x + y + 4z$$

We define the product **AX** to be the column matrix

$$\begin{bmatrix} 3x + 2y - 2z \\ 2x - y + 5z \\ -x + y + 4z \end{bmatrix}.$$

Now consider this matrix equation:

$$\begin{bmatrix} 3x + 2y - 2z \\ 2x - y + 5z \\ -x + y + 4z \end{bmatrix} = \begin{bmatrix} 4 \\ 3 \\ 7 \end{bmatrix}.$$

Equality for matrices is the same as for numbers, that is, a sentence such as $a = b$ says that a and b are two names for the same thing. Thus if the matrix equation above is true, the "two" matrices are really the same one. This means that $3x + 2y - 2z$ is 4, $2x - y + 5z$ is 3, and $-x + y + 4z$ is

21. Multiply:

$$\begin{bmatrix} 1 & 4 & 2 \\ -1 & 6 & 3 \\ 3 & 2 & -1 \\ 5 & 0 & 2 \end{bmatrix} \begin{bmatrix} 2 \\ 1 \\ 3 \end{bmatrix}.$$

7, or that

$$3x + 2y - 2z = 4,$$
$$2x - y + 5z = 3,$$
$$-x + y + 4z = 7.$$

Thus the matrix equation $\mathbf{AX} = \mathbf{B}$, or

$$\begin{bmatrix} 3 & 2 & -2 \\ 2 & -1 & 5 \\ -1 & 1 & 4 \end{bmatrix} \begin{bmatrix} x \\ y \\ z \end{bmatrix} = \begin{bmatrix} 4 \\ 3 \\ 7 \end{bmatrix},$$

is equivalent to the original system of equations.

Example 11 Multiply.

$$\begin{bmatrix} 3 & 1 & -1 \\ 1 & 2 & 2 \\ -1 & 0 & 5 \\ 4 & 1 & 2 \end{bmatrix} \begin{bmatrix} 1 \\ 2 \\ 1 \end{bmatrix} = \begin{bmatrix} 3 \cdot 1 + 1 \cdot 2 - 1 \cdot 1 \\ 1 \cdot 1 + 2 \cdot 2 + 2 \cdot 1 \\ -1 \cdot 1 + 0 \cdot 2 + 5 \cdot 1 \\ 4 \cdot 1 + 1 \cdot 2 + 2 \cdot 1 \end{bmatrix} = \begin{bmatrix} 4 \\ 7 \\ 4 \\ 8 \end{bmatrix}$$

◀

DO EXERCISE 21. _____

In all the examples discussed so far, the second matrix had only one column. If the second matrix has more than one column, we treat each of the columns in the same way when multiplying that we treated the single column. The product matrix will have as many columns as the second matrix.

22. Multiply:

$$\begin{bmatrix} 4 & 1 & 2 \\ -3 & 2 & 3 \\ 2 & 0 & 5 \\ 3 & 1 & 4 \end{bmatrix} \begin{bmatrix} 1 & 4 \\ 2 & 0 \\ -3 & 5 \end{bmatrix}.$$

Example 12 Multiply (compare with Example 11).

$$\underset{\mathbf{A}}{\begin{bmatrix} 3 & 1 & -1 \\ 1 & 2 & 2 \\ -1 & 0 & 5 \\ 4 & 1 & 2 \end{bmatrix}} \underset{\mathbf{B}}{\begin{bmatrix} 1 & 0 \\ 2 & 1 \\ 1 & 3 \end{bmatrix}}$$

$$= \begin{bmatrix} 4 & 3 \cdot 0 + 1 \cdot 1 + (-1)3 \\ 7 & 1 \cdot 0 + 2 \cdot 1 + 2 \cdot 3 \\ 4 & -1 \cdot 0 + 0 \cdot 1 + 5 \cdot 3 \\ 8 & 4 \cdot 0 + 1 \cdot 1 + 2 \cdot 3 \end{bmatrix} = \begin{bmatrix} 4 & -2 \\ 7 & 8 \\ 4 & 15 \\ 8 & 7 \end{bmatrix}$$

Same as in Example 11, the rows of $\mathbf{A}$ multiplied by the first column of $\mathbf{B}$

The rows of $\mathbf{A}$ multiplied by the second column of $\mathbf{B}$

◀

DO EXERCISE 22. _____

Example 13 Multiply.

$$\begin{bmatrix} 3 & 1 & -1 \\ 2 & 0 & 3 \end{bmatrix} \begin{bmatrix} 1 & 4 & 6 \\ 3 & -1 & 9 \\ 2 & 5 & 1 \end{bmatrix}$$

$$= \begin{bmatrix} 3\cdot1+1\cdot3-1\cdot2 & 3\cdot4+1\cdot(-1)-1\cdot5 & 3\cdot6+1\cdot9-1\cdot1 \\ 2\cdot1+0\cdot3+3\cdot2 & 2\cdot4+0\cdot(-1)+3\cdot5 & 2\cdot6+0\cdot9+3\cdot1 \end{bmatrix}$$

$$= \begin{bmatrix} 4 & 6 & 26 \\ 8 & 23 & 15 \end{bmatrix}$$ ◀

If matrix **A** has n columns and matrix **B** has n rows, then we can compute the product **AB**, regardless of other dimensions. The product will have as many rows as **A** and as many columns as **B**.

C A U T I O N ! Given any two matrices **A** and **B**, you may or may not be able to add, subtract, or multiply them. **A** + **B** and **A** − **B** exist only when the dimensions are the same. **AB** exists only when the number of columns in **A** is the same as the number of rows in **B**.

Consider the matrices

$$A = \begin{bmatrix} 3 & 1 & -1 \\ 2 & 0 & 3 \end{bmatrix} \quad \text{and} \quad B = \begin{bmatrix} 1 & 4 & 6 \\ 3 & -1 & 9 \\ 2 & 5 & 1 \end{bmatrix}.$$

The dimensions of **A** are 2×3 and the dimensions of **B** are 3×3. **A** + **B** and **A** − **B** do not exist because the dimensions of **A** and **B** are *not* the same. **AB** does exist because the number of columns in **A**, 3, is the same as the number of rows in **B**, 3. **AB** is given in Example 13.

$$\mathbf{A}_{2 \times 3} \qquad \mathbf{B}_{3 \times 3}$$

AB does exist: 3 = 3.
The dimensions of the
product are 2 × 3.

But **BA** does *not* exist because the number of columns in **B**, 3, is not the same as the number of rows in **A**, 2.

$$\mathbf{B}_{3 \times 3} \qquad \mathbf{A}_{2 \times 3}$$

BA does not exist: 3 ≠ 2.

In this context, it can also be pointed out that matrix multiplication is not commutative: **AB** exists and **BA** does not, so **AB** ≠ **BA**.

DO EXERCISES 23–26.

---

23. Find **AB** and **BA** if possible.

$$A = \begin{bmatrix} -2 & 4 & 0 \\ -3 & 0 & -8 \end{bmatrix},$$

$$B = \begin{bmatrix} -1 & -2 & -3 \\ 0 & 1 & 0 \\ 4 & 5 & 2 \end{bmatrix}$$

24. Multiply:

$$[4 \quad 1 \quad 0 \quad 2] \begin{bmatrix} 1 & 0 & 1 \\ 2 & -1 & 0 \\ 3 & 5 & 1 \\ 1 & 3 & 0 \end{bmatrix}.$$

25. Find **AB** and **BA** and compare.

$$A = \begin{bmatrix} -8 & 3 \\ -4 & 4 \end{bmatrix},$$

$$B = \begin{bmatrix} 1 & -4 \\ 2 & 0 \end{bmatrix}$$

26. Find **AI** and **IA**. Comment.

$$A = \begin{bmatrix} 3 & 2 \\ -1 & 5 \end{bmatrix},$$

$$I = \begin{bmatrix} 1 & 0 \\ 0 & 1 \end{bmatrix}$$

27. Write a matrix equation equivalent to this system of equations:

$$3x + 4y - 2z = 5,$$
$$2x - 2y + 5z = 3,$$
$$6x + 7y - z = 0.$$

 Equivalent Matrix Equations

For later purposes, it is important that we be able to write a matrix equation equivalent to a system of equations.

Example 14   Write a matrix equation equivalent to this system of equations:

$$4x + 2y - z = 3,$$
$$9x + z = 5,$$
$$4x + 5y - 2z = 1,$$
$$x + y + z = 0.$$

Solution   We write the coefficients on the left in a matrix. We write the product of that matrix and the column matrix containing the variables, and set the result equal to the column matrix containing the constants on the right:

$$\begin{bmatrix} 4 & 2 & -1 \\ 9 & 0 & 1 \\ 4 & 5 & -2 \\ 1 & 1 & 1 \end{bmatrix} \begin{bmatrix} x \\ y \\ z \end{bmatrix} = \begin{bmatrix} 3 \\ 5 \\ 1 \\ 0 \end{bmatrix}. \quad \blacktriangleleft$$

DO EXERCISE 27.

A Summary of Properties of Square Matrices

We now list a summary of some of the properties of square matrices of the same dimensions whose elements are real numbers. We restrict our discussion to square matrices so that all additions and multiplications are possible. Some of the proofs will be considered in the exercise set. Note that not all the field properties hold.

THEOREM 3

For any square matrices A, B, and C of the same dimensions, the following hold:

Commutativity. $A + B = B + A.$

Associativity. $A + (B + C) = (A + B) + C,$
$A(BC) = (AB)C.$

Identity.   There exists a unique matrix O, such that

$$A + O = O + A = A.$$

Inverses.   There exists a unique matrix $-A$, such that

$$A + (-A) = -A + A = O.$$

Distributivity. $A(B + C) = AB + AC,$
$(B + C)A = BA + CA.$

(continued)

For any square matrices **A** and **B** of the same dimensions and any real numbers k and m,

$$k(\mathbf{A} + \mathbf{B}) = k\mathbf{A} + k\mathbf{B},$$
$$(k + m)\mathbf{A} = k\mathbf{A} + m\mathbf{A},$$
$$(km)\mathbf{A} = k(m\mathbf{A}),$$

and

$$1\mathbf{A} = \mathbf{A}.$$

CAUTION! Note that, even with these restrictions, matrix multiplication is still *not* commutative. For example, let

$$\mathbf{A} = \begin{bmatrix} 1 & 0 \\ 2 & 0 \end{bmatrix} \quad \text{and} \quad \mathbf{B} = \begin{bmatrix} 3 & 4 \\ 0 & 0 \end{bmatrix}.$$

Then

$$\mathbf{AB} = \begin{bmatrix} 3 & 4 \\ 6 & 8 \end{bmatrix} \quad \text{and} \quad \mathbf{BA} = \begin{bmatrix} 11 & 0 \\ 0 & 0 \end{bmatrix},$$

so $\mathbf{AB} \neq \mathbf{BA}$.

● EXERCISE SET 9.4

A   Solve using matrices.

1. $4x + 2y = 11,$
$3x - y = 2$

2. $3x - 3y = 11,$
$9x - 2y = 5$

3. $x + 2y - 3z = 9,$
$2x - y + 2z = -8,$
$3x - y - 4z = 3$

4. $x - y + 2z = 0,$
$x - 2y + 3z = -1,$
$2x - 2y + z = -3$

5. $5x - 3y = -2,$
$4x + 2y = 5$

6. $3x + 4y = 7,$
$-5x + 2y = 10$

7. $4x - y - 3z = 1,$
$8x + y - z = 5,$
$2x + y + 2z = 5$

8. $3x + 2y + 2z = 3,$
$x + 2y - z = 5,$
$2x - 4y + z = 0$

9. $p + q + r = 1,$
$p + 2q + 3r = 4,$
$4p + 5q + 6r = 7$

10. $m + n + t = 9,$
$m - n - t = -15,$
$m + n + t = 3$

11. $-2w + 2x + 2y - 2z = -10,$
$w + x + y + z = -5,$
$3w + x - y + 4z = -2,$
$w + 3x - 2y + 2z = -6$

12. $-w + 2x - 3y + z = -8,$
$-w + x + y - z = -4,$
$w + x + y + z = 22,$
$-w + x - y - z = -14$

B   For Exercises 13–28, let

$$\mathbf{A} = \begin{bmatrix} 1 & 2 \\ 4 & 3 \end{bmatrix}, \qquad \mathbf{B} = \begin{bmatrix} -3 & 5 \\ 2 & -1 \end{bmatrix},$$

$$\mathbf{C} = \begin{bmatrix} 1 & -1 \\ -1 & 1 \end{bmatrix}, \qquad \mathbf{D} = \begin{bmatrix} 1 & 1 \\ 1 & 1 \end{bmatrix},$$

$$\mathbf{E} = \begin{bmatrix} 1 & 3 \\ 2 & 6 \end{bmatrix}, \qquad \mathbf{F} = \begin{bmatrix} 3 & 3 \\ -1 & -1 \end{bmatrix},$$

$$\mathbf{O} = \begin{bmatrix} 0 & 0 \\ 0 & 0 \end{bmatrix}, \quad \text{and} \quad \mathbf{I} = \begin{bmatrix} 1 & 0 \\ 0 & 1 \end{bmatrix}.$$

Find each of the following.

13. $\mathbf{A} + \mathbf{B}$

14. $\mathbf{B} + \mathbf{A}$

15. $\mathbf{E} + \mathbf{O}$

16. $2\mathbf{A}$

17. $3\mathbf{F}$

18. $(-1)\mathbf{D}$

19. $3\mathbf{F} + 2\mathbf{A}$

20. $\mathbf{A} - \mathbf{B}$

21. $\mathbf{B} - \mathbf{A}$

22. $\mathbf{AB}$

23. $\mathbf{BA}$

24. $\mathbf{OF}$

25. $\mathbf{CD}$

26. $\mathbf{EF}$

27. $\mathbf{AI}$

28. $\mathbf{IA}$

In Exercises 29–32, let

$$A = \begin{bmatrix} 1 & 0 & -2 \\ 0 & -1 & 3 \\ 3 & 2 & 4 \end{bmatrix}, \quad B = \begin{bmatrix} -1 & -2 & 5 \\ 1 & 0 & -1 \\ 2 & -3 & 1 \end{bmatrix},$$

$$C = \begin{bmatrix} -2 & 9 & 6 \\ -3 & 3 & 4 \\ 2 & -2 & 1 \end{bmatrix}, \quad \text{and} \quad I = \begin{bmatrix} 1 & 0 & 0 \\ 0 & 1 & 0 \\ 0 & 0 & 1 \end{bmatrix}.$$

Find each of the following.

29. AB   **30.** BA   **31.** CI   **32.** IC

Multiply.

33. $[-3 \quad 2]\begin{bmatrix} 4 \\ -2 \end{bmatrix}$ **34.** $[-2 \quad 0 \quad 4]\begin{bmatrix} 8 \\ -6 \\ \frac{1}{2} \end{bmatrix}$

35. $[-5 \quad 1 \quad 2]\begin{bmatrix} 1 & 3 \\ -1 & 0 \\ 4 & -2 \end{bmatrix}$ **36.** $\begin{bmatrix} -3 & 2 \\ 0 & 1 \\ -4 & 5 \end{bmatrix}\begin{bmatrix} 4 \\ 2 \end{bmatrix}$

 Write a matrix equation equivalent to the system of equations.

37. $3x - 2y + 4z = 17,$
$\quad 2x + y - 5z = 13$

38. $3x + 2y + 5z = 9,$
$\quad 4x - 3y + 2z = 10$

39. $x - y + 2z - 4w = 12,$
$\quad 2x - y - z + w = 0,$
$\quad x + 4y - 3z - w = 1,$
$\quad 3x + 5y - 7z + 2w = 9$

40. $2x + 4y - 5z + 12w = 2,$
$\quad 4x - y + 12z - w = 5,$
$\quad -x + 4y \qquad + 2w = 13,$
$\quad 2x + 10y + z \qquad = 5$

● SYNTHESIS

For Exercises 41–44, let

$$A = \begin{bmatrix} -1 & 0 \\ 2 & 1 \end{bmatrix} \quad \text{and} \quad B = \begin{bmatrix} 1 & -1 \\ 0 & 2 \end{bmatrix}.$$

41. Show that
$$(A + B)(A - B) \neq A^2 - B^2,$$
where
$$A^2 = AA \quad \text{and} \quad B^2 = BB.$$

42. Show that
$$(A + B)(A + B) \neq A^2 + 2AB + B^2.$$

43. Show that
$$(A + B)(A - B) = A^2 + BA - AB - B^2.$$

44. Show that
$$(A + B)(A + B) = A^2 + BA + AB + B^2.$$

● CHALLENGE

Let

$$A = \begin{bmatrix} a_{11} & a_{12} \\ a_{21} & a_{22} \end{bmatrix}, \quad B = \begin{bmatrix} b_{11} & b_{12} \\ b_{21} & b_{22} \end{bmatrix},$$

$$C = \begin{bmatrix} c_{11} & c_{12} \\ c_{21} & c_{22} \end{bmatrix}, \quad \text{and} \quad I = \begin{bmatrix} 1 & 0 \\ 0 & 1 \end{bmatrix}.$$

Prove each of the following.

45. $A + B = B + A$

46. $A + (B + C) = (A + B) + C$

47. $(k + m)A = kA + mA$ **48.** $k(A + B) = kA + kB$

49. $AI = IA = A$ **50.** $A(BC) = (AB)C$

OBJECTIVES

You should be able to:

A Evaluate determinants of square matrices.

B Use properties of determinants to simplify their evaluation.

C Factor certain determinants.

D Solve square systems of equations using Cramer's rule.

9.5

DETERMINANTS AND CRAMER'S RULE

A Determinants of Square Matrices

A matrix of m rows and n columns is called an $m \times n$ matrix (read "m by n"). If a matrix has the same number of rows and columns, it is called a **square matrix.** With every square matrix is associated a number called its **determinant,** defined as follows for 2×2 matrices.

DEFINITION   **Determinant of a 2 × 2 Matrix**

The determinant of the matrix $\begin{bmatrix} a & c \\ b & d \end{bmatrix}$ is denoted $\begin{vmatrix} a & c \\ b & d \end{vmatrix}$ and is defined as follows:

$$\begin{vmatrix} a & c \\ b & d \end{vmatrix} = ad - bc.$$

Evaluate.

1. $\begin{vmatrix} \sqrt{3} & -5 \\ -2 & -\sqrt{3} \end{vmatrix}$

Example 1   Evaluate: $\begin{vmatrix} \sqrt{2} & -3 \\ -4 & -\sqrt{2} \end{vmatrix}$.

Solution

$\begin{vmatrix} \sqrt{2} & -3 \\ -4 & -\sqrt{2} \end{vmatrix}$ The arrows indicate the products involved.

$$= \sqrt{2}(-\sqrt{2}) - (-4)(-3)$$
$$= -2 - 12$$
$$= -14$$

2. $\begin{vmatrix} 1 & 2 \\ 3 & 4 \end{vmatrix}$

DO EXERCISES 1–3.

We now consider a way to evaluate determinants of square matrices of dimensions 3 × 3 or higher. To do this, we need some new notation. The members, or elements, of a matrix will now be denoted by lower-case letters with two subscripts, as follows:

$$A = \begin{bmatrix} a_{11} & a_{12} & a_{13} \\ a_{21} & a_{22} & a_{23} \\ a_{31} & a_{32} & a_{33} \end{bmatrix}.$$

The element in the ith row and jth column is denoted a_{ij}.

3. $\begin{vmatrix} -2 & -3 \\ 4 & x \end{vmatrix}$

Example 2   Consider

$$[a_{ij}] = \begin{bmatrix} -8 & 0 & 6 \\ 4 & -6 & 7 \\ -1 & -3 & 5 \end{bmatrix}.$$

Find $a_{12}, a_{23},$ and a_{33}.

Solution

$a_{12} = 0$ 0 is the element in the first row and the second column.
$a_{23} = 7$ 7 is the element in the second row and the third column.
$a_{33} = 5$ 5 is the element in the third row and the third column.

4. For the matrix of Example 2, find $a_{11}, a_{13}, a_{22}, a_{31},$ and a_{32}.

DO EXERCISE 4.

5. For the matrix of Example 2, find the minors M_{22}, M_{32}, and M_{13}.

Minors

DEFINITION   **Minor**

The minor M_{ij} of an element a_{ij} is the determinant of the matrix found by deleting the ith row and the jth column.

Note that a minor is a certain determinant, hence is a number.

Example 3   In the matrix given in Example 2, find M_{11} and M_{23}.

Solution   To find M_{11}, we delete the first row and the first column:

$$\begin{bmatrix} -8 & 0 & 6 \\ 4 & -6 & 7 \\ -1 & -3 & 5 \end{bmatrix}.$$

We calculate the determinant of the matrix formed by the remaining elements:

$$M_{11} = \begin{vmatrix} -6 & 7 \\ -3 & 5 \end{vmatrix} = (-6) \cdot 5 - (-3) \cdot 7 = -30 - (-21)$$
$$= -30 + 21 = -9.$$

To find M_{23}, we delete the second row and the third column:

$$\begin{bmatrix} -8 & 0 & 6 \\ 4 & -6 & 7 \\ -1 & -3 & 5 \end{bmatrix}.$$

We calculate the determinant of the matrix formed by the remaining elements:

$$M_{23} = \begin{vmatrix} -8 & 0 \\ -1 & -3 \end{vmatrix}$$
$$= -8(-3) - (-1)0$$
$$= 24.$$

DO EXERCISE 5. _____

DEFINITION   **Cofactor**

The *cofactor* of an element a_{ij} is denoted A_{ij} and is given by

$$A_{ij} = (-1)^{i+j}M_{ij},$$

where M_{ij} is the minor of a_{ij}. In other words, to find the cofactor of an element, find its minor and multiply it by $(-1)^{i+j}$.

Note that $(-1)^{i+j}$ is 1 if $i + j$ is even and is -1 if $i + j$ is odd. Thus in calculating a cofactor, find the minor. Then add the number of the row and the number of the column. The sum is $i + j$. If this sum is odd, change the sign of the minor. If this sum is even, leave the minor as is.* Note too that the cofactor of an element is a number.

Example 4   In the matrix given in Example 2, find A_{11} and A_{23}.

Solution   In Example 3, we found that $M_{11} = -9$. In A_{11}, the sum of the subscripts, $1 + 1 = 2$, is even, so we do not change the sign of the minor:

$$A_{11} = -9.$$

In Example 3, we also found that $M_{23} = 24$. In A_{23}, the sum of the subscripts, $2 + 3 = 5$, is odd, so we do change the sign of the minor:

$$A_{23} = -24. \qquad \blacktriangleleft$$

DO EXERCISE 6.

Evaluating Determinants Using Cofactors

Consider the matrix **A** given by

$$\mathbf{A} = \begin{bmatrix} a_{11} & a_{12} & a_{13} \\ a_{21} & a_{22} & a_{23} \\ a_{31} & a_{32} & a_{33} \end{bmatrix}.$$

The determinant of the matrix, denoted $|\mathbf{A}|$, can be found as follows:

$$|\mathbf{A}| = a_{11}\mathbf{A}_{11} + a_{21}\mathbf{A}_{21} + a_{31}\mathbf{A}_{31}.$$

That is, multiply each element of the first column by its cofactor and add:

$$|\mathbf{A}| = a_{11} \cdot \begin{vmatrix} a_{22} & a_{23} \\ a_{32} & a_{33} \end{vmatrix} - a_{21} \cdot \begin{vmatrix} a_{12} & a_{13} \\ a_{32} & a_{33} \end{vmatrix} + a_{31} \cdot \begin{vmatrix} a_{12} & a_{13} \\ a_{22} & a_{23} \end{vmatrix}.$$

We have a minus sign with the second term since $2 + 1 = 3$, and 3 is odd. It can be shown that we can determine $|\mathbf{A}|$ by picking *any* row or column, multiplying each element by its cofactor, and adding. This is called *expanding* across a row or down a column. We just expanded down the first column. We now define the determinant function for square matrices of any dimensions.

> **DEFINITION**   **Determinant of Any Square Matrix**
>
> For any square matrix **A** of dimensions $n \times n$ $(n > 1)$, we define the *determinant* of **A**, denoted $|\mathbf{A}|$, as follows. Choose any row or column. Multiply each element in that row or column by its cofactor and add the results. The determinant of a 1×1 matrix is simply the element of the matrix. The value of a determinant will be the same no matter how it is evaluated.

---

* $(-1)^{i+j}$ can also be found by counting through the matrix horizontally and/or vertically, starting with a_{11} and $(+)$, saying $+, -, +, -$, and so on, until you come to a_{ij}.

Start here $(+)$

$$\begin{bmatrix} a_{11}^{+} & \to^{-} & \to^{+} & \to & \downarrow^{-} \\ & & & & \downarrow^{+} \\ & & & & \downarrow^{-} \\ & & & a_{ij}^{+} & \end{bmatrix} \qquad \text{The path does not matter.}$$

6. For the matrix of Example 2, find the cofactors A_{22}, A_{32}, and A_{13}.

7. Consider the matrix of Example 5. Find $|A|$ by expanding down the second column.

8. Consider the matrix of Example 5. Find $|A|$ by expanding down the third column.

Evaluate.

9. $\begin{vmatrix} 3 & 2 & 2 \\ -2 & 1 & 4 \\ 4 & -3 & 3 \end{vmatrix}$

10. $\begin{vmatrix} -5 & 0 & 0 \\ 4 & 2 & 0 \\ -3 & 5 & -6 \end{vmatrix}$

11. $\begin{vmatrix} x & 0 & x \\ 0 & x & 0 \\ 1 & 0 & x \end{vmatrix}$

Example 5   Evaluate $|A|$ by expanding across the third row.

$$A = \begin{bmatrix} -8 & 0 & 6 \\ 4 & -6 & 7 \\ -1 & -3 & 5 \end{bmatrix}$$

Solution

$$|A| = (-1)A_{31} + (-3)A_{32} + 5A_{33}$$

$$= (-1)(-1)^{3+1} \cdot \begin{vmatrix} 0 & 6 \\ -6 & 7 \end{vmatrix} + (-3)(-1)^{3+2} \cdot \begin{vmatrix} -8 & 6 \\ 4 & 7 \end{vmatrix}$$

$$+ 5(-1)^{3+3} \cdot \begin{vmatrix} -8 & 0 \\ 4 & -6 \end{vmatrix}$$

$$= (-1) \cdot 1 \cdot [0 \cdot 7 - (-6)6] + (-3)(-1)[-8 \cdot 7 - 4 \cdot 6]$$

$$+ 5 \cdot 1 \cdot [-8(-6) - 4 \cdot 0]$$

$$= -[36] + 3[-80] + 5[48]$$

$$= -36 - 240 + 240$$

$$= -36$$

The value of this determinant is -36 no matter how we evaluate it. That is, if we expand down the second column, we still get -36.   ◀

DO EXERCISES 7 AND 8.

Example 6   Evaluate. Expand down the first column.

$$\begin{vmatrix} -1 & 0 & 1 \\ -5 & 1 & -1 \\ 4 & 8 & 1 \end{vmatrix} = -1 \cdot \begin{vmatrix} 1 & -1 \\ 8 & 1 \end{vmatrix} - (-5) \cdot \begin{vmatrix} 0 & 1 \\ 8 & 1 \end{vmatrix} + 4 \cdot \begin{vmatrix} 0 & 1 \\ 1 & -1 \end{vmatrix}$$

$$= -1(1 + 8) + 5(-8) + 4(-1)$$

$$= -9 - 40 - 4 = -53$$ ◀

DO EXERCISES 9–11.

 Tips for Easy Evaluation of Certain Determinants

We can simplify the evaluation of certain determinants using the following properties.

THEOREM 4

If a row (or column) of a matrix A has all elements 0, then $|A| = 0$.

Proof.   Just evaluate by expanding across a row (or down a column) that has all 0's.

Examples Evaluate.

7. $\begin{vmatrix} 0 & 6 \\ 0 & 7 \end{vmatrix} = 0$

8. $\begin{vmatrix} 4 & 5 & -7 \\ 0 & 0 & 0 \\ -3 & 9 & 6 \end{vmatrix} = 0$ ◀

DO EXERCISES 12 AND 13.

THEOREM 5

If two rows (or columns) of a matrix **A** are interchanged to obtain a new matrix **B**, then $|A| = -|B|$.

Examples

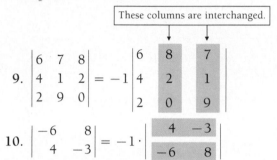

These columns are interchanged.

9. $\begin{vmatrix} 6 & 7 & 8 \\ 4 & 1 & 2 \\ 2 & 9 & 0 \end{vmatrix} = -1 \begin{vmatrix} 6 & 8 & 7 \\ 4 & 2 & 1 \\ 2 & 0 & 9 \end{vmatrix}$

10. $\begin{vmatrix} -6 & 8 \\ 4 & -3 \end{vmatrix} = -1 \cdot \begin{vmatrix} 4 & -3 \\ -6 & 8 \end{vmatrix}$ ◀

DO EXERCISES 14 AND 15.

THEOREM 6

If two rows (or columns) of a matrix **A** are the same, then $|A| = 0$.

Proof. Interchanging the rows (or columns) that are the same does not change **A**. Thus by Theorem 5, $|A| = -|A|$. This is possible only when $|A| = 0$.

Examples Evaluate.

11. $\begin{vmatrix} 6 & 7 & 8 \\ -2 & 6 & 5 \\ -2 & 6 & 5 \end{vmatrix} = 0$

12. $\begin{vmatrix} -5 & 4 & -5 \\ 3 & 7 & 3 \\ 0 & 12 & 0 \end{vmatrix} = 0$ ◀

DO EXERCISE 16.

Evaluate.

12. $\begin{vmatrix} 2 & 3 \\ 0 & 0 \end{vmatrix}$

13. $\begin{vmatrix} 2 & 3 & 0 \\ -3 & 7 & 0 \\ 2 & 4 & 0 \end{vmatrix}$

14. Given

$$A = \begin{bmatrix} 5 & -9 \\ -2 & 0 \end{bmatrix} \text{ and }$$

$$B = \begin{bmatrix} -2 & 0 \\ 5 & -9 \end{bmatrix}.$$

a) Find $|A|$ and $|B|$.
b) Why does $|A| = -|B|$?

15. Given

$$C = \begin{vmatrix} -2 & 3 & 4 \\ 4 & 5 & 6 \\ -9 & -1 & 0 \end{vmatrix} \text{ and }$$

$$D = \begin{vmatrix} 4 & 3 & -2 \\ 6 & 5 & 4 \\ 0 & -1 & -9 \end{vmatrix}.$$

a) Find $|C|$ and $|D|$.
b) Why does $|C| = -|D|$?

16. Evaluate:

$$\begin{vmatrix} -1 & 3 & -7 \\ -1 & 2 & -6 \\ -1 & 3 & -7 \end{vmatrix}.$$

Solve for x.

17. $\begin{vmatrix} -16 & 32 \\ -5 & -3 \end{vmatrix} = x \cdot \begin{vmatrix} 4 & -8 \\ -5 & -3 \end{vmatrix}$

18. $\begin{vmatrix} -3 & 12 & 2 \\ 5 & -6 & 3 \\ 0 & 18 & 5 \end{vmatrix}$

$= x \cdot \begin{vmatrix} -3 & 2 & 2 \\ 5 & -1 & 3 \\ 0 & 3 & 5 \end{vmatrix}$

9. Without expanding, evaluate $|A|$:

$A = \begin{bmatrix} 1 & 0 & -7 \\ -8 & 6 & -4 \\ 24 & -18 & 12 \end{bmatrix}$.

THEOREM 7

If all the elements of a row (or column) of a matrix **A** are multiplied by k, $|A|$ is multiplied by k. Or, if all the elements of a row (or column) of **A** have a common factor k, the k can be factored out of the determinant, $|A|$.

Examples

Factor out 3.

13. $\begin{vmatrix} 2 & 4 & 6 \\ -2 & 5 & 9 \\ 4 & -1 & -3 \end{vmatrix} = 3 \cdot \begin{vmatrix} 2 & 4 & 2 \\ -2 & 5 & 3 \\ 4 & -1 & -1 \end{vmatrix}$

14. $\begin{vmatrix} 10 & 25 \\ -4 & -7 \end{vmatrix} = 5 \cdot \begin{vmatrix} 2 & 5 \\ -4 & -7 \end{vmatrix}$

We leave it to the student to verify the truth of the equations in Examples 13 and 14 by evaluating each side. ◀

DO EXERCISES 17 AND 18.

Example 15 Without expanding, find $|A|$:

$A = \begin{bmatrix} -6 & 3 & 8 \\ 15 & -9 & -20 \\ -9 & -1 & 12 \end{bmatrix}$.

Solution

$|A| = (-3) \cdot \begin{vmatrix} 2 & 3 & 8 \\ -5 & -9 & -20 \\ 3 & -1 & 12 \end{vmatrix}$ Factoring -3 out of the first column

$= (-3)(4) \cdot \begin{vmatrix} 2 & 3 & 2 \\ -5 & -9 & -5 \\ 3 & -1 & 3 \end{vmatrix}$ Factoring 4 out of the third column

$= 0$ By Theorem 6, the first and the third columns are the same. ◀

DO EXERCISE 19.

THEOREM 8

If each element in a row (or column) is multiplied by a number k and each product is added to the corresponding element of another row (or column), the value of the determinant is not changed. That is, a multiple of any row (or column) can be added to any other row (or column) without changing the value of the determinant.

Example 16   Find a determinant having the same value as the one on the left by adding three times the second column to the first column.

$$\begin{vmatrix} 0 & 1 & 2 \\ 4 & 5 & 6 \\ 7 & 8 & 9 \end{vmatrix} = \begin{vmatrix} 0+3(1) & 1 & 2 \\ 4+3(5) & 5 & 6 \\ 7+3(8) & 8 & 9 \end{vmatrix} = \begin{vmatrix} 3 & 1 & 2 \\ 19 & 5 & 6 \\ 31 & 8 & 9 \end{vmatrix}$$

The student should check the truth of the equation by evaluating each determinant.   ◀

Example 17   Find a determinant having the same value as the one on the left by adding two times the third row to the first row.

$$\begin{vmatrix} 0 & 1 & 2 \\ 4 & 5 & 6 \\ 7 & 8 & 9 \end{vmatrix} = \begin{vmatrix} 0+2(7) & 1+2(8) & 2+2(9) \\ 4 & 5 & 6 \\ 7 & 8 & 9 \end{vmatrix}$$

The student should check the truth of the equation by evaluating each determinant.   ◀

DO EXERCISE 20.

We can use the properties of determinants to simplify their evaluation. We try to use the properties to find another determinant where in some row or column, one element is 1 and the rest are 0.

Example 18   Evaluate by first simplifying to a determinant where in one row or column, one element is 1 and the rest are 0:

$$\begin{vmatrix} 6 & 2 & 3 \\ 6 & -1 & 5 \\ -2 & 3 & 1 \end{vmatrix}.$$

Solution   We will try to get two 0's and a 1 in the third row. It already has a 1; that is why we chose the third row. We first factor a 2 out of column 1:

$$2 \cdot \begin{vmatrix} 3 & 2 & 3 \\ 3 & -1 & 5 \\ -1 & 3 & 1 \end{vmatrix}.\qquad \text{Theorem 7}$$

Now we multiply each element in column 3 by -3 and add the corresponding elements to column 2:

$$2 \cdot \begin{vmatrix} 3 & -7 & 3 \\ 3 & -16 & 5 \\ -1 & 0 & 1 \end{vmatrix}.\qquad \text{Theorem 8}$$

Next, we add the elements in column 3 to the corresponding elements in

20. Find a determinant equal to this one by adding twice the first row to the second row:

$$\begin{vmatrix} -2 & 3 & 4 \\ 1 & 4 & -3 \\ 0 & 9 & 7 \end{vmatrix}.$$

Evaluate by first simplifying to a determinant where in one row or column, one element is 1 and the rest are 0.

21. $\begin{vmatrix} 3 & -1 & 1 \\ 2 & 2 & -4 \\ 2 & 4 & 1 \end{vmatrix}$

22. $\begin{vmatrix} 5 & -4 & 2 & -2 \\ 3 & -3 & -4 & 7 \\ -2 & 3 & 2 & 4 \\ -8 & 9 & 5 & -5 \end{vmatrix}$

23. Factor:

$$\begin{vmatrix} a^2 & b^2 & c^2 \\ a & b & c \\ 1 & 1 & 1 \end{vmatrix}.$$

column 1 (Theorem 8):

$$2 \cdot \begin{vmatrix} 6 & -7 & 3 \\ 8 & -16 & 5 \\ 0 & 0 & 1 \end{vmatrix}.$$

Finally, we evaluate the determinant by expanding across the last row:

$$2 \cdot \left(0 - 0 + 1 \cdot \begin{vmatrix} 6 & -7 \\ 8 & -16 \end{vmatrix} \right) = 2 \cdot [6(-16) - 8(-7)] = -80.$$

DO EXERCISES 21 AND 22.

Factoring Certain Determinants

Example 19 Factor:

$$\begin{vmatrix} 1 & x & x^2 \\ 1 & y & y^2 \\ 1 & z & z^2 \end{vmatrix}.$$

Solution

$$\begin{vmatrix} 1 & x & x^2 \\ 1 & y & y^2 \\ 1 & z & z^2 \end{vmatrix} = \begin{vmatrix} 0 & x-y & x^2-y^2 \\ 1 & y & y^2 \\ 0 & z-y & z^2-y^2 \end{vmatrix}$$

By Theorem 8: Adding -1 times the second row to the first row, and -1 times the second row to the third row

$$= (x-y)(z-y) \cdot \begin{vmatrix} 0 & 1 & x+y \\ 1 & y & y^2 \\ 0 & 1 & z+y \end{vmatrix}$$

By Theorem 7: Factoring $x-y$ out of the first row and $z-y$ out of the third row

$$= (x-y)(z-y) \cdot \begin{vmatrix} 0 & 0 & x-z \\ 1 & y & y^2 \\ 0 & 1 & z+y \end{vmatrix}$$

By Theorem 8: adding -1 times the third row to the first row

$$= (x-y)(z-y)(x-z) \cdot \begin{vmatrix} 0 & 0 & 1 \\ 1 & y & y^2 \\ 0 & 1 & z+y \end{vmatrix}$$

By Theorem 7: Factoring $x-z$ out of the first row

$$= (x-y)(z-y)(x-z)$$

Expanding the determinant across the first row, we get 1.

DO EXERCISE 23.

Cramer's Rule: Square Systems

Determinants have many uses. One of these is in solving systems of linear equations in which the number of variables is the same as the number of equations—**square systems**—and in which the constants are not all 0. Let us consider a system of two equations:

$$a_1 x + b_1 y = c_1,$$
$$a_2 x + b_2 y = c_2.$$

Using the methods of the preceding sections, we can solve this system. We obtain

$$x = \frac{c_1 b_2 - c_2 b_1}{a_1 b_2 - a_2 b_1}, \qquad y = \frac{a_1 c_2 - a_2 c_1}{a_1 b_2 - a_2 b_1}.$$

The numerators and the denominators of the expressions for x and y can be written as determinants.

THEOREM 9   **Cramer's Rule: 2×2 Systems**

The solution of the system

$$a_1 x + b_1 y = c_1,$$
$$a_2 x + b_2 y = c_2,$$

if it is unique, is given by

$$x = \frac{\begin{vmatrix} c_1 & b_1 \\ c_2 & b_2 \end{vmatrix}}{\begin{vmatrix} a_1 & b_1 \\ a_2 & b_2 \end{vmatrix}}, \qquad y = \frac{\begin{vmatrix} a_1 & c_1 \\ a_2 & c_2 \end{vmatrix}}{\begin{vmatrix} a_1 & b_1 \\ a_2 & b_2 \end{vmatrix}}.$$

The equations above make sense only if the determinant in the denominator is not 0. If the denominator *is* 0, then one of two things happens.

1. If the denominator is 0 and the other two determinants in the numerators are also 0, then the system of equations is dependent.
2. If the denominator is 0 and at least one of the other determinants in the numerators is not 0, then the system is inconsistent.

To use this theorem, we compute the three determinants and then compute x and y as shown above. Note that the denominator in both cases contains the coefficients of x and y, in the same position as in the original equations. For x, the numerator is obtained by replacing the x-coefficients (the a's) by the c's. For y, the numerator is obtained by replacing the y-coefficients (the b's) by the c's.

Example 20   Solve using Cramer's rule:

$$2x + 5y = 7,$$
$$5x - 2y = -3.$$

Solution   We have

$$x = \frac{\begin{vmatrix} 7 & 5 \\ -3 & -2 \end{vmatrix}}{\begin{vmatrix} 2 & 5 \\ 5 & -2 \end{vmatrix}} = \frac{7(-2) - (-3)5}{2(-2) - 5 \cdot 5} = -\frac{1}{29},$$

$$y = \frac{\begin{vmatrix} 2 & 7 \\ 5 & -3 \end{vmatrix}}{\begin{vmatrix} 2 & 5 \\ 5 & -2 \end{vmatrix}} = \frac{2(-3) - 5 \cdot 7}{-29} = \frac{41}{29}.$$

The solution is $\left(-\frac{1}{29}, \frac{41}{29}\right)$.   ◀

Solve using Cramer's rule.

24. $2x - y = 5,$
    $x - 2y = 1$

25. $3x + 4y = -2,$
    $5x - 7y = 1$

26. $\sqrt{2}x - \pi y = 3,$
    $\pi x + \sqrt{2}y = 4$

DO EXERCISES 24–26.

THEOREM 10   **Cramer's Rule: 3×3 Systems**

The solution of the system

$$a_1 x + b_1 y + c_1 z = d_1,$$
$$a_2 x + b_2 y + c_2 z = d_2,$$
$$a_3 x + b_3 y + c_3 z = d_3$$

is found by considering the following determinants:

$$D = \begin{vmatrix} a_1 & b_1 & c_1 \\ a_2 & b_2 & c_2 \\ a_3 & b_3 & c_3 \end{vmatrix}, \qquad D_x = \begin{vmatrix} d_1 & b_1 & c_1 \\ d_2 & b_2 & c_2 \\ d_3 & b_3 & c_3 \end{vmatrix},$$

$$D_y = \begin{vmatrix} a_1 & d_1 & c_1 \\ a_2 & d_2 & c_2 \\ a_3 & d_3 & c_3 \end{vmatrix}, \qquad D_z = \begin{vmatrix} a_1 & b_1 & d_1 \\ a_2 & b_2 & d_2 \\ a_3 & b_3 & d_3 \end{vmatrix}.$$

The solution, if it is unique, is given by

$$x = \frac{D_x}{D}, \qquad y = \frac{D_y}{D}, \qquad z = \frac{D_z}{D}.$$

Note that we obtain the determinant D_x in the numerator for x from D by replacing the x-coefficients by d_1, d_2, and d_3. A similar thing happens with D_y and D_z. When $D = 0$, Cramer's rule cannot be used. If $D = 0$ and D_x, D_y, and D_z are 0, the system is dependent. If $D = 0$ and one of D_x, D_y, or D_z is not 0, then the system is inconsistent.

Example 21   Solve using Cramer's rule:

$$x - 3y + 7z = 13,$$
$$x + y + z = 1,$$
$$x - 2y + 3z = 4.$$

Solution   We have

$$D = \begin{vmatrix} 1 & -3 & 7 \\ 1 & 1 & 1 \\ 1 & -2 & 3 \end{vmatrix} = -10, \qquad D_x = \begin{vmatrix} 13 & -3 & 7 \\ 1 & 1 & 1 \\ 4 & -2 & 3 \end{vmatrix} = 20,$$

$$D_y = \begin{vmatrix} 1 & 13 & 7 \\ 1 & 1 & 1 \\ 1 & 4 & 3 \end{vmatrix} = -6, \qquad D_z = \begin{vmatrix} 1 & -3 & 13 \\ 1 & 1 & 1 \\ 1 & -2 & 4 \end{vmatrix} = -24.$$

Then

$$x = \frac{D_x}{D} = \frac{20}{-10} = -2,$$

$$y = \frac{D_y}{D} = \frac{-6}{-10} = \frac{3}{5},$$

$$z = \frac{D_z}{D} = \frac{-24}{-10} = \frac{12}{5}.$$

The solution is $(-2, \frac{3}{5}, \frac{12}{5})$. In practice, it is not necessary to evaluate D_z. When we have found values for x and y, we can substitute them into one of the equations and find z. ◀

DO EXERCISE 27.

27. Solve using Cramer's rule:
$$x - 3y - 7z = 6,$$
$$2x + 3y + z = 9,$$
$$4x + y \phantom{{}+{}z} = 7.$$

●EXERCISE SET 9.5

 A   Evaluate the determinant.

1. $\begin{vmatrix} -2 & -\sqrt{5} \\ -\sqrt{5} & 3 \end{vmatrix}$

2. $\begin{vmatrix} \sqrt{5} & -3 \\ 4 & 2 \end{vmatrix}$

3. $\begin{vmatrix} x & 4 \\ x & x^2 \end{vmatrix}$

4. $\begin{vmatrix} y^2 & -2 \\ y & 3 \end{vmatrix}$

5. $\begin{vmatrix} 3 & 1 & 2 \\ -2 & 3 & 1 \\ 3 & 4 & -6 \end{vmatrix}$

6. $\begin{vmatrix} 3 & -2 & 1 \\ 2 & 4 & 3 \\ -1 & 5 & 1 \end{vmatrix}$

7. $\begin{vmatrix} x & 0 & -1 \\ 2 & x & x^2 \\ -3 & x & 1 \end{vmatrix}$

8. $\begin{vmatrix} x & 1 & -1 \\ x^2 & x & x \\ 0 & x & 1 \end{vmatrix}$

Use the following matrix for Exercises 9–18:
$$A = \begin{bmatrix} 7 & -4 & -6 \\ 2 & 0 & -3 \\ 1 & 2 & -5 \end{bmatrix}.$$

9. Find a_{11}, a_{32}, and a_{22}.

10. Find a_{13}, a_{31}, and a_{23}.

11. Find M_{11}, M_{32}, and M_{22}.

12. Find M_{13}, M_{31}, and M_{23}.

13. Find A_{11}, A_{32}, and A_{22}.

14. Find A_{13}, A_{31}, and A_{23}.

15. Evaluate $|A|$ by expanding across the second row.

16. Evaluate $|A|$ by expanding down the second column.

17. Evaluate $|A|$ by expanding down the third column.

18. Evaluate $|A|$ by expanding across the first row.

Use the following matrix for Exercises 19–24:
$$A = \begin{bmatrix} 1 & 0 & 0 & -2 \\ 4 & 1 & 0 & 0 \\ 5 & 6 & 7 & 8 \\ -2 & -3 & -1 & 0 \end{bmatrix}.$$

19. Find M_{41} and M_{33}.

20. Find M_{12} and M_{44}.

21. Find A_{24} and A_{43}.

22. Find A_{22} and A_{34}.

23. Evaluate $|A|$ by expanding across the first row.

24. Evaluate $|A|$ by expanding down the third column.

Evaluate the determinant. Use any method of expanding.

25. $\begin{vmatrix} 5 & -4 & 2 & -2 \\ 3 & -3 & -4 & 7 \\ -2 & 3 & 2 & 4 \\ -8 & 9 & 5 & -5 \end{vmatrix}$

26. $\begin{vmatrix} x & p & q & r \\ 0 & y & s & t \\ 0 & 0 & z & u \\ 0 & 0 & 0 & w \end{vmatrix}$

B   Evaluate by first simplifying to a determinant where in one row or column, one element is 1 and the rest are 0.

27. $\begin{vmatrix} -4 & 5 \\ 6 & 10 \end{vmatrix}$

28. $\begin{vmatrix} 3 & -9 \\ -2 & 4 \end{vmatrix}$

29. $\begin{vmatrix} 2 & 1 & 1 \\ 2 & -3 & -1 \\ -4 & 5 & 2 \end{vmatrix}$

30. $\begin{vmatrix} 1 & 2 & 4 \\ 2 & 3 & 5 \\ 3 & 1 & 6 \end{vmatrix}$

31. $\begin{vmatrix} 11 & -15 & 20 \\ 16 & 24 & -8 \\ 6 & 9 & 15 \end{vmatrix}$

32. $\begin{vmatrix} 4 & -24 & 15 \\ -3 & 18 & -6 \\ 5 & -4 & 3 \end{vmatrix}$

33. $\begin{vmatrix} -3 & 0 & 2 & 6 \\ 2 & 4 & 0 & -1 \\ -1 & 0 & -5 & 2 \\ 0 & -1 & -2 & -3 \end{vmatrix}$

34. $\begin{vmatrix} -2 & 1 & 0 & 5 \\ 3 & 0 & -4 & -2 \\ 4 & -6 & -8 & -1 \\ 8 & 0 & -2 & -3 \end{vmatrix}$

Evaluate the determinant without expanding.

35. $\begin{vmatrix} x & y & z \\ 0 & 0 & 0 \\ p & q & r \end{vmatrix}$

36. $\begin{vmatrix} 5 & 5 & 5 \\ 3 & 3 & 3 \\ 2 & -7 & 8 \end{vmatrix}$

37. $\begin{vmatrix} 2a & t & -7a \\ 2b & u & -7b \\ 2c & v & -7c \end{vmatrix}$

38. $\begin{vmatrix} a & -1 & 4a \\ b & 2 & 4b \\ x & -3 & 4x \end{vmatrix}$

▶ Factor.

39. $\begin{vmatrix} x^2 & x & 1 \\ y^2 & y & 1 \\ z^2 & z & 1 \end{vmatrix}$

40. $\begin{vmatrix} 1 & 1 & 1 \\ a & b & c \\ a^2 & b^2 & c^2 \end{vmatrix}$

41. $\begin{vmatrix} x & x^2 & x^3 \\ y & y^2 & y^3 \\ z & z^2 & z^3 \end{vmatrix}$

42. $\begin{vmatrix} 1 & 1 & 1 \\ a & b & c \\ a^3 & b^3 & c^3 \end{vmatrix}$

▶ Solve using Cramer's rule.

43. $-2x + 4y = 3,$
$3x - 7y = 1$

44. $5x - 4y = -3,$
$7x + 2y = 6$

45. $\sqrt{3}x + \pi y = -5,$
$\pi x - \sqrt{3}y = 4$

46. $\pi x - \sqrt{5}y = 2,$
$\sqrt{5}x + \pi y = -3$

47. $3x + 2y - z = 4,$
$3x - 2y + z = 5,$
$4x - 5y - z = -1$

48. $3x - y + 2z = 1,$
$x - y + 2z = 3,$
$-2x + 3y + z = 1$

49. $\quad\quad 6y + 6z = -1,$
$8x \quad\quad + 6z = -1,$
$4x + 9y \quad\quad = 8$

50. $3x + 5y \quad\quad = 2,$
$2x \quad\quad - 3z = 7,$
$\quad\quad 4y + 2z = -1$

● **SYNTHESIS** _____

Solve.

51. $\begin{vmatrix} x & 5 \\ -4 & x \end{vmatrix} = 24$

52. $\begin{vmatrix} y & 2 \\ 3 & y \end{vmatrix} = y$

53. $\begin{vmatrix} x & -3 \\ -1 & x \end{vmatrix} \geq 0$

54. $\begin{vmatrix} y & -5 \\ -2 & y \end{vmatrix} < 0$

55. $\begin{vmatrix} x + 3 & 4 \\ x - 3 & 5 \end{vmatrix} = -7$

56. $\begin{vmatrix} m + 2 & -3 \\ m + 5 & -4 \end{vmatrix} = 3m - 5$

57. $\begin{vmatrix} 2 & x & 1 \\ 1 & 2 & -1 \\ 3 & 4 & -2 \end{vmatrix} = -6$

58. $\begin{vmatrix} x & 2 & x \\ 3 & -1 & 1 \\ 1 & -2 & 2 \end{vmatrix} = -10$

Rewrite each expression using a determinant. Answers may vary.

59. $2L + 2W$

60. $\pi r + \pi h$

61. $a^2 + b^2$

62. $\frac{1}{2}h(a + b)$

63. $2\pi r^2 + 2\pi rh$

64. $x^2y^2 - Q^2$

65. If a line contains the points (x_1, y_1) and (x_2, y_2), an equation of the line can be written as follows:

$$\begin{vmatrix} x & y & 1 \\ x_1 & y_1 & 1 \\ x_2 & y_2 & 1 \end{vmatrix} = 0.$$

Prove this.

66. Show that the points (x_1, y_1), (x_2, y_2), and (x_3, y_3) are collinear (on the same straight line) if and only if

$$\begin{vmatrix} x_1 & y_1 & 1 \\ x_2 & y_2 & 1 \\ x_3 & y_3 & 1 \end{vmatrix} = 0.$$

● **CHALLENGE** _____

67. Consider a triangle with vertices (x_1, y_1), (x_2, y_2), and (x_3, y_3). The area of this triangle is the absolute value of

$$\frac{1}{2} \cdot \begin{vmatrix} x_1 & y_1 & 1 \\ x_2 & y_2 & 1 \\ x_3 & y_3 & 1 \end{vmatrix}.$$

Prove this. (*Hint:* Look at the following figure. The area of triangle *ABC* is the area of trapezoid *ABDE* plus the area of trapezoid *AEFC* minus the area of trapezoid *BDFC*.)

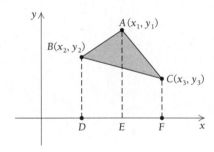

68. Prove that the lines $a_1x + b_1y = c_1$ and $a_2x + b_2y = c_2$ are parallel when

$$\begin{vmatrix} a_1 & b_1 \\ a_2 & b_2 \end{vmatrix} = 0$$

and either

$$\begin{vmatrix} c_1 & b_1 \\ c_2 & b_2 \end{vmatrix} \neq 0 \quad \text{or} \quad \begin{vmatrix} a_1 & c_1 \\ a_2 & c_2 \end{vmatrix} \neq 0.$$

9.6

INVERSES OF MATRICES

In this section, we learn the meaning of the *multiplicative inverse*, or simply *inverse*, of a square matrix. We also learn how to compute such inverses and how to use them to solve equations.

Identity Matrices

The letter **I** is used to represent square matrices such as

$$\mathbf{I} = \begin{bmatrix} 1 & 0 \\ 0 & 1 \end{bmatrix} \quad \text{and} \quad \mathbf{I} = \begin{bmatrix} 1 & 0 & 0 \\ 0 & 1 & 0 \\ 0 & 0 & 1 \end{bmatrix}.$$

These square matrices have 1's extending from the upper left down to the lower right along what is called the **main diagonal.** The rest of the elements are 0.

DO EXERCISE 1.

In Margin Exercise 1, you have proved the following theorem for the case $n = 2$.

THEOREM 11

For any square matrix **A** of dimensions $n \times n$,

  $\mathbf{AI} = \mathbf{IA} = \mathbf{A}$ (**I** is a multiplicative identity),

where **I** is the square matrix, defined above, of dimensions $n \times n$. For any matrix **A** with exactly n rows,

  $\mathbf{IA} = \mathbf{A}$,

where **I** is the square matrix of dimensions $n \times n$.

We usually say that **I** is an **identity matrix.** Suppose that for matrix **A** there is a matrix $\mathbf{A}^{-1}$ for which

$$\mathbf{A}^{-1} \cdot \mathbf{A} = \mathbf{I} = \mathbf{A} \cdot \mathbf{A}^{-1}.$$

Then $\mathbf{A}^{-1}$ is the **multiplicative inverse,** or simply **inverse,** of **A**. For example, for the matrix

$$\mathbf{A} = \begin{bmatrix} 5 & 3 \\ 3 & 2 \end{bmatrix},$$

we have

$$\mathbf{A}^{-1} = \begin{bmatrix} 2 & -3 \\ -3 & 5 \end{bmatrix}.$$

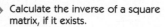

1. Consider

$$\mathbf{A} = \begin{bmatrix} a & b \\ c & d \end{bmatrix}, \quad \mathbf{X} = \begin{bmatrix} x \\ y \end{bmatrix},$$

and

$$\mathbf{I} = \begin{bmatrix} 1 & 0 \\ 0 & 1 \end{bmatrix}.$$

a) Find **AI**.

b) Find **IA**.

c) Compare **AI** and **IA**.

d) Find **IX**.

2. Let

$$A = \begin{bmatrix} 3 & 1 & 0 \\ 1 & -1 & 2 \\ 1 & 1 & 1 \end{bmatrix}$$

and

$$A^{-1} = \frac{1}{8} \begin{bmatrix} 3 & 1 & -2 \\ -1 & -3 & 6 \\ -2 & 2 & 4 \end{bmatrix}.$$

a) Find AA^{-1}.

b) Find $A^{-1}A$.

c) Compare AA^{-1} and $A^{-1}A$.

We can check that

$$A \cdot A^{-1} = \begin{bmatrix} 5 & 3 \\ 3 & 2 \end{bmatrix} \begin{bmatrix} 2 & -3 \\ -3 & 5 \end{bmatrix}$$

$$= \begin{bmatrix} 1 & 0 \\ 0 & 1 \end{bmatrix} = I.$$

We leave it to the student to verify that $A^{-1} \cdot A = I$.

DO EXERCISE 2. _____

A Calculating Matrix Inverses: The Gauss–Jordan Reduction Method

We now learn a way to find the inverse of a square matrix, if it exists. Consider the matrix

$$A = \begin{bmatrix} 2 & -1 & 1 \\ 1 & -2 & 3 \\ 4 & 1 & 2 \end{bmatrix}.$$

First we form a new (**augmented**) matrix consisting, on the left, of the matrix A and, on the right, of the corresponding identity matrix I:

This line is for clarity and can be omitted.

$$\begin{bmatrix} 2 & -1 & 1 & 1 & 0 & 0 \\ 1 & -2 & 3 & 0 & 1 & 0 \\ 4 & 1 & 2 & 0 & 0 & 1 \end{bmatrix}.$$

The matrix A   The identity matrix I

We now attempt to use the elimination method to transform A to an identity matrix, but whatever operations we perform, we do them on the entire augmented matrix. When we finish, we will get a matrix like the following:

$$\begin{bmatrix} 1 & 0 & 0 & a & b & c \\ 0 & 1 & 0 & d & e & f \\ 0 & 0 & 1 & g & h & i \end{bmatrix}.$$

The matrix on the right,

$$\begin{bmatrix} a & b & c \\ d & e & f \\ g & h & i \end{bmatrix},$$

will be A^{-1}.

Example 1 Find A^{-1}, where

$$A = \begin{bmatrix} 2 & -1 & 1 \\ 1 & -2 & 3 \\ 4 & 1 & 2 \end{bmatrix}.$$

Solution

a) We find the augmented matrix consisting of **A** and **I**:

$$\begin{bmatrix} 2 & -1 & 1 & 1 & 0 & 0 \\ 1 & -2 & 3 & 0 & 1 & 0 \\ 4 & 1 & 2 & 0 & 0 & 1 \end{bmatrix}.$$

Be sure to write the augmented matrix *before* doing any row operations. Otherwise, you may not obtain the inverse matrix.

b) We interchange the first and second rows so that the elements of the first column are multiples of the top number on the main diagonal:

$$\begin{bmatrix} 1 & -2 & 3 & 0 & 1 & 0 \\ 2 & -1 & 1 & 1 & 0 & 0 \\ 4 & 1 & 2 & 0 & 0 & 1 \end{bmatrix}. \begin{array}{l} \longrightarrow R_1' = R_2 \\ \longrightarrow R_2' = R_1 \end{array}$$

c) Next we obtain 0's in the rest of the first column. We multiply the first row by -2 and add it to the second row. Then we multiply the first row by -4 and add it to the third row:

$$\begin{bmatrix} 1 & -2 & 3 & 0 & 1 & 0 \\ 0 & 3 & -5 & 1 & -2 & 0 \\ 0 & 9 & -10 & 0 & -4 & 1 \end{bmatrix}. \begin{array}{l} \longrightarrow R_2' = -2R_1 + R_2 \\ \longrightarrow R_3' = -4R_1 + R_3 \end{array}$$

d) Next we move down the main diagonal to the number 3. We note that the number below it, 9, is a multiple of 3. Thus we multiply the second row by -3 and add it to the third:

$$\begin{bmatrix} 1 & -2 & 3 & 0 & 1 & 0 \\ 0 & 3 & -5 & 1 & -2 & 0 \\ 0 & 0 & 5 & -3 & 2 & 1 \end{bmatrix}. \longrightarrow R_3' = -3R_2 + R_3$$

e) Now we move down the main diagonal to the number 5. We check to see if each number above 5 in the third column is a multiple of 5. Since this is not the case, we multiply the first row by -5:

$$\begin{bmatrix} -5 & 10 & -15 & 0 & -5 & 0 \\ 0 & 3 & -5 & 1 & -2 & 0 \\ 0 & 0 & 5 & -3 & 2 & 1 \end{bmatrix}. \longrightarrow R_1' = -5R_1$$

f) Now we work back up. We add the third row to the second. We also multiply the third row by 3 and add it to the first:

$$\begin{bmatrix} -5 & 10 & 0 & -9 & 1 & 3 \\ 0 & 3 & 0 & -2 & 0 & 1 \\ 0 & 0 & 5 & -3 & 2 & 1 \end{bmatrix}. \begin{array}{l} \longrightarrow R_1' = 3R_3 + R_1 \\ \longrightarrow R_2' = R_3 + R_2 \end{array}$$

g) We move back to the number 3 on the main diagonal. We multiply the first row by -3, so the element on the top of the second column is a multiple of 3:

$$\begin{bmatrix} 15 & -30 & 0 & 27 & -3 & -9 \\ 0 & 3 & 0 & -2 & 0 & 1 \\ 0 & 0 & 5 & -3 & 2 & 1 \end{bmatrix}. \longrightarrow R_1' = -3R_1$$

Find A^{-1}. Use the Gauss–Jordan reduction method.

3. $A = \begin{bmatrix} 1 & 0 & 1 \\ 2 & 1 & 0 \\ 1 & -1 & 1 \end{bmatrix}$

h) We multiply the second row by 10 and add it to the first:

$$\begin{bmatrix} 15 & 0 & 0 & 7 & -3 & 1 \\ 0 & 3 & 0 & -2 & 0 & 1 \\ 0 & 0 & 5 & -3 & 2 & 1 \end{bmatrix} \longrightarrow R_1' = 10R_2 + R_1.$$

i) Finally, we get all 1's on the main diagonal. We multiply the first row by $\frac{1}{15}$, the second by $\frac{1}{3}$, and the third by $\frac{1}{5}$:

$$\begin{bmatrix} 1 & 0 & 0 & \frac{7}{15} & -\frac{1}{5} & \frac{1}{15} \\ 0 & 1 & 0 & -\frac{2}{3} & 0 & \frac{1}{3} \\ 0 & 0 & 1 & -\frac{3}{5} & \frac{2}{5} & \frac{1}{5} \end{bmatrix} \begin{aligned} &\longrightarrow R_1' = \tfrac{1}{15}R_1 \\ &\longrightarrow R_2' = \tfrac{1}{3}R_2 \\ &\longrightarrow R_3' = \tfrac{1}{5}R_3 \end{aligned}$$

We now have the matrix **I** on the left. Thus,

$$A^{-1} = \begin{bmatrix} \frac{7}{15} & -\frac{1}{5} & \frac{1}{15} \\ -\frac{2}{3} & 0 & \frac{1}{3} \\ -\frac{3}{5} & \frac{2}{5} & \frac{1}{5} \end{bmatrix}.$$

The student can always check by multiplying $A^{-1}A$ or AA^{-1}. If we cannot obtain the identity matrix on the left using the Gauss–Jordan reduction method, as would be the case when a system has no solution or infinitely many solutions, then A^{-1} does not exist. More specifically:

> If we obtain a row of all 0's in either of the two matrices in the augmented matrix, then A^{-1} does not exist.

4. $A = \begin{bmatrix} 3 & 5 \\ 1 & -2 \end{bmatrix}$

DO EXERCISES 3 AND 4.

B ▶ Solving Systems Using Inverses

We can use matrix inverses to solve certain kinds of systems. Consider the system

$$\begin{aligned} 3x + 5y &= -1, \\ x - 2y &= 4. \end{aligned}$$

We write a matrix equation equivalent to this system:

$$\begin{bmatrix} 3 & 5 \\ 1 & -2 \end{bmatrix} \begin{bmatrix} x \\ y \end{bmatrix} = \begin{bmatrix} -1 \\ 4 \end{bmatrix}.$$

Now we let

$$\begin{bmatrix} 3 & 5 \\ 1 & -2 \end{bmatrix} = A, \qquad \begin{bmatrix} x \\ y \end{bmatrix} = X, \quad \text{and} \quad \begin{bmatrix} -1 \\ 4 \end{bmatrix} = B.$$

Then we have the matrix equation

$$A \cdot X = B.$$

To solve a comparable equation $ax = b$ involving real numbers, we multiply on both sides by the real number a^{-1}, provided it exists. To solve the matrix equation $AX = B$, we multiply on both sides by the inverse, A^{-1}, of the coefficient matrix, A, provided the inverse exists. We found in Margin Exercise 4 that it does exist and is given by

$$A^{-1} = \begin{bmatrix} \frac{2}{11} & \frac{5}{11} \\ \frac{1}{11} & -\frac{3}{11} \end{bmatrix}, \quad \text{or} \quad \frac{1}{11}\begin{bmatrix} 2 & 5 \\ 1 & -3 \end{bmatrix}.$$

We solve the matrix equation $AX = B$ as follows:

$$A^{-1}(A \cdot X) = A^{-1} \cdot B \qquad \text{Multiplying by } A^{-1} \text{ on the left on each side}$$

$$(A^{-1} \cdot A) \cdot X = A^{-1} \cdot B \qquad \text{Associative law}$$

$$I \cdot X = A^{-1} \cdot B \qquad \text{Since } A^{-1} \cdot A = I$$

$$X = A^{-1} \cdot B. \qquad \text{Since } I \cdot X = X$$

Substituting, we now have

$$X = A^{-1} \cdot B$$

$$\begin{bmatrix} x \\ y \end{bmatrix} = \frac{1}{11}\begin{bmatrix} 2 & 5 \\ 1 & -3 \end{bmatrix}\begin{bmatrix} -1 \\ 4 \end{bmatrix}$$

$$= \frac{1}{11}\begin{bmatrix} 18 \\ -13 \end{bmatrix}$$

$$= \begin{bmatrix} \frac{18}{11} \\ -\frac{13}{11} \end{bmatrix}.$$

The solution of the system of equations is $(\frac{18}{11}, -\frac{13}{11})$, or $x = \frac{18}{11}$ and $y = -\frac{13}{11}$.

DO EXERCISE 5.

5. Consider the system

$$4x - 2y = -1,$$
$$x + 5y = 1.$$

a) Write a matrix equation equivalent to the system.

b) Find the coefficient matrix A.

c) Find A^{-1}.

d) Use the inverse of the coefficient matrix to solve the system.

●**EXERCISE SET** **9.6**

Find A^{-1}, if it exists. Use the Gauss–Jordan reduction method. Check your answers by calculating AA^{-1} and $A^{-1}A$.

1. $A = \begin{bmatrix} 3 & 2 \\ 5 & 3 \end{bmatrix}$

2. $A = \begin{bmatrix} 3 & 5 \\ 1 & 2 \end{bmatrix}$

3. $A = \begin{bmatrix} 11 & 3 \\ 7 & 2 \end{bmatrix}$

4. $A = \begin{bmatrix} 8 & 5 \\ 5 & 3 \end{bmatrix}$

5. $A = \begin{bmatrix} 4 & -3 \\ 1 & 2 \end{bmatrix}$

6. $A = \begin{bmatrix} 0 & -1 \\ 1 & 0 \end{bmatrix}$

7. $A = \begin{bmatrix} 3 & 1 & 0 \\ 1 & 1 & 1 \\ 1 & -1 & 2 \end{bmatrix}$

8. $A = \begin{bmatrix} 1 & 0 & 1 \\ 2 & 1 & 0 \\ 1 & -1 & 1 \end{bmatrix}$

9. $A = \begin{bmatrix} 1 & -1 & 2 \\ 0 & 1 & 3 \\ 2 & 1 & -2 \end{bmatrix}$

10. $A = \begin{bmatrix} 1 & -1 & 2 \\ 0 & 1 & 2 \\ 1 & -3 & -4 \end{bmatrix}$

11. $A = \begin{bmatrix} 1 & -4 & 8 \\ 1 & -3 & 2 \\ 2 & -7 & 10 \end{bmatrix}$

12. $A = \begin{bmatrix} -2 & 5 & 3 \\ 4 & -1 & 3 \\ 7 & -2 & 5 \end{bmatrix}$

13. $A = \begin{bmatrix} 1 & 2 & 3 & 4 \\ 0 & 1 & 3 & -5 \\ 0 & 0 & 1 & -2 \\ 0 & 0 & 0 & -1 \end{bmatrix}$

14. $A = \begin{bmatrix} -2 & -3 & 4 & 1 \\ 0 & 1 & 1 & 0 \\ 0 & 4 & -6 & 1 \\ -2 & -2 & 5 & 1 \end{bmatrix}$

15. $A = \begin{bmatrix} -2 & 5 & 3 \\ 4 & -1 & 3 \\ 4 & -10 & -6 \end{bmatrix}$

16. $A = \begin{bmatrix} -2 & 6 \\ -1 & 3 \end{bmatrix}$

In Exercises 17–20, a system of equations is given, together with the inverse of the coefficient matrix. Use the matrix inverse to solve the system.

17. $11x + 3y = -4,$
$7x + 2y = 5;$ $\quad A^{-1} = \begin{bmatrix} 2 & -3 \\ -7 & 11 \end{bmatrix}$

18. $8x + 5y = -6,$
$5x + 3y = 2;$ $\quad A^{-1} = \begin{bmatrix} -3 & 5 \\ 5 & -8 \end{bmatrix}$

19. $3x + y = 2,$
$2x - y + 2z = -5,$ $\quad A^{-1} = \dfrac{1}{9} \begin{bmatrix} 3 & 1 & -2 \\ 0 & -3 & 6 \\ -3 & 2 & 5 \end{bmatrix}$
$x + y + z = 5;$

20. $y - z = -4,$
$4x + y = -3,$ $\quad A^{-1} = -\dfrac{1}{2} \begin{bmatrix} 1 & -1 & -1 \\ -3 & 0 & 2 \\ -2 & 1 & 1 \end{bmatrix}$
$3x - y + 3z = 1;$

Write a matrix equation equivalent to the system and find the inverse of the coefficient matrix. Use the inverse of the coefficient matrix to solve each system. Show your work.

21. $4x - 3y = 2,$
$x + 2y = -1$

22. $3x + 5y = -4,$
$2x + 4y = -2$

23. $7x - 2y = -3,$
$9x + 3y = 4$

24. $5x + 3y = -2,$
$4x - y = 1$

25. $x + z = 1,$
$2x + y = 3,$
$x - y + z = 4$

26. $x + 2y + 3z = -1,$
$2x - 3y + 4z = 2,$
$-3x + 5y - 6z = 4$

27. $2w - 3x + 4y - 5z = 0,$
$3w - 2x + 7y - 3z = 2,$
$w + x - y + z = 1,$
$-w - 3x - 6y + 4z = 6$

28. $5w - 4x + 3y - 2z = -6,$
$w + 4x - 2y + 3z = -5,$
$2w - 3x + 6y - 9z = 14,$
$3w - 5x + 2y - 4z = -3$

● **SYNTHESIS**

29. Let
$$A = \begin{bmatrix} a & b & c \\ d & e & f \\ g & h & i \end{bmatrix} \quad \text{and} \quad I = \begin{bmatrix} 1 & 0 & 0 \\ 0 & 1 & 0 \\ 0 & 0 & 1 \end{bmatrix}.$$

Show that $AI = IA = A$.

State the conditions under which A^{-1} exists. Then find a formula for A^{-1}.

30. $A = [x]$

31. $A = \begin{bmatrix} x & 0 \\ 0 & y \end{bmatrix}$

32. $A = \begin{bmatrix} 0 & 0 & x \\ 0 & y & 0 \\ z & 0 & 0 \end{bmatrix}$

33. $A = \begin{bmatrix} x & 1 & 1 & 1 \\ 0 & y & 0 & 0 \\ 0 & 0 & z & 0 \\ 0 & 0 & 0 & w \end{bmatrix}$

● **CHALLENGE**

34. Consider
$$a_1 x + b_1 y = c_1,$$
$$a_2 x + b_2 y = c_2.$$

Find a general formula for the solution, and use it to prove Cramer's rule.

OBJECTIVES

You should be able to:

A. Determine whether an ordered pair of numbers is a solution of an inequality in two variables.

B. Graph linear inequalities.

C. Graph systems of linear inequalities.

D. Solve linear programming problems.

9.7

SYSTEMS OF INEQUALITIES AND LINEAR PROGRAMMING

A **graph** of an inequality is a drawing that represents its solutions. An inequality in one variable can be graphed on a number line. An inequality in two variables can be graphed on a coordinate plane.

 Solutions of Inequalities in Two Variables

The solutions of inequalities in two variables are ordered pairs.

Example 1 Determine whether $(-3, 2)$ is a solution of $5x - 4y \le 13$.

Solution   We replace x by -3 and y by 2.

$$
\begin{array}{c|c}
5x - 4y \le 13 \\
\hline
5(-3) - 4 \cdot 2 & 13 \\
-15 - 8 & \\
-23 & \text{TRUE}
\end{array}
$$

Since $-23 \le 13$ is true, $(-3, 2)$ is a solution.   ◀

Example 2   Determine whether $(6, -7)$ is a solution of $5x - 4y \le 13$.

Solution   We replace x by 6 and y by -7.

$$
\begin{array}{c|c}
5x - 4y \le 13 \\
\hline
5(6) - 4(-7) & 13 \\
30 + 28 & \\
58 & \text{FALSE}
\end{array}
$$

Since $58 \le 13$ is false, $(6, -7)$ is not a solution.   ◀

DO EXERCISES 1 AND 2.

B▷ Graphs of Linear Inequalities in Two Variables

A **linear inequality** is an inequality that is equivalent to

$$Ax + By < C, \quad Ax + By \le C, \quad Ax + By > C, \quad \text{or} \quad Ax + By \ge C.$$

That is, there is a first-degree polynomial on one side and a constant on the other. To graph a linear inequality, we first graph the related equation.

Example 3   Graph: $y < x$.

Solution   We first graph the line $y = x$ for comparison. Every solution of $y = x$ is an ordered pair like $(3, 3)$. The first and second coordinates are the same. The graph of $y = x$ is shown on the left below. We draw it dashed because the points on the line are *not* solutions of $y < x$.

 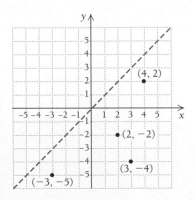

Now look at the graph on the right. Note that each of the ordered pairs plotted on the half-plane below $y = x$ is a solution of $y < x$. We can check a pair $(4, 2)$ as follows:

$$
\begin{array}{c|c}
y < x \\
\hline
2 & 4 \quad \text{TRUE}
\end{array}
$$

 1. Determine whether $(1, -4)$ is a solution of the inequality $4x - 5y \ge 12$.

2. Determine whether $(4, -3)$ is a solution of the inequality $3y - 2x < -60$.

It turns out that any point on the same side of $y = x$ as $(4, 2)$ is also a solution. Thus, if you know that one point in a half-plane is a solution, then all points in that half-plane are solutions. In this text, we will usually indicate those solutions by color shading. We shade the half-plane below $y = x$.

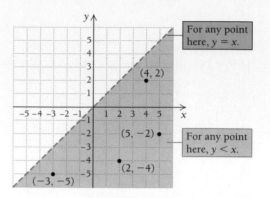

For any point here, $y = x$.

(4, 2)

(5, −2)

For any point here, $y < x$.

(2, −4)

(−3, −5)

Example 4   Graph: $8x + 3y \geq 24$.

Solution   First we sketch the line $8x + 3y = 24$. Points on the line $8x + 3y = 24$ are also in the graph of $8x + 3y \geq 24$, so we draw the line solid. This indicates that all points on the line are solutions. The rest of the solutions are either in the half-plane above the line or the half-plane below the line. To determine which, we select a point that is not on the line and determine whether it is a solution of $8x + 3y \geq 24$. We try $(-3, 4)$ as a test point:

$$
\begin{array}{c|c}
\multicolumn{2}{c}{8x + 3y \geq 24} \\
\hline
8(-3) + 3(4) & 24 \\
-24 + 12 & \\
-12 & \text{FALSE}
\end{array}
$$

We see that $-12 \geq 24$ is *false*. Since $(-3, 4)$ is not a solution, none of the points in the half-plane containing $(-3, 4)$ is a solution. Thus the points in the opposite half-plane are solutions. We shade that half-plane and obtain the graph:

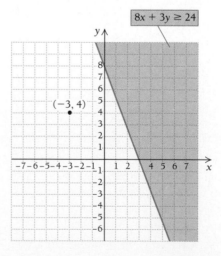

$8x + 3y \geq 24$

(−3, 4)

We graph linear inequalities as follows.

To Graph an Inequality in Two Variables

1. Replace the inequality symbol with an equals sign and graph this related equation.
2. If the inequality symbol is $<$ or $>$, draw the line dashed. If the inequality symbol is $\leq$ or $\geq$, draw the line solid.
3. The graph consists of a half-plane that is either above or below or left or right of the line and, if the line is solid, the line as well. To determine which half-plane to shade, choose a point not on the line as a test point. Substitute to determine whether that point is a solution. If it is, shade the half-plane containing that point. If not, shade the opposite half-plane.

Example 5   Graph: $6x - 2y < 12$.

Solution

1. We first graph the related equation $6x - 2y = 12$.
2. Since the inequality uses the symbol $<$, points on the line are not solutions of the inequality, so we draw a dashed line.
3. To determine which half-plane to shade, we consider a test point *not* on the line. We try $(0, 0)$ and substitute:

$$\frac{6x - 2y < 12}{\begin{array}{c|c} 6(0) - 2(0) & 12 \\ 0 - 0 & \\ 0 & \text{TRUE} \end{array}}$$

Since the inequality $0 < 12$ is *true,* the point $(0, 0)$ is a solution; each point in the half-plane containing $(0, 0)$ is a solution. Thus each point in the opposite half-plane is *not* a solution. The graph is shown below.

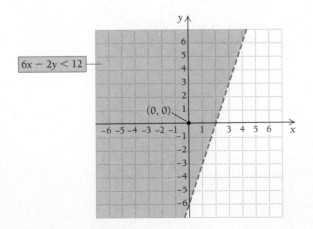

DO EXERCISES 3 AND 4.

Example 6   Graph $x > -3$ on a plane.

Solution   There is a missing variable in this inequality. If we graph the

Graph.

3. $6x - 3y < 18$

4. $4x + 3y \geq 12$

Graph on a plane.

5. $x < 3$

6. $y \geq -4$

inequality on a line, its graph is as follows:

However, we can also write this inequality as $x + 0y > -3$ and consider graphing it in the plane. We use the same technique that we have used with the other examples. We first graph the related equation $x = -3$ in the plane. We draw the boundary with a dashed line.

The graph is a half-plane to the right or left of the line $x = -3$. To determine which, we consider a test point, $(2, 5)$:

$$\frac{x + 0y > -3}{\begin{array}{c|c} 2 + 0(5) & -3 \\ 2 & \end{array}} \quad \text{TRUE}$$

Since $(2, 5)$ is a solution, all the pairs in the half-plane containing $(2, 5)$ are solutions. We shade that half-plane.

We see that the solutions of $x > -3$ are all those ordered pairs whose first coordinates are greater than -3. ◄

Example 7 Graph $y \leq 4$ on a plane.

Solution We first graph $y = 4$ using a solid line. We then use $(2, -3)$ as a test point and substitute:

$$\frac{0x + y \leq 4}{\begin{array}{c|c} 0(2) + (-3) & 4 \\ -3 & \end{array}} \quad \text{TRUE}$$

Since $(2, -3)$ is a solution, all points in the half-plane containing $(2, -3)$ are solutions. Note that this half-plane consists of all ordered pairs whose second coordinate is less than or equal to 4.

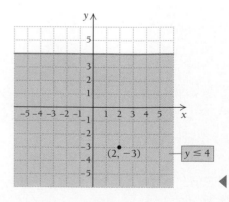

DO EXERCISES 5 AND 6.

Systems of Linear Inequalities

The following is an example of a system of two linear inequalities in two variables:

$$x + y \leq 4,$$
$$x - y < 4.$$

A **solution** of a system of linear inequalities is an ordered pair that is a solution of *both* inequalities. We now graph solutions of systems of linear inequalities. To do so, we graph each inequality and determine where the graphs overlap, or intersect.

Example 8   Graph the solutions of the system

$$x + y \leq 4,$$
$$x - y < 4.$$

Solution   We graph the inequality $x + y \leq 4$ by first graphing the equation $x + y = 4$ using a solid line. We consider $(0, 0)$ as a test point and find that it is a solution, so we shade all points on that side of the line using blue shading. The arrows at the ends of the line also indicate the half-plane that contains the solutions.

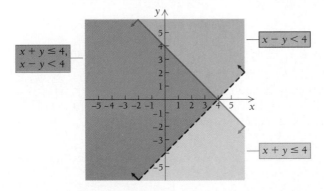

Next we graph $x - y < 4$. We begin by graphing the equation $x - y = 4$ using a dashed line and consider $(0, 0)$ as a test point. Again, $(0, 0)$ is a solution so we shade that side of the line using gray shading. The solution set of the system is the region that is shaded both blue and gray and part of the line $x + y = 4$.   ◀

DO EXERCISE 7.

Example 9   Graph: $-2 < x \leq 5$.

Solution   This is actually a conjunction, or system of inequalities:

$$-2 < x \quad and \quad x \leq 5.$$

We graph the equation $-2 = x$ and see that the graph of the first inequality is the half-plane to the right of the line $-2 = x$ (see the gray-shaded graph on the left below).

We graph the second inequality, starting with the line $x = 5$, and find that its graph is the line and also the half-plane to the left of it (see the blue-shaded graph on the right below).

7. Graph:

$$x + y \geq 1,$$
$$y - x \geq 2.$$

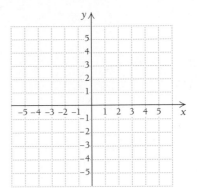

8. Graph:

$$-3 \le y < 4.$$

We shade the intersection of these graphs.

DO EXERCISE 8.

A system of inequalities may have a graph that consists of a polygon and its interior. In *linear programming,* which you will study later in this section, it is important to be able to find the vertices of such a polygon.

Example 10 Graph the following system of inequalities. Find the coordinates of any vertices formed.

$$6x - 2y \le 12, \tag{1}$$
$$y - 3 \le 0, \tag{2}$$
$$x + y \ge 0 \tag{3}$$

Solution We graph the lines $6x - 2y = 12$, $y - 3 = 0$, and $x + y = 0$ using solid lines. The regions for each inequality are indicated by the arrows at the ends of the lines. We then note where the regions overlap and shade the region of solutions using blue shading.

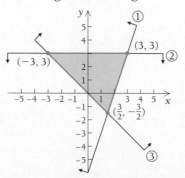

To find the vertices, we solve three different systems of equations. The system of equations from inequalities (1) and (2) is

$$6x - 2y = 12,$$
$$y - 3 = 0.$$

Solving, we obtain the vertex $(3, 3)$.

The system of equations from inequalities (1) and (3) is

$$6x - 2y = 12,$$
$$x + y = 0.$$

Solving, we obtain the vertex $(\frac{3}{2}, -\frac{3}{2})$.

The system of equations from inequalities (2) and (3) is

$$y - 3 = 0,$$
$$x + y = 0.$$

Solving, we obtain the vertex $(-3, 3)$.   ◀

DO EXERCISE 9.

Example 11   Graph the following system of inequalities. Find the coordinates of any vertices formed.

$$x + y \leq 16, \tag{1}$$
$$3x + 6y \leq 60, \tag{2}$$
$$x \geq 0, \tag{3}$$
$$y \geq 0 \tag{4}$$

Solution   We graph each inequality using solid lines. The regions for each inequality are indicated by the arrows at the ends of the lines. We then note where the regions overlap and shade the region of solutions using blue shading.

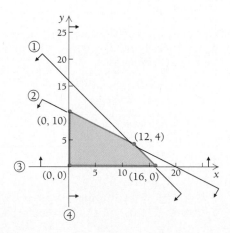

To find the vertices, we solve four different systems of equations. The system of equations from inequalities (1) and (2) is

$$x + y = 16,$$
$$3x + 6y = 60.$$

Solving, we obtain the vertex $(12, 4)$.

9. Graph the system of inequalities. Find the coordinates of any vertices formed.

$$5x + 6y \leq 30,$$
$$0 \leq y \leq 3,$$
$$0 \leq x \leq 4$$

10. Graph the system of inequalities. Find the coordinates of any vertices formed.

$$2x + 4y \leq 8,$$
$$x + y \leq 3,$$
$$x \geq 0,$$
$$y \geq 0$$

The system of equations from inequalities (1) and (4) is

$$x + y = 16,$$
$$y = 0.$$

Solving, we obtain the vertex $(16, 0)$.

The system of equations from inequalities (3) and (4) is

$$x = 0,$$
$$y = 0.$$

The vertex is obviously $(0, 0)$.

The system of equations from inequalities (2) and (3) is

$$3x + 6y = 60,$$
$$x = 0.$$

Solving, we obtain the vertex $(0, 10)$.

DO EXERCISE 10.

▷ An Application: Linear Programming

There are many problems in real life in which we want to find a greatest value (a maximum) or a least value (a minimum). For example, if you are in business, you would like to know how to make the *most* profit. Or you might like to know how to make your expenses the *least* possible. Many such problems can be solved using systems of inequalities and a branch of mathematics known as **linear programming**. The following is the basic theorem of linear programming.

THEOREM 12	**Linear Programming**

Suppose a linear function of two variables $F = ax + by + c$ is defined for the pairs (x, y) that are the solutions of a system of linear inequalities. Then maximum or minimum values of the function will occur at certain vertices. To find the maximum or minimum:

1. Graph the system of inequalities and find the vertices.
2. Compute the function values at the vertices. The largest and smallest of those values are the maximum and the minimum of the function.

This theorem was proven during World War II. Linear programming was developed then to deal with the complicated process of shipping personnel and supplies to Europe. Let us consider an example.

Example 12   You are taking a test in which items of type A are worth 10 points and items of type B are worth 15 points. It takes 3 minutes for each item of type A and 6 minutes for each item of type B. The total time allowed is 60 minutes and you are not allowed to answer more than 16 questions. Assuming that all your answers are correct, how many items of each type should you answer to get the best score?

Solution   Let x = the number of items of type A and y = the number of items of type B. The total score T is a linear function of the two variables x and y:

$$T = 10x + 15y.$$

This function has a *domain* that is a set of ordered pairs of numbers (x, y). This domain is determined by the following inequalities, also called *constraints:*

Total number of questions allowed, not more than 16	$x + y \leq 16,$
Total amount of time, not more than 60 min	$3x + 6y \leq 60,$
Numbers of items answered will not be negative	$x \geq 0,$
	$y \geq 0.$

We graphed this system of inequalities in Example 11. The coordinates of the vertices are $(0, 0)$, $(16, 0)$, $(12, 4)$, and $(0, 10)$. We know by Theorem 12 that the maximum and minimum values occur at certain vertices of the polygon. All we need do to find these values is to substitute the coordinates of the vertices in $T = 10x + 15y$.

Vertices (x, y)	Score $T = 10x + 15y$	
$(0, 0)$	$T = 10(0) + 15(0) = 0$	⟵ Minimum
$(16, 0)$	$T = 10(16) + 15(0) = 160$	
$(12, 4)$	$T = 10(12) + 15(4) = 180$	⟵ Maximum
$(0, 10)$	$T = 10(0) + 15(10) = 150$	

From the table, we see that the minimum value is 0 and the maximum is 180. To get this maximum, you must answer 12 items of type A and 4 items of type B.   ◀

Example 13   Find the maximum and minimum values of $F = 9x + 40y$ subject to the constraints

$$y - x \geq 1,$$
$$y - x \leq 3,$$
$$2 \leq x \leq 5.$$

Solution   We graph the system of inequalities, determine the vertices, and find the function values for those ordered pairs.

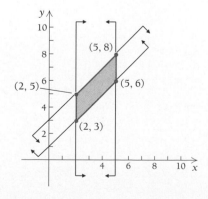

11. Find the maximum and minimum values of

$$F = 34x + 6y$$

subject to

$$x + y \leq 6,$$
$$x + y \geq 1,$$
$$1 \leq x \leq 3.$$

Vertices (x, y)	Score $F = 9x + 40y$	
(2, 3)	$F = 9(2) + 40(3) = 138$	◄──── Minimum
(2, 5)	$F = 9(2) + 40(5) = 218$	
(5, 6)	$F = 9(5) + 40(6) = 285$	
(5, 8)	$F = 9(5) + 40(8) = 365$	◄──── Maximum

The maximum value of F is 365 when $x = 5$ and $y = 8$. The minimum value of F is 138 when $x = 2$ and $y = 3$. ◄

DO EXERCISES 11 AND 12.

12. Find the maximum and minimum values of

$$G = 3x - 5y + 27$$

subject to

$$x + 2y \leq 8,$$
$$0 \leq y \leq 3,$$
$$0 \leq x \leq 6.$$

Example 14 A company manufactures motorcycles and bicycles. To stay in business, it must produce at least 10 motorcycles each month, but it does not have the facilities to produce more than 60 motorcycles. It also does not have the facilities to produce more than 120 bicycles. The total production of motorcycles and bicycles cannot exceed 160. The profit on a motorcycle is $134 and on a bicycle is $20. Find the number of each that should be manufactured in order to maximize profit.

Solution Let $x =$ the number of motorcycles to be produced and $y =$ the number of bicycles to be produced. The profit P is given by

$$P = \$134x + \$20y$$

subject to the constraints

$$10 \leq x \leq 60,$$
$$0 \leq y \leq 120,$$
$$x + y \leq 160.$$

We graph the system of inequalities, determine the vertices, and find the function values for those ordered pairs.

13. A college snack bar cooks and sells hamburgers and hot dogs during the lunch hour. To stay in business, it must sell at least 10 hamburgers but cannot cook more than 40. It must also sell at least 30 hot dogs but cannot cook more than 70. It cannot cook more than 90 sandwiches altogether. The profit is $1.83 on a hamburger and $1.21 on a hot dog. How many of each kind of sandwich should they sell in order to make the maximum profit? What is the maximum profit?

Vertices (x, y)	Score $P = \$134x + \$20y$
(10, 0)	$ 1,340
(60, 0)	$ 8,040
(60, 100)	$10,040
(40, 120)	$ 7,760
(10, 120)	$ 3,740

Thus the company will make a maximum profit of $10,040 by producing 60 motorcycles and 100 bicycles. ◀

DO EXERCISE 13 ON THE PRECEDING PAGE. _____

●EXERCISE SET **9.7**

A Determine whether the ordered pair is a solution of the inequality.

1. $(-4, 2)$; $2x + y > -5$

2. $(3, -6)$; $4x + 2y \leq 0$

3. $(8, 14)$; $2y - 3x \geq 5$

4. $(7, 20)$; $3x - y < -1$

B Graph the inequality on a plane, using graph paper.

5. $y > 2x$

6. $2y < x$

7. $y + x \geq 0$

8. $y + x < 0$

9. $y > x - 3$

10. $y \leq x + 4$

11. $x + y < 4$

12. $x - y \geq 5$

13. $3x - 2y \leq 6$

14. $2x - 5y < 10$

15. $3y + 2x \geq 6$

16. $2y + x \leq 4$

17. $3x - 2 \leq 5x + y$

18. $2x - 6y \geq 8 + 2y$

19. $x < -4$

20. $y \geq 5$

21. $y > -3$

22. $x \leq 5$

23. $-4 < y < -1$

24. $-1 < y < 4$

25. $-4 \leq x \leq 4$

26. $-3 \leq x \leq 3$

27. $y \geq |x|$

28. $y \leq |x|$

C Graph the system of inequalities. Find the coordinates of any vertices formed.

29. $y \leq x$,
       $y \geq 3 - x$

30. $y \geq x$,
       $y \leq x - 5$

31. $y \geq x$,
       $y \leq x - 4$

32. $y \geq x$,
       $y \leq 2 - x$

33. $y \geq -3$,
       $x \geq 1$

34. $y \leq -2$,
       $x \geq 2$

35. $x \leq 3$,
       $y \geq 2 - 3x$

36. $x \geq -2$,
       $y \leq 3 - 2x$

37. $x + y \leq 1$,
       $x - y \leq 2$

38. $y + 3x \geq 0$,
       $y + 3x \leq 2$

39. $2y - x \leq 2$,
       $y + 3x \geq -1$

40. $y \leq 2x + 1$,
       $y \geq -2x + 1$,
       $x \leq 2$

41. $x - y \leq 2$,
       $x + 2y \geq 8$,
       $y \leq 4$

42. $x + 2y \leq 12$,
       $2x + y \leq 12$,
       $x \geq 0$,
       $y \geq 0$

43. $4y - 3x \geq -12$,
       $4y + 3x \geq -36$,
       $y \leq 0$,
       $x \leq 0$

44. $8x + 5y \leq 40$,
       $x + 2y \leq 8$,
       $x \geq 0$,
       $y \geq 0$

45. $3x + 4y \geq 12$,
       $5x + 6y \leq 30$,
       $1 \leq x \leq 3$

46. $y - x \geq 1$,
       $y - x \leq 3$,
       $2 \leq x \leq 5$

D Find the maximum and minimum values of the function and the values of x and y for which they occur.

47. $P = 17x - 3y + 60$, subject to
       $6x + 8y \leq 48$,
       $0 \leq y \leq 4$,
       $0 \leq x \leq 7$.

48. $Q = 28x - 4y + 72$, subject to
       $5x + 4y \geq 20$,
       $0 \leq y \leq 4$,
       $0 \leq x \leq 3$.

49. $F = 5x + 36y$, subject to
       $5x + 3y \leq 34$,
       $3x + 5y \leq 30$,
       $x \geq 0$,
       $y \geq 0$.

50. $G = 16x + 14y$, subject to
       $3x + 2y \leq 12$,
       $7x + 5y \leq 29$,
       $x \geq 0$,
       $y \geq 0$.

51. The Hockeypuck Biscuit Factory makes two types of biscuits, Biscuit Jumbos and Mitimite Biscuits. The oven can cook at most 200 biscuits per day. Jumbos each require 2 oz of flour, Mitimites require 1 oz of flour, and there is at most 300 oz of flour available. The income from each Jumbo is $0.10 and from each Mitimite is $0.08. How many of each type of biscuit should be made in order to maximize income? What is the maximum income?

52. A student owns a car and a moped. The student has at most 12 gal of gasoline to be used between the car and the moped. The car's tank holds at most 10 gal and the moped's 3 gal. The mileage is 20 mpg for the car and 100 mpg for the moped. How many gallons of gasoline should each vehicle use if the student wants to travel as far as possible? What is the maximum number of miles?

53. You are about to take a test that contains questions of type A worth 10 points and questions of type B worth 25 points. You must do at least 3 questions of type A but time restricts doing more than 12. You must do at least 4 questions of type B but time restricts doing more than 15. You can do no more than 20 questions in total. How many of each type of question must you do to maximize your score? What is this maximum score?

54. You are about to take a test that contains questions of type A worth 4 points and questions of type B worth 7 points. You must do at least 5 questions of type A but time restricts doing more than 10. You must do at least 3 questions of type B but time restricts doing more than 10. You can do no more than 18 questions in total. How many of each type of question must you do to maximize your score? What is this maximum score?

55. A lumber company can convert logs into either lumber or plywood. In a given week, the mill can turn out 400 units of production, of which 100 units of lumber and 150 units of plywood are required by regular customers. The profit is $20 on a unit of lumber and $30 on a unit of plywood. How many units of each type should the mill produce in order to maximize the profit?

56. A farm consists of 240 acres of cropland. The farmer wishes to plant this acreage in corn or oats. Profit per acre in corn production is $40 and in oats is $30. An additional restriction is that the total number of hours of labor during the production period is 320. Each acre of land in corn production uses 2 hours of labor during the production period, while production of oats requires 1 hour per acre. Find how the land should be divided between corn and oats in order to give maximum profit.

57. A woman is planning to invest up to $40,000 in corporate or municipal bonds, or both. The least she is allowed to invest in corporate bonds is $6000, and she does not want to invest more than $22,000 in corporate bonds. She also does not want to invest more than $30,000 in municipal bonds. The interest is 8% on corporate bonds and $7\frac{1}{2}$% on municipal bonds. This is simple interest for one year. How much should she invest in each type of bond to maximize her income? What is the maximum income?

58. A man is planning to invest up to $22,000 in bank X or bank Y, or both. He wants to invest at least $2000 but no more than $14,000 in bank X. Bank Y does not insure more than a $15,000 investment, so he will invest no more than that in bank Y. The interest is 6% in bank X and $6\frac{1}{2}$% in bank Y. This is simple interest for one year. How much should he invest in each bank in order to maximize his income? What is the maximum income?

59. A pipe tobacco company has 3000 lb of English tobacco, 2000 lb of Virginia tobacco, and 500 lb of Latakia tobacco. To make one batch of Smello tobacco, it takes 12 lb of English tobacco and 4 lb of Latakia. To make one batch of Roppo tobacco, it takes 8 lb of English and 8 lb of Virginia tobacco. The profit is $10.56 per batch for Smello

and $6.40 for Roppo. How many batches of each kind of tobacco should be made to yield maximum profit? What is the maximum profit? (*Hint:* Organize the information in a table.)

60. It takes a tailoring firm 2 hr of cutting and 4 hr of sewing to make a knit suit. To make a worsted suit, it takes 4 hr of cutting and 2 hr of sewing. At most 20 hr per day are available for cutting and at most 16 hr per day are available for sewing. The profit is $34 on a knit suit and $31 on a worsted suit. How many of each kind of suit should be made in order to maximize profit?

● **SYNTHESIS** _____

Graph the system.

61. $y \geq x^2 - 2$,
$y \leq 2 - x^2$

62. $y < x + 1$,
$y \geq x^2$

63. *Hockey wins and losses.* A hockey team figures that it needs at least 60 points for the season to make the play-offs. A win w is worth 2 points and a tie t is worth 1 point. Find a system of inequalities that describes the situation. Graph the system.

64. *Elevators.* Many elevators have a capacity of 1 metric ton (1000 kg). An elevator contains c children, each weighing 35 kg, and a adults, each weighing 75 kg. Find a system of inequalities that asserts that the elevator is overloaded. Graph the system.

65. *Widths of a basketball floor.* Sizes of basketball floors vary due to building sizes and other constraints such as cost. The length L is to be at most 74 ft and the width is to be at most 50 ft. Find a system of inequalities that describes the perimeter of a basketball floor. Graph the system.

● **CHALLENGE** _____

66. $|x| + |y| \leq 1$

67. $|x + y| \leq 1$

68. $|x - y| > 0$

69. $|x| > |y|$

70. *Allocation of resources in a manufacturing process.* A furniture manufacturer produces chairs and sofas. The chairs require 20 ft of wood, 1 lb of foam rubber, and 2 sq yd of material. The sofas require 100 ft of wood, 50 lb of foam rubber, and 20 sq yd of material. The manufacturer has in stock 1900 ft of wood, 500 lb of foam rubber, and 240 sq yd of material. The chairs can be sold for $20 each and the sofas for $300 each. How many of each should be produced in order to maximize the income?

▓ **TECHNOLOGY CONNECTION** _____

Use a grapher to draw the graph of each linear inequality in the given system. Then use the trace and zoom features to find the coordinates of the vertices. Round to two decimal places.

71. $3.5x + 5.8y \leq 10$
$2.1x - 12.3y \geq 9$
$x \geq -2.3$

72. $20x - 45y < 105$
$12.5x + 13y > 10$
$0.7x - 7y > -35$

9.8
PARTIAL FRACTIONS

A There are situations in calculus in which it is helpful to do the reverse of adding rational expressions, that is, to **decompose** a rational expression into a sum of several rational expressions. Consider the following equation:

$$\frac{3}{x+2} + \frac{-2}{2x-3} = \frac{4x-13}{2x^2+x-6}.$$

Adding is straightforward. The present problem is to work backwards, taking the right-hand side and finding the rational expressions, or **partial fractions,** that were added to get it. This procedure is somewhat complicated and often makes use of systems of equations.

> **To Decompose a Rational Expression as a Sum of Partial Fractions**
>
> Consider any rational expression $P(x)/Q(x)$, with no common factor other than 1 or -1.
>
> *Case 1: The degree of $P(x)$ is greater than or equal to the degree of $Q(x)$. Then divide and carry out Case 2 with the remainder term.*
>
> *Case 2: The degree of $P(x)$ is less than the degree of $Q(x)$. Then use the following procedure.*
>
> a) *Find the factors of $Q(x)$.* Any polynomial can be factored into linear and quadratic factors. In practice, these factors may be hard to find, but any of the concepts discussed in Chapter 4 can be used. The factors then have the general forms $(px+q)^n$ and $(ax^2+bx+c)^m$. The quadratic factor ax^2+bx+c is presumed to be *irreducible*, meaning that it cannot be factored into linear factors with real coefficients.
>
> b) Assign to each linear factor $(px+q)^n$ the sum of n partial fractions:
>
> $$\frac{A_1}{(px+q)} + \frac{A_2}{(px+q)^2} + \cdots + \frac{A_n}{(px+q)^n}.$$
>
> c) Assign to each quadratic factor $(ax^2+bx+c)^m$ the sum of m partial fractions:
>
> $$\frac{B_1x+C_1}{ax^2+bx+c} + \frac{B_2x+C_2}{(ax^2+bx+c)^2} + \cdots + \frac{B_mx+C_m}{(ax^2+bx+c)^m}.$$

We then apply algebraic methods, as shown in the following examples. This procedure covers all situations, because any polynomial $Q(x)$ with real coefficients can be factored into linear and quadratic factors. We assume that $P(x)/Q(x)$ is simplified. For our first example, we decompose the rational expression in the opening paragraph.

Example 1 Decompose into partial fractions:

$$\frac{4x-13}{2x^2+x-6}.$$

1. Decompose into partial fractions:

$$\frac{13x + 5}{3x^2 - 7x - 6}.$$

Solution The degree of the numerator is less than the degree of the denominator. We begin by factoring the denominator: $(x + 2)(2x - 3)$. We know that there are constants A and B such that

$$\frac{4x - 13}{(x + 2)(2x - 3)} = \frac{A}{x + 2} + \frac{B}{2x - 3}.$$

To determine A and B, we add the expressions on the right:

$$\frac{4x - 13}{(x + 2)(2x - 3)} = \frac{A(2x - 3) + B(x + 2)}{(x + 2)(2x - 3)}.$$

Next, we equate the numerators:

$$4x - 13 = A(2x - 3) + B(x + 2).$$

Since the latter equation containing A and B is true for all x, we can substitute any value of x whatever and still have a true equation. We let $2x - 3 = 0$, or $x = \frac{3}{2}$. This gives us

$$4\left(\frac{3}{2}\right) - 13 = A\left(2 \cdot \frac{3}{2} - 3\right) + B\left(\frac{3}{2} + 2\right)$$

$$-7 = 0 + \frac{7}{2}B.$$

Solving, we obtain $B = -2$. Next we let $x + 2 = 0$, or $x = -2$, which gives us

$$4(-2) - 13 = A[2(-2) - 3] + B(-2 + 2).$$

Solving, we obtain $A = 3$.

The decomposition is as follows:

$$\frac{3}{x + 2} + \frac{-2}{2x - 3}, \quad \text{or} \quad \frac{3}{x + 2} - \frac{2}{2x - 3}.$$

To check, we can add to see if we get the original expression. ◀

The values A and B can also be determined with a system of equations resulting from equating the corresponding coefficients. The equation

$$4x - 13 = A(2x - 3) + B(x + 2)$$

can be written as the equation

$$4x - 13 = (2A + B)x + (2B - 3A).$$

Using this form, we can equate the coefficients and solve the resulting system:

$$4 = 2A + B \quad \text{and} \quad -13 = 2B - 3A.$$

We will consider this in more depth later.

DO EXERCISE 1.

Example 2 Decompose into partial fractions:

$$\frac{7x^2 - 29x + 24}{(2x - 1)(x - 2)^2}.$$

Solution The degree of the numerator is less than the degree of the denominator. The decomposition looks like the following:

$$\frac{A}{2x-1} + \frac{B}{x-2} + \frac{C}{(x-2)^2}.$$

As in Example 1, we add and equate the numerators. This gives us

$$7x^2 - 29x + 24 = A(x-2)^2 + B(2x-1)(x-2) + C(2x-1). \qquad (1)$$

Since the equation containing A, B, and C is true for all x, we can substitute any value of x whatever and still have a true equation. We let x be such that $2x - 1 = 0$, or $x = \frac{1}{2}$. This gives us

$$7\left(\frac{1}{2}\right)^2 - 29 \cdot \frac{1}{2} + 24 = A\left(\frac{1}{2} - 2\right)^2 + 0.$$

Solving, we obtain $A = 5$. Next, we let $x - 2 = 0$, or $x = 2$. Substituting gives us

$$7(2)^2 - 29(2) + 24 = 0 + C(2 \cdot 2 - 1).$$

Solving, we obtain $C = -2$.

To find B, we first simplify equation (1):

$$\begin{aligned} 7x^2 - 29x + 24 &= A(x^2 - 4x + 4) + B(2x^2 - 5x + 2) + C(2x - 1) \\ &= Ax^2 - 4Ax + 4A + 2Bx^2 - 5Bx + 2B + 2Cx - C \\ &= (A + 2B)x^2 + (-4A - 5B + 2C)x + (4A + 2B - C). \end{aligned}$$

Then we equate the coefficients of x^2:

$$7 = A + 2B.$$

Substituting 5 for A and solving for B gives us $B = 1$. The decomposition is as follows:

$$\frac{5}{2x-1} + \frac{1}{x-2} - \frac{2}{(x-2)^2}. \qquad \blacktriangleleft$$

DO EXERCISE 2.

Example 3   Decompose into partial fractions:

$$\frac{x^2 - 17x + 35}{(x^2 + 1)(x - 4)}.$$

Solution   The decomposition looks like the following:

$$\frac{Ax + B}{x^2 + 1} + \frac{C}{x - 4}.$$

Adding and equating numerators, we get

$$x^2 - 17x + 35 = (Ax + B)(x - 4) + C(x^2 + 1). \qquad (2)$$

Letting $x = 4$, we get

$$4^2 - 17 \cdot 4 + 35 = 0 + C(4^2 + 1).$$

Solving, we obtain $C = -1$.

2. Decompose into partial fractions:

$$\frac{3x^2 - 3x - 2}{(x + 1)(x - 1)^2}.$$

3. Decompose into partial fractions:

$$\frac{2x^2 + 4x + 5}{(x^2 + 1)(x + 2)}.$$

4. Decompose into partial fractions:

$$\frac{6x^3 + 29x^2 - 8x + 18}{2x^2 + 9x - 5}.$$

To find A and B, we first simplify equation (2):

$$x^2 - 17x + 35 = Ax^2 + Bx - 4Ax - 4B + Cx^2 + C$$
$$= (A + C)x^2 + (B - 4A)x + (-4B + C).$$

Equating the coefficients of x^2, we get $1 = A + C$. Since $C = -1$, we know that $A = 2$. Equating the constant terms, we get $35 = -4B + C$. This gives us $B = -9$. The decomposition is as follows:

$$\frac{2x - 9}{x^2 + 1} - \frac{1}{x - 4}. \quad \blacktriangleleft$$

DO EXERCISE 3. ────────────────────

Example 4   Decompose into partial fractions:

$$\frac{6x^3 + 5x^2 - 7}{3x^2 - 2x - 1}.$$

Solution   The degree of the numerator is greater than that of the denominator. Thus we divide and find an equivalent expression:

$$\require{enclose}
\begin{array}{r}
2x + 3 \\
3x^2 - 2x - 1 \enclose{longdiv}{6x^3 + 5x^2 - 7} \\
\underline{6x^3 - 4x^2 - 2x } \\
9x^2 + 2x - 7 \\
\underline{9x^2 - 6x - 3} \\
8x - 4
\end{array}$$

The original expression is thus equivalent to

$$2x + 3 + \frac{8x - 4}{3x^2 - 2x - 1}.$$

We decompose the fraction to get

$$\frac{8x - 4}{(3x + 1)(x - 1)} = \frac{5}{3x + 1} + \frac{1}{x - 1}.$$

The final result is

$$2x + 3 + \frac{5}{3x + 1} + \frac{1}{x - 1}. \quad \blacktriangleleft$$

DO EXERCISE 4. ────────────────────

We can also use systems of equations to decompose rational expressions into a sum of rational expressions. Let us reconsider Example 2.

Example 5   Decompose into partial fractions:

$$\frac{7x^2 - 29x + 24}{(2x - 1)(x - 2)^2}.$$

Solution   The decomposition looks like the following:

$$\frac{A}{2x - 1} + \frac{B}{x - 2} + \frac{C}{(x - 2)^2}.$$

We first add and equate the numerators:

$$7x^2 - 29x + 24 = A(x-2)^2 + B(2x-1)(x-2) + C(2x-1)$$
$$= A(x^2 - 4x + 4) + B(2x^2 - 5x + 2) + C(2x-1),$$

or

$$7x^2 - 29x + 24 = (A + 2B)x^2 + (-4A - 5B + 2C)x$$
$$+ (4A + 2B - C).$$

Then we equate corresponding coefficients:

$$7 = A + 2B, \qquad \text{The coefficients of the } x^2\text{-terms}$$
$$-29 = -4A - 5B + 2C, \qquad \text{The coefficients of the } x\text{-terms}$$
$$24 = 4A + 2B - C. \qquad \text{The constant terms}$$

We now have a system of three equations. We solve the system to obtain

$$A = 5, \qquad B = 1, \quad \text{and} \quad C = -2.$$

The decomposition is as follows:

$$\frac{5}{2x-1} + \frac{1}{x-2} - \frac{2}{(x-2)^2}.$$

DO EXERCISE 5.

Example 6   Decompose into partial fractions:

$$\frac{11x^2 - 8x - 7}{(2x^2 - 1)(x - 3)}.$$

Solution   The decomposition looks like the following:

$$\frac{Ax + B}{2x^2 - 1} + \frac{C}{x - 3}.$$

Adding and equating the numerators, we get

$$11x^2 - 8x - 7 = (Ax + B)(x - 3) + C(2x^2 - 1)$$
$$= Ax^2 - 3Ax + Bx - 3B + 2Cx^2 - C,$$

or

$$11x^2 - 8x - 7 = (A + 2C)x^2 + (-3A + B)x + (-3B - C).$$

We then equate corresponding coefficients:

$$11 = A + 2C, \qquad \text{The coefficients of the } x^2\text{-terms}$$
$$-8 = -3A + B, \qquad \text{The coefficients of the } x\text{-terms}$$
$$-7 = -3B - C. \qquad \text{The constant terms}$$

We solve this system of three equations and obtain

$$A = 3, \qquad B = 1, \quad \text{and} \quad C = 4.$$

The decomposition is as follows:

$$\frac{3x + 1}{2x^2 - 1} + \frac{4}{x - 3}.$$

DO EXERCISE 6.

5. Decompose into partial fractions:

$$\frac{x^2 + 13x + 7}{(x - 4)(x + 1)^2}.$$

6. Decompose into partial fractions:

$$\frac{5x^2 - 7x + 7}{(3x^2 + 2)(x - 1)}.$$

EXERCISE SET 9.8

A   Decompose into partial fractions.

1. $\dfrac{x+7}{(x-3)(x+2)}$

2. $\dfrac{2x}{(x+1)(x-1)}$

3. $\dfrac{7x-1}{6x^2-5x+1}$

4. $\dfrac{13x+46}{12x^2-11x-15}$

5. $\dfrac{3x^2-11x-26}{(x^2-4)(x+1)}$

6. $\dfrac{5x^2+9x-56}{(x-4)(x-2)(x+1)}$

7. $\dfrac{9}{(x+2)^2(x-1)}$

8. $\dfrac{x^2-x-4}{(x-2)^3}$

9. $\dfrac{2x^2+3x+1}{(x^2-1)(2x-1)}$

10. $\dfrac{x^2-10x+13}{(x^2-5x+6)(x-1)}$

11. $\dfrac{x^4-3x^3-3x^2+10}{(x+1)^2(x-3)}$

12. $\dfrac{10x^3-15x^2-35x}{x^2-x-6}$

13. $\dfrac{-x^2+2x-13}{(x^2+2)(x-1)}$

14. $\dfrac{26x^2+208x}{(x^2+1)(x+5)}$

15. $\dfrac{6+26x-x^2}{(2x-1)(x+2)^2}$

16. $\dfrac{5x^3+6x^2+5x}{(x^2-1)(x+1)^3}$

17. $\dfrac{6x^3+5x^2+6x-2}{2x^2+x-1}$

18. $\dfrac{2x^3+3x^2-11x-10}{x^2+2x-3}$

19. $\dfrac{2x^2-11x+5}{(x-3)(x^2+2x-5)}$

20. $\dfrac{3x^2-3x-8}{(x-5)(x^2+x-4)}$

Decompose into partial fractions using a system of equations.

21. $\dfrac{-4x^2-2x+10}{(3x+5)(x+1)^2}$

22. $\dfrac{26x^2-36x+22}{(x-4)(2x-1)^2}$

23. $\dfrac{36x+1}{12x^2-7x-10}$

24. $\dfrac{-17x+61}{6x^2+39x-21}$

25. $\dfrac{-4x^2-9x+8}{(3x^2+1)(x-2)}$

26. $\dfrac{11x^2-39x+16}{(x^2+4)(x-8)}$

SYNTHESIS

27. Decompose into partial fractions:

$$\frac{x}{x^4-a^4}.$$

28. Decompose into partial fractions:

$$\frac{9x^3-24x^2+48x}{(x-2)^4(x+1)}.$$

[*Hint:* Let the expression equal

$$\frac{A}{x+1}+\frac{P(x)}{(x-2)^4}$$

and find $P(x)$.]

Decompose into partial fractions and then graph by addition of ordinates (adding respective fraction values).

29. $f(x)=\dfrac{3x}{x^2+5x+4}$

30. $f(x)=\dfrac{x-1}{x^2-2x-3}$

CHALLENGE

Decompose into partial fractions.

31. $\dfrac{1+\ln x^2}{(\ln x+2)(\ln x-3)^2}$

32. $\dfrac{1}{e^{-x}+3+2e^x}$

SUMMARY AND REVIEW 9

TERMS TO KNOW

● REVIEW EXERCISES

Solve.

1. $5x - 3y = -4,$
$3x - y = -4$

2. $2x + 3y = 2$
$5x - y = -29$

3. $x + 5y = 12,$
$5x + 25y = 12$

4. $2x - 4y + 3z = -3,$
$-5x + 2y - z = 7,$
$3x + 2y - 2z = 4$

5. $x + 5y + 3z = 0,$
$3x - 2y + 4z = 0,$
$2x + 3y - z = 0$

6. $x - y = 5,$
$y - z = 6,$
$z - w = 7,$
$x + w = 8$

7. Classify each of the systems in Exercises 1–6 as consistent or inconsistent.

8. Classify each of the systems in Exercises 1–6 as dependent or independent.

Solve.

9. The value of 75 coins, consisting of nickels and dimes, is $5.95. How many of each kind are there?

10. A family invested $5000, part at 10% and the remainder at 10.5%. The annual income from both investments is $517. What is the amount invested at each rate?

11. In triangle ABC, the measure of angle B is three times that of angle A. The measure of angle C is $20°$ more than that of angle A. Find the angular measurements.

12. A student has a total of 225 on three tests. The sum of the scores on the first and second tests exceeds the third score by 61. The first score exceeds the second by 6. Find the three scores.

Solve using matrices. If there is more than one solution, list three of them.

13. $x + 2y = 5,$
$2x - 5y = -8$

14. $3x + 4y + 2z = 3,$
$5x - 2y - 13z = 3,$
$4x + 3y - 3z = 6$

15. $3x + 5y + z = 0,$
$2x - 4y - 3z = 0,$
$x + 3y + z = 0$

16. $w + x + y + z = -2,$
$-3w - 2x + 3y + 2z = 10,$
$2w + 3x + 2y - z = -12,$
$2w + 4x - y + z = 1$

17. Find numbers a, b, and c such that the function $f(x) = ax^2 + bx + c$ fits the data points $(0, 3)$, $(1, 0)$, and $(-1, 4)$. Then write the equation for the function.

Evaluate the determinant.

18. $\begin{vmatrix} 1 & -2 \\ 3 & 4 \end{vmatrix}$

19. $\begin{vmatrix} \sqrt{3} & -5 \\ -3 & -\sqrt{3} \end{vmatrix}$

20. $\begin{vmatrix} -2 & -3 \\ 4 & -x \end{vmatrix}$

21. $\begin{vmatrix} 2 & -1 & 1 \\ 1 & 2 & -1 \\ 3 & 4 & -3 \end{vmatrix}$

22. $\begin{vmatrix} -5.8 & 7.5 & 4.6 \\ 0 & 2.2 & 8.9 \\ 0 & 0 & 1.3 \end{vmatrix}$

23. $\begin{vmatrix} 3a & 3b & 3c \\ 5a & 5b & 5c \\ d & e & f \end{vmatrix}$

Solve for (x, y) using Cramer's rule.

24. $5x - 2y = 19,$
$7x + 3y = 15$

25. $ax - by = a^2,$
$bx + ay = ab$

26. Solve using Cramer's rule:
$3x - 2y + z = 5,$
$4x - 5y - z = -1,$
$3x + 2y - z = 4.$

For Exercises 27–34, let

$$A = \begin{bmatrix} 1 & -1 & 0 \\ 2 & 3 & -2 \\ -2 & 0 & 1 \end{bmatrix}, \quad B = \begin{bmatrix} -1 & 0 & 6 \\ 1 & -2 & 0 \\ 0 & 1 & -3 \end{bmatrix},$$

and

$$C = \begin{bmatrix} -2 & 0 \\ 1 & 3 \end{bmatrix}.$$

Find each of the following, if possible.

27. $A + B$

28. $-3A$

29. $-A$

30. AB

31. $B + C$

32. $A - B$

33. $2A - B$

34. $A + 3B$

Find A^{-1}, if it exists.

35. $A = \begin{bmatrix} -2 & 0 \\ 1 & 3 \end{bmatrix}$

36. $A = \begin{bmatrix} 0 & 0 & 3 \\ 0 & -2 & 0 \\ 4 & 0 & 0 \end{bmatrix}$

37. $A = \begin{bmatrix} 1 & 0 & 0 & 0 \\ 0 & 4 & -5 & 0 \\ 0 & 2 & 2 & 0 \\ 0 & 0 & 0 & 1 \end{bmatrix}$

38. Write a matrix equation equivalent to this system of equations:
$3x - 2y + 4z = 13,$
$x + 5y - 3z = 7,$
$2x - 3y + 7z = -8.$

Evaluate the determinant.

39. $\begin{vmatrix} -4 & \sqrt{3} \\ \sqrt{3} & 7 \end{vmatrix}$

40. $\begin{vmatrix} 1 & -1 & 2 \\ -1 & 2 & 0 \\ -1 & 3 & 1 \end{vmatrix}$

41. $\begin{vmatrix} 0 & a & b \\ -a & 0 & c \\ -b & -c & 0 \end{vmatrix}$

42. $\begin{vmatrix} 4 & -7 & 6 & 7 \\ 0 & -3 & 9 & -8 \\ 0 & 0 & -2 & 6 \\ 0 & 0 & 0 & 5 \end{vmatrix}$

43. Without expanding, show that

$$\begin{vmatrix} 5a & 5b & 5c \\ 3a & 3b & 3c \\ d & e & f \end{vmatrix} = 0.$$

Factor.

44. $\begin{vmatrix} 1 & a & bc \\ 1 & b & ac \\ 1 & c & ab \end{vmatrix}$

45. $\begin{vmatrix} 1 & x^2 & x^3 \\ 1 & y^2 & y^3 \\ 1 & z^2 & z^3 \end{vmatrix}$

46. $\begin{vmatrix} 1 & a & a^2 & a^3 \\ 1 & b & b^2 & b^3 \\ 1 & c & c^2 & c^3 \\ 1 & d & d^2 & d^3 \end{vmatrix}$

47. Write a matrix equation equivalent to this system and find the inverse of the coefficient matrix. Use the inverse of the coefficient matrix to solve each system. Show your work.

$$2x + 3y = 2,$$
$$5x - y = -29$$

48. Graph this system. Find the coordinates of any vertices formed.

$$2x + y \geq 9,$$
$$4x + 3y \geq 23,$$
$$x + 3y \geq 8,$$
$$x \geq 0,$$
$$y \geq 0$$

49. Maximize and minimize $T = 6x + 10y$ subject to

$$x + y \leq 10,$$
$$5x + 10y \geq 50,$$
$$x \geq 2,$$
$$y \geq 0.$$

50. You are about to take a test that contains questions of type A worth 7 points and questions of type B worth 12

points. The total number of questions worked must be at least 8. If you know that type A questions take 10 min and type B questions take 8 min and that the maximum time for the test is 80 min, how many of each type of question must you do in order to maximize your score? What is this maximum score?

51. Decompose into partial fractions:

$$\frac{5}{(x+2)^2(x+1)}.$$

● **SYNTHESIS** _____

52. One year, a person invested a total of $40,000, part at 12%, part at 13%, and the rest at $14\frac{1}{2}\%$. The total interest received on the investments was $5370. The interest received on the $14\frac{1}{2}\%$ investment was $1050 more than the interest received on the 13% investment. How much was invested at each rate?

Solve.

53. $\dfrac{2}{3x} + \dfrac{4}{5y} = 8,$

$\dfrac{5}{4x} - \dfrac{3}{2y} = -6$

54. $\dfrac{3}{x} - \dfrac{4}{y} + \dfrac{1}{z} = -2,$

$\dfrac{5}{x} + \dfrac{1}{y} - \dfrac{2}{z} = 1,$

$\dfrac{7}{x} + \dfrac{3}{y} + \dfrac{2}{z} = 19$

Graph.

55. $|x| - |y| \leq 1$

56. $|xy| > 1$

57. On the basis of Exercise 42, conjecture and prove a theorem regarding determinants.

● **THINKING AND WRITING** _____

1. Compare the procedure considered in Sections 9.1 and 9.2 for solving systems of equations with any you may have learned in earlier mathematics courses.

2. Discuss and compare as many methods for solving systems of equations as you can from this chapter.

3. Explain the difference between a square matrix and its determinant.

CHAPTER TEST 9

Solve.

1. $0.2x - 0.5y = -0.1,$
$0.01x + 0.1y = 0.12$

2. A boat travels 36 km downstream in 2 hr. It travels 48 km upstream in 4 hr. Find the speed of the boat and the speed of the stream.

3. A chemist has one solution of acid and water that is 15% acid and a second that is 75% acid. Find how many gallons of each should be mixed together in order to get 20 gallons of a solution that is 39% acid.

4. A factory has three machines A, B, and C. With all three working, they make 500 toothbrushes per day. With A and B working, they make 284 toothbrushes per day. With A and C working, they make 260 toothbrushes per day. What is the daily production of each machine?

Solve using matrices. If there is more than one solution, list three of them.

5. $8x - 2y = 1,$
$12x + 8y = 7$

6. $12x - 6y - 19z = 0,$
$3x - 2y - 6z = 0,$
$6x + 2y + 3z = 0$

7. $2x - 7y + z = 11,$
$3x + 2y - 4z = 15,$
$5x - 4y + 6z = 7$

Classify as consistent or inconsistent, dependent or independent.

8. $-x + y = -6,$
$-x - y = 6$

9. $x + 4y - 2z = 1,$
$2x + 6y - z = 0,$
$-3x - 10y + 3z = -1$

10. Find numbers a, b, and c such that the function $f(x) = ax^2 + bx + c$ fits the data points $(2, 1)$, $(0, 1)$, and $(-1, -5)$. Then write the equation for the function.

Evaluate the determinant.

11. $\begin{vmatrix} -3 & 1 \\ -4 & \frac{2}{3} \end{vmatrix}$

12. $\begin{vmatrix} 1 & 1 & -2 \\ 0 & 2 & -6 \\ 4 & 0 & 3 \end{vmatrix}$

Solve using Cramer's rule.

13. $7x - 3y = 31,$
$4x + 2y = 14$

14. $x - 2y + 7z = 11,$
$2x + y - 3z = -5,$
$6x + z = 1$

For Exercises 15–24, let

$$A = \begin{bmatrix} -1 & 3 \\ 0 & 4 \end{bmatrix}, \qquad B = \begin{bmatrix} -2 & 1 & 0 \\ 3 & 2 & 5 \end{bmatrix},$$

$$C = \begin{bmatrix} 4 & 0 \\ 1 & -1 \\ 2 & -3 \end{bmatrix}, \qquad D = \begin{bmatrix} 0 & 2 & -1 \\ 3 & -1 & 1 \\ 0 & 4 & 3 \end{bmatrix},$$

$$E = \begin{bmatrix} 5 & -1 & 3 \\ 0 & 4 & 2 \\ 1 & 0 & -6 \end{bmatrix}, \qquad F = [2 \quad -1 \quad -3],$$

$$G = \begin{bmatrix} -2 & -5 \\ 6 & -3 \end{bmatrix}, \quad O = \begin{bmatrix} 0 & 0 \\ 0 & 0 \end{bmatrix}, \quad I = \begin{bmatrix} 1 & 0 \\ 0 & 1 \end{bmatrix}.$$

Find each of the following, if possible.

15. GC

16. O + A

17. A + I

18. D + E

19. A − G

20. BC

21. B + C

22. −F

23. 2D − E

24. GI

Find A^{-1}, if it exists.

25. $A = \begin{bmatrix} -3 & 1 \\ 2 & 0 \end{bmatrix}$

26. $A = \begin{bmatrix} 3 & 4 & 3 \\ 1 & 0 & 1 \\ -2 & -5 & -2 \end{bmatrix}$

27. $A = \begin{bmatrix} 2 & -1 & 0 \\ 3 & 0 & 1 \\ -2 & 4 & 0 \end{bmatrix}$

28. Let

$$A = \begin{bmatrix} 3 & 1 & -2 \\ 2 & 0 & -1 \\ -5 & 0 & 4 \end{bmatrix}.$$

Find a_{12}, M_{12}, and A_{12}.

Evaluate the determinant.

29. $\begin{vmatrix} 1 & -1 & 3 \\ -2 & 4 & 2 \\ 1 & 2 & 1 \end{vmatrix}$

30. $\begin{vmatrix} -2 & 7 & -3 \\ 1 & 1 & 1 \\ 4 & -14 & 6 \end{vmatrix}$

31. $\begin{vmatrix} -2 & 3 & 4 & 6 \\ 1 & 0 & 0 & 0 \\ -1 & 1 & 2 & -3 \\ 0 & 5 & 1 & 3 \end{vmatrix}$

32. $\begin{vmatrix} 1 & -1 & -1 \\ 2 & -6 & 4 \\ -3 & 0 & 2 \end{vmatrix}$

33. Factor:

$$\begin{vmatrix} c^3 & b^3 & a^3 \\ c & b & a \\ 1 & 1 & 1 \end{vmatrix}.$$

34. Write a matrix equation equivalent to this system of equations and use the inverse of the coefficient matrix to solve the system. Show all your work.

$$2x - 3y = -9,$$
$$x + 4y = 1$$

35. Graph: $4x - 3y \geq 12$.

36. Maximize and minimize $T = 20x + 60y$ subject to
$$x + 2y \leq 16,$$
$$2 \leq x \leq 5,$$
$$y \leq 0.$$

37. You are about to take a test that contains questions of type A worth 5 points and type B worth 12 points. You must complete the test in 72 minutes. Type A questions take 4 minutes; type B questions take 8 minutes. The total number of problems worked must not exceed 12. If you are told that you must work at least 2 questions of type B, how many of each type of question must you do in order to maximize your score? What is this maximum score?

38. Decompose into partial fractions:

$$\frac{-8x + 23}{2x^2 + 5x - 12}.$$

● SYNTHESIS _____

39. Solve:

$$\frac{7}{x} - \frac{3}{y} = -16,$$
$$\frac{2}{x} + \frac{5}{y} = 13.$$

10

Conic Sections

In this chapter, we will study equations whose graphs are conic sections, which means that the curve is formed as a cross section of a cone. We have actually studied three other conic sections—*lines, circles,* and *parabolas*—in some detail in Chapters 3 and 4. • There are many applications of these equations of conic sections. We will consider many of them. •

10.1

CONIC SECTIONS: LINES AND ELLIPSES

In this chapter, we will study polynomial functions and relations that are graphs of second-degree equations of the type

$$Ax^2 + By^2 + Cx + Dy + E = 0.$$

Some of these equations define functions and some define relations that are not functions. Though we will not prove it here, most of these equations have graphs that are **conic sections,** meaning that their graphs are geometric figures formed by cross sections of cones. Examples are shown at the top of the following page.

The Greek mathematician Apollonius (c. 225 B.C.) did much to develop the conic sections. They have many applications, as we will see as we study this chapter.

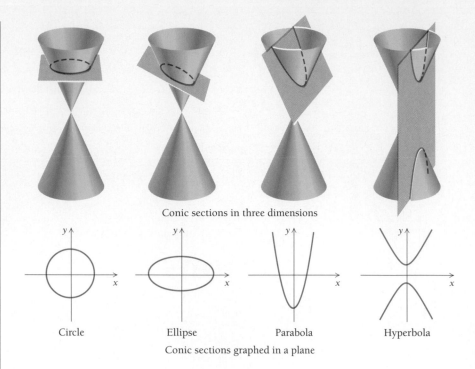

Conic sections in three dimensions

| Circle | Ellipse | Parabola | Hyperbola |

Conic sections graphed in a plane

Cones

Suppose that C is a circle with center O and that P is a point not in the same plane as C such that the line OP is perpendicular to the plane of the circle C. The set of points on all lines through P and a point of the circle form a **right circular cone** (or **conical surface**). Any line contained in the surface is called a **surface element** or a **generator line.** Note that there are two parts, or **nappes,** of a cone. Point P is called the **vertex** and line OP is called the **axis.**

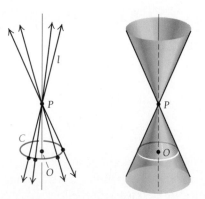

Conic Sections

The intersection of any plane with a cone is nonempty. That intersection is called a **conic section.** Some conic sections are shown at the top of the following page.

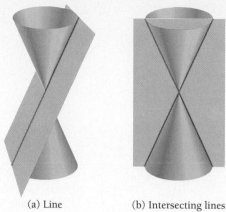

(a) Line   (b) Intersecting lines   (c) Single point

 Lines

Some second-degree equations have graphs consisting of a line. Some have graphs consisting of two lines. The graphs are conic sections except when the two lines are parallel.

Example 1   Graph: $3x^2 + 2xy - y^2 = 0$.

Solution   We factor and use the principle of zero products:

$$3x^2 + 2xy - y^2 = 0$$
$$(3x - y)(x + y) = 0$$
$$3x - y = 0 \quad \text{or} \quad x + y = 0$$
$$y = 3x \quad \text{or} \quad y = -x.$$

The graph consists of two intersecting lines.

Example 2   Graph: $(y - x)(y - x - 1) = 0$.

Solution   This is a second-degree equation:

$$(y - x)(y - x - 1) = 0$$
$$y - x = 0 \quad \text{or} \quad y - x - 1 = 0 \qquad \text{Using the principle of zero products}$$
$$y = x \quad \text{or} \quad y = x + 1.$$

The graph consists of two parallel lines. This equation does *not* represent a conic section.

Graph.

1. $x^2 - 4y^2 = 0$

2. $y^2 = 4$

3. $y^2 + 9x^2 = 6xy$

Example 3 Graph: $y^2 + x^2 = 2xy$.

Solution   We have

$$y^2 + x^2 = 2xy$$
$$y^2 - 2xy + x^2 = 0$$
$$(y - x)(y - x) = 0$$
$$y - x = 0 \quad \text{or} \quad y - x = 0$$
$$y = x \quad \text{or} \quad y = x.$$

The graph consists of a single line.

Generally, a second-degree equation with 0 on one side and a factorable expression on the other has a graph consisting of one or two lines. The equation $xy = 0$ has a graph that consists of the coordinate axes.

DO EXERCISES 1–3.

B   Single Points or No Points

When a plane intersects only the vertex of a cone, the result is a single point. The following is an equation for such a conic section.

Example 4 Graph: $x^2 + 4y^2 = 0$.

Solution   The expression $x^2 + 4y^2$ is not factorable in the real-number system. The only real-number solution of the equation is $(0, 0)$.

Example 5 Graph: $3x^2 + 7y^2 = -2$.

Solution   Since squares of numbers are never negative, the left side of the equation can never be negative. The equation has no real-number solutions, hence there are no points on the graph. This is a type of equation whose graph is *not* a conic section. Every plane intersects every cone. Thus an empty solution set cannot be found by the intersection of a plane and a cone.

DO EXERCISES 4 AND 5.

◉ Equations of Ellipses

Some equations of second degree have graphs that are circles. Circles are defined as follows.

> **DEFINITION**   **Circle**
>
> A *circle* is the locus or set of all points in a plane that are at a fixed distance from a fixed point in that plane.

When a plane intersects a cone perpendicular to the axis of the cone, as shown, a circle is formed. We will not prove that fact here.

Circle

We studied equations of circles in Section 3.2. You may wish to review that section before continuing. Recall the standard equation of a circle with radius r and center at the point (h, k).

> The equation, in standard form, of a circle with center (h, k) and radius r is
>
> $$(x - h)^2 + (y - k)^2 = r^2.$$

Some equations of second degree have graphs that are *ellipses*. Ellipses are defined as follows.

> **DEFINITION**   **Ellipse**
>
> An *ellipse* is the locus or set of all points P in a plane such that the sum of the distances from P to two fixed points F_1 and F_2 in the plane is constant. F_1 and F_2 are called *foci* (singular, *focus*) of the ellipse. The *center* of the ellipse is the midpoint of the segment joining F_1 and F_2.

When a plane intersects a cone with the plane not perpendicular to the axis of the cone, as shown, an ellipse may be formed. Again we will not prove this fact here.

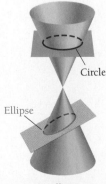

Circle

Ellipse

An ellipse in three dimensions

An ellipse in a plane

Graph.

4. $x^2 + 2y^2 = 0$

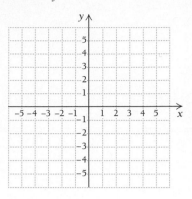

5. $x^2 + y^2 = -3$

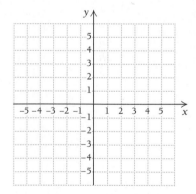

Here is a way to draw an ellipse. Stick two tacks in a piece of cardboard. These will be the foci F_1 and F_2. Attach a piece of string to the tacks. The length of the string will be the constant sum of the distances from the foci to any point on the ellipse. Take a pencil and pull the string tight. Now swing the pencil around, keeping the string tight.

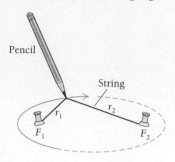

We first consider an equation of an ellipse whose center is at the origin and whose foci lie on one of the coordinate axes.

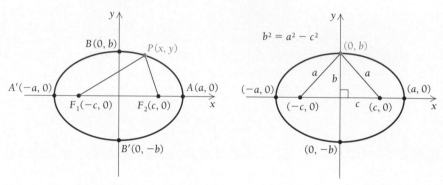

Suppose we have an ellipse with foci $F_1(-c, 0)$ and $F_2(c, 0)$. If $P(x, y)$ is a point on the ellipse, then F_1P is the distance from the first focus to the point $P(x, y)$ and F_2P is the distance from the second focus to the point $P(x, y)$. The sum $F_1P + F_2P$ is the given constant distance. We will call it $2a$:

$$F_1P + F_2P = 2a.$$

By the distance formula,

$$\sqrt{(x + c)^2 + y^2} + \sqrt{(x - c)^2 + y^2} = 2a,$$

or

$$\sqrt{(x + c)^2 + y^2} = 2a - \sqrt{(x - c)^2 + y^2}.$$

Squaring, we get

$$x^2 + 2cx + c^2 + y^2 = 4a^2 - 4a\sqrt{(x - c)^2 + y^2} + x^2 - 2cx + c^2 + y^2,$$

or

$$-4a^2 + 4cx = -4a\sqrt{(x - c)^2 + y^2}$$
$$-a^2 + cx = -a\sqrt{(x - c)^2 + y^2}.$$

Squaring again gives us

$$a^4 - 2a^2cx + c^2x^2 = a^2x^2 - 2cxa^2 + a^2c^2 + a^2y^2,$$

or

$$x^2(a^2 - c^2) + a^2y^2 = a^2(a^2 - c^2).$$

It follows from when P is at $(0, b)$ that $b^2 + c^2 = a^2$, or $b^2 = a^2 - c^2$. Substituting b^2 for $a^2 - c^2$ in the last equation, we have the equation of the ellipse $b^2x^2 + a^2y^2 = a^2b^2$, or the following.

> The equation, in standard form, of an ellipse with center at the origin is
>
> $$\frac{x^2}{a^2} + \frac{y^2}{b^2} = 1.$$
>
> When $a > b$, the foci are the points $(\pm c, 0)$, where $c^2 = a^2 - b^2$. When $b > a$, the foci are the points $(0, \pm c)$, where $c^2 = b^2 - a^2$.

We have proved that if a point is on the ellipse, then its coordinates satisfy this equation. We also need to know the converse, that is, if the coordinates of a point satisfy this equation, then the point is on the ellipse. The proof of the latter will be omitted here. In the above, the longer axis of symmetry $\overline{A'A}$ is called the **major axis.** The foci are always on the major axis. The shorter axis of symmetry $\overline{B'B}$ is called the **minor axis.** The intersection of these axes is called the **center.** The points A, A', B, and B' are called **vertices.** If the center of an ellipse is at the origin, the vertices are also the intercepts.

Ellipses as Stretched Circles

A **unit circle** is a circle centered at the origin with radius 1:

$$x^2 + y^2 = 1.$$

If we replace x by x/a and y by y/b, we get an equation of an ellipse:

$$\left(\frac{x}{a}\right)^2 + \left(\frac{y}{b}\right)^2 = 1, \quad \text{or} \quad \frac{x^2}{a^2} + \frac{y^2}{b^2} = 1.$$

An ellipse is a circle transformed by a stretch or shrink in the x-direction and in the y-direction. If $a = 2$, for example, the unit circle is stretched in the x-direction by a factor of 2. In any case, the x-intercepts become a and $-a$ and the y-intercepts become b and $-b$.

DO EXERCISE 6.

Example 6   For the ellipse $x^2 + 16y^2 = 16$, find the vertices and the foci. Then graph the ellipse.

Solution

a) We first multiply by $\frac{1}{16}$ to find standard form:

$$\frac{x^2}{16} + \frac{y^2}{1} = 1, \quad \text{or} \quad \frac{x^2}{4^2} + \frac{y^2}{1^2} = 1.$$

Thus, $a = 4$ and $b = 1$. Two of the vertices are $(-4, 0)$ and $(4, 0)$. These are also x-intercepts. The other vertices are $(0, 1)$ and $(0, -1)$. These are also y-intercepts. Since we know that $c^2 = a^2 - b^2$, we have $c^2 = 16 - 1$, so $c = \sqrt{15}$ and the foci are $(-\sqrt{15}, 0)$ and $(\sqrt{15}, 0)$.

b) We plot the vertices found in (a) and connect them with a smooth, oval-shaped curve. To be accurate, it is often helpful to find some other

6. Suppose a circle has been distorted to get the ellipse shown below.

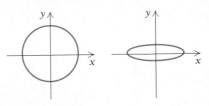

a) Has the circle been stretched or shrunk in the x-direction?

b) Has the circle been stretched or shrunk in the y-direction?

For each ellipse, find the vertices and the foci, and draw the graph using graph paper.

7. $x^2 + 9y^2 = 9$

8. $9x^2 + 25y^2 = 225$

9. $2x^2 + 4y^2 = 8$

points on the curve. We let $x = 2$ and solve for y:

$$x^2 + 16y^2 = 16$$
$$(2)^2 + 16y^2 = 16$$
$$4 + 16y^2 = 16$$
$$16y^2 = 12$$
$$y^2 = \frac{12}{16}$$
$$y = \pm\sqrt{\frac{12}{16}} = \pm\frac{\sqrt{3}}{2} \approx \pm 0.9.$$

Thus, $(2, 0.9)$ and $(2, -0.9)$ can also be plotted and used to draw the graph. Similarly, the points $(-2, 0.9)$ and $(-2, -0.9)$ can be computed and plotted.

DO EXERCISES 7–9.

 TECHNOLOGY CONNECTION

Drawing the graph of an ellipse on a grapher involves two steps:

1. Solve the equation for y. The result will include $\pm$ in front of a radical.
2. Graph both functions, the one including the $+$ sign and the one including the $-$ sign, on the same set of axes. Graphing both functions is necessary because ellipses are never functions.

For example, let's graph the ellipse

$$\frac{x^2}{4} + \frac{y^2}{16} = 1.$$

Solving for y, we get

$$y = \pm\sqrt{16 - 4x^2}.$$

To see the true shape of the ellipse, make sure the dimensions of the viewing box are such that one unit on the x-axis is exactly the same length as one unit on the y-axis. Then graph each of the functions

$$y = +\sqrt{16 - 4x^2} \quad \text{and} \quad y = -\sqrt{16 - 4x^2}$$

using the same set of axes. The result should look like the one shown here. (Your grapher may leave some "gaps" in the graph in the regions where the graph is nearly vertical.)

Graph each of the following ellipses.

TC 1. $10x^2 + 5y^2 = 10$

TC 2. $x^2 + 7y^2 = 70$

TC 3. $3.4x^2 + 6.7y^2 = 32.5$

Example 7 Graph this ellipse and its foci: $9x^2 + 2y^2 = 18$.

Solution

a) We first multiply by $\frac{1}{18}$:

$$\frac{x^2}{2} + \frac{y^2}{9} = 1, \quad \text{or} \quad \frac{x^2}{(\sqrt{2})^2} + \frac{y^2}{3^2} = 1.$$

Thus, $a = \sqrt{2}$ and $b = 3$, and the vertices are $(0, 3)$, $(0, -3)$, $(-\sqrt{2}, 0)$, and $(\sqrt{2}, 0)$.

b) Since $b > a$, the foci are on the y-axis and the major axis lies along the y-axis. To find c in this case, we proceed as follows:

$$c^2 = b^2 - a^2 = 9 - 2 = 7$$
$$c = \sqrt{7}.$$

The foci are $(0, \sqrt{7})$ and $(0, -\sqrt{7})$.

c) We plot the vertices and connect them with an oval-shaped curve as follows. Other points can be computed and plotted.

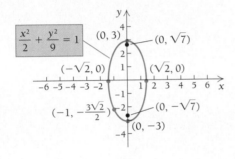

In summary, we have the following.

Equation	Vertices	Foci
$\dfrac{x^2}{a^2} + \dfrac{y^2}{b^2} = 1, \quad a > b$	$(\pm a, 0), (0, \pm b)$	$(\pm c, 0), \quad c^2 = a^2 - b^2$
$\dfrac{x^2}{a^2} + \dfrac{y^2}{b^2} = 1, \quad b > a$	$(\pm a, 0), (0, \pm b)$	$(0, \pm c), \quad c^2 = b^2 - a^2$

DO EXERCISES 10–12.

▷ Standard Form by Completing the Square

If the center of an ellipse is not at the origin but at some point (h, k), then we can think of the ellipse

$$\frac{x^2}{a^2} + \frac{y^2}{b^2} = 1$$

being translated h units left or right and k units up or down. The standard form of the equation is then as follows.

For each ellipse, find the center, the vertices, and the foci, and draw the graph.

10. $9x^2 + y^2 = 9$

11. $25x^2 + 9y^2 = 225$

12. $4x^2 + 2y^2 = 8$

An equation, in standard form, of an ellipse with center at (h, k) is

$$\frac{(x - h)^2}{a^2} + \frac{(y - k)^2}{b^2} = 1.$$

If $a > b$, the major axis and the foci are on the horizontal line $y = k$.
If $b > a$, the major axis and the foci are on the vertical line $x = h$.

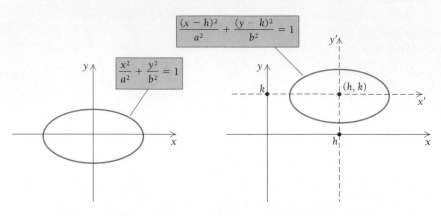

We can also think of such an ellipse under what is called **translation of axes** by considering the ellipse as being a graph of an equation

$$\frac{(x')^2}{a^2} + \frac{(y')^2}{b^2} = 1,$$

where the new axes are given by $x' = x - h$ and $y' = y - k$.

Example 8   For the ellipse

$$16x^2 + 4y^2 + 96x - 8y + 84 = 0,$$

find the center, the vertices, and the foci. Then graph the ellipse.

Solution

a) We first complete the square to get standard form:

$$16(x^2 + 6x +) + 4(y^2 - 2y +) = -84$$
$$16(x^2 + 6x + 9 - 9) + 4(y^2 - 2y + 1 - 1) = -84$$
$$16(x^2 + 6x + 9) + 4(y^2 - 2y + 1) = -84 + 144 + 4$$
$$16(x + 3)^2 + 4(y - 1)^2 = 64$$

$$\frac{1}{64}[16(x + 3)^2 + 4(y - 1)^2] = \frac{1}{64} \cdot 64$$

$$\frac{(x + 3)^2}{2^2} + \frac{(y - 1)^2}{4^2} = 1.$$

The center is $(-3, 1)$, $a = 2$, and $b = 4$.

b) The vertices of the ellipse $x^2/2^2 + y^2/4^2 = 1$ are $(2, 0), (-2, 0), (0, 4)$, and $(0, -4)$. Now $c^2 = 16 - 4 = 12$, so $c = 2\sqrt{3}$. Since the denominator of the y^2-term is larger, the major axis is on the y-axis. Thus the foci are $(0, 2\sqrt{3})$ and $(0, -2\sqrt{3})$.

c) Then the vertices and the foci of the translated ellipse are found by translation in the same way in which the center has been translated. Thus the vertices are

$$(-3 + 2, 1 + 0), \qquad (-3 - 2, 1 + 0),$$
$$(-3 + 0, 1 + 4), \qquad (-3 + 0, 1 - 4),$$

or

$$(-1, 1), \qquad (-5, 1), \qquad (-3, 5), \qquad (-3, -3).$$

The foci are $(-3, 1 + 2\sqrt{3})$ and $(-3, 1 - 2\sqrt{3})$.

d) The graph is as follows:

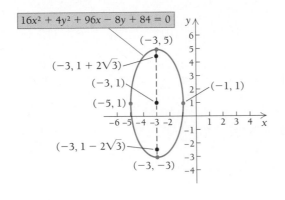

DO EXERCISES 13 AND 14.

For each ellipse, find the center, the vertices, and the foci, and draw the graph, using graph paper.

13. $25x^2 + 9y^2 + 150x$
$$- 36y + 260 = 0$$

14. $9x^2 + 25y^2 - 36x$
$$+ 150y + 260 = 0$$

 TECHNOLOGY CONNECTION

Ellipses whose centers are not at the origin can also be graphed on a grapher. They just require a bit more algebra to rewrite in the form $y = \pm f(x)$. For example, consider the ellipse in Example 8:

$$16x^2 + 4y^2 + 96x - 8y + 84 = 0.$$

In order to solve for y, we rewrite the ellipse as a quadratic in y:

$$4y^2 - 8y + (16x^2 + 96x + 84) = 0.$$

Using the quadratic equation

$$y = \frac{-b \pm \sqrt{b^2 - 4ac}}{2a}$$

gives us

$$y = \frac{-(-8) \pm \sqrt{(-8)^2 - 4(4)(16x^2 + 96x + 84)}}{2(4)},$$

which simplifies to

$$y = 1 \pm 2\sqrt{-x^2 - 6x - 5}.$$

When these two separate "pieces" of the ellipse are graphed using the same set of axes, the result is shown here (compare this with the graph in the text).

Graph each of the following ellipses.

TC 4. $4x^2 + 13y^2 + 6x + 4y - 10 = 0$

TC 5. $1.7x^2 + 4.3y^2 - 8.1x - 6.5y - 12.4 = 0$

Applications

Ellipses have many applications. For example, earth satellites travel in elliptical orbits. The planets travel around the sun in elliptical orbits with the sun at one focus.

Planetary orbit

An interesting attraction found in museums is the *whispering gallery*. It is elliptical. Persons with their feet at the foci can whisper and hear each other clearly, while persons at other positions cannot hear them. This happens because the sound waves emanating from one focus are reflected to the other focus, being concentrated there.

Whispering gallery

● **EXERCISE SET** 10.1

    Graph.

1. $x^2 - y^2 = 0$

2. $x^2 - 9y^2 = 0$

3. $3x^2 + xy - 2y^2 = 0$

4. $x^2 - xy - 2y^2 = 0$

5. $2x^2 + y^2 = 0$

6. $5x^2 + y^2 = -3$

C   Find the vertices and the foci, and draw the graph.

7. $\dfrac{x^2}{4} + \dfrac{y^2}{1} = 1$

8. $\dfrac{x^2}{1} + \dfrac{y^2}{4} = 1$

9. $16x^2 + 9y^2 = 144$

10. $9x^2 + 16y^2 = 144$

11. $2x^2 + 3y^2 = 6$

12. $5x^2 + 7y^2 = 35$

13. $4x^2 + 9y^2 = 1$

14. $25x^2 + 16y^2 = 1$

D   Find the center, the vertices, and the foci, and draw the graph.

15. $\dfrac{(x-1)^2}{4} + \dfrac{(y-2)^2}{1} = 1$

16. $\dfrac{(x-1)^2}{1} + \dfrac{(y-2)^2}{4} = 1$

17. $\dfrac{(x+3)^2}{25} + \dfrac{(y-2)^2}{16} = 1$

18. $\dfrac{(x-2)^2}{25} + \dfrac{(y+3)^2}{16} = 1$

19. $3(x+2)^2 + 4(y-1)^2 = 192$

20. $4(x-5)^2 + 3(y-5)^2 = 192$

21. $4x^2 + 9y^2 - 16x + 18y - 11 = 0$

22. $x^2 + 2y^2 - 10x + 8y + 29 = 0$

23. $4x^2 + y^2 - 8x - 2y + 1 = 0$

24. $9x^2 + 4y^2 + 54x - 8y + 49 = 0$

● **SYNTHESIS** _____

Find the equation of the ellipse with the following vertices. (*Hint:* Graph the vertices.)

25. $(2, 0), (-2, 0), (0, 3), (0, -3)$

26. $(1, 0), (-1, 0), (0, 4), (0, -4)$

27. $(1, 1), (5, 1), (3, 6), (3, -4)$

28. $(-1, -1), (-1, 5), (-3, 2), (1, 2)$

Find the equation of the ellipse satisfying the given conditions.

29. Center at $(-2, 3)$ with major axis of length 4 and parallel to the y-axis, minor axis of length 1

30. Vertices $(3, 0)$ and $(-3, 0)$ and containing the point $(2, \frac{22}{3})$

31. **a)** Graph $9x^2 + y^2 = 9$. Is this relation a function?
b) Solve $9x^2 + y^2 = 9$ for y.
c) Graph $y = 3\sqrt{1 - x^2}$ and determine whether it is a function. Find the domain and the range.
d) Graph $y = -3\sqrt{1 - x^2}$ and determine whether it is a function. Find the domain and the range.

32. Describe the graph of

$$\frac{x^2}{a^2} + \frac{y^2}{b^2} = 1$$

when $a^2 = b^2$.

33. ▇ Draw a large-scale precise graph of

$$\frac{x^2}{25} + \frac{y^2}{16} = 1$$

by calculating and plotting a large number of points.

34. The maximum distance of the earth from the sun is 9.3×10^7 miles. The minimum distance is 9.1×10^7 miles. The sun is at one focus of the elliptical orbit. Find the distance from the sun to the other focus.

35. The bridge support shown in this figure is the top half of an ellipse. Assuming that a coordinate system is superimposed on the drawing in such a way that the center of the ellipse is at point Q, find an equation of the ellipse.

36. In Washington, D.C., there is a large grassy area south of the White House known as the **Ellipse.** It is actually an ellipse with major axis of length 1048 ft and minor axis of length 898 ft. Assuming that a coordinate system is superimposed on the area in such a way that the center is at the origin and the major and minor axes are on the x- and y-axes of the coordinate system, find an equation of the ellipse.

37. Consider the figures below. The circle on the left has radius r. Suppose "each" radius shown is stretched or shrunk to form the ellipse on the right.

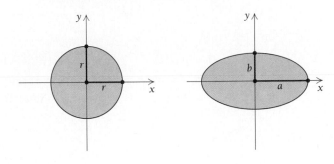

a) Conjecture a formula for the area of the ellipse:

$$\frac{x^2}{a^2} + \frac{y^2}{b^2} = 1.$$

(*Hint:* The area of the circle $x^2 + y^2 = r^2$ is $\pi \cdot r \cdot r$.)
b) Use the result of (a) to find the area of the ellipse

$$\frac{x^2}{16} + \frac{y^2}{25} = 1.$$

c) Use the result of (a) to find the area of the ellipse in Exercise 36.

38. The toy pictured here is called a "vacuum grinder." It consists of a rod hinged to two blocks A and B that slide in perpendicular grooves. One grasps the knob at C and grinds. Determine (and prove) whether or not the path of the handle C is an ellipse.

▨▧ **TECHNOLOGY CONNECTION** _____

Use a grapher to find the center and the vertices.

39. $4x^2 + 9y^2 - 16.025x + 18.0927y - 11.346 = 0$

40. $9x^2 + 4y^2 + 54.063x - 8.016y + 49.872 = 0$

10.2

CONIC SECTIONS: HYPERBOLAS

Some equations of second degree have graphs that are *hyperbolas*. Hyperbolas are defined as follows.

> **DEFINITION** **Hyperbola**
>
> A *hyperbola* is the locus or set of all points P in a plane such that the absolute value of the difference of the distances from P to two fixed points F_1 and F_2 in the plane is constant. The points F_1 and F_2 are called *foci* (singular, *focus*) and the midpoint of the segment joining them is called the *center*.

When a plane intersects a cone parallel to the axis of the cone as shown, a hyperbola is formed. Note that a hyperbola has two disconnected parts. These are called **branches.**

Hyperbola in three dimensions

Hyperbola in a plane

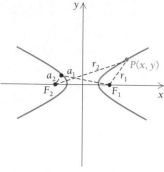

$|r_2 - r_1| = \text{constant} = |a_2 - a_1|$

A Equations of Hyperbolas

Now let us find equations for hyperbolas. We first consider an equation of a hyperbola whose center is at the origin and whose foci lie on one of the coordinate axes. Suppose we have a hyperbola, as shown, with foci $F_1(c, 0)$ and $F_2(-c, 0)$ on the x-axis. We consider a point $P(x, y)$ in the first quadrant. The proof for the other quadrants is similar to the proof that follows.

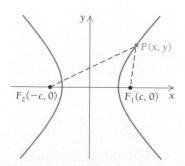

We know that $PF_2 > PF_1$, so $PF_2 - PF_1 > 0$ and $|PF_2 - PF_1| =$

$PF_2 - PF_1$. Let the constant difference be $2a$. Then

$$|PF_2 - PF_1| = PF_2 - PF_1 = 2a.$$

In the triangle F_2PF_1, $F_1F_2 + PF_1 > PF_2$ by the triangle inequality, so $PF_2 - PF_1 < F_1F_2$, or $2a < 2c$; therefore, $a < c$. Using the distance formula, we have

$$\sqrt{(x + c)^2 + y^2} - \sqrt{(x - c)^2 + y^2} = 2a,$$

or

$$\sqrt{(x + c)^2 + y^2} = 2a + \sqrt{(x - c)^2 + y^2}.$$

Squaring, we get

$$x^2 + 2xc + c^2 + y^2 = 4a^2 + 4a\sqrt{(x - c)^2 + y^2} + x^2 - 2xc + c^2 + y^2,$$

which simplifies to

$$4cx - 4a^2 = 4a\sqrt{(x - c)^2 + y^2}$$

or

$$cx - a^2 = a\sqrt{(x - c)^2 + y^2}.$$

Squaring again, we get

$$c^2x^2 - 2a^2cx + a^4 = a^2x^2 - 2a^2cx + a^2c^2 + a^2y^2,$$

or

$$x^2(c^2 - a^2) - a^2y^2 = a^2(c^2 - a^2).$$

Since $c > a$, $c^2 > a^2$, so $c^2 - a^2$ is positive. We represent $c^2 - a^2$ by b^2. The previous equation then becomes

$$x^2b^2 - a^2y^2 = a^2b^2,$$

or the following:

> The equation, in standard form, of a hyperbola with center at the origin and foci on the x-axis is
>
> $$\frac{x^2}{a^2} - \frac{y^2}{b^2} = 1.$$

We have shown that if a point is on the hyperbola, it satisfies this equation. We also need to know the converse: If a point satisfies the equation, then it is on the hyperbola. We omit the proof.

The following figure is a hyperbola.

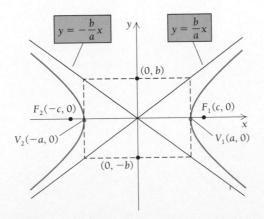

Points $V_1(a, 0)$ and $V_2(-a, 0)$ are called the **vertices,** and the line segment $\overline{V_1V_2}$ is called the **transverse axis.** The line segment from $(0, b)$ to $(0, -b)$ is called the **conjugate axis.** Note that when we replace x by $-x$ and/or y by $-y$, we get an equivalent equation. Thus the hyperbola is symmetric with respect to the origin, and the x- and y-axes are lines of symmetry.

The lines $y = (b/a)x$ and $y = -(b/a)x$ are **asymptotes.** They have slopes b/a and $-b/a$. The graph gets closer and closer to the asymptotes as x gets further away from 0.

Example 1   For the hyperbola $9x^2 - 16y^2 = 144$, find the vertices, the foci, and the asymptotes. Then graph the hyperbola.

Solution

a) We first multiply by $\frac{1}{144}$ to find the standard form:

$$\frac{x^2}{16} - \frac{y^2}{9} = 1.$$

Thus, $a = 4$ and $b = 3$. Next, we find the x-intercepts, or *vertices*. Let $y = 0$, and we see that $x^2/4^2 = 1$, so $x = \pm 4$. The vertices are $(4, 0)$ and $(-4, 0)$. Hyperbolas have only two vertices. Suppose we try to find vertices on the y-axis: the y-intercepts. If we set $x = 0$, we get $y^2/9 = -1$, and this equation has no real-number solutions. The foci are always on the same axis as the vertices. Since $b^2 = c^2 - a^2$,

$$c = \sqrt{a^2 + b^2} = \sqrt{4^2 + 3^2} = 5.$$

Thus the foci are $(5, 0)$ and $(-5, 0)$. The asymptotes are

$$y = \frac{b}{a}x = \frac{3}{4}x$$

and

$$y = -\frac{b}{a}x = -\frac{3}{4}x.$$

b) To graph the hyperbola, it is helpful to first graph the asymptotes. An easy way to do this is to draw the rectangle shown in the figure with vertices $(\pm a, \pm b)$, or $(\pm 4, \pm 3)$. Then draw the asymptotes out from the center to these vertices. The branches of the hyperbola are then drawn outward from the vertices of the hyperbola toward the asymptotes.

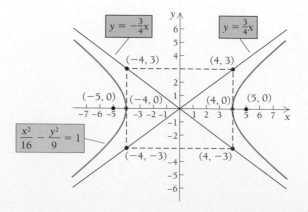

Why are $y = (b/a)x$ and $y = -(b/a)x$ asymptotes? To answer this, we solve $x^2/16 - y^2/9 = 1$ for y^2:

$$9x^2 - 16y^2 = 144 \qquad \text{Multiplying by 144}$$
$$-16y^2 = 144 - 9x^2$$
$$16y^2 = 9x^2 - 144$$
$$y^2 = \frac{1}{16}(9x^2 - 144) = \frac{9x^2 - 144}{16}.$$

From this last equation, we see that as $|x|$ gets larger, the term -144 is very small compared to $9x^2$, so y^2 gets close to $9x^2/16$. That is, when $|x|$ is large,

$$y^2 \approx \frac{9x^2}{16},$$

so

$$y \approx \left| \frac{3}{4}x \right| \quad \text{or} \quad y \approx \pm\frac{3}{4}x.$$

Thus the lines $y = \frac{3}{4}x$ and $y = -\frac{3}{4}x$ are asymptotes.

DO EXERCISES 1 AND 2.

The foci of a hyperbola can be on the y-axis. In that case, the equation is as follows.

> The equation, in standard form, of a hyperbola with center at the origin and foci on the y-axis is
>
> $$\frac{y^2}{b^2} - \frac{x^2}{a^2} = 1.$$

In this case, the slopes of the asymptotes are still $\pm b/a$ and it is still true that $b^2 = c^2 - a^2$. There are now y-intercepts, and they are $(0, \pm b)$.

In summary, we have the following.

Equation	Vertices	Foci
$\dfrac{x^2}{a^2} - \dfrac{y^2}{b^2} = 1$	$(-a, 0), (a, 0)$	$(-c, 0), (c, 0)$
$\dfrac{y^2}{b^2} - \dfrac{x^2}{a^2} = 1$	$(0, -b), (0, b)$	$(0, -c), (0, c)$
The asymptotes are $y = \pm\dfrac{b}{a}x.$		$c = \sqrt{a^2 + b^2}$

Example 2 For the hyperbola $25y^2 - 16x^2 = 400$, find the vertices, the foci, and the asymptotes. Then draw the graph.

Solution

a) We first multiply by $\frac{1}{400}$ to find the standard form:

$$\frac{y^2}{16} - \frac{x^2}{25} = 1.$$

For each hyperbola, find the vertices, the foci, and the asymptotes. Then draw the graph.

1. $4x^2 - 9y^2 = 36$

2. $x^2 - y^2 = 16$

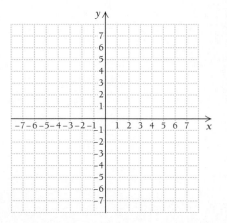

For each hyperbola, find the vertices, the foci, and the asymptotes. Then draw the graph using graph paper.

3. $9y^2 - 25x^2 = 225$

4. $y^2 - x^2 = 25$

Thus, $a = 5$ and $b = 4$. The vertices are $(0, 4)$ and $(0, -4)$ since the x^2-term is negative. Since $c = \sqrt{4^2 + 5^2} = \sqrt{41}$, the foci are $(0, \sqrt{41})$ and $(0, -\sqrt{41})$. The asymptotes are $y = \frac{4}{5}x$ and $y = -\frac{4}{5}x$.

b) The graph is as shown.

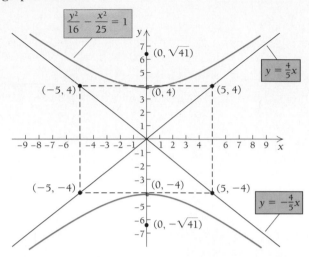

DO EXERCISES 3 AND 4.

B   Standard Form by Completing the Square

If a hyperbola with center at the origin is translated $|h|$ units left or right or $|k|$ units up or down, the center is at some point (h, k). The standard equation is then one of the following.

$$\frac{(x - h)^2}{a^2} - \frac{(y - k)^2}{b^2} = 1 \qquad \text{Hyperbola, transverse axis parallel to the } x\text{-axis}$$

$$\frac{(y - k)^2}{b^2} - \frac{(x - h)^2}{a^2} = 1 \qquad \text{Hyperbola, transverse axis parallel to the } y\text{-axis}$$

The asymptotes are $y - k = \pm \dfrac{b}{a}(x - h)$.

Example 3   For the hyperbola

$$4x^2 - y^2 + 24x + 4y + 28 = 0,$$

find the center, the vertices, the foci, and the asymptotes. Then draw the graph.

Solution

a) We complete the square to find standard form:

$$4(x^2 + 6x + \quad) - (y^2 - 4y + \quad) = -28$$
$$4(x^2 + 6x + 9 - 9) - (y^2 - 4y + 4 - 4) = -28$$
$$4(x^2 + 6x + 9) - (y^2 - 4y + 4) = -28 + 36 - 4$$
$$4(x + 3)^2 - (y - 2)^2 = 4$$
$$\frac{(x + 3)^2}{1} - \frac{(y - 2)^2}{4} = 1.$$

TECHNOLOGY CONNECTION

Hyperbolas can be graphed in a manner similar to the way we graphed ellipses earlier. First we solve for y. This is relatively simple for hyperbolas centered at the origin (like those in Examples 1 and 2). If the center is not at the origin, however, we treat the problem as a quadratic in y in order to solve for y, as in Example 3. Then we graph both functions using the same set of axes.

Graph each of the following hyperbolas.

TC 1. $5x^2 - 18y^2 - 23 = 0$

TC 2. $12y^2 - 2.5x^2 = 15$

TC 3. $4x^2 - 10y^2 - 6x + 4y = 10$

TC 4. $17.3x^2 - 4.3y^2 + 12.3x - 3.4y + 10.9 = 0$

The center is $(-3, 2)$.

b) Consider $x^2/1 - y^2/4 = 1$. We have $a = 1$ and $b = 2$. The vertices of this hyperbola are $(1, 0)$ and $(-1, 0)$. Also, $c = \sqrt{1^2 + 2^2} = \sqrt{5}$, so the foci are $(\sqrt{5}, 0)$ and $(-\sqrt{5}, 0)$. The asymptotes are $y = 2x$ and $y = -2x$.

c) The vertices, foci, and asymptotes of the translated hyperbola are found in the same way in which the center has been translated. The vertices are $(-3 + 1, 2 + 0)$, $(-3 - 1, 2 + 0)$, or $(-2, 2)$, $(-4, 2)$. The foci are $(-3 + \sqrt{5}, 2)$ and $(-3 - \sqrt{5}, 2)$, and the asymptotes are

$$y - 2 = 2(x + 3) \quad \text{and} \quad y - 2 = -2(x + 3).$$

d) The graph is as follows.

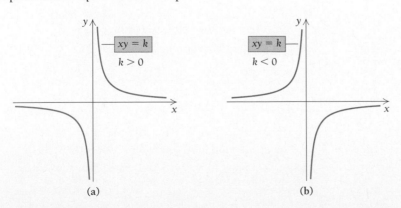

DO EXERCISES 5 AND 6.

Asymptotes on the Coordinate Axes

If a hyperbola has its center at the origin and its transverse axis at $45°$ to the coordinate axes, it has a simple equation. The coordinate axes are its asymptotes.

> $xy = k$, k a nonzero constant   Hyperbola, asymptotes the coordinate axes

If k is positive, the branches of the hyperbola lie in the first and third quadrants. If k is negative, the branches lie in the second and fourth quadrants. In either case, the asymptotes are the x-axis and the y-axis. We graphed these equations in Chapter 3.

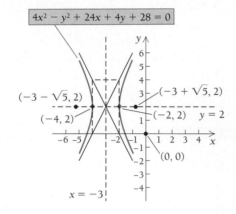

(a)   (b)

For each hyperbola, find the center, the vertices, the foci, and the asymptotes. Then draw the graph.

5. $4x^2 - 25y^2 - 8x$
$$- 100y - 196 = 0$$

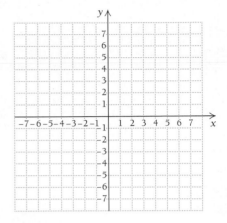

6. $16y^2 - 9x^2 - 64y$
$$- 18x - 89 = 0$$

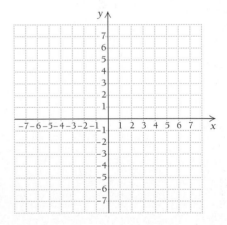

Graph using graph paper.

7. $xy = 3$

8. $xy = -12$

DO EXERCISES 7 AND 8.

Applications

Hyperbolas also have many applications. For example, a jet breaking the sound barrier creates a sonic boom whose wave front has the shape of a cone. The cone intersects the ground in one branch of a hyperbola.

Some comets travel in hyperbolic orbits. A cross section of an amphitheater may be half of one branch of a hyperbola.

● EXERCISE SET 10.2

A, B   Find the center, the vertices, the foci, and the asymptotes. Then draw the graph.

1. $\dfrac{x^2}{9} - \dfrac{y^2}{1} = 1$

2. $\dfrac{x^2}{1} - \dfrac{y^2}{9} = 1$

3. $\dfrac{(x-2)^2}{9} - \dfrac{(y+5)^2}{1} = 1$

4. $\dfrac{(x-2)^2}{1} - \dfrac{(y+5)^2}{9} = 1$

5. $\dfrac{(y+3)^2}{4} - \dfrac{(x+1)^2}{16} = 1$

6. $\dfrac{(y+3)^2}{25} - \dfrac{(x+1)^2}{16} = 1$

7. $x^2 - 4y^2 = 4$ **8.** $4x^2 - y^2 = 4$

9. $4y^2 - x^2 = 4$ **10.** $y^2 - 4x^2 = 4$

11. $x^2 - y^2 = 2$ **12.** $x^2 - y^2 = 3$

13. $x^2 - y^2 = \dfrac{1}{4}$ **14.** $x^2 - y^2 = \dfrac{1}{9}$

B   Find the center, the vertices, the foci, and the asymptotes. Then draw the graph.

15. $x^2 - y^2 - 2x - 4y - 4 = 0$

16. $4x^2 - y^2 + 8x - 4y - 4 = 0$

17. $36x^2 - y^2 - 24x + 6y - 41 = 0$

18. $9x^2 - 4y^2 + 54x + 8y + 45 = 0$

19. $9y^2 - 4x^2 - 18y + 24x - 63 = 0$

20. $x^2 - 25y^2 + 6x - 50y = 41$

21. $x^2 - y^2 - 2x - 4y = 4$

22. $9y^2 - 4x^2 - 54y - 8x + 41 = 0$

23. $y^2 - x^2 - 6x - 8y - 29 = 0$

24. $x^2 - y^2 = 8x - 2y - 13$

C   Graph.

25. $xy = 1$ **26.** $xy = -4$

27. $xy = -8$ **28.** $xy = 3$

●   SYNTHESIS

Find an equation of a hyperbola having:

29. Asymptotes $y = \frac{3}{2}x$ and $y = -\frac{3}{2}x$ and one vertex $(2, 0)$.

30. Vertices at $(1, 0)$ and $(-1, 0)$ and foci at $(2, 0)$ and $(-2, 0)$.

31. Transverse axis parallel to the y-axis and of length 11, conjugate axis of length 6, and center $(3, -8)$.

32. Vertices $(-9, 4)$ and $(-5, 4)$ and asymptotes $y = 3x + 25$ and $y = -3x - 17$.

33. Find an equation of the inverse of the relation of Exercise 32. Find the vertices and the center.

34. Show that the determinant equation

$$\begin{vmatrix} \dfrac{x-h}{a} & \dfrac{y-k}{b} \\[2mm] \dfrac{y-k}{b} & \dfrac{x-h}{a} \end{vmatrix} = 1$$

is an equation of a hyperbola with center (h, k).

35. a) Graph $x^2 - 4y^2 = 4$. Is this relation a function?
    b) Solve $x^2 - 4y^2 = 4$ for y.

c) Graph $y = \frac{1}{2}\sqrt{x^2 - 4}$ and determine whether it is a function. Find the domain and the range.

d) Graph $y = -\frac{1}{2}\sqrt{x^2 - 4}$ and determine whether it is a function. Find the domain and the range.

● **CHALLENGE** _____

36. In a navigation system called Loran, a radio transmitter at M (the *master* station) sends out pulses. Each pulse triggers another transmitter at S (the *slave* station), which then also transmits a pulse. A ship or airplane at A receives pulses from both M and S, and a device measures the difference in the time at which they arrive at A. Knowing this difference in time, the navigator can locate the vessel as being somewhere along a curve predrawn on a chart. What is the shape of that curve?

37. A rifle at A fires a bullet, which hits a target at B. A person at C hears the sound of the rifle shot and the sound of the bullet hitting the target simultaneously. Describe the set of all such points C.

TECHNOLOGY CONNECTION _____

Use a grapher to find the center, the vertices, and the asymptotes.

38. $5x^2 - 3.5y^2 + 14.6x - 6.7y + 3.4 = 0$

39. $x^2 - y^2 - 2.046x - 4.088y - 4.228 = 0$

10.3

CONIC SECTIONS: PARABOLAS

Some equations of second degree have graphs that are *parabolas*. Parabolas are defined as follows.

DEFINITION	**Parabola**

A *parabola* is the locus or set of all points P in a plane equidistant from a fixed line and a fixed point in the plane. The fixed line is called the *directrix* and the fixed point is called the *focus*.

When a plane intersects a cone parallel to an element of the cone, a parabola is formed.

Parabola in three
dimensions

Parabola in a
plane

OBJECTIVES

You should be able to:

A Given an equation of a parabola, find the vertex, the focus, and the directrix, and graph the parabola.

B Given the focus and the directrix of a parabola, find an equation of the parabola.

C Find the standard form of equations of parabolas by completing the square, if necessary, and graph the parabolas, having found the vertex, the focus, and the directrix.

D Classify an equation as having a graph that is a circle, an ellipse, a hyperbola, or a parabola.

⬛ Equations of Parabolas

Now let us find equations for parabolas. Given the focus F and the directrix l, we place the coordinate axes as shown. The y-axis contains F and is perpendicular to l. The x-axis is halfway between F and l. We call the distance from F to the x-axis p. Then F has coordinates $(0, p)$ and l has the equation $y = -p$.

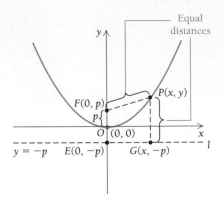

Let $P(x, y)$ be any point of the parabola and consider $\overline{PG}$ perpendicular to the line $y = -p$. The coordinates of G are $(x, -p)$. By definition of a parabola,

$$PF = PG.$$

Then using the distance formula, we have

$$\sqrt{(x - 0)^2 + (y - p)^2} = \sqrt{(x - x)^2 + (y + p)^2}.$$

Squaring, we get

$$x^2 + y^2 - 2py + p^2 = y^2 + 2py + p^2$$
$$x^2 = 4py.$$

Thus we have the following.

$$x^2 = 4py$$

Standard equation of a parabola with focus at $(0, p)$ and directrix $y = -p$. The vertex is $(0, 0)$ and the y-axis is the only line of symmetry.

We have shown that if $P(x, y)$ is on the parabola, then its coordinates satisfy this equation. The converse is also true, but we omit the proof.

Note that if $p > 0$, as above, the graph opens up. If $p < 0$, the graph opens down and the focus and the directrix exchange sides of the x-axis.

The inverse of the parabola above is described as follows.

$$y^2 = 4px$$

Standard equation of a parabola with focus at $(p, 0)$ and directrix $x = -p$. The vertex is $(0, 0)$ and the x-axis is the only line of symmetry.

We have already graphed parabolas in Chapter 4, but we review them here in the context of the new concepts of focus, directrix, and line of symmetry.

Example 1   For the parabola $y = x^2$, find the vertex, the focus, and the directrix, and draw the graph.

Solution   We first write $y = x^2$ in the form $x^2 = 4py$. We do this by re-writing the coefficient of y, 1, as $4(\frac{1}{4})$. Thus $p = \frac{1}{4}$ and

$$x^2 = 4py = y$$
$$x^2 = 4(\tfrac{1}{4})y.$$

Vertex: $(0, 0)$

Focus: $(0, \frac{1}{4})$

Directrix: $y = -\frac{1}{4}$

Example 2   For the parabola $y^2 = -12x$, find the vertex, the focus, and the directrix, and draw the graph.

Solution   We first write $y^2 = -12x$ in the form $y^2 = 4px$. We do this by rewriting the coefficient of x, -12, as $4(-3)$:

$$y^2 = 4px = -12x$$
$$y^2 = 4(-3)x.$$

Vertex: $(0, 0)$

Focus: $(-3, 0)$

Directrix: $x = -(-3)$
$\quad\quad\quad = 3$

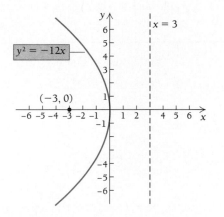

DO EXERCISES 1–3.

B   Focus and Directrix Known

Example 3   Find an equation of a parabola with focus $(5, 0)$ and directrix $x = -5$.

Solution   The focus is on the x-axis and $x = -5$ is the directrix, so the line of symmetry is the x-axis. Thus the equation is of the type

$$y^2 = 4px.$$

Since $p = 5$, the equation is $y^2 = 20x$.

For each parabola, find the vertex, the focus, and the directrix, and draw the graph.

1. $8y = x^2$

2. $y = 2x^2$

3. $y^2 = -6x$

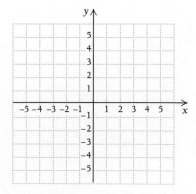

Find an equation of a parabola satisfying the given conditions.

4. Focus $(3, 0)$, directrix $x = -3$

5. Focus $(0, \frac{1}{2})$, directrix $y = -\frac{1}{2}$

6. Focus $(-6, 0)$, directrix $x = 6$

7. Focus $(0, -1)$, directrix $y = 1$

Example 4   Find an equation of a parabola with focus $(0, -7)$ and directrix $y = 7$.

Solution   The focus is on the y-axis and $y = 7$ is the directrix, so the line of symmetry is the y-axis. Thus the equation is of the type

$$x^2 = 4py.$$

Since $p = -7$, we obtain $x^2 = -28y.$

DO EXERCISES 4–7.

Standard Form by Completing the Square

If a parabola with vertex at the origin is translated h units left or right and k units up or down so that its vertex is (h, k) and its axis of symmetry is parallel to the y-axis, it has an equation as follows:

> $$(x - h)^2 = 4p(y - k),$$
>
> where the vertex is (h, k), the focus is $(h, k + p)$, and the directrix is $y = k - p$.

If a parabola is translated so that its vertex is (h, k) and its axis of symmetry is parallel to the x-axis, it has an equation as follows:

> $$(y - k)^2 = 4p(x - h),$$
>
> where the vertex is (h, k), the focus is $(h + p, k)$, and the directrix is $x = h - p$.

Example 5   For the parabola

$$x^2 + 6x + 4y + 5 = 0,$$

find the vertex, the focus, and the directrix, and draw the graph.

Solution   We complete the square:

$$x^2 + 6x \qquad\qquad = -4y - 5$$
$$x^2 + 6x + 9 - 9 = -4y - 5$$
$$x^2 + 6x + 9 = -4y + 4$$
$$(x + 3)^2 = -4(y - 1)$$
$$= 4(-1)(y - 1).$$

Vertex (h, k): $(-3, 1)$

Focus $(h, k + p)$: $(-3, 1 + (-1))$, or $(-3, 0)$

Directrix, $y = k - p$: $y = 1 - (-1)$, or
$$y = 2$$

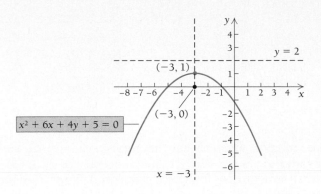

For each parabola, find the vertex, the focus, and the directrix, and draw the graph using graph paper.

8. $x^2 + 2x - 8y - 3 = 0$

9. $y^2 + 2y + 4x - 7 = 0$

Example 6   For the parabola

$$y^2 + 6y - 8x - 31 = 0,$$

find the vertex, the focus, and the directrix, and draw the graph.

Solution   We complete the square:

$$
\begin{aligned}
y^2 + 6y &= 8x + 31 \\
y^2 + 6y + 9 - 9 &= 8x + 31 \\
y^2 + 6y + 9 &= 8x + 40 \\
(y + 3)^2 &= 8(x + 5) = 4(2)(x + 5).
\end{aligned}
$$

Vertex (h, k): $(-5, -3)$
Focus $(h + p, k)$: $(-5 + 2, -3)$ or $(-3, -3)$
Directrix, $x = h - p$: $x = -5 - 2$, or $x = -7$

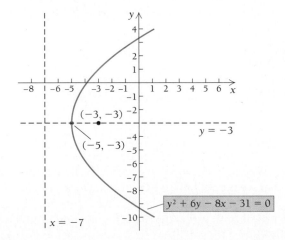

DO EXERCISES 8 AND 9.

Applications

Parabolas have many applications. For example, cross sections of headlights are parabolas. The bulb is located at the focus. All light from that point is reflected outward, parallel to the axis of symmetry.

TECHNOLOGY CONNECTION

Ellipses and hyperbolas are never functions. Parabolas may or may not be functions. Before graphing a parabola, we first determine whether it is a function.

- If the only y-term in the equation is a first-degree term, then the parabola is a function. We simply solve the equation for y, enter that function in the grapher, and draw the graph.

- If there is a second-degree y-term, then we solve for y the same way we did for ellipses and hyperbolas, and draw the two parts on the same axes.

Graph each of the following parabolas.

TC 1. $13x^2 - 8y - 9 = 0$

TC 2. $41x + 6y^2 = 12$

TC 3. $12x^2 - x + 3y - 14 = 0$

TC 4. $-6.5y^2 + 3.1x + 23.4y - 3 = 0$

(a)   (b)

Radar and radio antennas may have cross sections that are parabolas. Incoming radio waves are reflected and concentrated at the focus. Cables hung between structures to form suspension bridges form parabolas. When a cable supports only its own weight, it does not form a parabola, but rather a curve called a *catenary*.

▷ Classifying Equations

We said at the beginning of this chapter that certain equations of the type

$$Ax^2 + By^2 + Cx + Dy + E = 0$$

have graphs that are conic sections. For example, the equation $x^2 + y^2 = 4$ is a circle, but $x^2 + y^2 = -4$ has no real-number solutions, and it does *not* represent a conic section. How can we decide if the graph of an equation is a circle, an ellipse, a hyperbola, or a parabola? Certainly either A or B, or both, must be nonzero. Otherwise, we would not have a second-degree equation, and the graph is then probably a straight line, presuming either C or D is nonzero. The preceding shows that part of the answer rests in the value E. To know what the graph is, we would have to complete the square and try to find a standard form of one of the conic sections we have studied.

Let us consider some other examples.

Example 7   Classify each of the following equations as a circle, an ellipse, a hyperbola, or a parabola.

a) $6x^2 + 6y^2 = 24$
b) $y^2 + 6y = 8x + 31$
c) $4x^2 = 16y^2 + 64$
d) $16x^2 + 4y^2 + 96x = 8y - 84$

Solution

a) The fact that x and y are both squared tells us that we do not have a parabola. The fact that the squared terms both have positive coefficients tells us that we do not have a hyperbola. Do we have a circle? We need to get $x^2 + y^2$ by itself. Since both coefficients are the same, 6, we can factor 6 out of both terms on the left and then multiply by $\frac{1}{6}$:

$$6(x^2 + y^2) = 24$$
$$x^2 + y^2 = 4$$
$$x^2 + y^2 = 2^2.$$

The graph is a circle with center at the origin and radius 2.

b) Since only one of the variables is squared, we know that the graph is not a circle, an ellipse, or a hyperbola. We find the following equivalent equation:

$$(y + 3)^2 = 4(2)(x + 5).$$

This tells us that we have a parabola that opens to the right and whose axis of symmetry is parallel to the x-axis.

c) Both variables are squared, so the graph is not a parabola. We can obtain the equivalent equation:

$$\frac{x^2}{16} - \frac{y^2}{4} = 1.$$

The minus sign then tells us that the graph is a hyperbola.

d) Both variables are squared, so the graph is not a parabola. We have a plus sign between the squared terms, so the graph is not a hyperbola. If the coefficients of the squared terms were the same, we might have the graph of a circle, as in part (a), but they are not. We find the equivalent equation:

$$\frac{(x+3)^2}{4} + \frac{(y-1)^2}{16} = 1.$$

Thus the graph is an ellipse.

◀

We can also classify equations of the type

$$Ax^2 + By^2 + Cx + Dy + E = 0$$

by examining the coefficients, as follows.

Conic Section	Coefficients	Example
Ellipse	$A \neq B, AB > 0$	$16x^2 + 4y^2 = 64$
Circle	$A = B, A \neq 0, B \neq 0$	$x^2 + y^2 = 36$
Parabola	$A = 0 \text{ or } B = 0,$ but both cannot be 0.	$y^2 = 3(x-7),$ $(x+1)^2 = 6y + 11$
Hyperbola	$AB < 0$	$4x^2 = 16y^2 + 64$

DO EXERCISES 10–13.

Classify as a circle, an ellipse, a hyperbola, or a parabola.

10. $x^2 + 2x = 8y + 3$

11. $x^2 + y^2 + 4y = 14x + 11$

12. $4y^2 - 36 = 9x^2$

13. $36y = 25x^2 + 9y^2 + 150x + 260$

●**EXERCISE SET** 10.3

A Find the vertex, the focus, and the directrix, and draw the graph.

1. $x^2 = 20y$
2. $x^2 = 16y$
3. $y^2 = -6x$
4. $y^2 = -2x$
5. $x^2 - 4y = 0$
6. $y^2 + 4x = 0$
7. $y = 2x^2$
8. $y = \frac{1}{2}x^2$

B Find an equation of a parabola satisfying the given conditions.

9. Focus $(4, 0)$, directrix $x = -4$

10. Focus $(0, \frac{1}{4})$, directrix $y = -\frac{1}{4}$
11. Focus $(-\sqrt{2}, 0)$, directrix $x = \sqrt{2}$
12. Focus $(0, -\pi)$, directrix $y = \pi$
13. Focus $(3, 2)$, directrix $x = -4$
14. Focus $(-2, 3)$, directrix $y = -3$

C Find the vertex, the focus, and the directrix, and draw the graph.

15. $(x+2)^2 = -6(y-1)$
16. $(y-3)^2 = -20(x+2)$

17. $x^2 + 2x + 2y + 7 = 0$

18. $y^2 + 6y - x + 16 = 0$

19. $x^2 - y - 2 = 0$

20. $x^2 - 4x - 2y = 0$

21. $y = x^2 + 4x + 3$

22. $y = x^2 + 6x + 10$

23. $4y^2 - 4y - 4x + 24 = 0$

24. $4y^2 + 4y - 4x - 16 = 0$

▷ Classify as a circle, an ellipse, a hyperbola, or a parabola.

25. $x + 1 = 2y^2$

26. $10y + 40 = x^2 + y^2 + 8x$

27. $4y^2 + 25x^2 + 4 = 8y + 100x$

28. $9x^2 + 24y = 4y^2 + 36x + 36$

29. $2x + 13 + y^2 = 8y - x^2$

30. $x - \dfrac{5}{y} = 0$

31. $x = -16y + y^2 + 7$

32. $16x^2 + 5y^2 - 12x^2 + 8y^2 - 3x + 4y = 568$

33. $xy + 5x^2 = 9 + 7x^2 - 2x^2$

34. $56x^2 - 17y^2 = 234 - 13x^2 - 38y^2$

● **SYNTHESIS** _____

35. Graph each of the following using the same set of axes.

$$x^2 - y^2 = 0,$$
$$x^2 - y^2 = 1,$$
$$x^2 + y^2 = 1,$$

and

$$y = x^2.$$

36. Graph each of the following using the same set of axes.

$$x^2 - 4y^2 = 0,$$
$$x^2 - 4y^2 = 1,$$
$$x^2 + 4y^2 = 1,$$

and

$$x = 4y^2.$$

37. Find an equation of the following parabola: Line of symmetry parallel to the y-axis, vertex $(-1, 2)$, and passing through $(-3, 1)$.

38. a) Graph $(y - 3)^2 = -20(x + 1)$. Is this relation a function?

   b) In general, is $(y - k)^2 = 4p(x - h)$ a function?

39. Find an equation of a parabola containing the point $(-3, 5)$, symmetric with respect to a horizontal line, and with vertex $(-2, 1)$.

40. The cables of a suspension bridge are 50 ft above the roadbed at the ends of the bridge and 10 ft above it in the center of the bridge. The roadbed is 200 ft long. Vertical cables are to be spaced every 20 ft along the bridge. Calculate the lengths of these vertical cables.

41. Show that the determinant equation

$$\begin{vmatrix} y - k & x - h \\ 4p & y - k \end{vmatrix} = 0$$

is an equation of a parabola with vertex (h, k).

● **CHALLENGE** _____

42. Prove that when a cable supports a load distributed uniformly horizontally, it hangs in the shape of a parabola. (*Hint:* Proceed as follows.)

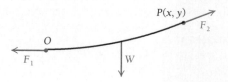

   a) Place a coordinate system as shown here, with the origin at the lowest point of the cable.

   b) For a point $P(x, y)$ on the cable, write an equation of rotational equilibrium. The forces involved are the tensions in the cable, F_1 and F_2, and the weight supported, W (which is a function of x). The weight of the cable is essentially neglected. Use point P as the center of rotation.

   c) Solve for y.

▨▧ **TECHNOLOGY CONNECTION** _____

Use a grapher to find the vertex, the focus, and the directrix.

43. $4.5x^2 - 7.8x + 9.7y = 0$

44. $134.1y^2 + 43.4x - 316.6y - 112.4 = 0$

10.4
SYSTEMS OF FIRST-DEGREE AND SECOND-DEGREE EQUATIONS

All the systems of equations that we have studied so far have been linear. We now consider systems of two equations in two variables in which at least one equation is not linear.

A ▸ Algebraic Solutions

We first consider systems of one first-degree and one second-degree equation. For example, the graphs may be a circle and a line. If so, there are three possibilities.

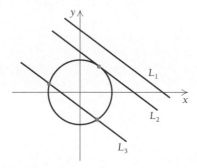

For L_1 there is no point of intersection, hence no solution in the set of real numbers. For L_2 there is one point of intersection, hence one real-number solution. For L_3 there are two points of intersection, hence two real-number solutions.

These systems can be solved graphically by finding the points of intersection. In solving algebraically, we use the substitution method.

Example 1 Solve the following system:

$$x^2 + y^2 = 25, \quad (1) \quad \text{(The graph is a circle.)}$$
$$3x - 4y = 0. \quad (2) \quad \text{(The graph is a line.)}$$

Solution We first solve the linear equation (2) for x:

$$x = \tfrac{4}{3}y. \quad (3)$$

We then substitute $\tfrac{4}{3}y$ for x in equation (1) and solve for y:

$$(\tfrac{4}{3}y)^2 + y^2 = 25$$
$$\tfrac{16}{9}y^2 + y^2 = 25$$
$$\tfrac{25}{9}y^2 = 25$$
$$y^2 = 9$$
$$y = \pm 3.$$

Now we substitute these numbers for y in equation (3) and solve for x:

$$x = \tfrac{4}{3}(3) = 4,$$
$$x = \tfrac{4}{3}(-3) = -4.$$

Solve. Sketch the graphs to confirm the solutions.

1. $x^2 + y^2 = 25,$
$y - x = -1$

2. $y = x^2 - 2x - 1,$
$y = x + 3$

Check: For $(4, 3)$:

$$\frac{x^2 + y^2 = 25}{\begin{array}{c|c} 4^2 + 3^2 & 25 \\ 16 + 9 & \\ 25 & \end{array}} \qquad \frac{3x - 4y = 0}{\begin{array}{c|c} 3(4) - 4(3) & 0 \\ 12 - 12 & \\ 0 & \end{array}}$$

For $(-4, -3)$:

$$\frac{x^2 + y^2 = 25}{\begin{array}{c|c} (-4)^2 - (-3)^2 & 25 \\ 16 + 9 & \\ 25 & \end{array}} \qquad \frac{3x - 4y = 0}{\begin{array}{c|c} 3(-4) - 4(-3) & 0 \\ -12 + 12 & \\ 0 & \end{array}}$$

The pairs $(4, 3)$ and $(-4, -3)$ check, so they are solutions. We can see the solutions in the graph. The graph of equation (1) is a circle, and the graph of equation (2) is a line. The graphs intersect at the points $(4, 3)$ and $(-4, -3)$.

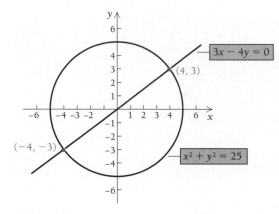

DO EXERCISES 1 AND 2.

Example 2   Solve the following system:

$$y + 3 = 2x, \qquad (1)$$
$$x^2 + 2xy = -1. \qquad (2)$$

Solution   We first solve the linear equation (1) for y:

$$y = 2x - 3. \qquad (3)$$

We then substitute $2x - 3$ for y in equation (2) and solve for x:

$$\begin{aligned} x^2 + 2x(2x - 3) &= -1 \\ x^2 + 4x^2 - 6x &= -1 \\ 5x^2 - 6x + 1 &= 0 \\ (5x - 1)(x - 1) &= 0 \qquad \text{Factoring} \\ 5x - 1 = 0 \quad \text{or} \quad x - 1 &= 0 \qquad \text{Using the principle of zero products} \\ x = \tfrac{1}{5} \quad \text{or} \qquad x &= 1. \end{aligned}$$

Now we substitute these numbers for x in equation (3) and solve for y:

$$y = 2(\tfrac{1}{5}) - 3 = -\tfrac{13}{5},$$
$$y = 2(1) - 3 = -1.$$

The pairs $(\tfrac{1}{5}, -\tfrac{13}{5})$ and $(1, -1)$ check, so they are solutions.

DO EXERCISE 3.

Example 3 Solve:

$$x + y = 5, \quad \text{(The graph is a line.)}$$
$$y = 3 - x^2. \quad \text{(The graph is a parabola.)}$$

Solution We substitute $3 - x^2$ for y in the first equation:

$$x + 3 - x^2 = 5$$
$$-x^2 + x - 2 = 0$$
$$x^2 - x + 2 = 0.$$

To solve this equation, we need the quadratic formula:

$$x = \frac{-b \pm \sqrt{b^2 - 4ac}}{2a}$$

$$= \frac{-(-1) \pm \sqrt{(-1)^2 - 4(1)(2)}}{2(1)}$$

$$= \frac{1 \pm \sqrt{1 - 8}}{2}$$

$$= \frac{1 \pm \sqrt{-7}}{2} = \frac{1}{2} \pm \frac{\sqrt{7}}{2} i.$$

Then solving the first equation for y, we obtain $y = 5 - x$. Substituting values for x gives us

$$y = 5 - \left(\frac{1}{2} + \frac{\sqrt{7}}{2} i\right) = \frac{9}{2} - \frac{\sqrt{7}}{2} i$$

and

$$y = 5 - \left(\frac{1}{2} - \frac{\sqrt{7}}{2} i\right) = \frac{9}{2} + \frac{\sqrt{7}}{2} i.$$

The solutions are

$$\left(\frac{1}{2} + \frac{\sqrt{7}}{2} i, \frac{9}{2} - \frac{\sqrt{7}}{2} i\right) \quad \text{and} \quad \left(\frac{1}{2} - \frac{\sqrt{7}}{2} i, \frac{9}{2} + \frac{\sqrt{7}}{2} i\right).$$

There are no real-number solutions. Note in the figure below that the graphs do not intersect. Getting only complex-number solutions tells us that the graphs do not intersect.

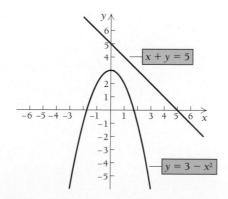

DO EXERCISE 4.

3. Solve:

$$y + 3x = 1,$$
$$x^2 - 2xy = 5.$$

4. Solve:

$$9x^2 - 4y^2 = 36,$$
$$5x + 2y = 0.$$

 TECHNOLOGY CONNECTION

Because the algebra is often difficult, finding the solution(s) to a system of nonlinear equations provides an excellent opportunity to use your grapher. No matter how complicated the equations, if they can be written in the form $y = f(x)$, then their intersections can be determined on the grapher. As with systems of linear equations, we simply zoom in as often as necessary and then use the trace feature to read the coordinates of the point(s) of intersection. Using a grapher restricts solutions to *real* numbers. Few graphers have the ability to give solutions that are complex numbers.

Solve each of the following systems using a grapher. Round to the nearest hundredth.

TC 1. $4xy - 7 = 0,$
$x - 3y - 2 = 0$

TC 2. $x^2 + y^2 = 14,$
$16x + 7y^2 = 0$

TC 3. $0.4x^2 + 0.6y^2 = 1,$
$y = 1.4x^2 + 1.4$

▶ Problem Solving

We now consider solving problems in which the translation is a system of equations, one quadratic and one linear.

Example 4   For a building at a community college, an architect wants to lay out a rectangular piece of ground that has a perimeter of 204 m and an area of 2565 m². Find the dimensions of the piece of ground.

Solution

1. **Familiarize.**   We draw a picture of the field, labeling the drawing. We let l = the length and w = the width.

$$\text{Perimeter} = 2l + 2w$$
$$= 204 \text{ m}$$

$$\text{Area} = lw$$
$$= 2565 \text{ m}^2$$

2. **Translate.**   We then have the following translation:

    Perimeter: $2w + 2l = 204;$

    Area: $lw = 2565.$

3. **Carry out.**   We solve the system

    $2w + 2l = 204,$ (The graph is a line.)

    $lw = 2565.$ (The graph is a hyperbola.)

    We solve the second equation for l and get $l = 2565/w$. Then we substitute $2565/w$ for l in the first equation and solve for w:

    $$2w + 2\left(\frac{2565}{w}\right) = 204$$

    $$2w^2 + 2(2565) = 204w \qquad \text{\small Multiplying by } w$$
    $$2w^2 - 204w + 2(2565) = 0 \qquad \text{\small Standard form}$$
    $$w^2 - 102w + 2565 = 0 \qquad \text{\small Multiplying by } \tfrac{1}{2}$$
    $$w = \frac{-(-102) \pm \sqrt{(-102)^2 - 4 \cdot 1 \cdot 2565}}{2 \cdot 1}$$

    Quadratic formula. Factoring could also be used, but the numbers are quite large.

    $$w = \frac{102 \pm \sqrt{144}}{2} = \frac{102 \pm 12}{2}$$
    $$w = 57 \quad \text{or} \quad w = 45.$$

    If $w = 57$, then $l = 2565/w = 2565/57 = 45$. If $w = 45$, then $l = 2565/w = 2565/45 = 57$. Since length is usually considered to be longer than width, we have the solution $l = 57$ and $w = 45$, or $(57, 45)$.

4. **Check.**   If $l = 57$ and $w = 45$, the perimeter is $2 \cdot 57 + 2 \cdot 45$, or 204. The area is $57 \cdot 45$, or 2565. The numbers check.

5. State.   The answer is that the length is 57 m and the width is 45 m.

DO EXERCISE 5.

5. The perimeter of a rectangular field is 34 m, and the length of a diagonal is 13 m. Find the dimensions of the field.

●**EXERCISE SET** 10.4

Ⓐ   Solve.

1. $x^2 + y^2 = 25,$
   $y - x = 1$

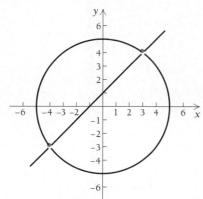

2. $x^2 + y^2 = 100,$
   $y - x = 2$

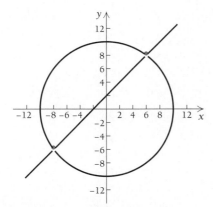

3. $4x^2 + 9y^2 = 36,$
   $3y + 2x = 6$

4. $9x^2 + 4y^2 = 36,$
   $3x + 2y = 6$

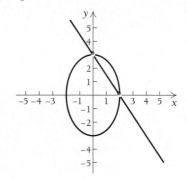

5. $y^2 - x^2 = 9,$
   $2x - 3 = y$

6. $x + y = -6,$
   $xy = -7$

7. $y^2 = x + 3,$
   $2y = x + 4$

8. $y = x^2,$
   $3x = y + 2$

9. $x^2 + 4y^2 = 25,$
   $x + 2y = 7$

10. $y^2 - x^2 = 16,$
   $2x - y = 1$

11. $x^2 - xy + 3y^2 = 27,$
   $x - y = 2$

12. $2y^2 + xy + x^2 = 7,$
   $x - 2y = 5$

13. $3x + y = 7,$
   $4x^2 + 5y = 56$

14. $2y^2 + xy = 5,$
   $4y + x = 7$

15. $a + b = 7,$
   $ab = 4$

16. $p + q = -4,$
   $pq = -5$

17. $2a + b = 1,$
   $b = 4 - a^2$

18. $4x^2 + 9y^2 = 36,$
   $x + 3y = 3$

19. $a^2 + b^2 = 89,$
   $a - b = 3$

20. $xy = 4,$
   $x + y = 5$

21. $x^2 + y^2 = 5,$
   $x - y = 8$

22. $4x^2 + 9y^2 = 36,$
   $y - x = 8$

Ⓑ   Solve.

23. The sum of two numbers is 12, and the sum of their squares is 90. What are the numbers?

24. The sum of two numbers is 15, and the difference of their squares is also 15. What are the numbers?

25. A rectangle has a perimeter of 28 cm, and the length of a diagonal is 10 cm. What are its dimensions?

26. A rectangle has a perimeter of 6 m, and the length of a diagonal is $\sqrt{5}$ m. What are its dimensions?

27. A rectangle has an area of 20 in² and a perimeter of 18 in. Find its dimensions.

28. A rectangle has an area of 2 yd² and a perimeter of 6 yd. Find its dimensions.

29. It will take 210 yd of fencing to enclose a rectangular field. The area of the field is 2250 yd². What are the dimensions?

30. The diagonal of a rectangle is 1 ft longer than the length of the rectangle and 3 ft longer than twice the width. Find the dimensions of the rectangle.

● **SYNTHESIS** _____

31. Find two numbers whose product is 2 and the sum of whose reciprocals is $\frac{33}{8}$.

32. Find an equation of a circle that passes through $(-2, 3)$ and $(-4, 1)$ and whose center is on the line $5x + 8y = -2$.

33. A piece of wire 100 cm long is to be cut into two pieces and those pieces are each to be bent to make a square. The area of one square is to be 144 cm² greater than that of the other square. How should the wire be cut?

34. The sum of two numbers is 1, and their product is 1. Find the sum of their cubes. There is a method to solve this problem that is easier than solving a system of one first-degree equation and one second-degree equation. Can you discover it?

35. Find an equation of an ellipse centered at the origin that passes through the points $(1, \sqrt{3}/2)$ and $(\sqrt{3}, 1/2)$.

36. Find an equation of a hyperbola of the type

$$\frac{x^2}{a^2} - \frac{y^2}{b^2} = 1$$

that passes through the points $(-3, -3\sqrt{5}/2)$, $(-3, 3\sqrt{5}/2)$, and $(-3/2, 0)$.

Solve for x and y.

37. $x - y = a + 2b,$
$x^2 - y^2 = a^2 + 2ab + b^2$

38. $\dfrac{x}{a - b} + \dfrac{y}{a + b} = 1,$
$x^2 - y^2 = (a - b)^2$

39. Given the area A and the perimeter P of a rectangle,

show that the length L and the width W are given by the formulas
$$L = \tfrac{1}{4}(P + \sqrt{P^2 - 16A}),$$
$$W = \tfrac{1}{4}(P - \sqrt{P^2 - 16A}).$$

40. Show that a hyperbola does not intersect its asymptotes. That is, solve the system

$$\frac{x^2}{a^2} - \frac{y^2}{b^2} = 1,$$

$$y = \frac{b}{a}x \quad \left(\text{or } y = -\frac{b}{a}x\right).$$

41. Find an equation of a circle that passes through the points $(2, 4)$ and $(3, 3)$ and whose center is on the line $3x - y = 3$.

42. Find an equation of a circle that passes through the points $(7, 3)$ and $(5, 5)$ and whose center is on the line $y - 4x = 1$.

● **CHALLENGE** _____

43. Solve:
$$x^3 + y^3 = 72,$$
$$x + y = 6.$$

44. Solve for h, k, and λ:

$$1 - 20\lambda k^2 = 0,$$
$$2 - 10\lambda hk = 0,$$
$$hk^2 - 640{,}000 = 0.$$

▤ **TECHNOLOGY CONNECTION** _____

Solve using a grapher.

45. $x^2 + y^2 = 19{,}380{,}510.36,$
$27{,}942.25x - 6.125y = 0$

46. $2x + 2y = 1660,$
$xy = 35{,}325$

47. $14.5x^2 - 13.5y^2 - 64.5 = 0,$
$5.5x - 6.3y - 12.3 = 0$

48. $13.5xy + 15.6 = 0,$
$5.6x - 6.7y - 42.3 = 0$

OBJECTIVES

You should be able to:

▲ Use the elimination method to solve systems of equations like
$$2x^2 + 5y^2 = 20,$$
$$3x^2 - y^2 = -1.$$
(continued)

10.5

SYSTEMS OF SECOND-DEGREE EQUATIONS

We now consider systems of two second-degree equations. The following figure shows the ways in which a circle and a hyperbola can intersect.

4 real solutions   3 real solutions   2 real solutions   1 real solution   0 real solutions

▶ Use the substitution method to solve systems of equations like

$$x^2 + 4y^2 = 20,$$

$$xy = 4.$$

▶ Solve applied problems involving systems of two second-degree equations.

A Algebraic Solutions by Elimination

To solve systems of two second-degree equations, we can use either the substitution method or the elimination method. The elimination method is generally used when each equation is of the form $Ax^2 + By^2 = C$. Then we can eliminate an x^2- or a y^2-term.

Example 1   Solve this system:

$$x^2 + y^2 = 25, \quad (1) \qquad \text{(The graph is a circle.)}$$
$$x^2 - y^2 = 25. \quad (2) \qquad \text{(The graph is a hyperbola.)}$$

Solution   We use the elimination method:

$$\begin{array}{r} x^2 + y^2 = 25 \\ x^2 - y^2 = 25 \\ \hline 2x^2 \quad\;\; = 50 \end{array} \qquad \text{Adding}$$

$$x^2 = 25$$
$$x = \pm 5.$$

If $x = 5$, $x^2 = 25$, and if $x = -5$, $x^2 = 25$, so substituting either 5 or -5 for x in equation (1) gives us

$$25 + y^2 = 25$$
$$y^2 = 0$$
$$y = 0.$$

Thus if $x = 5$, $y = 0$, and if $x = -5$, $y = 0$. The possible solutions are $(5, 0)$ and $(-5, 0)$.

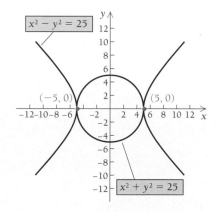

Check: Since $(5)^2 = 25$ and $(-5)^2 = 25$, we can do both checks at once.

1. Solve. Sketch graphs to confirm the solutions.

$$x^2 + y^2 = 4,$$
$$x^2 - y^2 = 4$$

Solve. Sketch graphs to confirm the solutions.

2. $x^2 + y^2 = 16,$
   $9x^2 + 16y^2 = 144$

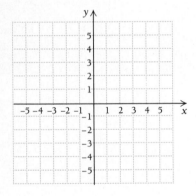

3. $x^2 + y^2 = 4,$
   $4x^2 + 25y^2 = 100$

$x^2 - y^2 = 25$	
$(\pm 5)^2 - (0)^2$	25
$25 - 0$	
25	

$x^2 + y^2 = 25$	
$(\pm 5)^2 + 0^2$	25
$25 + 0$	
25	

The solutions are $(5, 0)$ and $(-5, 0)$. We can see them as intersections of the graphs. ◀

DO EXERCISE 1.

Example 2 Solve this system:

$$2x^2 + 5y^2 = 22, \qquad (1)$$
$$3x^2 - y^2 = -1. \qquad (2)$$

Solution Here we multiply the second equation by 5 and use the elimination method:

$$
\begin{array}{ll}
2x^2 + 5y^2 = 22 & \\
\underline{15x^2 - 5y^2 = -5} & \text{Multiplying by 5} \\
17x^2 \qquad\quad = 17 & \text{Adding} \\
x^2 = 1 & \\
x = \pm 1. &
\end{array}
$$

If $x = 1$, $x^2 = 1$, and if $x = -1$, $x^2 = 1$. Thus substituting 1 or -1 for x in equation (2), we have

$$
\begin{aligned}
3 \cdot (\pm 1)^2 - y^2 &= -1 \\
3 - y^2 &= -1 \\
-y^2 &= -4 \\
y^2 &= 4 \\
y &= \pm 2.
\end{aligned}
$$

Thus, if $x = 1$, $y = 2$, or $y = -2$, and if $x = -1$, $y = 2$, or $y = -2$. The possible solutions are $(1, 2)$, $(1, -2)$, $(-1, 2)$, and $(-1, -2)$.

Check: Since $(2)^2 = 4$, $(-2)^2 = 4$, $(1)^2 = 1$, and $(-1)^2 = 1$, we can check all four pairs at one time.

$2x^2 + 5y^2 = 22$	
$2(\pm 1)^2 + 5(\pm 2)^2$	22
$2 + 20$	
22	

$3x^2 - y^2 = -1$	
$3(\pm 1)^2 - (\pm 2)^2$	-1
$3 - 4$	
-1	

The solutions are $(1, 2)$, $(1, -2)$, $(-1, 2)$, and $(-1, -2)$. ◀

DO EXERCISES 2–4. (EXERCISE 4 IS ON THE FOLLOWING PAGE.)

B The Substitution Method

When one equation contains a product of variables and the other equation is of the form $Ax^2 + By^2 = C$, we often solve for one of the variables in the equation with the product and then substitute in the other.

Example 3 Solve the system

$$x^2 + 4y^2 = 20, \qquad (1)$$
$$xy = 4. \qquad (2)$$

Solution Here we use the substitution method. First we solve equation (2) for y:

$$xy = 4$$

$$y = \frac{4}{x}.$$

Then we substitute $4/x$ for y in equation (1) and solve for x:

$$x^2 + 4y^2 = 20$$

$$x^2 + 4\left(\frac{4}{x}\right)^2 = 20$$

$$x^2 + \frac{64}{x^2} = 20$$

$$x^4 + 64 = 20x^2 \qquad \text{Multiplying by } x^2$$

$$x^4 - 20x^2 + 64 = 0$$

$$u^2 - 20u + 64 = 0 \qquad \text{Letting } u = x^2$$

$$(u - 16)(u - 4) = 0 \qquad \text{Factoring}$$

$$u = 16 \quad \text{or} \quad u = 4. \qquad \text{Principle of zero products}$$

Then $x^2 = 16$ or $x^2 = 4$, so $x = \pm 4$ or $x = \pm 2$. Since $y = 4/x$, if $x = 4$, $y = 1$; if $x = -4$, $y = -1$; if $x = 2$, $y = 2$; and if $x = -2$, $y = -2$. The solutions are $(4, 1)$, $(-4, -1)$, $(2, 2)$, and $(-2, -2)$. ◀

DO EXERCISE 5.

 Problem Solving

Example 4 The area of a rectangular Oriental rug is 300 yd², and the length of a diagonal is 25 yd. Find the dimensions of the rug.

Solution

1. **Familiarize.** We draw a picture and label it. We note that there is a right triangle in the figure. We let l = the length of the rectangle and w = the width of the rectangle.

2. **Translate.** We translate to a system of equations.

From the Pythagorean theorem: $l^2 + w^2 = 25^2$;   (The graph is a circle.)

From an area formula: $lw = 300$.   (The graph is a hyperbola.)

4. Solve:

$$2y^2 - 3x^2 = 6,$$
$$5y^2 + 2x^2 = 53.$$

5. Solve:

$$x^2 + xy + y^2 = 19,$$
$$xy = 6.$$

6. The area of a rectangle is $2\,\text{ft}^2$, and the length of a diagonal is $\sqrt{5}$ ft. Find the dimensions of the rectangle.

3. Carry out.   We solve the system

$$l^2 + w^2 = 625,$$
$$lw = 300$$

to get $(20, 15)$ and $(-20, -15)$.

4. Check.   Lengths of sides cannot be negative, so we need check only $(20, 15)$. In the right triangle, $20^2 + 15^2 = 400 + 225 = 625$, which is 25^2. The area is $20 \cdot 15 = 300$, so we have a solution.

5. State.   The answer is that the length is 20 yd and the width is 15 yd. ◄

DO EXERCISE 6. _____

• EXERCISE SET $\mathbf{10.5}$

Ⓐ, Ⓑ   Solve.

1. $x^2 + y^2 = 25,$
$\quad y^2 = x + 5$

2. $y = x^2,$
$\quad x = y^2$

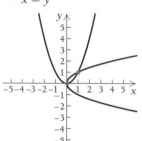

3. $x^2 + y^2 = 9,$
$\quad x^2 - y^2 = 9$

4. $y^2 - 4x^2 = 4,$
$\quad 4x^2 + y^2 = 4$

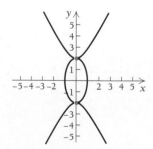

5. $x^2 + y^2 = 25,$
$\quad xy = 12$

6. $x^2 - y^2 = 16,$
$\quad x + y^2 = 4$

7. $x^2 + y^2 = 4,$
$\quad 16x^2 + 9y^2 = 144$

8. $x^2 + y^2 = 25,$
$\quad 25x^2 + 16y^2 = 400$

9. $x^2 + y^2 = 16,$
$\quad y^2 - 2x^2 = 10$

10. $x^2 + y^2 = 14,$
$\quad x^2 - y^2 = 4$

11. $x^2 + y^2 = 5,$
$\quad xy = 2$

12. $x^2 + y^2 = 20,$
$\quad xy = 8$

13. $x^2 + y^2 = 13,$
$\quad xy = 6$

14. $x^2 + 4y^2 = 20,$
$\quad xy = 4$

15. $x^2 + y^2 + 6y + 5 = 0,$
$\quad x^2 + y^2 - 2x - 8 = 0$

16. $2xy + 3y^2 = 7,$
$\quad 3xy - 2y^2 = 4$

17. $xy - y^2 = 2,$
$\quad 2xy - 3y^2 = 0$

18. $4a^2 - 25b^2 = 0,$
$\quad 2a^2 - 10b^2 = 3b + 4$

19. $m^2 - 3mn + n^2 + 1 = 0,$
$\quad 3m^2 - mn + 3n^2 = 13$

20. $ab - b^2 = -4,$
$\quad ab - 2b^2 = -6$

21. $a^2 + b^2 = 14,$
$\quad ab = 3\sqrt{5}$

22. $x^2 + xy = 5,$
$\quad 2x^2 + xy = 2$

23. $x^2 + y^2 = 25,$
$\quad 9x^2 + 4y^2 = 36$

24. $x^2 + y^2 = 1,$
$\quad 9x^2 - 16y^2 = 144$

Ⓒ

25. Find two numbers whose product is 156 if the sum of their squares is 313.

26. Find two numbers whose product is 60 if the sum of their squares is 136.

27. The area of a rectangle is $\sqrt{3}\,\text{m}^2$, and the length of a diagonal is 2 m. Find the dimensions.

28. The area of a rectangle is $\sqrt{2}\,\text{m}^2$, and the length of a diagonal is $\sqrt{3}$ m. Find the dimensions.

29. A garden contains two square peanut beds. Find the length of each bed if the sum of their areas is $832\,\text{ft}^2$ and the difference of their areas is $320\,\text{ft}^2$.

30. A certain amount of money saved for 1 yr at a certain interest rate yielded $7.50. If the principal had been $25 more and the interest rate 1% less, the interest would have been the same. Find the principal and the rate.

● SYNTHESIS _____

31. Find an equation of the circle that passes through the points $(4,6)$, $(-6,2)$, and $(1,-3)$.

32. Find an equation of the circle that passes through the points $(2,3)$, $(4,5)$, and $(0,-3)$.

33. The square of a certain number exceeds twice the square of another number by $\frac{1}{8}$. The sum of their squares is $\frac{5}{16}$. Find the numbers.

34. Four squares with sides 5 in. long are cut from the corners of a rectangular metal sheet that has an area of 340 in². The edges are bent up to form an open box with a volume of 350 in³. Find the dimensions of the box.

Solve for x and y.

35. $x^2 + xy = a,$
$y^2 + xy = b$

36. $x^2 - y^2 = a^2 - b^2,$
$x - y = a - b$

Solve.

37. $p^2 + q^2 = 13,$
$\dfrac{1}{pq} = -\dfrac{1}{6}$

38. $a + b = \dfrac{5}{6},$
$\dfrac{a}{b} + \dfrac{b}{a} = \dfrac{13}{6}$

39. $x^2 + y^2 = 4,$
$(x-1)^2 + y^2 = 4$

40. Solve for x and y:
$a^2b^2 - 2b^2x^2 - a^2y^2 = 0,$
$a^2b^2 - b^2x^2 - 2a^2y^2 = 0.$

41. Find k such that the following equations have two roots in common:
$(x-2)^4 - (x-2) = 0,$
$x^2 - kx + k = 0.$

Solve.

42. $10x^2 - xy + 4y^2 = 28,$
$2x^2 - 3xy - 2y^2 = 0$

43. $5^{x+y} = 100,$
$3^{2x-y} = 1000$

44. $e^x - e^{x+y} = 0,$
$e^y - e^{x-y} = 0$

45. The sum, the product, and the sum of the squares of two numbers are all the same. Find the numbers.

▨ TECHNOLOGY CONNECTION _____

Solve using a grapher.

46. ▤ $0.319x^2 + 2688.7y^2 = 56{,}548,$
$0.306x^2 - 2688.7y^2 = 43{,}452$

47. ▤ $18.465x^2 + 788.723y^2 = 6408,$
$106.535x^2 - 788.723y^2 = 2692$

SUMMARY AND REVIEW 10

● TERMS TO KNOW

Conic section, p. 634
Ellipse, p. 637
Foci, p. 637
Major axis, p. 639
Minor axis, p. 639

Vertices, p. 639
Hyperbola, p. 646
Transverse axis, p. 648
Conjugate axis, p. 648

Asymptote, p. 648
Parabola, p. 653
Directrix, p. 653
Line of symmetry, p. 654

● REVIEW EXERCISES

1. Graph: $2x^2 - 3xy - 2y^2 = 0$.

2. Find the center, the vertices, and the foci of the ellipse
$16x^2 + 25y^2 - 64x + 50y - 311 = 0.$
Then graph the ellipse.

3. Find an equation of the ellipse having vertices $(3,0)$ and $(0,4)$ and centered at the origin.

4. Find the center, the vertices, the foci, and the asymptotes of the hyperbola
$x^2 - 2y^2 + 4x + y - \frac{1}{8} = 0.$

Graph.

5. $xy = -2$

6. $4y^2 - x^2 = 16$

7. Find an equation of the parabola with directrix $y = \frac{3}{2}$ and focus $(0, -\frac{3}{2})$.

8. Find the focus, the vertex, and the directrix of the parabola $y^2 = -12x$.

9. Find the vertex, the focus, and the directrix of the parabola
$$x^2 + 10x + 2y + 9 = 0.$$

Classify the equation as a circle, an ellipse, a parabola, or a hyperbola.

10. $4x^2 + 4y^2 = 100 - 64x - 20y$

11. $9x^2 + 2y^2 = 18$

12. $y = -x^2 + 2x - 3 + 10y$

13. $\dfrac{y^2}{9} - \dfrac{x^2}{4} = 1$ **14.** $xy = 11$

15. $x = y^2 + 2y - 2 - 5x$ **16.** $xy = -3$

17. $x^2 + y^2 + 6x - 8y - 39 = 0$

Solve.

18. $x^2 - 16y = 0,$
$\quad x^2 - y^2 = 64$

19. $4x^2 + 4y^2 = 65,$
$\quad 6x^2 - 4y^2 = 25$

20. $x^2 - y^2 = 33,$
$\quad x + y = 11$

21. $x^2 - 2x + 2y^2 = 8,$
$\quad 2x + y = 6$

22. $x^2 - y = 3,$
$\quad 2x - y = 3$

23. $x^2 + y^2 = 25,$
$\quad x^2 - y^2 = 7$

24. $x^2 - y^2 = 3,$
$\quad y = x^2 - 3$

25. $x^2 + y^2 = 18,$
$\quad 2x + y = 3$

26. $x^2 + y^2 = 100,$
$\quad 2x^2 - 3y^2 = -120$

27. $x^2 + 2y^2 = 12,$
$\quad xy = 4$

28. The sides of a triangle are 8, 10, and 14. Find the altitude to the longest side.

29. The sum of two numbers is 11 and the sum of their squares is 65. Find the numbers.

30. A rectangle has a perimeter of 38 m and an area of 84 m². What are its dimensions?

31. Find two positive integers whose sum is 12 and the sum of whose reciprocals is $\frac{3}{8}$.

32. The perimeter of a square is 12 cm more than the perimeter of another square. Its area exceeds the area of the other by 39 cm². Find the perimeter of each square.

33. The sum of the area of two circles is 130π ft². The difference of the areas is 112π ft². Find the radius of each circle.

● **SYNTHESIS** _____

34. Find an equation of the ellipse that contains the point $(-1/2, 3\sqrt{3}/2)$ and two of whose vertices are $(-1, 0)$ and $(1, 0)$.

35. Find two numbers whose product is 4 and the sum of whose reciprocals is $\frac{65}{56}$.

36. Find an equation of the circle that passes through the points $(10, 7)$, $(-6, 7)$, and $(-8, 1)$.

● **THINKING AND WRITING** _____

1. Which, if any, of the conic sections studied in this chapter are functions?

2. Compare the number of x- and y-intercepts each type of conic section may have.

3. Explain the differences between the graph of a parabola and that of a hyperbola.

CHAPTER TEST 10

1. Graph: $3x^2 - 5xy - 2y^2 = 0$.

2. Find the center, the vertices, and the foci of the ellipse
$$9x^2 + y^2 - 36x - 8y + 43 = 0.$$
Then graph the ellipse.

3. Find an equation of the ellipse with vertices $(-7, 0)$, $(7, 0)$, $(0, -2)$, and $(0, 2)$.

4. Find the center, the vertices, the foci, and the asymptotes of the hyperbola
$$9y^2 - 25x^2 = 225.$$

5. Graph: $xy = 6$.

6. Find an equation of the parabola with directrix $y = 5$ and focus $(0, -5)$.

7. Find the vertex, the focus, and the directrix of the parabola
$$y^2 + 6y + 8x - 7 = 0.$$

Solve.

8. $x^2 + y^2 = 100,$
$\quad 2y + x = 20$

9. $x^2 + y^2 = 5,$
$\quad xy = 2$

10. The sum of two numbers is 4, and the difference of their squares is 32. What are the numbers?

11. The area of a rectangle is 10 yd^2, and the length of a diagonal is $\sqrt{101}$ yd. Find the dimensions.

12. In a fractional expression, the sum of the values of the numerator and the denominator is 23. The product of their values is 120. Find the values of the numerator and the denominator.

13. A rectangle with diagonal of length $5\sqrt{5}$ has an area of 22. Find the dimensions of the rectangle.

14. Two squares are such that the sum of their areas is 8 m^2 and the difference of their areas is 2 m^2. Find the length of a side of each square.

15. A rectangle has a diagonal of length 20 ft and a perimeter of 56 ft. Find the dimensions.

Classify the equation as a circle, an ellipse, a parabola, or a hyperbola.

16. $y = x^2 - 4x - 1 - 15y$

17. $x^2 + y^2 + 2x + 6y + 6 = 0$

18. $\dfrac{x^2}{9} - \dfrac{y^2}{4} = 1$

19. $16x^2 + 4y^2 = 64 - 128x$

20. $xy + 5 + 7x^2 = 7x^2$

21. $x = -y^2 + 8y + 16$

● **SYNTHESIS**

22. Find an equation of the ellipse with vertices $(-3, -4)$ and $(-3, 2)$ and containing the point $(-1, 1)$.

23. Find an equation of the circle through points $(-3, 8)$, $(4, 1)$, and $(-3, -4)$.

24. Find the point on the y-axis that is equidistant from $(-3, -5)$ and $(4, -7)$.

25. The sum of two numbers is 36 and the product is 4. Find the sum of the reciprocals of the numbers.

A desire to calculate odds in games of chance gave rise to the *theory of probability*, which today has many applications to business, medicine, sociology, psychology, and science. • The first part of this chapter is devoted to *sequences* and *series*. The idea of a sequence is a familiar one. For example, when a manager makes out a batting order, a sequence is being formed. Each batter is associated with a natural number; that is, a function is defined. When the members of a sequence are numbers, we can think of adding them. Such a sum is called a *series*. We also study a method of proof known as *mathematical induction,* which enables us to prove many important formulas, as well as other mathematical results. •

11

Sequences, Series, and Combinatorics

11.1
SEQUENCES AND SERIES

In this section, we discuss sets of numbers, considered in order, and their sums.

A Sequences

Suppose $1000 is invested at 8%, compounded annually. The amounts to which the account will grow after 1 year, 2 years, 3 years, 4 years, and so on, are as follows:

①   ②   ③   ④
↓   ↓   ↓   ↓
$1080.00 $1166.40 $1259.71 $1360.49, ...

Note that we can think of this as a function that maps 1 to the number $1080.00, 2 to the number $1166.40, 3 to the number $1259.71, 4 to the number $1360.49, and so on. A **sequence** is thus a *function*, where the domain is a set of consecutive positive integers.

If we keep computing the amounts in the account forever, we obtain an **infinite sequence:**

$1080.00, $1166.40, $1259.71, $1360.49, $1469.33, $1586.87, ...

OBJECTIVES

You should be able to:

A   Given a formula for one *n*th term (general term) of a sequence, find any term in the sequence, given a value for *n*.

B   Given a sequence, look for a pattern and try to guess a rule or formula for the general term.

C   Find and evaluate a series.

D   Convert between sigma (Σ) notation and other notation for a series.

E   Given a recursively defined sequence, construct its terms.

675

1. A sequence is given by $a(n) = 2n - 1$.

   a) Find the first 3 terms.

   b) Find the 34th term.

The three dots at the end indicate that the sequence goes on without stopping. If we stop after a certain number of years, we obtain a **finite sequence**:

$1080.00, $1166.40, $1259.71, $1360.49.

DEFINITION   **Sequence**

An *infinite sequence* is a function having for its domain the set of positive integers: $\{1, 2, 3, 4, 5, \ldots\}$.

A *finite sequence* is a function having for its domain the set of positive integers $\{1, 2, 3, 4, 5, \ldots, n\}$, for some positive integer n.

As another example, consider the sequence given by the formula

$$a(n) = 2^n, \quad \text{or} \quad a_n = 2^n.$$

The notation a_n means the same as $a(n)$ but is more commonly used with sequences. Some of the function values (also known as **terms** of the sequence) are as follows:

$$a_1 = 2^1 = 2,$$
$$a_2 = 2^2 = 4,$$
$$a_3 = 2^3 = 8,$$
$$a_6 = 2^6 = 64.$$

The first term of the sequence is a_1, the fifth term is a_5, and the nth term, or **general** term, is a_n. This sequence can also be denoted in the following ways:

a) $2, 4, 8, \ldots$;

b) $2, 4, 8, \ldots, 2^n, \ldots$.

2. A sequence is given by

$$a_n = \frac{(-1)^n}{n - 1}, \quad n \geq 2.$$

   a) Find the first 4 terms.

   b) Find the 47th term.

Example 1   Find the first 4 terms and the 57th term of the sequence whose general term is given by $a_n = (-1)^n/(n + 1)$.

Solution

$$a_1 = \frac{(-1)^1}{1 + 1} = -\frac{1}{2},$$

$$a_2 = \frac{(-1)^2}{2 + 1} = \frac{1}{3},$$

$$a_3 = \frac{(-1)^3}{3 + 1} = -\frac{1}{4},$$

$$a_4 = \frac{(-1)^4}{4 + 1} = \frac{1}{5},$$

$$a_{57} = \frac{(-1)^{57}}{57 + 1} = -\frac{1}{58}.$$

Note in Example 1 that the power $(-1)^n$ causes the signs of the terms to alternate between positive and negative, depending on whether n is even or odd.

DO EXERCISES 1 AND 2.

Finding the General Term

When a sequence is described merely by naming the first few terms, we do not know for sure what the general term is, but the student should look for a pattern and make a guess.

Examples   For each sequence, make a guess at the general term.

2. 1, 4, 9, 16, 25, . . .

These are squares of numbers, so the general term may be n^2.

3. $\sqrt{1}, \sqrt{2}, \sqrt{3}, \sqrt{4}, \ldots$

These are square roots of numbers, so the general term may be $\sqrt{n}$.

4. $-1, 2, -4, 8, -16, \ldots$

These are powers of 2 with alternating signs, so the general term may be $(-1)^n[2^{n-1}]$.

5. 2, 4, 8, . . .

If we see the pattern of powers of 2, we will see 16 as the next term and guess 2^n for the general term. We would then write the sequence with more terms as

2, 4, 8, 16, 32, 64, 128, ...

If we see that we can get the second term by adding 2, the third term by adding 4, and the next term by adding 6, and so on, we will see 14 as the next term. A general term for the sequence is then $n^2 - n + 2$, and we would then write the sequence with more terms as ◀

2, 4, 8, 14, 22, 32, 44, 58, ...

Example 5 illustrates that, in fact, you can never be certain about the general term unless you are playing some kind of guessing game in which someone knows a formula and you are to guess it.

DO EXERCISES 3–7.

 Sums and Series

> **DEFINITION**   **Series**
>
> Given the infinite sequence
>
> $$a_1, a_2, a_3, a_4, \ldots, a_n, \ldots,$$
>
> the sum of the terms
>
> $$a_1 + a_2 + a_3 + \cdots + a_n + \cdots$$
>
> is called an *infinite series*. A *partial sum* is the sum of the first n terms:
>
> $$a_1 + a_2 + a_3 + \cdots + a_n.$$
>
> A partial sum is also called a *finite series*, and is denoted S_n.

For each sequence, try to find a rule for finding the general term, or the nth term. Answers may vary.

3. 2, 4, 6, 8, 10, . . .

4. 1, 2, 3, 4, 5, 6, . . .

5. 1, 8, 27, 64, 125, . . .

6. $x, \dfrac{x^2}{2}, \dfrac{x^3}{3}, \dfrac{x^4}{4}, \dfrac{x^5}{5}, \ldots$

7. 1, 2, 4, 8, 16, 32, . . .

8. For the sequence $1, -1, 2, -2, 3, -3, 4, -4$, find each of the following.

 a) S_4

 b) S_7

Evaluate each sum.

9. $\displaystyle\sum_{k=1}^{3}\left(2+\frac{1}{k}\right)$

10. $\displaystyle\sum_{k=0}^{4} 5^k$

11. $\displaystyle\sum_{k=8}^{11} k^3$

Consider the sequence

$$3, 5, 7, 9, \ldots, 2n + 1.$$

We construct some partial sums:

$S_1 = 3,$	This is the first term of the given sequence.
$S_2 = 3 + 5 = 8,$	The sum of the first two terms
$S_3 = 3 + 5 + 7 = 15,$	The sum of the first three terms
$S_4 = 3 + 5 + 7 + 9 = 24.$	The sum of the first four terms

Note that if we write these partial sums in order, we create a new sequence:

$$3, 8, 15, 24, \ldots.$$

Example 6   For the sequence $-2, 4, -6, 8, -10, 12, -14$, find (a) S_3 and (b) S_5.

Solution

 a) $S_3 = -2 + 4 + (-6) = -4$

 b) $S_5 = -2 + 4 + (-6) + 8 + (-10) = -6$ ◀

DO EXERCISE 8. _____

▷ Sigma Notation

The Greek letter Σ (sigma) can be used to simplify notation when a series has a formula for the general term.

The sum of the first four terms of the sequence $3, 5, 7, 9, \ldots, 2k + 1$ can be named as follows, using what is called **sigma notation**, or **summation notation**:

$$\sum_{k=1}^{4} (2k + 1).$$

This is read "the sum as k goes from 1 to 4 of $(2k + 1)$." The letter k is called the **index of summation**. Sometimes the index of summation starts at a number other than 1.

Examples   Evaluate each sum.

7. $\displaystyle\sum_{k=1}^{5} k^2 = 1^2 + 2^2 + 3^2 + 4^2 + 5^2$
$$= 1 + 4 + 9 + 16 + 25 = 55$$

8. $\displaystyle\sum_{k=1}^{4} (-1)^k(2k)$
$$= (-1)^1(2 \cdot 1) + (-1)^2(2 \cdot 2) + (-1)^3(2 \cdot 3) + (-1)^4(2 \cdot 4)$$
$$= -2 + 4 - 6 + 8 = 4$$

9. $\displaystyle\sum_{k=0}^{3} (2^k + 5) = (2^0 + 5) + (2^1 + 5) + (2^2 + 5) + (2^3 + 5)$
$$= 6 + 7 + 9 + 13 = 35$$ ◀

DO EXERCISES 9–11. _____

Examples Write sigma notation for each sum.

10. $-1 + 3 - 5 + 7$

These are odd integers with alternating signs. Therefore, the general term is $(-1)^k(2k - 1)$, beginning with $k = 1$. Sigma notation is

$$\sum_{k=1}^{4} (-1)^k(2k - 1).$$

11. $3 + 9 + 27 + 81 + \cdots$

This is a sum of powers of 3, and it is also an infinite series. We use the symbol ∞ to represent infinity and name the infinite series with sigma notation, as follows:

$$\sum_{k=1}^{\infty} 3^k.$$

12. $x + x^2 + x^3 + x^4 + x^5 + x^6 = \sum_{k=1}^{6} x^k$

13. $\sin\dfrac{\pi}{2} - \sin\dfrac{3\pi}{2} + \sin\dfrac{5\pi}{2} - \sin\dfrac{7\pi}{2} + \cdots$

$$= \sum_{k=1}^{\infty} (-1)^{k+1} \sin\left(\dfrac{2k - 1}{2}\right)\pi \qquad \blacktriangleleft$$

DO EXERCISES 12–15.

Recursive Definitions

A sequence may be defined by **recursive definition.** Such a definition lists the first term, or the first few terms, and then tells how to get the rest of the terms from the given terms.

Example 14 Find the first 5 terms of the sequence defined by

$$a_1 = 5,$$
$$a_{k+1} = 2a_k - 3, \quad \text{for } k \geq 1.$$

Solution We have

$$a_1 = 5,$$
$$a_2 = 2a_1 - 3 = 2 \cdot 5 - 3 = 7,$$
$$a_3 = 2a_2 - 3 = 2 \cdot 7 - 3 = 11,$$
$$a_4 = 2a_3 - 3 = 2 \cdot 11 - 3 = 19,$$
$$a_5 = 2a_4 - 3 = 2 \cdot 19 - 3 = 35. \qquad \blacktriangleleft$$

DO EXERCISE 16.

Example 15 *The Fibonacci sequence.* One of the most famous recursively defined sequences is the *Fibonacci sequence.* So much mathematics has been derived from it that there is a journal, called the *Fibonacci Quarterly,* devoted to publishing the results. The Fibonacci sequence is defined as

---

Write sigma notation.

12. $2 + 4 + 6 + 8 + 10$

13. $1 + 8 + 27 + 64$

14. $x + \dfrac{x^2}{2} + \dfrac{x^3}{3} + \dfrac{x^4}{4} + \dfrac{x^5}{5} + \dfrac{x^6}{6} + \cdots$

15. $4 + 9 + 16 + 25 + 36 + \cdots$

16. Find the first 5 terms of this recursively defined sequence:

$$a_1 = -3,$$
$$a_{k+1} = (-1) \cdot a_k^2, \quad \text{for } k \geq 1.$$

17. Find the first 8 terms of the recursively defined sequence

$$a_1 = 5,$$
$$a_2 = 5,$$
$$a_{k+1} = a_k + a_{k-1}, \quad \text{for } k \geq 2.$$

follows:

$$a_1 = 1,$$
$$a_2 = 1,$$
$$a_{k+1} = a_k + a_{k-1}, \quad \text{for } k \geq 2.$$

Find the first 7 terms of the Fibonacci sequence.

Solution   We have

$$a_1 = 1,$$
$$a_2 = 1,$$
$$a_3 = a_2 + a_1 = 1 + 1 = 2,$$
$$a_4 = a_3 + a_2 = 2 + 1 = 3,$$
$$a_5 = a_4 + a_3 = 3 + 2 = 5,$$
$$a_6 = a_5 + a_4 = 5 + 3 = 8,$$
$$a_7 = a_6 + a_5 = 8 + 5 = 13.$$

DO EXERCISE 17.

● EXERCISE SET 11.1

A   In each of the following, the nth term of a sequence is given. In each case, find the first 4 terms, a_{10}, and a_{15}.

1. $a_n = 4n - 1$

2. $a_n = (n - 1)(n - 2)(n - 3)$

3. $a_n = \dfrac{n}{n-1}, n \geq 2$ **4.** $a_n = n^2 - 1$

5. $a_n = n^2 + 2n$ **6.** $a_n = \dfrac{n^2 - 1}{n^2 + 1}$

7. $a_n = n + \dfrac{1}{n}$ **8.** $a_n = \left(-\dfrac{1}{2}\right)^{n-1}$

9. $a_n = (-1)^n n^2$ **10.** $a_n = (-1)^n(n + 3)$

11. $a_n = (-1)^{n+1}(3n - 5)$ **12.** $a_n = (-1)^n(n^3 - 1)$

13. $a_n = \dfrac{n+2}{n+5}$ **14.** $a_n = \dfrac{2n-1}{3n-4}$

Find the indicated term of the given sequence.

15. $a_n = 5n - 6;\quad a_8$

16. $a_n = 3n + 10;\quad a_9$

17. $a_n = (3n - 4)(2n + 5);\quad a_7$

18. $a_n = (2n - 3)^2;\quad a_6$

19. $a_n = (-1)^{n-1}(4.6n - 18.3);\quad a_{12}$

20. $a_n = (-2)^{n-2}(54.76 - 1.3n);\quad a_{23}$

21. $a_n = 5n^2(4n - 100);\quad a_{11}$

22. $a_n = 4n^2(11n + 31);\quad a_{22}$

23. $a_n = \left(1 + \dfrac{1}{n}\right)^2;\quad a_{20}$ **24.** $a_n = \left(1 - \dfrac{1}{n}\right)^3;\quad a_{15}$

25. $a_n = \log 10^n;\quad a_{43}$ **26.** $a_n = \ln e^n;\quad a_{67}$

27. $a_n = 1 + \dfrac{1}{n^2};\quad a_{38}$ **28.** $a_n = 2 - \dfrac{1000}{n};\quad a_{100}$

B   For each sequence, find the general term, or nth term, a_n, or a rule for finding a_n. Answers may vary.

29. $1, 3, 5, 7, 9, \ldots$ **30.** $3, 9, 27, 81, 243, \ldots$

31. $-2, 6, -18, 54, \ldots$ **32.** $-2, 3, 8, 13, 18, \ldots$

33. $\dfrac{2}{3}, \dfrac{3}{4}, \dfrac{4}{5}, \dfrac{5}{6}, \dfrac{6}{7}, \ldots$

34. $\sqrt{2}, \sqrt{4}, \sqrt{6}, \sqrt{8}, \sqrt{10}, \ldots$

35. $\sqrt{3}, 3, 3\sqrt{3}, 9, 9\sqrt{3}, \ldots$

36. $1 \cdot 2, 2 \cdot 3, 3 \cdot 4, 4 \cdot 5, \ldots$

37. $-1, -4, -7, -10, -13, \ldots$

38. $\log 1, \log 10, \log 100, \log 1000, \ldots$

C   For each sequence, find the indicated partial sum.

39. $1, 2, 3, 4, 5, 6, 7, \ldots; S_7$

40. $1, -3, 5, -7, 9, -11, \ldots; S_8$

41. $2, 4, 6, 8, \ldots; S_5$

42. $1, \dfrac{1}{4}, \dfrac{1}{9}, \dfrac{1}{16}, \dfrac{1}{25}, \ldots; S_5$

▶ Evaluate each sum.

43. $\sum_{k=1}^{5} \dfrac{1}{2k}$

44. $\sum_{k=1}^{6} \dfrac{1}{2k+1}$

45. $\sum_{k=0}^{5} 2^k$

46. $\sum_{k=4}^{7} \sqrt{2k-1}$

47. $\sum_{k=7}^{10} \log k$

48. $\sum_{k=0}^{4} \pi k$

49. $\sum_{k=1}^{8} \dfrac{k}{k+1}$

50. $\sum_{k=1}^{4} \dfrac{k-1}{k+3}$

51. $\sum_{k=1}^{5} (-1)^k$

52. $\sum_{k=1}^{5} (-1)^{k+1}$

53. $\sum_{k=1}^{8} (-1)^{k+1}3k$

54. $\sum_{k=1}^{7} (-1)^k 4^{k+1}$

55. $\sum_{k=1}^{6} \dfrac{2}{k^2+1}$

56. $\sum_{k=1}^{10} k(k+1)$

57. $\sum_{k=0}^{5} (k^2-2k+3)$

58. $\sum_{k=0}^{5} (k^2-3k+4)$

59. $\sum_{k=1}^{10} \dfrac{1}{k(k+1)}$

60. $\sum_{k=1}^{10} \dfrac{2^k}{2^k+1}$

Write sigma notation.

61. $\dfrac{1}{2}+\dfrac{2}{3}+\dfrac{3}{4}+\dfrac{4}{5}+\dfrac{5}{6}+\dfrac{6}{7}$

62. $3+6+9+12+15$

63. $-2+4-8+16-32+64$

64. $\dfrac{1}{1^2}+\dfrac{1}{2^2}+\dfrac{1}{3^2}+\dfrac{1}{4^2}+\dfrac{1}{5^2}$

65. $4-9+16-25+\cdots+(-1)^n n^2$

66. $9-16+25+\cdots+(-1)^{n+1} n^2$

67. $5+10+15+20+25+\cdots$

68. $7+14+21+28+35+\cdots$

69. $\dfrac{1}{1\cdot 2}+\dfrac{1}{2\cdot 3}+\dfrac{1}{3\cdot 4}+\dfrac{1}{4\cdot 5}+\cdots$

70. $\dfrac{1}{1\cdot 2^2}+\dfrac{1}{2\cdot 3^2}+\dfrac{1}{3\cdot 4^2}+\dfrac{1}{4\cdot 5^2}+\cdots$

▶ Find the first 4 terms of each recursively defined sequence.

71. $a_1=4,\ a_{k+1}=1+\dfrac{1}{a_k}$

72. $a_1=256,\ a_{k+1}=\sqrt{a_k}$

73. $a_1=6561,\ a_{k+1}=(-1)^k\sqrt{a_k}$

74. $a_1=e^Q,\ a_{k+1}=\ln a_k$

75. $a_1=2,\ a_2=3,\ a_{k+1}=a_k+a_{k-1}$

76. $a_1=-10,\ a_2=8,\ a_{k+1}=a_k-a_{k-1}$

● **SYNTHESIS**

Find the first 5 terms of the sequence, and then find S_5.

77. $a_n=\dfrac{1}{2^n}\log 1000^n$

78. $a_n=i^n,\ i=\sqrt{-1}$

79. $a_n=\ln(1\cdot 2\cdot 3\cdots n)$

80. $a_n=\sin\dfrac{n\pi}{2}$

81. $a_n=\cos^{-1}(-1)^n$

82. $a_n=|\cos(x+\pi n)|$

83. a) Find the first few terms of the sequence
$$a_n=n^2-n+41.$$
  b) What pattern do you observe?
  c) Find the 41st term. Does the pattern you found in (b) still hold?

Find decimal notation, rounded to six decimal places, for the first 6 terms of each sequence.

84. ▪ $a_n=\left(1+\dfrac{1}{n}\right)^n$

85. ▪ $a_n=\sqrt{n+1}-\sqrt{n}$

86. ▪ $a_1=2,\ a_{k+1}=\sqrt{1+\sqrt{a_k}}$

87. ▪ $a_1=2,\ a_{k+1}=\dfrac{1}{2}\left(a_k+\dfrac{2}{a_k}\right)$

88. A single cell of bacteria divides into two every 15 min. Suppose that the same rate of division is maintained for 4 hr. Give a sequence that lists the number of cells after successive 15-min periods.

89. The value of an office machine is $5200. Its scrap value each year is 75% of its value the year before. Give a sequence that lists the scrap value of the machine for each year of a 10-year period.

90. A student gets $4.20 per hour for working in a warehouse for a publishing company. Each year, the student gets a $0.15 hourly raise. Give a sequence that lists the hourly salary of the student over a 10-year period.

● **CHALLENGE**

For each sequence, find a formula for S_n.

91. $a_n=\ln n$

92. $a_n=\dfrac{1}{n}-\dfrac{1}{n+1}$

11.2

ARITHMETIC SEQUENCES AND SERIES

In this section, we concentrate on what is called an *arithmetic sequence*. If we start with a particular first term and then add the same number successively, we obtain an **arithmetic sequence.** We will also study *arithmetic series*.

A Arithmetic Sequences

Consider this sequence:

$$2, 5, 8, 11, 14, 17, \ldots .$$

Note that adding 3 to any term produces the following term. In other words, the difference between any term and the preceding one is 3. This is an example of an arithmetic sequence.

DEFINITION   **Arithmetic Sequence**

A sequence is *arithmetic* if there exists a number d, called the *common difference*, such that $a_n = a_{n-1} + d$, or $a_n - a_{n-1} = d$, for any $n \geq 2$.

Arithmetic sequences are also called **arithmetic progressions.**

Examples   The following are arithmetic sequences. Identify the first term a_1 and the common difference d.

Sequence	First term, a_1	Common difference, d
1. 4, 9, 14, 19, 24, . . .	4	5
2. 34, 27, 20, 13, 6, −1, −8, . . .	34	−7
3. 2, $2\frac{1}{2}$, 3, $3\frac{1}{2}$, 4, $4\frac{1}{2}$, . . .	2	$\frac{1}{2}$

We obtain d for the first sequence by picking any term beyond the first—say, the second term, 9—and subtracting the preceding term from it: $9 - 4 = 5$. Then we can check by adding 5 to each term to see if we obtain the next:

$$a_1 = 4,$$
$$a_2 = 4 + 5 = 9,$$
$$a_3 = 9 + 5 = 14,$$
$$a_4 = 14 + 5 = 19, \quad \text{and so on.}$$

We now find a formula for the general, or *n*th, term of any arithmetic sequence. Let us denote the difference between successive terms (called the

common difference) by d, and write out the first few terms:

$$a_1,$$
$$a_2 = a_1 + d,$$
$$a_3 = a_2 + d = (a_1 + d) + d = a_1 + 2d,$$
$$a_4 = a_3 + d = (a_1 + 2d) + d = a_1 + 3d.$$

Note that the coefficient of d in each case is 1 less than the number of the term, n.

DO EXERCISE 1.

Generalizing, we obtain the following.

> ### THEOREM 1
>
> The nth term of an arithmetic sequence is given by
> $$a_n = a_1 + (n-1)d, \quad \text{for any } n \geq 1.$$

Example 4 Find the 14th term of the arithmetic sequence $4, 7, 10, 13, \ldots$.

Solution First note that $a_1 = 4$, $d = 3$, and $n = 14$. Then using the formula

$$a_n = a_1 + (n-1)d,$$

we obtain

$$a_{14} = 4 + (14 - 1) \cdot 3 = 4 + 13 \cdot 3 = 4 + 39 = 43.$$

The 14th term is 43. ◀

DO EXERCISE 2.

Example 5 In the sequence in Example 4, which term is 301? That is, what is n if $a_n = 301$?

Solution We substitute into the formula of Theorem 1 and solve for n:

$$a_n = a_1 + (n-1)d$$
$$301 = 4 + (n-1) \cdot 3$$
$$301 = 4 + 3n - 3$$
$$301 = 3n + 1$$
$$300 = 3n$$
$$100 = n.$$

The 100th term is 301. ◀

DO EXERCISE 3.

Given two terms and their places in an arithmetic sequence, we can construct the sequence.

---

1. In the following arithmetic sequence, identify the first term a_1 and the common difference d.

   $$3.1, 3.9, 4.7, 5.5, 6.3$$

2. Find the 13th term of the sequence $2, 6, 10, 14, \ldots$.

3. In the sequence given in Margin Exercise 2, which term is 298? That is, what is n if $a_n = 298$?

4. The 7th term of an arithmetic sequence is 79, and the 13th term is 151. Find a_1 and d and construct the sequence.

Example 6 The 3rd term of an arithmetic sequence is 8, and the 16th term is 47. Find a_1 and d and construct the sequence.

Solution We know that $a_3 = 8$ and $a_{16} = 47$. Thus we would have to add d thirteen times to get from 8 to 47. That is,

$$8 + 13d = 47.$$

Solving, we obtain

$$13d = 39$$
$$d = 3.$$

Since $a_3 = 8$, we subtract d twice to get to a_1. Thus,

$$a_1 = 8 - 2 \cdot 3 = 2.$$

The sequence is 2, 5, 8, 11, ...

DO EXERCISE 4. _____

B Sum of the First n Terms of an Arithmetic Sequence

Suppose we add the first 4 terms of the sequence

$$3, 5, 7, 9, 11, \ldots .$$

We get what is called an **arithmetic series:**

$$3 + 5 + 7 + 9, \quad \text{or} \quad 24.$$

The sum of the first n terms of a sequence is denoted S_n. Thus, for the preceding sequence, $S_4 = 24$. We want to find a formula for S_n when the sequence is arithmetic. We can denote an arithmetic sequence as

> This term is two terms back from the last. If you add d to this term, you get the next-to-last term, $a_n - d$.

$$a_1, \quad (a_1 + d), \quad (a_1 + 2d), \quad \ldots, \quad \overbrace{(a_n - 2d)}, \quad \underbrace{(a_n - d)}, \quad a_n.$$

> This is the next-to-last term. If you add d to this term, the result is a_n.

Then S_n is given by

$$S_n = a_1 + (a_1 + d) + (a_1 + 2d)$$
$$+ \cdots + (a_n - 2d) + (a_n - d) + a_n. \tag{1}$$

If we reverse the order of addition, we get

$$S_n = a_n + (a_n - d) + (a_n - 2d)$$
$$+ \cdots + (a_1 + 2d) + (a_1 + d) + a_1. \tag{2}$$

Suppose we add corresponding terms of each side of equations (1) and (2).

Then we get

$$2S_n = [a_1 + a_n] + [(a_1 + d) + (a_n - d)]$$
$$+ [(a_1 + 2d) + (a_n - 2d)] + \cdots + [(a_n - 2d) + (a_1 + 2d)]$$
$$+ [(a_n - d) + (a_1 + d)] + [a_n + a_1].$$

This simplifies to

$$2S_n = (a_1 + a_n) + (a_1 + a_n) + (a_1 + a_n) + \cdots + (a_1 + a_n).$$

Since there are n binomials $(a_1 + a_n)$ being added, it follows that

$$2S_n = n(a_1 + a_n),$$

from which we get the following formula.

THEOREM 2

The sum of the first n terms of an arithmetic sequence is given by

$$S_n = \frac{n}{2}(a_1 + a_n).$$

Example 7 Find the sum of the first 100 natural numbers.

Solution   The sum is

$$1 + 2 + 3 + \cdots + 99 + 100.$$

This is the sum of the first 100 terms of the arithmetic sequence for which

$$a_1 = 1, \qquad a_n = 100, \quad \text{and} \quad n = 100.$$

Then substituting into the formula

$$S_n = \frac{n}{2}(a_1 + a_n),$$

we get

$$S_{100} = \frac{100}{2}(1 + 100)$$

$$= 50(101) = 5050. \qquad \blacktriangleleft$$

DO EXERCISE 5.

The preceding formula is useful when we know both a_1 and a_n, the first and last terms, but it often happens that we do not know a_n. Therefore, we need a formula in terms of a_1, n, and d. We substitute the expression for a_n, which was given in Theorem 1, $a_n = a_1 + (n - 1)d$, into the formula of Theorem 2:

$$S_n = \frac{n}{2}(a_1 + [a_1 + (n - 1)d]).$$

This gives us the following.

5. Find the sum of the first 200 natural numbers.

6. Find the sum of the first 15 terms of the arithmetic sequence

$$1, 3, 5, 7, 9, \ldots.$$

THEOREM 3

The sum of the first n terms of an arithmetic sequence is given by

$$S_n = \frac{n}{2}[2a_1 + (n-1)d].$$

Example 8 Find the sum of the first 15 terms of the arithmetic sequence $4, 7, 10, 13, \ldots$.

Solution Note that

$$a_1 = 4, \qquad d = 3, \quad \text{and} \quad n = 15.$$

Here we use the formula of Theorem 3 since we do not know a_n:

$$S_n = \frac{n}{2}[2a_1 + (n-1)d].$$

We get

$$S_{15} = \frac{15}{2}[2 \cdot 4 + (15-1)3] = \frac{15}{2}[8 + 14 \cdot 3] = \frac{15}{2}[8 + 42]$$

$$= \frac{15}{2}[50] = 375.$$

◀

DO EXERCISE 6.

7. Find the sum:

$$\sum_{k=1}^{10} (9k - 4).$$

Example 9 Find the sum: $\displaystyle\sum_{k=1}^{13} (4k + 5)$.

Solution It is helpful to write out a few terms:

$$9 + 13 + 17 + \cdots.$$

We see that the sum is an arithmetic series coming from an arithmetic sequence with $a_1 = 9, d = 4$, and $n = 13$. We use the formula of Theorem 3:

$$S_n = \frac{n}{2}[2a_1 + (n-1)d]$$

$$S_{13} = \frac{13}{2}[2 \cdot 9 + (13-1)4] = \frac{13}{2}[18 + 12 \cdot 4] = \frac{13}{2} \cdot 66 = 429. \quad ◀$$

DO EXERCISE 7.

 Problem Solving

For some problem situations, the translations may involve sequences or series. We look at some examples.

Example 10 You take a job starting with an hourly rate of $14.25. You are promised a raise of 15¢ per hour every 2 months for 5 years. At the end of 5 years, what will be your hourly wage?

Solution One thing to do is write down your hourly wage for several two-month time periods. What appears is a *sequence of numbers*: 14.25, 14.40, 14.55, ... Is it an arithmetic sequence? Yes, because we add 0.15 each time in order to get the next term.

We ask ourselves what we know about arithmetic sequences. We recall, or if necessary look up, the pertinent formula(s). There are three:

$$a_n = a_1 + (n - 1)d, \qquad S_n = \frac{n}{2}(a_1 + a_n), \quad \text{and} \quad S_n = \frac{n}{2}[2a_1 + (n - 1)d].$$

In this case, we are not looking for a *sum,* so the first formula will give us our answer. We want to know the last term in a sequence. We will need to know a_1, n, and d. From our list above, we see that

$$a_1 = 14.25 \quad \text{and} \quad d = 0.15.$$

What is n? That is, how many terms are in the sequence? Each year, there are 6 raises, since you get a raise every 2 months. There are 5 years, so the total number of raises will be 5×6, or 30. There will be 31 terms: the original wage and 30 increased rates.

We want to find the 31st term of an arithmetic sequence, with $a_1 = 14.25$, $d = 0.15$, and $n = 31$. We substitute into the formula:

$$a_{31} = 14.25 + (31 - 1) \times 0.15$$
$$= \$18.75.$$

At the end of 5 years, your hourly wage will be $18.75. ◀

Example 10 is one in which the calculations or the translation could be done in a number of ways. Here is a point to remember: There is often a variety of ways in which a problem can be solved. In this chapter, however, we will concentrate on the use of sequences and series and their related formulas in problem solving.

DO EXERCISE 8.

Example 11 A stack of telephone poles has 30 poles in the bottom row. There are 29 poles in the second row, 28 in the next row, and so on. How many poles are in the stack?

Solution A picture will help in this case. The following figure shows the ends of the poles and the way in which they stack. There are 30 poles on the bottom, and we see that there will be one fewer in each succeeding row. How many rows will there be?

8. You take a job starting with an hourly rate of $16.75. You are promised a raise of 25¢ per hour every 2 months for 7 years. At the end of 7 years, what will be your hourly wage?

9. A stack of iron rods has 112 rods in the bottom row, 111 rods in the next row, and so on. The top row has 48 rods. How many rods are in the stack?

29 poles
30 poles

We go from 30 poles in a row, down to one pole in the top row, so there must be 30 rows.

We want the sum

$$30 + 29 + 28 + \cdots + 1.$$

Thus we want the sum of an arithmetic sequence. We recall the formula for the sum of an arithmetic sequence:

$$S_n = \frac{n}{2}(a_1 + a_n).$$

We want to find the sum of an arithmetic sequence, where $a_1 = 30$ and, since there are 30 terms, $n = 30$. There is just one pole on top, so $a_{30} = 1$. Substituting into the formula, we have

$$S_{30} = \frac{30}{2}(30 + 1)$$

$$= 465.$$

The answer is that there are 465 poles in the stack.   ◀

DO EXERCISE 9.

▷ **Arithmetic Means**

If p, m, and q form an arithmetic sequence, it can be shown (see Exercise 58) that $m = (p + q)/2$. We call m the **arithmetic mean** of p and q. Given two numbers p and q, if we find k other numbers $m_1, m_2, \ldots, m_k$ such that

$$p, m_1, m_2, \ldots, m_k, q$$

forms an arithmetic sequence, we say that we have "inserted k arithmetic means between p and q."

Example 12   Insert three arithmetic means between 4 and 13.

Solution   We look for numbers m_1, m_2, and m_3 such that 4, m_1, m_2, m_3, 13 is an arithmetic sequence. In this case, $a_1 = 4$, $n = 5$, and $a_5 = 13$. We use the formula of Theorem 1:

$$a_n = a_1 + (n - 1)d$$
$$13 = 4 + (5 - 1)d.$$

Then $d = 2\frac{1}{4}$, so we have

$$m_1 = a_1 + d = 4 + 2\frac{1}{4} = 6\frac{1}{4},$$
$$m_2 = m_1 + d = 6\frac{1}{4} + 2\frac{1}{4} = 8\frac{1}{2},$$
$$m_3 = m_2 + d = 8\frac{1}{2} + 2\frac{1}{4} = 10\frac{3}{4}.$$

DO EXERCISE 10.

10. Insert three arithmetic means between 4 and 16.

● **EXERCISE SET** **11.2**

A Find the first term and the common difference.

1. $3, 8, 13, 18, \ldots$

2. $1.08, 1.16, 1.24, 1.32, \ldots$

3. $9, 5, 1, -3, \ldots$

4. $-8, -5, -2, 1, 4, \ldots$

5. $\frac{3}{2}, \frac{4}{9}, 3, \frac{15}{4}, \ldots$

6. $\frac{3}{5}, \frac{1}{10}, -\frac{2}{5}, \ldots$

7. $\$1.07, \$1.14, \$1.21, \$1.28, \ldots$

8. $\$316, \$313, \$310, \$307, \ldots$

9. Find the 12th term of the arithmetic sequence
$2, 6, 10, \ldots.$

10. Find the 11th term of the arithmetic sequence
$0.07, 0.12, 0.17, \ldots.$

11. Find the 17th term of the arithmetic sequence
$7, 4, 1, \ldots.$

12. Find the 14th term of the arithmetic sequence
$3, \frac{7}{3}, \frac{5}{3}, \ldots.$

13. Find the 13th term of the arithmetic sequence
$\$1200, \$964.32, \$728.64, \ldots.$

14. Find the 10th term of the arithmetic sequence
$\$2345.78, \$2967.54, \$3589.30, \ldots.$

15. In the sequence of Exercise 9, what term is 106?

16. In the sequence of Exercise 10, what term is 1.67?

17. In the sequence of Exercise 11, what term is -296?

18. In the sequence of Exercise 12, what term is -27?

19. Find a_{17} when $a_1 = 5$ and $d = 6$.

20. Find a_{20} when $a_1 = 14$ and $d = -3$.

21. Find a_1 when $d = 4$ and $a_8 = 33$.

22. Find d when $a_1 = 8$ and $a_{11} = 26$.

23. Find n when $a_1 = 5$, $d = -3$, and $a_n = -76$.

24. Find n when $a_1 = 25$, $d = -14$, and $a_n = -507$.

25. In an arithmetic sequence, $a_{17} = -40$ and $a_{28} = -73$. Find a_1 and d. Write the first 5 terms of the sequence.

26. In an arithmetic sequence, $a_{17} = \frac{25}{3}$ and $a_{32} = \frac{95}{6}$. Find a_1 and d. Write the first 5 terms of the sequence.

B

27. Find the sum of the first 20 terms of the series
$$5 + 8 + 11 + 14 + \cdots.$$

28. Find the sum of the first 14 terms of the series
$$11 + 7 + 3 + \cdots.$$

29. Find the sum of the first 300 natural numbers.

30. Find the sum of the first 400 natural numbers.

31. Find the sum of the even numbers from 2 to 100, inclusive.

32. Find the sum of the odd numbers from 1 to 99, inclusive.

33. Find the sum of the multiples of 7 from 7 to 98, inclusive.

34. Find the sum of all multiples of 4 that are between 14 and 523.

35. If an arithmetic series has $a_1 = 2, d = 5$, and $n = 20$, what is S_n?

36. If an arithmetic series has $a_1 = 7$, $d = -3$, and $n = 32$, what is S_n?

C

37. A gardener is making a triangular planting, with 35 plants in the front row, 31 in the second row, 27 in the third row, and so on. If the pattern is consistent, how many plants will there be in the last row?

38. A formation of a marching band has 14 marchers in the front row, 16 in the second row, 18 in the third row, and so on, for 25 rows. How many marchers are in the last row? How many marchers are there altogether?

39. How many poles will be in a pile of telephone poles if there are 50 in the first layer, 49 in the second, and so on, until there is 1 in the last layer?

40. If 10¢ is saved on October 1, 20¢ on October 2, 30¢ on October 3, and so on, how much is saved during October? (October has 31 days.)

41. A family saves money in an arithmetic sequence. They save $600 the first year, $700 the second, and so on, for 20 years. How much do they save in all (disregarding interest)?

42. A student saves $30 on August 1, $50 on August 2, $70 on August 3, and so on. How much will she save in August? August has 31 days.

43. Theaters are often built with more seats per row as the rows move toward the back. Suppose that the main floor of a theater has 28 seats in the first row, 32 in the second, 36 in the third, and so on, for 50 rows. How many seats are on the main floor?

44. A person sets up an investment such that it will return $5000 the first year, $6125 the second year, $7250 the third year, and so on, for 25 years. How much in all is received from the investment?

45. Insert four arithmetic means between 4 and 13.

46. Insert three arithmetic means between −3 and 5.

● SYNTHESIS _____

47. Find a formula for the sum of the first n odd natural numbers:
$$1 + 3 + 5 + \cdots + (2n - 1).$$

48. Find a formula for the sum of the first n natural numbers:
$$1 + 2 + 3 + \cdots + n.$$

49. Find three numbers in an arithmetic sequence such that the sum of the first and third is 10 and the product of the first and second is 15.

50. Find the first term and the common difference for the arithmetic sequence where
$$a_2 = 40 - 3q \quad \text{and} \quad a_4 = 10p + q.$$

51. ▇ Find the first 10 terms of the arithmetic sequence for which $a_1 = \$8760$ and $d = -\$798.23$.

52. ▇ Find the sum of the first 10 terms of the sequence given in Exercise 51.

53. The zeros of this polynomial function form an arithmetic sequence. Find them.
$$f(x) = x^4 + 4x^3 - 84x^2 - 176x + 640$$

54. Insert enough arithmetic means between 1 and 50 so that the sum of the resulting series will be 459.

55. Suppose that the lengths of the sides of a right triangle form an arithmetic sequence. Prove that the triangle is similar to a right triangle whose sides have lengths 3, 4, and 5.

56. *Business: Straight-line depreciation.* A company buys an office machine for $5200 on January 1 of a given year. The machine is expected to last for 8 years, at the end of which time its **trade-in**, or **salvage, value** will be $1100. If the company figures the decline in value to be the same each year, then the **book values**, or **salvage values**, after t years, $0 \le t \le 8$, form an arithmetic sequence given by
$$a_t = C - t\left(\frac{C - S}{N}\right),$$
where C = the original cost of the item ($5200), N = the years of expected life (8), and S = the salvage value ($1100).

a) Find the formula for a_t for the straight-line depreciation of the office machine.

b) Find the salvage value after 0 years, 1 year, 2 years, 3 years, 4 years, 7 years, and 8 years.

● CHALLENGE _____

57. Prove that an expression for the general term of an arithmetic sequence defines a linear function.

58. Prove that if p, m, and q form an arithmetic sequence, then
$$m = \frac{p + q}{2}.$$

OBJECTIVES

You should be able to:

A Identify the common ratio of a geometric sequence. Use Theorem 4 for geometric sequences to find a given term.

B Use Theorem 5 to find the sum of the first n terms of a geometric sequence.

C Use Theorem 6 to find the sum of an infinite geometric series, if it exists.

D Use geometric sequences to solve problems.

11.3

GEOMETRIC SEQUENCES AND SERIES

For arithmetic sequences, we added a certain number to each term to get the next term. When we multiply each term by a certain number to get the next term, we get a **geometric sequence.** We also consider geometric series.

A Geometric Sequences

Consider the sequence
$$2, 6, 18, 54, 162, \ldots.$$

If we multiply each term by 3, we get the next term. Sequences in which each term can be multiplied by a certain number to get the next term are called **geometric.** We usually denote this number r. We refer to it as the

common ratio, because we can get r by dividing any term by the preceding term.

DEFINITION	Geometric Sequence

A sequence is *geometric* if there exists a number r, called the *common ratio,* such that

$$\frac{a_{n+1}}{a_n} = r, \quad \text{or} \quad a_{n+1} = a_n r, \quad \text{for any } n \geq 1.$$

1. $1, 5, 25, 125, \ldots$

A geometric sequence is also called a **geometric progression.**

2. $3, -9, 27, -81, \ldots$

Examples   The following are geometric sequences. Identify the common ratio.

Sequence	Common Ratio	
1. $3, 6, 12, 24, 48, 96, \ldots$	2	$6/3 = 2, 12/6 = 2,$ and so on
2. $3, -6, 12, -24, 48, -96, \ldots$	-2	$-6/3 = -2,$ $12/-6 = -2,$ and so on
3. \$5200, \$3900, \$2925, \$2193.75, \ldots	0.75	$\$3900/\$5200 = 0.75,$ $\$2925/\$3900 = 0.75$
4. \$1000, \$1080, \$1166.40, \ldots	1.08	$\$1080/\$1000 = 1.08$
5. $1, \frac{1}{2}, \frac{1}{4}, \frac{1}{8}, \ldots$	$\frac{1}{2}$	$\frac{1}{2}/1 = \frac{1}{2}, \frac{1}{4}/\frac{1}{2} = \frac{1}{2}$ ◀

3. \$6000, \$5100, \$4335, \$3684.75, \ldots

DO EXERCISES 1–5.

We now find a formula for the general, or nth, term of any geometric sequence. Let a_1 be the 1st term, and let r be the common ratio. We write out the first few terms as follows:

$$a_1,$$
$$a_2 = a_1 r,$$
$$a_3 = a_2 r = (a_1 r)r = a_1 r^2,$$
$$a_4 = a_3 r = (a_1 r^2)r = a_1 r^3.$$

4. \$100, \$109, \$118.81, \ldots

→ Note that the exponent is 1 less than the number of the term.

Generalizing, we obtain the following.

5. $1, \dfrac{1}{5}, \dfrac{1}{25}, \dfrac{1}{125}, \ldots$

THEOREM 4

The nth term of a geometric sequence is given by

$$a_n = a_1 r^{n-1}, \quad \text{for any } n \geq 1.$$

Example 6   Find the 7th term of the geometric sequence $4, 20, 100, \ldots$.

6. Find the 9th term of the geometric sequence

$$2, 4, 8, 16, \ldots.$$

Solution First note that

$$a_1 = 4 \quad \text{and} \quad n = 7.$$

To find the common ratio, we can divide any term by its predecessor, provided it has one. Since the second term is 20 and the first is 4, we get

$$r = \frac{20}{4}, \quad \text{or } 5.$$

Then using the formula

$$a_n = a_1 r^{n-1},$$

we have

$$a_7 = 4 \cdot 5^{7-1} = 4 \cdot 5^6 = 4 \cdot 15{,}625 = 62{,}500. \qquad \blacktriangleleft$$

DO EXERCISE 6. _____

Example 7 Find the 10th term of the geometric sequence

$$64, -32, 16, -8, \ldots.$$

Solution First note that

$$a_1 = 64, \qquad n = 10, \quad \text{and} \quad r = \frac{-32}{64}, \quad \text{or } -\frac{1}{2}.$$

Then using the formula

$$a_n = a_1 r^{n-1},$$

we have

7. Find the 6th term of the geometric sequence

$$-3, 1, -\frac{1}{3}, \frac{1}{9}, \ldots.$$

$$a_{10} = 64 \cdot \left(-\frac{1}{2}\right)^{10-1} = 64 \cdot \left(-\frac{1}{2}\right)^9 = 2^6 \cdot \left(-\frac{1}{2^9}\right) = -\frac{1}{2^3} = -\frac{1}{8}.$$

$\blacktriangleleft$

DO EXERCISE 7. _____

▶ B **Sum of the First *n* Terms of a Geometric Sequence**

We want to find a formula for the sum S_n of the first n terms of a geometric sequence

$$a_1, a_1 r, a_1 r^2, a_1 r^3, \ldots, a_1 r^{n-1}, \ldots.$$

A **geometric series** is given by

$$S_n = a_1 + a_1 r + a_1 r^2 + \cdots + a_1 r^{n-2} + a_1 r^{n-1}. \qquad (1)$$

We want to develop a formula that allows us to find this sum without a great amount of adding. If we multiply on both sides of equation (1) by r, we have

$$r S_n = a_1 r + a_1 r^2 + a_2 r^3 + \cdots + a_1 r^{n-1} + a_1 r^n. \qquad (2)$$

When we multiply on both sides of equation (1) by -1, we get

$$-S_n = -a_1 - a_1 r - a_1 r^2 - \cdots - a_1 r^{n-2} - a_1 r^{n-1}. \qquad (3)$$

Then, when we add corresponding sides of equations (2) and (3), certain terms have 0 as a sum, so we get

$$rS_n - S_n = a_1 r^n - a_1,$$

or

$$(r - 1)S_n = a_1(r^n - 1),$$

from which we obtain the following formula.

THEOREM 5

The sum of the first n terms of a geometric sequence is given by

$$S_n = \frac{a_1(r^n - 1)}{r - 1}, \quad \text{for any } r \neq 1.$$

Example 8   Find the sum of the first 7 terms of the geometric sequence

$$3, 15, 75, 375, \ldots.$$

Solution   First note that

$$a_1 = 3, \quad n = 7, \quad \text{and} \quad r = \tfrac{15}{3}, \quad \text{or } 5.$$

Then using the formula

$$S_n = \frac{a_1(r^n - 1)}{r - 1},$$

we have

$$S_7 = \frac{3(5^7 - 1)}{5 - 1} = \frac{3(78{,}125 - 1)}{4} = \frac{3(78{,}124)}{4} = 58{,}593. \quad \blacktriangleleft$$

DO EXERCISES 8 AND 9.

Example 9   Find the sum: $\displaystyle\sum_{k=1}^{11} (0.3)^k$.

Solution   This is a geometric series. The first term is 0.3, $r = 0.3$, and $n = 11$. Then

$$S_{11} = \frac{0.3[(0.3)^{11} - 1]}{0.3 - 1} = 0.42857\ldots. \quad \blacktriangleleft$$

DO EXERCISE 10.

 ## Infinite Geometric Series

Suppose we consider the sum of the terms of an infinite geometric sequence, such as $2, 4, 8, 16, 32, \ldots$. We get what is called an **infinite geometric series:**

$$2 + 4 + 8 + 16 + 32 + \cdots.$$

In this example, as n grows larger and larger, the sum of the first n terms,

---

8. Find the sum of the first 8 terms of the geometric sequence

$$4, 12, 36, 108, \ldots.$$

9. Find the sum of the first 10 terms of the geometric sequence

$$2, -1, \frac{1}{2}, -\frac{1}{4}, \ldots.$$

10. Find the sum:

$$\sum_{k=1}^{5} 3^k.$$

S_n, becomes larger and larger without bound. But there are infinite series for which the partial sums S_n get closer and closer to some specific number. Here is an example:

$$\frac{1}{2} + \frac{1}{4} + \frac{1}{8} + \frac{1}{16} + \cdots + \frac{1}{2^n} + \cdots.$$

Let's consider the partial sums S_n for some values of n:

$$S_1 = \tfrac{1}{2} = \tfrac{1}{2} = 0.5,$$
$$S_2 = \tfrac{1}{2} + \tfrac{1}{4} = \tfrac{3}{4} = 0.75,$$
$$S_3 = \tfrac{1}{2} + \tfrac{1}{4} + \tfrac{1}{8} = \tfrac{7}{8} = 0.875,$$
$$S_4 = \tfrac{1}{2} + \tfrac{1}{4} + \tfrac{1}{8} + \tfrac{1}{16} = \tfrac{15}{16} = 0.9375,$$
$$S_5 = \tfrac{1}{2} + \tfrac{1}{4} + \tfrac{1}{8} + \tfrac{1}{16} + \tfrac{1}{32} = \tfrac{31}{32} = 0.96875.$$

We see that we can write S_n in fractional notation in which the denominator of each term is the power 2^n, and the numerator is 1 less than the denominator:

$$S_n = \frac{2^n - 1}{2^n}.$$

Look at the decimal values above. Note that although the numerator is less than the denominator for all values of n, as n gets larger and larger, the values of S_n get closer and closer to 1. We say that 1 is the **sum of the infinite geometric series.** The sum of the infinite series, if it exists, is denoted S_∞. It can be shown (but we will not do it here) that the sum of the terms of an infinite geometric series exists if and only if $|r| < 1$ (that is, the absolute value of the common ratio is less than 1).

We want to find a formula for the sum of an infinite geometric series

$$a_1 + a_1 r + a_1 r^2 + \cdots.$$

We first consider the sum of the first n terms:

$$S_n = \frac{a_1(r^n - 1)}{r - 1} = \frac{a_1 - a_1 r^n}{1 - r}.$$

For $|r| < 1$, the values of r^n get closer and closer to 0 as n gets large. (Choose a number between -1 and 1 and check this by finding larger and larger powers on your calculator.) As r^n gets closer and closer to 0, so does $a_1 r^n$. Thus, S_n gets closer and closer to

$$\frac{a_1 - 0}{1 - r}, \quad \text{or} \quad \frac{a_1}{1 - r}.$$

THEOREM 6

When $|r| < 1$, the sum of an infinite geometric series

$$a_1 + a_1 r + a_1 r^2 + \cdots$$

is given by

$$S_\infty = \frac{a_1}{1 - r}.$$

When $|r| \geq 1$, an infinite geometric series does not have a sum.

Example 10 Determine whether this infinite geometric series has a sum. If so, find it.

$$1 + 3 + 9 + 27 + \cdots$$

Solution We have $|r| = |3| = 3$, and since $|r| > 1$, the series does *not* have a sum. ◀

Example 11 Determine whether this infinite geometric series has a sum. If so, find it.

$$1 - \tfrac{1}{2} + \tfrac{1}{4} - \tfrac{1}{8} + \tfrac{1}{16} - \cdots$$

Solution

a) $|r| = |-\tfrac{1}{2}| = \tfrac{1}{2}$, and since $|r| < 1$, the series does have a sum.
b) The sum is given by

$$S_\infty = \frac{1}{1 - (-\tfrac{1}{2})} = \frac{1}{\tfrac{3}{2}} = \frac{2}{3}.$$ ◀

DO EXERCISES 11–14.

Example 12 Find fractional notation for $0.63636363\ldots$, or $0.\overline{63}$.

Solution We can express this as

$$0.63 + 0.0063 + 0.000063 + \cdots.$$

This is an infinite geometric series, where $a_1 = 0.63$ and $r = 0.01$. Since $|r| < 1$, this series has a sum:

$$S_\infty = \frac{a_1}{1 - r} = \frac{0.63}{1 - 0.01} = \frac{0.63}{0.99} = \frac{63}{99}, \quad \text{or} \quad \frac{7}{11}.$$

Thus fractional notation for $0.63636363\ldots$ is $\tfrac{7}{11}$. ◀

DO EXERCISES 15 AND 16.

▷ Problem Solving

For some problem-solving situations, the translation may involve geometric sequences or series.

Example 13 Suppose someone offered you a job for the month of September (30 days) under the following conditions. You will be paid $0.01 for the first day, $0.02 for the second, $0.04 for the third, and so on, doubling your previous day's salary each day. How much would you earn? (Would you take the job? Make a guess before reading further.)

Solution You earn $0.01 the first day, $0.01(2)$ the second day, $0.01(2)(2)$ the third day, and so on. The amounts form a geometric sequence with $a_1 = \$0.01$, $r = 2$, and $n = 30$.
The amount earned is the geometric series

$$\$0.01 + \$0.01(2) + \$0.01(2^2) + \$0.01(2^3) + \cdots + \$0.01(2^{29}),$$

Determine whether the infinite geometric series has a sum. If so, find it.

11. $1 + 7 + 49 + 343 + \cdots$

12. $1 + (-1) + 1 + (-1) + \cdots$

13. $\dfrac{1}{2} + \dfrac{1}{4} + \dfrac{1}{8} + \dfrac{1}{16} + \dfrac{1}{32} + \cdots$

14. $625 + 250 + 100 + 40 + \cdots$

Find fractional notation.

15. $0.2\overline{22}$

16. $0.13\overline{13}$

17. Under the conditions of Example 13, how much would you make in October, which has 31 days?

where

$$a_1 = \$0.01, \quad n = 30, \quad \text{and} \quad r = 2.$$

Then using the formula

$$S_n = \frac{a_1(r^n - 1)}{r - 1},$$

we have

$$S_{30} = \frac{\$0.01(2^{30} - 1)}{2 - 1}$$

$$\approx \$0.01(1{,}074{,}000{,}000 - 1) \qquad \text{Using a calculator to approximate } 2^{30}$$

$$\approx \$0.01(1{,}074{,}000{,}000) \qquad 1{,}074{,}000{,}000 - 1 \approx 1{,}074{,}000{,}000$$

$$= \$10{,}740{,}000.$$

Since the salary for September is more than $10 million, most people would take the job.   ◀

DO EXERCISE 17.

18. In Example 14, suppose that 95% of the money will be spent again, and so on. What is the economic multiplier effect?

Example 14   *The economic multiplier.*   The NCAA finals have a tremendous effect on the economy of the host city. Recently, the finals were held in Dallas, Texas. Suppose that 20,000 people visited the city and spent $400 each while there. Then assume that 80% of that money is spent again in the city, and then 80% of that money is spent again, and so on. Find the total effect of this money on the economy. This is known as the **economic multiplier effect.**

Solution   According to certain economic theory, the money that this effectively puts into the economy can be calculated as the sum of an infinite geometric sequence as follows. The "initial effect" is 20,000 × $400, or $8,000,000. The total effect is

$$\$8{,}000{,}000 + \$8{,}000{,}000(0.80) + \$8{,}000{,}000(0.80)^2$$
$$+ \$8{,}000{,}000(0.80)^3 + \cdots.$$

Using the formula of Theorem 6, we find this amount to be

$$S_\infty = \frac{\$8{,}000{,}000}{1 - 0.80} = \$40{,}000{,}000.$$

Do you see why cities work so hard to be chosen to host the NCAA finals?   ◀

DO EXERCISE 18.

● EXERCISE SET 11.3

 Find the common ratio.

1. $2, 4, 8, 16, \ldots$

2. $18, -6, 2, -\frac{2}{3}, \ldots$

3. $-1, 1, -1, 1, \ldots$

4. $-8, -0.8, -0.08, -0.008, \ldots$

5. $\frac{1}{2}, -\frac{1}{4}, \frac{1}{8}, -\frac{1}{16}, \ldots$

6. $\frac{2}{3}, -\frac{4}{3}, \frac{8}{3}, -\frac{16}{3}, \ldots$

7. $75, 15, 3, \dfrac{3}{5}, \ldots$

8. $6.275, 0.6275, 0.06275, \ldots$

9. $\dfrac{1}{x}, \dfrac{1}{x^2}, \dfrac{1}{x^3}, \ldots$

10. $5, \dfrac{5m}{2}, \dfrac{5m^2}{4}, \dfrac{5m^3}{8}, \ldots$

11. $780, \$858, \$943.80, \$1038.18, \ldots$

12. $5600, \$5320, \$5054, \$4801.30, \ldots$

Find the indicated term.

13. $2, 4, 8, 16, \ldots$; the 6th term

14. $2, -10, 50, -250, \ldots$; the 9th term

15. $2, 2\sqrt{3}, 6, \ldots$; the 9th term

16. $1, -1, 1, -1, \ldots$; the 57th term

17. $\dfrac{8}{243}, \dfrac{8}{81}, \dfrac{8}{27}, \ldots$; the 10th term

18. $\dfrac{7}{625}, \dfrac{-7}{25}, \ldots$; the 23rd term

19. $\$1000, \$1080, \$1166.40, \ldots$; the 5th term

20. $\$1000, \$1070, \$1144.90, \ldots$; the 6th term

Find the nth, or general, term.

21. $1, 3, 9, \ldots$

22. $25, 5, 1, \ldots$

23. $1, -1, 1, -1, \ldots$

24. $2, 4, 8, \ldots$

25. $\dfrac{1}{x}, \dfrac{1}{x^2}, \dfrac{1}{x^3}, \ldots$

26. $5, \dfrac{5m}{2}, \dfrac{5m^2}{4}, \ldots$

27. Find the sum of the first 7 terms of the geometric series
$$6 + 12 + 24 + \cdots.$$

28. Find the sum of the first 6 terms of the geometric series
$$16 - 8 + 4 - \cdots.$$

29. Find the sum of the first 7 terms of the geometric series
$$\dfrac{1}{18} - \dfrac{1}{6} + \dfrac{1}{2} - \cdots.$$

30. Find the sum of the geometric series
$$-8 + 4 + (-2) + \cdots + \left(-\dfrac{1}{32}\right).$$

31. Find the sum of the first 8 terms of the series
$$1 + x + x^2 + x^3 + \cdots.$$

32. Find the sum of the first 10 terms of the series
$$1 + x^2 + x^4 + x^6 + \cdots.$$

33. Find the sum of the first 16 terms of the geometric sequence
$$\$200, \$200(1.06), \$200(1.06)^2, \ldots.$$

34. Find the sum of the first 23 terms of the geometric sequence
$$\$1000, \$1000(1.08), \$1000(1.08)^2, \ldots.$$

35. Find the sum:
$$\sum_{k=1}^{\infty} \left(\dfrac{1}{2}\right)^{k-1}.$$

36. Find the sum:
$$\sum_{k=1}^{\infty} 2^k.$$

Determine whether each of the following infinite geometric series has a sum. If so, find it.

37. $4 + 2 + 1 + \cdots$

38. $7 + 3 + \dfrac{9}{7} + \cdots$

39. $25 + 20 + 16 + \cdots$

40. $12 + 9 + \dfrac{27}{4} + \cdots$

41. $100 - 10 + 1 - \dfrac{1}{10} + \cdots$

42. $-6 + 18 - 54 + 162 - \cdots$

43. $8 + 40 + 200 + \cdots$

44. $-6 + 3 - \dfrac{3}{2} + \dfrac{3}{4} - \cdots$

45. $0.6 + 0.06 + 0.006 + \cdots$

46. $0.37 + 0.0037 + 0.000037 + \cdots$

47. $\$500(1.11)^{-1} + \$500(1.11)^{-2} + \$500(1.11)^{-3} + \cdots$

48. $\$1000(1.08)^{-1} + \$1000(1.08)^{-2} + \$1000(1.08)^{-3} + \cdots$

49. $\displaystyle\sum_{k=1}^{\infty} 16(0.1)^{k-1}$

50. $\displaystyle\sum_{k=1}^{\infty} 4(0.6)^{k-1}$

51. $\displaystyle\sum_{k=1}^{\infty} \dfrac{1}{2^{k-1}}$

52. $\displaystyle\sum_{k=1}^{\infty} \dfrac{8}{3}\left(\dfrac{1}{2}\right)^{k-1}$

Find fractional notation for each of the following infinite sums. (These are geometric series.)

53. $0.7\overline{77}$

54. $8.9\overline{999}$

55. $0.5\overline{333}$

56. $0.6\overline{444}$

57. $5.15\overline{15}$

58. $0.412\overline{5125}$

Problem Solving

59. A ping-pong ball is dropped from a height of 16 ft and always rebounds $\frac{1}{4}$ of the distance fallen. How high does it rebound the 6th time?

60. Approximate the total amount of the rebound heights of the ball in Exercise 59.

61. Gaintown has a population of 100,000 now, and the population is increasing by 3% each year. What will the population be in 15 years?

62. How long will it take for the population of Gaintown to double? (See Exercise 61.)

63. A student borrows $1200. The loan is to be repaid in 13 years at 12% interest, compounded annually. How much will be repaid at the end of 13 years?

64. A piece of paper is 0.01 in. thick. It is folded repeatedly in such a way that its thickness is doubled each time for 20 times. How thick is the result?

65. A superball dropped from the top of the Washington Monument (556 ft high) always rebounds $\frac{3}{4}$ of the distance fallen. How far (up and down) will the ball have traveled when it hits the ground for the 6th time?

66. Approximate the total distance that the ball of Exercise 65 will have traveled when it comes to rest.

67. Suppose someone offered you a job for the month of February (28 days) under the following conditions. You will be paid $0.01 the 1st day, $0.02 the 2nd, $0.04 the 3rd, and so on, doubling your previous day's salary each day. How much would you earn altogether?

68. *The amount of an annuity.* A person decides to save money in a savings account for retirement. At the beginning of each year, $1000 is invested at 11%, compounded annually. How much will be in the retirement fund at the end of 40 years?

69. *The economic multiplier.* The government is making a $13,000,000,000 expenditure for a new type of aircraft. If 85% of this gets spent again, and 85% of this gets spent again, and so on, what is the effect on the economy?

70. *Advertising effect.* A company is marketing a new product in a city of 5,000,000 people. They plan an advertising campaign that they think will induce 40% of the people to buy the product. They estimate that if those people like the product, they will induce 40% (of the 40% of 5,000,000) more to buy the product, and those will induce 40%, and so on. In all, how many people will buy the product as a result of the advertising campaign? What percentage of the population is this?

● **SYNTHESIS** _____

71. Prove that $\sqrt{3} - \sqrt{2}$, $4 - \sqrt{6}$, and $6\sqrt{3} - 2\sqrt{2}$ form a geometric sequence.

72. Consider the sequence $x + 3, x + 7, 4x - 2, \ldots$.

   a) If the sequence is arithmetic, find x and then determine each of the three terms and the 4th term.

b) If the sequence is geometric, find x and then determine each of the three terms and the 4th term.

73. Find the sum of the first n terms of
$$1 + x + x^2 + \cdots.$$

74. Find the sum of the first n terms of
$$x^2 - x^3 + x^4 - x^5 + \cdots.$$

For Exercises 75 and 76, assume that $a_1, a_2, a_3, \ldots$, is a geometric sequence.

75. Prove that $a_1^2, a_2^2, a_3^2, \ldots$, is a geometric sequence.

76. Prove that $\ln a_1, \ln a_2, \ln a_3, \ldots$, is an arithmetic sequence.

77. Prove that $5^{a_1}, 5^{a_2}, 5^{a_3}, \ldots$, is a geometric sequence, if $a_1, a_2, a_3, \ldots$, is an arithmetic sequence.

78. The sides of a square are 16 cm long. A second square is inscribed by joining the midpoints of the sides, successively. In the second square, we repeat the process, inscribing a third square. If this process is continued indefinitely, what is the sum of all the areas of all the squares? (*Hint:* Use an infinite geometric series.)

OBJECTIVES

You should be able to:

A List the statements of an infinite sequence that is defined by a formula.

B Do proofs by mathematical induction.

11.4

MATHEMATICAL INDUCTION

A Sequences of Statements

Infinite sequences of statements occur often in mathematics. In an infinite sequence of statements, there is, of course, a statement for each natural number. For example, consider the sentence

   For x between 0 and 1, $0 < x^n < 1$.

Let us think of this as $S(n)$, or S_n. Substituting natural numbers for n gives a sequence of statements. We list a few of them.*

Statement 1 (S_1):	For x between 0 and 1, $0 < x^1 < 1$.
Statement 2 (S_2):	For x between 0 and 1, $0 < x^2 < 1$.
S_3:	For x between 0 and 1, $0 < x^3 < 1$.
S_4:	For x between 0 and 1, $0 < x^4 < 1$.

_____

* Note that S_1, S_2, and so on do *not* represent sums in this context.

Example 1   List the first few statements in the sequence obtainable from $\log n < n$.

Solution   This time, S_n is "$\log n < n$."

S_1: $\log 1 < 1$
S_2: $\log 2 < 2$
S_3: $\log 3 < 3$ ◀

Many sequences of statements concern sums.

Example 2   List the first few statements in the sequence obtainable from
$$1 + 3 + 5 + \cdots + (2n - 1) = n^2.$$

Solution   This time, the entire equation is the statement S_n.

S_1: $1 = 1^2$
S_2: $1 + 3 = 2^2$
S_3: $1 + 3 + 5 = 3^2$
S_4: $1 + 3 + 5 + 7 = 4^2$ ◀

DO EXERCISES 1 AND 2. _____

 Proving Infinite Sequences of Statements

We now develop a method of proof, called **mathematical induction,** which we can use to try to prove that each statement in an infinite sequence of statements is true. The statements usually have the form:

For all natural numbers n, S_n,

where S_n is some sentence such as those in the preceding examples. Of course, we cannot prove each statement of an infinite sequence individually. Instead, we try to show that whenever S_k holds, then S_{k+1} must hold. We abbreviate this as $S_k \rightarrow S_{k+1}$. (This is read "S_k *implies* S_{k+1}.") If we can establish that this holds for all natural numbers k, we then have the following:

$S_1 \rightarrow S_2$ meaning whenever S_1 holds, S_2 must hold;
$S_2 \rightarrow S_3$ meaning whenever S_2 holds, S_3 must hold;
$S_3 \rightarrow S_4$ meaning whenever S_3 holds, S_4 must hold;
and so on, indefinitely.

Suppose we prove each of these statements. We do not yet know whether there is *any* k for which S_k holds. All we know is that if S_k holds, then S_{k+1} must hold. So we now show that S_k holds for some k, usually $k = 1$. We then have the following.

S_1 is true.   We have verified, or proved, this.

$S_1 \rightarrow S_2$ This means that whenever S_1 holds, S_2 must hold.
Therefore, S_2 is true.

$S_2 \rightarrow S_3$ This means that whenever S_2 holds, S_3 must hold.
Therefore, S_3 is true.

And so on.

We conclude that S_n is true for all natural numbers n.

1. List the first 5 statements in the sequence obtainable from
$$n^2 + 1 > n + 1.$$

2. List the first 5 statements in the sequence obtainable from
$$1 + 2 + \cdots + n = \frac{n(n + 1)}{2}.$$

We now state the principle of mathematical induction.

Principle of Mathematical Induction

We can prove an infinite sequence of statements S_n by showing the following.

a) S_1 is true. (This is called the *basis step*.)

b) For all natural numbers k, $S_k \rightarrow S_{k+1}$. (This is called the *induction step*.)

The situation with mathematical induction is analogous to lining up a sequence of dominoes. The induction step tells us that if any one domino is knocked over, then the one next to it will be hit and knocked over. The basis step tells us that the first domino can indeed be knocked over. Note that for all the dominoes to fall, *both* conditions must be satisfied.

When you are learning to do proofs by mathematical induction, it is helpful to first write out S_n, S_1, S_k, and S_{k+1}. This helps to identify what is to be assumed and what is to be deduced.

Example 3 Prove: For every natural number n,

$$1 + 3 + 5 + 7 + \cdots + (2n - 1) = n^2.$$

Proof. We first list S_n, S_1, S_k, and S_{k+1}.

S_n: $1 + 3 + 5 + \cdots + (2n - 1) = n^2$

S_1: $1 = 1^2$

S_k: $1 + 3 + 5 + \cdots + (2k - 1) = k^2$

S_{k+1}: $1 + 3 + 5 + \cdots + (2k - 1) + [2(k + 1) - 1] = (k + 1)^2$

a) *Basis step.* S_1, as listed, is obviously true.

b) *Induction step.* Let k be any natural number. We assume S_k as the hypothesis and try to show that it implies S_{k+1}. Now S_k is

$$1 + 3 + 5 + \cdots + (2k - 1) = k^2.$$

Starting with the left side of S_{k+1} and substituting k^2 for $1 + 3 +$

$\cdots + (2k - 1)$, we have

$$\underbrace{1 + 3 + \cdots + (2k - 1)} + [2(k + 1) - 1]$$
$$= k^2 + [2(k + 1) - 1] = k^2 + 2k + 1$$
$$= (k + 1)^2.$$

We have derived S_{k+1}. Thus we have shown that for all natural numbers k, $S_k \to S_{k+1}$. This completes the induction step and the proof is complete. We have proved that the equation is true for all natural numbers n. ◀

DO EXERCISE 3.

Example 4 Prove that

$$\sum_{p=1}^{n} \frac{1}{2^p} = \frac{2^n - 1}{2^n}$$

for all natural numbers n. In other words, prove that for all natural numbers n,

$$\frac{1}{2} + \frac{1}{4} + \frac{1}{8} + \cdots + \frac{1}{2^n} = \frac{2^n - 1}{2^n}.$$

Proof. We first list S_n, S_1, S_k, and S_{k+1}.

S_n: $\qquad \displaystyle\sum_{p=1}^{n} \frac{1}{2^p} = \frac{1}{2} + \frac{1}{4} + \frac{1}{8} + \cdots + \frac{1}{2^n} = \frac{2^n - 1}{2^n}$

S_1: $\qquad \displaystyle\frac{1}{2} = \frac{2^1 - 1}{2^1}$

S_k: $\qquad \displaystyle\sum_{p=1}^{k} \frac{1}{2^p} = \frac{1}{2} + \frac{1}{4} + \cdots + \frac{1}{2^k} = \frac{2^k - 1}{2^k}$

S_{k+1}: $\qquad \displaystyle\sum_{p=1}^{k+1} \frac{1}{2^p} = \frac{1}{2} + \frac{1}{4} + \frac{1}{8} + \cdots + \frac{1}{2^k} + \frac{1}{2^{k+1}} = \frac{2^{k+1} - 1}{2^{k+1}}$

a) *Basis step.* Since

$$\frac{2^1 - 1}{2^1} = \frac{2 - 1}{2} = \frac{1}{2},$$

S_1 is true.

b) *Induction step.* Let k be any natural number. Using S_k as the hypothesis, we have

$$\frac{1}{2} + \frac{1}{4} + \frac{1}{8} + \cdots + \frac{1}{2^k} = \frac{2^k - 1}{2^k}.$$

Starting with the left side of S_{k+1} and substituting

$$\frac{2^k - 1}{2^k}$$

for

$$\frac{1}{2} + \frac{1}{4} + \cdots + \frac{1}{2^k},$$

3. Consider

$$2 + 4 + 6 + \cdots + 2n = n(n + 1).$$

a) List S_1 and S_2.

b) List S_k.

c) List S_{k+1}.

d) Complete the basis step; that is, verify that S_1 is true.

e) Complete the proof that the formula holds for all n, by proving that S_k implies S_{k+1} for all natural numbers k.

4. Prove that

$$\sum_{p=1}^{n} (3p - 1) = \frac{n(3n + 1)}{2}$$

for all natural numbers n.

we have

$$\underbrace{\frac{1}{2} + \frac{1}{4} + \cdots + \frac{1}{2^k}} + \frac{1}{2^{k+1}}$$

$$= \frac{2^k - 1}{2^k} + \frac{1}{2^{k+1}} = \frac{2^k - 1}{2^k} \cdot \frac{2}{2} + \frac{1}{2^{k+1}}$$

$$= \frac{(2^k - 1) \cdot 2 + 1}{2^{k+1}}$$

$$= \frac{2^{k+1} - 1}{2^{k+1}}.$$

We have arrived at S_{k+1}. Thus we have shown that for all natural numbers k, $S_k \rightarrow S_{k+1}$. This completes the induction and the proof is complete. We have proved that the equation is true for all natural numbers n. ◀

DO EXERCISE 4. _____

Example 5 Prove: For every natural number n, $n < 2^n$.

Proof. We first list S_n, S_1, S_k, and S_{k+1}.

$\quad S_n: \qquad n < 2^n$
$\quad S_1: \qquad 1 < 2^1$
$\quad S_k: \qquad k < 2^k$
$\quad S_{k+1}: \qquad k + 1 < 2^{k+1}$

5. Prove that if x is a number greater than 1, then for any natural number n,

$$x \leq x^n.$$

a) *Basis step.* S_1, as listed, is obviously true.
b) *Induction step.* Let k be any natural number. We assume S_k as the hypothesis and try to show that it implies S_{k+1}. Now

$\quad k < 2^k \qquad$ By hypothesis (this is S_k)
$\quad 2k < 2 \cdot 2^k \qquad$ Multiplying on both sides by 2
$\quad 2k < 2^{k+1}. \qquad$ Adding exponents on the right

Now, since k is any natural number,

$\quad 1 \leq k$
$\quad k + 1 \leq k + k \qquad$ Adding k on both sides
$\quad k + 1 \leq 2k.$

Thus we have

$$k + 1 \leq 2k < 2^{k+1}$$

and

$$k + 1 < 2^{k+1}. \qquad \text{This is } S_{k+1}.$$

We have now shown that $S_k \rightarrow S_{k+1}$ for any natural number k. Thus the induction step and the proof are complete. We have proved that for every natural number n, $n < 2^n$. ◀

DO EXERCISE 5. _____

● EXERCISE SET 11.4

A   List the first 5 statements in the sequence obtainable from each of the following.

1. $n^2 < n^3$

2. $n^2 - n + 41$ is prime.

3. A polygon of n sides has $[n(n-3)]/2$ diagonals.

4. The sum of the angles of a polygon of n sides is $(n-2) \cdot 180°$.

B   Use mathematical induction to prove each of the following, for every natural number n.

5. $1 + 2 + 3 + \cdots + n = \dfrac{n(n+1)}{2}$

6. $4 + 8 + 12 + \cdots + 4n = 2n(n+1)$

7. $1 + 5 + 9 + \cdots + (4n-3) = n(2n-1)$

8. $3 + 6 + 9 + \cdots + 3n = \dfrac{3n(n+1)}{2}$

9. $\dfrac{1}{1 \cdot 2} + \dfrac{1}{2 \cdot 3} + \cdots + \dfrac{1}{n(n+1)} = \dfrac{n}{n+1}$

10. $2 + 4 + 8 + \cdots + 2^n = 2(2^n - 1)$

11. $1^3 + 2^3 + 3^3 + \cdots + n^3 = \dfrac{n^2(n+1)^2}{4}$

12. $\dfrac{1}{1 \cdot 2 \cdot 3} + \dfrac{1}{2 \cdot 3 \cdot 4} + \dfrac{1}{3 \cdot 4 \cdot 5} + \cdots$

$$+ \dfrac{1}{n(n+1)(n+2)} = \dfrac{n(n+3)}{4(n+1)(n+2)}$$

13. $n < n + 1$ 　　　　**14.** $2 \le 2^n$

15. $3^n < 3^{n+1}$ 　　　　**16.** $2n \le 2^n$

● SYNTHESIS _____

17. $\left(1 + \dfrac{1}{1}\right)\left(1 + \dfrac{1}{2}\right)\left(1 + \dfrac{1}{3}\right)\cdots\left(1 + \dfrac{1}{n}\right) = n + 1$

18. $a_1 + (a_1 + d) + (a_2 + d) + \cdots + [a_1 + (n-1)d]$

$$= \dfrac{n}{2}[2a_1 + (n-1)d]$$

(This is a formula for the sum of the first n terms of an *arithmetic* sequence (Theorem 3).)

19. $a_1 + a_1 r + a_1 r^2 + \cdots + a_1 r^{n-1} = \dfrac{a_1 - a_1 r^n}{1 - r}$

(This is a formula for the sum of the first n terms of a *geometric* sequence (Theorem 5).)

20. $\cos n\pi = (-1)^n$

(*Hint:* Use the identity for $\cos(\alpha + \beta)$.)

21. **DeMoivre's theorem.**   For any angle θ and any non-negative number r, and $i^2 = -1$,

$$[r(\cos\theta + i\sin\theta)]^n = r^n(\cos n\theta + i\sin n\theta).$$

22. $|\sin(nx)| \le n|\sin x|$

Prove using mathematical induction.

23. For every natural number $n \ge 2$,

$$\cos x \cdot \cos 2x \cdot \cos 4x \cdot \cdots \cdot \cos 2^{n-1}x = \dfrac{\sin 2^n x}{2^n \sin x}.$$

24. For every natural number $n \ge 2$,

$$\log_a(b_1 b_2 \ldots b_n) = \log_a b_1 + \log_a b_2 + \cdots + \log_a b_n.$$

25. For every natural number $n \ge 2$,

$$\left(1 - \dfrac{1}{2^2}\right)\left(1 - \dfrac{1}{3^2}\right)\cdots\left(1 - \dfrac{1}{n^2}\right) = \dfrac{n+1}{2n}.$$

Prove each of the following for any complex numbers $z_1, \ldots, z_n$, where $i^2 = -1$ and $\bar{z}$ is the conjugate of z (see Section 2.4).

26. $\overline{z^n} = \bar{z}^{\,n}$

27. $\overline{z_1 + z_2 + \cdots + z_n} = \bar{z}_1 + \bar{z}_2 + \cdots + \bar{z}_n$

28. $\overline{z_1 \cdot z_2 \cdot \cdots \cdot z_n} = \bar{z}_1 \cdot \bar{z}_2 \cdot \cdots \cdot \bar{z}_n$

29. i^n is either $1, -1, i,$ or $-i$.

For any integers a and b, b is a factor of a if there exists an integer c such that $a = bc$. Prove each of the following for any natural number n.

30. 3 is a factor of $n^3 + 2n$.

31. 2 is a factor of $n^2 + n$.

32. 5 is a factor of $n^5 - n$.

33. 3 is a factor of $n(n+1)(n+2)$.

● CHALLENGE _____

34. Use mathematical induction to prove that for every natural number $n \ge 2$,

$$\dfrac{1}{\sqrt{1}} + \dfrac{1}{\sqrt{2}} + \dfrac{1}{\sqrt{3}} + \cdots + \dfrac{1}{\sqrt{n}} > \sqrt{n}.$$

35. **The Tower of Hanoi problem.**   There are three pegs on a board. On one peg are n disks, each smaller than the one on which it rests. The problem is to move this pile of disks to another peg. The final order must be the same, but you can move only one disk at a time and you can never place a larger disk on a smaller one.

a) What is the *least* number of moves it takes to move 3 disks?

b) What is the *least* number of moves it takes to move 4 disks?

c) What is the *least* number of moves it takes to move 2 disks?

d) What is the *least* number of moves it takes to move 1 disk?

e) Conjecture a formula for the *least* number of moves it takes to move *n* disks. Prove it by mathematical induction.

36. Consider this statement: For every natural number n, $n = n + 1$.

a) Can you prove the basis step? If so, do it.

b) Can you prove the induction step? If so, do it.

c) Is the statement true? This illustrates the need for the basis step in an induction proof.

37. Find the error in this proof.

Statement.   Everyone is of the same sex.

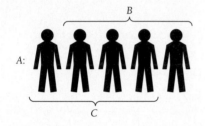

Proof.   Let S_n be the statement: If A is a set of n people, then all the people are of the same sex. Clearly, S_1 is true. Assume S_k. Let A be a set of $k + 1$ people. Then A is the union of two overlapping sets B and C, each containing k people. (Consider the illustration for $k = 5$.) By S_k, all the people in B are of the same sex. Since B and C overlap, all the people in A are of the same sex, and $S_k \rightarrow S_{k+1}$.

38. *Fractals.*   A **fractal**, simply stated, is an infinite recurring geometric figure. Fractals have application to the field of computer graphics and, subsequently, to physics and other fields of mathematics, such as topology and complex variables. Normally, if you make a drawing using a computer and zoom in on the drawing, you get to the place where you see the drawing made up of small squares called *pixels* (which means picture element). To avoid this happening, one uses a fractal technique, which in effect gives infinite resolution. Here is how a fractal might be constructed. You start with a basic figure, and at each step of an infinite sequence of steps, you build on the preceding figure.

Suppose we start with an equilateral triangle with side of length a, as in Figure (1). Then you add to each side another equilateral triangle, with side of length $a/3$, as in Figure (2). Then to each side of Figure (2) you add another equilateral triangle with side of length $a/9$, and so on, obtaining a design that looks very much like a snowflake.

a) Complete the following table and look for patterns.

Figure	Number of Sides	Perimeter	Area
(1)	3	$3a$	$\dfrac{a^2}{4}\sqrt{3}$
(2)			
(3)			
(4)			

b) Conjecture a formula for the number of sides of the nth figure. Prove it by mathematical induction.

c) Conjecture a formula for the perimeter of the nth figure. Prove it by mathematical induction.

d) Conjecture a formula for the area of the nth figure. Prove it by mathematical induction.

e) Find the limit of these areas.

OBJECTIVE

You should be able to:

A Evaluate factorial and permutation notation and solve related counting problems.

11.5

COMBINATORICS: PERMUTATIONS

In order to study probability, it is first necessary to study the theory of counting, called **combinatorics.** Such a study concerns itself with determining the number of ways in which a set can be arranged or combined, certain objects can be chosen, or a succession of events can occur.

The part of combinatorics that we will consider here is the study of *permutations*.

The study of *permutations* involves *order* and *arrangements*.

Example 1   How many 3-letter code symbols can be formed with the letters A, B, C, *without* repetition (that is, without allowing a letter to be repeated)?

Solution   Examples of such symbols are ABC, CBA, ACB, and so on. Consider placing the letters in these frames.

We can select any of the 3 letters for the first letter in the symbol. Once this letter has been selected, the second can be selected from the 2 remaining letters. The third letter is already determined, since only 1 possibility is left. The possibilities can be arrived at with a **tree diagram.**

| Tree Diagram | Outcomes |

1st pick   2nd pick   3rd pick

There are $3 \cdot 2 \cdot 1$, or 6, possibilities. The set of all of them is as follows:

{ABC, ACB, BAC, BCA, CAB, CBA}.

◀

Suppose we perform an experiment such as selecting letters (as in the preceding example), flipping a coin, or drawing a card. The results are called **outcomes.** An **event** is a set of outcomes. The following theorem concerns events that occur together, or are combined.

Fundamental Counting Principle

Given a combined action, or event, in which the first action can be performed in n_1 ways, the second action can be performed in n_2 ways, and so on, the total number of ways in which the combined action can be performed is the product

$$n_1 \cdot n_2 \cdot n_3 \cdot \; \cdots \; \cdot n_k.$$

Example 2   How many 3-letter code symbols can be formed with the letters A, B, and C, *with* repetition (that is, allowing letters to be repeated)?

Solution   There are 3 choices for the first letter and, since we allow repe-

1. How many 3-digit numbers can be named using all the digits 5, 6, 7 without repetition? with repetition?

2. **Zip codes in Canada.** A zip code for Montreal, Quebec, in Canada, is H2N 1M5. It consists of a letter in the first, third, and fifth places, and a number from 0 to 9 in the second, fourth, and sixth places.

   a) How many such zip codes are there?

   b) There are 25 million people in Canada. Can each person have his or her own unique zip code?

3. In how many ways can 5 different cars be parked in a row in a parking lot?

4. How many permutations are there of a set of 5 objects? Consider a set $\{A, B, C, D, E\}$.

Compute.

5. $_3P_3$

6. $_5P_5$

7. $_6P_6$

8. In how many different ways can 4 horses be lined up for a race?

9. In how many different ways can 6 people line up at a ticket window?

10. In how many different ways can the 9-person batting order of a baseball team be made up, if you assume that the pitcher bats last?

tition, 3 choices for the second and 3 for the third. Thus by the fundamental counting principle, there are $3 \cdot 3 \cdot 3$, or 27, choices. ◀

DO EXERCISES 1–3.

DEFINITION   **Permutation**

A *permutation* of a set of n objects is an ordered arrangement of all n objects.

Consider, for example, a set of 4 objects:

$$\{A, B, C, D\}.$$

To find the number of ordered arrangements of the set, we select a first letter: There are 4 choices. Then we select a second letter: There are 3 choices. Then we select a third letter: There are 2 choices. Finally, there is 1 choice for the last selection. Thus by the fundamental counting principle, there are $4 \cdot 3 \cdot 2 \cdot 1$, or 24, permutations of a set of 4 objects.

DO EXERCISE 4.

We can find a formula for the total number of permutations of all objects in a set of n objects. We have n choices for the first selection, $n - 1$ for the second, $n - 2$ for the third, and so on. For the nth selection, there is only 1 choice.

THEOREM 7

The total number of permutations of a set of n objects, denoted $_nP_n$, is given by

$$_nP_n = n(n - 1)(n - 2) \cdots (3)(2)(1).$$

Example 3   Find (a) $_4P_4$ and (b) $_7P_7$.

Solution

a) $_4P_4 = 4 \cdot 3 \cdot 2 \cdot 1 = 24$

b) $_7P_7 = 7 \cdot 6 \cdot 5 \cdot 4 \cdot 3 \cdot 2 \cdot 1 = 5040$ ◀

Example 4   In how many different ways can 9 different letters be placed in 9 mailboxes, one letter to a box?

Solution

$$_9P_9 = 9 \cdot 8 \cdot 7 \cdot 6 \cdot 5 \cdot 4 \cdot 3 \cdot 2 \cdot 1 = 362{,}880 \quad ◀$$

DO EXERCISES 5–10.

Factorial Notation

Products of successive natural numbers, such as $7 \cdot 6 \cdot 5 \cdot 4 \cdot 3 \cdot 2 \cdot 1$, are used so often that it is convenient to adopt a notation for them.

For the product $7 \cdot 6 \cdot 5 \cdot 4 \cdot 3 \cdot 2 \cdot 1$, we write 7!, read "7-factorial."

> **DEFINITION**   **Factorial Notation**
>
> $$n! = n(n-1)(n-2) \cdots (3)(2)(1)$$

Here are some examples.

$$
\begin{aligned}
7! &= 7 \cdot 6 \cdot 5 \cdot 4 \cdot 3 \cdot 2 \cdot 1 = 5040 \\
6! &= 6 \cdot 5 \cdot 4 \cdot 3 \cdot 2 \cdot 1 = 720 \\
5! &= 5 \cdot 4 \cdot 3 \cdot 2 \cdot 1 = 120 \\
4! &= 4 \cdot 3 \cdot 2 \cdot 1 = 24 \\
3! &= 3 \cdot 2 \cdot 1 = 6 \\
2! &= 2 \cdot 1 = 2 \\
1! &= 1 = 1
\end{aligned}
$$

DO EXERCISES 11 AND 12. _____

We also define 0! to be 1. We do this so that certain formulas and theorems can be stated concisely and with a consistent pattern.

We can now simplify the formula of Theorem 7 as follows:

> $$_nP_n = n!$$

DO EXERCISE 13. _____

Note that $8! = 8 \cdot 7!$. We can see this as follows. By definition of factorial notation,

$$
\begin{aligned}
8! &= 8 \cdot 7 \cdot 6 \cdot 5 \cdot 4 \cdot 3 \cdot 2 \cdot 1 \\
&= 8 \cdot (7 \cdot 6 \cdot 5 \cdot 4 \cdot 3 \cdot 2 \cdot 1) \\
&= 8 \cdot 7!.
\end{aligned}
$$

Generalizing, we get the following.

For any natural number n, $n! = n(n-1)!$.

By using this result repeatedly, we can further manipulate factorial notation.

Example 5   Rewrite 7! with a factor of 5!.

Solution

$$7! = 7 \cdot 6 \cdot 5!$$ ◀

DO EXERCISE 14. _____

11. Find 8!.

12. Find 9!.

13. Using factorial notation only, represent the number of permutations of 18 objects.

14. a) Rewrite 10! with a factor of 9!.
    b) Rewrite 20! with a factor of 15!.

Permutations of *n* Objects Taken *r* at a Time

Consider a set of 6 objects, say {A, B, C, D, E, F}. How many ordered arrangements are there having 3 members without repetition? We can select the first object in 6 ways. There are then 5 choices for the second and then 4 choices for the third. By the fundamental counting principle, there are then $6 \cdot 5 \cdot 4$ ways to construct the subset. In other words, there are $6 \cdot 5 \cdot 4$ permutations of a set of 6 objects taken 3 at a time. Note that

$$6 \cdot 5 \cdot 4 = \frac{6 \cdot 5 \cdot 4 \cdot 3 \cdot 2 \cdot 1}{3 \cdot 2 \cdot 1}, \quad \text{or} \quad \frac{6!}{3!}.$$

DEFINITION	Permutation: *n* Objects Taken *r* at a Time

A *permutation* of a set of *n* objects taken *r* at a time is an ordered arrangement of *r* objects taken from the set.

Consider a set of *n* objects and the selecting of an ordered arrangement of *r* objects. The first object can be selected in *n* ways. The second can be selected in $n - 1$ ways, and so on. The *r*th can be selected in $n - (r - 1)$ ways. By the fundamental counting principle, the total number of permutations is

$$n(n - 1)(n - 2) \cdots [n - (r - 1)].$$

We now multiply by 1:

$$n(n - 1)(n - 2) \cdots [n - (r - 1)] \frac{(n - r)!}{(n - r)!}$$

$$= \frac{n(n - 1)(n - 2)(n - 3) \cdots [n - (r - 1)](n - r)!}{(n - r)!}.$$

The numerator is now the product of all natural numbers from *n* to 1, hence is *n*!. Thus the total number of permutations is

$$\frac{n!}{(n - r)!}.$$

This gives us the following theorem.

THEOREM 8

The number of permutations of a set of *n* objects taken *r* at a time, denoted $_nP_r$, is given by

$$_nP_r = \underbrace{n(n - 1)(n - 2) \cdots [n - (r - 1)]}_{r \text{ factors}} \tag{1}$$

$$= \frac{n!}{(n - r)!}. \tag{2}$$

Formula (1) is most useful in application, but formula (2) will be important in a later development.

Example 6   Compute $_6P_4$ using both formulas of Theorem 8.

Solution   We do the calculation in two ways.
Using formula (1), we have

$$_6P_4 = \underbrace{6 \cdot 5 \cdot 4 \cdot 3}\qquad \text{Note that the 6 in } _6P_4 \text{ shows where to start and}$$

$$= 360. \qquad \text{the 4 in } _6P_4 \text{ shows how many factors there are.}$$

Using formula (2) of Theorem 8, we have

$$_6P_4 = \frac{6!}{(6-4)!} = \frac{6!}{2!} = \frac{6 \cdot 5 \cdot 4 \cdot 3 \cdot 2 \cdot 1}{2 \cdot 1} = 6 \cdot 5 \cdot 4 \cdot 3 = 360. \quad \blacktriangleleft$$

DO EXERCISES 15 AND 16.

Example 7   In how many ways can the letters of the set $\{A, B, C, D, E, F, G\}$ be arranged without repetition to form code words of (a) 7 letters? (b) 5 letters? (c) 4 letters? (d) 2 letters?

Solution

a) $_7P_7 = 7 \cdot 6 \cdot 5 \cdot 4 \cdot 3 \cdot 2 \cdot 1 = 5040$
b) $_7P_5 = 7 \cdot 6 \cdot 5 \cdot 4 \cdot 3 \qquad = 2520$
c) $_7P_4 = 7 \cdot 6 \cdot 5 \cdot 4 \qquad\quad = 840$
d) $_7P_2 = 7 \cdot 6 \qquad\qquad\quad = 42$ $\qquad \blacktriangleleft$

Example 8   A baseball manager arranges the batting order as follows: The 4 infielders will bat first, then the outfielders, catcher, and pitcher will follow, not necessarily in that order. How many different batting orders are possible?

Solution   The infielders can bat in 4! different ways; the rest in 5! different ways. Then by the fundamental counting principle, we have $_4P_4 \cdot _5P_5 = 4! \cdot 5!$, or 2880, possible batting orders. $\quad \blacktriangleleft$

DO EXERCISES 17–19.

Permutations of Sets with Nondistinguishable Objects

Consider a set of 7 marbles, 4 of which are blue and 3 of which are black. Although the marbles are all different, when they are lined up, one black marble will look just like any other black marble. In this sense, we say that the blue marbles are nondistinguishable and the black marbles are nondistinguishable.

We know that there are 7! permutations of this set. Many of them will look alike, however. We develop a formula for finding the number of distinguishable permutations.

---

15. Compute $_7P_3$.

16. Compute each of the following.
   a) $_{10}P_4$
   b) $_8P_2$
   c) $_{11}P_5$
   d) $_nP_1$
   e) $_nP_2$
   f) $_nP_0$

17. In how many ways can a 5-woman starting unit be selected from a 12-woman basketball squad and arranged in a straight line?

18. Many nations use flags consisting of three vertical stripes similar to the one shown here. For example, the flag of Ireland has its 1st stripe green, 2nd white, and 3rd gold. Suppose the following 9 colors are available: *black, yellow, red, blue, white, gold, orange, pink, purple*. How many different flags can be made up without repetition of colors? This assumes that the order in which a color appears as a stripe is considered.

19. How many 7-digit numbers can be named without repetition, using the digits 2, 3, 4, 5, 6, 7, and 8, if the even digits are listed first?

20. In how many distinguishable ways can the letters of the word MISSISSIPPI be arranged?

Consider a set of n objects in which n_1 are of one kind, n_2 are of a second kind, . . . , n_k are of the kth kind. By Theorem 7, the total number of permutations of the set is $n!$. Let P be the number of distinguishable permutations. For each of these P permutations, there are $n_1!$ actual permutations, obtained by permuting the objects of the first kind. For each of these $P \cdot n_1!$ permutations, there are $n_2!$ actual permutations, obtained by permuting the objects of the second kind, and so on. By the fundamental counting principle, the total number of actual permutations is

$$P \cdot n_1! \cdot n_2! \cdot \cdots \cdot n_k!.$$

Then we have $P \cdot n_1! \cdot n_2! \cdot \cdots \cdot n_k! = n!$. Solving for P, we obtain

$$P = \frac{n!}{n_1! n_2! \cdots n_k!}.$$

This proves the following theorem.

21. How many 6-digit numbers can be named with all the digits 3, 3, 3, 4, 4, and 5?

THEOREM 9

For a set of n objects in which n_1 are of one kind, n_2 are of another kind, . . . , n_k are of a kth kind, the number of distinguishable permutations is

$$\frac{n!}{n_1! \cdot n_2! \cdot \cdots \cdot n_k!}.$$

Example 9 In how many distinguishable ways can the letters of the word CINCINNATI be arranged?

Solution *Note:* There are 2 C's, 3 I's, 3 N's, 1 A, and 1 T, for a total of 10. Thus,

$$P = \frac{10!}{2! \cdot 3! \cdot 3! \cdot 1! \cdot 1!}, \quad \text{or} \quad 50{,}400. \qquad \blacktriangleleft$$

22. How many vertical signal-flag arrangements can be formed with 3 solid red, 3 solid green, and 2 solid yellow flags?

DO EXERCISES 20–22. _____

Repeated Use of the Same Object

Example 10 How many 5-letter code symbols can be formed with the letters A, B, C, and D if we allow a letter to occur more than once?

Solution We have five spaces:

We can select the first letter in 4 ways, the second in 4 ways, and so on. Thus there are 4^5, or 1024, arrangements. $\qquad \blacktriangleleft$

Generalizing, we have the following.

THEOREM 10

The number of distinct arrangements of n objects taken r at a time, allowing repetition, is n^r.

DO EXERCISE 23.

23. How many 5-letter code symbols can be formed by repeated use of the letters of the alphabet?

● **EXERCISE SET** 11.5

▷ Evaluate.

1. $_4P_3$

2. $_7P_5$

3. $_{10}P_7$

4. $_{10}P_3$

5. How many 5-digit numbers can be named using the digits 5, 6, 7, 8, and 9 without repetition? with repetition?

6. How many 4-digit numbers can be named using the digits 2, 3, 4, and 5 without repetition? with repetition?

7. In how many ways can 5 students be arranged in a straight line?

8. In how many ways can 7 athletes be arranged in a straight line?

9. In how many distinguishable ways can the letters of the word DIGIT be arranged?

10. In how many distinguishable ways can the letters of the word RABBIT be arranged?

11. How many 7-digit phone numbers can be formed with the digits 0, 1, 2, 3, 4, 5, 6, 7, 8, and 9, assuming that no digit is used more than once and the first digit is not 0?

12. A program is planned to have 5 rock numbers and 4 speeches. In how many ways can this be done if a rock number and a speech are to alternate and the rock numbers come first?

13. Suppose the expression $a^2b^3c^4$ is rewritten without exponents. In how many ways can this be done?

14. Suppose the expression a^3bc^2 is rewritten without exponents. In how many ways can this be done?

15. A penny, nickel, dime, quarter, and half dollar (if you have one) are arranged in a straight line.

   a) Considering just the coins, in how many ways can they be lined up?

   b) Considering the coins and heads and tails, in how many ways can they be lined up?

16. A penny, nickel, dime, and quarter are arranged in a straight line.

   a) Considering just the coins, in how many ways can they be lined up?

   b) Considering the coins and heads and tails, in how many ways can they be lined up?

17. ▮ Compute $_{52}P_4$.

18. ▮ Compute $_{50}P_5$.

19. A professor is going to grade her 24 students on a curve. She will give 3 A's, 5 B's, 9 C's, 4 D's, and 3 F's. In how many ways can she do this?

20. A professor is planning to grade his 20 students on a curve. He will give 2 A's, 5 B's, 8 C's, 3 D's, and 2 F's. In how many ways can he do this?

21. ▮ How many distinguishable code symbols can be formed from the letters of the word MATH? BUSINESS? PHILOSOPHICAL?

22. ▮ How many distinguishable code symbols can be formed from the letters of the word ORANGE? BIOLOGY? MATHEMATICS?

23. ▮ A state forms its license plates by first listing a number that corresponds to the county in which the car owner lives (the names of the counties are alphabetized and the number is its location in that order). Then the plate lists a letter of the alphabet, and this is followed by a number from 1 to 9999. How many such plates are possible if there are 80 counties?

24. How many code symbols can be formed using 4 out of 5 letters of A, B, C, D, E if the letters:

   a) are not repeated?

   b) can be repeated?

   c) are not repeated but must begin with D?

   d) are not repeated but must end with DE?

25. *Zip codes.* A zip code in Dallas, Texas, is 75247. A zip code in Cambridge, Massachusetts, is 02142.

   a) How many zip codes are possible if any of the digits 0 to 9 can be used?

   b) If each post office has its own zip code, how many possible post offices can there be?

26. *Zip codes.* Zip codes are sometimes given using a 9-digit number like 75247-5456, where the last 4 digits rep-

resent a post office box number.

a) How many 9-digit zip codes are possible?
b) There are 257 million people in the United States. If each person has a zip code and there were enough post office boxes, are there enough zip codes?

27. *Social security numbers.* A social security number is a 9-digit number like 293-36-0391.

a) How many social security numbers can there be?
b) There are 257 million people in the United States. Can each person have a social security number?

28. ▇ How "long" is 15!? You own 15 different books and decide to actually make up all possible arrangements of the books on a shelf. About how long, in years, would it take if you can make one arrangement per second?

● SYNTHESIS

Solve for n.

29. $_nP_5 = 7 \cdot {}_nP_4$

30. $_nP_4 = 8 \cdot {}_{n-1}P_3$

31. $_nP_5 = 9 \cdot {}_{n-1}P_4$

32. $_nP_4 = 8 \cdot {}_nP_3$

● CHALLENGE

33. In a single-elimination sports tournament consisting of n teams, a team is eliminated when it loses one game. How many games are required to complete the tournament?

34. In a double-elimination softball tournament consisting of n teams, a team is eliminated when it loses two games. At most, how many games are required to complete the tournament?

---

1. Consider the set $\{A, B, C, D, E\}$. How many combinations are there taken:

a) 5 at a time?
b) 4 at a time?
c) 2 at a time?
d) 1 at a time?
e) 0 at a time?

11.6

COMBINATORICS: COMBINATIONS

A If you play cards, you know that in most situations the *order* in which you hold cards *is not important!* It is just the contents of the hand, or set, of cards. We may sometimes make selections from a set *without regard to order.* Such selections are called **combinations.**

Permutation: *Combination:*
Order considered!   Order *not* considered!

Example 1 Find all the combinations of 3 elements taken from the set of 5 elements $\{A, B, C, D, E\}$. How many are there?

Solution The combinations are

$$\{A, B, C\}, \quad \{A, B, D\}, \quad \{A, B, E\}, \quad \{A, C, D\}, \quad \{A, C, E\},$$
$$\{A, D, E\}, \quad \{B, C, D\}, \quad \{B, C, E\}, \quad \{B, D, E\}, \quad \{C, D, E\}.$$

There are 10 combinations of 5 objects taken 3 at a time. ◀

When we find all the combinations of 5 objects taken 3 at a time, we are finding all the 3-element subsets. When we are naming a set, the order of the listing is *not* considered. Thus,

$$\{A, C, B\} \quad \text{is regarded as the same set as} \quad \{A, B, C\}.$$

DO EXERCISE 1.

DEFINITION   **Subset**

The set A is a *subset* of B, denoted $A \subseteq B$, if every element of A is an element of B.

$A \subseteq B$

A is a subset of B.

DEFINITION   **Combination of n Objects Taken r at a Time**

A *combination* containing r objects is a subset of a set that has n objects, $r \leq n$.

The elements of a subset are not ordered.

CAUTION!   When thinking of *combinations*, do *not* think about order!

Example 2   Find all the subsets of the set $\{A, B, C\}$. Identify these as combinations. How many subsets are there in all?

Solution

a) The empty set has 0 elements in it. It is denoted $\varnothing$. The empty set is a subset of every set. In this case, it is the combination of 3 objects taken 0 at a time. There is 1 such combination, $\varnothing$.

b) The following are all the one-element subsets of $\{A, B, C\}$:

$$\{A\}, \quad \{B\}, \quad \{C\}.$$

These are the combinations of 3 objects taken 1 at a time. There are 3 such combinations.

c) The following are all the two-element subsets of $\{A, B, C\}$:

$$\{A, B\}, \quad \{A, C\}, \quad \{B, C\}.$$

These are the combinations of 3 objects taken 2 at a time. There are 3 such combinations.

d) The following are all the three-element subsets of $\{A, B, C\}$:

$$\{A, B, C\}.$$

These are the combinations of 3 objects taken 3 at a time. There is only 1 such combination. A set is always a subset of itself.

The total number of subsets is $1 + 3 + 3 + 1$, or 8.   ◀

DO EXERCISES 2 AND 3.

2. Consider the set $\{A, B\}$.
   a) List all the subsets with 0 elements. How many such subsets are there?
   b) List all the subsets with 1 element. How many such subsets are there?
   c) List all the subsets with 2 elements. How many such subsets are there?
   d) How many subsets of the set $\{A, B\}$ are there in all?

3. Consider the set $\{A\}$.
   a) List all the subsets with 0 elements. How many such subsets are there?
   b) List all the subsets with 1 element. How many such subsets are there?
   c) How many subsets of the set $\{A\}$ are there in all?

We want to develop a formula for computing the number of combinations of n objects taken r at a time without actually listing the combinations, or subsets.

DEFINITION   **Combination Notation**

The number of combinations of n objects taken r at a time is denoted $_nC_r$.

We call $_nC_r$ **combination notation.** In Example 1 and Margin Exercise 1, we see that

$$_5C_5 = 1, \quad _5C_4 = 5, \quad _5C_3 = 10, \quad _5C_2 = 10, \quad _5C_1 = 5, \quad _5C_0 = 1$$

and that

The total number of subsets of a set of 5 objects

$$= {}_5C_5 + {}_5C_4 + {}_5C_3 + {}_5C_2 + {}_5C_1 + {}_5C_0$$
$$= 1 + 5 + 10 + 10 + 5 + 1 = 32.$$

We can derive some general results here. First, it is always true that $_nC_n = 1$, because a set with n objects has only 1 subset with n objects, the set itself. Second, $_nC_1 = n$ because a set with n objects has n subsets with 1 element each. Finally, $_nC_0 = 1$ because a set with n objects has only one subset with 0 elements, namely, the empty set $\varnothing$.

We want to derive a general formula for $_nC_r$, for any $r \leq n$. Let us return to Example 1 and compare the number of combinations with the number of permutations.

Combinations		*Permutations*					
$\{A, B, C\}$	$\longrightarrow$	ABC	BCA	CAB	CBA	BAC	ACB
$\{A, B, D\}$	$\longrightarrow$	ABD	BDA	DAB	DBA	BAD	ADB
$\{A, B, E\}$	$\longrightarrow$	ABE	BEA	EAB	EBA	BAE	AEB
$\{A, C, D\}$	$\longrightarrow$	ACD	CDA	DAC	DCA	CAD	ADC
$\{A, C, E\}$	$\longrightarrow$	ACE	CEA	EAC	ECA	CAE	AEC
$\{A, D, E\}$	$\longrightarrow$	ADE	DEA	EAD	EDA	DAE	AED
$\{B, C, D\}$	$\longrightarrow$	BCD	CDB	DBC	DCB	CBD	BDC
$\{B, C, E\}$	$\longrightarrow$	BCE	CEB	EBC	ECB	CBE	BEC
$\{B, D, E\}$	$\longrightarrow$	BDE	DEB	EBD	EDB	DBE	BED
$\{C, D, E\}$	$\longrightarrow$	CDE	DEC	ECD	EDC	DCE	CED

Note that each combination of 3 objects, say $\{A, C, E\}$, yields 3!, or 6, permutations, as shown above. It follows that

$$3! \cdot {}_5C_3 = 60 = {}_5P_3 = 5 \cdot 4 \cdot 3,$$

so

$$_5C_3 = \frac{_5P_3}{3!} = \frac{5 \cdot 4 \cdot 3}{3 \cdot 2 \cdot 1} = 10.$$

In general, the number of combinations of n objects taken r at a time, $_nC_r$, times the number of permutations of these r objects, $r!$, must equal the number of permutations of n objects taken r at a time:

$$r! \cdot {}_nC_r = {}_nP_r$$

$$_nC_r = \frac{_nP_r}{r!} = \frac{1}{r!} \cdot {}_nP_r = \frac{1}{r!} \cdot \frac{n!}{(n-r)!} = \frac{n!}{r!(n-r)!}.$$

This now gives us two formulas for computing $_nC_r$.

THEOREM 11

The total number of combinations of n objects taken r at a time, denoted $_nC_r$, is given by

$$_nC_r = \frac{n!}{r!(n-r)!}, \tag{1}$$

or

$$_nC_r = \frac{_nP_r}{r!} = \frac{n(n-1)(n-2)\cdots[n-(r-1)]}{r!}. \tag{2}$$

Another kind of notation used for $_nC_r$ is **binomial coefficient notation**. The reason for such terminology will be seen later.

DEFINITION   **Binomial Coefficient Notation**

$$\binom{n}{r} = {_nC_r}$$

You should be able to use either notation and either formula.

Example 3   Evaluate $\binom{7}{5}$, using formulas (1) and (2).

Solution

a) By formula (1),

$$\binom{7}{5} = \frac{7!}{5!2!} = \frac{7\cdot6\cdot5\cdot4\cdot3\cdot2\cdot1}{5\cdot4\cdot3\cdot2\cdot1\cdot2\cdot1} = \frac{7\cdot6\cdot5\cdot4\cdot3}{5\cdot4\cdot3\cdot2\cdot1} = \frac{7\cdot6}{2\cdot1} = 21.$$

b) By formula (2),

The 7 tells where to start.

$$\binom{7}{5} = \frac{7\cdot6\cdot5\cdot4\cdot3}{5\cdot4\cdot3\cdot2\cdot1} = \frac{7\cdot6}{2\cdot1} = 21.$$

The 5 tells us how many factors there are in both the numerator and the denominator and where to start the denominator.

CAUTION!

$$\binom{n}{r} \quad \text{does not mean} \quad n \div r \quad \text{or} \quad \frac{n}{r}.$$

DO EXERCISES 4–7.

Evaluate.

4. $\binom{10}{3}$

5. $\binom{10}{7}$

6. $_9C_4$

7. $_9C_5$

Evaluate.

8. $\dbinom{n}{1}$

The method in Example 3(b), using formula (2), is easier to carry out, but in some situations formula (1) does become useful.

Example 4   Evaluate $\dbinom{n}{0}$ and $\dbinom{n}{2}$.

Solution   We use formula (1) for the first expression and formula (2) for the second. Then

$$\binom{n}{0} = \frac{n!}{0!(n-0)!} = \frac{n!}{1 \cdot n!} = 1,$$

using formula (1), and

$$\binom{n}{2} = \frac{n(n-1)}{2!} = \frac{n(n-1)}{2}, \quad \text{or} \quad \frac{n^2-n}{2},$$

using formula (2).   ◀

DO EXERCISES 8 AND 9.

Note that

$$\binom{7}{2} = \frac{7 \cdot 6}{2 \cdot 1} = 21,$$

so that from Example 3,

$$\binom{7}{5} = \binom{7}{2}.$$

9. $\dbinom{n}{3}$

This says that the number of 5-element subsets of a set of 7 objects is the same as the number of 2-element subsets of a set of 7 objects. When 5 elements are chosen from a set, one also chooses *not* to include 2 elements. To see this, consider such a set:

Whenever we form a subset with 5 elements, we leave behind a subset with 2 elements, and vice versa.

Thus the numbers of each type of subset are the same. In general, we have the following.

THEOREM 12

$$\binom{n}{r} = \binom{n}{n-r} \quad \text{and} \quad {}_nC_r = {}_nC_{n-r}$$

The number of subsets of size r of a set with n objects is the same as the number of subsets of size $n-r$. The number of combinations of n objects taken r at a time is the same as the number of combinations of n objects taken $n-r$ at a time.

Theorem 12 provides an alternative way to compute. For example, it is much easier to compute $_{52}C_4$ than to compute $_{52}C_{48}$.

DO EXERCISES 10 AND 11. _____

We now solve problems involving combinations.

Example 5 **Michigan lotto.** The state of Michigan runs a 6-out-of-44-number lotto twice a week that pays at least $1.5 million. You purchase a card for $1 and pick any 6 numbers from 1 to 44. If your 6 numbers match those that the state draws, you win.

a) How many possible 6-number combinations are there for drawing?
b) Suppose it takes 10 minutes to pick your numbers and buy a ticket. How many tickets can you buy in 4 days?
c) How many people would you have to hire to buy all the tickets and ensure that you win?

Solution

a) No order is implied here. You pick any 6 numbers from 1 to 44. Thus the number of combinations is

$$_{44}C_6 = \binom{44}{6} = \frac{44 \cdot 43 \cdot 42 \cdot 41 \cdot 40 \cdot 39}{6 \cdot 5 \cdot 4 \cdot 3 \cdot 2 \cdot 1} = 7,059,052.$$

b) In four days, there are $4 \cdot 24 \cdot 60$, or 5760, minutes, so you could buy 5760/10, or 576, tickets in that entire time period.
c) You would need to hire 7,059,052/576, or about 12,256, people to buy all the tickets and ensure a win. (This presumes lottery tickets can be bought 24 hours a day, which is questionable.) ◀

DO EXERCISE 12. _____

Example 6 How many committees can be formed from a group of 5 governors and 7 senators if each committee contains 3 governors and 4 senators?

Solution The 3 governors can be selected in $_5C_3$ ways and the 4 senators can be selected in $_7C_4$ ways. If we use the fundamental counting principle, it follows that the number of possible committees is

$$_5C_3 \cdot {}_7C_4 = 10 \cdot 35 = 350.$$ ◀

DO EXERCISE 13. _____

10. a) Evaluate $\binom{8}{5}$ and $\binom{8}{3}$.

   b) Which seemed easier to compute?

11. Evaluate $\binom{100}{97}$.

12. An examination consists of 10 questions. A student is required to answer 8 of them. In how many different ways can the student choose 8 questions to answer? (*Hint:* Is the *order* in which the student answers the questions important, assuming that the answers themselves are numbered?)

13. A committee is to be formed from a group of 12 men and 8 women and is to consist of 3 men and 2 women. How many committees can be formed?

● **EXERCISE SET** **11.6**

Evaluate.

1. $_{13}C_2$ **2.** $_9C_6$ **3.** $\binom{13}{11}$ **4.** $\binom{9}{3}$ **5.** $\binom{7}{1}$ **6.** $\binom{8}{8}$

7. $\dfrac{_5P_3}{3!}$ **8.** $\dfrac{_{10}P_5}{5!}$ **9.** $\dbinom{6}{0}$

10. $\dbinom{6}{1}$ **11.** $\dbinom{6}{2}$ **12.** $\dbinom{6}{3}$

13. $_{12}C_{11}$ **14.** $_{12}C_{10}$ **15.** $_{12}C_9$

16. $_{12}C_8$ **17.** $\dbinom{m}{2}$ **18.** $\dbinom{t}{4}$

19. $\dbinom{p}{3}$ **20.** $\dbinom{m}{m}$

Find the sum.

21. $\dbinom{7}{0} + \dbinom{7}{1} + \dbinom{7}{2} + \dbinom{7}{3} + \dbinom{7}{4} + \dbinom{7}{5} + \dbinom{7}{6} + \dbinom{7}{7}$

22. $\dbinom{6}{0} + \dbinom{6}{1} + \dbinom{6}{2} + \dbinom{6}{3} + \dbinom{6}{4} + \dbinom{6}{5} + \dbinom{6}{6}$

In each of the following exercises, give an expression for the answer in terms of permutation notation, combination notation, factorial notation, or other products. Then evaluate.

23. There are 23 students in a fraternity. How many sets of 4 officers can be selected?

24. How many basketball games can be played in a 9-team league if each team plays all other teams once? twice?

25. On a test, a student is to select 6 out of 10 questions. In how many ways can he do this?

26. On a test, a student is to select 7 out of 11 questions. In how many ways can she do this?

27. How many lines are determined by 8 points, no 3 of which are collinear? How many triangles are determined by the same points?

28. How many lines are determined by 7 points, no 3 of which are collinear? How many triangles are determined by the same points?

29. Of the first 10 questions on a test, a student must answer 7. Of the second 5 questions, she must answer 3. In how many ways can this be done?

30. Of the first 8 questions on a test, a student must answer 6. Of the second 4 questions, he must answer 3. In how many ways can this be done?

31. Suppose the Senate of the United States consists of 58 Democrats and 42 Republicans. How many committees made up of 6 Democrats and 4 Republicans can be formed? You need not simplify the expression.

32. Suppose the Senate of the United States consists of 63 Republicans and 37 Democrats. How many committees made up of 8 Republicans and 12 Democrats can be formed? You need not simplify the expression.

33. How many 5-card poker hands consisting of 3 aces and 2 cards that are not aces are possible with a 52-card deck? (See Section 11.8 for a description of a 52-card deck.)

34. How many 5-card poker hands consisting of 2 kings and 3 cards that are not kings are possible with a 52-card deck?

35. Bresler's Ice Cream, a national firm, sells ice cream in 33 flavors.

  a) How many 3-dip cones are possible if order of flavors is to be considered and no flavor is repeated?

  b) How many 3-dip cones are possible if order is to be considered and a flavor can be repeated?

  c) How many 3-dip cones are possible if order is not considered and no flavor is repeated?

36. Baskin-Robbins Ice Cream, a national firm, sells ice cream in 31 flavors.

  a) How many 2-dip cones are possible if order of flavors is to be considered and no flavor is repeated?

  b) How many 2-dip cones are possible if order is to be considered and a flavor can be repeated?

  c) How many 2-dip cones are possible if order is not considered and no flavor is repeated?

37. Pizza Hut, a national pizza firm, has the following toppings for pizzas:

> cheese, pepperoni, sausage, mushroom, onion, green pepper, beef, Italian sausage, black olives, jalapeno peppers, ham, anchovies.

How many different kinds of pizza can Pizza Hut serve excluding size and thickness of pizzas?

38. Pizza Hut serves round pizzas in three sizes—9-inch, 13-inch, and 15-inch—and two thicknesses——thin-and-crispy and pan. Including all the toppings listed in Exercise 37, and considering sizes and thicknesses, how many different kinds of pizza can Pizza Hut serve?

●   **SYNTHESIS** _____

39. ▥ How many 5-card poker hands are possible with a 52-card deck?

40. ▥ How many 13-card bridge hands are possible with a 52-card deck?

41. There are 8 points on a circle. How many triangles can be inscribed with these points as vertices?

42. There are n points on a circle. How many quadrilaterals can be inscribed with these points as vertices?

43. A set of 5 parallel lines crosses another set of 8 parallel lines at angles that are not right angles. How many parallelograms are formed?

44. Prove: For any natural numbers n and $r \leq n$,
$$\binom{n}{r} = \binom{n}{n-r}.$$

45. How many games are played in a league with 8 teams if each team plays each other team once? twice?

46. How many games are played in a league with n teams if each team plays each other team once? twice?

Solve for n.

47. $\dbinom{n+1}{3} = 2 \cdot \dbinom{n}{2}$ **48.** $\dbinom{n}{n-2} = 6$

49. $\displaystyle \binom{n+2}{4} = 6 \cdot \binom{n}{2}$ **50.** $\displaystyle \binom{n}{3} = 2 \cdot \binom{n-1}{2}$

● CHALLENGE _____

51. How many line segments are determined by the 5 vertices of a pentagon? Of these, how many are diagonals?

52. How many line segments are determined by the 6 vertices of a hexagon? Of these, how many are diagonals?

53. How many line segments are determined by the n vertices of a n-gon? Of these, how many are diagonals? Use mathematical induction to prove the result for the diagonals.

54. Prove: For any natural numbers n and $r \leq n$,

$$\binom{n}{r-1} + \binom{n}{r} = \binom{n+1}{r}.$$

11.7
THE BINOMIAL THEOREM

OBJECTIVES

You should be able to:

A Expand a power of a binomial $(a+b)^n$ using Pascal's triangle.

B Expand a power of a binomial using factorial notation.

C Find a specific term of a binomial expansion.

D Find the total number of subsets of a set of **n objects.**

A Binomial Expansions Using Pascal's Triangle

Consider the following expanded powers of $(a+b)^n$, where $a+b$ is any binomial and n is a whole number. Look for patterns.

$$
\begin{aligned}
(a+b)^0 &= 1 \\
(a+b)^1 &= a+b \\
(a+b)^2 &= a^2 + 2ab + b^2 \\
(a+b)^3 &= a^3 + 3a^2b + 3ab^2 + b^3 \\
(a+b)^4 &= a^4 + 4a^3b + 6a^2b^2 + 4ab^3 + b^4 \\
(a+b)^5 &= a^5 + 5a^4b + 10a^3b^2 + 10a^2b^3 + 5ab^4 + b^5
\end{aligned}
$$

Each expansion is a polynomial. There are some patterns to be noted in the expansions.

1. In each term, the sum of the exponents is n.
2. The exponents of a start with n and decrease to 0. The last term, b^n, has no factor of a. The first term has no factor of b. The exponents of b start in the second term with 1 and increase to n, or we can think of them starting in the first term with 0 and increasing to n.
3. There is one more term than the power n. That is, there are $n+1$ terms in the expansion of $(a+b)^n$.
4. Now we consider the coefficients. The first and last coefficients are 1, and the coefficients have a symmetry to them. They start at 1 and increase through certain values about "half"-way and then decrease through these same values back to 1. Let us explore this further.

Suppose we want to find an expansion of $(a+b)^8$. If the patterns we have noticed were to continue, then we know that there are 9 terms in the expansion, which would be in the following form:

$$a^8 + c_1a^7b + c_2a^6b^2 + c_3a^5b^3 + c_4a^4b^4 + c_3a^3b^5 + c_2a^2b^6 + c_1ab^7 + b^8.$$

How can we determine these coefficients? We can answer this question in two different ways. Your instructor may direct you regarding which to learn. The first method is easier to describe, but it is not very efficient. It involves writing down the coefficients in a triangular array to get what is

1. Find as many patterns as you can in Pascal's triangle. See if you can use the patterns to write three more rows of the triangle.

known as **Pascal's triangle:**

$$(a + b)^0: \qquad\qquad\qquad 1$$
$$(a + b)^1: \qquad\qquad\quad 1 \qquad 1$$
$$(a + b)^2: \qquad\qquad 1 \qquad 2 \qquad 1$$
$$(a + b)^3: \qquad\quad 1 \qquad 3 \qquad 3 \qquad 1$$
$$(a + b)^4: \qquad 1 \qquad 4 \qquad 6 \qquad 4 \qquad 1$$
$$(a + b)^5: \quad 1 \qquad 5 \qquad 10 \qquad 10 \qquad 5 \qquad 1$$

There are many patterns in the triangle. Find as many as you can.

DO EXERCISE 1.

Perhaps you discovered a way to write the next row of numbers, given the numbers in the row above it. The end numbers of each row are both 1. Each remaining number is found by adding the two numbers above. This is shown as follows:

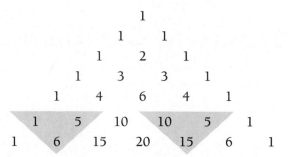

We see that

the 1st number is 1;

the 2nd number is $1 + 5$, or 6;

the 3rd number is $5 + 10$, or 15;

the 4th number is $10 + 10$, or 20;

the 5th number is $10 + 5$, or 15;

the 6th number is $5 + 1$, or 6; and

the 7th number is 1.

Thus the expansion of $(a + b)^6$ is

$$(a + b)^6 = a^6 + 6a^5b + 15a^4b^2 + 20a^3b^3 + 15a^2b^4 + 6ab^5 + b^6.$$

To find the expansion for $(a + b)^8$, we complete two more rows of Pascal's triangle:

$$1$$
$$1 \qquad 1$$
$$1 \qquad 2 \qquad 1$$
$$1 \qquad 3 \qquad 3 \qquad 1$$
$$1 \qquad 4 \qquad 6 \qquad 4 \qquad 1$$
$$1 \qquad 5 \qquad 10 \qquad 10 \qquad 5 \qquad 1$$
$$1 \qquad 6 \qquad 15 \qquad 20 \qquad 15 \qquad 6 \qquad 1$$
$$1 \qquad 7 \qquad 21 \qquad 35 \qquad 35 \qquad 21 \qquad 7 \qquad 1$$
$$1 \qquad 8 \qquad 28 \qquad 56 \qquad 70 \qquad 56 \qquad 28 \qquad 8 \qquad 1$$

Thus the expansion of $(a + b)^8$ is

$$(a + b)^8 = a^8 + 8a^7b + 28a^6b^2 + 56a^5b^3 + 70a^4b^4 + 56a^3b^5$$
$$+ 28a^2b^6 + 8ab^7 + b^8.$$

We can generalize our results as follows.

THEOREM 13   **The Binomial Theorem**

For any binomial $a + b$ and any natural number n,

$$(a + b)^n = c_0a^nb^0 + c_1a^{n-1}b + c_2a^{n-2}b^2 + \cdots$$
$$+ c_{n-1}a^1b^{n-1} + c_na^0b^n,$$

where the numbers $c_0, c_1, c_2, \ldots, c_n$ are from the $(n + 1)$st row of Pascal's triangle, and $c_0 = c_n = 1$.

DO EXERCISE 2.

Example 1   Expand: $(u - v)^5$.

Solution   Note that $a = u$, $b = -v$, and $n = 5$. We use the 6th row of Pascal's triangle:

$$1 \quad 5 \quad 10 \quad 10 \quad 5 \quad 1.$$

Then we have

$$(u - v)^5 = 1(u)^5 + 5(u)^4(-v)^1 + 10(u)^3(-v)^2$$
$$+ 10(u)^2(-v)^3 + 5(u)(-v)^4 + 1(-v)^5$$
$$= u^5 - 5u^4v + 10u^3v^2 - 10u^2v^3 + 5uv^4 - v^5.$$

Note that the signs of the terms alternate between $+$ and $-$. When the power of $-v$ is odd, the sign is $-$.   ◀

DO EXERCISE 3.

Example 2   Expand: $(2t + 3/t)^6$.

Solution   Note that $a = 2t$, $b = 3/t$, and $n = 6$. We use the 7th row of Pascal's triangle:

$$1 \quad 6 \quad 15 \quad 20 \quad 15 \quad 6 \quad 1.$$

Then we have

$$\left(2t + \frac{3}{t}\right)^6 = (2t)^6 + 6(2t)^5\left(\frac{3}{t}\right)^1 + 15(2t)^4\left(\frac{3}{t}\right)^2 + 20(2t)^3\left(\frac{3}{t}\right)^3$$

$$+ 15(2t)^2\left(\frac{3}{t}\right)^4 + 6(2t)^1\left(\frac{3}{t}\right)^5 + \left(\frac{3}{t}\right)^6$$

$$= 64t^6 + 6(32t^5)\left(\frac{3}{t}\right) + 15(16t^4)\left(\frac{9}{t^2}\right) + 20(8t^3)\left(\frac{27}{t^3}\right)$$

$$+ 15(4t^2)\left(\frac{81}{t^4}\right) + 6(2t)\left(\frac{243}{t^5}\right) + \frac{729}{t^6}$$

$$= 64t^6 + 576t^4 + 2160t^2 + 4320 + 4860t^{-2}$$
$$+ 2916t^{-4} + 729t^{-6}.$$ ◀

2. Write one more row of Pascal's triangle and use it to expand $(a + b)^9$.

3. Expand: $(x - y)^4$.

4. Expand: $\left(2t + \dfrac{1}{t}\right)^7$.

DO EXERCISE 4.

B   Binomial Expansion Using Factorial Notation

Suppose we want to find the expansion of $(a + b)^{11}$. The use of Pascal's triangle would yield the terms, but the computations would be inefficient. The method we now consider will allow us to find the rth terms of an expansion without computing any others. This method is also useful in more advanced courses such as finite mathematics and calculus, and uses, for obvious reasons, the **binomial coefficient notation** $\binom{n}{r}$ developed in Section 11.6.

We can restate the binomial theorem as follows.

THEOREM 14   **The Binomial Theorem**

For any binomial $a + b$ and any natural number n,

$$(a + b)^n = \binom{n}{0}a^n + \binom{n}{1}a^{n-1}b + \binom{n}{2}a^{n-2}b^2 + \cdots + \binom{n}{n}b^n.$$

The binomial theorem can be proved by mathematical induction, but we will not do so here.

Sigma notation for a binomial series is

$$(a + b)^n = \sum_{r=0}^{n} \binom{n}{r}a^{n-r}b^r.$$

Theorem 14 shows why $\binom{n}{r}$ is called a **binomial coefficient**. It should now be apparent why $0!$ is defined to be 1. In the binomial expansion, we want $\binom{n}{0}$ to equal 1 and we also want the definition

$$\binom{n}{r} = \frac{n!}{r!(n-r)!}$$

to hold for all whole numbers n and r, where $r \leq n$. Thus we must have

$$\binom{n}{0} = \frac{n!}{0!(n-0)!} = \frac{n!}{0!n!} = 1.$$

This will be satisfied if $0!$ is defined to be 1.

Example 3   Expand: $(x^2 - 2y)^5$.

Solution   Note that $a = x^2, b = -2y$, and $n = 5$. Then, using the binomial theorem, we have

$$(x^2 - 2y)^5 = \binom{5}{0}(x^2)^5 + \binom{5}{1}(x^2)^4(-2y) + \binom{5}{2}(x^2)^3(-2y)^2$$

$$+ \binom{5}{3}(x^2)^2(-2y)^3 + \binom{5}{4}x^2(-2y)^4 + \binom{5}{5}(-2y)^5$$

$$= \frac{5!}{0!5!}x^{10} + \frac{5!}{1!4!}x^8(-2y) + \frac{5!}{2!3!}x^6(-2y)^2$$

$$+ \frac{5!}{3!2!}x^4(-2y)^3 + \frac{5!}{4!1!}x^2(-2y)^4 + \frac{5!}{5!0!}(-2y)^5$$

$$= x^{10} - 10x^8y + 40x^6y^2 - 80x^4y^3 + 80x^2y^4 - 32y^5. \quad \blacktriangleleft$$

DO EXERCISES 5 AND 6.

Example 4   Expand: $(2/x + 3\sqrt{x})^4$.

Solution   Note that $a = 2/x$, $b = 3\sqrt{x}$, and $n = 4$. Then, using the binomial theorem, we have

$$\left(\frac{2}{x} + 3\sqrt{x}\right)^4 = \binom{4}{0}\cdot\left(\frac{2}{x}\right)^4 + \binom{4}{1}\cdot\left(\frac{2}{x}\right)^3(3\sqrt{x}) + \binom{4}{2}\cdot\left(\frac{2}{x}\right)^2(3\sqrt{x})^2$$

$$+ \binom{4}{3}\left(\frac{2}{x}\right)(3\sqrt{x})^3 + \binom{4}{4}(3\sqrt{x})^4$$

$$= \frac{4!}{0!4!}\cdot\frac{16}{x^4} + \frac{4!}{1!3!}\cdot\frac{8}{x^3}3\sqrt{x} + \frac{4!}{2!2!}\cdot\frac{4}{x^2}\cdot 9x$$

$$+ \frac{4!}{3!1!}\cdot\frac{2}{x}\cdot 27x^{3/2} + \frac{4!}{4!0!}\cdot 81x^2$$

$$= \frac{16}{x^4} + \frac{96}{x^{5/2}} + \frac{216}{x} + 216\sqrt{x} + 81x^2.$$ ◀

DO EXERCISES 7 AND 8.

   ## Finding a Specific Term

Suppose we want to determine only a particular term of an expansion. The method we have developed will allow us to find such a term without computing all the rows of Pascal's triangle or all the preceding coefficients.

THEOREM 15

The $(r + 1)$st term of $(a + b)^n$ is

$$\binom{n}{r}a^{n-r}b^r.$$

Example 5   Find the 5th term in the expansion of $(2x - 5y)^6$.

Solution   First, we note that $5 = 4 + 1$. Thus, $r = 4$, $a = 2x$, $b = -5y$, and $n = 6$. Then the 5th term of the expansion is

$$\binom{6}{4}(2x)^{6-4}(-5y)^4, \quad \text{or} \quad \frac{6!}{4!2!}(2x)^2(-5y)^4, \quad \text{or} \quad 37{,}500x^2y^4.$$ ◀

Example 6   Find the 8th term in the expansion of $(3x - 2)^{10}$.

Solution   First, we note that $8 = 7 + 1$. Thus, $r = 7$, $a = 3x$, $b = -2$, and $n = 10$. Then the 8th term of the expansion is

$$\binom{10}{7}(3x)^{10-7}(-2)^7, \quad \text{or} \quad \frac{10!}{7!3!}(3x)^3(-128), \quad \text{or} \quad -414{,}720x^3.$$ ◀

DO EXERCISES 9 AND 10.

Expand.

5. $(x + 5b)^5$

6. $(x^2 - 1)^5$

Expand.

7. $\left(2x + \dfrac{1}{y}\right)^4$

8. $(x - \sqrt{2})^6$

9. Find the 4th term of $(3a + 4b)^7$.

10. Find the 9th term of $(3x - 2)^{10}$.

11. How many subsets are there of the set $\{a, b, c, d, e, f\}$?

▷ Subsets

Suppose a set has n objects. The number of subsets containing r members is $\binom{n}{r}$, by Theorem 11. The total number of subsets of a set is the number with 0 elements, plus the number with 1 element, plus the number with 2 elements, and so on. The total number of subsets of a set with n members is

$$\binom{n}{0} + \binom{n}{1} + \binom{n}{2} + \cdots + \binom{n}{n}.$$

Now let us expand $(1 + 1)^n$:

$$(1 + 1)^n = \binom{n}{0} + \binom{n}{1} + \binom{n}{2} + \cdots + \binom{n}{n}.$$

Thus the total number of subsets is $(1 + 1)^n$, or 2^n. We have proved the following theorem.

THEOREM 16

The total number of subsets of a set with n members is 2^n.

12. How many subsets are there of the set of all states of the United States?

Example 7 The set $\{A, B, C, D, E\}$ has how many subsets?

Solution The set has 5 members, so the number of subsets is 2^5, or 32.   ◀

Example 8 Wendy's, a fast-food restaurant, advertised at one time that it made and sold hamburgers in 256 ways, using combinations of 8 seasonings. Show why.

Solution The total number of combinations is

$$\binom{8}{0} + \binom{8}{1} + \cdots + \binom{8}{8} = 2^8 = 256.$$ ◀

DO EXERCISES 11 AND 12.

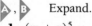 **EXERCISE SET 11.7**

A , B   Expand.

1. $(m + n)^5$
2. $(a - b)^4$
3. $(x - y)^6$
4. $(p + q)^7$
5. $(x^2 - 3y)^5$
6. $(3c - d)^7$
7. $(3c - d)^6$
8. $(t^{-2} + 2)^6$
9. $(x - y)^3$
10. $(x - y)^5$
11. $\left(\dfrac{1}{x} + y\right)^7$
12. $(2s - 3t^2)^3$
13. $\left(a - \dfrac{2}{a}\right)^9$
14. $\left(2x + \dfrac{1}{x}\right)^9$
15. $(1 - 1)^n$
16. $(1 + 3)^n$
17. $(\sqrt{3} - t)^4$
18. $(\sqrt{5} + t)^6$
19. $(\sqrt{2} + 1)^6 - (\sqrt{2} - 1)^6$
20. $(1 - \sqrt{2})^4 + (1 + \sqrt{2})^4$
21. $(x^{-2} + x^2)^4$
22. $\left(\dfrac{1}{\sqrt{x}} - \sqrt{x}\right)^6$

Find the indicated term of the binomial expression.

23. 3rd, $(a + b)^6$

24. 6th, $(x + y)^7$

25. 12th, $(a - 2)^{14}$

26. 11th, $(x - 3)^{12}$

27. 5th, $(2x^3 - \sqrt{y})^8$

28. 4th, $\left(\dfrac{1}{b^2} + \dfrac{b}{3}\right)^7$

29. Middle, $(2u - 3v^2)^{10}$

30. Middle two, $(\sqrt{x} + \sqrt{3})^5$

Determine the number of subsets of each of the following.

31. A set of 7 members

32. A set of 6 members

33. ■ The set of letters of the English alphabet, which contains 26 letters

34. ■ The set of letters of the Greek alphabet, which contains 24 letters

● **SYNTHESIS** _____

Expand.

35. $(\sqrt{2} - i)^4$, where $i^2 = -1$

36. $(1 + i)^6$, where $i^2 = -1$

37. $(\sin t - \csc t)^7$

38. $(\tan \theta + \cot \theta)^{11}$

39. Find a formula for
$$(a - b)^n.$$
Use sigma notation.

40. Expand and simplify:
$$\dfrac{(x + h)^n - x^n}{h}.$$
Use sigma notation.

Solve for x.

41. $\displaystyle\sum_{r=0}^{8} \binom{8}{r} x^{8-r} 3^r = 0$

42. $\displaystyle\sum_{r=0}^{4} \binom{4}{r} 5^{4-r} x^r = 64$

43. $\displaystyle\sum_{r=0}^{5} \binom{5}{r} (-1)^r x^{5-r} 3^r = 32$

44. $\displaystyle\sum_{r=0}^{9} \binom{9}{r} \sin^r x = 0$

45. ■ At one point in a recent season, Darryl Strawberry of the Los Angeles Dodgers had a batting average of 0.313. Suppose he came to bat 5 times in a game. The probability of his getting exactly 3 hits is the 3rd term of the binomial expansion of $(0.313 + 0.687)^5$. Find that term and use your calculator to estimate the probability.

46. ■ The probability that a woman will be either widowed or divorced is 85%. Suppose 8 women are interviewed. The probability that exactly 5 of them will be either widowed or divorced in her lifetime is the 6th term of the binomial expansion of $(0.15 + 0.85)^8$. Find that term and use your calculator to estimate the probability.

47. ■ In reference to Exercise 45, the probability that Strawberry will get at most 3 hits is found by adding the last 4 terms of the binomial expansion of $(0.313 + 0.687)^5$. Find these terms and use your calculator to estimate the probability.

48. ■ In reference to Exercise 46, the probability that at least 6 of the women interviewed will be widowed or divorced is found by adding the last 3 terms of the binomial expansion of $(0.15 + 0.85)^8$. Find these terms and use your calculator to estimate the probability.

49. Find the middle term of the expansion of $(8u + 3v^2)^{10}$.

50. Find the two middle terms of the expansion of $(\sqrt{x} - \sqrt{3})^5$.

51. Find the term of
$$\left(\dfrac{3x^2}{2} - \dfrac{1}{3x}\right)^{12}$$
that does not contain x.

52. Find the middle term of $(x^2 - 6y^{3/2})^6$.

53. Find the ratio of the 4th term of $(p^2 - \frac{1}{2} p \sqrt[3]{q})^5$ to the third term.

54. Find the term of $(\sqrt[3]{x} - 1/\sqrt{x})^7$ containing $1/x^{1/6}$.

55. What is the degree of $(x^5 + 3)^4$?

56. A money clip contains one each of the following bills: \$1, \$2, \$5, \$10, \$20, \$50, and \$100. How many different sums of money can be formed using the bills?

57. Find four consecutive integers such that the sum of the cubes of the three smallest of these is the cube of the fourth.

Find the sum.

58. $_{100}C_0 + _{100}C_1 + \cdots + _{100}C_{100}$

59. $_nC_0 + _nC_1 + \cdots + _nC_n$

● **CHALLENGE** _____

Simplify.

60. $\displaystyle\sum_{r=0}^{23} \binom{23}{r} (\log_a x)^{23-r} (\log_a t)^r$

61. $\ln\left[\displaystyle\sum_{r=0}^{15} \binom{15}{r} (\cos t)^{30-2r} (\sin t)^{2r}\right]$

62. Use mathematical induction and the property
$$\binom{n}{r-1} + \binom{n}{r} = \binom{n+1}{r}$$
to prove the binomial theorem.

11.8

PROBABILITY

We say that when a coin is tossed, we can reason that the chances that it will fall heads are 1 out of 2, or the **probability** that it will fall heads is $\frac{1}{2}$. Of course this does not mean that if a coin is tossed ten times, it will necessarily fall heads exactly five times. If the coin is tossed a great number of times, however, it will fall heads very nearly half of them.

Experimental and Theoretical Probability

If we toss a coin a great number of times, say 1000, and count the number of heads, we can determine the probability of getting a head. If there are 503 heads, we would calculate the probability of getting a head to be

$$\frac{503}{1000}, \quad \text{or} \quad 0.503.$$

This is an **experimental** determination of probability. Such a determination of probability is quite common. Here, for example, are some probabilities that have been determined *experimentally*:

1. If you kiss someone who has a cold, the probability of your catching a cold is 0.07.
2. A person who has just been released from prison has an 80% probability of returning.

If we consider a coin and reason that it is just as likely to fall heads as tails, we would calculate the probability to be $\frac{1}{2}$. This is a **theoretical** determination of probability. Here, for example, are some probabilities that have been determined *theoretically*:

1. If there are 30 people in a room, the probability that two or more of them have the same birthday (excluding year of birth) is 0.706.
2. You are on a vacation. You meet someone, and after a period of conversation, you discover that you have a common acquaintance. The typical reaction "It's a small world!" is actually not appropriate, because the probability of such an occurrence is quite high, just over 22%.

It is results like these that lend credence to the value of a study of probability. You might ask, "What is the *true* probability?" In fact, there is none. Experimentally, we can determine probabilities within certain limits. These may or may not agree with the probabilities that we obtain theoretically.

 Computing Probabilities

Experimental Probabilities

We first consider experimental determination of probability. The basic principle we use in computing such probabilities is as follows.

1. In reference to Example 1, what is the probability that a person is right-handed?

Principle P (Experimental)

An experiment is performed in which n observations are made. If a situation E, or event, occurs m times out of the n observations, then we say that the *experimental probability* of that event is given by

$$P(E) = \frac{m}{n}.$$

Example 1 *Sociological survey.* An actual experiment was conducted to determine the number of people who are left-handed, right-handed, or both. The results are shown in the graph.

a) Determine the probability that a person is left-handed.
b) Determine the probability that a person is ambidextrous (uses both hands equally well).

Solution

a) The number of people who are right-handed was 82, the number who are left-handed was 17, and there was 1 person who is ambidextrous. The total number of observations was $82 + 17 + 1$, or 100. Thus the probability that a person is left-handed is P, where

$$P = \frac{17}{100}.$$

b) The probability that a person is ambidextrous is P, where

$$P = \frac{1}{100}.$$

◀

DO EXERCISE 1.

Example 2 *Quality control.* It is very important to a manufacturer to produce as few defective products as possible. But since a company is producing thousands of products every day, it cannot afford to check every product to see if it is defective. To find out about the quality of its production, that is, what percentage of its products are defective, the company checks a smaller sample.

2. With another large batch of seeds, the company of Example 2 plants 500 seeds and 367 of them sprout. What is the probability that a seed will sprout? Did this batch of seeds pass government inspection?

The U.S. Department of Agriculture requires that 80% of the seeds that a company produces must sprout. To find out about the quality of the seeds it has produced, a company takes 500 seeds from those it has produced and plants them. It finds that 417 of the seeds sprout.

a) What is the probability that a seed will sprout?
b) Did the seeds pass government standards?

Solution

a) We know that 500 seeds were planted and 417 sprouted. The probability of a seed sprouting is P, where

$$P = \frac{417}{500} = 0.834, \quad \text{or} \quad 83.4\%.$$

b) Since the percentage of seeds exceeded the 80% requirement, the company deduces that it is producing quality seeds. ◀

DO EXERCISE 2.

3. In Example 3, what is the probability that a home was tuned to NBC? What is the probability that a home was tuned to a network other than CBS, ABC, or NBC, or was not tuned in at all?

Example 3 *TV ratings.* The major television networks and others such as cable TV are always concerned about the percentages of homes that have TVs and are watching their programs. It is too costly and unmanageable to contact every home in the country so a sample, or portion, of the homes are contacted. This is done by an electronic device attached to the TVs of about 1400 homes across the country. Viewing information is then fed into a computer. The following are the results of a recent survey.

Network	CBS	ABC	NBC	Other or not watching
Number of Homes Watching	258	231	206	705

What is the probability that a home was tuned to CBS during the time period considered? to ABC?

Solution The probability that a home was tuned to CBS is P, where

$$P = \frac{258}{1400} \approx 0.184 = 18.4\%.$$

The probability that a home was tuned to ABC is P, where

$$P = \frac{231}{1400} = 0.165 = 16.5\%.$$ ◀

DO EXERCISE 3.

The numbers that we found in Example 3 and in Margin Exercise 3 (18.4 for CBS, 16.5 for ABC, and 14.7 for NBC) are called the *ratings*.

Theoretical Probabilities

We need some terminology before we can continue. Suppose we perform an experiment such as flipping a coin, throwing a dart, drawing a card from a deck, or checking an item off an assembly line for quality. The results of an experiment are called **outcomes.** The set of all possible outcomes is called the **sample space.** An **event** is a set of outcomes, that is, a subset of the sample space. For example, for the experiment "throwing a dart," suppose the dartboard is as follows.

Then one event is

   {black},   (the outcome is "hitting black")

which is a subset of the sample space

   {black, white, gray},   (sample space)

assuming that the dart must hit the target somewhere.

   We denote the probability that an event E occurs as $P(E)$. For example, "getting a head" may be denoted by H. Then $P(H)$ represents the probability of getting a head. When all the outcomes of an experiment have the same probability of occurring, we say that they are *equally likely*. To see the distinction between events that are equally likely and those that are not, consider the dartboards shown below.

A   B

   For dartboard A, the events hitting *black, white,* and *gray* are equally likely, but for board B they are not. A sample space that can be expressed as a union of equally likely events can allow us to calculate probabilities of other events.

Principle P (Theoretical)

If an event E can occur m ways out of n possible equally likely outcomes of a sample space S, then the *theoretical probability* of that event is given by

$$P(E) = \frac{m}{n}.$$

   A die (pl., dice) is a cube, with six faces, each containing a number of dots from 1 to 6.

4. What is the probability of rolling a prime number on a die?

Example 4 What is the probability of rolling a 3 on a die?

Solution On a fair die, there are 6 equally likely outcomes and there is 1 way to get a 3. By Principle P, $P(3) = \frac{1}{6}$. ◄

Example 5 What is the probability of rolling an even number on a die?

Solution The event is getting an *even* number. It can occur in 3 ways (getting 2, 4, or 6). The number of equally likely outcomes is 6. By Principle P, $P(\text{even}) = \frac{3}{6}$, or $\frac{1}{2}$. ◄

DO EXERCISE 4.

5. Suppose we draw a card from a well-shuffled deck of 52 cards.

  a) What is the probability of drawing a king?

  b) What is the probability of drawing a spade?

  c) What is the probability of drawing a black card?

  d) What is the probability of drawing a jack or a queen?

We now use a number of examples related to a standard bridge deck of 52 cards. Such a deck is made up as shown in the following figure.

A DECK OF 52 CARDS

Example 6 What is the probability of drawing an ace from a well-shuffled deck of 52 cards?

Solution Since there are 52 outcomes (cards in the deck) and they are equally likely (from a well-shuffled deck) and there are 4 ways to obtain an ace, by Principle P we have

$$P(\text{drawing an ace}) = \frac{4}{52}, \quad \text{or} \quad \frac{1}{13}. ◄$$

6. Suppose we select, without looking, one marble from a bag containing 5 red marbles and 6 green marbles. What is the probability of selecting a green marble?

Example 7 Suppose we select, without looking, one marble from a bag containing 3 red marbles and 4 green marbles. What is the probability of selecting a red marble?

Solution There are 7 equally likely ways of selecting any marble, and since the number of ways of getting a red marble is 3,

$$P(\text{selecting a red marble}) = \frac{3}{7}. ◄$$

DO EXERCISES 5 AND 6.

The following are some results that follow from Principle *P*.

THEOREM 17

If an event *E* cannot occur, then $P(E) = 0$.

For example, in coin tossing, the event that a coin will land on its edge has probability 0.

THEOREM 18

If an event *E* is certain to occur (that is, every trial is a success), then $P(E) = 1$.

For example, in coin tossing, the event that a coin falls either heads or tails has probability 1.

In general:

THEOREM 19

The probability that an event *E* will occur is a number from 0 to 1:

$$0 \le P(E) \le 1.$$

DO EXERCISES 7 AND 8.

8. On a single roll of a die, what is the probability of getting a 1, 2, 3, 4, 5, or 6?

In the following examples, we use the combinatorics that we studied in Sections 11.5 and 11.6 to calculate theoretical probabilities.

Example 8 Suppose 2 cards are drawn from a well-shuffled deck of 52 cards. What is the probability that both of them are spades?

Solution The number of ways *n* of drawing 2 cards from a deck of 52 is $_{52}C_2$. Now 13 of the 52 cards are spades, so the number of ways *m* of drawing 2 spades is $_{13}C_2$. Thus,

$$P(\text{getting 2 spades}) = \frac{m}{n} = \frac{_{13}C_2}{_{52}C_2} = \frac{78}{1326} = \frac{1}{17}. \qquad \blacktriangleleft$$

Example 9 Suppose 2 people are selected at random from a group that consists of 6 men and 4 women. What is the probability that both of them are women?

Solution The number of ways of selecting 2 people from a group of 10 is $_{10}C_2$. The number of ways of selecting 2 women from a group of 4 is $_4C_2$. Thus the probability of selecting 2 women from the group of 10 is *P*, where

$$P = \frac{_4C_2}{_{10}C_2} = \frac{6}{45} = \frac{2}{15}. \qquad \blacktriangleleft$$

9. Suppose 3 cards are drawn from a well-shuffled deck of 52 cards. What is the probability that all 3 of them are spades?

10. Suppose 2 people are selected at random from a group that consists of 8 men and 5 women. What is the probability that both of them are women?

11. Suppose 3 people are selected at random from a group that consists of 8 men and 6 women. What is the probability that 2 men and 1 woman are selected?

12. What is the probability of getting a total of 7 on a roll of a pair of dice?

Example 10 Suppose 3 people are selected at random from a group that consists of 6 men and 4 women. What is the probability that 1 man and 2 women are selected?

Solution The number of ways of selecting 3 people from a group of 10 is $_{10}C_3$. One man can be selected in $_6C_1$ ways, and 2 women can be selected in $_4C_2$ ways. By the fundamental counting principle, the number of ways of selecting 1 man and 2 women is $_6C_1 \cdot _4C_2$. Thus the probability is

$$P = \frac{_6C_1 \cdot _4C_2}{_{10}C_3}, \quad \text{or} \quad \frac{3}{10}.$$

DO EXERCISES 9–11.

Example 11 What is the probability of getting a total of 8 on a roll of a pair of dice? (Assume that the dice are different, say one blue and one black.)

Solution On each die, there are 6 possible outcomes. The outcomes are paired so there are $6 \cdot 6$, or 36, possible ways in which the two can fall.

Blue die

6	(1,6)	(2,6)	(3,6)	(4,6)	(5,6)	(6,6)
5	(1,5)	(2,5)	(3,5)	(4,5)	(5,5)	(6,5)
4	(1,4)	(2,4)	(3,4)	(4,4)	(5,4)	(6,4)
3	(1,3)	(2,3)	(3,3)	(4,3)	(5,3)	(6,3)
2	(1,2)	(2,2)	(3,2)	(4,2)	(5,2)	(6,2)
1	(1,1)	(2,1)	(3,1)	(4,1)	(5,1)	(6,1)
	1	2	3	4	5	6 Black die

The pairs that total 8 are as shown. Thus there are 5 possible ways of getting a total of 8, so the probability is $\frac{5}{36}$.

DO EXERCISE 12.

Origin and Use of Probability

A desire to calculate odds in games of chance gave rise to the theory of probability. Today the theory of probability and its closely related field, mathematical statistics, have many applications, most of them not related to games of chance. Opinion polls, with such uses as predicting elections, are a familiar example. Quality control, in which a prediction about the percentage of faulty items manufactured is made without testing them all, is an important application, among many, in business. Still other applications are in the areas of genetics, medicine, and the kinetic theory of gases.

●EXERCISE SET 11.8

A

1. In an actual survey, 100 people were polled to determine the probability of a person wearing either glasses or contact lenses. Of those polled, 57 wore either glasses or contacts. What is the probability that a person wears either glasses or contacts? What is the probability that a person wears neither?

2. In another survey, 100 people were polled and asked to select a number from 1 to 5. The results are shown in the following table.

Number Choices	1	2	3	4	5
Number of People Who Selected that Number	18	24	23	23	12

What is the probability that the number selected is 1? 2? 3? 4? 5? What general conclusion might a psychologist make from this experiment?

Linguistics. An experiment was conducted to determine the relative occurrence of various letters of the English alphabet. A paragraph from a newspaper, one from a textbook, and one from a magazine were considered. In all, there was a total of 1044 letters. The number of occurrences of each letter of the alphabet is listed in the following table.

Letter	Number of Occ.	Letter	Number of Occ.
A	78	N	74
B	22	O	74
C	33	P	27
D	33	Q	4
E	140	R	67
F	24	S	67
G	22	T	95
H	63	U	31
I	60	V	10
J	2	W	22
K	9	X	8
L	35	Y	13
M	30	Z	1

Round answers to Exercises 3–6 to three decimal places.

3. What is the probability of the occurrence of the letter A? E? I? O? U?

4. What is the probability of a vowel occurring?

5. What is the probability of a consonant occurring?

6. What letter has the least probability of occurring? What is the probability of this letter not occurring?

Suppose we draw a card from a well-shuffled deck of 52 cards.

7. How many equally likely outcomes are there?

8. What is the probability of drawing a queen?

9. What is the probability of drawing a heart?

10. What is the probability of drawing a club?

11. What is the probability of drawing a 4?

12. What is the probability of drawing a red card?

13. What is the probability of drawing a black card?

14. What is the probability of drawing an ace or a deuce?

15. What is the probability of drawing a 9 or a king?

Suppose we select, without looking, one marble from a bag containing 4 red marbles and 10 green marbles.

16. What is the probability of selecting a red marble?

17. What is the probability of selecting a green marble?

18. What is the probability of selecting a purple marble?

19. What is the probability of selecting a white marble?

Suppose 4 cards are drawn from a well-shuffled deck of 52 cards.

20. What is the probability that all 4 are spades?

21. What is the probability that all 4 are hearts?

22. If 4 marbles are drawn at random all at once from a bag containing 8 white marbles and 6 black marbles, what is the probability that 2 will be white and 2 will be black?

23. From a group of 8 men and 7 women, a committee of 4 is chosen. What is the probability that 2 men and 2 women will be chosen?

24. What is the probability of getting a total of 6 on a roll of a pair of dice?

25. What is the probability of getting a total of 3 on a roll of a pair of dice?

26. What is the probability of getting snake eyes (a total of 2) on a roll of a pair of dice?

27. What is the probability of getting box-cars (a total of 12) on a roll of a pair of dice?

28. From a bag containing 5 nickels, 8 dimes, and 7 quarters, 5 coins are drawn at random, all at once. What is the probability of getting 2 nickels, 2 dimes, and 1 quarter?

29. From a bag containing 6 nickels, 10 dimes, and 4 quarters, 6 coins are drawn at random, all at once. What is the probability of getting 3 nickels, 2 dimes, and 1 quarter?

Roulette. A roulette wheel contains slots numbered 00, 0, 1, 2, 3, . . . , 35, 36. Eighteen of the slots numbered 1 through 36 are colored red and eighteen are colored black. The 00 and 0 slots are uncolored. The wheel is spun, and a ball is rolled around the rim until it falls into a slot. What is the probability that the ball falls in:

30. a black slot?   **31.** a red slot?

32. a red or black slot?   **33.** the 00 slot?

34. the 0 slot?

35. either the 00 or 0 slot? (Here the house always wins.)

36. an odd-numbered slot?

● **SYNTHESIS** _____

Five-card poker hands and probabilities. In part (a) of each problem, give a reasoned expression as well as the answer. Read all the problems before beginning.

37. ▦ How many 5-card poker hands can be dealt from a standard 52-card deck?

38. ▦ A *royal flush* consists of a 5-card hand with A-K-Q-J-10 of the same suit.

   a) How many royal flushes are there?
   b) What is the probability of getting a royal flush?

39. ▦ A *straight flush* consists of 5 cards in sequence in the same suit, but excludes royal flushes. An ace can be used low, before a two.

   a) How many straight flushes are there?
   b) What is the probability of getting a straight flush?

40. ▦ *Four of a kind* is a 5-card hand in which 4 of the cards are of the same denomination, such as J-J-J-J-6, 7-7-7-7-A, or 2-2-2-2-5.

   a) How many are there?
   b) What is the probability of getting four of a kind?

● **CHALLENGE** _____

41. ▦ A *full house* consists of a pair and three of a kind, such as Q-Q-Q-4-4.

   a) How many are there?
   b) What is the probability of getting a full house?

42. ▦ A *pair* is a 5-card hand in which just 2 of the cards are of the same denomination, such as Q-Q-8-A-3.

   a) How many are there?
   b) What is the probability of getting a pair?

43. ▦ *Three of a kind* is a 5-card hand in which exactly 3 of the cards are of the same denomination and the other 2 are *not* of the same denomination, such as Q-Q-Q-10-7.

   a) How many are there?
   b) What is the probability of getting three of a kind?

44. ▦ A *flush* is a 5-card hand in which all the cards are of the same suit, but not all in sequence (not a straight flush or royal flush).

   a) How many are there?
   b) What is the probability of getting a flush?

45. ▦ *Two pairs* is a hand like Q-Q-3-3-A.

   a) How many are there?
   b) What is the probability of getting two pairs?

46. ▦ A *straight* is any 5 cards in sequence, but not of the same suit—for example, 4 of spades, 5 of spades, 6 of diamonds, 7 of hearts, and 8 of clubs.

   a) How many are there?
   b) What is the probability of getting a straight?

— 11 —

SUMMARY AND REVIEW

● **TERMS TO KNOW**

● REVIEW EXERCISES

1. Find the 10th term in the arithmetic sequence $\frac{3}{4}, \frac{13}{12}, \frac{17}{12}, \ldots$.

2. Find the 6th term in the arithmetic sequence $a - b$, a, $a + b, \ldots$.

3. Find the sum of the first 18 terms of the arithmetic sequence $4, 7, 10, \ldots$.

4. Find the sum of the first 30 positive integers.

5. The first term of an arithmetic sequence is 5. The 17th term is 53. Find the 3rd term.

6. The common difference in an arithmetic sequence is 3. The 10th term is 23. Find the first term.

7. For a geometric sequence, $a_1 = -2$, $r = 2$, and $a_n = -64$. Find n and S_n.

8. For a geometric sequence, $r = \frac{1}{2}$, $n = 5$, and $S_n = \frac{31}{2}$. Find a_1 and a_n.

9. Determine whether this geometric sequence has a sum.

$$25, 27.5, 30.25, 33.275, \ldots$$

10. Determine whether this geometric sequence has a sum.

$$0.27, 0.0027, 0.000027, \ldots$$

11. Find this infinite sum. The series is geometric.

$$\frac{1}{2} - \frac{1}{6} + \frac{1}{18} - \cdots$$

12. Find fractional notation for $2.\overline{13}$.

13. Insert four arithmetic means between 5 and 9.

14. A golf ball is dropped from a height of 30 ft to the pavement, and the rebound is one fourth of the distance that it drops. If, after each descent, it continues to rebound one fourth of the distance dropped, what is the total distance that the ball has traveled when it reaches the pavement on its 10th descent?

15. You receive 10¢ on the first day of the year, 12¢ on the 2nd day, 14¢ on the 3rd day, and so on. How much will you receive on the 365th day? What is the sum of all these 365 gifts?

16. The present population of a city is 30,000. Its population is supposed to double every 10 yr. What will its population be at the end of 80 yr?

17. The sides of a square are each 16 in. long. A second square is inscribed by joining the midpoints of the sides, successively. In the second square, we repeat the process, inscribing a third square. If this process is continued indefinitely, what is the sum of the perimeters of all of the squares? (*Hint:* Use an infinite geometric series.)

18. A pendulum is moving back and forth in such a way that it traverses an arc 10 cm in length, and thereafter arcs are $\frac{4}{7}$ the length of the previous arc. What is the sum of the arc lengths that the pendulum traverses?

Use mathematical induction.

19. Prove: For all natural numbers n,

$$1 + 4 + 7 + \cdots + (3n - 2) = \frac{n(3n - 1)}{2}.$$

20. Prove: For all natural numbers n,

$$1 + 3 + 3^2 + \cdots + 3^{n-1} = \frac{3^n - 1}{2}.$$

21. Prove: For every natural number $n \geq 2$,

$$\left(1 - \frac{1}{2}\right)\left(1 - \frac{1}{3}\right)\cdots\left(1 - \frac{1}{n}\right) = \frac{1}{n}.$$

22. Find the first 4 terms of this recursively defined sequence.

$$a_1 = 5, \qquad a_{k+1} = 2a_k^2 + 1$$

23. Write Σ notation for this sequence.

$$0 + 3 + 8 + 15 + 24 + 35 + 48$$

24. In how many different ways can 6 books be arranged on a shelf?

25. If 9 different signal flags are available, how many different displays are possible using 4 flags in a row?

26. The winner of a contest can choose any 8 of 15 prizes. How many different selections can be made?

27. The Greek alphabet contains 24 letters. How many fraternity or sorority names can be formed using 3 different letters?

28. In how many distinguishable ways can the letters of the word TENNESSEE be arranged?

29. A manufacturer of houses has one floor plan but achieves variety by having 3 different colored roofs, 4 different ways of attaching the garage, and 3 different types of entrance. Find the number of different houses that can be produced.

30. How many code symbols can be formed using 5 out of 6 of the letters of G, H, I, J, K, L if the letters:
   a) cannot be repeated?
   b) can be repeated?
   c) cannot be repeated but must begin with K?
   d) cannot be repeated but must end with IGH?

31. Find the 4th term of $(a + x)^{12}$.

32. Find the 12th term of $(a + x)^{18}$. Do not multiply out the factorials.

Expand.

33. $(m + n)^7$

34. $(x^2 + 3y)^4$

35. $(5i + 1)^6$, where $i^2 = -1$

36. $(a + a^{-1})^8$

37. Before an election, a poll was conducted to see which candidate was favored. Three people were running for a particular office. During the polling, 86 favored A, 97 favored B, and 23 favored C. Assuming that the poll is a valid indicator of the election, what is the probability that the election will be won by A? B? C?

38. What is the probability of rolling a 10 on a roll of a pair of dice? on a roll of one die?

39. From a deck of 52 cards, 1 card is drawn. What is the probability that it is a club?

40. From a deck of 52 cards, 3 are drawn at random without replacement. What is the probability that 2 are aces and 1 is a king?

● **SYNTHESIS** _____

41. Explain why the following cannot be proved by mathematical induction: For every natural number n:

a) $3 + 5 + \cdots + (2n + 1) = (n + 1)^2$;

b) $1 + 3 + \cdots + (2n - 1) = n^2 + 3$.

42. Suppose a and b are geometric sequences. Prove that c is a geometric sequence, where $c_n = a_n b_n$.

43. Suppose a is an arithmetic sequence. Prove that c is a geometric sequence, where $c_n = b^{a_n}$, for some positive number b.

44. Suppose a is an arithmetic sequence. Under what conditions is b an arithmetic sequence where:

a) $b_n = |a_n|$?

b) $b_n = a_n + 8$?

c) $b_n = 7a_n$?

d) $b_n = \dfrac{1}{a_n}$?

e) $b_n = \log a_n$?

f) $b_n = a_n^3$?

45. The zeros of this polynomial form an arithmetic sequence. Find them.

$$x^4 - 4x^3 - 4x^2 + 16x$$

46. Write the first 3 terms of the infinite geometric sequence with $S_\infty = \frac{3}{11}$ and $r = 0.01$.

47. Write the first 3 terms of the infinite geometric sequence with $r = -\frac{1}{3}$ and $S_\infty = \frac{3}{8}$.

48. Simplify:

$$\sum_{r=0}^{10} (-1)^r \binom{10}{r} (\log x)^{10-r} (\log y)^r.$$

Solve for n.

49. $\dbinom{n}{n-1} = 36$

50. $26 \cdot \dbinom{n}{1} = \dbinom{n}{3}$

● **THINKING AND WRITING** _____

1. Chain letters have been outlawed by the government. Nevertheless, "chain" business deals still exist and they can be fraudulent. Suppose a saleperson is charged with the task of hiring four new salespersons. Each of them gives half of their profits to the person who hires them. Each of these people gets four new salespersons. Each of these gives half of his profits to the one who hired them. Half of these profits then go back to the original hiring person. Explain the lure of this business to someone who has managed several sequences of hirings. Explain the fallacy of such a business as well. Keep in mind that there are 257 million people in this country.

2. Examine the sequence of numbers given by 7^n. Then try to discover the one's digit in the number 7^{1000}.

CHAPTER TEST 11

1. Find the 18th term of the arithmetic sequence $\frac{1}{4}, 1, \frac{7}{4}, \frac{5}{2}, \ldots$.

2. The 2nd term of an arithmetic sequence is 9, and the 9th term is 37. Find the common difference.

3. Which term of the arithmetic sequence $1, \frac{3}{2}, 2, \frac{5}{2}, \ldots$ is $\frac{31}{2}$?

4. Insert three arithmetic means between 3 and 14.

5. Find the 8th term of the geometric sequence

$$0.2, 0.6, 1.8, \ldots.$$

6. Evaluate the sum

$$\sum_{k=1}^{5} \left(\frac{1}{2}\right)^{k+1}.$$

7. Which of the following infinite geometric sequences have sums?

a) $2, 0.2, 0.02, 0.002, \ldots$

b) $3, -6, 12, -24, 48, \ldots$

c) $\frac{1}{20}, \frac{1}{10}, \frac{1}{5}, \frac{2}{5}, \ldots$

8. Find the sum of the infinite geometric sequence

$$25, -5, 1, -\frac{1}{5}, \ldots.$$

9. A student made deposits in a savings account as follows: $20.50 the first month, $26 the second month, $31.50 the third month, and so on, for 2 years. What was the sum of the deposits?

10. A publishing company prints only $\frac{3}{5}$ as many books with each new printing of a book. If 100,000 copies of a book are printed originally, how many will be printed in the 5th printing?

11. Find fractional notation for $0.12\overline{8}$.

12. Use mathematical induction. Prove that for every natural number n,

$$5 + 10 + 15 + \cdots + 5n = \frac{5n(n + 1)}{2}.$$

13. Find the first 4 terms of this recursively defined sequence:

$$a_1 = 5, \qquad a_{k+1} = 4a_k + 3.$$

14. How many code symbols can be formed using 4 out of 6 of the letters of D, E, F, G, H, I if the letters:

a) can be repeated?

b) cannot be repeated?

c) cannot be repeated but must begin with FH?

15. On a test, a student must answer 4 out of 7 questions. In how many ways can this be done?

16. From a group of 20 seniors and 14 juniors, how many committees consisting of 3 seniors and 2 juniors are possible?

17. In how many distinguishable ways can the letters of the word ARKANSAS be arranged?

18. Determine the number of subsets of a set of 8 members.

19. Find the 3rd term of $(2a + b)^7$.

20. Expand $(x - \sqrt{2})^5$.

21. What is the probability of getting a total of 6 on a roll of a pair of dice?

22. From a deck of 52 cards, 1 card is drawn. What is the probability of drawing a 3 or a queen?

23. If 3 marbles are drawn at random all at once from a bag containing 5 green marbles, 7 red marbles, and 4 white marbles, what is the probability that 2 will be green and 1 will be white?

● **SYNTHESIS** _____

24. Find 4 numbers in an arithmetic sequence such that twice the second minus the fourth is 1 and the sum of the first and third is 14.

25. How many diagonals does a dodecagon (12-sided polygon) have?

26. Solve for n:

$$\binom{n}{6} = 3 \cdot \binom{n-1}{5}.$$

27. Solve for a:

$$\sum_{r=0}^{5} 9^{5-r} a^r = 0.$$

Tables

TABLE 1 Common Logarithms

x	0	1	2	3	4	5	6	7	8	9	x	0	1	2	3	4	5	6	7	8	9
1.0	.0000	.0043	.0086	.0128	.0170	.0212	.0253	.0294	.0334	.0374	5.5	.7404	.7412	.7419	.7427	.7435	.7443	.7451	.7459	.7466	.7474
1.1	.0414	.0453	.0492	.0531	.0569	.0607	.0645	.0682	.0719	.0755	5.6	.7482	.7490	.7497	.7505	.7513	.7520	.7528	.7536	.7543	.7551
1.2	.0792	.0828	.0864	.0899	.0934	.0969	.1004	.1038	.1072	.1106	5.7	.7559	.7566	.7574	.7582	.7589	.7597	.7604	.7612	.7619	.7627
1.3	.1139	.1173	.1206	.1239	.1271	.1303	.1335	.1367	.1399	.1430	5.8	.7634	.7642	.7649	.7657	.7664	.7672	.7679	.7686	.7694	.7701
1.4	.1461	.1492	.1523	.1553	.1584	.1614	.1644	.1673	.1703	.1732	5.9	.7709	.7716	.7723	.7731	.7738	.7745	.7752	.7760	.7767	.7774
1.5	.1761	.1790	.1818	.1847	.1875	.1903	.1931	.1959	.1987	.2014	6.0	.7782	.7789	.7796	.7803	.7810	.7818	.7825	.7832	.7839	.7846
1.6	.2041	.2068	.2095	.2122	.2148	.2175	.2201	.2227	.2253	.2279	6.1	.7853	.7860	.7868	.7875	.7882	.7889	.7896	.7903	.7910	.7917
1.7	.2304	.2330	.2355	.2380	.2405	.2430	.2455	.2480	.2504	.2529	6.2	.7924	.7931	.7938	.7945	.7952	.7959	.7966	.7973	.7980	.7987
1.8	.2553	.2577	.2601	.2625	.2648	.2672	.2695	.2718	.2742	.2765	6.3	.7993	.8000	.8007	.8014	.8021	.8028	.8035	.8041	.8048	.8055
1.9	.2788	.2810	.2833	.2856	.2878	.2900	.2923	.2945	.2967	.2989	6.4	.8062	.8069	.8075	.8082	.8089	.8096	.8102	.8109	.8116	.8122
2.0	.3010	.3032	.3054	.3075	.3096	.3118	.3139	.3160	.3181	.3201	6.5	.8129	.8136	.8142	.8149	.8156	.8162	.8169	.8176	.8182	.8189
2.1	.3222	.3243	.3263	.3284	.3304	.3324	.3345	.3365	.3385	.3404	6.6	.8195	.8202	.8209	.8215	.8222	.8228	.8235	.8241	.8248	.8254
2.2	.3424	.3444	.3464	.3483	.3502	.3522	.3541	.3560	.3579	.3598	6.7	.8261	.8267	.8274	.8280	.8287	.8293	.8299	.8306	.8312	.8319
2.3	.3617	.3636	.3655	.3674	.3692	.3711	.3729	.3747	.3766	.3784	6.8	.8325	.8331	.8338	.8344	.8351	.8357	.8363	.8370	.8376	.8382
2.4	.3802	.3820	.3838	.3856	.3874	.3892	.3909	.3927	.3945	.3962	6.9	.8388	.8395	.8401	.8407	.8414	.8420	.8426	.8432	.8439	.8445
2.5	.3979	.3997	.4014	.4031	.4048	.4065	.4082	.4099	.4116	.4133	7.0	.8451	.8457	.8463	.8470	.8476	.8482	.8488	.8494	.8500	.8506
2.6	.4150	.4166	.4183	.4200	.4216	.4232	.4249	.4265	.4281	.4298	7.1	.8513	.8519	.8525	.8531	.8537	.8543	.8549	.8555	.8561	.8567
2.7	.4314	.4330	.4346	.4362	.4378	.4393	.4409	.4425	.4440	.4456	7.2	.8573	.8579	.8585	.8591	.8597	.8603	.8609	.8615	.8621	.8627
2.8	.4472	.4487	.4502	.4518	.4533	.4548	.4564	.4579	.4594	.4609	7.3	.8633	.8639	.8645	.8651	.8657	.8663	.8669	.8675	.8681	.8686
2.9	.4624	.4639	.4654	.4669	.4683	.4698	.4713	.4728	.4742	.4757	7.4	.8692	.8698	.8704	.8710	.8716	.8722	.8727	.8733	.8739	.8745
3.0	.4771	.4786	.4800	.4814	.4829	.4843	.4857	.4871	.4886	.4900	7.5	.8751	.8756	.8762	.8768	.8774	.8779	.8785	.8791	.8797	.8802
3.1	.4914	.4928	.4942	.4955	.4969	.4983	.4997	.5011	.5024	.5038	7.6	.8808	.8814	.8820	.8825	.8831	.8837	.8842	.8848	.8854	.8859
3.2	.5051	.5065	.5079	.5092	.5105	.5119	.5132	.5145	.5159	.5172	7.7	.8865	.8871	.8876	.8882	.8887	.8893	.8899	.8904	.8910	.8915
3.3	.5185	.5198	.5211	.5224	.5237	.5250	.5263	.5276	.5289	.5307	7.8	.8921	.8927	.8932	.8938	.8943	.8949	.8954	.8960	.8965	.8971
3.4	.5315	.5328	.5340	.5353	.5366	.5378	.5391	.5403	.5416	.5428	7.9	.8976	.8982	.8987	.8993	.8998	.9004	.9009	.9015	.9020	.9025
3.5	.5441	.5453	.5465	.5478	.5490	.5502	.5514	.5527	.5539	.5551	8.0	.9031	.9036	.9042	.9047	.9053	.9058	.9063	.9069	.9074	.9079
3.6	.5563	.5575	.5587	.5599	.5611	.5623	.5635	.5647	.5658	.5670	8.1	.9085	.9090	.9096	.9101	.9106	.9112	.9117	.9122	.9128	.9133
3.7	.5682	.5694	.5705	.5717	.5729	.5740	.5752	.5763	.5775	.5786	8.2	.9138	.9143	.9149	.9154	.9159	.9165	.9170	.9175	.9180	.9186
3.8	.5798	.5809	.5821	.5832	.5843	.5855	.5866	.5877	.5888	.5899	8.3	.9191	.9196	.9201	.9206	.9212	.9217	.9222	.9227	.9232	.9238
3.9	.5911	.5922	.5933	.5944	.5955	.5966	.5977	.5988	.5999	.6010	8.4	.9243	.9248	.9253	.9258	.9263	.9269	.9274	.9279	.9284	.9289
4.0	.6021	.6031	.6042	.6053	.6064	.6075	.6085	.6096	.6107	.6117	8.5	.9294	.9299	.9304	.9309	.9315	.9320	.9325	.9330	.9335	.9340
4.1	.6128	.6138	.6149	.6160	.6170	.6180	.6191	.6201	.6212	.6222	8.6	.9345	.9350	.9355	.9360	.9365	.9370	.9375	.9380	.9385	.9390
4.2	.6232	.6243	.6253	.6263	.6274	.6284	.6294	.6304	.6314	.6325	8.7	.9395	.9400	.9405	.9410	.9415	.9420	.9425	.9430	.9435	.9440
4.3	.6335	.6345	.6355	.6365	.6375	.6385	.6395	.6405	.6415	.6425	8.8	.9445	.9450	.9455	.9460	.9465	.9469	.9474	.9479	.9484	.9489
4.4	.6435	.6444	.6454	.6464	.6474	.6484	.6493	.6503	.6513	.6522	8.9	.9494	.9499	.9504	.9509	.9513	.9518	.9523	.9528	.9533	.9538
4.5	.6532	.6542	.6551	.6561	.6571	.6580	.6590	.6599	.6609	.6618	9.0	.9542	.9547	.9552	.9557	.9562	.9566	.9571	.9576	.9581	.9586
4.6	.6628	.6637	.6646	.6656	.6665	.6675	.6684	.6693	.6702	.6712	9.1	.9590	.9595	.9600	.9605	.9609	.9614	.9619	.9624	.9628	.9633
4.7	.6721	.6730	.6739	.6749	.6758	.6767	.6776	.6785	.6794	.6803	9.2	.9638	.9643	.9647	.9652	.9657	.9661	.9666	.9671	.9675	.9680
4.8	.6812	.6821	.6830	.6839	.6848	.6857	.6866	.6875	.6884	.6893	9.3	.9685	.9689	.9694	.9699	.9703	.9708	.9713	.9717	.9722	.9727
4.9	.6902	.6911	.6920	.6928	.6937	.6946	.6955	.6964	.6972	.6981	9.4	.9731	.9736	.9741	.9745	.9750	.9754	.9759	.9763	.9768	.9773
5.0	.6990	.6998	.7007	.7016	.7024	.7033	.7042	.7050	.7059	.7067	9.5	.9777	.9782	.9786	.9791	.9795	.9800	.9805	.9809	.9814	.9818
5.1	.7076	.7084	.7093	.7101	.7110	.7118	.7126	.7135	.7143	.7152	9.6	.9823	.9827	.9832	.9836	.9841	.9845	.9850	.9854	.9859	.9863
5.2	.7160	.7168	.7177	.7185	.7193	.7202	.7210	.7218	.7226	.7235	9.7	.9868	.9872	.9877	.9881	.9886	.9890	.9894	.9899	.9903	.9908
5.3	.7243	.7251	.7259	.7267	.7275	.7284	.7292	.7300	.7308	.7316	9.8	.9912	.9917	.9921	.9926	.9930	.9934	.9939	.9943	.9948	.9952
5.4	.7324	.7332	.7340	.7348	.7356	.7364	.7372	.7380	.7388	.7396	9.9	.9956	.9961	.9965	.9969	.9974	.9978	.9983	.9987	.9991	.9996
x	0	1	2	3	4	5	6	7	8	9	x	0	1	2	3	4	5	6	7	8	9

TABLE 2 Exponential Functions

x	e^x	e^{-x}	x	e^x	e^{-x}	x	e^x	e^{-x}
0.00	1.0000	1.0000	0.55	1.7333	0.5769	3.6	36.598	0.0273
0.01	1.0101	0.9900	0.60	1.8221	0.5488	3.7	40.447	0.0247
0.02	1.0202	0.9802	0.65	1.9155	0.5220	3.8	44.701	0.0224
0.03	1.0305	0.9704	0.70	2.0138	0.4966	3.9	49.402	0.0202
0.04	1.0408	0.9608	0.75	2.1170	0.4724	4.0	54.598	0.0183
0.05	1.0513	0.9512	0.80	2.2255	0.4493	4.1	60.340	0.0166
0.06	1.0618	0.9418	0.85	2.3396	0.4274	4.2	66.686	0.0150
0.07	1.0725	0.9324	0.90	2.4596	0.4066	4.3	73.700	0.0136
0.08	1.0833	0.9231	0.95	2.5857	0.3867	4.4	81.451	0.0123
0.09	1.0942	0.9139	1.0	2.7183	0.3679	4.5	90.017	0.0111
0.10	1.1052	0.9048	1.1	3.0042	0.3329	4.6	99.484	0.0101
0.11	1.1163	0.8958	1.2	3.3201	0.3012	4.7	109.95	0.0091
0.12	1.1275	0.8869	1.3	3.6693	0.2725	4.8	121.51	0.0082
0.13	1.1388	0.8781	1.4	4.0552	0.2466	4.9	134.29	0.0074
0.14	1.1503	0.8694	1.5	4.4817	0.2231	5	148.41	0.0067
0.15	1.1618	0.8607	1.6	4.9530	0.2019	6	403.43	0.0025
0.16	1.1735	0.8521	1.7	5.4739	0.1827	7	1,096.6	0.0009
0.17	1.1853	0.8437	1.8	6.0496	0.1653	8	2,981.0	0.0003
0.18	1.1972	0.8353	1.9	6.6859	0.1496	9	8,103.1	0.0001
0.19	1.2092	0.8270	2.0	7.3891	0.1353	10	22,026	0.00005
0.20	1.2214	0.8187	2.1	8.1662	0.1225	11	59,874	0.00002
0.21	1.2337	0.8106	2.2	9.0250	0.1108	12	162,754	0.000006
0.22	1.2461	0.8025	2.3	9.9742	0.1003	13	442,413	0.000002
0.23	1.2586	0.7945	2.4	11.023	0.0907	14	1,202,604	0.0000008
0.24	1.2712	0.7866	2.5	12.182	0.0821	15	3,269,017	0.0000003
0.25	1.2840	0.7788	2.6	13.464	0.0743			
0.26	1.2969	0.7711	2.7	14.880	0.0672			
0.27	1.3100	0.7634	2.8	16.445	0.0608			
0.28	1.3231	0.7558	2.9	18.174	0.0550			
0.29	1.3364	0.7483	3.0	20.086	0.0498			
0.30	1.3499	0.7408	3.1	22.198	0.0450			
0.35	1.4191	0.7047	3.2	24.533	0.0408			
0.40	1.4918	0.6703	3.3	27.113	0.0369			
0.45	1.5683	0.6376	3.4	29.964	0.0334			
0.50	1.6487	0.6065	3.5	33.115	0.0302			

TABLE 3 **Values of Trigonometric Functions**

Degrees	Radians	Sin	Cos	Tan	Cot	Sec	Csc		
0° 00′	.0000	.0000	1.0000	.0000	—	1.000	—	1.5708	90° 00′
10	029	029	000	029	343.8	000	343.8	679	50
20	058	058	000	058	171.9	000	171.9	650	40
30	.0087	.0087	1.0000	.0087	114.6	1.000	114.6	1.5621	30
40	116	116	.9999	116	85.94	000	85.95	592	20
50	145	145	999	145	68.75	000	68.76	563	10
1° 00′	.0175	.0175	.9998	.0175	57.29	1.000	57.30	1.5533	89° 00′
10	204	204	998	204	49.10	000	49.11	504	50
20	233	233	997	233	42.96	000	42.98	475	40
30	.0262	.0262	.9997	.0262	38.19	1.000	38.20	1.5446	30
40	291	291	996	291	34.37	000	34.38	417	20
50	320	320	995	320	31.24	001	31.26	388	10
2° 00′	.0349	.0349	.9994	.0349	28.64	1.001	28.65	1.5359	88° 00′
10	378	378	993	378	26.43	001	26.45	330	50
20	407	407	992	407	24.54	001	24.56	301	40
30	.0436	.0436	.9990	.0437	22.90	1.001	22.93	1.5272	30
40	465	465	989	466	21.47	001	21.49	243	20
50	495	494	988	495	20.21	001	20.23	213	10
3° 00′	.0524	.0523	.9986	.0524	19.08	1.001	19.11	1.5184	87° 00′
10	553	552	985	553	18.07	002	18.10	155	50
20	582	581	983	582	17.17	002	17.20	126	40
30	.0611	.0610	.9981	.0612	16.35	1.002	16.38	1.5097	30
40	640	640	980	641	15.60	002	15.64	068	20
50	669	669	978	670	14.92	002	14.96	039	10
4° 00′	.0698	.0698	.9976	.0699	14.30	1.002	14.34	1.5010	86° 00′
10	727	727	974	729	13.73	003	13.76	981	50
20	756	756	971	758	13.20	003	13.23	952	40
30	.0785	.0785	.9969	.0787	12.71	1.003	12.75	1.4923	30
40	814	814	967	816	12.25	003	12.29	893	20
50	844	843	964	846	11.83	004	11.87	864	10
5° 00′	.0873	.0872	.9962	.0875	11.43	1.004	11.47	1.4835	85° 00′
10	902	901	959	904	11.06	004	11.10	806	50
20	931	929	957	934	10.71	004	10.76	777	40
30	.0960	.0958	.9954	.0963	10.39	1.005	10.43	1.4748	30
40	989	987	951	992	10.08	005	10.13	719	20
50	.1018	.1016	948	.1022	9.788	005	9.839	690	10
6° 00′	.1047	.1045	.9945	.1051	9.514	1.006	9.567	1.4661	84° 00′
10	076	074	942	080	9.255	006	9.309	632	50
20	105	103	939	110	9.010	006	9.065	603	40
30	.1134	.1132	.9936	.1139	8.777	1.006	8.834	1.4573	30
40	164	161	932	169	8.556	007	8.614	544	20
50	193	190	929	198	8.345	007	8.405	515	10
7° 00′	.1222	.1219	.9925	.1228	8.144	1.008	8.206	1.4486	83° 00′
10	251	248	922	257	7.953	008	8.016	457	50
20	280	276	918	287	7.770	008	7.834	428	40
30	.1309	.1305	.9914	.1317	7.596	1.009	7.661	1.4399	30
40	338	334	911	346	7.429	009	7.496	370	20
50	367	363	907	376	7.269	009	7.337	341	10
8° 00′	.1396	.1392	.9903	.1405	7.115	1.010	7.185	1.4312	82° 00′
10	425	421	899	435	6.968	010	7.040	283	50
20	454	449	894	465	6.827	011	6.900	254	40
30	.1484	.1478	.9890	.1495	6.691	1.011	6.765	1.4224	30
40	513	507	886	524	6.561	012	6.636	195	20
50	542	536	881	554	6.435	012	6.512	166	10
9° 00′	.1571	.1564	.9877	.1584	6.314	1.012	6.392	1.4137	81° 00′
		Cos	Sin	Cot	Tan	Csc	Sec	Radians	Degrees

(continued)

TABLE 3 (*continued*)

Degrees	Radians	Sin	Cos	Tan	Cot	Sec	Csc		
9° 00′	.1571	.1564	.9877	.1584	6.314	1.012	6.392	1.4137	81° 00′
10	600	593	872	614	197	013	277	108	50
20	629	622	868	644	084	013	166	079	40
30	.1658	.1650	.9863	.1673	5.976	1.014	6.059	1.4050	30
40	687	679	858	703	871	014	5.955	1.4021	20
50	716	708	853	733	769	015	855	992	10
10° 00′	.1745	.1736	.9848	.1763	5.671	1.015	5.759	1.3963	80° 00′
10	774	765	843	793	576	016	665	934	50
20	804	794	838	823	485	016	575	904	40
30	.1833	.1822	.9833	.1853	5.396	1.017	5.487	1.3875	30
40	862	851	827	883	309	018	403	846	20
50	891	880	822	914	226	018	320	817	10
11° 00′	.1920	.1908	.9816	.1944	5.145	1.019	5.241	1.3788	79° 00′
10	949	937	811	974	066	019	164	759	50
20	978	965	805	.2004	4.989	020	089	730	40
30	.2007	.1994	.9799	.2035	4.915	1.020	5.016	1.3701	30
40	036	.2022	793	065	843	021	4.945	672	20
50	065	051	787	095	773	022	876	643	10
12° 00′	.2094	.2079	.9781	.2126	4.705	1.022	4.810	1.3614	78° 00′
10	123	108	775	156	638	023	745	584	50
20	153	136	769	186	574	024	682	555	40
30	.2182	.2164	.9763	.2217	4.511	1.024	4.620	1.3526	30
40	211	193	757	247	449	025	560	497	20
50	240	221	750	278	390	026	502	468	10
13° 00′	.2269	.2250	.9744	.2309	4.331	1.026	4.445	1.3439	77° 00′
10	298	278	737	339	275	027	390	410	50
20	327	306	730	370	219	028	336	381	40
30	.2356	.2334	.9724	.2401	4.165	1.028	4.284	1.3352	30
40	385	363	717	432	113	029	232	323	20
50	414	391	710	462	061	030	182	294	10
14° 00′	.2443	.2419	.9703	.2493	4.011	1.031	4.134	1.3265	76° 00′
10	473	447	696	524	3.962	031	086	235	50
20	502	476	689	555	914	032	039	206	40
30	.2531	.2504	.9681	.2586	3.867	1.033	3.994	1.3177	30
40	560	532	674	617	821	034	950	148	20
50	589	560	667	648	776	034	906	119	10
15° 00′	.2618	.2588	.9659	.2679	3.732	1.035	3.864	1.3090	75° 00′
10	647	616	652	711	689	036	822	061	50
20	676	644	644	742	647	037	782	032	40
30	.2705	.2672	.9636	.2773	3.606	1.038	3.742	1.3003	30
40	734	700	628	805	566	039	703	974	20
50	763	728	621	836	526	039	665	945	10
16° 00′	.2793	.2756	.9613	.2867	3.487	1.040	3.628	1.2915	74° 00′
10	822	784	605	899	450	041	592	886	50
20	851	812	596	931	412	042	556	857	40
30	.2880	.2840	.9588	.2962	3.376	1.043	3.521	1.2828	30
40	909	868	580	994	340	044	487	799	20
50	938	896	572	.3026	305	045	453	770	10
17° 00′	.2967	.2924	.9563	.3057	3.271	1.046	3.420	1.2741	73° 00′
10	996	952	555	089	237	047	388	712	50
20	.3025	979	546	121	204	048	356	683	40
30	.3054	.3007	.9537	.3153	3.172	1.049	3.326	1.2654	30
40	083	035	528	185	140	049	295	625	20
50	113	062	520	217	108	050	265	595	10
18° 00′	.3142	.3090	.9511	.3249	3.078	1.051	3.236	1.2566	72° 00′
		Cos	Sin	Cot	Tan	Csc	Sec	Radians	Degrees

TABLE 3 (*continued*)

Degrees	Radians	Sin	Cos	Tan	Cot	Sec	Csc		
18° 00′	.3142	.3090	.9511	.3249	3.078	1.051	3.236	1.2566	**72° 00′**
10	171	118	502	281	047	052	207	537	50
20	200	145	492	314	018	053	179	508	40
30	.3229	.3173	.9483	.3346	2.989	1.054	3.152	1.2479	30
40	258	201	474	378	960	056	124	450	20
50	287	228	465	411	932	057	098	421	10
19° 00′	.3316	.3256	.9455	.3443	2.904	1.058	3.072	1.2392	**71° 00′**
10	345	283	446	476	877	059	046	363	50
20	374	311	436	508	850	060	021	334	40
30	.3403	.3338	.9426	.3541	2.824	1.061	2.996	1.2305	30
40	432	365	417	574	798	062	971	275	20
50	462	393	407	607	773	063	947	246	10
20° 00′	.3491	.3420	.9397	.3640	2.747	1.064	2.924	1.2217	**70° 00′**
10	520	448	387	673	723	065	901	188	50
20	549	475	377	706	699	066	878	159	40
30	.3578	.3502	.9367	.3739	2.675	1.068	2.855	1.2130	30
40	607	529	356	772	651	069	833	101	20
50	636	557	346	805	628	070	812	072	10
21° 00′	.3665	.3584	.9336	.3839	2.605	1.071	2.790	1.2043	**69° 00′**
10	694	611	325	872	583	072	769	1.2014	50
20	723	638	315	906	560	074	749	985	40
30	.3752	.3665	.9304	.3939	2.539	1.075	2.729	1.1956	30
40	782	692	293	973	517	076	709	926	20
50	811	719	283	.4006	496	077	689	897	10
22° 00′	.3840	.3746	.9272	.4040	2.475	1.079	2.669	1.1868	**68° 00′**
10	869	773	261	074	455	080	650	839	50
20	898	800	250	108	434	081	632	810	40
30	.3927	.3827	.9239	.4142	2.414	1.082	2.613	1.1781	30
40	956	854	228	176	394	084	595	752	20
50	985	881	216	210	375	085	577	723	10
23° 00′	.4014	.3907	.9205	.4245	2.356	1.086	2.559	1.1694	**67° 00′**
10	043	934	194	279	337	088	542	665	50
20	072	961	182	314	318	089	525	636	40
30	.4102	.3987	.9171	.4348	2.300	1.090	2.508	1.1606	30
40	131	.4014	159	383	282	092	491	577	20
50	160	041	147	417	264	093	475	548	10
24° 00′	.4189	.4067	.9135	.4452	2.246	1.095	2.459	1.1519	**66° 00′**
10	218	094	124	487	229	096	443	490	50
20	247	120	112	522	211	097	427	461	40
30	.4276	.4147	.9100	.4557	2.194	1.099	2.411	1.1432	30
40	305	173	088	592	177	100	396	403	20
50	334	200	075	628	161	102	381	374	10
25° 00′	.4363	.4226	.9063	.4663	2.145	1.103	2.366	1.1345	**65° 00′**
10	392	253	051	699	128	105	352	316	50
20	422	279	038	734	112	106	337	286	40
30	.4451	.4305	.9026	.4770	2.097	1.108	2.323	1.1257	30
40	480	331	013	806	081	109	309	228	20
50	509	358	001	841	066	111	295	199	10
26° 00′	.4538	.4384	.8988	.4877	2.050	1.113	2.281	1.1170	**64° 00′**
10	567	410	975	913	035	114	268	141	50
20	596	436	962	950	020	116	254	112	40
30	.4625	.4462	.8949	.4986	2.006	1.117	2.241	1.1083	30
40	654	488	936	.5022	1.991	119	228	054	20
50	683	514	923	059	977	121	215	1.1025	10
27° 00′	.4712	.4540	.8910	.5095	1.963	1.122	2.203	1.0996	**63° 00′**
		Cos	**Sin**	**Cot**	**Tan**	**Csc**	**Sec**	**Radians**	**Degrees**

TABLE 3 (*continued*)

Degrees	Radians	Sin	Cos	Tan	Cot	Sec	Csc		
27° 00′	.4712	.4540	.8910	.5095	1.963	1.122	2.203	1.0996	63° 00′
10	741	566	897	132	949	124	190	966	50
20	771	592	884	169	935	126	178	937	40
30	.4800	.4617	.8870	.5206	1.921	1.127	2.166	1.0908	30
40	829	643	857	243	907	129	154	879	20
50	858	669	843	280	894	131	142	850	10
28° 00′	.4887	.4695	.8829	.5317	1.881	1.133	2.130	1.0821	62° 00′
10	916	720	816	354	868	134	118	792	50
20	945	746	802	392	855	136	107	763	40
30	.4974	.4772	.8788	.5430	1.842	1.138	2.096	1.0734	30
40	.5003	797	774	467	829	140	085	705	20
50	032	823	760	505	816	142	074	676	10
29° 00′	.5061	.4848	.8746	.5543	1.804	1.143	2.063	1.0647	61° 00′
10	091	874	732	581	792	145	052	617	50
20	120	899	718	619	780	147	041	588	40
30	.5149	.4924	.8704	.5658	1.767	1.149	2.031	1.0559	30
40	178	950	689	696	756	151	020	530	20
50	207	975	675	735	744	153	010	501	10
30° 00′	.5236	.5000	.8660	.5774	1.732	1.155	2.000	1.0472	60° 00′
10	265	025	646	812	720	157	1.990	443	50
20	294	050	631	851	709	159	980	414	40
30	.5323	.5075	.8616	.5890	1.698	1.161	1.970	1.0385	30
40	352	100	601	930	686	163	961	356	20
50	381	125	587	969	675	165	951	327	10
31° 00′	.5411	.5150	.8572	.6009	1.664	1.167	1.942	1.0297	59° 00′
10	440	175	557	048	653	169	932	268	50
20	469	200	542	088	643	171	923	239	40
30	.5498	.5225	.8526	.6128	1.632	1.173	1.914	1.0210	30
40	527	250	511	168	621	175	905	181	20
50	556	275	496	208	611	177	896	152	10
32° 00′	.5585	.5299	.8480	.6249	1.600	1.179	1.887	1.0123	58° 00′
10	614	324	465	289	590	181	878	094	50
20	643	348	450	330	580	184	870	065	40
30	.5672	.5373	.8434	.6371	1.570	1.186	1.861	1.0036	30
40	701	398	418	412	560	188	853	1.0007	20
50	730	422	403	453	550	190	844	977	10
33° 00′	.5760	.5446	.8387	.6494	1.540	1.192	1.836	.9948	57° 00′
10	789	471	371	536	530	195	828	919	50
20	818	495	355	577	520	197	820	890	40
30	.5847	.5519	.8339	.6619	1.511	1.199	1.812	.9861	30
40	876	544	323	661	501	202	804	832	20
50	905	568	307	703	1.492	204	796	803	10
34° 00′	.5934	.5592	.8290	.6745	1.483	1.206	1.788	.9774	56° 00′
10	963	616	274	787	473	209	781	745	50
20	992	640	258	830	464	211	773	716	40
30	.6021	.5664	.8241	.6873	1.455	1.213	1.766	.9687	30
40	050	688	225	916	446	216	758	657	20
50	080	712	208	959	437	218	751	628	10
35° 00′	.6109	.5736	.8192	.7002	1.428	1.221	1.743	.9599	55° 00′
10	138	760	175	046	419	223	736	570	50
20	167	783	158	089	411	226	729	541	40
30	.6196	.5807	.8141	.7133	1.402	1.228	1.722	.9512	30
40	225	831	124	177	393	231	715	483	20
50	254	854	107	221	385	233	708	454	10
36° 00′	.6283	.5878	.8090	.7265	1.376	1.236	1.701	.9425	54° 00′
		Cos	Sin	Cot	Tan	Csc	Sec	Radians	Degrees

TABLE 3 (*continued*)

Degrees	Radians	Sin	Cos	Tan	Cot	Sec	Csc		
36° 00′	.6283	.5878	.8090	.7265	1.376	1.236	1.701	.9425	54° 00′
10	312	901	073	310	368	239	695	396	50
20	341	925	056	355	360	241	688	367	40
30	.6370	.5948	.8039	.7400	1.351	1.244	1.681	.9338	30
40	400	972	021	445	343	247	675	308	20
50	429	995	004	490	335	249	668	279	10
37° 00′	.6458	.6018	.7986	.7536	1.327	1.252	1.662	.9250	53° 00′
10	487	041	969	581	319	255	655	221	50
20	516	065	951	627	311	258	649	192	40
30	.6545	.6088	.7934	.7673	1.303	1.260	1.643	.9163	30
40	574	111	916	720	295	263	636	134	20
50	603	134	898	766	288	266	630	105	10
38° 00′	.6632	.6157	.7880	.7813	1.280	1.269	1.624	.9076	52° 00′
10	661	180	862	860	272	272	618	047	50
20	690	202	844	907	265	275	612	.9018	40
30	.6720	.6225	.7826	.7954	1.257	1.278	1.606	.8988	30
40	749	248	808	.8002	250	281	601	959	20
50	778	271	790	050	242	284	595	930	10
39° 00′	.6807	.6293	.7771	.8098	1.235	1.287	1.589	.8901	51° 00′
10	836	316	753	146	228	290	583	872	50
20	865	338	735	195	220	293	578	843	40
30	.6894	.6361	.7716	.8243	1.213	1.296	1.572	.8814	30
40	923	383	698	292	206	299	567	785	20
50	952	406	679	342	199	302	561	756	10
40° 00′	.6981	.6428	.7660	.8391	1.192	1.305	1.556	.8727	50° 00′
10	.7010	450	642	441	185	309	550	698	50
20	039	472	623	491	178	312	545	668	40
30	.7069	.6494	.7604	.8541	1.171	1.315	1.540	.8639	30
40	098	517	585	591	164	318	535	610	20
50	127	539	566	642	157	322	529	581	10
41° 00′	.7156	.6561	.7547	.8693	1.150	1.325	1.524	.8552	49° 00′
10	185	583	528	744	144	328	519	523	50
20	214	604	509	796	137	332	514	494	40
30	.7243	.6626	.7490	.8847	1.130	1.335	1.509	.8465	30
40	272	648	470	899	124	339	504	436	20
50	301	670	451	952	117	342	499	407	10
42° 00′	.7330	.6691	.7431	.9004	1.111	1.346	1.494	.8378	48° 00′
10	359	713	412	057	104	349	490	348	50
20	389	734	392	110	098	353	485	319	40
30	.7418	.6756	.7373	.9163	1.091	1.356	1.480	.8290	30
40	447	777	353	217	085	360	476	261	20
50	476	799	333	271	079	364	471	232	10
43° 00′	.7505	.6820	.7314	.9325	1.072	1.367	1.466	.8203	47° 00′
10	534	841	294	380	066	371	462	174	50
20	563	862	274	435	060	375	457	145	40
30	.7592	.6884	.7254	.9490	1.054	1.379	1.453	.8116	30
40	621	905	234	545	048	382	448	087	20
50	650	926	214	601	042	386	444	058	10
44° 00′	.7679	.6947	.7193	.9657	1.036	1.390	1.440	.8029	46° 00′
10	709	967	173	713	030	394	435	999	50
20	738	988	153	770	024	398	431	970	40
30	.7767	.7009	.7133	.9827	1.018	1.402	1.427	.7941	30
40	796	030	112	884	012	406	423	912	20
50	825	050	092	942	006	410	418	883	10
45° 00′	.7854	.7071	.7071	1.000	1.000	1.414	1.414	.7854	45° 00′
		Cos	Sin	Cot	Tan	Csc	Sec	Radians	Degrees

In the answer section, all graphs of curves described by specific equations have been computer-generated. Hundreds of *six-decimal* ordered pairs are calculated from the specific equation. These ordered pairs are plotted and connected by line segments. Because so many points are calculated, the points are close together and the short line segments connecting adjacent ordered pairs form a smooth curve. Thus the curve is *mathematically precise*, maximizing the accuracy of the graphs. Because of the printed line weight of the curves and the asymptotes and the high degree of accuracy provided by the computer-generated graphs, the curves and the asymptotes occasionally "appear" to be touching. Students will also observe this fact when using computers and graphing calculators.

Answers

CHAPTER 1

Margin Exercises, Section 1.1
1. $1, 19$ **2.** $0, 1, 19$ **3.** $-6, 0, 1, 19$ **4.** All of them
5. Rational   **6.** Rational   **7.** Rational   **8.** Rational
9. Irrational   **10.** Irrational   **11.** -12 **12.** -4.7
13. $-\frac{4}{5}$ **14.** -0.2 **15.** 5 **16.** 0 **17.** $-6, -6$
18. $8, 8$ **19.** $3.4, 3.4$ **20.** (a) 2; (b) -4 **21.** -24
22. $\frac{21}{25}$ **23.** 144 **24.** 1.3 **25.** 17 **26.** $-\frac{11}{5}$
27. -13 **28.** 4 **29.** -3 **30.** $-\frac{8}{3}$ **31.** 2 **32.** No
33. Yes   **34.** Yes   **35.** No

Exercise Set 1.1, pp. 9–10
1. $3, 14, \sqrt{16}$ **3.** $\sqrt{3}, -\sqrt{7}, \sqrt[3]{2}$
5. $-6, 0, 3, -2, 14, \sqrt{16}, -\sqrt[3]{8}$ **7.** Rational   **9.** Rational
11. Rational   **13.** Irrational   **15.** Irrational
17. Irrational   **19.** Rational   **21.** Irrational   **23.** $7, 7$
25. $-57, -57$ **27.** -87 **29.** -16 **31.** -10.3
33. $\frac{39}{10}$ **35.** 28 **37.** -49.2 **39.** 210 **41.** $-\frac{833}{5}$
43. 5 **45.** $-\frac{1}{7}$ **47.** $-\frac{3}{49}$ **49.** 25 **51.** -4 **53.** 18
55. -11.6 **57.** $-\frac{35}{8}$ **59.** (a) $1.96, 1.9881, 1.999396,$
$1.999962, 1.999990$; (b) $\sqrt{2}$ **61.** Identity $(+)$
63. Distributive   **65.** Commutative $(+)$
67. Associative $(\times)$ **69.** Inverse $(\times)$
71. Commutative $(+)$ **73.** Commutative $(+)$
75. $7 - 5 \neq 5 - 7$; $7 - 5 = 2, 5 - 7 = -2$
77. $16 \div (4 \div 2) \neq (16 \div 4) \div 2$; $16 \div (4 \div 2) = 8,$
$(16 \div 4) \div 2 = 2$ **79.** $\dfrac{3927}{1250}$ **81.** 1 **83.** $\dfrac{183,062}{9990}$
85. $(b + c)a = a(b + c) = ab + ac = ba + ca$

Margin Exercises, Section 1.2
1. 8^4 **2.** x^3 **3.** $(4y)^4$ **4.** $3 \cdot 3 \cdot 3 \cdot 3$, or 81
5. $5x \cdot 5x \cdot 5x \cdot 5x$, or $625 \cdot x \cdot x \cdot x \cdot x$
6. $(-5)(-5)(-5)(-5)$, or 625
7. $-[5 \cdot 5 \cdot 5 \cdot 5]$, or -625 **8.** 1 **9.** $25y^2$ **10.** $-8x^3$
11. 4^{-3} **12.** $\dfrac{1}{10^4}$, or $\dfrac{1}{10 \cdot 10 \cdot 10 \cdot 10}$, or $\dfrac{1}{10,000}$
13. $\dfrac{1}{4^3}, \dfrac{1}{4 \cdot 4 \cdot 4}, \dfrac{1}{64}$ **14.** 8^4 **15.** y^5 **16.** $-18x^{11}$
17. $-\dfrac{75}{x^{14}}$ **18.** $-\dfrac{10y^2}{x^{12}}$ **19.** $60y$ **20.** 4^3 **21.** 5^6
22. $\dfrac{1}{10^{14}}$ **23.** $\dfrac{1}{9^6}$ **24.** y^{11} **25.** $\dfrac{5}{y}$ **26.** $-\dfrac{2y^9}{x^3}$
27. 3^{49} **28.** $\dfrac{1}{8^{14}}$ **29.** $\dfrac{1}{y^{28}}$ **30.** $8x^3y^3$ **31.** $\dfrac{16y^{14}}{x^4}$
32. $\dfrac{z^{15}}{27x^{12}y^6}$ **33.** $\dfrac{8y^{45}z^3}{x^{30}}$ **34.** 4.65×10^5
35. 3.789×10^3 **36.** 1.45×10^{-4} **37.** 6.7×10^{-10}
38. 0.0000467 **39.** $7,894,000,000,000$

40. 8,166,000,000   **41.** 0.000001103   **42.** 3.1536×10^7
43. 6.308×10^{-10} **44.** 7.2×10^{-10} **45.** (a) 79; (b) 87
46. 27   **47.** (a) -13; (b) -13 **48.** 2   **49.** $\sqrt{3}$

50. 11.3   **51.** $\frac{3}{4}$ **52.** 20   **53.** 20   **54.** 4   **55.** 4

56. $6|a|\,|b|$ **57.** x^8 **58.** $10m^2n^2|n|$ **59.** $\dfrac{2x^2|x|}{y^2}$

Exercise Set 1.2, pp. 18–19

1. $\dfrac{1}{2}$ **3.** 1   **5.** 4^3 **7.** $6x^5$ **9.** $\dfrac{15b^5}{a}$ **11.** $72x^5$

13. $-18x^7yz$ **15.** b^3 **17.** $\dfrac{x^3}{y^3}$ **19.** 1   **21.** $3ab^2$

23. $\dfrac{4xy}{7z^5}$ **25.** $8a^3b^6$ **27.** $16x^{12}$ **29.** $-16x^{12}$

31. $36a^4b^6c^2$ **33.** $\dfrac{1}{25}c^2d^4$ **35.** 1   **37.** 32

39. $\dfrac{27a^8c^{18}}{4b^{10}}$ **41.** $\dfrac{3}{4}xy$ **43.** $\dfrac{32{,}768a^{20}c^{10}}{b^{25}}$ **45.** $-25, 25$

47. $-1.1664, 1.1664$ **49.** 5.8×10^7 **51.** 3.65×10^5
53. 2.7×10^{-6} **55.** 2.7×10^{-2} **57.** 9.11×10^{-28}
59. 3.664×10^9 **61.** 400,000   **63.** 0.0062
65. 7,690,000,000,000   **67.** 0.000000567
69. 9,460,000,000,000   **71.** 0.0000000769
73. 256,700,000   **75.** 1.395×10^3 **77.** 8.0×10^{-14}
79. 1.512×10^{10} ft^3 **81.** 10^{-9} sec   **83.** 1.47×10^{12} mi

85. 10   **87.** 3   **89.** 2048   **91.** $\dfrac{243}{8}$ **93.** 12   **95.** 47

97. $7|a|$ **99.** $8x^6$ **101.** $9|x|\,|y|$ **103.** $3a^2|b|$ **105.** x^{8t}
107. t^{8x} **109.** $(xy)^{ac+bc}$ **111.** $9x^{2a}y^{2b}$ **113.** $750.43
115. In $(x^3)^2$, exponents were added instead of multiplied; x^{10}.
117. In 2^3, the base 2 and the exponent 3 were multiplied. In $(x^{-4})^3$, the exponents were added. In $(y^6)^3$, the exponents were subtracted. In $(z^3)^3$, the exponents were added; $8x^{-12}y^{18}z^9$.

Margin Exercises, Section 1.3
1. 8, 6, 4, 9, 0; 9   **2.** 4, 4, 5, 6, 0; 6   **3.** $9x^3y^2 - 2x^2y^3$
4. $7xy^2 - 2x^2y$ **5.** $3x^4\sqrt{y} + 2$
6. $-4x^3 + 2x^2 - 4x - \frac{3}{2}$
7. $5p^2q^4 + p^2q^2 - 6pq^2 - 3q + 5$
8. $-(5x^2t^2 - 4xy^2t - 3xt + 6x - 5)$, $-5x^2t^2 + 4xy^2t + 3xt - 6x + 5$ **9.** $-(-3x^2y + 5xy - 7x + 4y + 2)$, $3x^2y - 5xy + 7x - 4y - 2$
10. $8xy^4 - 9xy^2 + 4x^2 + 2y - 7$
11. $3x^2y - 9x^3y^2 + 5x^2y^3 - x^2y^2 + 9y$

Exercise Set 1.3, p. 23
1. 4, 3, 2, 1, 0; 4   **3.** 3, 6, 6, 0; 6   **5.** 5, 6, 2, 1, 0; 6
7. $3x^2y - 5xy^2 + 7xy + 2$ **9.** $-10pq^2 - 5p^2q + 7pq - 4p + 2q + 3$ **11.** $3x + 2y - 2z - 3$
13. $5x\sqrt{y} - 4y\sqrt{x} - \frac{2}{5}$ **15.** $-(5x^3 - 7x^2 + 3x - 6)$, $-5x^3 + 7x^2 - 3x + 6$ **17.** $-2x^2 + 6x - 2$
19. $6a - 5b - 2c + 4d$ **21.** $x^4 - 3x^3 - 4x^2 + 9x - 3$
23. $9x\sqrt{y} - 3y\sqrt{x} + 9.1$
25. $-1.047p^2q - 2.479pq^2 + 8.879pq - 104.144$

Margin Exercises, Section 1.4
1. $3x^3y^2 + 4x^2y^2 - xy^2 + 6y^2$
2. $2p^4q^2 + 3p^3q^2 + 3p^2q^2 + 2q^2$
3. $2x^3y - 4xy + 3x^3 - 6x$ **4.** $15x^2 - xy - 6y^2$
5. $6xy - 2\sqrt{2}x + 3\sqrt{2}y - 2$ **6.** $16x^2 - 40xy + 25y^2$
7. $4y^4 + 24x^2y^3 + 36x^4y^2$ **8.** $16x^2 - 49$
9. $25x^4y^2 - 4y^2$ **10.** $16y^4 - 3$
11. $4x^2 + 12x + 9 - 25y^2$ **12.** $25t^2 - 4x^6y^4$
13. $x^3 + 3x^2 + 3x + 1$ **14.** $x^3 - 3x^2 + 3x - 1$
15. $t^6 - 9t^4b + 27t^2b^2 - 27b^3$
16. $8a^9 - 60a^6b^2 + 150a^3b^4 - 125b^6$

Exercise Set 1.4, pp. 26–27
1. $6x^3 + 4x^2 + 32x - 64$
3. $4a^3b^2 - 10a^2b^2 + 3ab^3 + 4ab^2 - 6b^3 + 4a^2b - 2ab + 3b^2$ **5.** $a^3 - b^3$ **7.** $4x^2 + 8xy + 3y^2$
9. $12x^3 + x^2y - \frac{3}{2}xy - \frac{1}{8}y^2$ **11.** $10p^3q^4 - 4p^2q^3r - 5r^2pq + 2r^3$ **13.** $4x^2 + 12xy + 9y^2$
15. $4x^4 - 12x^2y + 9y^2$ **17.** $4x^6 + 12x^3y^2 + 9y^4$
19. $\frac{1}{4}x^4 - \frac{3}{5}x^2y + \frac{9}{25}y^2$ **21.** $0.25x^2 + 0.70xy^2 + 0.49y^4$
23. $9x^2 - 4y^2$ **25.** $x^4 - y^2z^2$ **27.** $9x^4 - 2$
29. $4x^2 + 12xy + 9y^2 - 16$ **31.** $x^4 + 6x^2y + 9y^2 - y^4$
33. $x^4 - 1$ **35.** $16x^4 - y^4$
37. $0.002601x^2 + 0.00408xy + 0.0016y^2$
39. $2462.0358x^2 - 945.0214x - 38.908$
41. $y^3 + 15y^2 + 75y + 125$
43. $m^6 - 6m^4n + 12m^2n^2 - 8n^3$
45. $2x^3 - 2\sqrt{2}x^2y - \sqrt{2}xy^2 + 2y^3$ **47.** $a^{2n} - b^{2n}$
49. $x^{3m} - 3x^{2m}t^n + 3x^mt^{2n} - t^{3n}$ **51.** $x^6 - 1$
53. $16x^4 - 32x^3 + 16x^2$ **55.** $x^{a^2-b^2}$
57. $a^2 + b^2 + c^2 + 2ab + 2ac + 2bc$
59. $a^4 + 4a^3b + 6a^2b^2 + 4ab^3 + b^4$ **61.** $m^5 + t^5$
63. A term is missing—found by calculating twice the product of the terms; the first term is $9a^2$, not $3a^2$, where the 3 was not squared; $9a^2 + 6ab + b^2$
65. In step (1), $3x$ should be $6x$; the 2 and 3 were not multiplied. Similarly, -3 should be -12; the 4 and -3 were not multiplied. In step (2), $3x - 3$ is not x; $6x^2 + 6x - 12$.   **67.** $x^n - y^n$

Margin Exercises, Section 1.5
1. $4x^2y(5x + 3)$ **2.** $(p + q)(2x + y + 2)$
3. $(4x^2 - 3)(x + 5)$ **4.** $(x - 4)(x + 4)$
5. $(5y^2 + 4x)(5y^2 - 4x)$
6. $2(y^2 + 4x^2)(y - 2x)(y + 2x)$ **7.** $(x - \sqrt{3})(x + \sqrt{3})$
8. $(x + 5)(x + 1)$ **9.** $(x + 7)(x - 2)$
10. $(w^2 - 5)(w^2 - 2)$ **11.** $(3x + 2)(x + 1)$
12. $3(2x^2y^3 + 5)(x^2y^3 - 4)$ **13.** Not factorable
14. $(3y - 5)^2$ **15.** $(4x + 9y)^2$ **16.** $-3y^2(2x^2 - 5y^3)^2$
17. $(x - 2)(x^2 + 2x + 4)$ **18.** $(4 - t)(16 + 4t + t^2)$
19. $(3x + y)(9x^2 - 3xy + y^2)$
20. $(2m + 5t)(4m^2 - 10mt + 25t^2)$
21. $2y(4y^2 - 5x^2)(16y^4 + 20x^2y^2 + 25x^4)$
22. $(p + 2)(p - 2)(p^2 - 2p + 4)(p^2 + 2p + 4)$

Exercise Set 1.5, pp. 32–33
1. $(p + 4)(p + 2)$ **3.** $(2n - 7)(n + 8)$
5. $(y^2 - 7)(y^2 + 3)$ **7.** $3ab(6a - 5b)$
9. $(a + c)(b - 2)$ **11.** $(x^2 + 6)(x + 3)$

13. $(y + 2)(y - 2)(y - 3)$ **15.** $(3x - 5)(3x + 5)$
17. $4x(y^2 - z)(y^2 + z)$ **19.** $(y - 3)^2$ **21.** $(1 - 4x)^2$
23. $(2x - \sqrt{5})(2x + \sqrt{5})$ **25.** $(xy - 7)^2$
27. $4a(x + 7)(x - 2)$ **29.** $(a + b + c)(a + b - c)$
31. $(x + y - a - b)(x + y + a + b)$
33. $5(y^2 + 4x^2)(y - 2x)(y + 2x)$
35. $(x + 2)(x^2 - 2x + 4)$ **37.** $3(x - \frac{1}{2})(x^2 + \frac{1}{2}x + \frac{1}{4})$
39. $(x + 0.1)(x^2 - 0.1x + 0.01)$
41. $3(z - 2)(z^2 + 2z + 4)$
43. $(a - t)(a + t)(a^2 - at + t^2)(a^2 + at + t^2)$
45. $2ab(2a^2 + 3b^2)(4a^4 - 6a^2b^2 + 9b^4)$
47. $(x + 4.19524)(x - 4.19524)$
49. $37(x + 0.626y)(x - 0.626y)$ **51.** $(x - 4)(x + 8)$
53. $(3a - 3b - 2)(a - b + 4)$ **55.** $h(3x^2 + 3xh + h^2)$
57. $(y^2 + 12)(y^2 - 7)$ **59.** $(y + \frac{4}{7})(y - \frac{2}{7})$
61. $(t + 0.9)(t - 0.3)$ **63.** $(x^n + 8)(x^n - 3)$
65. $(x + a)(x + b)$ **67.** $(\frac{1}{2}t - \frac{2}{5})^2$
69. $(5y^m - x^n + 1)(5y^m + x^n - 1)$
71. $3(x^n - 2y^m)(x^{2n} + 2x^ny^m + 4y^{2m})$
73. $y(y - 1)^2(y - 2)$ **75.** $5(x - \frac{9}{5})$
77. **(a)** $x^5 + x - 1$; **(b)** $(x^2 - x + 1)(x^3 + x^2 - 1)$

Margin Exercises, Section 1.6
1. All real numbers except 5
2. All real numbers except -3 and -4
3. $\dfrac{(x + y)(x + y)}{(2x^2 - 1)(7x)}$ **4.** $\dfrac{(x - 2)(x + 4)}{(x + 2)(x + 2)}$ **5.** $\dfrac{x^2 + 5x + 6}{x^2 - 2x - 15}$;
all real numbers except 5; all real numbers except 5 and -3
6. $\dfrac{3x + 2}{x + 2}$; all real numbers except 0 and -2; all real
numbers except -2
7. $\dfrac{y + 2}{y - 1}$; all real numbers except 1 and -1; all real
numbers except 1   **8.** $\dfrac{3(x - y)}{x + y}$ **9.** $\dfrac{2ab(a + b)}{a - b}$
10. $x - y$ **11.** $\dfrac{3x^2 + 4x + 2}{x - 5}$ **12.** $\dfrac{2x^2 + 11}{x - 5}$
13. $\dfrac{4x^2 - xy + 4y^2}{2(2x - y)(x - y)}$ **14.** $\dfrac{x - 6}{(x + 4)(x + 6)}$ **15.** $\dfrac{1}{a - x}$
16. $\dfrac{a^2b^2}{b^2 - ab + a^2}$

Exercise Set 1.6, pp. 40–41
1. All real numbers except 0 and 1
3. All real numbers except -5, -2, and 2
5. $\dfrac{3}{x}$; all real numbers except 0
7. $\dfrac{7}{(x + 2)(x + 5)}$; all real numbers except -5 and -2
9. $\dfrac{5}{2}x$; all real numbers   **11.** $\dfrac{x - 2}{x + 2}$; all real numbers
except -2
13. $\dfrac{1}{x - y}$ **15.** $\dfrac{(x + 5)(2x + 3)}{7x}$ **17.** $\dfrac{a + 2}{a - 5}$
19. $m + n$ **21.** $\dfrac{3(x - 4)}{2(x + 4)}$ **23.** $\dfrac{1}{x + y}$ **25.** $\dfrac{x - y - z}{x + y + z}$

27. 1   **29.** $\dfrac{y - 2}{y - 1}$ **31.** $\dfrac{x + y}{2x - 3y}$ **33.** $\dfrac{3x - 4}{(x + 2)(x - 2)}$
35. $\dfrac{3y - 10}{(y - 5)(y + 4)}$ **37.** $\dfrac{4x - 8y}{(x + y)(x - y)}$ **39.** $\dfrac{3x - 4}{(x - 2)(x - 1)}$
41. $\dfrac{5a^2 + 10ab - 4b^2}{(a - b)(a + b)}$ **43.** $\dfrac{11x^2 - 18x + 8}{(2 + x)(2 - x)^2}$ **45.** 0
47. $\dfrac{x + y}{x}$ **49.** $\dfrac{a^2 - 1}{a^2 + 1}$ **51.** $\dfrac{c^2 - 2c + 4}{c}$ **53.** $\dfrac{xy}{x - y}$
55. $x - y$ **57.** $\dfrac{(x + y)(x - y)}{xy}$ **59.** $\dfrac{1 + a}{1 - a}$ **61.** $\dfrac{b + a}{b - a}$
63. $2x + h$ **65.** $3x^2 + 3xh + h^2$ **67.** x^5
69. Step (1) uses the wrong reciprocal. The reciprocal of a
sum is not the sum of the reciprocals. Step (2) would be
correct if step (1) had been. Step (3) would be correct if
steps (1) and (2) had been. Step (4) has an improper
simplification of the b in the denominator; $\dfrac{12a}{b(4a + 3b)}$.
71. $\dfrac{(n + 1)(n + 2)(n + 3)}{2 \cdot 3}$

Margin Exercises, Section 1.7
1. Yes, no   **2.** No, yes   **3.** Yes, yes   **4.** Yes, yes
5. $|x + 2|$ **6.** $|x| \cdot |y - 2|$, or $|x(y - 2)|$ **7.** $|x + 2|$
8. $|x + 4|$ **9.** $-4xy$ **10.** $\sqrt{133}$ **11.** $\sqrt{x^2 - 4y^2}$
12. $\sqrt[4]{81}$, or 3   **13.** $10\sqrt{3}$ **14.** $6|y|$ **15.** $|x + 1|\sqrt{2}$
16. $2\sqrt[3]{2}$ **17.** $(a + b)\sqrt[3]{a + b}$ **18.** $\dfrac{7}{8}$ **19.** $\dfrac{5}{|y|}$ **20.** $\dfrac{2}{3}$
21. $\dfrac{\sqrt[3]{7}}{5}$ **22.** 5   **23.** $\dfrac{|x|}{5}$ **24.** $\dfrac{2x}{y}$ **25.** 3^{10} **26.** 3^4
27. $-6\sqrt{5}$ **28.** $(10y + 7)\sqrt[3]{2y}$ **29.** $-4 - 9\sqrt{6}$
30. 37.42 mph   **31.** 15,65 m   **32.** $\dfrac{\sqrt[3]{15}}{5}$ **33.** $\dfrac{\sqrt{21xy}}{7y}$
34. $\dfrac{2a\sqrt[3]{18b^2}}{3b^3}$ **35.** $\dfrac{\sqrt{3} + \sqrt{5}}{-2}$ **36.** $\dfrac{x - 7\sqrt{x} + 10}{x - 4}$
37. $\dfrac{1}{\sqrt{a + 2} + \sqrt{a}}$ **38.** $\dfrac{x - 5}{x + 2\sqrt{5x} + 5}$

Exercise Set 1.7, pp. 49–50
1. No, yes   **3.** Yes, no   **5.** Yes, no   **7.** Yes, yes   **9.** 11
11. $4|x|$ **13.** $|b + 1|$ **15.** $-3x$ **17.** $|x - 2|$ **19.** 2
21. $6\sqrt{5}$ **23.** $3\sqrt[3]{2}$ **25.** $8\sqrt{2}|c|d^2$ **27.** $3\sqrt{2}$
29. $2x^2y\sqrt{6}$ **31.** $3x\sqrt[3]{4y}$ **33.** $2(x + 4)\sqrt[3]{(x + 4)^2}$
35. $\sqrt{7b}$ **37.** 2   **39.** $\dfrac{1}{2x}$ **41.** $\sqrt{a + b}$ **43.** $\dfrac{3a\sqrt{2b}}{4b}$
45. $\dfrac{y \cdot \sqrt[3]{20x^2z^2}}{5z^2}$ **47.** $8x^2\sqrt[3]{2}$ **49.** $ab^2x^2y\sqrt{a}$
51. $51\sqrt{2}$ **53.** $-12\sqrt{5} - 2\sqrt{2}$ **55.** $19\sqrt[3]{x^2} - 3x$
57. $4y\sqrt{3} - 2y\sqrt{6}$ **59.** 1   **61.** $4 + 2\sqrt{3}$
63. $t - 2x\sqrt{t} + x^2$ **65.** $10\sqrt{7}$ **67.** x
69. About 42.43 mph   **71.** About 13,709.5 ft
73. **(a)** $h = \dfrac{a}{2}\sqrt{3}$; **(b)** $A = \dfrac{a^2}{4}\sqrt{3}$ **75.** 8   **77.** $\dfrac{3(3 - \sqrt{5})}{2}$

79. $\dfrac{2\sqrt[3]{6}}{3}$ **81.** $\dfrac{8x - 20\sqrt{xy} - 6x\sqrt{y} + 15y\sqrt{x}}{4x - 25y}$

83. $\dfrac{187 + 75\sqrt{6}}{-23}$ **85.** $\dfrac{pq - p\sqrt{s} - q^2\sqrt{s} + qs}{q^2 - s}$

87. $\dfrac{2 - 5a}{6(\sqrt{2} - \sqrt{5a})}$ **89.** $\dfrac{x}{x + 2 - 2\sqrt{x + 1}}$

91. $\dfrac{a}{3(\sqrt{a + 3} + \sqrt{3})}$ **93.** $\dfrac{1}{\sqrt{x + h} + \sqrt{x}}$

95. $\dfrac{x - y}{x - 2\sqrt{xy} + y}$ **97.** $\dfrac{(2 + x^2)\sqrt{1 + x^2}}{1 + x^2}$

99. Let $a = 16$ and $b = 9$. Then $\sqrt{a + b} = 5$ and $\sqrt{a} + \sqrt{b} = 7$.   **101.** 0.0188 m
103. **(a)** 11,183 m/sec; **(b)** 4967 m/sec

Margin Exercises, Section 1.8

1. $n\sqrt{n}$ **2.** $\dfrac{1}{\sqrt[7]{y^6}}$, or $\dfrac{\sqrt[7]{y}}{y}$ **3.** 16   **4.** $\dfrac{1}{16}$

5. $(5ab)^{4/3}$, or $(5ab)\sqrt[3]{5ab}$ **6.** 8   **7.** $a^{2/3}$, or $\sqrt[3]{a^2}$
8. $2^{1/3}$, or $\sqrt[3]{2}$ **9.** $5^{11/6}$, or $5\sqrt[6]{5^5}$

10. $\sqrt[4]{a^5}$, or $a\sqrt[4]{a}$ **11.** $\dfrac{1}{\sqrt[5]{x^6}}$, or $\dfrac{1}{x\sqrt[5]{x}}$, or $\dfrac{\sqrt[5]{x^4}}{x^2}$

12. $\sqrt[4]{2^3} + \dfrac{1}{\sqrt[4]{2}}$, or $\dfrac{3\sqrt[4]{2^3}}{2}$ **13.** $\sqrt[6]{200}$ **14.** $\sqrt[6]{x^4 y^3 z^5}$

15. $\sqrt[4]{x + y}$ **16.** 5.98 ft   **17.** $\dfrac{p^{14} - q^{12}}{p^6 q^5}$

18. $\dfrac{5x + 4y^{3/4}}{x^{1/3} y^{1/4}}$ **19.** $\dfrac{5x^2(x + 3)}{(2x + 5)^{3/2}}$

Exercise Set 1.8, pp. 55–56

1. $\sqrt[4]{x^3}$ **3.** 8   **5.** $\dfrac{1}{5}$ **7.** $\dfrac{a}{b}\sqrt[4]{ab}$ **9.** $20^{2/3}$

11. $13^{5/4}$, or $13\sqrt[4]{13}$ **13.** $11^{1/6}$ **15.** $5^{5/6}$ **17.** 4

19. $2y^2$ **21.** $(a^2 + b^2)^{1/3}$ **23.** $3ab^3$ **25.** $\dfrac{m^2 n^4}{2}$

27. $8a^{4/2}$, or $8a^2$ **29.** $\dfrac{x^{-3}}{3^{-1} b^2}$, or $\dfrac{3}{x^3 b^2}$

31. $xy^{1/3}$, or $x\sqrt[3]{y}$ **33.** $\sqrt[5]{288}$ **35.** $\sqrt[12]{x^{11} y^7}$
37. $a\sqrt[6]{a^5}$ **39.** $(a + x)\sqrt[12]{(a + x)^{11}}$ **41.** 24.685
43. 43.138   **45.** 32.942   **47.** 5.56 ft   **49.** 7.07 ft
51. 34 hr, 16.3 hr, 14.5 hr, 13.2 hr, 8.2 hr, 6.2 hr

53. $\dfrac{b^{10} - a^5}{a^2 b^5}$ **55.** $\dfrac{5a + 2b}{a^{1/3} b^{1/2}}$ **57.** $\dfrac{y - x}{x^{1/3} y^{1/4}}$

59. $\dfrac{(3x - 2)(x + 1)^{1/4}}{(2x - 3)^3}$ **61.** $\dfrac{-2(14x + 19)}{(x + 1)^{1/2}(3x + 4)^{3/4}}$

63. $3(x^2 + 1)^2(7x^2 - 10x + 1)$ **65.** $\dfrac{-x^2 - 3}{x^4}$

67. $\dfrac{-x^2 - 2}{x^3(x^2 + 1)^{1/2}}$ **69.** $a^{a/2}$

Margin Exercises, Section 1.9

1. $18{,}600\,\dfrac{\text{m}}{\text{sec}}$ **2.** $0.5\,\dfrac{\text{m}}{\text{sec}}$ **3.** 62 ft   **4.** $\dfrac{23}{20}\,\text{kg}$

5. $105\,\dfrac{\text{cm}}{\text{sec}}$ **6.** 12 yd   **7.** 80 oz   **8.** $\dfrac{7}{10}$ **9.** $11.25\,\dfrac{\text{in.-lb}}{\text{hr}^2}$

10. $4\,\dfrac{\text{lb}^2}{\text{m}^2}$ **11.** 1224 in.   **12.** 58,080 ft   **13.** 18,000 sec

14. 20 yd   **15.** 36.96 km   **16.** 100 hr   **17.** $176\,\dfrac{\text{ft}}{\text{sec}}$

18. $0.36\,\text{m}^2$ **19.** $50\,\dfrac{\text{g}}{\text{cm}^3}$ **20.** $300\,\dfrac{\text{¢}}{\text{hr}}$

Exercise Set 1.9, pp. 59–60

1. 12 yd   **3.** 48 hr   **5.** 3 g   **7.** 8 m   **9.** $12\,\text{ft}^3$

11. $\dfrac{7\,\text{kg}^2}{10\,\text{m}^2}$ **13.** $720\,\dfrac{\text{lb-mi}^2}{\text{hr}^2\text{-ft}}$ **15.** $\dfrac{15}{2}\,\dfrac{\text{cm}^5\text{-kg}}{\text{sec}^3}$ **17.** 6 ft

19. 172,800 sec   **21.** $600\,\dfrac{\text{g}}{\text{cm}}$ **23.** $2{,}160{,}000\,\text{cm}^2$

25. $150\,\dfrac{\text{¢}}{\text{hr}}$ **27.** $6.228\,\dfrac{\text{L}}{\text{hr}}$ **29.** $5{,}865{,}696{,}000{,}000\,\dfrac{\text{mi}}{\text{yr}}$

31. $1621.8\,\dfrac{\text{m}}{\text{min}}$ **33.** $1664.9\,\text{km}^2$ **35.** 7.5 g, 1250 g

37. 1600 g   **39.** 15 moles   **41.** $4.4937 \times 10^{20}\,\dfrac{\text{g}\cdot\text{m}^2}{\text{sec}^3}$

43. 0.29979 m

Review Exercises: Chapter 1, pp. 61–62

1. [1.1] 12, -3, -1, -19, 31, 0   **2.** [1.1] 12, 31
3. [1.1] All except $\sqrt{7}$, $\sqrt[3]{10}$ **4.** [1.1] All
5. [1.1] $\sqrt{7}$, $\sqrt[3]{10}$ **6.** [1.1] 0, 12, 31
7. [1.1] -4 **8.** [1.2] -8 **9.** [1.1] -5
10. [1.1] 30   **11.** [1.1] -6 **12.** [1.1] 153
13. [1.1] -3000 **14.** [1.1] $-\frac{3}{16}$ **15.** [1.1] $\frac{31}{24}$
16. [1.2] 117   **17.** [1.2] -10 **18.** [1.2] 3,261,000
19. [1.2] 0.00041   **20.** [1.2] 277,000,000
21. [1.2] 0.0001009   **22.** [1.2] 1.432×10^{-2}
23. [1.2] 4.321×10^4 **24.** [1.2] 7.8125×10^{-22}

25. [1.2] 5.46×10^{-32} **26.** [1.2] $-\dfrac{14b^7}{a^2}$

27. [1.2] $\dfrac{6x^9 z^6}{y^6}$ **28.** [1.7] 3   **29.** [1.7] -2

30. [1.6] $\dfrac{b}{a}$ **31.** [1.6] $\dfrac{x + y}{xy}$

32. [1.7] -4 **33.** [1.7] $25x^4 - 10x^2\sqrt{2} + 2$
34. [1.7] $13\sqrt{5}$ **35.** [1.4] $x^3 + t^3$ **36.** [1.4] $125a^3 + 300a^2 b + 240ab^2 + 64b^3$ **37.** [1.3] $8xy^4 - 9xy^2 + 4x^2 + 2y - 7$ **38.** [1.5] $(x^2 - 3)(x + 2)$
39. [1.5] $3a(2a - 3b^2)(2a + 3b^2)$ **40.** [1.5] $(x + 12)^2$
41. [1.5] $x(9x - 1)(x + 4)$
42. [1.5] $(2x - 1)(4x^2 + 2x + 1)$
43. [1.5] $(3x^2 + 5y^2)(9x^4 - 15x^2 y^2 + 25y^4)$
44. [1.5] $6(x + 2)(x^2 - 2x + 4)$
45. [1.5] $(2x - 3)(2x + 3)(x - 1)$

46. [1.5] $(3x - 5)^2$ **47.** [1.5] $3(6x^2 - x + 2)$
48. [1.5] $(3x + y + 4)(3x + y - 1)$ **49.** [1.8] $y^3 \cdot \sqrt[6]{y}$
50. [1.8] $\sqrt[3]{(a+b)^2}$ **51.** [1.8] $\sqrt[5]{b^7}, b\sqrt[5]{b^2}$
52. [1.8] $\dfrac{m^4n^2}{3}$ **53.** [1.6] 3   **54.** [1.6] $\dfrac{x-5}{(x+3)(x+5)}$
55. [1.7] $\dfrac{x-y}{x + 2\sqrt{xy} + y}$ **56.** [1.7] $\dfrac{x - 2\sqrt{xy} + y}{x - y}$
57. [1.8] $\dfrac{2x - 3y}{x^{1/2}y^{3/4}}$ **58.** [1.8] $\dfrac{(4x+3)(3x+5)^{3/2}}{(x-2)^{3/4}}$
59. [1.7] 18.8 ft   **60.** [1.9] $\dfrac{500}{3}$, or $166\dfrac{2}{3}$ $\dfrac{\text{m}}{\text{min}}$
61. [1.1] Inverse $(+)$ **62.** [1.1] Distributive
63. [1.1] Associative $(\times)$ **64.** [1.1] Commutative $(\times)$
65. [1.4] $x^{2n} + 6x^n - 40$ **66.** [1.4] $t^{2a} + 2 + t^{-2a}$
67. [1.4] $y^{2b} - z^{2c}$
68. [1.4] $a^{3n} - 3a^{2n}b^m + 3a^nb^{2m} - b^{3m}$
69. [1.5] $(y^n + 8)^2$ **70.** [1.5] $(x^t - 7)(x^t + 4)$
71. [1.5] $m^{3n}(m^n - 1)(m^{2n} + m^n + 1)$
72. [1.6] $\dfrac{2xn^5}{(n+1)^5}$
73. [1.6] $\dfrac{(n-1)(n-2)(n-3)(n-4)}{-24}$

Test: Chapter 1, pp. 62–63
1. [1.1] 0, 233   **2.** [1.1] $\sqrt{8}, -\sqrt[3]{11}$ **3.** [1.1] All
4. [1.1] All except $\sqrt{8}, -\sqrt[3]{11}$ **5.** [1.1] 233
6. [1.1] $-14, -5, 0, 233$ **7.** [1.2] 0   **8.** [1.1] 2
9. [1.1] 12   **10.** [1.1] 8   **11.** [1.1] -3 **12.** [1.2] $\dfrac{726}{5}$
13. [1.2] 0.002834   **14.** [1.2] 470
15. [1.2] 4,450,000,000   **16.** [1.2] 0.0000445
17. [1.2] 8.16×10^{-4} **18.** [1.2] 4.8057×10^2
19. [1.2] 3.654×10^{-15} **20.** [1.2] 4.4×10^{-9}
21. [1.2] $-\dfrac{12x^9}{y^6}$ **22.** [1.2] $\dfrac{2p^4q^{14}}{3r^9}$ **23.** [1.7] -3
24. [1.7] 5   **25.** [1.7] 6   **26.** [1.6] $\dfrac{x+y}{x^2y}$
27. [1.4] $25a^4 - 20a^2b + 4b^2$ **28.** [1.3] $8x^2y - 5xy + y^2 + y - 12$ **29.** [1.4] $64y^3 - 144y^2 + 108y - 27$
30. [1.8] $\sqrt[20]{(c+d)^9}$ **31.** [1.8] $\sqrt[4]{t^2}$
32. [1.5] $4(2x + 3)^2$ **33.** [1.5] $(t - 7)(t^2 + 7t + 49)$
34. [1.5] $m^3(m - 3n)(m + 3n)$
35. [1.5] $3(4p^2 - 5)(p^2 + 2)$
36. [1.5] $(4 + 5a^2)(16 - 20a^2 + 25a^4)$
37. [1.5] $(3 + x)(x + 3y)$
38. [1.5] $a^2(a^2 - 2b^2)(a^4 + 2a^2b^2 + 4b^4)$
39. [1.6] $x - 5$ **40.** [1.6] $\dfrac{x+3}{(x+6)(x+2)}$
41. [1.7] $\dfrac{49 - 14\sqrt{x} + x}{49 - x}$ **42.** [1.4] $x^{3t} + 3x^t + 3x^{-t} + x^{-3t}$ **43.** [1.7] $\dfrac{49 - x}{49 + 14\sqrt{x} + x}$

44. [1.8] $\dfrac{b^{3/5}(b - a^2)}{a^{2/3}}$ **45.** [1.7] 90.8 ft
46. [1.8] $\dfrac{x^2(5x + 9)}{(2x + 3)^{3/2}}$ **47.** [1.9] $72\dfrac{\text{km}}{\text{hr}}$
48. [1.5] $(x^8 + 4)(x^4 + 2)(x^4 - 2)$
49. [1.2] $\dfrac{1}{2}\left|\dfrac{n(x - 2)}{n + 1}\right|$

CHAPTER 2

Margin Exercises, Section 2.1
1. $\{9\}$ **2.** $\{0, -1\}$ **3.** $\{\frac{4}{3}\}$ **4.** $\{\frac{17}{2}\}$ **5.** $\{-\frac{19}{8}\}$
6. $\varnothing$ **7.** All real numbers   **8.** $\{7, -\frac{3}{2}\}$ **9.** $\{5, -4\}$
10. $\{0, 5\}$ **11.** $\{-\frac{1}{5}\}$ **12.** $\{0, -\frac{1}{3}, 4\}$
13. $\{1, -1, -\frac{1}{5}\}$ **14.** $\{x | x > \frac{3}{2}\}$ **15.** $\{y | \frac{22}{13} \leq y\}$
16. $\{x | x < 5\}$ **17.** $\{x | x > \frac{5}{2}\}$ **18.** $\{y | y \geq -7\}$
19. $\{x | x^2 = 5\}$ **20.** $\{x | x \geq 2\}$ **21.** $\{x | x \geq -3\}$
22. $\{x | \frac{11}{2} \geq x\}$

Exercise Set 2.1, pp. 70–71
1. $\{12\}$ **3.** $\{-6\}$ **5.** $\{8\}$ **7.** $\{\frac{4}{5}\}$ **9.** $\{2\}$
11. $\{-\frac{3}{2}\}$ **13.** $\{-2\}$ **15.** $\{\frac{3}{2}, \frac{2}{3}\}$ **17.** $\{0, 1, -2\}$
19. $\{\frac{2}{3}, -1\}$ **21.** $\{4, 1\}$ **23.** $\{-1, -2\}$
25. $\{-\frac{5}{3}, 4, \frac{5}{2}\}$ **27.** $\{0, \frac{1}{4}, -\frac{1}{4}\}$ **29.** $\{0, 3\}$
31. $\{0, -\frac{1}{3}, 2\}$ **33.** $\{\frac{3}{2}, -\frac{2}{3}, 1\}$ **35.** $\{\frac{1}{2}, 0, -3\}$ **37.** $\varnothing$
39. $\{-1, -\frac{1}{7}, 1\}$ **41.** $\{-2, -1, 1\}$
43. All real numbers   **45.** $\{x | x > 3\}$ **47.** $\{x | x \geq -\frac{5}{12}\}$
49. $\{y | y \geq \frac{22}{13}\}$ **51.** $\{x | x \leq \frac{15}{34}\}$ **53.** $\{x | x < 1\}$
55. $\{x | x > 2.5\}$ **57.** $\{t | t^2 = 5\}$ **59.** $\{x | x \geq 3\}$
61. $\{x | \frac{3}{4} \geq x\}$ **63.** $\{0.7892\}$ **65.** $\{0, 2.1522\}$
67. $\{x | x < -0.7848\}$ **69.** $\{-2, 2\}$ **71.** $\{-5, -4, 5\}$

Margin Exercises, Section 2.2
1. Yes   **2.** Yes   **3.** No   **4.** Yes   **5.** No   **6.** No
7. Add $5x^2$.   **8.** Add $-5x^2$.   **9.** $\varnothing$ **10.** $\{4\}$
11. $\{-6, 6\}$ **12.** $\{\frac{16}{5}\}$ **13.** $\{10\}$

Exercise Set 2.2, pp. 76–77
1. Yes   **3.** No   **5.** No   **7.** $\{\frac{20}{9}\}$ **9.** $\varnothing$ **11.** $\{286\}$
13. $\varnothing$ **15.** $\{-2\}$ **17.** $\{6\}$ **19.** $\varnothing$ **21.** $\varnothing$
23. All real numbers except 0 and 6   **25.** $\{\frac{5}{3}\}$ **27.** $\varnothing$
29. $\{0.94656\}$ **31.** $\{-5\}$ **33.** $\{2\}$ **35.** $\{-\frac{19}{5}\}$
37. All real numbers except 0 and 9   **39.** All real numbers except 3   **41.** All real numbers except -2
43. (1) equivalent to (2); (2) not equivalent to (3); (3) equivalent to (4)   **45.** Identity   **47.** Identity
49. Not an identity

Margin Exercises, Section 2.3
1. $F = \frac{9}{5}C + 32$ **2.** $r_2 = \dfrac{Rr_1}{r_1 - R}$ **3.** \$2600
4. \$1920.60   **5.** \$1946.33   **6.** 20 ft   **7.** 36 km/h
8. 375 km   **9.** 50 mph, 60 mph   **10.** $2\frac{2}{9}$ hr
11. Laurie: 12 hr; Stacy: 6 hr

Exercise Set 2.3, pp. 87–89

1. $w = \dfrac{P - 2l}{2}$ **3.** $b = \dfrac{2A}{h}$ **5.** $r = \dfrac{d}{t}$ **7.** $I = \dfrac{E}{R}$

9. $T_1 = \dfrac{T_2 P_1 V_1}{P_2 V_2}$ **11.** $v_1 = \dfrac{H}{Sm} + v_2$ **13.** $p = \dfrac{Fm}{m - F}$

15. $x = \dfrac{5 + ab}{a - b}$ **17.** $x = -\dfrac{a}{9}$ **19.** $650

21. $26°, 130°, 24°$ **23.** 68 m, 93 m   **25.** 91%
27. 2 cm   **29.** $810,000$ **31.** 12 km/h
33. A: 46 mph; B: 58 mph   **35.** 98.3 mi   **37.** $1\frac{34}{71}$ hr
39. 6.21 hr   **41.** (a) $1087.50; (b) $1089.41; (c) $1090.41;
(d) $1091.43; **(e)** $1091.44   **43.** 32 mph   **45.** $53\frac{6}{23}$ mph
47. $51\frac{3}{7}$ mph   **49.** $10{:}38\frac{2}{11}$ **51.** (a) 1201.2 mi;
(b) less time to return to Los Angeles   **53.** $16

Margin Exercises, Section 2.4
1. $i\sqrt{6}$, or $\sqrt{6}i$ **2.** $-i\sqrt{10}$, or $-\sqrt{10}i$ **3.** $2i$
4. $-5i$ **5.** $-\sqrt{10}$ **6.** $\sqrt{11}$ **7.** $i\sqrt{7}$ **8.** $7i$ **9.** $3i$
10. $(\sqrt{17} + 3)i$ **11.** i **12.** -1 **13.** $-i$ **14.** $12 + i$
15. $5 - i$ **16.** $2 + 14i$ **17.** 8 **18.** $-6 + 8i$ **19.** $3i$
20. $(x + 2i)(x - 2i)$ **21.** $(3 + yi)(3 - yi)$ **22.** Yes
23. $x = -1, y = 2$ **24.** $7 - 2i$ **25.** $6 + 4i$ **26.** $5i$

27. $-3i$ **28.** -3 **29.** 8 **30.** $\dfrac{9}{13} + \dfrac{7}{13}i$

31. $\dfrac{4}{13} + \dfrac{7}{13}i$ **32.** $\dfrac{1}{3 + 4i}, \dfrac{3}{25} - \dfrac{4}{25}i$ **33.** $2 + 5i$

Exercise Set 2.4, pp. 96–97
1. $i\sqrt{15}$ **3.** $9i$ **5.** $-2i\sqrt{3}$ **7.** $9i$ **9.** $i(\sqrt{7} - \sqrt{10})$

11. $-\sqrt{55}$ **13.** $2\sqrt{5}$ **15.** $\sqrt{\dfrac{5}{2}}i$ **17.** $-\dfrac{3}{2}i$ **19.** -2

21. -1 **23.** $-i$ **25.** $-i$ **27.** -1 **29.** $6 + 5i$
31. 8 **33.** $2 + 4i$ **35.** $-4 - i$ **37.** $-5 + 5i$
39. $7 - i$ **41.** $-6 + 12i$ **43.** $-5 + 12i$

45. $(2x + 5yi)(2x - 5yi)$ **47.** Yes   **49.** $x = -\dfrac{3}{2}, y = 7$

51. $\dfrac{1}{2} + \dfrac{7}{2}i$ **53.** $\dfrac{1}{3} + \dfrac{2\sqrt{2}}{3}i$ **55.** $2 - 3i$ **57.** $\dfrac{1}{5} + \dfrac{2}{5}i$

59. $-\dfrac{1}{2} - \dfrac{1}{2}i$ **61.** $\dfrac{28}{65} - \dfrac{29}{65}i$ **63.** $-\dfrac{1}{2} + \dfrac{3}{2}i$

65. $\dfrac{5}{2} + \dfrac{13}{2}i$ **67.** $\dfrac{4}{25} - \dfrac{3}{25}i$ **69.** $\dfrac{5}{29} + \dfrac{2}{29}i$

71. $-i$ **73.** $\dfrac{1}{4}i$ **75.** $\dfrac{2}{5} + \dfrac{6}{5}i$ **77.** $\dfrac{8}{5} - \dfrac{9}{5}i$ **79.** $2 - i$

81. $\dfrac{11}{25} + \dfrac{2}{25}i$ **83.** For example, $\sqrt{-1}\sqrt{-1} = i^2 = -1$,
but $\sqrt{(-1)(-1)} = \sqrt{1} = 1$.   **85.** Let $z = a + bi$.
Then $z \cdot \bar{z} = (a + bi)(a - bi) = a^2 - b^2 i^2 = a^2 + b^2$. Since
a and b are real numbers, so is $a^2 + b^2$. Thus, $z \cdot \bar{z}$ is real.
87. Let $z = a + bi$ and $w = c + di$. Then $\overline{z + w} =$
$\overline{(a + bi) + (c + di)} = \overline{(a + c) + (b + d)i}$, by adding. We
now take the conjugate and obtain $(a + c) - (b + d)i$. Now
$\bar{z} + \bar{w} = \overline{(a + bi)} + \overline{(c + di)} = (a - bi) + (c - di)$, taking
the conjugates. We will now add to obtain $(a + c) - (b + d)i$,
the same result as before. Thus, $\overline{z + w} = \bar{z} + \bar{w}$.

89. By the definition of exponents, the conjugate of z^n is
the conjugate of the product of n factors of z. Using the
result of Exercise 88, we see that the conjugate of n factors
of z is the product of n factors of $\bar{z}$. Thus, $\overline{z^n} = \bar{z}^n$.

91. $3\bar{z}^5 - 4\bar{z}^2 + 3\bar{z} - 5$ **93.** $7 + \dfrac{8}{9}i$ **95.** $-bi$

97. $\sqrt{2} + i\sqrt{2}, -\sqrt{2} - i\sqrt{2}$ **99.** $\dfrac{ac + bd}{a^2 + b^2} + \dfrac{ad - bc}{a^2 + b^2}i$

Margin Exercises, Section 2.5

1. $\{\pm\sqrt{7}\}$ **2.** $\{0\}$ **3.** $\left\{\pm\sqrt{\dfrac{\pi}{3}}\right\}$ **4.** $\left\{\pm\sqrt{\dfrac{n}{m}}\right\}$

5. $\left\{\pm\dfrac{\sqrt{2}}{2}i\right\}$ **6.** $\{-4 \pm \sqrt{7}\}$ **7.** $\{5 \pm \sqrt{3}\}$

8. $\{-3, -7\}$ **9.** $4, (x + 2)^2$ **10.** $9, (x - 3)^2$

11. $\dfrac{25}{4}, \left(x + \dfrac{5}{2}\right)^2$ **12.** $\dfrac{49}{4}, \left(x - \dfrac{7}{2}\right)^2$ **13.** $\dfrac{9}{64}, \left(x + \dfrac{3}{8}\right)^2$

14. $\dfrac{1}{4}, \left(x - \dfrac{1}{2}\right)^2$ **15.** $\{-2 \pm \sqrt{7}\}$ **16.** $\{4, 2\}$

17. $\{2, 3\}$ **18.** $\left\{\dfrac{-1 \pm \sqrt{7}}{2}\right\}$ **19.** $\left\{-1, \dfrac{1}{4}\right\}$

20. $\left\{\dfrac{1}{2}, -4\right\}$ **21.** $\left\{\dfrac{4 \pm \sqrt{31}}{5}\right\}$ **22.** $\left\{\dfrac{1 \pm i\sqrt{7}}{2}\right\}$

23. Two real   **24.** One real   **25.** Two nonreal
26. $3x^2 + 7x - 20 = 0$ **27.** $x^2 + \sqrt{2}x - 4 = 0$
28. $x^2 + 25 = 0$

Exercise Set 2.5, pp. 104–105

1. $\{\pm 3\}$ **3.** $\{\pm i\}$ **5.** $\{\pm\sqrt{5}\}$ **7.** $\{0\}$ **9.** $\left\{\pm\dfrac{\sqrt{6}}{2}\right\}$

11. $\{\pm i\sqrt{7}\}$ **13.** $\left\{\pm\sqrt{\dfrac{b}{a}}\right\}$ **15.** $\{7 \pm \sqrt{5}\}$ **17.** $\left\{\pm\dfrac{3}{2}\right\}$

19. $\{h \pm \sqrt{a + 1}\}$ **21.** $\{-3 \pm \sqrt{5}\}$ **23.** $\{-10, 3\}$

25. $\left\{\dfrac{2 \pm \sqrt{14}}{5}\right\}$ **27.** $\left\{-5, \dfrac{3}{2}\right\}$ **29.** $\{-5, 1\}$

31. $\left\{-\dfrac{1}{2}, 2\right\}$ **33.** $\left\{\dfrac{-4 \pm \sqrt{7}}{3}\right\}$ **35.** $\{6 \pm \sqrt{33}\}$

37. $\left\{\dfrac{1 \pm i\sqrt{3}}{2}\right\}$ **39.** $\{2 \pm 3i\}$ **41.** $\left\{\dfrac{13 \pm \sqrt{509}}{10}\right\}$

43. $\{0.04 \pm 0.02\sqrt{79}\}$ **45.** $\left\{\dfrac{-19 \pm \sqrt{445}}{14}\right\}$

47. $\left\{\dfrac{-1 \pm i\sqrt{15}}{2}\right\}$ **49.** One real   **51.** Two nonreal
53. Two real   **55.** One real   **57.** Two nonreal
59. Two real   **61.** Two real   **63.** One real
65. $x^2 + 2x - 99 = 0$ $x^2 - 14x + 49 = 0$

69. $x^2 - \dfrac{4}{5}x - \dfrac{12}{25} = 0$, or $25x^2 - 20x - 12 = 0$

71. $x^2 - \left(\dfrac{c + d}{2}\right)x + \dfrac{cd}{4} = 0$ **73.** $x^2 - 4\sqrt{2}x + 6 = 0$

75. $x^2 + 9 = 0$ **77.** $\{1.1754, -0.4254\}$ **79.** $\left\{\dfrac{3}{2}, \dfrac{2}{3}\right\}$

81. $\left\{\dfrac{-1 \pm \sqrt{1 + 4\sqrt{2}}}{2}\right\}$ **83.** $\left\{-2, \dfrac{3}{4}\right\}$

85. $\left\{\dfrac{1 \pm \sqrt{113}}{2}\right\}$ **87.** $\{3 \pm \sqrt{5}\}$

89. (a) $\dfrac{-b + \sqrt{b^2 - 4ac}}{2a} + \dfrac{-b - \sqrt{b^2 - 4ac}}{2a} = \dfrac{-2b}{2a} = -\dfrac{b}{a}$;

(b) $\dfrac{-b + \sqrt{b^2 - 4ac}}{2a} \cdot \dfrac{-b - \sqrt{b^2 - 4ac}}{2a} =$

$\dfrac{b^2 - (b^2 - 4ac)}{4a^2} = \dfrac{4ac}{4a^2} = \dfrac{c}{a}$ **91. (a)** $k = -\dfrac{3}{5}$; **(b)** $-\dfrac{1}{3}$

93. (a) $k = 9 + 9i$; **(b)** $3 + 3i$ **95.** $x^2 - \sqrt{3}x + 8 = 0$

97. $h = -36, k = 15$ **99.** 2

Margin Exercises, Section 2.6

1. $r = \sqrt{\dfrac{3V}{\pi h}}$ **2.** $t = \dfrac{-v_0 + \sqrt{v_0^2 + 64S}}{32}$ **3.** 6.25%

4. $12 - 2\sqrt{22} \approx 2.6$ ft **5. (a)** 4.3 sec; **(b)** 1.9 sec;
(c) 44.9 m

Exercise Set 2.6, pp. 109–111

1. $d = \sqrt{\dfrac{kM_1M_2}{F}}$ **3.** $t = \sqrt{\dfrac{2S}{a}}$ **5.** $t = \dfrac{v_0 \pm \sqrt{v_0^2 - 64s}}{32}$

7. $n = \dfrac{3 + \sqrt{9 + 8d}}{2}$ **9.** $i = -1 + \sqrt{\dfrac{A}{P}}$ **11.** 8%

13. 5% **15.** 9 **17.** $24 - 4\sqrt{34} \approx 0.7$ ft **19.** 2 ft
21. 4.7 cm **23.** A: 15 mph; B: 20 mph
25. (a) 3.9 sec; **(b)** 1.9 sec; **(c)** 79.6 m **27.** 3.2 cm

29. $35 - 5\sqrt{33}$, or 6.3 ft **31.** $\dfrac{15 - \sqrt{115}}{2}$, or 2.1 cm

33. First part: $\dfrac{75 + 5\sqrt{185}}{2}$, or 71.5 mph;

second part: $\dfrac{55 + 5\sqrt{185}}{2}$, or 61.5 mph

35. $6 + 3\sqrt{5}$, or 12.7 mph **37.** 7 **39.** 12

41. $\left\{2, -\dfrac{3}{k}\right\}$ **43.** $\left\{\dfrac{1}{m+n}, \dfrac{-2}{m+n}\right\}$ **45.** 11.7%

47. $a_3 = \sqrt{a_1^2 + a_2^2}$ **49.** $x = 4$ cm, $y = 3$ cm

Margin Exercises, Section 2.7

1. $\varnothing$ **2.** $\{4\}$ **3.** $\left\{\dfrac{17}{3}\right\}$ **4.** $\{9\}$ **5.** $\{5\}$

6. $m = \sqrt{\dfrac{1}{1 - P^2}}$, or $\dfrac{\sqrt{1 - P^2}}{1 - P^2}$

Exercise Set 2.7, pp. 114–115

1. $\left\{\dfrac{5}{3}\right\}$ **3.** $\{\pm\sqrt{2}\}$ **5.** $\varnothing$ **7.** $\{7\}$ **9.** $\varnothing$

11. $\{-6\}$ **13.** $\{3, -1\}$ **15.** $\left\{\dfrac{80}{9}\right\}$ **17.** $\{1\}$ **19.** $\{7, 3\}$

21. $\{62.4459\}$ **23.** $\{-8\}$ **25.** $\{81\}$ **27.** $\left\{\dfrac{1}{64}\right\}$

29. $\{-125\}$ **31.** $L = \dfrac{gT^2}{4\pi^2}$, $g = \dfrac{4L\pi^2}{T^2}$ **33.** 208 mi

35. 14,400 ft **37.** $\{5 \pm 2\sqrt{2}\}$ **39.** $\left\{-\dfrac{8}{9}\right\}$ **41.** $\{2\}$

43. $\left\{\dfrac{-5 + \sqrt{61}}{18}\right\}$ **45.** $\{9\}$ **47.** $\{8\}$ **49.** $\{10\}$

Margin Exercises, Section 2.8
1. (a) 9; **(b)** $\sqrt{x} = 12 - x$; $(12 - x)^2 = x$;
$x = 144 - 24x + x^2$; $0 = x^2 - 25x + 144$;
$0 = (x - 9)(x - 16)$. The procedure in (a) was probably
easier, since the factoring was easier.

2. $\pm\sqrt{\dfrac{5 + \sqrt{3}}{2}}, \pm\sqrt{\dfrac{5 - \sqrt{3}}{2}}$ **3.** $\pm\sqrt{3}, 0$ **4.** $\{125, -8\}$

5. $\{\pm 3i, \pm 2\}$ **6.** 350.63 ft

Exercise Set 2.8, pp. 119–120
1. $\{1, 81\}$ **3.** $\{\pm\sqrt{5}\}$ **5.** $\{-27, 8\}$ **7.** $\{16\}$

9. $\{7, 5, -1, 1\}$ **11.** $\left\{1, 4, \dfrac{5 \pm \sqrt{37}}{2}\right\}$

13. $\{\pm\sqrt{2 + \sqrt{6}}, \pm\sqrt{2 - \sqrt{6}}\}$ **15.** $\left\{-\dfrac{1}{2}, \dfrac{1}{3}\right\}$

17. $\{-1, 2\}$ **19.** $\{\pm 5, \pm i\}$ **21.** $\left\{-1 \pm \sqrt{3}, \dfrac{9 \pm \sqrt{89}}{2}\right\}$

23. $\left\{\dfrac{100}{99}\right\}$ **25.** $\left\{-\dfrac{6}{7}\right\}$ **27.** $\left\{1 \pm \sqrt{2}, \dfrac{-1 \pm \sqrt{5}}{2}\right\}$

29. 132.7 ft **31.** 7% **33.** $\{2.05\}$ **35.** $\{1, 4\}$

37. $\{19\}$ **39.** $\left\{1, 3, \dfrac{-1 \pm i\sqrt{3}}{2}, \dfrac{-3 \pm 3i\sqrt{3}}{2}\right\}$

41. $\left\{-1, 6, \dfrac{5 \pm \sqrt{37}}{2}\right\}$ **43.** $\left\{-2 \pm \sqrt{2}, \dfrac{1 \pm i\sqrt{7}}{2}\right\}$

45. $\left\{\dfrac{-51 + 7\sqrt{61}}{194}\right\}$

Margin Exercises, Section 2.9
1. $y = 160x$ **2.** 4.5 kg **3.** 50 volts **4.** 176,250 tons

5. $y = \dfrac{6.4}{x}$ **6.** 7.5 hr **7.** $y = 3x^2$ **8.** $y = \dfrac{9}{x^2}$

9. $y = 7xz$ **10.** $y = 7\dfrac{xz}{w^2}$ **11.** 2 sec

12. (a) 128 lb; **(b)** 4000 mi

Exercise Set 2.9, pp. 124–125

1. $y = \dfrac{3}{2}x$ **3.** $y = \dfrac{4000}{x}$ **5.** $y = 5.375x$ **7.** $y = \dfrac{0.0015}{x^2}$

9. $y = \dfrac{xz}{w}$ **11.** $y = \dfrac{5}{4} \cdot \dfrac{xz}{w^2}$ **13.** y is doubled.

15. y is multiplied by $\dfrac{1}{n^2}$. **17.** 532,500 tons

19. L is multiplied by 16. **21.** 68.6 m

23. $624.24\,\text{m}^2$ **25.** 97

27. If p varies directly as q, then $p = kq$. Thus, $q = \dfrac{1}{k}p$,

so q varies directly as p. **29.** $\dfrac{\pi}{4}$ **31. (a)** $N = 0.001\dfrac{P_1 P_2}{d^2}$;

(b) 3834; **(c)** 1128 km; **(d)** Division by 0 is not defined.

Review Exercises: Chapter 2, pp. 126–127

1. [2.4] $-2\sqrt{10}i$ **2.** [2.4] $-4\sqrt{15}$ **3.** [2.4] $14 + 2i$

4. [2.4] $1 - 4i$ **5.** [2.4] $2 - i$ **6.** [2.4] $\dfrac{11}{10} + \dfrac{3}{10}i$

7. [2.4] No **8.** [2.4] $\dfrac{6}{85} + \dfrac{7}{85}i$ **9.** [2.4] $x = 2, y = -4$

10. [2.4] $\left\{-\dfrac{7}{15} + \dfrac{3}{5}i\right\}$ **11.** [2.2] $\{-1\}$

12. [2.1] $\left\{3, -\dfrac{2}{3}, -2\right\}$ **13.** [2.1] $\left\{\dfrac{4}{3}, -2\right\}$

14. [2.5] $\{1 \pm 3i\}$ **15.** [2.1] $\{-5, -2, 2\}$

16. [2.2] $\left\{\dfrac{27}{7}\right\}$ **17.** [2.8] $\left\{\pm\sqrt{\dfrac{3+\sqrt{5}}{2}}, \pm\sqrt{\dfrac{3-\sqrt{5}}{2}}\right\}$

18. [2.8] $\{1\}$ **19.** [2.8] $\{\pm\sqrt{3}, 0\}$ **20.** [2.8] $\{-8, 125\}$
21. [2.7] $\{5\}$ **22.** [2.7] $\{0, 3\}$ **23.** [2.5] $\{8, -2\}$
24. [2.1] $\{6, -3\}$ **25.** [2.1] $\{-20\}$
26. [2.1] $\{-5, 3\}$ **27.** [2.1] $\{-2, 1\}$
28. [2.1] $\{y\,|\,y > -2\}$ **29.** [2.1] $\{x\,|\,x \geq 5\}$
30. [2.1] $\{x\,|\,x \leq 4\}$ **31.** [2.5] Two nonreal solutions
32. [2.5] Two real solutions

33. [2.5] $x^2 + \dfrac{5}{2}x - \dfrac{3}{2} = 0$, or $2x^2 + 5x - 3 = 0$

34. [2.5] $x^2 - 2x + 5 = 0$ **35.** [2.7] $h = \dfrac{v^2}{2g}$

36. [2.3] $t = \dfrac{ab}{a+b}$ **37.** [2.3] 94% **38.** [2.3] $1\frac{1}{3}\,\text{hr}$

39. [2.3] $1\frac{1}{2}\,\text{hr}$ **40.** [2.6] 80 km/h

41. [2.6] $2 + 2\sqrt{2} \approx 4.8\,\text{km/h}$ **42.** [2.9] $y = \dfrac{0.5}{x^2}$

43. [2.9] $T = \dfrac{1}{180}\cdot\dfrac{x^2}{p}$ **44.** [2.9] \$4.96 per share

45. [2.9] $s = 16t^2$; $7\frac{1}{2}\,\text{sec}$ **46.** [2.2] No **47.** [2.2] No
48. [2.5] $-(a+c)$ **49.** [2.2] Yes

50. [2.9] $A = \dfrac{1}{4\pi}\cdot C^2$; $\dfrac{1}{4\pi}$ **51.** [2.7] $\{256\}$

52. [2.2] No **53.** [2.4] $x = 2 - i, y = -1 - 3i$

Test: Chapter 2, p. 127–128

1. [2.4] $-i$ **2.** [2.4] $4\sqrt{3}$ **3.** [2.4] $18 + 26i$

4. [2.4] $4 + 14i$ **5.** [2.4] $\dfrac{16}{13} + \dfrac{11}{13}i$ **6.** [2.4] 2

7. [2.4] $\dfrac{1}{5} - \dfrac{1}{10}i$ **8.** [2.4] $x = 2, y = -3$

9. [2.1] $\left\{\dfrac{9}{2}, -4, 5\right\}$ **10.** [2.8] $\{1, 36\}$

11. [2.1] $\left\{\dfrac{7}{2}, -3\right\}$ **12.** [2.5] $\left\{\dfrac{3 \pm \sqrt{89}}{8}\right\}$

13. [2.2] $\{11\}$ **14.** [2.5] $\left\{\dfrac{2 \pm 2i\sqrt{14}}{5}\right\}$ **15.** [2.7] $\{15\}$

16. [2.2] $\{0\}$ **17.** [2.5] $\{8, -4\}$
18. [2.1] $\{y\,|\,y > -2\}$ **19.** [2.1] $\{1, -1, 7\}$
20. [2.1] $\{0\}$ **21.** [2.3] 24 hr **22.** [2.3] 12 mph
23. [2.5] $\{2 \pm \sqrt{6}\}$ **24.** [2.5] Two real

25. [2.5] $x^2 + 25 = 0$ **26.** [2.3] $T_2 = \dfrac{S_1 T_1 W_2}{W_1 S_2}$

27. [2.6] 30 ft, 40 ft **28.** [2.9] $y = \dfrac{6xw}{z^2}$

29. [2.9] 16 cm **30.** [2.2] Yes **31.** [2.1] $\{x\,|\,2 \geq x\}$

32. [2.2], [2.5] $\left\{\dfrac{-1 \pm \sqrt{5}}{2}\right\}$. **33.** [2.7] $\{10 + 2\sqrt{21}\}$

CHAPTER 3

Margin Exercises, Section 3.1

1.

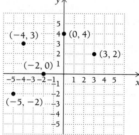

2. Yes **3.** No

4. No **5.** Yes

6.

7.

8. The shapes are the same, but this curve opens to the right instead of up.

9.

$xy = 1$

10. The shapes are the same, but this graph opens to the right instead of up.

$x = |y|$

11. The domain $= \{2, -4, -6\}$, and the range $= \{2, 3, 5, 7\}$.

12.

Range $= \{y| -4 \leq y \leq 5\}$

Domain $= \{x| -5 \leq x \leq 5\}$

13.

Range $= \{y| -1 \leq y \leq 1\}$

Domain $= \{x| 0 \leq x \leq 9\}$

14.

$y = x^2$

Range $= \{y| y \geq 0\}$

Domain $= \{x| x \text{ is a real number}\}$

Technology Connection, Section 3.1
TC 1.

$[-10, 10] \times [-10, 10]$

TC 2. $[-50, 50] \times [-60, 60]$; answers may vary.
TC 3. $[-1, 1] \times [-1, 1]$; answers may vary.

TC 4.   **TC 5.**

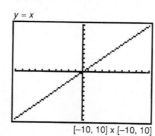

$y = x$

$[-10, 10] \times [-10, 10]$

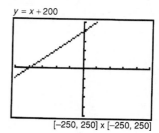

$y = x + 200$

$[-250, 250] \times [-250, 250]$

TC 6.   **TC 7.**

$y = x^2$

$[-10, 10] \times [-10, 10]$

$y = -4x^2$

$[-10, 10] \times [-10, 10]$

TC 8.   **TC 9.**

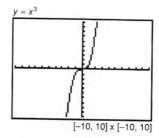

$y = x^3$

$[-10, 10] \times [-10, 10]$

$y = x^3 + 2x^2 - 4x - 13$

$[-10, 10] \times [-20, 10]$

Exercise Set 3.1, pp. 137–138
1.   **3.**

(6, 6)
(5, 4)
(4, 2)
(3, 0)

(3, 0)
(3, −1)
(3, −2)
(3, −3)
(3, −4)

5.

7.

33.

35.

9. Yes, no   **11.** No, no   **13.** Yes, no   **15.** Yes, no

17.

19.

37.

39.

21.

23.

41.

43.

45. Same graphs.

25.

27.

29.

31.

47. The domain $= \{x \mid -2 \le x \le 5\}$, and the range $= \{y \mid -2 \le y \le 4\}$.
49. The domain $= \{x \mid 2 \le x \le 6\}$, and the range $= \{y \mid 1 \le y \le 5\}$.
51. The domain $=$ the set of real numbers, and the range $= \{y \mid y \le 0\}$.
53. The domain $= \{x \mid x \ge 0\}$, and the range $=$ the set of real numbers.
55. The domain $= \{x \mid x \ge 0\}$, and the range $= \{y \mid y \ge 0\}$.
57. The domain $=$ the set of real numbers, and the range $= \{y \mid y \le 8\}$.

59.

$|x| + |y| = 1$

61.

$y = |x| + x$

$[-10, 10] \times [-10, 10]$

63.

$y = |x^2 - 4|$

$[-10, 10] \times [-5, 15]$

65.

$y = x^2|x|$

$[-10, 10] \times [-5, 15]$

67.

$y = \frac{1}{3}x^3 - x + \frac{2}{3}$

$[-10, 10] \times [-10, 10]$

69.

$y = 1 + \sqrt{2 - x}$

$[-10, 10] \times [-10, 10]$

Margin Exercises, Section 3.2

1. $\sqrt{149} \approx 12.207$ **2.** $6\sqrt{2} \approx 8.485$ **3.** 16   **4.** 8
5. Yes   **6.** No   **7.** $(\frac{3}{2}, -\frac{5}{2})$ **8.** $(9, -5)$
9. $x^2 + y^2 = 20$ **10.** $(x + 3)^2 + (y - 7)^2 = 25$
11. $(-1, 3), 2$

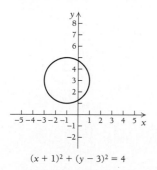

$(x + 1)^2 + (y - 3)^2 = 4$

12. $(7, -2), 8$
13. $(x + 1)^2 + (y - 4)^2 = 41$

Exercise Set 3.2, pp. 143–144

1. 5   **3.** $3\sqrt{2} \approx 4.243$ **5.** $\sqrt{a^2 + 64}$ **7.** $\sqrt{a^2 + b^2}$
9. $2\sqrt{a}$ **11.** 18.806   **13.** Yes   **15.** $(-\frac{1}{2}, -1)$
17. $(a, 0)$ **19.** $(-0.4485, -0.2733)$

21. $(0, 0), 6$

$x^2 + y^2 = 36$

23. $(3, 0), \sqrt{3}$

$(x - 3)^2 + y^2 = 3$

25. $(-1, -3), 2$

$(x + 1)^2 + (y + 3)^2 = 4$

27. $(8, -3), 2\sqrt{10}$ **29.** $(3, 0), \frac{1}{5}$
31. $(-4, 3), 2\sqrt{10}$ **33.** $(-3, 0), 3$
35. $(-4, 0), 10$ **37.** $\left(-\frac{21}{2}, -\frac{33}{2}\right), \frac{\sqrt{1462}}{2}$
39. $(-4.123, 3.174), 10.071$ **41.** $(0, 0), \frac{1}{3}$
43. $x^2 + y^2 = 25$ **45.** $(x + 4)^2 + (y - 1)^2 = 20$
47. $(5, 0)$ **49.** $(x - 2)^2 + (y - 4)^2 = 16$
51. $(x - 1)^2 + (y - 2)^2 = 41$
53. $(x + 8)^2 + (y - 5)^2 = 25$ **55.** $(8, 9), \frac{5\sqrt{2}}{2}$
57. Yes   **59.** Yes   **61.** Yes   **63.** No
65. We have the points $O(0, 0)$, $B(b, 0)$, $H(0, h)$, and
$P\left(\frac{b}{2}, \frac{h}{2}\right)$. Then each of the three distances PO, PB, PH is
$\frac{\sqrt{b^2 + h^2}}{2}$.
67. A circle with radius 2 centered at the origin

Margin Exercises, Section 3.3

1. Yes   **2.** Yes   **3.** No   **4.** Yes   **5.** B, C, D, E
6. $f(-1) = -5, f(0) = -3, f(5.6) = 8.2, f(10) = 17$
7. $f(0) = -1, f(1) = 5, f(-1) = -5,$
$f(2a) = 4a^2 + 10a - 1, f(a + 1) = a^2 + 7a + 5$
8. $f(16) = 4, f(3) = \sqrt{3}, f(-4)$ is not defined
9. $G(0) = 7, G(-3) = 7, G(\frac{1}{2}) = 7$; range: $\{7\}$
10. $6a + 3h$ **11.** $2a - 1 + h$
12. $\{x | x \neq -\frac{4}{3}$ and $x \neq -2\}$ **13.** $\{x | x \geq -2.5\}$
14. All real numbers

Technology Connection, Section 3.3

TC 1.   **TC 2.**

$y = 5x - 3$

$[-10, 10] \times [-10, 10]$

$y = x^2 - 4x + 3$

$[-10, 10] \times [-10, 10]$

TC 3.

$y = (x + 4)^2$

$[-10, 10] \times [-10, 10]$

TC 4. The domain $= \{x \mid x \neq -1, 1\}$, and the range $= \{y \mid y \neq 0\}$.
TC 5. The domain is all real numbers, and the range $= \{y \mid y \geq 0\}$.
TC 6. The domain $= \{x \mid x < 5\}$, and the range $= \{y \mid y > 0\}$.
TC 7. The domain $= \{x \mid x < 5\}$, and the range $= \{y \mid y \geq 0\}$.

Exercise Set 3.3, pp. 151–152
1. No   **3.** Yes   **5.** Yes   **7.** No   **9.** Yes
11. (a) 0; (b) 1; (c) 57; (d) $5t^2 + 4t$; (e) $5t^2 - 6t + 1$;
(f) $10a + 5h + 4$ **13.** (a) 5; (b) -2; (c) -4;
(d) $4|y| + 6y$; (e) $2|a + h| + 3a + 3h$;
(f) $\dfrac{2|a + h| + 3h - 2|a|}{h}$

15. (a) 1; (b) does not exist; (c) $-\dfrac{3}{5}$; (d) $-\dfrac{8}{9}$;

(e) $\dfrac{x + h}{2 - x - h}$; (f) $\dfrac{2}{(2 - x)(2 - x - h)}$

17. (a) $\dfrac{2}{3}$; (b) $\dfrac{10}{9}$; (c) 0; (d) does not exist;

(e) $\dfrac{h^2 - 3h}{2h^2 - 3h - 5}$;

(f) $\dfrac{a^2 + 2ab + b^2 - a - b - 2}{2a^2 + 4ab + 2b^2 - 5a - 5b - 3}$ **19.** $2a + h$

21. $f(0)$ does not exist as a real number; $f(2) = 2 + \sqrt{3}$;
$f(10) = 10 + 3\sqrt{11}$ **23.** All real numbers

25. $\{x \mid x \neq 0\}$ **27.** $\left\{x \mid x \geq -\dfrac{4}{7}\right\}$ **29.** $\{x \mid x \neq 2, -2\}$

31. $f(-1) = 2, f(7) = 9, f(5) = -6, f(-3) = 4$;
domain $= \{-1, -3, 5, 7\}$; range $= \{2, 4, -6, 9\}$

33. $-4 + 3i$ **35.** $\dfrac{-1}{x(x + h)}$ **37.** $\dfrac{1}{\sqrt{x + h} + \sqrt{x}}$

39. $\left\{x \mid x \neq -\dfrac{3}{4}, 2\right\}$ **41.** $\{x \mid x \neq 0, -2, 1\}$

43. $\{x \mid x \neq 2, -1 \text{ and } x \geq -3\}$ **45.** All real numbers

Margin Exercises, Section 3.4
1.   **2.**

3.   **4.**

5. $m = 2$ **6.** $m = -2$ **7.** $m = 6$ **8.** $m = -1$

9. 0   **10.** m is not defined   **11.** $-\dfrac{12}{41}$

12. $y = -3x - \dfrac{23}{4}$

13. $y = \dfrac{x}{4} - 9$ **14.** $y = -\dfrac{x}{2} + \dfrac{5}{2}$ **15.** $y = -3x + 7$

16. $y = -\dfrac{10x}{3} + 4$ **17.** $m = -7; b = 11$

18. $m = 0; b = -4$ **19.** (a) $y = \dfrac{2}{3}x + 2$; (b) $m = \dfrac{2}{3}; b = 2$

20.   **21.**

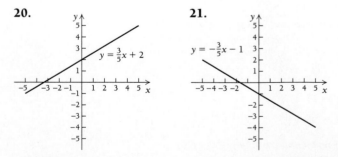

22. (a) $P(0) = 1, P(5) = 1\frac{5}{33}, P(10) = 1\frac{10}{33}, P(33) = 2$,
$P(200) = 7\frac{2}{33}$;

(b)

$$P(d) = 1 + \frac{1}{33}d$$

(c) Pressure can only be nonnegative. Thus the function is meaningful only for values of d for which $1 + \frac{1}{33}d \geq 0$. Thus $d \geq -33$. Since d also can only be nonnegative, the domain is all real numbers greater than or equal to zero.
23. No   **24.** Yes   **25.** Perpendicular   **26.** Neither
27. Parallel
28. Parallel: $y - 4 = -2(x - 3)$, or $y = -2x + 10$; perpendicular: $y - 4 = \frac{1}{2}(x - 3)$, or $y = \frac{1}{2}x + \frac{5}{2}$

Technology Connection, Section 3.4
TC 1.   **TC 2.**

TC 3.

Exercise Set 3.4, pp. 164–166
1.   **3.**

5.

7.

9. $\frac{1}{8}$ **11.** $\frac{1}{2}$ **13.** Not defined
15. Grade is 6.7%, $y = 0.067x$ **17.** $\frac{2}{5}$ ft, or 4.8 in.
19. $y = 4x - 10$ **21.** $y = 2x - 5$ **23.** $y = -\frac{2}{3}x + \frac{13}{3}$
25. $y = -8$ **27.** $y = \frac{1}{2}x + \frac{7}{2}$ **29.** $y = 6x + 17$
31. $y = 6$ **33.** $m = 2; b = 3$ **35.** $m = -3; b = 5$
37. $m = \frac{3}{4}; b = -3$ **39.** $m = 0; b = -\frac{10}{3}$
41.   **43.**

45.   **47.**

49. $y = 3.516x - 13.1602$
51. (a) $T(5) = 70°C$, $T(20) = 220°C$, $T(1000) = 10,020°C$;
(b)

$$T(d) = 10d + 20$$

(c) The domain is $\{d \mid 0 \leq d \leq 5600\}$.
53. (a) $V(0) = \$5200$, $V(1) = \$4687.50$, $V(2) = \$4175$, $V(3) = \$3662.50$, $V(8) = \$1100$;

(b)

$$V(t) = \$5200 - \$512.50t$$

55. Neither **57.** Perpendicular
59. $y = 3x + 3, \ y = -\frac{1}{3}x + 3$
61. $y = \frac{5}{2}x + \frac{5}{2}, \ y = -\frac{2}{5}x - \frac{31}{5}$ **63.** $x = 3, \ y = -3$
65. $f(x) = mx$ **67.** False **69.** False **71.** Yes
73. $\overline{AB}, \overline{DC}$: same slope; $\overline{BC}, \overline{AD}$: same slope
75. $F = \frac{9}{5}C + 32$
77. $P = mQ + b, \ m \neq 0$. Then we can solve for Q:

$Q = \dfrac{P}{m} - \dfrac{b}{m}.$ **79.** $y = -\frac{7}{3}x + \frac{22}{3}, \ y = \frac{3}{7}x - \frac{26}{7}$

81. $k = -\frac{31}{4}$ **83.** $y = \frac{5}{4}x + \frac{75}{8}$
85.

The midpoints are $X\left(\dfrac{a_1}{2}, \dfrac{a_2}{2}\right)$, $Y\left(\dfrac{a_1 + b_1}{2}, \dfrac{a_2 + b_2}{2}\right)$,

$Z\left(\dfrac{b_1 + c_1}{2}, \dfrac{b_2 + c_2}{2}\right)$, and $W\left(\dfrac{c_1}{2}, \dfrac{c_2}{2}\right)$. Then $m_{XY} = \dfrac{b_2}{b_1} = m_{ZW}$

and $\overline{XY} \| \overline{ZW}$. Also $m_{XW} = \dfrac{a_2 - c_2}{a_1 - c_1} = m_{YZ}$ and $\overline{XW} \| \overline{YZ}$. Thus

$XYZW$ is a parallelogram.

Margin Exercises, Section 3.5
1.

2.

3. (a), (b), (c), and (e) **4.** (a) $(-1, 3)$; (b) $(1, 4)$

5. (a) $(-2, 3)$; (b) $(0, 1)$; (c) $\left(-\dfrac{1}{4}, \sqrt{2}\right)$

6. (a) $[-1, 4]$; (b) $[-1, 4]$; (c) $(-1, 4)$; (d) $(-1, 4)$

7. (a) $\left[4, 5\frac{1}{2}\right]$; (b) $(-3, 0]$; (c) $\left[-\dfrac{1}{2}, \dfrac{1}{2}\right)$; (d) $(-\pi, \pi)$
8. (a) $(-\infty, 5]$; (b) $(4, \infty)$; (c) $(-\infty, 4.8)$; (d) $[3, \infty)$
9. (a) $[8, \infty)$; (b) $(-\infty, -7)$; (c) $(10, \infty)$; (d) $(-\infty, 0.78]$
10. (a) Increasing; (b) increasing; (c) decreasing;
(d) neither **11.** Increasing: $[-3, 0]$; decreasing: $[0, 3]$;
there are many answers.

12.

13.

14. (a) $4, -2, 1, 3, -3, -1$;

(b)

$f(x) = \text{INT}(x) + 1$

(c)

$f(x) = \text{INT}(x + 1)$

15. 1076.6 ft/sec, 1147.0 ft/sec, 1183.1 ft/sec
16. $V \approx 284.3$ in^3
17. $A(x) = x(40 - x) = 40x - x^2$, domain $= \{x | 0 < x < 40\}$
18. $d(t) = 40\sqrt{9t^2 + 100}$

Technology Connection, Section 3.5
TC 1. Increasing: $(-\infty, 1], [3, \infty)$; decreasing: $[1, 3]$
TC 2. Increasing: $(-\infty, -2], [-1.15, 1.15], [2, \infty)$;
decreasing: $[-2, -1.15], [1.15, 2]$
TC 3. **TC 4.**

Exercise Set 3.5, pp. 175–179

1.

3.

5.

7.

9.

11.

13.

15.

17. No **19.** Yes **21.** No **23.** Yes **25.** $(0, 5)$
27. $[-9, -4)$ **29.** $[x, x + h]$ **31.** (p, ∞) **33.** $[-3, 3]$
35. $[-14, -11)$ **37.** $(-\infty, -4]$ **39.** $(-\infty, 3.8)$
41. **(a)** Increasing; **(b)** neither; **(c)** decreasing; **(d)** neither

43.

45.

47.

49.

51.

$f(x) = \text{INT}(x - 2)$

53.

$f(x) = \text{INT}(x) + 2$

55. **(a)** \$4.74, \$5.05, \$5.81; **(b)** 2014
57. **(a)** 645 m above sea level; **(b)** at sea level
59. **(a)** $A(L) = L(L - 4)$, or $L^2 - 4L$;
(b) $A(W) = W(W + 4)$, or $W^2 + 4W$
61. $A(x) = x(17 - x)$, or $17x - x^2$
63. $A(x) = x\sqrt{256 - x^2}$

65. $V(x) = 5x^2 - \dfrac{1}{2}x^3$ **67.** $SA(x) = x^2 + \dfrac{432}{x}$

69. $d(s) = \dfrac{14}{s}$

71. **(a)** $V(r) = \dfrac{4}{3}\pi r^3 + 6\pi r^2$; **(b)** $S(r) = 4\pi r^2 + 12\pi r$

73. $C(x) = 3000(4 - x) + 5000(\sqrt{1 + x^2})$
75.

77. $\{x \mid 4 \leq x < 5\}$
79. **(a)** 182; **(b)** 171

81. No

$\text{INT}(y) = \text{INT}(x)$

83.

$f(x) = \begin{cases} |x|x - x^2|, & x < 0 \\ \dfrac{x^3}{1 + x}, & x \geq 0 \end{cases}$

DOT MODE

$[-5, 5] \times [-10, 10]$

Margin Exercises, Section 3.6

1. (a) $(-3, 2)$; **(b)** $(4, -5)$ **2. (a)** $(4, -3)$; **(b)** $(3, 5)$

3. x-axis: no; y-axis: yes   **4.** x-axis: yes; y-axis: no
5. x-axis: yes; y-axis: yes   **6.** x-axis: yes; y-axis: yes
7. a-axis: no; b-axis: no   **8.** p-axis: no; q-axis: no
9. (a) $(-3, -2)$; **(b)** $(4, -3)$; **(c)** $(5, 7)$

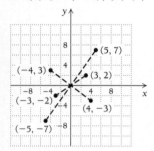

10. Yes   **11.** Yes   **12.** Yes   **13.** Yes   **14.** Yes
15. No   **16. (a)** Yes; **(b)** yes; **(c)** no; **(d)** yes
17. (a) No; **(b)** yes; **(c)** yes; **(d)** yes
18. (a) Odd; **(b)** odd; **(c)** even; **(d)** odd
19. (a) Neither; **(b)** even; **(c)** odd; **(d)** neither; **(e)** neither

Technology Connection, Section 3.6

1. y-axis   **2.** x-axis   **3.** Origin   **4.** x-axis, y-axis, origin
5. x-axis, y-axis, origin

Exercise Set 3.6, pp. 188–189

1. x-axis, no; y-axis, yes; origin, no
3. x-axis, no; y-axis, yes; origin, no   **5.** All yes   **7.** All yes
9. All no   **11.** All no   **13.** Yes   **15.** Yes   **17.** Yes
19. Yes   **21.** No   **23.** No   **25.** Yes   **27.** Yes
29. (a) Even; **(b)** even; **(c)** odd; **(d)** neither   **31.** Neither
33. Even   **35.** Neither   **37.** Even
39. Odd   **41.** Neither   **43.** Odd   **45.** Even and odd
47.   **49.**

     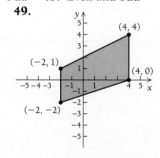

51. Let $E[x, y]$ represent an equation in x and y; let $\sim$
mean "is equivalent to"; and let w.r.t. mean "with respect to."

If the graph is symmetric w.r.t. the x-axis and origin,
i.e., if $E[x, -y] \sim E[x, y]$ and $E[-x, -y] \sim E[x, y]$, then
$E[x, -y] \sim E[-x, -y]$ which becomes, by symmetry w.r.t.
the origin, $E[-x, y] \sim E[x, y]$; hence the graph is symmetric
w.r.t. the y-axis.

   If the graph is symmetric w.r.t. the y-axis and origin,
i.e., if $E[-x, y] \sim E[x, y]$ and $E[-x, -y] \sim E[x, y]$, then
$E[-x, y] \sim E[-x, -y]$ which becomes, by symmetry w.r.t.
the origin, $E[x, -y] \sim E[x, y]$; hence the graph is symmetric
w.r.t. the x-axis.

   If the graph is symmetric w.r.t. both axes, i.e., if
$E[x, -y] \sim E[x, y]$ and $E[-x, y] \sim E[x, y]$, then
$E[x, -y] \sim E[-x, y]$; hence the graph is symmetric w.r.t.
the origin.
53. Odd   **55.** Neither   **57.** Even

Margin Exercises, Section 3.7

1. (a) $(f + g)(5) = \frac{1}{3}$, $(f + g)(2)$ does not exist;
(b) $(f - g)(7) = -\frac{119}{5}$, $(f - g)(-3) = \frac{79}{5}$;
(c) $fg(-3) = \frac{16}{5}$, $gg(1) = 576$; **(d)** $(f/g)(1) = \frac{1}{24}$,
$(g/f)(5) = 0$, $(f/g)(5)$ does not exist, $(f/g)(-5)$ does not
exist; **(e)** all reals except 2, all reals, all reals except 2,
all reals except 2, all reals except 2, all reals except 2, 5, and
-5 **2. (a)** $2x^2 - 6$; **(b)** 12; **(c)** $x^4 - 6x^2 - 27$;
(d) $\dfrac{x^2 + 3}{x^2 - 9}$; **(e)** $x^4 + 6x^2 + 9$; **(f)** all reals,

all reals, all reals, all reals except 3 and -3
3. (a) $P(x) = -0.5x^2 + 40x - 3$; **(b)** $R(40) = 1200$,
$C(40) = 403$, $P(40) = 797$ **4. (a)** $f \circ g(-3) = 13$,
$g \circ f(-3) = 3$; **(b)** $f \circ g(x) = x^2 + 4$,
$g \circ f(x) = x^2 + 10x + 24$ **5. (a)** $f \circ g(x) = \sqrt{x^2 + 3}$,
$g \circ f(x) = x + 3$; **(b)** domain of f: all reals greater than or
equal to -3; domain of g: all reals; domain of $f \circ g$: all reals
6. (a) $K \circ C(F) = \frac{5}{9}(F - 32) + 273$. This function
converts Fahrenheit temperature F to degrees in Kelvin
units; **(b)** $248°$ **7.** Answers may vary.
(a) $f(x) = \sqrt[3]{x + 1}$, $g(x) = x^2$;

(b) $f(x) = \dfrac{1}{x^4}$, $g(x) = x + 5$

Technology Connection, Section 3.7

1.

Exercise Set 3.7, pp. 196–197

1. (a) $(f + g)(x) = 2x + 1$, $(f - g)(x) = -7$,

$fg(x) = x^2 + x - 12$, $ff(x) = x^2 - 6x + 9$, $(f/g)(x) = \dfrac{x - 3}{x + 4}$,

$(g/f)(x) = \dfrac{x + 4}{x - 3}$, $f \circ g(x) = x + 1$, $g \circ f(x) = x + 1$;

(b) all reals, all reals, all reals, all reals, all reals, all reals, all reals except -4, all reals except 3, all reals, all reals

3. (a) $(f + g)(x) = x^3 + 2x^2 + 9x - 3$,

$(f - g)(x) = x^3 - 2x^2 - 9x + 3$, $fg(x) = 2x^5 + 9x^4 - 3x^3$,

$ff(x) = x^6$, $(f/g)(x) = \dfrac{x^3}{2x^2 + 9x - 3}$,

$(g/f)(x) = \dfrac{2x^2 + 9x - 3}{x^3}$,

$f \circ g(x) = (2x^2 + 9x - 3)^3$, $g \circ f(x) = 2x^6 + 9x^3 - 3$;

(b) all reals, all reals, all reals, all reals, all reals,

all reals except $\dfrac{-9 \pm \sqrt{105}}{4}$, all reals except 0, all reals,

all reals   **5.** -6 **7.** $x^2 - 2x - 9$ **9.** 55

11. Does not exist   **13.** $2x^3 + 5x^2 - 8x - 20$

15. $\dfrac{2x + 5}{x^2 - 4}$ **17.** $4x^2 + 20x + 21$ **19.** $4x + 15$

21. (a) $P(x) = -0.4x^2 + 57x - 13$; **(b)** $R(20) = 1040$,

$C(20) = 73$, $P(20) = 967$ **23.** $f \circ g(x) = x$, $g \circ f(x) = x$

25. $f \circ g(x) = x$, $g \circ f(x) = x$ **27.** $f \circ g(x) = x$,

$g \circ f(x) = x$ **29.** $f \circ g(x) = |x|$, $g \circ f(x) = x$

31. $f \circ g(x) = x$, $g \circ f(x) = x$ **33.** $f \circ g(x) = -6$,

$g \circ f(x) = 12$ **35.** $f(x) = x^5$, $g(x) = 4 - 3x$

37. $f(x) = \dfrac{1}{x^4}$, $g(x) = x - 1$ **39.** $f(x) = \dfrac{x - 1}{x + 1}$,

$g(x) = x^3$ **41.** $f(x) = x^6$, $g(x) = \dfrac{2 + x^3}{2 - x^3}$

43. $f(x) = \sqrt{x}$, $g(x) = \dfrac{x - 5}{x + 2}$

45. $f(x) = x^5 + x^4 + x^3 - x^2 + 4x$, $g(x) = x + 3$

47. (a) $a(t) = 250t$; **(b)** $P(a) = 300 + a$;

(c) $P \circ a(t) = 300 + 250t$. This function gives the distance
of the plane from the control tower in terms of the time t
that the plane travels.

49.

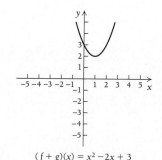

$(f + g)(x) = x^2 - 2x + 3$

51.

$(f + g)(x) = 1 - x^2 + \sqrt{x}$

53. $b = -2$ **55.** $f \circ f(x) = \dfrac{x - 1}{x}$, $f \circ f \circ f(x) = x$

57. Let $f(x)$ and $g(x)$ be even functions. Then by definition,
$f(x) = f(-x)$ and $g(x) = g(-x)$. Thus,
$(f + g)(x) = f(x) + g(x) = f(-x) + g(-x) = (f + g)(-x)$
and $f + g$ is even.   **59.** Let $f(x)$ and $g(x)$ be odd functions.
Then by definition, $-f(x) = f(-x)$, or $f(x) = -f(-x)$, and
$-g(x) = g(-x)$, or $g(x) = -g(-x)$. Thus,
$f \circ g(x) = f(g(x)) = f(-g(-x)) = -f(g(-x)) = -f(-g(x))$
and $f \circ g$ is odd.   **61.** Let f and g be increasing functions,
and thus by definition for all a and b in the domains of
f and g, if $a < b$, then $f(a) < f(b)$ and $g(a) < g(b)$. Since f
and g are each increasing, $g(a) < g(b)$ and $f(g(a)) < f(g(b))$,
or $f \circ g(a) < f \circ g(b)$. Thus, $f \circ g$ is increasing.
Using the addition property of inequalities, we get
$f(a) + g(a) < f(b) + g(b)$, or $(f + g)(a) < (f + g)(b)$
and $f + g$ is increasing.

63. $O(-x) = \dfrac{f(-x) - f(-(-x))}{2} = \dfrac{f(-x) - f(x)}{2}$,

$-O(x) = -\dfrac{f(x) - f(-x)}{2} = \dfrac{f(-x) - f(x)}{2}$. Thus,

$O(-x) = -O(x)$ and O is odd.

65.

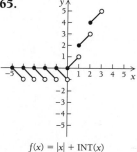

$f(x) = |x| + \text{INT}(x)$

Margin Exercises, Section 3.8

1.

2.

3.

$f(x) = 2x^2$
$f(x) = x^2$
$f(x) = 0.8x^2$

4.

$f(x) = -0.8x^2$
$f(x) = -x^2$
$f(x) = -2x^2$

Exercise Set 3.8, pp. 204–205

1.

$f(x) = |x| - 3$

3.

$f(x) = |x - 1|$

5.

$f(x) = x^2 - 4$
$f(x) = 4 - x^2$

6.

$y = \frac{1}{2}f(x)$

5.

$f(x) = -4|x|$

7.

$f(x) = \frac{1}{3}|x|$

7.

$y = -2f(x)$

8.

$y = f(2x)$

9.

$f(x) = |2x|$

11.

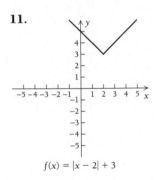

$f(x) = |x - 2| + 3$

9.

$y = f\left(-\frac{1}{2}x\right)$

13.

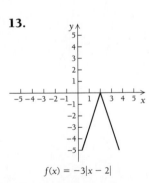

$f(x) = -3|x - 2|$

15.

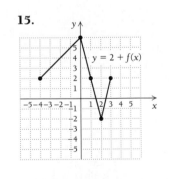

$y = 2 + f(x)$

10.

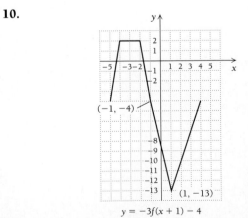

$(-1, -4)$
$(1, -13)$
$y = -3f(x + 1) - 4$

17.

$y = f(x - 1)$

19.

$y = -2f(x)$

21.

$y = \frac{1}{3}f(x)$

23.

$y = f(2x)$

39. The graph is translated 1.8 units to the left, stretched vertically by a factor of $\sqrt{2}$, and reflected across the *x*-axis.

41. $g(x) = x^3 - 3x^2 + 2$

43. $k(x) = (x + 1)^3 - 3(x + 1)^2$

45. .... (1) $y = f(x)$ is the given function.

---------- (2) $y = 3f(2x)$ is the result of shrinking (1) horizontally by a factor of $\frac{1}{2}$ and stretching vertically by a factor of 3.

_____ (3) $y = 3f[2(x + \frac{1}{4})]$, the required graph, is a translation of (2) to the left through $\frac{1}{4}$ unit.

25.

$y = f(-2x)$

27.

$y = f(\frac{x}{-2})$

(3) $y = 3f[2(x + \frac{1}{4})]$

(2) $y = 3f(2x)$

(1) $y = f(x)$

47.

$f(x) = x^2|x|$

$[-12, 12] \times [-30, 30]$

29.

$y = f(x - 2) + 3$

31.

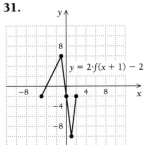

$y = 2 \cdot f(x + 1) - 2$

(a)

$f(x) = (4.7x)^2|4.7x|$

$[-12, 12] \times [-30, 30]$

(b)

$f(x) = -2.5x^2|x|$

$[-12, 12] \times [-30, 30]$

33.

$y = -\frac{1}{2}f(x - 3) + 2$

35.

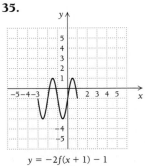

$y = -2f(x + 1) - 1$

(c)

$f(x) = x^2|x| - 11.5$

$[-12, 12] \times [-30, 30]$

(d)

$f(x) = (x + 9.9)^2|x + 9.9|$

$[-12, 12] \times [-30, 30]$

Review Exercises: Chapter 3, pp. 206–207

1. [3.3] No   **2.** [3.1] $\{3, 5, 7\}$

3. [3.1] $\{1, 3, 5, 7\}$

4. [3.1]

$x = |y|$

5. [3.1]

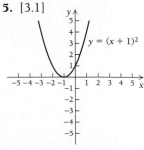

$y = (x + 1)^2$

37.

$y = \frac{5}{2}f(x - 3) - 2$

6. [3.5]

$g(x) = |x| - 2$

7. [3.5]

$f(x) = \sqrt{x}$

16. [3.4] $m = -2$; y-intercept: $(0, -7)$ **17.** [3.4] -1
18. [3.4] $y = 3x + 5$ **19.** [3.4] $y = \frac{1}{3}x - \frac{1}{3}$
20. [3.6] (b), (d), (f)   **21.** [3.6] (a), (b), (d), (g)
22. [3.6] (b), (c), (d), (h)
23. [3.7] $P(x) = -0.5x^2 + 105x - 6$ **24.** [3.5] (b)
25. [3.3] -3 **26.** [3.3] 9   **27.** [3.3] $2a + h - 1$
28. [3.3] 0   **29.** [3.3] 4   **30.** [3.3] $2\sqrt{a + 1}$
31. [3.3] $\{x | x \le \frac{7}{3}\}$ **32.** [3.3] $\{x | x \ne 1, 5\}$

33. [3.7] **(a)** $(f + g)(x) = 3 - 2x + \frac{4}{x^2}$,

$(f - g)(x) = \frac{4}{x^2} - 3 + 2x, fg(x) = \frac{12}{x^2} - \frac{8}{x}$,

$(f/g)(x) = \frac{4}{3x^2 - 2x^3}, f \circ g(x) = \frac{4}{(3 - 2x)^2}$,

$g \circ f(x) = 3 - \frac{8}{x^2}$; **(b)** all reals except 0, all reals,

all reals except 0, all reals except 0, all reals except 0,
all reals except 0 and $\frac{3}{2}$, all reals except $\frac{3}{2}$, all reals except 0

34. [3.7] **(a)** $(f + g)(x) = 3x^2 + 6x - 1$,
$(f - g)(x) = 3x^2 + 2x + 1, fg(x) = 6x^3 + 5x^2 - 4x$,

$(f/g)(x) = \frac{3x^2 + 4x}{2x - 1}, f \circ g(x) = 12x^2 - 4x - 1$,

$g \circ f(x) = 6x^2 + 8x - 1$; **(b)** all reals, all reals, all reals,
all reals, all reals, all reals except $\frac{1}{2}$, all reals, all reals

8. [3.5]

$f(x) = \sqrt{x - 2}$

9. [3.5]

$f(x) = 2\sqrt{x + 3}$

10. [3.5]

$f(x) = \frac{1}{2}\sqrt{x - 1} + 2$

11. [3.5]

$f(x) = \text{INT}(x)$

35. [3.8] $y = 1 + f(x)$

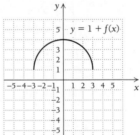

$y = 1 + f(x)$

36. [3.8] $y = \frac{1}{2}f(x)$

$y = \frac{1}{2}f(x)$

37. [3.8] $y = f(x + 1)$

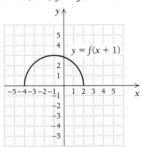

$y = f(x + 1)$

12. [3.5]

$f(x) = \text{INT}(x) - 3$

13. [3.5]

$f(x) = \text{INT}(x - 3)$

14. [3.4]

$5y - 2x = 10$

15. [3.4]

$y = -\frac{2}{3}x - 4$

38. [3.6] (a), (b), (c)
39. [3.6] (e), (f)
40. [3.6] (d)
41. [3.2] $\sqrt{34}$
42. [3.2] $(\frac{1}{2}, \frac{11}{2})$
43. [3.4] $y = -\frac{2}{3}x - \frac{1}{3}$
44. [3.4] $y = \frac{3}{2}x - \frac{5}{2}$
45. [3.4] Parallel   **46.** [3.4] Neither
47. [3.4] Perpendicular
48. [3.2] $(x + 2)^2 + (y - 6)^2 = 13$

49. [3.2] Center: $(-1, 3)$; radius: $\frac{3}{2}$;

$(x + 1)^2 + (y - 3)^2 = \frac{9}{4}$

50. [3.2] Center: $(5, -2)$; radius: $2\sqrt{2}$
51. [3.2] $(x - 3)^2 + (y - 4)^2 = 25$
52. [3.2] $(x - 2)^2 + (y - 4)^2 = 26$ **53.** [3.5] (c)
54. [3.5] (a)   **55.** [3.5] (b)   **56.** [3.5] $[-\pi, 2\pi]$
57. [3.5] $(0, 1]$ **58.** [3.5] $(-\infty, 14)$
59. [3.5]   **60.** [3.5]

61. [3.5] $d(t) = \sqrt{(55t)^2 + (50t)^2} = t\sqrt{5525} = 5t\sqrt{221}$
62. [3.5] $V(a) = 8\pi a^2 - 128$ **63.** [3.7] **(a)** $f(x) = \sqrt{x}$,

$g(x) = 5x + 2$; **(b)** $f(x) = \dfrac{x + 1}{x - 1}$, $g(x) = x^3$

64. [3.3] $\{x \mid x \neq 0, 3, -3\}$ **65.** [3.3] $\{x \mid x < 0\}$
66. [3.8] Graph $y = f(x)$. Then reflect the portion that lies below the x-axis, across the x-axis.
67. [3.1]

$|x - y| = 1$

Test: Chapter 3, pp. 207–209
1. [3.3] Yes   **4.** [3.5]
2. [3.1] $\{-2, 2, -7, 7\}$
3. [3.1] $\{-2, 0, 2, 7\}$

$f(x) = (x - 2)^2$

5. [3.1]

$x = |y + 1|$

6. [3.5]

$f(x) = 2\,\text{INT}(x)$

7. [3.6] (a), (d), (e)   **8.** [3.6] (b), (e)
9. [3.3] $\{x \mid x \neq -4, 4\}$
10. [3.7] $(f + g)(x) = x^2 + 2x + 5$,
$(f - g)(x) = -x^2 + 2x - 7$, $fg(x) = 2x^3 - x^2 + 12x - 6$,

$(f/g)(x) = \dfrac{2x - 1}{x^2 + 6}$, $f \circ g(x) = 2x^2 + 11$,

$g \circ f(x) = 4x^2 - 4x + 7$ **11.** [3.7] All reals, all reals, all reals, all reals, all reals, all reals, all reals, all reals
12. [3.5] (b)   **13.** [3.3] 4   **14.** [3.3] 2
15. [3.3] $a^2 - 3a + 4$ **16.** [3.3] $2a - 1 + h$
17. [3.7] $P(x) = -0.1x^2 + 100x - 20$ **18.** [3.4] $m = -\frac{3}{7}$,
y-intercept is $\frac{10}{7}$ **19.** [3.4] $y = 5x + 12$
20. [3.4] $y = 2x - 7$ **21.** [3.2] $\sqrt{13} \approx 3.606$
22. [3.2] $(\frac{11}{2}, -\frac{3}{2})$ **23.** [3.4] Perpendicular
24. [3.4]

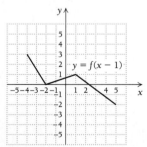

$y = \frac{3}{2}x - 1$

25. [3.4] $y = \frac{1}{3}x - \frac{14}{3}$
26. [3.2]
$(x + 2)^2 + (y - 4)^2 = 16$
27. [3.2] Center:
$(1, -3)$; radius: $\sqrt{5}$

28. [3.8] **(a)**   **(b)**

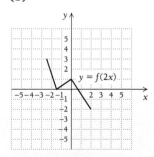

$y = f(x - 1)$ $y = f(2x)$

(c)   **29.** [3.6] (a), (c), (e)

$y = 3 + f(x)$

30. [3.6] (b), (f)
31. [3.6] (d)
32. [3.5] $(-7, 2)$
33. [3.5] $[-2, \infty)$
34. [3.5] (b)
35. [3.5] (c)

36. [3.5]

37. [3.5] $A(x) = 2x\sqrt{4 - x^2}$ **38.** [3.7] **(a)** $f(x) = \dfrac{1}{\sqrt{x}}$,
$g(x) = 7x + 2$; **(b)** $f(x) = 4x^2 + 9$, $g(x) = 5x - 1$
39. [3.7] Let $f(x)$ and $g(x)$ be odd functions.
Then by definition, $f(-x) = -f(x)$, or $f(x) = -f(-x)$, and
$g(-x) = -g(x)$, or $g(x) = -g(-x)$. Thus,
$(f + g)(x) = f(x) + g(x) = -f(-x) + [-g(-x)] = -[f(-x) + g(-x)] = -(f + g)(-x)$ and $f + g$ is odd.

CHAPTER 4

Margin Exercises, Section 4.1

1. (a)

(b) up; **(c)** y-axis, $x = 0$; **(d)** 0; **(e)** $(0, 0)$
2. (a)

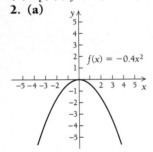

(b) down; **(c)** y-axis, $x = 0$; **(d)** 0; **(e)** $(0, 0)$
3. (a) and (b)

(c) $(2, 0)$; **(d)** $x = 2$; **(e)** 0; **(f)** up;
(g) horizontal translation to the right
4. (a) and (b)

(c) $(-2, 0)$; **(d)** $x = -2$;
(e) 0; **(f)** down
(g) horizontal translation to the left

5. (a) $(2, 4)$; **(b)** $x = 2$;
(c) no; **(d)** yes, 4
6. (a) $(-2, -1)$;
(b) $x = -2$; **(c)** yes, -1;
(d) no

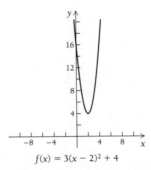

7. (a) $(5, \pi)$; **(b)** $x = 5$; **(c)** no; **(d)** yes, π
8. (a) $(5, 0)$; **(b)** $x = 5$; **(c)** yes, 0; **(d)** no
9. (a) $(-\frac{1}{4}, -6)$; **(b)** $x = -\frac{1}{4}$; **(c)** no; **(d)** yes, -6
10. (a) $(-9, 3)$; **(b)** $x = -9$; **(c)** yes, 3; **(d)** no
11. $f(x) = (x - 2)^2 + 3$ **12.** $f(x) = 3(x + 4)^2 - 38$
13. (a) $(\frac{3}{2}, -14)$, $x = \frac{3}{2}$; **(b)** -14 is a minimum
14. Vertex: $(-3, 34)$; maximum: 34
15.   **16.**

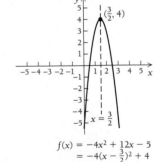

17. $(1 - \sqrt{6}, 0), (1 + \sqrt{6}, 0)$ **18.** $(-1, 0), (3, 0)$
19. $(-4, 0)$ **20.** None
21. Maximum height 12.4 m, at $t = \frac{2}{7}$. Reaches ground in 1.88 sec.
22. Maximum area: 400 ft²; length: 20 ft; width: 20 ft

Technology Connection, Section 4.1
TC 1. Vertex: $(-3.17, -28.57)$, $x = -3.17$
TC 2. Vertex: $(-1.30, 13.41)$, $x = -1.30$

TC 3. Vertex: $(-0.48, -9.90)$, $x = -0.48$
TC 4. $(1.13, 0)$, $(3.57, 0)$ **TC 5.** $(3.10, 0)$
TC 6. Do not exist   **TC 7.** 6.7 sec; 282.3 m; 14.2 sec
TC 8. 986.0 ft^2; 31.4 ft by 31.4 ft

Exercise Set 4.1, pp. 222–223
1. (a) $(0, 0)$; (b) $x = 0$; (c) 0 is a minimum
3. (a) $(9, 0)$; (b) $x = 9$; (c) 0 is a maximum
5. (a) $(1, -4)$; (b) $x = 1$; (c) -4 is a minimum
7. (a) $f(x) = -(x - 1)^2 + 4$;
(b) $(1, 4)$; (c) 4 is a maximum
9. (a) $f(x) = [x - (-\frac{3}{2})]^2 - \frac{9}{4}$;
(b) $(-\frac{3}{2}, -\frac{9}{4})$; (c) $-\frac{9}{4}$ is a minimum
11. (a) $f(x) = -\frac{3}{4}(x - 4)^2 + 12$;
(b) $(4, 12)$; (c) 12 is a maximum
13. (a) $f(x) = 3[x - (-\frac{1}{6})]^2 - \frac{49}{12}$;
(b) $(-\frac{1}{6}, -\frac{49}{12})$; (c) $-\frac{49}{12}$ is a minimum
15. Maximum: 0   **17.** Minimum: 5   **19.** Maximum: $-\frac{207}{16}$
21. Minimum: $-\frac{3}{4}$ **23.** Maximum: $\$\dfrac{435{,}625}{3}$

25.

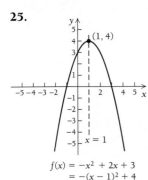

$f(x) = -x^2 + 2x + 3$
$= -(x - 1)^2 + 4$

27.

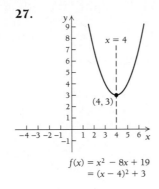

$f(x) = x^2 - 8x + 19$
$= (x - 4)^2 + 3$

29.

$f(x) = -\frac{1}{2}x^2 - 3x + \frac{1}{2}$
$= -\frac{1}{2}(x + 3)^2 + 5$

30.

$f(x) = 3x^2 - 24x + 50$
$= 3(x - 4)^2 + 2$

33. $(3, 0), (-1, 0)$ **35.** $(4 \pm \sqrt{11}, 0)$ **37.** None
39. 25 and 25   **41.** Base: 10 cm; height: 10 cm
43. 30 yd by 60 yd; maximum area $= 1800$ yd^2
45. 13.5 ft by 13.5 ft; 182.25 ft^2
47. 46 units; maximum profit $= \$1048$
49. 70 units; maximum profit $= \$19$
51. (a) 1662.5 m after 15 sec; (b) after 33.420 sec

53.

$f(x) = |x^2 - 1|$

55. Minimum, -6.95
57. $a = \frac{3}{4}$
59. $c = -\frac{945}{4}$
61. $\left(\dfrac{1}{q}, \dfrac{q^2 - 1}{q}\right)$
63. $c = 9$

65. $\$5.75$; 72,500   **67.** 25
69. The minimum occurs when $x = 24\pi/(x + 4)$, or approximately 10.6 in.
71. Vertex: $(0.9, -10.0)$; x-intercepts: $(-0.8, 0)$, $(2.6, 0)$
73. (a) 12.9 sec; (b) 920.6 m; (c) 26.6 sec

Margin Exercises, Section 4.2
1. $\{-3, 4\}$ **2.** $\{2, d, e\}$ **3.**

4.

5.

6. $\varnothing$; nothing to graph   **7.** $\{1, 2, 3, 4, 5\}$
8. $\{-3, -4, 1, 2, 3, 4, 8, 9, 11\}$ **9.** $\{1, 2, a, b, c, d, e\}$
10.   **11.**

12.   **13.**

14. $4 < x < 8$ **15.** $-3 \le x < 0$ **16.** $-5 \le x \le -2$
17. $-1 < x \le -\frac{1}{4}$ **18.** $-\frac{1}{2} < x$ and $x < 1$
19. $-\frac{17}{3} \le x$ and $x < -2$ **20.** $\frac{19}{4} \le x$ and $x \le \frac{37}{6}$
21. $\{x | -\frac{2}{3} < x \le 4\}$, or $(-\frac{2}{3}, 4]$

22. $\{x | 1 < x < 4\}$, or $(1, 4)$

23. $\{x | -\frac{1}{3} \le x \le \frac{1}{6}\}$, or $[-\frac{1}{3}, \frac{1}{6}]$

24. $\{x | x < -7$ or $x > -1\}$, or $(-\infty, -7) \cup (-1, \infty)$

25. $\{x | x \le -1$ or $x > 4\}$, or $(-\infty, -1] \cup (4, \infty)$

26. $\{x | x \ge \frac{5}{3}$ or $x \le 1\}$, or $(-\infty, 1] \cup [\frac{5}{3}, \infty)$

27. $\{x | x < -\frac{11}{4}$ or $x \ge \frac{1}{4}\}$, or $(-\infty, -\frac{11}{4}) \cup [\frac{1}{4}, \infty)$

Technology Connection, Section 4.2
TC 1. $(-8, 6]$ **TC 2.** $(4, 8)$ **TC 3.** $[-3.5, 1.5]$
TC 4. $(-\infty, 5) \cup (17, \infty)$ **TC 5.** $(-\infty, -2.5] \cup [2, \infty)$
TC 6. $(-\infty, 3] \cup (7, \infty)$

Exercise Set 4.2, p. 229
1. $\{3, 4, 5\}$ **3.** $\{0, 2, 4, 6, 8, 9\}$ **5.** $\{c\}$
7.   **9.**

11.   **13.**

15. $\varnothing$ **17.** $\{x | -3 \le x < 3\}$, or $[-3, 3)$
19. $\{x | 8 \le x \le 10\}$, or $[8, 10]$
21. $\{-7\}$ **23.** $\{x | -\frac{3}{2} < x < 2\}$, or $(-\frac{3}{2}, 2)$
25. $\{x | 1 < x \le 5\}$, or $(1, 5]$
27. $\{x | -\frac{11}{3} < x \le \frac{13}{3}\}$, or $(-\frac{11}{3}, \frac{13}{3}]$
29. $\{x | x \le -2 \text{ or } x > 1\}$, or $(-\infty, -2] \cup (1, \infty)$
31. $\{x | x \le -\frac{7}{2} \text{ or } x \ge \frac{1}{2}\}$, or $(-\infty, -\frac{7}{2}] \cup [\frac{1}{2}, \infty)$
33. $\{x | x < 9.6 \text{ or } x > 10.4\}$, or $(-\infty, 9.6) \cup (10.4, \infty)$
35. $\{x | x \le -\frac{57}{4} \text{ or } x \ge -\frac{55}{4}\}$, or $(-\infty, -\frac{57}{4}] \cup [-\frac{55}{4}, \infty)$
37. $\{w | 4.885 \text{ cm} < w < 53.67 \text{ cm}\}$
39. $\{S | 97\% \le S \le 100\%\}$; yes   **41.** $1981\frac{2}{5}° \le F < 4676°$
43. $\{d | 62\frac{2}{9} \text{ mi} < d < 136\frac{8}{27} \text{ mi}\}$ **45.** $\{1\}$
47. $\{x | x > -\frac{1}{5}\}$, or $(-\frac{1}{5}, \infty)$
49. $\{x | x \ge -\frac{3}{2}\}$, or $[-\frac{3}{2}, \infty)$ **51.** $\{x | x > 2\}$, or $(2, \infty)$
53. $(-5.2, 9]$ **55.** $(-\infty, -0.2) \cup [6.2, \infty)$

Margin Exercises, Section 4.3
1. $\{-5, 5\}$ **2.** $\{-3.5, 3.5\}$

3. $\{x | -5 < x < 5\}$, or $(-5, 5)$

4. $\{x | -3.5 \le x \le 3.5\}$, or $[-3.5, 3.5]$

5. $\{x | x \le -5 \text{ or } x \ge 5\}$, or $(-\infty, -5] \cup [5, \infty)$

6. $\{x | x < -3.5 \text{ or } x > 3.5\}$, or $(-\infty, -3.5) \cup (3.5, \infty)$

7. $\{-5, 3\}$ **8.** $\{x | -9 < x < -5\}$, or $(-9, -5)$
9. $\{x | x < 1 \text{ or } x > 5\}$, or $(-\infty, 1) \cup (5, \infty)$
10. $\{-1, \frac{11}{3}\}$ **11.** $\{x | -\frac{3}{10} < x < \frac{1}{10}\}$, or $(-\frac{3}{10}, \frac{1}{10})$
12. $\{x | x < -1 \text{ or } x > \frac{11}{3}\}$, or $(-\infty, -1) \cup (\frac{11}{3}, \infty)$
13. $\{-\frac{2}{3}, 8\}$ **14.** $\{-\frac{3}{2}\}$ **15.** $\{x | x \ge -\frac{5}{13}\}$, or $(-\frac{5}{13}, \infty)$
16. $\{x | x \le -\frac{5}{13}\}$, or $(-\infty, -\frac{5}{13}]$

Technology Connection, Section 4.3
TC 1. $\{-3, 15\}$ **TC 2.** $(-\infty, 1.5] \cup [4.5, \infty)$
TC 3. $(-5, 3)$

Exercise Set 4.3, p. 234
1. $\{-7, 7\}$;

3. $\{x | -7 < x < 7\}$, or $(-7, 7)$;

5. $\{x | x \le -4.5 \text{ or } x \ge 4.5\}$, or
$(-\infty, -4.5] \cup [4.5, \infty)$

7. $\{-3, 5\}$ **9.** $\{x | -17 < x < 1\}$, or $(-17, 1)$
11. $\{x | x \le -17 \text{ or } x \ge 1\}$, or $(-\infty, -17] \cup [1, \infty)$
13. $\{x | -\frac{1}{4} < x < \frac{3}{4}\}$, or $(-\frac{1}{4}, \frac{3}{4})$ **15.** $\{-\frac{1}{3}, \frac{3}{3}\}$
17. $\{-1, -\frac{1}{3}\}$ **19.** $\{x | -\frac{1}{3} < x < \frac{1}{3}\}$, or $(-\frac{1}{3}, \frac{1}{3})$
21. $\{x | -6 \le x \le 3\}$, or $[-6, 3]$
23. $\{x | x < 4.9 \text{ or } x > 5.1\}$, or $(-\infty, 4.9) \cup (5.1, \infty)$
25. $\{x | -\frac{7}{3} \le x \le 1\}$, or $[-\frac{7}{3}, 1]$
27. $\{x | -\frac{1}{2} \le x \le \frac{7}{2}\}$, or $[-\frac{1}{2}, \frac{7}{2}]$
29. $\{x | x < -8 \text{ or } x > 7\}$, or $(-\infty, -8) \cup (7, \infty)$
31. $\{x | x < -\frac{7}{4} \text{ or } x > -\frac{3}{2}\}$, or $(-\infty, -\frac{7}{4}) \cup (-\frac{3}{2}, \infty)$
33. $\{x | \frac{3}{8} \le x \le \frac{9}{8}\}$, or $[\frac{3}{8}, \frac{9}{8}]$ **35.** $\varnothing$ **37.** $\varnothing$
39. $\{x | x < -22.2182 \text{ or } x > -12.2158\}$, or
$(-\infty, -22.2182) \cup (-12.2158, \infty)$
41. $\{x | x \le -0.5746 \text{ or } x \ge 7.9277\}$, or
$(-\infty, -0.5746] \cup [7.9277, \infty)$ **43.** $\{\frac{5}{3}, 11\}$
45. $\{0, \frac{24}{7}\}$ **47.** $\{2\}$ **49.** $\{-\frac{3}{8}, \frac{7}{6}\}$
51. $\{x | \frac{3}{5} \le x \le 5\}$, or $[\frac{3}{5}, 5]$
53. $\{x | x \le \frac{3}{5} \text{ or } x \ge 5\}$, or $(-\infty, \frac{3}{5}] \cup [5, \infty)$ **55.** $\{\frac{4}{5}, 2\}$
57. $\{-4, 4\}$ **59.** $(-\infty, \frac{3}{2}]$ **61.** $(-\frac{9}{2}, \frac{11}{2})$
63. $(-\infty, -\frac{8}{3}) \cup (-2, \infty)$ **65.** $\{-\frac{3}{2}, \frac{5}{4}\}$
67. If $a = 0$, then $-|0| \le 0 \le |0|$. If $a < 0$, then $-|a| = a \le a \le |a|$, or $-|a| \le a \le |a|$. If $a > 0$, then
$-|a| = -a \le a \le |a|$, or $-|a| \le a \le |a|$.

69. By the triangle inequality, $|a + b| < |a| + |b| < \frac{e}{2} + \frac{e}{2}$;
hence $|a + b| < e$.
71.

Assume: $a \ne b$. Since x is the midpoint, $|x - a| = |x - b|$, or $x - a = \pm(x - b)$. The $+$ sign gives $a = b$,

not admissible; the $-$ sign gives $x = \dfrac{a + b}{2}$.

73.–81. See Exercises 55–63.

Margin Exercises, Section 4.4
1. $\{x | -3 < x < 1\}$, or $(-3, 1)$
2. $\{x | -3 \le x \le 1\}$, or $[-3, 1]$
3. $\{x | x < -3 \text{ or } x > 1\}$, or $(-\infty, -3) \cup (1, \infty)$

4. $\{x | x \le -3 \ or \ x \ge 1\}$, or $(-\infty, -3] \cup [1, \infty)$
5. $\{x | x < -1 \ or \ 0 < x < 1\}$, or $(-\infty, -1) \cup (0, 1)$
6. $\{x | 2 < x \le \frac{7}{2}\}$, or $(2, \frac{7}{2}]$
7. $\{x | x < 5 \ or \ x > 10\}$, or $(-\infty, 5) \cup (10, \infty)$

Technology Connection, Section 4.4
TC 1. $(-\infty, 1.13] \cup [3.57, \infty)$
TC 2. $(-2.36, -1.00) \cup (1.69, \infty)$
TC 3. $(-0.70, 2.81)$ **TC 4.** $[-0.8, 0)$
TC 5. $(-\infty, -5.36) \cup (3.36, 9)$ **TC 6.** $(-1.14, 0)$

Exercise Set 4.4, pp. 239–240
1. $(-\infty, -5) \cup (3, \infty)$ **3.** $[-2, 1]$ **5.** $(-2, 1)$
7. $(-\infty, -1] \cup [1, \infty)$ **9.** $(-\infty, -3] \cup [3, \infty)$
11. $(-\infty, \infty)$ **13.** $(2, 4)$ **15.** $(-3, \frac{5}{4})$
17. $\left(-\infty, \dfrac{-1 - \sqrt{41}}{4}\right) \cup \left(\dfrac{-1 + \sqrt{41}}{4}, \infty\right)$
19. $(-\infty, -2) \cup (0, 2)$ **21.** $(-3, -1) \cup (2, \infty)$
23. $(-\infty, -3) \cup (-2, 1)$ **25.** $(4, \infty)$ **27.** $(0, \frac{1}{3})$
29. $(-\infty, -\frac{2}{3}) \cup (3, \infty)$ **31.** $[-2, 0)$ **33.** $(\frac{3}{2}, 4]$
35. $(-\infty, -\frac{5}{2}] \cup (-2, \infty)$ **37.** $(-\infty, 0)$
39. $(-\infty, -\frac{11}{7})$ **41.** $(1, \infty)$ **43.** $(0, 2) \cup (2, \infty)$
45. $(-\infty, 0) \cup [1, \infty)$
47. $(-\infty, -3) \cup (-2, -1) \cup (2, \infty)$ **49.** $[-\sqrt{2}, \sqrt{2}]$
51. $(-\infty, \frac{5}{3}) \cup (11, \infty)$ **53.** $\varnothing$
55. $(-\infty, -\frac{1}{4}) \cup (\frac{1}{2}, \infty)$ **57.** $(-\infty, -5] \cup [5, \infty)$
59. $(-\infty, 0) \cup (0, \infty)$ **61.** $(-4, -2) \cup (-1, 1)$
63. $\{h | h > -2 + 2\sqrt{6} \text{ cm}\}$
65. **(a)** $\{x | 10 < x < 200\}$; **(b)** $\{x | 0 < x < 10 \ or \ x > 200\}$
67. **(a)** $10, 35$; **(b)** $\{x | 10 < x < 35\}$;
(c) $\{x | 0 < x < 10 \ or \ x > 35\}$
69. **(a)** $\{k | k > 2 \ or \ k < -2\}$; **(b)** $\{k | -2 < k < 2\}$
71. $\{x | -1 \le x \le 1\}$ **73.** $\{x | x \le -3 \ or \ x \ge 1\}$
75. Roots: $-2, 1, 3$; $f(x) < 0$: $(-\infty, -2) \cup (1, 3)$;
$f(x) > 0$: $(-2, 1) \cup (3, \infty)$
77. No roots; $f(x) < 0$: $(-\infty, 0)$; $f(x) > 0$: $(0, \infty)$
79. Roots: $-2, 1, 2, 3$; $f(x) < 0$: $(-2, 1) \cup (2, 3)$;
$f(x) > 0$: $(-\infty, -2) \cup (1, 2) \cup (3, \infty)$

Margin Exercises, Section 4.5
1.

$f(x) = (x - 1)^3$

2.

$f(x) = -\frac{1}{2}x^3 + 2$

3.

$f(x) = x^4$

4.

$f(x) = -\frac{1}{2}x^5$

5.

$f(x) = x^3 + 3x^2 - x - 3$

6.

$f(x) = x^4 - 10x^2 + 9$

7. **(a)** 165; **(b)**

$N(x) = \frac{1}{6}x^3 + \frac{1}{2}x^2 + \frac{1}{3}x$

Technology Connection, Section 4.5
TC 1.
$f(x) = x^3 - 4x - 2$

$[-10, 10] \times [-10, 10]$

TC 2.
$f(x) = -2x^4 + x^3 - x^2 + 1$

$[-5, 5] \times [-20, 5]$

TC 3.
$f(x) = -3x^3 + x^2 + x$

$[-5, 5] \times [-5, 5]$

Exercise Set 4.5, pp 246–247

1.

$f(x) = \frac{1}{3}x^6$

3.

$f(x) = -0.6x^5$

21.

$f(x) = x^3 - 3x^2 - 4x + 12$

23.

$f(x) = -x^4 - 3x^3 - 3x^2$

5.

$f(x) = (x + 1)^5 - 4$

7.

$f(x) = \frac{1}{4}(x + 1)^4$

25.

$f(x) = x(x - 2)(x + 1)(x + 3)$

9.

$f(x) = (x + 3)(x - 2)(x + 1)$

11.

$f(x) = 9x^2 - x^4$

27. (a) 0.0055, 0.4725, 1.8626, 16.1868, 38.8132, 55.4341, 76.2088;

(b)

$y = \frac{1}{13}x^3 - \frac{1}{14}x$

29. (a) 161.6 lb, 184.3 lb; **(b)** $W = 209.1$ lb. Therefore, he should watch his weight.

31. (a) and **(b)**

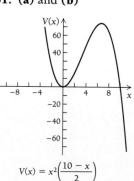

$$V(x) = x^2\left(\frac{10 - x}{2}\right)$$
$$= 5x^2 - \frac{1}{2}x^3$$

13.

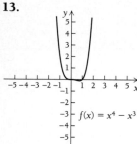

$f(x) = x^4 - x^3$

15.

$f(x) = x^3 - 4x$

17.

$f(x) = x^3 + x^2 - 2x$

19.

$f(x) = x^4 - 9x^2 + 20$

(c) $(-\infty, 0)$ and $(0, 10)$ **33.** 1, 4, 11, 12, 14, 19, 20
35. Every exponent of the polynomial must be even or $f(x) = c$.   **37.** Roots: $-3, -2, 1$
39. Roots: $-1.41421, -1, 1.41421, 2$

41. (a)

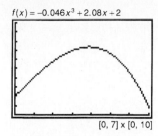

$f(x) = -0.046x^3 + 2.08x + 2$

$[0, 7] \times [0, 10]$

(b) Roots: $-6.17908, -0.98251, 7.16159$;
(c) approximately 7.2 hr

Margin Exercises, Section 4.6
1. (a) Yes; **(b)** no; **(c)** no   **2. (a)** Yes; **(b)** yes; **(c)** no; **(d)** no
3. (a) No; **(b)** no; **(c)** yes; **(d)** yes   **4. (a)** Yes; **(b)** no; **(c)** no
5. (a) Yes; **(b)** no   **6.** $Q(x) = x^2 + 5x + 10$, $R(x) = 24$,
$P(x) = (x - 3)(x^2 + 5x + 10) + 24$
7. (a) $P(3) = 24$, **(b)** same
8. $Q(x) = x^2 + 8x + 15$, $R(x) = 0$
9. $Q(x) = x^2 - 4x + 13$, $R(x) = -30$
10. $Q(y) = y^2 - y + 1$, $R(y) = 0$
11. (a) $P(10) = 73{,}120$; **(b)** $P(-8) = -37{,}292$
12. (a) Yes; **(b)** no; **(c)** yes
13. No   **14.** No   **15. (a)** Yes; **(b)** $x^2 + 8x + 15$;
(c) $(x - 2)(x + 5)(x + 3)$; **(d)** $2, -5, -3$

Technology Connection, Section 4.6
TC 1. $-2, 3, 7$

Exercise Set 4.6, pp. 252–253
1. 2 yes; 3 no, -1 no   **3. (a)** Yes; **(b)** no; **(c)** no
5. $Q(x) = x^2 + 8x + 15$, $R(x) = 0$,
$P(x) = (x - 2)(x^2 + 8x + 15) + 0$
7. $Q(x) = x^2 + 9x + 26$, $R(x) = 48$,
$P(x) = (x - 3)(x^2 + 9x + 26) + 48$
9. $Q(x) = x^2 - 2x + 4$, $R(x) = -16$,
$P(x) = (x + 2)(x^2 - 2x + 4) - 16$
11. $Q(x) = x^2 + 5$, $R(x) = 0$, $P(x) = (x^2 + 4)(x^2 + 5) + 0$
13. $P(x) = (2x^2 - x + 1) \cdot$
$$\left(\frac{5}{2}x^5 + \frac{5}{4}x^4 - \frac{5}{8}x^3 - \frac{39}{16}x^2 - \frac{29}{32}x + \frac{113}{64}\right) + \frac{171x - 305}{64}$$
15. $Q(x) = 2x^3 + x^2 - 3x + 10$, $R(x) = -42$
17. $Q(x) = x^2 - 4x + 8$, $R(x) = -24$
19. $Q(x) = x^3 + x^2 + x + 1$, $R(x) = 0$
21. $Q(x) = 2x^3 + x^2 + \frac{7}{2}x + \frac{7}{4}$, $R(x) = -\frac{1}{8}$
23. $Q(x) = x^3 + x^2y + xy^2 + y^3$, $R(x) = 0$
25. $P(1) = 0$, $P(-2) = -60$, $P(3) = 0$
27. $P(20) = 5{,}935{,}988$, $P(-3) = -772$
29. $P(2) = 0$, $P(-2) = 0$, $P(3) = 65$
31. -3 yes, 2 no   **33.** -3 no, $\frac{1}{2}$ no
35. $P(x) = (x - 1)(x + 2)(x + 3)$; $1, -2, -3$
37. $P(x) = (x - 2)(x - 5)(x + 1)$; $2, 5, -1$
39. $P(x) = (x - 2)(x - 3)(x + 4)$; $2, 3, -4$
41. $P(x) = (x - 1)(x - 2)(x - 3)(x + 5)$; $1, 2, 3, -5$
43. $0, 1$
45. 0 **47.** $-1, \frac{7}{6}$ **49.** $-5 < x < 1$ or $x > 2$

51. $\frac{14}{3}$ **53.** $k = 0$
55. $P(x) = (bx - r)Q(x) + R$ or
$$P(x) = \left(x - \frac{r}{b}\right) \cdot bQ(x) + R \text{ or } P(x) = \left(x - \frac{r}{b}\right)Q_1(x) + R.$$
Thus we would divide by $x - \dfrac{r}{b}$ and multiply the resulting

quotient $Q_1(x)$ by $\dfrac{1}{b}$ to obtain $Q(x)$. R is unchanged.

Margin Exercises, Section 4.7
1. 5, mult. 2; -6, mult. 1   **2.** -7, mult. 2; 3, mult. 1
3. -2, mult. 3; 3, mult. 1; -3, mult. 1   **4.** 4, mult. 2; 3,
mult. 2   **5.** 1, -1, each has mult. 1   **6.** $\pm 2i$,
$\pm\sqrt{3}$, each has mult. 1   **7.** 2, -2, $-\frac{1}{2}$, each has mult. 1
8. $x^3 - 6x^2 + 3x + 10$
9. $x^3 + (-1 + 5i)x^2 + (-2 - 5i)x - 10i$
10. $x^5 + 6x^4 + 12x^3 + 8x^2$ **11.** $x^4 + 2x^3 - 12x^2 + 14x - 5$
12. $7 + 2i$, $3 - \sqrt{5}$ **13.** $x^4 - 6x^3 + 11x^2 - 10x + 2$
14. $x^3 - 2x^2 + 4x - 8$ **15.** $-i, -2, 1$
16. (a) $3, -3, 1, -1$; **(b)** $2, -2, 1, -1$; **(c)** $\frac{3}{2}, -\frac{3}{2}, 3, -3$,
$\frac{1}{2}, -\frac{1}{2}, 1, -1$; **(d)** $\frac{1}{2}, -3$; **(e)** $3 + \sqrt{10}, 3 - \sqrt{10}$
17. (a) $3, -3, 1, -1$; **(b)** $1, -1$;
(c) same as for c (see part a); **(d)** all coefficients
positive; **(e)** -3; **(f)** $\dfrac{-3 \pm \sqrt{5}}{2}$ **18. (a)** $20, -20, 10$,
$-10, 5, -5, 4, -4, 2, -2, 1, -1$; **(b)** $1, -1$;
(c) same as for c; **(d)** all coefficients positive; **(e)** -5;
(f) $2i, -2i$ **19. (a)** All coefficients positive; **(b)** none

20. (a) None; **(b)** yes, $\dfrac{-3 \pm i\sqrt{3}}{2}$, quadratic formula

21. (a) $-\frac{1}{6}, -\frac{4}{3}, \frac{1}{6}, \frac{1}{3}$; **(b)** 6;
(c) $6P(x) = 6x^4 - x^3 - 8x^2 + x + 2$; **(d)** $1, -1, -\frac{1}{2}, \frac{2}{3}$;
(e) yes, $P(x) = 0$ and $6P(x) = 0$ are equivalent.

Exercise Set 4.7, pp. 261–262
1. -3, mult. 2; 1, mult. 1   **3.** 0, mult. 3; 1, mult. 2; -4,
mult. 1   **5.** $\pm\sqrt{3}, \pm 1$; each has mult. 1
7. $-3, -1, 1$; each has mult. 1   **9.** $x^3 - 6x^2 - x + 30$
11. $x^3 + 3x^2 + 4x + 12$ **13.** $x^3 - \sqrt{3}x^2 - 2x + 2\sqrt{3}$
15. $-3 - 4i, 4 + \sqrt{5}$ **17.** $x^3 - 4x^2 + 6x - 4$
19. $x^3 - 5x^2 + 16x - 80$ **21.** $x^4 + 4x^2 - 45$ **23.** $i, 2, 3$
25. $1 + 2i, 1 - 2i$ **27.** $1, -1$
29. $\pm(1, \frac{1}{3}, \frac{1}{5}, \frac{1}{15}, 2, \frac{2}{3}, \frac{2}{5}, \frac{2}{15})$
31. $-3, \sqrt{2}, -\sqrt{2}$; $(x + 3)(x - \sqrt{2})(x + \sqrt{2})$
33. $-2, 1$; $(x + 2)(x - 1)^2 = 0$ **35.** No rational roots
37. $-\frac{1}{5}, 1, 2i, -2i$; $(x + \frac{1}{5})(x - 1)(x - 2i)(x + 2i) = 0$
39. $-1, -2, 3 + \sqrt{13}, 3 - \sqrt{13}$;
$(x + 1)(x + 2)(x - 3 - \sqrt{13})(x - 3 + \sqrt{13})$
41. $2, 1 \pm \sqrt{3}$; $(x - 2)(x - 1 - \sqrt{3})(x - 1 + \sqrt{3}) = 0$
43. $-2, 1 \pm i\sqrt{3}$; $(x + 2)(x - 1 - i\sqrt{3})(x - 1 + i\sqrt{3})$
45. $\dfrac{1}{2}, \dfrac{1 \pm \sqrt{5}}{2}$; $\dfrac{1}{3}\left(x - \dfrac{1}{2}\right)\left(x - \dfrac{1 + \sqrt{5}}{2}\right)\left(x - \dfrac{1 - \sqrt{5}}{2}\right)$
47. None   **49.** None   **51.** None   **53.** $-2, 1, 2$

55. $4, i, -i$ **59.** $i, -i, 1 + \sqrt{2}, 1 - \sqrt{2}$ **59.** $-a$
61. $4\,\text{cm}$ **63.** $3\,\text{cm}, \dfrac{7 - \sqrt{33}}{2}\,\text{cm}$ **65.** Since $P(x)$
of odd degree n has n linear factors, of which an even
number corresponds to all the pairs of conjugate nonreal
roots, there is at least one other factor $(x - a)$ that gives
a real root a.   **67.** $\sqrt{5}$ is a root of $x^2 - 5 = 0$,
which has no rational roots (since ± 1 and ± 5 are not
roots). Thus $\sqrt{5}$ must be irrational.
69. $-\sqrt[3]{7}, \sqrt[3]{7}\left(\dfrac{1}{2} + \dfrac{\sqrt{3}}{2}i\right), \sqrt[3]{7}\left(\dfrac{1}{2} - \dfrac{\sqrt{3}}{2}i\right)$
71. **(a)** $1, 10, 20$; **(b)** $\{x \mid 1 < x < 10 \ or \ x > 20\}$;
(c) $\{x \mid 0 \le x < 1 \ or \ 10 < x < 20\}$

Margin Exercises, Section 4.8
1. $3; 2$ **2.** $2; 1$ **3.** Just $1; 2$ or 0 **4.** $5, 3,$ or $1;$ just 1
5. 3 or $1; 2$ or 0 **6.** 2 or $0; 0$ **7.** $[-4, 1]$
8. $[-2, 2]$ **9.** $[-1, 2]$ **10.** $-0.53, 0.65, 2.88$

Exercise Set 4.8, pp. 267–268
1. 3 or $1; 0$ **3.** $0; 3$ or 1 **5.** 2 or $0; 2$ or 0 **7.** $1; 1$
9. $1; 0$ **11.** 2 or $0; 2$ or 0 **13.** $3, -3$ **15.** $4, -1$
17. $5, -1$ **19.** $1, -3$ **21.** 3 or 1 positive;
1 negative; upper bound, 2; lower bound, -3
23. 1 positive; 1 negative; 2 nonreal; upper bound, 2;
lower bound, -2
25. 2 or 0 positive; 2 or 0 negative; upper bound, 4;
lower bound, -3
27. 0 positive, 0 negative   **29.** All roots are rational.
31. 2.2 **33.** No real roots   **35.** $-1.4, 1.4$
37. $-1.4, 1.4$ **39.** 0.79 **41.** -1.27
43. Let $P(x) = x^n - 1$. There is one variation of sign, so
there is just one positive root. Since n is even, $P(-x) = P(x)$.
Hence $P(-x)$ has just one variation of sign, and there is just
one negative root. Zero is not a root, so the total number of
real roots is two.
45. -1.3 **47.** $1.5, 5.7$ **49.** $7.16\,\text{hr}$

Margin Exercises, Section 4.9
1.

2.

3.

4.

5.

6.

7.

8.

9.

10.

11. $x = 0, x = -2, x = 3$ **12.** $x = 2, x = -2, x = -\frac{1}{2}$
13. (b) and (c)   **14.** $y = \frac{1}{2}$ **15.** $y = 3$ **16.** $y = 3x - 1$
17. $y = 5x$ **18.** $0, 3, -5$ **19.** $0, 1, -3$

20.

$$f(x) = \frac{1}{x^2 - 1}$$

21.

$$f(x) = \frac{1}{x^2 + 4}$$

13.

$$f(x) = \frac{x - 1}{x + 2}$$

15.

$$f(x) = \frac{3x}{x^2 + 5x + 4}$$

Technology Connection, Section 4.9

TC 1. x-intercept: $(2.46, 0)$; y-intercept: $(0, 0.45)$; vertical asymptotes: $x = -5.6$, $x = 3.4$; horizontal asymptote: $y = 0$

TC 2. x-intercepts: $(-1.15, 0)$, $(1.15, 0)$; y-intercept: $(0, -0.44)$; horizontal asymptote: $y = 1.5$

TC 3. y-intercept: $(0, 1.17)$

Exercise Set 4.9, pp. 278–279

1.

$$f(x) = \frac{1}{x - 3}$$

3.

$$f(x) = \frac{-2}{x - 5}$$

17.

$$f(x) = \frac{x^2 - 4}{x - 1}$$

19.

$$f(x) = \frac{x^2 + x - 2}{2x^2 + 1}$$

5.

$$f(x) = \frac{2x + 1}{x}$$

$(-\frac{1}{2}, 0)$

7.

$(0, \frac{1}{4})$

$$f(x) = \frac{1}{(x - 2)^2}$$

21.

$$f(x) = \frac{x - 1}{x^2 - 2x - 3}$$

23.

$$f(x) = \frac{x + 2}{(x - 1)^3}$$

9.

$$f(x) = \frac{2}{x^2}$$

11.

$(0, \frac{1}{3})$

$$f(x) = \frac{1}{x^2 + 3}$$

25.

$$f(x) = \frac{x^3 + 1}{x}$$

27.

$$f(x) = \frac{x^3 + 2x^2 - 15x}{x^2 - 5x - 14}$$

29.

$$f(x) = \frac{5x^4}{x^4 + 1}$$

31.

$$f(x) = \frac{x^2 - x - 2}{x + 2}$$

3. [4.1]

$$f(x) = 3x^2 + 6x + 1$$
$$= 3(x + 1)^2 - 2$$

33. **(a)** and **(b)**

$$t = \frac{500}{r}$$

35.

$$f(x) = \frac{x^3 + 4x^2 + x - 6}{x^2 - x - 2}$$

4. [4.1] None   **5.** [4.2] {5}
6. [4.2] {3, 4, 5, 7, 8, 9, 11, 12, 13}
7. [4.2]   **8.** [4.2]

-5 -4 -3 -2 -1 0 1 2 3 4 5   -5 -4 -3 -2 -1 0 1 2 3 4 5

9. [4.2] [2, 4]
10. [4.3] (1, 11)   **11.** [4.3] $(-\infty, -\frac{4}{3}) \cup (0, \infty)$
12. [4.3] [-20, 4]   **13.** [4.3] {2, -7}   **14.** [4.4] (-3, 3)
15. [4.4] $(-\infty, -\frac{1}{2}) \cup (2, \infty)$
16. [4.4] (-4, 1) ∪ (2, ∞)
17. [4.4] $(-\infty, -\frac{14}{3}) \cup (-3, \infty)$ **18.** [4.1] 20 × 20
19. [4.5]   **20.** [4.5]

37. **(a)** $T(w) = \dfrac{120,000}{40,000 - w^2}$; **(b)** 3.001876 hr, 3.007519 hr,

3.030303 hr; **(c)** domain = $\{w | 0 \le w < 200\}$,
range = $\{T | T \ge 3\}$;

(d)

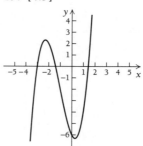

$$f(x) = x^3 + 3x^2 - 2x - 6$$

$$f(x) = x^4 - 3x^3 + 2x^2$$

21. [4.6] 0   **22.** [4.6] $Q = 2x^3 - 10x^2 + 27x - 59$, $R = 119$
23. [4.6] 88   **24.** [4.6] $(x - 1)(x + 3)(x + 5)$; 1, -3, -5
25. [4.6] No   **26.** [4.7] 0, mult. 2; 3, mult. 2; -4, mult. 3;
4, mult. 1   **27.** [4.7] $x^3 - 3x^2 + 2x$
28. [4.7] $(x^2 - 1)(x - 2)^2(x + 3)^3$ **29.** [4.7] ± 3, $-3i$
30. [4.7] $-8 + 7i$, $10 - \sqrt{5}$
31. [4.7] $\pm(1, 2, 3, 4, 6, 12, \frac{1}{2}, \frac{3}{2})$ **32.** [4.7] -3, 4, $\pm 3i$
33. [4.8] 2 or 0; 2 or 0   **34.** [4.8] 2, 0; 4, 2, or 0
35. [4.8] 2   **36.** [4.8] -2   **37.** [4.8] 1.41
38. [4.8] -0.9, 1.3, 2.5
39. [4.9]

39.

$$f(x) = \frac{x^4 + 3x^3 + 21x^2 - 50x + 80}{x^4 + 8x^3 - x^2 + 20x - 10}$$

[-20, 10] x [-20, 20]

41.

$$f(x) = \frac{x^3}{x^2 - 1}$$

[-5, 5] x [-5, 5]

$$f(x) = \frac{x^2 + x - 6}{x^2 - x - 20}$$

Review Exercises: Chapter 4, pp. 280–281
1. [4.1] **(a)** $f(x) = 3(x + 1)^2 - 2$; **(b)** (-1, -2); **(c)** $x = -1$;
(d) minimum: -2   **2.** [4.1] **(a)** $f(x) = -2(x + \frac{3}{4})^2 + \frac{57}{8}$;
(b) $(-\frac{3}{4}, \frac{57}{8})$; **(c)** $x = -\frac{3}{4}$; **(d)** maximum: $\frac{57}{8}$

40. [4.2] $[-2, \frac{1}{3})$　**41.** [4.3] $(-\infty, -\frac{1}{2}) \cup (\frac{1}{2}, \infty)$
42. [4.4] $(-\infty, 2)$　**43.** [4.3] $\{x | \frac{1}{3} \le x \le 1\}$
44. [4.3] $\{x | -1 < x < \frac{3}{7}\}$　**45.** [4.4] $\{n | n \ge 15\}$

46. [4.7] $(x - 1)\left(x + \frac{1}{2} + i\frac{\sqrt{3}}{2}\right)\left(x + \frac{1}{2} - i\frac{\sqrt{3}}{2}\right)$

47. [4.6] 7　**48.** [4.7] 4　**49.** [4.6] -4
50. [4.9]

$y = 1 - \dfrac{1}{x^2 + 4}$

Test: Chapter 4, pp. 281–282
1. [4.1] **(a)** $f(x) = 5(x - 1)^2 - 2$; **(b)** $(1, -2)$;
(c) minimum: -2　**2.** [4.1] **(a)** $f(x) = -4(x - \frac{3}{8})^2 - \frac{7}{16}$;
(b) $(\frac{3}{8}, -\frac{7}{16})$; **(c)** maximum: $-\frac{7}{16}$
3. [4.1]

$f(x) = 5x^2 - 10x + 3$
$= 5(x - 1)^2 - 2$

4. [4.1] $\left(\dfrac{1 - \sqrt{13}}{6}, 0\right), \left(\dfrac{1 + \sqrt{13}}{6}, 0\right)$

5. [4.2] $\{2, 3, 4, 6, 7, 8, 10, 11, 12\}$
6. [4.2]

7. [4.2] $(-5, -3]$　**8.** [4.3] $(-\infty, -\frac{3}{2}] \cup [3, \infty)$
9. [4.3] $(1, 5)$　**10.** [4.3] $\{-2, 3\}$
11. [4.4] $(-\infty, 2) \cup (6, \infty)$　**12.** [4.4] $(-\frac{3}{2}, \frac{1}{4})$
13. [4.4] $(-\infty, -7) \cup (-\frac{3}{2}, \infty)$　**14.** [4.1] $-10, -10$
15. [4.5]　**16.** [4.5]

$f(x) = x^4 - 5x^2$　　$f(x) = x^4 - 5x^2 + 6$

17. [4.6] 82　**18.** [4.6] Yes　**19.** [4.6] The quotient is $5x^2 - 6x + 10$. The remainder is -13.　**20.** [4.6] 2315
21. [4.6] $(x - 2)(x + 1)(x - 3)$; $-1, 2, 3$

22. [4.7] $x^4 - 4x^3 + x^2 + 16x - 20$
23. [4.7] $x(x + 1)(x - 2)^2(x - 1)^3$
24. [4.7] $x^3 - 8x^2 + 22x - 20$　**25.** [4.7] $\pm(\frac{1}{3}, \frac{2}{3}, 1, 2, 3, 6)$
26. [4.8] 3 or 1; 2 or 0　**27.** [4.8] $[-1, 3]$　**28.** [4.8] 1.4
29. [4.9]

$f(x) = \dfrac{x - 2}{x^2 - 2x - 15}$

30. [4.4] **(a)** After $2\frac{1}{2}$ seconds, the maximum height of 324 ft is attained. **(b)** 7 seconds; **(c)** between 2 and 3 seconds
31. [4.3] $(-\infty, -\frac{1}{8}) \cup (\frac{1}{4}, \infty)$
32. [4.4] $\{x | x \le -5 \text{ or } x \ge 2\}$
33. [4.9]

$f(x) = \left|2 - \dfrac{1}{3x}\right|$

34. [4.7] $i, -i, 1 - i, 1 + i$

CHAPTER 5

Margin Exercises, Section 5.1
1. $\{(4, -1), (5, 2), (-3, 0), (1, 5)\}$　**2.** **(a)** $x = 3y + 2$;
(b) $x = y$; **(c)** $y^2 + 3x^2 = 4$; **(d)** $x = 5y^2 + 2$;
(e) $x^2 = 4y - 5$; **(f)** $yx = 5$
3. **(a)** and **(b)**

● Relation P
× Inverse of P

Inverse of $P = \{(0, 5), (-2, 3), (-3, 0), (-4, 4)\}$

4.

5. (a)

12. The function is not one-to-one and does not have an inverse that is a function.
13. (a) and **(d)**   **14. (a)** Yes; **(b)** $f^{-1}(x) = 3 - x$

15. (a) Yes; **(b)** $g^{-1}(x) = \dfrac{x + 2}{3}$

16.

(b)

(c)

17. (a) The graph of $f(x) = x^3 + 1$ passes the horizontal-line test and thus has an inverse. **(b)** $f^{-1}(x) = \sqrt[3]{x - 1}$
(c)

6. WOMEN'S DRESS SIZES   ; yes

Domain (United States)	Range (France)
6	38
8	40
10	42
12	44
14	46
16	48
18	50

18. $f^{-1}(x) = \sqrt{x + 4},\ x \geq -4$

19. $f^{-1} \circ f(x) = f^{-1}\left(\dfrac{4}{x} - 3\right) = \dfrac{4}{\left(\dfrac{4}{x} - 3\right) + 3} = \dfrac{4}{\dfrac{4}{x}} = x,$

$f \circ f^{-1}(x) = f\left(\dfrac{4}{x + 3}\right) = \dfrac{4}{\dfrac{4}{x + 3}} - 3 = x + 3 - 3 = x$

7. SPORTS TEAMS   ; no

Domain	Range
Lakers	
Dodgers	Los Angeles
Rams	
Knickerbockers	
Yankees	New York
Giants	

20. $f(f^{-1}(1992)) = 1992,\ f^{-1}(f(-23,456)) = -23,456$

Technology Connection, Section 5.1
TC 1.

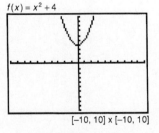

$[-10, 10] \times [-10, 10]$

not one-to-one

8. Assume that $f(a) = f(b)$ for any numbers a and b in the domain of f. Then:

$5a + 7 = 5b + 7$

$5a = 5b$

$a = b.$

Thus, if $f(a) = f(b)$, then $a = b$. This is true for any a and b in the domain of f, so f is one-to-one.
9. $-4 \neq 4$ and $g(-4) = g(4)$. Thus, g is not one-to-one.
10. The function is one-to-one and has an inverse that is also a function.
11. The function is not one-to-one and does not have an inverse that is a function.

TC 2.

$[-15, 15] \times [-10, 10]$

one-to-one,

$f^{-1}(x) = \sqrt[3]{x - 4}$

TC 3. one-to-one, $f^{-1}(x) = \dfrac{x+8}{2}$

$f(x) = 2x - 8$
$f^{-1}(x) = \dfrac{x+8}{2}$

[−15, 15] x [−10, 10]

TC 4.

$f(x) = 2\sqrt[3]{x}$
$f^{-1}(x) = \dfrac{x^3}{8}$

[−12, 12] x [−8, 8]

TC 5.

$f(x) = \dfrac{4}{x}$
$f^{-1}(x) = \dfrac{4}{x}$

[−12, 12] x [−8, 8]

TC 6.

$f(x) = 4x^3 + 2$
$f^{-1}(x) = \sqrt[3]{\dfrac{x-2}{4}}$

[−12, 12] x [−8, 8]

Exercise Set 5.1, pp. 293–294
1. $\{(1, 0), (6, 5), (-4, -2)\}$
3. $\{(8, 7), (8, -2), (-4, 3), (-8, 8)\}$ **5.** $x = 4y - 5$
7. $y^2 - 3x^2 = 3$ **9.** $x = 3y^2 + 2$ **11.** $yx = 7$
13

15.

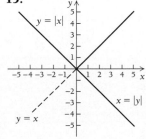

17. Yes **19.** No **21.** No **23.** No **25.** No **27.** Yes
29. (a) Yes; **(b)** $f^{-1}(x) = x - 4$ **31. (a)** Yes;
(b) $f^{-1}(x) = 5 - x$ **33. (a)** Yes; **(b)** $g^{-1}(x) = x + 3$
35. (a) Yes; **(b)** $f^{-1}(x) = \dfrac{1}{2}x$ **37. (a)** Yes;

(b) $g^{-1}(x) = \dfrac{x-5}{2}$ **39. (a)** Yes; **(b)** $h^{-1}(x) = \dfrac{4}{x} - 7$

41. (a) Yes; **(b)** $f^{-1}(x) = \dfrac{1}{x}$ **43. (a)** Yes;

(b) $f^{-1}(x) = \dfrac{4x-3}{2}$ **45. (a)** Yes; **(b)** $g^{-1}(x) = \dfrac{4+3x}{x-1}$
47. (a) Yes; **(b)** $f^{-1}(x) = \sqrt[3]{x+1}$ **49. (a)** Yes;
(b) $G^{-1}(x) = \sqrt[3]{x} + 4$
51. (a) Yes; **(b)** $f^{-1}(x) = x^3$
53. (a) Yes; **(b)** $f^{-1}(x) = \dfrac{\sqrt{x-3}}{2}, x > 39$ **55. (a)** Yes;

(b) $f^{-1}(x) = x^2 - 1, x \geq 0$ **57. (c)**
59.

61.

63.

65.

67.

69. $f^{-1} \circ f(x) =$
$f^{-1}(f(x)) = f^{-1}(\frac{7}{8}x) =$
$\frac{8}{7}(\frac{7}{8}x) = x, f \circ f^{-1}(x) =$
$f(f^{-1}(x)) = f(\frac{8}{7}x) =$
$\frac{7}{8}(\frac{8}{7}x) = x$

71. $f^{-1} \circ f(x) = f^{-1}(f(x)) = f^{-1}\left(\dfrac{1-x}{x}\right)$

$= \dfrac{1}{\dfrac{1-x}{x} + 1} = \dfrac{1}{\dfrac{1}{x}} = x,$

$f \circ f^{-1}(x) = f(f^{-1}(x)) = f\left(\dfrac{1}{x+1}\right)$

$= \dfrac{1 - \dfrac{1}{x+1}}{\dfrac{1}{x+1}} = \dfrac{\dfrac{x}{x+1}}{\dfrac{1}{x+1}} = x$

73. 3; -125 **75.** 12,053; $-17,243$
77. (a) 40, 42, 46, 50; **(b)** yes, $f^{-1}(x) = x - 32$;
(c) 8, 10, 14, 18
79. No; the function $f(x) = 5$ does not
pass the horizontal-line test.   **81.** No   **83.** No
85. Answers may vary. $f(x) = 2/x, f(x) = 4 - x, f(x) = x$
87. x-axis: no; y-axis: yes; origin: no

89. The inverse of the given relation $|x| - |y| = 1$ is
$|y| - |x| = 1$. Each of the two graphs is symmetric with
respect to the x-axis, the y-axis, and the origin.

91. Not one-to-one
93.

Margin Exercises, Section 5.2
1. (a) 1, 3, 9, 27, $\frac{1}{3}, \frac{1}{9}, \frac{1}{27}$;

(b)

2. (a) 1, $\frac{1}{3}, \frac{1}{9}, \frac{1}{27}$, 3, 9, 27;

(b)

3.

4.

5.

6.

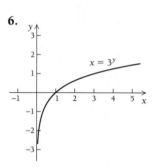

7. (a) $A(t) = 80,000(1.08)^t$; **(b)** $80,000, 108,839.12$,
$148,074.42, 172,714$; **(c)**

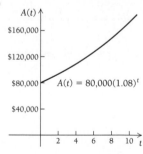

Technology Connection, Section 5.2
TC 1. $f(t) = 10,000(1.0675)^t$
TC 2.

$f(x) = 10000(1.0675)^t$

[0, 24] x [0, 60000]

TC 3. 10.6 yr
TC 4. 16.8 yr
TC 5. $39,400

Exercise Set 5.2, pp. 301–302

1.

$y = f(x) = 2^x$

3.

$y = 5^x$

5.

$y = 2^{x+1}$

7.

$y = 3^{x-2}$

9.

$y = 2^x - 3$

11.

$y = 5^{x+3}$

13.

$y = \left(\frac{1}{2}\right)^x$

15.

$y = \left(\frac{1}{5}\right)^x$

17.

$y = 2^{2x-1}$

19.

$y = 2^{x-1} - 3$

21.

$x = 2^y$

23.

$x = \left(\frac{1}{2}\right)^y$

25.

$x = 5^y$

27.

$x = \left(\frac{2}{3}\right)^y$

29.

$y = 2^x$
$x = 2^y$

31.

$y = \left(\frac{1}{2}\right)^x$
$x = \left(\frac{1}{2}\right)^y$

33. (a) 18.4 million, 45 million, 661.4 million, 58,320 million, 5,142,752.7 million;
(b)

$N(t) = 7.5(6)^{0.5t}$

Number of discs, in millions

Number of years (since 1985)

35. (a) $A(t) = \$50,000(1.09)^t$; **(b)** \$50,000, \$70,579.08, \$99,628.13, \$118,368.18;
(c)

$A(t) = 50,000(1.09)^t$

Amount

Years

37. (a) $5200, $3900, $2925, $1233.98, $292.83;
(b)

39. (a) 343; **(b)** 416.681217; **(c)** 450.409815;
(d) 451.287125; **(e)** 451.726421; **(f)** 451.805540
41. $\pi^{3.2}$
43.

45.

47.

49.

51.

53.

55. $\{x \mid x > 0\}$
59.

57.

Margin Exercises, Section 5.3
1. The domain is the set of positive real numbers. The range is the set of all real numbers.

2. $0 = \log_6 1$ **3.** $-3 = \log_{10} 0.001$ **4.** $0.25 = \log_{16} 2$
5. $T = \log_m P$ **6.** $2^5 = 32$ **7.** $10^3 = 1000$ **8.** $a^7 = Q$
9. $t^x = M$ **10.** 10,000   **11.** 3   **12.** $\frac{1}{4}$ **13.** 5
14. -4 **15.** 0   **16.** 0   **17.** 1   **18.** 1   **19.** 0

Exercise Set 5.3, p. 309
1.

3.

5.

7.

9.

11.

13.

$f(x) = 3^x$ $f^{-1}(x) = \log_3 x$

15. $3 = \log_{10} 1000$
17. $-3 = \log_5 \frac{1}{125}$
19. $\frac{1}{3} = \log_8 2$
21. $0.3010 = \log_{10} 2$
23. $3 = \log_e t$
25. $t = \log_Q x$
27. $3 = \log_e 20.0855$
29. $-1 = \log_e 0.3679$
31. $4^t = 7$ **33.** $2^5 = 32$

35. $10^{-1} = 0.1$ **37.** $10^{0.845} = 7$ **39.** $e^{3.4012} = 30$
41. $t^k = Q$ **43.** $e^{-0.9676} = 0.38$ **45.** $r^{-x} = M$
47. 1000   **49.** 1   **51.** 9   **53.** 4   **55.** $\frac{1}{2}$ **57.** 2   **59.** 3
61. -1 **63.** 0   **65.** 4   **67.** -2 **69.** 0   **71.** 0
73. $\frac{1}{2}$ **75.** 5

77.

$y = \left(\frac{2}{3}\right)^x$ $y = \log_{2/3} x$

79.

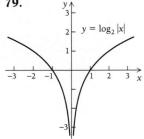

$y = \log_2 |x|$

81. 25   **83.** $(\sqrt{5})^{-3}$, or $5^{-3/2}$ **85.** 6   **87.** 1296
89. 3   **91.** 0   **93.** -8 **95.** All real numbers
97. $\{x \mid x \neq 0\}$ **99.** $\{x \mid x > \frac{4}{3}\}$ **101.** $\{x \mid 0 < x < 1\}$

Margin Exercises, Section 5.4
1. $\log_5 25 + \log_5 5$ **2.** $\log_b P + \log_b Q$ **3.** $\log_3 35$
4. $\log_a CABIN$ **5.** $5 \log_7 4$ **6.** $\frac{1}{2} \log_a 5$

7. $\log_b P - \log_b Q$ **8.** $\log_2 \dfrac{x}{25}$

9. $\frac{1}{2}(3 \log_a z - \log_a x - \log_a y)$
10. $2 \log_a x - 3 \log_a y - 1$

11. $3 \log_m a + 4 \log_m b - 5 - 9 \log_m n$ **12.** $\log_a \dfrac{x^5 \sqrt[4]{z}}{y}$

13. $-\frac{3}{2} \log_a b$ **14.** 0.602   **15.** 1   **16.** -0.398
17. 0.398   **18.** -0.699 **19.** $\frac{3}{2}$ **20.** 1.699   **21.** 1.204
22. 8   **23.** 4.3   **24.** 23   **25.** 3   **26.** x **27.** 42

Exercise Set 5.4, pp. 316–317
1. $\log_2 64 + \log_2 8$ **3.** $\log_4 32 + \log_4 64$
5. $\log_c Q + \log_c P$ **7.** $\log_b 720$ **9.** $\log_c PQ$
11. $4 \log_a x$ **13.** $5 \log_c y$ **15.** $-6 \log_b Q$
17. $\log_a 76 - \log_a 13$ **19.** $\log_b 5 - \log_b 4$ **21.** $\log_a \frac{18}{5}$
23. $3 \log_a x + 2 \log_a y + \log_a z$ **25.** $2 \log_b x + \log_b y - 3$
27. $\frac{1}{3}(4 \log_c x - 3 \log_c y - 2 \log_c z)$

29. $\frac{1}{4}(8 \log_a m + 12 \log_a n - 3 - 5 \log_a b)$ **31.** $\log_a \dfrac{x^{2/5}}{y^{1/3}}$

33. $\log_a \dfrac{2x^4}{y^3}$ **35.** $\frac{1}{2} - \log_a x$ **37.** 1.3023   **39.** -0.2457
41. -0.5283 **43.** $\frac{3}{2}$ **45.** 1.5283   **47.** 1.548   **49.** 11
51. $|x - 4|$ **53.** $4x$ **55.** Q **57.** 15   **59.** -7
61. False   **63.** True   **65.** False   **67.** False   **69.** True
71. $\log_a (x^6 - x^4 y^2 + x^2 y^4 - y^6)$
73. $\frac{1}{2}[\log_a (2 - x) + \log_a (2 + x)]$ **75.** $\frac{10}{3}$ **77.** $\{x \mid x > 0\}$
79. $\frac{9}{8}$ **81.** 3   **83.** $\{x \mid x > -2\}$ **85.** -2

87. $\log_a \dfrac{x + \sqrt{x^2 - 5}}{5} \cdot \dfrac{x - \sqrt{x^2 - 5}}{x - \sqrt{x^2 - 5}}$

$= \log_a \dfrac{5}{5(x - \sqrt{x^2 - 5})} = -\log_a (x - \sqrt{x^2 - 5})$

89. Let $\log_{a^{1/n}} x = M$. Then $(a^{1/n})^M = x$, or $(a^M)^{1/n} = x$, and $((a^M)^{1/n})^n = x^n$, or $a^M = x^n$, or $\log_a x^n = M$. Thus, $\log_{a^{1/n}} x = \log_a x^n$.

Margin Exercises, Section 5.5
1. 7.9913   **2.** -3.2097 **3.** 1.7251   **4.** -1.2711
5. (a) 3; (b) 1000; (c) -2; (d) 0.01   **6.** 11.3518
7. -6.7882 **8.** -0.9416 **9.** 7.3427
10. (a) 1.0986; (b) 3; (c) -0.7340; (d) 0.48
11. 1.1606   **12.** 5.7369

Technology Connection, Section 5.5
TC 1.
$y = 5^x$
$y = \log_5 x$

$[-6, 6] \times [-4, 4]$

TC 2.
$y = 1.2^x$
$y = \log_{1.2} x$

$[-9, 9] \times [-6, 6]$

TC 3.
$y = 0.3^x$
$y = \log_{0.3} x$

$[-3, 3] \times [-2, 2]$

Exercise Set 5.5, pp. 321–322
1. 0.4771   **3.** 0.9031   **5.** 0.3692   **7.** 1.8129
9. 1.7952   **11.** 2.7259   **13.** 4.1271   **15.** -0.2441
17. -1.2840 **19.** -2.0084 **21.** 0.0011
23. Does not exist   **25.** 1.0986   **27.** 2.0794
29. 4.4067   **31.** 9.0318   **33.** -5.1328 **35.** 2.8044
37. 0.0000027   **39.** 3.3219   **41.** 3.5850   **43.** -0.26144
45. -2.8074 **47.** -1.0959 **49.** 4.0229
51. $\ln x = \dfrac{\log x}{\log e} = \dfrac{1}{\log e} \cdot \log x = 2.3026 \log x$

53.

55.

57. 3   **59.** 253.3282   **61.** 2.9617

63. Let $a = e$, $b = 10$, and $M = e$. Substitute in the change-of-base formula:

$$\log e = \frac{\ln e}{\ln 10} = \frac{1}{\ln 10}.$$

65. Let $a = M$, $b = b$, and $M = M$. Substitute in the change-of-base formula:

$$\log_b M = \frac{\log_M M}{\log_M b} = \frac{1}{\log_M b}.$$

67. $\log_a (\log_a x) = \log_a \left(\dfrac{\log_b x}{\log_b a}\right)$

$\qquad\qquad\quad = \log_a (\log_b x) - \log_a (\log_b a)$

69. 4, 2.867972, 2.731999, 2.719642, 2.718418   **71.** $e^{\sqrt{\pi}}$

73.

$y = 4.5^x$
$y = \log_{4.5} x$

$[-6, 6] \times [-4, 4]$

75.

$y = 0.15^x$
$y = \log_{0.15} x$

$[-3, 3] \times [-2, 2]$

Margin Exercises, Section 5.6

1.

$f(x) = e^{2x}$

2.

$f(x) = e^{-2x}$

3.

$f(x) = e^{0.2x}$

4.

$f(x) = 1 - e^{-x}$

5.

$f(x) = 2 \ln x$

6.

$f(x) = \ln (x - 1)$

7. (a) \$59.63; **(b)**

$V(t) = \$45(1 - e^{-0.8t}) + \$15$

Technology Connection, Section 5.6

TC 1.

$f(x) = 2e^x - 3$

$[-5, 5] \times [-5, 5]$

TC 2.

$f(x) = \dfrac{4}{e^x + 1}$

$[-10, 10] \times [-5, 5]$

TC 3.

$f(x) = \dfrac{e^x}{x}$

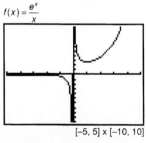

$[-5, 5] \times [-10, 10]$

TC 4.

$f(x) = 3e^{-1.7x}$

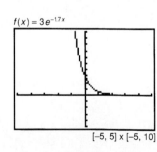

$[-5, 5] \times [-5, 10]$

TC 5.

$f(x) = \ln (x + 4)$

$[-6, 6] \times [-6, 6]$

TC 6.

$f(x) = \ln x + 4$

$[-6, 6] \times [-6, 6]$

TC 7.

$f(x) = 4 \ln x$

$[-6, 6] \times [-10, 10]$

TC 8.

$f(x) = \dfrac{1}{\ln(x + 4)}$

$[-6, 6] \times [-5, 5]$

17.

$f(x) = \ln(x - 2)$

19.

$f(x) = 2 - \ln x$

21. (a) 18.1%, 55.1%, 69.9%, 90.9%;

(b)

$P(t) = 1 - e^{-0.2t}$

Exercise Set 5.6, pp. 326–327

1.

$f(x) = 2e^x$

3.

$f(x) = e^{(1/2)x}$

5.

$f(x) = e^{x+1}$

7.

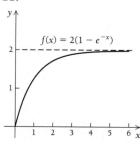

$f(x) = e^{2x} + 1$

23. (a) $58.69, $71.57, $77.29, $77.92, $77.99+;

(b)

$V(t) = \$58(1 - e^{-1.1t}) + \$20$

25.

$g(x) = e^{|x|}$

9.

$f(x) = 1 - e^{-0.01x}$

11.

$f(x) = 2(1 - e^{-x})$

27.

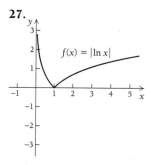

$f(x) = |\ln x|$

29.

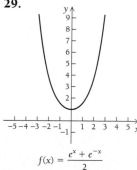

$f(x) = \dfrac{e^x + e^{-x}}{2}$

13.

$f(x) = 4 \ln x$

15.

$f(x) = \dfrac{1}{2} \ln x$

31. (a) 20, 42, 234, 602;   **(b)**

$N(t) = \dfrac{4800}{6 + 794e^{-0.4t}}$

33. (a)

$f(x) = x^2 e^{-x}$

$[-5, 5] \times [-5, 5]$

(b) 0; **(c)** minimum: 0

35. (a)

$f(x) = x^2 \ln x$

$[-1, 5] \times [-1, 5]$

(b) 1; **(c)** minimum: -0.2

37.–41. See answers to Exercises 25–29.

Margin Exercises, Section 5.7

1. 1   **2.** $\dfrac{3}{2}$ **3.** $\dfrac{\log 20}{\log 7} \approx 1.5395$ **4.** $\dfrac{\ln 80}{0.3} \approx 14.6068$

5. $x = \ln(t + \sqrt{t^2 + 1})$ **6.** 8   **7.** $\dfrac{35}{4}$ **8.** 5   **9.** 9

Technology Connection, Section 5.7

TC 1. 0.38   **TC 2.** -1.96 **TC 3.** 0.71
TC 4. $-10.00, 0.37$ **TC 5.** 343   **TC 6.** 7.06
TC 7. 2.44

Exercise Set 5.7, pp. 331–332

1. 5   **3.** 4   **5.** $\dfrac{3}{2}$ **7.** $\dfrac{3}{7}$ **9.** $\dfrac{\log 33}{\log 2} \approx 5.0444$

11. $\dfrac{\log 40}{\log 2} \approx 5.3219$ **13.** $\dfrac{5}{2}$ **15.** $-3, -1$

17. $\dfrac{\log 70}{\log 84} \approx 0.9589$ **19.** $\ln 1000 \approx 6.9078$

21. $-\ln 0.3 \approx 1.2040$ **23.** $\dfrac{\ln 0.08}{-0.03} \approx 84.1910$

25. $\dfrac{\log 2}{\log 2 - \log 3} \approx -1.7095$ **27.** $\dfrac{\log 48}{\log 3.9} \approx 2.8444$

29. $\dfrac{\log 250}{\log 1.87} \approx 8.8211$ **31.** $\dfrac{12}{5}$

33. $x = \ln\left(\dfrac{5t \pm \sqrt{25t^2 + 4}}{2}\right)$ **35.** $x = \dfrac{1}{2}\ln\dfrac{t+1}{t-1}$

37. 625   **39.** $\dfrac{1}{125}$ **41.** 100   **43.** 0.001   **45.** e

47. e^{-2} **49.** $-\dfrac{117}{7}$ **51.** 10   **53.** $\dfrac{1}{3}$ **55.** 3   **57.** $\dfrac{1}{3}$

59. 5   **61.** $1, 10^{16}$ **63.** $\pm 2\sqrt{6}$ **65.** $1, 100$ **67.** 4
69. $\varnothing$ **71.** $\pm\sqrt{58}$ **73.** 10^{1000} **75.** $-9, 9$
77. $\dfrac{1}{100}, 100$ **79.** $3, -8$ **81.** $\dfrac{1}{10}, \dfrac{1}{1000}$ **83.** $-\dfrac{8}{3}$

85. 1   **87.** $-\dfrac{1}{2}$ **89.** $t = \dfrac{\ln P - \ln P_0}{k}$

91. $t = -\dfrac{1}{k}\ln\dfrac{T - T_0}{T_1 - T_0}$ **93.** $\log_3 x = -3$

95. If $a > 1, x \geq 1$. If $0 < a < 1, 0 < x \leq 1$.

97. $x > \dfrac{\log 0.8}{\log 0.5} \approx 0.3219$ **99.** $1, 4$ **101.** $a = \dfrac{2}{3}b$

103.–121. See the answers to Exercises 69–87.

Margin Exercises, Section 5.8
1. (a) $A(t) = \$80,000(1.07)^t$; **(b)** 16.2 years; **(c)** 10.2 years
2. (a) 5000; **(b)** 5169;
(c)

(d) \$5011.87

3. (a) 5.6; **(b)** 1.6×10^{-6} **4.** 7.85   **5. (a)** 65 decibels;
(b) 140 decibels

Exercise Set 5.8, pp. 336–338
1. (a) 5.5 years; **(b)** 0.8 year   **3. (a)** $A(t) = \$50,000(1.09)^t$;
(b) 25.5 years; **(c)** 8.04 years   **5. (a)** 6.6 years; **(b)** 3.1 years
7. (a) 68%; **(b)** 54%, 40%;
(c)

(d) 6.9 months

9. (a) 1000; **(b)** 1140;
(c)   **(d)** \$23,988.33

11. 3.8   **13.** 4.2   **15.** 4.0×10^{-6} **17.** 1.6×10^{-5}
19. 8.25   **21.** 9.6   **23.** 34 decibels   **25.** 60 decibels
27. (a) 4.6 hr; (b) 9.2 hr   **29.** $I = 10^R I_0$
31. $[H^+] = 10^{-pH}$ **33.** (a) 3.0; (b) 436,515.8;
(c) 100 times

Margin Exercises, Section 5.9
1. (a) 3.1 ft/sec; (b) 1.3 ft/sec; (c) 2.4 ft/sec
2. 274 million, 334 million   **3.** (a) 33¢;
(b) in 110.5 yr from 1932, or in 2043   **4.** 6,010,389
5. (a) $k = 0.085$, or 8.5%; (b) $P(t) = \$10,000e^{0.085t}$;
(c) \$23,396.47; (d) 8.2 yr   **6.** 16.9 yr   **7.** 5.0%
8. 10.2%   **9.** 13,412

Exercise Set 5.9, pp. 346–349
1. 3.4 ft/sec   **3.** 1.5 ft/sec   **5.** (a) $C(t) = 0.05e^{0.097t}$;
(b) \$0.76, \$1.99; (c) in 47.5 years from 1962;
(d)   (e) 7.1 yr

7. (a) $P(t) = P_0e^{0.06t}$; (b) \$536.56; (c) \$724.27;
(d)

9. (a)

Yes; (b) $k = 0.135$, $C(t) = 80e^{0.135t}$; (c) \$3,505,000;
(d) in 26.8 yr from 1967; (e) 5.1 yr
11. (a) $P(t) = P_0e^{0.09t}$; (b) \$5470.87, \$5986.09; (c) 7.7 yr
13. 19.8 yr   **15.** 9.9%
17. (a) $k = 0.16$; $V(t) = 84,000e^{0.16t}$; (b) \$1,240,241,652;
(c) 4.3 yr; (d) 58.7 yr   **19.** In 17.3 yr from 1990

21. (a) $k = 0.007$; $P(t) = 2,812,000e^{0.007t}$; (b) 3,373,315
23. 5135   **25.** 23.1% per minute   **27.** 7.2 days
29. (a) 0.7%; (b) 81.1%   **31.** (a) 13.3 lb/in²;
(b) 7.24 lb/in²; (c) 46,052 ft; (d) 72,589 ft
33. (a) 24.7%, 1.5%, 0.09%, (3.98×10^{-29})%;
(b) 0.00008%   **35.** $R = e^{v/c}$
37. Measure the atmospheric pressure P at the top of the
building. Substitute that value in the equation
$P = 14.7e^{-0.00005a}$, and solve for the height, or altitude, a.
(*Note*: We assume that the base of the Empire State
Building is essentially at sea level.)
39. 9 A.M.

Review Exercises: Chapter 5, pp. 349–350
1. [5.1] $\{(5, -4), (-3, 2), (7, 1), (8, 8), (-4, 5)\}$
2. [5.1] $x = 3y^2 + 2y - 1$ **3.** [5.1] $x = \sqrt{y + 2}$
4. [5.1] (d)   **5.** [5.1] $f^{-1}(x) = (2x - 4)^2$
6. [5.1] $f^{-1}(x) = \sqrt[3]{x - 8}$ **7.** [5.1] a **8.** [5.1] t
9. [5.3]   **10.** [5.2]

11. [5.6]   **12.** [5.6]

13. [5.5] 2.0959   **14.** [5.5] 0.3869   **15.** [5.3] $8^{-2/3} = \frac{1}{4}$
16. [5.3] $\log_7 x = 2.3$ **17.** [5.4] $\log_b \dfrac{a^{1/2}c^{3/2}}{d^4}$
18. [5.4] $\frac{2}{3}\log M - \frac{1}{3}\log N$ **19.** [5.4] 1.255
20. [5.4] 0.544   **21.** [5.4] -0.602 **22.** [5.4] 0.2385
23. [5.4] $x^2 + 1$ **24.** [5.4] $\sqrt{9}$ **25.** [5.3] 4
26. [5.3] $\frac{1}{2}$ **27.** [5.3] 3   **28.** [5.7] $\frac{1}{5}$ **29.** [5.7] 4.3820
30. [5.7] 1   **31.** [5.7] 9   **32.** [5.7] 3   **33.** [5.7] 3
34. [5.8] 5.7 yr   **35.** [5.8] 30 decibels   **36.** [5.9] 6.25 g
37. [5.8] (a) 82%; (b) 50%; (c) 4.5 months
38. [5.9] (a) $k = 0.05$, $C(t) = 4.65e^{0.05t}$; (b) \$51.26;
(c) in 29.2 years since 1962; (d) 13.9 yr   **39.** [5.9] 3.9%
40. [5.9] 8.1 yr   **41.** [5.9] 2623   **42.** [5.8] 6.4
43. [5.8] 8   **44.** [5.5] -2.6655 **45.** [5.5] 6.1278
46. [5.5] 11.3780   **47.** [5.5] -10.4919 **48.** [5.7] 64, $\frac{1}{64}$
49. [5.7] 1

50. [5.3]

$y = |\log_3 x|$

51. [5.6]

$y = |e^x - 4|$

52. [5.7] $\{x|x > e^{6/5}\}$ **53.** [5.7] $\{x|x \neq \frac{1}{4}\ln 10\}$

Test: Chapter 5, p. 351
1. [5.1] $\{(-5,3), (-3,6), (-3,8), (3,-5), (-2.7,1.3)\}$
2. [5.1] $x = |y|$ **3.** [5.1] (b), (d)
4. [5.1] $f^{-1}(x) = x^2 + 6$ **5.** [5.1] 3
6. [5.3]   **7.** [5.6]

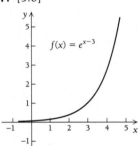

$y = \log_3 x$

$f(x) = e^{x-3}$

8. [5.5] 1.9482   **9.** [5.3] $(\sqrt{3})^4 = 9$
10. [5.3] $\log_x 0.03125 = 5$ **11.** [5.3] $3x$
12. [5.4] $\log_c \dfrac{x^3\sqrt{z}}{y^4}$ **13.** [5.4] $\dfrac{1}{4}\log w + \dfrac{3}{4}\log r$
14. [5.4] 0.477   **15.** [5.4] 1.699   **16.** [5.4] 0.233
17. [5.3] $0, \frac{1}{2}$ **18.** [5.3] 16   **19.** [5.7] 2   **20.** [5.7] 4
21. [5.7] 1.6094   **22.** [5.5] -2.2804 **23.** [5.5] 1.3507
24. [5.5] 4.9405   **25.** [5.5] -15.0534
26. [5.5] Not defined   **27.** [5.8] 6.1 yr   **28.** [5.8] 47 decibels
29. [5.9] 156,250   **30.** [5.8] (a) 2.5 ft/sec; (b) 437,502
31. [5.9] (a) $P(t) = 52e^{0.028t}$; (b) 155.0 million,
286.9 million; (c) in 62.6 years; (d) in 24.8 years
32. [5.9] 2.3%   **33.** [5.9] 8.1 yr   **34.** [5.9] 3984 yr
35. [5.8] 24 decibels   **36.** [5.8] 8.7   **37.** [5.4] False
38. [5.3], [5.6] $\{x|x > 1\}$ **39.** [5.3] 16

CHAPTER 6

Margin Exercises, Section 6.1

1. (a) $\dfrac{\pi}{2}$; (b) $\dfrac{3\pi}{4}$; (c) $\dfrac{3\pi}{2}$

2. (a) $\dfrac{\pi}{4}$; (b) $\dfrac{7\pi}{4}$; (c) $\dfrac{5\pi}{4}$

3. (a) $\dfrac{\pi}{6}$; (b) $\dfrac{4\pi}{3}$; (c) $\dfrac{11\pi}{6}$

4.

5.

6.

7. $N\left(\dfrac{\sqrt{3}}{2}, -\dfrac{1}{2}\right)$,

$P\left(-\dfrac{\sqrt{3}}{2}, -\dfrac{1}{2}\right)$,

$R\left(-\dfrac{\sqrt{3}}{2}, \dfrac{1}{2}\right)$

8. (a) $\left(\dfrac{3}{5}, \dfrac{4}{5}\right)$;   (b) $\left(-\dfrac{3}{5}, -\dfrac{4}{5}\right)$;   (c) $\left(-\dfrac{3}{5}, \dfrac{4}{5}\right)$;

(d) yes, by symmetry

9. (a) $\left(-\dfrac{\sqrt{35}}{6}, \dfrac{1}{6}\right)$; (b) $\left(\dfrac{\sqrt{35}}{6}, -\dfrac{1}{6}\right)$; (c) $\left(\dfrac{\sqrt{35}}{6}, \dfrac{1}{6}\right)$;

(d) yes, by symmetry

Exercise Set 6.1, pp. 358–359
1.

3.

5.

7. M: $\dfrac{2\pi}{3}$, $-\dfrac{4\pi}{3}$; N: $\dfrac{5\pi}{6}$,

$-\dfrac{7\pi}{6}$; P: $\dfrac{5\pi}{4}$, $-\dfrac{3\pi}{4}$; Q: $\dfrac{11\pi}{6}$,

$-\dfrac{\pi}{6}$

9. (a) $\left(-\dfrac{\sqrt{2}}{2}, -\dfrac{\sqrt{2}}{2}\right)$; (b) $\left(\dfrac{\sqrt{2}}{2}, \dfrac{\sqrt{2}}{2}\right)$; (c) $\left(\dfrac{\sqrt{2}}{2}, -\dfrac{\sqrt{2}}{2}\right)$

11. (a) $\left(\dfrac{2}{3}, -\dfrac{\sqrt{5}}{3}\right)$; (b) $\left(-\dfrac{2}{3}, \dfrac{\sqrt{5}}{3}\right)$; (c) $\left(-\dfrac{2}{3}, -\dfrac{\sqrt{5}}{3}\right)$

13. M(0, 1); N$\left(-\dfrac{\sqrt{2}}{2}, \dfrac{\sqrt{2}}{2}\right)$; P(0, −1); Q$\left(\dfrac{\sqrt{2}}{2}, -\dfrac{\sqrt{2}}{2}\right)$

15. $\left(\dfrac{\sqrt{3}}{2}, -\dfrac{1}{2}\right)$ **17.** $\left(\dfrac{3}{4}, \dfrac{\sqrt{7}}{4}\right)$

19. M: $\dfrac{8\pi}{3}$; N: $\dfrac{17\pi}{6}$; P: $\dfrac{13\pi}{4}$; Q: $\dfrac{23\pi}{6}$ **21.** $\pm\dfrac{2\sqrt{2}}{3}$

Margin Exercises, Section 6.2

1. (a) 0, (b) -1; (c) $\dfrac{\sqrt{2}}{2}$; (d) $-\dfrac{\sqrt{2}}{2}$; (e) $\dfrac{\sqrt{3}}{2}$; (f) $-\dfrac{1}{2}$

2. (a) -1; (b) 0; (c) $\dfrac{\sqrt{2}}{2}$; (d) $-\dfrac{\sqrt{2}}{2}$; (e) $-\dfrac{\sqrt{3}}{2}$; (f) $\dfrac{1}{2}$

3.

4. Yes **5.** The set of real numbers
6. The set of real numbers from -1 to 1, inclusive
7. (a) $(-a, -b)$; (b) $(-a, -b)$

8.

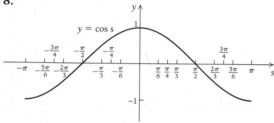

9. Yes **10.** The set of all real numbers
11. The set of all real numbers from -1 to 1, inclusive
12. $\cos y = d$, $\sin y = e$

Exercise Set 6.2, pp. 369–370

1. $\dfrac{\sqrt{2}}{2}$ **3.** $\dfrac{1}{2}$ **5.** -1 **7.** $\dfrac{\sqrt{3}}{2}$ **9.** $-\dfrac{\sqrt{2}}{2}$ **11.** $\dfrac{1}{2}$

13. 0 **15.** $-\dfrac{\sqrt{2}}{2}$ **17.** $\dfrac{\sqrt{2}}{2}$ **19.** $-\dfrac{\sqrt{3}}{2}$ **21.** $-\dfrac{\sqrt{3}}{2}$

23. $\dfrac{\sqrt{2}}{2}$ **25.** $-\dfrac{\sqrt{2}}{2}$ **27.** 0 **29.** $-\dfrac{\sqrt{2}}{2}$ **31.** 0

33. $\dfrac{\sqrt{2}}{2}$ **35.** 1 **37.** $-\dfrac{\sqrt{2}}{2}$ **39.** 0 **41.** $\dfrac{\sqrt{3}}{2}$ **43.** $-\dfrac{1}{2}$

45. $\cos x$ **47.** $-\sin x$ **49.** $-\cos x$ **51.** $\cos x$
53. $-\cos x$ **55.** (a) See Margin Exercise 3;

(b)

(c) same as (b);
(d) the same

57. (a) See Margin Exercise 3; **59.** (a) See Margin Exer. 8;
(b)

(c) same as (b);   (c) same as (b);
(d) the same   (d) the same

61. (a) $\dfrac{\pi}{2} + 2k\pi$, k any integer; (b) $\dfrac{3\pi}{2} + 2k\pi$, k any integer

63. $x = k\pi$, k any integer
65. $f \circ g(x) = \cos^2 x + 2\cos x$, $g \circ f(x) = \cos(x^2 + 2x)$
67.

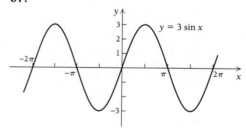

69. The domain is the set of all real numbers. The range is the set of all real numbers from 0 to 1 inclusive. The period is π. The amplitude is $\frac{1}{2}$.
71. The domain consists of the intervals

$$\left[-\dfrac{\pi}{2} + 2k\pi, \dfrac{\pi}{2} + 2k\pi \right], \ k \text{ any integer.}$$

73. The domain is the set of all real numbers except

$\dfrac{\pi}{2} + k\pi$ for any integer k.

75. The limit of $(\sin x)/x$ as x approaches 0 is 1.

Margin Exercises, Section 6.3
1. 0, not defined, -1, not defined **2.** -1, -1, $-\sqrt{2}$, $\sqrt{2}$

3. $\dfrac{\sqrt{3}}{3}$, $\sqrt{3}$, $\dfrac{2\sqrt{3}}{3}$, 2 **4.** $-\sqrt{3}$, $-\dfrac{\sqrt{3}}{3}$, 2, $-\dfrac{2\sqrt{3}}{3}$

5. Not defined, not defined, not defined

6. $-\dfrac{\pi}{2}, \dfrac{3\pi}{2}, -\dfrac{3\pi}{2}, \dfrac{5\pi}{2}, -\dfrac{5\pi}{2}$, and so on

7.

$$y = \tan x$$

8. π **9.** The set of all real numbers except $\pi/2 + k\pi$, k any integer   **10.** The set of all real numbers   **11.** Odd
12. $0, \pi, -\pi, 2\pi, -2\pi$, and so on
13.

$$y = \cot x$$

14. π **15.** The set of all real numbers except $k\pi$, k any integer   **16.** The set of all real numbers
17. Positive: I, III; negative: II, IV   **18.** Odd

19. (a) $\dfrac{\pi}{2}, -\dfrac{\pi}{2}, \dfrac{3\pi}{2}, -\dfrac{3\pi}{2}$, etc.; (b) $0, \pi, -\pi, 2\pi, -2\pi$, etc.

20.

$$y = \sec x$$

21.

$$y = \csc x$$

22. 2π; 2π **23.** The set of real numbers except $\pi/2 + k\pi$, k any integer; the set of real numbers except $k\pi$, k any integer
24. Its range consists of all real numbers 1 and greater, in addition to all real numbers -1 and less; its range is the same as that of the secant function.
25. Positive: I, IV; negative: II, III
26. Positive: I, II; negative: III, IV   **27.** Even   **28.** Odd

29.

Function	I	II	III	IV
sin, csc	+	+	−	−
cos, sec	+	−	−	+
tan, cot	+	−	+	−

30. $\tan s = \dfrac{\sin s}{\cos s} = \dfrac{\dfrac{1}{\cos s}}{\dfrac{1}{\sin s}} = \dfrac{1}{\cot s}$ **31.** $\cos s = \dfrac{1}{\dfrac{1}{\cos s}} = \dfrac{1}{\sec s}$

32. $\sin^2 s + \cos^2 s = 1$

$$\frac{\sin^2 s}{\cos^2 s} + \frac{\cos^2 s}{\cos^2 s} = \frac{1}{\cos^2 s}$$

$$\tan^2 s + 1 = \sec^2 s$$

33. $\sin^2 x = 1 - \cos^2 x$, $\sin x = \pm\sqrt{1 - \cos^2 x}$

34. $\cos s = -\dfrac{\sqrt{55}}{8}$; $\tan s = \dfrac{3}{\sqrt{55}}$; or $\dfrac{3\sqrt{55}}{55}$;

$\cot s = \dfrac{\sqrt{55}}{3}$; $\sec s = -\dfrac{8}{\sqrt{55}}$; or $-\dfrac{8\sqrt{55}}{55}$; $\csc s = -\dfrac{8}{3}$

35. $\sin s = \dfrac{3}{\sqrt{10}}$, or $\dfrac{3\sqrt{10}}{10}$; $\cos s = -\dfrac{1}{\sqrt{10}}$, or $-\dfrac{\sqrt{10}}{10}$;

$\cot s = -\dfrac{1}{3}$; $\sec s = -\sqrt{10}$; $\csc s = \dfrac{\sqrt{10}}{3}$

Technology Connection, Section 6.3

TC 1.

$y = \cos^2 x \sin x$

$[-2\pi, 2\pi] \times [-2, 2]$

TC 2.

$y = (\cos x + 1)(\tan x)$

$[-2\pi, 2\pi] \times [-2, 2]$

TC 3.

$y = \csc x + \cot x$

$[-2\pi, 2\pi] \times [-2, 2]$

TC 4.

$y = \sec x \cot x$

$[-2\pi, 2\pi] \times [-2, 2]$

Exercise Set 6.3, pp. 380–381

1. 1   **3.** $\dfrac{\sqrt{3}}{3}$ **5.** $\sqrt{2}$ **7.** Undefined   **9.** $-\sqrt{3}$

11. $\sqrt{2}$ **13.** $-\dfrac{\sqrt{3}}{3}$ **15.** cos, sec   **17.** sin, cos, sec, csc

19. Positive: I, III; negative: II, IV
21. Positive: I, IV; negative: II, III

23.

	$\pi/16$	$\pi/8$	$\pi/6$
sin	0.19509	0.38268	0.50000
cos	0.98079	0.92388	0.86603
tan	0.19891	0.41421	0.57735
cot	5.02737	2.41424	1.73206
sec	1.01959	1.08239	1.15469
csc	5.12584	2.61315	2.00000

	$\pi/4$	$3\pi/8$	$7\pi/16$
sin	0.70711	0.92388	0.98079
cos	0.70711	0.38268	0.19509
tan	1.00000	2.41424	5.02737
cot	1.00000	0.41421	0.19891
sec	1.41421	2.61315	5.12584
csc	1.41421	1.08239	1.01959

25. $\sin s = \dfrac{2\sqrt{2}}{3}$; $\tan s = 2\sqrt{2}$; $\cot s = \dfrac{1}{2\sqrt{2}}$, or

$\dfrac{\sqrt{2}}{4}$; $\sec s = 3$; $\csc s = \dfrac{3}{2\sqrt{2}}$, or $\dfrac{3\sqrt{2}}{4}$

27. $\sin s = -\dfrac{3}{\sqrt{10}}$, or $-\dfrac{3\sqrt{10}}{10}$; $\cos s = -\dfrac{1}{\sqrt{10}}$, or

$-\dfrac{\sqrt{10}}{10}$; $\cot s = \dfrac{1}{3}$; $\sec s = -\sqrt{10}$; $\csc s = -\dfrac{\sqrt{10}}{3}$

29. $\sin s = \dfrac{4}{5}$; $\cos s = -\dfrac{3}{5}$; $\tan s = -\dfrac{4}{3}$; $\cot s = -\dfrac{3}{4}$;

$\csc s = \dfrac{5}{4}$

31. $\cos s = -\dfrac{\sqrt{21}}{5}$; $\tan s = \dfrac{2}{\sqrt{21}}$, or $\dfrac{2\sqrt{21}}{21}$; $\cot s = \dfrac{\sqrt{21}}{2}$;

$\sec s = -\dfrac{5}{\sqrt{21}}$, or $-\dfrac{5\sqrt{21}}{21}$; $\csc s = -\dfrac{5}{2}$

33. The graph of $\sec(x - \pi)$ is like that of $\sec x$, moved π units to the right. The graph of $-\sec x$ is that of $\sec x$ reflected across the x-axis. The graphs are identical.
35. If the graph of $\tan x$ were reflected across the y-axis and then translated to the right a distance of $\pi/2$, the graph of $\cot x$ would be obtained. There are other ways to describe the relation.
37. The sine and tangent functions; the cosine and cotangent functions.
39.

$f(x) = |\tan x|$

41.

43. (a) $\triangle OPA \sim \triangle ODB$

Thus, $\dfrac{AP}{OA} = \dfrac{BD}{OB}$

$\dfrac{\sin \theta}{\cos \theta} = \dfrac{BD}{1}$

$\tan \theta = BD$;

(b) $\triangle OPA \sim \triangle ODB$

Thus, $\dfrac{OD}{OP} = \dfrac{OB}{OA}$

$\dfrac{OD}{1} = \dfrac{1}{\cos \theta}$

$OD = \sec \theta$;

(c) $\triangle OAP \sim \triangle ECO$

$\dfrac{OE}{PO} = \dfrac{CO}{AP}$

$\dfrac{OE}{1} = \dfrac{1}{\sin \theta}$

$OE = \csc \theta$;

(d) $\triangle OAP \sim \triangle ECO$

$\dfrac{CE}{AO} = \dfrac{CO}{AP}$

$\dfrac{CE}{\cos \theta} = \dfrac{1}{\sin \theta}$

$CE = \dfrac{\cos \theta}{\sin \theta}$

$CE = \cot \theta$

45. It is true that $\cos \theta \le \sec \theta$ for all θ in the intervals $\left(-\dfrac{\pi}{2} + k \cdot 2\pi, \dfrac{\pi}{2} + k \cdot 2\pi\right)$, k any integer.

Margin Exercises, Section 6.4
1. (a)

(b)

2. (a) 360°; **(b)** 180°; **(c)** 90°; **(d)** 45°; **(e)** 60°; **(f)** 30°
3. (a) I; **(b)** III; **(c)** IV; **(d)** III; **(e)** I; **(f)** IV; **(g)** I
4. Positive: 405°, 765°; negative: −315°, −675°. Answers may vary.   **5. (a)** Obtuse; **(b)** right; **(c)** acute; **(d)** straight

6. (a) $\dfrac{1}{60}$; **(b)** $\dfrac{1}{3600}$; **(c)** 56, 23, 11

7. Complement: 4°47′22″; supplement: 94°47′22″
8. Complement: 55.11°; supplement: 145.11°   **9.** 67.2211°
10. 37°27′   **11. (a)** III; **(b)** I; **(c)** IV; **(d)** III

12. (a) $\dfrac{5\pi}{4}$; **(b)** $\dfrac{7\pi}{4}$; **(c)** -4π

13. (a) 1.26; **(b)** 5.23; **(c)** −5.50
14. (a) 240°; **(b)** 450°; **(c)** −144°   **15.** 57.6 cm   **16.** 6
17. $\frac{3}{2}$ **18.** 377 cm/sec   **19.** r in cm, ω in radians/sec

20. km/yr   **21.** 7.2 radians/sec
22. 70.4 radians, or 11.21 revolutions

Exercise Set 6.4, pp. 393–395
1. I   **3.** III   **5.** I   **7.** II
9. Positive: 418°, 778°; negative: −302°, −662°.
Answers may vary.
11. Positive: 240°, 600°; negative: −480°, −840°.
Answers may vary.   **13.** 32°37′; 122°37′
15. 16°14′49″; 106°14′49″   **17.** 22.69°; 112.69°
19. 78.7656°; 168.7656°   **21.** 8°36′   **23.** 72°15′
25. 46°23′   **27.** 67°50′   **29.** 87°20′44″   **31.** 48°1′30″
33. 9.75°   **35.** 35.83°   **37.** 80.55°   **39.** 3.03°

41. 19.7897°   **43.** 31.9653°   **45.** $\dfrac{\pi}{6}$ **47.** $\dfrac{\pi}{3}$ **49.** $\dfrac{5\pi}{12}$

51. 0.2095π **53.** 1.1922π **55.** 2.094   **57.** 5.585
59. 3.491   **61.** 2.056   **63.** 0.024   **65.** 57.30°
67. 1440°   **69.** 135°   **71.** 74.66°   **73.** 135.576°

75. $30° = \dfrac{\pi}{6}$; $60° = \dfrac{\pi}{3}$, $135° = \dfrac{3\pi}{4}$, $180° = \pi$,

$225° = \dfrac{5\pi}{4}$, $270° = \dfrac{3\pi}{2}$, $315° = \dfrac{7\pi}{4}$ **77.** 1.1 radians, 63°

79. 5.236 radians   **81.** 16 m   **83.** 3150 cm/min
85. 52.4 cm/sec   **87.** 1047 mph   **89.** 68.75 radians/sec
91. 10 mph   **93.** 75.4 ft   **95.** 377 radians
97. **(a)** 53.33; **(b)** 170; **(c)** 25; **(d)** 142.86
99. 111.7 km, 69.8 mi   **101.** 0.03 radian
103. 2.094 radians/sec

105. **(a)** 395.35 rpm/sec, angular acc $= \dfrac{\Delta v}{t}$;

(b) 41.4 radians/sec²
107. 37.5°   **109.** 1.15 statute miles

Margin Exercises, Section 6.5

1. 0   **2.** 0   **3.** $\dfrac{\sqrt{2}}{2}$ **4.** 1   **5.** $-\dfrac{\sqrt{2}}{2}$

6. $\sin\theta = -\dfrac{3}{5}$; $\cos\theta = -\dfrac{4}{5}$; $\tan\theta = \dfrac{3}{4}$; $\cot\theta = \dfrac{4}{3}$;

$\sec\theta = -\dfrac{5}{4}$; $\csc\theta = -\dfrac{5}{3}$

7. $\sin\theta = \dfrac{\sqrt{3}}{2}$; $\cos\theta = -\dfrac{1}{2}$; $\tan\theta = -\sqrt{3}$; $\cot\theta = -\dfrac{1}{\sqrt{3}}$,

or $-\dfrac{\sqrt{3}}{3}$; $\sec\theta = -2$; $\csc\theta = \dfrac{2}{\sqrt{3}}$, or $\dfrac{2\sqrt{3}}{3}$

8. $\sin\theta = -\dfrac{2}{\sqrt{5}}$, or $-\dfrac{2\sqrt{5}}{5}$; $\cos\theta = -\dfrac{1}{\sqrt{5}}$, or $-\dfrac{\sqrt{5}}{5}$;

$\tan\theta = 2$ **9.** $\sin\theta$ and $\csc\theta$: +, +, −, −; $\cos\theta$ and
$\sec\theta$: +, −, −, +; $\tan\theta$ and $\cot\theta$: +, −, +, −
10. The cosine and secant are positive; the other four
function values are negative.
11. The sine and cosecant are positive; the other four
function values are negative.
12. The cosine and secant are positive; the other four
function values are negative.

13. $\sin 270° = -1$; $\cos 270° = 0$; $\tan 270°$: undefined;
$\cot 270° = 0$; $\sec 270°$: undefined; $\csc 270° = -1$

14. **(a)** $\sin\theta = \dfrac{12}{13}$, $\cos\theta = \dfrac{5}{13}$, $\tan\theta = \dfrac{12}{5}$, $\cot\theta = \dfrac{5}{12}$,

$\sec\theta = \dfrac{13}{5}$, $\csc\theta = \dfrac{13}{12}$; **(b)** $\sin\phi = \dfrac{5}{13}$, $\cos\phi = \dfrac{12}{13}$,

$\tan\phi = \dfrac{5}{12}$, $\cot\phi = \dfrac{12}{5}$, $\sec\phi = \dfrac{13}{12}$, $\csc\phi = \dfrac{13}{5}$

15. 127.3 ft

16. $\sin\theta = \dfrac{1}{\sqrt{10}}$, or $\dfrac{\sqrt{10}}{10}$; $\cos\theta = -\dfrac{3}{\sqrt{10}}$, or $-\dfrac{3\sqrt{10}}{10}$;

$\tan\theta = -\dfrac{1}{3}$; $\sec\theta = -\dfrac{\sqrt{10}}{3}$; $\csc\theta = \sqrt{10}$

17. $\sin\theta = -\dfrac{\sqrt{7}}{4}$; $\tan\theta = -\dfrac{\sqrt{7}}{3}$; $\cot\theta = -\dfrac{3}{\sqrt{7}}$,

or $-\dfrac{3\sqrt{7}}{7}$; $\sec\theta = \dfrac{4}{3}$; $\csc\theta = -\dfrac{4}{\sqrt{7}}$, or $-\dfrac{4\sqrt{7}}{7}$

Exercise Set 6.5, pp. 407–408

1. −1   **3.** $\dfrac{\sqrt{2}}{2}$ **5.** $-\dfrac{\sqrt{2}}{2}$ **7.** $-\dfrac{1}{\sqrt{3}}$, or $-\dfrac{\sqrt{3}}{3}$

9. $\sin\theta = \dfrac{5}{13}$; $\cos\theta = -\dfrac{12}{13}$; $\tan\theta = -\dfrac{5}{12}$; $\cot\theta = -\dfrac{12}{5}$;

$\sec\theta = -\dfrac{13}{12}$; $\csc\theta = \dfrac{13}{5}$

11. $\sin\theta = -\dfrac{3}{4}$; $\cos\theta = -\dfrac{\sqrt{7}}{4}$; $\tan\theta = \dfrac{3}{\sqrt{7}}$, or $\dfrac{3\sqrt{7}}{7}$;

$\cot\theta = \dfrac{\sqrt{7}}{3}$; $\sec\theta = -\dfrac{4}{\sqrt{7}}$, or $-\dfrac{4\sqrt{7}}{7}$; $\csc\theta = -\dfrac{4}{3}$

13. $\sin\theta = -\dfrac{2}{\sqrt{13}}$, or $-\dfrac{2\sqrt{13}}{13}$; $\cos\theta = \dfrac{3}{\sqrt{13}}$, or $\dfrac{3\sqrt{13}}{13}$;

$\tan\theta = -\dfrac{2}{3}$

15. $\sin\theta = \dfrac{5}{\sqrt{14}}$, or $\dfrac{5\sqrt{41}}{41}$; $\cos\theta = \dfrac{4}{\sqrt{41}}$, or $\dfrac{4\sqrt{41}}{41}$,

$\tan\theta = \dfrac{5}{4}$

17. The cosine and secant are positive; the other four
function values are negative.
19. The sine and cosecant are positive; the other four
function values are negative.
21. The sine and cosecant are positive; the other four
function values are negative.
23. The sine and cosecant are positive; the other four
function values are negative.
25. $\sin 90° = 1$; $\cos 90° = 0$; $\tan 90°$ is undefined;
$\cot 90° = 0$; $\sec 90°$ is undefined; $\csc 90° = 1$
27. $\sin(-180°) = 0$; $\cos(-180°) = -1$;
$\tan(-180°) = 0$; $\cot(-180°)$ is undefined;
$\sec(-180°) = -1$; $\csc(-180°)$ is undefined.

29. $\sin\theta = \dfrac{8}{17}$; $\cos\theta = \dfrac{15}{17}$; $\tan\theta = \dfrac{8}{15}$; $\cot\theta = \dfrac{15}{8}$;

$\sec\theta = \dfrac{17}{15}$; $\csc\theta = \dfrac{17}{8}$

31. $\sin\theta = \dfrac{3}{h}$; $\cos\theta = \dfrac{7}{h}$; $\tan\theta = \dfrac{3}{7}$; $\cot\theta = \dfrac{7}{3}$;

$\sec\theta = \dfrac{h}{7}$; $\csc\theta = \dfrac{h}{3}$

33. $\sin\theta = 0.8788$; $\cos\theta = 0.4771$; $\tan\theta = 1.8419$; $\cot\theta = 0.5429$; $\sec\theta = 2.0958$; $\csc\theta = 1.1379$

35.

θ	sin	cos	tan
45°	$\sqrt{2}/2$	$\sqrt{2}/2$	1
30°	$1/2$	$\sqrt{3}/2$	$1/\sqrt{3}$
60°	$\sqrt{3}/2$	$1/2$	$\sqrt{3}$

θ	cot	sec	csc
45°	1	$2/\sqrt{2}$	$2/\sqrt{2}$
30°	$\sqrt{3}$	$2/\sqrt{3}$	2
60°	$1/\sqrt{3}$	2	$2/\sqrt{3}$

37. 62.4 m

39. $\cos\theta = -\dfrac{2\sqrt{2}}{3}$; $\tan\theta = \dfrac{1}{2\sqrt{2}}$, or $\dfrac{\sqrt{2}}{4}$; $\cot\theta = 2\sqrt{2}$;

$\sec\theta = -\dfrac{3}{2\sqrt{2}}$, or $-\dfrac{3\sqrt{2}}{4}$; $\csc\theta = -3$

41. $\sin\theta = -\dfrac{4}{5}$; $\tan\theta = -\dfrac{4}{3}$; $\cot\theta = -\dfrac{3}{4}$;

$\sec\theta = \dfrac{5}{3}$; $\csc\theta = -\dfrac{5}{4}$

43. $\sin\theta = -\dfrac{1}{\sqrt{5}}$, or $-\dfrac{\sqrt{5}}{5}$, $\cos\theta = \dfrac{2}{\sqrt{5}}$, or

$\dfrac{2\sqrt{5}}{5}$; $\tan\theta = -\dfrac{1}{2}$; $\sec\theta = \dfrac{\sqrt{5}}{2}$; $\csc\theta = -\sqrt{5}$

45. $\cos\theta = -\dfrac{2\sqrt{2}}{3}$; $\tan\theta = -\dfrac{1}{2\sqrt{2}}$, or $-\dfrac{\sqrt{2}}{4}$;

$\cot\theta = -2\sqrt{2}$; $\sec\theta = -\dfrac{3}{2\sqrt{2}}$, or $-\dfrac{3\sqrt{2}}{4}$; $\csc\theta = 3$

47. $\cos\theta = \dfrac{7}{25}$; $\tan\theta = \dfrac{24}{7}$; $\cot\theta = \dfrac{7}{24}$; $\sec\theta = \dfrac{25}{7}$;

$\csc\theta = \dfrac{25}{24}$

49. $\sin\phi = \dfrac{2}{\sqrt{5}}$, or $\dfrac{2\sqrt{5}}{5}$; $\cos\phi = \dfrac{1}{\sqrt{5}}$, or $\dfrac{\sqrt{5}}{5}$; $\cot\phi = \dfrac{1}{2}$;

$\sec\phi = \sqrt{5}$; $\csc\phi = \dfrac{\sqrt{5}}{2}$

51.

$x = OQ = d_1 + d_2 =$
$R\cos\theta + \sqrt{L^2 - h^2} =$
$R\cos\theta + \sqrt{L^2 - (R\sin\theta)^2} =$
$R\cos\omega t + \sqrt{L^2 - R^2\sin^2\omega t}$

Margin Exercises, Section 6.6

1. $\sin 120° = \dfrac{\sqrt{3}}{2}$; $\cos 120° = -\dfrac{1}{2}$; $\tan 120° = -\sqrt{3}$;

$\cot 120° = -\dfrac{1}{\sqrt{3}}$, or $-\dfrac{\sqrt{3}}{3}$; $\sec 120° = -2$;

$\csc 120° = \dfrac{2}{\sqrt{3}}$, or $\dfrac{2\sqrt{3}}{3}$

2. $\sin 570° = -\dfrac{1}{2}$; $\cos 570° = -\dfrac{\sqrt{3}}{2}$; $\tan 570° = \dfrac{1}{\sqrt{3}}$, or

$\dfrac{\sqrt{3}}{3}$; $\cot 570° = \sqrt{3}$; $\sec 570° = -\dfrac{2}{\sqrt{3}}$, or $-\dfrac{2\sqrt{3}}{3}$;

$\csc 570° = -2$

3. $\sin(-945°) = \dfrac{\sqrt{2}}{2}$; $\cos(-945°) = -\dfrac{\sqrt{2}}{2}$;

$\tan(-945°) = -1$; $\cot(-945°) = -1$;

$\sec(-945°) = -\dfrac{2}{\sqrt{2}}$, or $-\sqrt{2}$; $\csc(-945°) = \dfrac{2}{\sqrt{2}}$, or $\sqrt{2}$

4. $\sin\dfrac{17\pi}{6} = \dfrac{1}{2}$; $\cos\dfrac{17\pi}{6} = -\dfrac{\sqrt{3}}{2}$; $\tan\dfrac{17\pi}{6} = -\dfrac{1}{\sqrt{3}}$, or

$-\dfrac{\sqrt{3}}{3}$; $\cot\dfrac{17\pi}{6} = -\sqrt{3}$; $\sec\dfrac{17\pi}{6} = -\dfrac{2}{\sqrt{3}}$, or $-\dfrac{2\sqrt{3}}{3}$;

$\csc\dfrac{17\pi}{6} = 2$

5. $\sin\left(-\dfrac{22\pi}{3}\right) = \dfrac{\sqrt{3}}{2}$; $\cos\left(-\dfrac{22\pi}{3}\right) = -\dfrac{1}{2}$;

$\tan\left(-\dfrac{22\pi}{3}\right) = -\sqrt{3}$; $\cot\left(-\dfrac{22\pi}{3}\right) = -\dfrac{1}{\sqrt{3}}$, or $-\dfrac{\sqrt{3}}{3}$;

$\sec\left(-\dfrac{22\pi}{3}\right) = -2$; $\csc\left(-\dfrac{22\pi}{3}\right) = \dfrac{2}{\sqrt{3}}$, or $\dfrac{2\sqrt{3}}{3}$

6. $\sin\dfrac{29\pi}{4} = -\dfrac{\sqrt{2}}{2}$; $\cos\dfrac{29\pi}{4} = -\dfrac{\sqrt{2}}{2}$; $\tan\dfrac{29\pi}{4} = 1$;

$\cot\dfrac{29\pi}{4} = 1$; $\sec\dfrac{29\pi}{4} = -\dfrac{2}{\sqrt{2}}$, or $-\sqrt{2}$; $\csc\dfrac{29\pi}{4} = -\dfrac{2}{\sqrt{2}}$,

or $-\sqrt{2}$ **7.** -0.754710 **8.** 0.416260 **9.** 0.292372
10. 1.683835 **11.** -0.379265 **12.** -1.143983
13.–15. The calculator was set in degree mode; the angle, or rotation, measure was in radians.   **16.** 0.4710
17. -0.2301 **18.** 1.7535 **19.** -0.88546
20. 0.36397 **21.** 0.83666 **22.** -0.27676
23. 0.65486 **24.** -5.84248 **25.** $24.80°$ **26.** $50°$
27. $45.50°$ **28.** $-13.67°$ **29.** 2.1788 **30.** $124.83°$
31. 1.5628 **32.** $304.83°$ **33.** $111.98°$ **34.** $193.67°$
35. $213.7°$

Exercise Set 6.6, pp. 417–419

1. $\dfrac{2}{\sqrt{2}}$, or $\sqrt{2}$ **3.** $\dfrac{1}{2}$ **5.** $\sqrt{3}$ **7.** -1 **9.** -1

11. $-\dfrac{1}{2}$ **13.** 1 **15.** $\dfrac{2}{\sqrt{2}}$, or $\sqrt{2}$ **17.** $\dfrac{\sqrt{2}}{2}$ **19.** -1

21. $-\dfrac{\sqrt{2}}{2}$ **23.** $-\sqrt{3}$

25. $\sin\dfrac{23\pi}{6} = -\dfrac{1}{2}$; $\cos\dfrac{23\pi}{6} = \dfrac{\sqrt{3}}{2}$; $\tan\dfrac{23\pi}{6} = -\dfrac{1}{\sqrt{3}}$,

or $-\dfrac{\sqrt{3}}{3}$; $\cot\dfrac{23\pi}{6} = -\sqrt{3}$; $\sec\dfrac{23\pi}{6} = \dfrac{2}{\sqrt{3}}$, or $\dfrac{2\sqrt{3}}{3}$;

$\csc\dfrac{23\pi}{6} = -2$

27. $\sin\left(-\dfrac{19\pi}{3}\right) = -\dfrac{\sqrt{3}}{2}$; $\cos\left(-\dfrac{19\pi}{3}\right) = \dfrac{1}{2}$;

$\tan\left(-\dfrac{19\pi}{3}\right) = -\sqrt{3}$; $\cot\left(-\dfrac{19\pi}{3}\right) = -\dfrac{1}{\sqrt{3}}$, or $-\dfrac{\sqrt{3}}{3}$;

$\sec\left(-\dfrac{19\pi}{3}\right) = 2$; $\csc\left(-\dfrac{19\pi}{3}\right) = -\dfrac{2}{\sqrt{3}}$, or $-\dfrac{2\sqrt{3}}{3}$

29. $\sin\dfrac{31\pi}{3} = \dfrac{\sqrt{3}}{2}$; $\cos\dfrac{31\pi}{3} = \dfrac{1}{2}$; $\tan\dfrac{31\pi}{3} = \sqrt{3}$;

$\cot\dfrac{31\pi}{3} = \dfrac{1}{\sqrt{3}}$, or $\dfrac{\sqrt{3}}{3}$; $\sec\dfrac{31\pi}{3} = 2$; $\csc\dfrac{31\pi}{3} = \dfrac{2}{\sqrt{3}}$, or $\dfrac{2\sqrt{3}}{3}$

31. $\sin(-750°) = -\dfrac{1}{2}$; $\cos(-750°) = \dfrac{\sqrt{3}}{2}$;

$\tan(-750°) = -\dfrac{1}{\sqrt{3}}$, or $-\dfrac{\sqrt{3}}{3}$; $\cot(-750°) = -\sqrt{3}$;

$\sec(-750°) = \dfrac{2}{\sqrt{3}}$, or $\dfrac{2\sqrt{3}}{3}$; $\csc(-750°) = -2$

33. 0.951057 **35.** 0.045410 **37.** 0.885664
39. 0.963241 **41.** 2.050197 **43.** 1.017124
45. -0.362184 **47.** -249.1107 **49.** -2.121903
51. -1.193718 **53.** 0.593653 **55.** 0.188238
57. -0.643538 **59.** -0.839072 **61.** 0
63. 0.680122 **65.** 1.236068 **67.** -4.493959 **69.** -1
71. -0.747022 **73.** $30.83°$, 0.5381 **75.** $49.37°$, 0.8616
77. $82.33°$, 1.4369 **79.** $9.17°$, 0.1600 **81.** $275.38°$
83. $200.15°$ **85.** $193.82°$ **87.** $162.51°$ **89.** 2.5270
91. 1.7839 **93.** 4.7648 **95.** 0.089321
97. 38.25 inches   **99.** 0.0670 radian
101. 0.47943 **103.** $865,000$ miles

105. (a) $t = \dfrac{V\sin\theta + \sqrt{V^2\sin^2\theta + 2gh}}{g}$;

(b) $x = V\cos\theta\,\dfrac{V\sin\theta + \sqrt{V^2\sin^2\theta + 2gh}}{g}$;

(c) 14.026 m/sec; **(d)** x is larger when h is larger; thus a taller shotputter can put the shot a greater distance; **(e)** $t \approx 2.57$ sec, $v \approx 124$ ft/sec

Margin Exercises, Section 6.7

1.

2. The amplitude is $\frac{2}{3}$.

3. The amplitude is 2.

4. The period is 4π.

5. The period is π.

6.

7.

Period π
Phase shift $-\frac{\pi}{2}$

8.

$y = 3 \cos 2x$

$y = 3 \cos (2x - \frac{\pi}{2})$
Amplitude 3
Period π
Phase shift $\frac{\pi}{4}$

9.

$y = -2 \sin (\pi x) + 2$
Amplitude 2
Period 2
Phase shift 0

10.

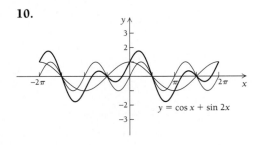

$y = \cos x + \sin 2x$

TC 5.

$f(x) = 5 \sin x$

$[-2\pi, 2\pi] \times [-6, 6]$

TC 6.

$f(x) = -5 \sin x$

$[-2\pi, 2\pi] \times [-6, 6]$

TC 7.

$f(x) = \frac{1}{2} \cos x$

$[-2\pi, 2\pi] \times [-1, 1]$

TC 8.

$f(x) = -\frac{1}{2} \cos x$

$[-2\pi, 2\pi] \times [-1, 1]$

TC 9.

$f(x) = \sin 4x$

$[-2\pi, 2\pi] \times [-2, 2]$

TC 10.

$f(x) = \sin (-4x)$

$[-2\pi, 2\pi] \times [-2, 2]$

Technology Connection, Section 6.7

TC 1.

$f(x) = \sin x + 4$

$[-2\pi, 2\pi] \times [-2, 6]$

TC 2.

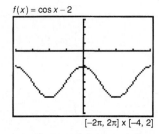

$f(x) = \cos x - 2$

$[-2\pi, 2\pi] \times [-4, 2]$

TC 3.

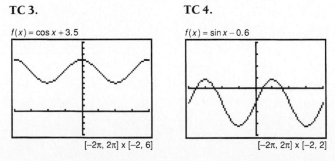

$f(x) = \cos x + 3.5$

$[-2\pi, 2\pi] \times [-2, 6]$

TC 4.

$f(x) = \sin x - 0.6$

$[-2\pi, 2\pi] \times [-2, 2]$

TC 11.

$f(x) = \cos (-\frac{1}{4} x)$

$[-2\pi, 2\pi] \times [-2, 2]$

TC 12.

$f(x) = \cos 2.5x$

$[-2\pi, 2\pi] \times [-2, 2]$

TC 13.

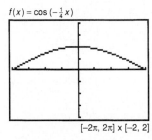

$f(x) = \cos (x - \frac{\pi}{4})$

$[-2\pi, 2\pi] \times [-2, 2]$

TC 14.

$f(x) = \cos (x + 135°)$

$[-360, 360] \times [-2, 2]$

TC 15.

$f(x) = \sin(x + \frac{3\pi}{2})$

[−2π, 2π] x [−2, 2]

TC 16.

$f(x) = \sin(x - 45°)$

[−360, 360] x [−2, 2]

TC 17.

$f(x) = 12.5 \cos(5x - 2.5\pi) - 3$

[−2π, 2π] x [−20, 15]

TC 18.

$f(x) = -20 \sin(-3x - 2\pi) + 1.5$

[−2π, 2π] x [−25, 25]

TC 19.

$f(x) = 0.25 \cos(4x + \frac{\pi}{3}) + 3.5$

[−2π, 2π] x [−1, 5]

TC 20.

$f(x) = 5 \cos x + \cos 5x$

[−2π, 2π] x [−10, 10]

TC 21.

$f(x) = \cos 2x - \sin\frac{1}{2}x$

[−2π, 2π] x [−3, 3]

TC 22.

$f(x) = 4 \cos x + \sin 3x$

[−2π, 2π] x [−5, 5]

Exercise Set 6.7, pp. 432–433

1. Amplitude: 1, period: 2π, phase shift: 0;

3. Amplitude: $\frac{1}{2}$, period: 2π, phase shift: 0;

5. Amplitude: 2, period: 2π, phase shift: 0;

7. Amplitude: $\frac{1}{2}$, period: 2π, phase shift: 0;

9. Amplitude: 1, period: π, phase shift: 0;

11. Amplitude: 1, period: π, phase shift: 0;

13. Amplitude: 1; period: 4π, phase shift: 0;

15. Amplitude: 1, period: 4π, phase shift: 0;

17. Amplitude: 1, period: π, phase shift: $\frac{\pi}{2}$;

19. Amplitude: 2, period: 4π, phase shift: π;

$$y = 2\cos\left(\tfrac{1}{2}x - \tfrac{\pi}{2}\right)$$

21. Amplitude: 3, period: $\dfrac{\pi}{2}$, phase shift: $\dfrac{\pi}{4}$;

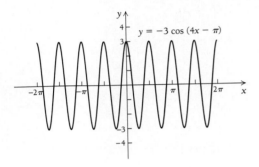

$$y = -3\cos(4x - \pi)$$

23. Amplitude: 3, period: 2, phase shift: $\dfrac{3}{\pi}$;

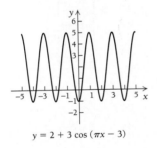

$$y = 2 + 3\cos(\pi x - 3)$$

25. Amplitude, 3; period, $\dfrac{2\pi}{3}$; phase shift, $\dfrac{\pi}{6}$

27. Amplitude, 5; period, $\dfrac{\pi}{2}$; phase shift, $-\dfrac{\pi}{12}$

29. Amplitude, $\dfrac{1}{2}$; period, 1; phase shift, $-\dfrac{1}{2}$

31.

$$y = 2\cos x + \cos 2x$$

33.

$$y = \sin x + \cos 2x$$

35.

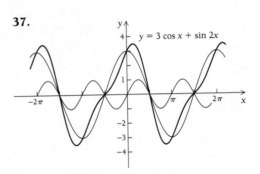

$$y = \sin x - \cos x$$

37.

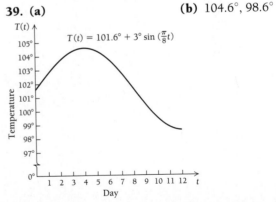

$$y = 3\cos x + \sin 2x$$

39. (a)   **(b)** $104.6°$, $98.6°$

$$T(t) = 101.6° + 3°\sin\left(\tfrac{\pi}{8}t\right)$$

41. (a)   **(b)** Amplitude: 3000, period: 90, phase shift: 10

$$y = 3000\left[\cos\tfrac{\pi}{45}(t - 10)\right]$$

43. $y = 3 - 8.6 \sin\left[\dfrac{2\pi}{5}(x + 11)\right]$; answers may vary.

45. $y = 2 + 16 \cos\left[6\left(x + \dfrac{2}{\pi}\right)\right]$

47.

49.

51.

$[-2\pi, 2\pi] \times [-10, 10]$

53.

$[-2\pi, 2\pi] \times [-8, 8]$

Margin Exercises, Section 6.8

1.

$y = 2 \cot\left(x + \dfrac{\pi}{4}\right)$

2.

$y = -\dfrac{2}{3} \csc\left(x - \dfrac{\pi}{2}\right)$

3.

Technology Connection, Section 6.8

TC 1.

$f(x) = 5 \tan(2x - 3\pi) + 6$

$[-\pi, \pi] \times [-15, 20]$

TC 2.

$f(x) = -4 \csc\left(-x - \dfrac{\pi}{2}\right) + 3.5$

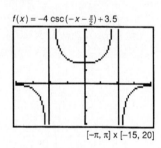

$[-\pi, \pi] \times [-15, 20]$

TC 3.

$f(x) = 2.5 \sec(-3x + 2\pi) - 4.5$

$[-\pi, \pi] \times [-20, 15]$

TC 4.

$f(x) = 0.4 \cot(8x + 2\pi) - 1.2$

$[-\pi/4, \pi/4] \times [-6, 4]$

TC 5.

$f(x) = e^{-0.6x} \cos x$

$[-2\pi, 2\pi] \times [-8, 3]$

TC 6.

$f(x) = 0.2 x^2 \sin x$

$[-2\pi, 2\pi] \times [-6, 6]$

TC 7.

$f(x) - 4^{-x} \cos x$

$[-2\pi, 2\pi] \times [-180, 20]$

Exercise Set 6.8, p. 438

1.

$y = -\tan x$

3.

$y = -\csc x$

5.

$y = \sec(-x)$

7.

$y = \cot(-x)$

9.

$y = -2 + \cot x$

11.

$y = -\frac{3}{2} \csc x$

13.

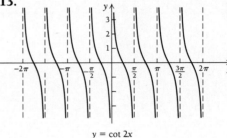

$y = \cot 2x$

15.

$y = \sec\left(-\frac{1}{2}x\right)$

17.

$y = 2 \sec(x - \pi)$

19.

$$y = 4 \tan \left(\tfrac{1}{4}x + \tfrac{\pi}{8}\right)$$

27.

21.

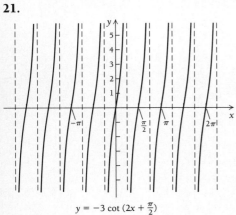

$$y = -3 \cot \left(2x + \tfrac{\pi}{2}\right)$$

29.

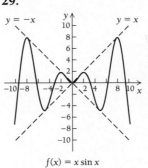

$$f(x) = x \sin x$$

31.

$$f(x) = 2^{-x} \sin x$$

23.

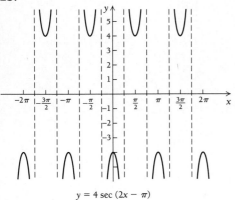

$$y = 4 \sec (2x - \pi)$$

33.

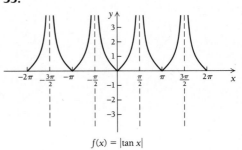

$$f(x) = |\tan x|$$

25.

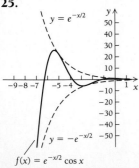

$$f(x) = e^{-x/2} \cos x$$

35.

$$f(x) = (\ln x)(\sin x)$$

37.

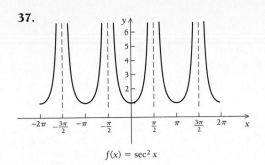

$$f(x) = \sec^2 x$$

39. (a)   **(b)** 4 inches

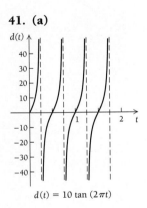

$$d(t) = 6e^{-0.8t} \cos(6\pi t) + 4$$

41. (a)

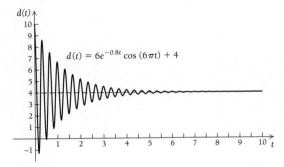

$$d(t) = 10 \tan(2\pi t)$$

(b) The function is undefined when the beam is parallel to the wall.

43. $-3.14, 3.14$ **45.** $-6.28, 6.28$ **47.** $-0.84, -0.41$

Review Exercises: Chapter 6, pp. 439–440

1. [6.1] u-axis: $\left(\frac{3}{5}, \frac{4}{5}\right)$; v-axis: $\left(-\frac{3}{5}, -\frac{4}{5}\right)$; origin: $\left(-\frac{3}{5}, \frac{4}{5}\right)$

2. [6.1]   **3.** [6.2], [6.3] I, IV

4. [6.2], [6.3]

	$\dfrac{\pi}{6}$	$\dfrac{\pi}{4}$	$\dfrac{\pi}{2}$	$\dfrac{2\pi}{3}$	$-\pi$	$-\dfrac{5\pi}{4}$
$\sin x$	$\dfrac{1}{2}$	$\dfrac{\sqrt{2}}{2}$	1	$\dfrac{\sqrt{3}}{2}$	0	$\dfrac{\sqrt{2}}{2}$
$\cos x$	$\dfrac{\sqrt{3}}{2}$	$\dfrac{\sqrt{2}}{2}$	0	$-\dfrac{1}{2}$	-1	$-\dfrac{\sqrt{2}}{2}$
$\cot x$	$\sqrt{3}$	1	0	$-\dfrac{\sqrt{3}}{3}$	Undefined	-1
$\csc x$	2	$\sqrt{2}$	1	$\dfrac{2\sqrt{3}}{3}$	Undefined	$\sqrt{2}$

5. [6.4] I, 0.483π, 1.52   **6.** [6.4] II, 0.806π, 2.53   **7.** [6.4] IV, -0.167π, -0.524 **8.** [6.4] 270°   **9.** [6.4] 171.89°

10. [6.4] Positive: 425°, 785°; negative: $-295°$, $-655°$

11. [6.4] $\dfrac{7\pi}{4}$, or 5.5 cm   **12.** [6.4] 2.25, 129°

13. [6.4] 1131 cm/min   **14.** [6.4] 146,215 radians/hr

15. [6.5] $\sin\theta = \dfrac{3}{\sqrt{13}}$, or $\dfrac{3\sqrt{13}}{13}$; $\cos\theta = -\dfrac{2}{\sqrt{13}}$, or $-\dfrac{2\sqrt{13}}{13}$; $\tan\theta = -\dfrac{3}{2}$; $\cot\theta = -\dfrac{2}{3}$; $\sec\theta = -\dfrac{\sqrt{13}}{2}$; $\csc\theta = \dfrac{\sqrt{13}}{3}$

16. [6.5] $\sin\theta = \dfrac{3}{\sqrt{73}}$, or $\dfrac{3\sqrt{73}}{73}$; $\cos\theta = \dfrac{8}{\sqrt{73}}$, or $\dfrac{8\sqrt{73}}{73}$; $\tan\theta = \dfrac{3}{8}$; $\cot\theta = \dfrac{8}{3}$; $\sec\theta = \dfrac{\sqrt{73}}{8}$; $\csc\theta = \dfrac{\sqrt{73}}{3}$

17. [6.6] $\dfrac{\sqrt{2}}{2}$ **18.** [6.6] 1   **19.** [6.6] $\sqrt{3}$ **20.** [6.6] $-\dfrac{\sqrt{3}}{2}$

21. [6.6] Undefined   **22.** [6.6] $-\dfrac{2\sqrt{3}}{3}$

23. [6.5] $\sin\theta = -\dfrac{2}{3}$; $\cos\theta = -\dfrac{\sqrt{5}}{3}$; $\cot\theta = \dfrac{\sqrt{5}}{2}$; $\sec\theta = -\dfrac{3\sqrt{5}}{5}$; $\csc\theta = -\dfrac{3}{2}$ **24.** [6.4] 22°12′

25. [6.4] 47.5575°   **26.** [6.6] 0.185095
27. [6.6] 0.993577   **28.** [6.6] 8.5796   **29.** [6.6] 187°
30. [6.6] 105.5°   **31.** [6.6] 0.2792   **32.** [6.6] 1.4108
33. [6.6] -0.9056 **34.** [6.6] -1.7013
35. [6.6] -1.7744 **36.** [6.6] Undefined
37. [6.6] -6.1685 **38.** [6.6] 0.2225
39. [6.2]

40. [6.3]

$y = \cot x$

41. [6.3]

$y = \csc x$

42. [6.2] Period: 2π; range: all reals from -1 to 1, inclusive

43. [6.2] Amplitude: 1; domain: all reals **44.** [6.3] Period: 2π; range: all real numbers 1 and greater, in addition to all real numbers -1 and less

45. [6.7]

$y = \sin\left(x + \frac{\pi}{2}\right)$

46. [6.7]

$y = 3 + \cos\left(x - \frac{\pi}{4}\right)$

47. [6.7] $\dfrac{\pi}{4}$ **48.** [6.7] 2π

49. [6.7]

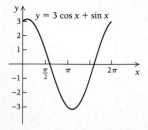

$y = 3\cos x + \sin x$

50. [6.8]

$y = 2\tan\left(x - \frac{\pi}{2}\right)$

51. [6.8]

$f(x) = e^{-0.7x}\cos x$

52. [6.2] No, $\sin x = \dfrac{7}{5}$, but sines are never greater than 1.

53. [6.2] All values
54. [6.7] Domain: set of all reals; range: $[-3, 3]$; period: 4π

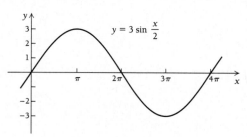

$y = 3\sin\dfrac{x}{2}$

55. [6.2] The domain consists of the intervals $\left(-\dfrac{\pi}{2} + 2k\pi, \dfrac{\pi}{2} + 2k\pi\right)$, k any integer

56. [6.5] $\cos x = -0.7890$; $\tan x = -0.7787$; $\cot x = -1.2842$; $\sec x = -1.2674$; $\csc x = 1.6276$

Test: Chapter 6, pp. 440–441
1. [6.1] Origin: $\left(\frac{4}{5}, \frac{2}{5}\right)$; u-axis: $\left(-\frac{4}{5}, \frac{2}{5}\right)$; v-axis: $\left(\frac{4}{5}, -\frac{2}{5}\right)$
2. [6.1]

3. [6.2]

4. [6.2] All reals from -1 to 1 inclusive   **5.** [6.2] 2π
6. [6.2] All reals   **7.** [6.2] 1
8. [6.2]

	$\pi/2$	$-4\pi/2$	$7\pi/2$
$\sin x$	1	0	-1
$\cos x$	0	1	0

9. [6.3]

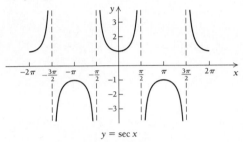

$y = \sec x$

10. [6.3] 2π

11. [6.3] All reals except $\dfrac{\pi}{2} + k\pi$ **12.** [6.2], [6.3] II, IV

13. [6.4] III, $-\dfrac{5\pi}{6}$, -2.62 **14.** [6.4] II, $\dfrac{13\pi}{20}$, 2.04

15. [6.4] III, $\dfrac{5\pi}{4}$, 3.93   **16.** [6.4] 315°   **17.** [6.4] 114.59°

18. [6.4] 10.47 cm   **19.** [6.4] $\dfrac{1}{2}$, 29°   **20.** [6.4] 2010

21. [6.4] 1250

22. [6.5] $\sin\theta = \dfrac{5}{\sqrt{34}}$, or $\dfrac{5\sqrt{34}}{34}$; $\cos\theta = -\dfrac{3}{\sqrt{34}}$,

or $-\dfrac{3\sqrt{34}}{34}$; $\tan\theta = -\dfrac{5}{3}$; $\cot\theta = -\dfrac{3}{5}$; $\sec\theta = -\dfrac{\sqrt{34}}{3}$;

$\csc\theta = \dfrac{\sqrt{34}}{5}$ **23.** [6.6] -1 **24.** [6.6] $-\dfrac{1}{2}$ **25.** [6.6] $\dfrac{1}{2}$

26. [6.5] $\cos\theta = -\dfrac{\sqrt{15}}{4}$; $\tan\theta = -\dfrac{\sqrt{15}}{15}$; $\cot\theta = -\sqrt{15}$;

$\sec\theta = -\dfrac{4\sqrt{15}}{15}$; $\csc\theta = 4$ **27.** [6.4] 19°15′

28. [6.4] 72.8°   **29.** [6.6] -0.830949 **30.** [6.6] 2.80201
31. [6.6] 0.614213   **32.** [6.6] 1.27396
33. [6.6] -3.73281 **34.** [6.6] 0.781286
35. [6.6] 60.62°, 1.0580   **36.** [6.6] 7.13°, 0.1244
37. [6.6] 154.50°   **38.** [6.6] -1.4834

39. [6.7]

$y = 2\cos\left(x - \dfrac{\pi}{4}\right)$

40. [6.7] 2π

41. [6.7] $\dfrac{\pi}{4}$

42. [6.7]

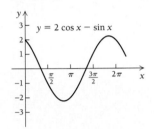

43. [6.5] **(a)** 0.99518; **(b)** 0.99518   **44.** [6.4] 132.5°

CHAPTER 7

Margin Exercises, Section 7.1

1. $\cos x + 1$ **2.** $\sin x$ **3.** $\sec\theta$ **4.** $\dfrac{1 + \sin x}{1 - \cot x}$

5. $\dfrac{5\sin x - \cos x}{\sin^2 x - \cos^2 x}$ **6.** $\dfrac{2}{\cos^2 x(\cot x - 2)}$

7. $\tan x \sin x \sqrt{\sin x}$ **8.** $\dfrac{\cos x}{\sqrt{3\cos x}}$

9. $\sqrt{4 - x^2} = 2\cos\theta$; $\cos\theta = \dfrac{\sqrt{4 - x^2}}{2}$; $\tan\theta = \dfrac{x}{\sqrt{4 - x^2}}$

Technology Connection, Section 7.1
TC 1. A, B, C   **TC 2.** A, B, C, E

Exercise Set 7.1, pp. 448–450
1. $\sin^2 x - \cos^2 x$ **3.** $\sin x - \sec x$ **5.** $\sin y + \cos y$
7. $\cot x - \tan x$ **9.** $1 - 2\sin y\cos y$
11. $2\tan x + \sec^2 x$ **13.** $\sin^2 y + \csc^2 y - 2$
15. $\cos^3 x - \sec^3 x$ **17.** $\cot^3 x - \tan^3 x$
19. $\cos^2 x$ **21.** $\cos x(\sin x + \cos x)$
23. $(\sin\theta - \cos\theta)(\sin\theta + \cos\theta)$ **25.** $\sin x(\sec x + 1)$
27. $(\sin x + \cos x)(\sin x - \cos x)$ **29.** $3(\cot y + 1)^2$
31. $(\csc^2 x + 5)(\csc x + 1)(\csc x - 1)$
33. $(\sin y + 3)(\sin^2 y - 3\sin y + 9)$
35. $(\sin v - \csc v)(\sin^2 v + 1 + \csc^2 v)$ **37.** $\tan x$
39. $\dfrac{2\cos^2 x}{9\sin x}$ **41.** $\cos x - 1$ **43.** $\cos\alpha + 1$

45. $\dfrac{2 \tan x + 1}{3 \sin x + 1}$ **47.** $\sec \theta$ **49.** 1

51. $\dfrac{1}{2 \cos x}$, or $\dfrac{1}{2} \sec x$ **53.** $\dfrac{3 \tan x}{\cos x - \sin x}$ **55.** $\dfrac{1}{3} \cot \gamma$

57. $\dfrac{1 - 2 \sin y + 2 \cos y}{\sin^2 y - \cos^2 y}$ **59.** -1 **61.** $\dfrac{5(\sin x - 3)}{3}$

63. $\sin x \cos x$ **65.** $\sqrt{\sin y}\,(\sin y + \cos y)$

67. $\sin x + \cos x$ **69.** $1 - \sin y$

71. $\sin x (\sqrt{2} + \sqrt{\cos x})$ **73.** $\dfrac{\sqrt{\sin x \cos x}}{\cos x}$ **75.** $\dfrac{\sqrt{\cos x}}{\cot x}$

77. $\dfrac{\sqrt{2} \cot x}{2}$ **79.** $\dfrac{\cos x}{1 - \sin x}$ **81.** $\dfrac{\sin x}{\sqrt{\cos x \sin x}}$

83. $\dfrac{\sin x}{\sqrt{\cos x}}$ **85.** $\dfrac{\cot x}{\sqrt{2}}$ **87.** $\dfrac{1 + \sin x}{\cos x}$

89. $\cos \theta = \dfrac{\sqrt{a^2 - x^2}}{a}$; $\tan \theta = \dfrac{x}{\sqrt{a^2 - x^2}}$

91. $\sin \theta = \dfrac{\sqrt{x^2 - 9}}{x}$; $\cos \theta = \dfrac{3}{x}$ **93.** $\sin \theta \tan \theta$

95. Let $x = \dfrac{\pi}{4}$. Then $\left(\sin \dfrac{\pi}{4} + \cos \dfrac{\pi}{4}\right)^2 = \left(\dfrac{\sqrt{2}}{2} + \dfrac{\sqrt{2}}{2}\right)^2 =$

$(\sqrt{2})^2 = 2$, and $\sin^2 \dfrac{\pi}{4} + \cos^2 \dfrac{\pi}{4} = \left(\dfrac{\sqrt{2}}{2}\right)^2 + \left(\dfrac{\sqrt{2}}{2}\right)^2 =$

$\dfrac{2}{4} + \dfrac{2}{4} = 1.$

97. Let $x = \dfrac{\pi}{6}$ and $y = \dfrac{\pi}{4}$. Then $\dfrac{\sin \dfrac{\pi}{6}}{\sin \dfrac{\pi}{4}} = \dfrac{\dfrac{1}{2}}{\dfrac{\sqrt{2}}{2}} = \dfrac{1}{\sqrt{2}}$, and

$\dfrac{x}{y} = \dfrac{\dfrac{\pi}{6}}{\dfrac{\pi}{4}} = \dfrac{2}{3}.$

99. Let $\alpha = \dfrac{\pi}{3}$ and $\beta = \dfrac{2\pi}{3}$. Then $\cos\left(\dfrac{\pi}{3} + \dfrac{2\pi}{3}\right) =$

$\cos \pi = -1$, and $\cos \dfrac{\pi}{3} + \cos \dfrac{2\pi}{3} = \dfrac{1}{2} + \left(-\dfrac{1}{2}\right) = 0.$

101. Let $\theta = \dfrac{\pi}{2}$. Then $\cos\left(2 \cdot \dfrac{\pi}{2}\right) = \cos \pi = -1$,

and $2 \cos \dfrac{\pi}{2} = 2 \cdot 0 = 0.$

103. Let $\theta = \dfrac{\pi}{2}$. Then $\ln\left(\sin \dfrac{\pi}{2}\right) = \ln(1) = 0$, and

$\sin\left(\ln \dfrac{\pi}{2}\right) = \sin(0.451583) = 0.436390.$

Margin Exercises, Section 7.2

1. $\dfrac{1}{2}$ **2.** $\dfrac{-\sqrt{6} + \sqrt{2}}{4}$, or $\dfrac{\sqrt{2}}{4}(1 - \sqrt{3})$ **3.** $-\dfrac{\sqrt{2} + \sqrt{6}}{4}$

4. $\dfrac{\sqrt{2} + \sqrt{6}}{4}$, or $\dfrac{\sqrt{2}}{4}(1 + \sqrt{3})$ **5.** $\sin(\alpha - \beta)$

6. $\tan(\alpha - \beta) = \dfrac{\sin(\alpha - \beta)}{\cos(\alpha - \beta)}$

$= \dfrac{\sin \alpha \cos \beta - \cos \alpha \sin \beta}{\cos \alpha \cos \beta + \sin \alpha \sin \beta} \cdot \dfrac{\dfrac{1}{\cos \alpha \cos \beta}}{\dfrac{1}{\cos \alpha \cos \beta}}$

$= \dfrac{\dfrac{\sin \alpha \cos \beta}{\cos \alpha \cos \beta} - \dfrac{\cos \alpha \sin \beta}{\cos \alpha \cos \beta}}{\dfrac{\cos \alpha \cos \beta}{\cos \alpha \cos \beta} + \dfrac{\sin \alpha \sin \beta}{\cos \alpha \cos \beta}}$

$= \dfrac{\tan \alpha - \tan \beta}{1 + \tan \alpha \tan \beta}$

7. $\dfrac{3 + \sqrt{3}}{3 - \sqrt{3}}$ **8.** $\cos \dfrac{7\pi}{12}$, or $\cos\left(-\dfrac{7\pi}{12}\right)$ **9.** $\cos 25°$

10. $\sin \dfrac{\pi}{6}$, or $\dfrac{1}{2}$ **11.** $\dfrac{3 + \sqrt{105}}{16}$

Exercise Set 7.2, pp. 455–457

1. $\dfrac{\sqrt{6} + \sqrt{2}}{4}$ **3.** $\dfrac{\sqrt{6} - \sqrt{2}}{4}$ **5.** $\dfrac{\sqrt{6} + \sqrt{2}}{4}$ **7.** $\dfrac{3 + \sqrt{3}}{3 - \sqrt{3}}$

9. $\sin 59°$, or 0.8572 **11.** $\tan 52°$, or 1.280 **13.** 1

15. 0   **17.** Undefined   **19.** 0.8448 **21.** $2 \sin \alpha \cos \beta$

23. $2 \cos \alpha \cos \beta$ **25.** $\cos u$ **27.** $0°$ **29.** $135°$

31. $22.83°$

33. $\dfrac{\cos(x + h) - \cos x}{h} = \dfrac{\cos x \cos h - \sin x \sin h - \cos x}{h}$

$= \dfrac{\cos x \cos h - \cos x}{h} - \dfrac{\sin x \sin h}{h}$

$= \cos x \left(\dfrac{\cos h - 1}{h}\right) - \sin x \left(\dfrac{\sin h}{h}\right)$

35. $\dfrac{1}{7}$ **37.** $\dfrac{1}{6}$ **39.** $2 \sin \theta \cos \theta$

41. $\cot(\alpha + \beta) = \dfrac{\cos(\alpha + \beta)}{\sin(\alpha + \beta)}$

$= \dfrac{\cos \alpha \cos \beta - \sin \alpha \sin \beta}{\sin \alpha \cos \beta + \cos \alpha \sin \beta} \cdot \dfrac{\dfrac{1}{\sin \alpha \sin \beta}}{\dfrac{1}{\sin \alpha \sin \beta}}$

$= \dfrac{\dfrac{\cos \alpha \cos \beta}{\sin \alpha \sin \beta} - 1}{\dfrac{\sin \alpha \cos \beta}{\sin \alpha \sin \beta} + \dfrac{\cos \alpha \sin \beta}{\sin \alpha \sin \beta}} = \dfrac{\cot \alpha \cot \beta - 1}{\cot \alpha + \cot \beta}$

43. $\sin\left(x + \dfrac{3\pi}{2}\right) = \sin x \cos \dfrac{3\pi}{2} + \cos x \sin \dfrac{3\pi}{2}$

$= (\sin x)(0) + (\cos x)(-1) = -\cos x$

45. $\cos\left(x + \dfrac{3\pi}{2}\right) = \cos x \cos \dfrac{3\pi}{2} - \sin x \sin \dfrac{3\pi}{2}$

$= (\cos x)(0) - (\sin x)(-1) = \sin x$

47. $\tan\left(x+\dfrac{\pi}{4}\right)=\dfrac{\tan x+\tan\dfrac{\pi}{4}}{1-\tan x\tan\dfrac{\pi}{4}}=\dfrac{1+\tan x}{1-\tan x}$

49. $\dfrac{\tan\alpha+\tan\beta}{1+\tan\alpha\tan\beta}=\dfrac{\dfrac{\sin\alpha}{\cos\alpha}+\dfrac{\sin\beta}{\cos\beta}}{1+\dfrac{\sin\alpha}{\cos\alpha}\cdot\dfrac{\sin\beta}{\cos\beta}}$

$=\dfrac{\dfrac{\sin\alpha\cos\beta+\cos\alpha\sin\beta}{\cos\alpha\cos\beta}}{\dfrac{\cos\alpha\cos\beta+\sin\alpha\sin\beta}{\cos\alpha\cos\beta}}$

$=\dfrac{\sin\alpha\cos\beta+\cos\alpha\sin\beta}{\cos\alpha\cos\beta+\sin\alpha\sin\beta}$

$=\dfrac{\sin(\alpha+\beta)}{\cos(\alpha-\beta)}$

51. $\sin(\alpha+\beta)+\sin(\alpha-\beta)=\sin\alpha\cos\beta+\cos\alpha\sin\beta+\sin\alpha\cos\beta-\cos\alpha\sin\beta=2\sin\alpha\cos\beta$

53. $\cos(\alpha+\beta)=\cos\alpha\cos\beta-\sin\alpha\sin\beta$

$=\cos\alpha\cos\beta-\cos\left(\dfrac{\pi}{2}-\alpha\right)\cos\left(\dfrac{\pi}{2}-\beta\right)$

Margin Exercises, Section 7.3

1. $\cot\left(\dfrac{\pi}{2}-\theta\right)=\dfrac{\cos\left(\dfrac{\pi}{2}-\theta\right)}{\sin\left(\dfrac{\pi}{2}-\theta\right)}=\dfrac{\sin\theta}{\cos\theta}=\tan\theta$

2. $\sec\left(\dfrac{\pi}{2}-\theta\right)=\dfrac{1}{\cos\left(\dfrac{\pi}{2}-\theta\right)}=\dfrac{1}{\sin\theta}=\csc\theta$

3. $\csc\left(\dfrac{\pi}{2}-\theta\right)=\dfrac{1}{\sin\left(\dfrac{\pi}{2}-\theta\right)}=\dfrac{1}{\cos\theta}=\sec\theta$

4. $\sin 15°=0.2588$; $\cos 15°=0.9659$; $\tan 15°=0.2679$; $\cot 15°=3.732$; $\sec 15°=1.035$; $\csc 15°=3.864$

5. (a) and **(b)**

(c)

(d) the same; **(e)** $\cos\left(x-\dfrac{\pi}{2}\right)=\sin x$

6. $\cos\left(x-\dfrac{\pi}{2}\right)=\cos x\cos\dfrac{\pi}{2}+\sin x\sin\dfrac{\pi}{2}$

$=\cos x\cdot 0+\sin x\cdot 1=\sin x$

7. $\sin\left(x-\dfrac{\pi}{2}\right)=\sin x\cos\dfrac{\pi}{2}-\cos x\sin\dfrac{\pi}{2}$

$=\sin x\cdot 0-\cos x\cdot 1=-\cos x$

8. (a) Translate the graph of $y=\cos x$ to obtain the graph of $y=\cos\left(x+\dfrac{\pi}{2}\right)$. Then graph $y=\sin x$ and reflect the graph to obtain $y=-\sin x$.

The graphs of $y=\cos\left(x+\dfrac{\pi}{2}\right)$ and $y=-\sin x$ are the same. Thus, $\cos\left(x+\dfrac{\pi}{2}\right)=-\sin x$.

(b) $\cos\left(x+\dfrac{\pi}{2}\right)=\cos x\cos\dfrac{\pi}{2}-\sin x\sin\dfrac{\pi}{2}$

$=\cos x\cdot 0-\sin x\cdot 1=-\sin x$

9. $\cot\left(x+\dfrac{\pi}{2}\right)=\dfrac{\cos\left(x+\dfrac{\pi}{2}\right)}{\sin\left(x+\dfrac{\pi}{2}\right)}=\dfrac{-\sin x}{\cos x}=-\tan x$

10. $\csc\left(x-\dfrac{\pi}{2}\right)=\dfrac{1}{\sin\left(x-\dfrac{\pi}{2}\right)}=\dfrac{1}{-\cos x}=-\sec x$

11. $\sin(x+\pi)=-\sin x$ **12.** $\tan(\theta-450°)=-\cot\theta$

TC 1. No   **TC 2.** Yes   **TC 3.** Yes   **TC 4.** Yes

Exercise Set 7.3, pp. 462–463

1. $\sin 25°=0.4226$; $\cos 25°=0.9063$; $\tan 25°=0.4663$; $\cot 25°=2.145$; $\sec 25°=1.103$; $\csc 25°=2.366$

3. (a) $\tan 22.5°=\dfrac{\sqrt{2-\sqrt{2}}}{\sqrt{2+\sqrt{2}}}$, $\cot 22.5°=\dfrac{\sqrt{2+\sqrt{2}}}{\sqrt{2-\sqrt{2}}}$,

$\sec 22.5°=\dfrac{2}{\sqrt{2+\sqrt{2}}}$, $\csc 22.5°=\dfrac{2}{\sqrt{2-\sqrt{2}}}$;

(b) $\sin 67.5° = \dfrac{\sqrt{2 + \sqrt{2}}}{2}$, $\cos 67.5° = \dfrac{\sqrt{2 - \sqrt{2}}}{2}$,

$\tan 67.5° = \dfrac{\sqrt{2 + \sqrt{2}}}{\sqrt{2 - \sqrt{2}}}$, $\cot 67.5° = \dfrac{\sqrt{2 - \sqrt{2}}}{\sqrt{2 + \sqrt{2}}}$,

$\sec 67.5° = \dfrac{2}{\sqrt{2 - \sqrt{2}}}$, $\csc 67.5° = \dfrac{2}{\sqrt{2 + \sqrt{2}}}$

5. $\tan\left(x - \dfrac{\pi}{2}\right) = \dfrac{\sin\left(x - \dfrac{\pi}{2}\right)}{\cos\left(x - \dfrac{\pi}{2}\right)} = \dfrac{-\cos x}{\sin x} = -\cot x$

7. $\csc\left(x + \dfrac{\pi}{2}\right) = \dfrac{1}{\sin\left(x + \dfrac{\pi}{2}\right)} = \dfrac{1}{\cos x} = \sec x$

9. $\sin(x - \pi) = \sin x \cos \pi - \cos x \sin \pi$
$= (\sin x)(-1) - (\cos x)(0) = -\sin x$

11. $\tan(\theta + 270°) = -\cot \theta$

13. $\sin(\theta + 450°) = \cos \theta$ **15.** $\sec(\pi - \theta) = -\sec \theta$

17. $\tan(x - 4\pi) = \tan x$ **19.** $\cos\left(\dfrac{9\pi}{2} + x\right) = -\sin x$

21. $\csc(540° - \theta) = \csc \theta$ **23.** $\cot\left(x - \dfrac{3\pi}{2}\right) = -\tan x$

25. (a) $\cos \theta = -\dfrac{2\sqrt{2}}{3}$; $\tan \theta = -\dfrac{1}{2\sqrt{2}}$, or $-\dfrac{\sqrt{2}}{4}$;

$\cot \theta = -2\sqrt{2}$; $\sec \theta = -\dfrac{3}{2\sqrt{2}}$, or $-\dfrac{3\sqrt{2}}{4}$; $\csc \theta = 3$;

(b) $\sin\left(\dfrac{\pi}{2} - \theta\right) = -\dfrac{2\sqrt{2}}{3}$, $\cos\left(\dfrac{\pi}{2} - \theta\right) = \dfrac{1}{3}$,

$\tan\left(\dfrac{\pi}{2} - \theta\right) = -2\sqrt{2}$, $\cot\left(\dfrac{\pi}{2} - \theta\right) = -\dfrac{\sqrt{2}}{4}$,

$\sec\left(\dfrac{\pi}{2} - \theta\right) = 3$, $\csc\left(\dfrac{\pi}{2} - \theta\right) = -\dfrac{3\sqrt{2}}{4}$;

(c) $\sin(\pi + \theta) = -\dfrac{1}{3}$, $\cos(\pi + \theta) = \dfrac{2\sqrt{2}}{3}$,

$\tan(\pi + \theta) = -\dfrac{\sqrt{2}}{4}$, $\cot(\pi + \theta) = -2\sqrt{2}$,

$\sec(\pi + \theta) = \dfrac{3\sqrt{2}}{4}$, $\csc(\pi + \theta) = -3$;

(d) $\sin(\pi - \theta) = \dfrac{1}{3}$, $\cos(\pi - \theta) = \dfrac{2\sqrt{2}}{3}$,

$\tan(\pi - \theta) = \dfrac{\sqrt{2}}{4}$, $\cot(\pi - \theta) = 2\sqrt{2}$,

$\sec(\pi - \theta) = \dfrac{3\sqrt{2}}{4}$, $\csc(\pi - \theta) = 3$;

(e) $\sin(2\pi - \theta) = -\dfrac{1}{3}$, $\cos(2\pi - \theta) = -\dfrac{2\sqrt{2}}{3}$,

$\tan(2\pi - \theta) = \dfrac{\sqrt{2}}{4}$, $\cot(2\pi - \theta) = 2\sqrt{2}$,

$\sec(2\pi - \theta) = -\dfrac{3\sqrt{2}}{4}$, $\csc(2\pi - \theta) = -3$

27. (a) $\cos 27° = 0.89101$, $\tan 27° = 0.50952$,
$\cot 27° = 1.9626$, $\sec 27° = 1.1223$, $\csc 27° = 2.2027$;
(b) $\sin 63° = 0.89101$, $\cos 63° = 0.45399$, $\tan 63° = 1.9626$,
$\cot 63° = 0.50952$, $\sec 63° = 2.2027$, $\csc 63° = 1.1223$
29. $\sin 128° = 0.78801$, $\cos 128° = -0.61566$,
$\tan 128° = -1.2799$, $\cot 128° = -0.78129$,
$\sec 128° = -1.6243$, $\csc 128° = 1.2690$
31. $\sin^2 x$ **33.** $\sin x(\cos y + \tan y)$

35. $-\cos x(1 + \cot x)$ **37.** $\cos x + 1$ **39.** $\dfrac{\cos x - 1}{1 + \tan x}$

41. $\cos x - 1$ **43.** $-\tan^2 y$

Margin Exercises, Section 7.4
1. $24/25$
2. II, $\cos 2\theta = -119/169$, $\tan 2\theta = -120/119$,
$\sin 2\theta = 120/169$ **3.** $\cos \theta - 4\sin^2 \theta \cos \theta$

4. $\sin^3 x = \sin x \dfrac{1 - \cos 2x}{2}$ **5.** $\dfrac{\sqrt{2 + \sqrt{3}}}{2}$ **6.** $\dfrac{1}{2 + \sqrt{3}}$,

or $2 - \sqrt{3}$

7. $\tan 2x$ **8.** $\dfrac{\frac{1}{2}(\cos^2 x - \sin^2 x)}{\sin x \cos x} = \cot 2x$

Technology Connection, Section 7.4
TC 1. B and D **TC 2.** A and D

Exercise Set 7.4, pp. 468–469

1. $\sin 2\theta = \dfrac{24}{25}$, $\cos 2\theta = -\dfrac{7}{25}$, $\tan 2\theta = -\dfrac{24}{7}$, II

3. $\sin 2\theta = \dfrac{24}{25}$, $\cos 2\theta = \dfrac{7}{25}$, $\tan 2\theta = \dfrac{24}{7}$, I

5. $\sin 2\theta = \dfrac{24}{25}$, $\cos 2\theta = -\dfrac{7}{25}$, $\tan 2\theta = -\dfrac{24}{7}$, II

7. $8\sin \theta \cos^3 \theta - 4\sin \theta \cos \theta$, or $4\sin \theta \cos^3 \theta - 4\sin^3 \theta \cos \theta$, or $4\sin \theta \cos \theta - 8\sin^3 \theta \cos \theta$

9. $\dfrac{3 - 4\cos 2\theta + \cos 4\theta}{8}$ **11.** $\dfrac{\sqrt{2 + \sqrt{3}}}{2}$ **13.** $2 + \sqrt{3}$

15. $\dfrac{\sqrt{2 + \sqrt{2}}}{2}$ **17.** $\dfrac{\sqrt{2 + \sqrt{2}}}{2}$ **19.** $\dfrac{\sqrt{2 + \sqrt{2}}}{2}$

21. $\dfrac{\sqrt{2 - \sqrt{2}}}{2}$ **23.** 0.6421 **25.** 0.9844 **27.** 0.1735

29. $\cos x$ **31.** $\cos x$ **33.** $\sin x$ **35.** $\cos x$ **37.** 1
39. 1 **41.** 8 **43.** $\sin 2x$
45. $(\sin x + \cos x)^2 = 1 + \sin 2x$

47. $\dfrac{2\cot x}{\cot^2 x - 1} = \tan 2x$ **49.** $2\sin^2 2x + \cos 4x = 1$

51. $\sin \theta = \sqrt{\dfrac{1}{2} + \dfrac{\sqrt{6}}{5}}$; $\cos \theta = \sqrt{\dfrac{1}{2} - \dfrac{\sqrt{6}}{5}}$;

$\tan \theta = \sqrt{\dfrac{5 + 2\sqrt{6}}{5 - 2\sqrt{6}}}$

53. $\sin \theta = -\dfrac{8}{17}$; $\cos \theta = \dfrac{15}{17}$; $\tan \theta = -\dfrac{8}{15}$

$\sec(2\pi - \theta) = -\dfrac{3\sqrt{2}}{4}$, $\csc(2\pi - \theta) = -3$

55. **(a)** 6062.76 ft; **(b)** 6097 ft;
(c) $N(\phi) = 6097 - 62 \cos^2 \phi$
57. Since $\cos^2 x - \sin^2 x = \cos 2x$, the graph of $f(x) = \cos^2 x - \sin^2 x$ is the same as the graph of $f(x) = \cos 2x$.

Margin Exercises, Section 7.5

1. *Method 1:*

Left side	Right side
$\dfrac{\csc t - 1}{t \csc t}$	$\dfrac{1 - \sin t}{t}$
$\dfrac{\dfrac{1}{\sin t} - 1}{t \cdot \dfrac{1}{\sin t}}$	
$\dfrac{\dfrac{1}{\sin t} - 1}{t \cdot \dfrac{1}{\sin t}} \cdot \dfrac{\sin t}{\sin t}$	
$\dfrac{1 - \sin t}{t}$	

2. *Method 1:*

Left side	Right side
$\sec \theta \csc \theta$	$\tan \theta + \cot \theta$
$\dfrac{1}{\cos \theta} \cdot \dfrac{1}{\sin \theta}$	
$\dfrac{1}{\cos \theta \sin \theta} \cdot \dfrac{\sin^2 \theta + \cos^2 \theta}{1}$	
	$\dfrac{\sin \theta}{\cos \theta} + \dfrac{\cos \theta}{\sin \theta}$
	$\tan \theta + \cot \theta$

3. *Method 1:*

Left side	Right side
$(\sin \theta + \cos \theta)^2$	$1 - \sin 2\theta$
	$1 - 2 \sin \theta \cos \theta$
	$(\sin^2 \theta + \cos^2 \theta) - 2 \sin \theta \cos \theta$
	$\sin^2 \theta - 2 \sin \theta \cos \theta + \cos^2 \theta$
	$(\sin \theta - \cos \theta)^2$

4. *Method 1:*

Left side	Right side
$(\sec u - \tan u)(1 + \sin u)$	$\cos u$
$\left(\dfrac{1}{\cos u} - \dfrac{\sin u}{\cos u}\right)(1 + \sin u)$	
$\dfrac{1}{\cos u} + \dfrac{\sin u}{\cos u} - \dfrac{\sin u}{\cos u} - \dfrac{\sin^2 u}{\cos u}$	
$\dfrac{1 - \sin^2 u}{\cos u}$	
$\dfrac{\cos^2 u}{\cos u}$	
$\cos u$	

5. *Method 1:*

Left side	Right side
$\dfrac{\sin t}{1 + \cos t}$	$\dfrac{1 - \cos t}{\sin t}$
$\dfrac{\sin t}{1 + \cos t} \cdot \dfrac{1 - \cos t}{1 - \cos t}$	
$\dfrac{(\sin t)(1 - \cos t)}{1 - \cos^2 t}$	
$\dfrac{(\sin t)(1 - \cos t)}{\sin^2 t}$	
$\dfrac{1 - \cos t}{\sin t}$	

6. *Method 2:*

Left side	Right side
$\cot^2 x - \cos^2 x$	$\cos^2 x \cot^2 x$
$\dfrac{\cos^2 x}{\sin^2 x} - \cos^2 x$	$\cos^2 x \cdot \dfrac{\cos^2 x}{\sin^2 x}$
$\dfrac{\cos^2 x - \sin^2 x \cos^2 x}{\sin^2 x}$	$\dfrac{\cos^4 x}{\sin^2 x}$
$\dfrac{\cos^2 x (1 - \sin^2 x)}{\sin^2 x}$	
$\dfrac{\cos^4 x}{\sin^2 x}$	

7. *Method 2:*

Left side	Right side
$\dfrac{\sin 2\theta + \sin \theta}{\cos 2\theta + \cos \theta + 1}$	$\tan \theta$
$\dfrac{2 \sin \theta \cos \theta + \sin \theta}{(2 \cos^2 \theta - 1) + \cos \theta + 1}$	$\dfrac{\sin \theta}{\cos \theta}$
$\dfrac{\sin \theta (2 \cos \theta + 1)}{2 \cos^2 \theta + \cos \theta}$	
$\dfrac{\sin \theta (2 \cos \theta + 1)}{\cos \theta (2 \cos \theta + 1)}$	
$\dfrac{\sin \theta}{\cos \theta}$	

8.

$\sin u \cdot \cos v$	$\frac{1}{2}[\sin (u + v) + \sin (u - v)]$
	$\frac{1}{2}[\sin u \cos v + \cos u \sin v$
	$\quad + \sin u \cos v - \cos u \sin v]$
	$\frac{1}{2}(2 \sin u \cos v)$
	$\sin u \cdot \cos v$

9. $2 \sin 7\theta \cos 5\theta = \sin 12\theta + \sin 2\theta$

Technology Connection, Section 7.5
TC 1. C   **TC 2.** D   **TC 3.** A   **TC 4.** B

Exercise Set 7.5, pp. 476–478

1.

$\csc x - \cos x \cot x$	$\sin x$
$\dfrac{1}{\sin x} - \cos x \dfrac{\cos x}{\sin x}$	
$\dfrac{1 - \cos^2 x}{\sin x}$	
$\dfrac{\sin^2 x}{\sin x}$	
$\sin x$	

3.

$\dfrac{1 + \cos \theta}{\sin \theta} + \dfrac{\sin \theta}{\cos \theta}$	$\dfrac{\cos \theta + 1}{\sin \theta \cos \theta}$
$\dfrac{1 + \cos \theta}{\sin \theta} \cdot \dfrac{\cos \theta}{\cos \theta} + \dfrac{\sin \theta \cdot \sin \theta}{\cos \theta \sin \theta}$	$\dfrac{1 + \cos \theta}{\sin \theta \cos \theta}$
$\dfrac{\cos \theta + \cos^2 \theta + \sin^2 \theta}{\sin \theta \cos \theta}$	
$\dfrac{1 + \cos \theta}{\sin \theta \cos \theta}$	

5.

$\dfrac{1 - \sin x}{\cos x}$	$\dfrac{\cos x}{1 + \sin x}$
$\dfrac{1 - \sin x}{\cos x} \cdot \dfrac{\cos x}{\cos x}$	$\dfrac{\cos x}{1 + \sin x} \cdot \dfrac{1 - \sin x}{1 - \sin x}$
$\dfrac{\cos x - \sin x \cos x}{\cos^2 x}$	$\dfrac{\cos x - \sin x \cos x}{1 - \sin^2 x}$
	$\dfrac{\cos x - \sin x \cos x}{\cos^2 x}$

7.

$\dfrac{1 + \tan \theta}{1 + \cot \theta}$	$\dfrac{\sec \theta}{\csc \theta}$
$\dfrac{1 + \dfrac{\sin \theta}{\cos \theta}}{1 + \dfrac{\cos \theta}{\sin \theta}}$	$\dfrac{\sin \theta}{\cos \theta}$
$\dfrac{\dfrac{\cos \theta + \sin \theta}{\cos \theta}}{\dfrac{\sin \theta + \cos \theta}{\sin \theta}}$	$\tan \theta$
$\dfrac{\cos \theta + \sin \theta}{\cos \theta} \cdot \dfrac{\sin \theta}{\sin \theta + \cos \theta}$	
$\dfrac{\sin \theta}{\cos \theta}$	
$\tan \theta$	

9.

$\dfrac{\sin x + \cos x}{\sec x + \csc x}$	$\dfrac{\sin x}{\sec x}$
$\dfrac{\sin x + \cos x}{\dfrac{1}{\cos x} + \dfrac{1}{\sin x}}$	
$\dfrac{\sin x + \cos x}{\dfrac{\sin x + \cos x}{\sin x \cos x}}$	
$(\sin x + \cos x) \cdot \dfrac{\sin x \cos x}{\sin x + \cos x}$	
$\sin x \cos x$	
$\dfrac{\sin x}{\sec x}$	

11.

$\dfrac{1 + \tan \theta}{1 - \tan \theta} + \dfrac{1 + \cot \theta}{1 - \cot \theta}$	0
$\dfrac{1 + \dfrac{\sin \theta}{\cos \theta}}{1 - \dfrac{\sin \theta}{\cos \theta}} + \dfrac{1 + \dfrac{\cos \theta}{\sin \theta}}{1 - \dfrac{\cos \theta}{\sin \theta}}$	
$\dfrac{\dfrac{\cos \theta + \sin \theta}{\cos \theta}}{\dfrac{\cos \theta - \sin \theta}{\cos \theta}} + \dfrac{\dfrac{\sin \theta + \cos \theta}{\sin \theta}}{\dfrac{\sin \theta - \cos \theta}{\sin \theta}}$	
$\dfrac{\cos \theta + \sin \theta}{\cos \theta} \cdot \dfrac{\cos \theta}{\cos \theta - \sin \theta} +$	
$\dfrac{\sin \theta + \cos \theta}{\sin \theta} \cdot \dfrac{\sin \theta}{\sin \theta - \cos \theta}$	
$\dfrac{\cos \theta + \sin \theta}{\cos \theta - \sin \theta} + \dfrac{\sin \theta + \cos \theta}{\sin \theta - \cos \theta}$	
$\dfrac{\cos \theta + \sin \theta}{\cos \theta - \sin \theta} - \dfrac{\cos \theta + \sin \theta}{\cos \theta - \sin \theta}$	
0	

13.

$\dfrac{1 + \cos 2\theta}{\sin 2\theta}$	$\cot \theta$
$\dfrac{1 + 2\cos^2 \theta - 1}{2 \sin \theta \cos \theta}$	$\dfrac{\cos \theta}{\sin \theta}$
$\dfrac{\cos \theta}{\sin \theta}$	

15.

$$
\begin{array}{c|c}
\sec 2\theta & \dfrac{\sec^2 \theta}{2 - \sec^2 \theta} \\[3mm]
\dfrac{1}{\cos 2\theta} & \dfrac{1 + \tan^2 \theta}{2 - 1 - \tan^2 \theta} \\[3mm]
\dfrac{1}{\cos^2 \theta - \sin^2 \theta} & \dfrac{1 + \tan^2 \theta}{1 - \tan^2 \theta} \\[3mm]
& \dfrac{1 + \dfrac{\sin^2 \theta}{\cos^2 \theta}}{1 - \dfrac{\sin^2 \theta}{\cos^2 \theta}} \\[5mm]
& \dfrac{\cos^2 \theta + \sin^2 \theta}{\cos^2 \theta - \sin^2 \theta} \\[3mm]
& \dfrac{1}{\cos^2 \theta - \sin^2 \theta}
\end{array}
$$

17.

$$
\begin{array}{c|c}
\dfrac{\sin (\alpha + \beta)}{\cos \alpha \cos \beta} & \tan \alpha + \tan \beta \\[3mm]
\dfrac{\sin \alpha \cos \beta + \cos \alpha \sin \beta}{\cos \alpha \cos \beta} & \dfrac{\sin \alpha}{\cos \alpha} + \dfrac{\sin \beta}{\cos \beta} \\[3mm]
& \dfrac{\sin \alpha \cos \beta + \cos \alpha \sin \beta}{\cos \alpha \cos \beta}
\end{array}
$$

19.

$$
\begin{array}{c|c}
1 - \cos 5\theta \cos 3\theta - \sin 5\theta \cos 3\theta & 2 \sin^2 \theta \\[2mm]
1 - [\cos 5\theta \cos 3\theta + \sin 5\theta \cos 3\theta] & 1 - \cos 2\theta \\[2mm]
1 - \cos (5\theta - 3\theta) & \\[2mm]
1 - \cos 2\theta &
\end{array}
$$

21.

$$
\begin{array}{c|c}
\dfrac{\tan \theta + \sin \theta}{2 \tan \theta} & \cos^2 \dfrac{\theta}{2} \\[4mm]
\dfrac{1}{2}\left[\dfrac{\dfrac{\sin \theta + \sin \theta \cos \theta}{\cos \theta}}{\dfrac{\sin \theta}{\cos \theta}}\right] & \dfrac{1 + \cos \theta}{2} \\[6mm]
\dfrac{1}{2}\left[\dfrac{\sin \theta (1 + \cos \theta)}{\cos \theta} \cdot \dfrac{\cos \theta}{\sin \theta}\right] & \\[4mm]
\dfrac{1 + \cos \theta}{2} &
\end{array}
$$

23.

$$
\begin{array}{c|c}
\cos^4 x - \sin^4 x & \cos 2x \\[2mm]
(\cos^2 x - \sin^2 x)(\cos^2 x + \sin^2 x) & \cos^2 x - \sin^2 x \\[2mm]
\cos^2 x - \sin^2 x &
\end{array}
$$

25.

$$
\begin{array}{c|c}
\dfrac{\tan 3\theta - \tan \theta}{1 + \tan 3\theta \tan \theta} & \dfrac{2 \tan \theta}{1 - \tan^2 \theta} \\[3mm]
\tan (3\theta - \theta) & \tan 2\theta \\[2mm]
\tan 2\theta &
\end{array}
$$

27.

$$
\begin{array}{c|c}
\dfrac{\cos^3 x - \sin^3 x}{\cos x - \sin x} & \dfrac{2 + \sin 2x}{2} \\[3mm]
\dfrac{(\cos x - \sin x)(\cos^2 x + \cos x \sin x + \sin^2 x)}{\cos x - \sin x} & \dfrac{2 + 2 \sin x \cos x}{2} \\[3mm]
1 + \cos x \sin x & 1 + \sin x \cos x
\end{array}
$$

29.

$$
\begin{array}{c|c}
\sin (\alpha + \beta) \sin (\alpha - \beta) & \sin^2 \alpha - \sin^2 \beta \\[2mm]
\left(\begin{array}{c}\sin \alpha \cos \beta + \\ \cos \alpha \sin \beta\end{array}\right)\left(\begin{array}{c}\sin \alpha \cos \beta - \\ \cos \alpha \sin \beta\end{array}\right) & \begin{array}{c}1 - \cos^2 \alpha - \\ (1 - \cos^2 \beta)\end{array} \\[4mm]
\sin^2 \alpha \cos^2 \beta - \cos^2 \alpha \sin^2 \beta & \cos^2 \beta - \cos^2 \alpha \\[2mm]
\begin{array}{c}\cos^2 \beta (1 - \cos^2 \alpha) - \\ \cos^2 \alpha (1 - \cos^2 \beta)\end{array} & \\[3mm]
\begin{array}{c}\cos^2 \beta - \cos^2 \alpha \cos^2 \beta - \\ \cos^2 \alpha + \cos^2 \alpha \cos^2 \beta\end{array} & \\[3mm]
\cos^2 \beta - \cos^2 \alpha &
\end{array}
$$

31.

$$
\begin{array}{c|c}
\cos (\alpha + \beta) + \cos (\alpha - \beta) & 2 \cos \alpha \cos \beta \\[2mm]
\begin{array}{c}\cos \alpha \cos \beta - \sin \alpha \sin \beta + \cos \alpha \cos \beta + \\ \sin \alpha \sin \beta\end{array} & \\[3mm]
2 \cos \alpha \cos \beta &
\end{array}
$$

33.

$$
\begin{array}{c|c}
\sin^2 x - \cos^2 x & 1 - 2 \cos^2 x \\[2mm]
-(\cos^2 x - \sin^2 x) & -(2 \cos^2 x - 1) \\[2mm]
-\cos 2x & -\cos 2x
\end{array}
$$

35.

$$
\begin{array}{c|c}
\sin^2 \theta & \cos^2 \theta (\sec^2 \theta - 1) \\[2mm]
& \cos^2 \theta \cdot \tan^2 \theta \\[2mm]
& \cos^2 \theta \cdot \dfrac{\sin^2 \theta}{\cos^2 \theta} \\[3mm]
& \sin^2 \theta
\end{array}
$$

37.

$$
\begin{array}{c|c}
\dfrac{\tan x}{\sec x - \cos x} & \dfrac{\sec x}{\tan x} \\[3mm]
\dfrac{\dfrac{\sin x}{\cos x}}{\dfrac{1}{\cos x} - \cos x} & \dfrac{\dfrac{1}{\cos x}}{\dfrac{\sin x}{\cos x}} \\[5mm]
\dfrac{\dfrac{\sin x}{\cos x}}{\dfrac{1 - \cos^2 x}{\cos x}} & \dfrac{1}{\cos x} \cdot \dfrac{\cos x}{\sin x} \\[5mm]
\dfrac{\sin x}{1 - \cos^2 x} & \dfrac{1}{\sin x} \\[3mm]
\dfrac{\sin x}{\sin^2 x} & \\[3mm]
\dfrac{1}{\sin x} &
\end{array}
$$

39.

$$\frac{\cos\theta + \sin\theta}{\cos\theta} \quad \Big| \quad 1 + \tan\theta$$

$$\frac{\cos\theta}{\cos\theta} + \frac{\sin\theta}{\cos\theta}$$

$$1 + \tan\theta \,\Big|$$

41.

$$\frac{\tan^2 x}{1 + \tan^2 x} \quad \Big| \quad \sin^2 x$$

$$\frac{\tan^2 x}{\sec^2 x}$$

$$\frac{\sin^2 x}{\cos^2 x}$$

$$\frac{1}{\cos^2 x}$$

$$\sin^2 x \,\Big|$$

43.

$$\frac{\cot x + \tan x}{\csc x} \quad \Big| \quad \sec x$$

$$\frac{\dfrac{\cos x}{\sin x} + \dfrac{\sin x}{\cos x}}{\dfrac{1}{\sin x}} \quad \Big| \quad \frac{1}{\cos x}$$

$$\frac{\cos^2 x + \sin^2 x}{\sin x \cos x} \cdot \frac{\sin x}{1}$$

$$\frac{1}{\cos x} \,\Big|$$

45.

$$\frac{\tan\theta + \cot\theta}{\csc\theta} \quad \Big| \quad \sec\theta$$

$$\frac{\dfrac{\sin\theta}{\cos\theta} + \dfrac{\cos\theta}{\sin\theta}}{\dfrac{1}{\sin\theta}} \quad \Big| \quad \frac{1}{\cos\theta}$$

$$\frac{\sin^2\theta + \cos^2\theta}{\cos\theta\sin\theta} \cdot \frac{\sin\theta}{1}$$

$$\frac{1}{\cos\theta} \,\Big|$$

47.

$$\frac{\sin x}{1 + \cos x} + \frac{1 + \cos x}{\sin x} \quad \Big| \quad 2\csc x$$

$$\frac{\sin^2 x + (1 + \cos x)^2}{(1 + \cos x)\sin x} \quad \Big| \quad \frac{2}{\sin x}$$

$$\frac{\sin^2 x + 1 + 2\cos x + \cos^2 x}{(1 + \cos x)\sin x}$$

$$\frac{2 + 2\cos x}{(1 + \cos x)\sin x}$$

$$\frac{2(1 + \cos x)}{(1 + \cos x)\sin x}$$

$$\frac{2}{\sin x} \,\Big|$$

49.

$$\cos\theta(1 + \csc\theta) - \cot\theta \quad \Big| \quad \cos\theta$$

$$\cos\theta\left(1 + \frac{1}{\sin\theta}\right) - \frac{\cos\theta}{\sin\theta}$$

$$\cos\theta + \frac{\cos\theta}{\sin\theta} - \frac{\cos\theta}{\sin\theta}$$

$$\cos\theta \,\Big|$$

51.

$$\frac{\tan x + \cot x}{\sec x + \csc x} \quad \Big| \quad \frac{1}{\cos x + \sin x}$$

$$\frac{\dfrac{\sin x}{\cos x} + \dfrac{\cos x}{\sin x}}{\dfrac{1}{\cos x} + \dfrac{1}{\sin x}}$$

$$\frac{\dfrac{\sin^2 x + \cos^2 x}{\cos x \sin x}}{\dfrac{\sin x + \cos x}{\cos x \sin x}}$$

$$\frac{\sin^2 x + \cos^2 x}{\sin x + \cos x}$$

$$\frac{1}{\cos x + \sin x} \,\Big|$$

53.

$$(\sec\theta - \tan\theta)(1 + \csc\theta) \quad \Big| \quad \cot\theta$$

$$\left(\frac{1}{\cos\theta} - \frac{\sin\theta}{\cos\theta}\right)\left(1 + \frac{1}{\sin\theta}\right) \quad \Big| \quad \frac{\cos\theta}{\sin\theta}$$

$$\left(\frac{1 - \sin\theta}{\cos\theta}\right)\left(\frac{\sin\theta + 1}{\sin\theta}\right)$$

$$\frac{1 - \sin^2\theta}{\cos\theta\sin\theta}$$

$$\frac{\cos^2\theta}{\cos\theta\sin\theta}$$

$$\frac{\cos\theta}{\sin\theta} \,\Big|$$

55.

$$(\sec x + \tan x)(1 - \sin x) \quad \Big| \quad \cos x$$

$$\left(\frac{1}{\cos x} + \frac{\sin x}{\cos x}\right)(1 - \sin x)$$

$$\left(\frac{1 + \sin x}{\cos x}\right)(1 - \sin x)$$

$$\frac{1 - \sin^2 x}{\cos x}$$

$$\frac{\cos^2 x}{\cos x}$$

$$\cos x \,\Big|$$

57.

$\cos^2\theta - \sin^2\theta$	$\cos^4\theta - \sin^4\theta$
	$(\cos^2\theta + \sin^2\theta)(\cos^2\theta - \sin^2\theta)$
	$\cos^2\theta - \sin^2\theta$

59.

$\dfrac{\cos\theta}{1-\cos\theta}$	$\dfrac{1+\sec\theta}{\tan^2\theta}$
	$\dfrac{1+\sec\theta}{\sec^2\theta - 1}$
	$\dfrac{1}{\sec\theta - 1}$
	$\dfrac{1}{\dfrac{1}{\cos\theta} - 1}$
	$\dfrac{1}{\dfrac{1-\cos\theta}{\cos\theta}}$
	$\dfrac{\cos\theta}{1-\cos\theta}$

61.

$\dfrac{\sin x - \cos x}{\cos^2 x}$	$\dfrac{\tan^2 x - 1}{\sin x + \cos x}$
	$\dfrac{\dfrac{\sin^2 x}{\cos^2 x} - 1}{\sin x + \cos x}$
	$\dfrac{\sin^2 x - \cos^2 x}{\cos^2 x} \cdot \dfrac{1}{\sin x + \cos x}$
	$\dfrac{\sin x - \cos x}{\cos^2 x}$

63.

$1 + \sec^4\theta$	$\tan^4\theta + 2\sec^2\theta$
$1 + (1 + \tan^2\theta)^2$	$\tan^4\theta + 2(1 + \tan^2\theta)$
$1 + 1 + 2\tan^2\theta + \tan^4\theta$	$\tan^4\theta + 2\tan^2\theta + 2$
$\tan^4\theta + 2\tan^2\theta + 2$	

65.

$\dfrac{1+\tan x}{1-\tan x}$	$\dfrac{\cot x + 1}{\cot x - 1}$
$\dfrac{1 + \dfrac{1}{\cot x}}{1 - \dfrac{1}{\cot x}}$	
$\dfrac{\dfrac{\cot x + 1}{\cot x}}{\dfrac{\cot x - 1}{\cot x}}$	
$\dfrac{\cot x + 1}{\cot x - 1}$	

67.

$\dfrac{\sin\theta \tan\theta}{\sin\theta + \tan\theta}$	$\dfrac{\tan\theta - \sin\theta}{\tan\theta \sin\theta}$
$\dfrac{\sin\theta \cdot \dfrac{\sin\theta}{\cos\theta}}{\sin\theta + \dfrac{\sin\theta}{\cos\theta}}$	$\dfrac{\dfrac{\sin\theta}{\cos\theta} - \sin\theta}{\dfrac{\sin\theta}{\cos\theta} \cdot \sin\theta}$
$\dfrac{\dfrac{\sin^2\theta}{\cos\theta}}{\sin\theta\cos\theta + \sin\theta}$	$\dfrac{\dfrac{\sin\theta - \sin\theta\cos\theta}{\cos\theta}}{\dfrac{\sin^2\theta}{\cos\theta}}$
$\dfrac{\dfrac{\sin^2\theta}{\cos\theta}}{\dfrac{\sin\theta\cos\theta + \sin\theta}{\cos\theta}}$	$\dfrac{\sin\theta - \sin\theta\cos\theta}{\sin^2\theta}$
$\dfrac{\sin\theta}{\cos\theta + 1}$	$\dfrac{1-\cos\theta}{\sin\theta}$
$\dfrac{\sin\theta}{\cos\theta + 1} \cdot \dfrac{1-\cos\theta}{1-\cos\theta}$	
$\dfrac{\sin\theta(1-\cos\theta)}{1-\cos^2\theta}$	
$\dfrac{\sin\theta(1-\cos\theta)}{\sin^2\theta}$	
$\dfrac{1-\cos\theta}{\sin\theta}$	

69.

$\dfrac{\cos x + 1}{\sin x} + \dfrac{\sin x}{\cos x + 1}$	$\dfrac{2}{\sin x}$
$\dfrac{(\cos x + 1)^2 + \sin^2 x}{\sin x (\cos x + 1)}$	
$\dfrac{\cos^2 x + 2\cos x + 1 + \sin^2 x}{\sin x (\cos x + 1)}$	
$\dfrac{2\cos x + 2}{\sin x (\cos x + 1)}$	
$\dfrac{2(\cos x + 1)}{\sin x (\cos x + 1)}$	
$\dfrac{2}{\sin x}$	

71.

$\sec\theta\csc\theta + \dfrac{\cot\theta}{\tan\theta - 1}$	$\dfrac{\tan\theta}{1 - \cot\theta} - 1$
$\dfrac{1}{\cos\theta}\cdot\dfrac{1}{\sin\theta} + \dfrac{\dfrac{\cos\theta}{\sin\theta}}{\dfrac{\sin\theta}{\cos\theta} - 1}$	$\dfrac{\dfrac{\sin\theta}{\cos\theta}}{1 - \dfrac{\cos\theta}{\sin\theta}} - 1$
$\dfrac{1}{\cos\theta\sin\theta} + \dfrac{\cos\theta}{\sin\theta}\cdot\dfrac{\cos\theta}{\sin\theta - \cos\theta}$	$\dfrac{\sin\theta}{\cos\theta}\cdot\dfrac{\sin\theta}{\sin\theta - \cos\theta} - 1$
$\dfrac{1}{\cos\theta\sin\theta} + \dfrac{\cos^2\theta}{\sin\theta(\sin\theta - \cos\theta)}$	$\dfrac{\sin^2\theta - \cos\theta(\sin\theta - \cos\theta)}{\cos\theta(\sin\theta - \cos\theta)}$
$\dfrac{\sin\theta - \cos\theta + \cos^3\theta}{\cos\theta\sin\theta(\sin\theta - \cos\theta)}$	$\dfrac{\sin^2\theta - \cos\theta\sin\theta + \cos^2\theta}{\cos\theta(\sin\theta - \cos\theta)}$
$\dfrac{\sin\theta - \cos\theta(1 - \cos^2\theta)}{\cos\theta\sin\theta(\sin\theta - \cos\theta)}$	$\dfrac{1 - \cos\theta\sin\theta}{\cos\theta(\sin\theta - \cos\theta)}$
$\dfrac{\sin\theta - \cos\theta\cdot\sin^2\theta}{\cos\theta\sin\theta(\sin\theta - \cos\theta)}$	
$\dfrac{1 - \cos\theta\sin\theta}{\cos\theta(\sin\theta - \cos\theta)}$	

73.

$\cot x + \csc x$	$\dfrac{\sin x}{1 - \cos x}$
$\dfrac{\cos x}{\sin x} + \dfrac{1}{\sin x}$	$\dfrac{\sin x}{1 - \cos x}\cdot\dfrac{1 + \cos x}{1 + \cos x}$
$\dfrac{\cos x + 1}{\sin x}$	$\dfrac{\sin x(1 + \cos x)}{1 - \cos^2 x}$
	$\dfrac{\sin x(1 + \cos x)}{\sin^2 x}$
	$\dfrac{1 + \cos x}{\sin x}$

75.

$\tan\theta + \cot\theta$	$\dfrac{1}{\cot\theta\sin^2\theta}$
$\dfrac{\sin\theta}{\cos\theta} + \dfrac{\cos\theta}{\sin\theta}$	$\dfrac{1}{\dfrac{\cos\theta}{\sin\theta}\cdot\sin^2\theta}$
$\dfrac{\sin^2\theta + \cos^2\theta}{\cos\theta\sin\theta}$	$\dfrac{1}{\cos\theta\sin\theta}$
$\dfrac{1}{\cos\theta\sin\theta}$	

77.

$2\cos^2 x - 1$	$\cos^4 x - \sin^4 x$
$2\cos^2 x - (\sin^2 x + \cos^2 x)$	$(\cos^2 x + \sin^2 x)(\cos^2 x - \sin^2 x)$
$\cos^2 x - \sin^2 x$	$\cos^2 x - \sin^2 x$

79.

$(\cos x + \sin x)(1 - \sin x\cos x)$	$\cos^3 x + \sin^3 x$
	$(\cos x + \sin x)(\cos^2 x - \cos x\sin x + \sin^2 x)$
	$(\cos x + \sin x)(1 - \sin x\cos x)$

81.

$\cos u\cdot\sin v$	$\frac{1}{2}[\sin(u + v) - \sin(u - v)]$
	$\frac{1}{2}[\sin u\cos v + \cos u\sin v - \sin u\cos v + \cos u\sin v]$
	$\frac{1}{2}(2\cos u\sin v)$
	$\cos u\cdot\sin v$

The other formulas follow similarly.

83. $\sin 3\theta - \sin 5\theta = 2\cos\dfrac{8\theta}{2}\sin\dfrac{-2\theta}{2} = -2\cos 4\theta\sin\theta$

85. $\sin 8\theta + \sin 5\theta = 2\sin\dfrac{13\theta}{2}\cos\dfrac{3\theta}{2}$

87. $\sin 7u\sin 5u = \frac{1}{2}(\cos 2u - \cos 12u)$

89. $7\cos\theta\sin 7\theta = \frac{7}{2}[\sin 8\theta - \sin(-6\theta)]$
   $= \frac{7}{2}(\sin 8\theta + \sin 6\theta)$

91. $\cos 55°\sin 25° = \frac{1}{2}(\sin 80° - \sin 30°) = \frac{1}{2}\sin 80° - \frac{1}{4}$

93.

$\sin 4\theta + \sin 6\theta$	$\cot\theta(\cos 4\theta - \cos 6\theta)$
$2\sin\dfrac{10\theta}{2}\cos\dfrac{-2\theta}{2}$	$\dfrac{\cos\theta}{\sin\theta}\left(2\sin\dfrac{10\theta}{2}\sin\dfrac{2\theta}{2}\right)$
$2\sin 5\theta\cos(-\theta)$	$\dfrac{\cos\theta}{\sin\theta}(2\sin 5\theta\sin\theta)$
$2\sin 5\theta\cos\theta$	$2\sin 5\theta\cos\theta$

95.

$\cot 4x \,(\sin x + \sin 4x + \sin 7x)$	$\cos x + \cos 4x + \cos 7x$
$\dfrac{\cos 4x}{\sin 4x}\left(\sin 4x + 2 \sin \dfrac{8x}{2} \cos \dfrac{-6x}{2}\right)$	$\cos 4x + 2 \cos \dfrac{8x}{2} \cdot \cos \dfrac{6x}{2}$
$\dfrac{\cos 4x}{\sin 4x}\,(\sin 4x + 2 \sin 4x \cos 3x)$	$\cos 4x + 2 \cos 4x \cdot \cos 3x$
$\cos 4x \,(1 + 2 \cos 3x)$	$\cos 4x \,(1 + 2 \cos 3x)$

97.

$\cot \dfrac{x+y}{2}$	$\dfrac{\sin y - \sin x}{\cos x - \cos y}$
$\dfrac{\cos \dfrac{x+y}{2}}{\sin \dfrac{x+y}{2}}$	$\dfrac{2 \cos \dfrac{x+y}{2} \sin \dfrac{y-x}{2}}{2 \sin \dfrac{x+y}{2} \sin \dfrac{y-x}{2}}$
	$\dfrac{\cos \dfrac{x+y}{2}}{\sin \dfrac{x+y}{2}}$

99.

$\tan \dfrac{\theta+\phi}{2}\,(\sin \theta - \sin \phi)$	$\tan \dfrac{\theta-\phi}{2}\,(\sin \theta + \sin \phi)$
$\dfrac{\sin \dfrac{\theta+\phi}{2}}{\cos \dfrac{\theta+\phi}{2}}\left(2 \cos \dfrac{\theta+\phi}{2} \sin \dfrac{\theta-\phi}{2}\right)$	$\dfrac{\sin \dfrac{\theta-\phi}{2}}{\cos \dfrac{\theta-\phi}{2}}\left(2 \sin \dfrac{\theta+\phi}{2} \cos \dfrac{\theta-\phi}{2}\right)$
$2 \sin \dfrac{\theta+\phi}{2} \cdot \sin \dfrac{\theta-\phi}{2}$	$2 \sin \dfrac{\theta+\phi}{2} \cdot \sin \dfrac{\theta-\phi}{2}$

101. $\ln |\sec x| = \ln \left|\dfrac{1}{\cos x}\right| = \ln \dfrac{1}{|\cos x|}$
$$= \ln (|\cos x|)^{-1} = -\ln |\cos x|$$

103. $\ln |\tan x| = \ln \left|\dfrac{\sin x}{\cos x}\right| = \ln \dfrac{|\sin x|}{|\cos x|}$
$$= \ln |\sin x| - \ln |\cos x|$$

105. $\ln e^{\sin t} = (\sin t)(\ln e) = \sin t \cdot 1 = \sin t$

107.

$\ln |\csc \theta - \cot \theta|$	$-\ln |\csc \theta + \cot \theta|$
$\ln \left|\dfrac{1}{\sin \theta} - \dfrac{\cos \theta}{\sin \theta}\right|$	$-\ln \left|\dfrac{1}{\sin \theta} + \dfrac{\cos \theta}{\sin \theta}\right|$
$\ln \left|\dfrac{1 - \cos \theta}{\sin \theta}\right|$	$-\ln \left|\dfrac{1 + \cos \theta}{\sin \theta}\right|$
$\ln \left|\dfrac{1 - \cos \theta}{\sin \theta} \cdot \dfrac{1 + \cos \theta}{1 + \cos \theta}\right|$	$\ln \left|\dfrac{1 + \cos \theta}{\sin \theta}\right|^{-1}$
$\ln \left|\dfrac{1 - \cos^2 \theta}{\sin \theta \,(1 + \cos \theta)}\right|$	$\ln \left|\dfrac{\sin \theta}{1 + \cos \theta}\right|$
$\ln \left|\dfrac{\sin^2 \theta}{\sin \theta \,(1 + \cos \theta)}\right|$	
$\ln \left|\dfrac{\sin \theta}{1 + \cos \theta}\right|$	

109. $\log (\cos x - \sin x) + \log (\cos x + \sin x)$
$$= \log [(\cos x - \sin x)(\cos x + \sin x)]$$
$$= \log (\cos^2 x - \sin^2 x) = \log \cos 2x$$

111. $\dfrac{1}{\omega C(\tan \theta + \tan \phi)}$

$$= \dfrac{1}{\omega C \left(\dfrac{\sin \theta}{\cos \theta} + \dfrac{\sin \phi}{\cos \phi}\right)}$$

$$= \dfrac{1}{\omega C \left(\dfrac{\sin \theta \cos \phi + \sin \phi \cos \theta}{\cos \theta \cos \phi}\right)}$$

$$= \dfrac{\cos \theta \cos \phi}{\omega C \sin (\theta + \phi)}$$

Margin Exercises, Section 7.6

1.

2.

3.

4.

5. $\dfrac{\pi}{3}$, or $60°$ **6.** $\dfrac{3\pi}{4}$, or $135°$ **7.** $\dfrac{3\pi}{4}$, or $135°$

8. $-\dfrac{\pi}{4}$, or $-45°$ **9.** $50.76°$ **10.** $-5.86°$ **11.** $53.38°$

12. $-87.45°$

Exercise Set 7.6, pp. 483–484

1. $\frac{\pi}{4}$, or 45°   **3.** $\frac{\pi}{4}$, or 45°   **5.** $-\frac{\pi}{4}$, or $-45°$

7. $\frac{3\pi}{4}$, or 135°   **9.** $\frac{\pi}{3}$, or 60°   **11.** $\frac{\pi}{4}$, or 45°

13. $-\frac{\pi}{6}$, or $-30°$ **15.** $\frac{3\pi}{4}$, or 135°   **17.** 0, or 0°

19. $\frac{\pi}{2}$, or 90°   **21.** $-\frac{\pi}{4}$, or $-45°$ **23.** $\frac{\pi}{3}$, or 60°

25. $-\frac{\pi}{3}$, or $-60°$ **27.** $\frac{2\pi}{3}$, or 120°   **29.** $\frac{\pi}{4}$, or 45°

31. $\frac{2\pi}{3}$, or 120°   **33.** 23°   **35.** $-38.31°$ **37.** 36.97°

39. 168.82°   **41.** 20.17°   **43.** 38.33°   **45.** 31.09°
47. $-170.83°$ **49.** 13.50°   **51.** $-39.50°$
53. $-62.70°$ **55.** $-22.23°$ **57.** 170.44°   **59.** 177.51°
61.

$y = \sec^{-1} x$

63. Let $x = \frac{\sqrt{2}}{2}$. Then $\sin^{-1}\frac{\sqrt{2}}{2} = \frac{\pi}{4} \approx 0.7854$, and

$$\left(\sin\frac{\sqrt{2}}{2}\right)^{-1} \approx (0.6496)^{-1} \approx 1.5393.$$

65. Let $x = 1$. Then $\tan^{-1} 1 = \frac{\pi}{4} \approx 0.7854$, and

$(\tan 1)^{-1} \approx (1.5574)^{-1} \approx 0.6421.$

67. Let $x = \frac{\sqrt{2}}{2}$. Then $\tan^{-1}\frac{\sqrt{2}}{2} \approx 0.6155$ and

$\dfrac{\sin^{-1}\frac{\sqrt{2}}{2}}{\cos^{-1}\frac{\sqrt{2}}{2}} = \dfrac{\pi/4}{\pi/4} = 1.$ **69.** $\tan\theta = \frac{50}{b}$, or $\theta = \tan^{-1}\frac{50}{b}$

71. (a) 3.1415927; **(b)** π

Margin Exercises, Section 7.7

1. $\frac{1}{2}$ **2.** 1   **3.** $\frac{\sqrt{2}}{2}$

4. The expression cannot be evaluated.   **5.** $\frac{2\pi}{3}$ **6.** $-\frac{\pi}{4}$

7. $\frac{\pi}{6}$ **8.** $\frac{2\pi}{3}$ **9.** $\frac{\sqrt{2}}{2}$ **10.** $\frac{\sqrt{3}}{2}$ **11.** $\frac{\pi}{3}$ **12.** $\frac{\pi}{2}$

13. 1   **14.** $\frac{3}{\sqrt{b^2+9}}$ **15.** $\frac{2}{\sqrt{5}}$ **16.** $\frac{t}{\sqrt{1-t^2}}$ **17.** 1

18. $\frac{1}{3}$

Exercise Set 7.7, pp. 489–490

1. 0.3   **3.** -4.2 **5.** $\frac{\pi}{3}$ **7.** $-\frac{\pi}{4}$ **9.** $\frac{\pi}{5}$ **11.** $-\frac{\pi}{3}$

13. $\frac{\sqrt{3}}{2}$ **15.** $\frac{1}{2}$ **17.** 1   **19.** $\frac{\pi}{6}$ **21.** $\frac{\pi}{3}$ **23.** $\frac{\pi}{2}$

25. $\frac{x}{\sqrt{x^2+4}}$ **27.** $\frac{\sqrt{x^2-9}}{3}$ **29.** $\frac{\sqrt{b^2-a^2}}{a}$ **31.** $\frac{3}{\sqrt{11}}$

33. $\frac{1}{3\sqrt{11}}$ **35.** $-\frac{1}{2\sqrt{6}}$ **37.** $\frac{1}{\sqrt{1+y^2}}$ **39.** $\frac{1}{\sqrt{1+t^2}}$

41. $\frac{\sqrt{1-y^2}}{y}$ **43.** $\sqrt{1-x^2}$ **45.** $\frac{p}{3}$ **47.** $\frac{1}{2}$

49. $\frac{\sqrt{2+\sqrt{3}}}{2}$ **51.** $\frac{24}{25}$ **53.** $\frac{119}{169}$ **55.** $\frac{3+4\sqrt{3}}{10}$

57. $-\frac{\sqrt{2}}{10}$ **59.** $xy + \sqrt{(1-x^2)(1-y^2)}$

61. $y\sqrt{1-x^2} - x\sqrt{1-y^2}$ **63.** 0.9861

65. $\sin\theta = x$; $\cos\theta = \sqrt{1-x^2}$; $\tan\theta = \frac{x}{\sqrt{1-x^2}}$;

$\cot\theta = \frac{\sqrt{1-x^2}}{x}$; $\sec\theta = \frac{1}{\sqrt{1-x^2}}$; $\csc\theta = \frac{1}{x}$

67. $\sin\theta = \frac{x}{\sqrt{1+x^2}}$; $\cos\theta = \frac{1}{\sqrt{1+x^2}}$; $\tan\theta = x$;

$\cot\theta = \frac{1}{x}$; $\sec\theta = \sqrt{1+x^2}$; $\csc\theta = \frac{\sqrt{1+x^2}}{x}$

69.

$\sin^{-1} x + \cos^{-1} x$	$\frac{\pi}{2}$
$\sin(\sin^{-1} x + \cos^{-1} x)$	$\sin\frac{\pi}{2}$
$[\sin(\sin^{-1} x)][\cos(\cos^{-1} x)] +$ $[\cos(\sin^{-1} x)][\sin(\cos^{-1} x)]$ $x \cdot x + \sqrt{1-x^2} \cdot \sqrt{1-x^2}$ $x^2 + 1 - x^2$	1
	1

71.

$\sin^{-1} x$	$\tan^{-1}\frac{x}{\sqrt{1-x^2}}$
$\sin(\sin^{-1} x)$	$\sin\left(\tan^{-1}\frac{x}{\sqrt{1-x^2}}\right)$
x	x

73.

$\arcsin x$	$\arccos\sqrt{1-x^2}$
$\sin(\arcsin x)$	$\sin(\arccos\sqrt{1-x^2})$
x	x

75. $\theta = \arctan\frac{y+h}{x} - \arctan\frac{y}{x}$

Margin Exercises, Section 7.8

1. $60° + 360°k, 300° + 360°k; \dfrac{\pi}{3} + 2k\pi, \dfrac{5\pi}{3} + 2k\pi$

2. $\dfrac{\pi}{6}, \dfrac{5\pi}{6}, \dfrac{7\pi}{6}, \dfrac{11\pi}{6}$; $30°, 150°, 210°, 330°$

3. $\dfrac{\pi}{6}, \dfrac{5\pi}{6}, \dfrac{7\pi}{6}, \dfrac{11\pi}{6}$ **4.** $40°, 320°$ **5.** $140°, 220°$

6. $120°, 240°, 75.52°, 284.48°$ **7.** $120°, 240°, 90°, 270°$
8. $241.66°, 118.34°$

Technology Connection, Section 7.8
TC 1. $-40.1°, 102.0°$ **TC 2.** $0.79, 0.98$ **TC 3.** ϕ
TC 4. $2.13, 3.85$

Exercise Set 7.8, p. 495

1. $\dfrac{\pi}{3} + 2k\pi, \dfrac{2\pi}{3} + 2k\pi$ **3.** $\dfrac{\pi}{4} + 2k\pi, -\dfrac{\pi}{4} + 2k\pi$

5. $20.17° + k \cdot 360°, 159.83° + k \cdot 360°$

7. $236.67°, 123.33°$ **9.** $\dfrac{4\pi}{3}, \dfrac{5\pi}{3}$ **11.** $123.69°, 303.69°$

13. $\dfrac{\pi}{6}, \dfrac{5\pi}{6}, \dfrac{7\pi}{6}, \dfrac{11\pi}{6}$ **15.** $\dfrac{\pi}{6}, \dfrac{5\pi}{6}, \dfrac{7\pi}{6}, \dfrac{11\pi}{6}$

17. $\dfrac{\pi}{6}, \dfrac{5\pi}{6}, \dfrac{3\pi}{2}$ **19.** 0 **21.** $0, \dfrac{\pi}{6}, \dfrac{5\pi}{6}, \pi, \dfrac{7\pi}{6}, \dfrac{11\pi}{6}$

23. $\dfrac{\pi}{6}, \dfrac{5\pi}{6}$ **25.** $109.47°, 250.53°, 120°, 240°$

27. $\dfrac{\pi}{6}, \dfrac{5\pi}{6}, \pi$ **29.** $0, \pi, \dfrac{\pi}{2}, \dfrac{3\pi}{2}$ **31.** $198.27°, 341.73°$

33. $70.12°, 128.31°, 250.12°, 308.31°$
35. $37.22°, 169.35°, 217.22°, 349.35°$
37. $60°, 120°, 240°, 300°$ **39.** $60°, 240°$
41. $30°, 60°, 120°, 150°, 210°, 240°, 300°, 330°$

43. $0, \pi$ **45.** $\dfrac{\pi}{2}$

47. (a)

The angle of elevation is equal to the angle of depression.
Thus, $\sin \theta = \dfrac{2000}{h}$. **(b)** $41.81°$

Margin Exercises, Section 7.9

1. $\dfrac{\pi}{2}, \pi$ **2.** $\dfrac{\pi}{2}, \pi$ **3.** $\dfrac{\pi}{4}, \dfrac{3\pi}{4}, \dfrac{5\pi}{4}, \dfrac{7\pi}{4}$ **4.** $\dfrac{\pi}{2}, \dfrac{3\pi}{2}, \dfrac{7\pi}{6}, \dfrac{11\pi}{6}$

5. $\dfrac{\pi}{12}, \dfrac{5\pi}{12}, \dfrac{13\pi}{12}, \dfrac{17\pi}{12}$ **6.** $\dfrac{\pi}{3}, \pi, \dfrac{5\pi}{3}$

Exercise Set 7.9, pp. 499–500

1. $0, \pi$ **3.** $\dfrac{3\pi}{4}, \dfrac{7\pi}{4}$ **5.** $\dfrac{\pi}{2}, \dfrac{3\pi}{2}, \dfrac{\pi}{6}, \dfrac{5\pi}{6}$

7. $\dfrac{\pi}{2}, \dfrac{3\pi}{2}, \dfrac{\pi}{4}, \dfrac{3\pi}{4}, \dfrac{5\pi}{4}, \dfrac{7\pi}{4}$ **9.** $0, \dfrac{\pi}{2}, \pi, \dfrac{3\pi}{2}$ **11.** 0

13. $0, \dfrac{\pi}{2}, \pi, \dfrac{3\pi}{2}$ **15.** $\dfrac{\pi}{6}, \dfrac{5\pi}{6}, \pi$ **17.** $\dfrac{\pi}{6}, \dfrac{5\pi}{6}, \dfrac{7\pi}{6}, \dfrac{11\pi}{6}$

19. $63.43°, 243.43°, 101.31°, 281.31°$

21. $\dfrac{\pi}{6}, \dfrac{\pi}{2}, \dfrac{5\pi}{6}, \dfrac{7\pi}{6}, \dfrac{3\pi}{2}, \dfrac{11\pi}{6}$ **23.** $\dfrac{2\pi}{3}, \dfrac{4\pi}{3}$ **25.** $\dfrac{\pi}{4}, \dfrac{7\pi}{4}$

27. $\dfrac{\pi}{12}, \dfrac{5\pi}{12}$ **29.** $\dfrac{\pi}{6}, \dfrac{3\pi}{2}$ **31.** $0.9669, 1.853, 4.108, 4.995$

33. $0.7297, 2.412, 3.665, 5.760$ **35.** $t \approx 1.24, 6.76$
37. $16.5°$ **39.** 1 **41.** $1.13, 5.66, -0.622, -0.516$, etc.
43. 0.1923

Review Exercises: Chapter 7, pp. 500–501
1. [7.3] $\cot (x - \pi) = \cot x$ **2.** [7.5] 1 **3.** [7.5] $\csc^2 x$
4. [7.3] $-\sin x$ **5.** [7.3] $\sin x$ **6.** [7.3] $-\cos x$
7. [7.1] $\tan x = \pm\sqrt{\sec^2 x - 1}$ **8.** [7.1] $\csc x$ **9.** [7.1] 1
10. [7.1] $\dfrac{\sqrt{\tan x \sec x}}{\sec x}$ **11.** [7.1] $\dfrac{\tan x}{\sqrt{\sec x \tan x}}$
12. [7.3] $\sin (90° - \theta) = 0.7314$; $\cos (90° - \theta) = 0.6820$;
$\tan (90° - \theta) = 1.0724$; $\cot (90° - \theta) = 0.9325$;
$\sec (90° - \theta) = 1.4663$; $\csc (90° - \theta) = 1.3673$

13. [7.2] $\cos x \cos \dfrac{3\pi}{2} - \sin x \sin \dfrac{3\pi}{2}$

14. [7.2] $\dfrac{\tan 45° - \tan 30°}{1 + \tan 45° \tan 30°}$

15. [7.2] $\cos (27° - 16°)$ or $\cos 11°$

16. [7.2] $\dfrac{-\sqrt{6} - \sqrt{2}}{4}$ **17.** [7.2] $2 - \sqrt{3}$

18. [7.4] $\sin 2\theta = \dfrac{24}{25}$, $\cos 2\theta = -\dfrac{7}{25}$, $\tan 2\theta = -\dfrac{24}{7}$, 2θ is

in quadrant II   **19.** [7.4] $\dfrac{1}{2}\sqrt{2 - \sqrt{2}}$ **20.** [7.4] $2 \cot \theta$

21.

$\tan 2\theta$	$\dfrac{2 \tan \theta}{1 - \tan^2 \theta}$
$\dfrac{\sin 2\theta}{\cos 2\theta}$	$\dfrac{2 \dfrac{\sin \theta}{\cos \theta}}{ }$
$\dfrac{2 \sin \theta \cos \theta}{\cos^2 \theta - \sin^2 \theta}$	$\dfrac{\cos^2 \theta}{\cos^2 \theta} - \dfrac{\sin^2 \theta}{\cos^2 \theta}$
	$\dfrac{2 \sin \theta}{\cos \theta} \cdot \dfrac{\cos^2 \theta}{\cos^2 \theta - \sin^2 \theta}$
	$\dfrac{2 \sin \theta \cos \theta}{\cos^2 \theta - \sin^2 \theta}$

22. [7.5]

$$\dfrac{\dfrac{\sec x - \cos x}{\tan x}}{} \quad \sin x$$

$$\dfrac{\dfrac{1}{\cos x} - \cos x}{\dfrac{\sin x}{\cos x}}$$

$$\dfrac{1 - \cos^2 x}{\cos x} \cdot \dfrac{\cos x}{\sin x}$$

$$\dfrac{1 - \cos^2 x}{\sin x}$$

$$\dfrac{\sin^2 x}{\sin x}$$

$$\sin x$$

23. [7.6] $\dfrac{\pi}{6}$　**24.** [7.6] $81°$　**25.** [7.6] $-45°$, or $-\dfrac{\pi}{4}$

26. [7.6] $150°$, or $\dfrac{5\pi}{6}$　**27.** [7.7] $\dfrac{7}{8}$　**28.** [7.7] $\dfrac{\pi}{3}$

29. [7.7] $\dfrac{\pi}{6}$　**30.** [7.7] $\dfrac{b}{\sqrt{b^2 + 25}}$　**31.** [7.8] $0, \pi$

32. [7.8] $\dfrac{\pi}{3}, \dfrac{5\pi}{3}$　**33.** [7.9] $\dfrac{\pi}{4}, \dfrac{3\pi}{4}, \dfrac{5\pi}{4}, \dfrac{7\pi}{4}$

34. [7.9] $0.5117, 1.845, 3.6533, 4.9861$

35. [7.8] $\dfrac{\pi}{6}, \dfrac{5\pi}{6}, \dfrac{7\pi}{6}, \dfrac{11\pi}{6}$　**36.** [7.8] No solution in $[0, 2\pi)$.

37. [7.4]

$y + 1 = 2\cos^2 x$

38. [7.6]

$f(x) = 2\sin^{-1}\left(x + \frac{\pi}{2}\right)$

Test: Chapter 7, pp. 501–502
1. [7.2] $\cos \pi \cos x + \sin \pi \sin x$

2. [7.2] $\dfrac{\tan 83° + \tan 15°}{1 - \tan 83° \tan 15°}$　**3.** [7.2] $\sin 35°$

4. [7.2] $\dfrac{\sqrt{2} - \sqrt{6}}{4}$　**5.** [7.2] $2 - \sqrt{3}$　**6.** [7.3] $\csc x$

7. [7.5] 1　**8.** [7.5] $\csc^2 x$　**9.** [7.3] $\cos x$　**10.** [7.3] $\sin x$
11. [7.3] $\sin x$　**12.** [7.1] $\csc x = \pm\sqrt{1 + \cot^2 x}$
13. [7.1] $\sec x$　**14.** [7.1] 1　**15.** [7.1] $\dfrac{\sqrt{\sec x \csc x}}{\csc x}$

16. [7.4] $\sin 2\theta = -\dfrac{120}{169}$; $\cos 2\theta = -\dfrac{119}{169}$; $\tan 2\theta = \dfrac{120}{119}$;

quadrant III　**17.** [7.4] $2 + \sqrt{3}$　**18.** [7.4] $2\sin 4x$

19. [7.5]

$$\dfrac{\dfrac{1 - \cos 2\theta}{\sin 2\theta}}{} \quad \tan \theta$$

$$\dfrac{1 - (1 - 2\sin^2 \theta)}{2\sin \theta \cos \theta} \quad \dfrac{\sin \theta}{\cos \theta}$$

$$\dfrac{2\sin^2 \theta}{2\sin \theta \cos \theta}$$

$$\dfrac{\sin \theta}{\cos \theta}$$

20. [7.6] $-\dfrac{\pi}{4}$　**21.** [7.6] $42.96°$　**22.** [7.6] $\dfrac{\pi}{6}$

23. [7.6] $\dfrac{3\pi}{4}$　**24.** [7.7] $\dfrac{\sqrt{3}}{2}$　**25.** [7.7] $\dfrac{\pi}{4}$

26. [7.8] $\dfrac{7\pi}{6}, \dfrac{11\pi}{6}$　**27.** [7.9] $\dfrac{\pi}{2}, \dfrac{3\pi}{2}, \dfrac{\pi}{4}, \dfrac{3\pi}{4}, \dfrac{5\pi}{4}, \dfrac{7\pi}{4}$

28. [7.8] $0.8730, 1.805, 4.015, 4.947$
29. [7.3] $\sin(90° - \theta) = 0.8910$, $\cos(90° - \theta) = 0.4540$,
$\tan(90° - \theta) = 1.963$, $\cot(90° - \theta) = 0.5095$,

$\sec(90° - \theta) = 2.203$, $\csc(90° - \theta) = 1.122$　**30.** [7.8] $\dfrac{\pi}{2}$

CHAPTER 8

Margin Exercises, Section 8.1

1. (a) $\dfrac{b}{c}$; (b) $\dfrac{a}{c}$; (c) $\dfrac{b}{a}$; (d) $\dfrac{c}{b}$; (e) $\dfrac{c}{a}$; (f) $\dfrac{a}{b}$　**2.** $42°$

3. $a = 14.7$, $b = 13.4$　**4.** $A = 41.79°$, $a = 6.4$, $c = 9.7$
5. $B = 52.2°$, $a = 6.69$, $c = 10.91$　**6.** (a) 76 m; (b) 142 m
7. $28.55°$　**8.** 35.7 ft　**9.** 240.3 ft　**10.** East: 136 km;
south: 63 km　**11.** 23.1 km　**12.** 15 mi

Exercise Set 8.1, pp. 511–513
1. $B = 60°$, $a = 3$, $b = 5$　**3.** $A = 22.7°$, $a = 53$, $c = 137$
5. $B = 47°38'$, $b = 25.4$, $c = 34.4$　**7.** $B = 53°48'$, $b = 37.2$,
$c = 46.1$　**9.** $A = 77.35°$, $a = 437.1$, $c = 448.0$
11. $B = 72.6°$, $a = 4.3$, $c = 14.3$　**13.** $B = 66°48'$, $b = 150$,
$c = 381$　**15.** $B = 42.42°$, $a = 35.7$, $b = 32.6$
17. $A = 7.7°$, $a = 0.132$, $b = 0.973$　**19.** $A = 34.05°$,
$B = 55.95°$, $c = 22.3$　**21.** $A = 53.13°$, $B = 36.87°$, $b = 12.0$
23. $A = 62.44°$, $B = 27.56°$, $a = 3.56$　**25.** 48.0 ft
27. 239 ft　**29.** 1.72°　**31.** 30.22°　**33.** 3.5 mi
35. 17,067 ft　**37.** 24 km　**39.** 25.9 cm　**41.** $\dfrac{25}{3}$ cm
43. 23 ft　**45.** 8 km　**47.** 3.4 km　**49.** 109 km
51. 3.3　**53.** Area $= \frac{1}{2} \cdot b \cdot a = \frac{1}{2}(c\cos A)(c\sin A) =$
$\frac{1}{2}c^2(\cos A \sin A) = \frac{1}{2}c^2(\frac{1}{2}\sin 2A) = \frac{1}{4}c^2\sin 2A$
55. $d = \sqrt{8000h + h^2}$, where d and h are in miles;
38.9 miles　**57.** Cut so that $\theta = 79.38°$.　**59.** 27°

Margin Exercises, Section 8.2

1. $C = 29°$, $c = 5.9$, $b = 11.1$ **2.** $\sin A = 2.8$; no solution
3. $A = 43°$, $C = 104°$, $c = 36$; or $A = 137°$, $C = 10°$, $c = 6$
4. $C = 18°$, $A = 124°$, $a = 27$ **5.** 22.5 in^2

Exercise Set 8.2, pp. 519–521

1. $C = 17°$, $a = 26$, $c = 11$ **3.** $A = 121°$, $a = 33$, $c = 14$
5. $B = 68.8°$, $a = 32.2 \text{ cm}$, $b = 32.3 \text{ cm}$
7. $A = 110.36°$, $a = 5 \text{ mi}$, $b = 3 \text{ mi}$
9. $C = 103°41'$, $a = 1804 \text{ km}$, $b = 5331 \text{ km}$
11. $B = 32°$, $a = 1752 \text{ in.}$, $c = 720 \text{ in.}$
13. $B = 57.4°$, $C = 86.1°$, $c = 40$, or $B = 122.6°$,
$C = 20.9°$, $c = 14$ **15.** $B = 19.0°$, $C = 44.7°$, $b = 6.2$
17. $A = 74°26'$, $B = 44°24'$, $a = 33.3$ **19.** No solution
21. $B = 14.7°$, $C = 135.0°$, $c = 28.04 \text{ cm}$ **23.** No solution
25. $B = 71°26'$, $A = 56°46'$, $a = 3.668 \text{ km}$; or $B = 108°34'$,
$A = 19°38'$, $a = 1.473 \text{ km}$ **27.** No solution
29. $C = 54°36'$, $B = 83°34'$, $b = 134.1 \text{ mi}$; or $C = 125°24'$,
$B = 12°46'$, $b = 29.8 \text{ mi}$ **31.** 8.2 ft^2 **33.** 12 yd^2
35. 596.98 ft^2 **37.** 76 m **39.** 51 ft **41.** 1467 km
43. From A: 35 mi; from B: 66 mi   **45.** 10.6 km or 2.4 km
47. See Student's Solution Manual or Instructor's Solution
Manual.   **49.**

$A = bh$, $h = a \sin \theta$, so $A = ab \sin \theta$
51. Area $\triangle ADB$ + Area $\triangle BDC$ = Area $\triangle ADC$. Using
Theorem 2, we have

$$\frac{1}{2}xy \sin \alpha + \frac{1}{2}yz \sin \beta = \frac{1}{2}xz \sin (\alpha + \beta).$$

Multiplying both sides by $\dfrac{2}{x \cdot y \cdot z}$ gives

$$\frac{\sin \alpha}{z} + \frac{\sin \beta}{x} = \frac{\sin (\alpha + \beta)}{y}.$$

Margin Exercises, Section 8.3

1. $a = 41$, $B = 21°$, $C = 37°$
2. $A = 108.21°$, $B = 22.33°$, $C = 49.46°$
3. Law of cosines; $A = 108.21°$, $B = 22.33°$, $C = 49.46°$
4. Law of sines; $C = 18°$, $A = 124°$, $a = 54$
5. Law of cosines; $a = 122$, $B = 21.9°$, $C = 35.7°$
6. Law of sines; $A = 43°$, $C = 104°$, $c = 107$
7. Law of sines; $B = 66°49'$, $C = 69°36'$, $c = 6.1$; or
$B = 113°11'$, $C = 23°14'$, $c = 2.6$
7. Law of sines; no solution

Exercise Set 8.3, pp. 525–527

1. $a = 15$, $B = 24°$, $C = 126°$
3. $b = 25$, $A = 20°$, $C = 27°$
5. $b = 75 \text{ m}$, $A = 94°51'$, $C = 12°29'$
7. $c = 51.2 \text{ cm}$, $A = 10.82°$, $B = 146.90°$
9. $a = 25.5 \text{ yd}$, $B = 38°2'$, $C = 45°45'$
11. $c = 45.17 \text{ mi}$, $A = 89.3°$, $B = 42.0°$
13. $A = 36.18°$, $B = 43.53°$, $C = 100.29°$
15. $A = 24.15°$, $B = 30.75°$, $C = 125.10°$ **17.** No solution
19. $A = 128.71°$, $B = 22.10°$, $C = 29.19°$
21. Law of sines; $C = 98°$, $a = 96.7$, $c = 101.9$
23. Law of cosines; $A = 73.71°$, $B = 51.75°$, $C = 54.54°$
25. Law of cosines; $c = 44$, $B = 42°$, $A = 91°$

27. Law of sines; $A = 61°20'$, $a = 5.633$, $b = 6.014$
29. 28 nautical mi, S 54°38′ E   **31.** 37 nautical mi
33. 59 ft   **35.** 67.98°, 67.98°, 44.04°   **37.** 46 ft
39. **(a)** 16 ft; **(b)** 121 ft^2 **41.** 3424 yd^2 **43.** 12 m, 24 m
45. 60.08°, 40.78°, 79.14°   **47.** 50 ft   **49.** 4.7 cm
51. See Student's Solution Manual or Instructor's Solution
Manual.   **53.** 64 in.   **55.** 162 ft
57. $A = \frac{1}{2}a^2 \sin \theta$; when $\theta = 90°$
59. Let $S^2 = a^2 + b^2 + c^2$. Then
$$a^2 = b^2 + c^2 - 2bc \cos A$$
$$b^2 = c^2 + a^2 - 2ca \cos B$$
$$(+) \quad \frac{c^2 = a^2 + b^2 - 2ab \cos C}{S^2 = S^2 + S^2 - 2(bc \cos A + ac \cos B + ab \cos C).}$$
$\therefore\ S^2 = a^2 + b^2 + c^2 = 2(bc \cos A + ac \cos B + ab \cos C)$
61. 10,106 ft

Margin Exercises, Section 8.4

1. 15.5 N, 20°   **2.** 75°, 171 km/h   **3.** E: 71, S: 71
4. 19 km/h from S 32° E   **5.** **(a)** 4 up, 21 left;
(b) 21.4, 11° with horizontal   **6.** $\langle 12, -8 \rangle$ **7.** $\langle 12, 4 \rangle$
8. $(8.5, 159°)$ **9.** $\langle 14.2, -4.9 \rangle$ **10.** **(a)** $\langle 1, 7 \rangle$;
(b) $\langle 7, -3 \rangle$; **(c)** $\langle -23, 21 \rangle$; **(d)** 17.1

Exercise Set 8.4, pp. 535–536

1. 57.0, 38°   **3.** 18.4, 37°   **5.** 20.9, 59°   **7.** 68.3, 18°
9. 13.6 N, 65°   **11.** 655 N, 21°   **13.** 22 ft/sec, 34°
15. 726 lb, 47°   **17.** 174 nautical mi, S 15° E
19. An angle of 12° upstream
21. Vertical 118, horizontal 92
23. S. 192 km/h, W. 161 km/h   **25.** 43 N, S 36° W
27. **(a)** N. 28, W. 7; **(b)** 28.9, N 14° W   **29.** $\langle 5, 16 \rangle$
31. $\langle 4.8, 13.7 \rangle$ **33.** $\langle -662, -426 \rangle$ **35.** $(5, 53°)$
37. $(18, 304°)$ **39.** $(5, 233°)$ **41.** $(18, 124°)$
43. $\langle 3.5, 2 \rangle$ **45.** $\langle -5.7, -8.2 \rangle$ **47.** $\langle 17.3, -10 \rangle$
49. $\langle 70.7, -70.7 \rangle$ **51.** $\langle 19, 36 \rangle$ **53.** $\langle 14, 8 \rangle$ **55.** 18
57. 17.9   **59.** 35.8   **61.** $\langle 0, 0 \rangle$
63. To show: If k is a scalar, then $k(\mathbf{u} + \mathbf{v}) = k\mathbf{u} + k\mathbf{v}$.
Proof: Let $\mathbf{u} = \langle a, b \rangle$ and $\mathbf{v} = \langle c, d \rangle$.
Then $k(\mathbf{u} + \mathbf{v}) = k(\langle a, b \rangle + \langle c, d \rangle)$
$\qquad\qquad = k\langle a + c, b + d \rangle$
$\qquad\qquad = \langle k(a + c), k(b + d) \rangle$
$\qquad\qquad = \langle ka + kc, kb + kd \rangle$
$\qquad\qquad = \langle ka, kb \rangle + \langle kc, kd \rangle$
$\qquad\qquad = k\langle a, b \rangle + k\langle c, d \rangle$
$\qquad\qquad = k\mathbf{u} + k\mathbf{v}$.

65.

If the vectors are *not* collinear
(or parallel), they form a triangle
as shown. From plane geometry
(Euclid), we know that (any)
one side of a triangle is *less than*
the sum of the other two sides.

If the vectors *are* collinear, then either

$|\mathbf{u} + \mathbf{v}| = |\mathbf{u}| + |\mathbf{v}|$

or

$|\mathbf{u} + \mathbf{v}| < |\mathbf{u}| + |\mathbf{v}|$.

Margin Exercises, Section 8.5

1.

2. Many answers possible. A: $(4, 30°)$, $(4, 390°)$, $(-4, 210°)$, etc.; B: $(5, -60°)$, $(5, 300°)$; C: $(2, 150°)$, $(2, -210°)$; D: $(3, 225°)$, $(3, -135°)$; E: $(5, 60°)$, $(-5, -120°)$ **3.** (a) $(3\sqrt{2}, 45°)$; (b) $(4, 270°)$;

(c) $(6, 120°)$; (d) $(4, 330°)$ **4.** (a) $\left(\frac{5}{2}\sqrt{3}, \frac{5}{2}\right)$;

(b) $(5, 5\sqrt{3})$; (c) $\left(-\frac{5}{\sqrt{2}}, -\frac{5}{\sqrt{2}}\right)$; (d) $(4\sqrt{3}, 4)$

5. $2r\cos\theta + 5r\sin\theta = 9$ **6.** $r^2 + 8r\cos\theta = 0$
7. $x^2 + y^2 = 49$ **8.** $y = 5$ **9.** $x^2 + y^2 - 3x = 5y$
10.

$r = 1 - \sin\theta$

11.

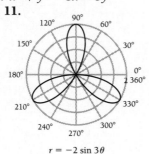

$r = -2\sin 3\theta$

Technology Connection, Section 8.5

TC 1.

$r = 8\sin 4\theta$

$[-15, 15] \times [-10, 10]$

TC 2.

$r = 10\tan\theta\sin\theta$

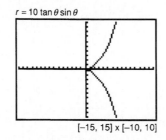

$[-15, 15] \times [-10, 10]$

TC 3.

$r = 4\sin 2\theta - 8$

$[-15, 15] \times [-10, 10]$

TC 4.

$r = 7\cos 3\theta + 4\sin 2\theta$

$[-15, 15] \times [-10, 10]$

Exercise Set 8.5, pp. 541–542

25. $(4\sqrt{2}, 45°)$ **27.** $(5, 90°)$ **29.** $(4, 0°)$
31. $(6, 60°)$ **33.** $(2, 30°)$ **35.** $(6, 30°)$
37. $\left(\frac{4}{\sqrt{2}}, \frac{4}{\sqrt{2}}\right)$ or $(2.8, 2.8)$ **39.** $(0, 0)$

41. $\left(-\frac{3}{\sqrt{2}}, -\frac{3}{\sqrt{2}}\right)$ **43.** $(3, -3\sqrt{3})$ **45.** $(5\sqrt{3}, 5)$

47. $\left(\frac{5\sqrt{3}}{2}, -\frac{5}{2}\right)$ **49.** $r(3\cos\theta + 4\sin\theta) = 5$

51. $r\cos\theta = 5$ **53.** $r^2 = 36$
55. $r^2(\cos^2\theta - 4\sin^2\theta) = 4$ **57.** $x^2 + y^2 = 25$
59. $y = x$ **61.** $y = 2$ **63.** $x^2 + y^2 = 4x$
65. $x^2 - 4y = 4$ **67.** $x^2 - 2x + y^2 - 3y = 0$
69.

$r = 4\cos\theta$

71.

$r = 1 - \cos\theta$

73.

$r = \sin 2\theta$

75.

$r = 2\cos 3\theta$

77.

$r\cos\theta = 4$

79.

$r = \dfrac{5}{1 + \cos\theta}$

81.

$r = \theta$

83.

$r^2 = \sin 2\theta$

7. $6 \operatorname{cis} 225°$ **8.** $20 \operatorname{cis} 55°$ **9.** $4 \operatorname{cis} \dfrac{5\pi}{4}$ **10.** $4 \operatorname{cis} 90°$

11. $2 \operatorname{cis} \dfrac{\pi}{4}$ **12.** $\sqrt{2} \operatorname{cis} 285°$

Exercise Set 8.6, pp. 548–549

1.

3.

85.

$r = e^{\theta/10}$

87. $y^2 = -4x + 4$

5.

7.

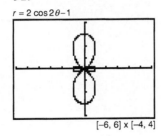

9. $\dfrac{3\sqrt{3}}{2} + \dfrac{3}{2}i$ **11.** $-10i$ **13.** $2 + 2i$ **15.** $-2 - 2i$

17. $\sqrt{2} \operatorname{cis} 315°$ **19.** $20 \operatorname{cis} 330°$ **21.** $5 \operatorname{cis} 180°$

23. $4 \operatorname{cis} 0°$, or 4 **25.** $8 \operatorname{cis} 120°$ **27.** $\operatorname{cis} 270°$, or $-i$

29. $2 \operatorname{cis} 270°$, or $-2i$ **31.** $z = a + bi$,
$|z| = \sqrt{a^2 + b^2}$; $-z = -a - bi$, $|-z| =$
$\sqrt{(-a)^2 + (-b)^2} = \sqrt{a^2 + b^2}$, $\therefore |z| = |-z|$

33. $|(a + bi)(a - bi)| = |a^2 + b^2| = a^2 + b^2$;
$|(a + bi)^2| = |a^2 + 2abi - b^2| = |a^2 - b^2 - 2abi| =$
$\sqrt{(a^2 - b^2)^2 + (2ab)^2} = \sqrt{a^4 + 2a^2b^2 + b^4} = a^2 + b^2$

35. $z \cdot w = (r_1 \operatorname{cis} \theta_1)(r_2 \operatorname{cis} \theta_2) = r_1 r_2 \operatorname{cis} (\theta_1 + \theta_2)$,
$|z \cdot w| = \sqrt{[r_1 r_2 \cos (\theta_1 + \theta_2)]^2 + [r_1 r_2 \sin (\theta_1 + \theta_2)]^2} =$
$\sqrt{(r_1 r_2)^2} = |r_1 r_2|$, $|z| = \sqrt{(r_1 \cos \theta_1)^2 + (r_1 \sin \theta_1)^2} =$
$\sqrt{r_1^2} = |r_1|$, $|w| = \sqrt{(r_2 \cos \theta_2)^2 + (r_2 \sin \theta_2)^2} = \sqrt{r_2^2} = |r_2|$.
Then $|z| \cdot |w| = |r_1| \cdot |r_2| = |r_1 r_2| = |z \cdot w|$

37.

39. $\cos \theta - i \sin \theta$

89.

$r = \sin \theta \tan \theta$

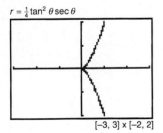

$[-3, -3] \times [-2, 2]$

91.

$r = 2 \cos 2\theta - 1$

$[-6, 6] \times [-4, 4]$

93.

$r = \frac{1}{4} \tan^2 \theta \sec \theta$

$[-3, 3] \times [-2, 2]$

Margin Exercises, Section 8.6

1.

2.

Margin Exercises, Section 8.7
1. $32 \operatorname{cis} 270°$, or $-32i$ **2.** $16 \operatorname{cis} 120°$, or $-8 + 8i\sqrt{3}$
3. $1 + i$, $-1 - i$ **4.** $\sqrt{5} + i\sqrt{5}$, $-\sqrt{5} - i\sqrt{5}$

3. (a) 5; **(b)** 13 **4.** $1 - i$ **5.** $\sqrt{3} - i$ **6.** $\sqrt{2} \operatorname{cis} 315°$

5. 1 cis 60°, 1 cis 180°, 1 cis 300°; or $\frac{1}{2} + \frac{\sqrt{3}}{2}i$, -1, $\frac{1}{2} - \frac{\sqrt{3}}{2}i$

Technology Connection, Section 8.7

TC 1. ± 1.495349, $\pm 1.495349i$

TC 2. -2.154435, $1.0772173 \pm 1.8657952i$

TC 3. ± 1, $\pm 0.80901699 \pm 0.58778525i$,
$\pm 0.30901699 \pm 0.95105652i$

TC 4. 1.1486984, $1.0493882 \pm 0.46721771i$,
$0.76862923 \pm 0.85364924i$, $0.35496731 \pm 1.0924771i$,
$-0.1200717 \pm 1.1424057i$, $-0.5743492 \pm 0.99480196i$,
$-0.9293165 \pm 0.67518795i$, $-1.123597 \pm 0.23882782i$

Exercise Set 8.7, p. 553

1. $8 \operatorname{cis} \pi$ **3.** $64 \operatorname{cis} \pi$ **5.** $8 \operatorname{cis} 270°$

7. $-8 - 8\sqrt{3}i$ **9.** $-8 - 8\sqrt{3}i$ **11.** i **13.** 1

15. $\operatorname{cis} 45°$, $\operatorname{cis} 225°$; or $\frac{\sqrt{2}}{2} + i\frac{\sqrt{2}}{2}$, $-\frac{\sqrt{2}}{2} - i\frac{\sqrt{2}}{2}$

17. $2 \operatorname{cis} 157.5°$, $2 \operatorname{cis} 337.5°$ **19.** $\sqrt{2} \operatorname{cis} 60°$,
$\sqrt{2} \operatorname{cis} 240°$; or $\frac{\sqrt{2}}{2} + \frac{\sqrt{6}}{2}i$, $-\frac{\sqrt{2}}{2} - \frac{\sqrt{6}}{2}i$

21. $\operatorname{cis} 30°$, $\operatorname{cis} 150°$, $\operatorname{cis} 270°$; or $\frac{\sqrt{3}}{2} + \frac{1}{2}i$, $-\frac{\sqrt{3}}{2} + \frac{1}{2}i$, $-i$

23. $2 \operatorname{cis} 0°$, $2 \operatorname{cis} 90°$, $2 \operatorname{cis} 180°$, $2 \operatorname{cis} 270°$; or 2, 2i,
-2, $-2i$

25. $\operatorname{cis} 36°$, $\operatorname{cis} 108°$, $\operatorname{cis} 180°$, $\operatorname{cis} 252°$, $\operatorname{cis} 324°$

27. $2 \operatorname{cis} 30°$, $2 \operatorname{cis} 120°$, $2 \operatorname{cis} 210°$, $2 \operatorname{cis} 300°$;
or $\sqrt{3} + i$, $-1 + \sqrt{3}i$, $-\sqrt{3} - i$, $1 - \sqrt{3}i$

29. $\sqrt[3]{4} \operatorname{cis} 110°$, $\sqrt[3]{4} \operatorname{cis} 230°$, $\sqrt[3]{4} \operatorname{cis} 350°$ **31.** $\operatorname{cis} 0°$,
$\operatorname{cis} 120°$, $\operatorname{cis} 240°$; or 1, $-\frac{1}{2} + \frac{\sqrt{3}}{2}i$, $-\frac{1}{2} - \frac{\sqrt{3}}{2}i$

33. $\operatorname{cis} 36°$, $\operatorname{cis} 108°$, $\operatorname{cis} 180°$, $\operatorname{cis} 252°$, $\operatorname{cis} 324°$

35. $\sqrt[5]{2} \operatorname{cis} 42°$, $\sqrt[5]{2} \operatorname{cis} 114°$, $\sqrt[5]{2} \operatorname{cis} 186°$, $\sqrt[5]{2} \operatorname{cis} 258°$,
$\sqrt[5]{2} \operatorname{cis} 330°$

37. $3 \operatorname{cis} 45°$, $3 \operatorname{cis} 135°$, $3 \operatorname{cis} 225°$, $3 \operatorname{cis} 315°$;
or $\frac{3\sqrt{2}}{2} + \frac{3\sqrt{2}}{2}i$, $-\frac{3\sqrt{2}}{2} + \frac{3\sqrt{2}}{2}i$, $-\frac{3\sqrt{2}}{2} - \frac{3\sqrt{2}}{2}i$,
$\frac{3\sqrt{2}}{2} - \frac{3\sqrt{2}}{2}i$

39. $-1.366 + 1.366i$, $0.366 - 0.366i$

41. ± 1.8612097, $\pm 1.8612097i$

43. ± 1.2311444, $\pm 0.99601675 \pm 0.72364853i$,
$\pm 0.38044455 \pm 1.1708879i$

Review Exercises: Chapter 8, p. 554

1. [8.1] $A = 58.09°$, $B = 31.91°$, $b = 4.5$

2. [8.1] $A = 38.83°$; $b = 37.9$; $c = 48.6$ **3.** [8.1] 1748 cm

4. [8.1] 14 ft **5.** [8.2] $A = 46°$, $C = 99°$, $c = 34$; or
$A = 134°$, $C = 11°$, $c = 7$

6. [8.2] $A = 34.2°$, $a = 0.622$, $c = 0.511$ **7.** [8.3] 8.0

8. [8.2] 419 cm **9.** [8.3] $A = 92.1°$, $B = 33.0°$, $C = 54.8°$

10. [8.2] 13.72 cm² **11.** [8.4] 76 mph; N 20° E

12. [8.2] 33.0 m² **13.** [8.4] 23 lb down, 17 lb left;
28.6 lb, 54° down from horizontal (to left)

14. [8.4] $\langle -18.8, 6.8 \rangle$ **15.** [8.4] (3.6, 124°)

16. [8.4] **(a)** $\langle 25, 0 \rangle$; **(b)** 7.1 **17.** [8.5] (4, 270°)

18. [8.5] (2, 330°) **19.** [8.5] $(-2, -2\sqrt{3})$

20. [8.5] $(-2, 2\sqrt{3})$ **21.** [8.5] $r + 2\cos\theta - 3\sin\theta = 0$

22. [8.5] $x^2 + y^2 = -4y$

23. [8.5]

$r = \cos 2\theta$

24. [8.5]

$r = 2\theta$

25. [8.5]

$r = 3(1 - \sin\theta)$

26. [8.6]

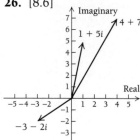

27. [8.6] $-\sqrt{2} + \sqrt{2}i$ **28.** [8.6] $\sqrt{2} \operatorname{cis} 45°$

29. [8.6] $70 \operatorname{cis} 50°$ **30.** [8.6] $\frac{\sqrt{2}}{2} \operatorname{cis} 15°$

31. [8.7] $4\sqrt{2} \operatorname{cis} 135°$ **32.** [8.7] $-\frac{81}{2} + \frac{81\sqrt{3}}{2}i$

33. [8.7] $\sqrt[6]{2} \operatorname{cis} 15°$; $\sqrt[6]{2} \operatorname{cis} 135°$; $\sqrt[6]{2} \operatorname{cis} 255°$

34. [8.7] 3, $\frac{3}{2}(-1 \pm \sqrt{3}i)$ **35.** [8.7] $2 \operatorname{cis} 45°$,
$2 \operatorname{cis} 225°$; or $\sqrt{2} + \sqrt{2}i$, $-\sqrt{2} - \sqrt{2}i$

36. [8.2] $50.52°$, $129.48°$ **37.** [8.4] $\langle \frac{36}{13}, \frac{15}{13} \rangle$

38. [8.5] $y^2 = 4x + 4$

Test: Chapter 8, p. 555

1. [8.1] $B = 67.6°$, $a = 7.6$, $c = 19.9$ **2.** [8.1]
$B = 66.87°$, $A = 23.13°$, $b = 26$ **3.** [8.1] 35°

4. [8.2] $A = 115°$, $a = 88.4$, $c = 31.8$ **5.** [8.3] 12
6. [8.2] 6 ft, 5 ft   **7.** [8.2] 3.3 cm^2 **8.** [8.4] 22.8 N, 61°
9. [8.4] (21.7, 12.5)   **10.** [8.4] (11.7, 149°)
11. [8.4] **(a)** $\langle -1, -14 \rangle$; **(b)** 8.1   **12.** [8.4]

(a) N. 7, W. 5; **(b)** 8.6, N 36° W   **13.** [8.5] $\left(-\dfrac{3\sqrt{3}}{2}, -\dfrac{3}{2} \right)$

14. [8.5] $x^2 + y^2 = 625$ **15.** [8.7] $2\sqrt{2}$ cis 45°,
$2\sqrt{2}$ cis 225°; or $2 + 2i$, $-2 - 2i$
16. [8.6]

17. [8.6] $\dfrac{3\sqrt{3}}{2} - \dfrac{3}{2}i$

18. [8.6] 2 cis 120°
19. [8.6] 10 cis 90°

20. [8.6] $\dfrac{1}{2}$ cis 80°

21. [8.7] 8 cis 270°

22. [8.7] 2 cis 60°, 2 cis 180°, 2 cis 300°;
or $1 + \sqrt{3}i$, -2, $1 - \sqrt{3}i$
23. [8.5]

$r = 4 \cos 3\theta$

24. [8.5]

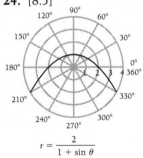

$r = \dfrac{2}{1 + \sin \theta}$

25. [8.5] $\left(-\dfrac{4}{\sqrt{53}}, \dfrac{14}{\sqrt{53}} \right)$

CHAPTER 9

Margin Exercises, Section 9.1
1. Yes   **2.** No   **3.** $(-2, 3)$
4. Infinitely many solutions   **5.** No solution   **6.** $(1, 2)$
7. $(4, -2)$ **8.** $(-2, 5)$ **9.** $(-3, 2)$ **10.** $(-\frac{1}{3}, \frac{1}{2})$
11. $(2, -1)$ **12.** $\frac{11}{2}, \frac{9}{2}$ **13.** $\frac{11}{2}, -\frac{9}{2}$
14. 10 km/h, 2 km/h   **15.** 5 L, 3 L

Technology Connection, Section 9.1
TC 1. (1.424, 1.334)   **TC 2.** Infinite number of solutions
TC 3. (1.614, -8.765)   **TC 4.** No solution

Exercise Set 9.1, pp. 565–567
1. No   **3.** $(-1, 3)$ **5.** No solution   **7.** $(\frac{39}{11}, -\frac{1}{11})$
9. $(-4, -2)$ **11.** $(-3, 0)$ **13.** $(10, 8)$ **15.** $(1, 1)$
17. $-\frac{11}{2}, -\frac{9}{2}$ **19.** 20 km/h, 3 km/h   **21.** 12.5 L, 7.5 L
23. 3 hr   **25.** 2 hr   **27.** 12 white, 28 printed
29. Paula is 32, Bob is 20   **31.** 76 m, 19 m

33. 137 m, 55 m   **35.** $6800 at 9%, $8200 at 10%
37. 4 dimes, 30 nickels   **39.** 5 lb of $4.05; 10 lb of $2.70
41. $(-12, 0)$ **43.** 4 km   **45.** 180   **47.** $\frac{7}{19}$
49. First train: 36 km/h; second train: 54 km/h
51. $96   **53.** 4 boys, 3 girls   **55.** $m = -\frac{4}{3}, b = \frac{1}{3}$
57. $(-\frac{1}{4}, -\frac{1}{2})$ **59.** $\{(5, 3), (-5, 3), (5, -3), (-5, -3)\}$
61. 40 mph   **63.** $(-4.699, 5.071)$ **65.** $(-0.210, 0.505)$

Margin Exercises, Section 9.2
1. (a) No; **(b)** yes   **2.** $(2, \frac{1}{2}, -2)$
3. A: 75; B: 84; C: 63   **4.** $f(x) = x^2 - 2x + 1$
5. $f(x) = \frac{5}{8}x^2 - 50x + 1150$; 510 accidents

Exercise Set 9.2, pp. 572–574
1. No   **3.** $(3, -2, 1)$ **5.** $(-3, 2, 1)$ **7.** $(\frac{1}{2}, \frac{2}{3}, -\frac{5}{6})$
9. $(1, -2, 4, -1)$ **11.** 8, 21, -3
13. $A = 30°$, $B = 90°$, $C = 60°$
15. 20 on Mon., 35 on Tues., 32 on Wed.
17. A: 2200; B: 2500; C: 2700   **19.** A: 10; B: 12; C: 15
21. Par-3: 4; par-4: 10; par-5: 4
23. 7%: $400; 8%: $500; 9%: $1600
25. $y = 2x^2 + 3x - 1$ **27. (a)** $E = -4t^2 + 40t + 2$;
(b) $98   **29. (a)** $f(x) = 104.5x^2 - 1501.5x + 6016$;
(b) 1682, 769, 1451   **31.** $(-1, \frac{1}{5}, -\frac{1}{2})$
33. A: 4 hr; B: 6 hr; C: 12 hr   **35.** 180°
37. $3x + 4y + 2z = 12$
39. Adults: 5; students: 1; children, 94

Margin Exercises, Section 9.3
1. No solution; inconsistent   **2.** $(-\frac{1}{2}, -2)$; consistent
3. Infinitely many solutions; consistent; dependent
4. $(2, 3)$; consistent; dependent
5. $\left(\dfrac{2y - 5}{3}, y \right)$ or $\left(x, \dfrac{3x + 5}{2} \right)$, $(1, 4)$, $(3, 7)$, $\left(0, \dfrac{5}{2} \right)$, etc.
6. $\left(x, \dfrac{1 - 2x}{5} \right)$ or $\left(\dfrac{1 - 5y}{2}, y \right)$, $(-7, 3)$, $(3, -1)$,
$\left(0, \dfrac{1}{5} \right)$, etc.
7. $(-2z + 7, 3z - 5, z)$; $(7, -5, 0)$, $(5, -2, 1)$,
$(3, 1, 2)$, etc.
8. $(-2z, 3z, z)$; $(-2, 3, 1)$, $(2, -3, -1)$, $(-4, 6, 2)$, etc.
9. $(0, 0, 0)$, only solution

Exercise Set 9.3, pp. 580–581
1. $\left(\dfrac{y + 5}{3}, y \right)$ or $(x, 3x - 5)$; $(0, -5)$, $(1, -2)$,
$(-1, -8)$, etc.   **3.** $\varnothing$
5. $\left(\dfrac{5 - 2y}{3}, y \right)$ or $\left(x, \dfrac{5 - 3x}{2} \right)$; $(3, -2)$, etc.
7. $\left(\dfrac{6y - 3}{4}, y \right)$ or $\left(x, \dfrac{4x + 3}{6} \right)$; $\left(1, \dfrac{7}{6} \right)$, etc.   **9.** $\varnothing$
11. $\left(\dfrac{10 + 11z}{9}, \dfrac{-11 + 5z}{9}, z \right)$; $\left(\dfrac{10}{9}, -\dfrac{11}{9}, 0 \right)$, etc.
13. $\left(\dfrac{11}{9}z, \dfrac{5}{9}z, z \right)$; $\left(\dfrac{11}{9}, \dfrac{5}{9}, 1 \right)$, etc.
15. $(4z - 5, -3z + 2, z)$; $(-1, -1, 1)$, etc.   **17.** $(0, 0, 0)$

19. Consistent: 1, 5, 7, 11, 13, 15, 17, the others are inconsistent; dependent: 1, 5, 7, 11, 13, 15, the others are independent

21. $\left(\dfrac{724y + 9160}{2013}, y\right)$ or $\left(x, \dfrac{2013x - 9160}{724}\right)$

23. (a) $\varnothing$; (b) inconsistent; (c) dependent **25.** $k = 2$
27. $(x, 18 - 2x, x)$ where $1 \le x \le 8$, x is an integer

Margin Exercises, Section 9.4

1. $\left(-\dfrac{63}{29}, -\dfrac{114}{29}\right)$ **2.** $(-1, 2, 3)$ **3.** 3×2

4. 2×2 **5.** 3×3 **6.** 1×2 **7.** 2×1 **8.** 1×1

9. 4, 5, 8 **10.** $A + B = \begin{bmatrix} -2 & -6 \\ 13 & 0 \end{bmatrix} = B + A$

11. $\begin{bmatrix} 1 & 1 & -10 \\ 2 & 4 & -3 \end{bmatrix}$

12. $A + O = \begin{bmatrix} 4 & -3 \\ 5 & 8 \end{bmatrix} = O + A = A$

13. $\begin{bmatrix} -1 & 4 & -7 \\ -2 & -4 & 8 \end{bmatrix}$ **14.** $\begin{bmatrix} -6 & 6 \\ 1 & -4 \\ -7 & 5 \end{bmatrix}$

15. $\begin{bmatrix} -2 & 1 & -5 \\ -6 & -4 & 3 \end{bmatrix}$ **16.** $\begin{bmatrix} 0 & 0 & 0 \\ 0 & 0 & 0 \end{bmatrix}$

17. $\begin{bmatrix} -1 & 4 & -7 \\ -2 & -4 & 8 \end{bmatrix}$ **18.** $\begin{bmatrix} 5 & -10 & 5x \\ 20 & 5y & 5 \\ 0 & -25 & 5x^2 \end{bmatrix}$

19. $\begin{bmatrix} t & -t & 4t & tx \\ ty & 3t & -2t & ty \\ t & 4t & -5t & ty \end{bmatrix}$ **20.** $[-13]$ **21.** $\begin{bmatrix} 12 \\ 13 \\ 5 \\ 16 \end{bmatrix}$

22. $\begin{bmatrix} 0 & 26 \\ -8 & 3 \\ -13 & 33 \\ -7 & 32 \end{bmatrix}$

23. $AB = \begin{bmatrix} 2 & 8 & 6 \\ -29 & -34 & -7 \end{bmatrix}$; BA not possible

24. $[8 \quad 5 \quad 4]$

25. $AB = \begin{bmatrix} -2 & 32 \\ 4 & 16 \end{bmatrix}$, $BA = \begin{bmatrix} 8 & -13 \\ -16 & 6 \end{bmatrix}$

26. $AI = \begin{bmatrix} 3 & 2 \\ -1 & 5 \end{bmatrix} = IA = A$

27. $\begin{bmatrix} 3 & 4 & -2 \\ 2 & -2 & 5 \\ 6 & 7 & -1 \end{bmatrix} \begin{bmatrix} x \\ y \\ z \end{bmatrix} = \begin{bmatrix} 5 \\ 3 \\ 0 \end{bmatrix}$

Exercise Set 9.4, pp. 591–592
1. $\left(\frac{3}{2}, \frac{5}{2}\right)$ **3.** $(-1, 2, -2)$ **5.** $\left(\frac{1}{2}, \frac{3}{2}\right)$
7. $\left(\frac{3}{2}, -4, 3\right)$ **9.** $(r - 2, 3 - 2r, r)$

11. $(1, -3, -2, -1)$ **13.** $\begin{bmatrix} -2 & 7 \\ 6 & 2 \end{bmatrix}$ **15.** $\begin{bmatrix} 1 & 3 \\ 2 & 6 \end{bmatrix}$

17. $\begin{bmatrix} 9 & 9 \\ -3 & -3 \end{bmatrix}$ **19.** $\begin{bmatrix} 11 & 13 \\ 5 & 3 \end{bmatrix}$ **21.** $\begin{bmatrix} -4 & 3 \\ -2 & -4 \end{bmatrix}$

23. $\begin{bmatrix} 17 & 9 \\ -2 & 1 \end{bmatrix}$ **25.** $\begin{bmatrix} 0 & 0 \\ 0 & 0 \end{bmatrix}$ **27.** $\begin{bmatrix} 1 & 2 \\ 4 & 3 \end{bmatrix}$ or A

29. $\begin{bmatrix} -5 & 4 & 3 \\ 5 & -9 & 4 \\ 7 & -18 & 17 \end{bmatrix}$ **31.** $\begin{bmatrix} -2 & 9 & 6 \\ -3 & 3 & 4 \\ 2 & -2 & 1 \end{bmatrix}$ or C

33. $[-16]$ **35.** $[2 \quad -19]$

37. $\begin{bmatrix} 3 & -2 & 4 \\ 2 & 1 & -5 \end{bmatrix} \begin{bmatrix} x \\ y \\ z \end{bmatrix} = \begin{bmatrix} 17 \\ 13 \end{bmatrix}$

39. $\begin{bmatrix} 1 & -1 & 2 & -4 \\ 2 & -1 & -1 & 1 \\ 1 & 4 & -3 & -1 \\ 3 & 5 & -7 & 2 \end{bmatrix} \begin{bmatrix} x \\ y \\ z \\ w \end{bmatrix} = \begin{bmatrix} 12 \\ 0 \\ 1 \\ 9 \end{bmatrix}$

41. $(A + B)(A - B) = \begin{bmatrix} -2 & 1 \\ 2 & -1 \end{bmatrix}$,

$A^2 - B^2 = \begin{bmatrix} 0 & 3 \\ 0 & -3 \end{bmatrix}$

43. $(A + B)(A - B) = \begin{bmatrix} -2 & 1 \\ 2 & -1 \end{bmatrix}$
$= A^2 + BA - AB - B^2$

45. $A + B = \begin{bmatrix} a_{11} + b_{11} & a_{12} + b_{12} \\ a_{21} + b_{21} & a_{22} + b_{22} \end{bmatrix}$

$= \begin{bmatrix} b_{11} + a_{11} & b_{12} + a_{12} \\ b_{21} + a_{21} & b_{22} + a_{22} \end{bmatrix}$

$= B + A$

47. $(k + m)A = \begin{bmatrix} (k + m)a_{11} & (k + m)a_{12} \\ (k + m)a_{21} & (k + m)a_{22} \end{bmatrix}$

$= \begin{bmatrix} ka_{11} + ma_{11} & ka_{12} + ma_{12} \\ ka_{21} + ma_{21} & ka_{22} + ma_{22} \end{bmatrix}$

$= \begin{bmatrix} ka_{11} & ka_{12} \\ ka_{21} & ka_{22} \end{bmatrix} + \begin{bmatrix} ma_{11} & ma_{12} \\ ma_{21} & ma_{22} \end{bmatrix}$

$= kA + mA$

49. Find AI, IA, and compare with A.

Margin Exercises, Section 9.5
1. -13 **2.** -2 **3.** $-2x + 12$
4. $a_{11} = -8$, $a_{13} = 6$, $a_{22} = -6$, $a_{31} = -1$, $a_{32} = -3$

5. $M_{22} = \begin{vmatrix} -8 & 6 \\ -1 & 5 \end{vmatrix} = -34$, $M_{32} = \begin{vmatrix} -8 & 6 \\ 4 & 7 \end{vmatrix} = -80$,

$M_{13} = \begin{vmatrix} 4 & -6 \\ -1 & -3 \end{vmatrix} = -18$

6. $A_{22} = -34$, $A_{32} = 80$, $A_{13} = -18$
7. $|A| = 0 \cdot A_{12} + (-6)A_{22} + (-3)A_{32} = -36$
8. $|A| = 6 \cdot A_{13} + 7A_{23} + 5A_{33} = -36$
9. 93   **10.** 60   **11.** $x^3 - x^2$ **12.** 0   **13.** 0
14. **(a)** $|A| = -18$, $|B| = 18$. **(b)** The rows are interchanged.
15. **(a)** $|C| = -10$, $|D| = 10$. **(b)** The first and third columns
are interchanged.   **16.** 0   **17.** $x = -4$ **18.** $x = 6$

19. 0   **20.** $\begin{vmatrix} -2 & 3 & 4 \\ -3 & 10 & 5 \\ 0 & 9 & 7 \end{vmatrix}$ **21.** 68   **22.** -195

23. $(b - a)(c - a)(b - c)$ **24.** $(3, 1)$

25. $\left(-\dfrac{10}{41}, -\dfrac{13}{41}\right)$

26. $\left(\dfrac{3\sqrt{2} + 4\pi}{2 + \pi^2}, \dfrac{4\sqrt{2} - 3\pi}{2 + \pi^2}\right)$

27. $(1, 3, -2)$

Exercise Set 9.5, pp. 603–604
1. -11 **3.** $x^3 - 4x$ **5.** -109 **7.** $-x^4 + x^2 - 5x$
9. $a_{11} = 7$, $a_{32} = 2$, $a_{22} = 0$
11. $M_{11} = 6$, $M_{32} = -9$, $M_{22} = -29$
13. $A_{11} = 6$, $A_{32} = 9$, $A_{22} = -29$ **15.** $|A| = -10$
17. $|A| = -10$ **19.** $M_{41} = -14$, $M_{33} = 20$
21. $A_{24} = 15$, $A_{43} = 30$ **23.** $|A| = 110$ **25.** -195
27. -70 **29.** -4 **31.** 9072   **33.** -153 **35.** 0
37. 0   **39.** $(x - y)(y - z)(x - z)$

41. $xyz(x - y)(y - z)(z - x)$ **43.** $\left(-\dfrac{25}{2}, -\dfrac{11}{2}\right)$

45. $\left(\dfrac{4\pi - 5\sqrt{3}}{3 + \pi^2}, \dfrac{4\sqrt{3} + 5\pi}{-3 - \pi^2}\right)$ **47.** $\left(\dfrac{3}{2}, \dfrac{13}{14}, \dfrac{33}{14}\right)$

49. $\left(\dfrac{1}{2}, \dfrac{2}{3}, -\dfrac{5}{6}\right)$ **51.** $2, -2$

53. $\{x | x \le -\sqrt{3} \text{ or } x \ge \sqrt{3}\}$ **55.** -34 **57.** 4

59. $\begin{vmatrix} L & -W \\ 2 & 2 \end{vmatrix}$ **61.** $\begin{vmatrix} a & b \\ -b & a \end{vmatrix}$ **63.** $\begin{vmatrix} 2\pi r & 2\pi r \\ -h & r \end{vmatrix}$

65. Evaluate the determinant and compare with the
two-point equation of a line.

67. $\dfrac{1}{2} \cdot \begin{vmatrix} x_1 & y_1 & 1 \\ x_2 & y_2 & 1 \\ x_3 & y_3 & 1 \end{vmatrix}$

$= \dfrac{1}{2}\left(x_1 \cdot \begin{vmatrix} y_2 & 1 \\ y_3 & 1 \end{vmatrix} - y_1 \cdot \begin{vmatrix} x_2 & 1 \\ x_3 & 1 \end{vmatrix} + 1 \cdot \begin{vmatrix} x_2 & y_2 \\ x_3 & y_3 \end{vmatrix}\right)$

$= \dfrac{1}{2}[x_1(y_2 - y_3) - y_1(x_2 - x_3) + (x_2 y_3 - x_3 y_2)]$

$= \dfrac{1}{2}[x_1 y_2 - x_1 y_3 - x_2 y_1 + x_3 y_1 + x_2 y_3 - x_3 y_2]$;

Area of triangle ABC

$= \dfrac{1}{2} \cdot (x_1 - x_2)(y_1 + y_2) + \dfrac{1}{2} \cdot (x_3 - x_1)(y_1 + y_3)$

$\quad - \dfrac{1}{2} \cdot (x_3 - x_2)(y_2 + y_3)$

$= \dfrac{1}{2}(x_1 y_1 + x_1 y_2 - x_2 y_1 - x_2 y_2 + x_3 y_1 + x_3 y_3 - x_1 y_1$

$\quad - x_1 y_3 - x_3 y_2 - x_3 y_3 + x_2 y_2 + x_2 y_3)$

$= \dfrac{1}{2}(x_1 y_2 - x_2 y_1 + x_3 y_1 - x_1 y_3 - x_3 y_2 + x_2 y_3)$

Margin Exercises, Section 9.6
1. **(a)** A; **(b)** A; **(c)** both equal A; **(d)** X

2. **(a)** $\begin{bmatrix} 1 & 0 & 0 \\ 0 & 1 & 0 \\ 0 & 0 & 1 \end{bmatrix} = I$; **(b)** I; **(c)** both equal I

3. $A^{-1} = \begin{bmatrix} -\frac{1}{2} & \frac{1}{2} & \frac{1}{2} \\ 1 & 0 & -1 \\ \frac{3}{2} & -\frac{1}{2} & -\frac{1}{2} \end{bmatrix}$

4. $A^{-1} = \dfrac{1}{11}\begin{bmatrix} 2 & 5 \\ 1 & -3 \end{bmatrix}$

5. **(a)** $\begin{bmatrix} 4 & -2 \\ 1 & 5 \end{bmatrix}\begin{bmatrix} x \\ y \end{bmatrix} = \begin{bmatrix} -1 \\ 1 \end{bmatrix}$; **(b)** $A = \begin{bmatrix} 4 & -2 \\ 1 & 5 \end{bmatrix}$;

(c) $A^{-1} = \dfrac{1}{22}\begin{bmatrix} 5 & 2 \\ -1 & 4 \end{bmatrix}$; **(d)** $x = -\dfrac{3}{22}$, $y = \dfrac{5}{22}$

Exercise Set 9.6, pp. 609–610
1. $A^{-1} = \begin{bmatrix} -3 & 2 \\ 5 & -3 \end{bmatrix}$ **3.** $A^{-1} = \begin{bmatrix} 2 & -3 \\ -7 & 11 \end{bmatrix}$

5. $A^{-1} = \begin{bmatrix} \frac{2}{11} & \frac{3}{11} \\ -\frac{1}{11} & \frac{4}{11} \end{bmatrix}$ **7.** $A^{-1} = \begin{bmatrix} \frac{3}{8} & -\frac{1}{4} & \frac{1}{8} \\ -\frac{1}{8} & \frac{3}{4} & -\frac{3}{8} \\ -\frac{1}{4} & \frac{1}{2} & \frac{1}{4} \end{bmatrix}$

9. $A^{-1} = \begin{bmatrix} \frac{1}{3} & 0 & \frac{1}{3} \\ -\frac{2}{5} & \frac{2}{5} & \frac{1}{5} \\ \frac{2}{15} & \frac{1}{5} & -\frac{1}{15} \end{bmatrix}$ **11.** A^{-1} does not exist.

13. $A^{-1} = \begin{bmatrix} 1 & -2 & 3 & 8 \\ 0 & 1 & -3 & 1 \\ 0 & 0 & 1 & -2 \\ 0 & 0 & 0 & -1 \end{bmatrix}$ **15.** A^{-1} does not exist.

17. $(-23, 83)$ **19.** $(-1, 5, 1)$

21. $\begin{bmatrix} 4 & -3 \\ 1 & 2 \end{bmatrix}\begin{bmatrix} x \\ y \end{bmatrix} = \begin{bmatrix} 2 \\ -1 \end{bmatrix}$, $A^{-1} = \dfrac{1}{11}\begin{bmatrix} 2 & 3 \\ -1 & 4 \end{bmatrix}$, $\left(\dfrac{1}{11}, -\dfrac{6}{11}\right)$

23. $\begin{bmatrix} 7 & -2 \\ 9 & 3 \end{bmatrix}\begin{bmatrix} x \\ y \end{bmatrix} = \begin{bmatrix} -3 \\ 4 \end{bmatrix}$, $A^{-1} = \dfrac{1}{39}\begin{bmatrix} 3 & 2 \\ -9 & 7 \end{bmatrix}$, $\left(-\dfrac{1}{39}, \dfrac{55}{39}\right)$

25. $\begin{bmatrix} 1 & 0 & 1 \\ 2 & 1 & 0 \\ 1 & -1 & 1 \end{bmatrix} \begin{bmatrix} x \\ y \\ z \end{bmatrix} = \begin{bmatrix} 1 \\ 3 \\ 4 \end{bmatrix}$,

$A^{-1} = \dfrac{1}{2} \begin{bmatrix} -1 & 1 & 1 \\ 2 & 0 & -2 \\ 3 & -1 & -1 \end{bmatrix}$, $(3, -3, -2)$

27. $(1, -1, 0, 1)$ **29.** Find **AI** and **IA** and compare with **A**.

31. A^{-1} exists if and only if $xy \neq 0$. $A^{-1} = \begin{bmatrix} x^{-1} & 0 \\ 0 & y^{-1} \end{bmatrix}$.

33. A^{-1} exists if and only if $xyzw \neq 0$.

$$A^{-1} = \begin{bmatrix} \dfrac{1}{x} & -\dfrac{1}{xy} & -\dfrac{1}{xz} & -\dfrac{1}{xw} \\ 0 & \dfrac{1}{y} & 0 & 0 \\ 0 & 0 & \dfrac{1}{z} & 0 \\ 0 & 0 & 0 & \dfrac{1}{w} \end{bmatrix}$$

Margin Exercises, Section 9.7
1. Yes   **2.** No
3.

4.

5.

6.

7.

8.

9. Vertices:
$(0, 0), (4, 0), (4, \frac{5}{3}),$
$(\frac{12}{5}, 3), (0, 3)$

10. Vertices:
$(0, 0), (3, 0), (2, 1), (0, 2)$

11. Maximum 120, when $x = 3$, $y = 3$; minimum 34, when $x = 1$, $y = 0$
12. Maximum 45, when $x = 6$, $y = 0$; minimum 12, when $x = 0$, $y = 3$
13. Maximum \$133.70, by selling 50 hot dogs and 40 hamburgers

Technology Connection, Section 9.7
TC 1.

TC 2.

TC 3.

TC 4. Vertices: $(0, 0),$
$(7.5, 0), (6.75, 0.75), (0, 3)$

Exercise Set 9.7, pp. 621–622
1. No   **3.** No

5.

7.

9.

11.

13.

15.

17.

19.

21.

23.

25.

27.

29.

31.

33.

35.

37.

39.

41.

43.

45.

47. The maximum value of P is 179 when $x = 7$ and $y = 0$. The minimum value of P is 48 when $x = 0$ and $y = 4$.
49. The maximum value of F is 216 when $x = 0$ and $y = 6$. The minimum value of F is 0 when $x = 0$ and $y = 0$.
51. Maximum income of $18 when 100 of each type of biscuit is made.
53. Maximum score of 425 when 5 questions of type A and 15 of type B are answered.
55. Maximum profit of $11,000 is achieved by producing 100 units of lumber and 300 units of plywood.
57. Maximum income of $3110 is achieved when $22,000 is invested in corporate bonds and $18,000 is invested in municipal bonds.
59. Maximum profit of $2520 when 125 batches of Smello and 187.5 batches of Roppo are made.

61.

63. $2w + t \geq 60,$
$\quad w \geq 0,$
$\quad t \geq 0$

65. $2L + 2W \leq 248,$
$\quad 0 \leq L \leq 74,$
$\quad 0 \leq W \leq 50$

67.

$|x + y| \leq 1$

69.

$|x| > |y|$

71. Vertices: $(-2.30, 3.11), (-2.30, -1.12), (3.17, -0.19)$

$3.5x + 5.8y \leq 10$
$2.1x - 12.3y \geq 9$
$x \geq -2.3$

$[-10, 10] \times [-10, 10]$

Margin Exercises, Section 9.8

1. $\dfrac{1}{3x + 2} + \dfrac{4}{x - 3}$ **2.** $\dfrac{1}{x + 1} + \dfrac{2}{x - 1} - \dfrac{1}{(x - 1)^2}$

3. $\dfrac{x + 2}{x^2 + 1} + \dfrac{1}{x + 2}$ **4.** $3x + 1 + \dfrac{4}{2x - 1} - \dfrac{3}{x + 5}$

5. $\dfrac{3}{x - 4} - \dfrac{2}{x + 1} + \dfrac{1}{(x + 1)^2}$ **6.** $\dfrac{2x - 5}{3x^2 + 2} + \dfrac{1}{x - 1}$

Exercise Set 9.8, p. 628

1. $\dfrac{2}{x - 3} + \dfrac{-1}{x + 2}$ **3.** $\dfrac{-4}{3x - 1} + \dfrac{5}{2x - 1}$

5. $\dfrac{-3}{x - 2} + \dfrac{2}{x + 2} + \dfrac{4}{x + 1}$ **7.** $\dfrac{-3}{(x + 2)^2} + \dfrac{-1}{x + 2} + \dfrac{1}{x - 1}$

9. $\dfrac{3}{x - 1} + \dfrac{-4}{2x - 1}$ **11.** $x - 2 + \dfrac{-\frac{11}{4}}{(x + 1)^2} + \dfrac{\frac{17}{16}}{x + 1} + \dfrac{-\frac{17}{16}}{x - 3}$

13. $\dfrac{3x + 5}{x^2 + 2} + \dfrac{-4}{x - 1}$ **15.** $\dfrac{-2}{x + 2} + \dfrac{10}{(x + 2)^2} + \dfrac{3}{2x - 1}$

17. $3x + 1 + \dfrac{2}{2x - 1} + \dfrac{3}{x + 1}$ **19.** $\dfrac{-1}{x - 3} + \dfrac{3x}{x^2 + 2x - 5}$

21. $\dfrac{5}{3x + 5} - \dfrac{3}{x + 1} + \dfrac{4}{(x + 1)^2}$ **23.** $\dfrac{8}{4x - 5} + \dfrac{3}{3x + 2}$

25. $\dfrac{2x - 5}{3x^2 + 1} - \dfrac{2}{x - 2}$ **27.** $\dfrac{-\frac{1}{2a^2}x}{x^2 + a^2} + \dfrac{\frac{1}{4a^2}}{x - a} + \dfrac{\frac{1}{4a^2}}{x + a}$

29. $\dfrac{-1}{x+1} + \dfrac{4}{x+4}$

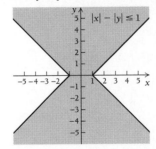

$$f(x) = \frac{3x}{x^2 + 5x + 4}$$

31. $-\dfrac{3}{25(\ln x + 2)} + \dfrac{3}{25(\ln x - 3)} + \dfrac{7}{5(\ln x - 3)^2}$

Review Exercises: Chapter 9, pp. 629–630
1. [9.1] $(-2, -2)$ **2.** [9.1] $(-5, 4)$ **3.** [9.3] $\varnothing$
4. [9.3] $\varnothing$ **5.** [9.2] $(0, 0, 0)$
6. [9.2] $(w, x, y, z) = (-5, 13, 8, 2)$
7. [9.3] Consistent: 1, 2, 5, 6; the others are inconsistent
8. [9.3] All are independent.   **9.** [9.1] 31 nickels, 44 dimes
10. [9.1] \$1600 at 10%, \$3400 at 10.5%
11. [9.2] A: $32°$; B: $96°$; C: $52°$
12. [9.2] A: 74.5; B: 68.5; C: 82   **13.** [9.4] $(1, 2)$
14. [9.4] $(-3, 4, -2)$
15. [9.3], [9.4] $\left(\dfrac{z}{2}, -\dfrac{z}{2}, z\right)$; $(0, 0, 0)$, $\left(\dfrac{1}{2}, -\dfrac{1}{2}, 1\right)$,
$(1, -1, 2)$, etc.   **16.** [9.4] $(-4, 1, -2, 3)$
17. [9.2] $f(x) = -x^2 - 2x + 3$ **18.** [9.5] 10
19. [9.5] -18 **20.** [9.5] $2x + 12$ **21.** [9.5] -6
22. [9.5] -16.588 **23.** [9.5] 0   **24.** [9.5] $(3, -2)$
25. [9.5] $(a, 0)$ **26.** [9.5] $\left(\dfrac{3}{2}, \dfrac{13}{14}, \dfrac{33}{14}\right)$

27. [9.4] $\begin{bmatrix} 0 & -1 & 6 \\ 3 & 1 & -2 \\ -2 & 1 & -2 \end{bmatrix}$ **28.** [9.4] $\begin{bmatrix} -3 & 3 & 0 \\ -6 & -9 & 6 \\ 6 & 0 & -3 \end{bmatrix}$

29. [9.4] $\begin{bmatrix} -1 & 1 & 0 \\ -2 & -3 & 2 \\ 2 & 0 & -1 \end{bmatrix}$ **30.** [9.4] $\begin{bmatrix} -2 & 2 & 6 \\ 1 & -8 & 18 \\ 2 & 1 & -15 \end{bmatrix}$

31. [9.4] Not possible   **32.** [9.4] $\begin{bmatrix} 2 & -1 & -6 \\ 1 & 5 & -2 \\ -2 & -1 & 4 \end{bmatrix}$

33. [9.4] $\begin{bmatrix} 3 & -2 & -6 \\ 3 & 8 & -4 \\ -4 & -1 & 5 \end{bmatrix}$ **34.** [9.4] $\begin{bmatrix} -2 & -1 & 18 \\ 5 & -3 & -2 \\ -2 & 3 & -8 \end{bmatrix}$

35. [9.6] $\begin{bmatrix} -\frac{1}{2} & 0 \\ \frac{1}{6} & \frac{1}{3} \end{bmatrix}$ **36.** [9.6] $\begin{bmatrix} 0 & 0 & \frac{1}{4} \\ 0 & -\frac{1}{2} & 0 \\ \frac{1}{3} & 0 & 0 \end{bmatrix}$

37. [9.6] $\begin{bmatrix} 1 & 0 & 0 & 0 \\ 0 & \frac{1}{9} & \frac{5}{18} & 0 \\ 0 & -\frac{1}{9} & \frac{2}{9} & 0 \\ 0 & 0 & 0 & 1 \end{bmatrix}$

38. [9.4] $\begin{bmatrix} 3 & -2 & 4 \\ 1 & 5 & -3 \\ 2 & -3 & 7 \end{bmatrix} \begin{bmatrix} x \\ y \\ z \end{bmatrix} = \begin{bmatrix} 13 \\ 7 \\ -8 \end{bmatrix}$ **39.** [9.5] -31

40. [9.5] -1 **41.** [9.7] 0   **42.** [9.5] 120

43. [9.5] $\begin{vmatrix} 5a & 5b & 5c \\ 3a & 3b & 3c \\ d & e & f \end{vmatrix} = 5(3) \begin{vmatrix} a & b & c \\ a & b & c \\ d & e & f \end{vmatrix} = 0$, since the first
two rows are the same.   **44.** [9.5] $(a - b)(b - c)(c - a)$
45. [9.5] $(x - y)(y - z)(z - x)(xy + yz + zx)$
46. [9.5] $(b - a)(c - a)(d - a)(c - b)(d - b)(d - c)$
47. [9.6] $(-5, 4)$ **48.** [9.7]

49. [9.7] Minimum $= 52$ at $(2, 4)$; maximum $= 92$ at $(2, 8)$
50. [9.7] Type A: 0; type B: 10; maximum score $= 120$ pts
51. [9.8] $\dfrac{5}{x + 1} - \dfrac{5}{x + 2} - \dfrac{5}{(x + 2)^2}$
52. [9.2] \$10,000 at 12%, \$12,000 at 13%, \$18,000 at $14\frac{1}{2}$%
53. [9.1] $(\frac{5}{18}, \frac{1}{7})$ **54.** [9.2] $(1, \frac{1}{2}, \frac{1}{3})$
55. [9.7]   **56.** [9.7]

57. [9.5] If a matrix has all 0's below the main diagonal, then its determinant is the product of the elements on the main diagonal. *Proof:* Expand down the first column.

Test: Chapter 9, pp. 630–631
1. [9.1] $(2, 1)$ **2.** [9.1] Boat: 15 km/h; stream: 3 km/h
3. [9.1] 12 gal of 15%, 8 gal of 75%
4. [9.2] A: 44; B: 240; C: 216   **5.** [9.4] $\left(\dfrac{1}{4}, \dfrac{1}{2}\right)$
6. [9.4] $\left(\dfrac{z}{3}, -\dfrac{5z}{2}, z\right)$: $(0, 0, 0)$, $\left(\dfrac{1}{3}, -\dfrac{5}{2}, 1\right)$, $\left(\dfrac{2}{3}, -5, 2\right)$,
etc. Answers may vary.   **7.** [9.4] $(3, -1, -2)$
8. [9.3] Consistent, independent
9. [9.3] Consistent, dependent
10. [9.2] $f(x) = -2x^2 + 4x + 1$ **11.** [9.5] 2
12. [9.5] -2 **13.** [9.5] $(4, -1)$ **14.** [9.5] $(0, -2, 1)$

15. [9.4] Not possible   **16.** [9.4] A or $\begin{bmatrix} -1 & 3 \\ 0 & 4 \end{bmatrix}$

17. [9.4] $\begin{bmatrix} 0 & 3 \\ 0 & 5 \end{bmatrix}$ **18.** [9.4] $\begin{bmatrix} 5 & 1 & 2 \\ 3 & 3 & 3 \\ 1 & 4 & -3 \end{bmatrix}$

19. [9.4] $\begin{bmatrix} 1 & 8 \\ -6 & 7 \end{bmatrix}$ **20.** [9.4] $\begin{bmatrix} -7 & -1 \\ 24 & -17 \end{bmatrix}$

21. [9.4] Not possible   **22.** [9.4] $[-2 \quad 1 \quad 3]$

23. [9.4] $\begin{bmatrix} -5 & 5 & -5 \\ 6 & -6 & 0 \\ -1 & 8 & 12 \end{bmatrix}$ **24.** [9.4] $\begin{bmatrix} -2 & -5 \\ 6 & -3 \end{bmatrix}$

25. [9.6] $\begin{bmatrix} 0 & \frac{1}{2} \\ 1 & \frac{3}{2} \end{bmatrix}$ **26.** [9.6] Does not exist

27. [9.6] $\begin{bmatrix} \frac{2}{3} & 0 & \frac{1}{6} \\ \frac{1}{3} & 0 & \frac{1}{3} \\ -2 & 1 & -\frac{1}{2} \end{bmatrix}$

28. [9.5] $a_{12} = 1$, $M_{12} = \begin{vmatrix} 2 & -1 \\ -5 & 4 \end{vmatrix} = 3$, $A_{12} = -3$

29. [9.5] -28 **30.** [9.5] 0   **31.** [9.5] 99
32. [9.5] 22
33. [9.5] $(a - b)(b - c)(c - a)(a + b + c)$
34. [9.6] $\begin{bmatrix} 2 & -3 \\ 1 & 4 \end{bmatrix}\begin{bmatrix} x \\ y \end{bmatrix} = \begin{bmatrix} -9 \\ 1 \end{bmatrix}$, $(-3, 1)$

35. [9.7]

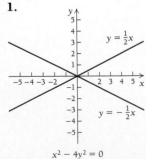

$4x - 3y \geq 12$

36. [9.7] Minimum 40 at $(2, 0)$, maximum 460 at $(2, 7)$
37. [9.7] Type A: 0; type B: 9; maximum score: 108

38. [9.8] $\dfrac{2}{2x - 3} - \dfrac{5}{x + 4}$ **39.** [9.1] $\left(-1, \dfrac{1}{3}\right)$

CHAPTER 10

Margin Exercises, Section 10.1

1.

$y = \frac{1}{2}x$

$y = -\frac{1}{2}x$

$x^2 - 4y^2 = 0$

2.

$y = 2$

$y = -2$

$y^2 = 4$

3.

$y = 3x$

$y^2 + 9x^2 = 6xy$

4. The graph consists just of $(0, 0)$.
5. There is no real-number solution, hence no graph.
6. (a) Stretched; (b) shrunk
7. $V: (-3, 0), (3, 0),$
$(0, -1), (0, 1)$;
$F: (-2\sqrt{2}, 0), (2\sqrt{2}, 0)$

8. $V: (-5, 0), (5, 0),$
$(0, -3), (0, 3)$;
$F: (-4, 0), (4, 0)$

$x^2 + 9y^2 = 9$

$9x^2 + 25y^2 = 225$

9. $V: (-2, 0), (2, 0),$
$(0, \sqrt{2}), (0, -\sqrt{2})$;
$F: (-\sqrt{2}, 0), (\sqrt{2}, 0)$

10. $C: (0, 0)$; $V: (-1, 0),$
$(1, 0), (0, 3), (0, -3)$;
$F: (0, 2\sqrt{2}), (0, -2\sqrt{2})$

$2x^2 + 4y^2 = 8$

$9x^2 + y^2 = 9$

11. $C: (0, 0)$; $V: (3, 0),$
$(-3, 0), (0, 5), (0, -5)$;
$F: (0, 4), (0, -4)$

12. $C: (0, 0)$; $V: (\sqrt{2}, 0),$
$(-\sqrt{2}, 0), (0, 2), (0, -2)$;
$F: (0, \sqrt{2}), (0, -\sqrt{2})$

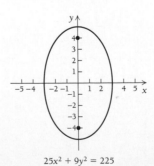

$25x^2 + 9y^2 = 225$

$4x^2 + 2y^2 = 8$

13. $C: (-3, 2)$;
$V: (-2\frac{4}{5}, 2), (-3\frac{1}{5}, 2),$
$(-3, 2\frac{1}{3}), (-3, 1\frac{2}{3})$;
$F: (-3, 2\frac{4}{15}), (-3, 1\frac{11}{15})$

$25x^2 + 9y^2 + 150x - 36y + 260 = 0$

14. $C: (2, -3)$;
$V: (2\frac{1}{3}, -3), (1\frac{2}{3}, -3),$
$(2, -2\frac{4}{5}), (2, -3\frac{1}{5})$;
$F: (2\frac{4}{15}, -3), (1\frac{11}{15}, -3)$

$9x^2 + 25y^2 - 36x + 150y + 260 = 0$

Technology Connection, Section 10.1

TC 1.

$10x^2 + 5y^2 = 10$

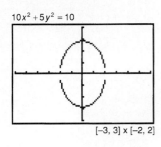

$[-3, 3] \times [-2, 2]$

TC 2.

$x^2 + 7y^2 = 70$

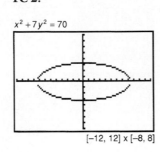

$[-12, 12] \times [-8, 8]$

TC 3.

$3.4x^2 + 6.7y^2 = 32.5$

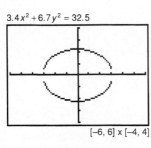

$[-6, 6] \times [-4, 4]$

TC 4.

$4x^2 + 13y^2 + 6x + 4y - 10 = 0$

$[-3, 3] \times [-2, 2]$

TC 5.

$1.7x^2 + 4.3y^2 - 8.1x - 6.5y - 12.4 = 0$

$[-9, 9] \times [-6, 6]$

Exercise Set 10.1, pp. 644–645

1.

$y = -x$ $y = x$

$x^2 - y^2 = 0$

3.

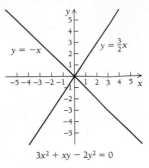

$y = -x$ $y = \frac{3}{2}x$

$3x^2 + xy - 2y^2 = 0$

5. The point $(0, 0)$

7. $V: (2, 0), (-2, 0),$
$(0, 1), (0, -1)$;
$F: (\sqrt{3}, 0), (-\sqrt{3}, 0)$

$\dfrac{x^2}{4} + \dfrac{y^2}{1} = 1$

9. $V: (-3, 0), (3, 0),$
$(0, 4), (0, -4)$;
$F: (0, \sqrt{7}), -(0, \sqrt{7})$

$16x^2 + 9y^2 = 144$

11. $V: (-\sqrt{3}, 0), (\sqrt{3}, 0),$
$(0, \sqrt{2}), (0, -\sqrt{2})$;
$F: (-1, 0), (1, 0)$

$2x^2 + 3y^2 = 6$

13. $V: (\pm\frac{1}{2}, 0), (0, \pm\frac{1}{3})$;
$F: \left(\dfrac{\pm\sqrt{5}}{6}, 0\right)$

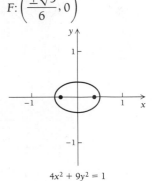

$4x^2 + 9y^2 = 1$

15. $C: (1, 2); V: (3, 2),$
$(-1, 2), (1, 3), (1, 1)$;
$F: (1 \pm \sqrt{3}, 2)$

$\dfrac{(x-1)^2}{4} + \dfrac{(y-2)^2}{1} = 1$

17. $C: (-3, 2); V: (2, 2),$
$(-8, 2), (-3, 6), (-3, -2)$;
$F: (0, 2), (-6, 2)$

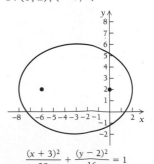

$\dfrac{(x+3)^2}{25} + \dfrac{(y-2)^2}{16} = 1$

19. $C: (-2, 1)$; $V: (-10, 1), (6, 1), (-2, 1 \pm 4\sqrt{3})$;
$F: (-6, 1), (2, 1)$

$3(x + 2)^2 + 4(y - 1)^2 = 192$

21. $C: (2, -1)$;
$V: (-1, -1), (5, -1)$,
$(2, 1), (2, -3)$;
$F: (2 \pm \sqrt{5}, -1)$

23. $C: (1, 1)$; $V: (0, 1)$,
$(2, 1), (1, 3), (1, -1)$;
$F: (1, 1 \pm \sqrt{3})$

$4x^2 + 9y^2 - 16x + 18y - 11 = 0$

$4x^2 + y^2 - 8x - 2y + 1 = 0$

25. $\dfrac{x^2}{4} + \dfrac{y^2}{9} = 1$

27. $\dfrac{(x - 3)^2}{4} + \dfrac{(y - 1)^2}{25} = 1$

29. $\dfrac{(x + 2)^2}{\frac{1}{4}} + \dfrac{(y - 3)^2}{4} = 1$

31. (a)

$9x^2 + y^2 = 9$

no; **(b)** $y = \pm 3\sqrt{1 - x^2}$;

(c)

$y = 3\sqrt{1 - x^2}$

yes;
domain $\{x | -1 \le x \le 1\}$,
range $\{y | 0 \le y \le 3\}$

(d)

$y = -3\sqrt{1 - x^2}$

yes; domain $\{x | -1 \le x \le 1\}$, range $\{y | -3 \le y \le 0\}$

35. $\dfrac{x^2}{50^2} + \dfrac{y^2}{12^2} = 1$, or $\dfrac{x^2}{2500} + \dfrac{y^2}{144} = 1$

37. (a) $A = \pi \cdot a \cdot b$; **(b)** 20π; **(c)** $739{,}141.4 \text{ ft}^2$

39. $C: (2.003125, -1.00515)$; $V: (5.0234304, -1.00515)$,
$(-1.0171804, -1.00515), (2.003125, -3.0186869)$,
$(2.003125, 1.0083869)$

Margin Exercises, Section 10.2

1. $V: (3, 0), (-3, 0)$;
$F: (\sqrt{13}, 0), (-\sqrt{13}, 0)$;
$A: y = \frac{2}{3}x, y = -\frac{2}{3}x$

2. $V: (4, 0), (-4, 0)$;
$F: (4\sqrt{2}, 0), (-4\sqrt{2}, 0)$;
$A: y = x, y = -x$

$4x^2 - 9y^2 = 36$

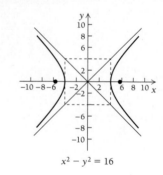

$x^2 - y^2 = 16$

3. $V: (0, 5), (0, -5)$;
$F: (0, \sqrt{34}), (0, -\sqrt{34})$;
$A: y = \frac{5}{3}x, y = -\frac{5}{3}x$

4. $V: (0, 5), (0, -5)$;
$F: (0, 5\sqrt{2}), (0, -5\sqrt{2})$;
$A: y = x, y = -x$

$9y^2 - 25x^2 = 225$

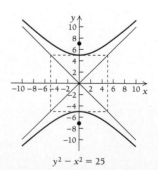

$y^2 - x^2 = 25$

5. C: $(1, -2)$;
V: $(6, -2)$, $(-4, -2)$;
F: $(1 + \sqrt{29}, -2)$,
$(1 - \sqrt{29}, -2)$;
A: $y = \frac{2}{5}x - \frac{12}{5}$,
$y = -\frac{2}{5}x - \frac{8}{5}$

$4x^2 - 25y^2 - 8x - 100y - 196 = 0$

6. C: $(-1, 2)$;
V: $(-1, 5)$, $(-1, -1)$;
F: $(-1, 7)$, $(-1, -3)$;
A: $y = \frac{3}{4}x + \frac{11}{4}$,
$y = -\frac{3}{4}x + \frac{5}{4}$

$16y^2 - 9x^2 - 64y - 18x - 89 = 0$

7.

$xy = 3$

8.

$xy = -12$

Exercise Set 10.2, pp. 652–653

1. C: $(0, 0)$;
V: $(-3, 0)$, $(3, 0)$;
F: $(-\sqrt{10}, 0)$, $(\sqrt{10}, 0)$;
A: $y = \frac{1}{3}x$, $y = -\frac{1}{3}x$

$\frac{x^2}{9} - \frac{y^2}{1} = 1$

3. C: $(2, -5)$:
V: $(-1, -5)$, $(5, -5)$;
F: $(2 - \sqrt{10}, -5)$,
$(2 + \sqrt{10}, -5)$;
A: $y = -\frac{x}{3} - \frac{13}{3}$,
$y = \frac{x}{3} - \frac{17}{3}$

$\frac{(x - 2)^2}{9} - \frac{(y + 5)^2}{1} = 1$

5. C: $(-1, -3)$;
V: $(-1, -1)$, $(-1, -5)$;
F: $(-1, -3 + 2\sqrt{5})$,
$(-1, -3 - 2\sqrt{5})$;
A: $y = \frac{1}{2}x - \frac{5}{2}$, $y = -\frac{1}{2}x - \frac{7}{2}$

$\frac{(y + 3)^2}{4} - \frac{(x + 1)^2}{16} = 1$

7. C: $(0, 0)$; V: $(-2, 0)$,
$(2, 0)$; F: $(-\sqrt{5}, 0)$,
$(\sqrt{5}, 0)$; A: $y = -\frac{1}{2}x$,
$y = \frac{1}{2}x$

$x^2 - 4y^2 = 4$

9. C: $(0, 0)$; V: $(0, 1)$,
$(0, -1)$; F: $(0, \sqrt{5})$,
$(0, -\sqrt{5})$; A: $y = -\frac{1}{2}x$,
$y = \frac{1}{2}x$

$4y^2 - x^2 = 4$

11. C: $(0, 0)$; V: $(-\sqrt{2}, 0)$,
$(\sqrt{2}, 0)$; F: $(-2, 0)$, $(2, 0)$;
A: $y = \pm x$

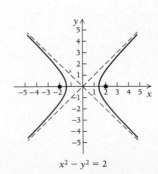

$x^2 - y^2 = 2$

Technology Connection, Section 10.2

TC 1.

$5x^2 - 18y^2 - 23 = 0$

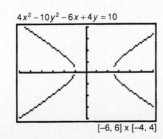

$[-9, 9] \times [-6, 6]$

TC 2.

$12y^2 - 2.5x^2 = 15$

$[-9, 9] \times [-6, 6]$

TC 3.

$4x^2 - 10y^2 - 6x + 4y = 10$

$[-6, 6] \times [-4, 4]$

TC 4.

$17.3x^2 - 4.3y^2 + 12.3x - 3.4y + 10.9 = 0$

$[-9, 9] \times [-6, 6]$

13. C: $(0, 0)$; V: $(-\frac{1}{2}, 0)$, $(\frac{1}{2}, 0)$; F: $\left(-\frac{\sqrt{2}}{2}, 0\right)$, $\left(\frac{\sqrt{2}}{2}, 0\right)$; A: $y = \pm x$

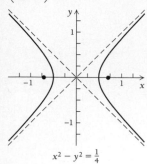

$x^2 - y^2 = \frac{1}{4}$

15. C: $(1, -2)$; V: $(0, -2)$, $(2, -2)$; F: $(1 - \sqrt{2}, -2)$, $(1 + \sqrt{2}, -2)$; A: $y = -x - 1$, $y = x - 3$

$x^2 - y^2 - 2x - 4y - 4 = 0$

17. C: $(\frac{1}{3}, 3)$; V: $(-\frac{2}{3}, 3)$, $(\frac{4}{3}, 3)$; F: $(\frac{1}{3} - \sqrt{37}, 3)$, $(\frac{1}{3} + \sqrt{37}, 3)$; A: $y = 6x + 1$, $y = -6x + 5$

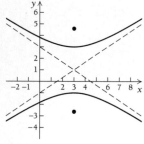

$36x^2 - y^2 - 24x + 6y - 41 = 0$

19. C: $(3, 1)$; V: $(3, 3)$, $(3, -1)$; F: $(3, 1 + \sqrt{13})$, $(3, 1 - \sqrt{13})$; A: $y = \frac{2}{3}x - 1$, $y = -\frac{2}{3}x + 3$

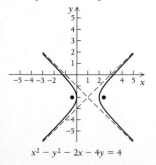

$9y^2 - 4x^2 - 18y + 24x - 63 = 0$

21. C: $(1, -2)$; V: $(2, -2)$, $(0, -2)$; F: $(1 + \sqrt{2}, -2)$, $(1 - \sqrt{2}, -2)$; A: $y = x - 3$, $y = -x - 1$

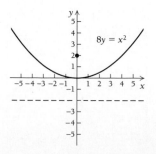

$x^2 - y^2 - 2x - 4y = 4$

23. C: $(-3, 4)$; V: $(-3, 10)$, $(-3, -2)$; F: $(-3, 4 + 6\sqrt{2})$, $(-3, 4 - 6\sqrt{2})$; A: $y = x + 7$, $y = -x + 1$

$y^2 - x^2 - 6x - 8y - 29 = 0$

25.

$xy = 1$

27.

$xy = -8$

29. $\dfrac{x^2}{4} - \dfrac{y^2}{9} = 1$ **31.** $\dfrac{(y + 8)^2}{\frac{121}{4}} - \dfrac{(x - 3)^2}{9} = 1$

33. $\dfrac{(y + 7)^2}{4} - \dfrac{(x - 4)^2}{36} = 1$; V: $(4, -5)$, $(4, -9)$; C: $(4, -7)$

35. (a) No; **(b)** $y = \pm\frac{1}{2}\sqrt{x^2 - 4}$;
(c) Yes, domain: $\{x | x \leq -2 \text{ or } x \geq 2\}$, range: $\{y | y \geq 0\}$;
(d) Yes, domain: $\{x | x \leq -2 \text{ or } x \geq 2\}$, range: $\{y | y \leq 0\}$
37. The (absolute value of the) difference of the distances from C to two fixed points A and B is a constant. By definition, such a locus is a hyperbola.
39. C: $(1.023, -2.044)$; V: $(2.07, -2.044)$, $(-0.024, -2.044)$; A: $y = x - 3.067$, $y = -x - 1.021$

Margin Exercises, Section 10.3
1. V: $(0, 0)$; F: $(0, 2)$; D: $y = -2$

$8y = x^2$

2. V: $(0, 0)$; F: $(0, \frac{1}{8})$; D: $y = -\frac{1}{8}$

$y = 2x^2$

3. $V: (0, 0)$; $F: (-\frac{3}{2}, 0)$; $D: x = \frac{3}{2}$

$y^2 = -6x$

4. $y^2 = 12x$ **5.** $x^2 = 2y$ **6.** $y^2 = -24x$ **7.** $x^2 = -4y$

8. $V: (-1, -\frac{1}{2})$; $F: (-1, \frac{3}{2})$; $D: y = -\frac{5}{2}$

9. $V: (2, -1)$; $F: (1, -1)$; $D: x = 3$

$x^2 + 2x - 8y - 3 = 0$

$y^2 + 2y + 4x - 7 = 0$

10. Parabola   **11.** Circle   **12.** Hyperbola   **13.** Ellipse

Technology Connection, Section 10.3

TC 1.

$13x^2 - 8y - 9 = 0$

$[-6, 6] \times [-4, 4]$

TC 2.

$41x + 6y^2 = 12$

$[-12, 12] \times [-8, 8]$

TC 3.

$12x^2 - x + 3y - 14 = 0$

$[-9, 9] \times [-6, 6]$

TC 4.

$-6.5y^2 + 3.1x + 23.4y - 3 = 0$

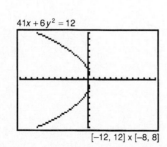

$[-9, 9] \times [-6, 6]$

Exercise Set 10.3, pp. 659–660

1. $V: (0, 0)$; $F: (0, 5)$; $D: y = -5$

$x^2 = 20y$

3. $V: (0, 0)$; $F: (-\frac{3}{2}, 0)$; $D: x = \frac{3}{2}$

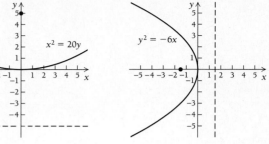

$y^2 = -6x$

5. $V: (0, 0)$; $F: (0, 1)$; $D: y = -1$

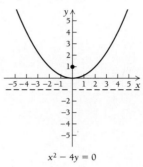

$x^2 - 4y = 0$

7. $V: (0, 0)$; $F: (0, \frac{1}{8})$; $D: y = -\frac{1}{8}$

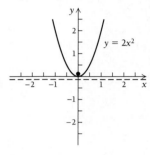

$y = 2x^2$

9. $y^2 = 16x$ **11.** $y^2 = -4\sqrt{2}x$

13. $(y - 2)^2 = 14(x + \frac{1}{2})$

15. $V: (-2, 1)$; $F: (-2, -\frac{1}{2})$; $D: y = \frac{5}{2}$

17. $V: (-1, -3)$; $F: (-1, -\frac{7}{2})$; $D: y = -\frac{5}{2}$

$(x + 2)^2 = -6(y - 1)$

$x^2 + 2x + 2y + 7 = 0$

19. $V: (0, -2)$; $F: (0, -1\frac{3}{4})$; $D: y = -2\frac{1}{4}$

21. $V: (-2, -1)$; $F: (-2, -\frac{3}{4})$; $D: y = -1\frac{1}{4}$

$x^2 - y - 2 = 0$

$y = x^2 + 4x + 3$

23. $V: \left(5\frac{3}{4}, \frac{1}{2}\right)$; $F: \left(6, \frac{1}{2}\right)$; $D: x = 5\frac{1}{2}$

$$4y^2 - 4y - 4x + 24 = 0$$

25. Parabola **27.** Ellipse **29.** Circle
31. Parabola **33.** Hyperbola
35. The graph of $x^2 - y^2 = 0$ is two lines. The others are, respectively, a hyperbola, a circle, and a parabola.
37. $(x + 1)^2 = -4(y - 2)$ **39.** $(y - 1)^2 = -16(x + 2)$
41. $(y - k)^2 - (x - h)(4p) = 0$; $(y - k)^2 = 4p(x - h)$
43. $V: (0.87, 0.35)$; $F: (0.87, -0.19)$; $D: y = 0.89$

Margin Exercises, Section 10.4
1. $(4, 3)$, $(-3, -4)$ **2.** $(4, 7)$, $(-1, 2)$
3. $\left(-\frac{5}{7}, \frac{22}{7}\right)$, $(1, -2)$ **4.** $\left(-\frac{3}{2}i, \frac{15}{4}i\right)$, $\left(\frac{3}{2}i, -\frac{15}{4}i\right)$
5. 5 m, 12 m

Technology Connection, Section 10.4
TC 1. $(3.5, 0.5)$, $(-1.5, -1.17)$
TC 2. $(-2.77, 2.52)$, $(-2.77, -2.52)$
TC 3. $\varnothing$

Exercise Set 10.4, pp. 665–666
1. $(-4, -3)$, $(3, 4)$ **3.** $(3, 0)$, $(0, 2)$ **5.** $(0, -3)$, $(4, 5)$
7. $(-2, 1)$ **9.** $\left(3, 2\right)$, $\left(4, \frac{3}{2}\right)$
11. $\left(\dfrac{5 + \sqrt{70}}{3}, \dfrac{-1 + \sqrt{70}}{3}\right)$, $\left(\dfrac{5 - \sqrt{70}}{3}, \dfrac{-1 - \sqrt{70}}{3}\right)$
13. $\left(\dfrac{15 + \sqrt{561}}{8}, \dfrac{11 - 3\sqrt{561}}{8}\right)$, $\left(\dfrac{15 - \sqrt{561}}{8}, \dfrac{11 + 3\sqrt{561}}{8}\right)$
15. $\left(\dfrac{7 - \sqrt{33}}{2}, \dfrac{7 + \sqrt{33}}{2}\right)$, $\left(\dfrac{7 + \sqrt{33}}{2}, \dfrac{7 - \sqrt{33}}{2}\right)$
17. $(3, -5)$, $(-1, 3)$ **19.** $(8, 5)$, $(-5, -8)$
21. $\left(\dfrac{8 + 3i\sqrt{6}}{2}, \dfrac{-8 + 3i\sqrt{6}}{2}\right)$, $\left(\dfrac{8 - 3i\sqrt{6}}{2}, \dfrac{-8 - 3i\sqrt{6}}{2}\right)$
23. 3, 9 **25.** 6 cm, 8 cm **27.** 4 in. by 5 in.
29. 75 yd by 30 yd **31.** $8, \frac{1}{4}$ **33.** 61.52 cm, 38.48 cm
35. $\dfrac{x^2}{4} + y^2 = 1$ **37.** $\left(\dfrac{2a^2 + 6ab + 5b^2}{2(a + 2b)}, \dfrac{-2ab - 3b^2}{2(a + 2b)}\right)$
39. $2(L + W) = P$, $L + W = \dfrac{P}{2}$, $LW = A$, $L = \dfrac{P}{2} - W$,
$W\left(\dfrac{P}{2} - W\right) = A$, $W^2 - \dfrac{WP}{2} + A = 0$,

$W = \dfrac{\dfrac{P}{2} \pm \sqrt{\left(\dfrac{P}{2}\right)^2 - 4A}}{2} = \dfrac{P}{4} \pm \dfrac{\sqrt{P^2 - 16A}}{4}$

$= \dfrac{1}{4}\left(P \pm \sqrt{P^2 - 16A}\right)$

41. $(x - 2)^2 + (y - 3)^2 = 1$ **43.** $(2, 4)$, $(4, 2)$
45. $(0.965, 4402.33)$, $(-0.965, -4402.33)$
47. $(2.11, -0.11)$, $(-13.04, -13.34)$

Margin Exercises, Section 10.5
1. $(2, 0)$, $(-2, 0)$ **2.** $(4, 0)$, $(-4, 0)$ **3.** $(0, 2)$, $(0, -2)$
4. $(2, 3)$, $(2, -3)$, $(-2, 3)$, $(-2, -3)$
5. $(3, 2)$, $(-3, -2)$, $(2, 3)$, $(-2, -3)$ **6.** 1 ft, 2 ft

Exercise Set 10.5, pp. 670–671
1. $(-5, 0)$, $(4, 3)$, $(4, -3)$ **3.** $(3, 0)$, $(-3, 0)$
5. $(4, 3)$, $(-4, -3)$, $(3, 4)$, $(-3, -4)$
7. $\left(\pm\dfrac{6\sqrt{21}}{7}, \pm\dfrac{4i\sqrt{35}}{7}\right)$
9. $(\sqrt{2}, \sqrt{14})$, $(-\sqrt{2}, \sqrt{14})$, $(\sqrt{2}, -\sqrt{14})$, $(-\sqrt{2}, -\sqrt{14})$
11. $(1, 2)$, $(-1, -2)$, $(2, 1)$, $(-2, -1)$
13. $(3, 2)$, $(-3, -2)$, $(2, 3)$, $(-2, -3)$
15. $\left(\dfrac{5 - 9\sqrt{15}}{20}, \dfrac{-45 + 3\sqrt{15}}{20}\right)$, $\left(\dfrac{5 + 9\sqrt{15}}{20}, \dfrac{-45 - 3\sqrt{15}}{20}\right)$ **17.** $(3, 2)$, $(-3, -2)$
19. $(2, 1)$, $(-2, -1)$, $(1, 2)$, $(-1, -2)$
21. $(3, \sqrt{5})$, $(-3, -\sqrt{5})$, $(\sqrt{5}, 3)$, $(-\sqrt{5}, -3)$
23. $\left(\dfrac{8\sqrt{5}}{5}i, \dfrac{3\sqrt{105}}{5}\right)$, $\left(\dfrac{8\sqrt{5}}{5}i, -\dfrac{3\sqrt{105}}{5}\right)$, $\left(-\dfrac{8\sqrt{5}}{5}i, \dfrac{3\sqrt{105}}{5}\right)$, $\left(-\dfrac{8\sqrt{5}}{5}i, -\dfrac{3\sqrt{105}}{5}\right)$
25. 13, 12 and -13, -12 **27.** 1 m by $\sqrt{3}$ m
29. 16 ft, 24 ft **31.** $\left(x + \dfrac{5}{13}\right)^2 + \left(y - \dfrac{32}{13}\right)^2 = \dfrac{5365}{169}$
33. $\left(\dfrac{1}{2}, \dfrac{1}{4}\right)$, $\left(\dfrac{1}{2}, -\dfrac{1}{4}\right)$, $\left(-\dfrac{1}{2}, \dfrac{1}{4}\right)$, $\left(-\dfrac{1}{2}, -\dfrac{1}{4}\right)$
35. $\left(\dfrac{a}{\sqrt{a + b}}, \dfrac{b}{\sqrt{a + b}}\right)$, $\left(\dfrac{-a}{\sqrt{a + b}}, \dfrac{-b}{\sqrt{a + b}}\right)$
37. $(3, -2)$, $(-3, 2)$, $(2, -3)$, $(-2, 3)$
39. $\left(\dfrac{1}{2}, \dfrac{\sqrt{15}}{2}\right)$, $\left(\dfrac{1}{2}, -\dfrac{\sqrt{15}}{2}\right)$ **41.** $k = 3$
43. $\left(\dfrac{2\log 3 + 3\log 5}{3(\log 3 \cdot \log 5)}, \dfrac{4\log 3 - 3\log 5}{3(\log 3 \cdot \log 5)}\right)$
45. $(0, 0)$, $\left(\dfrac{3 - i\sqrt{3}}{2}, \dfrac{3 + i\sqrt{3}}{2}\right)$, $\left(\dfrac{3 + i\sqrt{3}}{2}, \dfrac{3 - i\sqrt{3}}{2}\right)$
47. $(8.53, 2.53)$, $(8.53, -2.53)$, $(-8.53, 2.53)$, $(-8.53, -2.53)$

Review Exercises: Chapter 10, pp. 671–672

1. [10.1]

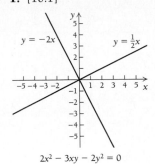

$2x^2 - 3xy - 2y^2 = 0$

2. [10.1]

$16x^2 + 25y^2 - 64x + 50y - 311 = 0$

C: $(2, -1)$; V: $(-3, -1)$, $(7, -1)$, $(2, 3)$, $(2, -5)$;

F: $(-1, -1)$, $(5, -1)$ **3.** [10.1] $\dfrac{x^2}{9} + \dfrac{y^2}{16} = 1$

4. [10.2] C: $\left(-2, \dfrac{1}{4}\right)$; V: $\left(0, \dfrac{1}{4}\right)$, $\left(-4, \dfrac{1}{4}\right)$;

F: $\left(-2 + \sqrt{6}, \dfrac{1}{4}\right)$, $\left(-2 - \sqrt{6}, \dfrac{1}{4}\right)$; A: $y - \dfrac{1}{4} = \pm\dfrac{\sqrt{2}}{2}(x + 2)$

5. [10.2]

6. [10.2]

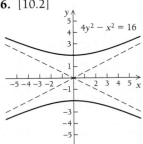

7. [10.3] $x^2 = -6y$

8. [10.3] F: $(-3, 0)$; V: $(0, 0)$; D: $x = 3$

9. [10.3] V: $(-5, 8)$; F: $\left(-5, \dfrac{15}{2}\right)$; D: $y = \dfrac{17}{2}$

10. [10.3] Circle   **11.** [10.3] Ellipse   **12.** [10.3] Parabola

13. [10.3] Hyperbola   **14.** [10.3] Hyperbola

15. [10.3] Parabola   **16.** [10.3] Hyperbola

17. [10.3] Circle   **18.** [10.5] $(-8\sqrt{2}, 8)$, $(8\sqrt{2}, 8)$

19. [10.5] $\left(3, \dfrac{\sqrt{29}}{2}\right)$, $\left(-3, \dfrac{\sqrt{29}}{2}\right)$, $\left(3, -\dfrac{\sqrt{29}}{2}\right)$,

$\left(-3, -\dfrac{\sqrt{29}}{2}\right)$ **20.** [10.4] $(7, 4)$

21. [10.4] $(2, 2)$, $\left(\dfrac{32}{9}, -\dfrac{10}{9}\right)$ **22.** [10.4] $(0, -3)$, $(2, 1)$

23. [10.5] $(4, 3)$, $(4, -3)$, $(-4, 3)$, $(-4, -3)$

24. [10.5] $(-\sqrt{3}, 0)$, $(\sqrt{3}, 0)$, $(-2, 1)$, $(2, 1)$

25. [10.4] $\left(-\dfrac{3}{5}, \dfrac{21}{5}\right)$, $(3, -3)$

26. [10.5] $(6, 8)$, $(6, -8)$, $(-6, 8)$, $(-6, -8)$

27. [10.5] $(2, 2)$, $(-2, -2)$, $(2\sqrt{2}, \sqrt{2})$, $(-2\sqrt{2}, -\sqrt{2})$

28. [10.5] $\dfrac{16\sqrt{6}}{7} \approx 5.6$ **29.** [10.4] 7, 4

30. [10.4] 12 by 7   **31.** [10.4] 4, 8

32. [10.4] 32 cm, 20 cm   **33.** [10.5] 11 ft, 3 ft

34. [10.1] $x^2 + \dfrac{y^2}{9} = 1$ **35.** [10.5] $\dfrac{8}{7}, \dfrac{7}{2}$

36. [10.1], [10.5] $(x - 2)^2 + (y - 1)^2 = 100$

Test: Chapter 10, pp. 672–673

1. [10.1]

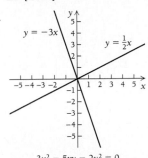

$3x^2 - 5xy - 2y^2 = 0$

2. [10.1] C: $(2, 4)$; V: $(1, 4)$, $(3, 4)$, $(2, 1)$, $(2, 7)$;

F: $(2, 4 + 2\sqrt{2})$, $(2, 4 - 2\sqrt{2})$

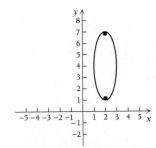

$9x^2 + y^2 - 36x - 8y + 43 = 0$

3. [10.1] $\dfrac{x^2}{49} + \dfrac{y^2}{4} = 1$ **4.** [10.2] C: $(0, 0)$; V: $(0, 5)$, $(0, -5)$;

F: $(0, \sqrt{34})$, $(0, -\sqrt{34})$; A: $y = \dfrac{5}{3}x$, $y = -\dfrac{5}{3}x$

5. [10.2]

6. [10.3] $x^2 = -20y$ **7.** [10.3] V: $(2, -3)$; F: $(0, -3)$;

D: $x = 4$ **8.** [10.4] $(8, 6)$, $(0, 10)$

9. [10.5] $(1, 2)$, $(2, 1)$, $(-2, -1)$, $(-1, -2)$

10. [10.4] $-2, 6$ **11.** [10.5] 10 yd by 1 yd

12. [10.4] Numerator 15, denominator 8; or numerator 8, denominator 15   **13.** [10.5] 11 by 2

14. [10.5] $\sqrt{5}$ m, $\sqrt{3}$ m   **15.** [10.4] 16 ft by 12 ft

16. [10.3] Parabola   **17.** [10.3] Circle
18. [10.3] Hyperbola   **19.** [10.3] Ellipse
20. [10.3] Hyperbola   **21.** [10.3] Parabola
22. [10.1] $\dfrac{5(x + 3)^2}{36} + \dfrac{(y + 1)^2}{9} = 1$
23. [10.1], [10.5] $(x + 2)^2 + (y - 2)^2 = 37$
24. [10.1] $(0, -\frac{31}{4})$ **25.** [10.4] 9

CHAPTER 11

Margin Exercises, Section 11.1
1. (a) 1, 3, 5; **(b)** 67   **2. (a)** $1, -\frac{1}{2}, \frac{1}{3}, -\frac{1}{4}$; **(b)** $\frac{1}{47}$

3. $2n$ **4.** n **5.** n^3 **6.** $\dfrac{x^n}{n}$ **7.** 2^{n-1}

8. (a) 0; **(b)** 4   **9.** $3 + 2\frac{1}{2} + 2\frac{1}{3} = 7\frac{5}{6}$
10. $5^0 + 5^1 + 5^2 + 5^3 + 5^4 = 781$

11. $8^3 + 9^3 + 10^3 + 11^3 = 3572$ **12.** $\sum\limits_{k=1}^{5} 2k$

13. $\sum\limits_{k=1}^{4} k^3$ **14.** $\sum\limits_{k=1}^{\infty} \dfrac{x^k}{k}$ **15.** $\sum\limits_{k=2}^{\infty} k^2$

16. $-3, -9, -81, -6561, -43{,}046{,}721$
17. $5, 5, 10, 15, 25, 40, 65, 105$

Exercise Set 11.1, pp. 680–681
1. $3, 7, 11, 15; 39, 59$ **3.** $2, \frac{2}{3}, \frac{4}{4}, \frac{5}{9}; \frac{10}{9}, \frac{15}{14}$
5. $3, 8, 15, 24; 120; 255$ **7.** $2, 2\frac{1}{2}, 3\frac{1}{3}, 4\frac{1}{4}; 10\frac{1}{10}; 15\frac{1}{15}$
9. $-1, 4, -9, 16; 100; -225$ **11.** $-2, -1, 4, -7; -25; 40$
13. $\frac{1}{2}, \frac{4}{7}, \frac{5}{8}, \frac{2}{3}; \frac{4}{5}; \frac{17}{20}$ **15.** 34   **17.** 323   **19.** -36.9

21. $-33{,}880$ **23.** $\dfrac{441}{400}$ **25.** 43   **27.** $1\dfrac{1}{1444}$

29. $2n - 1$ **31.** $(-1)^n \cdot 2 \cdot 3^{n-1}$ **33.** $\dfrac{n + 1}{n + 2}$

35. $(\sqrt{3})^n$ **37.** $(-1)(3n - 2)$ **39.** 28   **41.** 30
43. $\frac{1}{2} + \frac{1}{4} + \frac{1}{6} + \frac{1}{8} + \frac{1}{10} = \frac{137}{120}$
45. $2^0 + 2^1 + 2^2 + 2^3 + 2^4 + 2^5 = 63$
47. $\log 7 + \log 8 + \log 9 + \log 10$, or $\log (7 \cdot 8 \cdot 9 \cdot 10)$,
or $\log 5040$ **49.** $\frac{1}{2} + \frac{2}{3} + \frac{3}{4} + \frac{4}{5} + \frac{5}{6} + \frac{6}{7} + \frac{7}{8} + \frac{8}{9} = \frac{15{,}551}{2520}$
51. $-1 + 1 - 1 + 1 - 1 = -1$
53. $3 - 6 + 9 - 12 + 15 - 18 + 21 - 24 = -12$
55. $1 + \frac{2}{5} + \frac{1}{5} + \frac{2}{17} + \frac{1}{13} + \frac{2}{37} = \frac{75{,}581}{40{,}885}$
57. $3 + 2 + 3 + 6 + 11 + 18 = 43$
59. $\frac{1}{2} + \frac{1}{6} + \frac{1}{12} + \frac{1}{20} + \frac{1}{30} + \frac{1}{42} + \frac{1}{56} + \frac{1}{72} + \frac{1}{90} + \frac{1}{110} = \frac{10}{11}$

61. $\sum\limits_{k=1}^{6} \dfrac{k}{k + 1}$ **63.** $\sum\limits_{k=1}^{6} (-2)^k$ **65.** $\sum\limits_{k=2}^{n} (-1)^k k^2$

67. $\sum\limits_{k=1}^{\infty} 5k$ **69.** $\sum\limits_{k=1}^{\infty} \dfrac{1}{k(k + 1)}$ **71.** $4, 1\frac{1}{4}, 1\frac{4}{5}, 1\frac{5}{9}$

73. $6561, -81, 9i, -3\sqrt{i}$ **75.** $2, 3, 5, 8$
77. $\frac{3}{2}, \frac{3}{2}, \frac{9}{8}, \frac{3}{4}, \frac{15}{32}, \frac{171}{32}$ **79.** 0, 0.693, 1.792, 3.178, 4.787;
10.450   **81.** $\pi, 0, \pi, 0, \pi; 3\pi$ **83. (a)** 41, 43, 47, 53, 61, 71;
(b) all numbers are prime; **(c)** 1681; no
85. 0.414214, 0.317837, 0.267949, 0.236068,
0.213422, 0.196262
87. 2, 1.5, 1.416667, 1.414216, 1.414214, 1.414214
89. $3900, $2925, $2193.75, $1645.31, $1233.98, $925.49,
$694.12, $520.59, $390.44, $292.83   **91.** $\ln (1 \cdot 2 \cdot 3 \cdots \cdot n)$

Margin Exercises, Section 11.2
1. $a_1 = 3.1, d = 0.8$ **2.** 50   **3.** 75
4. $a_1 = 7, d = 12; 7, 19, 31, 43, \ldots$ **5.** 20,100
6. 225   **7.** 455   **8.** $27.25   **9.** 5200   **10.** 7, 10, 13

Exercise Set 11.2, pp. 689–690
1. $a_1 = 3, d = 5$ **3.** $a_1 = 9, d = -4$ **5.** $a_1 = \frac{3}{2}, d = \frac{3}{4}$
7. $a_1 = $1.07, d = 0.07 **9.** $a_{12} = 46$ **11.** $a_{17} = -41$
13. $a_{13} = -$1628.16$ **15.** 27th   **17.** 102nd
19. $a_{17} = 101$ **21.** $a_1 = 5$ **23.** $n = 28$ **25.** $a_1 = 8$;
$d = -3; 8, 5, 2, -1, -4$ **27.** 670   **29.** 45,150
31. 2550   **33.** 735   **35.** 990   **37.** 3   **39.** 1275
41. $31,000   **43.** 6300   **45.** $5\frac{4}{5}, 7\frac{3}{5}, 9\frac{2}{5}, 11\frac{1}{5}$
47. n^2 **49.** 3, 5, 7
51. $8760, $7961.77, $7163.54, $6365.31, $5567.08,
$4768.85, $3970.62, $3172.39, $2374.16, $1575.93
53. $-10, -4, 2, 8$ **55.** Sides are $a, a + d, a + 2d$; and

$a^2 + (a + d)^2 = (a + 2d)^2$. Solving, we get $d = \dfrac{a}{3}$. Thus the

sides are $a, \dfrac{4a}{3}$, and $\dfrac{5a}{3}$ in the ratio $3 : 4 : 5$.

57. $a_n = a_1 + (n - 1)d; a_n = f(n) = d \cdot n + (a_1 - d)$,
where $f(n)$ is a linear function of n with slope d and
y-intercept $(a_1 - d)$.

Margin Exercises, Section 11.3
1. 5   **2.** -3 **3.** 0.85   **4.** 1.09   **5.** $\frac{1}{5}$ **6.** $a_9 = 512$
7. $a_6 = \frac{1}{81}$ **8.** $S_8 = 13{,}120$ **9.** $S_{10} = \frac{341}{256}$ **10.** 363
11. No   **12.** No   **13.** Yes, 1   **14.** Yes, $\frac{3125}{3}$ **15.** $\frac{2}{9}$
16. $\frac{13}{99}$ **17.** $21,474,836.47$ **18.** $160,000,000$

Exercise Set 11.3, pp. 696–698
1. 2   **3.** -1 **5.** $-\dfrac{1}{2}$ **7.** $\dfrac{1}{5}$ **9.** $\dfrac{1}{x}$ **11.** 1.1   **13.** 64

15. 162   **17.** 648   **19.** $1360.49   **21.** 3^{n-1}

23. $(-1)^{n-1}$ **25.** $\dfrac{1}{x^n}$ **27.** 762   **29.** $\dfrac{547}{18}$ **31.** $\dfrac{x^8 - 1}{x - 1}$

33. $5134.51   **35.** 2   **37.** Yes, 8   **39.** Yes, 125
41. Yes, $\frac{1000}{11}$ **43.** No   **45.** Yes, $\frac{2}{3}$ **47.** Yes, $4545.45
49. Yes, $\frac{160}{9}$ **51.** Yes, 2   **53.** $\frac{7}{9}$ **55.** $\frac{8}{15}$ **57.** $\frac{510}{99}$, or $\frac{170}{33}$
59. $\frac{1}{256}$ ft   **61.** 155,797   **63.** $5236.19   **65.** 3100 ft
67. $2,684,355   **69.** $86,666,666,667

71. $(4 - \sqrt{6})/(\sqrt{3} - \sqrt{2}) = 2\sqrt{3} + \sqrt{2}$,
$(6\sqrt{3} - 2\sqrt{2})/(4 - \sqrt{6}) = 2\sqrt{3} + \sqrt{2}$; there exists a
common ratio, $2\sqrt{3} + \sqrt{2}$; thus the sequence is geometric.
73. $S_n = \dfrac{x^n - 1}{x - 1}$ **75.** $\dfrac{a_{n+1}}{a_n} = r$, so $\dfrac{(a_{n+1})^2}{(a_n)^2} = r^2$;
thus $a_1^2, a_2^2, \ldots$, is geometric, with ratio r^2.
77. Let the arithmetic sequence have common difference
$d = a_{n+1} - a_n$. Then the sequence $5^{a_1}, 5^{a_2}, 5^{a_3}, \ldots$, has
$\dfrac{5^{a_{n+1}}}{5^{a_n}} = 5^d$ for a common ratio and is therefore geometric.

Margin Exercises, Section 11.4
1. $1^2 + 1 > 1 + 1$ or $2 > 2; 2^2 + 1 > 2 + 1$ or $5 > 3$;
$3^2 + 1 > 3 + 1$ or $10 > 4; 4^2 + 1 > 4 + 1$ or $17 > 5$;
$5^2 + 1 > 5 + 1$ or $26 > 6$

2. $1 = \dfrac{1(1+1)}{2}$ or $1 = 1$; $1 + 2 = \dfrac{2(2+1)}{2}$, or

$1 + 2 = 3$; $1 + 2 + 3 = 6$; $1 + 2 + 3 + 4 = 10$;

$1 + 2 + 3 + 4 + 5 = 15$

3. (a) $2 \cdot 1 = 1(1+1)$, $2 + 4 = 2(2+1)$;

(b) $2 + 4 + \cdots + 2k = k(k+1)$;

(c) $2 + 4 + \cdots + 2(k+1) = (k+1)[(k+1)+1]$;

(d) $2 \cdot 1 = 1(1+1)$, $2 = 1 \cdot 2 = 2$;

(e) Assume S_k as hypothesis:

$$2 + 4 + \cdots + 2k = k(k+1).$$

Add $2(k+1)$ on both sides:

$$2 + 4 + \cdots + 2k + 2(k+1)$$
$$= k(k+1) + 2(k+1)$$
$$= (k+1)(k+2) \quad \text{(Simplifying)}$$

We now have

$$2 + 4 + \cdots + 2(k+1) = (k+1)(k+2)$$

or

$$(k+1)[(k+1)+1].$$

This is S_{k+1}. Hence $S_k \to S_{k+1}$ for all k. Finally, we conclude that $2 + 4 + \cdots + 2n = n(n+1)$ for all natural numbers n.

4. $S_1: 2 = \dfrac{1(3+1)}{2}$. True. $S_k: \displaystyle\sum_{p=1}^{k} (3p-1) = \dfrac{k(3k+1)}{2}$,

or $2 + 5 + 8 + \cdots + 3k - 1 = \dfrac{k(3k+1)}{2}$.

$S_{k+1}: \displaystyle\sum_{p=1}^{k+1} (3p-1) = \dfrac{(k+1)[3(k+1)+1]}{2}$, or

$2 + 5 + \cdots + 3k - 1 + 3(k+1) - 1 =$

$\dfrac{(k+1)[3(k+1)+1]}{2}$. Assume S_k. Then add $[3(k+1)-1]$

on both sides:

$$2 + 5 + \cdots + [3(k+1)-1]$$
$$= \frac{k(3k+1)}{2} + [3(k+1)-1]$$
$$= \frac{k(3k+1) + 2[3(k+1)-1]}{2}$$
$$= \frac{3k^2 + 7k + 4}{2} = \frac{(k+1)(3k+4)}{2}$$
$$= \frac{(k+1)[3(k+1)+1]}{2}$$

We have arrived at

$$\sum_{p=1}^{k+1} (3p-1) = \frac{(k+1)[3(k+1)-1]}{2}.$$

This is S_{k+1}. So $S_k \to S_{k+1}$ for all k. We conclude that

$$\sum_{p=1}^{n} (3p-1) = \frac{n(3n+1)}{2}$$

for *all* natural numbers n.

5. $S_1: x \le x$; $S_2: x \le x^2$. Both obviously true if $x > 1$. Thus the basis step is complete. $S_k: x \le x^k$; $S_{k+1}: x \le x^{k+1}$. Assume $S_k: x \le x^k$. We know by hypothesis that $1 < x$. Multiply the inequalities to get $x \cdot 1 \le x^k \cdot x$, or $x \le x^{k+1}$. We have arrived at S_{k+1}, hence have shown that $S_k \to S_{k+1}$ for all natural numbers k. We can now conclude that $x \le x^n$ for all natural numbers n.

Exercise Set 11.4, pp. 703–704

1. $1^2 < 1^3$, $2^2 < 2^3$, $3^2 < 3^3$, etc.

3. A polygon of 3 sides has $\dfrac{3(3-3)}{2}$ diagonals. A polygon

of 4 sides has $\dfrac{4(4-3)}{2}$ diagonals, etc.

5. S_n: $\quad 1 + 2 + 3 + \cdots + n = \dfrac{n(n+1)}{2}$

S_1: $\quad 1 = \dfrac{1(1+1)}{2}$

S_k: $\quad 1 + 2 + 3 + \cdots + k = \dfrac{k(k+1)}{2}$

S_{k+1}: $\quad 1 + 2 + 3 + \cdots + k + (k+1) = \dfrac{(k+1)(k+2)}{2}$

1. *Basis step:* S_1 true by substitution.

2. *Induction step:* Assume S_k. Deduce S_{k+1}. Starting with the left side of S_{k+1}, we have

$$\underbrace{1 + 2 + 3 + \cdots + k} + (k+1)$$
$$= \frac{k(k+1)}{2} + (k+1) \quad \text{(by } S_k)$$
$$= \frac{k(k+1) + 2(k+1)}{2} \quad \text{(adding)}$$
$$= \frac{(k+1)(k+2)}{2}. \quad \text{(distributive law)}$$

7. S_n: $\quad 1 + 5 + 9 + \cdots + (4n-3) = n(2n-1)$

S_1: $\quad 1 = 1(2 \cdot 1 - 1)$

S_k: $\quad 1 + 5 + 9 + \cdots + (4k-3) = k(2k-1)$

S_{k+1}: $\quad 1 + 5 + 9 + \cdots + (4k-3) + [4(k+1)-3]$
$$= (k+1)[2(k+1)-1]$$
$$= (k+1)(2k+1)$$

1. *Basis step:* S_1 true by substitution.

2. *Induction step:* Assume S_k. Deduce S_{k+1}. Starting with the left side of S_{k+1}, we have

$$\underbrace{1 + 5 + 9 + \cdots + (4k-3)} + [4(k+1)-3]$$
$$= k(2k-1) + [4(k+1)-3] \quad \text{(by } S_k)$$
$$= 2k^2 - k + 4k + 4 - 3$$
$$= (k+1)(2k+1).$$

9. S_n: $\quad \dfrac{1}{1 \cdot 2} + \dfrac{1}{2 \cdot 3} + \cdots + \dfrac{1}{n(n+1)} = \dfrac{n}{n+1}$

S_1: $\quad \dfrac{1}{1 \cdot 2} = \dfrac{1}{1+1}$

S_k: $\quad \dfrac{1}{1 \cdot 2} + \dfrac{1}{2 \cdot 3} + \cdots + \dfrac{1}{k(k+1)} = \dfrac{k}{k+1}$

S_{k+1}: $\quad \dfrac{1}{1 \cdot 2} + \dfrac{1}{2 \cdot 3} + \cdots + \dfrac{1}{k(k+1)} + \dfrac{1}{(k+1)(k+2)}$
$$= \frac{k+1}{k+2}$$

2. *Induction step:* Assume S_k. Deduce S_{k+1}. Add

$$\frac{1}{(k+1)(k+2)}$$

to both sides and simplify the right side.

11. S_1: $1^3 = \dfrac{1^2(1+1)^2}{4} = 1$

S_k: $1^3 + 2^3 + \cdots + k^3 = \dfrac{k^2(k+1)^2}{4}$

$1^3 + 2^3 + \cdots + (k+1)^3 = \dfrac{k^2(k+1)^2}{4} + (k+1)^3$

$= \dfrac{(k+1)^2}{4}[k^2 + 4(k+1)]$

$= \dfrac{(k+1)^2(k+2)^2}{4}$

13. *2. Induction step:*   Assume S_k. Deduce S_{k+1}. Now

$k < k+1$ (by S_k)

$k+1 < k+1+1$ (adding 1)

$\therefore k+1 < k+2$

15. S_1: $3^1 < 3^{1+1}$

S_k: $3^k < 3^{k+1}$

$3^k \cdot 3 < 3^{k+1} \cdot 3$

$3^{k+1} < 3^{(k+1)+1}$

17. S_1: $1 + \dfrac{1}{1} = 1 + 1$

S_k: $\left(1 + \dfrac{1}{1}\right)\cdots\left(1 + \dfrac{1}{k}\right) = k+1$

Multiply by $\left(1 + \dfrac{1}{k+1}\right)$:

$\left(1 + \dfrac{1}{1}\right)\cdots\left(1 + \dfrac{1}{k+1}\right) = (k+1)\left(1 + \dfrac{1}{k+1}\right)$

$= (k+1)\left(\dfrac{k+1+1}{k+1}\right)$

$= (k+1) + 1.$

19. S_1: $a_1 = \dfrac{a_1 - a_1 r}{1 - r} = \dfrac{a_1(1-r)}{1-r} = a_1$

S_k: $a_1 + \cdots + a_1 r^{k-1} = \dfrac{a_1 - a_1 r^k}{1-r}$

Add $a_1 r^k$:

$a_1 + \cdots + a_1 r^k = \dfrac{a_1 - a_1 r^k}{1-r} + a_1 r^k \dfrac{1-r}{1-r}$

$= \dfrac{a_1 - a_1 r^k + a_1 r^k - a_1 r^{k+1}}{1-r}$

$= \dfrac{a_1 - a_1 r^{k+1}}{1-r}.$

21. *2. Induction step:*   Assume S_k. Deduce S_{k+1}.
Starting with the left side of S_{k+1} we have

$[r(\cos\theta + i\sin\theta)]^{k+1}$

$= [r(\cos\theta + i\sin\theta)]^k[r(\cos\theta + i\sin\theta)]$

$= r^k[\cos k\theta + i\sin k\theta][r(\cos\theta + i\sin\theta)]$

$= r^{k+1}(\cos k\theta + i\sin k\theta)(\cos\theta + i\sin\theta)$

$= r^{k+1}[\cos k\theta \cos\theta + i\sin k\theta \cos\theta$

$\qquad\qquad + i\cos k\theta \sin\theta - \sin k\theta \sin\theta]$

(continued)

$= r^{k+1}[(\cos k\theta \cos\theta - \sin k\theta \sin\theta)$

$\qquad\qquad + i(\sin k\theta \cos\theta + \cos k\theta \sin\theta)]$

$= r^{k+1}[\cos(k\theta + \theta) + i\sin(k\theta + \theta)]$

$= r^{k+1}[\cos(k+1)\theta + i\sin(k+1)\theta].$

23. *2. Induction step:*   Assume S_k. Deduce S_{k+1}.
We start with the left side of S_{k+1}:

$\underbrace{\cos x \cdot \cos 2x \cdots \cos 2^{k-1}x} \cdot \cos 2^k x$

$= \dfrac{\sin 2^k x}{2^k \sin x} \cdot \cos 2^k x$

$= \dfrac{\sin 2^k x \cos 2^k x}{2^k \sin x}$

$= \dfrac{\frac{1}{2}\sin 2(2^k x)}{2^k \sin x}$ (since $\sin 2\theta = 2\sin\theta\cos\theta$)

$= \dfrac{\sin 2^{k+1} x}{2^{k+1} \sin x}.$

25. S_n: $\left(1 - \dfrac{1}{2^2}\right)\left(1 - \dfrac{1}{3^2}\right)\cdots\left(1 - \dfrac{1}{n^2}\right) = \dfrac{n+1}{2n}$

S_2: $1 - \dfrac{1}{2^2} = \dfrac{2+1}{2\cdot 2}$

S_k: $\left(1 - \dfrac{1}{2^2}\right)\left(1 - \dfrac{1}{3^2}\right)\cdots\left(1 - \dfrac{1}{k^2}\right) = \dfrac{k+1}{2k}$

S_{k+1}: $\left(1 - \dfrac{1}{2^2}\right)\left(1 - \dfrac{1}{3^2}\right)\cdots\left(1 - \dfrac{1}{k^2}\right)\left(1 - \dfrac{1}{(k+1)^2}\right)$

$= \dfrac{k+2}{2(k+1)}$

1. *Basis step:* S_2 is true by substitution.
2. *Induction step:*   Assume S_k. Deduce S_{k+1}.
Starting with the left side of S_{k+1} we have

$\underbrace{\left(1 - \dfrac{1}{2^2}\right)\left(1 - \dfrac{1}{3^2}\right)\cdots\left(1 - \dfrac{1}{k^2}\right)}\left(1 - \dfrac{1}{(k+1)^2}\right)$

$= \dfrac{k+1}{2k}\left(1 - \dfrac{1}{(k+1)^2}\right) = \dfrac{k+1}{2k} - \dfrac{1}{2k(k+1)}$

$= \dfrac{(k+1)(k+1) - 1}{2k(k+1)} = \dfrac{k^2 + 2k + 1 - 1}{2k(k+1)}$

$= \dfrac{k^2 + 2k}{2k(k+1)} = \dfrac{k(k+2)}{2k(k+1)} = \dfrac{k+2}{2(k+1)}$

27. S_2: $\overline{z_1 + z_2} = \overline{z_1} + \overline{z_2}$:

$\overline{(a+bi) + (c+di)} = \overline{(a+c) + (b+d)i}$

$= (a+c) - (b+d)i$

$\overline{(a+bi)} + \overline{(c+di)} = a - bi + c - di$

$= (a+c) - (b+d)i.$

S_k: $\overline{z_1 + z_2 + \cdots + z_k} = \overline{z_1} + \overline{z_2} + \cdots + \overline{z_k}.$

$\overline{(z_1 + z_2 + \cdots + z_k) + z_{k+1}}$

$= \overline{(z_1 + z_2 + \cdots + z_k)} + \overline{z_{k+1}}$ (by S_2).

$= \overline{z_1} + \overline{z_2} + \cdots + \overline{z_k} + \overline{z_{k+1}}$ (by S_k).

29. S_1: i is either i or -1 or $-i$ or 1.

S_k: i^k is either i or -1 or $-i$ or 1.

$i^{k+1} = i^k \cdot i$ is then $i \cdot i = -1$ or $-1 \cdot i = -i$ or $-i \cdot i = 1$ or $1 \cdot i = i$.

31. S_1:   2 is a factor of $1^2 + 1$.
    S_k: 2 is a factor of $k^2 + k$.

$$(k + 1)^2 + (k + 1) = k^2 + 2k + 1 + k + 1$$
$$= k^2 + k + 2(k + 1).$$

By S_k, 2 is a factor of $k^2 + k$; hence 2 is a factor of the right-hand side, so 2 is a factor of $(k + 1)^2 + (k + 1)$.
33. S_1:   3 is a factor of $1(1 + 1)(1 + 2)$
    S_k: 3 is a factor of $k(k + 1)(k + 2)$, or $k(k^2 + 3k + 2)$.

$$(k + 1)(k + 1 + 1)(k + 1 + 2)$$
$$= (k + 1)(k + 2)(k + 3)$$
$$= (k^2 + 3k + 2)(k + 3)$$
$$= k(k^2 + 3k + 2) + 3(k^2 + 3k + 2)$$

By S_k, 3 is a factor of $k(k^2 + 3k + 2)$; hence 3 is a factor of the right-hand side, so 3 is a factor of $(k + 1)(k + 2)(k + 3)$.
35. The least number of moves for

1 disk(s) is $1 = 2^1 - 1$,
2 disk(s) is $3 = 2^2 - 1$,
3 disk(s) is $7 = 2^3 - 1$,
4 disk(s) is $15 = 2^4 - 1$; etc.

Let P_n be the least number of moves for n disks. We conjecture and must show:
S_n: $P_n = 2^n - 1$.
1. *Basis step:* S_1 true by substitution.
2. *Induction step:*   Assume S_k for k disks: $P_k = 2^k - 1$. Show: $P_{k+1} = 2^{k+1} - 1$. Now suppose there are $k + 1$ disks on one peg. Move k of them to another peg in $2^k - 1$ moves (by S_k) and move the remaining disk to the free peg (1 move). Then move the k disks onto it in (another) $2^k - 1$ moves. Thus the total moves P_{k+1} is $2(2^k - 1) + 1 = 2^{k+1} - 1$: $P_{k+1} = 2^{k+1} - 1$.
37. B is a set of k people of the same sex (by S_k), and C is a set of k people of the same sex (by S_k). But if $k = 1$, B might be all men and C might be all women; and then the argument fails. Thus, $S_1 \not\rightarrow S_2$. In fact, $S_k \rightarrow S_{k+1}$ for $k > 1$.

Margin Exercises, Section 11.5
1. $3 \cdot 2 \cdot 1$, or 6; $3 \cdot 3 \cdot 3$, or 27
2. (a) $26 \cdot 10 \cdot 26 \cdot 10 \cdot 26 \cdot 10$, or 17,576,000; **(b)** no
3. $5 \cdot 4 \cdot 3 \cdot 2 \cdot 1$, or 120   **4.** $5 \cdot 4 \cdot 3 \cdot 2 \cdot 1$, or 120
5. $3 \cdot 2 \cdot 1$, or 6   **6.** $5 \cdot 4 \cdot 3 \cdot 2 \cdot 1$, or 120
7. $6 \cdot 5 \cdot 4 \cdot 3 \cdot 2 \cdot 1$, or 720   **8.** $4 \cdot 3 \cdot 2 \cdot 1$, or 24
9. $6 \cdot 5 \cdot 4 \cdot 3 \cdot 2 \cdot 1$, or 720
10. $8 \cdot 7 \cdot 6 \cdot 5 \cdot 4 \cdot 3 \cdot 2 \cdot 1$, or 40,320   **11.** 40,320
12. 362,880   **13.** 18!
14. (a) $10! = 10 \cdot 9!$; **(b)** $20! = 20 \cdot 19 \cdot 18 \cdot 17 \cdot 16 \cdot 15!$
15. $_7P_3 = 7 \cdot 6 \cdot 5 = 210$;
$$_7P_3 = \frac{7!}{4!} = \frac{7 \cdot 6 \cdot 5 \cdot 4 \cdot 3 \cdot 2 \cdot 1}{4 \cdot 3 \cdot 2 \cdot 1} = 7 \cdot 6 \cdot 5 = 210$$
16. (a) $_{10}P_4 = 10 \cdot 9 \cdot 8 \cdot 7 = 5040$; **(b)** $_8P_2 = 8 \cdot 7 = 56$;
(c) $_{11}P_5 = 11 \cdot 10 \cdot 9 \cdot 8 \cdot 7 = 55,440$; **(d)** $_nP_1 = n$;
(e) $_nP_2 = n(n - 1) = n^2 - n$; **(f)** 1
17. $_{12}P_5 = 12 \cdot 11 \cdot 10 \cdot 9 \cdot 8 = 95,040$
18. $_9P_3 = 9 \cdot 8 \cdot 7 = 504$ **19.** $_4P_4 \cdot _3P_3 = 4! \cdot 3! = 144$
20. $\dfrac{11!}{1!4!4!2!} = 34,650$ **21.** $\dfrac{6!}{3!2!1!} = 60$ **22.** $\dfrac{8!}{3!3!2!} = 560$
23. $26^5 = 11,881,376$

Exercise Set 11.5, pp. 711–712
1. $4 \cdot 3 \cdot 2$, or 24
3. $_{10}P_7 = 10 \cdot 9 \cdot 8 \cdot 7 \cdot 6 \cdot 5 \cdot 4$, or 604,800   **5.** 120; 3125
7. 120   **9.** $\dfrac{5!}{2!1!1!1!1!} = 5 \cdot 4 \cdot 3 = 60$
11. $9 \cdot 9 \cdot 8 \cdot 7 \cdot 6 \cdot 5 \cdot 4$, or 544,320   **13.** $\dfrac{9!}{2!3!4!} = 1260$
15. (a) 120; **(b)** 3840   **17.** $52 \cdot 51 \cdot 50 \cdot 49 = 6,497,400$
19. $\dfrac{24!}{3!5!9!4!3!} = 16,491,024,950,400$
21. $4! = 24$, $8! \div 3! = 6720$, $\dfrac{13!}{2!2!2!2!2!} = 194,594,400$
23. $80 \cdot 26 \cdot 9999 = 20,797,920$
25. (a) 10^5, or 100,000; **(b)** 100,000
27. (a) 10^9, or 1,000,000,000; **(b)** yes
29. 11   **31.** 9   **33.** $n - 1$

Margin Exercises, Section 11.6
1. (a) 1; **(b)** 5; **(c)** 10; **(d)** 5; **(e)** 1   **2. (a)** $\varnothing, 1$; **(b)** $\{A\}, \{B\}, 2$;
(c) $\{A, B\}$, 1; **(d)** 4   **3. (a)** $\varnothing$, 1; **(b)** $\{A\}$, 1; **(c)** 2   **4.** 120
5. 120   **6.** 126   **7.** 126   **8.** n **9.** $\dfrac{n(n - 1)(n - 2)}{6}$
10. (a) 56, 56; **(b)** $\dbinom{8}{3}$ **11.** 161,700   **12.** 45   **13.** 6160

Exercise Set 11.6, pp. 717–719
1. 78   **3.** 78   **5.** 7   **7.** 10   **9.** 1   **11.** 15   **13.** 12
15. 220   **17.** $\dfrac{m!}{2!(m - 2)!}$ **19.** $\dfrac{p!}{3!(p - 3)!}$ **21.** 128
23. $\dbinom{23}{4} = 8855$ **25.** $\dbinom{10}{6} = 210$
27. $\dbinom{8}{2} = 28$, $\dbinom{8}{3} = 56$ **29.** $\dbinom{10}{7} \cdot \dbinom{5}{3} = 1200$
31. $\dbinom{58}{6} \cdot \dbinom{42}{4}$ **33.** $\dbinom{4}{3} \cdot \dbinom{48}{2} = 4512$
35. (a) $_{33}P_3 = 32,736$; **(b)** $33^3 = 35,937$; **(c)** $_{33}C_3 = 5456$
37. 4096   **39.** $\dbinom{52}{2} = 2,598,960$ **41.** $\dbinom{8}{3} = 56$
43. $\dbinom{5}{2}\dbinom{8}{2} = 280$ **45.** 28, 56   **47.** 5   **49.** 7
51. $_5C_2 = 10$; $_5C_2 - 5 = 5$ **53. (a)** $_nC_2 = \dfrac{n(n - 1)}{2}$;
(b) $_nC_2 - n = \dfrac{n(n - 1)}{2} - \dfrac{2n}{2} = \dfrac{n(n - 3)}{2}$,
where $n = 4, 5, 6, \ldots$;
(c) Let D_n be the number of diagonals of an n-gon. We must prove S_n (below) using mathematical induction. We have

S_n: $D_n = \dfrac{n(n - 3)}{2}$, for $n = 4, 5, 6, \ldots$

S_4: $D_4 = \dfrac{4 \cdot 1}{2}$

S_k: $D_k = \dfrac{k(k - 3)}{2}$

S_{k+1}: $D_{k+1} = \dfrac{(k + 1)(k - 2)}{2}$

1. *Basis step:* S_4 is true (a quadrilateral has 2 diagonals).
2. *Induction step:* Assume S_k. Observe that when an additional vertex V_{k+1} is added to the k-gon, we gain k segments, 2 of which are sides [of the $(k+1)$-gon], and a former side $\overline{V_1V_k}$ becomes a diagonal. Thus the additional number of diagonals is $k - 2 + 1$, or $k - 1$. Then the new total of diagonals is $D_k + (k-1)$, or

$$D_{k+1} = D_k + (k-1)$$
$$- \frac{k(k-3)}{2} + (k-1) \qquad \text{(by } S_k\text{)}$$
$$= \frac{(k+1)(k-2)}{2}.$$

Margin Exercises, Section 11.7

1. 1 6 15 20 15 6 1;
1 7 21 35 35 21 7 1;
1 8 28 56 70 56 28 8 1
2. $a^9 + 9a^8b + 36a^7b^2 + 84a^6b^3 + 126a^5b^4 + 126a^4b^5 + 84a^3b^6 + 36a^2b^7 + 9ab^8 + b^9$
3. $x^4 - 4x^3y + 6x^2y^2 - 4xy^3 + y^4$
4. $128t^7 + 448t^5 + 672t^3 + 560t + 280t^{-1} + 84t^{-3} + 14t^{-5} + t^{-7}$
5. $x^5 + 25x^4b + 250x^3b^2 + 1250x^2b^3 + 3125xb^4 + 3125b^5$
6. $x^{10} - 5x^8 + 10x^6 - 10x^4 + 5x^2 - 1$
7. $16x^4 + 32x^3y^{-1} + 24x^2y^{-2} + 8xy^{-3} + y^{-4}$
8. $x^6 - 6\sqrt{2}x^5 + 30x^4 - 40\sqrt{2}x^3 + 60x^2 - 24\sqrt{2}x + 8$
9. $181{,}440a^4b^3$ **10.** $103{,}680x^2$ **11.** 2^6, or 64 **13.** 2^{50}

Exercise Set 11.7, pp. 724–725

1. $m^5 + 5m^4n + 10m^3n^2 + 10m^2n^3 + 5mn^4 + n^5$
3. $x^6 - 6x^5y + 15x^4y^2 - 20x^3y^3 + 15x^2y^4 - 6xy^5 + y^6$
5. $x^{10} - 15x^8y + 90x^6y^2 - 270x^4y^3 + 405x^2y^4 - 243y^5$
7. $729c^6 - 1458c^5d + 1215c^4d^2 - 540c^3d^3 + 135c^2d^4 - 18cd^5 + d^6$ **9.** $x^3 - 3x^2y + 3xy^2 - y^3$
11. $x^{-7} + 7x^{-6}y + 21x^{-5}y^2 + 35x^{-4}y^3 + 35x^{-3}y^4 + 21x^{-2}y^5 + 7x^{-1}y^6 + y^7$
13. $a^9 - 18a^7 + 144a^5 - 672a^3 + 2016a - 4032a^{-1} + 5376a^{-3} - 4608a^{-5} + 2304a^{-7} - 512a^{-9}$

15. $1 - n + \binom{n}{2} - \binom{n}{3} + \cdots + \binom{n}{n}(-1)^n$

17. $9 - 12\sqrt{3}t + 18t^2 - 4\sqrt{3}t^3 + t^4$ **19.** $140\sqrt{2}$
21. $x^{-8} + 4x^{-4} + 6 + 4x^4 + x^8$ **23.** $15a^4b^2$
25. $-745{,}472a^3$ **27.** $1120x^{12}y^2$ **29.** $-1{,}959{,}552u^5v^{10}$
31. 2^7, or 128 **33.** 2^{26}, or $67{,}108{,}864$ **35.** $-7 - 4\sqrt{2}i$
37. $\sin^7 t - 7\sin^5 t + 21\sin^3 t - 35\sin t + 35\csc t -$

$21\csc^3 t + 7\csc^5 t - \csc^7 t$ **39.** $\displaystyle\sum_{r=0}^{n}\binom{n}{r}(-1)^r a^{n-r}b^r$

41. -3 **43.** 5 **45.** $\binom{5}{2}(0.313)^3(0.687)^2 \approx 0.14473$

47. 0.96403 **49.** $2{,}006{,}581{,}248u^5v^{10}$ **51.** $\dfrac{55}{144}$

53. $-\dfrac{\sqrt[3]{q}}{2p}$ **55.** 20 **57.** $3, 4, 5, 6$ **59.** 2^n **61.** 0

Margin Exercises, Section 11.8

1. $\frac{82}{100}$ **2.** 73.4%; no **3.** $14.7\%, 50.4\%$ **4.** $\frac{1}{2}$

5. (a) $\frac{1}{13}$; **(b)** $\frac{1}{4}$; **(c)** $\frac{1}{2}$; **(d)** $\frac{2}{13}$ **6.** $\frac{6}{11}$ **7.** 0 **8.** 1
9. $\frac{11}{850}$ **10.** $\frac{5}{39}$ **11.** $\frac{6}{13}$ **12.** $\frac{1}{6}$

Exercise Set 11.8, pp. 733–734
1. $0.57, 0.43$ **3.** $0.075, 0.134, 0.057, 0.071, 0.030$
5. 0.633 **7.** 52 **9.** $\frac{1}{4}$ **11.** $\frac{1}{13}$ **13.** $\frac{1}{2}$ **15.** $\frac{2}{13}$

17. $\frac{5}{7}$ **19.** 0 **21.** $\dfrac{11}{4165}$ **23.** $\frac{28}{65}$ **25.** $\frac{1}{18}$ **27.** $\frac{1}{36}$

29. $\frac{30}{323}$ **31.** $\frac{9}{19}$ **33.** $\frac{1}{38}$ **35.** $\frac{1}{19}$ **37.** $2{,}598{,}960$
39. (a) 36; **(b)** 1.39×10^{-5}
41. (a) $(13 \cdot {}_4C_3) \cdot (12 \cdot {}_4C_2) = 3744$;

(b) $\dfrac{3744}{{}_{52}C_5} = \dfrac{3744}{2{,}598{,}960} \approx 0.00144$

43. (a) $13 \cdot \dbinom{4}{3} \cdot \dbinom{48}{2} - 3744 = 54{,}912$;

(b) $\dfrac{54{,}912}{{}_{52}C_5} = \dfrac{54{,}912}{2{,}598{,}960} \approx 0.0211$

45. (a) $\dbinom{13}{2}\dbinom{4}{2}\dbinom{4}{2}\dbinom{44}{1} = 123{,}552$;

(b) $\dfrac{123{,}552}{{}_{52}C_5} = \dfrac{123{,}552}{2{,}598{,}960} \approx 0.0475$

Review Exercises: Chapter 11, pp. 735–736
1. [11.2] $3\frac{3}{4}$ **2.** [11.2] $a + 4b$ **3.** [11.2] 531
4. [11.2] 465 **5.** [11.2] 11 **6.** [11.2] -4
7. [11.3] $n = 6, S_n = -126$ **8.** [11.3] $a_1 = 8, a_5 = \frac{1}{2}$
9. [11.3] No **10.** [11.3] Yes **11.** [11.3] $\frac{3}{8}$ **12.** [11.3] $\frac{211}{99}$
13. [11.2] $5\frac{4}{5}, 6\frac{3}{5}, 7\frac{2}{5}, 8\frac{1}{5}$ **14.** [11.3] ≈ 50 ft
15. [11.2] $\$7.38, \$1365.10$ **16.** [11.3] $7{,}680{,}000$

17. [11.3] $\dfrac{64\sqrt{2}}{\sqrt{2}-1}$ in., or $128 + 64\sqrt{2}$ in.
18. [11.3] $23\frac{1}{3}$ cm
19. [11.4]

$$S_n: \quad 1 + 4 + 7 + \cdots + (3n-2) = \frac{n(3n-1)}{2}$$

$$S_1: \quad 1 = \frac{1(3-1)}{2}$$

$$S_k: \quad 1 + 4 + 7 + \cdots + (3k-2) = \frac{k(3k-1)}{2}$$

$$S_{k+1}: \quad 1 + 4 + 7 + \cdots + [3(k+1)-2]$$
$$= 1 + 4 + 7 + \cdots + (3k-2) + (3k+1)$$
$$= \frac{(k+1)(3k+2)}{2}$$

1. *Basis step:* $1 = \dfrac{2}{2} = \dfrac{1(3-1)}{2}$ is true.

2. *Induction step:* Assume S_k. Add $(3k+1)$ to both sides.
$$1 + 4 + 7 + \cdots + (3k-2) + (3k+1)$$
$$= \frac{k(3k-1)}{2} + (3k+1)$$
$$= \frac{k(3k-1)}{2} + \frac{2(3k+1)}{2}$$

(continued)

$$= \frac{3k^2 - k + 6k + 2}{2}$$

$$= \frac{3k^2 + 5k + 2}{2}$$

$$= \frac{(k + 1)(3k + 2)}{2}$$

20. [11.4]

$$S_1: \quad 1 = \frac{3^1 - 1}{2}; S_2: 1 + 3 = \frac{3^2 - 1}{2}$$

$$S_k: \quad 1 + 3 + 3^2 + \cdots + 3^{k-1} = \frac{3^k - 1}{2}$$

2. *Induction step:*   Assume S_k. Add 3^k on both sides.

$$1 + 3 + \cdots + 3^{k-1} + 3^k$$

$$= \frac{3^k - 1}{2} + 3^k = \frac{3^k - 1}{2} + 3^k \cdot \frac{2}{2}$$

$$= \frac{3 \cdot 3^k - 1}{2} = \frac{3^{k+1} - 1}{2}$$

21. [11.4]

$$S_n: \quad \left(1 - \frac{1}{2}\right)\left(1 - \frac{1}{3}\right)\cdots\left(1 - \frac{1}{n}\right) = \frac{1}{n}$$

$$S_2: \quad \left(1 - \frac{1}{2}\right) = \frac{1}{2}$$

$$S_k: \quad \left(1 - \frac{1}{2}\right)\left(1 - \frac{1}{3}\right)\cdots\left(1 - \frac{1}{k}\right) = \frac{1}{k}$$

$$S_{k+1}: \quad \left(1 - \frac{1}{2}\right)\left(1 - \frac{1}{3}\right)\cdots\left(1 - \frac{1}{k}\right)\left(1 - \frac{1}{k+1}\right) = \frac{1}{k+1}$$

1. *Basis step:* S_2 is true by substitution.
2. *Induction step:*   Assume S_k. Deduce S_{k+1}.
   Starting with the left side of S_{k+1} we have:

$$\underbrace{\left(1 - \frac{1}{2}\right)\left(1 - \frac{1}{3}\right)\cdots\left(1 - \frac{1}{k}\right)}\left(1 - \frac{1}{k+1}\right)$$

$$= \frac{1}{k} \cdot \left(1 - \frac{1}{k+1}\right). \qquad \text{(by } S_k\text{)}$$

$$= \frac{1}{k} \cdot \left(\frac{k + 1 - 1}{k+1}\right)$$

$$= \frac{1}{k} \cdot \frac{k}{k+1}$$

$$= \frac{1}{k+1} \qquad \text{(simplifying)}$$

22. [11.1] 5; 51; 5203; 54,142,419

23. [11.1] $\sum_{n=1}^{7} (n^2 - 1)$ or $\sum_{n=0}^{6} n(n + 2)$

24. [11.5] $6! = 720$ **25.** [11.5] $9 \cdot 8 \cdot 7 \cdot 6 = 3024$

26. [11.6] $\binom{15}{8} = 6435$ **27.** [11.5] $24 \cdot 23 \cdot 22 = 12{,}144$

28. [11.5] $\frac{9!}{1!4!2!2!} = 3780$ **29.** [11.5] 36

30. [11.5] **(a)** $_6P_5 = 720$; **(b)** $6^5 = 7776$;

(c) $_5P_4 = 120$; **(d)** $_3P_2 = 6$ **31.** [11.7] $220a^9x^3$

32. [11.7] $\binom{18}{11}a^7x^{11}$

33. [11.7] $m^7 + 7m^6n + 21m^5n^2 + 35m^4n^3 + 35m^3n^4 + 21m^2n^5 + 7mn^6 + n^7$

34. [11.7] $x^8 + 12x^6y + 54x^4y^2 + 108x^2y^3 + 81y^4$

35. [11.7] $-6624 + 16{,}280i$

36. [11.7] $a^8 + 8a^6 + 28a^4 + 56a^2 + 70 + 56a^{-2} + 28a^{-4} + 8a^{-6} + a^{-8}$

37. [11.8] $\frac{86}{206} \approx 0.42$, $\frac{97}{206} \approx 0.47$, $\frac{23}{206} \approx 0.11$

38. [11.8] $\frac{1}{12}$, 0   **39.** [11.8] $\frac{1}{4}$ **40.** [11.8] $\frac{6}{5525}$

41. [11.4] S_1 fails.

42. [11.3] $\frac{a_{k+1}}{a_k} = r_1$, $\frac{b_{k+1}}{b_k} = r_2$,

so $\frac{a_{k+1}b_{k+1}}{a_k b_k} = r_1 r_2$ (constant)

43. [11.3] $a_{k+1} - a_k = d$,

so $\frac{c_{k+1}}{c_k} = \frac{b^{a_{k+1}}}{b^{a_k}} = b^{a_{k+1} - a_k} = b^d$ (constant)

44. [11.2] **(a)** a_n is all positive or all negative;
(b) always; **(c)** always; **(d)** $a_n = k$, k a constant;
(e) $a_n = k$, k a constant; **(f)** $a_n = k$, k a constant

45. [11.2] $-2, 0, 2, 4$ **46.** [11.3] 0.27, 0.0027, 0.000027

47. [11.3] $\frac{1}{2}, -\frac{1}{6}, \frac{1}{18}$ **48.** [11.6] $\left(\log\frac{x}{y}\right)^{10}$

49. [11.6] 36   **50.** [11.6] 14

Test: Chapter 11, pp. 736–737
1. [11.2] 13   **2.** [11.2] 4   **3.** [11.2] 30th

4. [11.2] $\frac{23}{4}, \frac{17}{2}, \frac{45}{4}$ **5.** [11.3] 437.4   **6.** [11.3] $\frac{31}{64}$

7. [11.3] a **8.** [11.3] $\frac{125}{6}$ **9.** [11.2] $2010

10. [11.3] 12,960   **11.** [11.3] $\frac{29}{225}$

12. [11.4]

$$S_n: \quad 5 + 10 + 15 + \cdots + 5n = \frac{5n(n+1)}{2}$$

$$S_1: \quad 5 = \frac{5 \cdot 1(1+1)}{2}$$

$$S_k: \quad 5 + 10 + 15 + \cdots + 5k = \frac{5k(k+1)}{2}$$

$$S_{k+1}: \quad 5 + 10 + 15 + \cdots + 5k + 5(k+1)$$

$$= \frac{5(k+1)(k+2)}{2}$$

1. *Basis step:* $\frac{5 \cdot 1(1+1)}{2} = 5$, so S_1 is true.

2. *Induction step:* Assume S_k. Then add $5(k + 1)$ on both sides.

$$5 + 10 + 15 + \cdots + 5k + 5(k + 1)$$

$$= \frac{5k(k + 1)}{2} + 5(k + 1)$$

$$= \frac{5k(k + 1)}{2} + \frac{10(k + 1)}{2}$$

$$= \frac{5k^2 + 15k + 10}{2}$$

$$= \frac{5(k^2 + 3k + 2)}{2}$$

$$= \frac{5(k + 1)(k + 2)}{2}$$

13. [11.1] 5, 23, 95, 383
14. [11.5] **(a)** 6^4, or 1296; **(b)** 360; **(c)** 12
15. [11.6] 35   16. [11.6] 103,740
17. [11.5] 3360
18. [11.7] 2^8, or 256
19. [11.7] $672a^5b^2$
20. [11.7] $x^5 - 5\sqrt{2}x^4 + 20x^3 - 20\sqrt{2}x^2 + 20x - 4\sqrt{2}$

21. [11.8] $\dfrac{5}{36}$ 22. [11.8] $\dfrac{2}{13}$

23. [11.8] $\dfrac{1}{14}$ 24. [11.2] 4, 7, 10, 13

25. [11.6] 54   26. [11.6] 18
27. [11.3] 9 cis 60°; 9 cis 120°; 9 cis 180°, or -9; 9 cis 240°; 9 cis 300°

Index

GEOMETRIC FUNCTIONS

Plane Geometry

Rectangle
Area: $A = lw$
Perimeter: $P = 2l + 2w$

Square
Area: $A = s^2$
Perimeter: $P = 4s$

Triangle
Area: $A = \frac{1}{2}bh$

Sum of Angle Measures:
$A + B + C = 180°$

Right Triangle
Pythagorean Theorem
(Equation):
$a^2 + b^2 = c^2$

Parallelogram
Area: $A = bh$

Trapezoid
Area: $A = \frac{1}{2}h(a + b)$

Circle
Area: $A = \pi r^2$
Circumference:
$C = \pi d = 2\pi r$
($\frac{22}{7}$ and 3.14 are different
approximations for π)

Solid Geometry

Rectangular Solid
Volume: $V = lwh$

Cube
Volume: $V = s^3$

Right Circular Cylinder
Volume: $V = \pi r^2 h$
Lateral Surface Area:
$L = 2\pi rh$
Total Surface Area:
$S = 2\pi rh + 2\pi r^2$

Right Circular Cone
Volume: $V = \frac{1}{3}\pi r^2 h$
Lateral Surface Area:
$L = \pi rs$
Total Surface Area:
$S = \pi r^2 + \pi rs$
Slant Height:
$s = \sqrt{r^2 + h^2}$

Sphere
Volume: $V = \frac{4}{3}\pi r^3$
Surface Area: $S = 4\pi r^2$

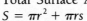

TRIGONOMETRIC FUNCTIONS

Real Numbers

Any Angle

Acute Angles

$$\sin \theta = v, \quad \csc \theta = \frac{1}{v},$$
$$\cos \theta = u, \quad \sec \theta = \frac{1}{u},$$
$$\tan \theta = \frac{v}{u}, \quad \cot \theta = \frac{u}{v}$$

$$\sin \theta = \frac{y}{r}, \quad \csc \theta = \frac{r}{y},$$
$$\cos \theta = \frac{x}{r}, \quad \sec \theta = \frac{r}{x},$$
$$\tan \theta = \frac{y}{x}, \quad \cot \theta = \frac{x}{y}$$

$$\sin \theta = \frac{\text{opp}}{\text{hyp}}, \quad \csc \theta = \frac{\text{hyp}}{\text{opp}},$$
$$\cos \theta = \frac{\text{adj}}{\text{hyp}}, \quad \sec \theta = \frac{\text{hyp}}{\text{adj}},$$
$$\tan \theta = \frac{\text{opp}}{\text{adj}}, \quad \cot \theta = \frac{\text{adj}}{\text{opp}}$$

BASIC TRIGONOMETRIC IDENTITIES

$$\sin (-x) = -\sin x,$$
$$\cos (-x) = \cos x,$$
$$\tan (-x) = -\tan x,$$

$$\tan x = \frac{\sin x}{\cos x},$$
$$\cot x = \frac{\cos x}{\sin x},$$

$$\csc x = \frac{1}{\sin x},$$
$$\sec x = \frac{1}{\cos x},$$
$$\cot x = \frac{1}{\tan x}$$

Pythagorean Identities

$$\sin^2 x + \cos^2 x = 1,$$
$$1 + \tan^2 x = \sec^2 x,$$
$$1 + \cot^2 x = \csc^2 x$$

Cofunction Identities

$$\sin \left(x \pm \frac{\pi}{2}\right) = \pm \cos x,$$
$$\cos \left(x \pm \frac{\pi}{2}\right) = \mp \sin x,$$
$$\tan \left(x \pm \frac{\pi}{2}\right) = -\cot x$$

Sum and Difference Identities

$$\sin (\alpha \pm \beta) = \sin \alpha \cos \beta \pm \cos \alpha \sin \beta,$$
$$\cos (\alpha \pm \beta) = \cos \alpha \cos \beta \mp \sin \alpha \sin \beta,$$
$$\tan (\alpha \pm \beta) = \frac{\tan \alpha \pm \tan \beta}{1 \mp \tan \alpha \tan \beta}$$

Double-Angle Identities

$$\sin 2x = 2 \sin x \cos x,$$
$$\cos 2x = \cos^2 x - \sin^2 x = 1 - 2 \sin^2 x$$
$$= 2 \cos^2 x - 1,$$
$$\tan 2x = \frac{2 \tan x}{1 - \tan^2 x},$$
$$\sin^2 x = \frac{1 - \cos 2x}{2},$$
$$\cos^2 x = \frac{1 + \cos 2x}{2}$$

Half-Angle Identities

$$\sin \frac{x}{2} = \pm \sqrt{\frac{1 - \cos x}{2}},$$
$$\cos \frac{x}{2} = \pm \sqrt{\frac{1 + \cos x}{2}},$$
$$\tan \frac{x}{2} = \pm \sqrt{\frac{1 - \cos x}{1 + \cos x}} = \frac{\sin x}{1 + \cos x} = \frac{1 - \cos x}{\sin x}$$